Reservoir Simulation: Problems and Solutions

Reservoir Simulation: Problems and Solutions

Society of Petroleum Engineers

Richardson, Texas, USA

Disclaimer

This book was prepared by members of the Society of Petroleum Engineers and their well-qualified colleagues from material published in the recognized technical literature and from their own individual experience and expertise. While the material presented is believed to be based on sound technical knowledge, neither the Society of Petroleum Engineers nor any of the authors or editors herein provide a warranty either expressed or implied in its application. Correspondingly, the discussion of materials, methods, or techniques that may be covered by letters patents implies no freedom to use such materials, methods, or techniques without permission through appropriate licensing. Nothing described within this book should be construed to lessen the need to apply sound engineering judgment nor to carefully apply accepted engineering practices in the design, implementation, or application of the techniques described herein.

ISBN 978-1-61399-693-5

First Printing 2019

Society of Petroleum Engineers
222 Palisades Creek Drive
Richardson, TX 75080-2040 USA

http://store.spe.org
service@spe.org 1.972.952.9393

Dedication

This book is dedicated to current and past petroleum-engineering students and practitioners of the world who have made petroleum engineering their career choice.

If I had an hour to solve a problem and my life depended on the solution, I would spend the first 55 minutes determining the proper question to ask. For once I know the proper question, I could solve the problem in less than 5 minutes.

—*Albert Einstein*

Preface

In solving a reservoir simulation problem, the very first step will involve showing a good understanding of the problem description. Upon successful problem identification, some prioritization of strategies toward the solution of the problem will be necessary. Once these two steps are correctly executed, the final step will involve the implementation of a formal solution protocol. To set the stage for solving a reservoir simulation problem, it will be necessary to ensure that a set of criteria is adopted to check the plausibility of the problem (well-posed problem). In many areas of reservoir simulation, as it happens in many areas of mathematics, a problem for the application of a certain numerical protocol is not considered to be well-formulated unless the existence of a unique solution is assured and the solution is stable. Once the solution is generated, it will be necessary to analyze and reflect on the solution.

As in any other engineering and scientific area, solving reservoir simulation problems is important because each time we solve a problem, we learn how to make decisions independently, which in turn generates self-confidence and builds self-esteem, both of which feed into developing creativity, persistence, and a proactive mind set to prepare ourselves for problems of the real world. But, reservoir simulation problem solving goes deeper than this because it involves powerful imagination, requiring the reservoir simulation problem solver to take all of what one has learned and incorporate it into seemingly unrelated actions. It is widely accepted that simulation problem solving has grounds in fact-learning, but merely knowing facts does not make one a good reservoir simulation practitioner. In other words, a good reservoir simulation engineer or scientist is one who assembles and puts to use all the tools at his/her disposal. As a result, solutions to reservoir simulation problems should come as a complete package, with components such as basic skills, concept understanding, literacy in mathematics, and problem-solving philosophy. Then, it can be said that solving a reservoir simulation problem is knowing what to do when you do not immediately know what to do.

Reservoir simulation has been in practice for more than 50 years and has been gaining significant momentum with its wider applications in increasingly more-complex reservoir systems. Accordingly, we believe that the timing of this new book is well in agreement with the developing industry practices and academic curricula. It is the authors' view that, most of the time, reservoir simulation technology has been treated in a prescriptive manner. We are hopeful that this new book will bring reservoir simulation methodology and its intricacies much closer to the readers, allowing them to grasp the ideas more effectively. In this book, we provide solutions to the exercises that were presented in the *Basic Applied Reservoir Simulation* textbook authored by Turgay Ertekin, Jamal H. Abou-Kassem, and Gregory R. King (SPE Textbook Series, Volume 7, 2001). Accordingly, the overall outline of the book follows the original outline of the textbook. While this book contains solutions to approximately 180 exercises from the original book, it also introduces a new set of 180 exercises and their solutions, all of which come from the homework and examination sets used in our courses taught at Penn State University.

We do not know of the existence of a similar "problems and solutions" book in the petroleum-engineering literature, and therefore believe that this new book will be instrumental in structuring effective solution strategies to a large spectrum of reservoir simulation problems and will fill an existing gap. Petroleum-engineering undergraduate and graduate students and more-recent petroleum-engineering graduates will benefit from this book, and it is our hope that it will help students, young engineers, and earth scientists become astute problem solvers of reservoir simulation applications.

As you start walking the way, the way appears. —Rumi, 13th Century mystic poet

Turgay Ertekin, Qian Sun, and Jian Zhang
Pennsylvania State University
University Park, Pennsylvania
May 2019

Acknowledgments

This book is a compendium of homework sets and midterm and final examinations of one senior-level and two graduate-level courses that we taught at Penn State University over a period of almost four decades. We owe a tremendous amount of gratitude to our students who have expended intense efforts to combine their creativity and understanding of the thematic topics of reservoir simulation to become better petroleum engineers. We also would like to extend our appreciation to many individuals from different parts of the world who have continuously encouraged us to write this book. We are indebted to the fine work and support of SPE Editorial Services Manager Ms. Jane Eden and SPE Editor Ms. Shashana Pearson-Hormilosa for guiding us, and perhaps more importantly, walking with us through the tortuous paths of producing a technical book of this kind. Finally, we thank the SPE Textbook Committee for inviting us to write this book and several colleagues of whom we do not know their names who have served as technical reviewers and provided much valuable feedback.

Table of Contents

Chapter 1

Introduction

Reservoir simulation is normally associated with modeling of bulk-fluid transport in porous media (macroscopic view), which is the subject of this book. With the recent inclusion of ultratight, nanoporous, and nanopermeability reservoirs in our resource portfolio, however, pore-scale molecular models, stochastic models, and models based on artificial intelligence are also of interest to the industry. One of the first questions that comes into a discussion about analytical methods and numerical methods is how do they compare when implemented into reservoir engineering applications? Perhaps a statement such as, "a numerical method provides an approximate solution to an exact problem, whereas an analytical method provides an exact solution to an approximate problem," describes the difference between the two formalisms in the most-effective way. However, one should realize that there is always some degree of exaggeration in the statement when the phrase "exact problem" is used. Mother Nature typically provides challenging and, to a certain extent, ill-posed problems (caused by our misunderstanding of certain processes). This is a result of the heterogeneous and anisotropic distribution of rock properties and nonlinear behavior of the fluid properties. Accordingly, in an analytical formulation it is necessary to remove some of these complexities, and to do so, we must incorporate some sweeping assumptions. As we do, we begin to deviate from the exact representation of a problem and end up with an approximate representation, which is a simplified version of the actual problem. Consider, for example, the classical well test model:

$$\frac{1}{r}\frac{\partial}{\partial r}\left(r\frac{\partial p}{\partial r}\right) = \frac{\phi\mu c}{k}\frac{\partial p}{\partial t}, \dots\dots\dots\dots\dots\dots\dots\dots\dots\dots\dots\dots\dots\dots\dots\dots \text{(1.1a)}$$

with typical boundary conditions stating constant flow rate and infinite-acting reservoir,

$$\left.r\frac{\partial p}{\partial r}\right|_{r=r_w} = \left.-\frac{q\mu}{2\pi kh}\right|_{r=r_w} \text{ for } t > 0, \text{ and as } r \to \infty, \ p \to p_i, \text{ for } t \geq 0, \dots\dots\dots\dots\dots\dots \text{(1.1b)}$$

and with the initial condition stating uniform pressure distribution over the reservoir domain to be tested,

$$p = p_i \text{ at } t = 0, \text{ for } r_w \leq r < \infty. \dots\dots\dots\dots\dots\dots\dots\dots\dots\dots\dots\dots\dots\dots\dots\dots \text{(1.1c)}$$

Eqs. 1.1a through 1.1c represent an analytical model of fluid flow in porous media. A close inspection of these equations reveals the following assumptions are already in place:

- Flow is taking place in a 1D domain in radial-cylindrical coordinates.
- The flow domain is infinitely large, and there is only one well that is produced at a constant rate.
- When the well is put on production at a constant rate of q, this rate is accomplished instantaneously.
- The reservoir has a uniform pressure distribution at $t = 0$ (before the well is put on production).
- The reservoir has homogeneous and isotropic property distribution.
- Fluid properties are assumed to be constant and a single-phase flow condition is assumed to be in place.
- Gravitational forces are ignored.
- Flow regime is laminar, and flowing fluid exhibits Newtonian behavior.
- Isothermal flow conditions exist.

It is obvious that with the inclusion of the assumptions listed, the problem is rather simplified so an analytical solution to this approximate problem can be developed. This solution is then going to be an exact representation of pressure distribution within the approximate representation of the flow domain and flow conditions.

The numerical formulation honors most of the challenges and complexities imposed by Mother Nature. One can incorporate all the heterogeneities of reservoir characteristics together with the anisotropic flow parameters and simultaneously incorporate not only gravitational forces but also capillary forces. In ultratight reservoirs with slow fluid velocities, however, subgrid-scale heterogeneities may be of interest, but cannot be captured in numerical models. One can, however, incorporate multimechanistic flow concepts, such as Darcian and Fickian flow components can be simultaneously taken into consideration (Ertekin et al. 1986). Of course, pressure and temperature dependency of all the fluids can also be incorporated wherever they exist. There is no doubt that such a representation will enable engineers and geoscientists to come much closer to the exact representation in problem descriptions. However, with all these complexities in place, an analytical-solution protocol will not be feasible any more, forcing us toward a numerical-solution technique that will produce a solution inherently approximate in nature.

1.1 Analysis of Reservoir Dynamics and Plausible Methodologies

In a typical reservoir engineering application, the reservoir engineer or scientist always seeks to find a solution to a problem, which can be represented by Eq. 1.2:

$$q_\psi(t) = M(\Phi, \psi), \dotfill (1.2)$$

where $q_\psi(t)$ is the production and/or pressure characteristics as a function of time (time series); $M(\Phi, \psi)$ is a mathematical representation (mathematical model) of a fluid flow problem; Φ is the flow domain characteristics; and ψ is the flow process, including all the thermodynamics of the reservoir system.

In solving Eq. 1.2, the basic approaches that can be used are

- Deterministic modeling: The right-hand side of the equation, M, Φ, and ψ, is fully prescribed.
- Stochastic modeling: The right-hand side of the equation, M, Φ, and ψ, is partially known.
- Neurosimulation (machine learning): The right-hand side of the equation, M, Φ, and ψ, is not known, but the left-hand side is available in the form of a time-series data set (production history and/or pressure history). A neurosimulation approach can be implemented in forward calculations, as well as in two different forms of inverse calculations.

In all of these applications, the goal is to generate tractable, robust, and low-cost solutions. In achieving such solutions, the engineer expends efforts to exploit imprecision and uncertainty.

1.1.1 Systems Analysis. Using an equation such as Eq. 1.2 can be considered an implementation of a systems-analysis technique. A systems-analysis problem can be schematically represented, as shown in **Fig. 1.1**.

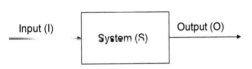

Input (I) → System (S) → Output (O)

Fig. 1.1—Systems analysis.

In Fig. 1.1, the system, S, represents the reservoir (with its pertinent characteristics) and wellbore and the related parameters it is in communication with. The input, I, represents the boundary conditions (external and internal), which can be considered a forcing function applied to the system S. The response of the system to the imposed forcing function is the output, O, which is a time-series representation of the pressure surface or production surface.

The Forward Problem (Forecasting). If the systems analysis is implemented in the following form,

$$[I] \times [S] \to [O], \dotfill (1.3)$$

the solution of Eq. 1.3 will yield the response of a reservoir system to the imposed boundary and initial conditions. This analysis typically is considered a prediction or forecasting.

The Inverse Problem I (History Matching/Characterization). In an inverse problem, the engineer makes an attempt to find S from the known I and O, which are imposed and measured entities, respectively. Then, the systems analysis is carried out as solving the equation

$$[O] / [I] \to [S]. \dotfill (1.4)$$

The Inverse Problem II (Design, Strategy Development). Consider the implementation of an improved-oil-recovery project in a reservoir with known characteristics. The system properties are well known and the expected response, which is considered a feasible response, is also determined through a feasibility analysis (economic analysis). Then, the problem reduces to the strategic design and implementation of the project so that it will yield the desired outcome. In other words,

$$[O] / [S] \to [I]. \dotfill (1.5)$$

In Eq. 1.4, [I] contains the required design- and project-implementation-related parameters. It should be recognized that the solution of inverse equations (Eqs. 1.4 and 1.5) can generate nonunique results, so ultimately it is the project engineer's responsibility to find the unique solution on which he/she is focusing.

It is also important to realize that in reservoir modeling, a phrase such as, "for every complex problem, there is always a simple solution," does not always hold up, because that simple solution may be the wrong solution. Accordingly, in formulating a modeling approach, the schematic representation shown in **Fig. 1.2** can be helpful in formulating the problem.

In Fig. 1.2, Quadrant I reminds us that as the problem complexity increases, the model sophistication accordingly should be increased. Quadrant III simply emphasizes the more-universal nature of the sophisticated models and their potential applications for simple problems. For example, a complex model can be used to generate solutions that can also be obtained by simple material-balance equations. Quadrant II simply states that for simple problems, one does not need sophisticated models. Quadrant IV underlines the impossibility of solving a complex problem with a simple model. Again, we should keep in mind the phrase, "for every complex problem, there is always a simple solution that is always wrong."

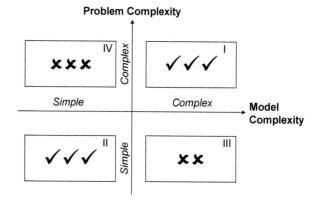

Fig. 1.2—Model complexity vs. problem complexity.

1.2 Problems and Solutions

Problem 1.2.1 (Ertekin et al. 2001)

What are the different ways a reservoir can be modeled?

Solution to Problem 1.2.1

- Traditional modeling approaches:
  - Analogical methods.
  - Experimental methods:
    - Analog models.
    - Physical models.
  - Mathematical methods:
    - Material-balance equations.
    - Decline-curve analysis.
    - Statistical approach.
    - Analytical approach.
- Reservoir simulation approach.

Problem 1.2.2 (Ertekin et al. 2001)

What are the differences between a mathematical, a numerical, and a computer model?

Solution to Problem 1.2.2

A numerical model is a class of mathematical models, and a computer model is an implementation of the numerical model for execution in a computational platform.

Problem 1.2.3 (Ertekin et al. 2001)

Match the items in the left column of **Table 1.1** with the corresponding item in the right-hand column.

Physical model	Simulator
Conceptual model	Partial-differential equations
Geological model	Material-balance equations
Mathematical model	Laboratory sandpacks
Computer model	Potentiometric model
Analog model	Depositional model
Numerical model	Empirical-correlation equations
Statistical model	Finite-difference equations

Table 1.1—Match the model on the left with the corresponding item on the right.

Solution to Problem 1.2.3

Table 1.2 displays the correct matches between models and their corresponding items.

Physical model	Laboratory sandpacks
Conceptual model	Partial-differential equations
Geological model	Depositional model
Mathematical model	Material-balance equations
Computer model	Simulator
Analog model	Potentiometric model
Numerical model	Finite-difference equations
Statistical model	Empirical-correlation equations

Table 1.2—Solution to Table 1.1: Models matched with their corresponding item.

Problem 1.2.4 (Ertekin et al. 2001)

To summarize the basic steps of a simulation study, put the following in sequential order:

- Define the study objectives.
- Prepare the data.
- Construct the geological model.
- History match.
- Predict performance.
- Analyze the results.
- Report.

Solution to Problem 1.2.4

1. Define the study objectives.
2. Construct the geological model.
3. Prepare the data.
4. Predict performance.
5. History match.
6. Analyze the results.
7. Report.

Problem 1.2.5 (Ertekin et al. 2001)

The equation $q = \dfrac{2\pi\beta_c kh\Delta p}{\mu \ln\left(\dfrac{r_e}{r_w}\right)}$ represents the steady-state radial flow of a fluid in a cylindrical porous medium. What are the analogous terms describing the heat and current flows in similar cylindrical systems? Identify the analogous terms and/or groups for

- Current flow: Ohm's law, $I = \dfrac{1}{R} \Delta E$.

- Heat Flow: $Q = \dfrac{KA\Delta T}{\Delta L}$.

Solution to Problem 1.2.5

Eqs. 1.6 through 1.8 express the analogous terms of Ohm's law:

$$I \sim q, \dotfill (1.6)$$

$$\frac{1}{R} \sim \frac{2\pi\beta_c hk}{\mu \ln\left(\dfrac{r_e}{r_w}\right)}, \dotfill (1.7)$$

$$\Delta E \sim \Delta p. \dotfill (1.8)$$

Eqs. 1.9 through 1.11 express the analogous terms of the heat flow equation:

$$Q \sim q, \dotfill (1.9)$$

$$\frac{KA}{\Delta L} \sim \frac{2\pi\beta_c hk}{\mu \ln\left(\dfrac{r_e}{r_w}\right)}, \dotfill (1.10)$$

$$\Delta T \sim \Delta p. \dotfill (1.11)$$

Problem in (Ertekin et al. 2001)

Comment on the accuracy of this statement: *The material-balance equation is considered a zero-dimensional model because time dependency is not incorporated into it.*

Solution to Problem 1.2.6

The statement is FALSE. The material-balance equation is a zero-dimensional model because it does not consider *spatial* and *temporal* variations of rock and fluid properties.

Problem 1.2.7

Comment on the accuracy of this statement: *Specifying constant flow rate as a boundary condition is equivalent to specifying the pressure gradient at the sandface.*

Solution to Problem 1.2.7

The statement is TRUE. A close inspection of Eq. 1.1b reveals that when flow rate q is specified at $r = r_w$, then

$$\left(\frac{\partial p}{\partial r}\right)_{r=r_w} = \frac{q\mu}{2\pi khr_w}. \dotfill (1.12)$$

Eq. 1.12 shows that pressure gradient at the wellbore is fixed because all the terms on the right-hand side of the equation are fixed.

Problem 1.2.8

Comment on the accuracy of this statement: *An infinite-acting assumption of classical well test analysis theory does not hold true for small reservoirs.*

Solution to Problem 1.2.8

The statement is not *necessarily* true. No matter how small the reservoir, there will be a period for a 'measurable' pressure transient to reach the boundary. Therefore, it is accurate to assume that during that period, the reservoir is infinite acting.

Nomenclature

A = cross-sectional area, L^2, ft^3

c = compressibility, Lt^2/m, psi^{-1}

h = thickness, L, ft

I = electrical current, q/t, A, and boundary conditions (external and internal)

k = permeability, L^2, darcies

K = heat conductivity, $ML/t^3/T$, Btu/hr/ft/°F

M = mathematical representation of a fluid flow problem

O = a time-series representation of the pressure surface of a production surface

p = pressure, m/Lt^2, psia

p_i = initial pressure, m/Lt^2, psia

q = production rate or flow rate, L^3/t, STB/D

q_ψ = production and/or pressure characteristics as function of time (time series)

Q = rate of heat transfer, m/t^3T, Btu/hr

r = distance from the wellbore, L, ft

r_e = radius of external boundary, L, ft

r_w = well radius, L, ft

R = electrical resistance, mL^3/tq^2, Ω

S = a forcing function applied to the system

t = time, days

β_c = transmissibility conversion factor, with a magnitude of 1.127

ΔE = voltage difference, V

ΔL = change of length, L, ft

Δp = pressure difference, m/Lt^2, psia

ΔT = temperature difference, °R

μ = viscosity, cp

ϕ = porosity, fraction

Φ = flow domain characteristics

ψ = flow process including all the thermodynamics of the reservoir system

References

Ertekin, T., Abou-Kassem, J. H., and King, G. R. 2001. *Basic Applied Reservoir Simulation*. Richardson, Texas, US: Society of Petroleum Engineers.

Ertekin, T., King, G. R., and Schwerer, F. S. 1986. Dynamic Gas Slippage: A Unique Dual-Mechanism Approach to the Flow of Gas in Tight Formations. *SPE Form Eval* **1**(01): 43–52. SPE-12045-PA. http://dx.doi.org/10.2118/12045-PA.

Chapter 2

Reservoir Rock and Fluid Properties and Basic Reservoir Engineering Concepts

A reservoir simulator's principal goal is to mimic and predict a reservoir's response to an excitation function imposed on it by the existing wells (producers and/or injectors). The magnitude of the reservoir's response is determined by the reservoir rock and fluid properties, as well as by certain structural features of the reservoir. Accordingly, while the reservoir rock and fluid properties can be considered bricks of a reservoir simulator, the reservoir engineering concepts work as the mortar that holds everything together in an orderly manner.

2.1 Reservoir Rock Properties
The most fundamental rock properties that appear in reservoir simulation equations are

- Porosity (ϕ), which is a property that defines the storage capacity of the rock.
- Permeability (k), which is a property that describes the reservoir's ability to transmit fluids.

While porosity is a dimensionless quality (dimensionless volume), the dimension of permeability is length squared (L^2). The permeability property of reservoir rock can exhibit directional dependency. If the permeability of a rock varies with direction, then the reservoir is anisotropic with respect to the permeability property. On the other hand, if the permeability of a reservoir is the same regardless of the direction in which it is measured, then that reservoir exhibits isotropic characteristics with respect to permeability. It should be noted that only those properties that are not volume based can exhibit directional dependency (in a 3D space). Permeability has the dimension of area, leaving one additional direction – which can vary – in expressing the area. However, porosity is a volume-based property; because it has all three dimensions, zero degrees of freedom are left in terms of directional variation. Therefore, it would be incorrect to call a reservoir's porosity distribution isotropic or anisotropic, because the porosity cannot exhibit any directional dependency.

Reservoir property distribution is also classified as homogeneous or heterogeneous. If a reservoir property is uniformly distributed in space, then the reservoir is homogeneous with respect to that property. In contrast, heterogeneous reservoir systems exhibit nonuniform spatial distribution. In orthogonal flow geometry, principal flow direction will be aligned along the maximum and minimum permeability values for an anisotropic medium. Therefore, it is important to place the coordinate system parallel to the principal flow directions so that the six off-diagonal entries of the permeability tensor can be eliminated for a 3D representation. Similarly, in a 2D system, two off-diagonal entries of the permeability tensor can be eliminated when principal flow dimensions and coordinate system directions coincide.

2.2 Reservoir Fluid Properties
The other important class of parameters in fluid flow dynamics in porous media is the fluid properties. Compared with rock properties, fluid properties exhibit much more dependency on pressure and temperature (in isothermal applications, temperature dependency does not come into the picture).

2.2.1 Gas Properties. Although a number of equations of state exists in calculating gas properties such as density, compressibility, and formation volume factor (FVF), the real-gas law is used as the basis. **Table 2.1** summarizes the equations we typically use in expressing gas-flow dynamics.

2.2.2 Oil Properties. In describing the flow dynamics of the liquid-hydrocarbon phase, properties such as density, FVF, viscosity, and the solubility of gas in oil will be needed. If the reservoir contains dead oil (no gas in solution), then the oil phase

Real-Gas Law	$pV = z_g nRT$
Gas Density	$\rho = \dfrac{pMW}{z_g RT}$
Gas Compressibility	$c = \dfrac{1}{p} - \dfrac{1}{z_g}\dfrac{\partial z_g}{\partial p}$
Compressibility Factor	$z_g = f(p,T)$
Formation Volume Factor	$B_g = \dfrac{z_g T p_{sc}}{T_{sc} p}$
Viscosity	$\mu = f(p,T)$

Table 2.1—Equations important in gas-property calculations.

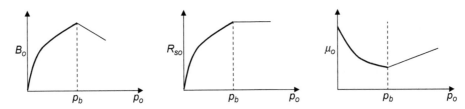

Fig. 2.1—Variation of oil properties with pressure.

can be treated as slightly compressible. Otherwise, when gas is dissolved in the liquid oil phase, it is necessary to use correlations and/or experimental data. **Fig. 2.1** shows typical variations of liquid-hydrocarbon properties.

Note that during a typical simulation study, a given section of the reservoir may traverse from undersaturated to saturated or from saturated to undersaturated conditions at different pressures. Most reservoir simulators implement variable bubblepoint protocols to handle such computationally challenging situations—challenging because of the discontinuity of the property at the bubblepoint pressure (i.e., a derivative of the property will have two values at the bubblepoint, depending on from which direction the bubblepoint is approached).

2.2.3 Water Properties. Again, pressure dependency of water properties is determined by the solubility of hydrocarbon gas in water. Because gas solubility in water is significantly smaller compared with oil, it is a common practice to assume that $B_w = 1.0$ res bbl/STB, $R_{sw} = 0.0$ scf/STB, and $\mu_w = $ the constant within the pressure ranges of interest.

2.3 Reservoir Rock and Fluid Interactions
2.3.1 Wettability and Interfacial Tension. When immiscible fluids such as oil and water co-exist in reservoir pore space, their individual interactions with the rock grains (i.e., the walls of the pore space) control their distribution as well as their movement in the porous structure. The two principal properties used to quantify these interactions are wettability (rock/fluid interactions) and interfacial tension (fluid/fluid interactions). In the case of rock/fluid interactions, it can be stated that in the presence of two different fluids, the fluid that adheres to the rock surface is called the wetting phase and is assigned a higher wettability index. The wetting affinity of a rock is measured through the contact angle, θ_c. The parameter that determines the wettability index is called adhesion tension. Interfacial tension is a measure of the surface energy per unit area of the interface between two immiscible fluids, and it has units of force per length.

2.3.2 Relative Permeability. When two or more immiscible fluids simultaneously flow in the same pore space, they compete against each other. By definition, relative permeability is the ratio of the effective permeability (i.e., the permeability to a fluid in the presence of another fluid). Because the effective permeability to a flowing fluid phase is directly related to the saturation of that phase, relative permeability is also a function of saturation. For example, for a two-phase system, $S_o + S_w = 1.00$:

$$k_{ro}(S_w) = \frac{k_{o_{effective}}(S_w)}{k_{abs}}, \dotfill (2.1a)$$

$$k_{rw}(S_w) = \frac{k_{w_{effective}}(S_w)}{k_{abs}}. \dotfill (2.1b)$$

Typically, $k_{ro} + k_{rw} < 1$ at any S_o and S_w combination, as shown in **Fig. 2.2.**

For three-phase-flow conditions (oil/water/gas), the relative permeability characteristics for each phase are described as follows (at any combination of S_o, S_w, and S_g):

$$k_{rw} = f(S_w), \dotfill (2.2a)$$

$$k_{rg} = f\left(S_g\right), \quad \dots\dots\dots\dots\dots\dots\dots \quad (2.2b)$$

$$k_{ro} = f\left(S_w, S_g\right). \quad \dots\dots\dots\dots\dots\dots \quad (2.2c)$$

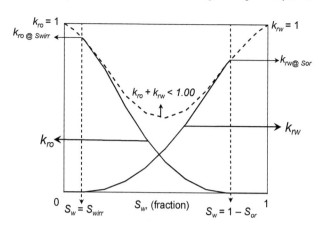

Note that Eq. 2.2c is not stated as $k_{ro} = f(S_o)$, because at the same S_o value there are infinite possible combinations of S_w and S_g values. Therefore, k_{ro} is described as the relative permeability to oil phase at particular water and gas saturations (in other words, at a particular value of oil saturation when $S_o = (1 - S_w - S_g)$.

2.3.3 Capillary Pressure. Capillary pressure exists whenever pores (capillaries) are saturated with two or more phases. In a two-phase system (oil/water), capillary pressure is, by definition,

Fig. 2.2—Typical relative permeability curves for two-phase (oil/water) flow.

$$P_{cow}\left(S_w\right) = p_o - p_w. \quad \dots\dots\dots\dots\dots\dots \quad (2.3a)$$

In Eq. 2.3a, it is assumed that water phase is the wetting phase, because the pressure in the nonwetting phase is larger than the pressure in the wetting phase, $p_o > p_w$, resulting in a positive p_{cow} value. Similarly, for a two-phase oil/gas system (oil is the wetting phase),

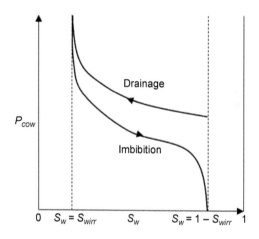

$$P_{cgo}(S_g) = p_g - p_o. \quad \dots\dots\dots\dots\dots\dots \quad (2.3b)$$

Fig. 2.3 shows the functional dependence of P_{cow} on water saturation, as well as saturation history (recall that imbibition is a process during which the wetting-phase saturation increases and, hence, in a drainage process, the wetting-phase saturation decreases).

Fig. 2.3—Capillary pressure in oil/water system (water-wet).

2.3.4 Microscopic Rock and Fluid Properties. Expensive and sophisticated measurement techniques and instrumentation have been developed for measuring and determining reservoir rock and fluid properties, including contact angle, interfacial tension, surface tension, specific surface area, and pore-throat diameter. It is noteworthy to recognize that all of these microscopic properties do not appear in transport equations of porous media that macroscopically describe the flow. The question that naturally comes to mind, then, is why so much investment is made to measure these rather exotic properties. In reality, the effects of these microscopic properties are carried into the macroscopic-transport equations of flow in porous media through some functional relationships, such as relative permeability characteristics and capillary pressure characteristics. In other words, effects of these microscopic properties are all captured by these two functions, and when they are implemented into the flow equations, their task is to transmit the relevant information to the transport equations of flow in porous media.

2.4 Basic Reservoir Engineering Concepts

2.4.1 Incompressible Fluid. For an incompressible fluid, density and viscosity are constant (independent of pressure), and FVF is equal to unity. This implies that compressibility of an incompressible fluid is zero ($c = 0$ psi^{-1}).

2.4.2 Slightly Compressible Fluid. For a slightly compressible fluid, density, viscosity, and FVF exhibit weak dependence on pressure. This implies that compressibility of a slightly compressible fluid within the pressure range of interest is assumed to be treated as constant (c is constant with values in the range of 10^{-7} to 10^{-8} psi^{-1}).

2.4.3 Compressible Fluid. For a compressible fluid, density, viscosity, FVF, compressibility, and compressibility factors are strong functions of pressure, and they introduce strong nonlinearities to the flow equations written for this class of fluids (real gases). For a real gas, Eq. 2.4 explicitly shows how compressibility of the gas changes with pressure:

$$c_g = \frac{1}{p} - \frac{1}{z_g}\frac{\partial z_g}{\partial p}. \quad \dots\dots\dots\dots\dots\dots\dots\dots\dots\dots\dots\dots\dots\dots\dots \quad (2.4)$$

2.4.4 Steady-State Flow. The steady-state flow regime is defined when pressure at every location within the reservoir domain does not change with time (note that this statement is valid for single-phase flow conditions). Mathematically, we can express this condition as Eq. 2.5:

$$\left(\frac{\partial p}{\partial t}\right)_{x,y,z} = 0. \quad\dotfill\quad (2.5)$$

Eq. 2.5 states that the rate of change of pressure, p, with respect to time, t, at any location is zero. For the steady-state flow condition to occur in a reservoir when a certain mass of fluid is taken out of the pore spaces, the same amount of mass must enter the reservoir and be instantaneously distributed over the entire pore space. In other words, the mass content of the pores at every point in the reservoir remains the same all the time.

2.4.5 Unsteady-State Flow. The unsteady-state flow, which is more commonly referred to as transient flow, implies that within the flow domain, the rate of change of pressure with respect to time, at any position in the reservoir, is not zero or uniform. In other words, change of pressure with time is expressed as

$$\left(\frac{\partial p}{\partial t}\right)_{x,y,z} = f(x,y,z,t). \quad\dotfill\quad (2.6)$$

2.4.6 Pseudosteady-State Flow. Pseudosteady-state flow regime takes place during the late-time period when the pressure transients contact a reservoir's all-no-flow outer boundaries (i.e., when there is a measurable pressure drop at the boundaries). Under pseudosteady-state conditions, the pressure throughout the reservoir decreases at the same constant rate. Such a condition can be expressed as Eq. 2.7:

$$\left(\frac{\partial p}{\partial t}\right)_{x,y,z} = \text{constant for all values of times.} \quad\dotfill\quad (2.7)$$

2.4.7 Flow-Potential Concept. The flow-potential concept defined by Hubbert (1940) is usually referred to as Hubbert's potential. **Fig. 2.4** shows the sign convention adopted to define flow potential. In this sign notation, positive direction is considered to be downward from the datum (usually sea level). The flow potential is defined as

$$\Phi = p - \frac{1}{144}\frac{g}{g_c}\rho G. \quad\dotfill\quad (2.8)$$

Eq. 2.8 is written in practical field units, and the $\frac{g}{g_c}$ coefficient can be taken as equal to unity (i.e., the ratio of local gravitational constant to universal gravitational constant) for simplification purposes. Accordingly, flow potential at Points 1 and 2 of the reservoir shown in Fig. 2.4 can be written in the forms of Eqs. 2.9 and 2.10, respectively (G_1 and G_2 are distances to Point 1 and Point 2, respectively, from the datum):

$$\Phi_1 = p_1 - \frac{1}{144}\frac{g}{g_c}\rho G_1, \quad\dotfill\quad (2.9)$$

$$\Phi_2 = p_2 - \frac{1}{144}\frac{g}{g_c}\rho(-G_2). \quad\dotfill\quad (2.10)$$

If pore-pressure values at Points 1 and 2 in Fig. 2.4 are equal to each other ($p_1 = p_2 = p$), then

$$\Phi_1 = p - \frac{1}{144}\frac{g}{g_c}\rho G_1, \quad\dots\quad (2.11)$$

$$\Phi_2 = p - \frac{1}{144}\frac{g}{g_c}\rho(-G_2). \quad\dots\quad (2.12)$$

Eqs. 2.11 and 2.12 clearly indicate that $\Phi_2 > \Phi_1$ when $p_1 = p_2$.

The gradient of flow potential along the arbitrary S direction [$S = S(x, y, z)$] can be written as Eq. 2.13a:

$$\frac{\partial \Phi}{\partial S} = \frac{\partial p}{\partial S} - \frac{1}{144}\rho\frac{\partial G}{\partial S}. \quad\dots\quad (2.13a)$$

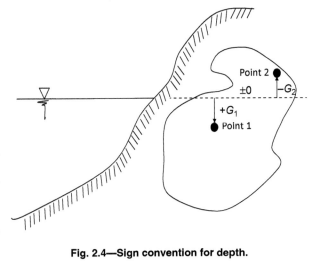

Fig. 2.4—Sign convention for depth.

Eq. 2.13a can be written in gradient form as Eq. 2.13b:

$$\nabla\Phi = \nabla p - \gamma\nabla G, \dots\dots\dots\dots\dots\dots\dots\dots\dots\dots\dots\dots\dots\dots\dots\dots\dots\dots\dots\dots\dots (2.13b)$$

where γ is the fluid density defined as pressure per distance in Eq. 2.13c:

$$\gamma = \frac{1}{144}\rho. \dots\dots\dots\dots\dots\dots\dots\dots\dots\dots\dots\dots\dots\dots\dots\dots\dots\dots\dots\dots\dots\dots\dots (2.13c)$$

2.4.8 Darcy's Law. Darcy's law (Darcy 1856) is a fundamental equation that describes the flow of a fluid through a porous medium. Darcy's law is central to the computations made in fields such as hydrogeology and petroleum and natural gas engineering. It essentially distinguishes the flow in porous media from flow dynamics in other flow fields. It serves as the constitutive equation that describes the relationship between the pressure field and the velocity field (Eq. 2.14):

$$v_s = \frac{q_s}{A_s} = -\frac{k_s}{\mu}\frac{\partial\Phi}{\partial S}. \dots\dots\dots\dots\dots\dots\dots\dots\dots\dots\dots\dots\dots\dots\dots\dots\dots\dots\dots (2.14)$$

In Eq. 2.14, S is an arbitrary direction. Darcy's law states that volumetric flow rate, q_s, is directly proportional to the area perpendicular to flow, to the permeability along the flow direction (k_S), and to the potential gradient along the flow direction $\frac{\partial\Phi}{\partial S}$, and it is inversely proportional to the viscosity of the fluid. Although, Henry Darcy empirically derived the relationship in 1856 (Darcy 1856), after one century it has been proved that Darcy's law can be derived from the Navier-Stokes equation (Hubbert 1956). **Table 2.2** summarizes the units of entries in Darcy's law for gas flow and liquid flow.

2.4.9 Conservation of Mass. Conservation of mass is a fundamental concept of physics, along with the conservation of energy and the conservation of momentum. The partial-differential equations (PDEs) governing the fluid flow in a continuum domain are fundamentally based on the conservation-of-mass principle under the assumption of isothermal conditions. The conservation-of-mass principle simply states in Eq. 2.15 that over a finite period of time,

$$(mass\ in) - (mass\ out) = (net\ change\ in\ mass). \dots\dots\dots\dots\dots\dots\dots\dots\dots\dots\dots\dots\dots\dots (2.15)$$

Darcy Units	Field Units (Gas Flow)	Field Units (Liquid Flow)
$q = -\dfrac{Ak}{\mu}\dfrac{\Delta p}{\Delta L}$	$q = -6.328\dfrac{Ak}{\mu}\dfrac{\Delta p}{\Delta L}$	$q = -1.127\dfrac{Ak}{\mu}\dfrac{\Delta p}{\Delta L}$
$v = -\dfrac{k}{\mu}\dfrac{\Delta p}{\Delta L}$	$v = -6.328\dfrac{k}{\mu}\dfrac{\Delta p}{\Delta L}$	$v = -1.127\dfrac{k}{\mu}\dfrac{\Delta p}{\Delta L}$
$q\left[\dfrac{L^3}{t}\right]$: cm³/sec	$q\left[\dfrac{L^3}{t}\right]$: ft³/D	$q\left[\dfrac{L^3}{t}\right]$: B/D
k [L²]: darcy	k [L²]: darcy	k [L²]: darcy
A [L²]: cm²	A [L²]: ft²	A [L²]: ft²
$\Delta p\left[\dfrac{m}{Lt^2}\right]$: atm	$\Delta p\left[\dfrac{m}{Lt^2}\right]$: psi	$\Delta p\left[\dfrac{m}{Lt^2}\right]$: psi
$\mu\left[\dfrac{m}{Lt}\right]$: cp	$\mu\left[\dfrac{m}{Lt}\right]$: cp	$\mu\left[\dfrac{m}{Lt}\right]$: cp
ΔL [L]: cm	ΔL [L]: ft	ΔL [L]: ft
$v\left[\dfrac{L}{t}\right]$: cm / sec	$v\left[\dfrac{L}{t}\right]$: ft / D	$v\left[\dfrac{L}{t}\right]$: B / D − ft²
Note (1): 1 darcy ≈ 10⁻⁸ cm²	Note (1): 1 darcy = 1.127 perms	Note (1): 1 darcy = 1.127 perms
	Note (2): if k is in perms,	Note (2): if k is in perms,
	$q = -5.615\dfrac{Ak}{\mu}\dfrac{\Delta p}{\Delta L}$	$q = -\dfrac{Ak}{\mu}\dfrac{\Delta p}{\Delta L}$
	$v = -5.615\dfrac{k}{\mu}\dfrac{\Delta p}{\Delta L}$	$v = -\dfrac{k}{\mu}\dfrac{\Delta p}{\Delta L}$

Table 2.2—Darcy's law and its units.

Note that in Eq. 2.15, we adopted a sign convention such that (*mass out* ≡ *production*) is a negative quantity, and (*mass in* ≡ *injection*) is a positive quantity. Accordingly, if net change in mass is negative, then it represents depletion of mass in the porous space, and if it is positive, we know that mass content is increased.

Fig. 2.5 represents a summary of the reservoir simulation, how reservoir models are assembled, and what kind of reservoir engineering concepts and reservoir rock and fluid properties are fused together to develop the reservoir simulation formalisms.

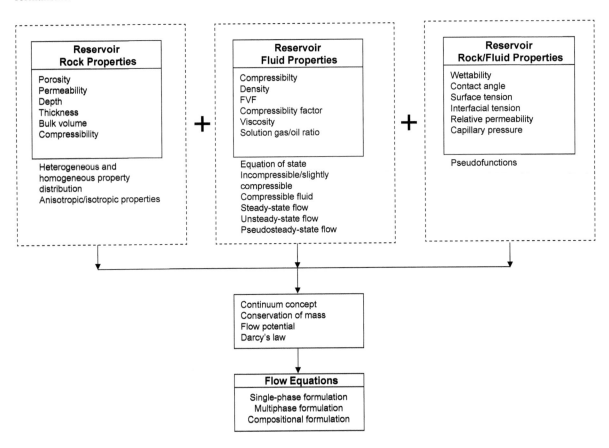

Fig. 2.5—Assembling a porous-media flow formulation.

2.5 Problems and Solutions

Problem 2.5.1 (Ertekin et al. 2001)

Consider the system shown in **Fig. 2.6.** Prove that equations $\Phi_A - \Phi_B = (p_A - p_B) - \gamma(G_A - G_B)$ and $\Phi - \Phi^0 = (p - p^0) - \gamma G$ hold true if the arbitrary new datum falls below the absolute datum.

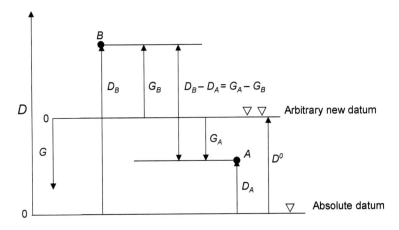

Fig. 2.6—Schematic representation of Problem 2.5.1.

Solution to Problem 2.5.1

Using the new datum as reference level,

$$\Phi_A - \Phi^0 = (p_A - p^0) - \gamma G_A, \dotfill (2.16)$$

$$\Phi_B - \Phi^0 = (p_B - p^0) - \gamma G_B, \dotfill (2.17)$$

which yields

$$\Phi_A - \Phi_B = (p_A - p_B) - \gamma(G_A - G_B), \dotfill (2.18)$$

which is true for any two depths. Thus, replacing Points A and B with any potential depth and new datum depth, one obtains

$$\Phi - \Phi^0 = (p - p^0) - \gamma G. \dotfill (2.19)$$

Problem 2.5.2 (Ertekin et al. 2001)

Fig. 2.7 shows a water reservoir dipping with a 15° angle. Points A, B, and C fall on the dip and are separated from each other, as indicated in the figure. The water density at reservoir conditions is 60 lbm/ft³. The pressure at Point B is 2,000 psia, and water is under hydrodynamic equilibrium at the time of discovery. What are the pressures at Points A and C?

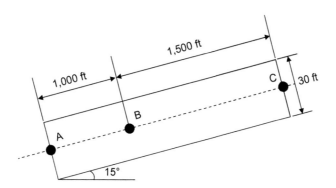

Fig. 2.7—Schematic representation of Problem 2.5.2.

Solution to Problem 2.5.2

Depth difference along the vertical directions:

$$G_{AB} = -1,000 \times \sin(15°) = -258.8 \text{ ft}, \dotfill (2.20)$$

$$G_{CB} = 1,500 \times \sin(15°) = 388.2 \text{ ft}. \dotfill (2.21)$$

Gravitational potential:

$$\gamma = \frac{\rho}{144 \, \text{lbm/ft}^2 - \text{psi}} = \frac{60}{144} = 0.41667 \frac{\text{psi}}{\text{ft}}. \dotfill (2.22)$$

Therefore,

$$p_A = p_B - \gamma G_{AB} = 2,000 - 0.4166 \times (-258.8) = 2,107.8 \text{ psia}, \dotfill (2.23)$$

$$p_C = p_B - \gamma G_{BC} = 2,000 - 0.4166 \times (388.2) = 1,838.3 \text{ psia}. \dotfill (2.24)$$

Problem 2.5.3 (Ertekin et al. 2001)

Consider the reservoir system in Problem 2.5.2: the water viscosity at reservoir conditions is 0.95 cp, the FVF of water is 1 res bbl/STB, and the permeability of the formation is 300 md. What is the rate of water flow between Points A and C if the formation width along the strike is 950 ft, and the pressures at Points A and C are maintained at 1,500 and 1,300 psia, respectively?

Solution to Problem 2.5.3

The depth difference is

$$G_{AC} = L \sin 15° = 2,500 \times 0.2588 = 647.05 \text{ ft.} \quad \dots \dots \dots \dots \dots \dots \dots \dots \dots \dots \dots \dots \dots \dots \dots \dots \dots (2.25)$$

To calculate the flow rate, we assume the direction from A to C is a positive direction; therefore,

$$q = -\frac{Ak}{\mu BL}\left(p_A - p_C - \frac{1}{144}\rho G_{AC}\right)$$

$$= -\frac{(30)(950)(300)(1.127 \times 10^{-3})}{(0.95)(1)(2,500)}\left[1,500 - 1,300 - \frac{60}{144}(647.05)\right] = 282.4 \text{ STB/D.} \quad \dots \dots \dots \dots \dots (2.26)$$

The sign change in this case indicates that the flow direction is from C to A.

Problem 2.5.4 (Ertekin et al. 2001)

A reservoir rock sample has a porosity of 0.18 measured at 14.7 psia. The rock compressibility is 0.77×10^{-5} psi^{-1}. What would the rock porosity be at 2,000 and at 3,000 psia? Plot the relationship between ϕ and p in a pressure range of 14.7 to 5,000 psia.

Solution to Problem 2.5.4

Knowing that $\phi = \phi^0\left[1 + c_\phi\left(p - p^0\right)\right]$, one can calculate the porosity values at 2,000 and 3,000 psi.

In **Fig. 2.8,** we show the porosity variation from 14.7 to 5,000 psi in a linear fashion, as dictated by the equation previously given:

$$\phi_{2,000} = 0.18\,[1 + 0.77 \times 10^{-5}\,(2,000 - 14.7)] = 0.18275, \dots \dots \dots \dots \dots \dots \dots \dots \dots \dots \dots \dots \dots \dots (2.27)$$

$$\phi_{2,000} = 0.18\,[1 + 0.77 \times 10^{-5}\,(3,000 - 14.7)] = 0.184114. \dots \dots \dots \dots \dots \dots \dots \dots \dots \dots \dots \dots \dots (2.28)$$

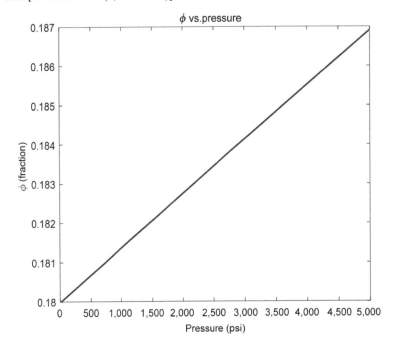

Fig. 2.8—Solution to Problem 2.5.4.

Problem 2.5.5 (Ertekin et al. 2001)

Obtain a flow-rate expression for the equivalent of a linear flow that takes place between Blocks 1 and 2 in **Fig. 2.9.** Note that the two blocks have different dimensions and permeabilities, the flow rates along the flow direction within Blocks 1 and 2 are equal, and $p_1 > p_2$.

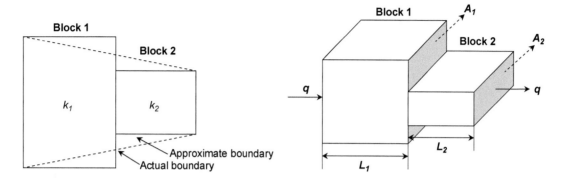

Fig. 2.9—Schematic representation of Problem 2.5.5.

Solution to Problem 2.5.5

On the basis of $q = \left(\dfrac{Ak}{L}\right)_{12} \dfrac{1}{\mu B} \Delta\Phi$, the term $\left.\dfrac{Ak}{L}\right|_{12}$ represents the overall conductivity of the system at the interface of of Blocks 1 and 2 in Fig. 2.9. Using the analogy between Ohm's law and Darcy's law, one can write the total resistivity of the system as

$$\frac{1}{\left.\dfrac{Ak}{L}\right|_{12}} = \frac{1}{\dfrac{A_1 k_1}{L_1}} + \frac{1}{\dfrac{A_2 k_2}{L_2}} = \frac{L_1}{A_1 k_1} + \frac{L_2}{A_2 k_2} = \frac{A_1 k_1 L_2 + A_2 k_2 L_1}{A_1 k_1 A_2 k_2}, \dotfill (2.29)$$

which simplifies to

$$\left.\frac{Ak}{L}\right|_{12} = \frac{A_1 k_1 A_2 k_2}{A_1 k_1 L_2 + A_2 k_2 L_1}. \dotfill (2.30)$$

Then, the final flow-rate equation will be

$$q = \frac{A_1 k_1 A_2 k_2}{A_1 k_1 L_2 + A_2 k_2 L_1} \frac{1}{\mu B} \Delta\Phi. \dotfill (2.31)$$

Problem 2.5.6 (Ertekin et al. 2001)

Use the Dranchuk and Abou-Kassem (1975) z-factor correlation to plot z and c_r vs. p_{pr} in the pseudoreduced pressure range of 1 to 10 for $T_{pr} = 1.1$ and 1.5.

Solution to Problem 2.5.6

The z-factor and c_r vs. p_{pr} figure can be found in **Fig. 2.10,** and the referenced tabulated data are listed in **Table 2.3.**

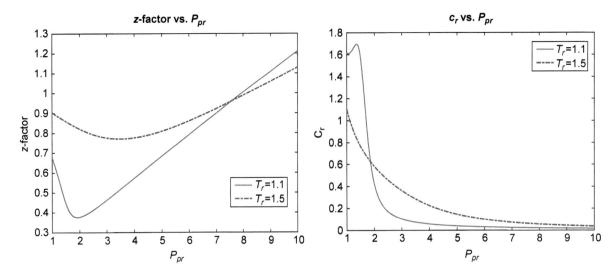

Fig. 2.10—Variation of z-factor and c_r with reduced pressure (Problem 2.5.6).

p_{pr}	z_g		c_r	
	$T_{pr}= 1.1$	$T_{pr}= 1.5$	$T_{pr}= 1.1$	$T_{pr}= 1.5$
1	0.67737	0.9034	1.61504	1.10229
2	0.37634	0.82147	0.41886	0.58194
3	0.46351	0.77613	0.10323	0.36124
4	0.57296	0.77626	0.05709	0.22515
5	0.68318	0.80913	0.03949	0.14558
6	0.79206	0.86059	0.03025	0.09986
7	0.89927	0.92207	0.02456	0.07266
8	1.00486	0.98895	0.0207	0.05561
9	1.10893	1.05072	0.01791	0.04433
10	1.2116	1.13002	0.0158	0.03649

Table 2.3—Tabulated data for Solution to Problem 2.5.6.

Problem 2.5.7 (Ertekin et al. 2001)

Table 2.4 lists the properties of gas and saturated oil at reservoir temperature. Calculate the oil- and gas-phase properties $(B, \mu, \text{and } \rho)$ at the following reservoir conditions when $\rho_{osc} = 49.098$ lbm/ft^3, gas molecular weight (MW) = 22.94 lbm/lbm mol, $c_o = 23.2 \times 10^{-6}$ psi^{-1} (compressibility to be used in density calculations), $c_\mu = 46 \times 10^{-6}$ psi^{-1} (compressibility to be used in viscosity calculations), $p_{sc} = 14.7$ psia, and $T_{sc} = 520$ °R:

1. $p = 3{,}014.7$ psia and $R_s = 930$ scf/STB.
2. $p = 3{,}014.7$ psia and $R_s = 800$ scf/STB.

Solution to Problem 2.5.7

1. $p = 3{,}014.7$ psia and $R_s = 930$ scf/STB: Oil phase is saturated. From Table 2.4, we get

$$B_o = 1.565 \frac{\text{res bbl}}{\text{STB}}, \; \mu_o = 0.594 \text{ cp}, \rho_o = \frac{\rho_{osc}}{B_o} = \frac{49.098}{1.565} = 31.37 \text{ lbm/ft}^3,$$

$$B_g = 0.00108 \text{ res bbl/scf}, \mu_g = 0.0228 \text{ cp},$$

$$\rho_g = \frac{\rho_{gsc}}{\alpha_c B_g} = \frac{p_{sc}(MW)}{RT_{sc}\alpha_c B_g} = \frac{(14.7)(22.94)}{(10.73)(520)(5.615)(0.00108)} = 9.966 \text{ lbm/ft}^3. \quad \dots \dots \dots \dots \dots (2.32)$$

2. $p = 3{,}014.7$ psia and $R_s = 800$ scf/STB: From Table 2.4, we get the same result as in Part 1 for the gas properties:

$$B_g = 0.00108 \text{ res bbl/scf}, \mu_g = 0.0228 \text{ cp},$$

$$\rho_g = \frac{\rho_{gsc}}{\alpha_c B_g} = \frac{p_{sc}(MW)}{RT_{sc}\alpha_c B_g} = \frac{(14.7)(22.94)}{(10.73)(520)(5.615)(0.00108)} = 9.966 \text{ lbm/ft}^3. \dots\dots\dots\dots\dots (2.33)$$

Because $R_s = 800$ scf/STB and is less than 930 scf/STB (data from the table), oil phase is undersaturated. One must apply linear interpolation to calculate the bubblepoint:

$$p_b = 2{,}014.7 + \frac{3{,}014.7 - 2{,}014.7}{930 - 636} \times (800 - 636) = 2{,}572.5 \text{ psia}, \dots\dots\dots\dots\dots\dots\dots (2.34)$$

$$B_b = 1.435 + \frac{1.565 - 1.435}{930 - 636} \times (800 - 636) = 1.5075 \text{ res bbl/STB}, \dots\dots\dots\dots\dots\dots (2.35)$$

$$\mu_b = 0.695 + \frac{0.594 - 0.695}{930 - 636} \times (800 - 636) = 0.639 \text{ cp}. \dots\dots\dots\dots\dots\dots\dots\dots (2.36)$$

Therefore,

$$B_o = B_b\left[1 - c_o(p - p_b)\right] = 1.5075\left[1 - 23.2 \times 10^{-6}(3{,}014.7 - 2{,}572.5)\right] = 1.492 \text{ res bbl/STB}, \dots\dots (2.37)$$

$$\mu_o = \frac{\mu_b}{\left[1 - c_\mu(p - p_b)\right]} = \frac{0.639}{\left[1 - 46 \times 10^{-6}(3{,}014.7 - 2{,}572.5)\right]} = 0.652 \text{cp}. \dots\dots\dots\dots (2.38)$$

$$\rho_o = \frac{\rho_{osc}}{B_o} = \frac{49.098}{1.492} = 32.91 \text{ lbm/ft}^3. \dots\dots\dots\dots\dots\dots\dots\dots\dots\dots\dots\dots\dots (2.39)$$

p (psia)	Oil				Gas	
	R_s (scf/STB)	B_o (res bbl/STB)	μ_o (cp)		B_g (res bbl/scf)	μ_g (cp)
14.70	1.00	1.062	1.040		1.66667	0.0080
514.70	180.00	1.207	0.910		0.00627	0.0112
1,014.70	371.00	1.295	0.830		0.00320	0.0140
2,014.70	636.00	1.435	0.695		0.00161	0.0189
3,014.70	930.00	1.565	0.594		0.00108	0.0228
4,014.70	1,270.00	1.695	0.510		0.00081	0.0268
5,014.70	1,618.00	1.827	0.449		0.00065	0.0309

Table 2.4—Properties of gas and saturated oil at reservoir temperature.

Problem 2.5.8 (Ertekin et al. 2001)

Using the information from Problem 2.5.7, calculate the oil- and gas-phase properties for the following reservoir conditions. Use linear interpolation for pressure entries not listed in Table 2.4 to obtain the values of the properties.

1. $p = 4{,}514.7$ psia and $R_s = 1{,}444$ scf/STB.
2. $p = 4{,}514.7$ psia and $R_s = 1{,}000$ scf/STB.

Solution to Problem 2.5.8

1. $p = 4{,}514.7$ psia and $R_s = 1{,}444$ scf/STB.

Applying linear interpolation, one obtains

$$R_s = 1{,}270 + \frac{1{,}618 - 1{,}270}{5{,}014.7 - 4{,}014.7} \times (4{,}514.7 - 4{,}014.7) = 1{,}444 \text{ scf/STB}, \dots\dots\dots\dots\dots (2.40)$$

which indicates that the oil phase is saturated.

From Table 2.4, for oil,

$$B_o = 1,695 + \frac{1,827 - 1,695}{5,014.7 - 4,014.7} \times (4,514.7 - 4,014.7) = 1,761 \text{ res bbl/STB,} \quad\dots\dots\dots\dots\dots\dots \text{(2.41)}$$

$$\mu_o = 0.51 + \frac{0.449 - 0.51}{5,014.7 - 4,014.7} \times (4,514.7 - 4,014.7) = 0.4795 \text{ cp,} \quad\dots\dots\dots\dots\dots\dots \text{(2.42)}$$

$$\rho_o = \frac{\rho_{osc}}{B_o} = \frac{49.098}{1.761} = 27.88 \text{ lbm/ft}^3; \quad\dots\dots\dots\dots\dots\dots\dots\dots \text{(2.43)}$$

and for gas,

$$B_g = 0.00081 + \frac{0.00065 - 0.00081}{5,014.7 - 4,014.7} \times (4,514.7 - 4,014.7) = 0.00073 \text{ res bbl/STB,} \quad\dots\dots\dots\dots \text{(2.44)}$$

$$\mu_g = 0.0268 + \frac{0.0309 - 0.0268}{5,014.7 - 4,014.7} \times (4,514.7 - 4,014.7) = 0.02885 \text{ cp,} \quad\dots\dots\dots\dots\dots \text{(2.45)}$$

$$\rho_g = \frac{\rho_{gsc}}{\alpha_c B_g} = \frac{p_{sc} MW}{RT_{sc} \alpha_c B_g} = \frac{(14.7)(22.94)}{(10.73)(520)(5.615)(0.00073)} = 14.742 \text{ lbm/ft}^3 . \quad\dots\dots\dots\dots \text{(2.46)}$$

2. $p = 4,514.7$ psia and $R_s = 1,000$ scf/STB: Because oil phase is undersaturated [$R_s = 1,000$ scf/STB, which is less than 1,444 scf/STB (bubblepoint data)], one must apply the following linear interpolation to calculate the bubblepoint:

$$p_b = 3,014.7 + \frac{4,014.7 - 3,014.7}{1,270 - 930} \times (1,000 - 930) = 3,220.6 \text{ psia,} \quad\dots\dots\dots\dots\dots\dots \text{(2.47)}$$

$$B_b = 1.565 + \frac{1.695 - 1.565}{1,270 - 930} \times (1,000 - 930) = 1.592 \text{ res bbl/scf,} \quad\dots\dots\dots\dots\dots\dots \text{(2.48)}$$

$$\mu_b = 0.594 + \frac{0.510 - 0.594}{1,270 - 930} \times (1,000 - 930) = 0.577 \text{ cp.} \quad\dots\dots\dots\dots\dots\dots\dots \text{(2.49)}$$

Thus,

$$B_o = B_b \left[1 - c_o (p - p_b) \right] = 1.592 \left[1 - 23.2 \times 10^{-6} (4,514.7 - 3,220.6) \right] = 1.544 \text{ res bbl/STB,} \quad\dots\dots\dots \text{(2.50)}$$

$$\mu_o = \frac{\mu_b}{\left[1 - c_\mu (p - p_b) \right]} = \frac{0.577}{\left[1 - 46 \times 10^{-6} (4,514.7 - 3,220.6) \right]} = 0.6135 \text{ cp,} \quad\dots\dots\dots\dots\dots \text{(2.51)}$$

$$\rho_o = \frac{\rho_{osc}}{B_o} = \frac{49.098}{1.544} = 30.84 \text{ lbm/ft}^3 . \quad\dots\dots\dots\dots\dots\dots\dots\dots\dots \text{(2.52)}$$

For gas properties, the same results can be obtained for B_g, μ_g, and ρ_g, as in Part 1.

Problem 2.5.9 (Ertekin et al. 2001)

Plot k_{row} vs. S_w in the range of $0 \le S_w \le 1$ using the following models:

- Corey model (Corey 1956; Corey et al. 1956): $k_{row} = (1 - S_{on})^2 \left[1 - (S_{on})^2 \right]$.

- Naar-Henderson model (Naar and Henderson 1961): $k_{row} = (1 - 2S_{on})^{3/2} \left[2 - (1 - 2S_{on})^{1/2} \right]$,

  where $S_{on} = \dfrac{S_w - S_{iw}}{1 - S_{iw}}$.

When $S_{iw} = 0.15$, examine the plots and provide comments on the prediction of k_{row} by the Corey and Naar-Henderson two-phase models.

Solution to Problem 2.5.9

The relative permeability curves generated using the Corey and Naar-Henderson methods are shown in **Fig. 2.11.** A set of tabulated results is listed in **Table 2.5.**

The k_{row} values predicted by both models are quite similar in the low-water-saturation region; however, disparity between the two models becomes bigger in the high-water-saturation region (the k_{row} values predicted by Corey's model are higher).

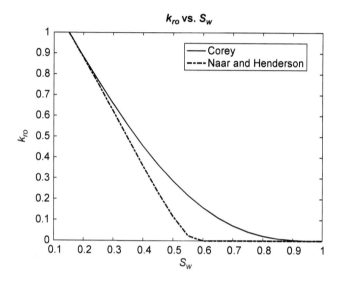

Fig. 2.11—Solution to Problem 2.5.9.

	k_{row} Calculation	
S_w	Corey	Naar-Henderson
0.15	1	1
0.25	0.7678	0.7527
0.35	0.5524	0.4901
0.45	0.3665	0.2325
0.55	0.2182	0.0251
0.65	0.1109	0
0.75	0.0434	0
0.85	0.0100	0
0.95	0.00040	0
1	0.00000	0

Table 2.5—Tabulated results from the Solution to Problem 2.5.9.

Problem 2.5.10 (Ertekin et al. 2001)

Coats et al. (1974) reported the two sets of two-phase relative permeability data given in **Table 2.6.** If, at a certain instant in time, phase saturations in a three-phase flow system are $S_w = 0.320$, $S_g = 0.250$, and $S_o = 0.430$, calculate the relative permeability for the three phases using Stones's three-phase Models 1 (Stone 1970) and 2 (Stone 1973).

	Oil/Water Data			Oil/Gas Data	
S_w	k_{rw}	k_{row}	S_g	k_{rg}	k_{rog}
0.130	0.0000	1.0000	0.000	0.0000	1.0000
0.191	0.0051	0.9990	0.101	0.0026	0.5169
0.250	0.0102	0.8000	0.150	0.0121	0.3373
0.294	0.0168	0.7241	0.195	0.0195	0.2919
0.357	0.0275	0.6206	0.250	0.0285	0.2255
0.414	0.0424	0.5040	0.281	0.0372	0.2100
0.490	0.0665	0.3170	0.337	0.0500	0.1764
0.557	0.0970	0.3029	0.386	0.0654	0.1433
0.630	0.1148	0.1555	0.431	0.0761	0.1172
0.673	0.1259	0.0956	0.485	0.0855	0.0883
0.719	0.1381	0.0576	0.567	0.1022	0.0461
0.789	0.1636	0.0000	0.605	0.1120	0.0294
			0.800	0.1700	0.0000

Table 2.6—Two-phase relative permeability data by Coats et al. (1974).

Solution to Problem 2.5.10

Knowing that $S_w = 0.320$ and $S_g = 0.250$,

$$k_{rw} = 0.0168 + \frac{0.0275 - 0.0168}{0.357 - 0.294} \times (0.320 - 0.294) = 0.0212, \dots\dots\dots\dots\dots\dots\dots\dots (2.53)$$

$$k_{row} = 0.7241 + \frac{0.6206 - 0.7241}{0.357 - 0.294} \times (0.320 - 0.294) = 0.6814. \dots\dots\dots\dots\dots\dots\dots (2.54)$$

From Table 2.6, one can obtain $k_{rg} = 0.0285$ and $k_{rog} = 0.2255$; the endpoint saturation and relative permeability data also can be obtained from the table: $S_{iw} = 0.13$, $S_{orw} = 1 - 0.789 = 0.211$, and $k_{rocw} = 1$. Employing Stone's Model 1, one can calculate the normalized saturations:

$$S_{wn} = \frac{S_w - S_{iw}}{1 - S_{iw} - S_{or}} = \frac{0.32 - 0.13}{1 - 0.13 - 0.211} = 0.2883, \dots\dots\dots\dots\dots\dots\dots (2.55)$$

$$S_{on} = \frac{S_o - S_{or}}{1 - S_{iw} - S_{or}} = \frac{0.43 - 0.211}{1 - 0.13 - 0.211} = 0.3323, \dots\dots\dots\dots\dots\dots\dots (2.56)$$

$$S_{gn} = 1 - S_{wn} - S_{on} = 1 - 0.2883 - 0.3323 = 0.3794, \dots\dots\dots\dots\dots\dots\dots (2.57)$$

$$\beta_w = \frac{k_{row}}{k_{rocw}} \frac{1}{(1 - S_{wn})} = \frac{0.6814}{1} \frac{1}{(1 - 0.2883)} = 0.9574, \dots\dots\dots\dots\dots\dots\dots (2.58)$$

$$\beta_g = \frac{k_{rog}}{k_{rocw}} \frac{1}{(1 - S_{gn})} = \frac{0.2255}{1} \frac{1}{(1 - 0.3794)} = 0.3633, \dots\dots\dots\dots\dots\dots\dots (2.59)$$

$$k_{ro} = k_{rocw} S_{on} \beta_w \beta_g = 1 \times 0.3323 \times 0.9574 \times 0.3633 = 0.1156. \dots\dots\dots\dots\dots\dots (2.60)$$

Using Stone's Model 2,

$$k_{ro} = k_{rocw} \left[\left(k_{rw} + \frac{k_{row}}{k_{rocw}} \right) \left(k_{rg} + \frac{k_{rog}}{k_{rocw}} \right) - \left(k_{rw} + k_{rg} \right) \right] \dots\dots\dots\dots\dots (2.61)$$

$$= 1 \times \left[\left(0.0212 + \frac{0.6814}{1} \right) \left(0.0285 + \frac{0.2255}{1} \right) - \left(0.0212 + 0.0285 \right) \right] = 0.1287.$$

Problem 2.5.11 (Ertekin et al. 2001)

Assuming the two-phase relative permeability data sets in Problem 2.5.10 are not given, what are the phase relative permeabilities predicted by the Naar-Henderson-Wygal model (Naar and Henderson 1961; Naar and Wygal 1961) if $S_{or} = 0.07$ (assuming $S_{iw} = 0.130$)?

Solution to Problem 2.5.11

Using the Naar-Henderson-Wygal relative permeability model, the following values are calculated:

$$k_{rw} = \left(\frac{S_w - S_{iw}}{1 - S_{iw}} \right)^4 = \left(\frac{0.32 - 0.13}{1 - 0.13} \right)^4 = 0.0023, \dots\dots\dots\dots\dots\dots\dots (2.62)$$

$$k_{rg} = \frac{S_g^3 \left(2 - S_g - 2S_{iw} \right)}{\left(1 - S_{iw} \right)^4} = \frac{(0.25)^3 \left(2 - 0.25 - 2 \times 0.13 \right)}{\left(1 - 0.13 \right)^4} = 0.0406, \dots\dots\dots\dots\dots (2.63)$$

$$k_{ro} = \frac{S_o^3 \left(1 - S_g + 2S_w - 3S_{iw} \right)}{\left(1 - S_{iw} \right)^4} = \frac{(0.43)^3 \left(1 - 0.25 + 2 \times 0.32 - 3 \times 0.13 \right)}{\left(1 - 0.13 \right)^4} = 0.1388. \dots\dots\dots (2.64)$$

Problem 2.5.12 (Ertekin et al. 2001)

Consider a thick reservoir represented in three dimensions in the shape of a perfect rectangular prism. If this reservoir is positioned parallel (along its lateral directions) to the datum plane, would it be possible to replace the potential gradients with the pressure gradients? Show your analysis.

Solution to Problem 2.5.12

No, the potential gradient along the vertical direction cannot be replaced with pressure gradient. Gravitational forces acting along the z-direction can create different flow-potential values, because the saturation distribution at a specific point above the lateral plane may be different compared with another point on the same plane.

Problem 2.5.13 (Ertekin et al. 2001)

Does $\dfrac{\partial B_g}{\partial p} = 0$ for an ideal gas? Justify your answer.

Solution to Problem 2.5.13

No, because the compressibility of gas is expressed as Eq. 2.65:

$$B_g = \frac{p_{sc}}{a_c T_{sc}} T \frac{z_g}{p}. \dotfill (2.65)$$

For ideal gas, $z_g = 1$: $B_g = \dfrac{p_{sc}}{a_c T_{sc}} T \dfrac{1}{p}$. Thus, $\dfrac{\partial B_g}{\partial p} = -\dfrac{p_{sc}}{a_c T_{sc}} T \dfrac{1}{p^2} \neq 0$.

Problem 2.5.14 (Ertekin et al. 2001)

Is compressibility of a real gas always less than that of an ideal gas at a constant temperature but of different pressures? Show your analysis. *Hint*: Think about the behavior of z with pressure at a constant temperature.

Solution to Problem 2.5.14

For a real gas, the compressibility can be expressed as Eq. 2.66:

$$c_{g,\text{real}} = \frac{1}{p} - \frac{1}{z_g} \frac{\partial z_g}{\partial p}. \dotfill (2.66)$$

For an ideal gas, $z_g = 1$, and thus, $c_{g,\text{ideal}} = \dfrac{1}{p}$.

At high pressure, $\dfrac{\partial z_g}{\partial p} < 0$, and $c_{g,\text{real}} > c_{g,\text{ideal}}$.

At low pressure, $\dfrac{\partial z_g}{\partial p} > 0$, and $c_{g,\text{real}} < c_{g,\text{ideal}}$.

At $p = 2{,}400$, $dz_g/dp = 0$ and $c_g = 1/p$.

The validity of the above three statements is shown in **Fig. 2.12.**

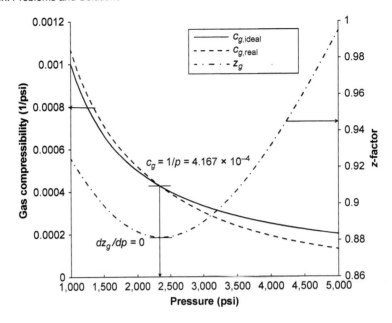

Fig. 2.12—Graphical explanation of the solution for Problem 2.5.14.

Problem 2.5.15 (Ertekin et al. 2001)

In a two-phase relative permeability data set, the summation of relative permeabilities goes through a minimum value. If a third phase is introduced into the system, how will the magnitude of this minimum value change? Explain.

Solution to Problem 2.5.15

The value will be equal to or less than before. If there is no capillary pressure between the third phase and the original two phases, then the value will not change; if there is capillary pressure, the additional energy lost between the phases will cause the summation of relative permeabilities to become even less.

Problem 2.5.16 (Ertekin et al. 2001)

Table 2.7 shows two-phase relative permeability data to be used in three-phase relative permeability characteristics:

1. On a ternary diagram, indicate the region in which possible saturation combinations can be encountered at any time during the life of the reservoir using the two-phase relative permeability characteristics from the table.
2. Using Stone's three-phase Models 1 and 2, calculate the k_{ro}, k_{rw}, and k_{rg} values when $S_o = 27\%$, $S_w = 50\%$, and $S_g = 23\%$.

Oil/Water Data			Oil/Gas Data		
S_w	k_{rw}	k_{row}	S_g	k_{rg}	k_{rog}
0.15	0	1	0.08	0	1
0.20	0.00001197	0.83	0.13	0.000669	0.613
0.25	0.0001197	0.677	0.23	0.0107	0.292
0.30	0.0009698	0.541	0.33	0.03849	0.134
0.40	0.007483	0.321	0.43	0.08926	0.05646
0.50	0.02874	0.169	0.53	0.167	0.02047
0.60	0.07855	0.075	0.63	0.276	0.005933
0.70	0.175	0.02555	0.73	0.419	0.001182
0.80	0.341	0.005409	0.83	0.6	0.000113
0.88	0.563	0	0.88	1	0

Table 2.7—Two-phase relative permeability data.

1. The answer is plotted in **Fig. 2.13.** Using data from Table 2.7, and that $S_w = 0.50$ and $S_g = 0.23$, one can read the following values directly from the table: $k_{rw} = 0.02874$, $k_{row} = 0.169$, $k_{rg} = 0.0107$, and $k_{rog} = 0.292$. The endpoint saturation and the relative permeability data also can be obtained from the table: $S_{iw} = 0.15$, $S_{orw} = 1 - 0.88 = 0.12$, $k_{rocw} = 1$, and $S_{gc} = 0.08$.

2. Employing Stone's Model 1, one can calculate the normalized saturations:

$$S_{wn} = \frac{S_w - S_{iw}}{1 - S_{iw} - S_{or}} = \frac{0.5 - 0.15}{1 - 0.15 - 0.12} = 0.4795, \quad \dots\dots\dots\dots\dots\dots\dots\dots\dots\dots\dots\dots\dots\dots\dots \quad (2.67)$$

$$S_{on} = \frac{S_o - S_{orw}}{1 - S_{iw} - S_{orw}} = \frac{0.27 - 0.12}{1 - 0.15 - 0.12} = 0.2055, \quad \dots\dots\dots\dots\dots\dots\dots\dots\dots\dots\dots\dots\dots\dots \quad (2.68)$$

$$S_{gn} = 1 - S_{wn} - S_{on} = 1 - 0.4795 - 0.2055 = 0.3151, \quad \dots\dots\dots\dots\dots\dots\dots\dots\dots\dots\dots\dots\dots\dots \quad (2.69)$$

$$\beta_w = \frac{k_{row}}{k_{rocw}} \frac{1}{(1 - S_{wn})} = \frac{0.1690}{1} \frac{1}{(1 - 0.4795)} = 0.3247, \quad \dots\dots\dots\dots\dots\dots\dots\dots\dots\dots\dots\dots\dots\dots \quad (2.70)$$

$$\beta_g = \frac{k_{rog}}{k_{rocw}} \frac{1}{(1 - S_{gn})} = \frac{0.2920}{1} \frac{1}{(1 - 0.3151)} = 0.4263, \quad \dots\dots\dots\dots\dots\dots\dots\dots\dots\dots\dots\dots\dots\dots \quad (2.71)$$

$$k_{ro} = k_{rocw} S_{on} \beta_w \beta_g = 1 \times 0.2055 \times 0.3247 \times 0.4263 = 0.0284. \quad \dots\dots\dots\dots\dots\dots\dots\dots\dots\dots\dots\dots \quad (2.72)$$

Using Stone's Model 2,

$$k_{ro} = k_{rocw} \left[\left(k_{rw} + \frac{k_{row}}{k_{rocw}} \right) \left(k_{rg} + \frac{k_{rog}}{k_{rocw}} \right) - \left(k_{rw} + k_{rg} \right) \right]$$

$$= 1 \times \left[(0.0287 + 0.1690) \left(0.0107 + \frac{0.2920}{1} \right) - (0.0287 + 0.0107) \right] = 0.0204. \quad \dots\dots\dots\dots\dots\dots\dots \quad (2.73)$$

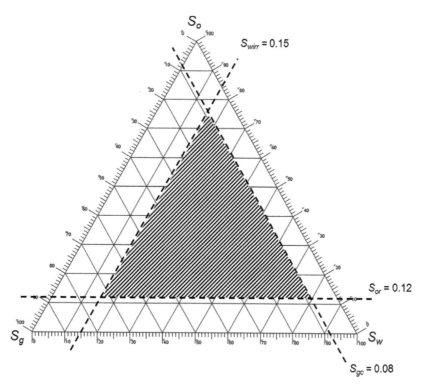

Fig. 2.13—Solution to Problem 2.5.16 (Part 1).

Problem 2.5.17 (Ertekin et al. 2001)

Comment on the accuracy of the statement, "A hydrocarbon reservoir that initially exhibits isotropic and homogeneous permeability characteristics will have anisotropic and heterogeneous permeability distribution as the reservoir is depleted, if permeability is a function of pressure."

Solution to Problem 2.5.17

When the reservoir becomes depleted, reservoir pressure values will start to vary from one location to another. Thus, the permeability values will become different (heterogeneous) as a function of the magnitude of pressure change at a specific location. Because pressure at a given location is the same in every direction (remember pressure is scalar and independent of direction), and permeability is isotropic for the system described, the magnitude of change in permeability in each direction will be the same, hence resulting in an isotropic permeability distribution as the reservoir pressure is depleted.

Problem 2.5.18 (Ertekin et al. 2001)

Is it possible to use the capillary pressure relationships $P_{cow}(S_w) = p_o - p_w$ and $P_{cgo}(S_g) = p_g - p_o$ for oil-wet reservoirs? Explain.

Solution to Problem 2.5.18

The equation $P_{cgo}(S_g) = p_g - p_o$ is still valid, but not $P_{cow}(S_w) = p_o - p_w$. Because in an oil-wet reservoir, $p_w > p_o$, and to end up with a positive capillary pressure between oil and water phases, $P_{cow}(S_w) = p_w - p_o$.

Problem 2.5.19 (Ertekin et al. 2001)

Derive the equation: $B_o = B_b\left[1 - c_o\left(p_o - p_b\right)\right]$ for the oil FVF above the bubblepoint pressure.

Solution to Problem 2.5.19

Recall the definition of compressibility written in terms of oil density:

$$c_o = \frac{1}{\rho_o}\frac{\partial \rho_o}{\partial p}. \dotfill (2.74)$$

Separating the variables and integrating both sides,

$$\int_{p_b}^{p_o} c_o \, dp = \int_{p_b}^{p_o} \frac{1}{\rho_o} d\rho_o, \dotfill (2.75)$$

one obtains

$$\left(p_o - p_b\right)c_o = \ln\frac{\rho_o}{\rho_b}. \dotfill (2.76)$$

Note that ρ_b and ρ_o values are densities at the bubblepoint pressure and at pressures higher than p_b, respectively. Now, Eq. 2.76 can be recast, as shown:

$$\exp\left[-\left(p_o - p_b\right)c_o\right] = \frac{\rho_b}{\rho_o}. \dotfill (2.77)$$

Invoking Taylor's expansion of the exponential function,

$$e^{-x} = 1 - x + \frac{x^2}{2!} - \frac{x^3}{3!} + \cdots. \dotfill (2.78)$$

The exponential function can be simplified by truncating after the second-order term:

$$\left[1 - \left(p_o - p_b\right)c_o\right] = \frac{\rho_b}{\rho_o}. \dotfill (2.79)$$

Recalling that $B_o = \rho_{osc}/\rho_o$ and $B_b = \rho_{osc}/\rho_b$,

thus

$$\frac{\rho_b}{\rho_o} = \frac{B_o}{B_b} = \left[1 - \left(p_o - p_b\right)c_o\right], \dots\dots\dots\dots\dots\dots\dots\dots\dots\dots\dots\dots\dots\dots\dots\dots (2.80)$$

which yields

$$B_o = B_b\left[1 - \left(p_o - p_b\right)c_o\right]. \dots\dots\dots\dots\dots\dots\dots\dots\dots\dots\dots\dots\dots\dots\dots (2.81)$$

Problem 2.5.20

Indicate whether the following statements are TRUE or FALSE, and justify your answer by providing explanations:

1. Pore-size distribution is not an important reservoir rock property; therefore, it does not appear in the flow equations.
2. Consider two two-phase (oil/water) systems, so that $\gamma_{(ow)1} = 35$ dynes/cm and $\gamma_{(ow)2} = 12$ dynes/cm. At any saturation combination, $(k_{ro} + k_{rw})_1 > (k_{ro} + k_{rw})_2$.
3. Consider two identical reservoirs; one of them is located 4,000 ft deeper than the other. The effect of the potential gradient for the deeper reservoir is expected to be more pronounced.
4. In a single-phase natural gas reservoir, the average gravity of the produced gas is always expected to remain constant.
5. Consider waterflooding of a single-phase oil reservoir. If oil and water phases are treated as incompressible, then the flow equations for this two-phase flow problem will describe a steady-state flow regime.
6. A compressible flow equation in one dimension for ideal gas can be treated as a linear equation.

Solution to Problem 2.5.20

1. FALSE. Pore-size distribution is incorporated into the flow equations by way of capillary pressure, relative permeability, and absolute permeability functions.
2. FALSE. As the interfacial tension between the phases decreases, the energy lost at the interface decreases and, thus, the summation of relative permeabilities at a given saturation combination increases.
3. FALSE. As shown in Eqs. 2.82 and 2.83, potential gradient is a function of the depth gradient, not of the depth:

$$\Phi = p - \frac{1}{144}\frac{g}{g_c}\rho G, \dots\dots\dots\dots\dots\dots\dots\dots\dots\dots\dots\dots\dots\dots\dots (2.82)$$

$$\frac{\partial \Phi}{\partial S} = \frac{\partial p}{\partial S} - \frac{1}{144}\frac{g}{g_c}\rho\frac{\partial G}{\partial S}. \dots\dots\dots\dots\dots\dots\dots\dots\dots\dots\dots\dots\dots (2.83)$$

4. FALSE. As the production continues, lighter components are stripped off from the reservoir. This results in progressively leaving higher-gravity gas in the reservoir.
5. FALSE. As the injection continues, the saturation values at a given point in the reservoir will change with time. Because saturation values are dependent variables of the problem in hand, and because they are changing with time, the problem is considered an unsteady-state problem.
6. FALSE. For a 1D gas reservoir, the governing flow equation is written as

$$\frac{\partial}{\partial x}\left(\frac{Ak}{\mu_g B_g}\frac{\partial p}{\partial x}\right)\Delta x + q = \frac{V_b \phi T_{sc}}{p_{sc} T}\frac{\partial p}{\partial t}, \dots\dots\dots\dots\dots\dots\dots\dots\dots\dots\dots (2.84)$$

and $B_g = \dfrac{p_{sc}T}{5.615 p T_{sc}}$.

Although $z_g = 1$, B_g, as shown in the preceding line, is still a nonlinear term (it is a function of pressure); therefore, the PDE is still a nonlinear equation.

Problem 2.5.21

Consider the slightly compressible flow in a 2D reservoir, as shown in **Fig. 2.14.1.** The reservoir is homogeneous and holds isotropic permeability distribution with uniform thickness. There is no depth gradient in the system. The reservoir is completely sealed, and a producer is located at the center of the system.

1. From the plot of sandface pressure vs. time shown in **Fig. 2.14.2,** plot production rate vs. time of this production plan, qualitatively.
2. From the plot of production rate vs. time shown in **Fig. 2.14.3,** plot sandface pressure of the well vs. time of this production plan, qualitatively.

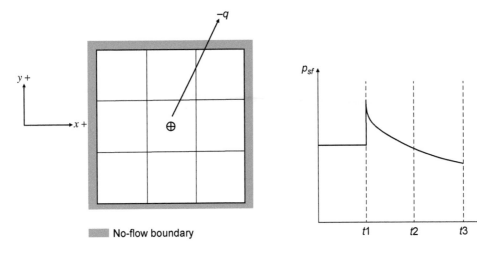

Fig. 2.14.1—2D Reservoir of Problem 2.5.21.   Fig. 2.14.2—Problem 2.5.21 (Part 1).

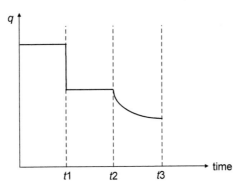

Fig. 2.14.3—Problem 2.5.21 (Part 2).

Solution to Problem 2.5.21

Figs. 2.14.4 and 2.14.5 show the solutions of the Part 1 and 2, respectively.

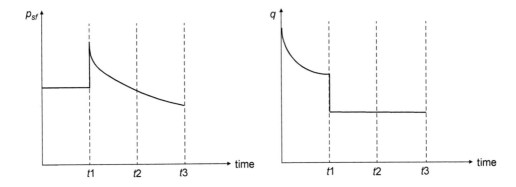

Fig. 2.14.4—Solution to Part 1 of Problem 2.5.21.

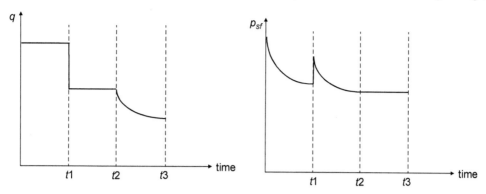

Fig. 2.14.5—Solution to Part 2 of Problem 2.5.21.

Problem 2.5.22

The two-phase relative permeability data shown in **Table 2.8** are given for the calculation of three-phase relative permeabilities. Use Stone's Model 1 to determine the relative permeability to oil of a system that reports k_{rw} to be 0.08 and k_{rg} to be 0.0067. Show all your calculations.

Oil/Water			Gas/Liquid		
S_w	k_{rw}	k_{row}	S_g	k_{rg}	k_{rog}
0.18	0	1	0.04	0	1
0.3	0.05	0.5	0.1	0.02	1
0.4	0.15	0.13	0.2	0.1	1
0.5	0.23	0.07	0.3	0.25	0.42
0.6	0.3	0.03	0.4	0.34	0.15
0.78	0.65	0.001	0.5	0.42	0.04
0.96	0.75	0	0.6	0.5	0.02
1	1	0	0.7	0.8	0.01
			0.78	1	0

Table 2.8—Two-phase relative permeability data for an oil/water system and a gas/liquid system.

Solution to Problem 2.5.22

Using the given k_{rw} and k_{rg} to interpolate the water and gas saturations when $k_{rw} = 0.08$,

$$S_w = \frac{(0.08 - 0.05)}{(0.15 - 0.05)}(0.4 - 0.3) + 0.3 = 0.33; \dots\dots\dots\dots\dots\dots\dots\dots\dots\dots\dots \text{(2.85)}$$

when $k_{rg} = 0.0067$,

$$S_g = \frac{(0.0067 - 0)}{(0.02 - 0)}(0.1 - 0.04) + 0.04 = 0.0601, \dots\dots\dots\dots\dots\dots\dots\dots\dots\dots \text{(2.86)}$$

$$S_o = 1 - 0.33 - 0.0601 = 0.6099. \dots\dots\dots\dots\dots\dots\dots\dots\dots\dots\dots\dots\dots \text{(2.87)}$$

Then, one can calculate k_{row} and k_{rog} values using the relative permeability table at the calculated saturation values:

$$k_{row}(S_w = 0.33) = \frac{(0.33 - 0.3)}{(0.4 - 0.3)}(0.13 - 0.5) + 0.5 = 0.389, \dots\dots\dots\dots\dots\dots\dots\dots \text{(2.88)}$$

$$k_{rog}(S_g = 0.0601) = 1.00, \text{ and } S_{wirr} = 0.18 \text{ and } S_{or} = 0.04, \dots\dots\dots\dots\dots\dots\dots\dots \text{(2.89)}$$

$$S_{on} = \frac{S_{on} - S_{or}}{1 - S_{wirr} - S_{or}} = \frac{(0.6099 - 0.04)}{(1 - 0.18 - 0.04)} = 0.7306, \dotfill \text{(2.90)}$$

$$S_{wn} = \frac{S_w - S_{wirr}}{1 - S_{wirr} - S_{or}} = \frac{(0.33 - 0.18)}{(1 - 0.18 - 0.04)} = 0.1923, \dotfill \text{(2.91)}$$

$$S_{gn} = \frac{S_g}{1 - S_{wirr} - S_{or}} = \frac{0.0601}{(1 - 0.18 - 0.04)} = 0.077, \dotfill \text{(2.92)}$$

$$\beta_w = \frac{k_{row}}{1 - S_{wn}} = \frac{0.389}{1 - 0.19} = 0.48, \dotfill \text{(2.93)}$$

$$\beta_g = \frac{k_{rog}}{1 - S_{gn}} = \frac{1}{1 - 0.077} = 1.08, \dotfill \text{(2.94)}$$

$$k_{ro} = S_{on}\,\beta_w\,\beta_g = (0.731)(0.48)(1.08) = 0.3813. \dotfill \text{(2.95)}$$

Problem 2.5.23

Match the following statements with Case 1 and Case 2 shown in **Fig. 2.15:**

1. Flow is from 1 to 2.
2. Flow is from 2 to 1.
3. Flow direction cannot be determined.

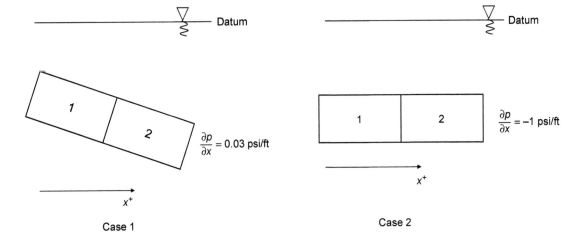

Fig. 2.15—Case 1 (on the left) and Case 2 (on the right) of Problem 2.5.23.

Solution to Problem 2.5.23

Case 1: Flow direction cannot be determined because the depth gradient and the density of the fluid are not given (potential gradient cannot be calculated).

Case 2: In this case, $\dfrac{\partial \Phi}{\partial x} = \dfrac{\partial p}{\partial x}$. Because the pressure is decreasing in the increasing direction of the x-axis, flow is from 1 to 2.

Problem 2.5.24

Match the statements for **Fig. 2.16:**

1. Nonwetting drainage.
2. Nonwetting imbibition.
3. Wetting imbibition.
4. Wetting drainage.

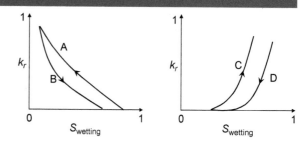

Fig. 2.16—Problem 2.5.24.

Solution to Problem 2.5.24

A. Nonwetting drainage.
B. Nonwetting imbibition.
C. Wetting imbibition.
D. Wetting drainage.

Problem 2.5.25

Compare the two reservoir rocks A and B, in terms of their pore-size distribution and permeability characteristics, on the basis of their capillary pressure curves, as shown in **Fig. 2.17.**

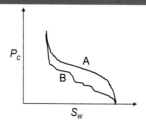

Fig. 2.17—Capillary pressure behavior of two reservoir rocks, A and B.

Solution to Problem 2.5.25

• Rock B has nonuniform pore-size distribution.
• Rock A has uniform pore-size distribution.
• Rock A is tighter than Rock B.

Problem 2.5.26

Consider a 1D single-phase flow in the following system, as shown in **Fig. 2.18:**

1. If the fluid in the system is water with a density of 62.4 lbm/ft^3, and the system is in hydrodynamic equilibrium, determine the pressure in Block 1.
2. Using the pressure of Block 1 obtained in Part 1, determine the flow direction if the fluid in the system is a gas with the following properties: MW = 16 lbm mol/mol, z-factor = 0.98, T = 150°F (an average temperature at both locations—isothermal reservoir).

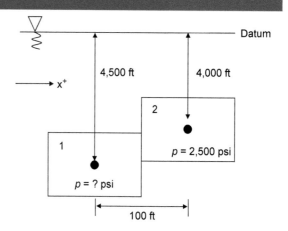

Fig. 2.18—Schematic representation of the 1D reservoir of Problem 2.5.26.

Solution to Problem 2.5.26

1. For a hydrodynamic equilibrium system, the flow potentials of Blocks 1 and 2 should be the same; therefore,

$$\Phi_1 = p_1 - \frac{1}{144}\frac{g}{g_c}\rho G_1, \quad\dotfill (2.96)$$

$$\Phi_2 = p_2 - \frac{1}{144}\frac{g}{g_c}\rho G_2. \quad\dotfill (2.97)$$

Because $\Phi_1 = \Phi_2$,

$$p_1 = p_2 - \frac{1}{144}\frac{g}{g_c}\rho(G_2 - G_1) = 2{,}716.67\,\text{psi}. \quad\dotfill (2.98)$$

2. Knowing that MW = 16 lbm/lbm mol, z-factor = 0.98, and $T = 150°F$,

$$\rho_1 = \frac{p_1 \text{MW}}{z_g RT} = \frac{(2{,}716.67)(16)}{(0.98)(10.73)(150+460)} = 6.78\,\text{lbm/ft}^3, \quad\dotfill (2.99)$$

$$\rho_2 = \frac{p_2 \text{MW}}{z_g RT} = \frac{(2{,}500)(16)}{(0.98)(10.73)(150+460)} = 6.24\,\text{lbm/ft}^3, \quad\dotfill (2.100)$$

$$\Phi_1 = p_1 - \frac{1}{144}\frac{g}{g_c}\rho_1 G_1 = 2{,}504.9\,\text{psi}, \quad\dotfill (2.101)$$

$$\Phi_2 = p_2 - \frac{1}{144}\frac{g}{g_c}\rho_2 G_2 = 2{,}329.78\,\text{psi}. \quad\dotfill (2.102)$$

Because $\Phi_1 > \Phi_2$, flow direction is from 1 to 2.

Problem 2.5.27

An incompressible fluid is flowing in an inclined reservoir, as shown in **Fig. 2.19.** Derive the expression of the contribution of the depth gradient to the total flow rate between Points A and B.

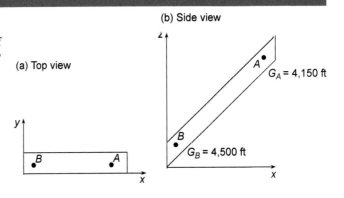

Fig. 2.19—Top and side views of the inclined reservoir of Problem 2.5.27.

Solution to Problem 2.5.27

Total flow rate is expressed under the presence of hydrostatic head and pressure in the reservoir, as shown in the parentheses of Eq. 2.103. Eq. 2.104 clearly describes the contribution of the depth gradient through the gravitational forces acting on the system:

$$q_t = -\frac{kA}{\mu B}\frac{\partial \Phi}{\partial x} = \frac{kA}{\mu B}\left(\frac{\partial p}{\partial x} - \frac{1}{144}\frac{g}{g_c}\rho\frac{\partial G}{\partial x}\right), \quad\dotfill (2.103)$$

$$q_d = -\frac{kA}{\mu B}\left(-\frac{1}{144}\frac{g}{g_c}\rho\frac{\partial G}{\partial x}\right). \quad\dotfill (2.104)$$

The ratio of the q_d to q_t (Eq. 2.105) provides the contribution of the depth gradient to the total flow rate between Points A and B:

$$\frac{q_d}{q_t} = \frac{\dfrac{1}{144}\dfrac{g}{g_c}\rho\dfrac{\partial G}{\partial x}}{\dfrac{\partial p}{\partial x} - \dfrac{1}{144}\dfrac{g}{g_c}\rho\dfrac{\partial G}{\partial x}}. \dots\dots\dots\dots\dots\dots\dots\dots\dots\dots\dots\dots\dots\dots\dots\dots\dots\dots\dots\dots\dots\dots\dots\dots\dots\dots\dots\dots (2.105)$$

Problem 2.5.28

Provide the governing flow equations in their simplest forms for the system described below. Also, give a full description of the characteristics of the system shown in **Fig. 2.20.**

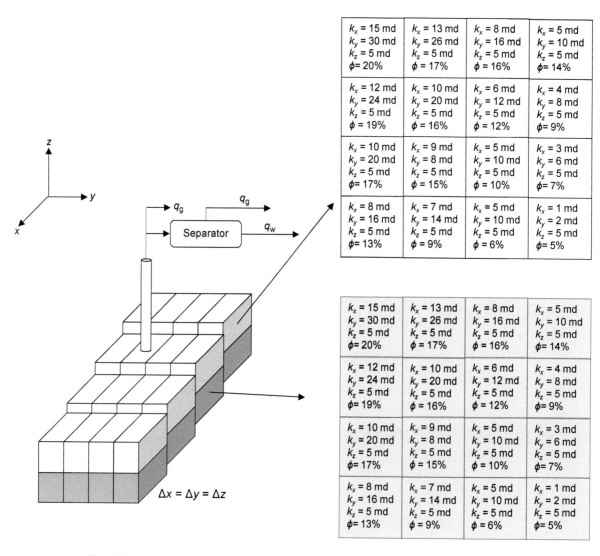

Fig. 2.20—Schematics of the reservoir of Problem 2.5.28 and its property characteristics.

Solution to Problem 2.5.28

One can write the governing equations representing the problem described in Fig. 2.20 in consideration of the following characteristics of the problem:

- Dimension: 3D, rectangular coordinates.
- Permeability: heterogeneous and anisotropic.

- Grid: uniform, depth gradient exists in x- and z-direction.
- Porosity: heterogeneous.
- Fluid: two-phase flow (water and gas).
- A production well exists.
- Water equation:

$$\frac{\partial}{\partial x}\left(\frac{A_x k_x}{\mu_w B_w}\frac{\partial \Phi_w}{\partial x}\right)\Delta x + \frac{\partial}{\partial y}\left(\frac{A_y k_y}{\mu_w B_w}\frac{\partial p_w}{\partial y}\right)\Delta y + \frac{\partial}{\partial z}\left(\frac{A_z k_z}{\mu_w B_w}\frac{\partial \Phi_w}{\partial z}\right)\Delta z + q_{wsc} = \frac{V_b}{5.615}\frac{\partial}{\partial t}\left(\frac{\phi S_w}{B_w}\right). \quad \ldots \ldots (2.106)$$

- Gas equation:

$$\frac{\partial}{\partial x}\left(\frac{A_x k_x}{\mu_g B_g}\frac{\partial \Phi_g}{\partial x} + R_{sw}\frac{A_x k_x}{\mu_w B_w}\frac{\partial \Phi_w}{\partial x}\right)\Delta x + \frac{\partial}{\partial y}\left(\frac{A_y k_y}{\mu_g B_g}\frac{\partial p_g}{\partial y} + R_{sw}\frac{A_y k_y}{\mu_w B_w}\frac{\partial p_w}{\partial y}\right)\Delta y$$

$$+ \frac{\partial}{\partial z}\left(\frac{A_z k_z}{\mu_g B_g}\frac{\partial \Phi_g}{\partial z} + R_{sw}\frac{A_z k_z}{\mu_w B_w}\frac{\partial \Phi_w}{\partial z}\right)\Delta z + q_{gsc} + R_{sw}\,q_{wsc} = \frac{V_b}{5.615}\frac{\partial}{\partial t}\left(\frac{\phi S_g}{B_g} + R_{sw}\frac{\phi S_w}{B_w}\right). \quad \ldots \ldots \ldots (2.107)$$

Problem 2.5.29

Specify whether each of the following statements is TRUE or FALSE. In each case, justify your answer:

1. The compressibility of a gas at pressures of 1,500 and 5,000 psia are 6.67×10^{-5} psi^{-1} and 4.85×10^{-5} psi^{-1}, respectively (use the z-factor chart for the gas, as shown in **Fig. 2.21.1,** in structuring your answer).
2. Flow in the system in **Fig. 2.21.2** is from Point B to Point A.
3. Consider the reservoir system in **Fig. 2.21.3.** If the flow is measured as 1,500 STB/D from Point B to Point A, the pressure difference between Point A and Point B can be calculated as 1,035.67 psi. Use $\rho = 64$ lbm/ft^3, $\mu = 1.3$ cp, $B = 1$ res bbl/STB, $k = 250$ md, and $A = 18,000$ ft^2.

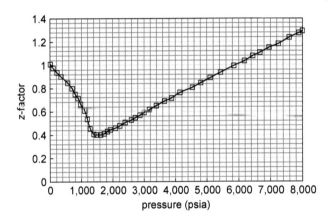

Fig. 2.21.1—Problem 2.5.29 (Part 1).

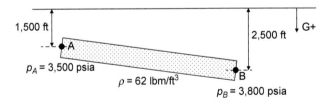

Fig. 2.21.2—Problem 2.5.29 (Part 2).

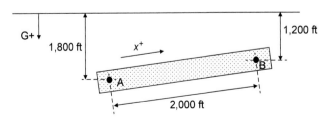

Fig. 2.21.3—Problem 2.5.29 (Part 3).

Solution to Problem 2.5.29

1. TRUE. As shown in **Fig. 2.21.4,** knowing that

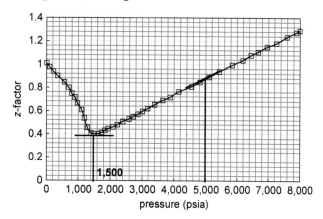

Fig. 2.21.4—Solution to Part 1 of Problem 2.5.29.

$$c_g = \frac{1}{p} - \frac{1}{z_g}\frac{\partial z_g}{\partial p}, \quad \dots\dots\dots\dots\dots\dots\dots\dots\dots\dots\dots\dots\dots\dots\dots\dots\dots\dots\dots\dots\dots\dots\dots \quad (2.108)$$

at $p = 1,500$ psi, $\dfrac{\partial z_g}{\partial p} = 0$, and $c_g = \dfrac{1}{1,500} - 0 \approx 6.67 \times 10^{-4}$ psi^{-1}, and at $p = 5,000$ psia, $\dfrac{\partial z_g}{\partial p} = \dfrac{0.16}{1,200} = 1.33 \times 10^{-4}$,

$$c_g = \frac{1}{5,000} - \frac{1}{0.88}\left(\frac{0.16}{1,200}\right) \approx 4.85 \times 10^{-5} \text{ psi}^{-1}.$$

2. FALSE.

$$\Phi_A = p_A - \frac{1}{144}\frac{g}{g_c}\rho G_A = 3,500 - \frac{1}{144}\frac{32.2}{32.2}(62)(1,500) = 2,854.17 \text{ psia}, \quad \dots\dots\dots\dots\dots\dots \quad (2.109)$$

$$\Phi_B = p_B - \frac{1}{144}\frac{g}{g_c}\rho G_B = 3,800 - \frac{1}{144}\frac{32.2}{32.2}(62)(2,500) = 2,723.61 \text{ psia.} \quad \dots\dots\dots\dots\dots\dots \quad (2.110)$$

Because $\Phi_A > \Phi_B$ from the preceding calculation results in Eqs. 2.109 and 2.110, the flow is from A to B.

3. FALSE. The flow equation between Points A and B can be written as

$$q = -1.127 \times 10^{-3}\frac{kA}{\mu}\left(\frac{\partial p}{\partial x} - \frac{1}{144}\frac{g}{g_c}\rho\frac{\partial G}{\partial x}\right) = -1.127 \times 10^{-3}\frac{kA}{\mu}\left(\frac{\Delta p}{\Delta x} - \frac{1}{144}\frac{g}{g_c}\rho\frac{G_B - G_A}{\Delta x}\right). \quad \dots\dots\dots \quad (2.111)$$

Solving for Δp, we get

$$\Delta p = \Delta x\left(\frac{q}{-1.127 \times 10^{-3}\dfrac{kA}{\mu}} + \frac{1}{144}\frac{g}{g_c}\rho\frac{G_B - G_A}{\Delta x}\right)$$

$$\dots\dots\dots\dots\dots \quad (2.112)$$

$$= (2,000)\left[\frac{-1,500}{-1.127 \times 10^{-3}\dfrac{(250)(18,000)}{1.3}} + \frac{1}{144}(64)\frac{(1,200 - 1,800)}{2,000}\right] = 502.34 \text{ psia.}$$

Problem 2.5.30

Consider the system shown in **Fig. 2.22:**

1. If the fluid in the system is in hydrodynamic equilibrium, determine the pressure at Point b.
2. For the pressure obtained in Part 1, if the fluid density is 62.0 lbm/ft^3, determine whether the system is in hydrodynamic equilibrium. If it is not in equilibrium, what is the flow direction between Points a and b?

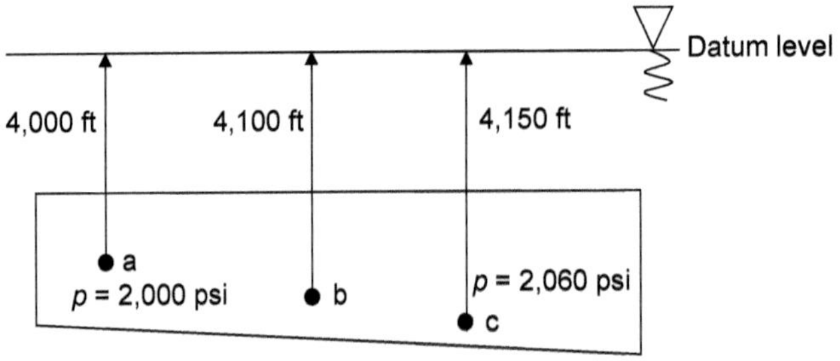

Fig. 2.22—Schematic of Problem 2.5.30.

Solution to Problem 2.5.30

1. Because the system is in hydrodynamic equilibrium, the hydrostatic gradient is the same for everywhere. Thus,

$$\gamma = \frac{p_c - p_a}{G_c - G_a} = \frac{2,060 - 2,000}{4,150 - 4,000} = 0.4 \, \text{psi/ft}, \quad \dots\dots\dots\dots\dots\dots\dots\dots\dots\dots\dots\dots\dots (2.113)$$

$$p_b = \gamma G_b = 0.4 \times 4,100 = 1,640 \, \text{psia}. \quad \dots\dots\dots\dots\dots\dots\dots\dots\dots\dots\dots (2.114)$$

2. If the fluid density is 62.0 lbm/ft³, the hydrostatic head is

$$\gamma' = \frac{\rho}{144} = \frac{62}{144} = 0.43 \, \text{psi/ft} \neq \gamma. \quad \dots\dots\dots\dots\dots\dots\dots\dots\dots\dots\dots (2.115)$$

So, if the system is not in equilibrium. The fluid potentials of Points a and b are

$$\Phi_a = p_a - \gamma' G_a = 2,000 - 0.43(4,000) = 280 \, \text{psia}, \quad \dots\dots\dots\dots\dots\dots\dots\dots (2.116)$$

$$\Phi_c = p_c - \gamma' G_c = 2,060 - 0.43(4,150) = 275.5 \, \text{psia}. \quad \dots\dots\dots\dots\dots\dots\dots (2.117)$$

Because $\Phi_a > \Phi_c$, flow is from Point a to c.

Problem 2.5.31

For a two-phase system (oil and water) with $S_{wirr} = 0.11$, k_{ro} predicted by Naar-Henderson's correlation is twice that predicted by Corey's correlation at $S_w = 0.34$. Is this TRUE or FALSE?

Solution to Problem 2.5.31

FALSE. $S_{wirr} = 0.11$ at $S_w = 0.34$; the normalized saturation is

$$S_{wn} = \frac{S_w - S_{wirr}}{1 - S_{wirr}} = \frac{0.34 - 0.11}{1 - 0.11} = 0.258. \quad \dots\dots\dots\dots\dots\dots\dots\dots\dots\dots (2.118)$$

- Corey model:

$$k_{ro_1} = (1 - S_{wn})^2 \, (1 - S_{wn}^2) = (1 - 0.258)^2 (1 - 0.258^2) = 0.514. \quad \dots\dots\dots\dots\dots (2.119)$$

- Naar-Henderson model:

$$k_{ro_2} = \frac{(1 - 2S_{wn})^{\frac{3}{2}}}{2 - (1 - 2S_{wn})^{\frac{1}{2}}} = \frac{[1 - 2(0.258)]^{\frac{3}{2}}}{2 - [1 - 2(0.258)]^{\frac{1}{2}}} = 0.258. \quad \dots\dots\dots\dots\dots\dots (2.120)$$

Therefore, $\dfrac{k_{ro_2}}{k_{ro_1}} = \dfrac{1}{2}$.

Problem 2.5.32

The two reservoirs in **Fig. 2.23** have the same rock and fluid properties, except permeability ($k_1 > k_2$). The fluid is considered to be slightly compressible.

   Compare p_1 and p_2, if the two wells are put on production at the same time and at the same constant rate.

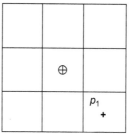

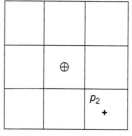

Reservoir 1 (permeability = k_1)   Reservoir 2 (permeability = k_2)

Fig. 2.23—Two reservoirs of Problem 2.5.32.

Solution to Problem 2.5.32

In this problem, p_1 would be greater than p_2. Higher permeability creates less pressure drop at any point in the reservoir of this problem for a constant-rate production. Because $k_1 > k_2$, $\Delta p_1 < \Delta p_2$ or $p_1 > p_2$.

Nomenclature

A	=	cross-sectional area, L^2, ft^2
A_S	=	cross-sectional area normal to the direction S, L^2, ft^2
A_x	=	cross-sectional area normal to the direction x, L^2, ft^2
A_y	=	cross-sectional area normal to the direction y, L^2, ft^2
A_z	=	cross-sectional area normal to the direction z, L^2, ft^2
B	=	FVF, L^3/L^3, res bbl/STB or res bbl/scf
B_b	=	FVF at the bubblepoint pressure, L^3/L^3, res bbl/STB or res bbl/scf
B_g	=	gas FVF, L^3/L^3, res bbl/scf
B_o	=	oil FVF, L^3/L^3, res bbl/STB
B_w	=	water FVF, L^3/L^3, res bbl//STB
c	=	compressibility, Lt^2/m, psi^{-1}
c_g	=	gas compressibility, Lt^2/m, psi^{-1}
$c_{g,\text{ideal}}$	=	gas compressibility for an ideal gas, Lt^2/m, psi^{-1}
$c_{g,\text{real}}$	=	gas compressibility for a real gas, Lt^2/m, psi^{-1}
c_o	=	oil compressibility, Lt^2/m, psi^{-1}
c_r	=	reduced compressibility
c_μ	=	compressibility to be used in viscosity calculation, Lt^2/m, psi^{-1}
c_ϕ	=	rock compressibility in terms of porosity, Lt^2/m, psi^{-1}
g	=	local gravitational acceleration, L/t^2
g_c	=	unit conversion factor in Newton's law or (gravitational acceleration on the equator), L/t^2
G	=	elevation with respect to absolute datum being positive upward, L, ft

k	=	permeability, L^2, darcies or perms
k_{abs}	=	absolute permeability, L^2, darcies or perms
$k_{o\text{effective}}$	=	effective permeability of oil phase, L^2, darcies or perms
k_{rg}	=	relative permeability to gas, dimensionless
k_{ro}	=	relative permeability to oil, dimensionless
k_{rocw}	=	relative permeability to gas at irreducible water saturation, dimensionless
k_{rog}	=	relative permeability to oil in oil/gas system, dimensionless
k_{row}	=	relative permeability to oil in oil/water system, dimensionless
k_{rw}	=	relative permeability to water
k_S	=	permeability along direction S, L^2, darcies or perms
k_x	=	permeability in the x-direction, L^2, darcies or perms
k_y	=	permeability in the y-direction, L^2, darcies or perms
k_z	=	permeability in the z-direction, L^2, darcies or perms
L	=	length dimension also length, L, ft
MW	=	molecular weight, lbm mol
p	=	pressure, m/Lt^2, psia
p_b	=	bubblepoint pressure, m/Lt^2, psia
p_g	=	gas pressure, m/Lt^2, psia
p_o	=	oil pressure, m/Lt^2, psia
p_{pr}	=	pseudoreduced pressure, dimensionless
p_{sc}	=	standard condition pressure, m/Lt^2, psia
p_w	=	water pressure, m/Lt^2, psia
P_{cgo}	=	gas/oil capillary pressure, m/Lt^2, psia
P_{cow}	=	oil/water capillary pressure, m/Lt^2, psia
q	=	volumetric flow rate, L^3/t, STB/D

q_d	=	volumetric flow rate induced by depth gradient, L^3/t, STB/D
q_{gsc}	=	gas volumetric flow rate at standard condition, L^3/t, STB/D
q_S	=	volumetric flow rate along direction S, L^3/t, STB/D
q_t	=	total flow rate, L^3/t, STB/D
q_{wsc}	=	water volumetric flow rate at standard condition, L^3/t, STB/D
R	=	universal gas constant
R_s	=	solution gas/oil ratio, L^3/L^3, scf/STB
R_{sw}	=	solution gas/water ratio, L^3/L^3, scf/STB
S	=	arbitrary spatial direction
S_g	=	gas saturation, fraction
S_{gn}	=	normalized gas saturation, fraction
S_{iw}	=	irreducible water saturation, fraction
S_o	=	oil saturation, fraction
S_{on}	=	normalized oil saturation, fraction
S_{or}	=	residual oil saturation, fraction
S_{orw}	=	residual oil saturation in oil/water system, fraction
S_w	=	water saturation, fraction
$S_{wetting}$	=	wetting-phase saturation, fraction
S_{wirr}	=	irreducible water saturation, fraction
S_{wn}	=	normalized water saturation, fraction
t	=	time, t, days
T	=	absolute temperature, T, °R
T_{sc}	=	standard condition temperature, T, °R
v_S	=	superficial velocity along direction S, L/t, bbl/ft²-D
V_b	=	bulk volume, L^3, ft³
x	=	distance in x-direction in the Cartesian coordinate system, L, ft
y	=	distance in y-direction in the Cartesian coordinate system, L, ft
z	=	distance in z-direction in the Cartesian coordinate system, L, ft
z_g	=	gas compressibility factor, dimensionless

α_c	=	unit conversion factor, equal to 5.615 ft³/bbl
β_g	=	intermediate variable in Stone's Model 1 for gas phase
β_w	=	intermediate variable in Stone's Model 1 for water phase
Δ	=	difference of a variable (independent or dependent)
∇	=	gradient operator
θ_c	=	contact angle, degrees
ρ	=	density, m/L^3, lbm/ft³
ρ_b	=	fluid density at bubblepoint pressure, m/L^3, lbm/ft³
ρ_g	=	gas density, m/L^3, lbm/ft³
ρ_{gsc}	=	gas density at the standard condition, m/L^3, lbm/ft³
ρ_o	=	oil density, m/L^3, lbm/ft³
ρ_{osc}	=	oil density at the standard condition, m/L^3, lbm/ft³
ρ_{sc}	=	density at the standard condition, m/L^3, lbm/ft³
μ	=	viscosity, cp
μ_b	=	viscosity at bubblepoint, cp
μ_g	=	gas viscosity, cp
μ_w	=	water viscosity, cp
μ_o	=	oil viscosity, cp
ϕ	=	porosity, fraction
Φ	=	flow potential, m/Lt^2, psia
Φ_g	=	flow potential of gas phase, m/Lt^2, psia
Φ_w	=	flow potential of water, m/Lt^2, psia

Superscripts

0	=	reference value of a variable

Subscripts

A	=	spatial location A
B	=	spatial location B
g	=	gas phase
o	=	oil phase
w	=	water phase

References

Coats, K. H., George, W. D., Chu, C. et al. 1974. Three-Dimensional Simulation of Steamflooding. *SPE J*, **14**(06): 573–592. SPE-4500-PA. http://dx.doi.org/10.2118/4500-PA.

Corey, A. T. 1956. The Interrelation Between Gas and Oil Relative Permeabilities. *Producer Monthly*, **19**(November): 38–41. https://www.discovery-group.com/pdfs/Corey_1954.pdf.

Corey, A. T., Rathjens, C. H., Henderson, J. H. et al. 1956. Three-Phase Relative Permeability. *J Pet Technol*, **8**(11): 63–65. SPE-737-G. http://dx.doi.org/10.2118/737-G.

Darcy, H. 1856. *Les Fontaines Publiques de la Ville de Dijon*, first edition. Paris: Victor Dalmont, ed; Libraire des Corps Impériaux des Ponts et Chaussées et des Mines.

Dranchuk, P. M. and Abou-Kassem, H. 1975. Calculation of Z Factor for Natural Gases Using Equation of State. *J Can Pet Technol*, **14**(3): 34. PETSOC-75-03-03. http://dx.doi.org/10.2118/75-03-03.

Ertekin, T., Abou-Kassem, J. H., and King, G. R. 2001. *Basic Applied Reservoir Simulation*. Richardson, Texas, US: Society of Petroleum Engineers.

Hubbert, M. K. 1940. The Theory of Ground-Water Motion. *The Journal of Geology*, **48**(08, Part 1): 785–944. https://www.jstor.org/stable/30057101.

Hubbert, M. K. 1956. Darcy's Law and the Field Equations of the Flow of Underground Fluids. In *Transactions of the Society of Petroleum Engineers*, Vol. 207: 222–239. Richardson, Texas: Society of Petroleum Engineers.

Naar, J. and Henderson, J. H. 1961. An Imbibition Model—Its Application to Flow Behavior and Prediction of Oil Recovery. *SPE J*, **1**(02): 61–70. SPE-1550-G. http://dx.doi.org/10.2118/1550-G.

Naar, J. and Wygal, R. J. 1961. Three-Phase Imbibition Relative Permeability. *SPE J*, **1**(04): 254–258. SPE-90-PA. http://dx.doi.org/10.2118/90-PA.

Stone, H. L. 1970. Probability Model for Estimating Three-Phase Relative Permeability. *J Pet Technol*, **22**(02): 214–218. SPE-2116-PA. http://dx.doi.org/10.2118/2116-PA.

Stone, H. L. 1973. Estimation of Three-Phase Relative Permeability and Residual Oil Data, *J Can Pet Technol*, **12**(4): 53. PETSOC-73-04-06. http://dx.doi.org/10.2118/73-04-06.

Chapter 3

Basic Mathematics of Reservoir Modeling

Partial-differential equations (PDEs) that describe fluid flow dynamics in porous media can be linear or nonlinear equations; however, they are always composed of only first-order and second-order derivatives. Therefore, good understanding of basic differential calculus forms the backbone of reservoir modeling of recovery processes.

3.1 Basic Differential Calculus

3.1.1 First-Order Derivative. The definition of a first-order derivative of a continuous function, f, is given by

$$\frac{df}{dx} = \lim_{\Delta x \to 0} \frac{f(x+\Delta x) - f(x)}{\Delta x}. \quad \text{..(3.1)}$$

If function f is a continuous function of the independent variables x and y and expressed as $f = f(x, y)$, then

$$\frac{df}{dx} = \lim_{\Delta x \to 0} \frac{f(x+\Delta x, y) - f(x, y)}{\Delta x}, \quad \text{..(3.2a)}$$

$$\frac{df}{dy} = \lim_{\Delta y \to 0} \frac{f(x, y+\Delta y) - f(x, y)}{\Delta y}. \quad \text{..(3.2b)}$$

Eqs. 3.2a and 3.2b represent the first-order partial derivatives of function $f(x, y)$ with regard to x and y, respectively.

3.1.2 Chain Rule. The chain rule gives the derivative of a function composed of a complex form of two functions. For example, porosity is a function of pressure, where pressure is a function of time:

$$\phi(t) = \phi[p(t)], \quad \text{..(3.3a)}$$

then,

$$\frac{\partial \phi}{\partial t} = \frac{\partial \phi}{\partial p} \frac{\partial p}{\partial t}. \quad \text{...(3.3b)}$$

3.1.3 Second-Order Derivative. The second derivative of a function f measures the concavity of the graph of function. If the second derivative of a function is positive, then the graph of the function will be concave upward (i.e., it holds water); and if the second derivative is negative, then the graph of the function will be concave downward (i.e., it spills water). If the second derivative is equal to zero, then the function goes through an inflection point.

3.1.4 Taylor-Series Expansion. The Taylor-series expansion of a function, such as $f = f(x, y)$, is a representation of the same function as an infinite sum of terms calculated from the values of the function's derivatives at a single point:

$$f(x, y) = f(x_0, y_0) + \frac{(x - x_0)}{1!} \frac{\partial f}{\partial x} + \frac{(y - y_0)}{1!} \frac{\partial f}{\partial y} + \frac{(x - x_0)^2}{2!} \frac{\partial^2 f}{\partial x^2} + \frac{(y - y_0)^2}{2!} \frac{\partial^2 f}{\partial y^2} + \cdots \quad \text{...................(3.4)}$$

3.1.5 Differential Equations. Differential equations are a special class of equations that relate to a function with one or more of its derivatives. A PDE is called an ordinary differential equation if there is only one independent variable (function itself is the dependent variable). PDEs contain dependent variables, which are functions of more than one independent variable. For example, while

$$\frac{d^2 y}{dx^2} + \frac{dy}{dx} + y(x) = 0. \qquad (3.5)$$

is an ordinary differential equation, the dependent variable, y, is a function of only one independent variable, which is x:

$$\frac{\partial^2 y}{\partial x^2} + \frac{\partial^2 y}{\partial y^2} + y(x, y, t) = \frac{\partial y}{\partial t}. \qquad (3.6)$$

Eq. 3.6 is a PDE (the dependent variable y is a function of three independent variables: x, y, t).

To solve a differential equation, the associated boundary and initial conditions must be specified. In general, each first-order derivative will require one specification. To solve Eq. 3.6 accordingly, it is necessary to have two boundary conditions along the x-direction, two boundary conditions along the y-direction, and one initial condition for the t dimension of the problem.

3.1.6 Leibniz's Rule. Leibniz's rule tells us that integration and differentiation are commutative (Eq. 3.7). For an arbitrary function $f(x,y)$, it states,

$$\frac{\partial}{\partial x} \int_a^b f(x, y)\, dy = \int_a^b \frac{\partial f(x, y)}{\partial x}\, dy. \qquad (3.7)$$

where $x, y, a, b \in \mathbb{R}$. It represents the fact that the interchange of a derivative and an integration is essentially equivalent.

The first-order derivatives in reservoir flow equations appear as either the first-order derivative in time or in the form of boundary conditions at the internal boundaries (e.g., wells) and external boundaries (e.g., structural boundaries of the reservoir). There are two typical boundary conditions specified for reservoir flow equations:

- Dirichlet-type boundary conditions: In this case, the value of the dependent variable is specified at the boundary (e.g., the specification of the sandface pressure at the well is an internal boundary).
- Neumann-type boundary condition: The gradient of the dependent variable is specified across the boundary (e.g., the specifying flow rate at the well). If the specified gradient is equal to zero, then it simply indicates no-flow conditions exist across the boundary where it is specified.

Note that the specification of a Dirichlet- and Neumann-type boundary condition at the same location concurrently is not allowed because such a specification will result in overspecified (ill-posed) problems.

In a PDE, if the coefficients of the derivatives are functions of independent variables (or constants), then such a differential equation is called a linear PDE. Accordingly, a nonlinear PDE will have coefficients that are functions of the dependent variables.

3.2 Finite-Difference Analogs of Derivatives: A Short Review of Finite-Difference Calculus

In finite-difference calculus, the specific notation shown in **Fig. 3.1** is widely used. The notation in Fig. 3.1 is for a 1D function along the x-direction. For a 2D function, the same notation will be $f_{i,j}$, and for a 3D function, we will have $f_{i,j,k}$. **Table 3.1** gives definitions of the most widely used finite-difference operators.

Note that finite-difference calculus operators obey commutative, associative, and distributive rules of the arithmetical operations listed in Eqs. 3.8a through 3.8d, respectively:

- Commutative Rule: $(\nabla + \delta) f_i = (\delta + \nabla) f_i$; $\qquad$ (3.8a)
- Associative Rule: $E \Delta f_i = \Delta E f_i$; $\qquad$ (3.8b)
- Distributive Rule: $\mu(\nabla + \delta) f_i = \mu \nabla f_i + \mu \delta f_i$; $\qquad$ (3.8c)
- Power Rule: $\Delta^{m+n} = \Delta^m \cdot \Delta^n$. $\qquad$ (3.8d)

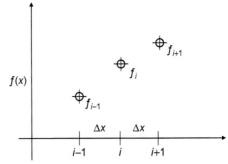

$f(x)$

Similar to trigonometric relationships, one can obtain many relationships between the operators. Eqs. 3.9a through 3.9e list a short summary of such relationships:

Fig. 3.1—Notation of finite-difference calculus.

$$\Delta = E - 1, \qquad (3.9a)$$

$$\nabla = 1 - E^{-1}, \qquad (3.9b)$$

Operation	Notation	Definition/Result
Forward-difference operator: $\Delta_x(f_i)$	Δ_x	$\Delta_x(f_i) = f_{i+1} - f_i$
Backward-difference operator: $\nabla_x(f_i)$	∇_x	$\nabla_x(f_i) = f_i - f_{i-1}$
Shift operator: $E_x(f_i)$	E_x	$E_x(f_i) = f_{i+1}$
Central-difference operator: $\delta_x(f_i)$	δ_x	$\delta_x(f_i) = f_{i+\frac{1}{2}} - f_{i-\frac{1}{2}}$
Average operator: $\mu_x(f_i)$	μ_x	$\mu_x(f_i) = \frac{1}{2}\left(f_{i+\frac{1}{2}} + f_{i-\frac{1}{2}}\right)$

Table 3.1—Common finite-difference operators.

$$\delta = E^{1/2} - E^{-1/2}, \quad (3.9c)$$

$$\mu\delta = \frac{1}{2}(E - E^{-1}), \quad (3.9d)$$

$$\mu^2 = \frac{\delta^2}{4} + 1. \quad (3.9e)$$

3.2.1 Relationship Between Derivatives and Difference Operators. A Taylor-series expansion can be written as

$$f_{i+1} = f_i + \Delta x \frac{\partial f}{\partial x}\Big|_i + \frac{(\Delta x)^2}{2!}\frac{\partial^2 f}{\partial x^2}\Big|_i + \frac{(\Delta x)^3}{3!}\frac{\partial^3 f}{\partial x^3}\Big|_i + \cdots \quad (3.10)$$

Recalling that $E_x(f_i) = f_{i+1}$, and letting $D_x(f_i) = \frac{\partial f}{\partial f}\Big|_i$, then Eq. 3.10 can be written as

$$E_x(f_i) = \left[1 + \Delta x\, D_x + \frac{(\Delta x)^2}{2!}D_x^2 + \frac{(\Delta x)^3}{2!}D_x^3 + \cdots\right](f_i), \quad (3.11)$$

which can be expressed as

$$E_x = 1 + \Delta x\, D_x + \frac{(\Delta x)^2}{2!}D_x^2 + \frac{(\Delta x)^3}{2!}D_x^3 + \cdots \quad (3.12)$$

Now, Eq. 3.12 can be expressed as

$$E_x = e^{\Delta x D_x}. \quad (3.13)$$

Recalling some of the relationships between the operators, one can write the following important relationships between the derivative operator and finite-difference operators:

$$\Delta x D_x = \ln(1 + \Delta_x), \quad (3.14a)$$

$$\Delta x D_x = \ln(1 - \nabla_x)^{-1}, \quad (3.14b)$$

$$\Delta x D_x = 2\sinh^{-1}\left(\frac{\delta}{2}\right). \quad (3.14c)$$

Now, one can write the relationships described by Eqs. 3.14a through 3.14c in an open form by writing the series expansions for the right-hand sides of these equations:

$$\Delta x D_x = \Delta_x - \frac{\Delta_x^2}{2} + \frac{\Delta_x^3}{3} - \frac{\Delta_x^4}{4} + \cdots, \quad (3.15)$$

or

$$D_x = \frac{1}{\Delta_x}\left(\Delta_x - \frac{\Delta_x^2}{2} + \frac{\Delta_x^3}{3} - \frac{\Delta_x^4}{4} + \cdots\right). \dots\dots\dots\dots\dots\dots\dots\dots (3.16)$$

For a first-order approximation,

$$D_x \cong \frac{1}{\Delta x}\Delta_x. \dots\dots\dots\dots\dots\dots\dots\dots\dots\dots\dots\dots\dots (3.17)$$

Eq. 3.17 says that

$$D_x(f_i) \cong \frac{1}{\Delta x}\Delta x(f_i), \dots\dots\dots\dots\dots\dots\dots\dots\dots\dots (3.18)$$

or

$$\left.\frac{df}{dx}\right|_i \cong \frac{f_{i+1} - f_i}{\Delta x}. \dots\dots\dots\dots\dots\dots\dots\dots\dots\dots\dots (3.19)$$

Eq. 3.19 gives the first-order approximation to the first derivative using a forward-difference operator. Following a similar procedure, one can write

$$\Delta x D_x = \nabla_x - \frac{\nabla_x^2}{2} + \frac{\nabla_x^3}{3} - \frac{\nabla_x^4}{4} + \cdots, \dots\dots\dots\dots\dots\dots\dots\dots (3.20)$$

which results in

$$D_x \cong \frac{1}{\Delta x}\nabla_x, \dots\dots\dots\dots\dots\dots\dots\dots\dots\dots\dots\dots\dots (3.21)$$

or

$$\left.\frac{df}{dx}\right|_i \cong \frac{f_i - f_{i-1}}{\Delta x}. \dots\dots\dots\dots\dots\dots\dots\dots\dots\dots\dots (3.22)$$

Eq. 3.22 gives another first-order approximation to the first derivative using a backward-difference operator. Using the Taylor-series expansion for $\sinh^{-1}(x)$, one can write

$$\Delta x D_x = \delta_x - \frac{\delta_x^3}{24} + \frac{\delta_x^5}{640} - \cdots, \dots\dots\dots\dots\dots\dots\dots\dots\dots (3.23)$$

or

$$D_x \cong \frac{1}{\Delta x}\delta_x, \dots\dots\dots\dots\dots\dots\dots\dots\dots\dots\dots\dots\dots (3.24)$$

and finally,

$$\left.\frac{df}{dx}\right|_i \cong \frac{f_{i+\frac{1}{2}} - f_{i-\frac{1}{2}}}{\Delta x}. \dots\dots\dots\dots\dots\dots\dots\dots\dots\dots (3.25)$$

Note that Eq. 3.25 can also be written at full nodes (because there is no discrete point at $i + 1/2$) as

$$\left.\frac{df}{dx}\right|_i \cong \frac{f_{i+1} - f_{i-1}}{2\Delta x}. \dots\dots\dots\dots\dots\dots\dots\dots\dots\dots (3.26)$$

Eqs. 3.19, 3.22, and 3.26 give the finite-difference approximations to the first-order derivative using forward-difference, backward-difference, and central-difference operators. Now, we can extend our discussion to second-order derivatives. From Eq. 3.24, one can write

$$D_x D_x \cong \frac{1}{\Delta x}\delta_x \frac{1}{\Delta x}\delta_x, \dots\dots\dots\dots\dots\dots\dots\dots\dots (3.27)$$

which can be reduced to

$$D_x^2 \cong \frac{1}{(\Delta x)^2} \delta_x^2, \dots\dots\dots\dots\dots\dots\dots\dots\dots\dots\dots\dots\dots\dots\dots\dots\dots\dots\dots\dots\dots \text{(3.28)}$$

and which can be used in approximating the second-order derivative as

$$\left.\frac{d^2 f}{dx^2}\right|_i \cong \frac{1}{(\Delta x)^2} \delta_x \left(f_{i+\frac{1}{2}} - f_{i-\frac{1}{2}} \right), \dots\dots\dots\dots\dots\dots\dots\dots\dots\dots\dots\dots \text{(3.29)}$$

or

$$\left.\frac{d^2 f}{dx^2}\right|_i \cong \frac{1}{(\Delta x)^2} \left[(f_{i+1} - f_i) - (f_i - f_{i-1}) \right], \dots\dots\dots\dots\dots\dots\dots\dots\dots\dots \text{(3.30)}$$

which finally gives

$$\left.\frac{d^2 f}{dx^2}\right|_i \cong \frac{f_{i+1} - 2f_i + f_{i-1}}{(\Delta x)^2}. \dots\dots\dots\dots\dots\dots\dots\dots\dots\dots\dots\dots\dots\dots\dots\dots \text{(3.31)}$$

Eq. 3.31 is the central-difference approximation to the second-order derivative. There is one more second-order derivative that needs to be addressed, which often appears in fluid flow equations of porous media:

$$\frac{\partial}{\partial x} \left[a(x) \frac{\partial f}{\partial x} \right]_i \cong ? \dots\dots\dots\dots\dots\dots\dots\dots\dots\dots\dots\dots\dots\dots\dots\dots\dots\dots\dots\dots \text{(3.32)}$$

The same second-order derivative can be written as Eq. 3.33:

$$\frac{\partial}{\partial x} \left[a(x) \frac{\partial f}{\partial x} \right]_i \cong \frac{\delta_x}{\Delta x} \left[a(x) \frac{\delta_x}{\Delta x} (f) \right]_i, \dots\dots\dots\dots\dots\dots\dots\dots\dots\dots\dots\dots\dots \text{(3.33)}$$

then

$$\frac{\partial}{\partial x} \left[a(x) \frac{\partial f}{\partial x} \right]_i \cong \frac{1}{(\Delta x)^2} \delta_x \left[a_i \left(f_{i+\frac{1}{2}} - f_{i-\frac{1}{2}} \right) \right]_i. \dots\dots\dots\dots\dots\dots\dots\dots\dots \text{(3.34)}$$

or

$$\frac{\partial}{\partial x} \left[a(x) \frac{\partial f}{\partial x} \right]_i \cong \frac{1}{(\Delta x)^2} \delta_x \left[a_{i+\frac{1}{2}} (f_{i+1} - f_i) - a_{i-\frac{1}{2}} (f_i - f_{i-1}) \right]. \dots\dots\dots\dots\dots\dots \text{(3.35)}$$

The use of Eq. 3.35 is critically important in reservoir simulation applications, because fluid-transport terms of the flow equations contain such derivatives as those found in heterogeneous reservoirs. It should also be noted that using the Taylor-series-expansion approximations given by Eqs. 3.19, 3.22, 3.26, and 3.35, can be obtained as shown:

$$f_{i+1} = f_i + \Delta x \left.\frac{\partial f}{\partial x}\right|_i + \frac{(\Delta x)^2}{2!} \left.\frac{\partial^2 f}{\partial x^2}\right|_i + \frac{(\Delta x)^3}{3!} \left.\frac{\partial^3 f}{\partial x^3}\right|_i + \cdots, \dots\dots\dots\dots\dots\dots \text{(3.36a)}$$

$$f_{i-1} = f_i - \Delta x \left.\frac{\partial f}{\partial x}\right|_i + \frac{(\Delta x)^2}{2!} \left.\frac{\partial^2 f}{\partial x^2}\right|_i - \frac{(\Delta x)^3}{3!} \left.\frac{\partial^3 f}{\partial x^3}\right|_i + \cdots. \dots\dots\dots\dots\dots\dots \text{(3.36b)}$$

From Eq. 3.36a, one can solve for $\left.\frac{\partial f}{\partial x}\right|_i$ by truncating the series expansion after the first-order derivative, which results in

$$f_{i+1} = f_i + \Delta x \left.\frac{\partial f}{\partial x}\right|_i. \dots\dots\dots\dots\dots\dots\dots\dots\dots\dots\dots\dots\dots\dots\dots\dots\dots\dots\dots \text{(3.37)}$$

or

$$\left.\frac{\partial f}{\partial x}\right|_i \cong \frac{f_{i+1} - f_i}{\Delta x}, \dots\dots\dots\dots\dots\dots\dots\dots\dots\dots\dots\dots\dots\dots\dots\dots\dots\dots\dots\dots\dots \text{(3.38)}$$

which is the same as Eq. 3.19 (forward-difference approximation). In a similar manner, from Eq. 3.36b one can obtain

$$\left.\frac{df}{dx}\right|_i \cong \frac{f_{i+1} - f_{i-1}}{\Delta x}, \quad \dotfill \quad (3.39)$$

which is the same as the backward-difference approximation given by Eq. 3.22. Now, by subtracting Eq. 3.36b from 3.36a, side-by-side, one obtains

$$f_{i+1} - f_{i-1} = 2\Delta x \left.\frac{\partial f}{\partial x}\right|_i + 2\frac{(\Delta x)^3}{3!}\left.\frac{\partial^3 f}{\partial x^3}\right|_i + \dots \quad \dotfill \quad (3.40)$$

Truncating after the first derivative, one obtains

$$\left.\frac{df}{dx}\right|_i \cong \frac{f_{i+1} - f_{i-1}}{2\Delta x}, \quad \dotfill \quad (3.41)$$

which is the same as the central-difference approximation given by Eq. 3.26. Note that the central-difference approximation is one order more accurate than either forward-difference or backward-difference operators. As in the central-difference approximation, the first term truncated is the third-order derivative, not the second-order derivative, as is the case for the other approximations. Finally, by adding Eqs. 3.36a and 3.36b side-by-side, one obtains

$$f_{i+1} + f_{i-1} = 2f_i + 2\frac{(\Delta x)^2}{2!}\left.\frac{\partial^2 f}{\partial x^2}\right|_i + \dots \quad \dotfill \quad (3.42)$$

From which the finite-difference approximation to the second-order derivative can be obtained as

$$\left.\frac{d^3 f}{dx^2}\right|_i \cong \frac{f_{i+1} - 2f_i + f_{i-1}}{(\Delta x)^2}, \quad \dotfill \quad (3.43)$$

which is the same as Eq. 3.31.

3.3 Matrix Representation of Algebraic Equations

In reservoir simulation, one of the concluding calculations involves solving a system of algebraic equations. Consider the $n \times n$ system of equations shown in Eq. 3.44:

$$
\begin{aligned}
a_{11}x_1 + a_{12}x_2 + a_{13}x_3 + \dots + a_{1m}x_m &= d_1 \\
a_{21}x_1 + a_{22}x_2 + a_{23}x_3 + \dots + a_{2m}x_m &= d_2 \\
a_{31}x_1 + a_{32}x_2 + a_{33}x_3 + \dots + a_{3m}x_m &= d_3 \\
\vdots \qquad \vdots \qquad \vdots \qquad \vdots \qquad \vdots & \\
a_{m1}x_1 + a_{m2}x_2 + a_{m3}x_3 + \dots + a_{mm}x_m &= d_m
\end{aligned}
\quad \dotfill \quad (3.44)
$$

The system of equations in Eq. 3.44 can be written in matrix notation, as follows:

$$
\begin{bmatrix}
a_{11} & a_{12} & a_{13} & \cdots & a_{1m} \\
a_{21} & a_{22} & a_{23} & \cdots & a_{2m} \\
a_{31} & a_{32} & a_{33} & \cdots & a_{3m} \\
\vdots & \vdots & \vdots & \ddots & \vdots \\
a_{m1} & a_{m2} & a_{m3} & \cdots & a_{m4}
\end{bmatrix}
\begin{bmatrix}
x_1 \\ x_2 \\ x_3 \\ \vdots \\ x_m
\end{bmatrix}
=
\begin{bmatrix}
d_1 \\ d_2 \\ d_3 \\ \vdots \\ d_m
\end{bmatrix}
\quad \dotfill \quad (3.45a)
$$

Eq. 3.45a can be written in shorthand notation as

$$[A][x] = [d]. \quad \dotfill \quad (3.45b)$$

In Eq. 3.45b, $[A]$ represents the coefficient matrix and $[x]$ and $[d]$ represent the unknown and right-hand-side vectors, respectively. Later, we will see that many properties of the coefficient matrix $[A]$ will make the solution process either more challenging or more straightforward.

3.4 Problems and Solutions

Problem 3.4.1 (Ertekin et al. 2001)

Show that $\dfrac{\partial}{\partial x}\left[a(p)\dfrac{\partial p}{\partial x}\right]$ can be transformed to $\dfrac{\partial^2 \psi}{\partial x^2}$ using the transformation $\psi = \displaystyle\int_0^p a(p)\,dp$.

Hint: Use the chain rule in conjunction with the Leibniz rule.

Solution to Problem 3.4.1

Applying the chain rule to $\dfrac{\partial p}{\partial x}$, one obtains

$$\frac{\partial p}{\partial x} = \frac{1}{\dfrac{\partial x}{\partial p}} = \frac{1}{\dfrac{\partial \psi}{\partial p}\dfrac{\partial x}{\partial \psi}}. \quad\dotfill\quad (3.46)$$

Note that

$$\frac{\partial \psi}{\partial p} = \frac{\partial}{\partial p}\int_0^p a(p)\,dp. \quad\dotfill\quad (3.47)$$

Using Leibniz rule,

$$\frac{\partial \psi}{\partial p} = \int_0^p \frac{\partial}{\partial p}\big[a(p)\big]\,dp = a(p), \quad\dotfill\quad (3.48)$$

one can write

$$\frac{\partial p}{\partial x} = \frac{1}{a(p)\dfrac{\partial x}{\partial \psi}} = \frac{1}{a(p)}\frac{\partial \psi}{\partial x}. \quad\dotfill\quad (3.49)$$

Thus,

$$\frac{\partial}{\partial x}\left[a(p)\frac{\partial p}{\partial x}\right] = \frac{\partial}{\partial x}\left[a(p)\frac{\partial \psi}{\partial x}\frac{1}{a(p)}\right] = \frac{\partial^2 \psi}{\partial x^2}. \quad\dotfill\quad (3.50)$$

Problem 3.4.2 (Ertekin et al. 2001)

Use the product rule to expand the expression $\dfrac{\partial}{\partial x}\left[\dfrac{A_x(x)\,k_x(x)}{\mu(p)B(p)}\dfrac{\partial p(x)}{\partial x}\right]$.

Solution to Problem 3.4.2

Applying the product rule to the original expression,

$$\frac{\partial}{\partial x}\left[\frac{A_x(x)\,k_x(x)}{\mu(p)B(p)}\frac{\partial p(x)}{\partial x}\right] = \frac{A_x(x)\,k_x(x)}{\mu(p)B(p)}\frac{\partial p^2(x)}{\partial x^2} + \frac{\partial}{\partial x}\left[\frac{A_x(x)\,k_x(x)}{\mu(p)B(p)}\right]\frac{\partial p(x)}{\partial x}, \quad\dotfill\quad (3.51)$$

one needs to further investigate the term $\dfrac{\partial}{\partial x}\left[\dfrac{A_x(x)\,k_x(x)}{\mu(p)B(p)}\right]$:

$$\frac{\partial}{\partial x}\left[\frac{A_x(x)\,k_x(x)}{\mu(p)B(p)}\right] = \frac{\left\{\mu(p)B(p)\dfrac{\partial\big[A_x(x)\,k_x(x)\big]}{\partial x} - A_x(x)\,k_x(x)\dfrac{\partial\big[\mu(p)B(p)\big]}{\partial x}\right\}}{\big[\mu(p)B(p)\big]^2} \quad\dotfill\quad (3.52)$$

$$= \frac{1}{\mu(p)B(p)}\frac{\partial\big[A_x(x)\,k_x(x)\big]}{\partial x} - \frac{A_x(x)\,k_x(x)}{\big[\mu(p)B(p)\big]^2}\frac{\partial\big[\mu(p)B(p)\big]}{\partial x}.$$

One can further extend two derivatives, $\dfrac{1}{\mu(p)B(p)}\dfrac{\partial\left[A_x(x)k_x(x)\right]}{\partial x}$ and $\dfrac{A_x(x)k_x(x)}{\left[\mu(p)B(p)\right]^2}\dfrac{\partial\left[\mu(p)B(p)\right]}{\partial x}$, as follows:

$$\frac{1}{\mu(p)B(p)}\frac{\partial\left[A_x(x)k_x(x)\right]}{\partial x}=\frac{A_x(x)}{\mu(p)B(p)}\frac{\partial\left[k_x(x)\right]}{\partial x}+\frac{k_x(x)}{\mu(p)B(p)}\frac{\partial\left[A_x(x)\right]}{\partial x},\ \dots\dots\dots\dots\dots\ (3.53)$$

and implementing chain rule into the second derivative,

$$\frac{A_x(x)k_x(x)}{\left[\mu(p)B(p)\right]^2}\frac{\partial\left[\mu(p)B(p)\right]}{\partial x}=\frac{\left[A_x(x)k_x(x)\right]}{\left[\mu(p)B(p)\right]^2}\frac{\partial p}{\partial x}\frac{\partial\left[\mu(p)B(p)\right]}{\partial p}$$
$$=\frac{\left[A_x(x)k_x(x)\right]}{\left[\mu(p)B(p)\right]^2}\frac{\partial p}{\partial x}\left\{\mu(p)\frac{\partial\left[B(p)\right]}{\partial p}+B(p)\frac{\partial\left[\mu(p)\right]}{\partial p}\right\}.\quad\dots\dots\dots\dots\ (3.54)$$

Substituting back into the original equation,

$$\frac{\partial}{\partial x}\left[\frac{A_x(x)k_x(x)}{\mu(p)B(p)}\frac{\partial p(x)}{\partial x}\right]=\frac{A_x(x)k_x(x)}{\mu(p)B(p)}\frac{\partial p^2(x)}{\partial x^2}+\frac{\partial}{\partial x}\left[\frac{A_x(x)k_x(x)}{\mu(p)B(p)}\right]\frac{\partial p(x)}{\partial x}$$
$$=\frac{A_x(x)k_x(x)}{\mu(p)B(p)}\frac{\partial p^2(x)}{\partial x^2}+\frac{A_x(x)}{\mu(p)B(p)}\frac{\partial\left[k_x(x)\right]}{\partial x}\frac{\partial p(x)}{\partial x}+\frac{k_x(x)}{\mu(p)B(p)}\quad\dots\dots\ (3.55)$$
$$\frac{\partial\left[Ak_x(x)\right]}{\partial x}\frac{\partial p(x)}{\partial x}-\frac{A_x(x)k_x(x)}{\left[\mu(p)B(p)\right]^2}\left(\frac{\partial p}{\partial x}\right)^2\left\{\mu(p)\frac{\partial\left[B(p)\right]}{\partial p}+B(p)\frac{\partial\left[\mu(p)\right]}{\partial p}\right\}.$$

Problem 3.4.3 (Ertekin et al. 2001)

Show that density of a liquid can be expressed as a function of pressure as $\rho=\rho_o\left[1+c(p-p_o)\right]$, where ρ_o = reference density at reference pressure (p_o) and reservoir temperature, and c = compressibility of the liquid (very small). *Hint:* Use the Taylor-series expansion.

Solution to Problem 3.4.3

One can write a Taylor-series expansion in the vicinity of ρ_o as follows:

$$\rho=\rho_o+(p-p_o)\frac{d\rho}{dp}+\frac{(p-p_o)^2}{2!}\frac{d^2\rho}{dp^2}+\cdots,\ \dots\dots\dots\dots\dots\dots\ (3.56)$$

where $\dfrac{1}{\rho_o}\dfrac{d\rho_o}{dp}=c$ and $\dfrac{d^n\rho}{dp^n}\approx 0$ for $n\geq 2$. Because $\dfrac{d\rho_o}{dp}=c\rho_o$, one obtains

$$\rho=\rho_o+c\rho_o(p-p_o)=\rho_o\left[1+c(p-p_o)\right].\ \dots\dots\dots\dots\ (3.57)$$

Problem 3.4.4 (Ertekin et al. 2001)

Derive an expression of $\dfrac{\partial B_g}{\partial p_g}$ for a real gas.

Solution to Problem 3.4.4

Recall that $B_g(p_g)=\dfrac{p_{sc}T}{\alpha_c T_{sc}}\dfrac{z_g}{p_g}$. Assuming isothermal reservoir conditions, the derivative of B_g in terms of pressure is

$$\frac{\partial B_g}{\partial p_g}=\frac{\partial}{\partial p_g}\left(\frac{p_{sc}T}{\alpha_c T_{sc}}\frac{z_g}{p_g}\right)=\frac{p_{sc}T}{\alpha_c T_{sc}}\frac{\partial}{\partial p_g}\left(\frac{z_g}{p_g}\right),\ \dots\dots\dots\dots\ (3.58)$$

because

$$\frac{\partial}{\partial p_g}\left(\frac{z_g}{p_g}\right) = \frac{p_g \dfrac{\partial z_g}{\partial p_g} - z_g}{p_g^2} = \frac{1}{p_g}\frac{\partial z_g}{\partial p_g} - \frac{z_g}{p_g^2}; \quad \dotfill \quad (3.59)$$

therefore,

$$\frac{\partial B_g}{\partial p_g} = \frac{\partial}{\partial p_g}\left(\frac{p_{sc}T}{\alpha_c T_{sc}}\frac{z_g}{p_g}\right) = \frac{p_{sc}T}{\alpha_c T_{sc}}\left(\frac{1}{p_g}\frac{\partial z_g}{\partial p_g} - \frac{z_g}{p_g^2}\right). \quad \dotfill \quad (3.60)$$

In addition, if one factors out $\dfrac{z_g}{p_g}$ from $\left(\dfrac{1}{p_g}\dfrac{\partial z_g}{\partial p_g} - \dfrac{z_g}{p_g^2}\right)$,

$$\left(\frac{1}{p_g}\frac{\partial z_g}{\partial p_g} - \frac{z_g}{p_g^2}\right) = \frac{z_g}{p_g}\left(\frac{1}{z_g}\frac{\partial z_g}{\partial p_g} - \frac{1}{p_g}\right) = -\frac{z_g}{p_g}c_g. \quad \dotfill \quad (3.61)$$

Recalling that $c_g = \dfrac{1}{p_g} - \dfrac{1}{z_g}\dfrac{\partial z_g}{\partial p_g}$, the original expression can be further simplified to

$$\frac{\partial B_g}{\partial p_g} = \frac{\partial}{\partial p_g}\left(\frac{p_{sc}T}{\alpha_c T_{sc}}\frac{z_g}{p_g}\right) = \frac{p_{sc}T}{\alpha_c T_{sc}}\frac{z_g}{p_g}c_g. \quad \dotfill \quad (3.62)$$

Problem 3.4.5 (Ertekin et al. 2001)

Identify the following equations as a linear or nonlinear PDEs:

1. $\dfrac{\partial}{\partial x}\left[C_1\dfrac{\partial p}{\partial x}\right] = 0$, where C_1 is a constant.

2. $\dfrac{\partial}{\partial x}\left[C_1(x)\dfrac{\partial p}{\partial x}\right] = 0$, where C_1 is a function of x.

3. $\dfrac{\partial}{\partial x}\left[C_1(p)\dfrac{\partial p}{\partial x}\right] = 0$, where $C_1(p)$ is a function of p.

Solution to Problem 3.4.5

1. Because C_1 is a constant, one can simplify the PDE to

$$\frac{\partial}{\partial x}\left[C_1\frac{\partial p}{\partial x}\right] = 0 \rightarrow C_1\frac{\partial^2 p}{\partial x^2} = 0, \quad \dotfill \quad (3.63)$$

which simplifies to

$$\left(\frac{\partial^2 p}{\partial x^2}\right)^1 = 0. \quad \dotfill \quad (3.64)$$

Because the derivative in Eq. 3.64 is raised to power one, the original equation is a linear PDE.

2. In this case, C_1 is a function of x, and one can simply expand the derivative by using the product rule:

$$\frac{\partial}{\partial x}\left[C_1(x)\frac{\partial p}{\partial x}\right] = 0 \rightarrow \left[\frac{\partial C_1(x)}{\partial x}\right]^1\left(\frac{\partial p}{\partial x}\right)^1 + C_1(x)\left(\frac{\partial^2 p}{\partial x^2}\right)^1 = 0. \quad \dotfill \quad (3.65)$$

Again, all of the derivatives shown in Eq. 3.65 are raised to power one. Therefore, the original PDE is linear.

3. When C_1 is a function of pressure (primary unknown of the original PDE), the original PDE can be simplified as

$$\frac{\partial}{\partial x}\left[C_1(p)\frac{\partial p}{\partial x}\right]=0 \rightarrow \frac{\partial C_1(p)}{\partial x}\frac{\partial p}{\partial x}+C_1(p)\frac{\partial^2 p}{\partial x^2}=0. \quad\dots\dots\dots (3.66)$$

One can apply the chain rule on term $\dfrac{\partial C_1(p)}{\partial x}$ and obtain

$$\left[\frac{\partial C_1(p)}{\partial p}\frac{\partial p}{\partial x}\right]\frac{\partial p}{\partial x}+C_1(p)\frac{\partial^2 p}{\partial x^2}=0 \rightarrow \frac{\partial C_1(p)}{\partial x}\left(\frac{\partial p}{\partial x}\right)^2+C_1(p)\left(\frac{\partial^2 p}{\partial x^2}\right)^1=0. \quad\dots (3.67)$$

In Eq. 3.67, the derivative $\dfrac{\partial p}{\partial x}$ is raised to power two, which signals that the original equation is a nonlinear PDE.

Problem 3.4.6 (Ertekin et al. 2001)

Consider a differential equation in the form of $k_x\dfrac{\partial^2 p}{\partial x^2}+k_y\dfrac{\partial^2 p}{\partial y^2}=0$. Use the transformations $X=\dfrac{x}{\sqrt{k_x}}$ and $Y=\dfrac{y}{\sqrt{k_y}}$ to transform the given differential equation into Laplace's equation, $\dfrac{\partial^2 p}{\partial X^2}+\dfrac{\partial^2 p}{\partial Y^2}=0$.

Solution to Problem 3.4.6

The suggested transformations, $X=\dfrac{x}{\sqrt{k_x}}$ and $Y=\dfrac{y}{\sqrt{k_y}}$, on the original PDE, $k_x\left(\dfrac{\partial^2 p}{\partial x^2}\right)+k_y\left(\dfrac{\partial^2 p}{\partial y^2}\right)=0$, will transform the coordinate (x-y) to another coordinate system (X-Y).

This implies that the transformation is taking place on the independent variables. For the first-order derivative of p in terms of x,

$$\frac{\partial p}{\partial x}=\frac{\partial p}{\partial X}\frac{\partial X}{\partial x}=\frac{\partial p}{\partial X}\frac{1}{\sqrt{k_x}}. \quad\dots\dots\dots (3.68)$$

For the second-order derivative,

$$\frac{\partial^2 p}{\partial x^2}=\frac{\partial}{\partial X}\left(\frac{\partial p}{\partial x}\right)\frac{\partial X}{\partial x}=\frac{\partial}{\partial X}\left(\frac{\partial p}{\partial X}\frac{1}{\sqrt{k_x}}\right)\frac{1}{\sqrt{k_x}}=\frac{1}{k_x}\frac{\partial^2 p}{\partial X^2}. \quad\dots (3.69)$$

In a similar fashion, we have also

$$\frac{\partial^2 p}{\partial x^2}=\frac{1}{k_y}\frac{\partial^2 p}{\partial Y^2}. \quad\dots\dots\dots (3.70)$$

Thus, the original PDE becomes

$$k_x\left(\frac{1}{k_x}\frac{\partial^2 p}{\partial X^2}\right)+k_y\left(\frac{1}{k_y}\frac{\partial^2 p}{\partial Y^2}\right)=0, \quad\dots\dots\dots (3.71)$$

or simply,

$$\frac{\partial^2 p}{\partial X^2}+\frac{\partial^2 p}{\partial Y^2}=0. \quad\dots\dots\dots (3.72)$$

Note that the original PDE was written for an anisotropic reservoir, whereas, with the help of the transformation, the new PDE is obtained for an isotropic reservoir. Therefore, once the solution on the (X-Y) plane is obtained as $p(X,Y)$, it needs to be brought back to the (x-y) plane by substituting for X and Y using $X=\dfrac{x}{\sqrt{k_x}}$ and $Y=\dfrac{y}{\sqrt{k_y}}$, respectively.

Problem 3.4.7 (Ertekin et al. 2001)

Use $x = \ln r$ and $y = \theta$ to transform $\dfrac{1}{r}\dfrac{\partial}{\partial r}\left(r\dfrac{\partial p}{\partial r}\right) + \dfrac{1}{r^2}\dfrac{\partial^2 p}{\partial \theta^2} = 0$ to $\dfrac{\partial^2 p}{\partial x^2} + \dfrac{\partial^2 p}{\partial y^2} = 0.$

Solution to Problem 3.4.7

The derivative of x in terms of r is $\dfrac{\partial x}{\partial r} = \dfrac{1}{r}$. The derivative of y in terms of θ is $\dfrac{\partial y}{\partial \theta} = 1$.

One can apply the chain rule to the original PDE:

$$\frac{1}{r}\frac{\partial}{\partial x}\left(r\frac{\partial p}{\partial x}\frac{\partial x}{\partial r}\right)\frac{\partial x}{\partial r} + \frac{1}{r^2}\frac{\partial p}{\partial y}\left(\frac{\partial p}{\partial y}\frac{\partial y}{\partial \theta}\right)\frac{\partial y}{\partial \theta} = 0. \quad\ldots\ldots\ldots\ldots\ldots\ldots (3.73)$$

Substituting $\dfrac{\partial x}{\partial r} = \dfrac{1}{r}$ and $\dfrac{\partial y}{\partial \theta} = 1$ into the Eq. 3.73 yields

$$\frac{\partial}{\partial x}\left(\frac{\partial p}{\partial x}\right)\frac{1}{r^2} + \frac{1}{r^2}\frac{\partial p}{\partial y}\left(\frac{\partial p}{\partial y}\right) = 0. \quad\ldots\ldots\ldots\ldots\ldots\ldots (3.74)$$

Finally,

$$\frac{\partial^2 p}{\partial x^2} + \frac{\partial^2 p}{\partial y^2} = 0. \quad\ldots\ldots\ldots\ldots\ldots\ldots\ldots\ldots\ldots\ldots\ldots\ldots\ldots (3.75)$$

Problem 3.4.8 (Ertekin et al. 2001)

Demonstrate the equivalence of the following operations:

1. $\Delta E = E\Delta$.
2. $\Delta b^x = b^x\left(b^{\Delta x} - 1\right)$.
3. $D = \dfrac{1}{\Delta x}\sinh^{-1}(\mu\delta)$, where $D = \dfrac{\partial}{\partial x}$.
4. $(E-2)(E-1)[f(x)] = -\Delta x$, where $f(x) = 2^{\frac{x}{\Delta x}} + x$.
5. $\Delta - \nabla = \delta^2$.
6. $\Delta\nabla = \delta^2$.
7. $\nabla = E^{-1}\Delta$.
8. $\Delta\nabla = \Delta\nabla$.
9. $\mu\delta = \dfrac{1}{2}(\Delta + \nabla)$.
10. $E = \left(\dfrac{\delta}{2} + \sqrt{1 + \dfrac{\delta^2}{4}}\right)^2$.

Solution to Problem 3.4.8

1. $\Delta E = E\nabla$: Applying the operator ΔE to an arbitrary function f_i yields

$$\Delta E(f_i) = \Delta(f_{i+1}) = f_{i+2} - f_{i+1}. \quad\ldots\ldots\ldots\ldots\ldots\ldots\ldots\ldots (3.76)$$

Similarly,

$$E\Delta(f_i) = E(f_{i+1} - f_i) = f_{i+2} - f_{i+1}. \quad\ldots\ldots\ldots\ldots\ldots\ldots\ldots (3.77)$$

Thus, $\Delta E = E\Delta$.

2. $\Delta b^x = b^x \left(b^{\Delta x} - 1 \right)$: Implementing Δ operator to function b^x,

$$\Delta(b^x) = b^{x+\Delta x} - b^x = b^x b^{\Delta x} - b^x = b^x \left(b^{\Delta x} - 1 \right), \quad\dots\dots\dots\dots\dots\dots\dots\dots\dots\dots\dots\dots \quad (3.78)$$

which is equivalent to the right-hand side of the original equation.

3. $D = \dfrac{1}{\Delta x} \sinh^{-1}(\mu\delta)$, where $D = \dfrac{\partial}{\partial x}$: Because $\Delta x D = \ln E$, therefore,

$$\sinh(\Delta x D) = \frac{e^{\ln E} - e^{-\ln E}}{2} = \frac{E^1 - E^{-1}}{2}. \quad\dots\dots\dots\dots\dots\dots\dots\dots\dots\dots\dots \quad (3.79)$$

Also, $\mu\delta = \dfrac{E^1 - E^{-1}}{2}$.

One can correlate the preceding two expressions, $\mu\delta = \sinh(\Delta x D)$, by taking the inverse of the hyperbolic sin function; and $\Delta x D = \sinh^{-1}(\mu\delta)$, which yields $D = \dfrac{1}{\Delta x} \sinh^{-1}(\mu\delta)$.

4. $(E-2)(E-1)[f(x)] = -\Delta x$, where $f(x) = 2^{\frac{x}{\Delta x}} + x$: Applying operator $(E-1)$ on function $f(x)$,

$$
\begin{aligned}
(E-1)[f(x)] &= \left(2^{\frac{x+\Delta x}{\Delta x}} + x + \Delta x - 2^{\frac{x}{\Delta x}} - x \right) \\
&= \left(2^{\frac{x}{\Delta x}} 2^{\frac{\Delta x}{\Delta x}} + x + \Delta x - 2^{\frac{x}{\Delta x}} - x \right). \quad\dots\dots\dots\dots\dots\dots\dots\dots \quad (3.80) \\
&= \left(2^{\frac{x}{\Delta x}} + \Delta x \right)
\end{aligned}
$$

Then, implementing operator $(E-2)$ on function $2^{\frac{x}{\Delta x}} + \Delta x$,

$$(E-2)\left(2^{\frac{x}{\Delta x}} + \Delta x \right) = 2^{\frac{x+\Delta x}{\Delta x}} + \Delta x - \left(2 \times 2^{\frac{x}{\Delta x}} + 2\Delta x \right) = -\Delta x. \quad\dots\dots\dots\dots\dots\dots \quad (3.81)$$

5. $\Delta - \nabla = \delta^2$: Applying operator $(\Delta - \nabla)$ on an arbitrary function f_i,

$$(\Delta - \nabla)f_i = \Delta f_i - \nabla f_i = (f_{i+1} - f_i) - (f_i - f_{i-1}). \quad\dots\dots\dots\dots\dots\dots\dots\dots\dots\dots \quad (3.82)$$

Similarly,

$$\delta^2 f_i = \delta\left(f_{i+\frac{1}{2}} - f_{i-\frac{1}{2}} \right) = (f_{i+1} - f_i) - (f_i - f_{i-1}). \quad\dots\dots\dots\dots\dots\dots\dots\dots\dots \quad (3.83)$$

Thus, $\Delta - \nabla = \delta^2$.

6. $\Delta\nabla = \delta^2$: Knowing that $\nabla f_i = (f_i - f_{i-1})$ and

$$\Delta\nabla f_i = \Delta(f_i - f_{i-1}) = (f_{i+1} - f_i) - (f_i - f_{i-1}). \quad\dots\dots\dots\dots\dots\dots\dots\dots\dots\dots \quad (3.84)$$

Moreover,

$$\delta^2 f_i = (f_{i+1} - f_i) - (f_i - f_{i-1}). \quad\dots\dots\dots\dots\dots\dots\dots\dots\dots\dots\dots\dots\dots\dots \quad (3.85)$$

The equivalency in Eq. 3.85 proves that $\Delta\nabla = \delta^2$.

7. $\nabla = E^{-1}\Delta$: Because $\nabla(f_i) = f_{i+1} - f_i$ and

$$E^{-1}\Delta(f_i) = E^{-1}(f_{i+1} - f_i) = (f_i - f_{i-1}), \text{ also, } \nabla(f_i) = -f_i - f_{i-1}. \quad\dots\dots\dots\dots\dots\dots\dots \quad (3.86)$$

Thus, $\nabla = E^{-1}\Delta$.

8. $\Delta\nabla = \Delta\nabla$: Because

$$\Delta\nabla f_i = \Delta(f_i - f_{i-1}) = (f_{i+1} - f_i) - (f_i - f_{i-1}), \quad \dots\dots\dots\dots\dots\dots\dots\dots\dots\dots\dots\dots\dots (3.87)$$

and

$$\Delta\nabla f_i = \nabla(f_{i+1} - f_i) = (f_{i+1} - f_i) - (f_i - f_{i-1}), \quad \dots\dots\dots\dots\dots\dots\dots\dots\dots\dots\dots\dots\dots (3.88)$$

therefore, $\Delta\nabla = \Delta\nabla$.

9. $\mu\delta = \dfrac{1}{2}(\Delta + \nabla)$: Because $\delta(f_i) = f_{i+\frac{1}{2}} - f_{i-\frac{1}{2}}$,

$$\mu[\delta(f_i)] = \mu\left(f_{i+\frac{1}{2}} - f_{i-\frac{1}{2}}\right) = \frac{1}{2}\left[(f_{i+1} - f_i) + (f_i - f_{i-1})\right], \quad \dots\dots\dots\dots\dots\dots\dots\dots (3.89)$$

$$\frac{1}{2}(\Delta + \nabla)(f_i) = \frac{1}{2}\left[\Delta(f_i) + \nabla(f_i)\right] = \frac{1}{2}\left[(f_{i+1} - f_i) + (f_i - f_{i-1})\right]. \quad \dots\dots\dots\dots\dots\dots (3.90)$$

Thus, $\mu\delta = \dfrac{1}{2}(\Delta + \nabla)$.

10. $E = \left(\dfrac{\delta}{2} + \sqrt{1 + \dfrac{\delta^2}{4}}\right)^2$: Knowing that $1 + \dfrac{\delta^2}{4} = \mu^2$ and $\left(\dfrac{\delta}{2} + \sqrt{1 + \dfrac{\delta^2}{4}}\right)^2 = \left(\dfrac{\delta}{2} + \mu\right)^2$;

also, substituting $\delta = E^{\frac{1}{2}} - E^{-\frac{1}{2}}$ and $\mu = \dfrac{E^{\frac{1}{2}} + E^{-\frac{1}{2}}}{2}$ into the expression, one ends up with

$$\left(\frac{\delta}{2} + \mu\right)^2 = \left(\frac{E^{\frac{1}{2}} - E^{-\frac{1}{2}}}{2} + \frac{E^{\frac{1}{2}} + E^{-\frac{1}{2}}}{2}\right)^2 = \left(E^{\frac{1}{2}}\right)^2 = E. \quad \dots\dots\dots\dots\dots\dots\dots\dots (3.91)$$

Problem 3.4.9 (Ertekin et al. 2001)

Although less common, finite differences can be used to represent integrals. If the operator J is defined as

$$J[f(x)] = \int_x^{x+h} f(t)\,dt, \text{ prove that } J[f(x)] = \frac{h\Delta f(x)}{\left(\Delta - \dfrac{\Delta^2}{2} + \dfrac{\Delta^3}{2} - \dfrac{\Delta^4}{2} + \cdots\right)}.$$

Solution to Problem 3.4.9

Because $J[f(x)] = \displaystyle\int_x^{x+h} f(t)\,dt$, one can take the first-order derivative in terms of x on both sides of the equation and obtain

$$\frac{\partial}{\partial x}\{J[f(x)]\} = f(x+h) - f(x) \text{ or } \frac{\partial}{\partial x}\{J[f(x)]\} = \Delta f(x), \quad \dots\dots\dots\dots\dots\dots\dots\dots (3.92)$$

which can be further simplified to

$$DJ[f(x)] = \Delta f(x). \quad \dots\dots\dots\dots\dots\dots\dots\dots\dots\dots\dots\dots\dots\dots\dots\dots\dots\dots\dots\dots\dots (3.93)$$

Thus,

$$J[f(x)] = \frac{\Delta}{D}[f(x)] = \frac{(\Delta_x)\Delta}{(\Delta_x)D}[f(x)] \text{ (note that in this case, } \Delta x = h). \quad \dots\dots\dots\dots\dots\dots (3.94)$$

Invoking that, $(\Delta_x)D = \ln(1+\Delta)$, $J[f(x)] = \dfrac{h\Delta[f(x)]}{\ln(1+\Delta)}$.

Recall the Taylor-series expansion:

$$\ln(1+\Delta) = \Delta - \frac{\Delta^2}{2} + \frac{\Delta^3}{3} - \frac{\Delta^4}{4} + \cdots \quad\quad\quad\quad\quad (3.95)$$

Therefore,

$$J\left[f(x_0)\right] = \frac{h\Delta\left[f(x_0)\right]}{\left(\Delta - \dfrac{\Delta^2}{2} + \dfrac{\Delta^3}{3} - \dfrac{\Delta^4}{4} + \cdots\right)}. \quad\quad\quad\quad (3.96)$$

Problem 3.4.10 (Ertekin et al. 2001)

Solve $\begin{bmatrix} 3 & 0 \\ 0 & 4 \end{bmatrix}\begin{bmatrix} x_1 \\ x_2 \end{bmatrix} = \begin{bmatrix} 6 \\ 16 \end{bmatrix}$ using the matrix-inversion method.

Solution to Problem 3.4.10

The inverse of the coefficient matrix is

$$\begin{bmatrix} 3 & 0 \\ 0 & 4 \end{bmatrix}^{-1} = \begin{bmatrix} \dfrac{1}{3} & 0 \\ 0 & \dfrac{1}{4} \end{bmatrix}, \quad\quad\quad\quad\quad (3.97)$$

and therefore,

$$\begin{bmatrix} x_1 \\ x_2 \end{bmatrix} = \begin{bmatrix} 3 & 0 \\ 0 & 4 \end{bmatrix}^{-1}\begin{bmatrix} 6 \\ 16 \end{bmatrix} = \begin{bmatrix} \dfrac{1}{3} & 0 \\ 0 & \dfrac{1}{4} \end{bmatrix}\begin{bmatrix} 6 \\ 16 \end{bmatrix} = \begin{bmatrix} 2 \\ 4 \end{bmatrix}. \quad\quad\quad (3.98)$$

Problem 3.4.11 (Ertekin et al. 2001)

What is the inverse of a diagonal matrix of the form $\begin{bmatrix} a_{11} & 0 & 0 & 0 \\ 0 & a_{22} & 0 & 0 \\ 0 & 0 & a_{33} & 0 \\ 0 & 0 & 0 & a_{44} \end{bmatrix}$?

Solution to Problem 3.4.11

The inverse of the diagonal matrix is shown in Eq. 3.99.

$$\begin{bmatrix} a_{11} & 0 & 0 & 0 \\ 0 & a_{22} & 0 & 0 \\ 0 & 0 & a_{33} & 0 \\ 0 & 0 & 0 & a_{44} \end{bmatrix}^{-1} = \begin{bmatrix} 1/a_{11} & 0 & 0 & 0 \\ 0 & 1/a_{22} & 0 & 0 \\ 0 & 0 & 1/a_{33} & 0 \\ 0 & 0 & 0 & 1/a_{44} \end{bmatrix}. \quad (3.99)$$

Problem 3.4.12 (Ertekin et al. 2001)

Calculate the coefficient of isothermal compressibility, c_g, in psi^{-1} for a real gas at 1,500 psia and 120°F, given that $z_g = \dfrac{577.1}{T} - 0.00015p$, where p is in psia and T is in °R.

Solution to Problem 3.4.12

Recall that the definition of compressibility for a real gas is given by $c_g = \dfrac{1}{p} - \dfrac{1}{z_g}\left(\dfrac{\partial z}{\partial p}\right)_T$, $\dfrac{\partial z_g}{\partial p} = -0.00015$, and

$$z_g = \frac{577.1}{120 + 459.67} - 0.00015(1,500) = 0.77. \quad\quad\quad\quad (3.100)$$

Then,

$$c_g = \frac{1}{p} - \frac{1}{z_g}\left(\frac{\partial z_g}{\partial p}\right)_T = \frac{1}{1500} - \frac{1}{0.77}(-0.00015) = 8.615 \times 10^{-4}\,\text{psi}^{-1}. \quad\text{................................ (3.101)}$$

Problem 3.4.13 (Ertekin et al. 2001)

The function $f(x) = 3x^2 + 6x - 5$ is given.

1. Calculate the value of $\left.\dfrac{\partial f}{\partial x}\right|_{x=1}$ using the forward-, backward-, and central-difference approximations with $\Delta x = 1$. Compare the results with the exact value of $\left.\dfrac{\partial f}{\partial x}\right|_{x=1}$.

2. Calculate the value of $\left.\dfrac{\partial^2 f}{\partial x^2}\right|_{x=1}$ using the central-difference approximation. Use $\Delta x = 1$ to compare the result against the exact value of $\left.\dfrac{\partial^2 f}{\partial x^2}\right|_{x=1}$.

Solution to Problem 3.4.13

1.
$$f(x)|_{x=1} = 3 \times 1^2 + 6 \times 1 - 5 = 4, \quad\text{... (3.102)}$$

$$f(x+\Delta x)|_{x=1,\Delta x=1} = 3 \times 2^2 + 6 \times 2 - 5 = 19, \quad\text{.. (3.103)}$$

$$f(x-\Delta x)|_{x=1,\Delta x=1} = 3 \times 0^2 + 6 \times 0 - 5 = -5. \quad\text{... (3.104)}$$

- Forward difference:

$$\frac{df}{dx} = \frac{f(x+\Delta x) - f(x)}{\Delta x} = \frac{19-4}{1} = 15. \quad\text{.. (3.105)}$$

- Backward difference:

$$\frac{df}{dx} = \frac{f(x) - f(x-\Delta x)}{\Delta x} = \frac{4-(-5)}{1} = 9. \quad\text{... (3.106)}$$

- Central difference:

$$\frac{df}{dx} = \frac{f(x+\Delta x) - f(x-\Delta x)}{2\Delta x} = \frac{19-(-5)}{2} = 12. \quad\text{..................................... (3.107)}$$

Analytical differentiation yields

$$\frac{\partial[f(x)]}{\partial x} = 6x + 6 = 6(1) + 6 = 12. \quad\text{.. (3.108)}$$

2. Central difference:

$$\frac{d^2 f}{dx^2} = \frac{f(x+\Delta x) + f(x-\Delta x) - 2f(x)}{(\Delta x)^2} = \frac{19 + (-5) - 2(4)}{1} = 6. \quad\text{............................. (3.109)}$$

Analytical differentiation yields

$$\left.\frac{d^2 f}{dx^2}\right|_{x=1} = 6. \quad\text{.. (3.110)}$$

Problem 3.4.14 (Ertekin et al. 2001)

Consider the function $f(x) = x^3 - 4x^2 + 6$.

1. Calculate the value of $\left.\dfrac{\partial f}{\partial x}\right|_{x=2}$ using backward-, forward-, and central-difference approximations (let $\Delta x = 1$).

2. Calculate the value of $\left.\dfrac{\partial^2 f}{\partial x^2}\right|_{x=2}$ using central-difference approximation (again, let $\Delta x = 1$).

3. Among the four approximations you calculated in Parts 1 and 2, which contains no error? Why?

Solution to Problem 3.4.14

1.

$$f(x)\big|_{x=2} = 2^3 - 4 \times 2^2 + 6 = -2, \dotfill (3.111)$$

$$f(x + \Delta x)\big|_{x=2, \Delta x=1} = 3^3 - 4 \times 3^2 + 6 = -3, \dotfill (3.112)$$

$$f(x - \Delta x)\big|_{x=2, \Delta x=1} = 1^3 - 4 \times 1^2 + 6 = 3. \dotfill (3.113)$$

- Backward-difference approximation:

$$\left.\frac{df}{dx}\right|_{x=2} = \frac{f(3) - f(2)}{1} = \frac{-3 - (-2)}{1} = -1. \dotfill (3.114)$$

- Forward-difference approximation:

$$\left.\frac{df}{dx}\right|_{x=2} = \frac{f(1) - f(1)}{1} = \frac{(-2) - (3)}{1} = -5. \dotfill (3.115)$$

- Central-difference approximation:

$$\left.\frac{df}{dx}\right|_{x=2} = \frac{f(3) - f(1)}{2 \times 1} = \frac{(-3) - (3)}{2 \times 1} = -3. \dotfill (3.116)$$

2.

$$\left.\frac{d^2 f}{dx^2}\right|_{x=2} = \frac{f(3) + f(1) - 2f(2)}{1^2} = \frac{(-3) + (3) - 2(-2)}{1^2} = 4. \dotfill (3.117)$$

3. Analytically,

$$\left.\frac{df}{dx}\right|_{x=2} = 3x^2 - 8x = 3 \times (2)^2 - 8 \times 2 = -4, \dotfill (3.118)$$

$$\left.\frac{d^2 f}{dx^2}\right|_{x=2} = 6x - 8 = 6 \times 2 - 8 = 4. \dotfill (3.119)$$

Central-difference approximation to the second-order derivative displays no error, because in the Taylor-series expansion, the first term truncated is a fourth-order term. The fourth-order derivative of the function given in this problem, $f(x) = x^3 - 4x^2 + 6$, vanishes, which is why no truncation error is introduced by way of truncating the Taylor-series expansion, starting with the fourth-order term.

Problem 3.4.15 (Ertekin et al. 2001)

Use the transformations $X = \left(\dfrac{k_y}{k_x}\right)^{1/4} x$ and $Y = \left(\dfrac{k_x}{k_y}\right)^{1/4} y$ on the PDE of the form $k_x \dfrac{\partial^2 p}{\partial x^2} + k_y \dfrac{\partial^2 p}{\partial y^2} = 0$. What important observation can be made on the basis of this transformation?

Solution to Problem 3.4.15

To apply the transformation, the second-order derivatives need to be expanded using the product rule:

$$\frac{\partial^2 p}{\partial x^2} = \frac{\partial}{\partial x}\left(\frac{\partial p}{\partial x}\right) = \frac{\partial}{\partial X}\left(\frac{\partial p}{\partial X}\frac{\partial X}{\partial x}\right)\frac{\partial X}{\partial x} = \frac{\partial^2 p}{\partial X^2}\sqrt{\frac{k_y}{k_x}}. \quad\ldots\ldots\ldots\ldots\ldots\ldots\ldots\ldots \text{(3.120)}$$

Similarly,

$$\frac{\partial^2 p}{\partial y^2} = \frac{\partial}{\partial y}\left(\frac{\partial p}{\partial y}\right) = \frac{\partial}{\partial Y}\left(\frac{\partial p}{\partial Y}\frac{\partial Y}{\partial y}\right)\frac{\partial Y}{\partial y} = \frac{\partial^2 p}{\partial Y^2}\sqrt{\frac{k_x}{k_y}}. \quad\ldots\ldots\ldots\ldots\ldots\ldots\ldots\ldots \text{(3.121)}$$

Thus, the original PDE becomes

$$k_x\frac{\partial^2 p}{\partial X^2}\sqrt{\frac{k_y}{k_x}} + k_y\frac{\partial^2 p}{\partial Y^2}\sqrt{\frac{k_x}{k_y}} = 0, \quad\ldots\ldots\ldots\ldots\ldots\ldots\ldots\ldots \text{(3.122)}$$

which simplifies to

$$(k_xk_y)^{\frac{1}{2}}\frac{\partial^2 p}{\partial X^2} + (k_xk_y)^{\frac{1}{2}}\frac{\partial^2 p}{\partial Y^2} = 0. \quad\ldots\ldots\ldots\ldots\ldots\ldots\ldots\ldots \text{(3.123)}$$

Observation: as seen in Eq. 3.123, an anisotropic-homogeneous domain is converted to an isotropic-homogeneous domain with permeabilities along the *X*- and *Y*-directions, because $k = (k_xk_y)^{\frac{1}{2}}$ (geometric mean of the original permeabilities along the *x*- and *y*-directions).

Problem 3.4.16 (Ertekin et al. 2001)

Use the differential equation known as Poisson's equation, $\frac{\partial^2 p}{\partial x^2} + \frac{\partial^2 p}{\partial y^2} + \frac{\partial^2 p}{\partial z^2} + C_1 = 0$, where C_1 is a constant. Implement the transformation $\psi(x,y,z) = p(x,y,z) - \frac{c}{2}x^2$ on the dependent variable *p*. What is the resulting PDE?

Solution to Problem 3.4.16

Using the transformation rule,

$$\frac{\partial^2 p}{\partial x^2} = \frac{\partial^2 \psi}{\partial x^2} - c, \quad\ldots\ldots\ldots\ldots\ldots\ldots\ldots\ldots \text{(3.124)}$$

$$\frac{\partial^2 p}{\partial y^2} = \frac{\partial^2 \psi}{\partial y^2}, \quad\ldots\ldots\ldots\ldots\ldots\ldots\ldots\ldots \text{(3.125)}$$

$$\frac{\partial^2 p}{\partial z^2} = \frac{\partial^2 \psi}{\partial z^2}. \quad\ldots\ldots\ldots\ldots\ldots\ldots\ldots\ldots \text{(3.126)}$$

Substituting the above transformation terms into the original PDE, yielding

$$\frac{\partial^2 \psi}{\partial x^2} + \frac{\partial^2 \psi}{\partial y^2} + \frac{\partial^2 \psi}{\partial z^2} = 0. \quad\ldots\ldots\ldots\ldots\ldots\ldots\ldots\ldots \text{(3.127)}$$

Problem 3.4.17

Consider functions $f(x) = x^2 - 2x + 3$ and $g(x) = x^4 - 6$. Calculate the values of $\left.\frac{\partial^2 f}{\partial x^2}\right|_{x=2}$ and $\left.\frac{\partial^2 g}{\partial x^2}\right|_{x=2}$ using central finite-difference approximations. Which one of the approximation contains fewer errors? Explain the reason. Use $(\Delta x = 1)$.

Solution to Problem 3.4.17

$$f(x+\Delta x) = f(2+1) = 3^2 - 2 \times 3 + 6, \quad \dots\dots\dots\dots\dots\dots\dots\dots \quad (3.128)$$

$$f(x-\Delta x) = f(2-1) = 1^2 - 2 \times 1 + 3 = 2, \quad \dots\dots\dots\dots\dots\dots\dots \quad (3.129)$$

$$f(x) = f(2) = 2^2 - 2 \times 2 + 3 = 3, \quad \dots\dots\dots\dots\dots\dots\dots\dots \quad (3.130)$$

$$g(x+\Delta x) = g(2+1) = 3^4 - 6 = 75, \quad \dots\dots\dots\dots\dots\dots\dots\dots \quad (3.131)$$

$$g(x-\Delta x) = g(2-1) = 1^4 - 6 = -5, \quad \dots\dots\dots\dots\dots\dots\dots\dots \quad (3.132)$$

$$g(x) = g(2) = 2^4 - 6 = 10. \quad \dots\dots\dots\dots\dots\dots\dots\dots\dots \quad (3.133)$$

Therefore,

$$\left.\frac{d^2 f}{dx^2}\right|_{x=2} = \frac{f(3) + f(1) - 2f(2)}{(\Delta x)(\Delta x)} = \frac{6 + 2 - 2 \times 3}{(1)(1)} = 2. \quad \dots\dots\dots \quad (3.134)$$

The exact solution is

$$\left.\frac{d^2 f}{dx^2}\right|_{x=2} = 2, \quad \dots\dots\dots\dots\dots\dots\dots\dots\dots\dots\dots \quad (3.135)$$

$$\left.\frac{d^2 g}{dx^2}\right|_{x=2} = \frac{g(3) + g(1) - 2g(2)}{(1)(1)} = \frac{75 - 55 - 2(40)}{1} = 50. \quad \dots\dots\dots \quad (3.136)$$

The exact solution is

$$\left.\frac{d^2 g}{dx^2}\right|_{x=2} = 12x^2 = 48. \quad \dots\dots\dots\dots\dots\dots\dots\dots\dots \quad (3.137)$$

Function $f(x)$ is a second-order function, and its third-order derivative vanishes. In writing the central-difference approximation for the second-order derivative, the truncated terms go to zero; therefore, they will not contribute to any error. Function $g(x)$ is a fourth-order function, and its third-order and fourth-order derivatives will not go to zero; therefore, they will contribute to the error observed (50 vs. 48).

Problem 3.4.18

Consider the function $y = 2x + 3$ within the interval of $0 < x < 2$. Using $\Delta x = 1$, as shown in **Fig. 3.2,** calculate the forward-, backward-, and central-difference approximation to $\left.\frac{dy}{dx}\right|_{x=1}$. Compare the results with the analytical derivative at the same point and comment.

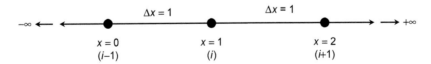

Fig. 3.2—Discretization of function $y(x)$ within the interval of $0 < x < 2$.

Solution to Problem 3.4.18

$$\left.y\right|_{x=1} = 2 \times (1) + 3 = 5, \quad \dots\dots\dots\dots\dots\dots\dots\dots\dots \quad (3.138)$$

$$\left.y\right|_{x+\Delta x} = 2 \times (1+1) + 3 = 7, \quad \dots\dots\dots\dots\dots\dots\dots\dots \quad (3.139)$$

$$\left.y\right|_{x-\Delta x} = 2 \times (1-1) + 3 = 3. \quad \dots\dots\dots\dots\dots\dots\dots\dots \quad (3.140)$$

- Forward-numerical difference:

$$\frac{dy}{dx}\bigg|_{x=1} = \frac{y|_{x+\Delta x} - y|_x}{\Delta x} = \frac{7-5}{1} = 2. \dotfill (3.141)$$

- Backward-numerical difference:

$$\frac{dy}{dx}\bigg|_{x=1} = \frac{y|_x - y|_{x-\Delta x}}{\Delta x} = \frac{5-3}{1} = 2. \dotfill (3.142)$$

- Central-numerical difference:

$$\frac{dy}{dx}\bigg|_{x=1} = \frac{y|_{x+\Delta x} - y|_{x-\Delta x}}{2\Delta x} = \frac{7-3}{2} = 2. \dotfill (3.143)$$

- Analytical solution: $\dfrac{dy}{dx} = 2.$

Because $y = 2x + 3$ is a linear function, the numerical differentiation and analytical differentiation yield the same values. Also, note that applying forward difference at $x = 0$ and backward difference at $x = 2$ will yield the same numerical value.

Problem 3.4.19

Given the following function, $f(x) = x^3 - 3x^2 + 4x - 6$, obtain approximate values of $\dfrac{\partial f}{\partial x}\bigg|_{x=5}$ and $\dfrac{\partial^2 f}{\partial x^2}\bigg|_{x=5}$ using the following:

1. Forward difference,
2. Backward difference,
3. Central difference,

with $\Delta x = 1$. Also, compare you results against the analytical solutions.

Solution to Problem 3.4.19

Knowing that $x = 5$ and $\Delta x = 1$,

$$f(5) = 5^3 - 3(5)^2 + 4 \times 5 - 6 = 64, \dotfill (3.144)$$

$$f(5+1) = 6^3 - 3(6)^2 + 4 \times 6 - 6 = 126, \dotfill (3.145)$$

$$f(5-1) = 4^3 - 3(4)^2 + 4 \times 4 - 4 = 26, \dotfill (3.146)$$

$$f'(x) = 3x^2 - 6x + 4, \dotfill (3.147)$$

$$f''(x) = 6x - 6. \dotfill (3.148)$$

At $x = 5$,

$$\frac{\partial f}{\partial x}\bigg|_{x=5} \approx 3(5)^2 - 6(5) + 4 = 49, \dotfill (3.149)$$

$$\frac{\partial^2 f}{\partial x^2}\bigg|_{x=5} \approx 6(5) - 6 = 24. \dotfill (3.150)$$

First-order derivative:

1. Forward difference:

$$\frac{\partial f}{\partial x}\bigg|_{x=5} \approx \frac{f_{i+1} - f_i}{\Delta x} = \frac{126 - 64}{1} = 62. \dotfill (3.151)$$

2. Backward difference:

$$\left.\frac{\partial f}{\partial x}\right|_{x=5} \approx \frac{f_i - f_{i-1}}{\Delta x} = \frac{64 - 26}{1} = 38. \dots\dots\dots\dots\dots\dots\dots\dots\dots (3.152)$$

3. Central difference:

$$\left.\frac{\partial f}{\partial x}\right|_{x=5} \approx \frac{f_{i+1} - f_{i-1}}{2\Delta x} = \frac{126 - 26}{2} = 50. \dots\dots\dots\dots\dots\dots\dots\dots (3.153)$$

Second-order derivative:

1. Forward difference:

$$\left.\frac{\partial^2 f}{\partial x^2}\right|_{x=5} \approx \frac{f_{i+2} - 2f_{i+1} + f_i}{(\Delta x)^2} = \frac{218 - 2(126) + 64}{(1)^2} = 30. \dots\dots\dots\dots (3.154)$$

2. Backward difference:

$$\left.\frac{\partial^2 f}{\partial x^2}\right|_{x=5} \approx \frac{f_i - 2f_{i-1} + f_{i-2}}{(\Delta x)^2} = \frac{64 - 2(26) + 6}{(1)^2} = 18. \dots\dots\dots\dots (3.155)$$

3. Central difference:

$$\left.\frac{\partial^2 f}{\partial x^2}\right|_{x=5} \approx \frac{f_{i+1} - 2f_i + f_{i-1}}{(\Delta x)^2} = \frac{26 - 2(64) + 126}{(1)^2} = 24. \dots\dots\dots\dots (3.156)$$

Nomenclature

a = constant

a_i = hypothetical coefficient

$a_{m,n}$ = the (m,n) element of matrix $[A]$

$[A]$ = coefficient matrix of a matrix equation

b = constant

B = formation volume factor, L^3/L^3, res bbl/STB or res bbl/scf

B_g = gas formation volume factor, L^3/L^3, res bbl/SCF

c = compressibility, Lt^2/m, psi^{-1}

c_g = gas compressibility, Lt^2/m, psi^{-1}

C = constant

$[d]$ = right-hand-side vector of a matrix equation

d_m = The mth element of vector $[d]$

D = elevation with respect to absolute datum being positive upward, L, ft

D_x = differential operator operating along the x-direction

E_x = shift operator operating along the x-direction

f = function

f_i = a tensional function-defined 1D domain

$f_{i,j}$ = a tensional function-defined 2D domain

$f_{i,j,k}$ = a tensional function-defined 3D domain

$g(x)$ = function of x

h = thickness, L, ft

J = an operator defined in Problem 3.4.9

k = permeability, L^2, darcies or perms

k_x = permeability in the x-direction, L^2, darcies or perms

k_y = permeability in the y-direction, L^2, darcies or perms

p = pressure, m/Lt^2, psia

p_0 = reference pressure, m/Lt^2, psia

p_g = gas pressure, m/Lt^2, psia

p_{sc} = standard condition pressure, m/Lt^2, psia

r = radial distance in both cylindrical and spherical coordinator system, L, ft

$\mathbb{R}$ = real number domain

t = time, t, days

T = absolute temperature, T, °R

T_{sc} = standard condition temperature, T, °R

x = distance in x-direction in the Cartesian coordinate system, L, ft; arbitrary constant

$[x]$ = the unknown vector of a matrix equation

x_m = The mth element of vector$[x]$

X = transformed x

y = distance in y-direction in the Cartesian coordinate system, L, ft; arbitrary constant

Y = transformed y

z = distance in z-direction in the Cartesian coordinate system, L, ft

z_g	=	gas compressibility factor, dimensionless
α_c	=	unit conversion factor, = 5.615 ft³/bbl
Δ	=	forward-difference operator
$\dot{\nabla}$	=	backward-difference operator
δ	=	central-difference operator
θ	=	angle in θ-direction in both cylindrical and spherical coordinator system, rad

μ	=	viscosity, cp
ρ	=	density, m/L³, lbm/ft³
ϕ	=	porosity, fraction
ψ	=	defined function

Superscripts

0	=	reference value of a variable

References

Ertekin, T., Abou-Kassem, J. H., and King, G. R. 2001. *Basic Applied Reservoir Simulation*. Richardson, Texas, US: Society of Petroleum Engineers.

Chapter 4

Flow Equations Describing Single-Phase Flow In Porous Media

In the formulation of basic equations for single-phase flow in porous media, one must consider the type of fluid and the flow geometry expected to be dominant. Flow equations are written in a certain geometry and in a certain number of dimensions in view of the reservoir's structural characteristics and well-completion techniques. For example, 1D radial-cylindrical flow geometry is most-widely used in single vertical-well applications. If the well is horizontal or a single well is hydraulically fractured with vertical fracture planes, an elliptical-cylindrical flow geometry is expected to develop and, hence, the use of an elliptical-cylindrical coordinate system is warranted. In field-scale simulation models with the presence of many wells located in the computational domain, rectangular flow geometry in one, two, or three dimensions typically is used. One important thing to keep in mind is the shape of the equipotential lines and streamlines. In an orthogonal-coordinate system, the elemental volume should conform to these two important elements of the flow problem. In doing so, it will also be possible to minimize the grid-orientation effects.

4.1 Flow Geometries and Flow Dimensions

To write the continuity equation as the first step, one has to determine what kind of flow geometry will be used to develop the flow equations. In writing the continuity equation, a volume element will be used as the control volume. The shape of the volume element depends on the coordinate system that will be used to describe the flow problem. The structural characteristics of the reservoir, together with the reservoir rock property distributions, will determine the number and nature of the flow dimensions that need to be included in developing the flow equation that most-closely describes the flow problem.

4.1.1 Flow Geometry. Most of the field-scale (multiwell) flow equations are studied in rectangular coordinates. The control volume in developing the flow equation in rectangular coordinates has to be a rectangular prism. The edges of the rectangular prism must be parallel to the orthogonal directions of the flow geometry shown in **Fig. 4.1.**

If flow is 1D (e.g., only in the x-direction), then streamlines will be parallel to the x-direction. If flow is in two dimensions (e.g., only in the x-y plane), then streamlines will be parallel to the x- and y-directions. Finally, if the flow is in three dimensions, then streamlines in all three principal directions will be considered (**Fig. 4.2**). It is necessary to visualize that the

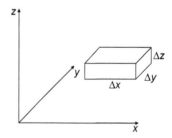

Fig. 4.1—Elemental volume in rectangular coordinates.

shape and orientation of the streamlines will not only depend on the dimensionality of the control volume, but also on the location and geometry of the source/sink (well). Originating in the perpendicular direction of the reservoir (control-volume) boundaries, the streamlines will converge in the direction of the pressure-sink point, which may result in straight

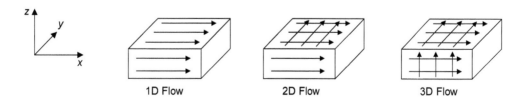

Fig. 4.2—Streamlines in 1D, 2D, and 3D flow in rectangular coordinates.

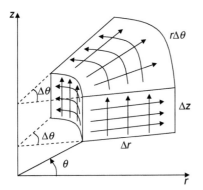

Fig. 4.3—Elemental volume in radial-cylindrical coordinates.

or curved streamline geometries. Depending on the shape of the streamline, to define it, one-, two-, or three-dimensional coordinates of each streamline point may be needed. However, it should be clearly visualized that at every given location, there is only one streamline point (corresponding to the velocity and direction of the fluid movement at this location). Accordingly, the representation (arrows) in Fig. 4.2 should not be interpreted as if there are streamlines going in multiple directions, which contradicts the fact that streamlines cannot intersect each other; this would imply that a fluid particle at the intersection point would have two different velocities and/or directions. The discussion presented here also applies to **Figs. 4.3 and 4.4.**

4.1.2 Radial-Cylindrical Flow Geometry. Radial-cylindrical flow geometry is particularly appealing for single-well problems. Fig. 4.3 shows the principal flow directions of the flow geometry and its elemental volume and orthogonal streamlines.

Similar to rectangular-flow geometry, 1D, 2D, and 3D flow geometries, and relevant streamlines in radial-cylindrical coordinates, are shown in Fig. 4.4.

Beyond rectangular and radial-cylindrical flow geometries, elliptical-cylindrical flow and spherical flow geometries may also become applicable. The elliptical-cylindrical flow geometry can be helpful in studying flow around a vertical well with a vertical hydraulic fracture plane or in formations with significantly large permeability contrast, with respect to the principal directions of flow. Spherical flow conditions may exist within the immediate vicinity of the perforations (See **Fig. 4.5** for a description of the spherical flow element and spherical flow components). Finally, ellipsoidal flow geometry, as shown in **Fig. 4.6,** can develop in three dimensions as well around horizontal wells completed in thick formations.

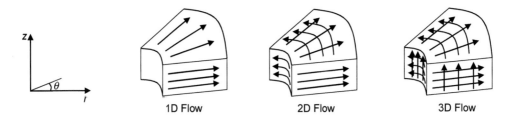

1D Flow   2D Flow   3D Flow

Fig. 4.4—Streamlines describing the flow in 1D, 2D, and 3D flow in radial-cylindrical coordinates.

$$A: \left(r-\frac{\Delta r}{2},\theta+\frac{\Delta\theta}{2},\varphi-\frac{\Delta\varphi}{2}\right)$$

$$B: \left(r-\frac{\Delta r}{2},\theta-\frac{\Delta\theta}{2},\varphi-\frac{\Delta\varphi}{2}\right)$$

$$C: \left(r-\frac{\Delta r}{2},\theta-\frac{\Delta\theta}{2},\varphi+\frac{\Delta\varphi}{2}\right)$$

$$D: \left(r-\frac{\Delta r}{2},\theta+\frac{\Delta\theta}{2},\varphi+\frac{\Delta\varphi}{2}\right)$$

$$E: \left(r+\frac{\Delta r}{2},\theta+\frac{\Delta\theta}{2},\varphi-\frac{\Delta\varphi}{2}\right)$$

$$F: \left(r+\frac{\Delta r}{2},\theta-\frac{\Delta\theta}{2},\varphi-\frac{\Delta\varphi}{2}\right)$$

$$G: \left(r+\frac{\Delta r}{2},\theta-\frac{\Delta\theta}{2},\varphi+\frac{\Delta\varphi}{2}\right)$$

$$H: \left(r+\frac{\Delta r}{2},\theta+\frac{\Delta\theta}{2},\varphi+\frac{\Delta\varphi}{2}\right)$$

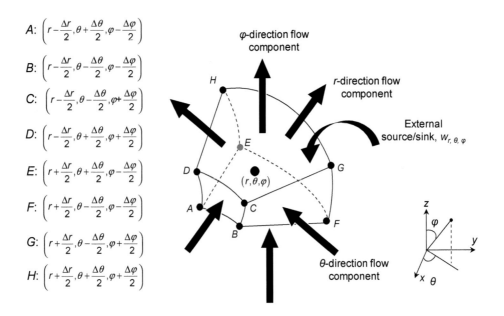

Fig. 4.5—Streamlines describing the flow in 1D, 2D, and 3D flow in radial-cylindrical coordinates.

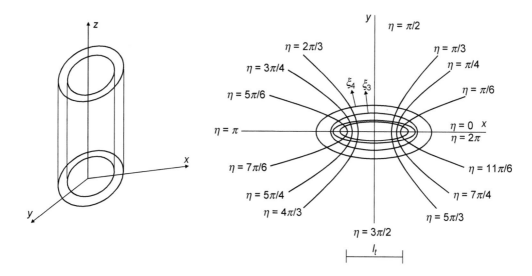

Fig. 4.6—Elliptic cylindrical flow geometry and representation of the streamlines (η) and equipotential (ξ) contours.

In all flow dynamics previously described, note that, ideally, streamlines and equipotential lines describe the shape of the elemental volume and, hence, the flow geometry. In other words, while rectangular flow geometry is most-extensively used, it is the geometry that differs most significantly from actual flow geometry, especially around the wellbore. The use of curvilinear coordinates, in which streamlines and equipotential lines are congruent with the principal directions of the coordinate system, will remove the grid-orientation effects. The radial-cylindrical, spherical, and ellipsoidal flow geometries can be considered as curvilinear coordinate systems. It is also possible to use hybrid coordinate systems, so that actual flow geometries can be more realistically captured, particularly around the wells and physical limits of the reservoir. For example, a hybrid flow geometry around a vertical well located in a rectangular block can be achieved by using the radial-cylindrical flow geometry to describe the flow geometry within the rectangular wellblock (Pedrosa and Aziz 1986). Also, in 3D flow, hemispherical flow geometry can be placed at the location of each perforation.

4.2 Single-Phase Flow Equations in Porous Media

The continuity equation is a mathematical expression that captures the principle of conservation of mass, and it can be developed by considering the flow of mass through a control volume (control volume is in rectangular coordinates):

$$\left(mass\ in\right) - \left(mass\ out\right) = \left(net\ change\ in\ mass\ content\right). \dots \dots \dots \dots \dots \dots \dots \dots \dots \text{(4.1)}$$

The general form of the single-phase flow equation in 3D rectangular coordinates can be written as (if Eq. 4.2 is written in field units with the conversion factor $\alpha_c = 5.615$)

$$-\frac{\partial}{\partial x}\left(\rho v_x A_x\right)\Delta x - \frac{\partial}{\partial y}\left(\rho v_y A_y\right)\Delta y - \frac{\partial}{\partial z}\left(\rho v_z A_z\right)\Delta z + \frac{q_m}{\alpha_c} = \frac{V_b}{\alpha_c}\frac{\partial}{\partial t}\left(\phi\rho\right). \dots \dots \dots \dots \text{(4.2)}$$

In Eq. 4.2, when a velocity term is introduced using Darcy's law as the constituent equation, then a fundamental law from the physics of flow in porous media is introduced into the formulation (note the use of potential gradients in Eq. 4.3):

$$v_s = -\frac{k_x}{\mu}\frac{\partial\Phi}{\partial s}, \text{ with } s = x,\ y, \text{ and } z. \dots \dots \dots \dots \dots \dots \dots \dots \dots \dots \dots \text{(4.3)}$$

In Eq. 4.3, permeability is expressed in perms (recall that 1 darcy = 1.127 perms).

After substituting Eq. 4.3, Eq. 4.2 becomes

$$\frac{\partial}{\partial x}\left(\rho\frac{A_x k_x}{\mu B}\frac{\partial\Phi}{\partial x}\right)\Delta x + \frac{\partial}{\partial y}\left(\rho\frac{A_y k_y}{\mu B}\frac{\partial\Phi}{\partial y}\right)\Delta y + \frac{\partial}{\partial z}\left(\rho\frac{A_z k_z}{\mu B}\frac{\partial\Phi}{\partial z}\right)\Delta z + \frac{q_m}{\alpha_c} = \frac{V_b}{\alpha_c}\frac{\partial}{\partial t}\left(\frac{\phi\rho}{B}\right). \dots \dots \dots \dots \text{(4.4)}$$

Eq. 4.4 is now written in surface conditions with the introduction of the formation-volume factor, B, which is expressed in reservoir volume/surface volume. Also, note that at this stage, q_m has the dimensions of $\left[\frac{m}{t}\right]$. Whether Eq. 4.4 represents the single-phase flow for an incompressible fluid, a slightly compressible fluid, or compressible fluid will be decided by how the density term is handled.

4.2.1 Single-Phase Incompressible Flow Equation. For an incompressible fluid, density is not a function of pressure, because it is constant. Dividing Eq. 4.4 by ρ and realizing that $q\,(\text{STB/D}) = \dfrac{q_m}{\alpha_c \rho}$, in which $\alpha_c = 5.615$, we end up with the incompressible flow equation for single-phase flow conditions (recall that in this case $B = 1$, and we also are assuming that ϕ is not a function of pressure):

$$\frac{\partial}{\partial x}\left(\frac{A_x k_x}{\mu B}\frac{\partial \Phi}{\partial x}\right)\Delta x + \frac{\partial}{\partial y}\left(\frac{A_y k_y}{\mu B}\frac{\partial \Phi}{\partial y}\right)\Delta y + \frac{\partial}{\partial z}\left(\frac{A_z k_z}{\mu B}\frac{\partial \Phi}{\partial z}\right)\Delta z + q = 0. \dots\dots\dots\dots\dots\dots\dots(4.5)$$

Eq. 4.5 represents the most-comprehensive form of the single-phase incompressible flow. Eq. 4.5 accommodates heterogeneous and anisotropic reservoir rock properties, allows the presence of the gravitational forces in three directions, and is written for reservoirs with some existing wells in the flow domain.

If the single-phase incompressible flow equation is written in radial-cylindrical coordinates in three dimensions, then one ends up with

$$\frac{1}{r}\frac{\partial}{\partial r}\left(r\frac{k_r}{\mu}\frac{\partial \Phi}{\partial r}\right) + \frac{1}{r^2}\frac{\partial}{\partial \theta}\left(\frac{k_\theta}{\mu}\frac{\partial \Phi}{\partial \theta}\right) + \frac{\partial}{\partial z}\left(\frac{k_z}{\mu}\frac{\partial \Phi}{\partial z}\right) = 0. \dots\dots\dots\dots\dots\dots\dots(4.6)$$

4.2.2 Single-Phase Slightly Compressible Flow Equation. As previously discussed, for a slightly compressible fluid, the density, viscosity, and formation volume factor (FVF) exhibit weak dependence on pressure. Furthermore, for a slightly compressible fluid, we assume that compressibility remains constant within the pressure range of interest (recall that $\alpha_c = 5.615$ and k is in perms):

$$\frac{\partial}{\partial x}\left(\frac{A_x k_x}{\mu B}\frac{\partial \Phi}{\partial x}\right)\Delta x + \frac{\partial}{\partial y}\left(\frac{A_y k_y}{\mu B}\frac{\partial \Phi}{\partial y}\right)\Delta y + \frac{\partial}{\partial z}\left(\frac{A_z k_z}{\mu B}\frac{\partial \Phi}{\partial z}\right)\Delta z + q = \frac{V_b \phi}{\alpha_c}\frac{\partial p}{\partial t}. \dots\dots\dots\dots\dots(4.7)$$

If we ignore the depth gradients, assume homogeneous and isotropic property distribution, and finally, treat μ and B as constants (because both μ and B are weak functions of pressure), then, assuming there is no existing well, Eq. 4.7 collapses to

$$\frac{\partial^2 p}{\partial x^2} + \frac{\partial^2 p}{\partial y^2} + \frac{\partial^2 p}{\partial z^2} = \frac{B}{\alpha_c}\frac{\phi \mu c}{k}\frac{\partial p}{\partial t}. \dots\dots\dots\dots\dots\dots\dots\dots\dots\dots(4.8)$$

In Eq. 4.8, the group that appears on the right-hand side, $\dfrac{\phi \mu c}{k}$, is called the inverse of the hydraulic-diffusivity constant or, in other terms,

$$\eta = \frac{k}{\phi \mu c}. \dots\dots\dots\dots\dots\dots\dots\dots\dots\dots\dots\dots\dots\dots\dots\dots\dots\dots\dots\dots\dots(4.9)$$

The hydraulic-diffusivity group has the dimensions of $\dfrac{\lfloor L^2 \rfloor}{[t]}$, and it scales the magnitude of pressure change at a given point in space over time. In other words, as η tends to be large, the right-hand side of Eq. 4.8 becomes smaller. In fact, in the limiting case in which $c \to 0$ for an incompressible fluid, $\eta \to \infty$ and the right-hand side of Eq. 4.8 goes to zero. This limiting case is true for incompressible fluids, because the right-hand side of the equation becomes zero. This indicates that the pressure surface over the reservoir is instantaneously generated, and it does not change with time.

One of the common uses of the slightly compressible equation is in well-test-analysis applications when it is written in 1D radial-cylindrical coordinates (r-direction only), as shown in Eq. 4.10:

$$\frac{1}{r}\frac{\partial}{\partial r}\left(r\frac{\partial p}{\partial x}\right) = \frac{\phi \mu c}{k_r}\frac{\partial p}{\partial t}. \dots\dots\dots\dots\dots\dots\dots\dots\dots\dots\dots\dots\dots\dots(4.10)$$

4.2.3 Single-Phase Compressible Flow Equation. Compressible fluid flow is the norm for gas reservoirs. Because the gravitational forces exerted by the gas columns are relatively small, it is customary to assume that $\dfrac{\partial \Phi}{\partial s} = \dfrac{\partial p}{\partial s}$ for $s = x, y, z$. The density of a gas is a function of pressure, and it can be represented using the real-gas law, such that

$$\rho = \frac{p\text{MW}}{z_g RT}. \dots\dots\dots\dots\dots\dots\dots\dots\dots\dots\dots\dots\dots\dots\dots\dots\dots\dots\dots\dots(4.11)$$

This time, it is appropriate to substitute Eq. 4.11 into Eq. 4.2. One can show that in this case, the single-phase compressible flow equation can be expressed as

$$\frac{\partial}{\partial x}\left(\frac{A_x k_x}{\mu_g B_g}\frac{\partial p}{\partial x}\right)\Delta x + \frac{\partial}{\partial y}\left(\frac{A_y k_y}{\mu_g B_g}\frac{\partial p}{\partial y}\right)\Delta y + \frac{\partial}{\partial z}\left(\frac{A_z k_z}{\mu_g B_g}\frac{\partial p}{\partial z}\right)\Delta z + q_{gsc} = \frac{V_b \phi}{\alpha_c}\frac{\partial}{\partial t}\left(\frac{1}{B_g}\right). \dots\dots\dots\dots(4.12)$$

Note that we arrived at Eq. 4.12 by using Eq. 4.13:

$$B_g = \frac{p_{sc} z_g T}{\alpha_c T_{sc} p}. \quad \dots\dots\dots\dots\dots\dots\dots\dots\dots\dots\dots\dots\dots\dots\dots\dots\dots\dots\dots\dots\dots\dots\dots \quad (4.13)$$

Eq. 4.12, like its previous counterpart Eq. 4.7, assumes that porosity is independent of pressure.

4.2.4 Practical Field Units of the Single-Phase Flow Equations. Table 4.1 gives a comprehensive summary of the practical fields used with single-phase flow equations.

Variable	Symbol(s)	Practical Field Unit	Equation Number
Length	Δx, Δy, Δz	ft	Eqs. 4.5, 4.7, and 4.12
Area	A_x, A_y, A_z	ft^2	Eqs. 4.5, 4.7, and 4.12
Permeability	k_x, k_y, k_z	perms	Eqs. 4.5, 4.7, and 4.12
Viscosity	μ	cp	Eqs. 4.5, 4.7, and 4.12
Gas FVF	B_g	res bbl/scf	Eqs. 4.12 and 4.13
Liquid FVF	B	res bbl/STB	Eqs. 4.5 and 4.7
Potential gradient	$\frac{\partial \Phi}{\partial s}(s=x,y,z)$	psi/ft	Eqs. 4.5 and 4.7
Pressure gradient	$\frac{\partial p}{\partial s}(s=x,y,z)$	psi/ft	Eqs 4.5, 4.7, and 4.12
Liquid flow rate	q_{sc}	STB/D	Eqs. 4.5 and 4.7
Gas flow rate	q_{gsc}	scf/D	Eq. 4.12
Gridblock bulk volume	V_b	ft^3	Eqs. 4.5, 4.7, and 4.12
Fluid density	ρ	lbm/ft^3	Eqs. 4.2, 4.3, 4.4, and 4.11
Compressibility	c	1/psi	Eq. 4.7
Absolute temperature	T	°R	Eq. 4.11
Porosity	ϕ	fraction	Eqs. 4.5, 4.7, and 4.12
Compressibility factor	z	dimensionless	Eqs. 4.11 and 4.13
Time	t	day	Eqs. 4.5, 4.7, and 4.12
Equation constant	α_c	dimensionless	Eqs. 4.7 and 4.12

Table 4.1—Variables and practical field units of the single-phase flow equations.

4.3 Problems and Solutions

Problem 4.3.1 (Ertekin et al. 2001)

Develop the single-phase flow equation in 3D spherical coordinates. To develop the mass balance, use the elementary volume shown in Fig. 4.5. (*Note:* Spherical-flow geometry is encountered around the perforations and also around wells that are partially penetrating a thick formation.)

Solution to Problem 4.3.1

The mass conservation of the displayed unit volume can be written using mass flux and accumulation:

$$\dot{m}_{r-\frac{\Delta r}{2},\theta,\varphi} - \dot{m}_{r+\frac{\Delta r}{2},\theta,\varphi} + \dot{m}_{r,\theta-\frac{\Delta\theta}{2},\varphi} - \dot{m}_{r,\theta+\frac{\Delta\theta}{2},\varphi} + \dot{m}_{r,\theta,\varphi-\frac{\Delta\varphi}{2}} - \dot{m}_{r,\theta,\varphi+\frac{\Delta\varphi}{2}} = \dot{m}_{r,\theta,\varphi}\Big|_{t+\Delta t} - \dot{m}_{r,\theta,\varphi}\Big|_{t}. \quad \dots\dots\dots\dots \quad (4.14)$$

And we have $m = \rho v$:

$$(\rho v)_{r-\frac{\Delta r}{2},\theta,\varphi} \Delta t - (\rho v)_{r+\frac{\Delta r}{2},\theta,\varphi} \Delta t + (\rho v)_{r,\theta-\frac{\Delta\theta}{2},\varphi} \Delta t - (\rho v)_{r,\theta+\frac{\Delta\theta}{2},\varphi} \Delta t$$
$$+ (\rho v)_{r,\theta,\varphi-\frac{\Delta\varphi}{2}} \Delta t - (\rho v)_{r,\theta,\varphi+\frac{\Delta\varphi}{2}} \Delta t = (\rho V_b \phi)_{r,\theta,\varphi}\Big|_{t+\Delta t} - (\rho V_b \phi)_{r,\theta,\varphi}\Big|_{t} \quad \dots\dots\dots\dots\dots\dots\dots\dots\dots \quad (4.15)$$

From Darcy's law,

$$v_r = \frac{\sin\varphi\Delta\varphi\Delta\theta r^2 k}{\mu}\frac{\partial p}{\partial r}, \quad\quad\quad\quad\quad (4.16)$$

$$v_\theta = \frac{\Delta r\,\Delta\varphi k}{\sin\varphi\mu}\frac{\partial p}{\partial\theta}, \quad\quad\quad\quad\quad (4.17)$$

$$v_\varphi = \frac{\Delta r\Delta\theta\sin\varphi k}{\mu}\frac{\partial p}{\partial\varphi}. \quad\quad\quad\quad\quad (4.18)$$

And we have $V_b = r^2\sin\varphi\Delta\varphi\Delta\theta dr$.

Thus,

$$
\begin{aligned}
&\sin\varphi\Delta\varphi\Delta\theta\left(r-\frac{\Delta r}{2}\right)^2\Delta t\frac{k\rho}{\mu}\frac{\partial p}{\partial r}\bigg|_{r-\frac{\Delta r}{2},\theta,\varphi} - \sin\varphi\Delta\varphi\Delta\theta\left(r+\frac{\Delta r}{2}\right)^2\Delta t\frac{k\rho}{\mu}\frac{\partial p}{\partial r}\bigg|_{r+\frac{\Delta r}{2},\theta,\varphi} \\
&+\frac{\Delta r\Delta\varphi}{\sin\varphi}\Delta t\frac{k\rho}{\mu}\frac{\partial p}{\partial\theta}\bigg|_{r,\theta-\frac{\Delta\theta}{2},\varphi} - \frac{\Delta r\Delta\varphi}{\sin\varphi}\Delta t\frac{k\rho}{\mu}\frac{\partial p}{\partial\theta}\bigg|_{r,\theta+\frac{\Delta\theta}{2},\varphi} \\
&+\Delta r\Delta\theta\sin\left(\varphi-\frac{\Delta\varphi}{2}\right)\Delta t\frac{k\rho}{\mu}\frac{\partial p}{\partial\theta}\bigg|_{r,\theta,\varphi-\frac{\Delta\varphi}{2}} - \Delta r\Delta\theta\sin\left(\varphi+\frac{\Delta\varphi}{2}\right)\Delta t\frac{k\rho}{\mu}\frac{\partial p}{\partial\theta}\bigg|_{r,\theta,\varphi+\frac{\Delta\varphi}{2}} \\
&=\left(\rho r^2\sin\varphi\Delta\varphi\Delta\theta dr\phi\right)_{r,\theta,\varphi}\bigg|_{t+\Delta t} - \left(\rho r^2\sin\varphi\Delta\varphi\Delta\theta dr\phi\right)_{r,\theta,\varphi}\bigg|_t .
\end{aligned} \quad (4.19)
$$

Divide $r^2\sin\varphi\Delta\varphi\Delta\theta\Delta r\Delta t$ by both sides of the equation:

$$
\begin{aligned}
&\frac{1}{r^2\,\Delta r}\left[\left(r-\frac{\Delta r}{2}\right)^2\frac{k\rho}{\mu}\frac{\partial p}{\partial r}\bigg|_{r-\frac{\Delta r}{2},\theta,\varphi} - \left(r+\frac{\Delta r}{2}\right)^2\frac{k\rho}{\mu}\frac{\partial p}{\partial r}\bigg|_{r+\frac{\Delta r}{2},\theta,\varphi}\right] + \frac{1}{r^2\sin^2\varphi\Delta\theta}\left(\frac{k\rho}{\mu}\frac{\partial p}{\partial\theta}\bigg|_{r,\theta-\frac{\Delta\theta}{2},\varphi} - \frac{k\rho}{\mu}\frac{\partial p}{\partial\theta}\bigg|_{r,\theta+\frac{\Delta\theta}{2},\varphi}\right) \\
&+\frac{1}{r^2\sin\varphi\Delta\varphi}\left[\sin\left(\varphi-\frac{\Delta\varphi}{2}\right)\frac{k\rho}{\mu}\frac{\partial p}{\partial\theta}\bigg|_{r,\theta,\varphi-\frac{\Delta\varphi}{2}} - \sin\left(\varphi+\frac{\Delta\varphi}{2}\right)\frac{k\rho}{\mu}\frac{\partial p}{\partial\theta}\bigg|_{r,\theta,\varphi+\frac{\Delta\varphi}{2}}\right] = \frac{1}{\Delta t}\left[\left(\rho\phi\right)_{r,\theta,\varphi}\bigg|_{t+\Delta t} - \left(\rho\phi\right)_{r,\theta,\varphi}\bigg|_t\right].
\end{aligned} \quad (4.20)
$$

Let $\Delta r \to 0, \Delta\theta \to 0, \Delta\varphi \to 0, \Delta t \to 0$, and divide both sides by ρ_{sc}, so we get

$$\frac{1}{r^2}\frac{\partial}{\partial r}\left(r^2\frac{k}{\mu B}\frac{\partial p}{\partial r}\right) + \frac{1}{r^2\sin^2\varphi}\frac{\partial}{\partial\theta}\left(\frac{k}{\mu B}\frac{\partial p}{\partial\theta}\right) + \frac{1}{r^2\sin\varphi}\frac{\partial}{\partial\varphi}\left(\sin\varphi\frac{k}{\mu B}\frac{\partial p}{\partial\varphi}\right) = \frac{\partial}{\partial t}\left(\frac{\phi}{B}\right). \quad (4.21)$$

Problem 4.3.2 (Ertekin et al. 2001)

Develop the 3D Laplace equation in the elliptic-cylindrical coordinates (ξ,η,z) shown in Fig. 4.6. (*Note:* The elliptic-cylindrical coordinate system is practical for analyzing fluid flow dynamics in porous media with a vertical fracture plane along the well. Furthermore, in distinctly anisotropic formations, equipotential lines are developed as confocal ellipses, rather than concentric circles.) Functional relationships between the rectangular- and elliptical-coordinate systems are $x = l_f\cosh\xi\cos\eta$, $y = l_f\sinh\xi\sin\eta$, and $z = z$, where $\xi\geq 0$, $0\leq\eta\leq 2\pi$, and $-\infty < z < +\infty$.

Solution to Problem 4.3.2

Invoking the Laplace equation written in a 3D Cartesian coordinate system,

$$\frac{\partial^2 p}{\partial x^2} + \frac{\partial^2 p}{\partial y^2} + \frac{\partial^2 p}{\partial z^2} = 0. \quad\quad\quad\quad\quad (4.22)$$

To apply the transformation, one needs to rewrite the equation to

$$\frac{\partial}{\partial x}\left(\frac{\partial p}{\partial x}\right) + \frac{\partial}{\partial y}\left(\frac{\partial p}{\partial y}\right) + \frac{\partial}{\partial z}\left(\frac{\partial p}{\partial z}\right) = 0. \quad\quad\quad\quad\quad (4.23)$$

The following transformation rule can be developed using the relationships, $x = l_f \cosh \xi \cos \eta$ and $y = l_f \sinh \xi \sin \eta$: $\dfrac{\partial x}{\partial \xi} = l_f \sinh \xi \cos \eta$ and $\dfrac{\partial y}{\partial \eta} = l_f \sinh \xi \cos \eta$.

Applying the chain rule, the original equation becomes

$$\frac{\partial}{\partial \xi}\left(\frac{\partial p}{\partial \xi}\frac{\partial \xi}{\partial x}\right)\frac{\partial \xi}{\partial x} + \frac{\partial}{\partial \eta}\left(\frac{\partial p}{\partial \eta}\frac{\partial \eta}{\partial y}\right)\frac{\partial \eta}{\partial y} + \frac{\partial}{\partial z}\left(\frac{\partial p}{\partial z}\right) = 0. \dots\dots\dots (4.24)$$

Substituting the transformation rule yields

$$\frac{\partial}{\partial \xi}\left(\frac{1}{l_f \sinh \xi \cos \eta}\frac{\partial p}{\partial \xi}\right)\frac{1}{l_f \sinh \xi \cos \eta} + \frac{\partial}{\partial \eta}\left(\frac{1}{l_f \sinh \xi \cos \eta}\frac{\partial p}{\partial \eta}\right)\frac{1}{l_f \sinh \xi \cos \eta} + \frac{\partial}{\partial z}\left(\frac{\partial p}{\partial z}\right) = 0. \dots\dots (4.25)$$

Simplifying,

$$\frac{\partial}{\partial \xi}\left(\frac{1}{\sinh \xi}\frac{\partial p}{\partial \xi}\right)\frac{1}{l_f^2 \sinh \xi \cos^2 \eta} + \frac{\partial}{\partial \eta}\left(\frac{1}{\cos \eta}\frac{\partial p}{\partial \eta}\right)\frac{1}{l_f^2 \sinh^2 \xi \cos \eta} + \frac{\partial}{\partial z}\left(\frac{\partial p}{\partial z}\right) = 0. \dots\dots\dots (4.26)$$

Multiply both side of the equation by $l_f^2 \sinh \xi \cos \eta$, and at last we can obtain

$$\frac{\partial}{\partial \xi}\left(\frac{1}{\sinh \xi}\frac{\partial p}{\partial \xi}\right)\frac{1}{\cos \eta} + \frac{\partial}{\partial \eta}\left(\frac{1}{\cos \eta}\frac{\partial p}{\partial \eta}\right)\frac{1}{\sinh \xi} + l_f^2 \sinh \xi \cos \eta \frac{\partial^2 p}{\partial z^2} = 0, \dots\dots\dots (4.27)$$

which is the 3D Laplace equation written in the elliptic-cylindrical coordinate system.

Problem 4.3.3 (Ertekin et al. 2001)

Derive Eq. 4.28 from Eq. 4.29 using coordinate transformation. Note that $x = r\cos\theta$, $y = r\sin\theta$, and $z = z$.

$$-\frac{1}{r}\frac{\partial}{\partial r}(r\rho v_r) - \frac{1}{r}\frac{\partial}{\partial \theta}(\rho v_\theta) - \frac{\partial}{\partial z}(\rho v_z) = \frac{1}{\alpha_c}\frac{\partial}{\partial t}(\phi\rho), \dots\dots\dots\dots\dots (4.28)$$

$$-\frac{\partial}{\partial x}(\rho v_x) - \frac{\partial}{\partial y}(\rho v_y) - \frac{\partial}{\partial z}(\rho v_z) = \frac{1}{\alpha_c}\frac{\partial}{\partial t}(\phi\rho). \dots\dots\dots\dots\dots (4.29)$$

Solution to Problem 4.3.3

As shown in **Fig. 4.7**, one can obtain the following geometric relationships between the velocity terms: $v_x = \cos\theta v_r$ and $v_y = \cos\theta v_\theta$. Start the transformation by modifying the component along the x-direction of Eq. 4.29:

$$-\frac{1}{r}\frac{\partial}{\partial x}(r\rho v_r) - \frac{\partial}{\partial y}(\rho v_y) - \frac{\partial}{\partial z}(\rho v_z) = \frac{1}{\alpha_c}\frac{\partial}{\partial t}(\phi\rho). \dots\dots (4.30)$$

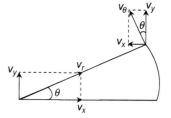

Then, applying the chain rule,

$$-\frac{1}{r}\frac{\partial(r\rho v_x)}{\partial r}\frac{\partial r}{\partial x} - \frac{\partial(\rho v_y)}{\partial \theta}\frac{\partial \theta}{\partial y} - \frac{\partial}{\partial z}(\rho v_z) = \frac{1}{\alpha_c}\frac{\partial}{\partial t}(\phi\rho). \dots\dots (4.31)$$

Fig. 4.7—Graphical representation of the geometric relationships of the velocity components in the x-y and r-θ planes of the rectangular and radial-cylindrical flow geometries.

Next, implementing the transformation rule, $\dfrac{\partial x}{\partial r} = \cos\theta$ and $\dfrac{\partial y}{\partial \theta} = r\cos\theta$, and substituting the velocity terms,

$$-\frac{1}{r}\frac{\partial}{\partial r}\left[(r\rho\cos\theta v_r)\frac{1}{\cos\theta}\right] - \frac{\partial}{\partial \theta}\left[(\cos\theta v_\theta)\frac{1}{r}\frac{1}{\cos\theta}\right] - \frac{\partial}{\partial z}(\rho v_z) = \frac{1}{\alpha_c}\frac{\partial}{\partial t}(\phi\rho). \dots\dots\dots\dots (4.32)$$

and finally resulting in Eq. 4.33:

$$-\frac{1}{r}\frac{\partial}{\partial r}(r\rho v_r) - \frac{1}{r}\frac{\partial}{\partial \theta}(\rho v_\theta) - \frac{\partial}{\partial z}(\rho v_z) = \frac{1}{\alpha_c}\frac{\partial}{\partial t}(\phi\rho). \dots\dots\dots\dots\dots (4.33)$$

Problem 4.3.4 (Ertekin et al. 2001)

Derive Eq. 4.28 from Eq. 4.29 in Problem 4.3.3, using the coordinate-transformation relationship: $x = r\cos\theta$, $y = r\sin\theta$, and $z = z$.

Solution to Problem 4.3.4

Applying the chain rule to Eq. 4.28,

$$-\frac{1}{r}\frac{\partial}{\partial x}\left(r\,\rho v_r\right)\frac{\partial x}{\partial r} - \frac{1}{r}\frac{\partial}{\partial y}\left(\rho v_\theta\right)\frac{\partial y}{\partial \theta} - \frac{\partial}{\partial z}\left(\rho v_z\right) = \frac{1}{\alpha_c}\frac{\partial}{\partial t}\left(\phi\rho\right). \quad\ldots\ldots\ldots\ldots\ldots\ldots\ldots (4.34)$$

Substituting the transformation rules, $\dfrac{\partial x}{\partial r} = \cos\theta$ and $\dfrac{\partial y}{\partial \theta} = r\cos\theta$, and the velocity relationships, $v_x = \cos\theta v_r$, and $v_y = \cos\theta v_\theta$, one may obtain the following:

$$-\frac{1}{r}\frac{\partial}{\partial x}\left(r\rho\,\frac{v_x}{\cos\theta}\right)\cos\theta - \frac{1}{r}\frac{\partial}{\partial y}\left(\rho\,\frac{v_y}{\cos\theta}\right)r\cos\theta - \frac{\partial}{\partial z}\left(\rho v_z\right) = \frac{1}{\alpha_c}\frac{\partial}{\partial t}\left(\phi\rho\right). \quad\ldots\ldots\ldots\ldots\ldots (4.35)$$

Such expression can be simplified to yield Eq. 4.29, as shown in Eq. 4.36:

$$-\frac{\partial}{\partial x}\left(\rho v_x\right) - \frac{\partial}{\partial y}\left(\rho v_y\right) - \frac{\partial}{\partial z}\left(\rho v_z\right) = \frac{1}{\alpha_c}\frac{\partial}{\partial t}\left(\phi\rho\right). \quad\ldots\ldots\ldots\ldots\ldots\ldots\ldots (4.36)$$

Problem 4.3.5 (Ertekin et al. 2001)

Give a complete mathematical description of the unsteady-state, single-phase, radial flow (1D) converging to a fully penetrating well with radius r_w producing at a constant rate q_{sc} for the following:

1. An infinitely large homogeneous reservoir.
2. A finite reservoir with constant pressure at the outer boundary.
3. A finite reservoir with no flow at the outer boundary.

Solution to Problem 4.3.5

Table 4.2 addresses the governing equations, boundary conditions, and initial conditions for each of the three different reservoirs.

	Governing Equation	Boundary Conditions	Initial Condition
1.	$\dfrac{1}{r}\dfrac{\partial}{\partial r}\left(r\dfrac{k_r}{\mu B}\dfrac{\partial p}{\partial r}\right) = \dfrac{\partial}{\partial t}\left(\dfrac{\phi}{B}\right)$	$\left.\dfrac{2\pi rhk_r}{\mu B}\dfrac{\partial p}{\partial r}\right|_{r_w} = q_{sc}$ $t\geq 0, \left.p\right|_{r\to r_e} \to p_i$	at $t=0$, $\left.p\right|_{r_w\leq r<\infty} = p_i$
2.	$\dfrac{1}{r}\dfrac{\partial}{\partial r}\left(r\dfrac{k_r}{\mu B}\dfrac{\partial p}{\partial r}\right) = \dfrac{\partial}{\partial t}\left(\dfrac{\phi}{B}\right)$	$\left.\dfrac{2\pi rhk_r}{\mu B}\dfrac{\partial p}{\partial r}\right|_{r_w} = q_{sc}$ $t\geq 0, \left.p\right|_{r=r_e} = p_i$	at $t=0$, $\left.p\right|_{r_w\leq r<r_e} = p_i$
3.	$\dfrac{1}{r}\dfrac{\partial}{\partial r}\left(r\dfrac{k_r}{\mu B}\dfrac{\partial p}{\partial r}\right) = \dfrac{\partial}{\partial t}\left(\dfrac{\phi}{B}\right)$	$\left.\dfrac{2\pi rhk_r}{\mu B}\dfrac{\partial p}{\partial r}\right|_{r_w} = q_{sc}$ $t\geq 0, \left.\dfrac{dp}{dr}\right|_{r\to r_e} \to 0$	at $t=0$, $\left.p\right|_{r_w\Delta r<r_e} = p_i$

Table 4.2—Governing equations, boundary conditions, and initial conditions for the reservoirs of Problem 4.3.5.

Problem 4.3.6 (Ertekin et al. 2001)

Consider the following 1D, horizontal, porous medium in which steady-state flow of oil is taking place in the positive x-direction. Note that there is no well in the system **(Fig. 4.8).** The permeability of the system varies according to the expression $k_x = \dfrac{1,000}{980+0.04\,x}$, in which x is in feet and k_x is in darcies. It is also known that at $x = 0$, $p = p_w$, and at $x = L$, $p = p_L$ (pressures are in psia). The viscosity of oil is μ_o cp, and the dimensions of the porous medium are h, b, and L. Give a complete mathematical statement for the reservoir described and obtain an expression that describes the pressure gradient at any point in the porous medium.

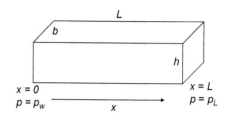

Fig. 4.8—Schematic representation of the 1D horizontal flow described in Problem 4.3.6.

Solution to Problem 4.3.6

The steady-state flow equation can be written as

$$\frac{\partial}{\partial x}\left(\frac{A_x k_x}{\mu_o B_o}\frac{\partial p}{\partial x}\right)\Delta x = 0. \dots\dots\dots\dots\dots\dots\dots\dots\dots\dots\dots\dots\dots\dots\dots (4.37)$$

Because $\dfrac{A_x}{\mu_o B_o}$ is constant, the equation can be simplified to

$$\frac{A_x}{\mu_o B_o}\frac{\partial}{\partial x}\left(k_x\frac{\partial p}{\partial x}\right)=0, \dots\dots\dots\dots\dots\dots\dots\dots\dots\dots\dots\dots\dots\dots\dots (4.38)$$

which means that $\dfrac{\partial}{\partial x}\left(k_x\dfrac{\partial p}{\partial x}\right)=0$. Therefore, $k_x\dfrac{\partial p}{\partial x}=C$, where C is a constant. The pressure gradient of the system can then be expressed as $\dfrac{\partial p}{\partial x}=\dfrac{C}{k_x}$.

To solve for the constant C, one needs to integrate on both sides of the equation $k_x\dfrac{\partial p}{\partial x}=C$:

$$\int_{p_w}^{p_L} dp = \int_0^L \frac{C}{k_x}dx, \dots\dots\dots\dots\dots\dots\dots\dots\dots\dots\dots\dots\dots\dots\dots\dots\dots\dots (4.39)$$

which can be written as

$$p_L - p_w = \frac{C}{1,000}\int_0^L (980+0.04x)dx = \frac{C}{1,000}\left(980L+0.02L^2\right). \dots\dots\dots\dots\dots\dots\dots\dots (4.40)$$

Solving for C,

$$C = \frac{1,000\left(p_L - p_w\right)}{\left(980L+0.02L^2\right)}. \dots\dots\dots\dots\dots\dots\dots\dots\dots\dots\dots\dots\dots\dots\dots\dots (4.41)$$

Substituting back to solve for the pressure gradient,

$$\frac{\partial p}{\partial x} = \frac{1,000\left(p_L - p_w\right)}{\left(980L+0.02L^2\right)}\frac{\left(980+0.04x\right)}{1,000} = \frac{\left(980+0.04x\right)\left(p_L - p_w\right)}{\left(980L+0.02L^2\right)}. \dots\dots\dots\dots\dots\dots (4.42)$$

Problem 4.3.7 (Ertekin et al. 2001)

The partial-differential equation (PDE), $\dfrac{\partial^2 p}{\partial x^2} = 400$, describes a specific fluid flow problem in porous media with the boundary conditions $p_{x=0\,\text{ft}} = 200\,\text{psi}$ and $p_{x=100\,\text{ft}} = 100\,\text{psi}$. After examining the given mathematical formulation, describe the flow problem to the fullest extent (depth gradient is ignored).

Solution to Problem 4.3.7

Invoking the governing equation for 1D incompressible flow system,

$$\frac{\partial}{\partial x}\left(\frac{A_x k_x}{\mu B}\frac{\partial p}{\partial x}\right)\Delta x + q = 0. \quad\dotfill(4.43)$$

If the reservoir is homogeneous and isotropic, Eq. 4.43 can be simplified as $\dfrac{\partial^2 p}{\partial x^2} + \dfrac{q\mu B}{k_x V_b} = 0$, which is $\dfrac{\partial^2 p}{\partial x^2} = -\dfrac{q\mu B}{k_x V_b}$.
Notably, such a governing equation displays a similar form as $\dfrac{\partial^2 p}{\partial x^2} = 400$ and $\dfrac{q\mu B}{k_x V_b} = -400$. Because μ, B, and V_b are all positive entities, $q < 0$. Therefore, the reservoir described by the PDE can be characterized as follows:

- The reservoir is a 1D system with single-phase incompressible fluid and homogeneous thickness and permeability distribution.
- Production well(s) are completed in the reservoir.
- Because $p\big|_{x=0\,\text{ft}} = 200\,\text{psi}$ is larger than $p\big|_{x=100\,\text{ft}} = 100\,\text{psi}$, one can conclude that there is an open boundary at $x = 0$, where fluid enters the reservoir.

Problem 4.3.8 (Ertekin et al. 2001)

Given $A_x = A_y = 100\,\text{ft}^2$, $\Delta x = 200\,\text{ft}$, $\Delta y = 100\,\text{ft}$, $k_y = 100\,\text{md}$, $\mu = 1\,\text{cp}$, and the PDE, $\dfrac{\partial^2 p}{\partial x^2} + \dfrac{1}{2}\dfrac{\partial^2 p}{\partial y^2} = 0$, that describes the fluid flow problem in a homogeneous system, what is the permeability of the system along the x-direction?

Solution to Problem 4.3.8

The generalized form of the incompressible flow equation in porous media in two dimensions is as follows (*Note:* there is no well in the reservoir, and the reservoir is positioned parallel to the datum plane along the x- and y-directions):

$$\frac{\partial}{\partial x}\left(\frac{A_x k_x}{\mu B}\frac{\partial p}{\partial x}\right)\Delta x + \frac{\partial}{\partial y}\left(\frac{A_y k_y}{\mu B}\frac{\partial p}{\partial y}\right)\Delta y = 0. \quad\dotfill (4.44)$$

Because the system is homogeneous, Eq. 4.44 becomes

$$\frac{\partial^2 p}{\partial x^2} + \frac{A_y k_y}{A_x k_x}\frac{\Delta y}{\Delta x}\frac{\partial^2 p}{\partial y^2} = 0. \quad\dotfill (4.45)$$

Thus,

$$\frac{A_y k_y}{A_x k_x}\frac{\Delta y}{\Delta x} = \frac{1}{2}. \quad\dotfill(4.46)$$

Solving for k_x,

$$k_x = 2\left(\frac{A_y k_y \Delta y}{A_x \Delta x}\right) = 2\left(\frac{100\times100\times100}{100\times200}\right) = 100\,\text{md}. \quad\dotfill(4.47)$$

Because $k_x = k_y = 100\,\text{md}$, the reservoir is characterized to be isotropic with respect to permeability.

Problem 4.3.9 (Ertekin et al. 2001)

Consider the following equation:

$$\left(1+\frac{a}{p}\right)\frac{\partial}{\partial x}\left(\frac{A_x}{\mu_g B_g}\frac{\partial p}{\partial x}\right)\Delta x-\frac{A_x}{\mu_g B_g}\frac{a}{p^2}\left(\frac{\partial p}{\partial x}\right)^2\Delta x=\frac{1}{\alpha_c}\frac{V_b}{k_\infty}\frac{\partial}{\partial t}\left(\frac{\phi}{B_g}\right).$$

Show that this PDE represents real-gas flow in a 1D porous system without wells, if the pressure dependency of the permeability is described by the Klinkenberg equation (Klinkenberg 1941), $k(p)=k_\infty\left(1+\frac{a}{p}\right)$, where k_∞ and a are constants.

Solution to Problem 4.3.9

Starting with the 1D gas-flow equation,

$$\frac{\partial}{\partial x}\left(\frac{A_x k_x}{\mu_g B_g}\frac{\partial p}{\partial x}\right)\Delta x=\frac{V_b}{\alpha_c}\frac{\partial}{\partial t}\left(\frac{\phi}{B_g}\right), \quad\dotfill(4.48)$$

and substituting k_x by the Klinkenberg equation, $k_x(p)=k_\infty\left(1+\frac{a}{p}\right)$,

$$\frac{\partial}{\partial x}\left[\frac{A_x k_\infty\left(1+\frac{a}{p}\right)}{\mu_g B_g}\frac{\partial p}{\partial x}\right]\Delta x=\frac{V_b}{\alpha_c}\frac{\partial}{\partial t}\left(\frac{\phi}{B_g}\right). \quad\dotfill(4.49)$$

Because k_∞ is constant, the Eq. 4.49 can be rewritten as follows:

$$\frac{\partial}{\partial x}\left[\frac{A_x\left(1+\frac{a}{p}\right)}{\mu_g B_g}\frac{\partial p}{\partial x}\right]\Delta x=\frac{1}{\alpha_c}\frac{V_b}{k_\infty}\frac{\partial}{\partial t}\left(\frac{\phi}{B_g}\right). \quad\dotfill(4.50)$$

Applying the product rule to the left-hand side of Eq. 4.50,

$$\left(1+\frac{a}{p}\right)\frac{\partial}{\partial x}\left(\frac{A_x}{\mu_g B_g}\frac{\partial p}{\partial x}\right)\Delta x+\frac{A_x}{\mu_g B_g}\frac{\partial p}{\partial x}\frac{\partial}{\partial x}\left(1+\frac{a}{p}\right)\Delta x=\frac{1}{\alpha_c}\frac{V_b}{k_\infty}\frac{\partial}{\partial t}\left(\frac{\phi}{B_g}\right). \quad\dotfill(4.51)$$

Applying the chain rule to the term $\frac{\partial}{\partial x}\left(1+\frac{a}{p}\right)$ yields

$$\frac{\partial}{\partial x}\left(1+\frac{a}{p}\right)=\frac{\partial}{\partial p}\left(1+\frac{a}{p}\right)\frac{\partial p}{\partial x}=-\frac{a}{p^2}\frac{\partial p}{\partial x}. \quad\dotfill(4.52)$$

Substituting Eq. 4.52 into the original equation,

$$\left(1+\frac{a}{p}\right)\frac{\partial}{\partial x}\left(\frac{A_x}{\mu_g B_g}\frac{\partial p}{\partial x}\right)\Delta x+\frac{A_x}{\mu_g B_g}\frac{\partial p}{\partial x}\frac{\partial p}{\partial x}\left(-\frac{a}{p^2}\right)\Delta x=\frac{1}{\alpha_c}\frac{V_b}{k_\infty}\frac{\partial}{\partial t}\left(\frac{\phi}{B_g}\right). \quad\dotfill(4.53)$$

Finally, we obtain

$$\left(1+\frac{a}{p}\right)\frac{\partial}{\partial x}\left(\frac{A_x}{\mu_g B_g}\frac{\partial p}{\partial x}\right)\Delta x-\frac{A_x}{\mu_g B_g}\frac{a}{p^2}\left(\frac{\partial p}{\partial x}\right)^2\Delta x=\frac{1}{\alpha_c}\frac{V_b}{k_\infty}\frac{\partial}{\partial t}\left(\frac{\phi}{B_g}\right). \quad\dotfill(4.54)$$

Problem 4.3.10 (Ertekin et al. 2001)

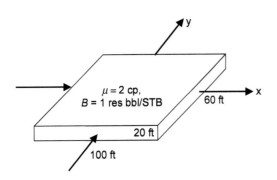

After examining the PDE, $\dfrac{\partial}{\partial x}\left(k_x\dfrac{\partial p}{\partial x}\right)+\dfrac{\partial^2 p}{\partial y^2}=-1$, and **Fig. 4.9,** quantitatively and qualitatively describe the flow problem to the fullest extent. (Ignore all gravitational forces.)

Fig. 4.9—Schematic representation of flow geometry described in Problem 4.3.10.

Solution to Problem 4.3.10

Qualitatively, the permeability of the system is characterized as anisotropic and heterogeneous along the x-direction and homogenous along y-direction. Starting with the original governing-flow equation of a 2D incompressible flow system in a porous medium with no depth gradients,

$$\frac{\partial}{\partial x}\left(\frac{A_x k_x}{\mu B}\frac{\partial p}{\partial x}\right)\Delta x+\frac{\partial}{\partial y}\left(\frac{A_y k_y}{\mu B}\frac{\partial p}{\partial y}\right)\Delta y+q=0. \qquad\qquad (4.55)$$

Because $A_x = 20$ ft $\times$ 60 ft $= 1{,}200$ ft^2 and $A_y = 100$ ft $\times$ 20 ft $= 2{,}000$ ft^2, and μ and B are constants, the equation can be simplified to $\dfrac{\partial}{\partial x}\left(k_x\dfrac{\partial p}{\partial x}\right)+k_y\dfrac{\partial^2 p}{\partial x^2}=-\dfrac{q\mu B}{V_b}$, knowing that $V_b = \Delta x\Delta y\Delta z = 120{,}000$ ft^3. If the simplified equation agrees with $\dfrac{\partial}{\partial x}\left(k_x\dfrac{\partial p}{\partial x}\right)+\dfrac{\partial^2 p}{\partial y^2}=-1$, and if—and only if—$k_y = 1$ perm $=\dfrac{1}{1.127\times10^{-3}}=887.311$ md and $-\dfrac{q\mu B}{V_b}=-1$, one can solve for the flow rate from the well ($\mu = 2$ cp and $B = 1$ res bbl/STB):

$$q=\frac{V_b}{\mu B}=6.0\times10^4 \text{ STB/D (injection well)} \qquad\qquad (4.56)$$

Problem 4.3.11 (Ertekin et al. 2001)

A fluid flow problem in a porous medium is expressed by the PDE (note that k_x is in darcies):

$$\frac{\partial}{\partial x}\left(1.127\,A_x k_x\frac{\partial p}{\partial x}\right)-0.0433\frac{\partial}{\partial x}\left(1.127\,A_x k_x\right)=0.$$

If properties of the flowing fluid are $\mu = 1$ cp, $\rho = 62.4$ lbm/ft^3, and $B = 1$ res bbl/STB, qualitatively and quantitatively describe the physical characteristics of the reservoir.

Solution to Problem 4.3.11

Observing the given PDE, one can state that it describes a 1D heterogeneous system with incompressible fluid flow. The PDE can be written using practical field units with $\mu = 1$ cp, $\rho = 62.4$ lbm/ft^3, $\dfrac{g}{g_c}=1$, and $B = 1$ bbl/STB:

$$\frac{\partial}{\partial x}\left(1.127\,A_x k_x\frac{\partial p}{\partial x}\right)\Delta x-\left(\frac{1}{144}\rho\frac{g}{g_c}\frac{\partial G}{\partial x}\right)\frac{\partial}{\partial x}\left(1.127\,A_x k_x\right)\Delta x=0. \qquad\qquad (4.57)$$

In this case, $-\dfrac{1}{144}\rho\dfrac{g}{g_c}\dfrac{\partial G}{\partial x}=-0.0433$ psi/ft, which suggests that $\dfrac{\partial G}{\partial x}=0.1$ ft/ft.

Problem 4.3.12 (Ertekin et al. 2001)

The differential equation $\dfrac{1}{r}\dfrac{\partial}{\partial r}\left(r\beta_c k_r \dfrac{\partial \Phi}{\partial r}\right)=0$ describes 1D single-phase incompressible flow in a radial-cylindrical, homogeneous reservoir. (*Note*: $\beta_c = 1.127$ and k_r is in darcies.)

1. For the boundary conditions shown in **Fig 4.10,** obtain an expression that gives the potential distribution as a function of radius.
2. Using the solution to Part 1, determine the flow rate into the well.
3. What is the flow rate across the outer boundary of the system?

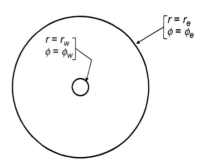

Fig. 4.10—Boundary conditions of the 1D flow described in Problem 4.3.12.

Solution to Problem 4.3.12

1. Because the reservoir is homogeneous, the equation becomes (k_r is constant)

$$\frac{1}{r}\frac{\partial}{\partial r}\left(r\frac{\partial \Phi}{\partial r}\right)=0 \text{ or } \frac{\partial}{\partial r}\left(r\frac{\partial \Phi}{\partial r}\right)=0. \dots\dots\dots\dots\dots\dots\dots\dots\dots\dots\dots(4.58)$$

Integrating both sides, one ends up with $r\dfrac{\partial \Phi}{\partial r}=C$, where C is a constant. Separating the variables and integrating both sides of the equation,

$$\int_{\Phi_w}^{\Phi_r} \partial \Phi = \int_{r_w}^{r} \frac{C}{r}dr. \dots\dots\dots\dots\dots\dots\dots\dots\dots\dots\dots\dots\dots\dots\dots\dots(4.59)$$

One then obtains the general form of the solution:

$$\Phi_r - \Phi_w = C \ln \frac{r}{r_w}. \dots\dots\dots\dots\dots\dots\dots\dots\dots\dots\dots\dots\dots\dots\dots(4.60)$$

Applying the boundary condition prescribed, $\Phi = \Phi_e$ at $r = r_e$,

$$\Phi_e - \Phi_w = C \ln \frac{r_e}{r_w}, \dots\dots\dots\dots\dots\dots\dots\dots\dots\dots\dots\dots\dots\dots\dots(4.61)$$

the constant C can be solved as

$$C = \frac{\Phi_e - \Phi_w}{\ln \dfrac{r_e}{r_w}}. \dots\dots\dots\dots\dots\dots\dots\dots\dots\dots\dots\dots\dots\dots\dots(4.62)$$

Thus, substituting for C in the general solution,

$$\Phi(r) = \frac{\Phi_e - \Phi_w}{\ln \dfrac{r_e}{r_w}} \ln \frac{r}{r_w} + \Phi_w. \dots\dots\dots\dots\dots\dots\dots\dots\dots\dots\dots\dots\dots(4.63)$$

2. Invoking Darcy's law expressed in the radial-cylindrical coordinate system,

$$q = \frac{2\pi r h \beta_c k_r}{\mu B}\frac{\partial \Phi}{\partial r}, \dots\dots\dots\dots\dots\dots\dots\dots\dots\dots\dots\dots\dots\dots(4.64)$$

and

$$\frac{\partial \Phi}{\partial r} = \frac{\Phi_e - \Phi_w}{\ln \dfrac{r_e}{r_w}}; \dots\dots\dots\dots\dots\dots\dots\dots\dots\dots\dots\dots\dots\dots\dots\dots(4.65)$$

therefore, the flow rate can be expressed as

$$q = \frac{2\pi h \beta_c k_r}{\mu B} \frac{\Phi_e - \Phi_w}{\ln \frac{r_e}{r_w}}. \quad \dots\dots\dots\dots (4.66)$$

3. Because this is a steady-state flow problem, the flow rate across the outer boundary is the same as that of the flow rate at the well.

Problem 4.3.13 (Ertekin et al. 2001)

Consider the incompressible fluid flow problem in the 1D porous body shown in **Fig. 4.11** (viscosity of the fluid is 2.0 cp, the formation volume factor is 1 res bbl/STB).

1. If the well in Block 2 is produced at a steady flow rate of –450 STB/D, and this rate creates a pressure distribution of $p_1 = 2,400$ psi, $p_2 = 800$ psi, and $p_3 = 800$ psi, what is the permeability of this linear, homogeneous system?
2. What kind of boundary condition would you assign to the extreme left end of the system? Quantify your answer.
3. If all the length dimensions of Blocks 1, 2, and 3 are doubled, what would the flow rate into the well be if the same pressure distribution is assigned to the new system?
4. Assume that the viscosity of the fluid of Part 1 is 4 cp and the wellblock pressure is maintained at 800 psia. What would the well-flow rate and the pressures in Blocks 1 and 3 be if the extreme ends of the system represent no-flow boundaries?

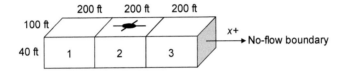

Fig. 4.11—Schematic representation of the 1D flow of Problem 4.3.13.

Solution to Problem 4.3.13

1. Because $p_2 = p_3$, the flow into the producer well comes from Block 1. Applying Darcy's law,

$$q = -\frac{Ak}{\mu B} \frac{\Delta p}{\Delta L}, \quad \dots\dots\dots\dots (4.67)$$

so that

$$k_x = -\frac{q \mu B \Delta L}{A \Delta p} = -\frac{(-450)(2)(1)(200)}{(100)(40)(2,400-800)(1.127\times 10^{-3})} \cong 25 \text{ md}. \quad \dots\dots\dots (4.68)$$

2. The system exhibits linear pressure distribution. There are two possibilities to characterize the boundary conditions on the left end of the system,
   • Dirichlet type:

$$p_{\text{boundary}} = p_1 + \frac{(p_1 - p_2)}{2} = 3,200 \text{ psia}. \quad \dots\dots\dots (4.69)$$

   • Neumann type:

$$\frac{\partial p}{\partial x} = \frac{\Delta p}{\Delta L} = \frac{800 - 2,400}{200} = -8 \frac{\text{psi}}{\text{ft}}. \quad \dots\dots\dots (4.70)$$

3. According to Darcy's law,

$$q = \frac{Ak}{\mu B} \frac{\Delta p}{\Delta L}, \quad \dots\dots\dots\dots (4.71)$$

when ΔL is doubled, flow rate will be halved: $q = \frac{-450}{2} = -225$ STB/D.

4. Because the fluid is incompressible, and when the outer boundaries are closed, there will be no flow into or out of the well.

Problem 4.3.14 (Ertekin et al. 2001)

Linearize the compressible flow equation,

$$\frac{\partial}{\partial x}\left(\frac{A_x k_x}{\mu_g B_g}\frac{\partial p}{\partial x}\right)\Delta x + \frac{\partial}{\partial y}\left(\frac{A_y k_y}{\mu_g B_g}\frac{\partial p}{\partial y}\right)\Delta y + \frac{\partial}{\partial z}\left(\frac{A_z k_z}{\mu_g B_g}\frac{\partial p}{\partial z}\right)\Delta z + q_{gsc} = \frac{V_b \phi T_{sc}}{p_{sc} T}\frac{\partial}{\partial t}\left(\frac{p}{z_g}\right),$$

by use of the transformation $m = \int_0^p \frac{p}{\mu z_g}\,dp$. *Hint:* Substitute the definition of B_g, $B_g\left(\frac{\text{res bbl}}{\text{scf}}\right) = \frac{p_{sc} T z_g}{\alpha_c T_{sc} p}$, into the flow equation before applying this transformation.

Solution to Problem 4.3.14

Recalling the definition of gas FVF, $B_g = \frac{p_{sc} T z_g}{\alpha_c T_{sc} p}$, and substituting B_g into the compressible flow equation, one obtains

$$\frac{\partial}{\partial x}\left(\alpha_c \frac{A_x k_x T_{sc} p}{\mu_g p_{sc} T z_g}\frac{\partial p}{\partial x}\right)\Delta x + \frac{\partial}{\partial y}\left(\alpha_c \frac{A_y k_y T_{sc} p}{\mu_g p_{sc} T z_g}\frac{\partial p}{\partial y}\right)\Delta y + \frac{\partial}{\partial z}\left(\alpha_c \frac{A_z k_z T_{sc} p}{\mu_g p_{sc} T z_g}\frac{\partial p}{\partial z}\right)\Delta z + q_{gsc} = \frac{V_b \phi T_{sc}}{p_{sc} T}\frac{\partial}{\partial t}\left(\frac{p}{z_g}\right). \quad \ldots (4.72)$$

Using

$$\frac{\partial m}{\partial p} = \frac{\partial}{\partial p}\left(\int_0^p \frac{p}{\mu z_g}\,dp\right) = \frac{p}{\mu z_g}, \quad \ldots\ldots\ldots\ldots\ldots\ldots\ldots\ldots\ldots\ldots\ldots\ldots\ldots\ldots\ldots (4.73)$$

and applying the chain rule to the spatial derivative along the x-direction,

$$\frac{\partial}{\partial x}\left(\alpha_c \frac{A_x k_x T_{sc} p}{\mu_g p_{sc} T z_g}\frac{\partial p}{\partial x}\right)\Delta x = \frac{\partial}{\partial x}\left(\alpha_c A_x k_x \frac{T_{sc} p}{\mu_g p_{sc} T z_g}\frac{\partial p}{\partial m}\frac{\partial m}{\partial x}\right)\Delta x$$

$$= \frac{\partial}{\partial x}\left(\alpha_c A_x k_x \frac{T_{sc} p}{\mu_g p_{sc} T z_g}\left(\frac{\mu z_g}{p}\right)\frac{\partial m}{\partial x}\right)\Delta x = \frac{\partial}{\partial x}\left(\frac{A_x k_x \alpha_c T_{sc}}{p_{sc} T}\frac{\partial m}{\partial x}\right)\Delta x. \quad \ldots\ldots\ldots\ldots\ldots (4.74)$$

Applying a similar transformation along y- and z-directions, one can finally obtain the following equation:

$$\frac{\partial}{\partial x}\left(\frac{A_x k_x \alpha_c T_{sc}}{p_{sc} T}\frac{\partial m}{\partial x}\right)\Delta x + \frac{\partial}{\partial y}\left(\frac{A_y k_y \alpha_c T_{sc}}{p_{sc} T}\frac{\partial m}{\partial y}\right)\Delta y + \frac{\partial}{\partial z}\left(\frac{A_z k_z \alpha_c T_{sc}}{p_{sc} T}\frac{\partial m}{\partial z}\right)\Delta z + q_{gsc} = \frac{V_b \phi T_{sc}}{p_{sc} T}\frac{\partial}{\partial t}\left(\mu\frac{\partial m}{\partial p}\right). \quad \ldots\ldots (4.75)$$

Problem 4.3.15 (Ertekin et al. 2001)

Does $\frac{\partial^2 p}{\partial x^2} = 0$ represent any fluid flow problem in porous media? If it does, provide a full description of the problem.

Solution to Problem 4.3.15

Starting with $\frac{\partial}{\partial x}\left(\frac{A_x k_x}{\mu B}\frac{\partial p}{\partial x}\right)\Delta x = 0$ to describe incompressible flow in a 1D homogeneous reservoir with no depth gradients and with no well, the preceding equation can be simplified to $\frac{\partial^2 p}{\partial x^2} = 0$.

Problem 4.3.16 (Ertekin et al. 2001)

Give the simplest form of the PDE that governs the flow of an incompressible fluid in a 2D homogeneous but anisotropic formation with no depth gradients. Assume there is no well in the reservoir.

Solution to Problem 4.3.16

Starting with the general form of the incompressible flow equation in a 2D system,

$$\frac{\partial}{\partial x}\left(\frac{A_x k_x}{\mu B}\frac{\partial p}{\partial x}\right)\Delta x + \frac{\partial}{\partial y}\left(\frac{A_y k_y}{\mu B}\frac{\partial p}{\partial y}\right)\Delta y + q = 0. \quad \ldots \ldots \ldots (4.76)$$

For a homogeneous and anisotropic system with an incompressible fluid and no well in the reservoir, Eq. 4.76 can be written as

$$k_x \frac{\partial^2 p}{\partial x^2} + k_y \frac{\partial^2 p}{\partial y^2} = 0. \quad \ldots \ldots \ldots (4.77)$$

Problem 4.3.17 (Ertekin et al. 2001)

Surface properties, such as surface tension, interfacial tension, and specific surface area, play important roles in fluid flow dynamics in porous media. Explain why these properties do not appear in the fluid flow equations.

Solution to Problem 4.3.17

The information about these properties are conveyed to the macroscopic flow equations by the relative permeability and capillary pressure functions.

Problem 4.3.18 (Ertekin et al. 2001)

Show that $-\dfrac{\partial}{\partial x}(\rho v_x)=\rho\phi(c+c_R)\dfrac{\partial p}{\partial t}$ represents the 1D form of the continuity equation for single-phase flow in which porosity is treated as a function of pressure. In this equation, c and c_R represent the fluid and pore compressibilities, respectively.

Solution to Problem 4.3.18

The 1D form of the continuity equation for single-phase flow is

$$-\frac{\partial}{\partial x}(\rho v_x)=\frac{\partial}{\partial t}(\phi\rho). \quad \ldots \ldots \ldots (4.78)$$

Expanding the right-hand side with the rule for the derivative of a product,

$$-\frac{\partial}{\partial x}(\rho v_x)=\phi\frac{\partial p}{\partial t}+\rho\frac{\partial\phi}{\partial t}. \quad \ldots \ldots \ldots (4.79)$$

Now, implementing the chain rule on both of the derivatives,

$$-\frac{\partial}{\partial x}(\rho v_x)=\phi\frac{\partial\rho}{\partial p}\frac{\partial p}{\partial t}+\rho\frac{\partial\phi}{\partial p}\frac{\partial p}{\partial t}. \quad \ldots \ldots \ldots (4.80)$$

Invoking the definition of compressibility in terms of porosity and fluid density,

$$c=\frac{1}{\rho}\frac{\partial\rho}{\partial p}, \quad \ldots \ldots \ldots (4.81)$$

and

$$c_R=\frac{1}{\phi}\frac{\partial\phi}{\partial p}; \quad \ldots \ldots \ldots (4.82)$$

therefore, $\rho c=\dfrac{\partial\rho}{\partial p}$ and $\phi c_R=\dfrac{\partial\phi}{\partial p}$, which results in

$$-\frac{\partial}{\partial x}(\rho v_x) = \phi\rho c \frac{\partial p}{\partial t} + \rho\phi c_R \frac{\partial p}{\partial t}, \quad\ldots\ldots\ldots\ldots\ldots\ldots\ldots\ldots\ldots\ldots\ldots\ldots\ldots\ldots\ldots\ldots\ldots\ldots\ldots\ldots\ldots\ldots\text{(4.83)}$$

and can be simplified to

$$-\frac{\partial}{\partial x}(\rho v_x) = \rho\phi(c + c_R)\frac{\partial p}{\partial t}. \quad\ldots\ldots\ldots\ldots\ldots\ldots\ldots\ldots\ldots\ldots\ldots\ldots\ldots\ldots\ldots\ldots\ldots\ldots\ldots\ldots\ldots\text{(4.84)}$$

Problem 4.3.19

Describe the porous-media flow problem represented by the following PDE in the fullest extent (*Note:* $k_x = 10$ md, $\mu = 2$ cp, $B = 1$ res bbl/STB, and $V_b = 10{,}000$ ft^3); ignore the depth gradient:

$$\frac{\partial^2 p}{\partial x^2} = -5.32.$$

Solution to Problem 4.3.19

Because there is no temporal derivative on the right-hand side of the equation, the PDE represents a 1D single-phase incompressible fluid system.

The general form of the governing-flow equation of such a problem (no depth gradient) can be written as,

$$\frac{\partial}{\partial x}\left(\frac{A_x k_x}{\mu B}\frac{\partial p}{\partial x}\right)\Delta x + q = 0, \quad\ldots\ldots\ldots\ldots\ldots\ldots\ldots\ldots\ldots\ldots\ldots\ldots\ldots\ldots\ldots\ldots\ldots\ldots\ldots\text{(4.85)}$$

which further simplifies to

$$\frac{\partial^2 p}{\partial x^2} = -\frac{q\mu B}{k_x V_b}, \quad\ldots\ldots\ldots\ldots\ldots\ldots\ldots\ldots\ldots\ldots\ldots\ldots\ldots\ldots\ldots\ldots\ldots\ldots\ldots\ldots\ldots\ldots\text{(4.86)}$$

and $-\dfrac{q\mu B}{k_x V_b} = -5.32$.

Then, one can solve for the flow rate:

$$q = \frac{5.32 k_x V_b}{\mu B} = \frac{5.32 \times 10 \times 1.127 \times 10^{-3} \times 10{,}000}{2 \times 1} = 299.78\text{ STB/D}. \quad\ldots\ldots\ldots\ldots\ldots\ldots\ldots\ldots\text{(4.87)}$$

The positive flow rate indicates that the well is an injector. Other reservoir characteristics can be summarized as follows:

- Homogeneous permeability.
- No depth gradient.
- Homogeneous thickness.
- Injection well(s) with a total rate of 299.78 STB/D exist.

Problem 4.3.20

Linearize the PDE of a single-phase liquid system with undersaturated oil at the bubblepoint. The formation permeability (k), oil viscosity (μ), and FVF (B) are all functions of pressure.

Solution to Problem 4.3.20

For a 1D system,

$$\frac{\partial}{\partial x}\left[\frac{Ak(p)}{\mu(p)B(p)}\frac{\partial p}{\partial x}\right]\Delta x + q = \frac{V_b\phi c}{5.615}\frac{\partial p}{\partial t}. \quad\ldots\ldots\ldots\ldots\ldots\ldots\ldots\ldots\ldots\ldots\ldots\ldots\ldots\ldots\text{(4.88)}$$

The nonlinear function, $\dfrac{k(p)}{\mu(p)B(p)}$, can be linearized by defining a transformation such as

$$m = \int_0^p \frac{k(p)}{\mu(p)B(p)}\,dp. \quad\dots\dots\dots\dots\dots\dots\dots\dots\dots\dots\dots\dots\dots\dots (4.89)$$

For the left-hand side of the equation,

$$\frac{\partial m}{\partial x} = \frac{\partial}{\partial x}\left[\int_0^p \frac{k(p)}{\mu(p)B(p)}\,dp\right]. \quad\dots\dots\dots\dots\dots\dots\dots\dots\dots\dots\dots (4.90)$$

Applying the chain rule,

$$\frac{\partial m}{\partial x} = \frac{\partial}{\partial p}\left[\int_0^p \frac{k(p)}{\mu(p)B(p)}\,dp\right]\frac{\partial p}{\partial x}. \quad\dots\dots\dots\dots\dots\dots\dots\dots\dots (4.91)$$

Using Leibniz's rule, it follows that

$$\frac{\partial m}{\partial x} = \left[\int_0^p \frac{\partial}{\partial p}\frac{k(p)}{\mu(p)B(p)}\,dp\right]\frac{\partial p}{\partial x}, \quad\dots\dots\dots\dots\dots\dots\dots\dots (4.92)$$

$$\frac{\partial m}{\partial x} = \int_0^p \partial\left[\frac{k(p)}{\mu(p)B(p)}\right]\frac{\partial p}{\partial x}, \quad\dots\dots\dots\dots\dots\dots\dots\dots\dots (4.93)$$

$$\frac{\partial m}{\partial x} = \frac{k(p)}{\mu(p)B(p)}\frac{\partial p}{\partial x}, \quad\dots\dots\dots\dots\dots\dots\dots\dots\dots\dots\dots\dots (4.94)$$

$$\frac{\partial p}{\partial x} = \frac{\mu(p)B(p)}{k(p)}\frac{\partial m}{\partial x}. \quad\dots\dots\dots\dots\dots\dots\dots\dots\dots\dots\dots\dots (4.95)$$

Focusing on the right-hand side,

$$\frac{\partial m}{\partial t} = \frac{k(p)}{\mu(p)B(p)}\frac{\partial p}{\partial t}, \quad\dots\dots\dots\dots\dots\quad\dots\dots\dots\dots\dots\dots\dots (4.96)$$

$$\frac{\partial p}{\partial t} = \frac{\mu(p)B(p)}{k(p)}\frac{\partial m}{\partial t}. \quad\dots\dots\dots\dots\dots\dots\dots\dots\dots\dots\dots\dots (4.97)$$

Substituting Eqs. 4.95 and 4.97 into Eq. 4.88, one obtains

$$\frac{\partial}{\partial x}\left[A\frac{\partial m}{\partial x}\right]\Delta x + q = \frac{V_b\phi c}{5.615}\frac{\mu B}{k}\bigg|_p \frac{\partial m}{\partial t}. \quad\dots\dots\dots\dots\dots\dots\dots\dots (4.98)$$

Problem 4.3.21

Provide the governing flow equation(s) and unknowns for the following systems:

1. A 2D single-phase slightly compressible fluid flow reservoir with homogeneous and anisotropic permeability distribution and heterogeneous and isotropic area distribution. Rock is incompressible, and depth gradients cannot be ignored. There are two wells in the system.
2. A 2D two-phase flow (oil and water) reservoir with homogeneous and isotropic permeability distribution and homogeneous and anisotropic area distribution. Depth gradients can be ignored. Porosity is changing with pressure, and there are wells in the system. Capillary pressure between the oil and water phase cannot be ignored.
3. A 2D two-phase flow (oil and water) reservoir with heterogeneous and isotropic permeability distribution. Gravitational forces cannot be ignored. Porosity is changing with pressure, and there are wells in the system. Capillary pressure between oil and water phase is not ignored.

4. A 2D two-phase flow (gas and water) reservoir with homogeneous and anisotropic permeability distribution. There is depth gradient in the system; however, the gravitational force for gas phase is negligible. There are both producers and injectors in the system. Porosity does not change throughout the operating period, and the solubility of gas in water is also taken into account. Capillary pressure between the gas and water phase is not ignored.

Solution to Problem 4.3.21

1. Because $A_x = A_y = A$ (but is heterogeneous) and $k_x \neq k_y$ (but is homogeneous), the following equation can be written:

$$k_x \frac{\partial}{\partial x}\left(\frac{A}{\mu_o B_o}\frac{\partial \Phi_o}{\partial x}\right)\Delta x + k_y \frac{\partial}{\partial y}\left(\frac{A}{\mu_o B_o}\frac{\partial \Phi_o}{\partial y}\right)\Delta y + q = \frac{V_b \phi c}{5.615}\frac{\partial p}{\partial t}. \quad \dots \dots \dots (4.99)$$

There is one primary unknown—pressure, which is embedded in the potential term of Eq. 4.99.

2. Because $k_x = k_y = k$ (and is homogeneous) and $A_x \neq A_y$ (but is homogeneous) in each direction,

- Oil equation:

$$A_x k \frac{\partial}{\partial x}\left(\frac{k_{ro}}{\mu_o B_o}\frac{\partial \Phi_o}{\partial x}\right)\Delta x + A_y k \frac{\partial}{\partial y}\left(\frac{k_{ro}}{\mu_o B_o}\frac{\partial \Phi_o}{\partial y}\right)\Delta y + q_o = \frac{V_b}{5.615}\frac{\partial}{\partial t}\left(\frac{\phi S_o}{B_o}\right), \quad \dots \dots \dots (4.100)$$

- Water equation:

$$A_x k \frac{\partial}{\partial x}\left(\frac{k_{rw}}{\mu_w B_w}\frac{\partial \Phi_w}{\partial x}\right)\Delta x + A_y k \frac{\partial}{\partial y}\left(\frac{k_{rw}}{\mu_w B_w}\frac{\partial \Phi_w}{\partial y}\right)\Delta y + q_w = \frac{V_b}{5.615}\frac{\partial}{\partial t}\left(\frac{\phi S_w}{B_w}\right), \quad \dots \dots \dots (4.101)$$

there are four primary unknowns: p_o, p_w, S_o, and S_w. The remaining two auxiliary equations are $S_o + S_w = 1$ and $P_{cow}(S_w) = p_o - p_w$ for a water-wet reservoir.

3. This time, $k_x = k_y = k$ (but is heterogeneously distributed).

- Oil equation:

$$\frac{\partial}{\partial x}\left(\frac{A_x k}{\mu_o B_o}\frac{\partial \Phi_o}{\partial x}\right)\Delta x + \frac{\partial}{\partial y}\left(\frac{A_y k}{\mu_o B_o}\frac{\partial \Phi_o}{\partial x}\right)\Delta y + q_o = \frac{V_b}{5.615}\frac{\partial}{\partial t}\left(\frac{\phi S_o}{B_o}\right). \quad \dots \dots \dots (4.102)$$

- Water equation:

$$\frac{\partial}{\partial x}\left(\frac{A_x k}{\mu_w B_w}\frac{\partial \Phi_w}{\partial x}\right)\Delta x + \frac{\partial}{\partial y}\left(\frac{A_y k}{\mu_w B_w}\frac{\partial \Phi_w}{\partial y}\right)\Delta y + q_w = \frac{V_b}{5.615}\frac{\partial}{\partial t}\left(\frac{\phi S_w}{B_w}\right). \quad \dots \dots \dots (4.103)$$

There are four primary unknowns: p_o, p_w, S_o, and S_w. The remaining two auxiliary equations are $S_o + S_w = 1$ and $P_{cow}(S_w) = p_o - p_w$ for a water-wet reservoir.

4. $k_x \neq k_y$, but both k_x and k_y show homogeneous distribution.

- Gas equation:

$$k_x \frac{\partial}{\partial x}\left(\frac{A_x}{\mu_g B_g}\frac{\partial p_g}{\partial x} + R_{sw}\frac{A_x}{\mu_w B_w}\frac{\partial \Phi_w}{\partial x}\right)\Delta x + k_y \frac{\partial}{\partial y}\left(\frac{A_y}{\mu_g B_g}\frac{\partial p_g}{\partial y} + R_{sw}\frac{A_y}{\mu_w B_w}\frac{\partial \Phi_w}{\partial y}\right)\Delta y + q_g$$

$$+ R_{sw} q_w = \frac{V_b \phi}{5.61S}\frac{\partial}{\partial t}\left(\frac{S_g}{B_g} + R_{sw}\frac{S_w}{B_w}\right). \quad \dots \dots \dots (4.104)$$

- Water equation:

$$k_x \frac{\partial}{\partial x}\left(\frac{A_x}{\mu_w B_w}\frac{\partial \Phi_w}{\partial x}\right)\Delta x + k_y \frac{\partial}{\partial y}\left(\frac{A_y}{\mu_w B_w}\frac{\partial \Phi_w}{\partial y}\right)\Delta y + q_w = \frac{V_b \phi}{5.615}\frac{\partial}{\partial t}\left(\frac{S_w}{B_w}\right). \quad \dots \dots \dots (4.105)$$

There are four primary unknowns: p_g, p_w, S_g, and S_w. The remaining two auxiliary equations are $S_g + S_w = 1$ and $P_{cgw}(S_g) = p_g - p_w$.

Problem 4.3.22

Give a full description of the reservoir flow systems represented by the following governing equations:

1.
- Oil Equation:

$$A\frac{\partial}{\partial x}\left(\frac{k_x k_{ro}}{\mu_o B_o}\frac{\partial p_o}{\partial x}\right)\Delta x + A\frac{\partial}{\partial y}\left(\frac{k_y k_{ro}}{\mu_o B_o}\frac{\partial p_o}{\partial y}\right)\Delta y - A\frac{\partial}{\partial x}\left(\frac{1}{144}\frac{g}{g_c}\frac{\rho_o k_x k_{ro}}{\mu_o B_o}\frac{\partial G}{\partial x}\right)\Delta x$$

$$-\frac{2\pi \bar{k} h k_{ro}}{\mu_o B_o\left[\ln\left(\frac{r_e}{r_w}\right)+s\right]}(p_o-200.0)=\frac{V_b}{5.615}\frac{\partial}{\partial t}\left(\frac{\phi s_o}{B_o}\right).\qquad \ldots\ldots\ldots\ldots (4.106)$$

- Gas Equation:

$$A\frac{\partial}{\partial x}\left(\frac{k_x k_{rg}}{\mu_g B_g}\frac{\partial p_g}{\partial x}+R_{so}\frac{k_x k_{ro}}{\mu_o B_o}\frac{\partial p_o}{\partial x}\right)\Delta x + A\frac{\partial}{\partial y}\left(\frac{k_y k_{rg}}{\mu_g B_g}\frac{\partial p_g}{\partial y}+R_{so}\frac{k_y k_{ro}}{\mu_o B_o}\frac{\partial p_o}{\partial y}\right)\Delta y$$

$$-A\frac{\partial}{\partial x}\left(\frac{1}{144}\frac{g}{g_c}\frac{R_{so}\rho_o k_x k_{ro}}{\mu_o B_o}\frac{\partial G}{\partial x}\right)\Delta x - \frac{2\pi \bar{k} h k_{rg}}{\mu_g B_g\left[\ln\left(\frac{r_e}{r_w}\right)+s\right]}(p_g-200.0)\qquad \ldots\ldots\ldots\ldots(4.107)$$

$$+\frac{2\pi R_{so}\bar{k} h k_{ro}}{\mu_o B_o\left[\ln\left(\frac{r_e}{r_w}\right)+s\right]}(p_o-200.0)=\frac{V_b}{5.615}\frac{\partial}{\partial t}\left(\frac{\phi S_g}{B_g}+\frac{R_{so}\phi S_o}{B_o}\right).$$

Knowing that, the wellbore model employed is

$$q_{fi,j,k}=-\frac{2\pi \bar{k} h k_{rf}}{\mu_f B_f\left[\ln\left(\frac{r_e}{r_w}\right)+s\right]}(p_f-p_{sf}). \quad \ldots\ldots\ldots\ldots\ldots\ldots\ldots\ldots (4.108)$$

2. $\dfrac{\partial^2 \Phi}{\partial x^2}+\dfrac{\partial^2 \Phi}{\partial y^2}=4.437\times 10^{-3}, \quad \ldots\ldots\ldots\ldots\ldots\ldots\ldots\ldots\ldots (4.109)$

where $k_y = 40$ md, $\mu = 1.0$ cp, $B = 1$ res bbl/STB, $\Delta x = 200$ ft, $\Delta y = 250$ ft, and thickness $= 50$ ft (uniform).

3. $\dfrac{\partial}{\partial x}\left(\dfrac{A_x}{\mu B}\dfrac{\partial p}{\partial x}\right)\Delta x + 2\dfrac{\partial}{\partial y}\left(\dfrac{A_y}{\mu B}\dfrac{\partial p}{\partial y}\right)\Delta y = -\dfrac{q}{k_x}+\dfrac{\partial}{\partial x}\left(\dfrac{A_x}{\mu B}\dfrac{\rho}{144}\dfrac{g}{g_c}\dfrac{\partial G}{\partial x}\right)\Delta x. \quad \ldots\ldots\ldots\ldots\ldots (4.110)$

4. $\dfrac{\partial}{\partial x}\left(\dfrac{k_{ro}}{\mu_o B_o}\dfrac{\partial p_o}{\partial x}\right)+\dfrac{\partial}{\partial y}\left(\dfrac{k_{ro}}{\mu_o B_o}\dfrac{\partial p_o}{\partial y}\right)-\left[\dfrac{\partial}{\partial x}\left(\dfrac{1}{144}\dfrac{g}{g_c}\rho_o\dfrac{k_{ro}}{\mu_o B_o}\dfrac{\partial G}{\partial x}\right)\right]+\dfrac{q_o}{kV_b}=\dfrac{\phi}{5.615k}\dfrac{\partial}{\partial t}\left(\dfrac{S_o}{B_o}\right), \quad \ldots\ldots\ldots\ldots (4.111)$

$$\dfrac{\partial}{\partial x}\left(\dfrac{k_{rg}}{\mu_g B_g}\dfrac{\partial p_g}{\partial x}+R_{so}\dfrac{k_{ro}}{\mu_o B_o}\dfrac{\partial p_o}{\partial x}\right)+\dfrac{\partial}{\partial y}\left(\dfrac{k_{rg}}{\mu_g B_g}\dfrac{\partial p_g}{\partial y}+R_{so}\dfrac{k_{ro}}{\mu_o B_o}\dfrac{\partial p_o}{\partial y}\right)-\left[\dfrac{\partial}{\partial x}\left(R_{so}\dfrac{1}{144}\dfrac{g}{g_c}\rho_o\dfrac{k_{ro}}{\mu_o B_o}\dfrac{\partial G}{\partial x}\right)\right]$$

$$\ldots\ldots (4.112)$$

$$+\dfrac{q_g}{kV_b}+R_{so}\dfrac{q_o}{kV_b}=\dfrac{\phi}{5.615k}\dfrac{\partial}{\partial t}\left(\dfrac{S_g}{B_g}+R_{so}\dfrac{S_o}{B_o}\right).$$

5. $\dfrac{\partial}{\partial x}\left(\dfrac{A_x}{\mu_g B_g}\dfrac{\partial p}{\partial x}\right)\Delta x+\dfrac{1}{3}\dfrac{\partial}{\partial y}\left(\dfrac{A_y}{\mu_g B_g}\dfrac{\partial p}{\partial y}\right)\Delta y+\dfrac{1}{10}\dfrac{\partial}{\partial z}\left(\dfrac{A_z}{\mu_g B_g}\dfrac{\partial p}{\partial z}\right)\Delta z-\dfrac{q_g}{k}=\dfrac{V_b\phi T_{sc}}{p_{sc}Tk}\dfrac{\partial}{\partial t}\left(\dfrac{p}{z_g}\right). \quad \ldots\ldots\ldots (4.113)$

6. $k_x\dfrac{\partial}{\partial x}\left(\dfrac{1}{\mu_w B_w}\dfrac{\partial p_w}{\partial x}\right)+\dfrac{\partial}{\partial y}\left(\dfrac{k_y}{\mu_w B_w}\dfrac{\partial \Phi_w}{\partial y}\right)+\dfrac{q_w}{2h(\Delta y)^2}=\dfrac{1}{5.615}\dfrac{\partial}{\partial t}\left(\dfrac{\phi}{B_w}\right). \quad \ldots\ldots\ldots\ldots\ldots (4.114)$

7. $k_x\dfrac{\partial}{\partial x}\left(\dfrac{A_x}{\mu_w B_w}\dfrac{\partial p_w}{\partial x}\right)\Delta x+k_y\dfrac{\partial}{\partial y}\left(\dfrac{A_y}{\mu_w B_w}\dfrac{\partial p_w}{\partial y}\right)\Delta y+q_w=\dfrac{V_b}{5.615}\dfrac{\partial}{\partial t}\left(\dfrac{\phi}{B_w}\right)$

$$\ldots\ldots\ldots\ldots\ldots\ldots\ldots (4.115)$$

$$+k_x\dfrac{\partial}{\partial x}\left(\dfrac{1}{144}\dfrac{g}{g_c}\dfrac{\rho_w A_x}{\mu_w B_w}\dfrac{\partial p_w}{\partial x}\right)\Delta x.$$

8. $\dfrac{\partial}{\partial x}\left(\dfrac{k_x}{\mu_g B_g}\dfrac{\partial p_g}{\partial x}\right)+\dfrac{\partial}{\partial y}\left(\dfrac{k_y}{\mu_g B_g}\dfrac{\partial p_g}{\partial y}\right)+k_z\dfrac{\partial}{\partial z}\left(\dfrac{1}{\mu_g B_g}\dfrac{\partial p_g}{\partial z}\right)+\dfrac{q_g}{V_b}=\dfrac{T_{sc}\phi}{p_{sc}T}\dfrac{\partial}{\partial t}\left(\dfrac{p}{z_g}\right).$ . . . . (4.116)

9. $\dfrac{\partial}{\partial x}\left(\dfrac{1}{\mu_o B_o}\dfrac{\partial p_o}{\partial x}\right)+\dfrac{\partial}{\partial y}\left(\dfrac{1}{\mu_o B_o}\dfrac{\partial p_o}{\partial y}\right)+\dfrac{q_o}{V_b k}=\dfrac{c_o\phi}{5.615 k}\dfrac{\partial p_o}{\partial t}.$ . . . . (4.117)

10. Include the grid system imposed over the flow domain,

$$\dfrac{\partial^2 p}{\partial x^2}+\dfrac{0.2}{\Delta x h}\dfrac{\partial}{\partial y}\left(A_y\dfrac{\partial p}{\partial y}\right)-\dfrac{0.2}{\Delta x h}\dfrac{\partial}{\partial y}\left(\dfrac{A_y}{144}\dfrac{g}{g_c}\rho\dfrac{\partial G}{\partial y}\right)-100\dfrac{\mu B}{V_b k_x}=\dfrac{\phi\mu c B}{5.615 k_x}\dfrac{\partial p}{\partial t}.$$ . . . . (4.118)

Solution to Problem 4.3.22

1.
- 2D flow in x-y domain.
- Homogeneous in thickness.
- Heterogeneous and anisotropic in permeability.
- Two-phase oil/gas flow.
- Deformable rock (i.e., porosity is a function of pressure).
- Depth gradients present along only the x-direction.
- Well with sandface pressure specification of $p_{sf}=200$ psi.

2.
- 2D flow in x-y domain.
- Homogeneous in thickness.
- Homogeneous and isotropic permeability distribution.
- Single-phase, incompressible fluid flow.
- With depth gradients along x- and y-directions.
- There is a well in the system. The flow rate can be calculated using

$$\dfrac{\partial^2 \Phi}{\partial x^2}+\dfrac{\partial^2 \Phi}{\partial y^2}=\dfrac{-q\mu B}{V_b k};$$ . . . . (4.119)

therefore, $\dfrac{-q\mu B}{V_b k}=4.437\times10^{-3}$. The flow rate can be solved as

$$q=-\dfrac{4.437\times10^{-3}\,kV_b}{\mu B}=-\dfrac{4.437\times10^{-3}\times40\times1.127\times10^{-3}\times200\times250\times50}{1\times1}=-500\,\text{STB/D}.$$ . . . . (4.120)

The negative sign indicates that the well is a production well.

3.
- 2D flow in x-y domain.
- Homogeneous and anisotropic in permeability ($k_y=2k_x$).
- Nonuniform formation thickness.
- Depth gradient is present only along the x-direction.
- Single-phase incompressible fluid flow.
- There is a well in the system.

4.
- 2D flow in x-y domain.
- Homogeneous and isotropic permeability.
- Two-phase (oil/gas) flow, with gas dissolved in the oleic phase.
- Homogeneous in thickness.
- Rock is nondeformable (i.e., porosity is not a function of pressure).
- Depth gradient only exists along the x-direction.
- There is a well in the system.

5.
- 3D flow domain represented in Cartesian coordinate.
- Single-phase, compressible fluid flow.
- Nonuniform thickness distribution.
- Well(s) exist in the system.
- Rock is nondeformable (i.e., porosity is not a function of pressure).
- Homogeneous and anisotropic in permeability, and $k_x=3k_y$ and $k_x=10k_z$.

6.
- 2D flow domain captured in Cartesian coordinates.
- Slightly compressible fluid flow.
- $\Delta x = 2\Delta y$.
- Anisotropic in permeability (homogeneous k_x and heterogeneous k_y distributions).
- Homogeneous in thickness.
- Rock is deformable (i.e., porosity is a function of pressure).
- Depth gradients exist along only the y-direction.
- There is a well in the system.

7.
- 2D flow domain represented in x-y plane of Cartesian coordinates.
- Single-phase, slightly compressible fluid flow.
- Anisotropic and homogeneous in permeability.
- Nonuniform thickness distribution.
- Depth gradient is only in x-direction.
- Rock is deformable (i.e., porosity is a function of pressure).
- There is a well in the system.

8.
- 3D flow domain captured in Cartesian coordinates.
- Single-phase compressible fluid flow.
- Anisotropic in permeability, k_x and k_y are heterogeneously distributed, whereas, k_z distribution is homogenous.
- Homogeneous in thickness.
- Rock is nondeformable (i.e., porosity is not a function of pressure).
- There is a well in the system.

9.
- 2D flow domain captured in Cartesian coordinates.
- Single-phase, slightly compressible fluid flow.
- Isotropic and homogenous in permeability.
- Uniform gridblock dimensions.
- Depth gradients are negligible.
- Rock is nondeformable (i.e., porosity is not a function of pressure).
- There is a well in the system.

10.
- The fluid is slightly compressible.
- The fluid flow is in two dimensions (x and y).
- Reservoir rock and fluid properties are not considered dependent on pressure.
- Homogeneous, but anisotropic $\left(\dfrac{k_y}{k_x} = \dfrac{1}{5} \right)$, permeability distribution.
- Δy and thickness values are uniform.
- There is no depth gradient along the x-direction.
- There is depth gradient along the y-direction.
- A producer is located in the block with a flow rate of 100 STB/D.

Problem 4.3.23

State whether the following assertions involving the incompressible, slightly compressible, and compressible fluid flow problems are TRUE or FALSE. In each case, justify your answer.

1. Consider the incompressible fluid flow systems shown in **Fig. 4.12.1**; the two reservoirs have the same rock and fluid properties. The injection rates of Reservoirs 1 and 2 are q_1 and q_2, respectively. Note that producers are operated at constant sandface pressures and injectors are operated at constant flow rates. Is the following statement TRUE or FALSE?

$$\text{If } |q_2| > |q_1|, \text{ then } (p_2 > p_1).$$
(*Note*: p_1 and p_2 are block pressures.)

2. Consider the incompressible fluid flow systems shown in **Fig. 4.12.2**; the two reservoirs have the same rock and fluid properties, with the exception that Reservoir 1 is more permeable than Reservoir 2. The injection rates of the two reservoirs are 100 STB/D. The production wells of the two systems are operated at the same constant sandface pressure. Is the following statement TRUE or FALSE?

The two systems have the same production rates; therefore, p_1 is always equal to p_2.
(*Note*: p_1 and p_2 are wellblock pressures.)

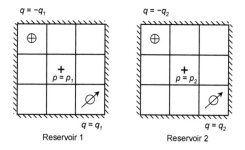

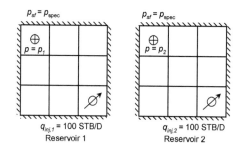

Fig. 4.12.1—Presentation of the flow problem described in Part 1 of Problem 4.3.23.

Fig. 4.12.2—Presentation of the flow problem described in Part 2 of Problem 4.3.23.

3. Consider the slightly compressible fluid flow systems shown in **Fig. 4.12.3**; there is a production well located at the center of the reservoir. The two systems have the same initial conditions and rock and fluid properties, except permeability. The two systems have homogeneous and isotropic permeability distribution. The two producers are producing with a constant flow rate at 100 STB/D. Is the following statement TRUE or FALSE?

$$\text{If } (k_1 > k_2), \text{ then } (p_1 > p_2 \text{ at any time}).$$
(*Note*: p_1 and p_2 are block pressures.)

4. Consider the slightly compressible fluid flow systems shown in **Fig. 4.12.4**; the two reservoirs have the same initial conditions and rock and fluid properties. In each system, there is a producer located in Block 1 and an injector located in Block 9. In both systems, the producers are operated at a constant sandface pressure of 20 psia, and the injectors are operated at a constant injection rate of 100 STB/D. The southern boundary of Block 7 is a constant pressure-gradient boundary. Is the following statement TRUE or FALSE?

$$\text{If } c_1 \text{ and } c_2 > 0 \text{ and } c_1 > c_2, \text{ then } (q_2 > q_1) \text{ at any time.}$$

5. Consider the compressible fluid flow systems shown in **Fig. 4.12.5**; there is a production well located at the center of the reservoir. Both wells are operated at constant sandface pressures. The two systems have the same initial conditions and rock and fluid properties: $\Delta x = \Delta y = 500$ ft, $h = 100$ ft, $k = 100$ md, $\phi = 0.2$, no depth gradient, $r_w = 0.2$ ft, and skin factor $= 0$. Is the following statement TRUE or FALSE?

$$\text{If } \left(p_{sf_1} = p_{sf_2} \right), \text{ then } \left(|q_1| = |q_2| \right) \text{ at any time.}$$

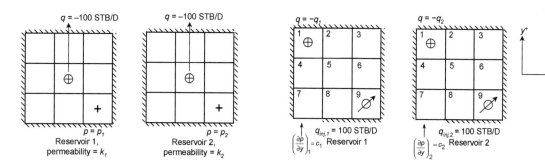

Fig. 4.12.3—Presentation of the flow problem described in Part 3 of Problem 4.3.23.

Fig. 4.12.4—Presentation of the flow problem described in Part 4 of Problem 4.3.23.

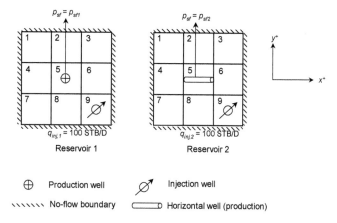

Fig. 4.12.5—Description of Part 5 of Problem 4.3.23.

Solution to Problem 4.3.23

1. TRUE. Because $q_2 > q_1$, and the sandface pressure at the producers is the same, the block pressure at the producer block is higher in Reservoir 2, and the difference between the pressure in the center block and the pressure at the producer block is bigger. Thus, p_2 has to be larger than p_1 to support the large flow rate into the producer.

2. FALSE. Invoking Peaceman's equation (Peaceman 1978, 1983):

$$q = \frac{2\pi \bar{k} h}{\mu B \left[\ln\left(\frac{r_e}{r_w}\right) + s \right]} \left(p - p_{sf} \right).$$

   Because p_{sf} and q are the same, one can conclude that $k_1 > k_2$ and $p_1 < p_2$.

3. TRUE. The more permeable the reservoir is, the less pressure difference needed between the sandface and wellblock to support the same flow rate.

4. TRUE. Because (c_1 and $c_2 > 0$), the flow at the southern boundary would leave the systems. When $c_1 > c_2$, there is more flow coming out of Reservoir 1; hence, the pressure drop in Reservoir 1 would be more rapid. This implies that pressure values in Reservoir 2 will always be larger than that in Reservoir 2 at the respective plots; hence, $q_2 > q_1$.

5. FALSE. We know that $q = -\Omega \left(p - p_{sf} \right)$ and the productivity indices for the vertical well in Reservoir 1 and for the horizontal well in Reservoir 2 will be as follows:

   - For the vertical producer in Reservoir 1,

$$\Omega_1 = \frac{2\pi k h}{\mu B \left(\ln \frac{r_{e_1}}{r_w} + s \right)}. \quad \dots\dots\dots\dots\dots\dots\dots\dots\dots\dots\dots\dots\dots\dots\dots\dots \text{(4.121)}$$

   - For the horizontal producer in Reservoir 2,

$$\Omega_2 = \frac{2\pi k \Delta x}{\mu B \left(\ln \frac{r_{e_2}}{r_w} + s \right)}. \quad \dots\dots\dots\dots\dots\dots\dots\dots\dots\dots\dots\dots\dots\dots\dots\dots \text{(4.122)}$$

   Because $\Delta x = \Delta y \neq h \rightarrow r_{e1} \neq r_{e2} \rightarrow \Omega_1 \neq \Omega_2$, the flow rate would be different for the vertical and horizontal wells.

Problem 4.3.24

Consider the fluid flow in the reservoir shown in **Fig. 4.13,** with properties listed in **Table 4.3.**

1. Describe the reservoir system.
2. Write the governing flow equation.
3. Calculate the injection rate.
4. Can we determine the pressure at each block? Why?

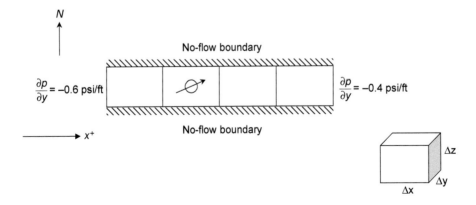

Fig. 4.13—Boundary conditions of the 1D flow domain as described in Problem 4.3.24.

k_x = 100 md	B = 1 res bbl/STB
Δx = 150 ft	r_w = 0.25 ft
Δy = 100 ft	s = –0.5
h = 100 ft	c = 0 psi^{-1}
μ = 1.0 cp	

Table 4.3—Fluid flow reservoir properties for the flow problem described in Fig. 4.13.

Solution to Problem 4.3.24

1.
- 1D reservoir expressed by Cartesian coordinates.
- Single-phase and incompressible fluid flow.
- Homogeneous permeability distribution.
- Uniform formation thickness.
- Uniform gridblock dimensions.
- No depth gradient.
- There is a well in the system.

2. Starting with the general form of the governing equation, $\dfrac{\partial}{\partial x}\left(\dfrac{A_x k_x}{\mu B}\dfrac{\partial p}{\partial x}\right)\Delta x + q = 0$ can be reduced to

$$\frac{\partial^2 p}{\partial x^2} + \frac{q\mu B}{V_b k_x} = 0, \quad\ldots\ldots\ldots\ldots\ldots\ldots\ldots\ldots\ldots\ldots\ldots\ldots\ldots\ldots\ldots\ldots\ldots\ldots\ldots\ldots\ldots\ldots\ldots\ldots\ldots (4.123)$$

after implementing the conditions indicated in Part 1.

3. As shown in Fig. 4.13, the flow is into the reservoir and out of the system across the western and eastern boundaries of the reservoir, respectively. Therefore, the total fluid flow into the reservoir by way of the boundary can be calculated as

$$q_{\text{in}} = -\frac{\Delta y h k_x}{\mu B}\frac{\partial p}{\partial x} = \frac{(100)(100)(100)(1.127\times 10^{-3})}{(1)(1)}(0.6 - 0.4) = -225.4\,\text{STB/D}. \quad\ldots\ldots\ldots\ldots\ldots (4.124)$$

Because the fluid is incompressible, excess fluid moving into the reservoir fluid should flow out of the well. Hence, the well is actually a production well, and the production rate is –225.4 STB/D.

4. No. Infinite number of systems with parallel pressure surfaces will generate the same flow rate because of the lack of a specified pressure anywhere within the flow domain.

Problem 4.3.25

For the system shown in **Fig. 4.14,** and using the rock and fluid properties found in **Table 4.4,** it is impossible to produce fluid without having injectors in the system. The system boundaries are completely sealed. Is this TRUE or FALSE?

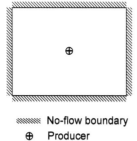

▨▨▨▨ No-flow boundary
⊕ Producer

Fig. 4.14—Physical boundaries and well position described in Problem 4.3.25.

ϕ = 0.2
$k_x = k_y$ = 50 md
h = 50 ft
c_f = 0 psi^{-1}

Table 4.4—Rock and fluid properties for the flow problem described in Fig. 4.14.

Solution to Problem 4.3.25

TRUE. Because $c_f = 0$ psi^{-1}, the fluid in the system is incompressible. And because no fluid is entering across the boundary of the system, no fluid can leave the reservoir.

Problem 4.3.26

State whether the following three problems are well posed or not (mathematically and/or physically). In either case, justify your answer.

1. Use **Fig. 4.15.1** and the reservoir rock/fluid properties found in **Table 4.5.1** to determine your answer.
2. Use **Fig. 4.15.2** and the reservoir rock/fluid properties found in **Table 4.5.2** to determine your answer.
3. Use **Fig. 4.15.3** and the reservoir rock/fluid properties found in **Table 4.5.3** to determine your answer.

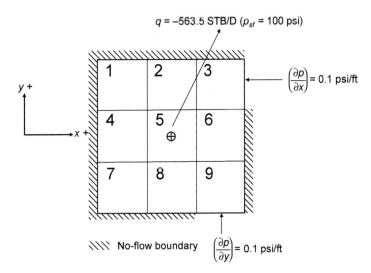

Fig. 4.15.1—Schematic representation of Part 1 of Problem 4.3.26.

$\Delta x = \Delta y = 500$ ft
$h = 50$ ft
$k_x = k_y = 100$ md
$B = 1$ res bbl/STB
$\mu = 1$ cp
$\phi = 0.2$
$c_f = 0$ psi^{-1}
No depth gradient

Table 4.5.1—Reservoir rock and fluid properties for the flow problem described in Fig. 4.15.1.

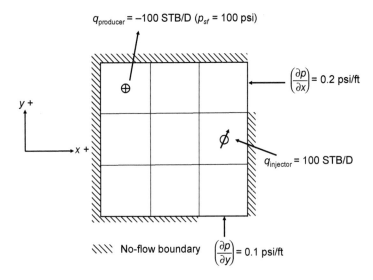

Fig. 4.15.2—Schematic representation of Part 2 of Problem 4.3.26.

$\Delta x = 600$ ft
$\Delta y = 300$ ft
$h = 50$ ft
$k_x = k_y = 100$ md
$B = 1$ res bbl/STB
$\mu = 1$ cp
$\phi = 0.2$
$c_f = 0$ psi^{-1}

Table 4.5.2—Reservoir rock and fluid properties for the flow problem described in Fig. 4.15.2.

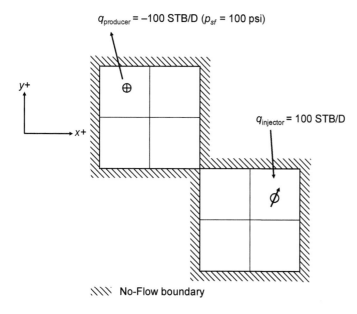

$q_{producer} = -100$ STB/D ($p_{sf} = 100$ psi)

$y+$

$x+$

$q_{injector} = 100$ STB/D

⌀

⧆

\\\\ No-Flow boundary

Fig. 4.15.3—Schematic representation of Part 3 of Problem 4.3.26.

$\Delta x = 500$ ft
$\Delta y = 300$ ft
$h = 50$ ft
$k_x = k_y = 100$ md
$B = 1$ res bbl/STB
$\mu = 1$ cp
$\phi = 0.2$
$c_f = 0$ psi^{-1}

Table 4.5.3—Reservoir rock and fluid properties for the flow problem described in Fig. 4.15.3.

Solution to Problem 4.3.26

1. Not well-posed. Because $c_f = 0\,\text{psi}^{-1}$, the fluid is incompressible. Because the reservoir is homogeneous and isotropic, the flow leaving the system across the southern boundary is equal to the flow going into the system across the eastern boundary. Therefore, the flow rate at the well must be equal to zero.
2. Well-posed. Because $c_f = 0\,\text{psi}^{-1}$, the fluid is incompressible. Because the reservoir is homogeneous and isotropic, the flow leaving the system across the southern boundary is equal to the flow going into the system across the eastern boundary. Therefore, the injector and producer rates of both of the wells must be equal to each other.
3. Not well-posed. There are two disconnected reservoir systems. The reservoir with the producer is isolated from the reservoir with the injector. The flow rates of both wells should be zero because of no-flow boundaries surrounding both reservoirs.

Problem 4.3.27

Identify the characterizations of the single-phase reservoirs on the basis of the following descriptions. If the governing flow equation in a 3D reservoir with single-phase fluid flow is a Laplace equation with dependent variable Φ,

1. Write the flow equation of the problem.
2. Characterize this reservoir flow system to the fullest extent.

Solution to Problem 4.3.27

1. Laplace equation: $\nabla^2\Phi=0$,

$$\frac{\partial^2 \Phi}{\partial x^2} + \frac{\partial^2 \Phi}{\partial y^2} + \frac{\partial^2 \Phi}{\partial z^2} = 0. \quad\ldots\ldots\ldots\ldots\ldots\ldots\ldots\ldots\ldots\ldots\ldots\ldots\ldots\ldots\ldots\ldots\ldots\ldots\ldots\text{(4.125)}$$

2. See **Table 4.6** for a characterization of the reservoir fluid system.

Properties	Description		
Fluid	incompressible		
Well	does not exist		
Thickness	uniform		
	isotropic ($k_x = k_y = k_z = k$)		
Permeability	x-direction	y-direction	z-direction
	homogeneous	homogeneous	homogeneous
Rock	nondeformable		
Depth Gradient	x-direction	y-direction	z-direction
	yes	yes	yes

Table 4.6—Characterization of the reservoir fluid system found in Part 2 of the Solution to Problem 4.3.27.

Problem 4.3.28

Show that a real-gas flow equation written in terms of pseudopressures at high pressures collapses to a form analogous to the slightly compressible fluid flow equation. Explain why this is expected.

Solution to Problem 4.3.28

Define the following:

$$\Psi(p) = \int_0^p \frac{\partial}{\partial p}\left(\frac{p}{\mu_g z_g}\right) dp \ \text{ or } \ \frac{\partial \Psi(p)}{\partial p} = \frac{p}{\mu_g z_g}. \qquad (4.126)$$

For gas flow in a 1D porous media,

$$\frac{\partial}{\partial x}\left(\frac{A_x k_x}{\mu_g B_g}\frac{\partial p}{\partial x}\right)\Delta x + q_g = \frac{V_b \phi T_{sc}}{p_{sc} T}\frac{\partial}{\partial t}\left(\frac{p}{z_g}\right), \qquad (4.127)$$

$$B_g = \frac{p_{sc} z_g T}{5.615 T_{sc} p}. \qquad (4.128)$$

Therefore,

$$\frac{\partial}{\partial x}\left(A_x k_x \frac{\partial \Psi}{\partial x}\right)\Delta x + \frac{p_{sc} T}{5.615 T_{sc}}q_g = \frac{V_b \phi}{5.615}\frac{\partial}{\partial t}\left(\frac{p}{z_g}\right), \qquad (4.129)$$

$$\frac{\partial}{\partial t}\left(\frac{p}{z_g}\right) = \frac{\partial\left(\frac{p}{z_g}\right)}{\partial p}\frac{\partial p}{\partial t}, \qquad (4.130)$$

$$\frac{\partial\left(\frac{p}{z_g}\right)}{\partial p} = \frac{z_g - p\frac{\partial z_g}{\partial p}}{z_g^2} = \frac{p}{z_g}\left(\frac{1}{p} - \frac{1}{z_g}\frac{\partial z_g}{\partial p}\right), \qquad (4.131)$$

and

$$c_g = \left(\frac{1}{p} - \frac{1}{z_g}\frac{\partial z_g}{\partial p}\right), \partial p = \frac{\mu_g z_g}{p}\partial \Psi. \qquad (4.132)$$

Thus,

$$\frac{\partial}{\partial x}\left(A_x k_x \frac{\partial \Psi}{\partial x}\right)\Delta x + \frac{p_{sc} T}{5.615 T_{sc}}q_g = \frac{V_b \phi \mu_g c_g}{5.615}\frac{\partial \Psi}{\partial t}. \qquad (4.133)$$

Eq. 4.132 is analogous to the single-phase liquid equation:

$$\frac{\partial}{\partial x}\left(\frac{A_x k_x}{\mu B}\frac{\partial p}{\partial x}\right)\Delta x + q_o = \frac{V_b \phi c_t}{5.615}\frac{\partial p}{\partial t}. \dots\dots\dots\dots\dots\dots\dots\dots\dots\dots\dots\dots(4.134)$$

The pseudopressure-transformation application to the gas equation is simply a linearization treatment that eliminates the nonlinear terms of the left-hand side of the original gas equation. Eqs. 4.133 and 4.134 are expected to be analogous, simply because at high pressures, gases will behave like liquids. Note that the gas equations in pseudopressure form and the single-phase slightly compressible flow equations still have nonlinear terms (e.g., c_g, μ_g, and c_t) on their respective right-hand sides. This is why any equation in that form is referred to as a quasilinear equation.

Problem 4.3.29

Consider the incompressible fluid flow problem in the 1D porous body shown in **Fig. 4.16.** Viscosity of the fluid is $\mu = 1.5$ cp.

1. If the well in Block 2 is produced at a steady-state flow rate of 600 STB/D, and this rate creates a pressure distribution of $p_1 = 1,500$ psia, $p_2 = 700$ psia, and $p_3 = 700$ psia, what is the permeability of this linear, homogeneous system?
2. What kind of boundary condition would you assign to the extreme left end of the system? Quantify your answer.

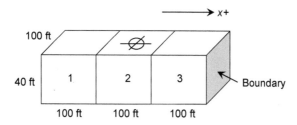

Fig. 4.16—Schematic representation of the flow domain described in Problem 4.3.29.

Solution to Problem 4.3.29

1. Because the pressure of Block 1 is higher than the pressure in Block 2, the flow direction is from Block 1 to 2. Because Blocks 2 and 3 are at identical pressures, no flow takes place between Blocks 2 and 3.
   Applying Darcy's law between Blocks 1 and 2,

$$q = -\frac{0.001127 kA}{\mu B}\frac{\partial p}{\partial x}. \dots\dots\dots\dots\dots\dots\dots\dots\dots\dots\dots\dots\dots\dots\dots\dots(4.135)$$

Knowing that, $A = 100\,\text{ft} \times 40\,\text{ft} = 4,000\,\text{ft}^2$.

For an incompressible fluid flow, B can be assumed to be equal to 1:

$$\frac{\partial p}{\partial x} = \left(\frac{700 - 1,500}{100}\right) = -8\,\frac{\text{psia}}{\text{ft}}. \dots\dots\dots\dots\dots\dots\dots\dots\dots\dots\dots\dots(4.136)$$

The permeability of the system can be solved:

$$-600\,\text{STB/D} = \frac{0.001127 \times k \times 4,000\,\text{ft}^2}{1.5} \times (-8)\left(\frac{\text{psia}}{\text{ft}}\right). \dots\dots\dots\dots\dots\dots\dots\dots(4.137)$$

Therefore, $k = 24.95 \cong 25\,\text{md}$.

2. Either a constant-pressure support or constant-flow support across the left-extreme boundary is possible:

   - $p = 2,300$ (presence of aquifer or some form of constant-pressure support).
   - Flow rate $q = -600$ STB/D from the left boundary or $\frac{\partial p}{\partial x} = -8\,\frac{\text{psia}}{\text{ft}}$.

Problem 4.3.30

A fluid flow problem is described by the following PDE:

$$\frac{1}{2}\frac{\partial}{\partial x}\left(k_x\frac{\partial p}{\partial x}\right)+\frac{\partial^2 p}{\partial y^2}=-1, \text{ with } \Delta x = 150\,\text{ft},\ \Delta y = 78\,\text{ft},\ \Delta z = 25\,\text{ft, and }\mu = 1.1\,\text{cp}.$$

1. Quantitatively and qualitatively describe the flow problem to the fullest extent. Ignore the gravitational effects on the fluid flow dynamics and assume that permeability has a unit of md.
2. Calculate the flow rate of the external sink/source without solving the equation.

Solution to Problem 4.3.30

1. The PDE represents a fluid flow problem with the following characteristics:

   - 2D single-phase flow.
   - Incompressible fluid.
   - Anisotropic, with respect to permeability.
   - Homogenous in y-direction, with respect to permeability.
   - Heterogonous in x-direction, with respect to permeability.
   - There is an injection well.
   - Areas perpendicular to the flow directions are individually homogenous.

2. Reservoir flow equation states that

$$\frac{\partial}{\partial x}\left(\frac{A_x k_x}{\mu B}\frac{\partial p}{\partial x}\right)\Delta x+\frac{\partial}{\partial y}\left(\frac{A_y k_y}{\mu B}\frac{\partial p}{\partial y}\right)\Delta y+q=0. \quad\dotfill (4.138)$$

On simplification,

$$\frac{\partial}{\partial x}\left(k_x\frac{\partial p}{\partial x}\right)+\frac{\partial}{\partial y}\left(k_y\frac{\partial p}{\partial y}\right)+\frac{q\mu B}{V_b(0.001127)}=0; \quad\dotfill (4.139)$$

when Eq. 4.139 is compared with the original equation,

$$\frac{\partial}{\partial x}\left(k_x\frac{\partial p}{\partial x}\right)+2\frac{\partial^2 p}{\partial y^2}+\frac{q\mu B}{V_b(0.001127)}=0, \quad\dotfill (4.140)$$

one obtains $k_y = 2$ md and $\dfrac{q\mu B}{V_b(0.001127)}=-2$.

Assuming $B = 1$ res bbl/STB, one can calculate $V_b = 25\times150\times78 = 292,500\,\text{ft}^3$, and therefore,

$$q=-\frac{2(292,500)(0.001127)}{1.1}=-600\,\text{STB/D}.$$

Nomenclature

a	= constant	B_w	= water FVF, L³/L³, res bbl/STB
A	= cross-sectional area, L², ft²	c	= compressibility, Lt²/m, psi⁻¹
A_x	= cross-sectional area normal to the direction x, L², ft²	c_f	= fluid compressibility, Lt²/m, psi⁻¹
		c_R	= formation-rock compressibility, Lt²/m, psi⁻¹
A_y	= cross-sectional area normal to the direction y, L², ft²	c_t	= total compressibility, Lt²/m, psi⁻¹
		C	= constant
A_z	= cross-sectional area normal to the direction z, L², ft²	G	= acceleration of gravity, L/t², ft/s²
B	= FVF, L³/L³, res bbl/STB or res bbl/scf; constant	g_c	= unit conversion factor in Newton's law
B_f	= FVF of composition f ($f = o,w,g$), L³/L³, res bbl/STB or res bbl/scf	G	= elevation with respect to absolute datum being positive upward, L, ft
B_g	= gas FVF, L³/L³, res bbl/scf	h	= thickness, L, ft
B_o	= oil FVF, L³/L³, res bbl/STB	k	= permeability, L², darcies

k_r = permeability in the r-direction, L^2, darcies

k_{rf} = relative permeability to phase f ($f = o, w, g$), dimensionless

k_{rg} = relative permeability to gas, dimensionless

k_{ro} = relative permeability to oil, dimensionless

k_{rw} = relative permeability to water

k_x = permeability in the x-direction, L^2, darcies or perms

k_y = permeability in the y-direction, L^2, darcies or perms

k_z = permeability in the z-direction, L^2, darcies or perms

k_Θ = permeability in the θ-direction of the radial-cylindrical coordinate system, darcies or perms (1 darcy = 1.127 perms)

k_∞ = absolute permeability, L^2, darcies or perms

l_f = distance between two focuses of an ellipse

L = length, L, ft

M = mass per unit volume of porous media, m/L^3, lbm/ft^3

$\dot{m}$ = mass flux vector, m/L^{2}t, lbm/D–ft^2

MW = molecular weight, lbm mol

p = pressure, m/Lt^2, psia

P_{cgw} = gas/water capillary pressure, m/Lt^2, psia

P_{cow} = oil/water capillary pressure, m/Lt^2, psia

p_f = phase f ($f = o, w, g$) pressure, m/Lt^2, psia

p_g = gas pressure, m/Lt^2, psia

p_i = initial pressure, m/Lt^2, psia

p_L = pressure measured at a spatial location of $x = L$, m/Lt^2, psia

p_o = oil pressure, m/Lt^2, psia

p_{sc} = standard condition pressure, m/Lt^2, psia

p_{sf} = sandface pressure, m/Lt^2, psia

p_w = water pressure, m/Lt^2, psia

q = flow rate, L^3/t, STB/D for liquid and scf/D for gas

q_f = flow rate of phase f ($f = o, w, g$), L^3/t, STB/D for liquid and scf/D for gas

$q_{fi,j,k}$ = fluid flow rate from well located in block i,j,k, L^3/t, STB/D for liquid and scf/D for gas

q_{gsc} = gas flow rate at standard conditions, L^3/t, scf/D

q_{in} = flow rate entering the system, L^3/t, STB/D for liquid and scf/D for gas

$q_{injector}$ = injection rate of an injection well, L^3/t, res bbl/D

q_m = mass fluid flow rate of the fluid, M/t, lbm/D

$q_{producer}$ = production rate of a production well, L^3/t, STB/D for liquid and scf/D for gas

q_{sc} = liquid flow rate at standard conditions, L^3/t, STB/D

r = radial distance in both cylindrical and spherical coordinate system, L, ft

r_e = radius of the external boundary, L, ft

r_w = well radius, L, ft

R = universal gas constant

R_{so} = solution gas/oil ratio, L^3/L^3, scf/STB

R_{sw} = solution gas/water ratio, L^3/L^3, scf/STB

s = skin factor, dimensionless

S_g = gas saturation, fraction

S_o = oil saturation, fraction

S_w = water saturation, fraction

t = time, t, days

T = absolute temperature, T, °R

T_{sc} = standard condition temperature, T, °R

v = superficial velocity, L/t, res bbl/(D–ft^2)

v_r = superficial velocity component along r-direction, L/t, res bbl/(D–ft^2)

v_s = superficial velocity component along arbitrary s-direction, L/t, res bbl/(D–ft^2)

v_x = superficial velocity component along x-direction, L/t, res bbl/(D–ft^2)

v_y = superficial velocity component along y-direction, L/t, res bbl/(D–ft^2)

v_z = superficial velocity component along z-direction, L/t, res bbl/(D–ft^2)

v_θ = superficial velocity component along θ-direction, L/t, res bbl/(D–ft^2)

v_φ = superficial velocity component along φ-direction, L/t, res bbl/(D–ft^2)

V_b = bulk volume, L^3, ft^3

x = distance in x-direction in the Cartesian coordinate system, L, ft

y = distance in y-direction in the Cartesian coordinate system

z = distance in z-direction in the Cartesian coordinate system

z_g = gas compressibility factor, dimensionless

α_c = unit conversion factor between ft^3 and bbl, with a magnitude of 5.615

β_c = transmissibility conversion factor, with a magnitude of 1.127

∇ = gradient operator

Δ = difference of parameters

ΔL = position differential (distance), L, ft

Δp = pressure differential, m/Lt^2, psia

Δr = control volume dimension along the r-direction, L, ft

$\Delta\theta$	=	control volume dimension along the θ-direction, radian
$\Delta\varphi$	=	control volume dimension along the φ-direction, radian
Δt	=	timestep size, t, D
Δx	=	control-volume dimension along the x-direction, L, ft
Δy	=	control-volume dimension along the y-direction, L, ft
Δz	=	control-volume dimension along the z-direction, L, ft
η	=	the η coordinate of elliptical-cylindrical coordinate system and hydraulic diffusivity factor
θ	=	angle in θ-direction in both cylindrical and spherical coordinate systems, rad
μ	=	viscosity, m/Lt, cp
μ_f	=	viscosity of phase $f(f = o,w,g)$, m/Lt, cp
μ_g	=	gas viscosity, m/Lt, cp
μ_o	=	oil viscosity, m/Lt, cp
μ_w	=	water viscosity, m/Lt, cp
ξ	=	the ξ coordinator in the elliptic-cylindrical coordinate system
ρ	=	density, m/L^3, lbm/ft^3

ρ_o	=	oil density, m/L^3, lbm/ft^3
Φ	=	flow potential, m/Lt^2, psia
Φ_e	=	flow potential at external radius, m/Lt^2, psia
Φ_o	=	flow potential of oil phase, m/Lt^2, psia
Φ_r	=	flow potential at radius r, m/Lt^2, psia
Φ_w	=	flow potential at the well radius, or flow potential of the water phase, m/Lt^2, psia
ϕ	=	porosity, fraction
ψ	=	pseudopressure, m/Lt^3, psi^2/cp
φ	=	the φ coordinate of elliptical-cylindrical coordinate system

Subscripts

boundary	=	on the boundary
g	=	gas phase
o	=	oil phase
r	=	r-direction (cylindrical and spherical coordinates)
t	=	time dimension
w	=	water phase
θ	=	θ-direction in cylindrical and spherical coordinates

References

Ertekin, T., Abou-Kassem, J. H., and King, G. R. 2001. *Basic Applied Reservoir Simulation*. Richardson, Texas, US: Society of Petroleum Engineers.

Klinkenberg, L. J. 1941. The Permeability Of Porous Media To Liquids And Gases. Paper presented at the API Drilling and Production Practice conference, New York, New York, 1 January. API-41-2OO. http://dx.doi.org/10.5510/OGP20120200114.

Peaceman, D. W. 1978. Interpretation of Well-Block Pressures in Numerical Reservoir Simulation (includes associated paper 6988). *SPE J* **18**(03): 183–194. SPE-6893-PA. http://dx.doi.org/10.5510/6893-PA.

Peaceman, D.W. 1983. Interpretation of Well-Block Pressures in Numerical Reservoir Simulation with Nonsquare Grid Blocks and Anisotropic Permeability. *SPE J* **23**(3): 531–543. SPE-10528-PA. https://dx.doi.org/10.2118/10528-PA.

Pedrosa, O. A. Jr. and Aziz, K. 1986. Use of a Hybrid Grid in Reservoir Simulation. *SPE Res Eng* **1**(06): 611–621; In *Transactions of the Society of Petroleum Engineers*, Vol. 281, SPE-13507-PA. http://dx.doi.org/10.2118/13507-PA.

Chapter 5

Gridding Methods and Finite-Difference Representation of Single-Phase Flow Equations

In previous chapters, we have learned finite-difference analogs of first-order and second-order derivatives. In Chapter 4, we learned the partial-differential equations (PDEs) that describe the single-phase flow of incompressible, slightly compressible, and compressible flow dynamics in porous media. Because these PDEs are all composed of first-order and second-order derivatives, writing the finite-difference representation of them is quite straightforward, using their respective finite-difference analogs. Because the finite-difference equations are implemented at preselected discrete points in the reservoir, a discretized representation of the reservoir also will be necessary. The discretized representation of the reservoir is achieved by superimposing gridblocks over the reservoir; this process is called gridding. In this chapter, we first will briefly review the different types of grids, and then we will learn how to couple the finite-difference approximations and the discretized (gridded) reservoir architecture.

5.1 Grid Types and Grid Design

There typically are two types of grids used in reservoir simulation: block centered (i.e., body centered) and mesh centered (i.e., point distributed). The finite-difference approximation for implementations with either of the grid systems will be the same, as discussed later in this chapter.

5.1.1 Block-Centered (Body-Centered) Grids. In body-centered grids the discrete points in the reservoir are located at the center of each gridblock, and there will be no discrete points on the external boundaries of the reservoir (see **Fig. 5.1**). The stepwise procedure for placing a body-centered grid over the reservoir can be summarized as follows:

1. Determine the principal flow directions and orient the coordinate directions accordingly.
2. Draw the smallest rectangle that will encompass the reservoir in its entirety. The edges of the rectangle should be placed parallel to the principal axes of the coordinate system.
3. Place the gridblocks that will host the wells; try to locate the wells at the center of these blocks. There should not be more than one well in a given block. If you cannot avoid having more than one well in a block (if you have two, for example), combine them into one hypothetical well located at the center of the block.
4. Place the remaining major grid lines to cover the entire reservoir domain. To more-accurately capture the irregularities of the external boundaries, use a necessary nonuniform spacing between the grid lines.

5.1.2 Mesh-Centered (Point-Distributed) Grids. In mesh-centered or point-distributed grids, the discrete points are located at the intersections of the major grid lines. Contrary to the body-centered grids, in mesh-centered grids the discrete points are not necessarily located in the center of the blocks they are representing; furthermore, there are discrete points located on the external boundaries of the reservoir. The stepwise procedure for placing a mesh-centered grid over the reservoir of Fig. 5.1 is described below in conjunction with **Fig. 5.2:**

1. Determine the principal flow directions and orient the coordinate directions accordingly.
2. Draw the smallest rectangle that will encompass the external boundaries of the reservoir. The edges of the rectangle should be placed parallel to the principal axes of the coordinate system.
3. Place the major grid lines so that they intersect at the well locations (well locations will coincide with the discrete points).
4. Place the remaining major grid lines to cover the entire reservoir domains. The optimum spacing of the grid lines must be determined to more-effectively capture the irregular aspects of the physical boundaries.

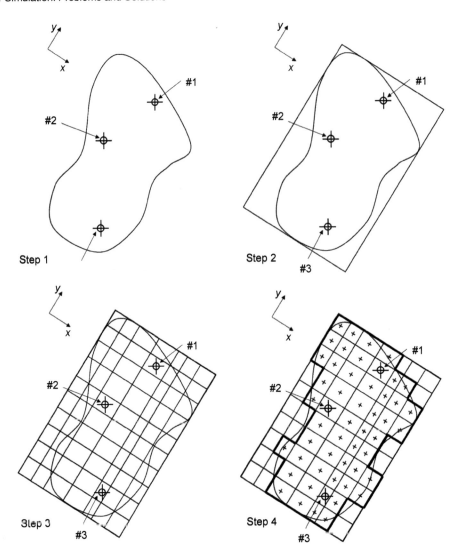

Fig. 5.1—Laying a body-centered grid over a reservoir with three existing wells (Note that major grid lines traverse across the entire dimensions of the outermost rectangle).

5.1.3 Types of Boundaries. In reservoir simulation, two different types of boundaries are encountered: internal and external. Internal boundaries are commonly placed at the well locations. At the internal boundaries there will be two options to specify the boundary conditions. These two options include the specification of flow rate (Neumann-type boundary condition) or the specification of pressure at the sandface. Recall that the flow-rate specification is synonymous with the specification of the pressure gradient at the wellbore; it can easily be seen in the following equation:

$$r \left. \frac{\partial p}{\partial r} \right|_{r=r_w} = - \left. \frac{q \mu B}{2 \pi k h} \right|_{r=r_w} \quad \dots \dots \dots \dots \dots \dots \dots \dots \dots \dots \dots \dots \dots \dots \dots \dots \dots \dots \dots \dots \dots \dots \dots \dots (5.1)$$

Eq. 5.1 is Darcy's law as it is written in the immediate vicinity of the wellbore. By specifying (fixing) the flow rate, q, the value of $\left. \frac{\partial p}{\partial r} \right|_{r=r_w}$ is fixed (the Neumann-type boundary condition) for a given set of rock and fluid properties. It should be emphasized that Eq. 5.1 is also consistent with the sign convention adopted throughout this book ($q < 0$ indicates production and $q > 0$ indicates injection). In Eq. 5.1 if q is negative, then the well is a producer and $\frac{dp}{dr} > 0$, indicating that pressure will increase in the increasing direction of r (away from the wellbore). Similarly, if the well of Eq. 5.1 is an injector, then $\frac{dp}{dr}$ will become negative, showing that pressure is decreasing as one moves away from the wellbore. As a final cautionary note, remember that it is not physically realistic to specify a pressure and a pressure gradient at the same location and at the same

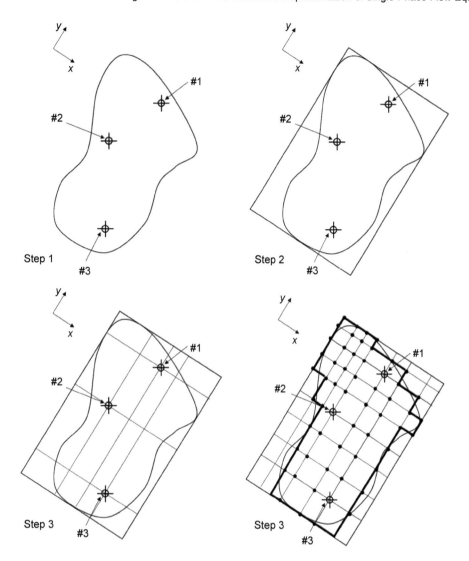

Fig. 5.2—Laying a mesh-centered grid over a reservoir with three existing wells (Note the existence of discrete points on the boundaries).

time on the boundary. For example, a well cannot produce under both imposed sandface and flow rate specifications. **Fig. 5.3** displays the pressure distribution along the r-direction when the Neumann-type inner-boundary condition is specified.

The external boundaries typically are the physical (i.e., structural) limits of the reservoir. Again, it is possible to specify either the pressure gradient across the boundary or the value of the pressure on the boundary. The specification of the pressure gradient across the physical limits of an external boundary must be made with regard to the positive directions of the coordinate system considered in conjunction with the principal flow directions. We use **Fig. 5.4** and **Table 5.1** to illustrate the definitions of positive and negative directions of the pressure gradients on the basis of prescribed $x+$ an $y+$ orientations. This representation clearly shows that we have adopted positive values of flow rate for injection and negative values for production.

5.2 Finite-Difference Representation of Single-Phase Flow Equations

5.2.1 Incompressible Flow Equation. The PDE describing the single-phase flow of an incompressible fluid in porous media is repeated here (ignoring the depth gradients):

$$\frac{\partial}{\partial x}\left(\frac{A_x k_x}{\mu B}\frac{\partial \Phi}{\partial x}\right)\Delta x + \frac{\partial}{\partial y}\left(\frac{A_y k_y}{\mu B}\frac{\partial \Phi}{\partial y}\right)\Delta y + \frac{\partial}{\partial z}\left(\frac{A_z k_z}{\mu B}\frac{\partial \Phi}{\partial x}\right)\Delta z + q = 0. \dots\dots\dots\dots\dots\dots\dots\dots\dots\dots\dots\dots\dots (5.2)$$

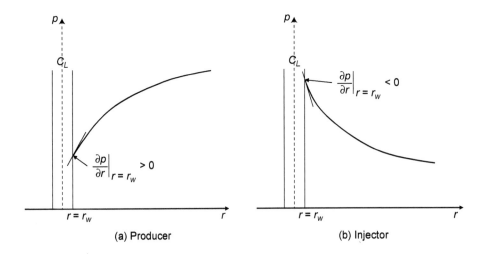

Fig. 5.3—Neumann-type specification at the well (note that C_L denotes the centerline of the well).

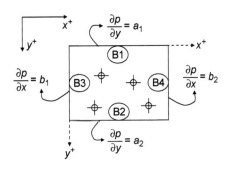

Fig. 5.4—Implications of the sign of pressure gradient on the nature of the boundary in conjunction with the specified positive-coordinate-system directions. For example, across Boundary 2 if a_2 has a negative value, then flow is toward the outside, across Boundary 2. Similarly, if b_1 is negative, flow is from the outside into the reservoir across the Boundary 3.

Physical Boundary	Sign of Gradient	Nature of Boundary
	$a_1 > 0$	Flow is out across B1
B1	$a_1 = 0$	No flow across B1
	$a_1 < 0$	Flow is in across B1
	$a_2 > 0$	Flow is in across B2
B2	$a_2 = 0$	No flow across B2
	$a_2 < 0$	Flow is out across B2
	$a_3 > 0$	Flow is out across B3
B3	$a_3 = 0$	No flow across B3
	$a_3 < 0$	Flow is in across B3
	$a_4 > 0$	Flow is in across B4
B4	$a_4 = 0$	No flow across B4
	$a_4 < 0$	Flow is out across B4

Table 5.1—Physical boundary, sign of the gradient, and nature of the boundary, as depicted in Fig. 5.4.

In Eq. 5.2, because of the incompressible nature of fluid, the viscosity term, μ, is constant, and the formation volume factor, B, is equal to unity. In an effort to not lose the generality of Eq. 5.2, we will keep these two terms in their respective locations. Using a five-point central-difference approximation for the second-order space derivatives, and ignoring depth gradients ($\Phi = p$), the finite-difference approximation of Eq. 5.2 can be written as follows (refer to Table 2.2 from Chapter 2 and Table 4.1 from Chapter 4 for units of the terms):

$$\left.\frac{A_x k_x}{\mu B \Delta x}\right|_{i+\frac{1}{2},j,k}\left(p_{i+1,j,k}-p_{i,j,k}\right)-\left.\frac{A_x k_x}{\mu B \Delta x}\right|_{i-\frac{1}{2},j,k}\left(p_{i,j,k}-p_{i-1,j,k}\right)+\left.\frac{A_y k_y}{\mu B \Delta y}\right|_{i,j+\frac{1}{2},k}\left(p_{i,j+1,k}-p_{i,j,k}\right)$$

$$-\left.\frac{A_y k_y}{\mu B \Delta y}\right|_{i,j-\frac{1}{2},k}\left(p_{i,j,k}-p_{i,j-1,k}\right)+\left.\frac{A_z k_z}{\mu B \Delta z}\right|_{i,j,k+\frac{1}{2}}\left(p_{i,j,k+1}-p_{i,j,k}\right)-\left.\frac{A_z k_z}{\mu B \Delta z}\right|_{i,j,k-\frac{1}{2}}\left(p_{i,j,k}-p_{i,j,k-1}\right) \dots \dots \dots \dots \dots (5.3)$$

$$+q_{i,j,k}=0.$$

Eq. 5.3 has seven principal unknowns that are the block-pressure values, including the pressure of the central block (i,j,k) and the six pressure values of the surrounding six blocks $(i, j, k–1)$, $(i, j–1, k)$, $(i–1, j, k)$, $(i, j, k+1)$, $(i, j+1, k)$, and $(i+1, j, k)$. Also note that the source/sink term, $q_{i,j,k}$, will bring an eighth unknown; therefore, it is necessary to have one more equation to close the system. The eighth equation is known as the wellbore equation, and typically it is expressed as

$$q_{i,j,k}=\left.\frac{2\pi r_w \overline{k} h}{\mu B\left[\ln \dfrac{r_{eq}}{r_w}+s\right]}\right|_{i,j,k}\left(p_{i,j,k}-p_{sf_{i,j,k}}\right), \dots \dots \dots \dots \dots \dots \dots \dots \dots \dots \dots \dots \dots \dots \dots \dots \dots (5.4a)$$

or through identifying the productivity index group $\Omega_{i,j,k}$ and enforcing the sign convention we have adopted in Eq. 5.4a:

$$q_{i,j,k} = -\Omega_{i,j,k}\left(p_{i,j,k} - p_{sf_{i,j,k}}\right). \quad\dots\dots\dots\dots\dots\dots\dots\dots\dots\dots\dots\dots\dots\dots\dots \quad (5.4b)$$

Eq. 5.4 has two additional variables, which are $q_{i,j,k}$ and $p_{sf\,i,j,k}$. As previously discussed, one of these two variables will be specified as the boundary condition, and then Eq. 5.4 will provide the remaining additional equation needed to close the system (to make the number of equations and number of the unknowns equal to each other to end up with a well-posed problem). The $\overline{k}$ term in Eq. 5.4a is the geometric mean of the permeabilities in the principal directions of the x-y plane (assuming the well is parallel to the z-direction and, hence, perpendicular to the x-y plane). Also note that the $\Omega_{i,j,k}$ term in Eq. 5.4b is the productivity index term with units (STB/D-psi). Eq. 5.4 is an important equation and will be discussed in more detail in the next chapter.

5.2.2 Slightly Compressible Flow Equation. The equation that describes the flow of a slightly compressible fluid was previously developed (for a reservoir with no depth gradients) as

$$\frac{\partial}{\partial x}\left(\frac{A_x k_x}{\mu B}\frac{\partial p}{\partial x}\right)\Delta x + \frac{\partial}{\partial y}\left(\frac{A_y k_y}{\mu B}\frac{\partial p}{\partial y}\right)\Delta y + \frac{\partial}{\partial z}\left(\frac{A_z k_z}{\mu B}\frac{\partial p}{\partial z}\right)\Delta z + q = \frac{V_b \phi c}{5.615}\frac{\partial p}{\partial t}. \quad\dots\dots\dots\dots\dots\dots\dots\dots \quad (5.5)$$

Contrary to Eq. 5.2, Eq. 5.5 represents an unsteady-state flow problem. This implies that pressure values at reservoir locations will change as a function of time. The finite-difference representation of Eq. 5.5 can be written as shown in Eq. 5.6. Notably, the right-hand side of Eq. 5.5 is discretized by a specific time interval (Δt) at the current timestep level; however, the Δt term may have different values for different timestep levels.

$$\left.\frac{A_x k_x}{\mu B \Delta x}\right|^n_{i+\frac{1}{2},j,k}\left(p^{n+1}_{i+1,j,k} - p^{n+1}_{i,j,k}\right) - \left.\frac{A_x k_x}{\mu B \Delta x}\right|^n_{i-\frac{1}{2},j,k}\left(p^{n+1}_{i,j,k} - p^{n+1}_{i-1,j,k}\right)$$

$$+ \left.\frac{A_y k_y}{\mu B \Delta y}\right|^n_{i,j+\frac{1}{2},k}\left(p^{n+1}_{i,j+1,k} - p^{n+1}_{i,j,k}\right) - \left.\frac{A_y k_y}{\mu B \Delta y}\right|^n_{i,j-\frac{1}{2},k}\left(p^{n+1}_{i,j,k} - p^{n+1}_{i,j-1,k}\right)$$

$$\dots\dots\dots\dots\dots\dots\dots\dots\dots\dots\dots\dots\dots \quad (5.6)$$

$$+ \left.\frac{A_z k_z}{\mu B \Delta z}\right|^n_{i,j,k+\frac{1}{2}}\left(p^{n+1}_{i,j,k+1} - p^{n+1}_{i,j,k}\right) - \left.\frac{A_z k_z}{\mu B \Delta z}\right|^n_{i,j,k-\frac{1}{2}}\left(p^{n+1}_{i,j,k} - p^{n+1}_{i,j,k-1}\right)$$

$$+ q^{n+1}_{i,j,k} = \left(\frac{V_b \phi c}{5.615}\right)^n_{i,j,k}\frac{p^{n+1}_{i,j,k} - p^n_{i,j,k}}{\Delta t}.$$

Similarly, the wellbore model will take the following form:

$$q^{n+1}_{i,j,k} = -\left.\frac{2\pi r_w \overline{k} h}{\mu B\left[\ln\dfrac{r_{eq}}{r_w} + s\right]}\right|^n_{i,j,k}\left(p^{n+1}_{i,j,k} - p^{n+1}_{sf_{i,j,k}}\right) \quad\dots\dots\dots\dots\dots\dots\dots\dots\dots\dots\dots\dots\dots \quad (5.7a)$$

or

$$q^{n+1}_{i,j,k} = -\Omega^n_{i,j,k}\left(p^{n+1}_{i,j,k} - p^{n+1}_{sf_{i,j,k}}\right). \quad\dots\dots\dots\dots\dots\dots\dots\dots\dots\dots\dots\dots\dots\dots\dots \quad (5.7b)$$

In Eqs. 5.6 and 5.7, the superscript n and superscript $n+1$ refer to old-time and new-time levels, respectively. Because a slightly compressible flow equation is a weakly nonlinear PDE, it does not introduce significant errors to keep the transmissibility coefficients at the old-time level, n.

5.2.3 Compressible Flow Equation. The compressible flow equation is a strongly nonlinear PDE and can be expressed as Eq. 5.8 (note that units are consistent, as described in Table 4.1 from Chapter 4):

$$\frac{\partial}{\partial x}\left(\frac{A_x k_x}{\mu B}\frac{\partial p}{\partial x}\right)\Delta x + \frac{\partial}{\partial y}\left(\frac{A_y k_y}{\mu B}\frac{\partial p}{\partial y}\right)\Delta y + \frac{\partial}{\partial z}\left(\frac{A_z k_z}{\mu B}\frac{\partial p}{\partial z}\right)\Delta z + q = \frac{V_b \phi T_{sc}}{p_{sc} T}\frac{\partial}{\partial t}\left(\frac{p}{z_g}\right). \quad\dots\dots\dots\dots\dots\dots \quad (5.8)$$

In Eq. 5.8, it is perfectly reasonable to ignore the depth gradients caused by small values of the density of gas (flow potential and the pressure gradients are very close to each other). Accordingly, the finite-difference representation of Eq. 5.8 can be written in the following form [the superscripts (k) and ($k+1$) indicate the former and the current iteration levels, respectively. The superscripts $n+1$ and n indicate the current time level and previous time level, respectively]:

$$\frac{A_x k_x}{\mu_g B_g \Delta x}\Bigg|_{i+\frac{1}{2},j,k}^{(k)\,n+1} \left(\overset{(k+1)}{p_{i+1,j,k}^{n+1}} - \overset{(k+1)}{p_{i,j,k}^{n+1}} \right) - \frac{A_x k_x}{\mu_g B_g \Delta x}\Bigg|_{i-\frac{1}{2},j,k}^{(k)\,n+1} \left(\overset{(k+1)}{p_{i,j,k}^{n+1}} - \overset{(k+1)}{p_{i-1,j,k}^{n+1}} \right)$$

$$+\frac{A_y k_y}{\mu_g B_g \Delta y}\Bigg|_{i,j+\frac{1}{2},k}^{(k)\,n+1} \left(\overset{(k+1)}{p_{i,j+1,k}^{n+1}} - \overset{(k+1)}{p_{i,j,k}^{n+1}} \right) - \frac{A_y k_y}{\mu_g B_g \Delta y}\Bigg|_{i,j-\frac{1}{2},k}^{(k)\,n+1} \left(\overset{(k+1)}{p_{i,j,k}^{n+1}} - \overset{(k+1)}{p_{i,j-1,k}^{n+1}} \right)$$

$$+\frac{A_z k_z}{\mu_g B_g \Delta z}\Bigg|_{i,j,k+\frac{1}{2}}^{(k)\,n+1} \left(\overset{(k+1)}{p_{i,j,k+1}^{n+1}} - \overset{(k+1)}{p_{i,j,k}^{n+1}} \right) - \frac{A_z k_z}{\mu_g B_g \Delta z}\Bigg|_{i,j,k-\frac{1}{2}}^{(k)\,n+1} \left(\overset{(k+1)}{p_{i,j,k}^{n+1}} - \overset{(k+1)}{p_{i,j,k-1}^{n+1}} \right)$$

$$+\overset{(k+1)}{q_{i,j,k}^{n+1}} = \left(\frac{V_b \phi T_{sc}}{p_{sc} T} \right)_{i,j,k}^{n+1} \frac{1}{\Delta t}\left(\frac{\overset{(k+1)}{p_{i,j,k}^{n+1}}}{\overset{(k+1)}{z_{gi,j,k}^{n+1}}} - \frac{p_{i,j,k}^{n}}{z_{gi,j,k}^{n}} \right). \qquad\qquad (5.9)$$

The wellbore model, then, can be written as

$$\overset{(k+1)}{q_{i,j,k}^{n+1}} = -\frac{2\pi r_w \bar{k} h}{\mu_g B_g \left(\ln\dfrac{r_{eq}}{r_w} + s \right)}\Bigg|_{i,j,k}^{(k)\,n} \left(\overset{(k+1)}{p_{i,j,k}^{n+1}} - \overset{(k+1)}{p_{sf_{i,j,k}}^{n+1}} \right) \qquad\qquad\qquad (5.10a)$$

or

$$\overset{(k+1)}{q_{i,j,k}^{n+1}} = -\overset{(k)}{\Omega_{i,j,k}^{n}} \left(\overset{(k+1)}{p_{i,j,k}^{n+1}} - \overset{(k+1)}{p_{sf_{i,j,k}}^{n+1}} \right). \qquad\qquad\qquad (5.10b)$$

In Eqs. 5.9 and 5.10, a second set of superscripts, k and $k+1$, appears. The superscript k refers to the old iteration level, and the superscript $k+1$ refers to the new iteration level. The existing nonlinearities in the compressible fluid flow equation forces us to embed a simple iterative protocol into the solution process so that nonlinear coefficients can be allowed to follow one iteration level behind.

5.2.4 Strongly Implicit Pressure (SIP) Notation. At this junction, as a shorthand notation, we introduce the SIP notation. With the help of the SIP notation, Eqs. 5.3, 5.6, and 5.9 can be written in the following forms:

- Incompressible fluid flow equation in SIP notation:

$$B_{i,j,k} p_{i,j,k-1} + S_{i,j,k} p_{i,j-1,k} + W_{i,j,k} p_{i-1,j,k} + C_{i,j,k} p_{i,j,k} + E_{i,j,k} p_{i+1,j,k} + N_{i,j,k} p_{i,j+1,k} + A_{i,j,k} p_{i,j,k+1} = Q_{i,j,k}. \qquad (5.11)$$

- Slightly compressible fluid flow equation in SIP notation:

$$B_{i,j,k}^n p_{i,j,k-1}^{n+1} + S_{i,j,k}^n p_{i,j-1,k}^{n+1} + W_{i,j,k}^n p_{i-1,j,k}^{n+1} + C_{i,j,k}^n p_{i,j,k}^{n+1} + E_{i,j,k}^n p_{i+1,j,k}^{n+1} + N_{i,j,k}^n p_{i,j+1,k}^{n+1} + A_{i,j,k}^n p_{i,j,k+1}^{n+1} = Q_{i,j,k}^{n+1}. \qquad (5.12)$$

- Compressible fluid flow equation in SIP notation:

$$\overset{(k)}{B_{i,j,k}^{n+1}} \overset{(k+1)}{p_{i,j,k-1}^{n+1}} + \overset{(k)}{S_{i,j,k}^{n+1}} \overset{(k+1)}{p_{i,j-1,k}^{n+1}} + \overset{(k)}{W_{i,j,k}^{n+1}} \overset{(k+1)}{p_{i-1,j,k}^{n+1}} + \overset{(k)}{C_{i,j,k}^{n+1}} \overset{(k+1)}{p_{i,j,k}^{n+1}} + \overset{(k)}{E_{i,j,k}^{n+1}} \overset{(k+1)}{p_{i+1,j,k}^{n+1}} + \overset{(k)}{N_{i,j,k}^{n+1}} \overset{(k+1)}{p_{i,j+1,k}^{n+1}} + \overset{(k)}{A_{i,j,k}^{n+1}} \overset{(k+1)}{p_{i,j,k+1}^{n+1}} = \overset{(k+1)}{Q_{i,j,k}^{n+1}}. \qquad (5.13)$$

5.2.5 Definitions of the SIP Coefficients. Coefficients such as $B_{i,j,k}$, $S_{i,j,k}$, $W_{i,j,k}$, $E_{i,j,k}$, $N_{i,j,k}$, and $A_{i,j,k}$ in Eqs. 5.11 through 5.13 are commonly called the transmissibility coefficients, because they describe the volumetric rate of fluid exchange between the central block (i,j,k) for which the equation is written and the surrounding blocks, $(i, j, k-1)$, $(i, j-1, k)$, $(i-1, j, k)$, $(i, j, k+1)$, $(i, j+1, k)$, and $(i+1, j, k)$. These interactions are schematically represented in **Fig. 5.5.**
Accordingly, definitions for the transmissibility coefficients are as follows:

$$B_{i,j,k} = \frac{A_z k_z}{\mu B \Delta z}\Bigg|_{i,j,k-\frac{1}{2}}, \qquad\qquad\qquad\qquad\qquad (5.14a)$$

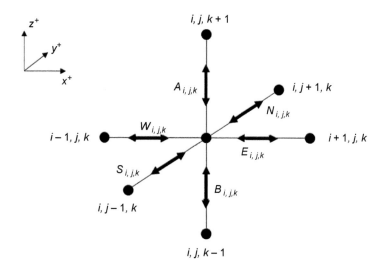

Fig. 5.5—Transmissibility coefficients of a 3D representation in cartesian coordinates.

$$S_{i,j,k} = \left.\frac{A_y k_y}{\mu B \Delta y}\right|_{i,j-\frac{1}{2},k}, \quad \dots\dots\dots\dots\dots\dots\dots\dots\dots\dots\dots\dots\dots\dots\dots\dots\dots\dots\dots (5.14b)$$

$$W_{i,j,k} = \left.\frac{A_x k_x}{\mu B \Delta x}\right|_{i-\frac{1}{2},j,k}, \quad \dots\dots\dots\dots\dots\dots\dots\dots\dots\dots\dots\dots\dots\dots\dots\dots\dots\dots\dots (5.14c)$$

$$E_{i,j,k} = \left.\frac{A_x k_x}{\mu B \Delta x}\right|_{i+\frac{1}{2},j,k}, \quad \dots\dots\dots\dots\dots\dots\dots\dots\dots\dots\dots\dots\dots\dots\dots\dots\dots\dots\dots (5.14d)$$

$$N_{i,j,k} = \left.\frac{A_y k_y}{\mu B \Delta y}\right|_{i,j+\frac{1}{2},k}, \quad \dots\dots\dots\dots\dots\dots\dots\dots\dots\dots\dots\dots\dots\dots\dots\dots\dots\dots\dots (5.14e)$$

$$A_{i,j,k} = \left.\frac{A_z k_z}{\mu B \Delta z}\right|_{i,j,k+\frac{1}{2}}. \quad \dots\dots\dots\dots\dots\dots\dots\dots\dots\dots\dots\dots\dots\dots\dots\dots\dots\dots\dots (5.14f)$$

Transmissibility coefficients, as defined in Eqs. 5.14a through 5.14f, require the use of harmonic averaging techniques, $H_{ave} = \frac{1}{2}\left(\frac{1}{a_1}+\frac{1}{a_2}\right)$, for different groups of terms in transmissibility. These terms will be further discussed in Chapter 8.

There are two more SIP coefficients, $C_{i,j,k}$ and $Q_{i,j,k}$, that we have not yet defined because they require special attention. Let us consider the following situations:

- For a block that does not host a well ($q_{i,j,k} = 0$):
  - For incompressible flow,

$$C_{i,j,k} = -\left(B_{i,j,k} + S_{i,j,k} + W_{i,j,k} + E_{i,j,k} + N_{i,j,k} + A_{i,j,k}\right), \quad \dots\dots\dots\dots\dots\dots\dots\dots\dots\dots (5.15a)$$

$$Q_{i,j,k} = 0. \quad \dots\dots\dots\dots\dots\dots\dots\dots\dots\dots\dots\dots\dots\dots\dots\dots\dots\dots\dots\dots\dots\dots\dots (5.15b)$$

  - For slightly compressible flow,

$$C_{i,j,k}^n = -\left(B_{i,j,k}^n + S_{i,j,k}^n + W_{i,j,k}^n + E_{i,j,k}^n + N_{i,j,k}^n + A_{i,j,k}^n + \Gamma_{i,j,k}^n\right), \quad \dots\dots\dots\dots\dots\dots\dots\dots (5.16a)$$

$$Q_{i,j,k}^{n+1} = -\Gamma_{i,j,k}^n p_{i,j,k}^n, \quad \dots\dots\dots\dots\dots\dots\dots\dots\dots\dots\dots\dots\dots\dots\dots\dots\dots\dots\dots (5.16b)$$

where

$$\Gamma_{i,j,k}^n = \left(\frac{V_b \phi c}{5.615 \Delta t}\right)_{i,j,k}^n. \quad \dots\dots\dots\dots\dots\dots\dots\dots\dots\dots\dots\dots\dots\dots\dots\dots\dots\dots (5.16c)$$

○ For compressible flow,

$$\overset{(k)}{C_{i,j,k}^{n+1}} = -\left(\overset{(k)}{B_{i,j,k}^{n+1}} + \overset{(k)}{S_{i,j,k}^{n+1}} + \overset{(k)}{W_{i,j,k}^{n+1}} + \overset{(k)}{E_{i,j,k}^{n+1}} + \overset{(k)}{N_{i,j,k}^{n+1}} + \overset{(k)}{A_{i,j,k}^{n+1}} + \overset{(k)}{\Gamma_{i,j,k}^{n+1}} \right), \quad \dots \dots \dots \dots \dots \dots \quad (5.17a)$$

$$\overset{(k)}{Q_{i,j,k}^{n+1}} = -\Gamma_{i,j,k}^{n} P_{i,j,k}^{n}, \quad \dots \dots \dots \dots \dots \dots \dots \dots \dots \dots \dots \dots \dots \dots \dots \dots \dots \dots \dots \dots \dots \dots \dots \quad (5.17b)$$

where

$$\overset{(k)}{\Gamma_{i,j,k}^{n+1}} = \left(\frac{V_b \phi T_{sc}}{p_{sc} T z_g \Delta t} \right)^{\overset{(k)}{n+1}}_{i,j,k}, \quad \dots \dots \dots \dots \dots \dots \dots \dots \dots \dots \dots \dots \dots \dots \dots \dots \dots \quad (5.17c)$$

$$\Gamma_{i,j,k}^{n} = \left(\frac{V_b \phi T_{sc}}{p_{sc} T z_g \Delta t} \right)^{n}_{i,j,k}. \quad \dots \dots \dots \dots \dots \dots \dots \dots \dots \dots \dots \dots \dots \dots \dots \dots \dots \dots \quad (5.17d)$$

Note that Eq. 5.17d is written at the former time level; therefore, no iteration level is assigned.

- For a block that hosts a well, it is necessary to consider the type of boundary condition specified at the wellpoint:
  ○ If $q_{i,j,k}$ is specified ($q_{i,j,k} = q_{spec}$), then for incompressible flow,

$$C_{i,j,k} = -\left(B_{i,j,k} + S_{i,j,k} + W_{i,j,k} + E_{i,j,k} + N_{i,j,k} + A_{i,j,k} \right), \quad \dots \dots \dots \dots \dots \dots \quad (5.18a)$$

$$Q_{i,j,k} = -q_{spec}. \quad \dots \dots \dots \dots \dots \dots \dots \dots \dots \dots \dots \dots \dots \dots \dots \dots \dots \dots \dots \dots \dots \dots \dots \quad (5.18b)$$

  ○ If $q_{i,j,k}$ is specified ($q_{i,j,k} = q_{spec}$), then for slightly compressible flow,

$$C_{i,j,k}^{n} = -\left(B_{i,j,k}^{n} + S_{i,j,k}^{n} + W_{i,j,k}^{n} + E_{i,j,k}^{n} + N_{i,j,k}^{n} + A_{i,j,k}^{n} + \Gamma_{i,j,k}^{n} \right), \quad \dots \dots \dots \dots \dots \quad (5.19a)$$

$$Q_{i,j,k}^{n+1} = -q_{spec}^{n+1} - \Gamma_{i,j,k}^{n} P_{i,j,k}^{n}. \quad \dots \dots \dots \dots \dots \dots \dots \dots \dots \dots \dots \dots \dots \dots \dots \dots \quad (5.19b)$$

  ○ If $q_{i,j,k}$ is specified ($q_{i,j,k} = q_{spec}$), then for compressible flow,

$$\overset{(k)}{C_{i,j,k}^{n+1}} = -\left(\overset{(k)}{B_{i,j,k}^{n+1}} + \overset{(k)}{S_{i,j,k}^{n+1}} + \overset{(k)}{W_{i,j,k}^{n+1}} + \overset{(k)}{E_{i,j,k}^{n+1}} + \overset{(k)}{N_{i,j,k}^{n+1}} + \overset{(k)}{A_{i,j,k}^{n+1}} + \overset{(k)}{\Gamma_{i,j,k}^{n+1}} \right), \quad \dots \dots \dots \dots \quad (5.20a)$$

$$\overset{(k)}{Q_{i,j,k}^{n+1}} = q_{spec}^{n+1} - \Gamma_{i,j,k}^{n} P_{i,j,k}^{n}. \quad \dots \dots \dots \dots \dots \dots \dots \dots \dots \dots \dots \dots \dots \dots \dots \dots \dots \quad (5.20b)$$

- If flowing sandface pressure, $p_{sf\,i,j,k}$, is specified ($p_{sf\,i,j,k} = p_{sf\,spec}$), then it is necessary to substitute the well equation for the $q_{i,j,k}$ term. In this way, the specified sandface pressure is introduced into the transport equation using Eqs. 5.4b, 5.7b, and 5.10b for incompressible, slightly compressible, and compressible flow systems, respectively. Such a substitution will provide the following definitions:
  ○ For incompressible flow,

$$C_{i,j,k} = -\left(B_{i,j,k} + S_{i,j,k} + W_{i,j,k} + E_{i,j,k} + N_{i,j,k} + A_{i,j,k} + \Omega_{i,j,k} \right), \quad \dots \dots \dots \dots \dots \quad (5.21a)$$

$$Q_{i,j,k} = -\Omega_{i,j,k} P_{sf_{spec}}. \quad \dots \dots \dots \dots \dots \dots \dots \dots \dots \dots \dots \dots \dots \dots \dots \dots \dots \dots \dots \quad (5.21b)$$

  ○ For slightly compressible flow,

$$C_{i,j,k}^{n} = -\left(B_{i,j,k}^{n} + S_{i,j,k}^{n} + W_{i,j,k}^{n} + E_{i,j,k}^{n} + N_{i,j,k}^{n} + A_{i,j,k}^{n} + \Gamma_{i,j,k}^{n} + \Omega_{i,j,k}^{n} \right), \quad \dots \dots \dots \dots \quad (5.22a)$$

$$Q_{i,j,k}^{n+1} = -\Omega_{i,j,k}^{n+1} P_{sf_{spec}} - \Gamma_{i,j,k}^{n} P_{i,j,k}^{n}. \quad \dots \dots \dots \dots \dots \dots \dots \dots \dots \dots \dots \dots \dots \dots \quad (5.22b)$$

  ○ For compressible flow,

$$\overset{(k)}{C_{i,j,k}^{n+1}} = -\left(\overset{(k)}{B_{i,j,k}^{n+1}} + \overset{(k)}{S_{i,j,k}^{n+1}} + \overset{(k)}{W_{i,j,k}^{n+1}} + \overset{(k)}{E_{i,j,k}^{n+1}} + \overset{(k)}{N_{i,j,k}^{n+1}} + \overset{(k)}{A_{i,j,k}^{n+1}} + \overset{(k)}{\Gamma_{i,j,k}^{n+1}} + \overset{(k)}{\Omega_{i,j,k}^{n+1}} \right), \quad \dots \dots \quad (5.23a)$$

$$\overset{(k)}{Q_{i,j,k}^{n+1}} = -\overset{(k)}{\Omega_{i,j,k}^{n+1}} P_{sf_{spec}} - \Gamma_{i,j,k}^{n} P_{i,j,k}^{n}. \quad \dots \dots \dots \dots \dots \dots \dots \dots \dots \dots \dots \dots \dots \dots \dots \quad (5.23b)$$

If depth gradients are not ignored, then finite-difference approximation expressions representing the derivatives involving the depth gradients need to be introduced into the $Q_{i,j,k}$ terms. For example, for the incompressible flow equation, the $Q_{i,j,k}$ term becomes (for a block with no wells)

$$Q_{i,j,k} = \begin{bmatrix} \left.\dfrac{1}{144}\dfrac{g}{g_c}\rho\dfrac{A_x k_x}{\mu B \Delta x}\right|_{i+\frac{1}{2},j,k}\left(G_{i+1,j,k}-G_{i,j,k}\right)+\left.\dfrac{1}{144}\dfrac{g}{g_c}\rho\dfrac{A_x k_x}{\mu B \Delta x}\right|_{i-\frac{1}{2},j,k}\left(G_{i-1,j,k}-G_{i,j,k}\right) \\ \left.\dfrac{1}{144}\dfrac{g}{g_c}\rho\dfrac{A_y k_y}{\mu B \Delta y}\right|_{i,j+\frac{1}{2},k}\left(G_{i+1,j,k}-G_{i,j,k}\right)+\left.\dfrac{1}{144}\dfrac{g}{g_c}\rho\dfrac{A_y k_y}{\mu B \Delta y}\right|_{i,j-\frac{1}{2},k}\left(G_{i,j-1,k}-G_{i,j,k}\right) \\ \left.\dfrac{1}{144}\dfrac{g}{g_c}\rho\dfrac{A_z k_z}{\mu B \Delta z}\right|_{i,j,k+\frac{1}{2}}\left(G_{i,j,k+1}-G_{i,j,k}\right)+\left.\dfrac{1}{144}\dfrac{g}{g_c}\rho\dfrac{A_z k_z}{\mu B \Delta z}\right|_{i,j,k-\frac{1}{2}}\left(G_{i,j,k-1}-G_{i,j,k}\right) \end{bmatrix}. \quad\quad (5.24)$$

Note that the G terms in Eq. 5.24 represent distances to the center of seven blocks from the datum level (positive downwards).

Similar changes must be incorporated to Eqs. 5.16b, 5.18b, 5.19b, 5.20b, 5.21b, and 5.22b. Note that for the $Q_{i,j,k}$ values associated with the compressible flow equation (Eqs. 5.17b, 5.20b, and 5.23b), we are not recommending any changes because of the assumption that $\Phi \approx p$. Ignoring gravitational forces in compressible flow equations is a common strategy because of the low density of the gas.

5.3 Truncation Error, Stability, and Consistency Analysis of Finite-Difference Schemes

5.3.1 Truncation Error Analysis. The finite-difference representations previously discussed are approximations of the original PDEs. In other words, the finite-difference forms are obtained by way of truncating the high-order terms from Taylor expansions of the PDEs. Hence, the truncation error is one cause of the approximate (inexact) nature of the solution; therefore, it is necessary to address the magnitude of the truncation error and discuss the strategies used in decreasing it.

The local truncation error can be expressed as the difference between the finite-difference representation (f_{fd}) and original differential-equation (f_d) representation of function f:

$$\varepsilon_{L_i}^n = \left(f_{fd}\right)_i^n - \left(f_d\right)_i^n. \quad\quad\quad (5.25)$$

For example, consider the diffusivity equation in 1D form:

$$\frac{\partial^2 p}{\partial x^2} = \frac{1}{D}\frac{\partial p}{\partial t}. \quad\quad\quad (5.26)$$

Its corresponding local-truncation error, hence, is described as

$$\varepsilon_{L_i}^n = \left[\frac{p_{i-1}^n - 2p_i^n + p_{i+1}^n}{(\Delta x)^2} - \frac{1}{D_i}\frac{p_i^{n+1}-p_i^n}{\Delta t}\right] - \left(\frac{\partial^2 p}{\partial x^2} - \frac{1}{D}\frac{\partial p}{\partial t}\right)_i^n. \quad\quad\quad (5.27)$$

5.3.2 Stability Analysis. Because truncation errors exist, the solution of the finite-difference representation of the problem will not converge to the exact solution of the PDE, even if the grid dimensions are small. Even still, the convergence is not guaranteed—the round-off error generated by computers may accumulate and, finally, dominate the desired solution, leading to divergence. Stability analysis helps in determining whether a numerical scheme is capable of controlling the growth of this error and generating stable solutions.

Several procedures were developed to analyze the stability of a given finite-difference approximation. It is essential to determine the stability criteria (unconditionally stable, conditionally stable, or unconditionally unstable) to ensure that the solution is stable. Here, two commonly used procedures are introduced.

Fourier-Series Method. The initial error in the finite-difference form can be represented by the Fourier-series form:

$$\sum_n A_n e^{\left(\frac{In\pi x}{l}\right)}, \quad\quad\quad (5.28)$$

in which $I = \sqrt{-1}$ and l = interval through which the function is defined. Hence, the finite-difference approximation can be expressed as a product of space-dependent (γ) and time-dependent (ξ) terms. For example, consider a 1D problem:

$$p_i^n = \xi^n e^{I(\gamma_1 x)}, \quad\quad\quad (5.29a)$$

$$p_{i+1}^n = \xi^n e^{I[\gamma_1(x+\Delta x)]}, \quad\quad\quad (5.29b)$$

$$p_i^{n+1} = \xi^{n+1} e^{I(\gamma_1 x)}. \quad\quad\quad (5.29c)$$

A scheme is stable as long as the amplification factor, μ_{max}, is less than one:

$$\mu_{max} = \left| \frac{\xi^{n+1}}{\xi^n} \right|_{max} < 1. \dots\dots\dots\dots\dots\dots\dots\dots\dots\dots (5.30)$$

In other words, the error generated in the new timestep (ξ^{n+1}) should be less than that of the old timestep (ξ^n) to guarantee convergence.

Matrix Method. The Fourier-series method ignores the effect of the boundary conditions on stability. The matrix method can be applied if boundary conditions are taken into consideration. In this method, the finite-difference equations are expressed in matrix form and the eigenvalues of the resulting iteration matrix are examined. To achieve convergence, the spectral radius (absolute value of the maximum eigenvalue) of the matrix needs to be less than one. This method is computationally expensive in calculating eigenvalues. In addition, in less-complex cases, the amplification factor is very close to or equal to that of the spectral radius. Hence, the matrix method is rarely used in the practical applications.

5.3.3 Consistency Analysis.
The consistency (or compatibility) analysis of a finite-difference scheme is to determine whether the approximation will collapse to the original form of the PDE when the mesh size in the limit is approaching zero. For example, for the 1D problem, $\dfrac{\partial^2 p}{\partial x^2} = \dfrac{1}{D} \dfrac{\partial p}{\partial t}$, with $D = 1$, one can write,

$$\frac{p_{i+1}^n - \left(p_i^n + p_i^{n-1} \right) + p_{i-1}^n}{(\Delta x)^2} = \frac{p_i^{n+1} - p_i^{n-1}}{2\Delta t}. \dots\dots\dots\dots\dots\dots\dots\dots (5.31)$$

The truncation error is expressed as the difference between the original PDE and its finite-difference analog:

$$\varepsilon_{L_i}^n = \left(\frac{\partial p}{\partial t} - \frac{\partial^2 p}{\partial x^2} \right)_i^n + \left[\frac{(\Delta t)^2}{6} \frac{\partial^3 p}{\partial t^3} \bigg|_i^n - \frac{(\Delta x)^2}{12} \frac{\partial^4 p}{\partial t^4} \bigg|_i^n + \frac{(\Delta t)^2}{(\Delta x)^2} \frac{\partial^2 p}{\partial t^2} \bigg|_i^n \right]. \dots\dots (5.32)$$

Please note that $\left[\dfrac{(\Delta t)^2}{6} \dfrac{\partial^3 p}{\partial t^3} \bigg|_i^n - \dfrac{(\Delta x)^2}{12} \dfrac{\partial^4 p}{\partial t^4} \bigg|_i^n + \dfrac{(\Delta t)^2}{(\Delta x)^2} \dfrac{\partial^2 p}{\partial t^2} \bigg|_i^n \right]$ terms are obtained when the Taylor-series expansions of dependent variables in Eq. 5.31 are substituted into the finite-difference analog.

Let us investigate the following two cases:

- When $\Delta t = r \Delta x$ (where r is a positive constant),

$$\varepsilon_{L_i}^n = \left(\frac{\partial p}{\partial t} - \frac{\partial^2 p}{\partial x^2} \right)_i^n + \left[\frac{r^2 (\Delta t)^2}{6} \frac{\partial^3 p}{\partial t^3} \bigg|_i^n - \frac{(\Delta x)^2}{12} \frac{\partial^4 p}{\partial t^4} \bigg|_i^n + r^2 \frac{\partial^2 p}{\partial t^2} \bigg|_i^n \right], \dots\dots (5.33)$$

in the limit when mesh size $\Delta x \to 0$,

$$\varepsilon_{L_i}^n = \left(\frac{\partial p}{\partial t} - \frac{\partial^2 p}{\partial x^2} \right)_i^n + r^2 \frac{\partial^2 p}{\partial t^2} \bigg|_i^n \neq 0. \dots\dots\dots\dots\dots\dots\dots (5.34)$$

Hence, the finite-difference scheme is not consistent with the PDE.
- When $\Delta t = r(\Delta x)^2$,

$$\varepsilon_{L_i}^n = \left(\frac{\partial p}{\partial t} - \frac{\partial^2 p}{\partial x^2} \right)_i^n + \left[\frac{r^2 (\Delta t)^2}{6} \frac{\partial^3 p}{\partial t^3} \bigg|_i^n - \frac{(\Delta x)^2}{12} \frac{\partial^4 p}{\partial t^4} \bigg|_i^n + r^2 (\Delta x)^2 \frac{\partial^2 p}{\partial t^2} \bigg|_i^n \right], \dots\dots (5.35)$$

in the limit when mesh size $\Delta x \to 0$,

$$\varepsilon_{L_i}^n = \left(\frac{\partial p}{\partial t} - \frac{\partial^2 p}{\partial x^2} \right)_i^n = 0. \dots\dots\dots\dots\dots\dots\dots\dots\dots (5.36)$$

Hence, the finite-difference scheme is consistent with the original PDE.

5.4 Problems and Solutions

Problem 5.4.1 (Ertekin et al. 2001)

Which grid systems (rectangular, cylindrical, spherical, corner-point, hybrid) can be used in the following reservoir engineering applications:

1. Single-well simulation.
2. Full-field simulation.
3. Pattern simulation.
4. Cross-sectional simulation.

Solution to Problem 5.4.1

1. Single-well simulation → Radial-cylindrical grid system.
2. Full-field simulation → Rectangular, corner-point grid system.
3. Pattern simulation → Hybrid grid system.
4. Cross-sectional simulation → Rectangular grid system.

Problem 5.4.2 (Ertekin et al. 2001)

Derive the identity of $\Delta x_{i+1/2} = x_{i+1} - x_i = 1/2(\Delta x_{i+1} + \Delta x_i)$ for block-centered grids.

Solution to Problem 5.4.2

Because $\Delta x_i = \Delta x_{i+1/2} - \Delta x_{i-1/2}$, substituting i with $i+1/2$ and $i-1/2$, one obtains

$$\Delta x_{i+1/2} = x_{i+1} - x_i, \dotfill (5.37)$$

and

$$\Delta x_{i-1/2} = x_i - x_{i-1}. \dotfill (5.38)$$

Because $\Delta x_i = \dfrac{1}{2}\left(\Delta x_{i+1/2} + \Delta x_{i-1/2}\right)$, substituting i with $i+1/2$,

$$\Delta x_{i+\frac{1}{2}} = \frac{1}{2}\left(\Delta x_{i+1} + \Delta x_i\right), \dotfill (5.39)$$

which confirms the identity.

Problem 5.4.3 (Ertekin et al. 2001)

A 2D slightly compressible fluid transport equation is given as (*Note*: $\beta_c = 1.127$)

$$\frac{\partial}{\partial x}\left(\beta_c \frac{A_x k_x}{\mu B}\frac{\partial p}{\partial x}\right)\Delta x + \frac{\partial}{\partial y}\left(\beta_c \frac{A_y k_y}{\mu B}\frac{\partial p}{\partial y}\right)\Delta y + q = \frac{V_b \phi c_t}{\alpha_c B_l^0}\frac{\partial p}{\partial t}:$$

1. Write the forward-difference approximation to this equation.
2. Write the backward-difference approximation to this equation.

Solution to Problem 5.4.3

1. Forward-difference approximation:

$$\frac{\partial}{\partial x}\left(\beta_c \frac{A_x k_x}{\mu B}\frac{\partial p}{\partial x}\right)\Bigg|_i \approx \frac{1}{\Delta x_i}\left[\left(\beta_c \frac{A_x k_x}{\mu B}\right)_{i+1}\left(\frac{\partial p}{\partial x}\right)_{i+1} - \left(\beta_c \frac{A_x k_x}{\mu B}\right)_i\left(\frac{\partial p}{\partial x}\right)_i\right], \quad\dots\dots\dots\dots (5.40)$$

$$\left(\frac{\partial p}{\partial x}\right)_{i+1} = \frac{p_{i+2}-p_{i+1}}{\Delta x_{i+1}}, \quad\dots\dots\dots\dots\dots\dots\dots\dots\dots\dots\dots\dots\dots\dots\dots (5.41)$$

$$\left(\frac{\partial p}{\partial x}\right)_i = \frac{p_{i+1}-p_i}{\Delta x_i}. \quad\dots\dots\dots\dots\dots\dots\dots\dots\dots\dots\dots\dots\dots\dots\dots\dots (5.42)$$

Therefore,

$$\frac{\partial}{\partial x}\left(\beta_c \frac{A_x k_x}{\mu B}\frac{\partial p}{\partial x}\right)\Delta x = \left(\beta_c \frac{A_x k_x}{\mu B \Delta x}\right)_{i+1,j}\left(p_{i+2,j}-p_{i+1,j}\right)-\left(\beta_c \frac{A_x k_x}{\mu B \Delta x}\right)_{i,j}\left(p_{i+1,j}-p_{i,j}\right), \quad\dots\dots\dots\dots (5.43)$$

and similarly,

$$\frac{\partial}{\partial x}\left(\beta_c \frac{A_y k_y}{\mu B}\frac{\partial p}{\partial y}\right)\Delta y = \left(\beta_c \frac{A_y k_y}{\mu B \Delta y}\right)_{i,j+1}\left(p_{i,j+2}-p_{i,j+1}\right)-\left(\beta_c \frac{A_y k_y}{\mu B \Delta y}\right)_{i,j}\left(p_{i,j}-p_{i+1,j}\right). \quad\dots\dots\dots\dots (5.44)$$

Then, the finite-difference representation of the original equation using the forward-difference approximation becomes (written at node i, j, and time level n)

$$\left(\beta_c \frac{A_x k_x}{\mu B \Delta x}\right)_{i+1,j}^n \left(p_{i+2,j}^n - p_{i+1,j}^n\right) - \left(\beta_c \frac{A_x k_x}{\mu B \Delta x}\right)_{i,j}^n \left(p_{i+1,j}^n - p_{i,j}^n\right)$$
$$+\left(\beta_c \frac{A_y k_y}{\mu B \Delta y}\right)_{i,j+1}^n \left(p_{i,j+2}^n - p_{i,j+1}^n\right) - \left(\beta_c \frac{A_y k_y}{\mu B \Delta y}\right)_{i,j}^n \left(p_{i,j+1}^n - p_{i+1,j}^n\right) + q_i^n = \left(\frac{V_b \phi c_t}{\alpha_c B_l^o}\right)_i^n \frac{p_i^{n+1}-p_i^n}{\Delta t}. \quad\dots\dots (5.45)$$

2. Backward-difference approximation:

$$\frac{\partial}{\partial x}\left(\beta_c \frac{A_x k_x}{\mu B}\frac{\partial p}{\partial x}\right)\Bigg|_i \approx \frac{1}{\Delta x_i}\left[\left(\beta_c \frac{A_x k_x}{\mu B}\right)_i\left(\frac{\partial p}{\partial x}\right)_i - \left(\beta_c \frac{A_x k_x}{\mu B}\right)_{i-1}\left(\frac{\partial p}{\partial x}\right)_{i-1}\right], \quad\dots\dots\dots\dots (5.46)$$

$$\left(\frac{\partial p}{\partial x}\right)_i = \frac{p_i-p_{i-1}}{\Delta x_i}, \quad\dots\dots\dots\dots\dots\dots\dots\dots\dots\dots\dots\dots\dots\dots\dots\dots (5.47)$$

$$\left(\frac{\partial p}{\partial x}\right)_{i-1} = \frac{p_{i-1}-p_{i-2}}{\Delta x_{i-1}}. \quad\dots\dots\dots\dots\dots\dots\dots\dots\dots\dots\dots\dots\dots\dots\dots (5.48)$$

Therefore,

$$\frac{\partial}{\partial x}\left(\beta_c \frac{A_x k_x}{\mu B}\frac{\partial p}{\partial x}\right)\Delta x = \left(\beta_c \frac{A_x k_x}{\mu B \Delta x}\right)_{i,j}\left(p_{i,j}-p_{i-1,j}\right)-\left(\beta_c \frac{A_x k_x}{\mu B \Delta x}\right)_{i-1,j}\left(p_{i-1,j}-p_{i-2,j}\right), \quad\dots\dots\dots\dots (5.49)$$

and similarly,

$$\frac{\partial}{\partial x}\left(\beta_c \frac{A_y k_y}{\mu B}\frac{\partial p}{\partial y}\right)\Delta y = \left(\beta_c \frac{A_y k_y}{\mu B \Delta y}\right)_{i,j}\left(p_{i,j}-p_{i,j-1}\right)-\left(\beta_c \frac{A_y k_y}{\mu B \Delta y}\right)_{i,j-1}\left(p_{i,j-1}-p_{i,j-2}\right). \quad\dots\dots\dots\dots (5.50)$$

The backward-difference representation of the original PDE becomes (written at node i, j, and time level n)

$$\left(\beta_c \frac{A_x k_x}{\mu B \Delta x}\right)_{i,j}^n \left(p_{i,j}^{n+1} - p_{i-1,j}^{n+1}\right) - \left(\beta_c \frac{A_x k_x}{\mu B \Delta x}\right)_{i-1,j}^n \left(p_{i-1,j}^{n+1} - p_{i-2,j}^{n+1}\right)$$
$$+\left(\beta_c \frac{A_y k_y}{\mu B \Delta y}\right)_{i,j}^n \left(p_{i,j}^{n+1} - p_{i,j-1}^{n+1}\right) - \left(\beta_c \frac{A_y k_y}{\mu B \Delta y}\right)_{i,j-1}^n \left(p_{i,j-1}^{n+1} - p_{i,j-2}^{n+1}\right) + q_i^{n+1} = \left(\frac{V_b \phi c_t}{\alpha_c B_l^o}\right)_i^n \frac{p_i^{n+1}-p_i^n}{\Delta t}. \quad\dots\dots (5.51)$$

Problem 5.4.4 (Ertekin et al. 2001)

Use the results of Problem 5.4.3:

1. Write the 2D forward-difference approximation in a format suitable for an explicit calculation method (i.e., solve the equation in Part 1 of Problem 5.4.3 for $p_{i,j}^{n+1}$).
2. Write the 2D backward-difference approximation in a format suitable for an implicit calculation method (i.e., write the equation in Part 2 of Problem 5.4.3 with $p_{i+1,j}^{n+1}, p_{i,j+1}^{n+1}, p_{i,j}^{n+1}, p_{i,j-1}^{n+1}$, and $p_{i-1,j}^{n+1}$ on the left side, and $a_{sc,i,j}$ and $p_{i,j}^{n}$ on the right side).

Solution to Problem 5.4.4

1.

$$\left(\beta_c \frac{A_x k_x}{\mu B \Delta x}\right)_{i+1,j}^n \left(p_{i+2,j}^n - p_{i+1,j}^n\right) - \left(\beta_c \frac{A_x k_x}{\mu B \Delta x}\right)_{i,j}^n \left(p_{i+1,j}^n - p_{i,j}^n\right)$$

$$+\left(\beta_c \frac{A_y k_y}{\mu B \Delta y}\right)_{i,j+1}^n \left(p_{i,j+2}^n - p_{i,j+1}^n\right) - \left(\beta_c \frac{A_y k_y}{\mu B \Delta y}\right)_{i,j}^n \left(p_{i,j+1}^n - p_{i,j+1}^n\right) + q_i^{n+1} = \left(\frac{V_b \phi c_t}{\alpha_c B_l^o}\right)_i^n \frac{p_{i,j}^{n+1} - p_{i,j}^n}{\Delta t}. \qquad \ldots \ldots (5.52)$$

Rearranging,

$$p_{i,j}^{n+1} = \left[\begin{array}{l} \left(\beta_c \dfrac{A_x k_x}{\mu B \Delta x}\right)_{i+1,j}^n \left(p_{i+2,j}^n - p_{i+1,j}^n\right) - \left(\beta_c \dfrac{A_x k_x}{\mu B \Delta x}\right)_{i,j}^n \left(p_{i+1,j}^n - p_{i,j}^n\right) + \\[3mm] \left(\beta_c \dfrac{A_y k_y}{\mu B \Delta y}\right)_{i,j+1}^n \left(p_{i,j+2}^n - p_{i,j+1}^n\right) - \left(\beta_c \dfrac{A_y k_y}{\mu B \Delta y}\right)_{i,j}^n \left(p_{i,j+1}^n - p_{i,j+1}^n\right) + q_i^{n+1} \end{array}\right] \left(\frac{\Delta t \alpha_c B_l^o}{V_b \phi c_t}\right)_i^n + p_{i,j}^n. \quad \ldots (5.53)$$

2.

$$\left(\beta_c \frac{A_x k_x}{\mu B \Delta x}\right)_{i,j} \left(p_{i,j}^{n+1} - p_{i-1,j}^{n+1}\right) - \left(\beta_c \frac{A_x k_x}{\mu B \Delta x}\right)_{i-1,j} \left(p_{i-1,j}^{n+1} - p_{i-2,j}^{n+1}\right) + \left(\beta_c \frac{A_y k_y}{\mu B \Delta y}\right)_{i,j} \left(p_{i,j}^{n+1} - p_{i,j-1}^{n+1}\right)$$

$$-\left(\beta_c \frac{A_y k_y}{\mu B \Delta y}\right)_{i,j-1} \left(p_{i,j-1}^{n+1} - p_{i,j-2}^{n+1}\right) - \left(\frac{V_b \phi c_t}{\alpha_c B_l^o \Delta t}\right)_i p_{i,j}^{n+1} = -\left(\frac{V_b \phi c_t}{\alpha_c B_l^o \Delta t}\right)_i p_{i,j}^n - q_i^{n+1}. \qquad \ldots \ldots \ldots (5.54)$$

Note that in Parts 1 and 2, it is assumed that flow rate is specified; in other words, $q_i^{n+1} = q_{spec}$.

Problem 5.4.5 (Ertekin et al. 2001)

Use **Fig. 5.6,** where the reservoir has no-flow boundary: $\Delta x = 1,000$ ft, $\Delta y = 750$ ft, $\Delta z = 50$ ft, $k_x = 50$ md, $\phi = 0.27$, $c = 1.6 \times 10^{-6}$ psi^{-1}, $\mu = 30$ cp, $p_i = 4,500$ psi, $B = 1$ res bbl/STB, and $B^0 = 1$ res bbl/STB.

1. Calculate a stable timestep size for the explicit formulation.
2. Perform the explicit calculation for several timesteps.

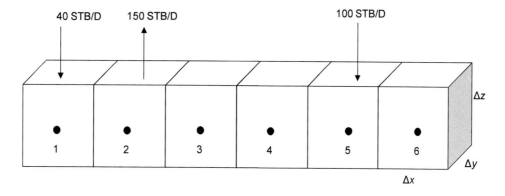

Fig. 5.6—Schematic representation of the reservoir system described in Problem 5.4.5.

Solution to Problem 5.4.5

1. The 1D slightly compressible flow equation for homogeneous reservoir properties can be written as

$$\frac{\partial^2 p}{\partial x^2} = \frac{\mu B \phi c_t}{5.615 k_x} \frac{\partial p}{\partial t}. \quad\dotfill (5.55)$$

A forward finite-difference analog of the PDE can be written as

$$\frac{p_{i+1}^n - 2p_i^n + p_{i-1}^n}{(\Delta x)^2} = \frac{1}{\left(\dfrac{5.615 k_x}{\mu B \phi c_t}\right)} \left(\frac{p_i^{n+1} - p_i^n}{\Delta t}\right). \quad\dotfill (5.56)$$

Let $r = \dfrac{\left(\dfrac{5.615 k_x}{\mu B \phi c_t}\right)(\Delta t)}{(\Delta x)^2}$, and the original equation becomes

$$r\left(p_{i+1}^n - 2p_i^n + p_{i-1}^n\right) = p_i^{n+1} - p_i^n. \quad\dotfill (5.57)$$

Substituting the spatial terms using the Fourier series (note that a product of time- and space-dependent groups are used): $p_{i+1}^n = \xi^n e^{I\gamma(x+\Delta x)}$, $p_{i-1}^n = \xi^n e^{I\gamma(x-\Delta x)}$, $p_i^{n+1} = \xi^{n+1} e^{I\gamma x}$, and $p_i^n = \xi^n e^{I\gamma x}$, the original equation becomes
$r\left[\xi^n e^{I\gamma(x+\Delta x)} - 2\xi^n e^{I\gamma x} + \xi^n e^{I\gamma(x-\Delta x)}\right] = \xi^{n+1} e^{I\gamma x} - \xi^n e^{I\gamma x}$.

Dividing both sides of the equation by $e^{I\gamma x}$ yields

$$\xi^n \left[1 + r\left(e^{I\gamma\Delta x} + e^{-I\gamma\Delta x}\right) - 2r\right] = \xi^{n+1}. \quad\dotfill (5.58)$$

Using the Euler's identity, $e^{I\gamma\pm\Delta x} = \cos(\gamma\Delta x) \pm I \sin(\gamma\Delta x)$ and $\dfrac{\xi^{n+1}}{\xi^n} = \left\{1 + r\left[2\cos(\gamma\Delta x) - 2\right]\right\}$.

Stability analysis requires that $\left(\dfrac{\xi^{n+1}}{\xi^n}\right)_{max} = \left|1 + 2r\left[\cos(\gamma\Delta x) - 1\right]\right| \leq 1$. This implies that $-1 \leq \left\{1 + 2r\left[\cos(\gamma\Delta x) - 1\right]\right\} \leq 1$. Knowing that $\left\{1 + 2r\left[\cos(\gamma\Delta x) - 1\right]\right\} \leq 1$ is always true because $\cos(\gamma\Delta x)$ is a quantity less than 1, one therefore needs to investigate the expression

$$\left\{1 + 2r\left[\cos(\gamma\Delta x) - 1\right]\right\} \geq -1. \quad\dotfill (5.59)$$

Solving the inequality for r gives $r \leq \dfrac{1}{1 - \cos\theta}$. Hence, $r = \dfrac{\left(\dfrac{5.615 k_x}{\mu B \phi c_t}\right)(\Delta t)}{(\Delta x)^2} \leq \dfrac{1}{2}$. Therefore, the timestep should satisfy

$$\Delta t \leq \frac{1}{2}\frac{(\Delta x)^2 \mu B \phi c_t}{5.615 k_x} = \frac{1}{2}\frac{(1{,}000)^2 (30)(1)(0.27)(1.6\times10^{-6})}{5.615(1.127\times10^{-3})(50)} = 20.45. \quad\dotfill (5.60)$$

Thus, any timestep less than 20.45 days can be used for the solution. We will use $\Delta t = 1$ day to illustrate the solution process.

2.

$$p_{i+1}^n + p_{i-1}^n - 2p_i^n + \frac{q\mu B \Delta x}{k_x h \Delta y} = \frac{(\Delta x)^2 \mu B \phi c_t}{5.615 k_x \Delta t}\left(p_i^{n+1} - p_i^n\right). \quad\dotfill (5.61)$$

In an explicit form,

$$p_i^{n+1} = \frac{5.615 k_x \Delta t}{(\Delta x)^2 \mu B \phi c_t}\left(p_{i+1}^n + p_{i-1}^n - 2p_i^n\right) + \frac{5.165\Delta t}{(\Delta x)^2 \phi c_t}\frac{q\Delta x}{h\Delta y} + p_i^n. \quad\dotfill (5.62)$$

The tabulated solution for the first 10 timesteps is listed in **Table 5.2.**

Timestep/Nodes	1	2	3	4	5	6
0	4,500	4,500	4,500	4,500	4,500	4,500
1	4,510.40	4,461.01	4,500.00	4,500.00	4,526.00	4,500
2	4,519.59	4,424.17	4,499.05	4,500.63	4,550.72	4,500.635
3	4,527.66	4,389.34	4,497.26	4,501.82	4,574.27	4,501.857
4	4,534.68	4,356.36	4,494.74	4,503.48	4,596.73	4,503.625
5	4,540.72	4,325.09	4,491.57	4,505.54	4,618.18	4,505.898
6	4,545.86	4,295.43	4,487.85	4,507.95	4,638.68	4,508.64
7	4,550.14	4,267.25	4,483.64	4,510.65	4,658.31	4,511.814
8	4,553.63	4,240.44	4,479.02	4,513.59	4,677.12	4,515.391
9	4,556.39	4,214.92	4,474.04	4,516.74	4,695.18	4,519.339
10	4,558.45	4,190.59	4,468.75	4,520.06	4,712.52	4,523.632

Table 5.2—Tabulated solution for first 10 timesteps from Part 2 of the Solution to Problem 5.4.5.

Problem 5.4.6 (Ertekin et al. 2001)

Use the case given in Problem 5.4.5 to complete the following:

1. Write the system of equations for the first timestep of the implicit formulation (do not attempt to solve the 6×6 matrix equation). Use the same timestep size ($\Delta t = 1$ day) as that of the stable-explicit-scheme timestep size.
2. Lump the gridblocks to form three 2,000×750×50-ft gridblocks.
3. Write the system of equation for the new grid system.
4. Solve the 3×3 matrix equation for the new grid system.

Solution to Problem 5.4.6

1. Recalling the governing-flow equation,

$$\frac{\partial^2 p}{\partial x^2} + \frac{q\mu B}{k_x V_b} = \frac{\mu B \phi c_t}{k 5.615} \frac{\partial p}{\partial t}. \quad \dots\dots\dots\dots\dots\dots\dots\dots\dots\dots\dots\dots\dots \text{(5.63)}$$

and its finite-difference analog,

$$p_{i-1}^{n+1} - 2p_i^{n+1} + p_{i+1}^{n+1} + \left(\frac{q\mu B \Delta x}{k_x h \Delta y}\right)_i^{n+1} = \frac{(\Delta x)^2 \mu B \phi c_t}{5.615 \Delta t k_x}\left(p_i^{n+1} - p_i^n\right); \quad \dots\dots\dots\dots\dots\dots\dots \text{(5.64)}$$

Eq. 5.64 can be rearranged as follows:

$$p_{i-1}^{n+1} - \left(2 + \frac{(\Delta x)^2 \mu B \phi c_t}{5.615 k_x \Delta t}\right)p_i^{n+1} + p_{i+1}^{n+1} = -\left(\frac{q\mu B \Delta x}{k_x h \Delta y}\right)_i^{n+1} - \frac{(\Delta x)^2 \mu B \phi c_t}{5.615 k_x \Delta t} p_i^n. \quad \dots\dots\dots\dots \text{(5.65)}$$

The intermediate variable,

$$\frac{(\Delta x)^2 \mu B \phi c_t}{5.615 \Delta t k_x} = \frac{(1,000)(1,000)(30)(1)(0.27)\left(1.6 \times 10^{-6}\right)}{5.615(1)(50)\left(1.127 \times 10^{-3}\right)} = 40.96, \quad \dots\dots\dots\dots\dots\dots\dots \text{(5.66)}$$

and

$$\left(\frac{\mu B \Delta x}{k h \Delta y}\right)_i^{n+1} = \frac{(30)(1)(1,000)}{(50)\left(1.127 \times 10^{-3}\right)(50)(750)} = 14.20. \quad \dots\dots\dots\dots\dots\dots\dots\dots \text{(5.67)}$$

Substituting the variables to the finite-difference equations for each block (recall that, in agreement with the adopted sign notation, injection rates and production rates have positive and negative signs, respectively):

- Equation for Block 1 ($i = 1$):

$$-41.96 p_1^{n+1} + p_2^{n+1} = -184,888.5. \quad \dots\dots\dots\dots\dots\dots\dots\dots\dots\dots\dots\dots\dots\dots\dots \text{(5.68)}$$

- Equation for Block 2 ($i = 2$):

$$p_1^{n+1} - 42.96 p_2^{n+1} + p_3^{n+1} = -182,191.04. \quad \dots\dots\dots\dots\dots\dots\dots\dots\dots\dots\dots\dots\dots \text{(5.69)}$$

- Equation for Block 3 ($i = 3$):

$$p_2^{n+1} - 42.96 p_3^{n+1} + p_4^{n+1} = -184,321. \dots\dots\dots\dots\dots\dots (5.70)$$

- Equation for Block 4 ($i = 4$):

$$p_3^{n+1} - 42.96 p_4^{n+1} + p_5^{n+1} = -184,321. \dots\dots\dots\dots\dots\dots (5.71)$$

- Equation for Block 5 ($i = 5$):

$$p_4^{n+1} - 42.96 p_5^{n+1} + p_6^{n+1} = -184,321. \dots\dots\dots\dots\dots\dots (5.72)$$

- Equation for Block 6 ($i = 6$):

$$p_5^{n+1} - 42.96 p_6^{n+1} = -185,740.3. \dots\dots\dots\dots\dots\dots (5.73)$$

This system of equations can be expressed in the following matrix form:

$$
\begin{bmatrix}
-41.96 & 1 & 0 & 0 & 0 & 0 \\
1 & -42.96 & 1 & 0 & 0 & 0 \\
0 & 1 & -42.96 & 1 & 0 & 0 \\
0 & 0 & 1 & -42.96 & 1 & 0 \\
0 & 0 & 0 & 1 & -42.96 & 1 \\
0 & 0 & 0 & 0 & 1 & -41.96
\end{bmatrix}
\begin{bmatrix}
p_1^{n+1} \\
p_2^{n+1} \\
p_3^{n+1} \\
p_4^{n+1} \\
p_5^{n+1} \\
p_6^{n+1}
\end{bmatrix}
=
\begin{bmatrix}
-184,888.5 \\
-182,191 \\
-184,321 \\
-184,321 \\
-184,321 \\
-185,740.3
\end{bmatrix}
\dots\dots\dots\dots (5.74)
$$

2. **Fig. 5.7** shows the reduced system with $\Delta x = 2,000$ ft. In the new grid system, the intermediate variable,

$$\frac{(\Delta x)^2 \mu B \phi c_t}{5.615 \Delta t k} = \frac{(2,000)(2,000)(30)(1)(0.27)(1.6 \times 10^{-6})}{5.615(1)(50)(1.127 \times 10^{-3})} = 163.84, \dots\dots\dots\dots\dots\dots (5.75)$$

$$\left(\frac{\mu B \Delta x}{k h \Delta y} \right)_i^{n+1} = \frac{(30)(1)(2,000)}{(50)(1.127 \times 10^{-3})(50)(750)} = 28.39. \dots\dots\dots\dots\dots\dots (5.76)$$

- Equation for Block 1 ($i = 1$):

$$-164.84 p_1^{n+1} + p_2^{n+1} = -734,158.99. \dots\dots\dots\dots\dots\dots (5.77)$$

- Equation for Block 2 ($i = 2$):

$$p_1^{n+1} - 165.84 p_2^{n+1} + p_3^{n+1} = -737,282.3302. \dots\dots\dots\dots\dots\dots (5.78)$$

- Equation for Block 3 ($i = 3$):

$$p_2^{n+1} - 164.84 p_3^{n+1} = -740,121.7268. \dots\dots\dots\dots\dots\dots (5.79)$$

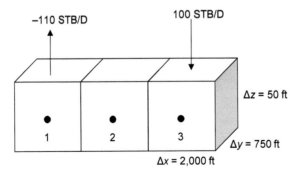

Fig. 5.7—Schematic representation of the reservoir system of Problem 5.4.6, Part 2.

Such systems of equations can be expressed in the following matrix form:

$$\begin{bmatrix} -164.84 & 1 & 0 \\ 1 & -165.84 & 1 \\ 0 & 1 & -164.84 \end{bmatrix} \begin{bmatrix} p_1^{n+1} \\ p_2^{n+1} \\ p_3^{n+1} \end{bmatrix} = \begin{bmatrix} -734,158.99 \\ -737,282.3302 \\ -740,121.7268 \end{bmatrix} \quad \ldots\ldots\ldots\ldots\ldots\ldots\ldots (5.80)$$

3. Solving for the pressures, $p_1^{n+1} = 4,481.05$ psi, $p_2^{n+1} = 4,499.99$ psi, $p_3^{n+1} = 4,517.23$ psi.

A detailed discussion on material-balance checks is presented in Chapter 8, but in the meantime, implementing a material-balance check is an important way to validate results (i.e., mass content at time level $n + 1$ – mass content at time level n = total fluid accumulation + total fluid depletion):

- Total fluid expansion:

$$\Delta V = cV_p \sum_{i=1}^{3} \Delta p_i = 1.6 \times 10^{-6} \times (360,6411.3\,\mathrm{bbl})$$
$$\times [17.23 + (-0.0104) + (-18.95)] = -10 \text{ STB.} \quad \ldots\ldots\ldots\ldots\ldots\ldots\ldots\ldots\ldots (5.81)$$

- Total flow in and out,

$$Q_{\text{total}} = (q_{\text{in}} + q_{\text{out}})\Delta t = 1\,\text{day} \times (-110 + 100) \text{ STB/D} = -10 \text{ STB.} \quad \ldots\ldots\ldots\ldots\ldots (5.82)$$

Such results indicate a good material balance.

Problem 5.4.7 (Ertekin et al. 2001)

Consider the single-phase incompressible fluid flow problem taking place in a 1D homogeneous porous medium composed of four uniform gridblocks shown in **Fig. 5.8.1.** Use the rock and fluid properties listed in **Table 5.3** to respond to the following:

1. Write the PDE that governs the flow problem described.
2. Give a finite-difference form of the PDE, and put it into a characteristic form.
3. Generate the system of equations that can be used to solve the pressure distribution.
4. Can you offer an educated guess for the flow rate of the well?

k_x =	40 md
Δx =	600 ft
Δy =	300 ft
h =	140 ft
μ =	1.4 cp
B =	1 res bbl/STB
r_w =	0.24 ft
s =	0
p_{sf} =	1,240 psi

Table 5.3—Rock and fluid properties found in Fig. 5.8.1 (Problem 5.4.7).

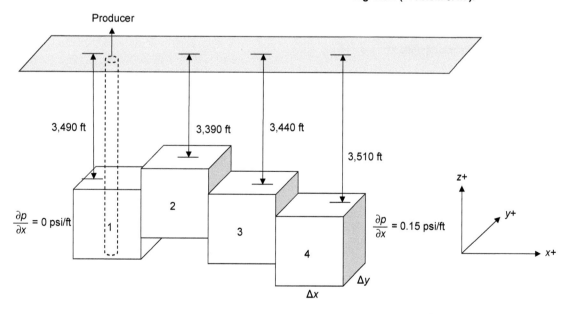

Fig. 5.8.1—Schematic representation of the reservoir system of Problem 5.4.7, Part 1.

Solution to Problem 5.4.7

1. The governing equation for incompressible fluid flow of a linear reservoir with depth gradient is

$$\frac{\partial}{\partial x}\left(\frac{A_x k_x}{\mu B}\frac{\partial p}{\partial x}\right)\Delta x + q = \frac{\partial}{\partial x}\left(\frac{1}{144}\frac{g}{g_c}\rho\frac{A_x k_x}{\mu B}\frac{\partial G}{\partial x}\right)\Delta x. \quad\ldots\ldots\ldots\ldots\ldots\ldots\ldots\ldots\ldots (5.83)$$

For homogeneous and isotropic reservoir properties, Eq. 5.82 can be simplified to

$$\frac{\partial^2 p}{\partial x^2} + \frac{q\mu B}{V_b k} = \frac{1}{144}\rho\frac{\partial^2 G}{\partial x^2}. \quad\ldots\ldots\ldots\ldots\ldots\ldots\ldots\ldots\ldots\ldots\ldots\ldots\ldots (5.84)$$

2. The characteristic form of the finite-difference equation can be written as

$$p_{i-1} - 2p_i + p_{i+1} + \left(\frac{q\mu B\Delta x}{\Delta ykh}\right)_i = \frac{1}{144}\rho\left(G_{i-1} - 2G_i + G_{i+1}\right). \quad\ldots\ldots\ldots\ldots\ldots\ldots\ldots (5.85)$$

3.
- Equation for Block 1 ($i = 1$): The gravitational term is

$$\frac{1}{144}\rho\left(G_2 - G_1\right) = \frac{1}{144}(56)(3,390 - 3,490) = -38.89. \quad\ldots\ldots\ldots\ldots\ldots\ldots\ldots\ldots (5.86)$$

The coefficient associated with the q term is

$$\frac{\mu B\Delta x}{\Delta ykh} = \frac{(1.4)(1)(600)}{(300)(40)(1.127\times10^{-3})(140)} = 0.4437.$$

Calling on Peaceman's wellbore model (Peaceman 1985), the details of which can be found in Chapter 6, $q_1 = -\Omega\left(p_1 - p_{sf1}\right)$, with $r_{eq} = 0.14\sqrt{\left(\Delta x^2 + \Delta y^2\right)} = 93.91$ ft (the equivalent wellblock radius equation is described in Chapter 6 of this book),

$$\Omega = \frac{2\pi kh}{\mu B\left[\ln\left(\dfrac{r_{eq}}{r_w}\right)+s\right]} = \frac{2\pi(40)\left(1.127\times10^{-3}\right)(140)}{(1.4)(1)\left[\ln\left(\dfrac{93.91}{0.24}\right)+0\right]} - 4.7425 \text{ STB/D-psi.} \quad\ldots\ldots\ldots\ldots (5.87)$$

Substituting these coefficients into the finite-difference equation yields

$$p_2 - p_1 - 2.1042\left(p_1 - 1,240\right) = -38.89, \quad\ldots\ldots\ldots\ldots\ldots\ldots\ldots\ldots\ldots\ldots\ldots 5.88)$$

which can be simplified to

$$-3.1042 p_1 + p_2 = -2,648.16. \quad\ldots\ldots\ldots\ldots\ldots\ldots\ldots\ldots\ldots\ldots\ldots\ldots\ldots (5.89)$$

- Equation for Block 2 ($i = 2$): The gravitational term is

$$\frac{1}{144}\rho\left(G_1 - 2G_2 + G_3\right) = \frac{1}{144}(56)(3,490 - 2\times3,390 + 3,440) = 58.33. \quad\ldots\ldots\ldots\ldots (5.90)$$

One then obtains

$$p_1 - 2p_2 + p_3 = 58.33. \quad\ldots\ldots\ldots\ldots\ldots\ldots\ldots\ldots\ldots\ldots\ldots\ldots\ldots\ldots\ldots (5.91)$$

- Equation for Block 3 ($i = 3$): The gravitational term is

$$\frac{1}{144}\rho(G_1 - 2G_2 + G_3) = \frac{1}{144}(56)(3,390 - 2\times3,440 + 3,510) = 7.78, \quad\ldots\ldots\ldots\ldots\ldots (5.92)$$

$$p_2 - 2p_3 + p_4 = 7.78. \quad\ldots\ldots\ldots\ldots\ldots\ldots\ldots\ldots\ldots\ldots\ldots\ldots\ldots\ldots\ldots\ldots (5.93)$$

• Equation for Block 4 ($i = 4$): The gravitational term is

$$\frac{1}{144} \rho \left(G_3 - G_4 \right) = -27.23.$$

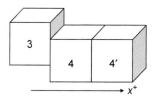

For Block 4, there is a Neumann-type boundary condition specified at the right side of the block. As shown in **Fig. 5.8.2,** an imaginary Block 4′ is used to treat the boundary condition.

In this case, $p_3 - 2p_4 + p_4' = -27.33$. The boundary condition fulfills $\frac{p_4' - p_4}{600} = 0.15$, which states that $p_4' = 90 + p_4$. Therefore, the original equation for Block 4 can be rewritten as

Fig. 5.8.2—Handling of the boundary condition across the Eastern boundary (Problem 5.4.7, Part 3).

$$p_3 - p_4 = -117.23. \dots\dots\dots\dots\dots\dots\dots\dots\dots\dots\dots\dots\dots (5.94)$$

4. For an incompressible fluid system, the production rate of the well can be estimated by flow entering the reservoir across the eastern external boundary:

$$q = -\frac{kA}{\mu B} \frac{\partial p}{\partial x} = \frac{\left(1.127 \times 10^{-3}\right)(40)(300)(140)}{(1.4)(1)} (0.15) = -202.86 \text{ STB/D}. \dots\dots\dots\dots\dots (5.95)$$

Problem 5.4.8 (Ertekin et al. 2001)

Determine the truncation error involved in the following approximation of the differential equation:

$$\frac{\partial^2 p}{\partial x^2} = \left(\frac{1}{D}\right)\left(\frac{\partial p}{\partial t}\right), \quad \frac{p_{i+1}^n - \left(p_i^{n+1} + p_i^{n-1}\right) + p_{i-1}^n}{(\Delta x)^2} = \frac{p_i^{n+1} - p_i^{n-1}}{2D\Delta t}.$$

This three-time-level approximation is known as the DuFort-Frankel approximation (Dufort and Frankel 1953).

Solution to Problem 5.4.8

Use the Taylor expansion to determine the truncation errors.

• Spatial terms:

$$p_{i+1}^n = p_i^n + \Delta x \left.\frac{\partial p}{\partial x}\right|_i^n + \frac{(\Delta x)^2}{2!} \left.\frac{\partial^2 p}{\partial x^2}\right|_i^n + \frac{(\Delta x)^3}{3!} \left.\frac{\partial^3 p}{\partial x^3}\right|_i^n + \cdots, \dots\dots\dots\dots\dots\dots\dots\dots (5.96)$$

$$p_{i-1}^n = p_i^n - \Delta x \left.\frac{\partial p}{\partial x}\right|_i^n + \frac{(\Delta x)^2}{2!} \left.\frac{\partial^2 p}{\partial x^2}\right|_i^n - \frac{(\Delta x)^3}{3!} \left.\frac{\partial^3 p}{\partial x^3}\right|_i^n + \cdots. \dots\dots\dots\dots\dots\dots\dots\dots (5.97)$$

• Temporal terms:

$$p_i^{n+1} = p_i^n + \Delta t \left.\frac{\partial p}{\partial t}\right|_i^n + \frac{(\Delta t)^2}{2!} \left.\frac{\partial^2 p}{\partial t^2}\right|_i^n + \frac{(\Delta t)^3}{3!} \left.\frac{\partial^3 p}{\partial t^3}\right|_i^n + \cdots, \dots\dots\dots\dots\dots\dots\dots (5.98)$$

$$p_i^{n-1} = p_i^n - \Delta t \left.\frac{\partial p}{\partial t}\right|_i^n + \frac{(\Delta t)^2}{2!} \left.\frac{\partial^2 p}{\partial t^2}\right|_i^n + \frac{(\Delta t)^3}{3!} \left.\frac{\partial^3 p}{\partial t^3}\right|_i^n + \cdots. \dots\dots\dots\dots\dots\dots\dots (5.99)$$

The truncation error is expressed as

$$\varepsilon_{L_i}^n = \left[\frac{p_{i+1}^n - \left(p_i^{n+1} + p_i^{n-1}\right) + p_{i-1}^n}{(\Delta x)^2} - \frac{p_i^{n+1} - p_i^{n+1}}{2D\Delta t}\right] - \left(\frac{\partial^2 p}{\partial x^2} - \frac{1}{D}\frac{\partial p}{\partial x}\right). \dots\dots\dots\dots\dots\dots (5.100)$$

Substituting the Taylor's expansion terms,

$$\frac{p_{i+1}^n - \left(p_i^{n+1} + p_i^{n-1}\right) + p_{i-1}^n}{(\Delta x)^2} = \frac{\partial^2 p}{\partial x^2}\bigg|_i^n + \frac{(\Delta x)^2}{12}\frac{\partial^4 p}{\partial x^4}\bigg|_i^n - \frac{(\Delta t)^2}{(\Delta x)^2}\frac{\partial^2 p}{\partial t^2}\bigg|_i^n - \frac{(\Delta t)^4}{12(\Delta x)^2}\frac{\partial^4 p}{\partial t^4}\bigg|_i^n + \cdots, \quad\quad\dots\dots\dots\dots\dots \text{(5.101)}$$

$$\frac{p_i^{n+1} - p_i^{n-1}}{2D\Delta t} = \frac{1}{D}\frac{\partial p}{\partial t}\bigg|_i^n + \frac{(\Delta t)^2}{6D}\frac{\partial^3 p}{\partial t^3}\bigg|_i^n + \cdots, \quad\quad\dots\dots\dots\dots\dots\dots\dots\dots\dots\dots\dots\dots\dots \text{(5.102)}$$

$$\varepsilon_{L_i}^n = \left[\frac{\partial^2 p}{\partial x^2}\bigg|_i^n + \frac{(\Delta x)^2}{12}\frac{\partial^4 p}{\partial t^4}\bigg|_i^n - \frac{(\Delta t)^2}{(\Delta x)^2}\frac{\partial^2 p}{\partial t^2}\bigg|_i^n - \frac{(\Delta t)^4}{12(\Delta x)^2}\frac{\partial^4 p}{\partial t^4}\bigg|_i^n + \cdots - \frac{1}{D}\frac{\partial p}{\partial t}\bigg|_i^n - \frac{(\Delta t)^4}{6D}\frac{\partial^3 p}{\partial t^3}\bigg|_i^n - \cdots\right]^n$$

$$- \left(\frac{\partial^2 p}{\partial x^2} - \frac{1}{D}\frac{\partial p}{\partial t}\right)\bigg|_i^n . \quad\quad\dots\dots\dots\dots \text{(5.103)}$$

Rewriting Eq. 5.103, one obtains

$$\varepsilon_{L_i}^n = \frac{(\Delta x)^2}{12}\frac{\partial^4 p}{\partial x^4}\bigg|_i^n - \frac{(\Delta t)^2}{(\Delta x)^2}\frac{\partial^2 p}{\partial t^2}\bigg|_i^n - \frac{(\Delta t)^4}{12(\Delta x)^2}\frac{\partial^4 p}{\partial t^4}\bigg|_i^n + \cdots - \frac{(\Delta t)^4}{6D}\frac{\partial^3 p}{\partial t^3}\bigg|_i^n - \cdots. \quad\dots\dots\dots\dots\dots\dots \text{(5.104)}$$

The main part of the truncation error is

$$\varepsilon_{L_i}^n = \frac{(\Delta x)^2}{12}\frac{\partial^4 p}{\partial x^4}\bigg|_i^n - \frac{(\Delta t)^2}{(\Delta x)^2}\frac{\partial^2 p}{\partial t^2}\bigg|_i^n, \quad\quad\dots\dots\dots\dots\dots\dots\dots\dots\dots\dots\dots\dots \text{(5.105)}$$

which can be expressed as

$$O\left[(\Delta x)^2\right] + O\left[\left(\frac{\Delta t}{\Delta x}\right)^2\right]. \quad\quad\dots\dots\dots\dots\dots\dots\dots\dots\dots\dots\dots\dots\dots\dots\dots \text{(5.106)}$$

Recall the original equation,

$$\frac{\partial^2 p}{\partial x^2} - \frac{1}{D}\frac{\partial p}{\partial t} = 0, \quad\quad\dots\dots\dots\dots\dots\dots\dots\dots\dots\dots\dots\dots\dots\dots\dots\dots\dots\dots\dots \text{(5.107)}$$

where p can be treated as a general function. Substituting $p = \dfrac{\partial^2 p}{\partial x^2}$ into Eq. 5.106 one obtains

$$\frac{\partial p}{\partial x^2}\left(\frac{\partial^2 p}{\partial x^2}\right) - \frac{1}{D}\frac{\partial}{\partial t}\left(\frac{\partial^2 p}{\partial x^2}\right) = 0, \quad\quad\dots\dots\dots\dots\dots\dots\dots\dots\dots\dots\dots\dots\dots\dots\dots\dots \text{(5.108)}$$

then

$$\frac{\partial^4 p}{\partial x^4} - \frac{1}{D^2}\frac{\partial^2 p}{\partial t^2} = 0, \quad\quad\dots\dots\dots\dots\dots\dots\dots\dots\dots\dots\dots\dots\dots\dots\dots\dots\dots\dots\dots \text{(5.109)}$$

which leads to

$$\frac{\partial^4 p}{\partial x^4} - \frac{1}{D^2}\frac{\partial^2 p}{\partial t^2}. \quad\quad\dots\dots\dots\dots\dots\dots\dots\dots\dots\dots\dots\dots\dots\dots\dots\dots\dots\dots\dots\dots \text{(5.110)}$$

The truncation error is expressed as

$$\varepsilon_{L_i}^n = \frac{(\Delta x)^2}{12}\frac{\partial^4 p}{\partial x^4}\bigg|_i^n - \frac{(\Delta t)^2}{(\Delta x)^2}\frac{\partial^2 p}{\partial t^2}\bigg|_i^n = 0, \quad\quad\dots\dots\dots\dots\dots\dots\dots\dots\dots\dots\dots\dots \text{(5.111)}$$

$$\varepsilon_{L_i}^n \cong \frac{(\Delta x)^2}{12}\frac{1}{D^2}\frac{\partial^2 p}{\partial x^2}\bigg|_i^n - \frac{(\Delta t)^2}{(\Delta x)^2}\frac{\partial^2 p}{\partial t^2}\bigg|_i^n = 0, \quad\quad\dots\dots\dots\dots\dots\dots\dots\dots\dots\dots\dots \text{(5.112)}$$

$$\frac{(\Delta x)^2}{12D^2} - \frac{(\Delta t)^2}{(\Delta x)} = 0, \quad\dots\dots\dots\dots\dots\dots\dots\dots\dots\dots\dots\dots\dots\dots\dots\dots\dots\dots\dots (5.113)$$

or

$$(\Delta t)^2 = \frac{(\Delta x)^4}{12D^2}. \quad\dots\dots\dots\dots\dots\dots\dots\dots\dots\dots\dots\dots\dots\dots\dots\dots\dots\dots\dots\dots\dots (5.114)$$

Therefore, to make the truncation error approximately equal to 0, $\Delta t \le \dfrac{(\Delta x)^2}{(2\sqrt{3})D}$.

Problem 5.4.9 (Ertekin et al. 2001)

Investigate the stability of the DuFort-Frankel approximation (DuFort and Frankel 1953), as applied to the diffusivity equation in Problem 5.4.8, using the Fourier-series analysis method.

Solution to Problem 5.4.9

The DuFort-Frankel scheme requires the following:

$$\frac{p_{i+1}^n - \left(p_i^{n+1} + p_i^{n-1}\right) + p_{i-1}^n}{(\Delta x)^2} = \frac{p_i^{n+1} - p_i^{n-1}}{2D\Delta t}. \quad\dots\dots\dots\dots\dots\dots\dots\dots\dots\dots\dots\dots\dots (5.115)$$

Rearranging the DuFort-Frankel approximation, we write

$$\frac{2D\Delta t}{(\Delta x)^2}\left[p_{i+1}^n - \left(p_i^{n+1} + p_i^{n-1}\right) + p_{i-1}^n\right] = p_i^{n+1} - p_i^{n-1}. \quad\dots\dots\dots\dots\dots\dots\dots\dots\dots\dots (5.116)$$

Let $r = \dfrac{2D\Delta t}{(\Delta x)^2}$, thus Eq. 5.115 becomes

$$r\left[p_{i+1}^n - \left(p_i^{n+1} + p_i^{n-1}\right) + p_{i-1}^n\right] = p_i^{n+1} - p_i^{n-1}. \quad\dots\dots\dots\dots\dots\dots\dots\dots\dots\dots\dots\dots (5.117)$$

Substituting the spatial terms using the Fourier series, $p_{i+1}^n = \xi^n e^{I\gamma(x+\Delta x)}$, $p_{i-1}^n = \xi^n e^{I\gamma(x-\Delta x)}$, $p_i^{n+1} = \xi^{n+1} e^{I\gamma x}$, and $p_i^{n-1} = \xi^{n-1} e^{I\gamma x}$, the original equation becomes

$$r\left[\xi^n e^{I\gamma(x+\Delta x)} - \left(\xi^{n+1} e^{I\gamma x} + \xi^{n-1} e^{I\gamma x}\right) + \xi^n e^{I\gamma(x-\Delta x)}\right] = \xi^{n+1} e^{I\gamma x} - \xi^{n-1} e^{I\gamma x}. \quad\dots\dots\dots\dots\dots (5.118)$$

Dividing Eq. 5.118 by $\xi^n e^{I\gamma x}$, and knowing that $\mu_{amp} = \dfrac{\xi^{n+1}}{\xi^n} = \dfrac{\xi^n}{\xi^{n-1}}$ (here, μ_{amp} is known as the amplification factor),

$$r\left[e^{I\gamma\Delta x} - \left(\mu_{amp} + \frac{1}{\mu_{amp}}\right) + e^{I\gamma(-\Delta x)}\right] = \mu_{amp} - \frac{1}{\mu_{amp}}. \quad\dots\dots\dots\dots\dots\dots\dots\dots\dots\dots (5.119)$$

On the basis of the Euler's identity, $e^{I\gamma\pm\Delta x} = \cos(\gamma\Delta x) \pm I\sin(\gamma\Delta x)$,

$$r\left[2\cos(\gamma\Delta x) - \left(\mu_{amp} + \frac{1}{\mu_{amp}}\right)\right] = \mu_{amp} - \frac{1}{\mu_{amp}}, \quad\dots\dots\dots\dots\dots\dots\dots\dots\dots\dots (5.120)$$

or

$$r\left[2\mu_{amp}\cos(\gamma\Delta x) - \mu_{amp}^2 - 1\right] = \mu_{amp}^2 - 1, \quad\dots\dots\dots\dots\dots\dots\dots\dots\dots\dots\dots\dots\dots (5.121)$$

which is

$$(r+1)\mu_{amp}^{2} - 2r\cos(\gamma\Delta x)\mu_{amp} + (r-1) = 0. \qquad \text{(5.122)}$$

The roots of the preceding quadratic equation are

$$\mu_{amp1,2} = \frac{2r\cos(\gamma\Delta x) \pm \sqrt{4r^{2}\cos^{2}(\gamma\Delta x) - 4(r+1)(r-1)}}{2(r+1)}. \qquad \text{(5.123)}$$

To satisfy the stability criteria, the larger root should be smaller than or equal to 1:

$$\frac{2r\cos(\gamma\Delta x) + \sqrt{4r^{2}\cos^{2}(\gamma\Delta x) - 4(r+1)(r-1)}}{2(r+1)} \le 1. \qquad \text{(5.124)}$$

Simplify to get $r(1+r)\cos(\gamma\Delta x) \le r^{2}+r+1$, which is always true, because $0 \le \cos(\gamma\Delta x) \le 1$. Therefore, the DuFort-Frankel approximation is unconditionally stable.

Problem 5.4.10 (Ertekin et al. 2001)

Use the Fourier stability-analysis method to investigate the stability of the following finite-difference scheme for the PDE, $\partial^{2}p/\partial x^{2} = (\partial p/\partial t)$:

$$p_{i}^{n+1} = \frac{1}{6}\left(p_{i-1}^{n} + p_{i+1}^{n}\right) + \frac{2}{3}p_{i}^{n}.$$

Solution to Problem 5.4.10

The proposed finite-difference scheme can be written as follows:

$$6p_{i}^{n+1} - \left(p_{i-1}^{n} + p_{i+1}^{n}\right) - 4p_{i}^{n} = 0. \qquad \text{(5.125)}$$

Substituting for the dependent terms using the Fourier series,

$$p_{i+1}^{n} = \xi^{n}e^{I\gamma(x+\Delta x)}, \; p_{i-1}^{n} = \xi^{n}e^{I\gamma(x-\Delta x)}, \; p_{i}^{n+1} = \xi^{n+1}e^{I\gamma x}, \text{ and } p_{i}^{n-1} = \xi^{n-1}e^{I\gamma x},$$

one obtains

$$6\xi^{n+1}e^{I\gamma x} - \xi^{n}e^{I\gamma(x-\Delta x)} - \xi^{n}e^{I\gamma(x+\Delta x)} - 4\xi^{n}e^{I\gamma x} = 0. \qquad \text{(5.126)}$$

Dividing the entire equation by $\xi^{n}e^{I\gamma x}$, and knowing that $\mu_{amp} = \dfrac{\xi^{n+1}}{\xi^{n}}$,

$$6\mu_{amp} - e^{I\gamma(-\Delta x)} - e^{I\gamma\Delta x} - 4 = 0. \qquad \text{(5.127)}$$

On the basis of the Euler's identity, $e^{I\gamma\pm\Delta x} = \cos(\gamma\Delta x) \pm I\sin(\gamma\Delta x)$,

$$6\mu_{amp} - 2\cos(\gamma\Delta x) - 4 = 0. \qquad \text{(5.128)}$$

Solving for μ_{amp} from Eq. 5.127 yields

$$\mu_{amp} = \frac{\cos(\gamma\Delta x) + 2}{3}. \qquad \text{(5.129)}$$

Because $0 \le \cos(\gamma\Delta x) \le 1$, $\mu_{amp} = \dfrac{\cos(\gamma\Delta x) + 2}{3} \le 1$; therefore, the finite-difference scheme is unconditionally stable.

Problem 5.4.11 (Ertekin et al. 2001)

The equation $\partial^2 p / \partial x^2 = (\partial p / \partial t)$ is approximated by the equation

$$\frac{p_{i+1}^n - 2p_i^n + p_{i-1}^n}{(\Delta x)^2} = 3\left(\frac{p_i^{n+1} - p_i^{n-1}}{2\Delta t}\right) - \left(\frac{p_i^n - p_i^{n-1}}{2\Delta t}\right).$$

Investigate the stability of the proposed scheme.

Solution to Problem 5.4.11

The proposed scheme can be rearranged as follows:

$$\frac{2\Delta t}{(\Delta x)^2}\left(p_{i+1}^n - 2p_i^n + p_{i-1}^n\right) = 3\left(p_i^{n+1} - p_i^n\right) - \left(p_i^n - p_i^{n-1}\right).$$

Letting $r = \dfrac{2\Delta t}{(\Delta x)^2}$, one obtains

$$r\left(p_{i+1}^n - 2p_i^n + p_{i-1}^n\right) = 3\left(p_i^{n+1} - p_i^n\right) - \left(p_i^n - p_i^{n-1}\right). \qquad \ldots\ldots (5.130)$$

Substituting for the dependent terms using the Fourier series: $p_{i+1}^n = \xi^n e^{I\gamma(x+\Delta x)}$, $p_{i-1}^n = \xi^n e^{I\gamma(x-\Delta x)}$, $p_i^{n+1} = \xi^{n+1} e^{I\gamma x}$, and $p_i^{n-1} = \xi^{n-1} e^{I\gamma x}$, one obtains

$$r\left[\xi^n e^{I\gamma(x+\Delta x)} - 2\xi^n e^{I\gamma x} + \xi^n e^{I\gamma(x-\Delta x)}\right] = 3\left(\xi^{n+1} e^{I\gamma x} - \xi^n e^{I\gamma x}\right) - \left(\xi^n e^{I\gamma x} - \xi^{n-1} e^{I\gamma x}\right). \qquad \ldots\ldots (5.131)$$

Dividing the entire equation by $\xi^n e^{I\gamma x}$, and knowing that $\mu_{amp} = \dfrac{\xi^{n+1}}{\xi^n} = \dfrac{\xi^n}{\xi^{n-1}}$,

$$r\left[e^{I\gamma\Delta x} - 2 + e^{I\gamma(-\Delta x)}\right] = 3\left(\mu_{amp} - 1\right) - \left(1 - \frac{1}{\mu_{amp}}\right). \qquad \ldots\ldots (5.132)$$

Based on the Euler identity, $e^{I\gamma \pm \Delta x} = \cos(\gamma\Delta x) \pm I\sin(\gamma\Delta x)$,

$$r\left[2\cos(\gamma\Delta x) - 2\right] = 3\left(\mu_{amp} - 1\right) - \left(1 - \frac{1}{\mu_{amp}}\right). \qquad \ldots\ldots (5.133)$$

Let

$$M = r\left[2\cos(\gamma\Delta x) - 2\right]. \qquad \ldots\ldots (5.134)$$

Thus,

$$3\mu_{amp}^2 - (4 + M)\mu_{amp} + 1 = 0. \qquad \ldots\ldots (5.135)$$

Solving for μ_{amp},

$$\mu_{amp1,2} = \frac{(4+M) \pm \sqrt{(4+M)^2 - 12}}{6}. \qquad \ldots\ldots (5.136)$$

Checking if the larger root is smaller than or equal to 1,

$$\mu_{amp1} = \frac{(4+M) + \sqrt{(4+M)^2 - 12}}{6} \leq 1. \qquad \ldots\ldots (5.137)$$

Simplifying,

$$16 + 8M + M^2 \leq 16 - 4M + M^2. \dotfill (5.138)$$

One can solve the inequality that yields

$$M = r\left[2\cos\left(\gamma\Delta x\right) - 2\right] \leq 0. \dotfill (5.139)$$

Because r is always larger than 0, and $0 \leq \cos\left(\gamma\Delta x\right) \leq 1$, $M \leq 0$ is always true, implying that the finite-difference scheme is unconditionally stable.

Problem 5.4.12 (Ertekin et al. 2001)

For the 1D block-centered grid shown in **Fig. 5.9,** determine the pressure distribution for 90 days using $\Delta t = 30$ days. Implement explicit finite-difference scheme in the solution. The initial reservoir pressure is 6,000 psia; $\Delta x = 1,000$ ft; $\Delta y = 1,000$ ft; $\Delta z = 75$ ft; $c_l = 3.5 \times 10^{-6}$ psi^{-1}; $k_x = 15$ md; $\phi = 0.18$; and $\mu_l = 10$ cp. B_l varies according to the relationship $B_l = \dfrac{B_l^0}{1 + c_l\left(p - p^0\right)}$, in which $p^0 = 6,000$ psia. *Hint:* After each timestep, use the expression given here for B_l and recalculate the transmissibility terms $T_{lxi+1/2}$ and $T_{lxi-1/2}$ before entering the new timestep calculations.

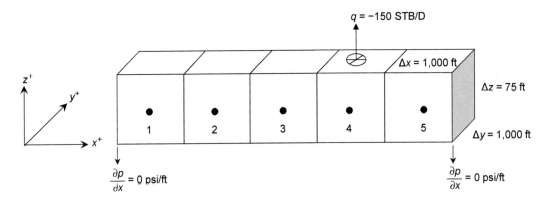

Fig. 5.9—1D block-centered gridblock representation of Problem 5.4.12.

Solution to Problem 5.4.12

The governing flow equation of the described system can be written as

$$\frac{\partial^2 p}{\partial x^2} + \frac{q\mu B}{k_x V_b} = \frac{\mu B \phi c_t}{5.615 k_x}\frac{\partial p}{\partial t}. \dotfill (5.140)$$

The finite-difference analog can be expressed as

$$p_{i-1}^n - 2p_i^n + p_{i+1}^n + \frac{q\mu B}{kh} = \frac{(\Delta x)^2 \mu B \phi c_t}{5.615 k \Delta t}\left(p_i^{n+1} - p_i^n\right). \dotfill (5.141)$$

Solving for the pressure in timestep $n + 1$,

$$p_i^{n+1} = \frac{5.615 \Delta t}{(\Delta x)^2 \mu B \phi c_t}\left(p_{i-1}^n - 2p_i^n + p_{i+1}^n\right) + \frac{5.165 \Delta t}{(\Delta x)^2 \phi c_t}\frac{q}{kh} + p_i^n. \dotfill (5.142)$$

Table 5.4 lists the computational procedure of the problem, which involves a marching scheme along the system.

Timestep/Nodes		1	2	3	4	5
$t = 0$	p	6,000	6,000	6,000	6,000	6,000
	$B =$	1	1	1	1	1
$t = 30$	$\dfrac{5.615\Delta t}{(\Delta x)^2 \mu B \phi c_t} =$	0.45201	0.45201	0.45201	0.45201	0.45201
	p	6,000.00	6,000.00	6,000.00	5,465.24	6,000.00
	$B =$	1	1	1	1.00188	1
$t = 60$	$\dfrac{5.615\Delta t}{(\Delta x)^2 \mu B \phi c_t} =$	0.4520	0.4520	0.4520	0.4512	0.4520
	p	6,000.00	6,000.00	5,758.28	5,413.00	5,758.28
	$B =$	1	1	1.000847	1.00206	1.000847
$t = 90$	$\dfrac{5.615\Delta t}{(\Delta x)^2 \mu B \phi c_t} =$	0.45201	0.45201	0.45163	0.45108	0.45163
	p	6,000.00	5,890.74	5,711.51	5,189.74	5,602.35

Table 5.4—Computational procedure and results for Problem 5.4.12.

Problem 5.4.13

Write the single-phase incompressible flow equation and its characteristic finite-difference approximation for the homogeneous 1D system shown in **Fig. 5.10**.

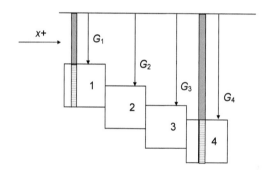

Fig. 5.10—Schematic representation of the dipping reservoir of Problem 5.4.13.

Solution to Problem 5.4.13

The generalized form of the equation is

$$\frac{\partial}{\partial x}\left(\frac{A_x k_x}{\mu B}\frac{\partial p}{\partial x}\right)\Delta x + q = \frac{\partial}{\partial x}\left(\frac{1}{144}\frac{g}{g_c}\rho\frac{A_x k_x}{\mu B}\frac{\partial G}{\partial x}\right)\Delta x. \quad\quad\quad (5.143)$$

For a homogeneous system, such an equation can be simplified to

$$\frac{\partial^2 p}{\partial x^2} + \frac{q\mu B}{k_x V_b} = \frac{1}{144}\frac{g}{g_c}\rho\frac{\partial^2 G}{\partial x^2}. \quad\quad\quad (5.144)$$

In finite-difference form,

$$p_{i-1} - 2p_i + p_{i+1} + \left(\frac{q\mu B}{hk}\right)_i = \frac{1}{144}\frac{g}{g_c}\rho\left(G_{i-1} - 2G_i + G_{i+1}\right). \quad\quad\quad (5.145)$$

Problem 5.4.14

Consider the 2D, homogeneous and isotropic reservoir shown in **Fig. 5.11.1.** The fluid in the reservoir is single phase and incompressible. The reservoir does not have any depth gradients and has the boundary conditions shown in the figure. A uniform, body-centered grid system is used in the discretization of the reservoir.

1. Write the governing flow equation for the system.
2. Write the corresponding characteristic finite-difference approximation.
3. Generate a system of equations that will provide the pressure distribution when it is solved.

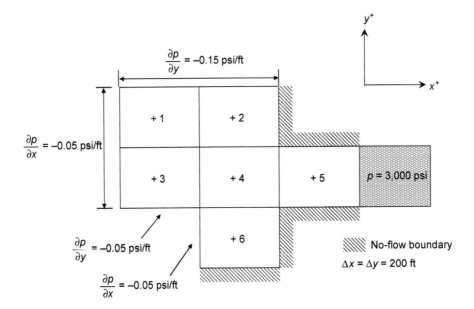

Fig. 5.11.1—Schematic representation of the reservoir system and the mixed boundary conditions of Problem 5.4.14.

Solution to Problem 5.4.14

1. For a 2D incompressible reservoir with homogeneous and isotropic property distributions and no well, the governing flow equation can be written as

$$\frac{\partial^2 p}{\partial x^2} + \frac{\partial^2 p}{\partial y^2} = 0. \quad\ldots\ldots\ldots\ldots\ldots\ldots\ldots\ldots\ldots\ldots\ldots\ldots\ldots\ldots\ldots\ldots\ldots\ldots\ldots (5.146)$$

2. Because $\Delta x = \Delta y$, the finite-difference approximation for an arbitrary block (i,j) is

$$p_{i,j-1} + p_{i-1,j} - 4p_{i,j} + p_{i+1,j} + p_{i,j+1} = 0. \quad\ldots\ldots\ldots\ldots\ldots\ldots\ldots\ldots\ldots\ldots\ldots\ldots\ldots\ldots (5.147)$$

3. To treat the boundary conditions, p'^+ is used to represent the reflection block along the x-direction and p''^+ to represent the reflection block along the y-direction. For instance, **Fig. 5.11.2** illustrates the notations used in the reflection nodes of Block 1, the + indicates that the reflection block is along the positive direction of the y-axis, while the – indicates that the reflection block is along the negative direction of the x-axis. The detailed results of the finite-difference equations and the treatments of the boundary conditions of each block are included in **Table 5.5.**

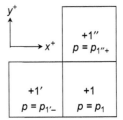

Fig. 5.11.2—Handling of the boundary conditions around Block 1 of Problem 5.4.14, Part 3.

Block	Original Finite-Difference Analog Equation	Treatment of Boundary Condition	Simplified Form of Finite-Difference Analog Equation
1	$p_3 + p_{1'_-} - 4p_1 + p_2 + p_{1''_+} = 0$	$\dfrac{p_1 - p_{1'_-}}{\Delta x} = -0.05 \rightarrow p_{1'_-} = p_1 + 0.05\Delta x = p_1 + 10$ $\dfrac{p_{1''_+} - p_1}{\Delta y} = -0.15 \rightarrow p_{1''_+} = p_1 - 0.15\Delta y = p_1 - 30$	$-2p_1 + p_2 + p_3 = 20$
2	$p_4 + p_1 - 4p_2 + p_{2'_+} + p_{2''_+} = 0$	$\dfrac{p_{2''_+} - p_2}{\Delta y} = -0.15 \rightarrow p_{2''_+} = p_2 - 0.15\Delta y = p_2 - 30$ $p_{2'_+} = p_2$	$p_1 - 2p_2 + p_4 = 30$
3	$p_{3''_-} + p_{3'_-} - 4p_3 + p_4 + p_1 = 0$	$\dfrac{p_3 - p_{3'_-}}{\Delta x} = -0.05 \rightarrow p_{3'_-} = p_3 + 0.05\Delta x = p_3 + 10$ $\dfrac{p_3 - p_{3''_-}}{\Delta y} = -0.1 \rightarrow p_{3''_-} = p_3 + 0.1\Delta y = p_3 + 20$	$p_1 - 2p_3 + p_4 = -30$
4	$p_1 - 2p_3 + p_4 = -30$	N/A	$p_1 - 2p_3 + p_4 = -30$
5	$p_{5''_-} + p_4 - 4p_5 + p_{5'_+} + p_{5''_+} = 0$	$p_{5''_-} = p_5, p_{5''_+} = p_5, p_{5'_+} = 3{,}000$	$p_4 - 2p_5 = -3{,}000$
6	$p_{6''_-} + p_{6'_-} - 4p_6 + p_{6'_+} + p_4 = 0$	$\dfrac{p_6 - p_{6'_-}}{\Delta x} = -0.1 \rightarrow p_{6'_-} = p_6 + 0.1\Delta x = p_6 + 20$ $p_{6''_-} = p_6, p_{6''_+} = p_6$	$p_4 - p_6 = -20$

Table 5.5—Detailed results of the finite-difference equations and the treatments of the boundary conditions of each block (Problem 5.4.14, Part 3).

Problem 5.4.15

For the incompressible system shown in **Fig. 5.12,** determine the nature of the boundary condition using the data in **Table 5.6.** Ignore all the gravitational forces.

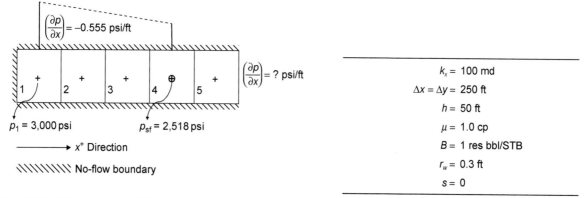

Fig. 5.12—1D Incompressible reservoir and its known and unknown boundary conditions of Problem 5.4.15.

k_x =	100 md
$\Delta x = \Delta y$ =	250 ft
h =	50 ft
μ =	1.0 cp
B =	1 res bbl/STB
r_w =	0.3 ft
s =	0

Table 5.6—Data used to determine boundary conditions in Fig. 5.12 of Problem 5.4.15.

Solution to Problem 5.4.15

Using the linear pressure gradient, calculate the block pressure of Block 4:

$$p_4 = 3{,}000 - 3 \times 250 \times 0.555 = 2{,}584.0. \quad\quad\quad\quad (5.148)$$

Because $p_4 > p_{sf}$, the well in Block 4 can produce

$$\Omega_4 = \frac{2\pi \bar{k} h}{\mu B\left[\ln\left(\dfrac{r_{eq}}{r_w}\right) + s\right]} = \frac{2\pi(100)\left(1.127 \times 10^{-3}\right)(50)}{(1)(1)\left[\ln\left(\dfrac{50}{0.3}\right) + 0\right]} = 6.92 \text{ STB/D-psi}, \quad\quad\quad (5.149)$$

$$q_{well} = -\Omega_4(p_4 - p_{sf}) = -6.92(2{,}584 - 2{,}518) = -455 \text{ STB/D.} \quad \dotfill (5.150)$$

For an incompressible fluid reservoir, the production rate of Well 4 is contributed by the flow from Block 3 to Block 4, and from Block 5 to Block 4. Mass flow from Block 3 to 4 is

$$q_{3to4} = \frac{kA}{\mu B}\frac{dp}{dx} = \frac{(1.127\times10^{-3})(50)(50)(250)}{(1)(1)}(0.555) = -390.1 \text{ STB/D.} \quad \dotfill (5.151)$$

Mass flow from Block 5 to 4 is

$$q_{5to4} = q_{well} - q_{3\,to\,4} = -455 - (-391.0) = -64.0 \text{ STB/D.} \quad \dotfill (5.152)$$

Note that

$$q_{5to4} = -\frac{kA}{\mu B}\frac{dp}{dx}\bigg|_{unknown} = -64.0 \text{ STB/D.} \quad \dotfill (5.153)$$

Thus, the unknown pressure gradient to the right of Block 5 is calculated as

$$\frac{dp}{dx}\bigg|_{unknown} = -\frac{\mu B q_{5to4}}{kA} = +0.09 \text{ psi/ft.} \quad \dotfill (5.154)$$

A positive pressure gradient on the right-most reservoir boundary indicates flow into the reservoir.

Problem 5.4.16

Consider the single-phase incompressible fluid flow in a 1D homogeneous porous medium. The plan view of the reservoir is shown in **Fig. 5.13.** Depths to the centers of Blocks 1, 2, and 3 are G_1, G_2, and G_3, respectively. Complete statements below using one of the following symbols:

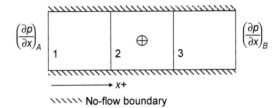

< less than   = equal to
> greater than   U undefined

1. If $\left(\dfrac{\partial p}{\partial x}\right)_A = 0$ and $G_1 < G_2$, then: $p_1\,(\ )\,p_2$.

2. If $\left(\dfrac{\partial p}{\partial x}\right)_A > 0$ and $G_1 = G_2$, then: $p_1\,(\ )\,p_2$.

3. If $\left(\dfrac{\partial p}{\partial x}\right)_A < 0$ and $G_1 = G_2$, then: $p_1\,(\ )\,p_2$.

4. If $\left(\dfrac{\partial p}{\partial x}\right)_B < 0$ and $G_2 = G_3$, then: $p_2\,(\ )\,p_3$.

5. If $\left(\dfrac{\partial p}{\partial x}\right)_B < 0$ and $G_2 > G_3$, then: $p_2\,(\ )\,p_3$.

Fig. 5.13—Schematic representation of the reservoir system of Problem 5.4.16.

Solution to Problem 5.4.16

Considering a homogeneous 1D reservoir with depth gradient.

- At Block 1:

$$p_2 - p_1 - \left(\frac{\partial p}{\partial x}\right)_A \Delta x = \frac{1}{144}\frac{g}{g_c}\rho(G_2 - G_1). \quad \dotfill (5.155)$$

- At Block 2:

$$p_3 - 2p_1 + p_1 + \frac{\mu B \Delta x}{A_x k_x}q = \frac{1}{144}\frac{g}{g_c}\rho(G_3 - 3G_2 + G_1). \quad \dotfill (5.156)$$

- At Block 3:

$$p_2 - p_3 + \left(\frac{\partial p}{\partial x}\right)_B \Delta x = \frac{1}{144} \frac{g}{g_c} \rho \left(G_2 - G_3\right). \dots\dots\dots\dots\dots\dots\dots\dots\dots\dots\dots\dots\dots\dots (5.157)$$

1. If $\left(\frac{\partial p}{\partial x}\right)_A = 0$ and $G_1 < G_2$, then $p_2 - p_1 = \frac{1}{144} \frac{g}{g_c} \rho \left(G_2 - G_1\right) > 0$.
   Hence, $p_1 < p_2$.

2. If $\left(\frac{\partial p}{\partial x}\right)_A > 0$ and $G_1 < G_2$, then $\left(\frac{\partial p}{\partial x}\right)_A = \frac{p_1 - p_1'}{\Delta x} > 0$ indicates that $p_1 > p_1'$ (flow is out of the reservoir), and
   then $p_1 < p_2$.

3. Because $p_2 - p_1 = \left(\frac{\partial p}{\partial x}\right)_A \Delta x < 0$, hence, $p_1 > p_2$.

4. If $\left(\frac{\partial p}{\partial x}\right)_B < 0$ and $G_2 = G_3$, then $p_2 - p_3 = -\left(\frac{\partial p}{\partial x}\right)_B \Delta x < 0$.
   Hence, $p_2 < p_3$.

5. If $\left(\frac{\partial p}{\partial x}\right)_B < 0$ and $G_2 > G_3$, then $p_2 - p_3 = -\left(\frac{\partial p}{\partial x}\right)_B \Delta x + \frac{1}{144} \frac{g}{g_c} \rho \left(G_2 - G_3\right) > 0$.
   Hence, $p_2 > p_3$.

Problem 5.4.17

A brine formation (incompressible fluid) is discretized by a 4×4×4 uniform block-centered grid system ($\Delta x = \Delta y = \Delta z$), as displayed in **Fig. 5.14.1.** In this problem, all gravitational forces can be neglected. The reservoir is homogeneous in porosity and thickness. Permeability is isotropic and homogeneous along all directions. The boundary conditions are specified as indicated:

- There is a well completed at Block (2, 2, 2) injecting brine at a constant rate of q.
- There is a well completed at Block (3, 3, 3) producing brine at a constant pressure of p_{sf}.
- All blocks on the outer surface of the system are kept at constant pressure.

For the system described, answer the following questions:

1. Write the PDE that describes this problem. To solve the problem, how many finite-difference equations do you need to write?
2. Can you solve the same problem as a reduced system? In that case, how many unknowns will you have? If you cannot, explain why.

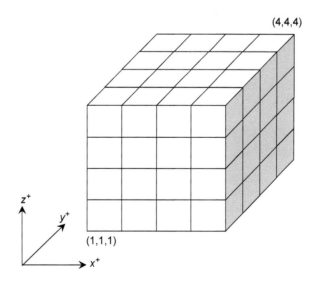

Fig. 5.14.1—Reservoir of Problem 5.4.17.

Solution to Problem 5.4.17

1. The PDE describing this system is

$$\frac{\partial^2 p}{\partial x^2} + \frac{\partial^2 p}{\partial y^2} + \frac{\partial^2 p}{\partial z^2} + \frac{q\mu B}{V_b k} = 0. \quad \dots\dots\dots\dots\dots\dots\dots\dots\dots\dots\dots\dots\dots\dots \text{(5.158)}$$

In view of the boundary conditions and dimensions of the grid shown in **Fig. 5.14.2,** the unknowns of the problem could be reduced to the eight block pressures of a 2×2×2 system, the flow rate of the production well, and the sandface pressure of the injection well. Strictly speaking, eight finite-difference equations are required to solve for this problem.

2. Possible flow paths:

    *   (2,2,2)-- > (3,2,2)-- >(3,3,2)-- > (3,3,3)
        (2,2,2)-- > (3,2,2)-- >(3,2,3)-- > (3,3,3)
    *   (2,2,2)-- > (2,2,3)-- >(2,3,3)-- > (3,3,3)
        (2,2,2)-- > (2,2,3)-- >(3,2,3)-- > (3,3,3)
    *   (2,2,2)-- > (2,3,2)-- >(3,3,2)-- > (3,3,3)
        (2,2,2)-- > (2,3,2)-- >(2,3,3)-- > (3,3,3)

The reservoir is homogeneous and isotropic in permeability and homogeneous in thickness. The used grid system is uniform ($\Delta x = \Delta y = \Delta z$), and all gravitational force are ignored. Considering the fluid paths and symmetry, Blocks (2,2,3), (3,2,2), and (2,3,2) are at the same pressure, and Blocks (3,3,2), (2,3,3), and (3,2,3) are also at the same pressure. Thus, the number of finite-difference equations can be reduced to 4.

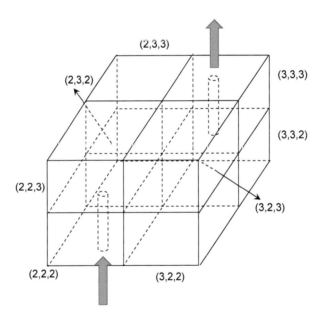

Fig. 5.14.2—Use of symmetry planes in simplifying the Problem 5.4.17, Part 2.

Problem 5.4.18

Fig. 5.15 represents a portion of a body-centered grid imposed on a 2D reservoir. The geological evidence indicates that all of the external boundaries of the reservoir that are oriented parallel to the y-direction have a Neumann-type specification with $\partial p/\partial x = -0.1$ psi/ft. The remaining external boundaries, which are parallel to the x-direction, are no-flow boundaries.

Consider the Blocks numbered 1 through 7. For the boundary blocks write the equations that implement the boundary conditions previously specified.

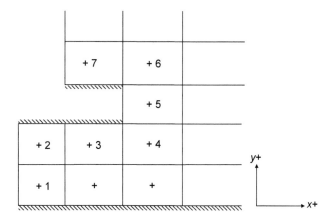

Fig. 5.15—Part of a body-centered grid system imposed on the reservoir of Problem 5.4.18.

Solution to Problem 5.4.18

Boundary blocks are Blocks 1, 2, 3, 5, and 7.

- Block 1: $p_1' = p_1 + 0.1\Delta x$, because $(p_1 - p_1')/\Delta x = 0.1$ psi/ft.
- Block 2: $p_2' = p_2 + 0.1\Delta x$, because $(p_2 - p_2')/\Delta x = 0.1$ psi/ft.
- Block 3: $p_3' = p_3$.
- Block 5: $p_5' = p_5 + 0.1\Delta x$, because $(p_5 - p_5')/\Delta x = 0.1$ psi/ft.
- Block 7: $p_7' = p_7 + 0.1\Delta x$, because $(p_7 - p_7')/\Delta x = 0.1$ psi/ft.

In the preceding equations, p_1', p_2', p_3', p_5', and p_7' represent the pressures of the reflection nodes across the respective boundaries.

Problem 5.4.19

Consider a 2D, single-phase flow in the reservoir shown in **Fig. 5.16.** Block 3 is connected to an infinitely large aquifer, and the pressure of the aquifer is 3,000 psia. It is known that the eastern boundary of Block 5 is leaking. All other boundaries are known to be sealed. The well is producing at its maximum capacity of 8,000 STB/D. Provide a quantitative analysis of the eastern boundary of Block 5 in Fig. 5.16, using the reservoir properties in **Table 5.7.**

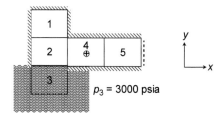

Fig. 5.16—2D reservoir of Problem 5.4.19.

$\mu = 1.0$ cp
$B = 1$ res bbl/STB
$k_x = 50$ md (uniform)
$k_y = 100$ md (uniform)
$\Delta x = \Delta y = 100$ ft (uniform)
$h = 100$ ft (uniform)
$r_w = 0.25$ ft
$s = 0$
$c = 0$ psi^{-1}

Table 5.7—Reservoir properties for Fig. 5.16 (Problem 5.4.19).

Solution to Problem 5.4.19

Because $c = 0$ psi^{-1}, the fluid is incompressible. Block 1 is an inactive block, so we have $p_1 = p_2$. Because the well is producing at its maximum capacity, the sandface pressure of Block 4 must be 14.7 psia:

$$\Omega = \frac{2\pi\sqrt{k_x k_y}\,h}{\mu B\left[\ln\left(\dfrac{r_{eq}}{r_w}\right)+s\right]} = \frac{2(3.14)\sqrt{(50)(100)}\left(1.127\times10^{-3}\right)(100)}{(1)(1)\left[\ln\left(\dfrac{20}{0.25}\right)+0\right]} = 11.42 \text{ STB/D-psi,} \dots\dots\dots\dots\dots (5.159)$$

where $r_{eq} = 0.2\Delta x = 20$ ft,

$$p_4 = -\frac{q_4}{\Omega} + p_{sf} = -\frac{(-8,000)}{11.42} + 14.7 = 715.225 \text{ psia,} \dots\dots\dots\dots\dots\dots\dots\dots\dots\dots\dots (5.160)$$

and at Block 2,

$$k_y p_3 - \left(k_x + k_y\right)p_2 + k_x p_4 = 0. \dots\dots\dots\dots\dots\dots\dots\dots\dots\dots\dots\dots\dots (5.161)$$

The block pressure can be solved:

$$p_2 = \frac{k_y p_3 + k_x p_4}{k_x + k_y} = \frac{(100)(3,000)+(50)(715.225)}{100+50} = 2,238.408 \text{ psia.} \dots\dots\dots\dots\dots (5.162)$$

Fluid flow rate encroaching into the reservoir from the aquifer is

$$q_a = \frac{hk_y}{\mu B}(p_3 - p_2) = \frac{(100)(100)(1.127\times10^{-3})}{(1)(1)}(3,000 - 2,238.408) = 8,583.14 \text{ STB/D.} \dots\dots\dots\dots (5.163)$$

So, the leakage across the eastern boundary of Block 5 is

$$q_a - q_4 = 8,583.14 - 8,000 = 583.14 \text{ STB/D.} \dots\dots\dots\dots\dots\dots\dots\dots\dots\dots\dots (5.164)$$

Problem 5.4.20

1. Consider the homogeneous reservoir and the specified boundary conditions at Blocks 1 and 4 shown in **Fig. 5.17**; provide a qualitative description of the type of flow at each of these boundaries.
2. Write the PDE that governs the incompressible fluid flow problem described in Fig. 5.17. Reduce the equation to its simplest form.
3. Estimate the flow rate of the well knowing the sandface pressure of the well in Block 3 is 800 psi.
4. Write the finite-difference approximation for this problem.
5. Generate the system of equations for each block in the figure.

k_x = 50 md
$\Delta x = \Delta y$ = 200 ft
h = 100 ft
μ = 1.5 cp
B = 1 res bbl/STB
c = 0 psi^{-1}
r_w = 0.25 ft
s = 0

Table 5.8—Data used in Fig. 5.17 (Problem 5.4.20).

Note: Use the data provided in **Table 5.8** for the calculations.

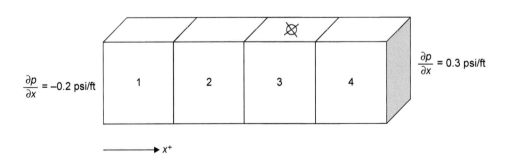

Fig. 5.17—Schematic representation of the reservoir system of Problem 5.4.20.

Solution to Problem 5.4.20

1. At the boundary to the left in Fig. 5.17, where $\partial p/\partial x = -0.2$, flow is into the system, because the pressure outside the reservoir is greater than the pressure inside. The same is also true across the right-hand-side boundary (consistent with the positive x-direction).

2. Starting with the general form of the PDE written in one dimension with no depth gradients,

$$\frac{\partial}{\partial x}\left(\frac{A_x k_x}{\mu B}\frac{\partial p}{\partial x}\right)\Delta x + q = 0. \dots\dots\dots\dots\dots\dots\dots\dots\dots\dots\dots\dots\dots\dots\dots\dots\dots\dots(5.165)$$

For a homogeneous system,

$$\frac{\partial^2 p}{\partial x^2} + \frac{q\mu B}{kV_b} = 0. \dots\dots\dots\dots\dots\dots\dots\dots\dots\dots\dots\dots\dots\dots\dots\dots\dots\dots\dots(5.166)$$

3. For an incompressible flow reservoir, the production rate from the well must equal the fluid entering across the external boundaries. One can calculate the fluid entering the reservoir across the western and eastern boundaries as follows:

$$q_1 = \frac{kA}{\mu B}\frac{dp}{dx}\bigg|_1 = \frac{(1.127\times10^{-3})\times(50)\times(200)\times(100)}{1.5\times1}(0.2) = 150 \text{ STB/D}, \dots\dots\dots\dots\dots(5.167)$$

$$q_2 = \frac{kA}{\mu B}\frac{dp}{dx}\bigg|_2 = \frac{(1.127\times10^{-3})\times(50)\times(200)\times(100)}{1.5\times1}(0.3) = 225 \text{ STB/D}. \dots\dots\dots\dots\dots(5.168)$$

Total inflow into the system is 375 STB/D. Then, the production rate from the well for an incompressible system must be equal to 375 STB/D.

4. The finite-difference approximation can be written as

$$p_{i-1} - 2p_i + p_{i+1} + \frac{q\mu B}{hk_x} = 0. \dots\dots\dots\dots\dots\dots\dots\dots\dots\dots\dots\dots\dots\dots\dots\dots\dots(5.169)$$

5.
- Block 1:

$$p_1' - 2p_1 + p_2 = 0. \dots\dots\dots\dots\dots\dots\dots\dots\dots\dots\dots\dots\dots\dots\dots\dots\dots\dots\dots(5.170)$$

Boundary condition:

$$\frac{p_1 - p_1'}{\Delta x} = -0.2 \rightarrow p_1' = p_1 + 40. \dots\dots\dots\dots\dots\dots\dots\dots\dots\dots\dots\dots\dots\dots\dots(5.171)$$

Substituting into the finite-difference equation,

$$-p_1 + p_2 = -40. \dots\dots\dots\dots\dots\dots\dots\dots\dots\dots\dots\dots\dots\dots\dots\dots\dots\dots\dots(5.172)$$

- Block 2:

$$p_1 - 2p_2 + p_3 = 0. \dots\dots\dots\dots\dots\dots\dots\dots\dots\dots\dots\dots\dots\dots\dots\dots\dots\dots\dots(5.173)$$

- Block 3:

$$p_2 - 2p_3 + p_4 + \frac{q\mu B}{hk} = 0, \dots\dots\dots\dots\dots\dots\dots\dots\dots\dots\dots\dots\dots\dots\dots\dots\dots(5.174)$$

$$\frac{q\mu B}{hk} = \frac{(-375)(1.5)(1)}{(100)(50)(0.001127)} = -100, \dots\dots\dots\dots\dots\dots\dots\dots\dots\dots\dots\dots\dots\dots(5.175)$$

$$p_2 - 2p_3 + p_4 = 100. \dots\dots\dots\dots\dots\dots\dots\dots\dots\dots\dots\dots\dots\dots\dots\dots\dots\dots\dots(5.176)$$

- Block 4:

$$p_3 - 2p_4 + p_4' = 0. \dots\dots\dots\dots\dots\dots\dots\dots\dots\dots\dots\dots\dots\dots\dots\dots\dots (5.177)$$

Boundary condition:

$$\frac{p_4' - p_4}{\Delta x} = 0.3 \rightarrow p_4' = p_4 + 60. \dots\dots\dots\dots\dots\dots\dots\dots\dots\dots\dots\dots\dots (5.178)$$

Thus,

$$p_3 - p_4 = -60. \dots\dots\dots\dots\dots\dots\dots\dots\dots\dots\dots\dots\dots\dots\dots\dots\dots\dots (5.179)$$

Problem 5.4.21

Consider the reservoir shown in **Fig. 5.18** with the characteristics described in **Table 5.9:**

Fig. 5.18—Reservoir of Problem 5.4.21.

- 2D reservoir.
- Single-phase fluid flow.
- Incompressible fluid.
- Homogeneous thickness.
- Anisotropic reservoir characteristics.
- No-flow outer boundaries.
- No depth gradients along the flow direction.

1. Write the general flow equation that describes the reservoir.
2. Reduce the generalized form to its simplest form.
3. Write the finite-difference approximation for this reservoir.
4. Given that pressure values in the blocks are $p_1 = 1,500$ psi, $p_2 = 2,000$ psi, $p_3 = 1,500$ psi, and $p_4 = 500$ psi, calculate the flow rate of the wellbore.

$$k_x = 25 \text{ md}$$
$$k_y = 100 \text{ md}$$
$$\Delta x = \Delta y = 500 \text{ ft}$$
$$h = 50 \text{ ft}$$
$$\mu = 2 \text{ cp}$$
$$B = 1 \text{ res bbl/STB}$$
$$c = 0 \text{ psi}^{-1}$$
$$r_w = 0.25 \text{ ft}$$
$$s = 0$$

Table 5.9—Data used for Fig. 5.18 in Problem 5.4.21.

Solution to Problem 5.4.21

1. The general flow equation can be written as

$$\frac{\partial}{\partial x}\left(\frac{A_x k_x}{\mu B}\frac{\partial p}{\partial x}\right)\Delta x + \frac{\partial}{\partial y}\left(\frac{A_y k_y}{\mu B}\frac{\partial p}{\partial y}\right)\Delta y + q = 0. \dots\dots\dots\dots\dots\dots\dots\dots (5.180)$$

2. Eq. 5.180 can be simplified as

$$k_x \frac{\partial^2 p}{\partial x^2} + k_y \frac{\partial^2 p}{\partial y^2} + \frac{q\mu B}{V_b} = 0. \dots\dots\dots\dots\dots\dots\dots\dots\dots\dots\dots\dots\dots (5.181)$$

3. The finite-difference approximation for this reservoir is

$$p_{i-1,j} - 2p_{i,j} + p_{i+1,j} + \frac{k_y}{k_x}\left(p_{i,j-1} - 2p_{i,j} + p_{i,j+1}\right) + \frac{q\mu B}{hk_x} = 0 \dots\dots\dots\dots\dots\dots (5.182)$$

and $\dfrac{k_y}{k_x} = \dfrac{100}{25} = 4$.

Rearranging,

$$4p_{i,j-1} + p_{i-1,j} - 10p_{i,j} + p_{i+1,j} + 4p_{i,j+1} + \frac{q\mu B}{hk_x} = 0. \quad \dotfill \quad (5.183)$$

4. One should write the finite-difference equation in Block 4 to calculate the flow rate at the sandface:

$$4p_{4''} + p_1 - 10p_4 + p_{4'} + 4p_3 + \frac{q\mu B}{hk_x} = 0. \quad \dotfill \quad (5.184)$$

Implementing the boundary conditions, $p_{4''} = p_4$ and $p_{4'} = p_4$ (because the outer boundaries are no-flow boundaries):

$$p_1 - 5p_4 + 4p_3 + \frac{q\mu B}{hk_x} = 0. \quad \dotfill \quad (5.185)$$

Solving for q,

$$q = -\frac{hk_x}{\mu B}\left(p_1 - 5p_4 + 4p_3\right) = -\frac{(50)(25)\left(1.127 \times 10^{-3}\right)}{(2)(1)}\left(1,500 - 5 \times 500 + 4 \times 1,500\right) \quad \dotfill \quad (5.186)$$

$$= -3,521.88 \text{ STB/D}.$$

The negative sign indicates that the well is a production well.

Problem 5.4.22

Discretize the reservoir shown in **Fig. 5.19.1** using no more than 30 active blocks. Note that the principle flow directions are known to be in the northeast-southwest direction.

1. Using a body-centered grid system, identify the gridblocks for the system that you have discretized.
2. Using a mesh-centered grid system, identify the gridblocks for the system that you have discretized.

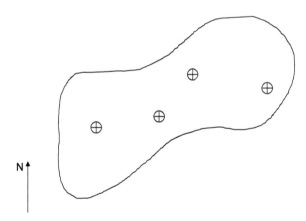

Fig. 5.19.1—2D reservoir of Problem 5.4.22.

Solution to Problem 5.4.22

Answers to Parts 1 and 2 are displayed in **Figs. 5.19.2 and 5.19.3,** respectively.

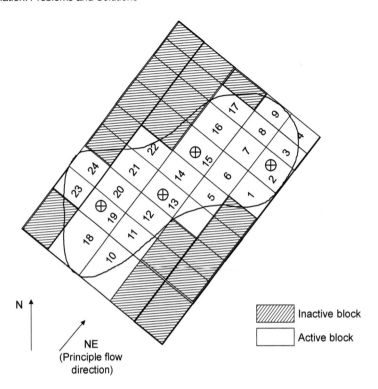

Fig. 5.19.2—Body-centered grid system imposed on the reservoir of Problem 5.4.2, Part 1.

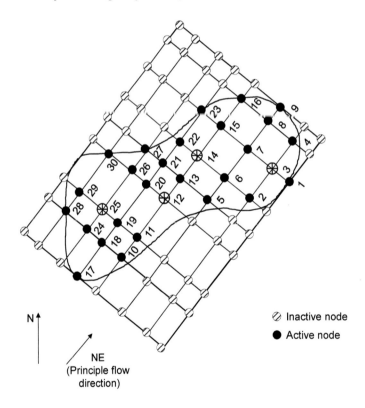

Fig. 5.19.3—Mesh-centered grid system imposed on the reservoir of Problem 5.4.22, Part 2.

Problem 5.4.23

1. Generate an isopach map using the information provided in **Figs. 5.20.1 and 5.20.2.**
2. A reservoir engineer places a grid using a body-centered grid system, as displayed in **Fig. 5.20.3.** Provide a cross-sectional view along the line A-A′, showing the respective cross sections of the blocks located on the same line.
3. Document the following information for the gridblock hosting the well:

   • Block dimensions.
   • Depth from the datum level.

Solution to Problem 5.4.23

Using the structural map provided, design a body-centered grid with the total number of blocks not exceeding 100.

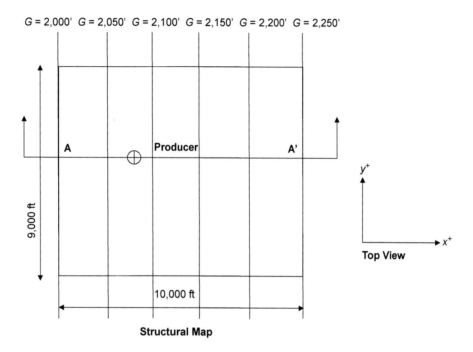

Fig. 5.20.1—Structural map of reservoir of Problem 5.4.23 (*Note:* the intervals between the isodepth lines are uniform and the well is located in the center between line *G* = 2,050 and *G* = 2,100).

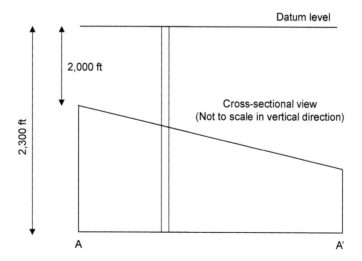

Fig. 5.20.2—Cross-sectional view of the reservoir of Problem 5.4.23.

1. **Fig. 5.20.4** is the generated isopach map for the reservoir.
2. Cross-sectional view along line A-A′, **(Fig. 5.20.5).**
3. Information for the gridblock hosting the well:

   - Block dimensions: $\Delta x = 625$ ft, $\Delta y = 715$ ft, $h = 225$ ft.
   - Depth from the datum level: $2{,}075+225/2 = 2{,}187.5$ ft.

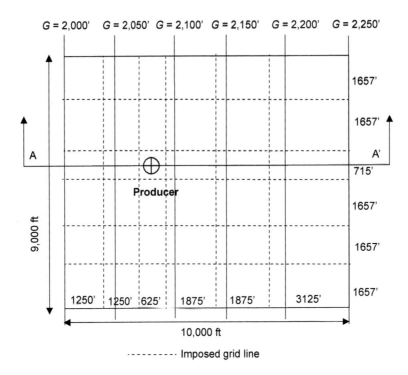

Fig. 5.20.3—Reservoir of Problem 5.4.23 with an overlaid body-centered grid.

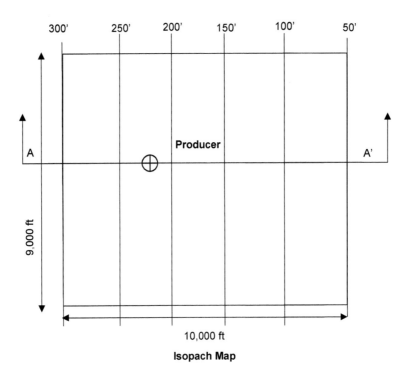

Isopach Map

Fig. 5.20.4—Solution to Problem 5.4.23, Part 1.

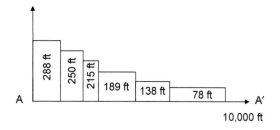

Fig. 5.20.5—Solution to Problem 5.4.23, Part 2.

Problem 5.4.24

Consider the 2D flow of an incompressible liquid in the reservoir displayed in **Fig. 5.21.** Assume that the reservoir exhibits homogeneous and isotropic property distribution; there are no depth gradients; and the injection and production wells located in the corner blocks are operated at the same specified injection and production rates, respectively.

1. Give a mathematical formulation of the single-phase problem described.
2. Obtain a characteristic finite-difference approximation of the governing PDE for the uniform grid spacing imposed on the system shown in Fig. 5.21.
3. Generate the system of equations that will be used in solving for the pressure distribution in the system.

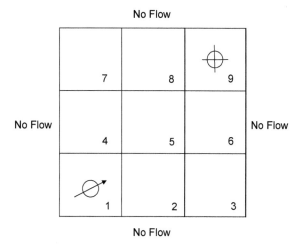

Fig. 5.21—Schematic representation of the reservoir system of Problem 5.4.24.

Solution to Problem 5.4.24

1. General form of the single-phase incompressible flow equation is

$$\frac{\partial}{\partial x}\left(\frac{A_x k_x}{\mu B}\frac{\partial p}{\partial x}\right)\Delta x + \frac{\partial}{\partial y}\left(\frac{A_y k_y}{\mu B}\frac{\partial p}{\partial y}\right)\Delta y + q = 0. \dots\dots\dots\dots\dots\dots\dots\dots\dots\dots\dots\dots (5.187)$$

For homogeneous and isotropic conditions ($k_x = k_y$ and $A_x = A_y$), it reduces to

$$\frac{\partial^2 p}{\partial x^2} + \frac{\partial^2 p}{\partial y^2} + \frac{q\mu B}{kV_b} = 0. \dots\dots\dots\dots\dots\dots\dots\dots\dots\dots\dots\dots\dots\dots\dots\dots\dots (5.188)$$

2. Using the finite-difference analog to the second-order derivatives, we can write

$$\frac{p_{i-1,j} - 2p_{i,j} + p_{i+1,j}}{(\Delta x)_{i,j}^2} + \frac{p_{i,j-1} - 2p_{i,j} + p_{i,j+1}}{(\Delta y)_{i,j}^2} + \frac{q\mu B}{V_b k} = 0. \dots\dots\dots\dots\dots\dots\dots\dots (5.189)$$

Because $\Delta x = \Delta y$ and $V_b = \Delta x \Delta y h = (\Delta x)^2 h$,

$$p_{i,j-1} + p_{i-1,j} - 4p_{i,j} + p_{i+1,j} + p_{i,j+1} = -\frac{q\mu B}{hk}. \quad \dots\dots\dots\dots\dots\dots\dots\dots\dots\dots\dots (5.190)$$

3. The system of equation of every block can be written as

  • Block 1:

$$-2p_1 + p_2 + p_4 = -\frac{q_1\mu}{hk} \text{ (with a positive } q_1 \text{ value, } q_1 > 0). \quad \dots\dots\dots\dots\dots\dots (5.191)$$

  • Block 2:

$$p_1 - 3p_2 + p_3 + p_5 = 0. \quad \dots\dots\dots\dots\dots\dots\dots\dots\dots\dots\dots\dots (5.192)$$

  • Block 3:

$$p_2 - 2p_3 + p_6 = 0. \quad \dots\dots\dots\dots\dots\dots\dots\dots\dots\dots\dots\dots\dots (5.193)$$

  • Block 4:

$$p_1 - 3p_4 + p_5 + p_7 = 0. \quad \dots\dots\dots\dots\dots\dots\dots\dots\dots\dots\dots (5.194)$$

  • Block 5:

$$p_2 + p_4 - 4p_5 + p_6 + p_8 = 0. \quad \dots\dots\dots\dots\dots\dots\dots\dots\dots (5.195)$$

  • Block 6:

$$p_3 + p_5 - 3p_6 + p_8 = 0. \quad \dots\dots\dots\dots\dots\dots\dots\dots\dots\dots (5.196)$$

  • Block 7:

$$p_4 - 2p_7 + p_8 = 0. \quad \dots\dots\dots\dots\dots\dots\dots\dots\dots\dots\dots\dots (5.197)$$

  • Block 8:

$$p_5 + p_7 - 3p_8 + p_9 = 0. \quad \dots\dots\dots\dots\dots\dots\dots\dots\dots\dots (5.198)$$

  • Block 9:

$$p_6 + p_8 - 2p_9 = -\frac{q_9\mu}{hk} \text{ (with a negative } q_9 \text{ value, } q_9 < 0). \quad \dots\dots\dots\dots (5.199)$$

Problem 5.4.25

In an incompressible system, as shown in **Fig. 5.22,** Block 1 is connected to a large aquifer, which maintains its pressure at 1,000 psi. Other boundaries are completely sealed. Knowing the reservoir properties listed in **Table 5.10,** determine the pressure distribution of the system, without solving the system of equations.

$k_x =$	1,000 md
$\Delta x =$	300 ft
$\Delta y =$	300 ft
$h =$	200 ft
$\mu =$	1.127 cp
$B =$	1 res bbl/STB
$r_w =$	0.5 ft
$s =$	0
Ignore the depth gradient.	

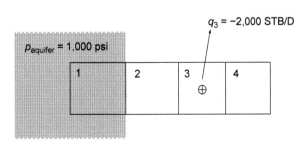

Fig. 5.22—1D reservoir of Problem 5.4.25.

Table 5.10—Reservoir properties for Fig. 5.22 (Problem 5.4.25).

Solution to Problem 5.4.25

For an incompressible fluid reservoir, the fluid produced from the production well has to be equal to the fluid that enters from the aquifer.

Mass flux from Block 1 to 2:

$$q_{1\text{to}2} = -\frac{kh}{\mu B}(p_1 - p_2) = q_{\text{well}} = -2,000 \text{ STB/D}, \dots\dots\dots\dots\dots\dots\dots\dots\dots\dots\dots\dots\dots\dots (5.200)$$

and

$$\frac{kh}{\mu B} = \frac{(0.001127)(1,000)(200)}{(1.127)(1)} = 200. \dots\dots\dots\dots\dots\dots\dots\dots\dots\dots\dots\dots\dots\dots\dots\dots (5.201)$$

Therefore, $(p_1 - p_2) = \dfrac{-2,000}{-200} = 10$, solve for p_2, $p_2 = p_1 - 10 = 990$ psi.

Similarly, $p_3 = 980$ psi. In this case, Block 4 is an inactive block and thus, $p_4 = p_3 = 980$ psi.

Problem 5.4.26

For the following incompressible fluid system displayed in **Fig. 5.23.1,** the number of unknowns to be solved is 18. The wells are both production wells with rates of 1,000 STB/D. Is this TRUE or FALSE?

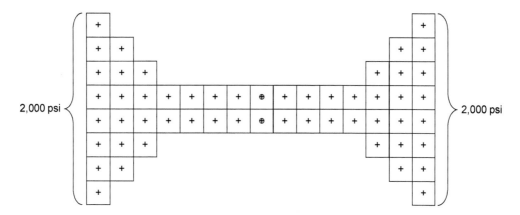

Fig. 5.23.1—Schematic representation of the reservoir of Problem 5.4.26.

Solution to Problem 5.4.26

FALSE. The number of unknown block pressures is 10 (using the existing symmetry), shown by the crosshatched blocks in **Fig. 5.23.2.** If sandface pressure is also to be calculated as another unknown, then the number of unknowns is 11.

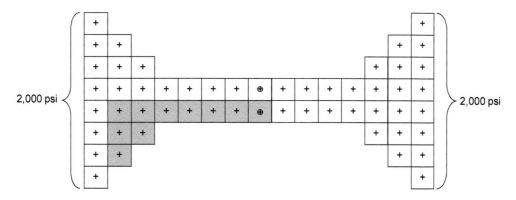

Fig. 5.23.2—Identification of the symmetrical nodes with unknown pressures of Problem 5.4.26.

Problem 5.4.27

Fig. 5.24 represents a 2D (no depth gradients) homogeneous isotropic reservoir, in which flow of a single-phase incompressible fluid is taking place. The reservoir is surrounded by no-flow boundaries, except on the east side of the reservoir, which is charged by an infinitely large aquifer. The production well located in Block 1 is put on production at a constant sandface pressure of 450 psia, and the production well in Block 6 is put on production at a constant flow rate of 500 STB/D.

Calculate the flow rate from the well located in Block 1, and determine the pressure distribution in this system without solving the system of equations. Use the rock and fluid properties listed in **Table 5.11** for the calculation.

$k_x = k_y = 100$ md

$\Delta x = \Delta y = 1{,}000$ ft

$h = 100$ ft

$\mu = 1$ cp

$B = 1$ res bbl/STB

$r_w = 0.3$ ft

$r_{eq}^* = 198$ ft

$s = 0$

*Calculation of equivalent wellblock radius r_{eq} is discussed in Chapter 6

Table 5.11—Rock and fluid properties for Fig. 5.24 (Problem 5.4.27).

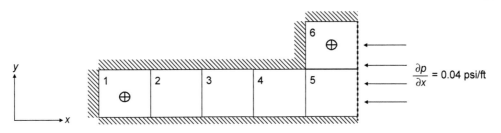

Fig. 5.24—Schematic representation of the reservoir system of Problem 5.4.27.

Solution to Problem 5.4.27

Total flow into the reservoir is

$$q_{in} = 2 \times \frac{kA}{\mu B} \frac{\partial p}{\partial x} = (2)\frac{(1{,}000)(100)(0.1)(1.127)}{(1)(1)}(0.04) = 901.6 \text{ STB/D.} \quad \ldots \ldots \quad (5.202)$$

If $q_{in} + q_{out} = 0$, then $901.6 - 500 + q_1 = 0$. Thus, $q_1 = -401.6$ STB/D. Calculating the productivity index,

$$\Omega = \frac{2\pi kh}{\mu B\left[\ln\left(\dfrac{r_{eq}}{r_w}\right) + s\right]} = \frac{2\pi(0.1)(1.127)(100)}{(1)(1)\left(\ln\dfrac{198}{0.3}\right)} = 9.67 \text{ STB/D-psi.} \quad \ldots \ldots \quad (5.203)$$

The block pressure of Block 1 can be calculated from

$$q_1 = -\Omega_1\left(p_1 - p_{sf_1}\right) \quad \ldots \ldots \quad (5.204)$$

or

$$p_1 = \frac{401.6}{11.5} + 450 = 485 \text{ psi.} \quad \ldots \ldots \quad (5.205)$$

For a homogeneous and isotropic system, the PDE can be written as

$$\frac{\partial^2 p}{\partial x^2} + \frac{\partial^2 p}{\partial y^2} + \frac{q\mu B}{kV_b} = 0. \quad \ldots \ldots \quad (5.206)$$

The finite-difference approximation of Eq. 5.203 is

$$p_{i,j-1} + p_{i-1,j} - 4p_{i,j} + p_{i+1,j} + p_{i,j+1} + \frac{q\mu B}{hk} = 0. \quad\text{...} (5.207)$$

- Block 1:

$$p_2 - 2p_1 + p_1 - 401.6\frac{(1)(1)}{(100)(1.127)(0.1)} = 0. \quad\text{...} (5.208)$$

Because p_1 is already determined ($p_1 = 485$ psi), $p_2 = 520.6$ psi.
- Block 2:

$$p_3 - 2p_2 + p_1 = 0. \quad\text{...} (5.209)$$

The only unknown in Eq. 5.209 is p_3; therefore, $p_3 = 556$ psi.
- Block 3:

$$p_4 - 2p_3 + p_2 = 0. \quad\text{...} (5.210)$$

Thus, $p_4 = 591$ psi.
- Block 4: In a similarly progressive way, one can solve for p_5 as

$$p_5 - 2p_4 + p_3 = 0, \quad\text{...} (5.211)$$

Hence, $p_5 = 626$ psi.
- Block 5: Finally, p_6 can be solved from

$$p_5 + p_4 - 4p_5 + p_5 + 40 + p_6 = 0, \quad\text{...} (5.212)$$

So, $p_6 = 621$ psi.

To check the material balance, now we can write the equation for Block 6 with all the other known block pressures:

$$p_5 + p_6 - 4p_6 + p_6 + 40 + p_6 - 500\frac{(1)(1)}{(100)(1.127)(0.1)} = 0. \quad\text{...} (5.213)$$

Material balance check: $626 - 621 - 40 - 45 = 0$, which gives a perfect material-balance check.

Problem 5.4.28

Fig. 5.25 represents a 2D (no depth gradients) homogeneous isotropic reservoir, in which flow of a single-phase incompressible fluid is taking place. Which of the following boundary conditions are PLAUSIBLE? In each case, justify your answer. Use the rock and fluid properties listed in **Table 5.12** for the calculation.

1. $c_1 = c_2$, production rate from the well in Block 1 is equal to the injection rate of the well in Block 3.
2. $c_1 = c_2$, wells in Blocks 1 and 3 are producing at a constant flow rate of 500 STB/D.
3. $c_1 = 0$ and $c_2 = 0.1$ psi/ft. The total combined production rate from the wells located in Blocks 1 and 3 is 11,270 STB/D.

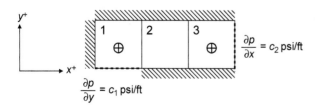

$k_x = k_y =$ 100 md
$\Delta x = \Delta y =$ 1,000 ft
$h =$ 50 ft
$\mu =$ 1 cp
$B =$ 1 res bbl/STB
$r_w =$ 0.3 ft
$s = 0$

Fig. 5.25—Schematic representation of the reservoir system of Problem 5.4.28.

Table 5.12—Rock and fluid properties for Fig. 5.25 (Problem 5.4.28).

Solution to Problem 5.4.28

1. PLAUSIBLE. For an incompressible system, flow into the system has to be equal to flow out of the system. Because $c_1 = c_2$, fluid entering the reservoir across the external boundaries should be equal to each other. Therefore, the injection and production rates from the wells must be identical.
2. NOT PLAUSIBLE. If $c_1 = c_2$, the wells cannot produce. For example, if c_1 and c_2 are negative and equal to each other, the flow rate representing the fluid entering the reservoir in Block 1 is equal to the fluid leaving the reservoir from Block 3. Thus, if both wells are producers, they cannot produce any volume.
3. PLAUSIBLE. The flow across the eastern boundary is

$$q = \frac{kA}{\mu B} c_2 = \frac{(1,000)(100)(1.127)(0.1)}{(1)(1)} = 11,270 \text{ STB/D}. \dots \dots \dots \dots \dots \dots \dots \dots \dots \dots \dots (5.214)$$

For an incompressible fluid reservoir, the total production rate equals the flow rate into the reservoir across the boundary, which is 11,270 STB/D.

Problem 5.4.29

Consider a single-phase, incompressible flow reservoir system with heterogeneous and anisotropic permeability distribution (**Fig. 5.26**). There is one producer and one injector in the system. The permeability in the z-direction is zero. $\Delta x = \Delta y = \Delta z$ for all gridblocks.

1. Provide a governing flow equation for the system.
2. Write a finite-difference approximation of the differential equation you wrote in Part 1.

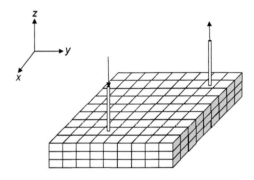

Fig. 5.26—3D representation of the reservoir system of Problem 5.4.29.

Solution to Problem 5.4.29

1. The generalized form of the PDE governing the flow is

$$\frac{\partial}{\partial x}\left(\frac{A_x k_x}{\mu B}\frac{\partial p}{\partial x}\right)\Delta x + \frac{\partial}{\partial y}\left(\frac{A_y k_y}{\mu B}\frac{\partial p}{\partial y}\right)\Delta y + \frac{\partial}{\partial z}\left(\frac{A_z k_z}{\mu B}\frac{\partial p}{\partial z}\right)\Delta z + q = 0, \dots \dots \dots \dots \dots (5.215)$$

which can be simplified as

$$\frac{\partial}{\partial x}\left(k_x \frac{\partial p}{\partial x}\right) + \frac{\partial}{\partial y}\left(k_y \frac{\partial p}{\partial y}\right) + \frac{\partial}{\partial z}\left(k_z \frac{\partial p}{\partial z}\right) + \frac{q\mu B}{V_b} = 0. \dots \dots \dots \dots \dots \dots \dots \dots (5.216)$$

2.

$$\left[\frac{k_x}{(\Delta x)^2}\right]_{i+\frac{1}{2},j,k}\left(p_{i+1,j,k} - p_{i,j,k}\right) + \left[\frac{k_x}{(\Delta x)^2}\right]_{i-\frac{1}{2},j,k}\left(p_{i-1,j,k} - p_{i,j,k}\right)$$
$$+ \left[\frac{k_y}{(\Delta y)^2}\right]_{i,j+\frac{1}{2},k}\left(p_{i,j+1,k} - p_{i,j,k}\right) + \left[\frac{k_y}{(\Delta y)^2}\right]_{i,j-\frac{1}{2},k}\left(p_{i,j-1,k} - p_{i,j,k}\right) \quad \dots \dots \dots \dots (5.217)$$
$$+ \left[\frac{k_z}{(\Delta z)^2}\right]_{i,j,k+\frac{1}{2}}\left(p_{i,j,k+1} - p_{i,j,k}\right) + \left[\frac{k_z}{(\Delta z)^2}\right]_{i,j,k-\frac{1}{2}}\left(p_{i,j,k-1} - p_{i,j,k}\right) + \frac{q_{i,j,k}\mu B}{V_{b_{i,j,k}}} = 0.$$

Note: Because the permeability along the z-direction is given as zero, all terms that have k_z in Eqs. 5.215 through 5.217 can be dropped.

Problem 5.4.30

State whether or not the following are well-posed problems (mathematically and/or physically). In each case, justify your answer.

1. Use **Fig. 5.27.1** and the information in **Table 5.13.1** to determine your answer.
2. Use **Fig. 5.27.2** and the information in **Table 5.13.2** to determine your answer.
3. Use **Fig. 5.27.3** and the information in **Table 5.13.3** to determine your answer.
4. Use **Fig. 5.27.4** and the information in **Table 5.13.4** to determine your answer.
5. Use **Fig. 5.27.5** and the information in **Table 5.13.5** to determine your answer.

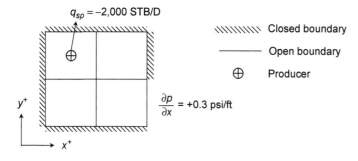

Fig. 5.27.1—Problem 5.4.30, Part 1.

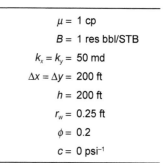

μ = 1 cp	
B = 1 res bbl/STB	
$k_x = k_y$ = 50 md	
$\Delta x = \Delta y$ = 200 ft	
h = 200 ft	
r_w = 0.25 ft	
ϕ = 0.2	
c = 0 psi^{-1}	

Table 5.13.1—Reservoir properties for Fig. 5.27.1 (Problem 5.4.30, Part 1).

Fig. 5.27.2—Problem 5.4.30, Part 2.

μ = 1 cp
B = 1 res bbl/STB
$k_x = k_y$ = 50 md
$\Delta x = \Delta y$ = 200 ft
h = 200 ft
r_w = 0.25 ft
ϕ = 0.2
c = 7x10^{-6} psi^{-1}
p_i = 2,000 psia

Table 5.13.2—Reservoir properties for Fig. 5.27.2 (Problem 5.4.30, Part 2).

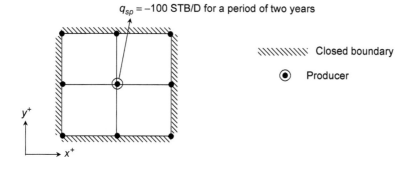

Fig. 5.27.3—Problem 5.4.30, Part 3.

μ = 1 cp
B = 1 res bbl/STB
$k_x = k_y$ = 50 md
$\Delta x = \Delta y$ = 200 ft
h = 200 ft
r_w = 0.25 ft
ϕ = 0.2
c = 0 psi^{-1}
p_i = 2,000 psia

Table 5.13.3—Reservoir properties for Fig. 5.27.3 (Problem 5.4.30, Part 3).

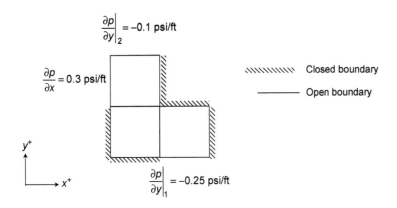

$$\frac{\partial p}{\partial y}\bigg|_2 = -0.1 \text{ psi/ft}$$

$$\frac{\partial p}{\partial x} = 0.3 \text{ psi/ft}$$

\\\\\\\\\\ Closed boundary

———— Open boundary

y^+

x^+

$$\frac{\partial p}{\partial y}\bigg|_1 = -0.25 \text{ psi/ft}$$

Fig. 5.27.4—Problem 5.4.30, Part 4.

$\mu = 1$ cp
$B = 1$ res bbl/STB
$k_x = k_y = 50$ md
$\Delta x = \Delta y = 200$ ft
$h = 200$ ft
$r_w = 0.25$ ft
$\phi = 0.2$
$c = 0$ psi^{-1}

Table 5.13.4—Reservoir properties for Fig. 5.27.4 (Problem 5.4.30, Part 4).

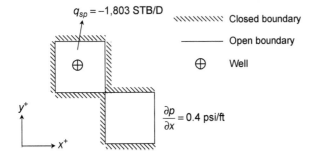

$q_{sp} = -1,803$ STB/D

\\\\\\\\\\ Closed boundary

———— Open boundary

⊕ Well

y^+

x^+

$$\frac{\partial p}{\partial x} = 0.4 \text{ psi/ft}$$

Fig. 5.27.5—Problem 5.4.30, Part 5.

$\mu = 1$ cp
$B = 1$ res bbl/STB
$k_x = k_y = 50$ md
$\Delta x = \Delta y = 200$ ft
$h = 200$ ft
$r_w = 0.25$ ft
$\phi = 0.2$
$c = 0$ psi^{-1}
$p_i = 2,000$ psia

Table 5.13.5—Reservoir properties for Fig. 5.27.5 (Problem 5.4.30, Part 5).

Solution to Problem 5.4.30

1. The system is ill conditioned. The reason is because we have an incompressible fluid, and the flow rate from the well must be equal to the fluid that is entering across the open boundary:

$$q_{\text{in}} = -\frac{kA}{\mu B}\frac{\partial p}{\partial x} = -\frac{1.127 \times 0.050 \times 200 \times 200 \times 0.3}{1 \times 1} = -676.2 \text{ STB/D.} \quad \dots\dots\dots\dots\dots\dots\dots(5.218)$$

We can see that the –676.2 STB/D is not in agreement with the flow rate that is specified.

2. To check the validity of the specified flow rate at the well, we need to find the maximum oil that could be produced from the reservoir with the expansion of the fluid. The minimum pressure that can be reached over the entire field is 14.7 psia. The total oil production at the specified rate is

$$\text{Cumulative} = 100 \text{ STB/D} \times 2 \text{ years} \times 365 \text{ days/year} = 73,000 \text{ STB.} \quad \dots\dots\dots\dots\dots\dots(5.219)$$

The original oil in place (OOIP) is given by

$$\text{OOIP} = \frac{V_p}{5.615} = \frac{400 \times 400 \times 200 \times 0.2}{5.615} = 1,140 \text{ MSTB.} \quad \dots\dots\dots\dots\dots\dots\dots(5.220)$$

Using the definition of compressibility, the total fluid expansion = $1,140,000 \times (2,000 - 14.7) \times (7 \times 10^{-6})$ = 15,840 STB. This is the maximum expected production from the reservoir, and it tells us that it is not possible to produce for two years at a specified flow rate of 100 STB/D. Note that the total fluid expansion when the entire field pressure drops to 14.7 psia is only 15,840 STB.

3. We need to check if the specifications are consistent. We have an incompressible fluid, and flow rate from the well must be equal to the total fluid that is entering and leaving the reservoir across the open boundary:

$$q = -\frac{1.127 \times 0.050 \times 200 \times 200}{1 \times 1} \times (0.4 - 0.6) = 451 \text{ STB/D}. \quad \dots\dots\dots\dots\dots\dots\dots\dots\dots\dots(5.221)$$

We see that the 451 STB/D is in agreement with the flow rate that is specified. However, if one needs to solve for the pressure distribution, the problem is not well posed.

4. Because the fluid is incompressible, the total flow into the reservoir must be equal to the total flow leaving the reservoir (no accumulation or depletion can take place). Note that ($k_y = 2k_x$),

$$-\left.\frac{\partial p}{\partial x}\right|_{\text{out}} - 2\left.\frac{\partial p}{\partial y}\right|_{1-\text{in}} + 2\left.\frac{\partial p}{\partial y}\right|_{2-\text{out}} = -0.3 + 2 \times 0.25 - 2 \times 0.1 = -0.3 + 0.5 - 0.2 = 0. \quad \dots\dots\dots\dots\dots (5.222)$$

There is a balance between the flow entering across the boundary and the flow leaving the reservoir.

5. Physically not possible for the following reasons:

- No communication between the blocks (two separate reservoirs).
- The fluid is incompressible, and one of the blocks is producing without any encroachment.
- The fluid is incompressible, and there is a block with a boundary, across which some fluid is entering while no fluid is leaving the system.

Nomenclature

a	= constant	$E_{i,j,k}$	= transmissibility coefficient using SIP notation between blocks (i, j, k) and ($i+1, j, k$), L^{4}t/m, STB/D-psia for liquid, scf/D-psia for gas
A	= cross-sectional area, L^2, ft^2		
$A_{i,j,k}$	= transmissibility coefficient using SIP notation between blocks (i, j, k) and ($i, j, k+1$), L^{4}t/m, STB/D-psia for liquid, scf/D-psia for gas	f	= function
		f_f	= original differential-equation representation of function f
A_n	= n^{th} coefficient of Fourier series determined from initial and boundary condition	f_{fd}	= finite-difference representation of function f
A_x	= cross-sectional area normal to the direction x, L^2, ft^2	g	= acceleration of gravity, L/t^2, ft/s^2
		g_c	= unit conversion factor in Newton's law
A_y	= cross-sectional area normal to the direction y, L^2, ft^2	G	= elevation with respect to absolute datum being positive upward, L, ft
A_z	= cross-sectional area normal to the direction z, L^2, ft^2	h	= thickness, L, ft
b	= constant	H_{ave}	= harmonic average
B	= formation volume factor, L^3/L^3, res bbl/STB or res bbl/scf,	I	= $\sqrt{-1}$
		k	= permeability, L^2, darcies or perms (1.127 perms = 1 darcy)
$B_{i,j,k}$	= transmissibility coefficient using SIP notation between blocks (i, j, k) and ($i, j, k-1$), L^{4}t/m, STB/D-psia for liquid, scf/D-psia for gas	$\bar{k}$	= geometric mean of the permeabilities in the plane perpendicular to the well, L^2, darcies or perms (1.127 perms = 1 darcy)
B_l	= liquid formation volume factor, L^3/L^3, res bbl/STB	k_x	= permeability in the x-direction, L^2, darcies or perms (1.127 perms = 1 darcy)
B_o	= oil formation volume factor, L^3/L^3, res bbl/STB	k_y	= permeability in the y-direction, L^2, darcies or perms (1.127 perms = 1 darcy)
c	= constant; compressibility, Lt^2/m, psi^{-1}		
c_l	= liquid compressibility, Lt^2/m, psi^{-1}	k_z	= permeability in the z-direction, L^2, darcies or perms (1.127 perms = 1 darcy)
C_L	= marker for centerline (as in Fig. 5.3)		
c_t	= total compressibility, Lt^2/m, psi^{-1}	$N_{i,j,k}$	= transmissibility coefficient using SIP notation between blocks (i, j, k) and ($i, j+1, k$), L^{4}t/m, STB/D-psia for liquid, scf/D-psia for gas
$C_{i,j,k}$	= transmissibility coefficient using SIP notation of central block (i, j, k), L^{4}t/m, STB/D-psia for liquid, scf/D-psia for gas		
		M	= intermediate variable used in Problem 5.4.11
D	= diffusivity coefficient, L^2/t, ft^2/D	O	= order of

p = pressure, m/Lt^2, psia

$p_{aquifer}$ = aquifer pressure, m/Lt^2, psia

p_{sc} = standard condition pressure, m/Lt^2, psia

p_{sf} = well sandface pressure, m/Lt^2, psia

$p_{sf,spec}$ = well sandface pressure specification, m/Lt^2, psia

q = production or flow rate, L^3/t, STB/D for liquid and scf/D for gas

q_{in} = flow rate entering the system, L^3/t, STB/D for liquid and scf/D for gas

q_{spec} = flow rate specification, L^3/t, STB/D for liquid and scf/D for gas

q_{well} = flow rate of a well, L^3/t, STB/D for liquid and scf/D for gas

$Q_{i,j,k}$ = right-hand-side coefficient of governing flow equation using SIP notation for block (i, j, k) L^3/t, STB/D–psia for liquid, scf/D for gas

Q_{total} = total flow rate, L^3/t, STB/D for liquid and scf/D for gas

r = radial direction in both cylindrical and spherical coordinate systems

r_{eq} = equivalent well block radius, L, ft

r_w = well radius, L, ft

R = universal gas constant

s = skin factor, dimensionless

$S_{i,j,k}$ = transmissibility coefficient using SIP notation between blocks (i, j, k) and $(i, j-1, k)$, L^{4}t/m, STB/D-psia for liquid, scf/D-psia for gas

t = time, t, days

T = transmissibility, L^{4}t/m, STB/D-psia for liquid, scf/D–psia for gas

T_{sc} = standard condition temperature, T, °R

V = volumetric quantity, L^3, ft^3

V_b = bulk volume, L^3, ft^3

V_p = pore volume, L^3, ft^3

ΔV = incremental volumetric quantity, L^3, ft^3

$W_{i,j,k}$ = transmissibility coefficient using SIP notation between blocks (i, j, k) and $(i–1, j, k)$, L^{4}t/m, STB/D–psia for liquid, scf/D-psia for gas

x = distance in x-direction in the Cartesian coordinate system, L, ft

y = distance in y-direction in the Cartesian coordinate system

z = distance in z-direction in the Cartesian coordinate system

z_g = gas compressibility factor, dimensionless

α_c = unit conversion factor between ft^3 and bbl, with a magnitude of 5.615

β_c = transmissibility conversion factor, with a magnitude of 1.127

Γ = accumulation coefficient, L^3/t, ft^3/D

γ = space-dependent term of Fourier-series

Δ = difference of a parameter

Δx = control volume dimension along the x-direction, L, ft

Δy = control volume dimension along the y-direction, L, ft

Δz = control volume dimension along the z-direction, L, ft

δ = central difference operator

ε_L = local truncation error

θ = angle in θ-direction in both cylindrical and spherical coordinator system, rad

μ = viscosity, cp

μ_g = gas viscosity, cp

μ_l = liquid viscosity, cp

μ_{amp} = amplification factor

μ_{max} = maximum amplification factor, dimensionless

ξ = time-dependent terms of Fourier-Series

ρ = density, m/L^3, lbm/ft^3

ϕ = porosity, fraction

Φ = flow potential, m/Lt^2, psia

Ω = well productivity index group used in the SIP notation, STB/D–psi

Superscripts

0 = physical parameters measured at reference condition

g = gaseous phase

k = previous iteration level

k+1 = current iteration level

n = previous timestep level

n+1 = current timestep level

o = oleic phase

w = aqueous phase

Subscripts

i = pertaining to central block i in a 1D representation (x-direction only)

j = pertaining to central block j in a 1D representation (y-direction only)

k = pertaining to central block k in a 1D representation (z-direction only)

i,j = pertaining to central block i,j in a 2D representation (x- and y-directions only)

i,j,k = pertaining to central block i,j,k in a 3D representation

l = liquid

x = indicator for the x-direction

References

Dufort, E.C. and Frankel, S. P. 1953. Stability Conditions in the Numerical Treatment of Parabolic Differential Equations. *Mathematical Tables and Other Aids to Computation,* **7**(43): 135–152. http://dx.doi.org/10.2307/2002754.

Ertekin, T., Abou-Kassem, J. H., and King, G. R. 2001. *Basic Applied Reservoir Simulation.* Richardson, Texas, US: Society of Petroleum Engineers.

Peaceman, D.W. 1983. Interpretation of Well-Block Pressures in Numerical Reservoir Simulation with Nonsquare Grid Blocks and Anisotropic Permeability. *SPE J* **23**(3): 531–543. SPE-10528-PA. https://dx.doi.org/10.2118/10528-PA.

Chapter 6

Coupling of the Well Equation with the Reservoir Flow Equations

Wells represent the internal boundaries of the reservoir systems. At the well location, like any other boundary, a boundary condition must be specified. An equation representing the relationship between the pressure of the block hosting the well, the sandface pressure, and the flow rate at the sandface is known as the well reservoir coupling. In this chapter, we will review how a well equation is incorporated into reservoir simulation models.

6.1 Boundary-Condition Specification at the Well

At the wellpoint (the sandface), there are two options to constrain the well: either the pressure or the flow rate at the sandface is specified. While pressure specification at the sandface is a Dirichlet-type boundary condition, flow-rate specification at the sandface is a Neumann-type boundary condition. **Fig. 6.1** shows the three important dependent variables—pressure of the block hosting the well ($p_{i,j,k}$), sandface pressure of the well ($p_{sf_{i,j,k}}$), and the flow rate of the well ($q_{sc_{i,j,k}}$)—that coexist in a wellblock.

The well model describes the relationship among the three variables: $p_{i,j,k}$, $p_{sf_{i,j,k}}$, and $q_{sc_{i,j,k}}$. The wellblock pressure, $p_{i,j,k}$, is an unknown, like any other block pressure. Because it is impossible to control the wellblock pressure ($p_{i,j,k}$), it is not realistic to specify the value of block pressure as a constraint (although it is mathematically possible). From the remaining two dependent variables, only one can be specified; the other one is left as an additional unknown. Because every block enters the computations with its block equation, there is one extra unknown introduced by the well equation. Therefore, it is necessary to have one more equation so that the problem becomes a well-posed problem. This additional equation is known as the wellbore model, and in this book, Peaceman's wellbore equation (Peaceman 1983) is the one that is more extensively included in the applications.

6.2 Wellbore Models

6.2.1 van Poollen et al. Model. The model developed by van Poollen et al. (1968) was one of the earliest wellbore models to calculate the equivalent wellbore radius and the production/injection rate of the wells. Although the model is rarely used in current reservoir simulation models, it is considered an insightful progression in the reservoir simulation history. The van Poollen et al. model employed the steady-state radial flow assumption shown in Eq. 6.1:

$$q_{sc} = -\frac{2\pi kh\left(p_e - p_{sf}\right)}{\mu B\left[\ln\left(\dfrac{r_{eq}}{r_w}\right) + s\right]} \quad \dotfill \quad (6.1)$$

At an arbitrary location of the reservoir, pressure function can be expressed as

$$p = p_{sf} - \frac{q_{sc}\mu B}{2\pi kh}\left[\ln\left(\frac{r}{r_w}\right) + s\right] \quad \dotfill \quad (6.2)$$

The average reservoir pressure is defined using an areal average over the domain interest:

$$\bar{p} = \frac{\int_0^{V_p} p\,dV_p}{V_p} \quad \dotfill \quad (6.3)$$

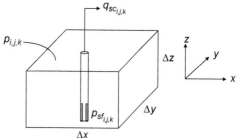

Fig. 6.1—Three dependent variables of a wellblock ($p_{i,j,k}$, $p_{sf_{i,j,k}}$ and $p_{i,j,k}$).

Note that

$$dV_p = 2\pi r\phi h dr, \dotfill (6.4)$$

because

$$V_p = \pi\left(r^2 - r_w^2\right)\phi h. \dotfill (6.5)$$

Substituting Eq. 6.2 and Eq. 6.4 into Eq. 6.3, yields

$$\bar{p} = \frac{\int_0^{V_p} p_{sf} 2\pi rh\phi dr}{V_p} - \frac{q_{sc}\mu B}{2\pi khV_p}\int_0^{V_p}\left[\ln\left(\frac{r}{r_w}\right) + s\right]2\pi rh\phi dr. \dotfill (6.6)$$

Employing Eq. 6.4, one can simplify

$$\frac{\int_0^{V_p} p_{sf} 2\pi rh\phi dr}{V_p} = p_{sf}, \dotfill (6.7)$$

and

$$\frac{q_{sc}\mu B}{2\pi khV_p}\int_0^{V_p}\left[\ln\left(\frac{r}{r_w}\right) + s\right]2\pi rh\phi dr = \frac{q_{sc}\mu B}{\pi k r_{eq}^2 h}\left[\int_0^{r_{eq}}\ln\left(\frac{r}{r_w}\right)dr + s\int_0^{r_{eq}} rdr\right]. \dotfill (6.8)$$

Here we assume that $r_{eq} \gg r_w$.
   Further simplifying,

$$\frac{q_{sc}\mu B}{2\pi khV_p}\int_0^{V_p}\left[\ln\left(\frac{r}{r_w}\right) + s\right]2\pi rh\phi dr = \frac{q_{sc}\mu B}{2\pi kh}\left[s - \ln r_w\right] - \frac{q_{sc}\mu B}{\pi k h r_{eq}^2}\int_0^{r_{eq}} r\ln r\, dr. \dotfill (6.9)$$

Implementing integration by parts,

$$\int_0^{r_{eq}} r\ln r\, dr = \frac{r_{eq}^2}{2}\left[\ln r_{eq} - \frac{1}{2}\right]. \dotfill (6.10)$$

Finally, Eq. 6.6 can be written as

$$\bar{p} = p_{sf} - \frac{q_{sc}\mu B}{2\pi kh}\left[\ln\left(\frac{r}{r_w}\right) + s - \frac{1}{2}\right]. \dotfill (6.11)$$

or

$$q_{sc} = -\frac{2\pi \bar{k} h\left(\bar{p} - p_{sf}\right)}{\mu B\left[\ln\left(\frac{r_{eq}}{r_w}\right) + s - \frac{1}{2}\right]}, \dotfill (6.12)$$

where

$$\bar{k} = \sqrt{k_x k_y}. \dotfill (6.13)$$

The equivalent wellbore radius is calculated from

$$\pi\left(r_{eq}\right)^2 = \Delta x\Delta y, \dotfill (6.14)$$

which yields

$$r_{eq} = \sqrt{\frac{\Delta x\Delta y}{\pi}}. \dotfill (6.15)$$

With the advance of reservoir simulation technology, more robust and practical wellbore models were later introduced by Peaceman (1983, 1993).

6.2.2 Peaceman Model. The Peaceman wellbore model effectively establishes the relationships between well flow rate, well sandface pressure, and wellblock pressure, and is given by

$$q_{sc_{i,j,k}}^{n+1} = \left[\frac{2\pi \bar{k} \Delta L}{\mu B \left(\ln \frac{r_{eq}}{r_w} + s \right)} \right]_{i,j,k}^{n+1} \left(p_{i,j,k}^{n+1} - p_{sf_{i,j,k}}^{n+1} \right). \quad \text{...} (6.16)$$

The critical assumption made by the Peaceman model is that the wellblock pressure is the same as the pressure at the equivalent wellblock radius (r_{eq}). Thus, the dependent variables, such as flow rate, sandface pressure, and wellblock pressure, can be related by the inflow-performance relation demonstrated in Eq. 6.16.

As previously discussed, in Eq. 6.16 either $q_{sc_{i,j,k}}^{n+1}$ or $p_{sf_{i,j,k}}^{n+1}$ must be specified. If $q_{sc_{i,j,k}}^{n+1}$ is specified, the remaining unknowns in the wellblock will be $p_{i,j,k}^{n+1}$ and $p_{sf_{i,j,k}}^{n+1}$. On the other hand, if the well is constrained by the $p_{sf_{i,j,k}}^{n+1}$ specification, then $p_{i,j,k}^{n+1}$ and $q_{sc_{i,j,k}}^{n+1}$ will be the two remaining unknowns. **Fig. 6.2** shows the boundary-condition specification and the remaining unknowns in the wellblock. (*Note*: $p_{i,j,k}^{n+1}$ is always an unknown.)

It is an interesting issue to discuss what happens if a well has been producing with an atmospheric pressure specification at the sandface ($p_{sf} = p_{atm}$). This is only possible when no fluid accumulation is allowed inside the wellbore (a situation only achievable if a downhole pump is located across the sandface). The well will continue to produce, albeit in small amounts, until the wellbore pressure approaches the atmospheric pressure and, at the same time, the pressure gradient at the sandface goes to zero. If a well is being produced at a constant rate, as shown in the top-left portion of **Fig. 6.2**, the sandface pressure will be continuously declining (top right). If the downhole pump is placed at a certain elevation above the sandface, at a certain time again, the sandface pressure and the back pressure exerted by the fluid column below the intake valve of the pump will equilibrate (i.e., pressure at the sandface cannot decrease any further), and if the pump is not further lowered, the well will cease production. At that specific time, by lowering the pump farther down toward the sandface, the well will continue to operate at a constant flow-rate specification. When the pump cannot be further lowered (e.g., when it is located across the sandface), the sandface pressure is equal to the atmospheric pressure and the well switches to the constant sandface-pressure specification. Thus, flow rate will start to decline. This switching of well constraints is schematically shown in **Fig. 6.3**.

Now we can return to Eq. 6.16, which is usually written in shorthand notation as follows:

$$q_{sc_{i,j,k}}^{n+1} = -\Omega_{i,j,k}^{n+1} \left(p_{i,j,k}^{n+1} - p_{sf_{i,j,k}} \right). \quad \text{...} (6.17)$$

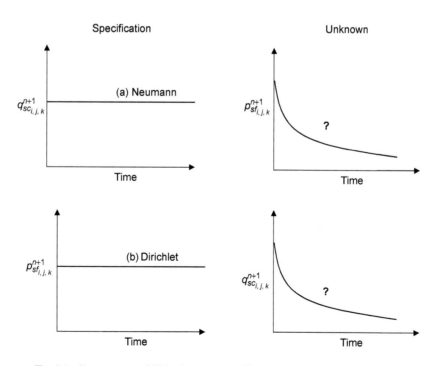

Fig. 6.2—Neumann- and Dirichlet-type specifications at the production well.

where

$$\Omega_{i,j,k}^{n+1} = \left[\frac{2\pi \bar{k} \Delta L}{\mu B \left(\ln \frac{r_{eq}}{r_w} + s \right)} \right]_{i,j,k}^{n+1} \quad \dots\dots\dots\dots\dots\dots (6.18)$$

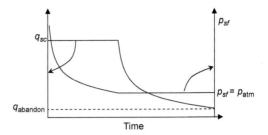

Fig. 6.3—Switching of wellbore constraint from flow-rate specification to sandface-pressure specification (in a wellbore in which the downhole pump is placed across the formation at the bottom of the well to allow no fluid accumulation in the wellbore).

The $\Omega_{i,j,k}^{n+1}$ term is known as the productivity index (PI) for a producing well and as the injectivity index (II) for an injector.

Before we review the terms that appear in the definition of $\Omega_{i,j,k}^{n+1}$, it is important to once more discuss the sign notation we adopted. Note that the value of $\Omega_{i,j,k}^{n+1}$ is always positive. Therefore, Eq. 6.17 will generate a negative $q_{sc_{i,j,k}}^{n+1}$ if and only if $p_{sf_{i,j,k}}^{n+1} < p_{i,j,k}^{n+1}$, which is always true for a production well. In a production well, for flow to take place from within the block toward the wellbore, wellblock pressure must be larger than the flowing sandface pressure. On the other hand, in an injection well $p_{sf_{i,j,k}}^{n+1} > p_{i,j,k}^{n+1}$, resulting in a positive $q_{sc_{i,j,k}}^{n+1}$ in Eq. 6.17. Note that these statements are consistent with the sign convention we adopted in this book ($q > 0$ for injection and $q < 0$ for production).

Now, we can examine the terms that appear in the definition of $\Omega_{i,j,k}^{n+1}$ (Eq. 6.18). We are already familiar with terms such as viscosity, μ; formation volume factor, B; wellbore radius, r_w; and skin factor, s. The skin factor is an indication of the near-wellbore productivity or injectivity environment, which typically is employed to represent damage, stimulation, perforation, or partial penetration effects (Ertekin et al. 2001). The skin factor is a dimensionless pressure group signifying the magnitude of the additional pressure drop lost and/or gained as a result of the presence of skin. A positive skin represents a damaged near-wellbore region, while a negative skin indicates a stimulated condition. Introducing the skin factor to the well index formulation results in higher or lower pressure drops, respectively, for the damaged or stimulated wells. Now, let us examine other terms that make up the $\Omega_{i,j,k}^{n+1}$.

$\Delta L_{i,j,k}$. This term represents the length of the wellbore in the wellblock. The well shown in Fig. 6.1 is a vertical well, drilled perpendicular to the x-y plane. Therefore, $\Delta L = \Delta z$ (if this is a 2D reservoir, note that $\Delta z = h$). In other words, a vertical well is completed by penetrating the wellblock. This observation introduces a good degree of flexibility for horizontal wells. For example, if the block i,j,k of Fig. 6.1 contains a horizontal well drilled perpendicular to the y-z plane, then $\Delta L = \Delta x$. Finally, if the same well is drilled perpendicular to the x-z plane (i.e., parallel to the y-direction), then $\Delta L = \Delta y$. It should be noted that, especially in the case of a horizontal well, a large number of blocks will be penetrated. In Eq. 6.18, ΔL always maintains the length of the wellblock edge parallel to the direction the well is drilled (penetrating the length of the horizontal well segment is equal to the length of the block in the direction the well is drilled).

$\bar{k}_{i,j,k}$. This term represents the geometric mean of the directional permeabilities in the plane to which the wellbore trajectory is orthogonal. In other words, for the vertical well drilled perpendicular to the x-y plane,

$$\bar{k}_{i,j,k} = \left(k_{x_{i,j,k}} k_{y_{i,j,k}} \right)^{1/2}. \quad \dots\dots\dots\dots\dots\dots\dots\dots\dots\dots\dots\dots\dots\dots\dots\dots\dots\dots (6.19a)$$

Similarly, for a horizontal well segment drilled perpendicular to the y-z plane,

$$\bar{k}_{i,j,k} = \left(k_{x_{i,j,k}} k_{z_{i,j,k}} \right)^{1/2}. \quad \dots\dots\dots\dots\dots\dots\dots\dots\dots\dots\dots\dots\dots\dots\dots\dots\dots\dots (6.19b)$$

Finally, if the horizontal well is oriented parallel to the y-direction (i.e., orthogonal to the x-z plane),

$$\bar{k}_{i,j,k} = \left(k_{x_{i,j,k}} k_{z_{i,j,k}} \right)^{1/2}. \quad \dots\dots\dots\dots\dots\dots\dots\dots\dots\dots\dots\dots\dots\dots\dots\dots\dots\dots (6.19c)$$

$r_{eq_{i,j,k}}$. Peaceman (1983) carried out numerical experiments that calculate the radius of an area (r_{eq}), at which the steady-state reservoir pressure is equal to the wellblock pressure (Peaceman 1983). In his 1983 paper, Peaceman shows that for a single-phase, five-spot-pattern flow problem, the equivalent wellblock radius for a nonsquare wellblock with anisotropic permeability is defined as follows:

- For a vertical well drilled perpendicular to x-y plane,

$$r_{eq_{i,j,k}} = 0.28 \frac{\left[\left(\frac{k_{y_{i,j,k}}}{k_{x_{i,j,k}}} \right)^{1/2} \left(\Delta x_{i,j,k} \right)^2 + \left(\frac{k_{x_{i,j,k}}}{k_{y_{i,j,k}}} \right)^{1/2} \left(\Delta y_{i,j,k} \right)^2 \right]^{1/2}}{\left(\frac{k_{y_{i,j,k}}}{k_{x_{i,j,k}}} \right)^{1/4} + \left(\frac{k_{x_{i,j,k}}}{k_{y_{i,j,k}}} \right)^{1/4}} \quad \dots\dots\dots\dots\dots\dots\dots\dots\dots\dots\dots\dots\dots\dots (6.20a)$$

- For a horizontal well drilled perpendicular to the y-z plane,

$$r_{eq_{i,j,k}} = 0.28 \frac{\left[\left(\frac{k_{z_{i,j,k}}}{k_{y_{i,j,k}}} \right)^{\frac{1}{2}} \left(\Delta y_{i,j,k} \right)^2 + \left(\frac{k_{y_{i,j,k}}}{k_{z_{i,j,k}}} \right)^{\frac{1}{2}} \left(\Delta z_{i,j,k} \right)^2 \right]^{\frac{1}{2}}}{\left(\frac{k_{z_{i,j,k}}}{k_{y_{i,j,k}}} \right)^{\frac{1}{4}} + \left(\frac{k_{y_{i,j,k}}}{k_{z_{i,j,k}}} \right)^{\frac{1}{4}}} \quad \dots \dots \dots \dots \dots \dots \dots \dots \dots \quad (6.20\text{b})$$

- Finally, for a horizontal well drilled perpendicular to the x-z plane,

$$r_{eq_{i,j,k}} = 0.28 \frac{\left[\left(\frac{k_{z_{i,j,k}}}{k_{x_{i,j,k}}} \right)^{\frac{1}{2}} \left(\Delta x_{i,j,k} \right)^2 + \left(\frac{k_{x_{i,j,k}}}{k_{z_{i,j,k}}} \right)^{\frac{1}{2}} \left(\Delta z_{i,j,k} \right)^2 \right]^{\frac{1}{2}}}{\left(\frac{k_{z_{i,j,k}}}{k_{x_{i,j,k}}} \right)^{\frac{1}{4}} + \left(\frac{k_{x_{i,j,k}}}{k_{z_{i,j,k}}} \right)^{\frac{1}{4}}} \quad \dots \dots \dots \dots \dots \dots \dots \dots \dots \quad (6.20\text{c})$$

Before we conclude our discussion on the PI, we can make some observations to simplify the definition of $r_{e_{i,j,k}}$. This is especially helpful because it is quite probable that in the lateral x-y plane, some reservoirs can exhibit isotropic permeability (i.e., $k_x = k_y$), so then, Eq. 6.20a collapses to

$$r_{eq_{i,j,k}} = 0.14 \left[\left(\Delta x_{i,j,k} \right)^2 + \left(\Delta y_{i,j,k} \right)^2 \right]^{\frac{1}{2}} \quad \dots \dots \dots \dots \dots \dots \dots \dots \dots \dots \dots \dots \dots \quad (6.21)$$

Furthermore, if the block hosting the well is a square block ($\Delta x = \Delta y$), then

$$r_{eq_{i,j,k}} = 0.198 \Delta x. \quad \dots \dots \dots \dots \dots \dots \dots \dots \dots \dots \dots \dots \dots \dots \dots \dots \dots \quad (6.22)$$

Such simplifications are realistic for the horizontal wells, because typically the vertical permeability, k_z, is always much less than the lateral permeabilities, k_x and k_y. The practical field units, as provided and used in earlier chapters, are still applicable for the Peaceman wellbore model entries.

6.3 Abou-Kassem and Aziz Model for Off-Centered Wells

Abou-Kassem and Aziz (1985) developed another equivalent wellblock radius equation applicable to wells located at off-center locations in a square or rectangular gridblock, with an aspect ratio of $\Delta y/\Delta x$ in the range of ½ to 2. The equivalent wellblock radius is defined as

$$r_{eq} = \left\{ \exp(-2\pi f) \prod_i \left[r_{i,1}^{T_i} \prod_j \left(\frac{r_{i,j}}{a_j} \right)^{T_i} \right] \right\}^b \quad \dots \dots \dots \dots \dots \dots \dots \dots \dots \dots \quad (6.23)$$

The gridblock that hosts a well is numbered as Block 0, and the well is numbered as 1. A block that hosts a well is defined as an interior block if it is located inside the reservoir, which means that none of the block edges is adjacent to any of the boundaries of the reservoir. Otherwise, the block hosting a well is treated as boundary block. For a boundary block, the well creates an image well by reflecting all the physical properties across the reservoir boundary. One or more image wells will exist, depending on the number of reservoir boundaries that fall on the edges of the wellblock. Notably, the image well would not exist for an interior block. These descriptions easily can be visualized by reviewing **Figs. 6.4 and 6.5.**

The distance between Well 1 and its image well, Well j, is referred to as a_j in Eq. 6.23. $\prod_j$ is the product over all existing well images, and $\prod_i$ is the product over all existing surrounding gridpoints. The factor f equals the fraction of well flow from the wellblock:

- $f = 1$ for an interior well.
- $f = \frac{1}{2}$ for a well on one boundary (on the edge).
- $f = \frac{1}{4}$ for a well on two boundaries (on the corner).

Finally, the exponent $b = \dfrac{1}{\sum_i T_i}$ includes the summation over all existing surrounding gridpoints. Fig. 6.5 shows Block 0 (wellblock) and its surrounding blocks in a point-distributed grid and provides the definitions of the entries of Eq. 6.23 for point-distributed grids.

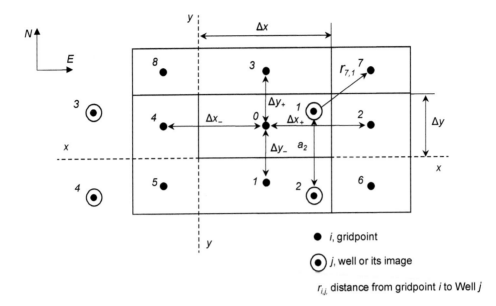

Fig. 6.4—Wellblock 0 and its surrounding gridblocks in a block-centered grid.

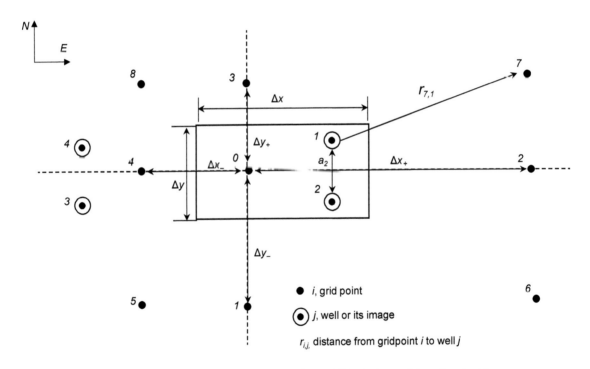

Fig. 6.5— Wellblock 0 and its surrounding gridblocks in a point-centered grid.

Table 6.1 summarizes the equations for the interface-transmissibility factor, T_i, between a surrounding Block, i, and the wellblock.

Note that for a uniform square grid, $\Delta x = \Delta y$, the transmissibility factors simplify to $T_5 = T_6 = T_7 = T_8 = 0$ (for the five-point difference scheme) and 1/6 (for the nine-point difference scheme); and $T_1 = T_2 = T_3 = T_4 = 1$ (for the five-point difference scheme) and 2/3 (for the nine-point difference scheme).

It is possible to construct the entries in Table 6.1 to use in Eq. 6.23 (note that for an interior block, there are no image wells),

with $b = \dfrac{1}{\sum\limits_i T_i} = \dfrac{1}{4}$ and $f = 1$. Then, simplify the equivalent wellblock radius equation to

$$r_{eq} = \left[\exp(-2\pi) \prod_i \left(r_{i,1}^{T_i} \right) \right]^{1/4} = \left[\exp(-2\pi)(\Delta x)^4 \right]^{1/4} = e^{(-\pi/2)} \Delta x. \quad \ldots (6.24)$$

6.4 Babu and Odeh's Model for Horizontal Wells

The Babu and Odeh (1989) inflow-performance equation for a horizontal well is

$$q_{sc} = \frac{2\pi \beta_c d\bar{k} \left(p_{i,j,k} - p_{sf_{i,j,k}} \right)}{\mu B \left[\ln\left(\dfrac{A^{1/2}}{r_w} \right) + \ln(C_H) + s - \dfrac{3}{4} \right]}. \quad \ldots \ldots (6.25)$$

Compared with the Peaceman model, this development is more suitable to horizontal wells that are completely offset from the center of the wellblock. Moreover, it considers the skin factors of multiple damage or stimulation mechanisms. Because Eq. 6.25 is written for a horizontal well placed parallel to the y-direction, Eq. 6.19c gives the definition of the permeability entry, $\bar{k}$.

$$\ln(C_H) = 6.28 \frac{c}{h} \sqrt{\frac{k_z}{k_x}} \left[\frac{1}{3} - \frac{X_0}{c} + \left(\frac{X_0}{c} \right)^2 \right] - \ln\left(\sin \frac{\pi Z_0}{h} \right) - 0.5 \ln\left[\left(\frac{c}{h} \right) \sqrt{\frac{k_z}{k_x}} \right] - 1.088. \quad \ldots \ldots \ldots \ldots (6.26)$$

New entries in Eq. 6.26 are defined as $d = \Delta y$, $A = \Delta x \Delta z$, $c = \Delta x$, and X_0 and Z_0 are the coordinates of the center of the well in the vertical plane. As shown in **Fig. 6.6**, (X_0, y_1, Z_0) and (X_0, y_2, Z_0) represent the coordinates of the beginning and end of the well, respectively. The approximation of the entry $\ln(C_H)$ described in Eq. 6.26 is valid when Eq. 6.27 is fulfilled (Babu and Odeh 1989):

$$C_H \geq 0.75 h \sqrt{\frac{k_x}{k_z}}, \quad \ldots \ldots \ldots \ldots \ldots (6.27)$$

and when a minimum distance between the well and the boundaries exists. This minimum distance is defined as

$$\min\left(X_0, c - X_0 \right) \geq 0.75 h \sqrt{\frac{k_x}{k_z}}. \quad \ldots \ldots \ldots \ldots \ldots \ldots \ldots \ldots \ldots \ldots \ldots \ldots \ldots \ldots (6.28)$$

Because $h \ll c$, Eqs. 6.27 and 6.28 are commonly satisfied in practice. The skin factor, s, in Eq. 6.25 is a composite skin that includes both the mechanical skin, s_m, and the skin resulting from partial penetration, s_p. In other words,

$$s = s_p + s_m. \quad \ldots \ldots \ldots \ldots \ldots \ldots \ldots \ldots \ldots \ldots \ldots \ldots \ldots \ldots \ldots \ldots \ldots \ldots \ldots \ldots \ldots \ldots \ldots \ldots \ldots \ldots \ldots \ldots (6.29)$$

If the horizontal well fully penetrates the wellblock, $L = y_1 - y_2 = d$, then $s_p = 0$. For the computation of s_p, two cases must be considered:

- Case 1— $\left(\dfrac{C}{\sqrt{k_x}} \right) \geq 0.75 \left(\dfrac{d}{\sqrt{k_y}} \right) \geq 0.75 \left(\dfrac{h}{\sqrt{k_z}} \right)$,

  and if $L < d$,

  $$s_p = P_{xyz} + P'_{xy}, \quad \ldots \ldots \ldots \ldots \ldots \ldots \ldots \ldots \ldots \ldots \ldots \ldots \ldots \ldots \ldots \ldots \ldots \ldots \ldots \ldots \ldots \ldots \ldots \ldots \ldots \ldots (6.30)$$

  where

  $$P_{xyz} = \left(\frac{d}{L} - 1 \right) \left[\ln\left(\frac{h}{r_w} \right) + 0.25 \ln\left(\frac{k_x}{k_z} \right) - \ln\left(\sin \frac{\pi Z_0}{h} \right) - 1.84 \right], \quad \ldots \ldots \ldots \ldots \ldots \ldots \ldots (6.31)$$

  and

$$T_1 = \left(\frac{\Delta x}{\Delta y_-} \right) - (T_5 + T_6) \quad T_5 = \frac{1}{3} \left[\frac{(\Delta x_-)(\Delta y_-)}{(\Delta x_-)^2 + (\Delta y_-)^2} \right]$$

$$T_2 = \left(\frac{\Delta y}{\Delta x_+} \right) - (T_6 + T_7) \quad T_6 = \frac{1}{3} \left[\frac{(\Delta x_+)(\Delta y_-)}{(\Delta x_+)^2 + (\Delta y_-)^2} \right]$$

$$T_3 = \left(\frac{\Delta x}{\Delta y_+} \right) - (T_7 + T_8) \quad T_7 = \frac{1}{3} \left[\frac{(\Delta x_+)(\Delta y_+)}{(\Delta x_+)^2 + (\Delta y_+)^2} \right]$$

$$T_4 = \left(\frac{\Delta y}{\Delta x_-} \right) - (T_5 + T_8) \quad T_8 = \frac{1}{3} \left[\frac{(\Delta x_-)(\Delta y_+)}{(\Delta x_-)^2 + (\Delta y_-)^2} \right]$$

Note: For a uniform square grid, $\Delta x = \Delta y$, the transmissibility factors simplify to $T_5 = T_6 = T_7 = T_8 = 0$ (for the five-point difference scheme) and 1/6 (for the nine-point difference scheme); and $T_1 = T_2 = T_3 = T_4 = 1$ (for the five-point difference scheme) and 2/3 (for the nine-point difference scheme.

Table 6.1—Interface transmissibility factors between a neighboring gridblock, *i*, and the wellblock.

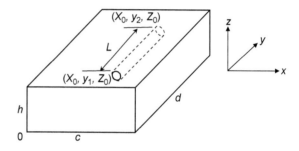

Fig. 6.6—Schematic of a horizontal well depicted by the Babu and Odeh (1989) model.

$$P'_{xy} = \frac{2d^2}{Lh}\sqrt{\frac{k_z}{k_y}} \times \left\{ F\left(\frac{L}{2d}\right) + 0.5\left[F\left(\frac{4y_{mid}+L}{2}\right) - F\left(\frac{4y_{mid}-L}{2}\right)\right]\right\}, \quad \dots\dots\dots\dots\dots\dots (6.32)$$

where y_{mid} = midpoint location of the well [$y_{mid} = 0.5\,(y_1 + y_2)$]. The function, F, in Eq. 6.32 describes the effects of the well location in the horizontal plane. **Table 6.2** shows the evaluation of function F for different arguments.

Argument	Function F
$\dfrac{L}{2d}$	$-\left(\dfrac{L}{2d}\right)\left[0.145 + \ln\left(\dfrac{L}{2d}\right) - 0.137\left(\dfrac{L}{2d}\right)^2\right]$
$\dfrac{4y_{mid}+L}{2d} \le 1$	$-\left(\dfrac{4y_{mid}+L}{2d}\right)\left[0.145 + \ln\left(\dfrac{4y_{mid}+L}{2d}\right) - 0.137\left(\dfrac{4y_{mid}+L}{2d}\right)^2\right]$
$\dfrac{4y_{mid}+L}{2d} > 1$	$\left[2-\left(\dfrac{4y_{mid}+L}{2d}\right)\right]\left\{0.145 + \ln\left[2-\left(\dfrac{4y_{mid}+L}{2d}\right)\right] - 0.137\left[2-\left(\dfrac{4y_{mid}+L}{2d}\right)\right]^2\right\}$
$\dfrac{4y_{mid}-L}{2d} \le 1$	$-\left(\dfrac{4y_{mid}-L}{2d}\right)\left[0.145 + \ln\left(\dfrac{4y_{mid}-L}{2d}\right) - 0.137\left(\dfrac{4y_{mid}-L}{2d}\right)^2\right]$
$\dfrac{4y_{mid}-L}{2d} > 1$	$\left[2-\left(\dfrac{4y_{mid}-L}{2d}\right)\right]\left\{0.145 + \ln\left[2-\left(\dfrac{4y_{mid}-L}{2d}\right)\right] - 0.137\left[2-\left(\dfrac{4y_{mid}-L}{2d}\right)\right]^2\right\}$

Table 6.2—Formulas to calculate function *F*.

- Case 2—$\left(\dfrac{d}{\sqrt{k_y}}\right) > 1.33\left(\dfrac{c}{\sqrt{k_x}}\right)\left(\dfrac{h}{\sqrt{k_z}}\right)$ and $L < d$, then

$$s_p = P_{xyz} + P_y + P_{xy}. \quad \dots\dots\dots\dots\dots\dots (6.33)$$

P_{xyz} is the same as that given by Eq. 6.31. P_y is calculated from

$$P_y = \frac{6.28d^2}{ch}\frac{\sqrt{k_x k_z}}{k_y}\left\{\left[\frac{1}{3} - \frac{y_{mid}}{d} + \left(\frac{y_{mid}}{d}\right)^2\right] + \frac{L}{24d}\left(\frac{L}{d}-3\right)\right\}, \quad \dots\dots\dots\dots\dots\dots (6.34)$$

where, again, y_{mid} = midpoint coordinate of the well. Finally, the P_{xy} component is defined by

$$P_{xy} = \left(\frac{d}{L}-1\right)\left(\frac{6.28c}{h}\sqrt{\frac{k_z}{k_x}}\right)\left[\frac{1}{3} - \frac{X_0}{c} + \left(\frac{X_0}{c}\right)^2\right]. \quad \dots\dots\dots\dots\dots\dots (6.35)$$

6.5 Vertical Multilayer Well Models

In the previous section, we did not consider a vertical well going through a number of layers and creating a column of wellblocks. For the segment of the vertical well in the kth layer, the Peaceman's equation can be written as

$$q_{sc,k} = -\Omega_k\left(p_k - p_{sf,k}\right), \quad \dots\dots\dots\dots\dots\dots (6.36)$$

with

$$\Omega = \left[\frac{2\pi \bar{k}\,\Delta z}{\mu B\left(\ln\dfrac{r_{eq}}{r_w} + s\right)}\right]_k. \quad \dots\dots\dots\dots\dots\dots (6.37)$$

The total flow rate for the well will be the summation of the flow rates for all layers:

$$q_{sc,well} = -\sum_k \Omega_k\left(p_k - p_{sf,k}\right). \quad \dots\dots\dots\dots\dots\dots (6.38)$$

In the following subsections, we discuss various treatments of wells with multilayer completions. A critical concept (i.e., the reference layer) is employed to anchor the internal boundary conditions of the wells. In this book, we assume that the topmost perforated layer is the reference layer. **Fig. 6.7** displays an illustration of multilayer well completion (in some of the developments, we make references to multiphase flow conditions, a topic which is discussed in much more depth in Chapter 9).

6.5.1 Explicit Treatment of Multilayer Wells. The explicit treatment indicates that the primary unknowns, such as pressure and saturation values, calculated in the previous timestep level, n, is employed to determine the well-productivity index of the individual layer of the current timestep level, $n + 1$.

Pressure-Specified Wells. For a pressure-specified well, where $p_{sf,ref} = p_{sf,sp}$, the wellbore pressure of each individual layer can be calculated by

$$p_{sf,k} = p_{sf,ref} + \int_{H_{ref}}^{H_k} \gamma_{wb} \, dH. \dots\dots\dots\dots\dots\dots\dots\dots\dots (6.39)$$

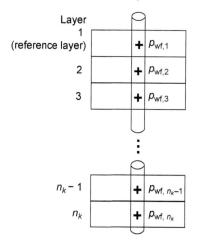

Fig. 6.7—A vertical well penetrating through several layers.

The integration term can be expressed using the average wellbore hydrostatic pressure gradient ($\overline{\gamma}_{wb}$, psi/ft):

$$\int_{H_{ref}}^{H_k} \gamma_{wb} \, dH = \overline{\gamma}_{wb} \left(H_k - H_{ref} \right). \dots\dots\dots\dots\dots\dots\dots\dots\dots\dots (6.40)$$

Note that $p_{sf,ref}$ and H_{ref} are the sandface pressure and depth of the reference layer, respectively. The term $\overline{\gamma}_{wb}$ is defined by

$$\overline{\gamma}_{wb} = \frac{1}{144} \frac{g}{g_c} \overline{\rho}_l. \dots\dots\dots\dots\dots\dots\dots\dots\dots\dots\dots\dots\dots\dots\dots (6.41)$$

In this discussion, the friction and inertial loss are assumed negligible compared to the hydrostatic gradient. For single-phase fluid, or multiphase flow when in phase l ($l = o, w,$ and g), the average liquid density $\overline{\rho}_l$ can be calculated using the formation volume factor (FVF), flow rates, and fluid densities at standard condition:

$$\overline{\gamma}_{wb} = \left(\frac{\rho_{osc} q_{osc} + \rho_{wsc} q_{wsc} + \dfrac{\rho_{gsc} q_{gsc}}{5.615}}{B_o q_{osc} + B_w q_{wsc} + B_g q_{fgsc}} \right). \dots\dots\dots\dots\dots\dots\dots\dots (6.42)$$

Note that the flow rate used in Eq. 6.42 is determined from the previous timestep level, and q_{fgsc} is the gas flow rate of the free-gas phase in three-phase flow, which is

$$q_{fgsc} = q_{gsc} - R_s q_{osc}. \dots\dots\dots\dots\dots\dots\dots\dots\dots\dots\dots\dots\dots\dots\dots (6.43)$$

It is important to emphasize that the FVF values must be evaluated at the average wellbore pressure condition,

$$B_l = B_l \left(\overline{p}_{wf} \right), \dots\dots\dots\dots\dots\dots\dots\dots\dots\dots\dots\dots\dots\dots\dots\dots (6.44)$$

where $l = o, w,$ and g and

$$\overline{p}_{wf} = \frac{p_{sf,ref} + p_{sf,n_k}}{2}, \dots\dots\dots\dots\dots\dots\dots\dots\dots\dots\dots\dots\dots\dots (6.45)$$

and $p_{sf,nk}$ is the sandface pressure at the lower-most perforated layer.

The following algorithm steps are used to estimate the average wellbore hydrostatic gradient:

1. Start the algorithm by assuming $\overline{p}_{wf} = p_{sf,sp}$.
2. Evaluate the FVF, B_l ($l = o, w,$ or g) at $\overline{p}_{wf}$.
3. Calculate $\overline{\gamma}_{wb}$ using Eq. 6.42.
4. Calculate p_{sf1} through p_{sf,n_k} using Eq. 6.39.
5. Re-evaluate $\overline{p}_{wf}$ from Eq. 6.45.
6. Check the convergence on $\overline{p}_{wf}$ and go back to Step 2 for iteration if the convergence criterion is not satisfied.

When the variable pressure gradient is considered for each layer, the algorithm can be implemented with the following external loop:

1. Start from Layer 2, which is the first layer beneath the reference layer (recall that $p_{wf1} = p_{sf, sp}$), and use the algorithm to go through the iteration from the reference layer (Layer 1) to Layer 2. Note that the average wellbore pressure gradient of Layer 2 must be calculated with the flow rate contributed from Layer 2 using Eq. 6.42, and the average wellbore pressure will be the average between the sandface pressure of Layers 1 and 2.
2. Once a convergence is achieved at Layer 2, repeat the same procedure for Layers 2 and 3.
3. Continue applying the loop for all completed layers.

Once the sandface pressure of all the layers is calculated, the flow rate, $q_{sc,k}$, can be calculated using Eq. 6.38 and substituted into the finite-difference equation.

Rate-Specified Wells. When the production or injection rate of the multilayer well is specified ($q_{sc} = q_{sc,sp}$), the calculation is even more complex, because it cannot directly be used as the sink or source term in the finite-difference equation. One must determine how to distribute the specified flow rate to the perforated layers. In this discussion, we introduce two methods: the potential method and the productivity-index-weighted method.

Potential Method. The idea behind the potential method is to distribute the specified flow rate to each perforated layer by means of dividing Eq. 6.36 by Eq. 6.38:

$$q_{sc,k} = \frac{\Omega_k \left(p_k - p_{sf,k} \right)}{\sum_k^{n_k} \Omega_k \left(p_k - p_{sf,k} \right)} q_{sc,sp} . \quad\dots\dots\dots\dots \text{(6.46)}$$

In the explicit treatment, the productivity indices (Ω_k) and block pressure (p_k) are calculated using the properties of the previous timestep level (n). But the sandface pressure of each layer ($p_{sf,k}$) still needs to be determined. Substituting Eq. 6.39 into Eq. 6.38, the expression of reference sandface pressure can be obtained:

$$p_{sf,ref} = \frac{\sum_k^{n_k} \left\{ \Omega_k \left[p_k - \overline{\gamma}_{wb} \left(H_k - H_{ref} \right) \right] \right\} + q_{sc,sp}}{\sum_k^{n_k} \Omega_k} . \quad\dots\dots\dots\dots \text{(6.47)}$$

Eq. 6.47 can be employed by an iterative algorithm to find out the sandface pressure values:

1. Assume a value of $\overline{\gamma}_{wb}$ to initialize the iteration; for instance, the $\overline{\gamma}_{wb}$ value from the previous timestep can be used.
2. Calculate $p_{sf,ref}$ from Eq. 6.47.
3. Calculate the sandface pressure of all layers using Eq. 6.39.
4. Calculate the average wellbore pressure using Eq. 6.45.
5. Update the pressure gradient using Eq. 6.42.
6. Check the convergence on the pressure gradient, if the convergence criterion is not satisfactory, go back to Step 1.

Note that a similar external loop, described in the treatment of the rate-specified well, can be employed here to consider the variable wellbore pressure gradient of different layers.

Productivity-Index-Weighted Method. This approach assumes that the pressure drawdown and buildup are constant for the production and injection wells, respectively. Therefore, Eq. 6.46 can be simplified as

$$q_{sc,k} = \frac{\Omega_k}{\sum_k^{n_k} \Omega_k} q_{sc,sp} . \quad\dots\dots\dots\dots\dots\dots\dots\dots \text{(6.48)}$$

Eq. 6.48 provides an effective way to allocate the specified flow rate to each perforated layer. Again, in the explicit treatment, the productivity indices are calculated using properties evaluated from the previous timestep.

6.5.2 Implicit Treatment of Multilayer Wells.
The explicit treatment of the multilayer completions may generate unphysical solutions, especially when small changes to the primary unknowns (pressure and saturation values) introduce large variations into the results. For example, when water breakthrough occurs in water-flooding projects, the implicit treatment aims at improving the stability of the finite-difference formulation, and the parameters involved in the well-injection/production-rate calculation are calculated at the current timestep ($n+1$) and iteration level ($v+1$). The flow rate at layer k can be expanded using Taylor-series expansion between the previous (v) and current iteration level ($v+1$):

$$q_{sc,k}^{(v+1)^{n+1}} = q_{sc,k}^{(v)^{n+1}} + \frac{\partial q_{sc,k}}{\partial S_{w,k}} \bigg|^{(v)^{n+1}} \left[S_{w,k}^{(v+1)^{n+1}} - S_{w,k}^{(v)^{n+1}} \right] + \frac{\partial q_{sc,k}}{\partial S_{g,k}} \bigg|^{(v)^{n+1}} \left[S_{g,k}^{(v+1)^{n+1}} - S_{g,k}^{(v)^{n+1}} \right] + \frac{\partial q_{sc,k}}{\partial p_{o,k}} \bigg|^{(v)^{n+1}}$$

$$\left[p_{o,k}^{(v+1)^{n+1}} - p_{o,k}^{(v)^{n+1}} \right] + \frac{\partial q_{sc,k}}{\partial p_{sf,ref}} \bigg|^{(v)^{n+1}} \left[p_{sf,ref}^{(v+1)^{n+1}} - p_{sf,ref}^{(v)^{n+1}} \right] . \quad\dots\dots\dots\dots \text{(6.49)}$$

Eq. 6.49 considers the oil pressure (p_o), water saturation (S_w), and gas saturation (S_g) as primary unknowns. The partial derivatives, with the respective to the unknowns, are expressed as

$$\frac{\partial q_{sc,k}}{\partial S_{w,k}} = -\frac{\partial \Omega_k}{\partial S_{w,k}}\left(p_{o,k} - p_{sf,k}\right), \dots\dots\dots\dots\dots\dots\dots\dots\dots\dots\dots\dots\dots\dots\dots\dots\dots\dots\dots\dots\dots\dots\dots\dots\dots (6.50)$$

$$\frac{\partial q_{sc,k}}{\partial S_{g,k}} = -\frac{\partial \Omega_k}{\partial S_{g,k}}\left(p_{o,k} - p_{sf,k}\right), \dots\dots\dots\dots\dots\dots\dots\dots\dots\dots\dots\dots\dots\dots\dots\dots\dots\dots\dots\dots\dots\dots\dots\dots\dots (6.51)$$

$$\frac{\partial q_{sc,k}}{\partial p_{o,k}} = -\frac{\partial \Omega_k}{\partial p_{o,k}}\left(p_{o,k} - p_{sf,k}\right) - \Omega_k, \dots\dots\dots\dots\dots\dots\dots\dots\dots\dots\dots\dots\dots\dots\dots\dots\dots\dots\dots\dots\dots\dots (6.52)$$

$$\frac{\partial q_{sc,k}}{\partial p_{sf,\text{ref}}} = \Omega_k \frac{\partial p_{sf,k}}{\partial p_{sf,\text{ref}}}. \dots\dots\dots\dots\dots\dots\dots\dots\dots\dots\dots\dots\dots\dots\dots\dots\dots\dots\dots\dots\dots\dots\dots\dots\dots\dots\dots\dots (6.53)$$

And, from Eq. 6.24 one obtains

$$\frac{\partial p_{sf,k}}{\partial p_{sf,\text{ref}}} = 1 + \frac{\partial \overline{\gamma}_{wb}}{\partial p_{sf,\text{ref}}}\left(H_k - H_{\text{ref}}\right). \dots\dots\dots\dots\dots\dots\dots\dots\dots\dots\dots\dots\dots\dots\dots\dots\dots\dots\dots\dots (6.54)$$

For single-phase problems, all derivatives with respect to saturation become zero. Eq. 6.49 contains three primary unknowns: $q_{sc,k}^{n+1}$, $p_{o,k}^{n+1}$, and $p_{sf,\text{ref}}^{n+1}$. To solve for these unknowns, we have Eq. 6.54 and one material-balance equation.

For a three-phase problem, seven unknowns are considered in Eq. 6.49: $q_{osc,k}^{n+1}$, $q_{wsc,k}^{n+1}$, $q_{gsc,k}^{n+1}$, $S_{w,k}^{n+1}$, $S_{g,k}^{n+1}$, $p_{o,k}^{n+1}$, and $p_{sf,\text{ref}}^{n+1}$. To solve these unknowns, Eq. 6.39 is written for gas, water, and oil phases and three material-balance equations. All the unknowns in single and multiphase problems are at the new iteration level, $v+1$. Therefore, in the fully implicit approach, we must eliminate one primary unknown by applying the well specifications to solve the system of equations.

Pressure-Specified Wells. For pressure-specified wells, because the sandface pressure at the reference layer is specified,

$$\overset{(v+1)}{p_{sf,\text{ref}}^{n+1}} = \overset{(v)}{p_{sf,\text{ref}}^{n+1}} = p_{sf,sp}. \dots\dots\dots\dots\dots\dots\dots\dots\dots\dots\dots\dots\dots\dots\dots\dots\dots\dots\dots\dots\dots\dots\dots\dots\dots\dots (6.55)$$

Substituting Eq. 6.40 into Eq. 6.34 yields

$$\overset{(v+1)}{q_{sc,k}^{n+1}} = \overset{(v)}{q_{sc,k}^{n+1}} + \left.\frac{\partial q_{sc,k}}{\partial S_{w,k}}\right|^{\overset{(v)}{n+1}}\left[\overset{(v+1)}{S_{w,k}^{n+1}} - \overset{(v)}{S_{w,k}^{n+1}}\right] + \left.\frac{\partial q_{sc,k}}{\partial S_{g,k}}\right|^{\overset{(v)}{n+1}}\left[\overset{(v+1)}{S_{g,k}^{n+1}} - \overset{(v)}{S_{g,k}^{n+1}}\right] + \left.\frac{\partial q_{sc,k}}{\partial p_{o,k}}\right|^{\overset{(v)}{n+1}}\left[\overset{(v+1)}{p_{o,k}^{n+1}} - \overset{(v)}{p_{o,k}^{n+1}}\right]. \dots\dots\dots (6.56)$$

In this way, we successfully eliminate terms $\overset{(v+1)}{p_{sf,\text{ref}}^{n+1}}$ and $\overset{(v)}{p_{sf,\text{ref}}^{n+1}}$ from Eq. 6.49. To explain the treatment more effectively, we introduce a single-phase $2 \times 2 \times 2$ system with a vertical well completed in two layers, shown in **Fig. 6.8**. The coefficient matrix of the described system, with the pressure specification at the well, is given in **Fig. 6.9**.

Rate-Specified Wells. For rate-specified wells, the total production or injection rates are specified. To implement the specification, one must calculate the summation of Eq. 6.49 when it is written for each perforated layer, as shown:

$$\sum_k^{n_k} \overset{(v+1)}{q_{sc,k}^{n+1}} = \sum_k^{n_k} \overset{(v)}{q_{sc,k}^{n+1}} + \sum_k^{n_k}\left\{\left.\frac{\partial q_{sc,k}}{\partial S_{w,k}}\right|^{\overset{(v)}{n+1}}\left[\overset{(v+1)}{S_{w,k}^{n+1}} - \overset{(v)}{S_{w,k}^{n+1}}\right] + \left.\frac{\partial q_{sc,k}}{\partial S_{g,k}}\right|^{\overset{(v)}{n+1}}\left[\overset{(v+1)}{S_{g,k}^{n+1}} - \overset{(v)}{S_{g,k}^{n+1}}\right] + \right.$$

$$\left. \left.\frac{\partial q_{sc,k}}{\partial p_{o,k}}\right|^{\overset{(v)}{n+1}}\left[\overset{(v+1)}{p_{o,k}^{n+1}} - \overset{(v)}{p_{o,k}^{n+1}}\right] + \left.\frac{\partial q_{sc,k}}{\partial p_{sf,\text{ref}}}\right|^{\overset{(v)}{n+1}}\left[\overset{(v+1)}{p_{sf,\text{ref}}^{n+1}} - \overset{(v)}{p_{sf,\text{ref}}^{n+1}}\right]\right\}. \dots\dots\dots\dots\dots\dots (6.57)$$

In Eq. 6.57, because the well production/injection rate is specified,

$$\sum_k^{n_k} \overset{(v+1)}{q_{sc,k}^{n+1}} = \sum_k^{n_k} \overset{(v)}{q_{sc,k}^{n+1}} = q_{sc,sp}. \dots\dots\dots\dots\dots\dots\dots\dots\dots\dots\dots\dots\dots\dots\dots\dots\dots\dots\dots\dots\dots\dots\dots\dots\dots (6.58)$$

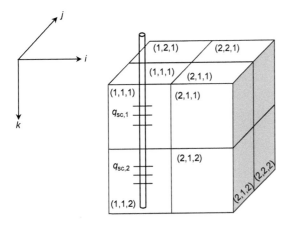

Fig. 6.8—Schematic representation of a two-layer reservoir system with a multilayer vertical-well completion.

	$p_{1,1,1}$	$p_{2,1,1}$	$p_{1,2,1}$	$p_{2,2,1}$	$p_{1,1,2}$	$p_{2,1,2}$	$p_{1,2,2}$	$p_{2,2,2}$	$q_{sc,1}$	$q_{sc,2}$
	X	X	X	O	X	O	O	O	X	O
	X	X	O	X	O	X	O	O	O	O
	X	O	X	X	O	O	X	O	O	O
	O	X	X	X	O	O	O	X	O	O
	X	O	O	O	X	X	X	O	O	X
	O	X	O	O	X	X	O	X	O	O
	O	O	X	O	X	O	X	X	O	O
	O	O	O	X	O	X	X	X	O	O
	X	O	O	O	O	O	O	O	X	O
	O	O	O	O	X	O	O	O	O	X

(Top headers: Unknown pressure | Unknown flow rates; left-side row groups: Material-balance equations, Well equations)

Fig. 6.9—Coefficient matrix of the 2 × 2 × 2 reservoir well system of Fig. 6.8 for a pressure-specified well (X represents nonzero entries).

Substituting Eq. 6.58 into Eq. 6.57, one obtains

$$\sum_{k}^{n_k}\left\{\left.\frac{\partial q_{sc,k}}{\partial S_{w,k}}\right|_{n+1}^{(v)}\left[\overset{(v+1)}{S_{w,k}^{n+1}}-\overset{(v)}{S_{w,k}^{n+1}}\right]+\left.\frac{\partial q_{sc,k}}{\partial S_{g,k}}\right|_{n+1}^{(v)}\left[\overset{(v+1)}{S_{g,k}^{n+1}}-\overset{(v)}{S_{g,k}^{n+1}}\right]+\left.\frac{\partial q_{sc,k}}{\partial p_{o,k}}\right|_{n+1}^{(v)}\right.$$

$$\left.\left[\overset{(v+1)}{p_{o,k}^{n+1}}-\overset{(v)}{p_{o,k}^{n+1}}\right]+\left.\frac{\partial q_{sc,k}}{\partial p_{sf,\text{ref}}}\right|_{n+1}^{(v)}\left[\overset{(v+1)}{p_{sf,\text{ref}}^{n+1}}-\overset{(v)}{p_{sf,\text{ref}}^{n+1}}\right]\right\}=0. \quad\quad (6.59)$$

The sandface pressure at the current timestep and iteration level $\left(\overset{(v+1)}{p_{wf,\text{ref}}^{n+1}}\right)$ can be depicted as

$$\overset{(v+1)}{p_{sf,\text{ref}}^{n+1}}=\overset{(v)}{p_{sf,\text{ref}}^{n+1}}-\frac{\sum_{k}^{n_k}\left\{\left.\frac{\partial q_{sc,k}}{\partial S_{w,k}}\right|_{n+1}^{(v)}\left[\overset{(v+1)}{S_{w,k}^{n+1}}-\overset{(v)}{S_{w,k}^{n+1}}\right]+\left.\frac{\partial q_{sc,k}}{\partial S_{g,k}}\right|_{n+1}^{(v)}\left[\overset{(v+1)}{S_{g,k}^{n+1}}-\overset{(v)}{S_{g,k}^{n+1}}\right]+\left.\frac{\partial q_{sc,k}}{\partial p_{v,k}}\right|_{n+1}^{(v)}\left[\overset{(v+1)}{p_{o,k}^{n+1}}-\overset{(v)}{p_{o,k}^{n+1}}\right]\right\}}{\left.\frac{\partial q_{sc,k}}{\partial p_{sf,\text{ref}}}\right|_{n+1}^{(v)}}. \quad\quad (6.60)$$

Eq. 6.60 is considered the additional equation to solve for $\overset{(v+1)}{p_{sf,\text{ref}}^{n+1}}$, illustrated by the coefficient matrix shown in **Fig. 6.10**.

Let us now summarize the coupling of the well equation with the reservoir flow equation. As previously explained through the definition of $C_{i,j,k}$ and $Q_{i,j,k}$ coefficients, such a coupling is achieved through the $q_{sc_{i,j,k}}$ term in the reservoir flow equation written for the wellblock in the finite-difference form. If flow rate is specified, then the value of $q_{sc_{i,j,k}}$ is known, and it can be moved to the right-hand side of the equation, which changes the definition of the $Q_{i,j,k}$. However, if the wellbore constraint is introduced by

	$p_{1,1,1}$	$p_{2,1,1}$	$p_{1,2,1}$	$p_{2,2,1}$	$p_{1,1,2}$	$p_{2,1,2}$	$p_{1,2,2}$	$p_{2,2,2}$	$p_{sf,ref}$
	X	X	X	O	X	O	O	O	X
	X	X	O	X	O	X	O	O	O
	X	O	X	X	O	O	X	O	O
	O	X	X	X	O	O	O	X	O
	X	O	O	O	X	X	X	O	X
	O	X	O	O	X	X	O	X	O
	O	O	X	O	X	O	X	X	O
	O	O	O	X	O	X	X	X	O
	X	O	O	O	O	O	O	O	X

(Top headers: Unknown pressure | Unknown sandface pressure; left-side row groups: Material-balance equations, Well equations)

Fig. 6.10—Coefficient matrix of the 2 × 2 × 2 reservoir well system of Fig. 6.8 for a rate-specified well (X represents nonzero entries).

specifying the $p_{sf_{i,j,k}}$, it is necessary to substitute the well equation, $q_{sc_{i,j,k}}=-\Omega_{i,j,k}\left(p_{i,j,k}-p_{i,j,k}\right)$, with the $q_{sc_{i,j,k}}$ term of the macroscopic flow equation. The $-\Omega_{i,j,k}p_{i,j,k}$ product remains on the left-hand side, because $p_{i,j,k}$ is one of the unknowns of the problem. This changes the definition of $C_{i,j,k}$. At the same time, the $\Omega_{i,j,k}p_{sf_{i,j,k}}$ product is carried to the right-hand side, changing the definition of the $Q_{i,j,k}$ term.

6.6 Problems and Solutions

Problem 6.6.1 (Ertekin et al. 2001)

Compare the flow rates generated by the models of van Poollen et al. (1968) and Peaceman (1983) for the following data: $\Delta x = \Delta y = 600$ ft; $h = 40$ ft; $k_x = k_y = 100$ md; $p_{wf} = 350$ psia; $p_0 = \overline{p} = 1{,}600$ psia; $\mu = 2$ cp; $B = 1.1$ res bbl/STB; $s = 0$; $r_w = 0.25$ ft.

[Note that in the van Poollen et al. (1968) model, $r_{eq} = \sqrt{\dfrac{\Delta x \Delta y}{\pi}}$].

Solution to Problem 6.6.1

The van Poollen model for $\Delta x = \Delta y$ gives $r_{eq} = 0.5642\Delta x = 338.52$ ft, 335.52 ft,

$$\Omega = \frac{2\pi kh}{\mu B\left[\ln\left(\dfrac{r_{eq}}{r_w}\right) + s - 0.5\right]} = \frac{2\pi 100\left(1.127\times 10^{-3}\right)(40)}{(1.1)(2)\left[\ln\left(\dfrac{338.52}{0.25}\right) - 0.5\right]} = 1.9185 \text{ STB/D-psi.} \quad \dots\dots\dots\dots\dots\dots\dots (6.61)$$

On the other hand, the Peaceman model states: $r_{eq} = 0.198\Delta x = 118.8$ ft,

$$\Omega = \frac{2\pi kh}{\mu B\left[\ln\left(\dfrac{r_{eq}}{r_w}\right) + s\right]} = \frac{2\pi 100\left(1.127\times 10^{-3}\right)(40)}{(1.1)(2)\left[\ln\left(\dfrac{118.8}{0.25}\right)\right]} = 2.088 \text{ STB/D-psi.} \quad \dots\dots\dots\dots\dots\dots\dots\dots (6.62)$$

Problem 6.6.2 (Ertekin et al. 2001)

Repeat Problem 6.6.1 for $\Delta x = \Delta y = 200$, 400, 800, 1000, and 1200 ft, and state your observation. Repeat the calculation with various $\Delta x = \Delta y$ values.

Solution to Problem 6.6.2

Table 6.3 summarizes the results of Problem 6.6.2 in a tabulated form. As seen in the last column of the table, the dimensions of the square block become larger, the radial flow regime develops in a more-pronounced manner, and the radial flow assumptions employed by the van Poollen model in a square block becomes increasingly more accurate. The differences in well-productivity indices calculated using the van Poollen model and the Peaceman model is a result of the different formulations employed to compute the equivalent wellblock radius. Note that PI calculated using the van Poollen model is always lower than that of the Peaceman model with the same Δx and Δy values.

$\Delta x = \Delta y$	PI, STB/D–psi (Peaceman Model)	PI, STB/D–psi (van Poollen Model)	% Difference
200	2.541882	2.294058	9.75%
400	2.235903	2.041874	8.68%
600	2.088818	1.918505	8.15%
800	1.995673	1.839643	7.82%
1,000	1.928953	1.7828	7.58%
1,200	1.877663	1.738899	7.39%

Table 6.3—Tabulated results for Problem 6.6.2.

Problem 6.6.3 (Ertekin et al. 2001)

Compare and contrast the equivalent wellblock radii calculated with the Peaceman model and the van Poollen model for different aspect ratios of $\Delta y/\Delta x$.

Solution to Problem 6.6.3

Using $\Delta x = 100$ ft for the calculation, **Table 6.4** lists the calculated r_{eq} values using various aspect ratios listed in the table and displayed in **Fig. 6.11.** The r_{eq} calculated using the van Poollen model is larger than that of the Peaceman model, until the aspect ratio ($\Delta y/\Delta x$) becomes extremely large (to approximately 16).

Δy	Aspect Ratio ($\Delta y/\Delta x$)	Peaceman Model	van Poollen Model	% Difference
10	0.1	14.0698	17.8412	26.80%
50	0.5	15.6525	39.8942	154.87%
100	1	19.7990	56.4190	184.96%
200	2	31.3050	79.7885	154.87%
300	3	44.2719	97.7205	120.73%
400	4	57.7235	112.8379	95.48%
500	5	71.3863	126.1566	76.72%
1,000	10	140.6983	178.4124	26.80%
5,000	50	700.14	398.94	43.02%

Table 6.4—Tabulated results for Problem 6.6.3.

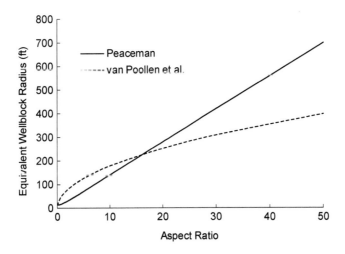

Fig. 6.11—Variation of equivalent wellblock radius with aspect ratios of a rectangular block in Peaceman and van Poollen models (Solution to Problem 6.6.3).

Problem 6.6.4 (Ertekin et al. 2001)

Calculate the equivalent wellblock radius, r_{eq}, of a well in a corner block in the block-centered grid shown in **Fig. 6.12.1** using the Abou-Kassem and Aziz (1985) model given:

$$r_{eq} = \left\{ \exp(-2\pi f) \prod_i \left[r_{i,1}^{T_i} \prod_j \left(\frac{r_{i,j}}{a_j} \right)^{T_i} \right] \right\}^b .$$

Solution to Problem 6.6.4

In this case, the wellblock is a boundary block surrounded by two no-flow boundaries. Therefore, there are three image wells, which are marked as nodes 4, 5, and 6 in **Fig. 6.12.2.**

To use the Abou-Kassem and Aziz model, one needs to calculate the distances between the well marked in Node 1 and reflection Nodes 4, 5, and 6: $a_4 = \Delta x + 2x_w$, $a_6 = \Delta y + 2y_w$, and $a_5 = \sqrt{\left(\Delta x + 2x_w\right)^2 + \left(\Delta y + 2y_w\right)^2}$.

The distances between the surrounding nodes and the actual well are $r_{2,1} = \sqrt{\left(\Delta x - x_w\right)^2 + \left(y_w\right)^2}$ and $r_{3,1} = \sqrt{\left(\Delta y - y_w\right)^2 + \left(x_w\right)^2}$.

The distances between the surrounding nodes and all image wells are

$$r_{2,4} = \sqrt{\left(2\Delta x + x_w\right)^2 + \left(y_w\right)^2}, \ldots\ldots\ldots (6.63)$$

$$r_{2,5} = \sqrt{\left(2\Delta x + x_w\right)^2 + \left(\Delta y + 2y_w\right)^2}, \ldots\ldots (6.64)$$

$$r_{2,6} = \sqrt{\left(\Delta x - x_w\right)^2 + \left(\Delta y + y_w\right)^2}, \ldots\ldots\ldots (6.65)$$

$$r_{3,4} = \sqrt{\left(\Delta x + x_w\right)^2 + \left(\Delta y - y_w\right)^2}, \ldots\ldots\ldots (6.66)$$

$$r_{3,5} = \sqrt{\left(2\Delta y + y_w\right)^2 + \left(\Delta x + x_w\right)^2}, \ldots\ldots (6.67)$$

$$r_{3,6} = \sqrt{\left(x_w\right)^2 + \left(2\Delta y + y_w\right)^2}. \ldots\ldots\ldots\ldots (6.68)$$

In this case, $f = \frac{1}{4}$, $\sum_i T_i = \dfrac{\Delta x}{\Delta y} + \dfrac{\Delta y}{\Delta x} = \dfrac{\Delta x^2 + \Delta y^2}{\Delta x \Delta y}$ and $b = \dfrac{1}{\sum_i T_i} = \dfrac{\Delta x \Delta y}{\Delta x^2 + \Delta y^2}$.

In the Abou-Kassem and Aziz model,

$$r_{eq} = \left\{ \exp\left(-2\pi f\right) \prod_i \left[r_{i,1}^T \prod_j \left(\frac{r_{i,j}}{a_j} \right)^{T_i} \right] \right\}^b \ldots (6.69)$$

For the product term,

$$\prod_i \left[r_{i,j}^{T_i} \prod_j \left(\frac{r_{i,j}}{a_j} \right)^{T_i} \right] = \left\{ r_{2,1}^{T_1} \left[\left(\frac{r_{2,4}}{a_4} \right) \left(\frac{r_{2,5}}{a_5} \right) \left(\frac{r_{2,6}}{a_6} \right) \right]^{T_1} \right\}$$
$$\left\{ r_{3,1}^{T_2} \left[\left(\frac{r_{3,4}}{a_4} \right) \left(\frac{r_{3,5}}{a_5} \right) \left(\frac{r_{3,6}}{a_6} \right) \right]^{T_2} \right\}, \ldots\ldots (6.70)$$

$$r_{eq} = \left[\exp\left(-\pi/2\right) \left\{ r_{2,1}^{T_1} \left[\left(\frac{r_{2,4}}{a_4} \right) \left(\frac{r_{2,5}}{a_5} \right) \left(\frac{r_{2,6}}{a_6} \right) \right] \right\} \left\{ r_{3,1}^{T_2} \left[\left(\frac{r_{3,4}}{a_4} \right) \left(\frac{r_{3,5}}{a_5} \right) \left(\frac{r_{3,6}}{a_6} \right) \right]^{T_2} \right\} \right]^{\frac{\Delta x \Delta y}{\Delta x^2 + \Delta y^2}} \ldots\ldots\ldots\ldots\ldots (6.71)$$

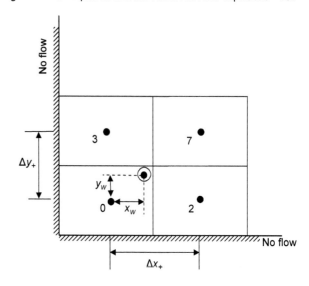

Fig 6.12.1—Schematic representation of Problem 6.6.4.

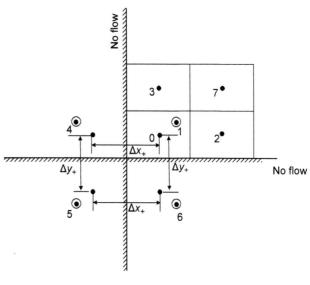

Fig. 6.12.2—Three image well locations in Problem 6.6.4.

Problem 6.6.5 (Ertekin et al. 2001)

Calculate the equivalent wellblock radius, r_{eq}, using the Abou-Kassem and Aziz model for a well in a corner block in the point-distributed grid shown in **Fig. 6.13.1.**

Solution to Problem 6.6.5

In this case, the wellblock is a boundary block surrounded by two no-flow boundaries. Therefore, there are three image wells, which are marked Nodes 4, 5, and 6 in **Fig. 6.13.2.** To use the Abou-Kassem and Aziz model, one needs to calculate the distances between the well, as marked in Node 1 and reflection Nodes 4, 5, and 6: $a_4 = 2x_w$, $a_6 = 2y_w$, and $a_5 = \sqrt{\left(2x_w\right)^2 + \left(2y_w\right)^2}$.

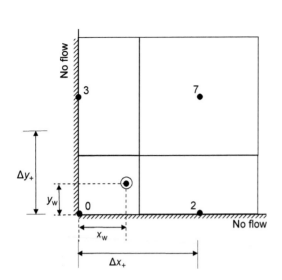

Fig. 6.13.1—Schematic representation of Problem 6.6.5.

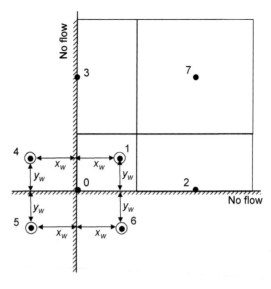

Fig. 6.13.2—Three image well locations (Nodes 4, 5, and 6) of Problem 6.6.5.

The distances between the surrounding nodes and the actual well are $r_{2,1} = \sqrt{(\Delta x - x_w)^2 + (y_w)^2}$ and $r_{3,1} = \sqrt{(\Delta y - y_w)^2 + (x_w)^2}$.

The distances between the surrounding nodes and all image wells are

$$r_{2,4} = \sqrt{(\Delta x + x_w)^2 + (y_w)^2}, \dotfill (6.72)$$

$$r_{2,5} = \sqrt{(\Delta x + x_w)^2 + (y_w)^2}, \dotfill (6.73)$$

$$r_{2,6} = \sqrt{(\Delta x - x_w)^2 + (y_w)^2}, \dotfill (6.74)$$

$$r_{3,4} = \sqrt{(x_w)^2 + (\Delta y - y_w)^2}, \dotfill (6.75)$$

$$r_{3,5} = \sqrt{(\Delta y + y_w)^2 + (x_w)^2}, \dotfill (6.76)$$

$$r_{3,6} = \sqrt{(x_w)^2 + (\Delta y + y_w)^2}. \dotfill (6.77)$$

In this case, $f = 1/4$, and $\sum_i T_i = \dfrac{\Delta x}{\Delta y} + \dfrac{\Delta y}{\Delta x} = \dfrac{\Delta x^2 + \Delta y^2}{\Delta x \Delta y}$ and $b = \dfrac{1}{\sum_i T_i} = \dfrac{\Delta x \Delta y}{\Delta x^2 + \Delta y^2}$.

In the Abou-Kassem and Aziz model,

$$r_{eq} = \left\{ \exp(-2\pi f) \prod_i \left[r_{i,j}^{T_i} \prod_j \left(\frac{r_{i,j}}{a_j} \right)^{T_i} \right] \right\}^b. \dotfill (6.78)$$

For the product term,

$$\prod_i \left[r_{i,j}^{T_i} \prod_j \left(\frac{r_{i,j}}{a_j} \right)^{T_i} \right] = \left\{ r_{2,1}^{T_1} \left[\left(\frac{r_{2,4}}{a_4} \right) \left(\frac{r_{2,5}}{a_5} \right) \left(\frac{r_{2,6}}{a_6} \right) \right]^{T_1} \right\} \left\{ r_{3,1}^{T_1} \left[\left(\frac{r_{3,4}}{a_4} \right) \left(\frac{r_{3,5}}{a_5} \right) \left(\frac{r_{3,6}}{a_6} \right) \right]^{T2} \right\}, \dotfill (6.79)$$

$$r_{eq} = \left[\exp(-\pi/2) \left\{ r_{2,1}^{T_1} \left[\left(\frac{r_{2,4}}{a_4} \right) \left(\frac{r_{2,5}}{a_5} \right) \left(\frac{r_{2,6}}{a_6} \right) \right]^{T_1} \right\} r_{3,1}^{T_2} \left[\left(\frac{r_{3,4}}{a_4} \right) \left(\frac{r_{3,5}}{a_5} \right) \left(\frac{r_{3,6}}{a_6} \right) \right]^{T_2} \right]. \dotfill (6.80)$$

Problem 6.6.6 (Ertekin et al. 2001)

Consider the multiblock well completion shown in **Fig. 6.14.** At the middle of the top layer, flowing sandface pressure is specified as 1,643 psia.

1. Assuming incompressible fluid behavior, calculate the flowing sandface pressures at the midpoints of layers 2 through 4 (use a fluid density of 53 lbm/ft^3).

2. For the flowing sandface pressure calculated in Part 1, a numerical simulator calculates the wellblock pressures for Blocks 1 through 4 as 2,400; 2,443; 2,479; and 2,503 psia, respectively. First, calculate the PI for each wellblock and calculate the total flow rate out of the well. Use Peaceman's equivalent wellblock radius concept. Assume that $\mu = 1.3$ cp, $B = 1$ res bbl/STB, $s = 0$ for all wellblocks, and $r_w = 0.3$ ft.

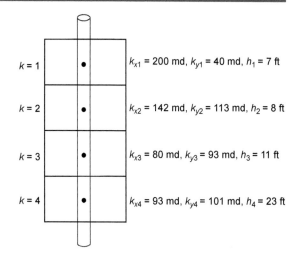

$k = 1$ $k_{x1} = 200$ md, $k_{y1} = 40$ md, $h_1 = 7$ ft

$k = 2$ $k_{x2} = 142$ md, $k_{y2} = 113$ md, $h_2 = 8$ ft

$k = 3$ $k_{x3} = 80$ md, $k_{y3} = 93$ md, $h_3 = 11$ ft

$k = 4$ $k_{x4} = 93$ md, $k_{y4} = 101$ md, $h_4 = 23$ ft

Fig. 6.14—Schematic representation of multiblock well completion described in Problem 6.6.6.

Solution to Problem 6.6.6

1. The sandface of an arbitrary layer n can be calculated using the hydrostatic-head relationship to the reference layer:

$$p_{wfn} = p_{wf,\text{ref}} + \gamma\left(h_n - h_{\text{ref}}\right), \dots\dots\dots\dots\dots\dots\dots\dots\dots\dots\dots\dots\dots\dots\dots (6.81)$$

where $\gamma = 52/144 = 0.368$ psi/ft. Thus, the sandface pressure of each layer can be calculated by way of Eqs. 6.82 through 6.84:

$$p_{wf,2} = 1,643 + 0.368\left(\frac{7+8}{2}\right) = 1,645.76 \text{ psi}, \dots\dots\dots\dots\dots\dots\dots\dots\dots\dots\dots\dots (6.82)$$

$$p_{wf,3} = 1,643 + 0.368\left(\frac{7}{2} + 8 + \frac{11}{2}\right) = 1,649.26 \text{ psi}, \dots\dots\dots\dots\dots\dots\dots\dots\dots\dots (6.83)$$

$$p_{wf,4} = 1,643 + 0.368\left(\frac{7}{2} + 8 + 11 + \frac{23}{2}\right) = 1,655.51 \text{ psi.} \dots\dots\dots\dots\dots\dots\dots\dots (6.84)$$

2. **Table 6.5** lists the calculated well PI and flow rate for each layer. The total production rate is

$$Q_{\text{total}} = q_1 + q_2 + q_3 + q_4 = 4,518.8 \text{ STB.}$$

Layer	Ω, STB/D–psi	Flow Rate, STB/D
1	0.683521	517.43
2	1.152931	919.16
3	1.09247	906.47
4	2.561175	2,175.74

Table 6.5—Calculated well PI values and flow rates for each layer (Problem 6.6.6, Part 2).

Problem 6.6.7 (Ertekin et al. 2001)

Investigate the effect of well length, L, and the ratio of well length to penetration distance, L/d, using Babu and Odeh's PI model.

Solution to Problem 6.6.7

For a given system, the well length impacts the skin factor as a result of the partial penetration effects. For the convenience of analysis, assume the reservoir properties listed in **Table 6.6**.

The horizontal wellbore is completed at the center of the x-z plane, which indicates that $X_0 = 400$ and $Z_0 = 20$ ft. **Table 6.7** summarizes the calculated results for various y_1 and y_2 values using the Babu and Odeh model.

Observe that as the length of the wellbore decreases, the s_p increases, while the PI (J) and production rate decrease accordingly. A sample calculation for $y_1 = 150$ and $y_2 = 650$, this case satisfies $\left(c / \sqrt{k_x} \right) \geq 0.75 \left(c / \sqrt{k_y} \right) \gg 0.75 \left(h / \sqrt{k_z} \right)$ and $L < d$:

$$P_{xyz} = \left(\frac{d}{L} - 1 \right) \left[\ln \left(\frac{h}{r_w} \right) + 0.25 \ln \left(\frac{k_x}{k_z} \right) - \ln \left(\sin \frac{\pi Z_0}{h} \right) - 1.84 \right] = 2.0045 \text{ psi,} \quad \text{...........................} \quad (6.85)$$

$$P'_{xy} = \frac{2d^2}{Lh} \sqrt{\frac{k_z}{k_y}} \left\{ F \left(\frac{L}{2d} \right) + 0.5 \left[F \left(\frac{4 y_{mid} + L}{2d} \right) \right] - F \left(\frac{4 y_{mid} - L}{2d} \right) \right\} = 2.4270 \text{ psi,} \quad \text{.....................} \quad (6.86)$$

$$s_p = P_{xyz} + P'_{xyz} = 4.4315 \text{ psi.} \quad \text{.....................................} \quad (6.87)$$

$k_x = k_y$ =	100 md
k_z =	10 md
Δx =	800 ft
Δy =	800 ft
h =	40 ft
μ =	1 cp
B =	1 res bbl/STB
r_w =	0.4 ft
ϕ =	18%
s_m =	-2
c =	0 psi^{-1}
p_{block} =	1,500 psi
p_{sf} =	1,000 psi

Table 6.6—Reservoir properties for Problem 6.6.7.

y_1	y_2	L, ft	L/d	s_p	s	J, STB/D–psi	q, STB/D
0	800	800	1	0.0000	-2.0000	38.50	19,251.38
50	750	700	0.875	1.1024	-0.8976	31.13	15,563.72
100	700	600	0.75	2.5318	0.5318	24.93	12,467.28
150	650	500	0.625	4.4315	2.4315	19.72	9,860.03
200	600	400	0.5	7.0762	5.0762	15.27	7,636.71
250	550	300	0.375	11.0732	9.0732	11.39	5,695.73
300	500	200	0.25	18.1345	16.1345	7.86	3,930.73
350	450	100	0.125	36.2176	34.2176	4.38	2,191.58

Table 6.7—Tabulated results for Problem 6.6.7.

Problem 6.6.8 (Ertekin et al. 2001)

What are the effects of well location and drainage volume on the productivity of a horizontal wellbore? Use the Babu and Odeh model in your investigation.

Solution to Problem 6.6.8

For calculations, the properties found in **Table 6.8** are used. To investigate the effects of wellbore location against the PI, various X_0 and Z_0 values are used to calculate the PI. **Fig. 6.15** displays the PI as a function of X_0 and Z_0 values. The tabulated values are listed in **Table 6.9**.

One can observe that the PI reaches a maximum value at the center of the formation, where $X_0 = 0.5 \Delta x$ and $Z_0 = 0.5h$. **Table 6.10** summarizes the PI for various V_b values when different Δx and Δy values are used.

X_0	Z_0	J, STB/D–psi
200	10	17.686
250	12	20.211
300	14	22.508
350	16	24.187
400	18	24.877
450	20	24.407
500	22	22.903
550	24	20.715
600	26	18.241
650	28	15.789
700	30	13.539
750	32	11.561

$k_x = k_y =$	100 md
$k_z =$	10 md
$\Delta x =$	800 ft
$\Delta y =$	800 ft
$h =$	40 ft
$\mu =$	1 cp
$B =$	1 res bbl/STB
$r_w =$	0.4 ft
$s_m =$	–2
$c =$	0 psi^{-1}

Table 6.8—Reservoir properties for Problem 6.6.8.

Table 6.9—PI as a function of X_0 and Z_0 values (Problem 6.6.8).

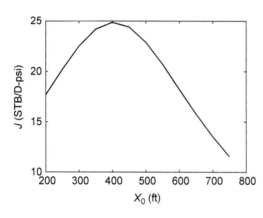
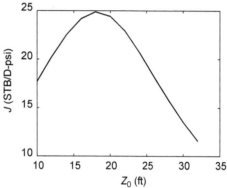

Fig. 6.15—Solution to Problem 6.6.8.

Δx	Δy	V_b, MM ft^3	J, STB/D–psi
800	800	25.6	24.935
1,000	1,000	40	19.482
1,200	1,200	57.6	16.250
1,400	1,400	78.4	13.964
1,600	1,600	102.4	12.403
1,800	1,800	129.6	11.270
2,000	2,000	160	10.409

Table 6.10—PI for various V_b values when different Δx and Δy values are used (Problem 6.6.8).

Problem 6.6.9 (Ertekin et al. 2001)

Compare and contrast the PIs when using the Babu and Odeh model and the Peaceman model for a given anisotropic-porous medium. Under what conditions do these two equations generate similar values and when do they deviate? Note that to make a meaningful comparison, you must write both equations for the same orientation of the horizontal well. Furthermore, assume that the horizontal well fully penetrates the wellblock for both equations.

Solution to Problem 6.6.9

Use the following reservoir properties to carry out the formulation, assume the well is completed perpendicular to the x-z plane and at the center of the formation. Use the reservoir properties in **Table 6.11** to complete the calculations.

Table 6.12 summarizes the PI calculated using the Babu and Odeh model and the Peaceman model. From the numerical experiments conducted, results appear to be more in agreement at low k_z/k_x ratios; however, they start to part from each other as the k_z/k_x ratio increases.

k_z/k_x	PI, STB/D–psi (Babu and Odeh Model)	PI, STB/D–psi (Peaceman Model)	% Difference
0.05	21.63	27.23	20.57%
0.1	26.93	36.45	26.12%
0.2	32.27	49.10	34.28%
0.3	35.26	58.62	39.85%
0.4	37.26	66.56	44.02%
0.5	38.73	73.51	47.31%
0.6	39.88	79.77	50.01%
0.7	40.81	85.50	52.27%
0.8	41.58	90.82	54.22%
0.9	42.24	95.81	55.91%
1	42.81	100.51	57.41%

Reservoir properties:

$k_x = k_y =$ 100 md

$\Delta x =$ 800 ft

$\Delta y =$ 800 ft

$h =$ 40 ft

$\mu =$ 1 cp

$B =$ 1 res bbl/STB

$r_w =$ 0.4 ft

$s_m =$ 0

$c =$ 0 psi^{-1}

Table 6.11—Reservoir properties for Problem 6.6.9.

Table 6.12—PI calculated using different models with various permeability aspect ratios (Problem 6.6.9).

Problem 6.6.10 (Ertekin et al. 2001)

Consider single-phase incompressible fluid flow taking place in the two identical 1D homogeneous reservoirs of **Fig. 6.16.** As indicated in the figure, the boundary conditions imposed on both systems are identical. The only difference is the values of skin encountered at the well locations. How would you compare the expected flow rates from these two wells? Explain.

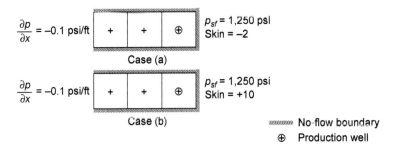

Case (a): $\dfrac{\partial p}{\partial x} = -0.1$ psi/ft, $p_{sf} = 1{,}250$ psi, Skin $= -2$

Case (b): $\dfrac{\partial p}{\partial x} = -0.1$ psi/ft, $p_{sf} = 1{,}250$ psi, Skin $= +10$

No-flow boundary

⊕ Production well

Fig. 6.16—Schematic representation of the two reservoirs of Problem 6.6.10.

Solution to Problem 6.6.10

For an incompressible system, the production rate of the well would be the same regardless of the skin factor, because the external boundary conditions are identical. However, Case (b) would yield lower block pressure to support the production rate as a result of the high positive skin factor (damage), as seen in the flow-rate expression:

$$q = -\frac{2\pi kh}{\mu B\left[\ln\left(\dfrac{r_{eq}}{r_w}\right) + s\right]}\left(p_{block} - p_{sf}\right). \qquad (6.88)$$

Problem 6.6.11 (Ertekin et al. 2001)

Consider the body-centered-grid representation of a 1D reservoir shown in **Fig. 6.17.** The reservoir has homogeneous property distribution and is 100% saturated with water (assume zero compressibility for water). Blocks 1 and 4 are kept at 3,500 and 3,000 psia, respectively, by strong edgewater drives, and the well in Block 2 ($r_w = 0.5$ ft) is produced at a flowing sandface pressure of $p_{wf} = 2{,}000$ psia. Calculate the pressure distribution in the reservoir and the production rate from the well. In the calculation, use the following gridblock information and reservoir rock and fluid properties: $h = 100$ ft; $\Delta x = 800$ ft; $A_x = 2{,}000$ ft^2; $k_x = 36$ md; $\phi = 17\%$, $s = -2$, $c = 0$ psi^{-1}; $B = 1$ res bbl/STB; and $\mu = 1.0$ cp. Ignore all gravitational forces.

Fig. 6.17—Schematic representation the reservoir of Problem 6.6.11.

Solution to Problem 6.6.11

The general form of the incompressible flow equation in one dimension is

$$\frac{\partial}{\partial x}\left(\frac{A_x k_x}{\mu B}\frac{\partial p}{\partial x}\right)\Delta x + q = 0. \quad\dotfill (6.89)$$

For a homogeneous system,

$$\frac{\partial^2 p}{\partial x^2} + \frac{q\mu B}{kV_b} = 0. \quad\dotfill (6.90)$$

The finite-difference analog of the previous equation is

$$p_{i-1} - 2p_i + p_{i+1} + \frac{q\mu B\Delta x}{\Delta y h k} = 0. \quad\dotfill (6.91)$$

Knowing that $A_x = 2,000$ ft^2 and $h = 100$ ft, $\Delta y = A_x/h = 20$ ft,

$$r_{eq} = 0.14\sqrt{(\Delta x)^2 + (\Delta y)^2} = 112.04 \text{ ft.}, \quad\dotfill (6.92)$$

$$\Omega = \frac{2\pi \bar{k} h}{\mu B\left(\ln\frac{r_{eq}}{r_w} + s\right)} = \frac{2\pi(36)(1.127\times 10^{-3})(100)}{(1)(1)\left(\ln\frac{112.04}{0.5} - 2\right)} = 7.47 \text{ STB/D-psi.} \quad\dotfill (6.93)$$

Knowing that pressures of Blocks 1 and 4 are kept at 3,500 and 3,000 psi, accordingly, the equation solving for pressure at Block 2 is

$$p_1 - 2p_2 + p_3 - 73.62(p_2 - 2,000) = 0, \quad\dotfill (6.94)$$

$$-75.62p_2 + p_3 = -150,745. \quad\dotfill (6.95)$$

Solving for Block 3 is

$$p_2 - 2p_3 + p_4 = 0, \quad\dotfill (6.96)$$

$$p_2 - 2p_3 = -3,000. \quad\dotfill (6.97)$$

Solving for unknown block pressures, $p_2 = 2,026.62$ psi and $p_3 = 2,513.31$ psi. Flow rate is

$$q = -\Omega\left(p_2 - p_{sf}\right) = -73.62(2,026.62 - 2,000) = -198.81 \text{ STB/D}.$$

Problem 6.6.12 (Ertekin et al. 2001)

Investigate the change in PI of a horizontal well as a function of the formation thickness. Use the Babu and Odeh model in your analysis.

Solution to Problem 6.6.12

In this problem, assume the rock and fluid properties found in **Table 6.13.** Assume that the horizontal well is completed at the center of the formation. Use various formation thicknesses to investigate the sensitivity of PI using the Babu and Odeh model.

As displayed in **Fig. 6.18** and in **Table 6.14,** PI increases as the formation thickness increases; however, the rate of the increments decreases (i.e., dJ/dh becomes progressively smaller).

$k_x = k_y =$	100 md
$k_z =$	10 md
$\Delta x =$	800 ft
$\Delta y =$	800 ft
$L_w =$	800 ft
$\mu =$	1 cp
$B =$	1 res bbl/STB
$r_w =$	0.4 ft
$S_m =$	0
$c =$	0 psi^{-1}

Table 6.13—Reservoir properties for Problem 6.6.12.

h, ft	J, STB/D–psi
10	12.755
20	23.052
30	31.061
40	37.259
50	42.076
60	45.846
70	48.822
80	51.189
90	53.084
100	54.613
200	60.422

Table 6.14—Tabulated results for Problem 6.6.12.

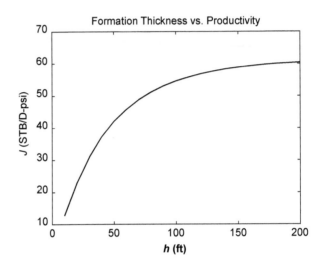

Fig. 6.18—Solution to Problem 6.6.12.

Problem 6.6.13

Consider the incompressible fluid flow through the 2D porous media shown in **Fig. 6.19.** The reservoir has homogeneous and anisotropic property distribution with no-flow boundaries. An injector and a producer are completed in Blocks 6 and 4, respectively. The injection rate is + 600 STB/D and the production well produces at a constant sandface pressure of 1,000 psi. Both wells are stimulated and have skin factors of $s = -2$. Blocks 13, 14, 15, and 16 are maintained at 3,000 psia. Ignore all gravitational forces. Use the reservoir properties listed in **Table 6.15** to do the following:

1. Write the differential equation that describes the problem, and identify the unknown blocks and unknowns of the problem.
2. Write the finite-difference equations for Blocks 4, 6, 7, and 12 using the calculated numerical values of the coefficients.

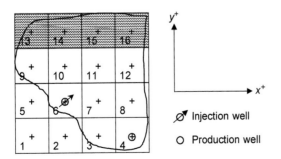

Fig. 6.19—Schematic representation of the reservoir of Problem 6.6.13.

$k_x =$	300 md
$k_y =$	200 md
$\Delta x =$	1,000 ft
$\Delta y =$	1,000 ft
$h =$	20 ft
$\mu =$	1.127 cp
$B =$	1 res bbl/STB
$r_w =$	0.2 ft
$s =$	–2

Table 6.15—Reservoir properties for Problem 6.6.13.

Solution to Problem 6.6.13

1. For a 2D incompressible system, the governing flow equation is

$$\frac{\partial}{\partial x}\left(\frac{A_x k_x}{\mu B}\frac{\partial p}{\partial x}\right)\Delta x+\frac{\partial}{\partial y}\left(\frac{A_y k_y}{\mu B}\frac{\partial p}{\partial y}\right)\Delta y+q=0. \dots\dots (6.98)$$

The reservoir properties display homogeneous distribution with anisotropic permeability values:

$$k_x\frac{\partial^2 p}{\partial x^2}+k_y\frac{\partial^2 p}{\partial y^2}+\frac{q\mu B}{V_b}=0. \dots\dots (6.99)$$

The finite-difference analog of the previous equation is

$$\frac{\frac{k_x}{k_y}\left(p_{i-1,j}-2p_{i,j}+p_{i+1,j}\right)}{(\Delta x)^2}+\frac{\left(p_{i,j-1}-2p_{i,j}+p_{i,j+1}\right)}{(\Delta y)^2}+\frac{q\mu B}{V_b k_y}=0. \dots\dots (6.100)$$

Because $\frac{k_x}{k_y}=\frac{300}{200}=1.5$ and $\Delta x=\Delta y$, the previous equation can be simplified as

$$p_{i,j-1}+1.5p_{i-1,j}-5p_{i,j}+1.5p_{i+1,j}+p_{i,j+1}+\frac{q\mu B}{hk_y}=0. \dots\dots (6.101)$$

The blocks with unknown pressure values are Blocks 9, 10, 11, 12, 6, 7, 8, 3, and 4. Therefore, the unknowns of the problem are $p_9, p_{10}, p_{11}, p_{12}, p_6, p_7, p_8, p_3$, and p_4. The two additional unknowns are p_{sf6} and q_4.

2.

$$\Omega=\frac{2\pi\bar{k}h}{\mu B\left(\ln\frac{r_{eq}}{r_w}+s\right)}=\frac{2\pi(244.9)(1.127\times10^{-3})(20)}{(1)(1.2)\left(\ln\frac{199.00}{0.3}-2\right)}=6.42 \text{ STB/D-psi}, \dots\dots (6.102)$$

where

$$r_{eq}=\frac{0.28\left[\left(\frac{k_y}{k_x}\right)^{\frac{1}{2}}(\Delta x)^2+\left(\frac{k_x}{k_y}\right)^{\frac{1}{2}}(\Delta y)^2\right]^{\frac{1}{2}}}{\left(\frac{k_y}{k_x}\right)^{\frac{1}{4}}+\left(\frac{k_x}{k_y}\right)^{\frac{1}{4}}}=\frac{0.28\left[\left(\frac{200}{300}\right)^{\frac{1}{2}}(1,000)^2+\left(\frac{300}{200}\right)^{\frac{1}{2}}(1,000)^2\right]^{\frac{1}{2}}}{\left(\frac{200}{300}\right)^{\frac{1}{4}}+\left(\frac{300}{200}\right)^{\frac{1}{4}}}=199.00 \text{ ft, } (6.103)$$

and

$$\bar{k}=\sqrt{k_x k_y}=\sqrt{200\times300}=244.95 \text{ md}. \dots\dots (6.104)$$

- Block 4: Because Block 4 hosts a production well, the rate-associated coefficient is

$$\frac{\mu B}{hk_y}=\frac{(1.127)(1)}{(20)(200)(1.127\times10^{-3})}=0.25. \dots\dots (6.105)$$

Substitute the flow-rate term by the Peaceman equation:

$$q_4=-\Omega_4\left(p_4-p_{sf}\right). \dots\dots (6.106)$$

Therefore, the finite-difference equation for Block 4 is

$$1.5p_3-4.11p_4+p_8=1,605.0. \dots\dots (6.107)$$

- Block 6:

$$p_2 + 1.5p_5 - 5p_6 + 1.5p_7 + p_{10} + (600)(0.25) = 0. \quad\ldots\ldots\ldots\ldots\ldots\ldots\ldots\ldots\ldots \quad (6.108)$$

Because Blocks 2 and 5 are inactive blocks ($p_2 = p_5 = p_6$), the previous equation can be rewritten as

$$-2.5p_6 + 1.5p_7 + p_{10} = -150. \quad\ldots\ldots\ldots\ldots\ldots\ldots\ldots\ldots\ldots\ldots\ldots\ldots \quad (6.109)$$

- Block 7:

$$p_3 + 1.5p_6 - 5p_7 + 1.5p_8 + p_{11} = 0. \quad\ldots\ldots\ldots\ldots\ldots\ldots\ldots\ldots\ldots\ldots\ldots \quad (6.110)$$

- Block 12:

$$p_8 + 1.5p_{11} - 5p_{12} + 1.5p_{12} + p_{16} = 0. \quad\ldots\ldots\ldots\ldots\ldots\ldots\ldots\ldots\ldots\ldots \quad (6.111)$$

Because $p_{16} = 3{,}000$ psi, the previous equation can be reduced to

$$p_8 + 1.5p_{11} - 3.5p_{12} = -3{,}000. \quad\ldots\ldots\ldots\ldots\ldots\ldots\ldots\ldots\ldots\ldots\ldots \quad (6.112)$$

Problem 6.6.14

Consider the flow of a single-phase, slightly compressible fluid in a horizontal reservoir shown in **Fig. 6.20.** All external boundaries are no-flow boundaries. Initial pressure of the reservoir is 3,000 psia. Earlier reservoir simulation studies in this hypothetical case indicate that a vertical well would produce at an average rate of 217.5 STB/D for the first 30 days. The horizontal-drilling department claims that instead of drilling a vertical well in Block 1, drilling a horizontal well will cause the additional cost of horizontal drilling to break even in 30 days as a result of the expected increase in production rate. In this scenario, the additional cost of horizontal drilling is USD 250,000 (use $20/STB as crude oil market price). Assume both vertical and horizontal wells are operated at 1,000-psia sandface pressures, and the wellbore radius is 0.25 ft in both cases. Determine the feasibility of drilling the horizontal well. Perform a material-balance check (MBC) and comment on the results. Use the reservoir properties listed in **Table 6.16.**

Solution to Problem 6.6.14

The governing equation for a 1D slightly compressible fluid flow is

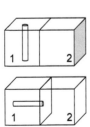

Fig. 6.20—Schematic representation of the vertical and horizontal wells of Problem 6.6.14.

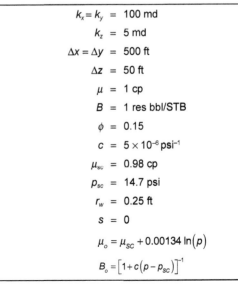

$k_x = k_y$	=	100 md
k_z	=	5 md
$\Delta x = \Delta y$	=	500 ft
Δz	=	50 ft
μ	=	1 cp
B	=	1 res bbl/STB
ϕ	=	0.15
c	=	5×10^{-6} psi^{-1}
μ_{sc}	=	0.98 cp
p_{sc}	=	14.7 psi
r_w	=	0.25 ft
s	=	0
$\mu_o = \mu_{sc} + 0.00134\ln(p)$		
$B_o = \left[1 + c(p - p_{sc})\right]^{-1}$		

Table 6.16—Reservoir properties for Problem 6.6.14.

$$\frac{\partial}{\partial x}\left(\frac{A_x k_x}{\mu B}\frac{\partial p}{\partial x}\right)\Delta x + q = \frac{V_b \phi c}{5.615}\frac{\partial p}{\partial t}. \dots \dots \dots \dots \dots \dots \dots \dots \dots \dots \dots \dots \dots (6.113)$$

For a homogeneous system, the previous equation can be reduced to

$$\frac{\partial^2 p}{\partial x^2} + \frac{q\mu B}{kV_b} = \frac{\phi c \mu B}{5.615k}\frac{\partial p}{\partial t}. \dots \dots \dots \dots \dots \dots \dots \dots \dots \dots \dots \dots (6.114)$$

In the finite-difference form,

$$p_{i-1}^{n+1} - 2p_i^{n+1} + p_{i+1}^{n+1} - \frac{\mu B}{kh}\Omega\left(p_i^{n+1} - p_{sf}^{n+1}\right) = \frac{(\Delta x)^2 \phi c \mu B}{5.615k\Delta t}\left(p_i^{n+1} - p_i^{n}\right). \dots \dots \dots \dots \dots (6.115)$$

Calculating the viscosity and formation volume factor (at $p = 3,000$ psia),

$$\mu_o = \mu_{sc} + 0.00134 \ln (p) = 0.98 + 0.00134 \ln (3,000) = 0.99 \text{ cp}, \dots \dots \dots \dots \dots \dots (6.116)$$

$$B_o = \left[1 + 5\times 10^{-6}\times(3,000 - 14.7)\right]^{-1} = 0.985 \text{ res bbl/STB}, \dots \dots \dots \dots \dots \dots (6.117)$$

$$\frac{(\Delta x)^2 \phi c \mu B}{5.615k\Delta t} = \frac{(0.15)(5\times 10^{-6})(0.99)(0.985)(500)(500)}{(1.127\times 10^{-3})(100)(5.615)(30)} = 9.63\times 10^{-3}. \dots \dots \dots \dots \dots \dots (6.118)$$

Now, one can calculate the PI for the horizontal well: $\bar{k} = \sqrt{k_x k_y} = 22.36$ md,

$$r_{eq} = \frac{0.28\left[\left(\frac{k_y}{k_z}\right)^{\frac{1}{2}}(\Delta z)^2 + \left(\frac{k_z}{k_y}\right)^{\frac{1}{2}}(\Delta y)^2\right]^{\frac{1}{2}}}{\left(\frac{k_y}{k_z}\right)^{\frac{1}{4}} + \left(\frac{k_z}{k_y}\right)^{\frac{1}{4}}} = 28.026 \text{ ft}, \dots \dots \dots \dots \dots \dots (6.119)$$

$$\Omega = \frac{2\pi \bar{k}h}{\mu B\left(\ln\frac{r_{eq}}{r_w} + s\right)} = \frac{2\pi(22.36)(1.127\times 10^{-3})(500)}{(0.99)(0.985)\left(\ln\frac{28.026}{0.25}\right)} = 17.19 \text{ STB/D-psi}. \dots \dots \dots \dots \dots (6.120)$$

- Block 1:

$$p_2^{n+1} - p_1^{n+1} - 2.97\left(p_1^{n+1} - p_{sf}\right) = 9.63\times 10^{-3}\left(p_1^{n+1} - p_1^{n}\right), \dots \dots \dots \dots \dots \dots (6.121)$$

$$p_2^{n+1} - 3.987 p_1^{n+1} = -3,005. \dots \dots \dots \dots \dots \dots \dots \dots \dots \dots \dots \dots (6.122)$$

- Block 2:

$$p_1^{n+1} - 1.00963 p_2^{n+1} = -28.89. \dots \dots \dots \dots \dots \dots \dots \dots \dots \dots \dots \dots (6.123)$$

Solving, $p_1 = 1,012.37$ psi and $p_2 = 1,031.33$ psi, one can calculate the production rate of the well:

$$q = -\Omega\left(p_1 - p_{sf}\right) = -17.19\times(1,012.37 - 1,000) = 212.64 \text{ STB/D}. \dots \dots \dots \dots \dots \dots (6.124)$$

The total production during the first 30 days is

$$Q_{total} = q\Delta t = 6379.321 \text{ STB}. \dots \dots \dots \dots \dots \dots \dots \dots \dots \dots \dots \dots (6.125)$$

Employing the MBC to validate the results,

$$\text{MBC} = \frac{\left|\dfrac{\phi V_b}{5.615}\sum\limits_{i=1}^{2}\left[\left(\dfrac{1}{B}\right)^{n+1}-\left(\dfrac{1}{B}\right)^{n}\right]\right|}{Q_{\text{total}}} = \left|\frac{6,420}{6,379}\right| = 0.999. \dotfill (6.126)$$

However, the profit obtained is $\$6,379.21 \times 20 = \$127,584 < \$250,000$, indicating that the plan for drilling a horizontal well is not a feasible way to break even with costs involved at the end of the first 30-day production period.

Problem 6.6.15

Consider an incompressible, 2D homogeneous anisotropic horizontal reservoir shown in **Fig. 6.21.** The reservoir is sealed and will be used for a filtration project. The lithology of the reservoir is known to have the capability of filtering some contaminates that exist in the injection water. The contaminated water will be injected through the vertical well located in Block 1. The injection pressure is at 3,150 psia. The filtered water will be produced through a horizontal well, which is located in Block 3. Note that a downhole pump is installed at the horizontal well, such that the horizontal well can be produced at a sandface pressure of 14.7 psia. Ignore all gravitational forces.

Using the reservoir properties given in **Table 6.17,** complete the following:

1. Solve the pressure distribution of the reservoir using the Gauss-Seidel iterative procedure (use a set of initial guesses of $p_1 = 2,000$ psi; $p_2 = 1,800$ psi; and $p_3 = 500$ psi).
2. Calculate the flow rates at the injector and producer.
3. Perform an MBC and comment on the accuracy of the solutions.
4. Think of another field-development strategy, if one exists, in which you can increase the injection rate without changing the specified operating conditions at the injection well. Explain in detail why it is possible or not possible.

$k_x =$	25 md
$k_y =$	100 md
$k_z =$	20 md
$\Delta x =$	500 ft
$\Delta y =$	500 ft
$h =$	100 ft
$\mu =$	1.0 cp
$B =$	1.0 res bbl/STB
$r_w =$	0.25 ft
$s =$	0

Table 6.17—Reservoir properties for Problem 6.6.15.

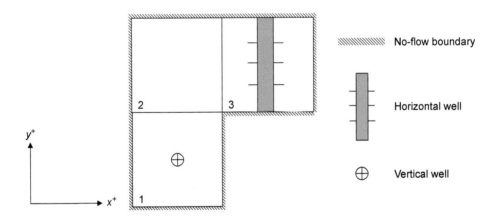

Fig. 6.21—Schematic representation of the 2D reservoir of Problem 6.6.15.

Solution to Problem 6.6.15

1. For an incompressible-flow equation in a 2D horizontal reservoir with no depth gradients, the governing equation can be written as

$$\frac{\partial}{\partial x}\left(\frac{A_x k_x}{\mu B}\frac{\partial p}{\partial x}\right)\Delta x + \frac{\partial}{\partial y}\left(\frac{A_y k_y}{\mu B}\frac{\partial p}{\partial y}\right)\Delta y + q = 0. \dotfill (6.127)$$

For a homogeneous and anisotropic system,

$$k_x \frac{\partial^2 p}{\partial x^2} + k_y \frac{\partial^2 p}{\partial y^2} + \frac{q\mu B}{V_b} = 0. \quad \dots\dots\dots\dots\dots\dots\dots\dots\dots\dots\dots\dots\dots\dots\dots\dots \quad (6.128)$$

Because $k_y = 4k_x$,

$$\frac{\partial^2 p}{\partial x^2} + 4 \frac{\partial^2 p}{\partial y^2} + \frac{q\mu B}{k_x V_b} = 0, \quad \dots\dots\dots\dots\dots\dots\dots\dots\dots\dots\dots\dots\dots\dots \quad (6.129)$$

the finite-difference equation can be written as

$$\left[\frac{p_{i-1,j} - 2p_{i,j} + p_{i+1,j}}{(\Delta x)^2} \right] + 4 \left[\frac{p_{i,j-1} - 2p_{i,j} + p_{i,j-1}}{(\Delta y)^2} \right] + \frac{q\mu B}{V_b k} = 0, \quad \dots\dots\dots\dots\dots \quad (6.130)$$

$$p_{i-1,j} + p_{i+1,j} - 10p_{i,j} + 4p_{i,j-1} + 4p_{i,j+1} + \frac{q\mu B}{hk_x} = 0. \quad \dots\dots\dots\dots\dots\dots\dots \quad (6.131)$$

For a vertical well,

$$r_{eq} = \frac{0.28 \left[\left(\frac{k_y}{k_x} \right)^{\frac{1}{2}} (\Delta x)^2 + \left(\frac{k_x}{k_y} \right)^{\frac{1}{2}} (\Delta y)^2 \right]^{\frac{1}{2}}}{\left(\frac{k_y}{k} \right)^{\frac{1}{4}} + \left(\frac{k_x}{k_y} \right)^{\frac{1}{4}}} = 104.35 \text{ ft,} \quad \dots\dots\dots\dots\dots\dots\dots\dots \quad (6.132)$$

where $\bar{k} = \sqrt{k_x k_y} = 50$ md,

$$\Omega_1 = \frac{2\pi \bar{k} h}{\mu B \left[\ln \left(\frac{r_{eq}}{r_w} \right) + s \right]} = \frac{2\pi(50)(1.127 \times 10^{-3})(100)}{(1)(2) \left(\ln \frac{104.35}{0.25} \right)} = 5.852 \text{ STB/D-psi.} \quad \dots\dots\dots\dots\dots \quad (6.133)$$

For a horizontal well (drilled perpendicular to the x-z plane),

$$r_{eq} = \frac{0.28 \left[\left(\frac{k_x}{k_z} \right)^{\frac{1}{2}} (\Delta z)^2 + \left(\frac{k_z}{k_x} \right)^{\frac{1}{2}} (\Delta x)^2 \right]^{\frac{1}{2}}}{\left(\frac{k_x}{k_z} \right)^{\frac{1}{4}} + \left(\frac{k_z}{k_x} \right)^{\frac{1}{4}}} = 67.73 \text{ ft,} \quad \dots\dots\dots\dots\dots\dots\dots\dots \quad (6.134)$$

where $\bar{k} = \sqrt{k_x k_z} = 22.36$ md,

$$\Omega_3 = \frac{2\pi \bar{k} h}{\mu B \left[\ln \left(\frac{r_{eq}}{r_w} \right) + s \right]} = \frac{2\pi(22.36)(1.127 \times 10^{-3})(500)}{(1)(1) \left(\ln \frac{67.73}{0.25} \right)} = 14.076 \text{ STB/D-psi.} \quad \dots\dots\dots\dots \quad (6.135)$$

- Block 1:

$$4p_1 + p_1 - 10p_1 + p_1 + 4p_2 + \frac{-\Omega_1(p_1 - p_{sf})\mu B}{hk} = 0, \quad \dots\dots\dots\dots\dots\dots\dots\dots \quad (6.136)$$

which simplifies to

$$-6.09p_1 + 4p_2 = -6,583.5. \quad \dots\dots\dots\dots\dots\dots\dots\dots\dots\dots\dots\dots\dots\dots\dots\dots \quad (6.137)$$

- Block 2:

$$4p_1 + p_2 - 10p_2 + p_3 + 4p_2 = 0, \quad \dots\dots\dots\dots\dots\dots\dots\dots\dots\dots\dots\dots\dots\dots \quad (6.138)$$

which simplifies to

$$4p_1 - 5p_2 + p_3 = 0. \quad\dotfill \quad (6.139)$$

- Block 3:

$$4p_3 + p_2 - 10p_3 + p_3 + 4p_3 + \frac{-\Omega_3(p_3 - p_{sf})\mu B}{hk} = 0, \quad\dotfill\quad (6.140)$$

which simplifies to

$$p_2 - 6.027p_3 = -73.90. \quad\dotfill\quad (6.141)$$

The Gauss-Seidel iterative algorithm is used to solve the system of equations with a pressure tolerance of 1 psi, and the intermediate results are displayed in **Table 6.18.**

Iteration #	p_1	p_2	p_3
0	2,000	1,800	500
1	2,263.3	1,910.64	329.275
2	2,335.971	1,934.632	333.2556
3	2,351.728	1,948.034	335.4793
4	2,360.531	1,955.521	336.7216
5	2,365.449	1,959.703	337.4155
6	2,368.196	1,962.04	337.8032
7	2,369.731	1,963.345	338.0198
8	2,370.588	1,964.074	338.1408

Table 6.18—Gauss-Seidel iterative process in the solution of Problem 6.6.15.

2. Flow rate of the wells is as follows:

$$q_1 = -\Omega_1(p_1 - p_{sf,1}) = -5.852(2,370.6 - 3,150) = 4,561 \text{ STB/D}, \quad\dotfill\quad (6.142)$$

$$q_3 = -\Omega_3(p_3 - p_{sf,3}) = -14.076(338.1 - 14.7) = -4,552 \text{ STB/D}. \quad\dotfill\quad (6.143)$$

3. Validation of the results using the MBC is as follows:

$$\text{MBC} = \left|\frac{q_1}{q_3}\right| = \left|\frac{4,561}{-4,552}\right| = 1.002. \quad\dotfill\quad (6.144)$$

The MBC shows a good level of accuracy, but it can be further improved by using a finer convergence criterion when the Gauss-Seidel iteration is carried out.

4. Because $k_y > k_x$, the injection rate is expected to significantly increase by completing the horizontal injection well parallel to the x-direction (perpendicular to the y-z plane). For a horizontal injection well completed parallel to the x-direction,

$$r_{eq} = \frac{0.28\left[\left(\frac{k_y}{k_z}\right)^{\frac{1}{2}}(\Delta z)^2 + \left(\frac{k_z}{k_y}\right)^{\frac{1}{2}}(\Delta y)^2\right]^{\frac{1}{2}}}{\left(\frac{k_y}{k_z}\right)^{\frac{1}{4}} + \left(\frac{k_z}{k_y}\right)^{\frac{1}{4}}} = 47.39 \text{ ft.,} \quad\dotfill\quad (6.145)$$

where $\overline{k} = \sqrt{k_y k_z} = 44.72$ md,

$$\Omega_3 = \frac{2\pi \bar{k} h}{\mu B \left[\ln \left(\dfrac{r_{eq}}{r_w} \right) + s \right]} = \frac{2\pi (500)(1.127 \times 10^{-3})(44.72)}{(1)(1) \left(\ln \dfrac{47.39}{0.25} \right)} = 30.174 \text{ STB/D-psi.} \quad \dots \dots \dots \dots \quad (6.146)$$

The equation of Block 3 can be changed to

$$p_2 - 11.782 p_3 = -158.5. \quad \dots \dots \dots \dots \dots \dots \dots \dots \dots \dots \dots \dots \dots \dots \dots \dots \quad (6.147)$$

Solving for the unknown block pressures gives $p_1 = 2,327$ psi; $p_2 = 1,897$ psi; and $p_3 = 174$ psi,

$$q_1 = -\Omega_1 (p_1 - p_{sf,1}) = -5.852(2,327 - 3,150) = 4,816.2 \text{ STB/D}, \quad \dots \dots \dots \dots \dots \dots \quad (6.148)$$

$$q_3 = -\Omega_3 (p_3 - p_{sf,3}) = -30.19(174 - 14.7) = -4,809.27 \text{ STB/D}. \quad \dots \dots \dots \dots \dots \dots \quad (6.149)$$

MBC gives

$$\text{MBC} = \left| \frac{q_1}{q_3} \right| = \left| \frac{4,816.2}{-4,809.27} \right| = 1.0015. \quad \dots \dots \dots \dots \dots \dots \dots \dots \dots \dots \dots \dots \dots \quad (6.150)$$

Problem 6.6.16

Consider the slightly compressible flow in a 2D reservoir shown in **Fig 6.22**. The reservoir is homogeneous and exhibits anisotropic permeability distribution. There is no depth gradient in the reservoir. The reservoir is completely sealed. A producer is located at the center of the system, and it operates with a specified flow rate of–500 STB/D (production). There are two different production strategies, Case (a) and Case (b).

Using the properties listed in **Table 6.19,** answer the following questions:

1. If one plots the block pressure of Block 9 vs. time for both strategies, do you expect that the two plots will be the same? Justify your answer.
2. How would you compare the sandface pressures of the two cases?
3. Which production strategy would you recommend for maximizing the production? Justify your answer.

$k_x =$	200 md
$k_y =$	100 md
$k_z =$	50 md
$\Delta x = \Delta y =$	500 ft
$h =$	50 ft
$\mu =$	1 cp
$B =$	1 res bbl/STB
$c_f =$	1×10^{-5} psi^{-1}
$\phi =$	0.2
$r_w =$	0.25 ft
$s =$	0

Table 6.19—Reservoir properties for Problem 6.6.16.

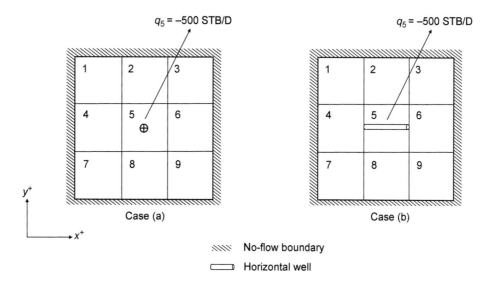

Fig. 6.22—Schematic representation of Problem 6.6.16.

Solution to Problem 6.6.16

1. The production wells are producing at the same rate for the horizontal and vertical wells. Therefore, governing flow equations, when written for both cases, will generate the same system of equations:

$$\frac{\partial}{\partial x}\left(\frac{A_x k_x}{\mu B}\frac{\partial p}{\partial x}\right)\Delta x + \frac{\partial}{\partial y}\left(\frac{A_y k_y}{\mu B}\frac{\partial p}{\partial y}\right)\Delta y + q = \frac{V_b \phi c_t}{5.615}\frac{\partial p}{\partial t}, \dots\dots\dots\dots\dots\dots\dots (6.151)$$

where $q = -500$ STB/D for both of the cases.

Considering the discretization scheme imposed on the reservoir, the horizontal and vertical wells are completed at the same gridblock (Block 5). With identical constant-rate specifications ($q = -500$ STB/D) for both cases, the solved pressure distribution will be the same, regardless of the well completion schemes (simply because Block 5 is to be treated as the same sink term for the reservoir systems). However, the productivity index of the horizontal well is higher than that of the vertical well; therefore, the sandface pressure required to sustain the specified flow rate for the horizontal well is larger than that of the vertical well. In other words, the horizontal well completion can sustain the −500 STB/D for a longer period of time than the vertical well completion.

2.

$$r_{eq}^a = \frac{0.28\left[\left(\frac{k_y}{k_x}\right)^{\frac{1}{2}}(\Delta x)^2 + \left(\frac{k_x}{k_y}\right)^{\frac{1}{2}}(\Delta y)^2\right]^{\frac{1}{2}}}{\left(\frac{k_y}{k_x}\right)^{\frac{1}{4}} + \left(\frac{k_x}{k_y}\right)^{\frac{1}{4}}} = \frac{0.28\left[\left(\frac{100}{200}\right)^{\frac{1}{2}}(500)^2 + \left(\frac{200}{100}\right)^{\frac{1}{2}}(500)^2\right]^{\frac{1}{2}}}{\left(\frac{100}{200}\right)^{\frac{1}{4}} + \left(\frac{200}{100}\right)^{\frac{1}{4}}} = 100.4 \text{ ft}, \dots (6.152)$$

$$r_{eq}^b = \frac{0.28\left[\left(\frac{k_y}{k_z}\right)^{\frac{1}{2}}(\Delta z)^2 + \left(\frac{k_z}{k_y}\right)^{\frac{1}{2}}(\Delta y)^2\right]^{\frac{1}{2}}}{\left(\frac{k_y}{k_z}\right)^{\frac{1}{4}} + \left(\frac{k_z}{k_y}\right)^{\frac{1}{4}}} = \frac{0.28\left[\left(\frac{100}{50}\right)^{\frac{1}{2}}(50)^2 + \left(\frac{50}{100}\right)^{\frac{1}{2}}(500)^2\right]^{\frac{1}{2}}}{\left(\frac{100}{50}\right)^{\frac{1}{4}} + \left(\frac{50}{100}\right)^{\frac{1}{4}}} = 58.6 \text{ ft}, \dots (6.153)$$

$$\frac{\Omega_a}{\Omega_b}\frac{\sqrt{200}}{\sqrt{50}}\frac{h}{\Delta x}\frac{\ln\left(\frac{r_{eq}^b}{r_w}\right)}{\ln\left(\frac{r_{eq}^a}{r_w}\right)} = 0.2\frac{\ln\left(\frac{58.6}{r_w}\right)}{\ln\left(\frac{100}{r_w}\right)} < 1. \dots\dots\dots\dots\dots\dots\dots (6.154)$$

Therefore, the PI of the horizontal well is larger than that of vertical well. Thus, the sandface pressure of the horizontal well should be larger than that of vertical well at the same production rate.

3. The horizontal well could produce for a longer period of time at −500 STB/D than the vertical well, until the sandface pressure drops to 14.7 psia. Thus, the development strategy with a horizontal well yields a higher productivity than a vertical well. However, drilling and completion of a horizontal well is expected to be more expensive, and a comprehensive project-economic analysis should be conducted to make a final decision.

Problem 6.6.17

The reservoir represented in **Fig. 6.23** is known to be homogeneous with no depth gradient. It contains a single-phase, slightly compressible fluid. All boundaries are sealed. Two wells are drilled at the beginning, but during the first five days, only the well in Block 2 was produced at a constant rate of −100 STB/D. At day five, the well was shut in, and on the tenth day, the following pressures were recorded: $p_1 = 3,298.17$ psi and $p_2 = 3,298.08$ psi. Unfortunately, initial pressure of the reservoir was not recorded when the wells were completed. Use the reservoir properties listed in **Table 6.20** to complete the following:

1. Calculate the sandface pressure of Well 2 after five days.
2. Assuming a different scenario, if both wells were put on production from the beginning but $q_2 = -200$ STB/D, what would the sandface pressure of Well 2 be at the end of the fifth day?

$\Delta x = \Delta y =$	200 ft
$k =$	100 md
$h =$	50 ft
$\phi =$	0.2
$\mu =$	1.1 cp
$c =$	5×10^{-6} psi^{-1}
$B =$	1.2 res bbl/STB
$r_w =$	0.25 ft
$s =$	0

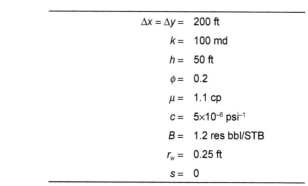

Fig. 6.23— Reservoir system of Problem 6.6.17.   **Table 6.20—Reservoir properties for Problem 6.6.17.**

Solution to Problem 6.6.17

1. The finite-difference analog of the reservoir described in this problem is

$$p_{i+1}^{n+1} - 2p_i^{n+1} + p_{i-1}^{n+1} + \frac{q\mu B}{hk} = \frac{(\Delta x)^2 \phi c \mu B}{5.615 \Delta tk} \left(p_i^{n+1} - p_i^n \right), \dots\dots\dots\dots\dots\dots\dots\dots\dots\dots (6.155)$$

where

$$\frac{(\Delta x)^2 \phi c \mu B}{5.615 \Delta tk} = \frac{(200)^2 (0.2)(5 \times 10^{-6})(1.2)(1.1)}{(5.615)(100)(1.127 \times 10^{-3})(5)} = 0.01669, \dots\dots\dots\dots\dots\dots\dots (6.156)$$

and

$$\frac{q\mu B}{hk} = \frac{(-100)(1.1)(1.2)}{(50)(1.127 \times 10^{-3})(100)} = -23.425. \dots\dots\dots\dots\dots\dots\dots\dots\dots\dots\dots (6.157)$$

The PI of the well is

$$\Omega = \frac{2\pi kh}{\mu B \left[\ln\left(\frac{r_{eq}}{r_w} \right) + s \right]} = \frac{2\pi (100)(0.001127)(50)}{(1.1)(1.2)\left[\ln\left(\frac{39.6}{0.25} \right) \right]} = 5.28 \text{ STB/D-psi}, \dots\dots\dots\dots\dots\dots (6.158)$$

where $r_{eq} = 0.198 \Delta x = 39.6$ ft.
Because the pressure at the end of the tenth day is recorded,

- From $t = 5$ to 10 for Block 1 (n at $t = 5$ and $n+1$ at $t = 10$):

$$p_2^{n+1} - p_1^{n+1} = (0.01669)\left(p_1^{n+1} - p_1^n \right), \dots\dots\dots\dots\dots\dots\dots\dots\dots\dots\dots\dots\dots (6.159)$$

$$p_1^{t=5} = p_1^{t=10} - \frac{p_2^{t=10} - p_1^{t=10}}{0.01669} = 3,309.933 \text{ psi}. \dots\dots\dots\dots\dots\dots\dots\dots\dots\dots (6.160)$$

- From $t = 5$ to 10 for Block 2:

$$p_1^{n+1} - p_2^{n+1} = (0.01669)\left(p_2^{n+1} - p_2^n \right), \dots\dots\dots\dots\dots\dots\dots\dots\dots\dots\dots\dots\dots (6.161)$$

$$p_2^{t=5} = p_2^{t=10} - \frac{p_1^{t=10} - p_2^{t=10}}{0.01669} = 3,292.317 \text{ psi}. \dots\dots\dots\dots\dots\dots\dots\dots\dots\dots (6.162)$$

Because $q_2 = -\Omega_2 \left(p_2 - p_{sf,2} \right)$,

$$p_{sf2} = p_2 + \frac{q}{\Omega} = 3,292.317 + \frac{(-100)}{5.2824} = 3,273.42 \text{ psi}. \dots\dots\dots\dots\dots\dots\dots\dots\dots\dots (6.163)$$

- From $t = 0$ to 5 for Block 1,

$$p_2^{n+1} - p_1^{n+1} = (0.01669)\left(p_1^{n+1} - p_1^{n}\right), \quad \dots \dots \dots \dots \dots \dots \dots \dots \dots \dots \dots \dots (6.164)$$

$$p_1^{t=0} = p_1^{t=5} - \frac{p_2^{t=5} - p_1^{t=5}}{0.01669} = 4{,}000 \text{ psi}. \quad \dots \dots \dots \dots \dots \dots \dots \dots \dots (6.165)$$

- From $t = 0$ to 5 for Block 2,

$$p_1^{n+1} - p_2^{n+1} + (-23.425) = (0.01669)\left(p_2^{n+1} - p_2^{n}\right), \quad \dots \dots \dots \dots \dots \dots \dots (6.166)$$

$$p_2^{t=0} = p_2^{t=5} - \frac{p_1^{t=5} - p_2^{t=5}}{0.01669} = 4{,}000 \text{ psi}. \quad \dots \dots \dots \dots \dots \dots \dots \dots (6.167)$$

Therefore, the initial reservoir pressure is 4,000 psi.

2. If both wells are producing,

- Block 1:

$$-1.016686\, p_1^{n+1} + p_2^{n+1} = -43.325. \quad \dots \dots \dots \dots \dots \dots \dots \dots \dots \dots \dots (6.168)$$

- Block 2:

$$p_1^{n+1} - 1.016686\, p_2^{n+1} = -19.90. \quad \dots \dots \dots \dots \dots \dots \dots \dots \dots \dots \dots \dots (6.169)$$

Solving two equations in two unknowns, one obtains $p_1^{n+1} = 1{,}900.18$ psi and $p_2^{n+1} = 1{,}888.57$ psi.
The sandface pressure can be calculated as

$$p_{sf2} = p_2 + \frac{q}{\Omega} = 1{,}888.57 + \frac{-200}{5.2824} = 1{,}850.78 \text{ psi}. \quad \dots \dots \dots \dots \dots \dots (6.170)$$

MBC: The total volume expanded in Blocks 1 and 2 should be equal to the total production at the end of five days:

$$\Delta V = \Delta V_1 + \Delta V_2$$

$$= -cV_p(\Delta p_1 + \Delta p_2) = 5 \times 10^{-6} \frac{(200)(200)(50)(0.2)}{5.615}\left[(2{,}099.82) + (2{,}111.43)\right] = 1{,}500 \text{ STB}. \quad \dots \dots (6.171)$$

The total production within the five-day period is

$$Q_{total} = 5 \text{ days} \times (-100-200) \text{ STB/D} = -1{,}500 \text{ STB}, \quad \dots \dots \dots \dots \dots \dots \dots \dots (6.172)$$

$$\text{MBC} = \left|\frac{\Delta V}{Q_{total}}\right| = \left|\frac{1{,}500}{-1{,}500}\right| = 1.00, \quad \dots \dots \dots \dots \dots \dots \dots \dots \dots \dots \dots (6.173)$$

which indicates a good MBC.

Problem 6.6.18

Fig 6.24 represents a 1D, homogeneous reservoir in which a single-phase slightly compressible fluid resides in hydrodynamic equilibrium. The reservoir is surrounded by no-flow boundaries A production well is drilled and completed in Block 1. Before the well is put on production, a pressure measurement is conducted in this well and gives a bottomhole pressure of 3,000 psia. When the well starts to produce, it produces at a constant sandface-pressure constraint of $p_{sf1} = 2{,}000$ psi.

A colleague in your office suggests that if the well had been located in Block 2 instead of Block 1, the initial production rate from the well would have been higher. Quantitatively show how the flow rate initially would differ if the well were completed in Block 2 instead of in Block 1. In both cases, the constant sandface-pressure constraint of $p_{sf2} = 2{,}000$ psi should be applied. What is the flow rate from the well after 30 days of production? Use a timestep size of 30 days, and the properties listed in **Table 6.21** for your calculations.

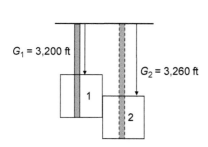

$k =$	100 md
$\Delta x = \Delta y =$	1,000 ft
$h =$	100 ft
$\phi =$	20%
$\mu =$	0.98 cp
$B =$	0.97 res bbl/STB
$c =$	5×10^{-6} psi^{-1}
$r_w =$	0.3 ft
$s =$	0
$p_{sf} =$	2,000 psi
$\rho =$	60 lbm/ft^3

Fig. 6.24—Reservoir system of Problem 6.6.18.   Table 6.21—Reservoir properties for Problem 6.6.18.

Solution to Problem 6.6.18

For the hydrodynamic equilibrium system of this problem, $\Phi_1 = \Phi_2$, where

$$\Phi_1 = p_1 - \frac{1}{144} \frac{g}{g_c} \rho G_1, \quad\dots\dots\dots\dots\dots\dots\dots\dots\dots\dots\dots\dots\dots\dots \text{(6.174)}$$

and

$$\Phi_2 = p_2 - \frac{1}{144} \frac{g}{g_c} \rho G_2. \quad\dots\dots\dots\dots\dots\dots\dots\dots\dots\dots\dots\dots\dots\dots \text{(6.175)}$$

Because $p_1 = 3{,}000$ psi, one can solve for p_2:

$$p_2 = p_1 + \frac{1}{144} \frac{g}{g_c} \rho(G_2 - G_1) = 3{,}025 \text{ psi}. \quad\dots\dots\dots\dots\dots\dots\dots\dots\dots\dots\dots \text{(6.176)}$$

The productivity index is

$$\Omega = \frac{2\pi k h}{\mu B \left[\ln\left(\dfrac{r_{eq}}{r_w}\right) + s \right]} = \frac{2\pi (0.1)(1.127)(100)}{(0.98)(0.97)\left(\ln \dfrac{198}{0.3} \right)} = 11.5 \text{ STB/D-psi}, \quad\dots\dots\dots\dots\dots\dots\dots\dots \text{(6.177)}$$

where $r_{eq} = 0.198\Delta x = 198$ ft.

The governing macroscopic flow equation for the reservoir described in this problem is

$$\frac{\partial}{\partial x}\left(\frac{A_x k_x}{\mu B} \frac{\partial p}{\partial x} \right)\Delta x - \frac{\partial}{\partial x}\left(\frac{1}{144} \frac{g}{g_c} \rho \frac{A_x k_x}{\mu B} \frac{\partial G}{\partial x} \right)\Delta x + q = \frac{V_b \phi c_t}{5.615}\left(\frac{\partial p}{\partial t} \right). \quad\dots\dots\dots\dots\dots \text{(6.178)}$$

For a homogeneous system, the previous equation reduces to

$$\frac{\partial^2 p}{\partial x^2} - \frac{1}{144} \frac{g}{g_c} \rho \frac{\partial^2 G}{\partial x^2} + \frac{q\mu B}{V_b k} = \frac{\phi c_t \mu B}{5.615 k}\left(\frac{\partial p}{\partial t} \right). \quad\dots\dots\dots\dots\dots\dots\dots\dots\dots\dots \text{(6.179)}$$

The characteristic finite-difference equation will be

$$p_{i-1}^{n+1} - 2p_i^{n+1} + p_{i+1}^{n+1} + \frac{q\mu B}{hk} = \frac{\mu B \phi c_t \Delta x^2}{5.615 k \Delta t}\left(p_i^{n+1} - p_i^n \right) + \frac{1}{144} \frac{g}{g_c} \rho(G_{i-1} - 2G_i + G_{i+1}), \quad\dots\dots\dots\dots \text{(6.180)}$$

where

$$\frac{\mu B \phi c_t (\Delta x)^2}{5.165 k \Delta t} = \frac{(0.97)(0.98)(5 \times 10^{-6})(0.2)(1000)^2}{5.165(100)(1.127 \times 10^{-3})(30)} = 0.05. \quad\dots\dots\dots\dots\dots\dots\dots\dots\dots \text{(6.181)}$$

Next we examine cases for the different well completion blocks.

If the well is completed at Block 1:

- For Block 1,

$$-2.017 p_1^{n+1} + p_2^{n+1} = -2,059.84, \dots\dots\dots\dots\dots\dots\dots\dots\dots\dots\dots\dots\dots\dots \quad (6.182)$$

- For Block 2,

$$p_1^{n+1} - 1.05 p_2^{n+1} = -176.25, \dots\dots\dots\dots\dots\dots\dots\dots\dots\dots\dots\dots\dots\dots \quad (6.183)$$

One can solve for two unknowns in two equations, $p_1^{n+1} = 2,091.7$ psi and $p_2^{n+1} = 2,160.2$ psi:

$$q_1 = -\Omega\left(p_1 - p_{sf,1}\right) = -11.5(2,091.70 - 2,000) = -1,054.55 \text{ STB/D}. \dots\dots\dots\dots\dots\dots \quad (6.184)$$

MBC: The total volume expanded in Block 1 and 2 should be equal to the total production at the end of 30 days:

$$\Delta V = \Delta V_1 + \Delta V_2$$

$$= -cV_p(\Delta p_1 + \Delta p_2) = 5 \times 10^{-6} \frac{(1,000)(1,000)(100)(0.2)}{5.615}\left[(908.3) + (1,864.8)\right]$$

$$= 31,577.9 \text{ STB}. \dots\dots\dots\dots\dots\dots\dots\dots\dots\dots\dots\dots\dots\dots\dots\dots\dots\dots\dots \quad (6.185)$$

The total production within the 30-day period is

$$Q_{\text{total}} = 30 \text{ days} \times -1,054.55 \text{ STB/D} = -31,636.5 \text{ STB}, \dots\dots\dots\dots\dots\dots\dots\dots \quad (6.186)$$

and

$$\text{MBC} = \left|\frac{Q_{\text{total}}}{\Delta V}\right| = \left|\frac{-31,636.5}{31,5779}\right| = 0.999, \dots\dots\dots\dots\dots\dots\dots\dots\dots\dots\dots \quad (6.187)$$

which indicates a good material balance.

If the well is completed at Block 2, the system of equations needs to be written as

- For Block 1,

$$-1.05 p_1^{n+1} + p_2^{n+1} = -125.219. \dots\dots\dots\dots\dots\dots\dots\dots\dots\dots\dots\dots\dots\dots \quad (6.188)$$

- For Block 2,

$$p_1^{n+1} - 2.017 p_2^{n+1} = -2,111.09. \dots\dots\dots\dots\dots\dots\dots\dots\dots\dots\dots\dots\dots\dots \quad (6.189)$$

One can solve for two unknowns in the two equations ($p_1^{n+1} = 2,113.47$ psi and $p_2^{n+1} = 2,094.08$ psi):

$$q_2 = -\Omega\left(p_2 - p_{sf,2}\right) = -11.5(2,094.08 - 2,000) = -1,078.92 \text{ STB/D}. \dots\dots\dots\dots\dots \quad (6.190)$$

MBC: The total fluid volume expanded in Blocks 1 and 2 should be equal to the total production at the end of 30 days:

$$\Delta V = \Delta V_1 + \Delta V_2$$

$$= cV_p(\Delta p_1 + \Delta p_2) = 5 \times 10^{-6} \frac{(1,000)(1,000)(100)(0.2)}{5.615}\left[(3,000 - 213,048) + (3,025 - 2,094.08)\right]$$

$$= -32,367.73 \text{ STB.}$$

$$\dots\dots\dots\dots\dots\dots\dots\dots\dots\dots\dots\dots\dots\dots\dots\dots\dots\dots\dots\dots\dots\dots \quad (6.191)$$

Total production within the 30-day period is

$$Q_{\text{total}} = 30 \text{ days} \times 1,078.92 \text{ STB/D} = 32,367.73 \text{ STB}. \dots\dots\dots\dots\dots\dots\dots\dots \quad (6.192)$$

$$\text{MBC} = \left|\frac{Q_{\text{total}}}{\Delta V}\right| = \left|\frac{-32,367.73}{32,367.73}\right| = 1.000, \dots\dots\dots\dots\dots\dots\dots\dots\dots\dots\dots \quad (6.193)$$

which indicates a good material balance.

Comparing the production rates of Case 1 and Case 2 with the well completed in Block 1 and Block 2, respectively, one obtains $q_1 < q_2$. The simulation results agree with the colleague's recommendation.

Problem 6.6.19

Fig. 6.25 represents the flow of a single-phase, slightly compressible fluid in a linear, horizontal, and homogeneous reservoir represented by the body-centered grid system. All outer boundaries are no-flow boundaries.

First, the well in Block 3 has been drilled and put on production at a constant rate of –1,000 STB/D. Thirty days later, the well in Block 1 has been drilled and put on production at a constant rate of –1,000 STB/D, along with the well producing in Block 3. During the drilling operation, both wells have been damaged as a result of the invasion of drilling mud flowing into the formation. Consequently, wellbore-stimulation techniques have been applied to minimize the damage. As a result of the stimulation process, the damage is successfully treated and well test data showed both wells having a skin factor of –2.

Using the previous information and the data provided in **Table 6.22**, determine whether both wells can support –1,000 STB/D production for up to 60 days. Conduct an MBC and comment on the accuracy of your calculations.

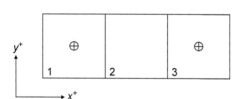

$\Delta x =$	1,000 ft
$\Delta y =$	1,000 ft
$h =$	100 ft
$k =$	10 md
$\phi =$	0.15
$c_t =$	5×10^{-6} psi^{-1}
$\mu =$	2.0 cp
$B =$	1 res bbl/STB
$r_w =$	0.25 ft
$s =$	–2
$p_i =$	3,000 psi

Fig. 6.25—Schematic representation of the reservoir of Problem 6.6.19.

Table 6.22—Reservoir properties for Problem 6.6.19.

Solution to Problem 6.6.19

For a 1D, slightly compressible reservoir system, the generalized flow equation is

$$\frac{\partial}{\partial x}\left(\frac{A_x k_x}{\mu B}\frac{\partial p}{\partial x}\right)\Delta x + q = \frac{V_b \phi c}{5.615}\frac{\partial p}{\partial t}. \quad\quad\quad (6.194)$$

For a homogeneous reservoir, it can be reduced to

$$\frac{\partial^2 p}{\partial x^2} + \left(\frac{q\mu B}{kV_b}\right) = \frac{\phi c\mu B}{5.615k}\frac{\partial p}{\partial t}. \quad\quad\quad (6.195)$$

Writing the finite-difference analog,

$$p_{i+1}^{n+1} - 2p_i^{n+1} + p_{i-1}^{n+1} + \frac{q\mu B}{hk} = \frac{(\Delta x)^2 \phi c\mu B}{5.615\Delta tk}\left(p_i^{n+1} - p_i^n\right), \quad\quad\quad (6.196)$$

where

$$\frac{(\Delta x)^2 \phi c\mu B}{5.615\Delta tk} = \frac{(1,000)^2 (0.15)(5 \times 10^{-6})(2)(1)}{(5.615)(30)(1.127 \times 10^{-3})(10)} = 0.79, \quad\quad\quad (6.197)$$

and

$$\frac{q\mu B}{hk} = \frac{(-1,000)(2)(1)}{(100)(1.127 \times 10^{-3})(10)} = -1,774.623. \quad\quad\quad (6.198)$$

The PI is $\Omega = \dfrac{2\pi kh}{\mu B \left[\ln\left(\dfrac{r_{eq}}{r_w} \right) + s \right]} = 0.7574$ STB/D-psi, where $r_{eq} = 0.198$, $\Delta x = 198$ ft.

From 0 to 30 days,

- Block 1: $-1.79 p_1 + p_2 = -2{,}370.38$. . . . . (6.199)
- Block 2: $p_1 - 2.79 p_2 + p_3 = -2{,}370.38$. . . . . (6.200)
- Block 3: $p_2 - 1.79 p_3 = -595.755$. . . . . (6.201)

Solving three equations in three unknowns, one obtains $p_1 = 2{,}668.966$, $p_2 = 2{,}407.408$, and $p_3 = 1{,}677.626$:

$$q_3 = -\Omega\left(p_3 - p_{sf} \right) = -1{,}000 \text{ STB/D}. \quad \text{............................} \quad (6.202)$$

Solving for p_{sf}, one obtains 357.36 psi. Because the calculated sandface pressure is larger than 14.7 psi, we can conclude that Well 3 can support the production rate of 1,000 STB/D for the first 30 days.

Validating the results using the MBC,

$$\Delta V_{t0\,to\,30} = \Delta V_1 + \Delta V_2 + \Delta V_3 = -cV_p \left(\Delta p_1 + \Delta p_2 + \Delta p_3 \right)$$

$$= 5\times 10^{-6} \left[\frac{(1{,}000)(1{,}000)(100)(0.15)}{5.615} \right] \left[(3{,}000 - 2{,}668.966) + (3{,}000 - 2{,}407.408) + (3{,}000 - 1{,}677.626) \right],$$

$$= 30{,}000 \text{ STB} \quad \text{...} \quad (6.203)$$

$$\text{MBC} = \left| \frac{\Delta V}{Q_{total}\,(t=30)} \right| = \left| \frac{30{,}000}{30 \times 1{,}000} \right| = 1.00000. \quad \text{............................} \quad (6.204)$$

From 30 to 60 days,

- Block 1: $-1.79 p_1 + p_2 = -333.14$. . . . . (6.205)
- Block 2: $p_1 - 2.79 p_2 + p_3 = -1{,}901.81$. . . . . (6.206)
- Block 3: $p_2 - 1.79 p_3 = -448.33$. . . . . (6.207)

Solving, $p_1 = 799.8$ psia, $p_2 = 1{,}098.5$ psia, and $p_3 = 363.22$ psia. Knowing that

$$q = -\Omega\left(p - p_{sf} \right), \quad \text{....................................} \quad (6.208)$$

and

$$p_{sf} = p + \frac{q}{\Omega}. \quad \text{...} \quad (6.209)$$

When p_{sf1} and p_{sf2} are computed,

$$p_{sf,1} = p_1 + \frac{q}{\Omega} = 799.8 - \frac{1{,}000}{0.7574} = -520.51 \text{ psi} \quad \text{....................} \quad (6.210)$$

and

$$p_{sf,2} = p_2 + \frac{q}{\Omega} = 363.22 - \frac{1{,}000}{0.7574} = -957.09 \text{ psi}, \quad \text{....................} \quad (6.211)$$

the negative sandface pressures imply that neither of the wells can sustain the production rate during the second 30-day period.

Problem 6.6.20

Fig. 6.26 describes the flow of a single-phase, slightly compressible fluid in a horizontal, homogeneous, and isotropic reservoir represented by the body-center grid system. All outer boundaries are no-flow boundaries.

During the drilling operation, the well in Block 2 has been damaged as a result of the invasion of drilling mud into the formation. A common way to include the wellbore damage (or stimulation) in reservoir models is to represent the damage (or stimulation) as skin factor, s, and add the skin factor into the wellbore flow equation. A well test analysis has been conducted to find the skin factor. Test results show that skin factor, $s = +5$, which indicates a significant damage for the well in Block 2.

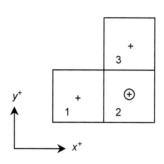

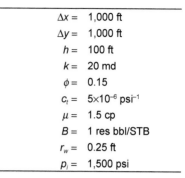

$\Delta x =$	1,000 ft
$\Delta y =$	1,000 ft
$h =$	100 ft
$k =$	20 md
$\phi =$	0.15
$c_t =$	5×10^{-6} psi^{-1}
$\mu =$	1.5 cp
$B =$	1 res bbl/STB
$r_w =$	0.25 ft
$p_i =$	1,500 psi

Fig. 6.26—Schematic representation of the 2D reservoir of Problem 6.6.20.

Table 6.23—Reservoir properties for Problem 6.6.20.

Using the data provided in **Table 6.23,** complete the following:

1. Determine the pressure distribution, and conduct an MBC after 20 days of production with a rate of –1,000 STB/D.
2. Determine if it is realistic for the well to support the production for 20 days.
3. Repeat the analysis for Part 2 if $s = -2$.

Solution to Problem 6.6.20

Considering the existing symmetry in the system, one can state that the pressure of Block 1 is equal to that of Block 3. Therefore, the primary unknowns of the problem are p_1 and p_2:

$$\frac{\partial}{\partial x}\left(\frac{A_x k_x}{\mu B}\frac{\partial p}{\partial x}\right)\Delta x + \frac{\partial}{\partial x}\left(\frac{A_y k_y}{\mu B}\frac{\partial p}{\partial y}\right)\Delta y + q = \frac{V_b \phi c_t}{5.615}\left(\frac{\partial p}{\partial t}\right). \quad\quad (6.212)$$

For an isotropic and homogeneous system,

$$\frac{\partial^2 p}{\partial x^2} + \frac{\partial^2 p}{\partial y^2} + \frac{q\mu B}{V_b k} = \frac{\phi c_t \mu B}{5.615 k}\left(\frac{\partial p}{\partial t}\right). \quad\quad (6.213)$$

The finite-difference equation for Block 1 is

$$p_2^{n+1} - p_1^{n+1} + \frac{q\mu B}{hk} = \frac{\mu B \phi c_t \Delta x^2}{5.615 k \Delta t}\left(p_1^{n+1} - p_1^{n}\right). \quad\quad (6.214)$$

The finite-difference equation for Block 2 is

$$p_1^{n+1} - p_2^{n+1} + \frac{q\mu B}{hk} = \frac{\mu B \phi c_t \Delta x^2}{5.615 k \Delta t}\left(p_2^{n+1} - p_2^{n}\right), \quad\quad (6.215)$$

where $\frac{q\mu B}{hk} = -665.484$ and $\frac{\mu B \phi c_t \Delta x^2}{5.615 k \Delta t} = 4.44$.

Writing the system of equations in a matrix form yields

$$\begin{bmatrix} -1.44 & 1 \\ 2 & -2.44 \end{bmatrix}\begin{bmatrix} p_1^{n+1} \\ p_2^{n+1} \end{bmatrix} = \begin{bmatrix} -666.669 \\ -1.18519 \end{bmatrix}, \quad\quad (6.216)$$

and solving for p_1^{n+1} and p_2^{n+1}, $p_2^{n+1} = 1,065.291$ psi, and $p_1^{n+1} = p_3^{n+1} = 872.08$ psi.

Validating the results by conducting an MBC,

$$\Delta V_{t0 to 30} = \Delta V_1 + \Delta V_2 + \Delta V_3 = -c V_p \left(\Delta p_1 + \Delta p_2 + \Delta p_3\right)$$

$$= 6 \times 10^{-7}\left[\frac{(1,000)(1,000)(100)(0.15)}{5.615}\right]\left[(1,500 - 872.08) + (1,500 - 2,1065.29) + (1,500 - 872.08)\right]$$

$$= 20,000 \text{ STB} \cdot \quad\quad (6.217)$$

The MBC yields

$$\text{MBC} = \frac{\Delta V}{Q_{\text{total}}(t=20)} = \frac{20{,}000}{1{,}000\ \text{STB/D} \times 20} = 1.00000. \quad\text{(6.218)}$$

To determine whether the well is competent to support the prescribed production rate of 1,000 STB/D for 20 days, one needs to calculate the sandface pressure. Knowing that $r_{eq} = 0.2\Delta x = 800$ ft, if $s = 0$,

$$\Omega = \frac{2\pi kh}{\mu B\left[\ln\left(\dfrac{r_{eq}}{r_w}\right)+s\right]} = 0.8083\ \text{STB/D-psi}, \quad\text{(6.219)}$$

$$q = -\Omega\left(p - p_{sf}\right) = -1{,}000\ \text{STB/D}, \quad\text{(6.220)}$$

$$p_{sf} = p + \frac{q}{\Omega} = -365.079\ \text{psi}. \quad\text{(6.221)}$$

The negative sandface pressure indicates that the well cannot support $q = -1{,}000$ STB/D for 20 days.
If $s = -2$,

$$\Omega = \frac{2\pi kh}{\mu B\left[\ln\left(\dfrac{r_{eq}}{r_w}\right)+s\right]} = 2.0188\ \text{STB/D-psi}, \quad\text{(6.222)}$$

$$p_{sf} = p + \frac{q}{\Omega} = 569.65\ \text{psi}. \quad\text{(6.223)}$$

The positive sandface pressure indicates that the well is able to produce at a rate of $q = -1{,}000$ STB/D for 20 days.

Problem 6.6.21

As shown in **Fig. 6.27,** can the well completed in Block 2 sustain a production rate of −1,000 STB/D? Use the data provided in **Table 6.24** for your analysis.

Solution to Problem 6.6.21

For steady-state flow, the characteristic finite-difference equation is

$$p_{i-1} - 2p_i + p_{i+1} = -\frac{q\mu B}{hk}. \quad\text{(6.224)}$$

When the previous equation is written for Block 2,

$$p_1 - 2p_2 + p_3 = +\frac{q\mu B}{hk} = 0. \quad\text{(6.225)}$$

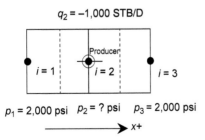

$q_2 = -1{,}000$ STB/D

$p_1 = 2{,}000$ psi $p_2 = ?$ psi $p_3 = 2{,}000$ psi

$x+$

Fig. 6.27—Reservoir system for Problem 6.6.21.

$k_x =$	36 md
$\Delta x =$	200 ft
$\Delta y =$	200 ft
$h =$	100 ft
$\mu =$	1 cp
$B =$	1 res bbl/STB
$r_w =$	0.5 ft
$s =$	−2
$c =$	0 psi^{-1}

Table 6.24—Reservoir properties for Problem 6.6.21.

Solving for p_2,

$$p_2 = \dfrac{\dfrac{q\mu B}{hk} + p_1 + p_3}{2} = 1,876.76 \text{ psi.} \quad\dots\dots\dots\dots\dots\dots\dots\dots\dots\dots\dots\dots\dots\dots\dots\dots (6.226)$$

The PI of the well is

$$\Omega_2 = \dfrac{2\pi kh}{\mu B\left[\ln\left(\dfrac{r_{eq}}{r_w}\right) + s\right]} = \dfrac{2\pi(36)(0.001127)(100)}{(1)(1)\left[\ln\left(\dfrac{115.45}{0.5}\right) - 2\right]} = 7.4063 \text{ STB/D-psi,} \quad\dots\dots\dots\dots\dots\dots\dots\dots\dots (6.227)$$

where $r_{eq} = 0.14\sqrt{(\Delta x)^2 + (\Delta y)^2} = 0.14\sqrt{(800)^2 + (200)^2} = 115.45$ ft,

$$p_{sf,2} = p_2 + \dfrac{q}{\Omega_2} = 1,741.67 \text{ psi} > 14.7 \text{ psi.} \quad\dots\dots\dots\dots\dots\dots\dots\dots\dots\dots\dots\dots\dots\dots\dots (6.228)$$

Because the calculated sandface pressure is larger than the atmospheric pressure, it indicates that the well can sustain the prescribed production rate. To validate the results, one can flow fluid from Node 1 to 2 and from Node 3 to 2:

$$q_{12} = \dfrac{kA}{\mu B}\dfrac{(p_1 - p_2)}{\Delta x} = \dfrac{(0.001127)(36)(200)(100)}{(1)(1)}\dfrac{(2,000 - 1,876.76)}{200} = 500 \text{ STB/D,} \quad\dots\dots\dots\dots\dots\dots (6.229)$$

$$q_{32} = \dfrac{kA}{\mu B}\dfrac{(p_3 - p_2)}{\Delta x} = \dfrac{(0.001127)(36)(200)(100)}{(1)(1)}\dfrac{(2,000 - 1,876.76)}{200} = 500 \text{ STB/D,} \quad\dots\dots\dots\dots\dots\dots (6.230)$$

$$\text{MBC} = \dfrac{|q_{12} + q_{32}|}{q_{\text{well}}} = \left|\dfrac{1,000}{-1,000}\right| = 1.00000, \text{ which confirms the validity of the results.}$$

Problem 6.6.22

Fig. 6.28.1 represents a 2D homogeneous but anisotropic reservoir in which flow of a single-phase incompressible liquid is taking place. Note that the nature of the northern boundary around Block 3 indicates a nonsealing fault. All other boundaries are no-flow boundaries. Reservoir block and fluid properties are provided in **Table 6.25.** Ignore gravitational forces.

1. The well in Block 1 is producing at a constant flowing sandface pressure of 1,450 psia. Compare the magnitudes of pressure differences between Blocks 3 and 2 and also between Blocks 2 and 1. To confirm the accuracy of your solution, perform an MBC.
2. If the skins around the well were removed, could the well have produced at a higher rate?
3. If the well was located in Block 2, would the pressure of Blocks 1 and 2 be the same?
4. If k_x were identical to k_y, can you treat the system as a 1D system?

$k_x =$	92 md
$k_y =$	37 md
$\Delta x =$	500 ft
$\Delta y =$	500 ft
$h =$	73 ft
$\mu =$	1.2 cp
$B =$	1 res bbl/STB
$r_w =$	0.3 ft
$s =$	+2

Fig. 6.28.1—Reservoir system for Problem 6.6.22.   Table 6.25—Reservoir properties for Problem 6.6.22.

Solution to Problem 6.6.22

1. The governing equation for a 2D, incompressible reservoir flow system is (assuming no depth gradients):

$$\frac{\partial}{\partial x}\left(\frac{A_x k_x}{\mu B}\frac{\partial p}{\partial x}\right)\Delta x + \frac{\partial}{\partial y}\left(\frac{A_y k_y}{\mu B}\frac{\partial p}{\partial y}\right)\Delta y + q = 0. \quad\dotfill (6.231)$$

For a homogeneous and anisotropic reservoir, the previous equation can be reduced to

$$k_x\frac{\partial^2 p}{\partial x^2} + k_y\frac{\partial^2 p}{\partial y^2} + \frac{q\mu B}{V_b} = 0. \quad\dotfill (6.232)$$

Mass flux from the northern boundary into the reservoir $\left(\frac{\partial p}{\partial y} > 0\right)$,

$$q = \frac{k_y A_y}{\mu B}\frac{\partial p}{\partial y} = \frac{(1.127\times10^{-3})(37)(500)(73)}{(1)(1.2)}(1.577) = 2{,}000\ \text{STB/D}. \quad\dotfill (6.233)$$

For the incompressible system of this problem, the mass that enters the boundary needs to be produced from the well. Thus, $q_w = -2{,}000$ STB/D,

$$q = -\Omega\left(p_1 - p_{sf}\right), \quad\dotfill (6.234)$$

where

$$\Omega = \frac{2\pi\bar{k}h}{\mu B\left[\ln\left(\frac{r_{eq}}{r_w}\right) + s\right]} = \frac{2\pi(58.3438)(1.127\times10^{-3})(73)}{(1)(1.2)\left(\ln\frac{101.45}{0.3} + 2\right)} = 3.211\ \text{STB/D-psi}, \quad\dotfill (6.235)$$

where $\bar{k} = \sqrt{k_x k_y} = 58.34$ md, and $r_{eq} = \dfrac{0.28\left[\left(\frac{k_y}{k_x}\right)^{\frac{1}{2}}(\Delta x)^2 + \left(\frac{k_x}{k_y}\right)^{\frac{1}{2}}(\Delta y)^2\right]^{\frac{1}{2}}}{\left(\frac{k_y}{k_x}\right)^{\frac{1}{4}} + \left(\frac{k_x}{k_y}\right)^{\frac{1}{4}}} = 101.45\ \text{ft},$

$$p_1 = p_{sf} - \frac{q}{\Omega} = 2{,}072.86\ \text{psi}. \quad\dotfill (6.236)$$

The pressure of other blocks can be calculated in a progressive manner:

- For mass transfer between Blocks 1 and 2,

$$p_2 = p_1 - \frac{q\mu B}{k_x h} = 2{,}389.95\ \text{psi}. \quad\dotfill (6.237)$$

- For mass transfer between Blocks 2 and 3,

$$p_3 = p_2 - \frac{q\mu B}{k_y h} = 3{,}178.38\ \text{psi}. \quad\dotfill (6.238)$$

- Across the northern interface of Block 2,

$$\frac{\partial p}{\partial y} = \frac{p_3 - p_2}{\Delta y} = 1.576\ \text{psi/ft}. \quad\dotfill (6.239)$$

The calculated pressure gradient is close to the given pressure gradient (1.577 psi/ft), indicating a satisfactory material balance:

$$p_2 - p_3 = -788.43 \text{ psi}, \dots\dots\dots\dots\dots\dots\dots\dots\dots\dots\dots\dots\dots\dots\dots\dots \quad (6.240)$$

$$p_1 - p_2 = -317.09 \text{ psi}. \dots\dots\dots\dots\dots\dots\dots\dots\dots\dots\dots\dots\dots\dots\dots\dots \quad (6.241)$$

2. The production rate of the well should be the same because the mass flux from the external boundary is sustained; however, the block pressure will be different.
3. If the well is located in Block 2, Block 1 becomes an inactive block, indicating that $p_1 = p_2$.
4. For an isotropic and homogeneous property distribution, the reservoir can be treated as a 1D system (because $\Delta x = \Delta y$), as shown in **Fig. 6.28.2.**

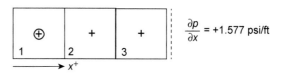

Fig. 6.28.2—Equivalent reservoir system for Problem 6.6.22.

Problem 6.6.23

Consider the two reservoir systems, A and B, shown in **Fig. 6.29.1.** Both reservoirs have identical fluid properties and initial reservoir pressures; however, the permeability of Reservoir A is greater than that of Reservoir B. At initial conditions, the wells in both reservoirs are filled with fluid, and the wellhead pressure gauges read 1,500 psia. A downhole pump is placed in the middle of the formation in both cases. Estimate the initial reservoir pressure if the fluid density is 60 lbm/ft³.

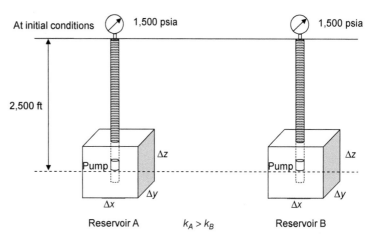

Fig. 6.29.1—Two reservoir systems of Problem 6.6.23.

1. As shown in **Fig. 6.29.2**, both reservoirs are put on production at the same flow rate and at the same time. Is it true that the liquid level in the wellbore of the well in Reservoir A is expected to be higher than that in Reservoir B all the time? Justify your answer.

2. Consider the two identical reservoir systems in **Fig. 6.29.3.** Both systems are set to produce at the same constant flow rate; however, downhole pumps are to be installed in both wells at different depths, as shown in the figure. Draw the expected production profiles and sandface pressure profiles for both cases. Which case is expected to produce more? Why?

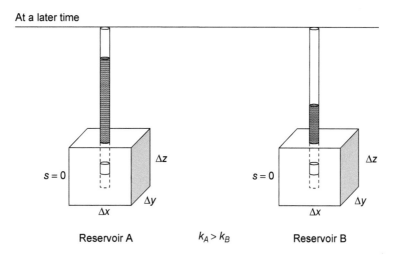

Fig. 6.29.2—Schematic representation of Part 1 of Problem 6.6.23.

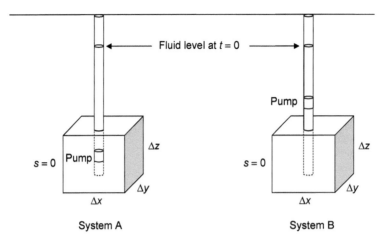

Fig. 6.29.3—Schematic representation of Part 2 of Problem 6.6.23.

Solution to Problem 6.6.23

1. The initial reservoir pressure can be obtained by adding the hydrostatic pressure of the fluid column and the wellhead pressure:

$$p_i = p_{surface} + \frac{\rho}{144}G = 1,500 + \frac{60}{144}2,500 = 2,541.67 \text{ psia.} \quad \dots\dots\dots\dots\dots\dots\dots\dots\dots\dots\dots\dots (6.242)$$

Because $k_A > k_B$,

$$\Omega_A = \frac{2\pi k_A h}{\mu B\left[\ln\left(\dfrac{r_{eq}}{r_w}\right)+s\right]} \text{ and } \Omega_B = \frac{2\pi k_B h}{\mu B\left[\ln\left(\dfrac{r_{eq}}{r_w}\right)+s\right]}, \quad \dots\dots\dots\dots\dots\dots\dots\dots\dots\dots\dots (6.243)$$

$$p_{sfA} = p_i + \frac{q}{\Omega_A}, \quad \dots\dots\dots\dots\dots\dots\dots\dots\dots\dots\dots\dots\dots\dots\dots\dots\dots\dots\dots\dots\dots (6.244)$$

$$p_{sfB} = p_i + \frac{q}{\Omega_B}. \quad \dots\dots\dots\dots\dots\dots\dots\dots\dots\dots\dots\dots\dots\dots\dots\dots\dots\dots\dots\dots\dots (6.245)$$

One can conclude that $p_{sfA} > p_{sfB}$ because $\Omega_A > \Omega_B$. The hydrostatic head of the wellbore gives $p_{sf} = 0.433\gamma H$, where γ is the specific gravity of the fluid and H (in ft) is the fluid-column height in the wellbore. The fluid-column height can be calculated as

$$H = \frac{p_{sf}}{0.433\gamma} \quad \dots\dots\dots\dots\dots (6.246)$$

The fluid densities of Reservoir A and B are identical as $\rho = 60 \text{ lbm/ft}^3$. Thus, the inequality $p_{sfA} > p_{sfB}$ yields $\dfrac{p_{sfA}}{0.433\gamma_A} > \dfrac{p_{sfB}}{0.433\gamma_B}$. Combining with the fluid-column-height calculation, one can observe that $H_A > H_B$.

2. As shown in **Fig. 6.29.4,** Reservoir A will produce more, because Reservoir B will stop producing when the fluid column is lower than the elevation level of the pump.

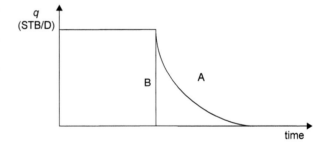

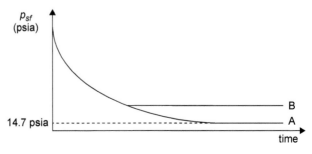

Fig. 6.29.4—Solution to Part 2 of Problem 6.6.23.

Problem 6.6.24

Consider the horizontal reservoir shown in **Fig. 6.30.** The reservoir boundaries are totally sealed. The slightly compressible fluid is produced through the production well in Block 1 at a rate of −1,000 STB/D. The initial reservoir pressure is 3,000 psi.

1. Determine the sandface pressure of the production well at the end of 30 and 60 days using the properties in **Table 6.26.** Perform incremental and cumulative MBCs at the end of each timestep to validate the results.
2. If the reservoir under study is a single-phase gas reservoir with a reservoir temperature of 120°F, the well in Block 1 is produced at a rate of 10 MMscf/D. Initial reservoir pressure is 4,000 psia. At the given pressure and temperature, z-factor is 0.98 and gas viscosity is 0.02 cp. Assume that the pressure drop in Block 1 at the end of first day of production is not significant. Determine the sandface pressure at the end of day 1 if

   - $s = +7$.
   - $s = 0$.
   - $s = -4$.

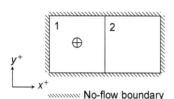

Fig. 6.30—Schematic representation of the reservoir of Problem 6.6.24.

$k_1 =$	200 md
$k_2 =$	100 md
$\Delta x = \Delta y =$	600 ft
$h =$	100 ft
$\mu =$	0.9 cp
$B =$	1 res bbl/STB
$c =$	1×10^{-5} psi^{-1}
$\phi =$	25%
$r_w =$	0.25 ft
$s =$	0

Table 6.26—Reservoir properties for Problem 6.6.24.

Solution to Problem 6.6.24

1. For a 1D, slightly compressible system,

$$\frac{\partial}{\partial x}\left(\frac{A_x k_x}{\mu B}\frac{\partial p}{\partial x}\right)\Delta x + q = \frac{V_b \phi c}{5.615}\frac{\partial p}{\partial t}. \quad \text{...} \quad (6.247)$$

For a system with homogeneous thickness, but heterogeneous permeability,

$$\frac{\partial}{\partial x}\left(k_x \frac{\partial p}{\partial x}\right)\Delta x + \frac{q\mu B}{V_b} = \frac{\phi c \mu B}{5.615}\frac{\partial p}{\partial t}. \quad \text{.............................} \quad (6.248)$$

In finite-difference form,

$$\left[k_{i+\frac{1}{2}}\left(p_{i+1}^{n+1} - p_i^{n+1}\right) - k_{i-\frac{1}{2}}\left(p_i^{n+1} - p_{i-1}^{n+1}\right)\right] + \frac{q\mu B}{h} = \frac{(\Delta x)^2 \phi c \mu B}{5.615\Delta t}\left(p_i^{n+1} - p_i^n\right). \quad \text{..........}(6.249)$$

Using harmonic averaging to calculate the permeability between Blocks 1 and 2,

$$k_{1\text{to}2} = \frac{2k_1 k_2}{k_1 + k_2} = \frac{2(0.2254)(0.1127)}{0.2254 + 0.1127} = 0.1503 \text{ perms.} \quad \text{.................................} \quad (6.250)$$

Also,

$$\frac{(\Delta x)^2 \phi c \mu B}{5.615\Delta t} = \frac{(600)^2 (0.25)(1\times10^{-5})(0.9)(1)}{(5.615)(30)} = 0.0048. \quad \text{.................................} \quad (6.251)$$

At the end of 30 days, the results were as follows:

- At Block 1,

$$-0.1551 p_1 + 0.1503 p_2 = -5.426. \quad \text{...} \quad (6.252)$$

- At Block 2,

$$-0.1503p_1 + 0.1551p_2 = -14.4. \quad \dots\dots\dots\dots\dots\dots\dots\dots\dots\dots\dots\dots\dots \text{(6.253)}$$

Solving $p_1 = 2,050.5$ psi and $p_2 = 2,080$ psi,

$$\Omega = \frac{2\pi kh}{\mu B\left[\ln\left(\dfrac{r_{eq}}{r_w}\right)+s\right]} = \frac{2\pi(100)(0.2254)}{(0.9)(1)\left(\ln\dfrac{120}{0.25}\right)} = 25.47 \text{ STB/D-psi}, \quad \dots\dots\dots\dots\dots\dots \text{(6.254)}$$

$$p_{sf} = -\frac{q_1}{\Omega} + p_1 = 2,050.5 - \frac{1,000}{25.47} \text{ psi} = 2,011 \text{ psi}. \quad \dots\dots\dots\dots\dots\dots\dots\dots\dots \text{(6.255)}$$

Incremental MBC (IMBC),

$$\Delta V = \Delta V_1 + \Delta V_2 = -cV_p\left(\Delta p_1 + \Delta p_2\right)$$

$$= -1\times10^{-6}\,\frac{(600)(600)(100)(0.25)}{5.615}\left[(2,050.5-3,000)+(2,080-3,000)\right]$$

$$= 29,965 \text{ STB}, \quad \dots\dots\dots\dots\dots\dots\dots\dots\dots\dots\dots\dots\dots\dots \text{(6.256)}$$

which yields,

$$\text{IMBC} = \frac{\Delta V}{Q_{\text{total}}\left(t=30\right)} = 0.999. \quad \dots\dots\dots\dots\dots\dots\dots\dots\dots\dots\dots\dots \text{(6.257)}$$

At the end of the first timestep, the cumulative material balance equals the incremental material balance. At the end of 60 days, the results were as follows:

- At Block 1,

$$-0.1551p_1 + 0.1503p_2 = -0.8424. \quad \dots\dots\dots\dots\dots\dots\dots\dots\dots\dots\dots\dots \text{(6.258)}$$

- At Block 2,

$$-0.1503p_1 + 0.1551p_2 = -9.984. \quad \dots\dots\dots\dots\dots\dots\dots\dots\dots\dots\dots\dots \text{(6.259)}$$

Solving $p_1 = 1,112.78$ psi and $p_2 = 1,142.72$ psi,

$$p_{sf} = -\frac{q_1}{\Omega} + p_1 = 1,112.78 - \frac{1,000}{25.47} \text{ psi} = 1,073.6 \text{ psi}. \quad \dots\dots\dots\dots\dots\dots \text{(6.260)}$$

IMBC,

$$\Delta V_{t\,30\,\text{to}\,60} = \Delta V_1 + \Delta V_2 = -cV_p\left(\Delta p_1 + \Delta p_2\right)$$

$$= 1\times10^{-6}\,\frac{(600)(600)(100)(0.25)}{5.615}\left[(2,050.5-1,112.78)+(2,080-1,142.72)\right]$$

$$= 30,053 \text{ STB}, \quad \dots\dots\dots\dots\dots\dots\dots\dots\dots\dots\dots\dots\dots\dots \text{(6.261)}$$

which yields,

$$\text{IMBC} = \frac{\Delta V}{Q_{\text{total}}\left(t=30\right)} = 1.002. \quad \dots\dots\dots\dots\dots\dots\dots\dots\dots\dots\dots\dots \text{(6.262)}$$

Cumulative MBC (CMBC),

$$\Delta V_{t\,0\,\text{to}\,60} = \Delta V_1 + \Delta V_2 = -cV_p\left(\Delta p_1 + \Delta p_2\right)$$

$$= -1\times10^{-6}\,\frac{(600)(600)(100)(0.25)}{5.615}\left[(3,000-1,112.78)+(3,000-1,142.72)\right]$$

$$= 60,018.7 \text{ STB}, \quad \dots\dots\dots\dots\dots\dots\dots\dots\dots\dots\dots\dots\dots\dots \text{(6.263)}$$

which yields,

$$CMBC = \frac{\Delta V}{Q_{total}(t=60)} = 1.000. \dotfill (6.264)$$

2. For the gas flow, the gas viscosity and FVF are strong functions of pressure. In this problem, we assume that the pressure change is not significant. Therefore, one can use the gas viscosity and FVF evaluated at the initial pressure of 4,000 psi. One can calculate the FVF of gas in res bbl/scf:

$$B_g = -\frac{z_g T p_{sc}}{5.615 T_{sc} p} = \frac{(0.98)(580)(14.7)}{(5.165)(520)(4,000)} \, 7.2 \times 10^{-4} \text{ res bbl/scf.} \dotfill (6.265)$$

- For $s = +7$:

$$\Omega = \frac{2\pi kh}{\mu_g B_g \left[\ln\left(\frac{r_{eq}}{r_w}\right) + s \right]} = \frac{2\pi(100)(0.2254)}{(0.02)(7.2\times10^{-4})\left(\ln\frac{120}{0.25} + 7 \right)} = 74.63 \text{ Mscf/D-psi.} \dotfill (6.266)$$

Solving for the sandface pressure,

$$p_{sf} = -\frac{q_1}{\Omega} + p_1 = 3,866 \text{ psi.} \dotfill (6.267)$$

- For $s = 0$:

$$\Omega = \frac{2\pi kh}{\mu_g B_g \left[\ln\left(\frac{r_{eq}}{r_w}\right) + s \right]} = \frac{2\pi(100)(0.2254)}{(0.02)(7.2\times10^{-4})\left(\ln\frac{120}{0.25} + 0 \right)} = 158.73 \text{ Mscf/D-psi.} \dotfill (6.268)$$

Solving for the sandface pressure,

$$p_{sf} = -\frac{q_1}{\Omega} + p_1 = 3,937 \text{ psi.} \dotfill (6.269)$$

- For $s = -4$:

$$\Omega = \frac{2\pi kh}{\mu_g B_g \left[\ln\left(\frac{r_{eq}}{r_w}\right) + s \right]} = \frac{2\pi(100)(0.2254)}{(0.02)(7.2\times10^{-4})\left(\ln\frac{120}{0.25} + -4 \right)} = 454.55 \text{ Mscf/D-psi.} \dotfill (6.270)$$

Solving for the sandface pressure,

$$p_{sf} = -\frac{q_1}{\Omega} + p_1 = 3,978 \text{ psi.} \dotfill (6.271)$$

Problem 6.6.25

Consider the following 2D problem for incompressible fluid flow in the homogeneous and isotropic reservoir shown in **Fig. 6.31.** There is an observation well located in Block 4 with a pressure gauge reading of 2,200 psia. The reservoir has two open boundaries in Blocks 1 and 3. Determine the following using the reservoir properties listed in **Table 6.27:**

1. The flow rate for the well located in Block 5.
2. The pressure at Blocks 1, 2, 3, and 5.
3. Whether the well is damaged or stimulated.

$\mu =$	1.1 cp
$B =$	1 res bbl/STB
$k_x = k_y =$	100 md
$\Delta x = \Delta y =$	1,000 ft
$h =$	50 ft
$r_w =$	0.25 ft
p_{sf}(Well 5) =	1,800 psia

Table 6.27—Reservoir properties for Problem 6.6.25.

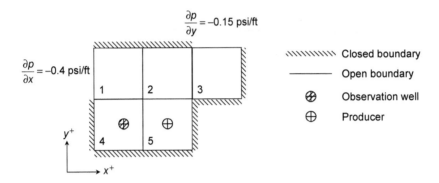

Fig 6.31—Schematic representation of the reservoir of Problem 6.6.25.

Solution to Problem 6.6.25

1. Flow rate of Well 5 must be equal to that of the total fluid entering the reservoir:

$$q_{well-5} = -\frac{kA}{\mu B}\left(\frac{\partial p}{\partial x} + \frac{\partial p}{\partial y}\right) = -\frac{0.1127 \times 1,000 \times 50}{1.1 \times 1}(0.4 - 0.15) = -1,280.68 \text{ STB/D.} \quad \dots\dots\dots\dots \text{(6.272)}$$

2. Flow equations by node (*Note*: $p_4 = p_{sf4} = 2,200$ psi in a shut-in well, sandface pressure is equal to the block pressure).

- Node 1:

$$-4p_1 + p_2 + p_4 + p_1 + p_1 + 0.4\Delta x = 0, \quad \dots\dots\dots\dots\dots\dots\dots\dots\dots\dots\dots\dots\dots\dots \text{(6.273)}$$

which can be simplified as

$$-2p_1 + p_2 = -2,600. \quad \dots\dots\dots\dots\dots\dots\dots\dots\dots\dots\dots\dots\dots\dots\dots\dots \text{(6.274)}$$

- Node 2:

$$p_1 - 3p_2 + p_3 + p_5 = 0. \quad \dots\dots\dots\dots\dots\dots\dots\dots\dots\dots\dots\dots\dots\dots\dots\dots \text{(6.275)}$$

- Node 3:

$$p_2 - p_3 + 0.15\Delta y = 0, \quad \dots\dots\dots\dots\dots\dots\dots\dots\dots\dots\dots\dots\dots\dots\dots\dots \text{(6.276)}$$

$$p_2 - p_3 = 150, \quad \dots\dots\dots\dots\dots\dots\dots\dots\dots\dots\dots\dots\dots\dots\dots\dots\dots\dots \text{(6.277)}$$

- Node 5:

$$p_2 - 2p_5 + 2,200 + \frac{q\mu B}{hk} = 0, \quad \dots\dots\dots\dots\dots\dots\dots\dots\dots\dots\dots\dots\dots \text{(6.278)}$$

where

$$\frac{q\mu B}{hk} = \frac{-1,280.68 \times 1.1 \times 1}{50 \times 0.1127} = -250.0. \quad \dots\dots\dots\dots\dots\dots\dots\dots\dots\dots\dots\dots \text{(6.279)}$$

Thus, Eq. 6.277 can be simplified:

$$p_2 - 2p_5 = -1,950. \quad \dots\dots\dots\dots\dots\dots\dots\dots\dots\dots\dots\dots\dots\dots\dots\dots \text{(6.280)}$$

In matrix notation,

$$\begin{bmatrix} -2 & 1 & 0 & 0 \\ 1 & -3 & 1 & 1 \\ 0 & 1 & -1 & 0 \\ 0 & 1 & 0 & -2 \end{bmatrix} \begin{bmatrix} p_1 \\ p_2 \\ p_3 \\ p_5 \end{bmatrix} = \begin{bmatrix} -2,600 \\ 0 \\ 150 \\ -1,950 \end{bmatrix}. \quad \dots\dots\dots\dots\dots\dots\dots\dots \text{(6.281)}$$

The solution of the previous system of equations gives $p_1 = 2,362.5$ psia, $p_2 = 2,125$ psia, $p_3 = 1,975$ psia, and $p_5 = 2,037.5$ psia.

3. The expression of the skin factor can be derived from the Peaceman wellbore model:

$$s = -\ln\left(\frac{r_{eq}}{r_w}\right) + \frac{2\pi\bar{k}h}{\mu Bq}\left(p_5 - p_{sf}\right), \quad \dots\dots\dots\dots\dots\dots\dots\dots\dots\dots \text{(6.282)}$$

where $r_{eq} = 0.2 \times \Delta x = 0.2 \times 1,000$ ft = 200 ft.
One can solve for the skin factor,

$$s = -\ln\left(\frac{200}{0.25}\right) + \frac{2 \times 3.14 \times 1.127 \times 0.1 \times 50}{1.1 \times 1 \times 1280.68}(2,037.5 - 1,800) = -0.72. \quad \dots\dots\dots\dots\dots \text{(6.283)}$$

Therefore, the well is stimulated.

Nomenclature

a	=	constant
a_j	=	distance between a well and its image well (Fig. 6.4), L, ft
A	=	cross-sectional area, L^2, ft^2
A_x	=	cross-sectional area perpendicular to flow along the x-direction, L^2, ft^2
b	=	constant
B	=	FVF, L^3/L^3, res bbl/STB or res bbl/scf
B_g	=	gas FVF, L^3/L^3, res bbl/scf
B_o	=	oil FVF, L^3/L^3, res bbl/STB
c	=	compressibility, Lt^2/m, psi^{-1}
$C_{i,j,k}$	=	transmissibility coefficient of block i,j,k in SIP notation, L^4t/m, STB/(D–psi)
C_H	=	constant defined by the Babu and Odeh model
d	=	distance between two points, such as penetration distance, L, ft
f	=	fraction of well flow associated with the wellblock defined by the Abou-Kassem and Aziz model (Fig. 6.4)
g	=	acceleration of gravity, L/t^2, ft/s^2
g_c	=	unit conversion factor in Newton's law
G	=	elevation with respect to absolute datum being positive upward, L, ft
h	=	thickness, L, ft

H	=	fluid-column height in the wellbore, L, ft
i	=	node number for the blocks surrounding the well (Fig. 6.4)
j	=	well number (Fig. 6.4)
J	=	productivity or injectivity index, L^4t/m, STB/D–psi
k	=	permeability, L^2, perms (1.127 perms = 1 darcy)
$\bar{k}$	=	the geometric mean of the directional permeabilities in the plane to which the well is orthogonal, L^2, perms
k_x	=	permeability in the x-direction, L^2, perms
k_y	=	permeability in the y-direction, L^2, perms
k_z	=	permeability in the z-direction, L^2, perms
L	=	the length of the wellbore in the wellblock, L, ft
p	=	pressure, m/Lt^2, psia
p_{atm}	=	atmospheric pressure, m/Lt^2, psia
p_{block}	=	block pressure, m/Lt^2, psia
p_e	=	pressure at the external radius, m/Lt^2, psia
p_i	=	initial pressure, m/Lt^2, psia
p_{sc}	=	standard condition pressure, m/Lt^2, psia
p_{sf}	=	sandface pressure, m/Lt^2, psia

$p_{surface}$	=	surface pressure, m/Lt², psia
p_{wf}	=	well bottomhole flowing pressure, m/Lt², psia
$\bar{p}_{wf}$	=	average well bottomhole flowing pressure, m/Lt², psia
$\bar{p}$	=	average reservoir pressure, m/Lt², psia
p'_{xy}	=	argument defined by Babu and Odeh's (1989) horizontal wellbore model
P_{xy}	=	argument defined by Babu and Odeh's (1989) horizontal wellbore model
P_{xyz}	=	argument defined by Babu and Odeh's (1989) horizontal wellbore model
P_y	=	argument defined by Babu and Odeh's (1989) horizontal wellbore model
q	=	production or flow rate, L³/t, STB/D
$q_{abandon}$	=	abandonment flow rate, L³/t, STB/D
q_{sc}	=	volumetric flow rate at standard condition, L³/t, B/D for liquid and scf/D for gas
$Q_{i,j,k}$	=	right-hand-side of the SIP equation for block i,j,k, L³/t, STB/D
Q_{total}	=	L³, STB for liquid and scf for gas
R_s	=	solution gas oil ratio, L³/L³, scf/STB
r	=	radial distance in cylindrical and spherical coordinator systems, L, ft
$r_{i,j}$	=	distance from Node i to Well j (Fig. 6.4), L, ft
r_{eq}	=	equivalent wellblock radius, L, ft
r_w	=	wellblock radius, L, ft
s	=	skin factor, dimensionless
s_m	=	mechanical skin factor, dimensionless
s_p	=	partial penetration skin factor, dimensionless
S_g	=	gas saturation, fraction
S_o	=	oil saturation, fraction
S_w	=	water saturation, fraction
t	=	time, t, days
T	=	absolute temperature, T, °R,
T_i	=	interface transmissibility fraction between a surrounding block and the wellblock as defined by the Abou-Kassem and Aziz model (Fig. 6.4), dimensionless
V	=	total fluid volume expanded, L³, STB
V_b	=	bulk volume, L³, ft³
V_p	=	pore volume, L³, ft³
X_0	=	x coordinate of the center of the horizontal well in the x-z plane

x	=	distance in x-direction in the Cartesian coordinate system, L, ft
x_w	=	well coordinate in the x-direction with center at wellblock point, L, ft
y	=	distance in y-direction in the Cartesian coordinate system
y_{mid}	=	midpoint location of the well
y_w	=	well coordinate in the y-direction with center at wellblock point, L, ft
z	=	distance in z-direction in the Cartesian coordinate system, L, ft
z_g	=	gas compressibility factor, dimensionless
Z_0	=	z coordinate of the center of the horizontal well in the x-z plane
β_c	=	transmissibility conversion factor, with a magnitude of 1.127
γ	=	fluid gravity, m/L²t², psi/ft
$\bar{\gamma}_{wb}$	=	average wellbore hydrostatic pressure gradient, m/L²t², psi/ft
Δ	=	difference of a variable (independent or dependent)
Δx	=	control volume dimension along the x-direction, L, ft
Δy	=	control volume dimension along the y-direction, L, ft
Δz	=	control volume dimension along the z-direction, L, ft
μ	=	viscosity, cp
μ_o	=	oil viscosity, m/Lt, cp
ρ	=	density, m/L³, lbm/ft³
Φ	=	flow potential, m/Lt², psia
ϕ	=	porosity, fraction
Ω	=	well productivity/injection, L⁴t/m, STB/D–psia for liquid, scf/D–psia for gas

Superscripts

a	=	Case a
b	=	Case b
$n+1$	=	current (new) time level
n	=	previous (old) time level
$v+1$	=	current (new) iteration level
v	=	previous (old) iteration level

Subscripts

f	=	fluid; free gas in Eq. 6.43
g	=	gas-related characteristic
i,j,k	=	central block i, j, k
k	=	index of perforated layer

l	=	phase (l = o, w, g)		sf	=	sandface
n	=	layer number		sp	=	specified
o	=	oil-related characteristic		t	=	total
ref	=	physical parameter at reference condition		w	=	water-related characteristic
sc	=	standard conditions		well	=	pertaining to a well

References

Abou-Kassem, J. H. and Aziz, K. 1985. Analytical Well Models for Reservoir Simulation. *SPE J.* **25**(4): 573–579. SPE-11719-PA. http://dx.doi.org/10.2118/11719-PA.

Babu, D. K. and Odeh, A. S. 1989. Productivity of a Horizontal Well. *SPE Res Eval & Eng* **4**(04): 417–421. SPE-18298-PA. http://dx.doi.org/10.2118/18298-PA.

Ertekin, T., Abou-Kassem, J. H., and King, G. R. 2001. *Basic Applied Reservoir Simulation*. Richardson, Texas, US: Society of Petroleum Engineers.

Peaceman, D. W. 1983. Interpretation of Well-Block Pressures in Numerical Reservoir Simulation with Nonsquare Grid Blocks and Anisotropic Permeability. *SPE J* **23**(3): 531–543. SPE-10528-PA. https://dx.doi.org/10.2118/10528-PA.

Peaceman, D. W. 1993. Representation of a Horizontal Well In Numerical Reservoir Simulation. *SPE Advanced Technology Series* **1**(01): 7–16. SPE-21217-PA. https://dx.doi.org/10.2118/21217-PA.

van Poollen, H. K., Breitenbach, E. A., and Thurnau, D. H. 1968. Treatment of Individual Wells and Grids in Reservoir Modeling. *SPE J.* **8**(04): 341–346. SPE-2022-PA. http://dx.doi.org/10.2118/2022-PA.

Chapter 7

Solution Methods for Systems of Linear Algebraic Equations

In the previous chapters, we have discussed the physical aspects and pertinent formalisms that must be considered in studying fluid flow in porous media. We developed the governing partial-differential equations (PDEs) and learned about the associated boundary conditions and initial conditions used in describing the flow problem. Next, we learned how to represent the PDE and its associated boundary conditions using finite-difference analogs. In addition to discretization of the different forms of boundary conditions of the governing PDEs, we discussed how to achieve a similar-level discretization on the structural representation, as well as accommodation of the wells of a reservoir using different types of gridding techniques. At this stage, we have a characteristic finite-difference equation representing the flow problem at preselected points in the reservoir. This characteristic finite-difference equation, when written at each node where the dependent variables are not known, will generate a system of linear or nonlinear algebraic equations that need to be simultaneously solved. In this chapter, we discuss the overall merits of direct- and iterative-solution techniques as they are applied to systems of linear algebraic equations generated from the characteristic finite-difference approximations.

7.1 Solution Methods
The solution methods applicable to systems of linear algebraic equations are classified under two general groups: direct methods and iterative methods.

7.1.1 Direct Methods. A direct method is capable of producing an exact solution on a computing machine that has no limitation on character length in its results, so there are no round-off errors. **Fig. 7.1** is a schematic representation of a direct solver. In a direct-solver application, the solution is obtained after carrying out a fixed number of operations. Therefore, the computational costs involved with their use are predictable. Direct methods are also much more easily transportable from one application to another because they most often appear in the computational workflow as an external subroutine subprogram. Direct solvers have the following advantages and disadvantages:

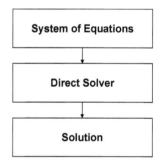

Fig. 7.1—Direct-solution process.

- More-accurate solutions are generated.
- Computational costs are predictable.
- Long and skinny reservoir applications are as well suited for direct solvers, because the bandwidth of the coefficient matrix is smaller.
- Larger storage is required.
- Computational overhead is demanding.

The most popular direct methods are as follows (for addition discussion of these methods, see Ertekin et al. 2001):

- Gaussian elimination.
- Gauss-Jordan reduction.
- Matrix factorization (Crout reduction).

In the application of Gaussian elimination for a problem, such as

$$[A][P] = [Q], \quad \dots\dots\dots\dots\dots\dots\dots\dots\dots\dots\dots\dots\dots\dots\dots\dots\dots\dots\dots (7.1)$$

where $[A]$ is the coefficient matrix, $[P]$ is the unknown vector, and $[Q]$ is the right-hand-side vector, the first step involves converting the coefficient matrix to an upper-triangular matrix so that Eq. 7.1 becomes

$$[A'][P] = [Q']. \quad \dots\dots\dots\dots\dots\dots\dots\dots\dots\dots\dots\dots\dots\dots\dots\dots\dots\dots\dots (7.2)$$

where $[A']$ is the new form of the coefficient matrix and its structure is upper triangular. In an upper-triangular matrix, all the elements of the matrix below the main diagonal are zero. The original matrix equation (Eq. 7.1) is converted to an upper-triangular-matrix equation (Eq. 7.2) through use of a process called forward elimination. During the forward-elimination process, unknowns are systematically removed from the rows of the matrix equation. In carrying out the elimination stage, the first equation is divided by the coefficient of the first unknown; then, the first unknown is eliminated from the succeeding equations. In the next step of the forward elimination, the modified second equation is divided by the coefficient of the second unknown, and the result is used to eliminate the second unknown from the remaining equations. This process is continued until $[A']$ becomes an upper-triangular matrix. When this stage is complete, the last equation in the upper-triangular-matrix equation will have one unknown, which can be solved explicitly. After solving for the last unknown by back substitution, the remaining unknowns are systematically solved in a successive manner, moving from bottom to top.

The Gauss-Jordan method is a variation of Gaussian elimination, and it accomplishes the effect of back substitution (the second stage of Gaussian elimination) at the same time of the forward elimination, with the reduction of the elements below the main diagonal. Again, we start with the original problem indicated in Eq. 7.1: $[A][P] = [Q]$, and form the augmented matrix, $[A \vdots Q \vdots I]$. This time, we perform normalization on this augmented matrix until we end with an augmented matrix, $[I \vdots P \vdots A^{-1}]$. The column represented by $\vdots P \vdots$ in this augmented matrix will be the solution vector.

Matrix factorization is sometimes referred to as an $[L][U]$ factorization of the coefficient matrix. The algorithmic process basically is the same as the Gaussian-elimination algorithm, but has the advantage of not operating on the right-hand-side vector using the factorization. In this process, the equation to be solved is again Eq. 7.1: $[A][P] = [Q]$.

The goal is to factorize $[A]$ into upper-$[A]$ and lower-$[A]$ triangular matrices, or

$$[A] = [L][U]. \dotfill (7.3)$$

The original problem of Eq. 7.1 can be rewritten as

$$[L]\{[U][P]\} = [Q], \dotfill (7.4)$$

or using an intermediate vector of $[r]$ first, the following problem is solved:

$$[L][r] = [Q]. \dotfill (7.5)$$

Eq. 7.5 is easy to solve because $[L]$ is a lower-triangular matrix. Once the intermediate vector $[r]$ is obtained, then in the second stage, the unknown vector $[P]$ can be obtained from

$$[U][P] = [r]. \dotfill (7.6)$$

Again, Eq. 7.6 is easy to solve because this time $[U]$ is an upper-triangular matrix. The solution of Eq. 7.6 starts from the last unknown and by back substitution moves from the bottom of $[U]$ to the top row of $[U]$ when the first unknown of the problem is finally solved. As expected, most of the computational overhead is encountered in the factorization of the coefficient matrix $[A]$.

In this chapter, we include the procedures to implement the following direct solvers (Price and Coats 1974; Ertekin et al. 2001):

- Gaussian elimination.
- Gauss-Jordan reduction.
- Crout reduction.
- Thomas' algorithm.

Consider the following form of the linear system of equations with coefficient matrix $[A]$, unknown vector $[x]$, and right-hand-side vector $[d]$:

$$\begin{bmatrix} a_{11} & a_{12} & \cdots & \cdots & \cdots & a_{1n} \\ a_{21} & a_{22} & \cdots & \cdots & \cdots & a_{2n} \\ a_{31} & a_{32} & \cdots & \cdots & \cdots & a_{3n} \\ \vdots & \cdots & \ddots & \cdots & \cdots & \vdots \\ \vdots & \cdots & \cdots & \ddots & \cdots & \vdots \\ a_{n1} & a_{n2} & \cdots & \cdots & \cdots & a_{nn} \end{bmatrix} \begin{bmatrix} x_1 \\ x_2 \\ x_3 \\ \vdots \\ \vdots \\ x_n \end{bmatrix} = \begin{bmatrix} d_1 \\ d_2 \\ d_3 \\ \vdots \\ \vdots \\ d_n \end{bmatrix}. \dotfill (7.7)$$

Gaussian Elimination Algorithm. For $i = 1, 2, \ldots, n$, set $d_i^{(0)} = d_i$ and $a_{i,j}^{(0)} = a_{i,j}$ for $j = 1, 2, \ldots, n$;

- Set $d_i^{(i)} = \dfrac{d_i^{(i-1)}}{a_{i,i}^{(i-1)}}$ for $i = 1, 2, 3, \ldots, n$; \dotfill (7.8)

- Set $a_{i,j}^{(i)} = \dfrac{a_{i,j}^{(i-1)}}{a_{i,i}^{(i-1)}}$ for $j = i + 1, i + 2, ..., n$ .... (7.9)

  and $a_{i,i}^{(i)} = 1.0$;

- Set $d_k^{(i)} = d_k^{(i-1)} - d_i^{(i)} a_{k,i}^{(i-1)}$ for $k = i + 1, i + 2, ..., n$; .... (7.10)

- For $j = i + 1, i + 2, ..., n$; $a_{k,j}^{(i)} = a_{k,j}^{(i-1)} - d_i^{(i)} a_{k,i}^{(i-1)}$ .... (7.11)

  and $a_{k,i}^{(i)} = 0$;

- Set $x_n = d_n^{(n)}$, for $i = n - 1, n - 2, ..., 1$, set

$$x_i = d_i^{(n)} - \sum_{j=i+1}^{n} a_{i,j}^{(n)} x_j. \qquad (7.12)$$

Gauss-Jordan Reduction. For $i = 1, 2, ..., n$, set $d_i^{(0)} = d_i$ and $a_{i,j}^{(0)} = a_{i,j}$ for $j = 1, 2, ..., n$;

- Set $d_i^{(i)} = \dfrac{d_i^{(i-1)}}{a_{i,i}^{(i)}}$ for $i = 1, 2, 3, ..., n$; .... (7.13)

- Set $a_{i,j}^{(i)} = \dfrac{a_{i,j}^{(i-1)}}{a_{i,i}^{(i-1)}}$ for $j = i + 1, i + 2, ..., n$ .... (7.14)

  and $a_{i,i}^{(i)} = 1.0$;

- Set $d_k^{(i)} = d_k^{(i-1)} - d_i^{(i)} a_{k,i}^{(i-1)}$ for $k = i + 1, i + 2, ..., n$ .... (7.15)

  and $k \neq i$;

- For $j = i + 1, i + 2, ..., n$, $a_{k,j}^{(i)} = a_{k,j}^{(i-1)} - a_{i,j}^{(i)} a_{k,i}^{(i-1)}$ .... (7.16)

  and $a_{k,i}^{(i)} = 0$;

- For $i = 1, 2, ..., n$, set $x_i = d_i^{(n)}$. .... (7.17)

Crout Reduction. For $i = 1, 2, ..., n$ set $l_{ii} = 1$;

- For $j = 1, 2, ... n$, perform the following two computational steps.

  1. For $i \leq j$, $u_{i,j} = a_{i,j}$ for $i = 1$.

     for $i = 2, 3, ..., j$, $u_{i,j} = a_{i,j} - \sum_{k=1}^{i-1} l_{i,k} u_{k,j}$. .... (7.18)

  2. For $i > j$, for $j = 1$: $l_{i,1} = \dfrac{a_{i,1}}{u_{11}}$, ... (7.19)

     and for $i = j + 1, j + 2, ..., n$,

     and for $j \geq 2$: $l_{i,j} = \dfrac{1}{u_{i,j}} \left(a_{i,j} - \sum_{k=1}^{j-1} l_{i,k} u_{k,j} \right)$ .... (7.20)

- Set $y_1 = d_1$, then for $i = 2, 3, ..., n$: $y_i = d_i - \sum_{j=1}^{i-1} l_{i,j} y_j$. .... (7.21)

- Set $x_n = \dfrac{y_n}{u_{nm}}$ then, for $i = n - 2, n - 2, ..., 1$:

$$x_i = \frac{1}{u_{i,j}} \left(y_i - \sum_{j=i+1}^{n} u_{i,j} x_j \right). \qquad (7.22)$$

Thomas' Algorithm. Thomas' algorithm is a linear solver, which is capable of solving the linear system of algebraic equations with a tridiagonal coefficient matrix structure:

$$
\begin{bmatrix}
b_1 & c_1 & \cdots & \cdots & \cdots & 0 \\
a_2 & b_2 & c_2 & \cdots & \cdots & 0 \\
0 & a_3 & b_3 & c_3 & \cdots & 0 \\
\vdots & \cdots & \ddots & \cdots & \cdots & \vdots \\
\vdots & \cdots & \cdots & \ddots & \cdots & \vdots \\
\vdots & \cdots & \cdots & 0 & a_m & b_m
\end{bmatrix}
\begin{bmatrix}
x_1 \\ x_2 \\ x_3 \\ \vdots \\ \vdots \\ x_m
\end{bmatrix}
=
\begin{bmatrix}
d_1 \\ d_2 \\ d_3 \\ \vdots \\ \vdots \\ d_m
\end{bmatrix}
. \quad \dots \dots \dots \dots \dots \dots \dots \dots \dots \dots \dots \dots \dots (7.23)
$$

The algorithmic process to implement Thomas' algorithm is

- For $i = 1$, define

$$
\omega_1 = \frac{c_1}{b_1} \quad \dots \dots \dots \dots \dots \dots \dots \dots \dots \dots \dots \dots \dots \dots \dots \dots \dots \dots (7.24)
$$

and

$$
g_1 = \frac{d_1}{b_1}. \quad \dots \dots \dots \dots \dots \dots \dots \dots \dots \dots \dots \dots \dots \dots \dots \dots (7.25)
$$

- For $i = 2, 3, \dots m - 1$,

$$
w_i = \frac{c_i}{b_i - a_i w_{i-1}}. \quad \dots \dots \dots \dots \dots \dots \dots \dots \dots \dots \dots \dots \dots (7.26)
$$

- For $i = 2, 3, \dots m$,

$$
g_i = \frac{d_i - a_i g_i}{b_i - a_i w_{i-1}}, \quad \dots \dots \dots \dots \dots \dots \dots \dots \dots \dots \dots (7.27)
$$

then $x_m = g_m$.

- For $i = m - 1, m - 2, \dots 3, 2, 1$,

$$
x_i = g_i - w_i x_{i+1}. \quad \dots \dots \dots \dots \dots \dots \dots \dots \dots \dots \dots \dots \dots \dots (7.28)
$$

7.1.2 Iterative Methods. For large systems of equations, direct methods, most of the time, will not be applicable in terms of the nature of the computational overhead that will be realized. For large problems, iterative processes typically are preferred. **Fig. 7.2** gives a schematic representation of an iterative process.

The major features of iterative methods are as follows:

- Asymptotically converging solutions (approximate in nature).
- Unpredictable computer costs.
- Lower storage requirements.
- Faster processing than direct methods for large problems.
- Less transportability than direct solvers.
- Efficacy of the solution depends on the users' experiences; experience in using iterative methods is important.

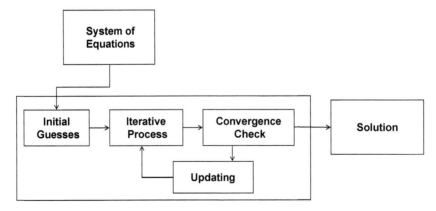

Fig. 7.2—Iterative-solution process.

In reservoir simulation, a powerful solver—whether a direct method or iterative method—is an absolute necessity for these reasons: to provide accurate material-balance checks (MBCs), to ensure stability over time, and to build confidence in the use of the model for prediction. In the next subsection, we briefly review some of the more widely used iterative-solution protocols.

Systematic Iterative Methods. Within the systematic iterative methods, we explore Jacobi's iteration and the Gauss-Seidel procedure (for additional information on these procedures, see Ertekin et al. 2001), and illustrate each of these protocols by way of the incompressible equation, as written with the strongly implicit pressure (SIP) notation, which is repeated here as Eq. 7.29:

$$B_{i,j,k}p_{i,j,k-1} + S_{i,j,k}p_{i,j-1,k} + W_{i,j,k}p_{i-1,j,k} + C_{i,j,k}p_{i,j,k} + E_{i,j,k}p_{i+1,j,k} + N_{i,j,k}p_{i,j+1,k} + A_{i,j,k}p_{i,j,k+1} = Q_{i,j,k}. \quad \dots \dots \dots \dots (7.29)$$

When written at each discrete point of a reservoir represented by a 3D grid structure, Eq. 7.29 generates a system of equations and block pressures that will be the unknowns of the problem. (For example, a 5×6×3 reservoir model will generate 90 blocks with 90 unknown block pressures.)

Jacobi's Iteration Equations. Eq. 7.30 gives the Jacobi's iterative method, as applied to Eq. 7.29:

$$p_{i,j,k}^{(k+1)} = \frac{1}{C_{i,j,k}}\Big[Q_{i,j,k} - B_{i,j,k}p_{i,j,k-1}^{(k)} - S_{i,j,k}p_{i,j-1,k}^{(k)} - W_{i,j,k}p_{i-1,j,k}^{(k)} -$$
$$E_{i,j,k}p_{i+1,j,k}^{(k)} - N_{i,j,k}p_{i,j+1,k}^{(k)} - A_{i,j,k}p_{i,j,k+1}^{(k)}\Big]. \quad \dots \dots \dots \dots (7.30)$$

Eq. 7.30 is executed for $i = 1, \dots, N_x, j = 1, \dots, N_y$ and $k = 1, \dots, N_z$. Once all block pressures at the new iteration level are calculated, a convergence check is implemented:

$$\left|p_{i,j,k}^{(k+1)} - p_{i,j,k}^{(k)}\right| \le \varepsilon_p \text{ for } i = 1, \dots, N_x, j = 1, \dots, N_y \text{ and } k = 1, \dots, N_z. \quad \dots \dots \dots \dots (7.31)$$

In Eq. 7.31, ε_p is the convergence criterion. During the convergence check, even if only one block violates the convergence criterion, a new level of iteration must be conducted.

Gauss-Seidel Iterative Equation. In the Gauss-Seidel iterative process, Jacobi's iteration is accelerated by using block pressures at the new iteration level, as soon as they become available. Accordingly, as block pressures are calculated in the increasing orders of i, j, and k, the algorithmic equation for the implementation of the Gauss-Seidel procedure will be in the following form:

$$p_{i,j,k}^{(k+1)} = \frac{1}{C_{i,j,k}}\Big[Q_{i,j,k} - B_{i,j,k}p_{i,j,k-1}^{(k+1)} - S_{i,j,k}p_{i,j-1,k}^{(k+1)} - W_{i,j,k}p_{i-1,j,k}^{(k+1)} -$$
$$E_{i,j,k}p_{i+1,j,k}^{(k)} - N_{i,j,k}p_{i,j+1,k}^{(k)} - A_{i,j,k}p_{i,j,k+1}^{(k)}\Big]. \quad \dots \dots \dots \dots (7.32)$$

The convergence criterion will be implemented in the same way as Jacobi's iteration. The Gauss-Seidel procedure converges two times faster than Jacobi's iteration.

Successive Over-Relaxation Methods. In successive over-relaxation (SOR) methods, the objective is to accelerate the convergence rate by modifying the value of the unknowns calculated at a given iteration level. If one adds and subtracts $p_{i,j,k}^{(k)}$ to the right-hand side of the Gauss-Seidel equation, then it will be possible to see that in Gauss-Seidel iterations what is achieved can be simply expressed as

$$p_{i,j,k}^{(k+1)} = \frac{1}{C_{i,j,k}}\Big[Q_{i,j,k} - B_{i,j,k}p_{i,j,k-1}^{(k+1)} - S_{i,j,k}p_{i,j-1,k}^{(k+1)} - W_{i,j,k}p_{i-1,j,k}^{(k+1)} -$$
$$E_{i,j,k}p_{i+1,j,k}^{(k)} - N_{i,j,k}p_{i,j+1,k}^{(k)} - C_{i,j,k}p_{i,j,k}^{(k)} - A_{i,j,k}p_{i,j,k+1}^{(k)}\Big] + p_{i,j,k}^{(k)}. \quad \dots \dots \dots \dots (7.33a)$$

A careful examination of Eq. 7.33a shows that $p_{i,j,k}^{(k+1)} - p_{i,j,k}^{(k)}$ represents improvement in the solution between iteration levels (k) and $(k+1)$:

$$\text{Improvement} = \frac{1}{C_{i,j,k}}\Big[Q_{i,j,k} - B_{i,j,k}p_{i,j,k-1}^{(k+1)} - S_{i,j,k}p_{i,j-1,k}^{(k+1)} - W_{i,j,k}p_{i-1,j,k}^{(k+1)} -$$
$$E_{i,j,k}p_{i+1,j,k}^{(k+1)} - N_{i,j,k}p_{i,j+1,k}^{(k)} - C_{i,j,k}p_{i,j,k}^{(k)} - A_{i,j,k}p_{i,j,k+1}^{(k)}\Big]. \quad \dots \dots \dots \dots (7.33b)$$

Now, Eq. 7.33a can be written as,

$$p_{i,j,k}^{(k+1)} = p_{i,j,k}^{(k)} + \text{Improvement}. \quad \dots \dots \dots \dots (7.33c)$$

In Eq. 7.33b, the correction term refers to the improvement made by each convergent Gauss-Seidel iteration. Therefore, if a common factor is found that uniformly can be used for each individual unknown, and if by multiplying that factor, the correction term makes the improvement for each of the unknowns even larger, then the Gauss-Seidel iteration will converge faster. This common factor, ω_{opt}, is known as the acceleration parameter, and its value is dependent on certain characteristics of the coefficient matrix.

Pointwise-Successive Over-Relaxation (PSOR). The algorithmic protocol for the PSOR implementation can be written as

$$p_{i,j,k}^{(k+1)} = \frac{\omega_{opt}}{C_{i,j,k}} \Big[Q_{i,j,k} - B_{i,j,k} p_{i,j,k-1}^{(k+1)} - S_{i,j,k} p_{i,j-1,k}^{(k+1)} - W_{i,j,k} p_{i-1,j,k}^{(k+1)} -$$
$$E_{i,j,k} p_{i+1,j,k}^{(k+1)} - N_{i,j,k} p_{i,j+1,k}^{(k)} - A_{i,j,k} p_{i,j,k+1}^{(k)} \Big] - \left(\omega_{opt} - 1 \right) p_{i,j,k}^{(k)}. \quad\dots\dots\dots\dots\dots\dots\dots (7.34)$$

For $i = 1, \dots, N_x, j = 1, \dots, N_y$ and $k = 1, \dots, N_z$, the convergence check will be conducted as usual:

$$\left| p_{i,j,k}^{(k+1)} - p_{i,j,k}^{(k)} \right| \le \varepsilon_p \text{ for } i = 1,\dots,N_x, j = 1,\dots,N_y \text{ and } k = 1,\dots,N_z.$$

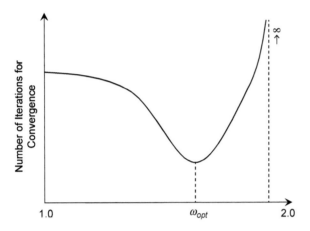

Fig. 7.3—Impact of acceleration parameter on the number of iterations needed for convergence.

The impact of the ω_{opt} in accelerating the convergence rate will depend on how close the value of the acceleration parameter is to its optimum value. The value of ω cannot be less than one, because any value between 0 and 1 will decelerate the convergence rate rather than accelerate it. When $\omega_{opt} = 1$, Eq. 7.34 collapses to the Gauss-Seidel equation (Eq. 7.32). It is also important to realize that as ω_{opt} approaches 2, the number of iterations will approach infinity, as shown in **Fig. 7.3**.

Typically, calculation of ω_{opt} is done as prescribed in the following steps:

- Start the algorithm (Eq. 7.34) with $\omega_{opt} = 1$ (Gauss-Seidel).
- After each iteration is completed, calculate the maximum improvement achieved among all the unknowns.
- Look for a stabilized value for the ratio of two successive maximum improvements (in absolute values), as shown by Eq. 7.35:

$$\rho_{GS} = \frac{\left| \text{Maximum improvement} \right|^{(k+1)}}{\left| \text{Maximum improvement} \right|^{(k)}} \quad\dots\dots\dots\dots\dots\dots\dots\dots\dots\dots\dots\dots\dots\dots\dots\dots\dots\dots (7.35)$$

- Iterate until the value of ρ_{GS} stabilizes at the second or third decimal point; such a stabilized value may be reached within the first 10 Gauss-Seidel iterations. The stabilized value of ρ_{GS} is known as the spectral radius (maximum eigenvalue) of the Gauss-Seidel iteration matrix.
- Calculate the value of ω_{opt} from (Eq. 7.36):

$$\omega_{opt} = \frac{2}{1 + \sqrt{1 - \rho_{GS}}}. \quad\dots\dots\dots\dots\dots\dots\dots\dots\dots\dots\dots\dots\dots\dots\dots\dots\dots\dots\dots\dots\dots\dots (7.36)$$

- Once the value of ω_{opt} is calculated, continue with the iterations found in Eq. 7.34 for the value of ω_{opt}, as calculated from Eq. 7.36.

Note that calculated ω_{opt} does not represent the exact value of the ω_{opt}, because the ρ_{GS} value used in Eq. 7.34 is not the exact spectral radius of the iteration matrix. One option to improve accuracy is to continue with the Gauss-Seidel iterations until ρ_{GS} stabilizes with higher levels of accuracy, and then calculate the ω_{opt} with this more-accurate value of ρ_{GS} to obtain a ω_{opt} value, which is closer to the true ω_{opt}. However, keep in mind that such a strategy will increase the number of slow-converging iterations of the Gauss-Seidel protocol. If the ω_{opt} is calculated as prescribed above, the PSOR convergence rate will be found to be 2X faster than the Gauss-Seidel convergence rate on the same problem.

Linewise-Successive Over-Relaxation (LSOR). In LSOR implementation, unknown block pressures for blocks located on the same line are simultaneously solved. For example, if we consider a 2D incompressible flow problem, the characteristic finite-difference equation will take the following form:

$$S_{i,j} p_{i,j-1} + W_{i,j} p_{i-1,j} + C_{i,j} p_{i,j} + E_{i,j} p_{i+1,j} + N_{i,j} p_{i,j+1} = Q_{i,j}. \quad\dots\dots\dots\dots\dots\dots\dots\dots\dots\dots\dots (7.37)$$

If one simultaneously solves the unknowns located on the line parallel to the x-direction, then the characteristic LSOR equation for line j will be written as

$$W_{i,j} p_{i-1,j}^{(k+1)} + C_{i,j} p_{i,j}^{(k+1)} + E_{i,j} p_{i+1,j}^{(k+1)} = Q_{i,j} - S_{i,j} p_{i,j-1}^{(k+1)} - N_{i,j} p_{i,j+1}^{(k)}. \quad\dots\dots\dots\dots\dots\dots\dots\dots\dots (7.38)$$

Eq. 7.38 will generate a system of equations that will be solved for each line j. In other words, if there are N_y lines parallel to the x-direction, Eq. 7.38 will be solved for $j = 1, 2,..., N_y$. When the $j = N_y$ line solution is obtained, one LSOR iteration will be complete. The search for ω_{opt} is similar to the ω_{opt} search for PSOR, as previously described. The LSOR scheme can be accelerated at the end of each completed LSOR iteration using Eq. 7.39:

$$p_{i,j}^{(k+1)}\,(\text{accelerated}) = p_{i,j}^{(k)} + \omega_{opt}\left[p_{i,j}^{(k+1)}\,(\text{unaccelerated}) - p_{i,j}^{(k)}\right]. \dots\dots\dots\dots\dots (7.39)$$

The LSOR procedure is approximately 2X faster than the PSOR procedure.

Blockwise Successive Over-Relaxation (BSOR). A protocol similar to the LSOR process can be implemented by solving a block of unknowns located on the same plane together. Here, the objective is to simultaneously solve more of the unknowns. A good application of BSOR is in the 3D grid representations. Again, let us consider the single-phase incompressible flow equation in its characteristic finite-difference form. If one simultaneously solves all the unknowns located on the x-y planes, layer by layer, then one will iterate through the following equation:

$$S_{i,j,k}p_{i,j-1,k}^{(k+1)} + W_{i,j,k}p_{i-1,j,k}^{(k+1)} + C_{i,j,k}p_{i,j,k}^{(k+1)} + E_{i,j,k}p_{i+1,j,k}^{(k+1)} + N_{i,j,k}p_{i,j+1,k}^{(k+1)}$$
$$= Q_{i,j,k} - B_{i,j}p_{i,j,k-1}^{(k+1)} - A_{i,j}p_{i,j,k+1}^{(k+1)}. \dots\dots\dots\dots\dots\dots (7.40)$$

Eq. 7.40 will be solved for each plane for $k = 1, 2, ..., N_z$. When all the planes are exposed to Eq. 7.40, a BSOR iteration will be complete. Searching for ω_{opt} is similar to the procedure discussed for PSOR. The solution at the end of each BSOR iteration is accelerated using Eq. 7.39.

Advanced Iterative Methods. All the iterative solution methods discussed thus far are quite limited in terms of their convergence speed. Schemes that excessively converge in a gradual manner can be problematic, especially in large-scale problems in which the number of gridblocks with unknowns will be large. In this subsection, we briefly review three additional iterative solvers that are much more efficient in terms of their convergence speeds.

Iterative Alternating-Direction Implicit Procedure (ADIP). We discuss the iterative ADIP method using the finite-difference representation of the 2D single-phase, slightly compressible flow equation written using SIP notation:

$$S_{i,j}^{n}p_{i,j-1}^{n+1} + W_{i,j}^{n}p_{i-1,j}^{n+1} + C_{i,j}^{n}p_{i,j}^{n+1} + E_{i,j}^{n}p_{i+1,j}^{n+1} + N_{i,j}^{n}p_{i,j+1}^{n+1} = Q_{i,j}^{n+1}. \dots\dots\dots (7.41)$$

In solving the system of equations generated by Eq. 7.41, a two-stage process is used. We describe the central coefficient $C_{i,j}$ as follows:

$$C_{i,j}^{n} = C_{xi,j}^{n} + C_{yi,j}^{n}, \dots\dots\dots\dots\dots\dots\dots\dots (7.42a)$$

where

$$C_{xi,j}^{n} = -\left(W_{i,j}^{n} + E_{i,j}^{n}\right), \dots\dots\dots\dots\dots\dots\dots\dots (7.42b)$$

and

$$C_{yi,j}^{n} = -\left(S_{i,j}^{n} + N_{i,j}^{n}\right). \dots\dots\dots\dots\dots\dots\dots\dots (7.42c)$$

At the current iteration level, our goal during the first stage of a given timestep computation is to bring the solution to an intermediate level, $p_{i,j}^{n+1\,(k)} \to p_{i,j}^{n+1\,(*)}$, and then during the second stage, the iteration will be complete, $p_{i,j}^{n+1\,(*)} \to p_{i,j}^{n+1\,(k+1)}$. Accordingly, during Stage 1, if we select the x-direction terms as implicit, then

$$W_{i,j}^{n}p_{i-1,j}^{n+1\,(*)} + C_{xi,j}^{n}p_{i,j}^{n+1\,(*)} + E_{i,j}^{n}p_{i+1,j}^{n+1\,(*)} =$$
$$Q_{i,j}^{n+1} - S_{i,j}^{n}p_{i,j-1}^{n+1\,(k)} - C_{yi,j}^{n}p_{i-1,j}^{n+1\,(k)} - N_{i,j}^{n}p_{i,j+1}^{n+1\,(k)} + \alpha_m\sum_{1}^{4}T_{i,j}^{n}\left(p_{i,j}^{n+1\,(*)} - p_{i,j}^{n+1\,(k)}\right) + \Gamma_{i,j}^{n}p_{i,j}^{n+1\,(*)} - \Gamma_{i,j}^{n}p_{i,j}^{n}. \dots\dots\dots\dots (7.43a)$$

The group of terms under the summation sign in Eq. 7.43a is known as the normalization of the iteration parameters. By collecting all unknown terms on the left-hand side of Eq. 7.43a, we obtain the final form of the first-stage calculations:

$$W_{i,j}^{n}p_{i-1,j}^{n+1\,(*)} + \left(C_{xi,j}^{n} - \Gamma_{i,j}^{n} - \alpha_m\sum_{1}^{4}T_{i,j}^{n}\right)p_{i,j}^{n+1\,(*)} + E_{i,j}^{n}p_{i+1,j}^{n+1\,(*)} =$$
$$Q_{i,j}^{n+1} - S_{i,j}^{n}p_{i,j-1}^{n+1\,(k)} - \left(C_{yi,j}^{n} + \alpha_m\sum_{1}^{4}T_{i,j}^{n}\right)p_{i,j}^{n+1\,(k)} - N_{i,j}^{n}p_{i,j+1}^{n+1\,(k)} - \Gamma_{i,j}^{n}p_{i,j}^{n}. \dots\dots\dots\dots (7.43b)$$

When Eq. 7.43b is solved for all unknowns, all block pressures will be brought to the $\left[\substack{(*)\\n+1}\right]$ level. In the second stage, the

goal is to bring the solution from $\left[\substack{(*)\\n+1}\right]$ level to $\left[\substack{(k+1)\\n+1}\right]$ level to complete the iteration. Accordingly, the second-stage equations will be written in the alternative direction (y–direction):

$$S_{i,j}^n \overset{(k+1)}{p_{i,j-1}^{n+1}} + C_{yi,j}^n \overset{(k+1)}{p_{i,j}^{n+1}} + N_{i,j}^n \overset{(k+1)}{p_{i,j+1}^{n+1}} =$$

$$Q_{i,j}^{n+1} - W_{i,j}^n \overset{(*)}{p_{i-1,j}^{n+1}} - E_{i,j}^n \overset{(*)}{p_{i+1,j}^{n+1}} - C_{xi,j}^n \overset{(*)}{p_{i,j}^{n+1}} + \alpha_m \sum_1^4 T_{i,j}^n \left(\overset{(k+1)}{p_{i,j}^{n+1}} - \overset{(*)}{p_{i,j}^{n+1}} \right) + \Gamma_{i,j}^n \overset{(*)}{p_{i,j}^{n+1}} - \Gamma_{i,j}^n p_{i,j}^n . \quad \dots \dots (7.44a)$$

Once again, collecting all the unknowns on the left-hand side of Eq. 7.44a (unknowns of this stage are the block pressures with superscripts, $\left[\substack{(k+1)\\n+1}\right]$), one ends up with the final form of the second-stage equation:

$$S_{i,j}^n \overset{(k+1)}{p_{i,j-1}^{n+1}} + \left(C_{yi,j}^n - \Gamma_{i,j}^n - \alpha_m \sum_1^4 T_{i,j}^n \right) \overset{(k+1)}{p_{i,j}^{n+1}} + N_{i,j}^n \overset{(k+1)}{p_{i,j+1}^{n+1}} =$$

$$Q_{i,j}^{n+1} - W_{i,j}^n \overset{(*)}{p_{i-1,j}^{n+1}} - \left(C_{xi,j}^n + \alpha_m \sum_1^4 T_{i,j}^n \right) \overset{(*)}{p_{i,j}^{n+1}} - E_{i,j}^n \overset{(*)}{p_{i+1,j}^{n+1}} - \Gamma_{i,j}^n p_{i,j}^n . \quad \dots \dots (7.44b)$$

When the second-stage equations generated by Eq. 7.44b are solved, then we will have all blocks with unknown pressure, this time with superscripts $\left[\substack{(k+1)\\n+1}\right]$ calculated. This completes the second stage of calculation; this solution is now carried from iteration level k to iteration level $k+1$.

ADIP Iteration Parameters. In Eqs. 7.43a through 7.44b, we see that iteration parameters, α_m, are externally introduced. Subscript m indicates the number of iteration parameters to be cyclically used after they are lined up in a circular manner.

The determination of these parameters can be demanding, but they can be approached by trial-and-error techniques. The simulation engineer needs to establish the upper and lower limits of the parameter range. Some highlights of α_m are listed below:

- $\alpha_m \geq 0$ forces convergence.
- $\alpha_m < 0$ forces divergence.
- The same α_m should be used for all stages.
- α_m consists of a series of parameters cyclically used.
- The iteration parameters should be geometrically spaced: If a total of m parameters are chosen per cycle, then

$$\alpha_m \big/ \alpha_1 = \beta^{m-1}. \quad \dots \dots (7.45)$$

In other words,

$$\frac{\alpha_2}{\alpha_1} = \frac{\alpha_3}{\alpha_2} = \frac{\alpha_4}{\alpha_3} = \dots = \frac{\alpha_{m-1}}{\alpha_{m-2}} = \frac{\alpha_m}{\alpha_{m-1}} = \beta. \quad \dots \dots (7.46)$$

- If α_1 is the minimum and α_m is the maximum parameter, then, β, may be calculated from

$$\ln \beta = \frac{\ln\left(\dfrac{\alpha_{max}}{\alpha_{min}}\right)}{k-1} = \frac{\ln\left(\dfrac{\alpha_m}{\alpha_1}\right)}{k-1}. \quad \dots \dots (7.47)$$

In Eq. 7.47, k is the number of parameters to be used in a cycle. Usually, k is 4 or 5 for a small range on α_m (e.g., 0.01 to 2) and 6 to 8 for a larger range on α_m (e.g., 0.001 to 2). The minimum iteration parameter α_{min} is calculated from Eq. 7.48 by finding the minimum of two scalar groups placed within the rectangular brackets:

- $$\alpha_{min} = \alpha_1 = \min\left[\frac{\pi^2}{2(N_x)^2}\left(\frac{1}{1+\dfrac{T_y}{T_x}}\right) ; \frac{\pi^2}{2(N_y)^2}\left(\frac{1}{1+\dfrac{T_x}{T_y}}\right) \right]. \quad \dots \dots (7.48)$$

- Also, if $T_x \neq T_y$, then $\alpha_{\max} = 1$; if $T_x \gg T_y$, then $\alpha_{\max} = 2$.

Strongly Implicit Procedure (SIP). This method is an approximate factorization method. We will summarize the SIP protocol using a 2D flow problem written in finite-difference form, as previously expressed in Eq. 7.15: $S_{i,j} p_{i,j-1} + W_{i,j} p_{i-1,j} + C_{i,j} p_{i,j} + E_{i,j} p_{i+1,j} + N_{i,j} p_{i,j+1} = Q_{i,j}$.

The previous problem can be written as $[A][P] = [Q]$. Our goal here is to find another coefficient matrix $[A']$ that easily can be factorized into lower-triangular $[L']$ and upper-triangular $[U']$ matrices, such that

$$[A'] = [L'][U'], \dotfill (7.49)$$

and at the same time, to ensure that $[A']$ is somehow close to $[A]$. Then it becomes necessary to find the $b_{i,j}$, $c_{i,j}$, and $d_{i,j}$ entries of the $[L']$, as well as the $e_{i,j}$ and $f_{i,j}$ of $[U']$ (main diagonal entries of $[U']$ have only ones). Stone's SIP factorization algorithm (Stone 1968) will provide definitions of the lowercase coefficients, as follows:

$$b_{i,j} = \frac{S_{i,j}}{\left(1 + \alpha_k e_{i,j-1}\right)}, \dotfill (7.50a)$$

$$c_{i,j} = \frac{W_{i,j}}{\left(1 + \alpha_k f_{i-1,j}\right)}, \dotfill (7.50b)$$

$$d_{i,j} = C_{i,j} + \alpha_k b_{i,j} e_{i,j-1} + \alpha_k c_{i,j} f_{i-1,j} - b_{i,j} f_{i,j-1} - c_{i,j} e_{i-1,j}, \dotfill (7.50c)$$

$$e_{i,j} = \frac{E_{i,j} - \alpha_k b_{i,j} e_{i,j-1}}{d_{i,j}}, \dotfill (7.50d)$$

$$f_{i,j} = \frac{N_{i,j} - \alpha_k e_{i,j} f_{i-1,j}}{d_{i,j}}. \dotfill (7.50e)$$

Note that Eqs. 7.50a through 7.50e must be solved sequentially in increasing order of i and j. Furthermore, recall that $S_{i,1}$ and $W_{1,i}$ are always equal to zero. After solving for all lower-case coefficients, the forward equation is solved for the intermediate vector $\left[v_{i,j}^{(k+1)}\right]$ from

$$b_{i,j} v_{i,j-1}^{(k+1)} + c_{i,j} v_{i-1,j}^{(k+1)} + d_{i,j} v_{i,j}^{(k+1)} = \alpha_k R_{i,j}^{(k)}. \dotfill (7.51)$$

In Eq. 7.51, α_k represents the SIP iteration parameters and $R_{i,j}^{(k)}$ is the residual of the original equation at iteration level k.

Once the intermediate vector $v_{i,j}^{(k+1)}$ is obtained using the forward equation (Eq. 7.51), then the backward equation given in Eq. 7.52 can be solved to obtain $\delta_{i,j}^{(k+1)}$:

$$\delta_{i,j}^{(k+1)} + e_{i,j} \delta_{i+1,j}^{(k+1)} + f_{i,j} \delta_{i,j+1}^{(k+1)} = v_{i,j}^{(k+1)}. \dotfill (7.52)$$

Once all $\delta_{i,j}^{(k+1)}$ values are obtained, the SIP iteration is completed by evaluating

$$p_{i,j}^{(k+1)} = p_{i,j}^{(k)} + \delta_{i,j}^{(k+1)}, \dotfill (7.53)$$

where $\delta_{i,j}^{(k+1)}$ represents the improvement in a converging SIP iteration. Accordingly, when $\delta_{i,j}^{(k+1)}$ becomes smaller than the convergence criterion, then the solution is obtained:

$$\left|\delta_{i,j}^{(k+1)}\right| \leq \varepsilon_p \text{ for } i = 1, \dots, N_x, j = 1, \dots, N_y. \dotfill (7.54)$$

SIP Iteration Parameters. The highlights of SIP iteration parameter are as follows:

- Varying α_k through a cycle of values is critically important.
- The minimum value of α_k is not critical and can be assigned a positive value in the vicinity of zero.
- The maximum value of α_k is critically important and can be calculated from

$$1 - \alpha_{\max} = \min_{i,j} \left\{ \frac{\pi^2}{2(N_x)^2} \left[\frac{1}{1 + \frac{S_{i,j}}{W_{i,j}}} \right]; \frac{\pi^2}{2(N_y)^2} \left[\frac{1}{1 + \frac{W_{i,j}}{S_{i,j}}} \right] \right\}. \dotfill (7.55)$$

- The different values of $(1-\alpha_k)$ in the cycle are chosen to be geometrically spaced between 1 and $(1-\alpha_{max})$, which can be expressed as

$$1-\alpha_r = (1-\alpha_{max})^{\frac{\gamma}{s-1}} \text{ for } r = 0, 1, 2, \ldots s-1. \quad\ldots\ldots\ldots\ldots\ldots\ldots\ldots\ldots\ldots\ldots\ldots\ldots\ldots\ldots\ldots\ldots (7.56)$$

In Eq. 7.56, s is the number of iteration parameters to be used in a given cycle.

Conjugate Gradient Method (CGM). The SOR, ADIP, and SIP methods discussed in the previous subsections of this chapter are, to a certain extent, handicapped, because finding the right set of iteration parameters uses empirically generated rules of thumb. CGM can accelerate finding the solution without recourse to externally introduced parameters. The convergence of the CGM is guaranteed. Theoretically, after N number of iterations, the convergence will be achieved where N is the number of unknowns in N number of equations to be solved. **Fig. 7.4** represents the basic premise of the CGM, such that instead of finding the zero (root) of a function in the modified problem, the minimum of the integrated form of the derivative function is searched.

The CGMs are also known as "the marble in the salad bowl," because they resemble the tracking of the marble from the edges of a salad bowl to the global minimum (in **Fig. 7.5,** the pressure surface map is considered analogous to the salad bowl).

There are different versions of the CGMs:

- Basic CGM applies to positive-definite symmetric $n \times n$ coefficient-matrix systems.
- Preconditioned CGM applies to a positive-definite symmetric $n \times n$ coefficient-matrix system. Through preconditioning, it is possible to obtain early convergence. The solution vector obtained at the end of the first iteration from the preconditioned conjugate gradient will be much closer to the exact solution obtained from the basic CGMs.
- Biconjugate gradient method applies to positive-definite nonsymmetric $n \times n$ coefficient-matrix systems.

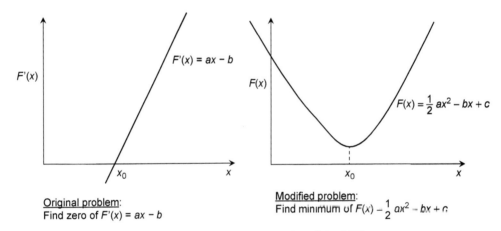

Fig. 7.4—Schematic representation of the CGM.

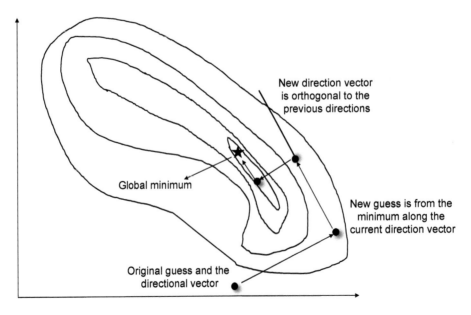

Fig. 7.5—Graphical representation of CGM for system of equations (marble in a bowl).

- Preconditioned biconjugate gradient method applies to positive-definite nonsymmetric systems, and it is 2X faster than the biconjugate gradient method without preconditioning.
- Preconditioned biconjugate gradient-stabilized method applies to positive-definite nonsymmetric matrix systems; its convergence rate is the fastest.

The following algorithms are presented to show the implementation processes for the aforementioned versions of CGMs to solve a system of equations, as indicated in Eq. 7.7, $[A][x] = [b]$:

- Basic conjugate gradient algorithm (Hestenes and Stiefel 1952):

  - x_0 = initial guess.

  - $k = 0$.

  - $r_0 = b - Ax_0,$ .... (7.57)

    $k = k + 1$, if $k = 1$,

    $$p_1 = r_0,$$ .... (7.58)

    or else,

    $$\beta_k = \frac{r_{k-1}^T r_{k-1}}{r_{k-2}^T r_{k-2}},$$ .... (7.59)

    $$p_k = r_{k-1} + \beta_k p_{k-1};$$ .... (7.60)

    end if

    $$g_k = AP_k,$$ .... (7.61)

    $$\alpha_k = \frac{r_{k-1}^T r_{k-1}}{p_k^T g_k},$$ .... (7.62)

    $$x_k = x_{k-1} + \alpha_k p_k,$$ .... (7.63)

    $$r_k = r_{k-1} - \alpha_k g_k.$$ .... (7.64)

    Check the convergence, and continue if necessary.

Where α_k and β_k are scalars, A is a positive-definite symmetric $n \times n$ matrix, and g_k, r_k, p_k, and x_k are vectors.

- Preconditioned conjugate gradient algorithm (Barrett et al. 1994):

  - x_0 = initial guess.

  - $k = 0$.

  - $r_0 = b - Ax_0,$ .... (7.65)

    $k = k + 1$.

    Solve for z_{k-1} from equation $Mz_{k-1} = r_{k-1},$ .... (7.66)

    $$\rho_k = r_{k-1}^T z_{k-1},$$ .... (7.67)

    if $k = 1$;

    $$p_k = z_{k-1},$$ .... (7.68)

    or else

    $$\beta_k = \frac{\rho_k}{\rho_{k-1}},$$ .... (7.69)

    $$p_k = z_{k-1} + \beta_k p_{k-1};$$ .... (7.70)

    end if

    $$g_k = Ap_k,$$ .... (7.71)

    $$\alpha_k = \frac{\rho_k}{p_k^T g_k},$$ .... (7.72)

    $$x_k = x_{k-1} + \alpha_k p_k,$$ .... (7.73)

    $$r_k = r_{k-1} - \alpha_k g_k.$$ .... (7.74)

    Check the convergence, and continue if necessary.

ρ_k is a scalar; M is the preconditioner used by the solution, which is a positive-definite symmetric $n \times n$ matrix; and z_k is a vector. The remainder of the variables is the same as those previously defined.

- Biconjugate gradient algorithm (Barrett et al. 1994):

  ○ $x_0 = $ initial guess.
  ○ $k = 0$.
  ○ $r_0 = b - Ax_0$, ... (7.75)

  $k = k + 1$,
  if $k = 1$,

  $$p_1 = r_0, \qquad\qquad\qquad\qquad\qquad\qquad\qquad\qquad\qquad\qquad (7.76)$$

  $$\tilde{p}_1 = p_1, \qquad\qquad\qquad\qquad\qquad\qquad\qquad\qquad\qquad\qquad (7.77)$$

  or else

  $$\tilde{p}_k = A^T r_{k-1}, \qquad\qquad\qquad\qquad\qquad\qquad\qquad\qquad\qquad (7.78)$$

  $$\beta_k \frac{\tilde{p}_k^T \tilde{p}_k}{\tilde{p}_{k-1}^T \tilde{p}_{k-1}}, \qquad\qquad\qquad\qquad\qquad\qquad\qquad\qquad (7.79)$$

  $$p_k = \tilde{p}_k + \beta_k p_{k-1}; \qquad\qquad\qquad\qquad\qquad\qquad\qquad\qquad (7.80)$$

  end if

  $$g_k = Ap_k, \qquad\qquad\qquad\qquad\qquad\qquad\qquad\qquad\qquad\qquad (7.81)$$

  $$\alpha_k \frac{\tilde{p}_k^T \tilde{p}_k}{g_k^T g_k}, \qquad\qquad\qquad\qquad\qquad\qquad\qquad\qquad\qquad (7.82)$$

  $$x_k = x_{k-1} + \alpha_k p_k, \qquad\qquad\qquad\qquad\qquad\qquad\qquad\qquad (7.83)$$

  $$r_k = r_{k-1} - \alpha_k g_k. \qquad\qquad\qquad\qquad\qquad\qquad\qquad\qquad (7.84)$$

  Check the convergence, and continue if necessary.

- Preconditioned biconjugate gradient algorithm (Knyazev 1991):

  ○ $x_0 = $ initial guess.
  ○ $k = 0$.
  ○ $r_0 = b - Ax_0$, ... (7.85)

  $k = k + 1$.

  Solve $Mz_{k-1} = r_{k-1}$, ... (7.86)

  for z_{k-1}.

  Solve $M^T \tilde{z}_{k-1} = \tilde{r}_{k-1}$, ... (7.87)

  for $\tilde{z}_{k-1}$,

  $$\rho_k = z_{k-1}^T \tilde{r}_{k-1}; \qquad\qquad\qquad\qquad\qquad\qquad\qquad\qquad (7.88)$$

  if $k = 1$,

  $$p_k = z_{k-1}, \qquad\qquad\qquad\qquad\qquad\qquad\qquad\qquad\qquad (7.89)$$

  $$\tilde{p}_k = \tilde{z}_{k1}, \qquad\qquad\qquad\qquad\qquad\qquad\qquad\qquad\qquad (7.90)$$

  or else

  $$\beta_k \frac{\rho_k}{\rho_{k-1}}, \qquad\qquad\qquad\qquad\qquad\qquad\qquad\qquad\qquad (7.91)$$

  $$p_k = z_{k-1} + \beta_k p_{k-1}, \qquad\qquad\qquad\qquad\qquad\qquad\qquad (7.92)$$

  $$\tilde{p}_k = \tilde{z}_{k-1} + \beta_k \tilde{p}_{k-1}; \qquad\qquad\qquad\qquad\qquad\qquad\qquad (7.93)$$

  end if

  $$g_k = Ap_k, \qquad\qquad\qquad\qquad\qquad\qquad\qquad\qquad\qquad\qquad (7.94)$$

$$\tilde{g}_k = A^T \tilde{p}_k, \dotfill (7.95)$$

$$\alpha_k = \frac{\rho_k}{\tilde{p}_k^T g_k}, \dotfill (7.96)$$

$$x_k = x_{k-1} + \alpha_k p_k, \dotfill (7.97)$$

$$r_k = r_{k-1} - \alpha_k g_k, \dotfill (7.98)$$

$$\tilde{r}_k = \tilde{r}_{k-1} - \alpha_k \tilde{g}_k. \dotfill (7.99)$$

Check the convergence, and continue if necessary.

Where $\tilde{r}_k$, $\tilde{z}_k$, and $\tilde{g}_k$ are vectors, and the remainder of the variables are the same as previously defined.

- Preconditioned biconjugate gradient-stabilized algorithm (van der Vorst 1992):

  ○ $x_0 = $ initial guess.
  ○ $k = 0$.
  ○ $r_0 = b - Ax_0, \dotfill (7.100)$

  $k = k + 1$,

  $$\rho_k = r_0 \tilde{r}_{k-1}, \dotfill (7.101)$$

  and if $\rho_k = 0$, the method fails.

  If $k = 1$

  $$p_k = r_{k-1}, \dotfill (7.102)$$

  or else

  $$\beta_k = \left(\frac{\rho_k}{\rho_{k-1}}\right) \times \left(\frac{\alpha_{k-1}}{\omega_{k-1}}\right), \dotfill (7.103)$$

  $$p_k = r_{k-1} + \beta_k \left(p_{k-1} - \omega_{k-1} v_{k-1}\right); \dotfill (7.104)$$

  end if.

  Solve for $\tilde{p}_k$ from $M\tilde{p}_k = p_k, \dotfill (7.105)$

  $$v_k = A\tilde{p}_k, \dotfill (7.106)$$

  $$\alpha_k = \frac{\rho_k}{\tilde{r}_{k-1}^T v_k}, \dotfill (7.107)$$

  $$S_k = r_{k-1} - a_k v_k. \dotfill (7.108)$$

  Solve $M\tilde{s}_k = s_k$ for $\tilde{s}_k$,

  $$t_k = A\tilde{s}_k, \dotfill (7.109)$$

  $$\omega_k = \left(\frac{t_k^T s_k}{t_k^T t_k}\right), \dotfill (7.110)$$

  $$x_k = x_{k-1} + \alpha_k \tilde{p}_k + \omega_k \tilde{s}_k, \dotfill (7.111)$$

  $$r_k = s_k - \omega_k t_k. \dotfill (7.112)$$

Check the convergence, and continue if necessary.

Where v_k, s_k, $\tilde{s}_k$, and t_k are vectors, and the remainder of the variables are the same as previously defined.

7.2 Ordering of Gridblocks

There is always one-to-one correspondence between the re-ordering of matrix elements and the renumbering of gridblocks. Cartesian number schemes result in less-stiff matrix structures, thereby making the solution process much smoother, typically by decreasing the bandwidth of the coefficient matrix. The most common ordering schemes are

- Standard ordering by rows (**Fig. 7.6.1**).
- Standard ordering by columns (**Fig. 7.6.2**).
- Cyclic-2 (checkerboard) ordering (**Fig. 7.6.3**).
- D4 (checkerboard) ordering (**Fig. 7.6.4**).
- D2 ordering (**Fig. 7.6.5**).

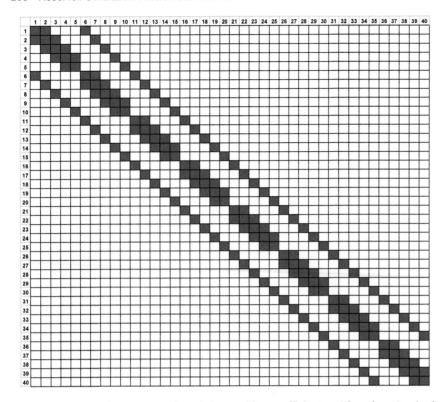

Fig. 7.6.1—Schematic representation of the resulting coefficient matrix using standard ordering by rows (with a bandwidth of 11).

1	2	3	4	5
6	7	8	9	10
11	12	13	14	15
16	17	18	19	20
21	22	23	24	25
26	27	28	29	30
31	32	33	34	35
36	37	38	39	40

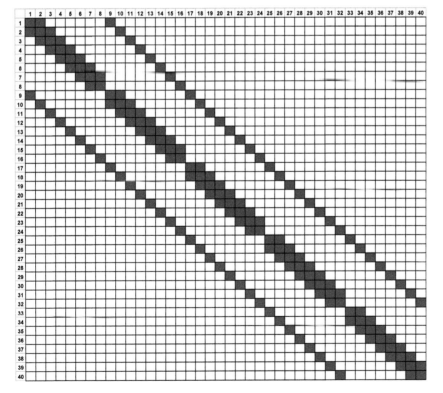

Fig. 7.6.2—Schematic representation of the resulting coefficient matrix using standard ordering by columns (with a bandwidth of 17).

1	9	17	25	33
2	10	18	26	34
3	11	19	27	35
4	12	20	28	36
5	13	21	29	37
6	14	22	30	38
7	15	23	31	39
8	16	24	32	40

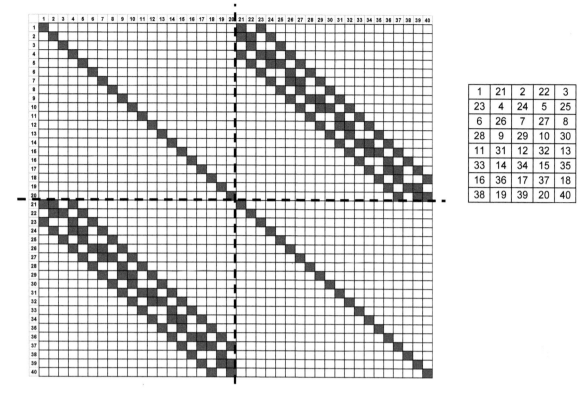

1	21	2	22	3
23	4	24	5	25
6	26	7	27	8
28	9	29	10	30
11	31	12	32	13
33	14	34	15	35
16	36	17	37	18
38	19	39	20	40

Fig. 7.6.3—Schematic representation of the resulting coefficient matrix using cyclic-2 ordering (note that there are two diagonal and two additional partitioned matrices with bandwidths of 6).

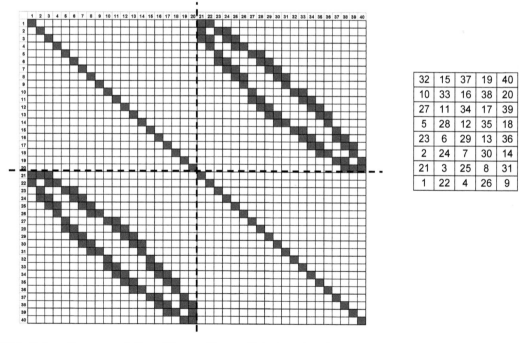

32	15	37	19	40
10	33	16	38	20
27	11	34	17	39
5	28	12	35	18
23	6	29	13	36
2	24	7	30	14
21	3	25	8	31
1	22	4	26	9

Fig. 7.6.4—Schematic representation of the resulting coefficient matrix using D4 ordering.

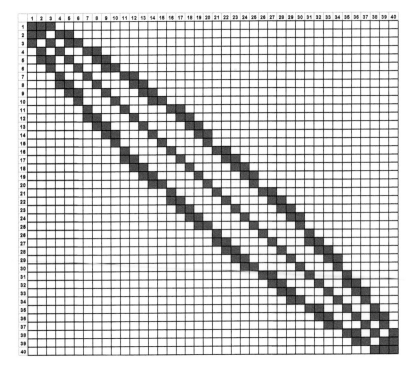

Fig. 7.6.5—Schematic representation of the resulting coefficient matrix using D2 ordering.

Figs. 7.6.1 through 7.6.5 show the schematic representation of these ordering schemes using a 5×8 system. Note that in these figures, the filled-in squares represent the nonzero entries of the coefficient matrix. **Table 7.1** provides a comparison of the computational work and storage requirements for different ordering schemes.

Ordering of Grid Blocks	Computational Storage	Computational Work
Standard ordering	$(N_x)(N_y)^2$	$(N_x)(N_y)^3$
Cyclic-2 ordering	$\dfrac{(N_x)(N_y)^2}{2}$	$\dfrac{(N_x)(N_y)^3}{3}$
D4 ordering	$\dfrac{(N_x)(N_y)^2}{2} - \dfrac{(N_y)^3}{6}$	$\dfrac{(N_x)(N_y)^3}{2} - \dfrac{(N_y)^2}{4}$
D2 ordering	$(N_x)(N_y)^2 - \dfrac{(N_y)^3}{6}$	$(N_x)(N_y)^3 - \dfrac{(N_y)^4}{2}$

Table 7.1—Computational storage and work requirements for different ordering of gridblocks (*Note*: it is assumed that $N_y \ll N_x$).

7.3 Problems and Solutions

Problem 7.3.1 (Ertekin et al. 2001)

Fig. 7.7.1 shows a 2D reservoir with regular boundaries represented by 30 uniform blocks. Use the following schemes to order the reservoir gridblocks, and show the structure of the resulting coefficient matrix (mark nonzero entries) for each ordering scheme on engineering paper.

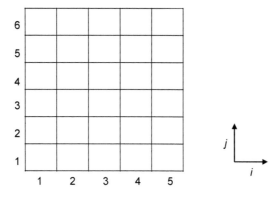

Fig. 7.7.1—Reservoir system for Problem 7.3.1.

1. Standard ordering by rows.
2. Standard ordering by columns.
3. D2 ordering.
4. Cyclic-2 ordering.
5. D4 ordering.

Solution to Problem 7.3.1

In Parts 1 through 5, the filled-in squares represent the nonzero entries of the coefficient matrix, and the blank squares represent the zero entries.

1. Standard ordering by rows is shown in **Fig. 7.7.2.**
2. Standard ordering by columns is shown in **Fig. 7.7.3.**
3. D2 ordering is shown in **Fig. 7.7.4.**
4. Cyclic-2 ordering is shown in **Fig. 7.7.5.**
5. D4 ordering is shown in **Fig. 7.7.6.**

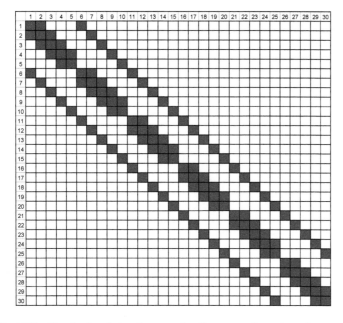

Fig. 7.7.2—Standard ordering by rows for the reservoir system shown and the structure of the resulting coefficient matrix.

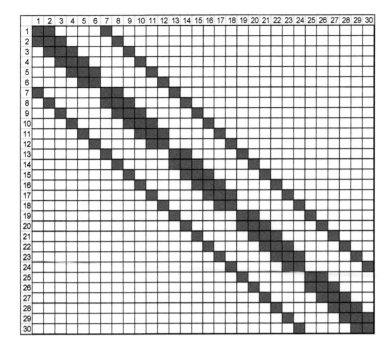

	1	2	3	4	5
1	1	7	13	19	25
2	2	8	14	20	26
3	3	9	15	21	27
4	4	10	16	22	28
5	5	11	17	23	29
6	6	12	18	24	30

Fig. 7.7.3—Standard ordering by columns for the reservoir system shown and the structure of the resulting coefficient matrix.

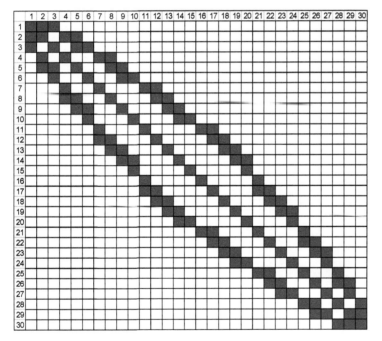

	1	2	3	4	5
1	16	21	25	28	30
2	11	17	22	26	29
3	7	12	18	23	27
4	4	8	13	19	24
5	2	5	9	14	20
6	1	3	6	10	15

Fig. 7.7.4—D2 ordering for the reservoir system shown and the structure of the resulting coefficient matrix.

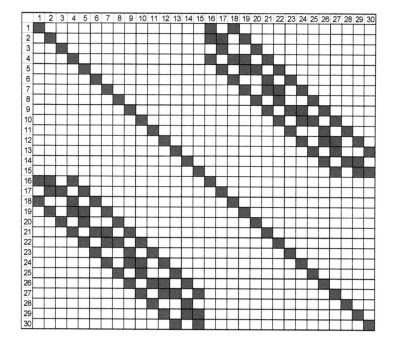

	1	2	3	4	5
1	1	16	2	17	3
2	18	4	19	5	20
3	6	21	7	22	8
4	23	9	24	10	25
5	11	26	12	27	13
6	28	14	29	15	30

Fig. 7.7.5—Cyclic-2 ordering for the reservoir system shown and the structure of the resulting coefficient matrix.

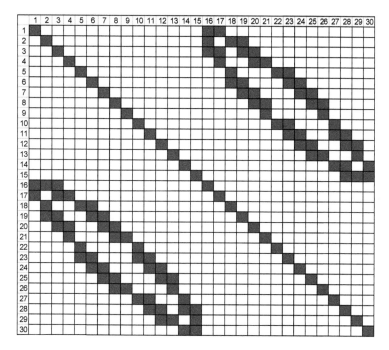

	1	2	3	4	5
1	22	10	27	14	30
2	5	23	11	28	15
3	18	6	24	12	29
4	2	19	7	25	13
5	16	3	20	8	26
6	1	17	4	21	9

Fig. 7.7.6—D4 ordering for the reservoir system shown and the structure of the resulting coefficient matrix.

Problem 7.3.2 (Ertekin et al. 2001)

Answer all questions in Problem 7.3.1 if the reservoir has irregular boundaries, as shown in **Fig. 7.8.1** (*Note:* number only the active blocks).

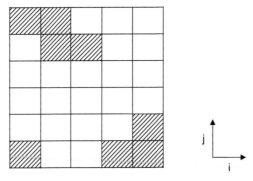

Fig. 7.8.1—Reservoir system for Problem 7.3.2.

Solution to Problem 7.3.2

1. Standard ordering by rows is shown in **Fig. 7.8.2.**
2. Standard ordering by rows is shown in **Fig. 7.8.3.**
3. D2 ordering is shown in **Fig. 7.8.4.**
4. Cyclic-2 ordering is shown in **Fig. 7.8.5.**
5. D4 ordering is shown in **Fig. 7.8.6.**

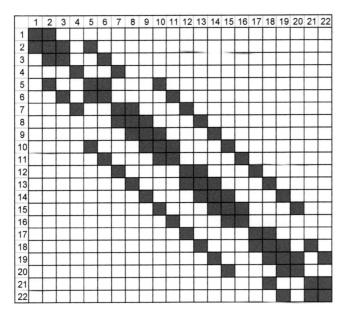

Fig. 7.8.2—Standard ordering by rows.

	1	2	3	4	5
1			1	2	3
2	4			5	6
3	7	8	9	10	11
4	12	13	14	15	16
5	17	18	19	20	
6		21	22		

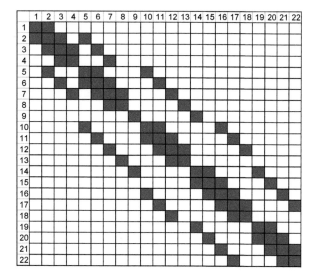

Fig. 7.8.3—Standard ordering by columns.

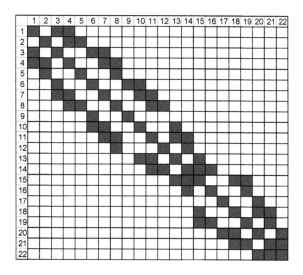

Fig. 7.8.4—D2 ordering.

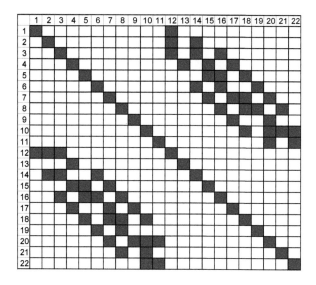

Fig. 7.8.5—Cyclic-2 ordering.

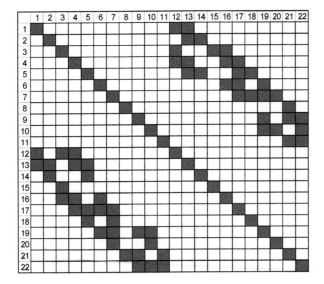

Fig. 7.8.6—D4 ordering.

Problem 7.3.3 (Ertekin et al. 2001)

Use 1) Jacobi's iteration, and 2) Gauss-Seidel's algorithm to solve the following systems of equations:

- $5x_1 + 3x_2 + 4x_3 = 12$.
- $3x_1 + 6x_2 + 4x_3 = 13$.
- $4x_1 + 4x_2 + 5x_3 = 13$.

Use an initial guess of zero for all unknowns, and comment on the results.

Solution to Problem 7.3.3

1. In this case, Jacobi's iteration fails to solve the systems of equations. **Table 7.2** displays ten Jacobi iterations, which indicate a divergence of the improvement.
2. Gauss-Seidel's algorithm for this symmetric matrix is

$$x_1^{(k+1)} = \frac{1}{5}\left[12 - 3x_2^{(k)} - 4x_3^{(k)}\right], \quad \dots\dots\dots\dots\dots\dots\dots\dots\dots\dots\dots\dots \text{(7.113)}$$

$$x_2^{(k+1)} = \frac{1}{6}\left[13 - 3x_1^{(k+1)} - 4x_3^{(k)}\right], \quad \dots\dots\dots\dots\dots\dots\dots\dots\dots\dots\dots \text{(7.114)}$$

$$x_3^{(k+1)} = \frac{1}{5}\left[13 - 4x_2^{(k+1)} - 4x_1^{(k+1)}\right]. \quad \dots\dots\dots\dots\dots\dots\dots\dots\dots\dots \text{(7.115)}$$

Using $x_1 = 0$, $x_2 = 0$, and $x_3 = 0$ as the initial guess,

$$x_1^{(1)} = \frac{1}{5}(1) = 2.4, \quad \dots\dots\dots\dots\dots\dots\dots\dots\dots\dots\dots\dots\dots\dots\dots \text{(7.116)}$$

$$x_2^{(1)} = \frac{1}{6}(13 - 3 \times 2.4) = 0.967, \quad \dots\dots\dots\dots\dots\dots\dots\dots\dots\dots\dots \text{(7.117)}$$

$$x_3^{(1)} = \frac{1}{5}\left[13 - 4 \times x_1^{(1)} - 4 \times x_2^{(1)}\right] = -0.0933. \quad \dots\dots\dots\dots\dots\dots \text{(7.118)}$$

The process of the Gauss-Seidel iteration is shown in **Table 7.3.**

Iteration Level	x_1	x_2	x_3	$\max\left\|x_i^{(k+1)} - x_i^{(k)}\right\|$
0	0.000	0.000	0.000	—
1	2.400	2.167	2.600	2.600
2	−0.980	−0.767	−1.053	3.653
3	3.703	3.359	3.997	5.051
4	−2.813	−2.350	−3.049	7.047
5	6.249	5.606	6.730	9.779
6	−6.348	−5.445	−6.884	13.614
7	11.174	9.930	12.034	18.918
8	−13.185	−11.443	−14.283	26.317
9	20.693	18.282	22.303	36.586
10	−26.411	−23.048	−28.579	50.882

Table 7.2—Ten Jacobi iterations indicating a divergence of the improvement.

Iteration Level	x_1	x_2	x_3	$\max\left\|x_i^{(k+1)} - x_i^{(k)}\right\|$
0	0	0	0	—
1	2.4	0.967	−0.093	2.4
2	1.895	1.282	0.059	0.505
3	1.584	1.335	0.265	0.311
4	1.387	1.297	0.453	0.197
5	1.260	1.235	0.604	0.151
6	1.176	1.176	0.719	0.114
7	1.119	1.128	0.802	0.083
8	1.082	1.091	0.862	0.060
9	1.056	1.064	0.904	0.042
10	1.038	1.045	0.934	0.029
11	1.026	1.031	0.954	0.021
12	1.018	1.022	0.968	0.014
13	1.012	1.015	0.978	0.010

Table 7.3—Process of the Gauss-Seidel iteration.

Problem 7.3.4 (Ertekin et al. 2001)

Use the CGM to solve the following system of equations:

$$\begin{bmatrix} -1.021 & 1 & 0 & 0 \\ 1 & -2.021 & 1 & 0 \\ 0 & 1 & -2.021 & 1 \\ 0 & 0 & 2 & -2.021 \end{bmatrix} \begin{bmatrix} p_1^{n+1} \\ p_2^{n+1} \\ p_3^{n+1} \\ p_4^{n+1} \end{bmatrix} = \begin{bmatrix} -84 \\ 116 \\ -84 \\ -84 \end{bmatrix}. \quad\ldots\ldots\ldots\ldots\ldots\ldots\ldots\ldots\ldots\ldots\ldots\ldots\ldots (7.119)$$

Start with an initial guess of 4,000 psia. *Hint:* Observe the symmetry of the matrix and select a suitable version of the CGM.

Solution to Problem 7.3.4

Because the coefficient matrix is a nonsymmetrical matrix, a basic CGM may be used to solve the system of equations in the solution with a tolerance of 1 psi. Results of the end of successive iteration levels are displayed in **Table. 7.4.**

	$p^{(0)}$	$p^{(1)}$	$p^{(2)}$	$p^{(3)}$	$p^{(4)}$	$p^{(5)}$
p_1	4,000	4,018.27	4,025.99	4,001.34	1,279.49	1,279.49
p_2	4,000	3,963.08	3,949.15	3,940.49	1,222.36	1,222.36
p_3	4,000	4,018.27	4,007.07	4,017.17	1,306.90	1,306.90
p_4	4,000	4,000.00	4,016.88	4,032.43	1,334.88	1,334.88

Table 7.4—Results at the end of five successive iterations.

Problem 7.3.5 (Ertekin et al. 2001)

Use 1) Jacobi's iteration, 2) Gauss-Seidel's algorithm, and 3) the PSOR protocol with ω_{opt} to solve the following system of equations:

$$\begin{bmatrix} -1.021 & 1 & 0 & 0 \\ 1 & -2.021 & 1 & 0 \\ 0 & 1 & -2.021 & 1 \\ 0 & 0 & 2 & -2.021 \end{bmatrix} \begin{bmatrix} p_1^{(k+1)} \\ p_2^{(k+1)} \\ p_3^{(k+1)} \\ p_4^{(k+1)} \end{bmatrix} = \begin{bmatrix} -84 \\ 116 \\ -84 \\ -84 \end{bmatrix}. \quad \dots\dots\dots\dots\dots\dots\dots\dots\dots \text{(7.120)}$$

Begin with an initial guess of 1,200 psia for all unknowns and use a convergence tolerance of 0.5 psi.

Solution to Problem 7.3.5

1. Jacobi's iteration is shown in the equations that follow and in **Table 7.5.**

$$p_1^{(k+1)} = -\frac{1}{1.021}\left[-84 - p_2^{(k)}\right], \quad \dots\dots\dots\dots\dots\dots\dots\dots\dots\dots\dots\dots \text{(7.121)}$$

$$p_2^{(k+1)} = -\frac{1}{2.021}\left[116 - p_1^{(k)} - p_3^{(k)}\right], \quad \dots\dots\dots\dots\dots\dots\dots\dots\dots\dots \text{(7.122)}$$

$$p_3^{(k+1)} = \frac{1}{2.021}\left[-84 - p_2^{(k)} - p_4^{(k)}\right], \quad \dots\dots\dots\dots\dots\dots\dots\dots\dots\dots \text{(7.123)}$$

$$p_4^{(k+1)} = -\frac{1}{2.021}\left[-84 - 2p_3^{(k)}\right]. \quad \dots\dots\dots\dots\dots\dots\dots\dots\dots\dots\dots \text{(7.124)}$$

Iteration Level	p_1	p_2	p_3	p_4
0	1,200	1,200	1,200	1,200
1	1,257.59	1,130.13	1,229.09	1,229.09
2	1,189.16	1,173.03	1,208.92	1,257.89
3	1,231.17	1,129.18	1,244.39	1,237.92
4	1,188.23	1,167.52	1,212.82	1,273.02
5	1,225.78	1,130.65	1,249.16	1,241.78
...				
355	1,278.70	1,221.06	1,306.09	1,333.57
356	1,278.21	1,221.57	1,305.60	1,334.09
357	1,278.72	1,221.09	1,306.11	1,333.60
358	1,278.24	1,221.59	1,305.63	1,334.11
359	1,278.74	1,221.12	1,306.13	1,333.63

Table 7.5—Jacobi's iterations from the Solution to Problem 7.3.5.

2. Gauss-Seidel's algorithm is shown in the equations that follow and in **Table 7.6.**

$$p_1^{(k+1)} = -\frac{1}{1.021}\left[-84 - p_2^{(k)}\right], \dotfill (7.125)$$

$$p_2^{(k+1)} = -\frac{1}{2.021}\left[116 - p_1^{(k+1)} - p_3^{(k)}\right], \dotfill (7.126)$$

$$p_3^{(k+1)} = -\frac{1}{2.021}\left[-84 - p_2^{(k+1)} - p_4^{(k)}\right], \dotfill (7.127)$$

$$p_4^{(k+1)} = -\frac{1}{2.021}\left[-84 - 2p_3^{(k+1)}\right]. \dotfill (7.128)$$

Iteration Level	p_1	p_2	p_3	p_4
0	1,200	1,200	1,200	1,200
1	1,257.59	1,158.63	1,208.62	1,237.63
2	1,217.07	1,142.85	1,219.43	1,248.33
3	1,201.61	1,140.55	1,223.59	1,252.44
4	1,199.36	1,141.49	1,226.09	1,254.91
5	1,200.28	1,143.18	1,228.15	1,256.96
...				
61	1,258.61	1,201.55	1,286.24	1,314.44
62	1,259.11	1,202.05	1,286.73	1,314.93
63	1,259.60	1,202.54	1,287.22	1,315.41

Table 7.6—Gauss-Seidel iterations from the Solution to Problem 7.3.5.

3. PSOR with ω_{opt} is shown in the equations that follow and in **Table 7.7.**

$$p_1^{(k+1)} = -\frac{\omega_{opt}}{1.021}\left(-84 - p_2^{(k)}\right) - \left(\omega_{opt} - 1\right)p_1^{(k)}, \dotfill (7.129)$$

$$p_2^{(k+1)} = -\frac{\omega_{opt}}{2.021}\left(116 - p_1^{(k+1)} - p_1^{(k)}\right) - \left(\omega_{opt} - 1\right)p_2^{(k)}, \dotfill (7.130)$$

$$p_3^{(k+1)} = -\frac{\omega_{opt}}{2.021}\left(-84 - p_2^{(k+1)} - p_4^{(k)}\right) - \left(\omega_{opt} - 1\right)p_3^{(k)}, \dotfill (7.131)$$

$$p_4^{(k+1)} = -\frac{1}{2.021}\left(-84 - 2p_3^{(k+1)}\right) - \left(\omega_{opt} - 1\right)p_3^{(k)}, \dotfill (7.132)$$

where $\omega_{opt} = 1.736$.

Note: Because 0.5 psi represents a relatively large tolerance in this example, we see that the Gauss-Seidel and PSOR algorithms converge faster than expected.

Iteration Level	p_1	p_2	p_3	p_4
0	1,200	1,200	1,200	1,200
1	1,299.98	1,164.59	1,220.10	1,285.04
2	1,166.18	1,092.98	1,216.83	1,216.83
3	1,142.90	1,122.89	1,186.34	1,214.66
...				
21	1,277.15	1,220.45	1,305.15	1,333.35
22	1,277.97	1,220.95	1,305.66	1,333.89
23	1,278.22	1,221.24	1,305.99	1,334.06

Table 7.7—PSOR iterations with ω_{opt} from the Solution to Problem 7.3.5.

Problem 7.3.6 (Ertekin et al. 2001)

Use the 1) Gaussian elimination algorithm, 2) Thomas' algorithm, and 3) Crout reduction algorithm to solve the system of equations in Problem 7.3.5.

Solution to Problem 7.3.6

1. Gaussian elimination algorithm is shown in the following matrices:

$$\begin{bmatrix} 1 & -0.979 & 0 & 0 \\ 0 & -1.042 & 1 & 0 \\ 0 & 1 & -2.021 & 1 \\ 0 & 0 & 2 & -2.021 \end{bmatrix} \begin{bmatrix} p_1 \\ p_2 \\ p_3 \\ p_4 \end{bmatrix} = \begin{bmatrix} 82.27 \\ 33.73 \\ -84 \\ -84 \end{bmatrix}, \quad \dots\dots\dots\dots\dots \quad (7.133)$$

$$\begin{bmatrix} 1 & -0.979 & 0 & 0 \\ 0 & 1 & -0.960 & 0 \\ 0 & 0 & -1.061 & 1 \\ 0 & 0 & 2 & -2.021 \end{bmatrix} \begin{bmatrix} p_1 \\ p_2 \\ p_3 \\ p_4 \end{bmatrix} = \begin{bmatrix} 82.27 \\ -32.38 \\ -51.62 \\ -84 \end{bmatrix}, \quad \dots\dots\dots\dots\dots \quad (7.134)$$

$$\begin{bmatrix} 1 & -0.979 & 0 & 0 \\ 0 & 1 & -0.960 & 0 \\ 0 & 0 & 1 & -0.943 \\ 0 & 0 & 0 & -0.136 \end{bmatrix} \begin{bmatrix} p_1 \\ p_2 \\ p_3 \\ p_4 \end{bmatrix} = \begin{bmatrix} 82.27 \\ -32.38 \\ -48.65 \\ -181.31 \end{bmatrix}. \quad \dots\dots\dots\dots\dots \quad (7.135)$$

The backward-solution equations are as follows:

- $p_4 = -181.31 \,/\, -0.136 = 1{,}334.88 \text{ psi},$ $\dots\dots\dots\dots\dots$ (7.136)
- $p_3 = 48.65 + 0.943 p_4 = 1{,}306.90 \text{ psi},$ $\dots\dots\dots\dots\dots$ (7.137)
- $p_2 = -32.38 + 0.960 p_3 = 1{,}222.40 \text{ psi},$ $\dots\dots\dots\dots\dots$ (7.138)
- $p_1 = -0.979 p_2 + 82.27 = 1{,}279.50 \text{ psi}.$ $\dots\dots\dots\dots\dots$ (7.139)

2. Thomas' algorithm is shown in the matrices that follow:

   - Forward solution (refer to Eqs. 7.24 and 7.26 for the definitions of ω):

$$\vec{w} = \begin{bmatrix} -0.97943 & -0.96009 & -0.94259 & 0 \end{bmatrix}, \quad \dots\dots\dots\dots\dots \quad (7.140)$$

$$\vec{g} = \begin{bmatrix} 82.27228 & -32.3817 & 48.6548 & 1{,}334.882 \end{bmatrix}. \quad \dots\dots\dots\dots\dots \quad (7.141)$$

   - Backward solution:

$$\vec{p} = \begin{bmatrix} 1{,}279.491 & 1{,}222.36 & 1{,}306.899 & 1{,}334.882 \end{bmatrix}. \quad \dots\dots\dots\dots\dots \quad (7.142)$$

3. Crout reduction algorithm is shown in the matrices that follow:

$$[L] = \begin{bmatrix} 1 & 0 & 0 & 0 \\ -0.979 & 1 & 0 & 0 \\ 0 & -0.960 & 1 & 0 \\ 0 & 0 & -1.885 & 1 \end{bmatrix}, \quad \dots\dots\dots\dots\dots \quad (7.143)$$

$$[U] = \begin{bmatrix} -1.021 & 1 & 0 & 0 \\ 0 & -1.042 & 1 & 0 \\ 0 & 0 & -1.061 & 1 \\ 0 & 0 & 0 & -0.136 \end{bmatrix}. \quad \dots\dots\dots\dots\dots \quad (7.144)$$

Solving for y,

$$\begin{bmatrix} 1 & 0 & 0 & 0 \\ -0.979 & 1 & 0 & 0 \\ 0 & -0.960 & 1 & 0 \\ 0 & 0 & -1.885 & 1 \end{bmatrix} \begin{bmatrix} y_1 \\ y_2 \\ y_3 \\ y_4 \end{bmatrix} = \begin{bmatrix} -84 \\ 116 \\ -84 \\ -84 \end{bmatrix}, \quad\dots\dots\dots\dots\dots\dots\dots\dots\dots\dots\dots\dots\dots (7.145)$$

$$\vec{y} = \begin{bmatrix} -84 \\ 33.73 \\ -51.62 \\ -181.31 \end{bmatrix}. \quad\dots\dots\dots\dots\dots\dots\dots\dots\dots\dots\dots\dots\dots\dots\dots\dots\dots (7.146)$$

Solving for x,

$$\begin{bmatrix} -1.021 & 1 & 0 & 0 \\ 0 & -1.042 & 1 & 0 \\ 0 & 0 & -1.061 & 1 \\ 0 & 0 & 0 & -0.136 \end{bmatrix} \begin{bmatrix} x_1 \\ x_2 \\ x_3 \\ x_4 \end{bmatrix} = \begin{bmatrix} -84 \\ 33.73 \\ -51.62 \\ -181.31 \end{bmatrix}, \quad\dots\dots\dots\dots\dots\dots\dots\dots\dots\dots\dots (7.147)$$

$$\vec{x} = \begin{bmatrix} 1,279.491 \\ 1,222.36 \\ 1,306.899 \\ 1,334.882 \end{bmatrix} \quad\dots\dots\dots\dots\dots\dots\dots\dots\dots\dots\dots\dots\dots\dots\dots\dots\dots\dots\dots (7.148)$$

Problem 7.3.7 (Ertekin et al. 2001)

Consider the following three equations in three unknowns:

- $x_1 + 1.0139\, x_2 - 0.8124\, x_3 = 3.0778$.

- $1.0139\, x_1 + 2.0611\, x_2 + 0.4054\, x_3 = 2.7350$.

- $0.8124\, x_1 + 0.4054\, x_2 + 2.8143\, x_3 = 1.3113$.

Use the 1) Gaussian elimination algorithm, 2) Gauss-Jordan reduction algorithm (for additional discussion on this algorithm, see Ertekin et al. 2001), and 3) Crout reduction algorithm to solve the system of equations.

Solution to Problem 7.3.7

1. Gaussian elimination method:

$$\begin{bmatrix} 1 & 0.324458 & -0.25998 \\ 0 & 1.732132 & 0.66899 \\ 0 & 0.14181 & 3.025505 \end{bmatrix} \begin{bmatrix} x_1 \\ x_2 \\ x_3 \end{bmatrix} = \begin{bmatrix} 0.984928 \\ 1.736382 \\ 0.511145 \end{bmatrix}, \quad\dots\dots\dots\dots\dots\dots\dots\dots\dots\dots (7.149)$$

$$\begin{bmatrix} 1 & 0.324458 & -0.25998 \\ 0 & 1 & 0.386224 \\ 0 & 0 & 2.970734 \end{bmatrix} \begin{bmatrix} x_1 \\ x_2 \\ x_3 \end{bmatrix} = \begin{bmatrix} 0.984928 \\ 1.002454 \\ 0.368987 \end{bmatrix}. \quad\dots\dots\dots\dots\dots\dots\dots\dots\dots\dots (7.150)$$

Solving for $x_3 = 0.124$, $x_2 = 0.9545$, and $x_1 = 0.7075$.

2. Gauss-Jordan reduction algorithm:

$$\begin{bmatrix} 1 & 0.324458 & -0.25998 \\ 0 & 1.732132 & 0.66899 \\ 0 & 0.14181 & 3.025505 \end{bmatrix} \begin{bmatrix} x_1 \\ x_2 \\ x_3 \end{bmatrix} = \begin{bmatrix} 0.984928 \\ 1.736382 \\ 0.511145 \end{bmatrix}, \quad\dots\dots\dots\dots\dots\dots\dots\dots\dots\dots (7.151)$$

$$\begin{bmatrix} 1 & 0 & -0.38529 \\ 0 & 1 & 0.386224 \\ 0 & 0 & 2.970734 \end{bmatrix} \begin{bmatrix} x_1 \\ x_2 \\ x_3 \end{bmatrix} = \begin{bmatrix} 0.659673 \\ 1.002454 \\ 0.368987 \end{bmatrix}, \quad\quad\quad\quad (7.152)$$

$$\begin{bmatrix} 1 & 0 & 0 \\ 0 & 1 & 0 \\ 0 & 0 & 1 \end{bmatrix} \begin{bmatrix} x_1 \\ x_2 \\ x_3 \end{bmatrix} = \begin{bmatrix} 0.707529 \\ 0.954482 \\ 0.124207 \end{bmatrix}. \quad\quad\quad\quad (7.153)$$

Thus, $x_3 = 0.124$, $x_2 = 0.9545$, and $x_1 = 0.7075$.

3. Crout reduction algorithm:

$$L = \begin{bmatrix} 1 & 0 & 0 \\ 0.324458 & 1 & 0 \\ 0.259976 & 0.08187 & 1 \end{bmatrix}, \quad\quad\quad\quad (7.154)$$

$$U = \begin{bmatrix} 3.1249 & 1.0139 & -0.8124 \\ 0 & 1.732132 & 0.66899 \\ 0 & 0 & 2.970734 \end{bmatrix}. \quad\quad\quad\quad (7.155)$$

Solving for y,

$$\begin{bmatrix} 1 & 0 & 0 \\ 0.324458 & 1 & 0 \\ 0.259976 & 0.08187 & 1 \end{bmatrix} \begin{bmatrix} y_1 \\ y_2 \\ y_3 \end{bmatrix} = \begin{bmatrix} 3.0778 \\ 2.735 \\ 1.3113 \end{bmatrix}, \quad\quad\quad\quad (7.156)$$

$$\vec{y} = \begin{bmatrix} 3.078 \\ 1.736 \\ 0.369 \end{bmatrix}. \quad\quad\quad\quad (7.157)$$

Solving for x,

$$\begin{bmatrix} 3.1249 & 1.0139 & -0.8124 \\ 0 & 1.732132 & 0.66899 \\ 0 & 0 & 2.970734 \end{bmatrix} \begin{bmatrix} x_1 \\ x_2 \\ x_3 \end{bmatrix} = \begin{bmatrix} 3.078 \\ 1.736 \\ 0.369 \end{bmatrix}, \quad\quad\quad\quad (7.158)$$

$$\vec{x} = \begin{bmatrix} 0.7075 \\ 0.9545 \\ 0.1240 \end{bmatrix}. \quad\quad\quad\quad (7.159)$$

Problem 7.3.8 (Ertekin et al. 2001)

Use 1) Jacobi's iteration, 2) Gauss-Seidel's iteration, 3) the PSOR method, and 4) CGM to solve the following system of equations:

$$\begin{bmatrix} 6 & -4 & 1 & 0 & 0 \\ -4 & 6 & -4 & 1 & 0 \\ 1 & -4 & 6 & -4 & 1 \\ 0 & 1 & -4 & 6 & -4 \\ 0 & 0 & 1 & -4 & 6 \end{bmatrix} \begin{bmatrix} x_1 \\ x_2 \\ x_3 \\ x_4 \\ x_5 \end{bmatrix} = \begin{bmatrix} 0 \\ 0 \\ 0 \\ 0 \\ 0 \end{bmatrix}. \quad\quad\quad\quad (7.160)$$

Solution to Problem 7.3.8

1. Jacobi's iteration: In this case, Jacobi's iteration fails to converge. **Table 7.8** shows the diverging iterations of Jacobi's method.
2. **Table 7.9** shows the Gauss-Seidel iteration.
3. **Table 7.10** shows the PSOR with $\omega_{opt} = 1.581$.
4. **Table 7.11** shows the CGM.

Iteration Level	x_1	x_2	x_3	x_4	x_5
0	0.0000	0.0000	0.0000	0.0000	0.0000
1	0.0000	0.0000	0.0000	0.0000	0.1667
2	0.0000	0.0000	−0.0278	0.1111	0.1667
3	0.0046	−0.0370	0.0463	0.0926	0.2454
4	−0.0324	0.0185	−0.0046	0.2006	0.2207
5	0.0131	−0.0581	0.1147	0.1409	0.3012
6	−0.0579	0.0617	0.0028	0.2870	0.2415
7	0.0407	−0.0845	0.2018	0.1526	0.3575
8	−0.0900	0.1363	−0.0210	0.3870	0.2348
9	0.0943	−0.1385	0.3247	0.1198	0.4281

Table 7.8—Diverging iterations of Jacobi's method from the Solution to Problem 7.3.8.

Iteration Level	x_1	x_2	x_3	x_4	x_5
0	0.00000	0.00000	0.00000	0.000000	0.000000
1	0.00000	0.00000	0.00000	0.000000	0.166667
2	0.00000	0.00000	−0.02778	0.092593	0.233025
3	0.00463	−0.03086	0.001543	0.161523	0.274091
4	−0.02083	−0.03978	0.038952	0.215325	0.303725
...					
117	0.178496	0.428424	0.642690	0.714157	0.535657
118	0.178501	0.428434	0.642702	0.714166	0.535661
119	0.178506	0.428444	0.642713	0.714175	0.535664
120	0.178511	0.428453	0.642723	0.714183	0.535668
121	0.178515	0.428461	0.642732	0.71419	0.535671

Table 7.9—The Gauss-Seidel iteration from the Solution to Problem 7.3.8.

Iteration Level	x_1	x_2	x_3	x_4	x_5
0	0.00000	0.00000	0.00000	0.00000	0.000000
1	0.00000	0.00000	0.00000	0.00000	0.263557
2	0.00000	0.00000	−0.06946	0.20462	0.344364
3	0.018307	−0.10786	0.046806	0.321855	0.390336
4	−0.13669	−0.11688	0.12203	0.383847	0.409139
...					
33	0.17855	0.428538	0.642829	0.714271	0.53571
34	0.178556	0.428549	0.642839	0.714277	0.535712
35	0.178561	0.428557	0.642847	0.714281	0.535714

Table 7.10—PSOR with ω_{opt} = 1.581 from the Solution to Problem 7.3.8.

Iteration Level	$x^{(0)}$	$x^{(1)}$	$x^{(2)}$	$x^{(3)}$	$x^{(4)}$	$x^{(5)}$
x_1	0	0.00000	0.00000	0.00698	−0.0748	0.17857
x_2	0	0.00000	0.00000	−0.0558	−0.026	0.42857
x_3	0	0.00000	−0.033	0.04303	0.16499	0.64286
x_4	0	0.00000	0.13204	0.2467	0.36773	0.71429
x_5	0	0.16667	0.26019	0.32396	0.38432	0.53571

Table 7.11—CGM from the Solution to Problem 7.3.8.

Problem 7.3.9 (Ertekin et al. 2001)

Consider the following three equations:

- $-8x_1 + 8x_2 = -1$.
- $2x_1 - 8x_2 + 4x_3 = -1$.
- $4x_2 - 8x_3 = -1$.

Use the Gauss-Seidel iteration to solve the system of equations. Start with an initial guess of $x_1^{(0)} = x_2^{(0)} = x_3^{(0)} = 1.0000$, and use a tolerance of 0.001 for the convergence check.

Solution to Problem 7.3.9

Gauss-Seidel iteration equations are

$$x_1^{(k+1)} = \frac{1}{(-8)}\left(-1 - 8x_2^{(k)}\right), \quad \dots\dots\dots\dots\dots\dots\dots\dots\dots\dots\dots\dots \quad (7.161)$$

$$x_2^{(k+1)} = \frac{1}{(-8)}\left(-1 - 2x_1^{(k+1)} - 4x_3^{(k)}\right), \quad \dots\dots\dots\dots\dots\dots\dots\dots\dots \quad (7.162)$$

$$x_3^{(k+1)} = \frac{1}{(-8)}\left(-1 - 4x_2^{(k)}\right). \quad \dots\dots\dots\dots\dots\dots\dots\dots\dots\dots\dots\dots \quad (7.163)$$

The iteration results using the Gauss-Seidel method are shown in **Table 7.12.**

Iteration Level	x_1	x_2	x_3
0	1.00000	1.00000	1.00000
1	1.12500	0.90625	0.35156
2	1.03125	0.55859	0.26465
3	0.68359	0.42822	0.23206
4	0.55322	0.37933	0.21983
5	0.50433	0.36100	0.21525
6	0.48600	0.35413	0.21353
7	0.47913	0.35155	0.21289
8	0.47655	0.35058	0.21265
9	0.47558	0.35022	0.21255

Table 7.12—Iteration results using the Gauss-Seidel method from the Solution to Problem 7.3.9.

Problem 7.3.10 (Ertekin et al. 2001)

Consider the 2D, steady-state flow of water in a porous medium with the associated boundary conditions shown in **Fig. 7.9.** Generate the LSOR equations by sweeping the system parallel to the i-direction. Express each line equation in matrix form so that Thomas' algorithm is readily applicable.

Solution to Problem 7.3.10

The governing flow equation of the 2D steady-state flow problem is

$$\frac{\partial}{\partial x}\left(\frac{A_x k_x}{\mu B}\frac{\partial p}{\partial x}\right)\Delta x + \frac{\partial}{\partial y}\left(\frac{A_y k_y}{\mu B}\frac{\partial p}{\partial y}\right)\Delta y + q = 0, \quad \dots\dots\dots\dots\dots\dots\dots \quad (7.164)$$

which can be written in a finite-difference analog as

$$T_{i,j-1}p_{i,j-1}^{(k+1)} + T_{i-1,j}p_{i-1,j}^{(k+1)} - \left(T_{i,j-1} + T_{i-1,j} + T_{i+1,j} + T_{i,j+1}\right)p_{i,j}^{(k+1)} + T_{i+1,j}p_{i+1,j}^{(k+1)} + T_{i,j+1}p_{i,j+1}^{(k+1)} = 0, \quad \dots\dots\dots\dots\dots \quad (7.165)$$

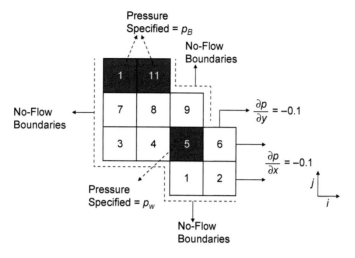

Fig. 7.9—Schematic representation of the reservoir system for Problem 7.3.10.

where T represents the transmissibility between two adjacent blocks.

- Line 1 (Blocks 1 and 2):

  - Block 1: $T_{12}p_2^{(k+1)} - \left(T_{15} + T_{12}\right)p_1^{(k+1)} = -T_{15}p_5^{(k)}$. . . . .(7.166)

  - Block 2: $T_{12}p_1^{(k+1)} - \left(T_{12} + T_{26}\right)p_2^{(k+1)} = 0.1\Delta x\, T_{2x}' - T_{26}p_6^{(k)}$. . . . . (7.167)

  In matrix form,

  $$\begin{bmatrix} -\left(T_{15} + T_{12}\right) & T_{12} \\ T_{12} & -\left(T_{12} + T_{26}\right) \end{bmatrix} \begin{bmatrix} p_1^{(k+1)} \\ p_2^{(k+1)} \end{bmatrix} = \begin{bmatrix} -T_{15}p_5^{(k)} \\ 0.1\Delta x T_{2x}' - T_{26}p_6^{(k)} \end{bmatrix}. \quad\text{. . . . . (7.168)}$$

- Line 2 (Blocks 3, 4, and 6):

  - Block 3: $T_{34}p_4^{(k+1)} - \left(T_{37} + T_{34}\right)p_3^{(k+1)} = -T_{37}p_7^{(k)}$, ... (7.169)

  - Block 4: $T_{34}p_4^{(k+1)} - \left(T_{34} + T_{48} + T_{45}\right)p_4^{(k+1)} = -T_{48}p_8^{(k)} - T_{45}p_w$, ... (7.170)

  - Block 6: $-\left(T_{56} + T_{26}\right)p_6^{(k+1)} = 0.1\Delta x T_{6x}' + 0.1\Delta y T_{6y}' - T_{56}p_w - T_{26}p_2^{(k)}$. . . . . (7.171)

  In matrix form,

  $$\begin{bmatrix} -\left(T_{37} + T_{34}\right) & T_{34} & 0 \\ T_{34} & -\left(T_{34} + T_{48} + T_{45}\right) & 0 \\ 0 & 0 & -\left(T_{56} + T_{26}\right) \end{bmatrix} \begin{bmatrix} p_3^{(k+1)} \\ p_4^{(k+1)} \\ p_6^{(k+1)} \end{bmatrix} = \begin{bmatrix} -T_{37}p_7^{(k)} \\ -T_{48}p_8^{(k)} - T_{45}p_w \\ 0.1\Delta x T_{6x}' + 0.1\Delta y T_{6y}' - T_{56}p_w - T_{26}p_2^{(k)} \end{bmatrix} \quad\text{. . . (7.172)}$$

- Line 3 (Blocks 7, 8, and 9):

  - Block 7: $T_{78}p_8^{(k+1)} - \left(T_{7,10} + T_{37} + T_{78}\right)p_7^{(k+1)} = -T_{7,10}p_B - T_{37}p_3^{(k)}$. . . . . (7.173)

  - Block 8: $T_{78}p_7^{(k+1)} - \left(T_{8,11} + T_{89} + T_{78} + T_{48}\right)p_8^{(k+1)} + T_{89}p_9^{(k+1)} = -T_{8,11}p_B - T_{48}p_4^{(k)}$. . . . . (7.174)

  - Block 9: $T_{89}p_8^{(k+1)} - \left(T_{89} + T_{59}\right)p_9^{(k+1)} = -T_{59}p_w$. . . . . (7.175)

  In matrix form,

  $$\begin{bmatrix} -\left(T_{7,10} + T_{37} + T_{78}\right) & T_{78} & 0 \\ T_{78} & -\left(T_{8,11} + T_{89} + T_{78} + T_{48}\right) & T_{89} \\ 0 & T_{89} & -\left(T_{89} + T_{59}\right) \end{bmatrix} \begin{bmatrix} p_3^{(k+1)} \\ p_4^{(k+1)} \\ p_6^{(k+1)} \end{bmatrix} = \begin{bmatrix} -T_{7,10}p_B - T_{37}p_3^{(k)} \\ -T_{8,11}p_B - T_{48}p_4^{(k)} \\ -T_{59}p_w \end{bmatrix}. \quad\text{. . . . . .(7.176)}$$

Problem 7.3.11 (Ertekin et al. 2001)

Consider the single-phase, steady-state flow of oil in the horizontal, 2D isotropic system in **Fig. 7.10**. Observe that there is a well in the system with a flow-rate specification ($q = q_{spec}$). Apply the LSOR method to solve for the pressure distribution in the system. Write the equations along the j-direction to obtain the iterative equations of iteration level $(k + 1)$ for the unknown blocks in the system. Assume $\omega = 1$, and remember to sweep the system parallel to the j-direction.

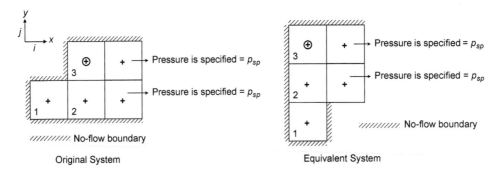

Fig. 7.10—Reservoir system and its equivalent representation (Problem 7.3.11).

Solution to Problem 7.3.11

For an isotropic and homogeneous system with $\Delta x = \Delta y$, the original system can be transformed to a 1D system, as shown on the right side of Fig. 7.10. In this case,

- Block 1: $T_{12} p_2^{(k+1)} - T_{12} p_1^{(k+1)} = 0.$ . . . . (7.177)

- Block 2: $T_{12} p_1^{(k+1)} - \left(T_{12} + T_{23} + T_2'\right) p_2^{(k+1)} + T_{23} p_3^{(k+1)} = -T_2' p_{sp}.$ . . . . (7.178)

- Block 3: $T_{23} p_{23}^{(k+1)} - \left(T_{23} + T_3'\right) p_3^{(k+1)} = q - T_3' p_{sp}.$ . . . . (7.179)

Problem 7.3.12 (Ertekin et al. 2001)

Consider the 3D homogeneous finite-difference body of **Fig. 7.11.1.** A production well at (3,3,3) is completed and kept at a constant pressure. On the outer surfaces of the system, a uniform Dirichlet-type boundary condition is specified. Ignoring gravitational effects, answer the following questions:

1. What is the total number of unknowns?

2. Consider the symmetry: What is the minimum number of equations you need to solve for the pressures? Indicate the nodes for which you will write equations.

3. If $k_x = k_y \neq k_z$, how many unknowns would the system have? What is the minimum number of equations you need to solve for the unknowns of the system?

4. If the same nonzero Neumann-type boundary condition is specified on ($y = 0$ and $y = 4\Delta y$) and ($x = 0$ and $x = 4\Delta x$) surfaces, and if all the other surfaces of the system are represented by the no-flow boundary conditions, how many unknowns does the system have? What is the minimum number of equations you need to consider solving for the unknowns of this problem if $k_x = k_y = k_z$?

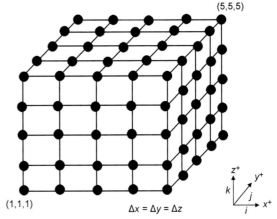

Fig. 7.11.1—Reservoir system and its point-distributed grid for Problem 7.3.12.

5. If the entire surface of the body except the lowermost plane is closed to flow and the nodes on the lowermost plane are kept at a constant pressure, how many unknowns does the system have? What is the minimum number of equations you must consider solving for the unknowns if this problem is $k_x = k_y = k_z$?

Solution to Problem 7.3.12

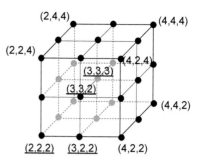

Fig. 7.11.2—The nodes with unknown pressures for the isotropic reservoir system of Problem 7.3.12.

All unknowns for this structure are located under its outer surfaces, as shown in **Fig. 7.11.2.**

1. With a Dirichlet boundary condition specified on the external boundaries, the total number of unknowns is 28, without considering the symmetry of the system. In this case, we have 27 node pressures and one flow rate to calculate.

2. In view of the symmetry of the system, the number of unknown nodes that needs to be solved is four. For a 3×3×3 cube and a well located in its center, equations must be written for a corner node, (2,2,2); a side node, (3,2,2); a face-center node, (3,2,3); and the well node (3,3,3). The nodes with unknown pressures are underlined in Fig. 7.11.2.

3. For the case $k_x = k_y \neq k_z$, the total number of unknowns is 28, with 27 node pressures and one flow rate. Considering the symmetry, there are seven unknowns, which include six node pressures and the flow rate of the well. The node pressures to be solved are at (2,2,2), (3,2,2), (3,3,2), (2,2,3), (3,2,3), and (3,3,3). These nodes with unknown pressures are underlined in **Fig. 7.11.3.**

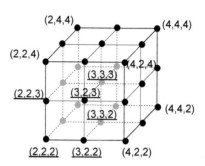

Fig. 7.11.3—The nodes with unknown pressures when $k_x = k_y \neq k_z$ (Problem 7.3.12).

4. All the nodes represent the point-pressure values that need to be calculated. There are 125 node pressures and one flow rate, making the total number of unknowns 126. Considering the symmetry along the diagonal plane, and considering the symmetry with respect to the $z = 3$ plane, there will be 42 node pressures and one flow rate to be solved.

5. Nodes located on the lowermost x-y plane ($z = 0$) are nodes with known pressure values. All the remaining 100 nodes located above the $z = 0$ plane, plus the flow rate of the well are the unknowns (a total of 101 unknowns). Considering the symmetrically imposed boundary conditions, there are 25 node pressures and one well flow rate to be determined.

Problem 7.3.13 (Ertekin et al. 2001)

Determine the steady-state temperature distribution in the 3D structure in **Fig. 7.12.1,** which is obtained by the intersection ($z = 0$) of three finite slabs. Assume homogeneous and isotropic property distribution. The lowermost plane ($z = 0$) is kept at 100°C, and the uppermost plane ($z = 12$) is kept at 0°C. All other outer surfaces of the structure are completely insulated.

Use conventional seven-point, finite-difference approximations to solve the problem. In solving finite-difference equations, use the LSOR algorithm with optimum acceleration parameter. For the solution of the line equation, have the PSOR and Thomas' algorithm available.

Note: Do not use the rule of symmetry to solve this problem; instead use the existing symmetry to check the validity of your solution.

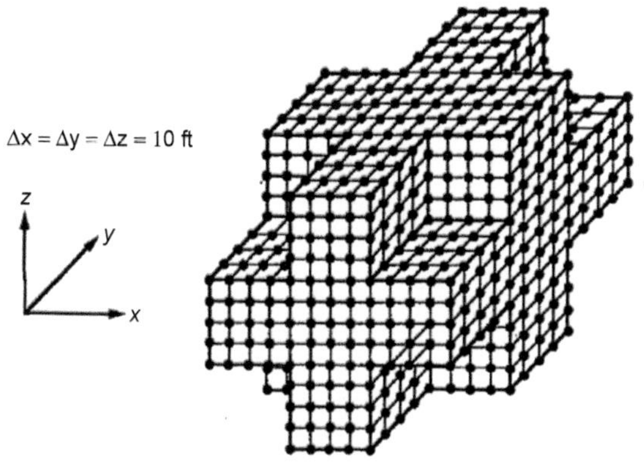

$\Delta x = \Delta y = \Delta z = 10$ ft

Fig. 7.12.1—Reservoir system for Problem 7.3.13.

Solution to Problem 7.3.13

For a 3D steady-state system, the governing equation can be written as

$$\frac{\partial^2 T}{\partial x^2} + \frac{\partial^2 T}{\partial y^2} + \frac{\partial^2 T}{\partial z^2} = 0. \quad \quad (7.180)$$

In a finite-difference form,

$$T_{i+1,j,k} + T_{i-1,j,k} + T_{i,j+1,k} + T_{i,j-1,k} + T_{i,j,k+1} + T_{i,j,k-1} - 6T_{i,j,k} = 0. \quad \quad (7.181)$$

In this case, the 13×13×13 nodes of a cube, with some inactive nodes, need to be solved. Because the plane at $z = 0$ is kept at 100°C and plane $z = 12$ is kept at 0°C, temperature nodes located on the remaining planes between $z = 1$ and

$z = 11$ (inclusive) need to be computed. In **Figs. 7.12.2 through 7.12.12,** results of these calculations are presented for each plane between $z = 1$ and $z = 11$. As can be seen from the temperature-distribution figures, nodes that are symmetrically located with respect to each other exhibit symmetrical temperature values, and as one moves from plane $z = 0$ to plane $z = 12$, temperature decreases in a linear fashion along the z-direction of any two nodes located on top of each other:

				90.9497	90.9978	91.0235	90.9978	90.9497				
				90.9786	91.0246	91.0492	91.0246	90.9786				
				91.0323	91.0744	91.0968	91.0744	91.0323				
				91.0986	91.1375	91.1574	91.1375	91.0986				
90.9497	90.9786	91.0323	91.0986	91.1506	91.1976	91.2167	91.1976	91.1506	91.0986	91.0323	90.9786	90.9497
90.9978	91.0246	91.0744	91.1375	91.1976	91.2435	91.2609	91.2435	91.1976	91.1375	91.0744	91.0246	90.9978
91.0235	91.0492	91.0968	91.1574	91.2167	91.2608	91.2772	91.2608	91.2167	91.1574	91.0968	91.0492	91.0235
90.9978	91.0246	91.0744	91.1375	91.1976	91.2435	91.2608	91.2435	91.1976	91.1375	91.0744	91.0246	90.9978
90.9497	90.9786	91.0323	91.0986	91.1506	91.1976	91.2167	91.1976	91.1506	91.0986	91.0323	90.9786	90.9497
				91.0986	91.1375	91.1574	91.1375	91.0986				
				91.0323	91.0744	91.0968	91.0744	91.0323				
				90.9786	91.0246	91.0492	91.0246	90.9786				
				90.9497	90.9978	91.0235	90.9978	90.9497				

Fig. 7.12.2—Temperature distribution on z = 1 plane.

				81.822	81.991	82.073	81.991	81.822				
				81.886	82.048	82.126	82.048	81.886				
				82.01	82.155	82.225	82.155	82.01				
				82.172	82.297	82.356	82.297	82.172				
81.822	81.886	82.01	82.172	82.311	82.437	82.487	82.437	82.311	82.172	82.01	81.886	81.822
81.991	82.048	82.155	82.297	82.437	82.544	82.584	82.544	82.437	82.297	82.155	82.048	81.991
82.073	82.126	82.225	82.356	82.487	82.584	82.62	82.584	82.487	82.356	82.225	82.126	82.073
81.991	82.048	82.155	82.297	82.437	82.544	82.584	82.544	82.437	82.297	82.155	82.048	81.991
81.822	81.886	82.01	82.172	82.311	82.437	82.487	82.437	82.311	82.172	82.01	81.886	81.822
				82.172	82.297	82.356	82.297	82.172				
				82.01	82.155	82.225	82.155	82.01				
				81.886	82.048	82.126	82.048	81.886				
				81.822	81.991	82.073	81.991	81.822				

Fig. 7.12.3—Temperature distribution on z = 2 plane.

				72.463	73.015	73.233	73.015	72.463				
				72.574	73.103	70.011	73.103	72.574				
				72.803	73.278	73.463	73.278	72.803				
				73.145	73.524	73.671	73.524	73.145				
72.463	72.574	72.803	73.145	73.499	73.786	73.889	73.786	73.499	73.145	72.803	72.574	72.463
73.015	73.103	73.278	73.524	73.786	73.979	74.049	73.979	73.786	73.524	73.278	73.103	73.015
73.233	73.311	73.463	73.671	73.889	74.049	74.107	74.049	73.889	73.671	73.463	73.311	73.233
73.015	73.103	73.278	73.524	73.786	73.979	74.049	73.979	73.786	73.524	73.278	73.103	73.015
72.463	72.574	72.803	73.145	73.499	73.786	73.889	73.786	73.499	73.145	72.803	72.574	72.463
				73.145	73.524	73.671	73.524	73.145				
				72.803	73.278	73.463	73.278	72.803				
				72.574	73.103	73.311	73.103	72.574				
				72.463	73.015	73.233	73.015	72.463				

Fig. 7.12.4—Temperature distribution on z = 3 plane.

52.631	53.14	54.396	57.019	62.44	64.284	64.751	64.284	62.44	57.019	54.396	53.14	52.631
53.14	53.619	54.812	57.328	62.613	64.392	64.839	64.392	62.613	57.328	54.812	53.619	53.14
54.396	54.812	55.854	58.105	63.009	64.617	65.016	64.617	63.009	58.105	55.854	54.812	54.396
57.019	57.328	58.105	59.803	63.727	64.966	65.272	64.966	63.727	59.803	58.105	57.328	57.019
62.44	62.613	63.009	63.727	64.819	65.387	65.555	65.387	64.819	63.727	63.009	62.613	62.44
64.284	64.392	64.617	64.966	65.387	65.663	65.755	65.663	65.387	64.966	64.617	64.392	64.284
64.751	64.839	65.016	65.272	65.555	65.755	65.824	65.755	65.555	65.272	65.016	64.839	64.751
64.284	64.392	64.617	64.966	65.387	65.663	65.755	65.663	65.387	64.966	64.617	64.392	64.284
62.44	62.613	63.009	63.727	64.819	65.387	65.555	65.387	64.819	63.727	63.009	62.613	62.44
57.019	57.328	58.105	59.803	63.727	64.966	65.272	64.966	63.727	59.803	58.105	57.328	57.019
54.396	54.811	55.854	58.105	63.009	64.617	65.016	64.617	63.009	58.105	55.854	54.811	54.396
53.14	53.619	54.811	57.328	62.613	64.392	64.839	64.392	62.613	57.328	54.811	53.619	53.14
52.631	53.14	54.396	57.019	62.44	64.284	64.751	64.284	62.44	57.019	54.396	53.14	52.631

Fig. 7.12.5—Temperature distribution on z = 4 plane.

51.614	51.913	52.616	53.91	55.822	56.823	57.114	56.823	55.822	53.91	52.616	51.913	51.614
51.913	52.194	52.859	54.094	55.935	56.894	57.172	56.894	55.935	54.094	52.859	52.194	51.913
52.616	52.859	53.439	54.53	56.19	57.043	57.289	57.043	56.19	54.53	53.439	52.859	52.616
53.91	54.094	54.53	55.351	56.621	57.27	57.457	57.27	56.621	55.351	54.53	54.094	53.91
55.822	55.935	56.19	56.621	57.188	57.533	57.642	57.533	57.188	56.621	56.19	55.935	55.822
56.823	56.894	57.043	57.27	57.533	57.713	57.774	57.713	57.533	57.27	57.043	56.894	56.823
57.113	57.172	57.289	57.457	57.642	57.774	57.82	57.774	57.642	57.457	57.289	57.172	57.113
56.823	56.894	57.043	57.27	57.533	57.713	57.774	57.713	57.533	57.27	57.043	56.894	56.823
55.822	55.935	56.19	56.621	57.188	57.533	57.642	57.533	57.188	56.621	56.19	55.935	55.822
53.91	54.094	54.53	55.351	56.621	57.27	57.457	57.27	56.621	55.351	54.53	54.094	53.91
52.616	52.859	53.439	54.53	56.19	57.043	57.289	57.043	56.19	54.53	53.439	52.859	52.616
51.913	52.194	52.859	54.094	55.935	56.894	57.172	56.894	55.935	54.094	52.859	52.194	51.913
51.614	51.913	52.616	53.91	55.822	56.823	57.113	56.823	55.822	53.91	52.616	51.913	51.614

Fig. 7.12.6—Temperature distribution on z = 5 plane.

50.000	50.000	50.000	50.000	50.000	50.000	50.000	50.000	50.000	50.000	50.000	50.000	50.000
50.000	50.000	50.000	50.000	50.000	50.000	50.000	50.000	50.000	50.000	50.000	50.000	50.000
50.000	50.000	50.000	50.000	50.000	50.000	50.000	50.000	50.000	50.000	50.000	50.000	50.000
50.000	50.000	50.000	50.000	50.000	50.000	50.000	50.000	50.000	50.000	50.000	50.000	50.000
50.000	50.000	50.000	50.000	50.000	50.000	50.000	50.000	50.000	50.000	50.000	50.000	50.000
50.000	50.000	50.000	50.000	50.000	50.000	50.000	50.000	50.000	50.000	50.000	50.000	50.000
50.000	50.000	50.000	50.000	50.000	50.000	50.000	50.000	50.000	50.000	50.000	50.000	50.000
50.000	50.000	50.000	50.000	50.000	50.000	50.000	50.000	50.000	50.000	50.000	50.000	50.000
50.000	50.000	50.000	50.000	50.000	50.000	50.000	50.000	50.000	50.000	50.000	50.000	50.000
50.000	50.000	50.000	50.000	50.000	50.000	50.000	50.000	50.000	50.000	50.000	50.000	50.000
50.000	50.000	50.000	50.000	50.000	50.000	50.000	50.000	50.000	50.000	50.000	50.000	50.000
50.000	50.000	50.000	50.000	50.000	50.000	50.000	50.000	50.000	50.000	50.000	50.000	50.000
50.000	50.000	50.000	50.000	50.000	50.000	50.000	50.000	50.000	50.000	50.000	50.000	50.000

Fig. 7.12.7—Temperature distribution on z = 6 plane.

48.386	48.088	47.385	46.09	44.179	43.178	42.887	43.178	44.179	46.09	47.385	48.088	48.386
48.088	47.806	47.141	45.907	44.065	43.106	42.829	43.106	44.065	45.907	47.141	47.806	48.088
47.385	47.141	46.562	45.471	43.81	42.957	42.712	42.957	43.81	45.471	46.562	47.141	47.385
46.09	45.906	45.471	44.649	43.379	42.73	42.543	42.73	43.379	44.649	45.471	45.906	46.09
44.179	44.065	43.81	43.379	42.812	42.467	42.358	42.467	42.812	43.379	43.81	44.065	44.179
43.178	43.106	42.957	42.73	42.467	42.287	42.226	42.287	42.467	42.73	42.957	43.106	43.178
42.887	42.829	42.712	42.543	42.358	42.226	42.18	42.226	42.358	42.543	42.712	42.829	42.887
43.178	43.106	42.957	42.73	42.467	42.287	42.226	42.287	42.467	42.73	42.957	43.106	43.178
44.179	44.065	43.81	43.379	42.812	42.467	42.358	42.467	42.812	43.379	43.81	44.065	44.179
46.09	45.906	45.471	44.649	43.379	42.73	42.543	42.73	43.379	44.649	45.471	45.906	46.09
47.385	47.141	46.562	45.471	43.81	42.957	42.711	42.957	43.81	45.471	46.562	47.141	47.385
48.087	47.806	47.141	45.906	44.065	43.106	42.829	43.106	44.065	45.906	47.141	47.806	48.087
48.386	48.087	47.385	46.09	44.179	43.178	42.887	43.178	44.179	46.09	47.385	48.087	48.386

Fig. 7.12.8—Temperature distribution on z = 7 plane.

47.369	46.86	45.604	42.982	37.56	35.716	35.25	35.716	37.56	42.982	45.604	46.86	47.369
46.86	46.381	45.189	42.672	37.387	35.609	35.161	35.609	37.387	42.672	45.189	46.381	46.86
45.604	45.189	44.146	41.895	36.991	35.383	34.984	35.383	36.991	41.895	44.146	45.189	45.604
42.982	42.672	41.895	40.197	36.273	35.034	34.728	35.034	36.273	40.197	41.895	42.672	42.982
37.56	37.387	36.991	36.273	35.181	34.613	34.445	34.613	35.181	36.273	36.991	37.387	37.56
35.716	35.609	35.383	35.034	34.613	34.338	34.246	34.338	34.613	35.034	35.383	35.609	35.716
35.25	35.161	34.984	34.728	34.445	34.246	34.176	34.246	34.445	34.728	34.984	35.161	35.25
35.716	35.609	35.383	35.034	34.613	34.338	34.246	34.338	34.613	35.034	35.383	35.609	35.716
37.56	37.387	36.991	36.273	35.181	34.613	34.445	34.613	35.181	36.273	36.991	37.387	37.56
42.982	42.672	41.895	40.197	36.273	35.034	34.728	35.034	36.273	40.197	41.895	42.672	42.982
45.604	45.189	44.146	41.895	36.991	35.383	34.984	35.383	36.991	41.895	44.146	45.189	45.604
46.86	46.381	45.189	42.672	37.387	35.609	35.161	35.609	37.387	42.672	45.189	46.381	46.86
47.369	46.86	45.604	42.981	37.56	35.716	35.25	35.716	37.56	42.981	45.604	46.86	47.369

Fig. 7.12.9—Temperature distribution on z = 8 plane.

				27.538	26.986	26.767	26.986	27.538				
				27.427	26.897	26.689	26.897	27.427				
				27.197	26.723	26.537	26.723	27.197				
				26.855	26.476	26.329	26.476	26.855				
27.537	27.427	27.197	26.855	26.501	26.214	26.111	26.214	26.501	26.855	27.197	27.427	27.537
26.986	26.897	26.723	26.476	26.214	26.021	25.951	26.021	26.214	26.476	26.723	26.897	26.986
26.767	26.689	26.537	26.329	26.111	25.951	25.894	25.951	26.111	26.329	26.537	26.689	26.767
26.986	26.897	26.723	26.476	26.214	26.021	25.951	26.021	26.214	26.476	26.723	26.897	26.986
27.537	27.427	27.197	26.855	26.501	26.214	26.111	26.214	26.501	26.855	27.197	27.427	27.537
				26.855	26.476	26.329	26.476	26.855				
				27.197	26.723	26.537	26.723	27.197				
				27.427	26.897	26.689	26.897	27.427				
				27.537	26.986	26.767	26.986	27.537				

Fig. 7.12.10—Temperature distribution on z = 9 plane.

				18.178	18.009	17.927	18.009	18.178				
				18.114	17.953	17.875	17.953	18.114				
				17.99	17.845	17.775	17.845	17.99				
				17.828	17.703	17.644	17.703	17.828				
18.178	18.114	17.99	17.828	17.689	17.563	17.513	17.563	17.689	17.828	17.99	18.114	18.178
18.009	17.953	17.845	17.703	17.563	17.456	17.416	17.456	17.563	17.703	17.845	17.953	18.009
17.927	17.875	17.775	17.644	17.513	17.416	17.38	17.416	17.513	17.644	17.775	17.875	17.927
18.009	17.952	17.845	17.703	17.563	17.456	17.416	17.456	17.563	17.703	17.845	17.952	18.009
18.178	18.114	17.99	17.828	17.689	17.563	17.513	17.563	17.689	17.828	17.99	18.114	18.178
				17.828	17.703	17.644	17.703	17.828				
				17.99	17.845	17.775	17.845	17.99				
				18.114	17.952	17.875	17.952	18.114				
				18.178	18.009	17.927	18.009	18.178				

Fig. 7.12.11—Temperature distribution on z = 10 plane.

				9.05	9.002	8.977	9.002	9.05				
				9.021	8.975	8.951	8.975	9.021				
				8.968	8.926	8.903	8.926	8.968				
				8.901	8.863	8.843	8.863	8.901				
9.05	9.021	8.968	8.901	8.849	8.802	8.783	8.802	8.849	8.901	8.968	9.021	9.05
9.002	8.975	8.926	8.863	8.802	8.757	8.739	8.757	8.802	8.863	8.926	8.975	9.002
8.977	8.951	8.903	8.843	8.783	8.739	8.723	8.739	8.783	8.843	8.903	8.951	8.977
9.002	8.975	8.926	8.863	8.802	8.757	8.739	8.757	8.802	8.863	8.926	8.975	9.002
9.05	9.021	8.968	8.901	8.849	8.802	8.783	8.802	8.849	8.901	8.968	9.021	9.05
				8.901	8.863	8.843	8.863	8.901				
				8.968	8.926	8.903	8.926	8.968				
				9.021	8.975	8.951	8.975	9.021				
				9.05	9.002	8.977	9.002	9.05				

Fig. 7.12.12—Temperature distribution on z = 11 plane.

If one repeats the computations by considering the existing symmetry, the number of unknowns to be solved will drastically decrease. Such a numerical exercise provides the temperature-distribution solutions on each plane (note that this time we are solving 1/8) as shown in **Figs. 7.12.13 through 7.12.23** for the temperature distribution of planes z = 1 through z = 11:

		91.1506	91.0985	91.0322	90.9786	90.9497
	91.2435	91.1975	91.1374	91.0744	91.0245	90.9977
91.2772	91.2608	91.2166	91.1573	91.0967	91.0491	91.0234

Fig. 7.12.13—Temperature distribution on z = 1 plane by considering the existing symmetry.

		82.3113	82.1723	82.0098	81.8863	81.8224
	82.5442	82.4370	82.2969	82.1552	82.0475	81.9909
82.6200	82.5841	82.4866	82.3557	82.2252	82.1255	82.0727

Fig. 7.12.14—Temperature distribution on z = 2 plane by considering the existing symmetry.

		73.4987	73.1450	72.8030	72.5734	72.4625
	73.9793	73.7858	73.5238	73.2775	73.1025	73.0144
74.1065	74.0487	73.8889	73.6712	73.4630	73.3108	73.2325

Fig. 7.12.15—Temperature distribution on z = 3 plane by considering the existing symmetry.

						52.6312
					53.6191	53.1397
				55.8541	54.8112	54.3962
			59.8029	58.1047	57.3282	57.0185
		64.8192	63.7272	63.0091	62.6129	62.4398
	65.6625	65.387	64.9662	64.6173	64.3915	64.2837
65.824	65.7544	65.5552	65.272	65.0161	64.8386	64.7504

Fig. 7.12.16—Temperature distribution on z = 4 plane by considering the existing symmetry.

						51.614
					52.1935	51.9125
				53.4384	52.8586	52.6154
			55.3506	54.5294	54.0935	53.9097
		57.1879	56.6209	56.1896	55.9352	55.8214
	57.7127	57.5333	57.2701	57.0432	56.8939	56.8225
57.8197	57.7736	57.6421	57.4571	57.2885	57.1714	57.1133

Fig. 7.12.17—Temperature distribution on z = 5 plane by considering the existing symmetry.

						50
					50	50
				50	50	50
			50	50	50	50
		50	50	50	50	50
	50	50	50	50	50	50
50	50	50	50	50	50	50

Fig. 7.12.18—Temperature distribution on z = 6 plane by considering the existing symmetry.

						48.3857
					47.8062	48.0872
				46.5614	47.1411	47.3843
			44.6492	45.4703	45.9062	46.09
		42.8119	43.3789	43.8102	44.0646	44.1784
	42.2871	42.4665	42.7297	42.9566	43.1059	43.1774
42.1801	42.2263	42.3578	42.5427	42.7113	42.8285	42.8865

Fig. 7.12.19—Temperature distribution on z = 7 plane by considering the existing symmetry.

						47.3685
					46.3806	46.86
				44.1457	45.1885	45.6035
			40.1969	41.895	42.6716	42.9813
		35.1807	36.2726	36.9907	37.3869	37.56
	34.3374	34.6128	35.0336	35.3826	35.6083	35.7161
34.1759	34.2454	34.4446	34.7278	34.9837	35.1612	35.2495

Fig. 7.12.20—Temperature distribution on z = 8 plane by considering the existing symmetry.

		26.5012	26.8549	27.1969	27.4264	27.5373
	26.0206	26.2141	26.4761	26.7224	26.8973	26.9854
25.8934	25.9512	26.1110	26.3287	26.5368	26.6891	26.7673

Fig. 7.12.21—Temperature distribution on z = 9 plane by considering the existing symmetry.

		17.6886	17.8276	17.9901	18.1136	18.1775
	17.4558	17.5629	17.7030	17.8447	17.9524	18.0090
17.3799	17.4158	17.5134	17.6442	17.7747	17.8744	17.9272

Fig. 7.12.22—Temperature distribution on z = 10 plane by considering the existing symmetry.

		8.8494	8.9014	8.9677	9.0214	9.0503
	8.7565	8.8024	8.8625	8.9256	8.9754	9.0022
8.7228	8.7392	8.7833	8.8426	8.9032	8.9508	8.9765

Fig. 7.12.23—Temperature distribution on z = 11 plane by considering the existing symmetry.

Problem 7.3.14 (Ertekin et al. 2001)

Solve the equation, $\dfrac{\partial^2 u}{\partial x^2} + \dfrac{\partial^2 u}{\partial y^2} = 0$, for the 2D region shown in **Fig. 7.13.1,** with the indicated boundary conditions for u. Use the Gaussian elimination method and Crout reduction algorithms. Compare the computational time required for both algorithms.

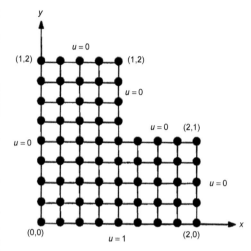

Fig. 7.13.1—Reservoir system for Problem 7.3.14.

Solution to Problem 7.3.14

The PDE can be written in a finite-difference form using five-point approximation:

$$u_{i+1,j} + u_{i-1,j} + u_{i,j+1} + u_{i,j-1} - 4u_{i,j} = 0. \quad \ldots\ldots\ldots\ldots (7.182)$$

Implementing Crout reduction and Gaussian elimination yields the same result. The solution is provided in **Fig. 7.13.2,** which shows Crout reduction and Gaussian elimination generating the same results, as expected. Gaussian elimination takes 2.5 milliseconds of central-processing-unit (CPU) time, while Crout reduction takes 1.3 milliseconds of CPU time. It should be noted that the Crout reduction is approximately 2X faster than Gaussian elimination for a problem this size.

0.00000	0.00000	0.00000	0.00000	0.00000	0.00000	0.00000	0.00000	0.00000
0.00000	0.00657	0.00961	0.00712	0.00000	0.00000	0.00000	0.00000	0.00000
0.00000	0.01666	0.02476	0.01885	0.00000	0.00000	0.00000	0.00000	0.00000
0.00000	0.03532	0.05391	0.04354	0.00000	0.00000	0.00000	0.00000	0.00000
0.00000	0.07073	0.11200	0.10139	0.00000	0.00000	0.00000	0.00000	0.00000
0.00000	0.13559	0.22199	0.25002	0.22780	0.20648	0.16808	0.10026	0.00000
0.00000	0.24964	0.39034	0.44889	0.45471	0.43003	0.36559	0.23298	0.00000
0.00000	0.47262	0.64086	0.70047	0.71213	0.69335	0.63125	0.46606	0.00000
0.00000	1.00000	1.00000	1.00000	1.00000	1.00000	1.00000	1.00000	1.00000

Fig. 7.13.2—Solution of Problem 7.3.14.

Problem 7.3.15 (Ertekin et al. 2001)

Solve Laplace's 3D equation approximately, as given by

$$\frac{\partial^2 u}{\partial x^2} + \frac{\partial^2 u}{\partial y^2} + \frac{\partial^2 u}{\partial z^2} = 0.$$

Inside a cube **(Fig 7.14.1)** whose edge is of unit length, if $u = u(x,y,z) = 1$ on one of the faces and 0 on the remaining faces, use $\Delta x = \Delta y = \Delta z = 0.1$, with a mesh-centered grid. Use the Gauss-Seidel algorithm. (*Hint:* Use the rule of symmetry to reduce the size of the problem).

Solution to Problem 7.3.15

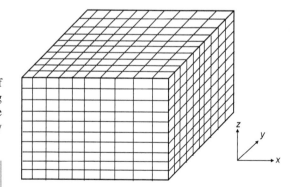

Fig. 7.14.1—Illustration of the cube-shaped reservoir and its gridblocks for Problem 7.3.15.

Discretizing the Laplace equation in finite-difference form, one obtains

$$u_{i+1,j,k} + u_{i-1,j,k} + u_{i,j+1,k} + u_{i,j-1,k} + u_{i,j,k+1} + u_{i,j,k-1} - 6u_{i,j,k} = 0. \quad \ldots\ldots\ldots\ldots\ldots\ldots\ldots\ldots\ldots\ldots (7.183)$$

Using the symmetry of the system, the linear system of equations can be written for 1/8 of the cube, which includes 135 unknowns. Assuming that at plane $z = 1$, $u = 1$, and at the remaining faces $u = 0$, the solutions on surface $z = 0$ through $z = 1.0$ are shown in **Figs. 7.14.2 through 7.14.12.**

0	–	–	–	–	–	–
0	0	–	–	–	–	–
0	0	0	–	–	–	–
0	0	0	0	–	–	–
0	0	0	0	0	–	–
0	0	0	0	0	0	–
0	0	0	0	0	0	0

Fig. 7.14.2—Solution on surface z = 0.

0	-	-	-	-	-	-
0	0.017708	-	-	-	-	-
0	0.016877	0.016085	-	-	-	-
0	0.014435	0.013758	0.011769	-	-	-
0	0.01056	0.010065	0.008611	0.006302	-	-
0	0.005583	0.005322	0.004554	0.003333	0.001763	-
0	0	0	0	0	0	0

Fig. 7.14.3—Solution on surface z = 0.1.

0	-	-	-	-	-	-
0	0.03876	-	-	-	-	-
0	0.03696	0.03525	-	-	-	-
0	0.03167	0.0302	0.02588	-	-	-
0	0.02322	0.02215	0.01898	0.01392	-	-
0	0.0123	0.01173	0.01006	0.00738	0.00391	-
0	0	0	0	0	0	0

Fig. 7.14.4—Solution on surface z = 0.2.

0	-	-	-	-	-	-
0	0.06703	-	-	-	-	-
0	0.064	0.0611	-	-	-	-
0	0.055	0.05252	0.04516	-	-	-
0	0.0405	0.03868	0.03327	0.02453	-	-
0	0.02154	0.02057	0.0177	0.01306	0.00695	-
0	0	0	0	0	0	0

Fig. 7.14.5—Solution on surface z = 0.3.

0	-	-	-	-	-	-
0	0.10746	-	-	-	-	-
0	0.1028	0.09835	-	-	-	-
0	0.08884	0.08501	0.07352	-	-	-
0	0.0659	0.06308	0.05459	0.04058	-	-
0	0.03528	0.03378	0.02926	0.02178	0.0117	-
0	0	0	0	0	0	0

Fig. 7.14.6—Solution on surface z = 0.4.

0	-	-	-	-	-	-
0	0.16658	-	-	-	-	-
0	0.15984	0.15339	-	-	-	-
0	0.13933	0.13376	0.11675	-	-	-
0	0.10464	0.1005	0.08785	0.06625	-	-
0	0.05672	0.0545	0.04771	0.03606	0.01968	-
0	0	0	0	0	0	0

Fig. 7.14.7—Solution on surface $z = 0.5$.

0	-	-	-	-	-	-
0	0.25268	-	-	-	-	-
0	0.24358	0.23484	-	-	-	-
0	0.21517	0.20757	0.18376	-	-	-
0	0.1649	0.15921	0.1413	0.10909	-	-
0	0.09137	0.0883	0.07859	0.06095	0.03426	-
0	0	0	0	0	0	0

Fig. 7.14.8—Solution on surface $z = 0.6$.

0	-	-	-	-	-	-
0	0.37525	-	-	-	-	-
0	0.36411	0.3534	-	-	-	-
0	0.32809	0.31869	0.28809	-	-	-
0	0.25981	0.25269	0.22932	0.18376	-	-
0	0.15003	0.14614	0.13327	0.10773	0.06397	-
0	0	0	0	0	0	0

Fig. 7.14.9—Solution on surface $z = 0.7$.

0	-	-	-	-	-	-
0	0.5424	-	-	-	-	-
0	0.53097	0.51994	-	-	-	-
0	0.49211	0.48234	0.44877	-	-	-
0	0.41046	0.40295	0.37682	0.31938	-	-
0	0.25672	0.25255	0.23784	0.20445	0.13407	-
0	0	0	0	0	0	0

Fig. 7.14.10—Solution on surface $z = 0.8$.

0	-	-	-	-	-	-
0	0.75528	-	-	-	-	-
0	0.74733	0.73962	-	-	-	-
0	0.71847	0.71158	0.6862	-	-	-
0	0.64822	0.64284	0.62264	0.57	-	-
0	0.47472	0.47168	0.45997	0.42766	0.33157	-
0	0	0	0	0	0	0

Fig. 7.14.11—Solution on surface $z = 0.9$.

1	-	-	-	-	-	-
1	1	-	-	-	-	-
1	1	1	-	-	-	-
1	1	1	1	-	-	-
1	1	1	1	1	-	-
1	1	1	1	1	1	-
1	1	1	1	1	1	1

Fig. 7.14.12—Solution on surface z = 1.0.

Problem 7.3.16 (Ertekin et al. 2001)

Poisson's equation is $\dfrac{\partial^2 u}{\partial x^2} + \dfrac{\partial^2 u}{\partial y^2} = \rho(x, y)$, where $\rho(x,y)$ is a given function of x and y. Approximately solve this equation for the case in which $\rho(x,y) = 100/(1 + x^2 + y^2)$, if the region and boundary conditions are the same as those in Fig. 7.13.1 from Problem 7.3.14. Use the Crout reduction algorithm.

Solution to Problem 7.3.16

Poisson's equation can be written in finite-difference form:

$$\frac{u_{i+1,j} - 2u_{i,j} + u_{i-1,j}}{(\Delta x)^2} + \frac{u_{i,j+1} - 2u_{i,j} + u_{i,j-1}}{(\Delta y)^2} = 100 / (1 + x^2 + y^2). \quad\ldots\ldots\ldots\ldots\ldots\ldots\ldots\ldots\ldots\ldots\ldots\ldots (7.184)$$

Because $\Delta x = \Delta y$,

$$u_{i+1,j} + u_{i-1,j} + u_{i,j+1} + u_{i,j-1} - 4u_{i,j} = (\Delta x)^2 \times 100 / (1 + x^2 + y^2). \quad\ldots\ldots\ldots\ldots\ldots\ldots\ldots\ldots\ldots\ldots\ldots\ldots (7.185)$$

In this system, $\Delta x = 0.25$. Solving the system of equations using Crout reduction produces **Fig. 7.15.**

0.000	0.000	0.000	0.000	0.000	0.000	0.000	0.000	0.000
0.000	−1.485	−1.908	−1.419	0.000	0.000	0.000	0.000	0.000
0.000	−2.517	−3.279	−2.418	0.000	0.000	0.000	0.000	0.000
0.000	−3.417	−4.487	−3.334	0.000	0.000	0.000	0.000	0.000
0.000	−4.282	−5.694	−4.432	0.000	0.000	0.000	0.000	0.000
0.000	−4.988	−6.798	−6.262	−4.306	−3.171	−2.269	−1.326	0.000
0.000	−5.026	−6.799	−6.571	−5.351	−4.110	−2.938	−1.685	0.000
0.000	−3.554	−4.634	−4.423	−3.641	−2.759	−1.902	−1.026	0.000
0.000	1.000	1.000	1.000	1.000	1.000	1.000	1.000	1.000

Fig. 7.15—Solution of Problem 7.3.16.

Problem 7.3.17 (Ertekin et al. 2001)

Obtain the steady-state temperature distribution in a square slab with the dimensions of unity aligned with the x and y axes. The boundaries $x = 1$, $y = 0$, and $y = 1$ are kept at zero temperature, while the boundary $x = 0$ is left at a temperature of 100°C.

1. Solve the problem using conventional five-point, finite-difference approximation, letting $\Delta x = \Delta y = 0.05$.
2. Re-solve the same problem using the higher-accuracy nine-point, finite-difference approximation:

$$\left[\frac{\partial^2 u}{\partial x^2} + \frac{\partial^2 u}{\partial y^2}\right]_{i,j} = \frac{1}{6h^2}\left[\begin{array}{l} u_{i+1,j+1} + 4u_{i,j+1} + u_{i-1,j+1} + 4u_{i+1,j} \\ -20u_{i,j} + 4u_{i-1,j} + u_{i+1,j-1} + u_{i,j-1} + u_{i-1,j-1} \end{array}\right] + 0\left(h^4\right), \text{ where } h = \Delta x = \Delta y.$$

3. Compare your results with the exact solution provided here. Also compare the computational time, storage requirements, and number of iterations. Tabulate your results. The analytical solution is

$$T_{(x,y)} = \sum_{n=1}^{\infty} T_n(x,y), \text{ where } T_n = E_n \sin(n\pi x)\left[\sinh(n\pi x) - \tanh(n\pi x)\cosh(n\pi x)\right], \text{ and}$$

$$E_n = -\frac{2}{\tanh(n\pi)} \int_0^1 100 \sin(n\pi y)\, dy.$$

Solution to Problem 7.3.17

1. The five-point finite-difference approximation yields the temperature distribution shown in **Fig. 7.16.1.**
2. The nine-point finite-difference approximation yields the temperature distribution shown in **Fig. 7.16.2.**
3. The analytical solution for this problem yields the results shown in **Fig. 7.16.3.**

The maximum relative error encountered against the analytical solution of the five-point finite-difference approximation is 2.47%, while the maximum error of nine-point finite-difference approximation is 0.0824%.

100.000	0.000	0.000	0.000	0.000	0.000	0.000	0.000	0.000	0.000	0.000	0.000	0.000	0.000	0.000	0.000	0.000	0.000	0.000	0.000	0.000
100.000	49.726	29.690	20.133	14.793	11.421	9.093	7.382	6.066	5.020	4.168	3.461	2.864	2.353	1.910	1.519	1.170	0.852	0.556	0.274	0.000
100.000	69.212	48.903	36.047	27.619	21.798	17.571	14.369	11.861	9.845	8.191	6.810	5.642	4.639	3.767	2.998	2.309	1.681	1.097	0.542	0.000
100.000	78.220	60.661	47.533	37.839	30.582	25.022	20.661	17.165	14.309	11.941	9.949	8.253	6.793	5.520	4.396	3.387	2.467	1.610	0.795	0.000
100.000	83.008	67.989	55.584	45.622	37.667	31.276	26.089	21.827	18.287	15.313	12.791	10.630	8.761	7.126	5.678	4.378	3.189	2.082	1.028	0.000
100.000	85.824	72.703	61.192	51.400	43.187	36.327	30.590	25.769	21.697	18.234	15.271	12.715	10.494	8.544	6.813	5.255	3.830	2.501	1.235	0.000
100.000	87.585	75.805	65.082	55.598	47.353	40.256	34.173	28.964	24.497	20.657	17.342	14.467	11.955	9.744	7.775	6.000	4.375	2.858	1.411	0.000
100.000	88.713	77.850	67.735	58.554	50.373	43.172	36.883	31.415	26.671	22.554	18.975	15.854	13.117	10.700	8.543	6.596	4.811	3.143	1.552	0.000
100.000	89.415	79.146	69.453	60.512	52.413	45.175	38.771	33.143	28.217	23.912	20.151	16.857	13.960	11.394	9.103	7.030	5.129	3.351	1.655	0.000
100.000	89.802	79.866	70.420	61.628	53.590	46.345	39.884	34.169	29.140	24.728	20.860	17.463	14.470	11.816	9.442	7.294	5.322	3.478	1.718	0.000
100.000	89.925	80.097	70.732	61.991	53.975	46.729	40.252	34.509	29.448	25.000	21.096	17.666	14.641	11.957	9.556	7.383	5.387	3.520	1.739	0.000
100.000	89.802	79.866	70.420	61.628	53.590	46.345	39.884	34.169	29.140	24.728	20.860	17.463	14.470	11.816	9.442	7.294	5.322	3.478	1.718	0.000
100.000	89.415	79.146	69.453	60.512	52.413	45.175	38.771	33.143	28.217	23.912	20.151	16.857	13.960	11.394	9.103	7.030	5.129	3.351	1.655	0.000
100.000	88.713	77.850	67.735	58.554	50.373	43.172	36.883	31.415	26.671	22.554	18.975	15.854	13.117	10.700	8.543	6.596	4.811	3.143	1.552	0.000
100.000	87.585	75.805	65.082	55.598	47.353	40.256	34.173	28.964	24.497	20.657	17.342	14.467	11.955	9.744	7.775	6.000	4.375	2.858	1.411	0.000
100.000	85.824	72.703	61.192	51.400	43.187	36.327	30.590	25.769	21.697	18.234	15.271	12.715	10.494	8.544	6.813	5.255	3.830	2.501	1.235	0.000
100.000	83.008	67.989	55.584	45.622	37.667	31.276	26.089	21.827	18.287	15.313	12.791	10.630	8.761	7.126	5.678	4.378	3.189	2.082	1.028	0.000
100.000	78.220	60.661	47.533	37.839	30.582	25.022	20.661	17.165	14.309	11.941	9.949	8.253	6.793	5.520	4.396	3.387	2.467	1.610	0.795	0.000
100.000	69.212	48.903	36.047	27.619	21.798	17.571	14.369	11.861	9.845	8.191	6.810	5.642	4.639	3.767	2.998	2.309	1.681	1.097	0.542	0.000
100.000	49.726	29.690	20.133	14.793	11.421	9.093	7.382	6.066	5.020	4.168	3.461	2.864	2.353	1.910	1.519	1.170	0.852	0.556	0.274	0.000
100.000	0.000	0.000	0.000	0.000	0.000	0.000	0.000	0.000	0.000	0.000	0.000	0.000	0.000	0.000	0.000	0.000	0.000	0.000	0.000	0.000

Fig. 7.16.1—Temperature distribution solved using the five-point finite-difference approximation.

50.000	0.000	0.000	0.000	0.000	0.000	0.000	0.000	0.000	0.000	0.000	0.000	0.000	0.000	0.000	0.000	0.000	0.000	0.000	0.000	0.000
100.000	49.726	28.950	19.675	14.541	11.278	9.007	7.327	6.029	4.994	4.149	3.447	2.853	2.345	1.904	1.515	1.166	0.849	0.554	0.274	0.000
100.000	69.956	48.906	35.810	27.394	21.630	17.452	14.285	11.802	9.802	8.159	6.786	5.623	4.624	3.755	2.989	2.302	1.676	1.094	0.540	0.000
100.000	78.683	60.908	47.540	37.746	30.470	24.923	20.581	17.102	14.260	11.902	9.918	8.229	6.774	5.505	4.383	3.378	2.460	1.606	0.793	0.000
100.000	83.267	68.228	55.696	45.634	37.628	31.220	26.031	21.776	18.242	15.275	12.759	10.604	8.739	7.108	5.663	4.366	3.181	2.077	1.025	0.000
100.000	85.976	72.887	61.326	51.466	43.203	36.314	30.562	25.736	21.663	18.203	15.242	12.690	10.471	8.525	6.797	5.242	3.821	2.495	1.232	0.000
100.000	87.682	75.943	65.208	55.685	47.403	40.275	34.172	28.950	24.477	20.634	17.319	14.444	11.934	9.725	7.759	5.987	4.364	2.851	1.408	0.000
100.000	88.779	77.955	67.845	58.647	50.439	43.212	36.902	31.419	26.663	22.540	18.958	15.835	13.098	10.682	8.527	6.583	4.800	3.136	1.549	0.000
100.000	89.466	79.232	69.551	60.604	52.487	45.228	38.804	33.159	28.219	23.906	20.139	16.841	13.943	11.378	9.087	7.017	5.118	3.344	1.652	0.000
100.000	89.845	79.941	70.510	61.718	53.667	46.404	39.925	34.192	29.149	24.726	20.851	17.450	14.454	11.800	9.427	7.281	5.312	3.471	1.715	0.000
100.000	89.966	80.169	70.820	62.079	54.053	46.790	40.295	34.535	29.459	25.000	21.088	17.653	14.626	11.942	9.541	7.370	5.377	3.513	1.736	0.000
100.000	89.845	79.941	70.510	61.718	53.667	46.404	39.925	34.192	29.149	24.726	20.851	17.450	14.454	11.800	9.427	7.281	5.312	3.471	1.715	0.000
100.000	89.466	79.232	69.551	60.604	52.487	45.228	38.804	33.159	28.219	23.906	20.139	16.841	13.943	11.378	9.087	7.017	5.118	3.344	1.652	0.000
100.000	88.779	77.955	67.845	58.647	50.439	43.212	36.902	31.419	26.663	22.540	18.958	15.835	13.098	10.682	8.527	6.583	4.800	3.136	1.549	0.000
100.000	87.682	75.943	65.208	55.685	47.403	40.275	34.172	28.950	24.477	20.634	17.319	14.444	11.934	9.725	7.759	5.987	4.364	2.851	1.408	0.000
100.000	85.976	72.887	61.326	51.466	43.203	36.314	30.562	25.736	21.663	18.203	15.242	12.690	10.471	8.525	6.797	5.242	3.821	2.495	1.232	0.000
100.000	83.267	68.228	55.696	45.634	37.628	31.220	26.031	21.776	18.242	15.275	12.759	10.604	8.739	7.108	5.663	4.366	3.181	2.077	1.025	0.000
100.000	78.683	60.908	47.540	37.746	30.470	24.923	20.581	17.102	14.260	11.902	9.918	8.229	6.774	5.505	4.383	3.378	2.460	1.606	0.793	0.000
100.000	69.956	48.906	35.810	27.394	21.630	17.452	14.285	11.802	9.802	8.159	6.786	5.623	4.624	3.755	2.989	2.302	1.676	1.094	0.540	0.000
100.000	49.726	28.950	19.675	14.541	11.278	9.007	7.327	6.029	4.994	4.149	3.447	2.853	2.345	1.904	1.515	1.166	0.849	0.554	0.274	0.000
50.000	0.000	0.000	0.000	0.000	0.000	0.000	0.000	0.000	0.000	0.000	0.000	0.000	0.000	0.000	0.000	0.000	0.000	0.000	0.000	0.000

Fig. 7.16.2—Temperature distribution solved using the nine-point finite-difference approximation.

50.000	0.000	0.000	0.000	0.000	0.000	0.000	0.000	0.000	0.000	0.000	0.000	0.000	0.000	0.000	0.000	0.000	0.000	0.000	0.000	0.000
100.000	49.726	28.974	19.679	14.542	11.278	9.007	7.327	6.029	4.994	4.149	3.447	2.853	2.345	1.904	1.515	1.166	0.849	0.554	0.274	0.000
100.000	69.932	48.906	35.813	27.395	21.630	17.452	14.285	11.802	9.802	8.159	6.786	5.623	4.624	3.755	2.989	2.302	1.676	1.094	0.540	0.000
100.000	78.679	60.904	47.540	37.746	30.471	24.924	20.581	17.102	14.260	11.902	9.918	8.229	6.774	5.505	4.383	3.378	2.460	1.606	0.793	0.000
100.000	83.266	68.226	55.695	45.634	37.628	31.220	26.031	21.776	18.242	15.275	12.759	10.604	8.739	7.108	5.663	4.366	3.181	2.077	1.025	0.000
100.000	85.975	72.886	61.325	51.466	43.203	36.314	30.562	25.736	21.663	18.203	15.242	12.690	10.471	8.525	6.797	5.242	3.821	2.495	1.232	0.000
100.000	87.681	75.942	65.207	55.685	47.402	40.275	34.172	28.950	24.477	20.634	17.319	14.444	11.934	9.725	7.759	5.987	4.364	2.851	1.408	0.000
100.000	88.779	77.955	67.845	58.647	50.439	43.212	36.902	31.419	26.663	22.540	18.958	15.835	13.098	10.682	8.527	6.583	4.800	3.136	1.549	0.000
100.000	89.466	79.232	69.551	60.604	52.487	45.228	38.804	33.159	28.219	23.906	20.139	16.841	13.943	11.378	9.087	7.017	5.118	3.344	1.652	0.000
100.000	89.845	79.941	70.510	61.718	53.667	46.404	39.925	34.192	29.149	24.726	20.851	17.450	14.454	11.800	9.427	7.281	5.312	3.471	1.715	0.000
100.000	89.966	80.169	70.819	62.079	54.053	46.790	40.294	34.535	29.459	25.000	21.088	17.653	14.626	11.942	9.541	7.370	5.377	3.513	1.736	0.000
100.000	89.845	79.941	70.510	61.718	53.667	46.404	39.925	34.192	29.149	24.726	20.851	17.450	14.454	11.800	9.427	7.281	5.312	3.471	1.715	0.000
100.000	89.466	79.232	69.551	60.604	52.487	45.228	38.804	33.159	28.219	23.906	20.139	16.841	13.943	11.378	9.087	7.017	5.118	3.344	1.652	0.000
100.000	88.779	77.955	67.845	58.647	50.439	43.212	36.902	31.419	26.663	22.540	18.958	15.835	13.098	10.682	8.527	6.583	4.800	3.136	1.549	0.000
100.000	87.681	75.942	65.207	55.685	47.402	40.275	34.172	28.950	24.477	20.634	17.319	14.444	11.934	9.725	7.759	5.987	4.364	2.851	1.408	0.000
100.000	85.975	72.886	61.325	51.466	43.203	36.314	30.562	25.736	21.663	18.203	15.242	12.690	10.471	8.525	6.797	5.242	3.821	2.495	1.232	0.000
100.000	83.266	68.226	55.695	45.634	37.628	31.220	26.031	21.776	18.242	15.275	12.759	10.604	8.739	7.108	5.663	4.366	3.181	2.077	1.025	0.000
100.000	78.679	60.904	47.540	37.746	30.471	24.924	20.581	17.102	14.260	11.902	9.918	8.229	6.774	5.505	4.383	3.378	2.460	1.606	0.793	0.000
100.000	69.932	48.906	35.813	27.395	21.630	17.452	14.285	11.802	9.802	8.159	6.786	5.623	4.624	3.755	2.989	2.302	1.676	1.094	0.540	0.000
100.000	49.726	28.974	19.679	14.542	11.278	9.007	7.327	6.029	4.994	4.149	3.447	2.853	2.345	1.904	1.515	1.166	0.849	0.554	0.274	0.000
50.000	0.000	0.000	0.000	0.000	0.000	0.000	0.000	0.000	0.000	0.000	0.000	0.000	0.000	0.000	0.000	0.000	0.000	0.000	0.000	0.000

Fig. 7.16.3—Temperature distribution solved using the analytical solution.

Problem 7.3.18 (Ertekin et al. 2001)

Obtain the steady-state temperature distribution in a cube with the dimensions of unity aligned with x, y, and z axes. The external surface $z = 1$ is kept at 100°C, while all other external surfaces are kept at 0°C.

1. Solve the temperature distribution using conventional seven-point, finite-difference approximations, and the LSOR algorithm with $\omega_{opt} = 1.0$. In the solution of the line equation, use the PSOR algorithm with ω_{opt}.

2. Develop an algorithm of plane-successive over-relaxation (BSOR). Test your algorithm with $\omega_{opt} = 1.0$. In the solution of plane equations, use LSOR with $\omega_{opt} = 1.0$. In the solution of the line equation, use PSOR algorithm with ω_{opt}.

3. Compare the resulting solution with the exact solution provided here. Also, compare the computational time, storage requirements, and number of iterations. Tabulate your results. The analytical solution for this system is

$$T_{(x,y,z)} = 400 \sum_{m=1}^{\infty} \sum_{n=1}^{\infty} d_{mn} \frac{\sinh\left(\alpha_{mn} z\right)}{\sinh\left(\alpha_{mn}\right)} \sin\left(m\pi x\right) \sin\left(n\pi y\right), \text{ where } \alpha_{mn} = \pi\left(m^2 + n^2\right)^{1/2} \text{ and}$$

$$d_{mn} = \int_0^1 \int_0^1 \sin\left(m\pi x\right) \sin\left(n\pi y\right) dx\, dy.$$

Solution to Problem 7.3.18

The governing equation of the problem can be written as

$$\frac{\partial^2 T}{\partial x^2} + \frac{\partial^2 T}{\partial y^2} + \frac{\partial^2 T}{\partial z^2} = 0. \quad\dotfill (7.186)$$

1. With the finite-difference solution with seven-point approximation, the equation can be written in a finite-difference form:

$$T_{i+1,j,k} + T_{i-1,j,k} + T_{i,j+1,k} + T_{i,j-1,k} + T_{i,j,k+1} + T_{i,j,k-1} - 6T_{i,j,k} = 0. \quad\dotfill (7.187)$$

If LSOR is implemented, the line equation can be written as follows (assuming the sweeping direction is along the j axis):

$$T_{i+1,j,k}^{(k+1)} + T_{i-1,j,k}^{(k+1)} - 6T_{i,j,k}^{(k+1)} = -\left[T_{i,j+1,k}^{(k)} + T_{i,j-1,k}^{(k+1)} + T_{i,j,k-1}^{(k+1)} + T_{i-1,j,k+1}^{(k)} \right]. \quad\dotfill (7.188)$$

Figs. 7.17.1 through 7.17.11 show the temperature on $z = 0$ through $z = 1$ planes.

0	0	0	0	0	0	0	0	0	0	0
0	0	0	0	0	0	0	0	0	0	0
0	0	0	0	0	0	0	0	0	0	0
0	0	0	0	0	0	0	0	0	0	0
0	0	0	0	0	0	0	0	0	0	0
0	0	0	0	0	0	0	0	0	0	0
0	0	0	0	0	0	0	0	0	0	0
0	0	0	0	0	0	0	0	0	0	0
0	0	0	0	0	0	0	0	0	0	0
0	0	0	0	0	0	0	0	0	0	0
0	0	0	0	0	0	0	0	0	0	0

Fig. 7.17.1—Temperature distribution on $z = 0$ plane.

0	0	0	0	0	0	0	0	0	0	0
0	0.17653	0.33375	0.45602	0.533	0.55921	0.533	0.45602	0.33375	0.17653	0
0	0.33375	0.63106	0.8624	1.0081	1.05772	1.0081	0.8624	0.63106	0.33375	0
0	0.45602	0.8624	1.17875	1.37809	1.44601	1.37809	1.17875	0.8624	0.45602	0
0	0.53299	1.0081	1.37809	1.61132	1.69079	1.61132	1.37809	1.0081	0.53299	0
0	0.55921	1.05772	1.44601	1.69079	1.77421	1.69079	1.446	1.05772	0.55921	0
0	0.53299	1.0081	1.37809	1.61132	1.69079	1.61132	1.37809	1.0081	0.53299	0
0	0.45601	0.86239	1.17875	1.37809	1.446	1.37809	1.17875	0.86239	0.45601	0
0	0.33374	0.63106	0.86239	1.00809	1.05772	1.00809	0.86239	0.63106	0.33374	0
0	0.17653	0.33374	0.45601	0.53299	0.55921	0.53299	0.45601	0.33374	0.17653	0
0	0	0	0	0	0	0	0	0	0	0

Fig. 7.17.2—Temperature distribution on $z = 0.1$ plane.

0	0	0	0	0	0	0	0	0	0	0
0	0.39169	0.73886	1.00696	1.17465	1.23153	1.17465	1.00696	0.73886	0.39169	0
0	0.73886	1.39408	1.90045	2.21737	2.32492	2.21737	1.90045	1.39408	0.73886	0
0	1.00696	1.90045	2.59152	3.02438	3.17133	3.02438	2.59152	1.90045	1.00696	0
0	1.17465	2.21737	3.02438	3.53014	3.70189	3.53014	3.02438	2.21737	1.17465	0
0	1.23153	2.32492	3.17133	3.70189	3.88209	3.70189	3.17133	2.32492	1.23153	0
0	1.17464	2.21737	3.02437	3.53013	3.70189	3.53013	3.02437	2.21737	1.17464	0
0	1.00696	1.90044	2.59151	3.02437	3.17132	3.02437	2.59151	1.90044	1.00696	0
0	0.73886	1.39408	1.90044	2.21737	2.32491	2.21736	1.90044	1.39408	0.73886	0
0	0.39169	0.73886	1.00696	1.17464	1.23153	1.17464	1.00695	0.73886	0.39169	0
0	0	0	0	0	0	0	0	0	0	0

Fig. 7.17.3—Temperature distribution on $z = 0.2$ plane.

0	0	0	0	0	0	0	0	0	0	0
0	0.6959	1.30669	1.77178	2.05903	2.15576	2.05903	1.77178	1.30669	0.6959	0
0	1.30669	2.45482	3.33034	3.87174	4.05418	3.87174	3.33034	2.45482	1.30669	0
0	1.77178	3.33033	4.52071	5.25781	5.50639	5.25781	4.52071	3.33033	1.77178	0
0	2.05902	3.87173	5.25781	6.11695	6.40685	6.11695	5.25781	3.87173	2.05902	0
0	2.15575	4.05417	5.50638	6.40685	6.71075	6.40685	5.50638	4.05417	2.15575	0
0	2.05902	3.87173	5.2578	6.11694	6.40684	6.11694	5.2578	3.87173	2.05902	0
0	1.77177	3.33033	4.5207	5.2578	5.50638	5.2578	4.5207	3.33033	1.77177	0
0	1.30669	2.45481	3.33033	3.87173	4.05417	3.87173	3.33032	2.45481	1.30669	0
0	0.6959	1.30669	1.77177	2.05902	2.15575	2.05902	1.77177	1.30669	0.6959	0
0	0	0	0	0	0	0	0	0	0	0

Fig. 7.17.4—Temperature distribution on $z = 0.3$ plane.

0	0	0	0	0	0	0	0	0	0	0
0	1.17034	2.17878	2.92766	3.38023	3.53077	3.38023	2.92766	2.17878	1.17034	0
0	2.17878	4.06075	5.46251	6.31169	6.59453	6.31169	5.46251	4.06075	2.17878	0
0	2.92766	5.46251	7.35642	8.50668	8.89033	8.50668	7.35642	5.46251	2.92765	0
0	3.38022	6.31169	8.50668	9.84224	10.28815	9.84224	8.50668	6.31169	3.38022	0
0	3.53076	6.59452	8.89032	10.28815	10.75501	10.28815	8.89032	6.59452	3.53076	0
0	3.38022	6.31169	8.50667	9.84224	10.28814	9.84223	8.50667	6.31168	3.38022	0
0	2.92765	5.46251	7.35641	8.50667	8.89032	8.50667	7.35641	5.4625	2.92765	0
0	2.17878	4.06074	5.4625	6.31168	6.59451	6.31168	5.4625	4.06074	2.17878	0
0	1.17033	2.17878	2.92765	3.38022	3.53076	3.38022	2.92765	2.17878	1.17033	0
0	0	0	0	0	0	0	0	0	0	0

Fig. 7.17.5—Temperature distribution on z = 0.4 plane.

0	0	0	0	0	0	0	0	0	0	0
0	1.96855	3.60726	4.77263	5.4522	5.67385	5.4522	4.77263	3.60726	1.96855	0
0	3.60726	6.62707	8.78822	10.05447	10.46848	10.05447	8.78822	6.62707	3.60726	0
0	4.77263	8.78822	11.67943	13.38157	13.93953	13.38157	11.67943	8.78821	4.77263	0
0	5.4522	10.05446	13.38157	15.34683	15.99221	15.34683	13.38156	10.05446	5.4522	0
0	5.67384	10.46847	13.93952	15.9922	16.66672	15.9922	13.93952	10.46847	5.67384	0
0	5.4522	10.05446	13.38156	15.34683	15.9922	15.34683	13.38156	10.05446	5.4522	0
0	4.77263	8.78821	11.67942	13.38156	13.93951	13.38156	11.67942	8.78821	4.77262	0
0	3.60725	6.62706	8.78821	10.05445	10.46847	10.05445	8.78821	6.62706	3.60725	0
0	1.96855	3.60725	4.77262	5.45219	5.67384	5.45219	4.77262	3.60725	1.96855	0
0	0	0	0	0	0	0	0	0	0	0

Fig. 7.17.6—Temperature distribution on z = 0.5 plane.

0	0	0	0	0	0	0	0	0	0	0
0	3.42644	6.09651	7.86045	8.83203	9.13942	8.83203	7.86045	6.09651	3.42644	0
0	6.09651	10.91073	14.13318	15.92463	16.49403	15.92463	14.13318	10.91073	6.09651	0
0	7.86044	14.13318	18.38057	20.76246	21.52300	20.76246	18.38057	14.13318	7.86044	0
0	0.00203	15.02462	20.76246	23.49119	24.36517	23.49119	20.76246	15.92462	8.83203	0
0	9.13942	16.49403	21.52299	24.36517	25.27647	24.36517	21.52299	16.49402	9.13942	0
0	8.83203	15.92462	20.76245	23.49118	24.36517	23.49118	20.76245	15.92462	8.83203	0
0	7.86044	14.13317	18.38056	20.76245	21.52298	20.76245	18.38056	14.13317	7.86044	0
0	6.09651	10.91072	14.13317	15.92461	16.49402	15.92461	14.13317	10.91072	6.09651	0
0	3.42644	6.09651	7.86044	8.83203	9.13941	8.83203	7.86044	6.09651	3.42644	0
0	0	0	0	0	0	0	0	0	0	0

Fig. 7.17.7—Temperature distribution on z = 0.6 plane.

0	0	0	0	0	0	0	0	0	0	0
0	6.39705	10.77421	13.32831	14.61551	15.00458	14.61551	13.32831	10.77421	6.39705	0
0	10.77421	18.37790	22.93448	25.27157	25.98402	25.27157	22.93448	18.37790	10.77421	0
0	13.32831	22.93448	28.81269	31.87381	32.81431	31.87381	28.81269	22.93448	13.32830	0
0	14.61551	25.27157	31.87381	35.34503	36.41697	35.34503	31.87381	25.27156	14.61550	0
0	15.00458	25.98402	32.81430	36.41697	37.53143	36.41696	32.81430	25.98402	15.00457	0
0	14.61550	25.27156	31.87380	35.34502	36.41696	35.34502	31.87380	25.27156	14.61550	0
0	13.32830	22.93448	28.81269	31.87380	32.81430	31.87380	28.81268	22.93448	13.32830	0
0	10.77421	18.37789	22.93448	25.27156	25.98401	25.27156	22.93448	18.37789	10.77420	0
0	6.39705	10.77420	13.32830	14.61550	15.00457	14.61550	13.32830	10.77420	6.39705	0
0	0	0	0	0	0	0	0	0	0	0

Fig. 7.17.8—Temperature distribution on z = 0.7 plane.

0	0	0	0	0	0	0	0	0	0	0
0	13.40747	20.44547	23.78518	25.25655	25.67300	25.25655	23.78518	20.44547	13.40747	0
0	20.44547	31.93928	37.68325	40.29695	41.04808	40.29695	37.68325	31.93928	20.44547	0
0	23.78518	37.68325	44.87899	48.23680	49.21422	48.23680	44.87899	37.68325	23.78518	0
0	25.25655	40.29695	48.23680	51.99743	53.10083	51.99743	48.23679	40.29695	25.25655	0
0	25.67300	41.04807	49.21422	53.10083	54.24421	53.10083	49.21422	41.04807	25.67300	0
0	25.25655	40.29695	48.23679	51.99742	53.10083	51.99742	48.23679	40.29695	25.25654	0
0	23.78518	37.68325	44.87899	48.23679	49.21421	48.23679	44.87899	37.68325	23.78518	0
0	20.44547	31.93927	37.68325	40.29695	41.04807	40.29695	37.68325	31.93927	20.44547	0
0	13.40747	20.44547	23.78518	25.25654	25.67300	25.25654	23.78518	20.44547	13.40747	0
0	0	0	0	0	0	0	0	0	0	0

Fig. 7.17.9—Temperature distribution on z = 0.8 plane.

0	0	0	0	0	0	0	0	0	0	0
0	33.15681	42.76669	45.99752	47.16864	47.47226	47.16864	45.99752	42.76669	33.15681	0
0	42.76669	57.00032	62.26462	64.28548	64.82330	64.28548	62.26462	57.00032	42.76669	0
0	45.99752	62.26462	68.62116	71.15937	71.84850	71.15937	68.62116	62.26462	45.99752	0
0	47.16864	64.28548	71.15937	73.96427	74.73473	73.96427	71.15937	64.28548	47.16863	0
0	47.47226	64.82330	71.84850	74.73473	75.53052	74.73473	71.84850	64.82330	47.47226	0
0	47.16863	64.28548	71.15937	73.96427	74.73473	73.96427	71.15937	64.28548	47.16863	0
0	45.99752	62.26462	68.62116	71.15937	71.84850	71.15937	68.62116	62.26462	45.99752	0
0	42.76669	57.00031	62.26462	64.28548	64.82330	64.28548	62.26462	57.00031	42.76668	0
0	33.15681	42.76668	45.99752	47.16863	47.47226	47.16863	45.99752	42.76668	33.15681	0
0	0	0	0	0	0	0	0	0	0	0

Fig. 7.17.10—Temperature distribution on z = 0.9 plane.

100	100	100	100	100	100	100	100	100	100	100
100	100	100	100	100	100	100	100	100	100	100
100	100	100	100	100	100	100	100	100	100	100
100	100	100	100	100	100	100	100	100	100	100
100	100	100	100	100	100	100	100	100	100	100
100	100	100	100	100	100	100	100	100	100	100
100	100	100	100	100	100	100	100	100	100	100
100	100	100	100	100	100	100	100	100	100	100
100	100	100	100	100	100	100	100	100	100	100
100	100	100	100	100	100	100	100	100	100	100
100	100	100	100	100	100	100	100	100	100	100

Fig. 7.17.11—Temperature distribution on z = 1 plane.

2. In implementing the plane-successive over-relaxation, the linear equation will be written as

$$T_{i+1,j,k}^{(m+1)} + T_{i-1,j,k}^{(m+1)} + T_{i,j+1,k}^{(m+1)} + T_{i,j-1,k}^{(m+1)} - 6T_{i,j,k}^{(m+1)} = -\left[T_{i,j,k-1}^{(m+1)} + T_{i-1,j,k+1}^{(m)}\right]. \quad \dots\dots\dots\dots\dots\dots\dots\dots\dots\dots\dots (7.189)$$

Thus, the temperature at a certain surface is simultaneously solved. The same temperature distribution can be obtained as that in Part 1.

3. The analytical solution is shown in **Figs. 7.17.12 through 7.17.22** for temperature on z = 0 through z = 1.0 planes.

0	0	0	0	0	0	0	0	0	0	0
0	0	0	0	0	0	0	0	0	0	0
0	0	0	0	0	0	0	0	0	0	0
0	0	0	0	0	0	0	0	0	0	0
0	0	0	0	0	0	0	0	0	0	0
0	0	0	0	0	0	0	0	0	0	0
0	0	0	0	0	0	0	0	0	0	0
0	0	0	0	0	0	0	0	0	0	0
0	0	0	0	0	0	0	0	0	0	0
0	0	0	0	0	0	0	0	0	0	0
0	0	0	0	0	0	0	0	0	0	0

Fig. 7.17.12—Temperature on z = 0 plane.

0	0	0	0	0	0	0	0	0	0	0
0	0.17032	0.32280	0.44233	0.51812	0.54403	0.51812	0.44233	0.32280	0.17032	0
0	0.32280	0.61182	0.83841	0.98211	1.03126	0.98211	0.83841	0.61182	0.32280	0
0	0.44233	0.83841	1.14898	1.34600	1.41339	1.34600	1.14898	0.83841	0.44233	0
0	0.51812	0.98211	1.34600	1.57688	1.65585	1.57688	1.34600	0.98211	0.51812	0
0	0.54403	1.03126	1.41339	1.65585	1.73879	1.65585	1.41339	1.03126	0.54403	0
0	0.51812	0.98211	1.34600	1.57688	1.65585	1.57688	1.34600	0.98211	0.51812	0
0	0.44233	0.83841	1.14898	1.34600	1.41339	1.34600	1.14898	0.83841	0.44233	0
0	0.32280	0.61182	0.83841	0.98211	1.03126	0.98211	0.83841	0.61182	0.32280	0
0	0.17032	0.32280	0.44233	0.51812	0.54403	0.51812	0.44233	0.32280	0.17032	0
0	0	0	0	0	0	0	0	0	0	0

Fig. 7.17.13—Temperature distribution on $z = 0.1$ plane.

0	0	0	0	0	0	0	0	0	0	0
0	0.37764	0.71465	0.97746	1.14328	1.19980	1.14328	0.97746	0.71465	0.37764	0
0	0.71465	1.35253	1.85012	2.16416	2.27123	2.16416	1.85012	1.35253	0.71465	0
0	0.97746	1.85012	2.53111	2.96105	3.10766	2.96105	2.53111	1.85012	0.97746	0
0	1.14328	2.16416	2.96105	3.46430	3.63594	3.46430	2.96105	2.16416	1.14328	0
0	1.19980	2.27123	3.10766	3.63594	3.81614	3.63594	3.10766	2.27123	1.19980	0
0	1.14328	2.16416	2.96105	3.46430	3.63594	3.46430	2.96105	2.16416	1.14328	0
0	0.97746	1.85012	2.53111	2.96105	3.10766	2.96105	2.53111	1.85012	0.97746	0
0	0.71465	1.35253	1.85012	2.16416	2.27123	2.16416	1.85012	1.35253	0.71465	0
0	0.37764	0.71465	0.97746	1.14328	1.19980	1.14328	0.97746	0.71465	0.37764	0
0	0	0	0	0	0	0	0	0	0	0

Fig. 7.17.14—Temperature distribution on $z = 0.2$ plane.

0	0	0	0	0	0	0	0	0	0	0
0	0.66960	1.26322	1.72129	2.00731	2.10419	2.00731	1.72129	1.26322	0.66960	0
0	1.26322	2.38356	3.24865	3.78914	3.97229	3.78914	3.24865	2.38356	1.26322	0
0	1.72129	3.24865	4.42894	5.16692	5.41710	5.16692	4.42894	3.24865	1.72129	0
0	2.00731	3.78914	5.16692	6.02891	6.32123	6.02891	5.16692	3.78914	2.00731	0
0	2.10419	3.97229	5.41710	6.32123	6.62788	6.32123	5.41710	3.97229	2.10419	0
0	2.00731	3.78914	5.16692	6.02891	6.32123	6.02891	5.16692	3.78914	2.00731	0
0	1.72129	3.24865	4.42894	5.16692	5.41710	5.16692	4.42894	3.24865	1.72129	0
0	1.26322	2.38356	3.24865	3.78914	3.97229	3.78914	3.24865	2.38356	1.26322	0
0	0.66960	1.26322	1.72129	2.00731	2.10419	2.00731	1.72129	1.26322	0.66960	0
0	0	0	0	0	0	0	0	0	0	0

Fig. 7.17.15—Temperature distribution on $z = 0.3$ plane.

0	0	0	0	0	0	0	0	0	0	0
0	1.12127	2.10287	2.84543	3.30022	3.45248	3.30022	2.84543	2.10287	1.12127	0
0	2.10287	3.94558	5.34161	6.19782	6.48472	6.19782	5.34161	3.94558	2.10287	0
0	2.84543	5.34161	7.23599	8.39978	8.79013	8.39978	7.23599	5.34161	2.84543	0
0	3.30022	6.19782	8.39978	9.75427	10.20892	9.75427	8.39978	6.19782	3.30022	0
0	3.45248	6.48472	8.79013	10.20892	10.68530	10.20892	8.79013	6.48472	3.45248	0
0	3.30022	6.19782	8.39978	9.75427	10.20892	9.75427	8.39978	6.19782	3.30022	0
0	2.84543	5.34161	7.23599	8.39978	8.79013	8.39978	7.23599	5.34161	2.84543	0
0	2.10287	3.94558	5.34161	6.19782	6.48472	6.19782	5.34161	3.94558	2.10287	0
0	1.12127	2.10287	2.84543	3.30022	3.45248	3.30022	2.84543	2.10287	1.12127	0
0	0	0	0	0	0	0	0	0	0	0

Fig. 7.17.16—Temperature distribution on $z = 0.4$ plane.

0	0	0	0	0	0	0	0	0	0	0
0	1.87137	3.47094	4.63722	5.32716	5.55347	5.32716	4.63722	3.47094	1.87137	0
0	3.47094	6.44460	8.62031	9.91150	10.33580	9.91150	8.62031	6.44460	3.47094	0
0	4.63722	8.62031	11.54603	13.28863	13.86248	13.28863	11.54603	8.62031	4.63722	0
0	5.32716	9.91150	13.28863	15.30559	15.97085	15.30559	13.28863	9.91150	5.32716	0
0	5.55347	10.33580	13.86248	15.97085	16.66667	15.97085	13.86248	10.33580	5.55347	0
0	5.32716	9.91150	13.28863	15.30559	15.97085	15.30559	13.28863	9.91150	5.32716	0
0	4.63722	8.62031	11.54603	13.28863	13.86248	13.28863	11.54603	8.62031	4.63722	0
0	3.47094	6.44460	8.62031	9.91150	10.33580	9.91150	8.62031	6.44460	3.47094	0
0	1.87137	3.47094	4.63722	5.32716	5.55347	5.32716	4.63722	3.47094	1.87137	0
0	0	0	0	0	0	0	0	0	0	0

Fig. 7.17.17—Temperature distribution on z = 0.5 plane.

0	0	0	0	0	0	0	0	0	0	0
0	3.21871	5.84381	7.63072	8.62513	8.94023	8.62513	7.63072	5.84381	3.21871	0
0	5.84381	10.63765	13.92796	15.77209	16.35874	15.77209	13.92796	10.63765	5.84381	0
0	7.63072	13.92796	18.28824	20.75107	21.53796	20.75107	18.28824	13.92796	7.63072	0
0	8.62513	15.77209	20.75107	23.57906	24.48549	23.57906	20.75107	15.77209	8.62513	0
0	8.94023	16.35874	21.53796	24.48549	25.43131	24.48549	21.53796	16.35874	8.94023	0
0	8.62513	15.77209	20.75107	23.57906	24.48549	23.57906	20.75107	15.77209	8.62513	0
0	7.63072	13.92796	18.28824	20.75107	21.53796	20.75107	18.28824	13.92796	7.63072	0
0	5.84381	10.63765	13.92796	15.77209	16.35874	15.77209	13.92796	10.63765	5.84381	0
0	3.21871	5.84381	7.63072	8.62513	8.94023	8.62513	7.63072	5.84381	3.21871	0
0	0	0	0	0	0	0	0	0	0	0

Fig. 7.17.18—Temperature distribution on z = 0.6 plane.

0	0	0	0	0	0	0	0	0	0	0
0	5.93302	10.31799	12.93760	14.25146	14.64579	14.25146	12.93760	10.31799	5.93302	0
0	10.31799	18.06358	22.78527	25.19050	25.91807	25.19050	22.78527	18.06358	10.31799	0
0	12.93760	22.78527	28.90440	32.07096	33.03686	32.07096	28.90440	22.78527	12.93760	0
0	14.25146	25.19050	32.07096	35.66938	36.77342	35.66938	32.07096	25.19050	14.25146	0
0	14.64579	25.91807	33.03686	36.77342	37.92218	36.77342	33.03686	25.91807	14.64579	0
0	14.25146	25.19050	32.07096	35.66938	36.77342	35.66938	32.07096	25.19050	14.25146	0
0	12.93760	22.78527	28.90440	32.07096	33.03686	32.07096	28.90440	22.78527	12.93760	0
0	10.31799	18.06358	22.78527	25.19050	25.91807	25.19050	22.78527	18.06358	10.31799	0
0	5.93302	10.31799	12.93760	14.25146	14.64579	14.25146	12.93760	10.31799	5.93302	0
0	0	0	0	0	0	0	0	0	0	0

Fig. 7.17.19—Temperature distribution on z = 0.7 plane.

0	0	0	0	0	0	0	0	0	0	0
0	12.49619	19.82136	23.22922	24.68275	25.08777	24.68275	23.22922	19.82136	12.49619	0
0	19.82136	31.98081	37.92631	40.54423	41.28418	40.54423	37.92631	31.98081	19.82136	0
0	23.22922	37.92631	45.40104	48.78973	49.76135	48.78973	45.40104	37.92631	23.22922	0
0	24.68275	40.54423	48.78973	52.59588	53.69760	52.59588	48.78973	40.54423	24.68275	0
0	25.08777	41.28418	49.76135	53.69760	54.84066	53.69760	49.76135	41.28418	25.08777	0
0	24.68275	40.54423	48.78973	52.59588	53.69760	52.59588	48.78973	40.54423	24.68275	0
0	23.22922	37.92631	45.40104	48.78973	49.76135	48.78973	45.40104	37.92631	23.22922	0
0	19.82136	31.98081	37.92631	40.54423	41.28418	40.54423	37.92631	31.98081	19.82136	0
0	12.49619	19.82136	23.22922	24.68275	25.08777	24.68275	23.22922	19.82136	12.49619	0
0	0	0	0	0	0	0	0	0	0	0

Fig. 7.17.20—Temperature distribution on z = 0.8 plane.

0	0	0	0	0	0	0	0	0	0	0
0	33.16301	43.24028	46.25636	47.31189	47.58460	47.31189	46.25636	43.24028	33.16301	0
0	43.24028	58.31616	63.37371	65.24517	65.73930	65.24517	63.37371	58.31616	43.24028	0
0	46.25636	63.37371	69.53324	71.91908	72.56219	71.91908	69.53324	63.37371	46.25636	0
0	47.31189	65.24517	71.91908	74.57244	75.29712	74.57244	71.91908	65.24517	47.31189	0
0	47.58460	65.73930	72.56219	75.29712	76.04734	75.29712	72.56219	65.73930	47.58460	0
0	47.31189	65.24517	71.91908	74.57244	75.29712	74.57244	71.91908	65.24517	47.31189	0
0	46.25636	63.37371	69.53324	71.91908	72.56219	71.91908	69.53324	63.37371	46.25636	0
0	43.24028	58.31616	63.37371	65.24517	65.73930	65.24517	63.37371	58.31616	43.24028	0
0	33.16301	43.24028	46.25636	47.31189	47.58460	47.31189	46.25636	43.24028	33.16301	0
0	0	0	0	0	0	0	0	0	0	0

Fig. 7.17.21—Temperature distribution on z = 0.9 plane.

100	100	100	100	100	100	100	100	100	100	100
100	100	100	100	100	100	100	100	100	100	100
100	100	100	100	100	100	100	100	100	100	100
100	100	100	100	100	100	100	100	100	100	100
100	100	100	100	100	100	100	100	100	100	100
100	100	100	100	100	100	100	100	100	100	100
100	100	100	100	100	100	100	100	100	100	100
100	100	100	100	100	100	100	100	100	100	100
100	100	100	100	100	100	100	100	100	100	100
100	100	100	100	100	100	100	100	100	100	100
100	100	100	100	100	100	100	100	100	100	100

Fig. 7.17.22—Temperature distribution on z = 1 plane.

The CPU time in the implementation of the analytical solution is 45.87 seconds.

The numerical solution takes 0.78 seconds of CPU time using the LSOR algorithm, and it takes 94 iterations to achieve a convergence tolerance of 1×10^{-5} °C.

The numerical solution takes 0.64 seconds of CPU time using the plane-successive over-relaxation (BSOR) algorithm, and it takes 94 iterations to achieve a convergence tolerance of 1×10^{-5} °C.

Problem 7.3.19 (Ertekin et al. 2001)

Obtain the steady-state temperature distribution in a hollow finite cylinder 100 ft in length, with outer and inner radii of 10 and 1 ft, respectively. The cylinder is kept at 0°C on the surfaces $r = 1$ ft, $r = 10$ ft, and $z = 0$ ft, while the surface $z = 100$ ft is kept at a temperature of 100°C.

Note that the 3D, steady-state heat-flow equation in cylindrical coordinates is given as

$$\frac{\partial^2 T}{\partial r^2} + \frac{1}{r}\frac{\partial T}{\partial r} + \frac{1}{r^2}\frac{\partial^2 T}{\partial \theta^2} + \frac{\partial^2 T}{\partial z^2} = 0.$$

Use equal Δr and $\Delta \theta$ spacing in r and θ directions. To solve the finite-difference equations, use the PSOR algorithm with optimum acceleration parameters. Check the convergence rate of PSOR against that of Gauss-Seidel.

Solution to Problem 7.3.19

The PDE can be discretized into a finite-difference form:

$$\frac{T_{i+1,j,k} - 2T_{i,j,k} + T_{i-1,j,k}}{(\Delta r)^2} + \frac{1}{r_i}\frac{T_{i+1,j,k} - T_{i,j,k}}{(\Delta r)} + \frac{1}{(r_i)^2}\frac{T_{i,j+1,k} - 2T_{i,j,k} + T_{i,j-1,k}}{(\Delta \theta)^2} + \frac{T_{i,j,k+1} - 2T_{i,j,k} + T_{i,j,k-1}}{(\Delta z)^2} = 0. \quad \ldots\ldots(7.190)$$

In this case, i, j, and k represent block indices along the r-, θ-, and z-directions, respectively. We present the tabulated data showing the temperature distribution at $z = 90$ ft (**Table 7.13**) and $z = 80$ ft (**Table 7.14**) for validation purposes.

For a convergence tolerance of 1×10^{-5} °C, the solution is achieved by conducting 94 Gauss-Seidel and 48 PSOR iterations with $\omega_{opt} = 1.5004$.

r/θ	$\pi/3$	$2\pi/3$	π	$4\pi/3$	$5\pi/3$
1	0	0	0	0	0
2	4.6140	4.6140	4.6140	4.6140	4.6140
3	7.0826	7.0826	7.0826	7.0826	7.0826
4	8.2866	8.2866	8.2866	8.2866	8.2866
5	8.5775	8.5775	8.5775	8.5775	8.5775
6	8.1243	8.1243	8.1243	8.1243	8.1243
7	7.0131	7.0131	7.0131	7.0131	7.0131
8	5.2843	5.2843	5.2843	5.2843	5.2843
9	2.9496	2.9496	2.9496	2.9496	2.9496
10	0	0	0	0	0

Table 7.13—Temperature distribution at plane $z = 90$ ft.

r/θ	$\pi/3$	$2\pi/3$	π	$4\pi/3$	$5\pi/3$
1	0	0	0	0	0
2	0.3214	0.3214	0.3214	0.3214	0.3214
3	0.5091	0.5091	0.5091	0.5091	0.5091
4	0.6041	0.6041	0.6041	0.6041	0.6041
5	0.6231	0.6231	0.6231	0.6231	0.6231
6	0.5775	0.5775	0.5775	0.5775	0.5775
7	0.4784	0.4784	0.4784	0.4784	0.4784
8	0.3383	0.3383	0.3383	0.3383	0.3383
9	0.1726	0.1726	0.1726	0.1726	0.1726
10	0	0	0	0	0

Table 7.14—Temperature distribution at plane $z = 80$ ft.

Problem 7.3.20 (Ertekin et al. 2001)

Find the steady-state temperature over a set of discrete points of the semi-infinite region that is shaded in **Fig. 7.18.1,** if the temperatures are maintained as indicated. Note that in polar coordinates, the analytical solution is

$$T(r,\theta) = \frac{60}{\pi} \tan^{-1}\left[\frac{(r^2 - 1)\sin\theta}{(r^2 + 1)\cos\theta - 2r}\right] - \frac{60}{\pi} \tan^{-1}\left[\frac{(r^2 - 1)\sin\theta}{(r^2 + 1)\cos\theta - 2r}\right].$$

Use equal $\Delta\theta$ spacing and variable Δr spacing by increasing Δr as you move away from the physical boundaries. By use of a trial-and-error procedure, determine the number of nodes you would need in the r-direction to place a hypothetical boundary at a certain r distance, while ensuring that the approximate solution is within ±5% of the exact solution. Use the Gauss-Seidel algorithm to solve the finite-difference equations.

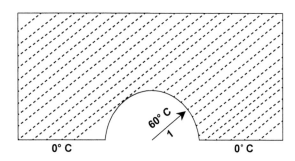

Fig. 7.18.1—Semi-infinite region of Problem 7.3.20.

Solution to Problem 7.3.20

The PDE can be discretized into a finite-difference form:

$$\frac{T_{i+1,j} - 2T_{i,j} + T_{i-1,j}}{(\Delta r)^2} + \frac{1}{r_i}\frac{T_{i+1,j} - T_{i,j}}{(\Delta r)} + \frac{1}{(r_i)^2}\frac{T_{i,j+1} - 2T_{i,j} + T_{i,j-1}}{(\Delta\theta)^2} = 0. \quad\dots\dots\dots\dots\dots\dots\dots\dots\dots\dots (7.191)$$

In this case, i and j are the block indices in r- and θ-directions, respectively.
Boundary conditions are as follows:

- $T = 60°C$ at $r = 1$ and $0 < \theta < \pi$.
- $T = 0°C$ at $r > 1$ and $\theta = \pi$ or $\theta = 0$.

Figs. 7.18.2 and 7.18.3 show the solved temperature profiles at $r_e = 5$ and $r_e = 10$, respectively.

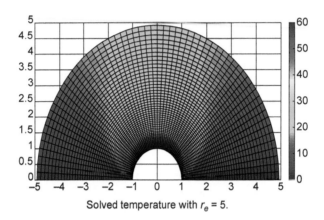

Solved temperature with $r_e = 5$.

Fig. 7.18.2—Solved temperature distribution with $r_e = 5$.

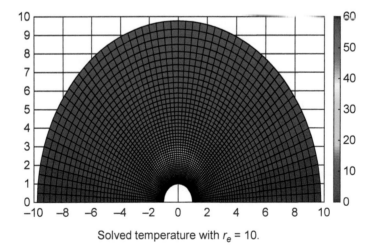

Solved temperature with $r_e = 10$.

Fig. 7.18.3—Solved temperature distribution with $r_e = 10$.

Problem 7.3.21 (Ertekin et al. 2001)

An infinite wedge-shaped region, *ABDE* of angle $\pi/4$ in **Fig. 7.19.1** has one of its sides (*AB*) maintained at a constant temperature, T_1. The other side, *BDE* has part *BD* insulated, while the remaining part (*DE*) is maintained at a constant temperature, T_2. Find the temperature distribution over a set of discrete points. In polar coordinates, the analytical solution is

$$T = T_1 + \frac{2(T_1 - T_1)}{\pi} \times \sin^{-1}\left[\frac{1}{2}(r^4 + 2r^2 \cos 2\theta + 1)^{\frac{1}{2}} - \frac{1}{2}(r^4 - 2r^2 \cos 2\theta + 1)^{\frac{1}{2}}\right].$$

Compare the accuracy of the numerical solution with the exact solution. *Hint:* the 2D, steady-state heat-flow equation in radial coordinates is

$$\frac{\partial^2 T}{\partial r^2} + \frac{1}{r}\frac{\partial T}{\partial r} + \frac{1}{r^2}\frac{\partial^2 T}{\partial \theta^2} = 0.$$

Use equal Δr and $\Delta\theta$ spacing in the r- and θ-directions, respectively. By trial and error, find out the number of nodes you need to approximate the exact solution within ±1%.

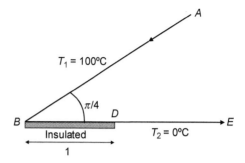

Fig. 7.19.1—Infinitely large wedge-shaped region for Problem 7.3.21.

Solution to Problem 7.3.21

Use the PSOR algorithm with ω_{opt}. Repeat with the Gauss-Seidel algorithm. The PDE can be discretized into a finite-difference form:

$$\frac{T_{i+1,j} - 2T_{i,j} + T_{i-1,j}}{(\Delta r)^2} + \frac{1}{r_i}\frac{T_{i+1,j} - T_{i,j}}{(\Delta r)} + \frac{1}{(r_i)^2}\frac{T_{i,j+1} - 2T_{i,j} + T_{i,j-1}}{(\Delta\theta)^2} = 0. \quad\ldots\ldots\ldots\ldots\ldots\ldots\ldots\ldots(7.192)$$

In this case, i and j represent r- and θ-directions, respectively.
Boundary conditions are as follows:

- $T = 100°C$ at $\theta = \pi/4$.
- $T = 0°C$ at $r > 1$ and $\theta = 0$.

Use $r_e = 20$, $\Delta\theta = 3°$, and $\Delta r = 0.1$; the absolute maximum error is 0.9% comparing against the analytical solution. **Figs. 7.19.2 and 7.19.3** show the temperature distributions solved by numerical and analytical solutions, respectively. It takes 12,248 iterations and 1,706 iterations, respectively, to achieve a prescribed temperature tolerance of 1×10^{-5} °C using the Gauss-Seidel algorithm and the PSOR protocol with $\omega_{opt} = 1.3345$.

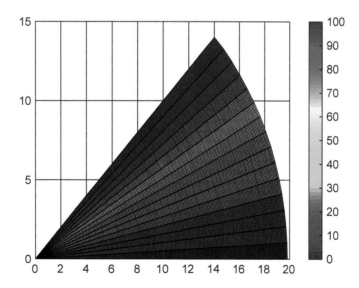

Fig. 7.19.2—Temperature profile generated by the finite-difference solution.

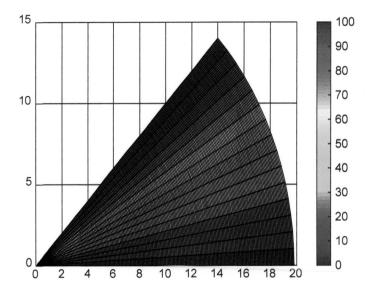

Fig. 7.19.3—Temperature generated by the analytical solution.

For the purpose of validation, a low-resolution solution is listed in **Tables 7.15 and 7.16** for $r_e = 20$, $\Delta\theta = 15°$, and $\Delta r = 1$.

r/θ	0	$\pi/12$	$\pi/6$	$\pi/4$
1	85.27718	85.27718	91.33677	100
2	0	41.07172	72.34286	100
3	0	35.43357	68.46842	100
4	0	34.12146	67.39036	100
5	0	33.69372	67.00885	100
6	0	33.52138	66.84849	100
7	0	33.44116	66.77204	100
8	0	33.3997	66.73195	100
9	0	33.37652	66.70933	100
10	0	33.36274	66.69581	100
11	0	33.35415	66.68734	100
12	0	33.3486	66.68185	100
13	0	33.34491	66.67819	100
14	0	33.34241	66.67571	100
15	0	33.34071	66.67402	100
16	0	33.33957	66.67289	100
17	0	33.33883	66.67215	100
18	0	33.33839	66.67172	100
19	0	33.3382	66.67152	100
20	0	33.3382	66.67152	100

Table 7.15—Solution generated by finite-difference approximations.

r/θ	0	$\pi/12$	$\pi/6$	$\pi/4$
1	50	50	76.14367	100
2	0	34.21494	67.50796	100
3	0	33.50428	66.83604	100
4	0	33.38725	66.72043	100
5	0	33.3554	66.68871	100
6	0	33.34397	66.6773	100
7	0	33.33907	66.67241	100
8	0	33.3367	66.67003	100
9	0	33.33543	66.66877	100
10	0	33.33471	66.66804	100
11	0	33.33427	66.66761	100
12	0	33.334	66.66733	100
13	0	33.33382	66.66715	100
14	0	33.33369	66.66703	100
15	0	33.33361	66.66694	100
16	0	33.33354	66.66688	100
17	0	33.3335	66.66683	100
18	0	33.33346	66.6668	100
19	0	33.33344	66.66677	100
20	0	33.33344	66.66677	100

Table 7.16—Solution generated by analytical solution.

Problem 7.3.22 (Ertekin et al. 2001)

Consider **Fig. 7.20,** which represents a portion of a larger grid system with two prescribed boundary conditions. Furthermore, consider that $\Delta x = 800$ ft, $\Delta y = 200$ ft, $h = 100$ ft, and $k_x = k_y = 36$ md. Assume you are dealing with an incompressible liquid, so that $B = 1$ res bbl/STB and $\mu = 1$ cp. The final form of the finite-difference equation for mesh point 4 is $Ap_4 + Bp_5 + Cp_7 = D$.

Determine the coefficients A, B, and C, and the right-side entry, D, for this system.

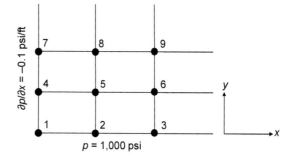

Fig. 7.20—Grid system for Problem 7.3.22.

Solution to Problem 7.3.22

For a 2D incompressible system,

$$\frac{\partial}{\partial x}\left(\frac{A_x k_x}{\mu B}\frac{\partial p}{\partial x}\right)\Delta x + \frac{\partial}{\partial y}\left(\frac{A_y k_y}{\mu B}\frac{\partial p}{\partial y}\right)\Delta y = 0. \quad\dotfill (7.193)$$

For a homogeneous and anisotropic system,

$$\frac{\partial^2 p}{\partial x^2} + \frac{\partial^2 p}{\partial y^2} = 0. \quad\dotfill (7.194)$$

Using five-point finite-difference analog,

$$\frac{p_{i+1,j} - 2p_{i,j} + p_{i-1,j}}{(\Delta x)^2} + \frac{p_{i,j+1} - 2p_{i,j} + p_{i,j-1}}{(\Delta y)^2} = 0. \quad\dotfill (7.195)$$

Because $\Delta x = 800$ ft and $\Delta y = 200$ ft,

$$p_{i+1,j} + p_{i-1,j} - 2p_{i,j} + 16\left(p_{i,j+1} + p_{i,j-1} - 2p_{i,j}\right) = 0. \quad\dotfill (7.196)$$

At the boundary,

$$\frac{p_5 - p_5'}{2\Delta x} = -0.1 \quad \dots\dots\dots\dots\dots\dots\dots\dots\dots\dots\dots\dots\dots\dots\dots\dots\dots\dots \quad (7.197)$$

and

$$p_5' = p_5 + 160; \quad \dots\dots\dots\dots\dots\dots\dots\dots\dots\dots\dots\dots\dots\dots\dots\dots\dots\dots \quad (7.198)$$

therefore, $p_5 + 160 + p_5 - 34p_4 + 16p_7 + 16(1,000) = 0.$ $\dots\dots\dots\dots\dots\dots\dots\dots\dots\dots\dots\dots \quad (7.199)$

Simplifying: $17p_4 - p_5 - 8p_7 = 8,080;$ $\dots\dots\dots\dots\dots\dots\dots\dots\dots\dots\dots\dots\dots\dots \quad (7.200)$

therefore, $A = 17$, $B = -1$, $C = -8$, and $D = 8,080$.

Problem 7.3.23 (Ertekin et al. 2001)

Consider the 2D, body-centered grid with the boundary conditions shown in **Fig. 7.21.1.** Show all reflection nodes and the respective written equations needed to solve this problem.

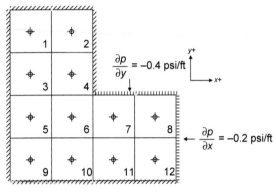

///////// No-flow boundary

|||||||||| Constant pressure gradient specification

Fig. 7.21.1—Reservoir system for Problem 7.3.23.

Solution to Problem 7.3.23

The reflection nodes can be implemented into nodes 7, 8, and 12 **(Fig. 7.21.2).** For the blocks located along the no-flow boundaries, the reflection nodes and the actual node will be at the same pressure; for example, $p_3' = p_3$. The equations that should be written for the reflection nodes are across the nonzero pressure gradients:

$$\frac{p_{8x}' - p_8}{\Delta x} = -0.2, \quad \dots\dots\dots\dots\dots\dots\dots \quad (7.201a)$$

$$\frac{p_{12x}' - p_{12}}{\Delta x} = -0.2, \quad \dots\dots\dots\dots\dots\dots\dots \quad (7.201b)$$

$$\frac{p_{8y}' - p_8}{\Delta y} = -0.4, \quad \dots\dots\dots\dots\dots\dots\dots \quad (7.201c)$$

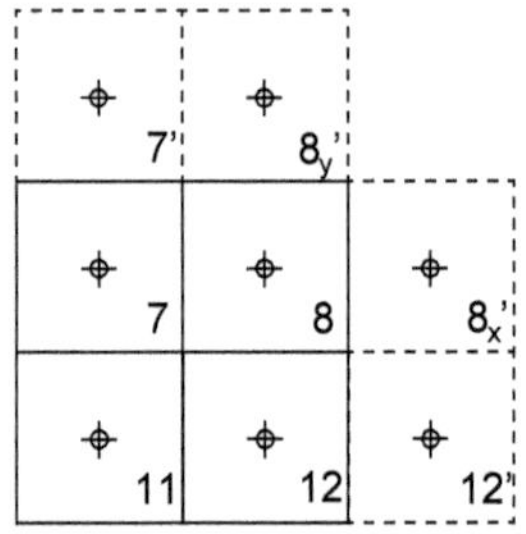

Fig. 7.21.2—Reflection nodes implemented into nodes 7, 8, and 12.

$$\frac{p_7' - p_7}{\Delta y} = -0.4, \quad \dots\dots\dots\dots\dots\dots\dots\dots\dots\dots\dots\dots\dots\dots\dots\dots \quad (7.201d)$$

$$p_{8x}' = p_8 - 0.2\Delta x, \quad \dots\dots\dots\dots\dots\dots\dots\dots\dots\dots\dots\dots\dots\dots\dots\dots \quad (7.201e)$$

$$p_{12}' = p_{12} - 0.2\Delta x, \quad \dots\dots\dots\dots\dots\dots\dots\dots\dots\dots\dots\dots\dots\dots\dots\dots \quad (7.201f)$$

$$p_{8y}' = p_8 - 0.4\Delta y, \quad \dots\dots\dots\dots\dots\dots\dots\dots\dots\dots\dots\dots\dots\dots\dots\dots \quad (7.201g)$$

$$p_7' = p_7 - 0.4\Delta y. \quad \dots\dots\dots\dots\dots\dots\dots\dots\dots\dots\dots\dots\dots\dots\dots\dots \quad (7.201h)$$

Problem 7.3.24 (Ertekin et al. 2001)

Construct the coefficient matrix, the unknown vector, and the right-side vector for the 2D incompressible, homogeneous, isotropic reservoir shown in **Fig. 7.22**. Assume you are simultaneously solving for all unknowns.

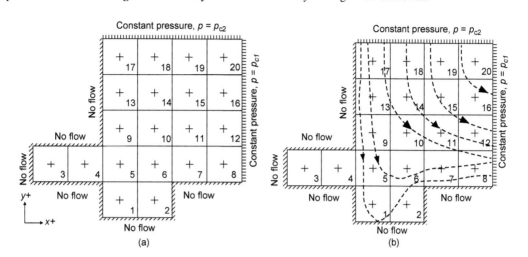

Fig. 7.22—Reservoir system for Problem 7.3.24.

Solution to Problem 7.3.24

If the fluid is incompressible fluid, then

- For a homogeneous and anisotropic system,

$$\frac{\partial^2 p}{\partial x^2}+\frac{\partial^2 p}{\partial y^2}=0. \dotfill (7.202)$$

- In a finite-difference form,

$$p_{i+1,j} + p_{i-1,j} + p_{i,j+1} + p_{i,j-1} - 4p_{i,j} = 0. \dotfill (7.203)$$

The streamlines roughly will look like those shown in Fig. 7.22 (b), if $p_{c2} > p_{c1}$.

In this problem $p_3 = p_4 = p_5$, because p_3 and p_4 are inactive blocks. Thus, there are 18 nodes with unknown pressures:

$$
\begin{bmatrix}
-2 & 1 & 0 & 0 & 1 & 0 & 0 & 0 & 0 & 0 & 0 & 0 & 0 & 0 & 0 & 0 & 0 & 0 \\
1 & -2 & 0 & 0 & 0 & 1 & 0 & 0 & 0 & 0 & 0 & 0 & 0 & 0 & 0 & 0 & 0 & 0 \\
1 & 0 & -3 & 1 & 0 & 0 & 1 & 0 & 0 & 0 & 0 & 0 & 0 & 0 & 0 & 0 & 0 & 0 \\
0 & 1 & 1 & -4 & 1 & 0 & 0 & 1 & 0 & 0 & 0 & 0 & 0 & 0 & 0 & 0 & 0 & 0 \\
0 & 0 & 0 & 1 & -3 & 1 & 0 & 0 & 1 & 0 & 0 & 0 & 0 & 0 & 0 & 0 & 0 & 0 \\
0 & 0 & 0 & 0 & 1 & -3 & 0 & 0 & 0 & 1 & 0 & 0 & 0 & 0 & 0 & 0 & 0 & 0 \\
0 & 0 & 1 & 0 & 0 & 0 & -3 & 1 & 0 & 0 & 1 & 0 & 0 & 0 & 0 & 0 & 0 & 0 \\
0 & 0 & 0 & 1 & 0 & 0 & 1 & -4 & 1 & 0 & 0 & 1 & 0 & 0 & 0 & 0 & 0 & 0 \\
0 & 0 & 0 & 0 & 1 & 0 & 0 & 1 & -4 & 1 & 0 & 0 & 1 & 0 & 0 & 0 & 0 & 0 \\
0 & 0 & 0 & 0 & 0 & 1 & 0 & 0 & 1 & -4 & 0 & 0 & 0 & 1 & 0 & 0 & 0 & 0 \\
0 & 0 & 0 & 0 & 0 & 0 & 1 & 0 & 0 & 0 & -3 & 1 & 0 & 0 & 1 & 0 & 0 & 0 \\
0 & 0 & 0 & 0 & 0 & 0 & 0 & 1 & 0 & 0 & 1 & -4 & 1 & 0 & 0 & 1 & 0 & 0 \\
0 & 0 & 0 & 0 & 0 & 0 & 0 & 0 & 1 & 0 & 0 & 1 & -4 & 1 & 0 & 0 & 1 & 0 \\
0 & 0 & 0 & 0 & 0 & 0 & 0 & 0 & 0 & 1 & 0 & 0 & 1 & -4 & 0 & 0 & 0 & 1 \\
0 & 0 & 0 & 0 & 0 & 0 & 0 & 0 & 0 & 0 & 1 & 0 & 0 & 0 & -3 & 1 & 0 & 0 \\
0 & 0 & 0 & 0 & 0 & 0 & 0 & 0 & 0 & 0 & 0 & 1 & 0 & 0 & 1 & -4 & 1 & 0 \\
0 & 0 & 0 & 0 & 0 & 0 & 0 & 0 & 0 & 0 & 0 & 0 & 1 & 0 & 0 & 1 & -4 & 1 \\
0 & 0 & 0 & 0 & 0 & 0 & 0 & 0 & 0 & 0 & 0 & 0 & 0 & 1 & 0 & 0 & 1 & -4
\end{bmatrix}
\begin{bmatrix}
p_1 \\ p_2 \\ p_5 \\ p_6 \\ p_7 \\ p_8 \\ p_9 \\ p_{10} \\ p_{11} \\ p_{12} \\ p_{13} \\ p_{14} \\ p_{15} \\ p_{16} \\ p_{17} \\ p_{18} \\ p_{19} \\ p_{20}
\end{bmatrix}
=
\begin{bmatrix}
0 \\ 0 \\ 0 \\ 0 \\ 0 \\ -p_{c1} \\ 0 \\ 0 \\ 0 \\ -p_{c1} \\ 0 \\ 0 \\ 0 \\ -p_{c1} \\ -p_{c2} \\ -p_{c2} \\ -p_{c2} \\ -p_{c2}-p_{c1}
\end{bmatrix}.
$$

$$\dotfill (7.204)$$

Problem 7.3.25 (Ertekin et al. 2001)

Consider the slightly compressible fluid flow through the 2D porous medium shown in **Fig. 7.23.1.** The porous medium has homogeneous and isotropic property distribution, and all external boundaries are no-flow boundaries. The well is produced at a constant flow rate of q_{sc} STB/D.

Use the implicit finite-difference approximation and the LSOR procedure to solve for pressure distribution. While sweeping the system parallel to the x-direction, use Thomas' algorithm to solve the line equations.

1. Construct the coefficient matrix for the equations of line $j = 1$ at the (k+1) LSOR iteration level.
2. In completing one LSOR iteration across the entire reservoir, how many times is Thomas' algorithm called for?
3. Can the answer to Part 2 be six? Why? Justify your answer explicitly by constructing a coefficient matrix.
4. Write the algorithm equation you would use to solve the line equation with Jacobi's iteration, but not with Thomas' algorithm.

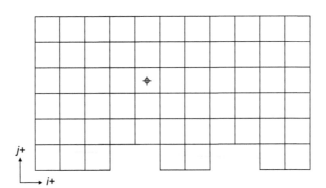

Fig. 7.23.1—Reservoir system for Problem 7.3.25.

Solution to Problem 7.3.25

For a 2D slightly compressible flow problem, the PDE can be written as

$$\frac{\partial}{\partial x}\left(\frac{A_x k_x}{\mu B}\frac{\partial p}{\partial x}\right)\Delta x + \frac{\partial}{\partial y}\left(\frac{A_y k_y}{\mu B}\frac{\partial p}{\partial y}\right)\Delta y + q = \frac{V_b \phi c_t}{5.615}\frac{\partial p}{\partial t}. \quad \dots \dots \dots \dots (7.205)$$

Simplifying the equation for a homogeneous and isotropic system,

$$\frac{\partial^2 p}{\partial x^2} + \frac{\partial^2 p}{\partial y^2} + \frac{q\mu B}{kV_b} = \frac{\mu B \phi c_t}{5.615 V_b}\frac{\partial p}{\partial t}. \quad \dots \dots \dots \dots (7.206)$$

Writing the equation in a finite-difference form,

$$p_{i,j-1}^{n+1} + p_{i-1,j}^{n+1} - 4p_{i,j}^{n+1} + p_{i+1,j}^{n+1} + p_{i,j+1}^{n+1} + \frac{q\mu B}{kh} = \frac{(\Delta x)^2 \mu B \phi c_t}{5.615\Delta t}\left(p_{i,j}^{n+1} - p_{i,j}^n\right). \quad \dots \dots \dots (7.207)$$

1. At line $j = 1$, there are seven active blocks. The coefficient matrix would be an 11×11 tridiagonal matrix; therefore, in constructing the $j = 1$ line equations, Blocks 4, 5, 8, and 9 can be considered fictitious blocks not connected to the surrounding blocks. This is suggested to preserve the tridiagonal structure of the coefficient matrix. The general form of the equation along line $j = 1$ would be of the following form:

$$p_{i,j-1}^{(k+1)\,n+1} + p_{i-1,j}^{(k+1)\,n+1} - \left[4 + \frac{(\Delta x)^2 \mu B \phi c_t}{5.615\Delta t}\right]p_{i,j}^{(k+1)\,n+1} + p_{i+1,j}^{(k+1)\,n+1} + \frac{q\mu B}{kh} = \frac{(\Delta x)^2 \mu B \phi c_t}{5.615\Delta t}p_{i,j+1}^{(k)\,n+1} - p_{i,j}^n. \quad \dots \dots (7.208)$$

Along line $j = 1$, all the $p_{i,j-1}^{(k+1)\,n+1} = p_{i,j}^{(k+1)\,n+1}$, because the external boundaries are no-flow boundaries. There is no well positioned on line $j = 1$. Thus, the above equation can be simplified to

$$p_{i-1,j}^{(k+1)\,n+1} - \left[3 + \frac{(\Delta x)^2 \mu B \phi c_t}{5.615\Delta t}\right]p_{i,j}^{(k+1)\,n+1} + p_{i+1,j}^{(k+1)\,n+1} = -\frac{(\Delta x)^2 \mu B \phi c_t}{5.615\Delta t}p_{i,j+1}^{(k)\,n+1} - p_{i,j}^n. \quad \dots \dots \dots (7.209)$$

2. To complete the LSOR iteration for the entire reservoir, Thomas' algorithm needs to be called for several times to solve a line equation. Depending on the sweeping direction, the times of calling for the algorithm could be different; if the sweeping direction is along the x axis, Thomas' algorithm is called for 11 times, because the system contains 11 columns.

3. If the sweeping direction is along the y axis, Thomas' algorithm is called for six times, because the system contains six rows. Note that four hypothetical equations need to be included for the four blocks that do not exist on $j = 1$ (see the four main diagonal entries highlighted with boxes in **Fig. 7.23.2**).

4. If the sweeping direction is along the x axis, and if Jacobi's iteration is implemented to solve column equations,

$$\overset{(k+1)}{\underset{(m+1)}{p}}{}_{i,j}^{n+1} = \frac{1}{-4-\Gamma}\left[-\frac{q\mu B}{kh} - \Gamma p_{i,j}^{n} - \overset{(k)}{\underset{(m)}{p}}{}_{i+1,j}^{n+1} - \overset{(k)}{\underset{(m+1)}{p}}{}_{i-1,j}^{n+1} - \left(\overset{(k)}{\underset{(m+1)}{p}}{}_{i,j+1}^{n+1} + \overset{(k)}{\underset{(m+1)}{p}}{}_{i,j-1}^{n+1} \right) \right] \quad\dots\dots\dots\dots\dots\dots\dots\text{(7.210)}$$

If the sweeping direction is along the y axis, and if Jacobi's iteration is implemented to solve row equations,

$$\overset{(k+1)}{\underset{(m+1)}{p}}{}_{i,j}^{n+1} = \frac{1}{-4-\Gamma}\left[-\frac{q\mu B}{kh} - \Gamma p_{i,j}^{n} - \overset{(k)}{\underset{(m)}{p}}{}_{i,j+1}^{n+1} - \overset{(k)}{\underset{(m)}{p}}{}_{i,j-1}^{n+1} - \left(\overset{(k)}{\underset{(m+1)}{p}}{}_{i+1,j}^{n+1} + \overset{(k)}{\underset{(m+1)}{p}}{}_{i-1,j}^{n+1} \right) \right], \quad\dots\dots\dots\dots\dots\dots\dots\text{(7.211)}$$

where $\Gamma = \dfrac{(\Delta x)^2 \, \mu B \phi c_t}{5.615 \Delta t}$, k is the Jacobi's iteration level and m is the LSOR iteration level.

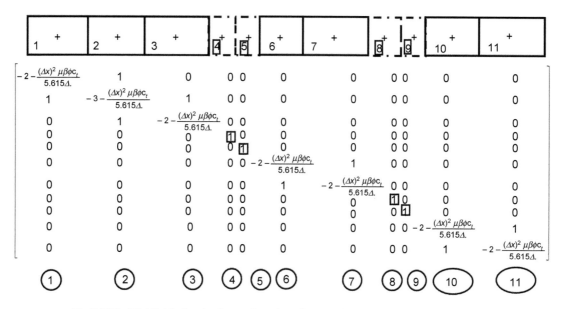

Fig. 7.23.2—Highlighted main diagonal entries of the coefficient matrix of Problem 7.3.25.

Problem 7.3.26 (Ertekin et al. 2001)

The coefficient matrix in **Fig. 7.24.1** was constructed to solve for the pressure distribution for the single-phase fluid flow taking place within the system shown in **Fig. 7.24.2**. However, the numbering scheme used for this reservoir is not given. Determine the numbering of the gridblocks that generated the given coefficient matrix.

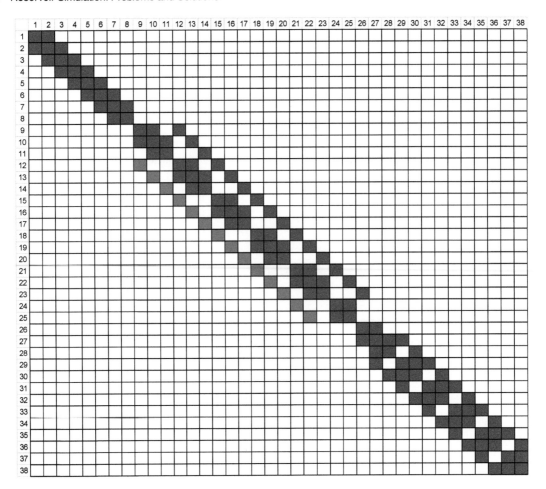

Fig. 7.24.1—Coefficient matrix of Problem 7.3.26.

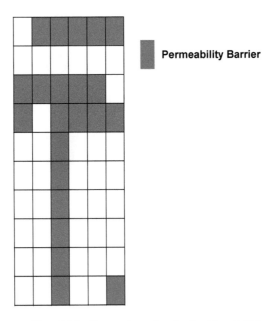

Permeability Barrier

Fig. 7.24.2—Reservoir system for Problem 7.3.26.

Solution to Problem 7.3.26

Fig. 7.24.3 illustrates that the entire reservoir is divided into three sections by the permeability barriers. More importantly, it can be observed that the bandwidth of Section 1 < Section 3 < Section 2. Because Section 1, Section 2, and Section 3 are not in communication, they therefore can be independently solved. This becomes obvious if one observes the resulting coefficient matrix structures shown in **Fig. 7.24.4.** Finally, the resulting numbering scheme is shown in **Fig. 7.24.5.**

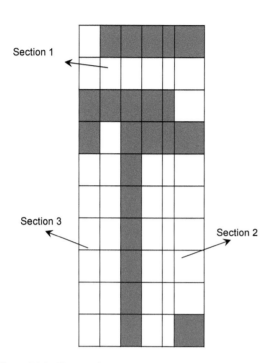

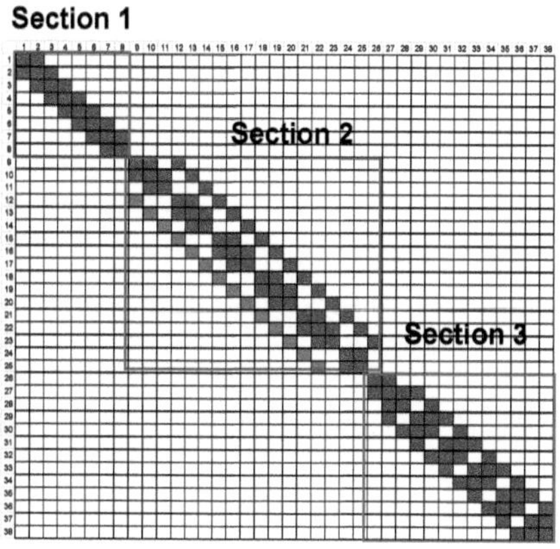

Fig. 7.24.4—Coefficient matrix with sections marked for Problem 7.3.26.

Fig. 7.24.3—Reservoir system with sections marked for Problem 7.3.26.

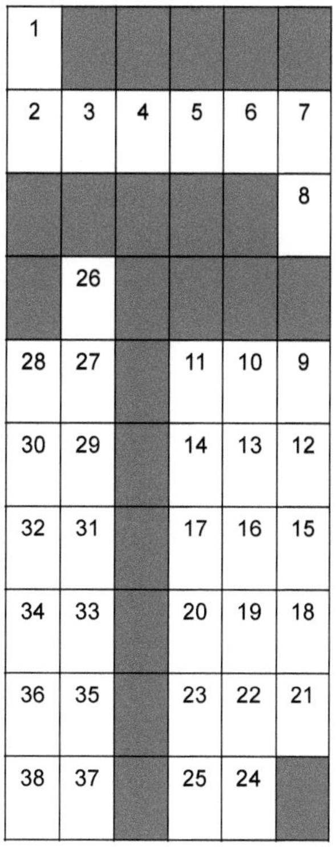

Fig. 7.24.5—Solution for Problem 7.3.26.

Problem 7.3.27

Consider the flow of a single-phase, incompressible liquid in the 2D heterogeneous, anisotropic porous medium shown in **Fig. 7.25.1.** As you can see, there are two wells in the reservoir. The system is surrounded by no-flow boundaries. At the production well, sandface pressure is specified, and at the injection well, the injection rate is specified. Ignore the depth gradient.

1. Provide a list of reservoir rock and fluid properties data that will be used in constructing a mathematical model for this problem.
2. Give a complete mathematical formulation of the problem.
3. Generate a characteristic finite-difference representation of the problem using SIP notation.
4. Express the system of equation in matrix form using SIP notation.
5. Redo Part 4 if the system is numbered as shown in **Fig. 7.25.2.**

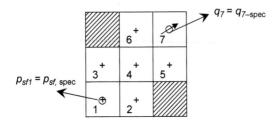

Fig. 7.25.1—Reservoir system and numbering of blocks for Problem 7.3.27.

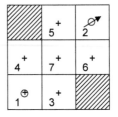

Fig. 7.25.2—Reservoir system for Problem 7.3.27 with the different numbering scheme suggested.

Solution to Problem 7.3.27

1. Reservoir rock and fluid properties data used to construct the mathematical model:
   - Thickness distribution, h.
   - Permeability distributions for k_x and k_y.
   - Fluid properties, viscosity (μ), and formation volume factor ($B = 1$).
   - Grid dimensions, Δx and Δy.
2. The governing equation for the incompressible flow for a heterogeneous and anisotropic reservoir system is

$$\frac{\partial}{\partial x}\left(\frac{A_x k_x}{\mu B}\frac{\partial p}{\partial x}\right)\Delta x + \frac{\partial}{\partial y}\left(\frac{A_y k_y}{\mu B}\frac{\partial p}{\partial y}\right)\Delta y + q = 0. \quad\quad\quad (7.212)$$

Wellbore model,

$$q = -\Omega(p - p_{sf}), \quad\quad\quad (7.213)$$

where

$$\Omega = \frac{2\pi \bar{k} h}{\mu B\left[\ln\left(\dfrac{r_e}{r_w}\right) + s\right]}, \qu\quad\quad\quad (7.214)$$

$$\bar{k} = \sqrt{k_x k_y}, r_e = \frac{0.28\left[\left(\dfrac{k_y}{k_x}\right)^{\frac{1}{2}}(\Delta x)^2 + \left(\dfrac{k_x}{k_y}\right)^{\frac{1}{2}}(\Delta y)^2\right]^{\frac{1}{2}}}{\left(\dfrac{k_y}{k_x}\right)^{\frac{1}{4}} + \left(\dfrac{k_x}{k_y}\right)^{\frac{1}{4}}}. \qu\quad\quad (7.215)$$

3. The characteristic finite-difference equation written using SIP notation is

$$S_{i,j} p_{i,j-1} + W_{i,j} p_{i-1,j} + C_{i,j} p_{i,j} + E_{i,j} p_{i+1,j} + N_{i,j} p_{i,j+1} = Q_{i,j}. \qu\quad\quad (7.216)$$

4. See **Fig. 7.25.3** for the matrix form of the SIP notation.
5. See **Fig. 7.25.4** for results.

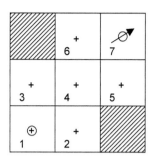

$$\begin{array}{ccccccc} 1 & 2 & 3 & 4 & 5 & 6 & 7 \end{array}$$
$$\begin{bmatrix} C_1 & E_1 & N_1 & 0 & 0 & 0 & 0 \\ W_2 & C_2 & 0 & N_2 & 0 & 0 & 0 \\ S_3 & 0 & C_3 & E_3 & 0 & 0 & 0 \\ 0 & S_4 & W_4 & C_4 & E_4 & N_4 & 0 \\ 0 & 0 & 0 & W_5 & C_5 & 0 & N_5 \\ 0 & 0 & 0 & S_6 & 0 & C_6 & E_6 \\ 0 & 0 & 0 & 0 & S_7 & W_7 & C_7 \end{bmatrix} \begin{bmatrix} p_1 \\ p2 \\ p_3 \\ p_4 \\ p_5 \\ p_6 \\ p_7 \end{bmatrix} = \begin{bmatrix} Q_1 \\ Q_2 \\ Q_3 \\ Q_4 \\ Q_5 \\ Q_6 \\ Q_7 \end{bmatrix}$$

Fig. 7.25.3—System of equations of Problem 7.3.27.4 in matrix form.

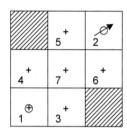

$$\begin{array}{ccccccc} 1 & 2 & 3 & 4 & 5 & 6 & 7 \end{array}$$
$$\begin{bmatrix} C_1 & 0 & E_1 & N_1 & 0 & 0 & 0 \\ 0 & C_2 & 0 & 0 & W_2 & S_2 & 0 \\ W_3 & 0 & C_3 & 0 & 0 & 0 & N_3 \\ S_4 & 0 & 0 & C_4 & 0 & 0 & E_4 \\ 0 & E_5 & 0 & 0 & C_5 & 0 & S_5 \\ 0 & N_6 & 0 & 0 & 0 & C_6 & W_6 \\ 0 & 0 & S_7 & W_7 & N_7 & E_7 & C_7 \end{bmatrix} \begin{bmatrix} p_1 \\ p_2 \\ p_3 \\ p_4 \\ p_5 \\ p_6 \\ p_7 \end{bmatrix} = \begin{bmatrix} Q_1 \\ Q_2 \\ Q_3 \\ Q_4 \\ Q_5 \\ Q_6 \\ Q_7 \end{bmatrix}$$

Fig. 7.25.4—System of equations of Problem 7.3.27.5 in matrix form.

Problem 7.3.28

1. Is the resulting coefficient matrix of the system ordered, as shown in **Fig. 7.26.1,** a pentadiagonal matrix?
2. Is the resulting coefficient matrix of the system ordered, as shown in **Fig. 7.26.2,** a pentadiagonal matrix?

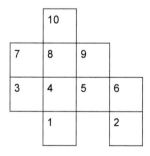

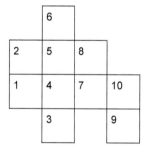

Fig. 7.26.1—Reservoir system for Problem 7.3.28, Part 1.   **Fig. 7.26.2—Reservoir system for Problem 7.3.28, Part 2.**

Solution to Problem 7.3.28

1. No, it is not; the structure of the coefficient matrix is shown in **Fig. 7.26.3.**
2. Yes, the pentadiagonal structure is as shown in **Fig. 7.26.4.**

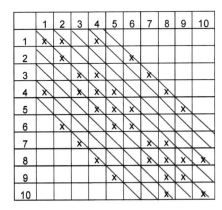

Fig. 7.26.3—Solution for Problem 7.3.28, Part 1.

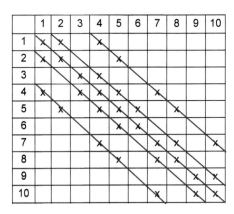

Fig. 7.26.4—Solution for Problem 7.3.28, Part 2.

Problem 7.3.29

State whether the following statements are TRUE or FALSE, and justify your answer.

1. Consider two different reservoirs with different property distributions. The pressure distributions of both reservoirs are obtained using the same iterative solution technique. The pressure convergence criteria used in obtaining both of the solutions are 0.1 psi. The level of accuracy for the solutions of both reservoirs is expected to be the same.
2. When PSOR and Gauss-Seidel solution procedures are applied to the same reservoir with the same convergence criterion, the level of accuracy for PSOR will be better than that of Gauss-Seidel.
3. For the same reservoir, PSOR and Gauss-Seidel procedures will generate solutions of the same level of accuracy, if the same number of iterations is performed with each protocol.
4. When a reservoir system is treated as incompressible and slightly compressible, the computational work involved will be the same if the same solver is used in solving the system of equations.
5. Thomas' algorithm cannot be used to solve 3D single-phase, slightly compressible fluid flow problems when an LSOR iterative scheme is used.
6. In compressible fluid flow simulation, the iteration parameter α used in SIP applications should be updated every timestep.

Solution to Problem 7.3.29

1. FALSE. For reservoirs with varying property distributions, the convergence path during the iteration process could be different, depending on the magnitude of entries of the coefficient matrix. Therefore, using 0.1 psi as convergence criteria cannot guarantee that the solution is at the same level of accuracy.
2. FALSE. For the same convergence criterion, solving the system of equations using PSOR and Gauss-Seidel will yield solutions at the same level of accuracy. However, PSOR converges to that particular solution faster than the Gauss-Seidel.
3. FALSE. Because the convergence rate of PSOR is faster than the Gauss-Seidel within the same number of iterations, solutions provided by PSOR will be closer to the results generated by Gauss-Seidel.
4. FALSE. In solving a slightly compressible system, the resulting linear system of equations needs to be solved for each timestep, and the equation solver will be used multiple times, based on the number of timesteps involved in the project. However, the incompressible problem does not have any time level; therefore, the solver is called for only once. Thus, the computational overhead for solving a slightly compressible system is much larger.
5. FALSE. The overall idea of LSOR is to solve the gridblocks line by line. Therefore, in 3D application, it generates a series of systems of equations that have coefficient matrices with a tridiagonal structure. Thomas' algorithm is designed to solve tridiagonal matrices; therefore, it can be used.
6. FALSE. For compressible fluid flow problems, transmissibility terms are updated every iteration, yielding a new system of equations with different coefficient matrices. Therefore, iteration parameters need to be updated to solve the new system of equations for each iteration.

Problem 7.3.30

Consider the single-phase incompressible fluid flow taking place in the 2D homogeneous and isotropic porous media shown in **Fig. 7.27,** using the reservoir parameters outlined in **Table 7.17.** The reservoir is surrounded by no-flow boundaries, except one side of Block 1 and the right-hand side of Block 2, which are connected to constant-pressure-gradient boundaries.

There is one well located in Block 2 with a constant sandface-pressure specification of 3,000 psia. Note that there is no depth gradient in the reservoir.

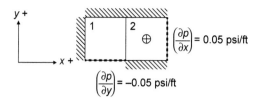

1. Write the differential equation representing the fluid flow dynamics in this reservoir.
2. Obtain the finite-difference approximation of the governing differential equation.
3. Calculate the pressure of each block using the PSOR iterative procedure with ω_{opt}. Begin the PSOR iteration with a uniform initial-guess pressure of 3,000 psia. Use a pressure tolerance of 1 psia for the convergence check. For spectral radius calculation, use a tolerance of 0.01.
4. Calculate the flow rate from the well in Block 2, and describe the type of well (injector or producer).
5. Perform an MBC to verify your results.

Fig. 7.27—Reservoir system for Problem 7.3.30.

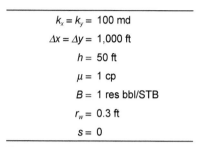

Table 7.17—Reservoir properties for Problem 7.3.30.

Solution to Problem 7.3.30

1. The governing equation for a homogeneous incompressible flow system is

$$\frac{\partial^2 p}{\partial x^2} + \frac{q\mu B}{kV_b} = 0. \qquad\qquad\qquad (7.217)$$

2. The finite-difference approximation is

$$p_{i-1} - 2p_i + p_{i+1} + \frac{q\mu B}{hk} = 0. \qquad\qquad\qquad (7.218)$$

3. The pressure of each block is as follows:
   • Block 1:

$$p_1' - 2p_1 + p_2 = 0. \qquad\qquad\qquad (7.219)$$

   Boundary condition:

$$\frac{p_1 - p_1'}{\Delta y} = -0.05, \qquad\qquad\qquad (7.220)$$

   which is

$$p_1' = p_1 + 0.05\Delta y. \qquad\qquad\qquad (7.221)$$

   Thus, the original equation can be written as

$$-p_1 + p_2 = -50. \qquad\qquad\qquad (7.222)$$

   • Block 2:

$$p_2' - 2p_2 + p_1 + \frac{q\mu B}{hk} = 0. \qquad\qquad\qquad (7.223)$$

   Boundary conditions:

$$q = -\frac{2\pi kh\left(p_1 - p_{sf}\right)}{\mu B\left[\ln\left(\dfrac{r_e}{r_w}\right) + s\right]}, \qquad\qquad\qquad (7.224)$$

   where $r_e = 0.198\,\Delta x = 198$ ft and

$$\frac{p_2' - p_2}{\Delta x} = 0.05, \qquad\qquad\qquad (7.225)$$

which gives $p_2' = p_2 + 0.05\Delta x.$ .... (7.226)

Substituting into the original equation yields,

$$p_1 + 0.05\Delta x - p_2 + \frac{\mu B}{hk}\left\{-\frac{2\pi kh\left(p_1 - p_{sf}\right)}{\mu B\left[\ln\left(\frac{r_e}{r_w}\right)+s\right]}\right\} = 0, \dots\dots\dots\dots\dots\dots\dots\dots\dots\dots (7.227)$$

which can be simplified to

$$p_1 - 1.9678\,p_2 = -2{,}953.4. \dots\dots\dots\dots\dots\dots\dots\dots\dots\dots\dots\dots\dots\dots\dots (7.228)$$

Implementing PSOR,

$$p_1^{(k+1)} = p_1^{(k)} + \frac{\omega_{opt}}{-1}\left[-50 + p_1^{(k)} - p_2^{(k)}\right], \dots\dots\dots\dots\dots\dots\dots\dots\dots\dots\dots (7.229)$$

$$p_2^{(k+1)} = p_2^{(k)} + \frac{\omega_{opt}}{-1.9678}\left[-2{,}953.4 - p_1^{(k+1)} + 1.9678\,p_2^{(k)}\right], \dots\dots\dots\dots\dots\dots (7.230)$$

and

$$\omega_{opt} = \frac{2}{1 + \sqrt{1 - \rho_g}} = 1.1756. \dots\dots\dots\dots\dots\dots\dots\dots\dots\dots\dots\dots\dots (7.231)$$

Table 7.18 summarizes the results at each iteration level. The solution is: $p_1 = 3{,}153.24$ psi and $p_2 = 3{,}103.3$ psi.

k	p_1	p_2	d_{max}	$\left\|\dfrac{d_{max}^{(k+1)}}{d_{max}^{(k)}}\right\|$	ω_{opt}
0	3,000	3,000	—	—	—
1	3,050.00	3,050.82	50.82	0.00	1.00
2	3,100.82	3,076.64	50.82	0.00	1.00
3	3,126.64	3,089.77	25.82	0.508	1.00
4	3,139.77	3,096.44	13.12	0.508	1.1756
5	3,147.61	3,101.12	7.84		1.1756
6	3,151.74	3,102.76	4.13		1.1756
7	3,152.94	3,103.20	1.21		1.1756
8	3,153.24	3,103.30	0.30		1.1756

Table 7.18—Summary of results from Solution to Problem 7.3.30.

4. Peaceman wellbore flow equation yields

$$q_2 = -\frac{2\pi kh\left(p_2 - p_{sf}\right)}{\mu B\left[\ln\left(\frac{r_e}{r_w}\right)+s\right]} = -565 \text{ STB/D}. \dots\dots\dots\dots\dots\dots\dots\dots\dots\dots\dots (7.232)$$

5. Fluid flow from the boundary condition:

$$q = \left(\frac{\partial p}{\partial x} + \frac{\partial p}{\partial y}\right)\left(\frac{kA}{\mu B}\right) = (0.05 + 0.05)\frac{(1{,}000)(50)(100)\left(1.127 \times 10^{-3}\right)}{(1)(1)} = 563.5 \text{ STB/D}. \dots\dots\dots (7.233)$$

MBC:

$$\left|\frac{q_{in}}{q_{well}}\right| = \left|\frac{565}{563.5}\right| = 1.003. \dots\dots\dots\dots\dots\dots\dots\dots\dots\dots\dots\dots\dots\dots\dots\dots\dots (7.234)$$

Problem 7.3.31

Consider the single-phase compressible flow problem in 2D rectangular coordinates. If you are set to solve this problem using the simple iteration method, write the algorithmic equation for the Gauss-Seidel procedure using the SIP notation that will lead you to the solution of the system of equations.

Solution to Problem 7.3.31

The characteristic governing equation written in SIP notation is

$$S_{i,j}^{n+1}\overset{(k)}{\underset{p_{i,j-1}^{n+1}}{}}\overset{(k+1)}{} + W_{i,j}^{n+1}\overset{(k)}{\underset{p_{i-1,j}^{n+1}}{}}\overset{(k+1)}{} + C_{i,j}^{n+1}\overset{(k)}{\underset{p_{i,j}^{n+1}}{}}\overset{(k+1)}{} + E_{i,j}^{n+1}\overset{(k)}{\underset{p_{i+1,j}^{n+1}}{}}\overset{(k+1)}{} + N_{i,j}^{n+1}\overset{(k)}{\underset{p_{i,j+1}^{n+1}}{}}\overset{(k+1)}{} = Q_{i,j}. \dots (7.235)$$

Eq. 7.236 expresses the formulation that solves the system of equations:

$$\overset{(m+1)}{\underset{(k+1)}{p_{i,j}^{n+1}}} = \frac{1}{\overset{(k)}{C_{i,j}^{n+1}}}\left\{Q_{i,j} - \left[\overset{(k)}{S_{i,j}^{n+1}}\overset{(m+1)}{\underset{(k)}{p_{i,j-1}^{n+1}}} + \overset{(k)}{W_{i,j}^{n+1}}\overset{(m+1)}{\underset{(k+1)}{p_{i-1,j}^{n+1}}} + \overset{(k)}{E_{i,j}^{n+1}}\overset{(m)}{\underset{(k+1)}{p_{i+1,j}^{n+1}}} + \overset{(k)}{N_{i,j}^{n+1}}\overset{(m)}{\underset{(k+1)}{p_{i,j+1}^{n+1}}}\right]\right\}. \dots (7.236)$$

Problem 7.3.32

Fig. 7.28 shows a 1D, heterogeneous reservoir in which flow of a single-phase compressible fluid is taking place. The reservoir is surrounded by no-flow boundaries. The production well in Block 1 is put on production at a constant flow rate of q_g in scf/D. Formulate the problem and provide an algorithmic expression that solves the system of equations using Jacobi's iteration. Write your equations using SIP notation and provide definitions of the coefficients.

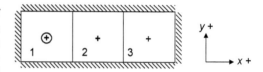

Fig. 7.28—Reservoir system for Problem 7.3.32.

Solution to Problem 7.3.32

The finite-difference equation in SIP notation is

$$\overset{(k)}{W_i^{n+1}}\overset{(k+1)}{p_{i-1}^{n+1}} + \overset{(k)}{C_i^{n+1}}\overset{(k+1)}{p_i^{n+1}} + \overset{(k)}{E_i^{n+1}}\overset{(k+1)}{p_{i+1}^{n+1}} = Q_i. \dots (7.237)$$

Solving the system of equations using Jacobi's iteration (superscript m represents the Jacobi iteration levels, and superscript k denotes the iterations conducted in conjunction with the simple iteration method, as applied to the compressible flow equation):

$$\overset{(m+1)}{\underset{(k+1)}{p_i^{n+1}}} = \frac{1}{\overset{(k)}{C_i^{n+1}}}\left\{Q_i - \left[\overset{(k)}{W_i^{n+1}}\overset{(m)}{\underset{(k+1)}{p_{i-1}^{n+1}}} + \overset{(k)}{E_i^{n+1}}\overset{(m)}{\underset{(k+1)}{p_{i+1}^{n+1}}}\right]\right\}. \dots (7.238)$$

The definitions of the SIP coefficients are

$$W_{i,j}^{n+1} = A_x \left.\frac{k_x}{\Delta x}\right|_{i-\frac{1}{2}} \frac{\frac{1}{2}\left(p_i^{n+1} + p_{i-1}^{n+1}\right)}{\frac{1}{2}\left[\mu_g\left(p_i^{n+1}\right) + \mu_g\left(p_{i-1}^{n+1}\right)\right] \times \frac{1}{2}\left[\mu_g\left(p_i^{n+1}\right) + \mu_g\left(p_{i+1}^{n+1}\right)\right]}, \dots (7.239)$$

$$E_{i,j}^{n+1} = A_x \left.\frac{k_x}{\Delta x}\right|_{i+\frac{1}{2}} \frac{\frac{1}{2}\left(p_i^{n+1} + p_{i+1}^{n+1}\right)}{\frac{1}{2}\left[\mu_g\left(p_i^{n+1}\right) + \mu_g\left(p_{i+1}^{n+1}\right)\right] \times \frac{1}{2}\left[\mu_g\left(p_i^{n+1}\right) + \mu_g\left(p_{i+1}^{n+1}\right)\right]}, \dots (7.240)$$

$$C_{i,j}^{n+1} = -\left\{\left[\frac{\phi V_b T_{sc}}{p_{sc} T z_g\left(p_i^{n+1}\right)}\right]\frac{1}{\Delta t} + W_{i,j}^{n+1} + E_{i,j}^{n+1}\right\}, \dots (7.241)$$

$$Q_i = -\frac{p_i^n}{z_g\left(p_i^n\right)}\left(\frac{\phi V_b T_{sc}}{p_{sc}T}\right)\frac{1}{\Delta t} - q_g. \quad\dotfill \quad (7.242)$$

Problem 7.3.33

Find the pressure distribution in a reservoir represented by the 2D mesh-centered grid system shown in **Fig. 7.29.1.** Fluid in the reservoir is considered incompressible. The reservoir is surrounded by a strong aquifer providing constant pressure support of 2,000 psi. Assume homogeneous and isotropic property distributions. Two wells are completed with constant sandface-pressure specifications of 1,650 psi, and the productivity indices of the wells are 7.7 STB/D-psi.

1. If the coefficient matrix is required to have a pentadiagonal structure, as shown in **Fig. 7.29.2,** how would you number the gridblocks?
2. Calculate the entries of the coefficient matrix. Do not attempt to solve the system of equations.

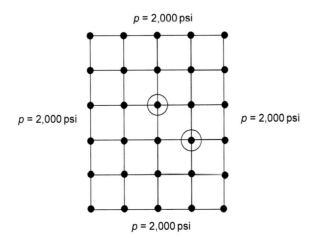

Fig. 7.29.1—Reservoir system for Problem 7.3.33.

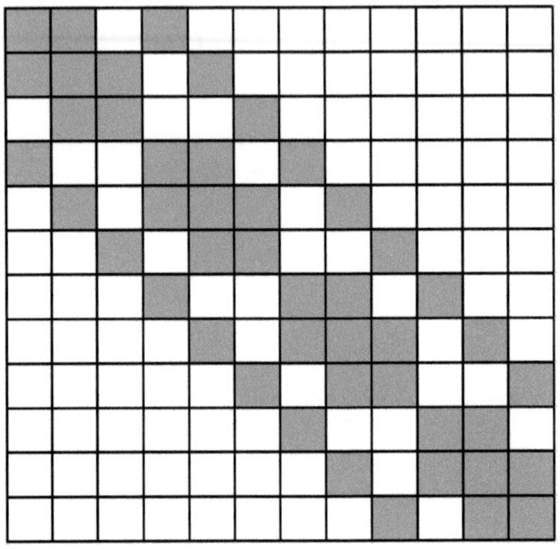

Fig. 7.29.2—Coefficient matrix with a pentadiagonal structure (Problem 7.3.33).

Solution to Problem 7.3.33

1. The numbering scheme is depicted in **Fig. 7.29.3.**

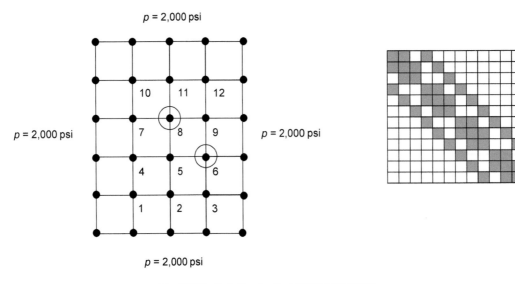

Fig. 7.29.3—Solution for Problem 7.3.33, Part 1.

2. For a 2D incompressible reservoir flow system,

$$\frac{\partial}{\partial x}\left(\frac{A_x k_x}{\mu B}\frac{\partial p}{\partial x}\right)\Delta x + \frac{\partial}{\partial y}\left(\frac{A_y k_y}{\mu B}\frac{\partial p}{\partial y}\right)\Delta y + q = 0, \dots\dots\dots\dots\dots\dots\dots\dots\dots\dots\dots (7.243)$$

which, with homogeneous and isotropic permeability, reduces the homogeneous thickness distributions to the following form:

$$\frac{\partial^2 p}{\partial x^2} + \frac{\partial^2 p}{\partial y^2} + \frac{q\mu B}{kV_b} = 0. \dots\dots\dots\dots\dots\dots\dots\dots\dots\dots\dots\dots\dots\dots\dots (7.244)$$

The finite-difference analog can be written as

$$p_{i,j-1} + p_{i-1,j} - 4p_{i,j} + p_{i+1,j} + p_{i,j+1} - \Omega\left(p_{i,j} - p_{sfi,j}\right)\frac{\mu B}{kh} = 0, \dots\dots\dots\dots\dots\dots\dots (7.245)$$

in which $\Omega = 7.70$ STB/D-psi. Thus, the coefficient matrix $[A]$ can be expressed as

$$[A] = \begin{bmatrix} -4 & 1 & 0 & 1 & 0 & 0 & 0 & 0 & 0 & 0 & 0 & 0 \\ 1 & -4 & 1 & 0 & 1 & 0 & 0 & 0 & 0 & 0 & 0 & 0 \\ 0 & 1 & -4 & 0 & 0 & 1 & 0 & 0 & 0 & 0 & 0 & 0 \\ 1 & 0 & 0 & -4 & 1 & 0 & 1 & 0 & 0 & 0 & 0 & 0 \\ 0 & 1 & 0 & 1 & -4 & 1 & 0 & 1 & 0 & 0 & 0 & 0 \\ 0 & 0 & 1 & 0 & 1 & -11.7 & 0 & 0 & 1 & 0 & 0 & 0 \\ 0 & 0 & 0 & 1 & 0 & 0 & -4 & 1 & 0 & 1 & 0 & 0 \\ 0 & 0 & 0 & 0 & 1 & 0 & 1 & -11.7 & 1 & 0 & 1 & 0 \\ 0 & 0 & 0 & 0 & 0 & 1 & 0 & 1 & -4 & 0 & 0 & 1 \\ 0 & 0 & 0 & 0 & 0 & 0 & 1 & 0 & 0 & -4 & 1 & 0 \\ 0 & 0 & 0 & 0 & 0 & 0 & 0 & 1 & 0 & 1 & -4 & 1 \\ 0 & 0 & 0 & 0 & 0 & 0 & 0 & 0 & 1 & 0 & 1 & -4 \end{bmatrix} \dots\dots\dots\dots (7.246)$$

Problem 7.3.34

Consider the single-phase, incompressible fluid flow problem shown in **Fig. 7.30.** The specific gravity of the reservoir fluid is 1 and the reservoir is homogeneous and isotropic, and there are no depth gradients. The reservoir external boundary is completely sealed and the well completed in Block 5 is operated at a constant sandface pressure, which can be calculated using the hydrostatic head of the fluid column in the well.

1. Calculate the production rate of the well using the reservoir parameters in **Table 7.19.**
2. Perform an MBC on the solution obtained.

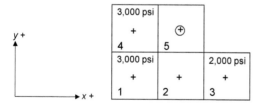

Fig. 7.30—Reservoir system for Problem 7.3.34.

$k_x = k_y = 180$ md	
$\Delta x = 4,000$ ft	
$\Delta y = 4,000$ ft	
$h = 100$ ft	
$\mu = 1$ cp	
$B = 1$ res bbl/STB	
$r_w = 0.2$ ft	
$s = 0$	
Depth $= 4,000$ ft	

Table 7.19—Reservoir properites for Problem 7.3.34.

Solution to Problem 7.3.34

1. The sandface pressure of the well can be calculated using the hydrostatic head of the well:

$$p_{sf} = 0.433 \times 4,000 + 14.7 = 1,746.70 \text{ psi}, \dots\dots\dots\dots\dots\dots\dots\dots\dots\dots\dots\dots\dots (7.247)$$

where $r_e = 0.2\Delta x = 800$ ft and $\Omega = \dfrac{2\pi kh}{\mu B \left[\ln\left(\dfrac{r_e}{r_w}\right) + s \right]} = 15.38$ STB/D-psi,

$$p_{i,j-1} + p_{i-1,j} - 4p_{i,j} + p_{i+1,j} + p_{i,j+1} + \frac{q\mu B}{kh} = 0. \quad\ldots\ldots\ldots\ldots\ldots\ldots\ldots (7.248)$$

- Block 2: $-3p_2 + p_5 = -5,000.$ $\ldots\ldots\ldots\ldots\ldots\ldots\ldots\ldots\ldots\ldots\ldots\ldots\ldots\ldots$ (7.249)
- Block 5: $p_2 - 2.758p_5 = -4,324.163.$ $\ldots\ldots\ldots\ldots\ldots\ldots\ldots\ldots\ldots\ldots\ldots\ldots$ (7.250)

Production rates using the Gauss-Seidel method are laid out in **Table 7.20:**

$$q = -\Omega\left(p_5 - p_{sf}\right) = -15.3787\left(2,470.791 - 1,746.70\right) = -11,135.6 \text{ STB/D.} \quad\ldots\ldots\ldots (7.251)$$

Iteration Level	p_2	p_5
0	2,500	2,500
1	2,500	2,474.316
2	2,491.439	2,471.212
3	2,490.404	2,470.836
4	2,490.279	2,470.791

Table 7.20—Production rates using the Gauss-Seidel method in the Solution to Problem 7.3.34.

2. The MBC:

$$q_{in} = \frac{kh}{\mu B}\left(p_4 - p_5 + p_1 - p_5\right) = 21,075.74 \text{ STB/D.} \quad\ldots\ldots\ldots\ldots\ldots\ldots\ldots\ldots (7.252)$$

$$q_{out} = \frac{kh}{\mu B}\left(p_2 - p_3\right) = 9,945.8 \text{ STB/D.} \quad\ldots\ldots\ldots\ldots\ldots\ldots\ldots\ldots\ldots\ldots\ldots (7.253)$$

$$\text{MBC} = \frac{|q_{in} - q_{out}|}{|q_{well}|} = \frac{11,129.94}{11,135.6} = 0.999. \quad\ldots\ldots\ldots\ldots\ldots\ldots\ldots\ldots\ldots (7.254)$$

Problem 7.3.35

A brine reservoir with an incompressible fluid is discretized using a body-centered grid system $(\Delta x \neq \Delta y)$, as shown in **Fig. 7.31.** The Neumann-type boundary conditions are specified at Blocks 1 and 9. Note that there is a well located in Block 8 that is producing brine at a constant bottomhole-pressure specification.

The reservoir characteristics are known to be as follows:

- Permeability distribution is isotropic, but heterogeneous.
- Thickness is uniform.
- No depth gradient is present.

Can you solve the system of equations generated from the finite-difference approximation using Thomas' algorithm? If yes, write the system of equations in matrix form using the SIP notation. If no, explain why not.

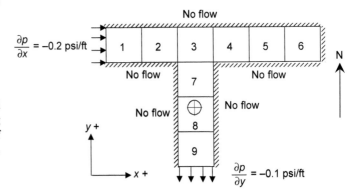

Fig. 7.31—Reservoir system for Problem 7.3.35.

Solution to Problem 7.3.35

Yes, this problem can be solved using Thomas' algorithm. Blocks 4, 5, and 6 are inactive blocks ($p_3 = p_4 = p_5 = p_6$). We can write the system of equations as follows:

$$
\begin{matrix} 1 & 2 & 3 & 7 & 8 & 9 \end{matrix}
$$
$$
\begin{bmatrix}
C_1 & E_1 & 0 & 0 & 0 & 0 \\
W_2 & C_2 & E_2 & 0 & 0 & 0 \\
0 & W_3 & C_3 & S_3 & 0 & 0 \\
0 & 0 & N_7 & C_7 & S_7 & 0 \\
0 & 0 & 0 & N_8 & C_8 & S_8 \\
0 & 0 & 0 & 0 & N_9 & C_9
\end{bmatrix}
\begin{bmatrix} p_1 \\ p_2 \\ p_3 \\ p_7 \\ p_8 \\ p_9 \end{bmatrix}
=
\begin{bmatrix} Q_1 \\ Q_2 \\ Q_3 \\ Q_7 \\ Q_8 \\ Q_9 \end{bmatrix}
. \quad\dots\dots\dots\dots\dots\dots\dots\dots\dots\dots\dots\dots (7.255)
$$

As seen, a tridiagonal coefficient matrix will be generated.

Problem 7.3.36

Consider the reservoir shown in **Fig. 7.32.** All outside boundaries are closed. The reservoir fluid is single-phase and incompressible. If the LSOR algorithmic protocol is used to solve the pressure distribution in this reservoir, what is the minimum number of line equations that need to be solved during one LSOR iteration? Justify your answer with a detailed explanation.

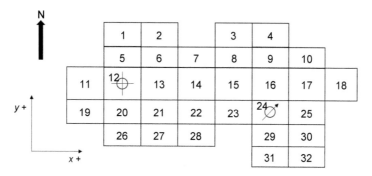

Fig. 7.32—Reservoir system for Problem 7.3.36.

Solution to Problem 7.3.36

The best way to use LSOR is to generate the line equations along the longer edge of the grid (x-direction). In this problem, the minimum number of line equations that need to be solved in one LSOR iteration is six. When applying Thomas' algorithm to solve the pressures for each line, nonexistent blocks may cause the problem of distorting the tridiagonal structure required to obtain the solution (especially the nonexisting blocks between Blocks 28 and 29 and between Blocks 2 and 3). This issue, however, can be solved by assigning the number 1 on the diagonal of the coefficient matrix and leaving all other entries of that row 0.

For example, for the top row, one can write the line equation as follows:

$$
\begin{matrix} 1 \\ 2 \\ 2' \\ 3 \\ 4 \end{matrix}
\begin{bmatrix}
C_1 & E_1 & 0 & 0 & 0 \\
W_2 & C_2 & 0 & 0 & 0 \\
0 & 0 & 1 & 0 & 0 \\
0 & 0 & 0 & C_3 & E_3 \\
0 & 0 & 0 & W_4 & C_4
\end{bmatrix}_{5\times5}
\begin{bmatrix} p_1 \\ p_2 \\ p_{2'} \\ p_3 \\ p_5 \end{bmatrix}_{5\times1}
=
\begin{bmatrix} -S_1 \\ -S_2 \\ \text{any value} \\ -S_3 \\ -S_5 \end{bmatrix}_{5\times1}
. \quad\dots\dots\dots\dots\dots\dots\dots (7.256)
$$

Problem 7.3.37

A 3D single-phase incompressible fluid flow reservoir is displayed in **Fig. 7.33.1.** All external boundaries are completely sealed. The reservoir has isotropic and homogeneous permeability distribution. One well is completed in Block (1,1,1) to inject brine at a rate of 2,000 STB/D, and one well is producing brine in Block (2,2,2). Assume that the pressure of the block hosting the production well is known to be 100 psia. Use the following rock and fluid properties in your calculations: $k = 100$ md, $\mu = 1$ cp, $B = 1$ res bbl/STB. The grid system has uniform block dimensions of $\Delta x = \Delta y = \Delta z = 200$ ft. Ignore all gravitational forces.

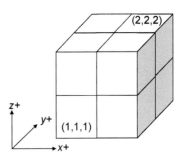

Fig. 7.33.1—Reservoir system for Problem 7.3.37.

1. Write the linear system of equations to solve the pressure distribution in the reservoir in matrix form.
2. Calculate the pressure distribution of the system using the PSOR method with a pressure tolerance of 5 psia. Use iteration parameter $\omega_{opt} = 1.36$ and an initial guess of 100 psia.

Solution to Problem 7.3.37

Numbering the blocks with unknown pressures by considering the existing symmetry in the system will result in the representation depicted in **Fig. 7.33.2.**

Considering the symmetry of the system shown in **Fig. 7.33.3,** pressure values at Blocks identified as 1, 2, and 3 need to be solved by knowing that $p_4 = 100$ psia. For a system with homogeneous and isotropic permeability and $\Delta x = \Delta y = \Delta z$, the PDE can be simplified to

$$\frac{\partial^2 p}{\partial x^2} + \frac{\partial^2 p}{\partial y^2} + \frac{\partial^2 p}{\partial z^2} + \frac{q\mu B}{kV_b} = 0. \quad \dots\dots\dots (7.257)$$

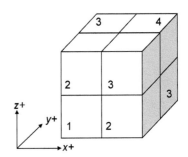

Fig. 7.33.2—Numbering of the blocks with unknown pressures using the existing symmetry (Problem 7.3.37).

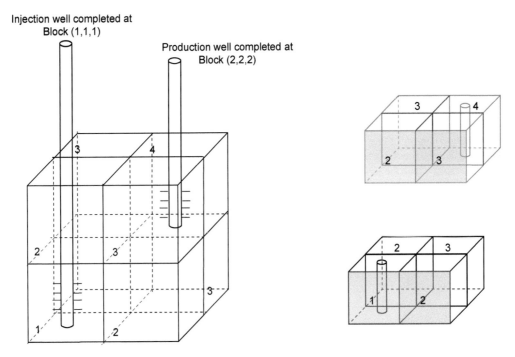

Fig. 7.33.3—Symmetry of the system (Problem 7.3.37).

The finite-difference analog of the PDE is

$$\frac{p_{i-1,j,k} - 2p_{i,j,k} + p_{i+1,j,k}}{(\Delta x)^2} + \frac{p_{i,j-1,k} - 2p_{i,j,k} + p_{i,j+1,k}}{(\Delta y)^2} + \frac{p_{i,j,k-1} - 2p_{i,j,k} + p_{i,j,k+1}}{(\Delta z)^2} + \frac{q\mu B}{kV_b} = 0 \cdot \quad \dots\dots\dots\dots\dots (7.258)$$

Because $\Delta x = \Delta y = \Delta z$, the above equation can be further simplified to the following form:

$$p_{i,j,k-1} + p_{i,j-1,k} + p_{i-1,j,k} - 6p_{i,j,k} + p_{i+1,j,k} + p_{i,j+1,k} + p_{i,j,k+1} + \frac{q\mu B}{kh} = 0 \cdot \quad \dots\dots\dots\dots\dots\dots (7.259)$$

- Block 1:

$$3p_2 - 3p_1 + \frac{q\mu B}{k\Delta x} = 0, \quad \dots\dots\dots\dots\dots\dots\dots\dots\dots\dots\dots\dots\dots\dots\dots (7.260)$$

where

$$\frac{q\mu B}{k\Delta x} = \frac{(2,000)(1)(1)}{(100)(1.127 \times 10^{-3})(200)} = 88.73 \cdot \quad \dots\dots\dots\dots\dots\dots\dots\dots\dots\dots\dots\dots (7.261)$$

Thus, the original equation can be rewritten as $p_2 - p_1 = -29.58$. $\quad \dots\dots\dots\dots\dots\dots\dots\dots\dots (7.262)$

- Block 2:

$$p_1 - 3p_2 + 2p_3 = 0 \cdot \quad \dots\dots\dots\dots\dots\dots\dots\dots\dots\dots\dots\dots\dots\dots\dots\dots\dots (7.263)$$

- Block 3:

$$2p_2 - 3p_3 = -100 \cdot \quad \dots\dots\dots\dots\dots\dots\dots\dots\dots\dots\dots\dots\dots\dots\dots\dots\dots (7.264)$$

The linear system of equations can be expressed in a matrix form:

$$\begin{bmatrix} -1 & 1 & 0 \\ 1 & -3 & 2 \\ 0 & 2 & -3 \end{bmatrix} \begin{bmatrix} p_1 \\ p_2 \\ p_3 \end{bmatrix} = \begin{bmatrix} -29.577 \\ 0 \\ -100 \end{bmatrix}. \quad \dots\dots\dots\dots\dots\dots\dots\dots\dots\dots\dots\dots (7.265)$$

The system of equations can be solved using PSOR; see **Table 7.21** for iteration results. Note that for this problem, it is a given that $\omega_{opt} = 1.36$:

$$p_1^{(k+1)} = p_1^{(k)} + \frac{\omega_{opt}}{-1}\left[-29.577 + p_1^{(k)} - p_2^{(k)}\right], \quad \dots\dots (7.266)$$

$$p_2^{(k+1)} = p_2^{(k)} + \frac{\omega_{opt}}{-3}\left[0 - p_1^{(k+1)} + 3p_2^{(k)} - 2p_3^{(k)}\right], \quad \dots\dots (7.267)$$

$$p_3^{(k+1)} = p_3^{(k)} + \frac{\omega_{opt}}{-3}\left[-100 - 2p_2^{(k+1)} + 3p_3^{(k)}\right]. \quad \dots\dots (7.268)$$

Iteration Level	p_1	p_2	p_3
0	100	100	100
1	140.22	118.24	116.53
2	150.54	131.34	122.46
3	164.65	138.39	126.72
4	169.16	141.76	128.24

Table 7.21—Iteration results for system of equations found in Solution to Problem 7.3.37.

An MBC is carried out to validate the result. Total fluid from adjacent blocks to the production block (Block 4) is

$$q_{production} = 3 \times \frac{kA}{\mu B}\frac{(p_3 - p_4)}{\Delta L} = 3 \times \frac{(100)(1.127 \times 10^{-3})(200)(200)}{(1)(1)}\frac{(128.24 - 100)}{(200)} = 1,910 \text{ STB/D.} \quad \dots\dots (7.269)$$

Total injection rate = 2,000 STB/D:

$$\text{MBC} = \left|\frac{q_{production}}{q_{injection}}\right| = \left|\frac{1,910}{2,000}\right| = 0.95. \quad \dots\dots\dots\dots\dots\dots\dots\dots\dots\dots\dots\dots\dots\dots (7.270)$$

The MBC error is caused by the relatively coarse-prescribed tolerance of the PSOR iteration (5 psia). If the PSOR iteration continues until the pressure improvement is less than 0.1 psi, a higher-resolution solution can be generated: $p_1 = 173.94$ psi, $p_2 = 144.37$ psi, and $p_3 = 129.58$ psi.

This pressure distribution will generate a flow rate of

$$q_{production} = 3 \times \frac{kA}{\mu B} \frac{(p_3 - p_4)}{\Delta L} = 3 \times \frac{(100)(1.127 \times 10^{-3})(200)(200)}{(1)(1)} \frac{(129.58 - 100)}{(200)} = 2,000 \, \text{STB/D}, \quad \ldots \ldots \ldots (7.271)$$

which gives a perfect MBC, two digits after the decimal point.

Problem 7.3.38

A single-phase, slightly compressible reservoir with heterogeneous property distributions is discretized using the grid system shown in **Fig. 7.34.1.** A reservoir engineer would like to solve the problem with SIP algorithm.

1. Write the system of equations in matrix form with a minimum number of unknowns to be solved using SIP notation.
2. Clearly specify how the blocks are numbered.

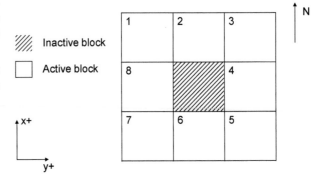

Fig. 7.34.1—Reservoir system for Problem 7.3.38.

Solution to Problem 7.3.38

1. The following pentadiagonal coefficient matrix will be generated:

$$
\begin{array}{c}
\\ 1 \\ 2 \\ 3 \\ 4 \\ 5 \\ 6 \\ 7 \\ 8
\end{array}
\begin{bmatrix}
C_1 & E_1 & 0 & 0 & 0 & 0 & 0 & S_1 \\
W_2 & C_2 & E_2 & 0 & 0 & 0 & 0 & 0 \\
0 & W_3 & C_3 & S_3 & 0 & 0 & 0 & 0 \\
0 & 0 & N_4 & C_4 & S_4 & 0 & 0 & 0 \\
0 & 0 & 0 & N_5 & C_5 & W_5 & 0 & 0 \\
0 & 0 & 0 & 0 & W_6 & C_6 & E_6 & 0 \\
0 & 0 & 0 & 0 & 0 & E_7 & C_7 & N_7 \\
N_8 & 0 & 0 & 0 & 0 & 0 & S_8 & C_8
\end{bmatrix}
\begin{bmatrix}
p_1 \\ p_2 \\ p_3 \\ p_4 \\ p_5 \\ p_6 \\ p_7 \\ p_8
\end{bmatrix}
=
\begin{bmatrix}
Q_1 \\ Q_2 \\ Q_3 \\ Q_4 \\ Q_5 \\ Q_6 \\ Q_7 \\ Q_8
\end{bmatrix} \quad \ldots\ldots\ldots\ldots\ldots\ldots\ldots (7.272)
$$

2. In generating the coefficient matrix, the numbering scheme in **Fig. 7.34.2** is used.

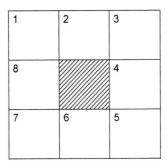

Fig. 7.34.2—Numbering scheme to be used to generate a system of equations with minimum number of unknowns (Problem 7.3.38).

Problem 7.3.39

Consider the single-phase, slightly compressible reservoir shown in **Fig. 7.35.1.** Can the system of equations within each timestep be solved using Thomas' algorithm? If so, write the system of equations in matrix form using the SIP notation and the numbering scheme of the blocks. If not, explain why not.

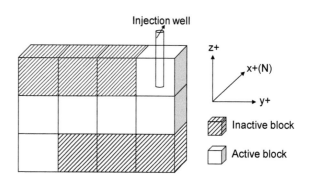

Fig. 7.35.1—Reservoir system for Problem 7.3.39.

Solution to Problem 7.3.39

To solve the problem using Thomas' algorithm, a tridiagonal coefficient matrix needs to be constructed by numbering the active blocks, as shown in **Fig. 7.35.2.**

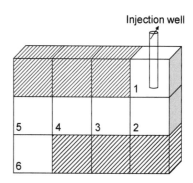

Fig. 7.35.2—Numbering of the blocks of Problem 7.3.39.

A tridiagonal coefficient matrix and the corresponding right-hand-side vector are generated using the numbering scheme shown:

$$\begin{bmatrix} C_1 & B_1 & 0 & 0 & 0 & 0 \\ A_2 & C_2 & W_2 & 0 & 0 & 0 \\ 0 & E_3 & C_3 & W_3 & 0 & 0 \\ 0 & 0 & E_4 & C_4 & W_1 & 0 \\ 0 & 0 & 0 & E_5 & C_5 & B_5 \\ 0 & 0 & 0 & 0 & A_6 & C_6 \end{bmatrix} \begin{bmatrix} p_1 \\ p_2 \\ p_3 \\ p_4 \\ p_5 \\ p_6 \end{bmatrix} = \begin{bmatrix} Q_1 \\ Q_2 \\ Q_3 \\ Q_4 \\ Q_5 \\ Q_6 \end{bmatrix} \dots\dots\dots\dots\dots\dots\dots\dots\dots\dots\dots\dots\dots\dots\dots (7.273)$$

Problem 7.3.40

An incompressible fluid reservoir is discretized using a mesh-centered grid system with constant pressure boundaries, as displayed in **Fig. 7.36.** Wells in Nodes 1 and 6 are producing at constant rates of 5,000 STB/D. Use the reservoir properties shown in Fig. 7.36 and in **Table 7.22** to do the following:

1. Write the generalized PDE in its simplest form for the previously described problem.
2. Calculate the pressures at Nodes 1 through 6. Use any linear solver to solve the system of equations. Use a pressure tolerance of 1 psi for convergence, if you are using an iterative solver.
3. Calculate the sandface pressures of the wells at Nodes 1 and 6.

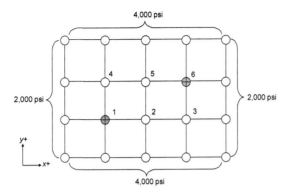

4,000 psi

2,000 psi

2,000 psi

4,000 psi

Fig. 7.36—Reservoir system for Problem 7.3.40.

k_x =	300 md, homogeneous
k_y =	100 md, homogeneous
h =	50 ft, homogeneous
$\Delta x = \Delta y$ =	60 ft, uniform
μ =	0.98 cp
B =	1.0 res bbl/STB
Skin factor =	0
r_w =	0.25 ft, for both wells
Ignore gravitational forces	

Table 7.22—Reservoir properties for Problem 7.3.40.

Solution to Problem 7.3.40

1. Starting with the general form of the PDE,

$$\frac{\partial}{\partial x}\left(\frac{A_x k_x}{\mu B}\frac{\partial p}{\partial x}\right)\Delta x + \frac{\partial}{\partial y}\left(\frac{A_y k_y}{\mu B}\frac{\partial p}{\partial y}\right)\Delta y + q = 0 \cdot \dots \dots \dots \dots \dots \dots \dots (7.274)$$

For a homogeneous and anisotropic system, the previous equation can be further simplified to

$$k_x \frac{\partial^2 p}{\partial x^2} + k_y \frac{\partial^2 p}{\partial y^2} + \frac{q\mu B}{V_b} = 0 \cdot \dots \dots \dots \dots \dots \dots \dots \dots \dots \dots \dots (7.275)$$

2. Considering the symmetry in place, $p_1 = p_6$, $p_3 = p_4$, and $p_2 = p_5$. Generate the system of equations by implementing the finite-difference approximation only at Nodes 1, 2, and 4. Because $k_x = 300$ md and $k_y = 100$ md, the equation can be further simplified by substituting $k_x = 3k_y$:

$$\frac{k_x}{k_y}\frac{p_{i-1,j} - 2p_{i,j} + p_{i+1,j}}{(\Delta x)^2} + \frac{p_{i,j-1} - 2p_{i,j} + p_{i,j+1}}{(\Delta y)^2} + \frac{q\mu B}{k_y V_b} = 0. \dots \dots \dots \dots \dots \dots (7.276)$$

Because $\Delta x = \Delta y$, the previous equation can be rewritten as

$$\frac{k_x}{k_y}\left(p_{i-1,j} - 2p_{i,j} + p_{i+1,j}\right) + \left(p_{i,j-1} - 2p_{i,j} + p_{i,j+1}\right) + \frac{q\mu B}{k_y V_b} = 0. \dots \dots \dots \dots (7.277)$$

- Node 1: $3\left(2,000 - 2p_1 + p_2\right) + \left(p_4 - 2p_1 + 4,000\right) + \frac{q\mu B}{hk_y} = 0. \dots \dots \dots \dots \dots \dots (7.278)$

- Node 2: $3\left(p_1 - 2p_2 + p_4\right) + \left(p_2 - 2p_2 + 4,000\right) = 0. \dots \dots \dots \dots \dots \dots \dots \dots (7.279)$

- Node 4: $3\left(2,000 - 2p_4 + p_2\right) + \left(p_1 - 2p_4 + 4,000\right) = 0. \dots \dots \dots \dots \dots \dots \dots (7.280)$

Knowing that $k_y = 100$ md, $q = -5,000$ STB/D, $\mu = 0.98$ cp, and $B = 1$ res bbl/STB,

- $-8p_1 + 3p_2 + p_4 = -9,130. \dots \dots \dots \dots \dots \dots \dots \dots \dots \dots \dots \dots \dots (7.281)$
- $p_1 + 3p_2 - 8p_4 = -10,000. \dots \dots \dots \dots \dots \dots \dots \dots \dots \dots \dots \dots (7.282)$
- $3p_1 - 7p_2 + 3p_4 = -4,000. \dots \dots \dots \dots \dots \dots \dots \dots \dots \dots \dots \dots (7.283)$

Solving three equations in three unknowns,

- $p_1 = 2,498.6$ psi.
- $p_2 = 2,754.52$ psi.
- $p_4 = 2,595.26$ psi.

MBCs show

- Flow in from the pressure boundary with 4,000 psi:

$$q_{\text{in}} = \frac{k_y h}{\mu B}\left(3 \times 4,000 - p_1 - p_2 - p_3\right) = -23,871.81 \text{ STB/D}. \dots \dots \dots \dots \dots \dots (7.284)$$

- Flow out from the pressure boundary with 2,000 psi:

$$q_{\text{out}} = \frac{k_x h}{\mu B}\left(p_1 + p_4 - 2 \times 2,000\right) = 18,869.1 \text{ STB/D}, \dots \dots \dots \dots \dots \dots \dots (7.285)$$

$$q_{\text{in}} + q_{\text{out}} = -5,002.73 \text{ STB/D}. \dots \dots \dots \dots \dots \dots \dots \dots \dots \dots \dots \dots (7.286)$$

3. Calculate the sandface pressures of the wells at Nodes 1 and 6:

$$r_e = \frac{0.28\sqrt{\left(\frac{k_x}{k_y}\right)^{\frac{1}{2}}(\Delta y)^2 + \left(\frac{k_y}{k_x}\right)^{\frac{1}{2}}(\Delta x)^2}}{\left(\frac{k_x}{k_y}\right)^{\frac{1}{4}} + \left(\frac{k_y}{k_x}\right)^{\frac{1}{4}}} = 12.3 \text{ ft},\ \dots\dots\dots\dots\dots\dots\dots\dots \quad (7.287)$$

where $\bar{k} = \sqrt{k_x k_y} = 173.2$ md.

Using the Peaceman wellbore equation,

$$\Omega = \frac{2\pi \bar{k} h}{\mu B \ln\left(\frac{r_e}{r_w}\right)} = 16.05 \text{ STB/(psi}-\text{D)},\ \dots\dots\dots\dots\dots\dots\dots\dots\dots \quad (7.288)$$

$$p_{sf} = p_1 + \frac{q_1}{\Omega} = 2{,}498.6 \text{ psi} - \frac{5{,}000 \text{ STB / D}}{16.05 \text{STB / (psi}-\text{D)}} = 2{,}187.07 \text{ psi}.\ \dots\dots\dots\dots\dots \quad (7.289)$$

Problem 7.3.41

A 2D homogeneous, anisotropic horizontal reservoir is shown in **Fig. 7.37.** The reservoir is known to be sealed, except along the constant-pressure boundary represented by Block 4. This constant pressure of 5,000 psia in Block 4 is supported by a strong water drive. A single-phase, incompressible fluid is produced through a production well in Block 2 at a rate of –1,400 STB/D.

1. Solve for the pressure distribution in the reservoir using the Gauss-Seidel iterative procedure and the reservoir properties found in **Table 7.23.** Use 4,700 psi as initial guess and 1 psia as the pressure tolerance.
2. Calculate the producing sandface pressure of the well in Block 2.
3. Perform an MBC and comment on the accuracy of the solutions.

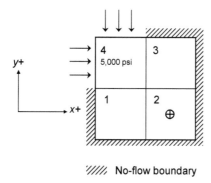

//// No-flow boundary

Fig. 7.37—Reservoir system for Problem 7.3.41.

k_x	= 100 md
k_y	= 50 md
Δx	= 500 ft
Δy	= 500 ft
h	= 50 ft
μ	= 1 cp
B	= 1.0 res bbl/STB
r_w	= 0.25 ft
s	= 0

Table 7.23—Reservoir properties for Problem 7.3.41.

Solution to Problem 7.3.41

For a 2D incompressible system, the governing PDE is:

$$\frac{\partial}{\partial x}\left(\frac{A_x k_x}{\mu B}\frac{\partial p}{\partial x}\right)\Delta x + \frac{\partial}{\partial y}\left(\frac{A_y k_y}{\mu B}\frac{\partial p}{\partial y}\right)\Delta y + q = 0.\ \dots\dots\dots\dots\dots\dots\dots\dots \quad (7.290)$$

The reservoir properties exhibit homogeneous distributions, but permeability displays anisotropic values, then Eq. 290 can be simplified to

$$k_x \frac{\partial^2 p}{\partial x^2} + k_y \frac{\partial^2 p}{\partial y^2} + \frac{q\mu B}{V_b} = 0.\ \dots\dots\dots\dots\dots\dots\dots\dots\dots\dots\dots \quad (7.291)$$

In a finite-difference form,

$$k_x \frac{p_{i-1,j} - 2p_{i,j} + p_{i+1,j}}{(\Delta x)^2} + k_y \frac{p_{i,j-1} - 2p_{i,j} + p_{i,j+1}}{(\Delta y)^2} + \frac{q\mu B}{V_b} = 0.\ \dots\dots\dots\dots \quad (7.292)$$

Because $\Delta x = \Delta y$, after dividing the entire equation by k_y, the previous equation can be rewritten as

$$\frac{k_x}{k_y}(p_{i-1,j} - 2p_{i,j} + p_{i+1,j}) + (p_{i,j-1} - 2p_{i,j} + p_{i,j-1}) + \frac{q\mu B}{hk_y} = 0, \dots\dots\dots\dots\dots\dots\dots \quad (7.293)$$

$$2(p_{i-1,j} + p_{i+1,j} - 2p_{i,j}) + (p_{i,j-1} + p_{i,j-1} - 2p_{i,j}) + \frac{q\mu B}{hk_y} = 0, \dots\dots\dots\dots\dots\dots \quad (7.294)$$

$$2p_{i-1,j} + 2p_{i+1,j} - 6p_{i,j} + p_{i,j-1} + p_{i,j-1} + \frac{q\mu B}{hk_y} = 0. \dots\dots\dots\dots\dots\dots\dots \quad (7.295)$$

- For Block 1:

$$p_1 + 2p_1 - 6p_1 + 2p_2 + 5,000 = 0, \dots\dots\dots\dots\dots\dots\dots\dots\dots\dots\dots \quad (7.296)$$

$$-3p_1 + 2p_2 = -5,000. \dots\dots\dots\dots\dots\dots\dots\dots\dots\dots\dots\dots\dots \quad (7.297)$$

- For Block 2:

$$p_3 + 2p_1 - 6p_3 + 2p_2 + \frac{(-1,500)(1)(1)}{(50)(1.127\times10^{-3})(50)} = 0, \dots\dots\dots\dots\dots\dots \quad (7.298)$$

$$2p_1 - 3p_2 + p_3 = 532.4. \dots\dots\dots\dots\dots\dots\dots\dots\dots\dots\dots\dots \quad (7.299)$$

- For Block 3, see **Table 7.24** and the following equations:

$$p_2 + 2(5,000) - 6p_3 + 2p_3 + p_3 = 0, \dots\dots\dots\dots\dots\dots\dots\dots\dots\dots \quad (7.300)$$

$$p_2 - 3p_3 = -10,000, \dots\dots\dots\dots\dots\dots\dots\dots\dots\dots\dots\dots\dots \quad (7.301)$$

$$p_1^{(k+1)} = \frac{1}{3}(5,000 + 2p_2^{(k)}), \dots\dots\dots\dots\dots\dots\dots\dots\dots\dots\dots \quad (7.302)$$

$$p_2^{(k+1)} = \frac{1}{3}(2p_1^{(k+1)} + p_3^{(k)} - 532.4), \dots\dots\dots\dots\dots\dots\dots\dots\dots \quad (7.303)$$

$$p_3^{(k+1)} = \frac{1}{3}(10,000 + p_2^{(k+1)}). \dots\dots\dots\dots\dots\dots\dots\dots\dots\dots \quad (7.304)$$

Therefore, $p_1 = 4,733.01$ psi, $p_2 = 4,600.09$ psi, $p_3 = 4,866.70$ psi.

Iteration Level	p_1	p_2	p_3
0	4,700.00	4,700.00	4,700.00
1	4,800.00	4,589.20	4,863.07
2	4,726.13	4,594.31	4,864.77
3	4,729.54	4,597.15	4,865.72
4	4,731.43	4,598.73	4,866.24
5	4,732.49	4,599.60	4,866.53
6	4,733.07	4,600.09	4,866.70

Table 7.24—Iterations for Block 3 in the Solution to Problem 7.3.41.

Calculating the sandface pressure,

$$q_2 = -\Omega(p_2 - p_{sf}), \dots\dots\dots\dots\dots\dots\dots\dots\dots\dots\dots\dots\dots \quad (7.305)$$

$$\Omega = \frac{2\pi \bar{k}h}{\mu B\left[\ln\left(\frac{r_e}{r_w}\right) + s\right]} = \frac{2\pi(70.71)(1.127\times10^3)(50)}{(1)(1)\left(\ln\frac{100}{0.25}\right)} = 4.18 \text{ STB}/(\text{psi}-\text{D}), \dots\dots \quad (7.306)$$

where $r_e = \dfrac{0.28\left[\left(\frac{k_y}{k_x}\right)^{\frac{1}{2}}(\Delta x)^2 + \left(\frac{k_x}{k_y}\right)^{\frac{1}{2}}(\Delta y)^2\right]^{\frac{1}{2}}}{\left(\frac{k_y}{k_x}\right)^{\frac{1}{4}} + \left(\frac{k_x}{k_y}\right)^{\frac{1}{4}}} = 100.44$ ft and $\bar{k} = \sqrt{k_x k_y} = 70.71$ md.

Thus,

$$p_{sf} = -\frac{q_2}{\Omega} + p_2 = 4,600 - \frac{1,500}{4.18} \text{ psi} = 4,241 \text{ psi}. \dots\dots\dots\dots\dots\dots\dots\dots\dots (7.307)$$

MBC results should indicate that flow into the system must be equal to the summation of flows from Block 4 to Block 3 and from Block 4 to Block 1:

$$q_{4to3} = \frac{kA}{\mu B}\frac{\partial p}{\partial x} = \frac{(1.127\times10^{-3})(100)(500)(50)}{(1)(1)}\left[\frac{p_4 - p_3}{\Delta x}\right] = 2,817.5\left[\frac{5,000 - 4,867}{500}\right], \dots\dots\dots\dots (7.308)$$

$$= 749.5 \text{ STB / D}$$

$$q_{4to1} = \frac{kA}{\mu B}\frac{\partial p}{\partial x} = \frac{(1.127\times10^{-3})(50)(500)(50)}{(1)(1)}\left[\frac{p_4 - p_1}{\Delta y}\right] = 2,254\left[\frac{5,000 - 4,733}{500}\right] = 752.3 \text{ STB / D,} \dots\dots (7.309)$$

$$q_{in} = q_{4to3} + q_{4to1} = 749.5 + 752.3 = 1,501.8 \text{ STB / D}. \dots\dots\dots\dots\dots\dots\dots\dots\dots\dots (7.310)$$

Then, the MBC shows that $q_{out} = 1,500$ STB/D and

$$\text{MBC} = \left|\frac{q_{in}}{q_{out}}\right| = \left|\frac{1,501.8}{1,500}\right| = 1.001. \dots\dots\dots\dots\dots\dots\dots\dots\dots\dots\dots\dots (7.311)$$

Problem 7.3.42

1. Construct the resulting coefficient matrix of the 2D system shown in **Fig. 7.38.1** using the SIP notation. What type of matrix does this result in?
2. Determine whether the following statement is TRUE or FALSE. If the statement is false, write the correct equation:
   The general expression to solve the slightly compressible flow equation using the PSOR algorithm can be written as

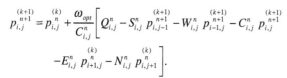

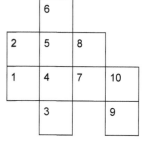

Fig. 7.38.1—Reservoir system for Problem 7.3.42, Part 1.

3. Is the following statement TRUE or FALSE? Justify your answer.
   Consider two different reservoirs with different property distributions. The pressure distribution of both reservoirs is obtained using the same iterative solution technique, and the pressure convergence criteria in obtaining both solutions are 0.01 psia. The level of accuracy of both reservoirs is expected to be the same.
4. Is the following statement TRUE or FALSE? Justify your answer.
   When poor MBCs are obtained, improvements to the solution can be made by increasing the number of blocks in the reservoir.

Solution to Problem 7.3.42

1. As displayed in **Fig. 7.38.2,** a pentadiagonal coefficient matrix is formed.

	1	2	3	4	5	6	7	8	9	10
1	C_1	N_1		E_1						
2	S_2	C_2		E_2						
3			C_3	N_3						
4	N_4		S_4	C_4	N_4		E_4			
5		W_5		S_5	C_5	N_5		E_5		
6					S_6	C_6				
7				W_7			C_7	N_8		E_8
8					W_8		S_8	C_8		
9									C_9	N_{10}
10							W_{10}		S_{10}	C_{10}

Fig. 7.38.2—Solution for Problem 7.3.42, Part 1.

2. FALSE. The pressures all should be at the time level $n+1$:

$$\overset{(k+1)}{p_{i,j}^{n+1}} = \overset{(k)}{p_{i,j}^{n}} + \frac{\omega_{opt}}{C_{i,j}^{n}}\left[Q_{i,j}^{n} - S_{i,j}^{n}\,\overset{(k+1)}{p_{i,j-1}^{n+1}} - W_{i,j}^{n}\,\overset{(k+1)}{p_{i-1,j}^{n+1}} - C_{i,j}^{n}\,\overset{(k+1)}{p_{i,j}^{n+1}} \right.$$
$$\left. - E_{i,j}^{n}\,\overset{(k)}{p_{i+1,j}^{n+1}} - N_{i,j}^{n}\,\overset{(k)}{p_{i,j+1}^{n+1}} \right] \qquad\dots\dots\dots\dots\dots\dots\dots\dots\dots (7.312)$$

3. FALSE. Because the property distributions are different for each reservoir, the coefficient matrix also will be different. Thus, even applying the same solver and convergence criteria, the resulting accuracy levels may be different if diagonal dominance of each coefficient matrix is significantly different.
4. TRUE. A finer (high-resolution) grid system will increase the accuracy of the computation.

Problem 7.3.43

Consider a single-phase, slightly compressible fluid flow system with homogeneous permeability distribution, as presented in **Fig. 7.39**. There is a well located in Block 1. The well is put on production at a flow rate of q. Fluid viscosity, formation volume factor, and density values are treated as constants throughout the system.

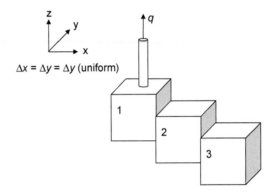

$\Delta x = \Delta y = \Delta y$ (uniform)

1. Develop a finite-difference approximation of the governing flow equation using a backward-difference scheme.
2. Set up the finite-difference equations in a matrix form such that Thomas' algorithm is readily applicable.

Fig. 7.39—Reservoir system for Problem 7.3.43.

Solution to Problem 7.3.43

1. The governing flow equation can be written as

$$\frac{\partial^2 p}{\partial x^2} - \frac{1}{144}\frac{g}{g_c}\rho\frac{\partial^2 G}{\partial x^2} + \frac{q\mu B}{V_b k_x} = \frac{\phi c_t \mu B}{5.615 k_x}\frac{\partial p}{\partial t}. \qquad\dots\dots\dots\dots\dots\dots\dots\dots (7.313)$$

The backward-difference scheme can be written as

$$\frac{p_{i+1}^{n+1} - 2p_i^{n+1} + p_{i-1}^{n+1}}{(\Delta x)^2} - \frac{1}{144}\frac{g}{g_c}\rho\frac{G_{i+1} - 2G_i + G_{i-1}}{(\Delta x)^2} + \frac{q\mu B}{V_b k_x} = \frac{\phi c_t \mu B}{5.615 k_x}\left(\frac{p_i^{n+1} - p_i^{n}}{\Delta t}\right). \qquad\dots\dots\dots\dots (7.314)$$

2. The characteristic finite-difference equation can be rearranged as

$$\frac{1}{(\Delta x)^2}p_{i-1}^{n+1} - \frac{1}{(\Delta x)^2}\left(2 + \frac{\phi c_t \mu B}{5.615 k_x \Delta t}\right)p_i^{n+1} + \frac{1}{(\Delta x)^2}p_{i+1}^{n+1} = \frac{1}{144}\frac{g}{g_c}\rho\frac{G_{i+1} - 2G_i + G_{i-1}}{(\Delta x)^2}$$
$$- \frac{q\mu B}{V_b k_x} - \frac{\phi c_t \mu B}{5.615 k_x \Delta t}p_i^{n} \qquad\dots\dots\dots\dots\dots (7.315)$$

Now, writing the equations for each block,

- Block 1:

$$\frac{1}{(\Delta x)^2}p_1^{n+1} - \frac{1}{(\Delta x)^2}\left(2 + \frac{\phi c_t \mu B}{5.615 k_x \Delta t}\right)p_1^{n+1} + \frac{1}{(\Delta x)^2}p_2^{n+1} =$$
$$\frac{1}{144}\frac{g}{g_c}\rho\frac{G_1 - 2G_1 + G_2}{(\Delta x)^2} - \frac{q\mu B}{V_b k_x} - \frac{\phi c_t \mu B}{5.615 k_x \Delta t}p_1^{n} - \frac{1}{(\Delta x)^2}\left(1 + \frac{\phi c_t \mu B}{5.615 k_x \Delta t}\right)p_1^{n+1} \qquad\dots\dots\dots\dots (7.316)$$
$$+ \frac{1}{(\Delta x)^2}p_2^{n+1} = \frac{1}{144}\frac{g}{g_c}\rho\frac{-G_1 + G_2}{(\Delta x)^2} - \frac{q\mu B}{V_b k_x} - \frac{\phi c_t \mu B}{5.615 k_x \Delta t}p_1^{n}.$$

- Block 2:

$$\frac{1}{(\Delta x)^2} p_1^{n+1} - \frac{1}{(\Delta x)^2}\left(2+\frac{\phi c_t \mu B}{5.615 k_x \Delta t}\right) p_2^{n+1} + \frac{1}{(\Delta x)^2} p_3^{n+1}$$

$$= \frac{1}{144}\frac{g}{g_c}\rho\frac{G_1 - 2G_2 + G_3}{(\Delta x)^2} - \frac{q\mu B}{V_b k_x} - \frac{\phi c_t \mu B}{5.615 k_x \Delta t} p_2^{n}. \qquad\qquad \dotfill (7.317)$$

- Block 3:

$$\frac{1}{(\Delta x)^2} p_2^{n+1} - \frac{1}{(\Delta x)^2}\left(1+\frac{\phi c_t \mu B}{5.615 k_x \Delta t}\right) p_3^{n+1} = \frac{1}{144}\frac{g}{g_c}\rho\frac{G_2 - G_3}{(\Delta x)^2} - \frac{q\mu B}{V_b k_x} - \frac{\phi c_t \mu B}{5.615 k_x \Delta t} p_3^{n}. \quad \dotfill (7.318)$$

Equations for Blocks 1, 2, and 3 can be assembled in matrix form:

$$\begin{bmatrix} -\frac{1}{(\Delta x)^2}\left(1+\frac{\phi c_t \mu B}{5.615 k_x \Delta t}\right) & \frac{1}{(\Delta x)^2} & 0 \\[2ex] \frac{1}{(\Delta x)^2} & -\frac{1}{(\Delta x)^2}\left(2+\frac{\phi c_t \mu B}{5.615 k_x \Delta t}\right) & \frac{1}{(\Delta x)^2} \\[2ex] 0 & \frac{1}{(\Delta x)^2} & -\frac{1}{(\Delta x)^2}\left(1+\frac{\phi c_t \mu B}{5.615 k_x \Delta t}\right) \end{bmatrix} \begin{bmatrix} p_1^{n+1} \\[1ex] p_2^{n+1} \\[1ex] p_3^{n+1} \end{bmatrix} =$$

$$\begin{bmatrix} \frac{1}{144}\frac{g}{g_c}\rho\frac{-G_1 + G_2}{(\Delta x)^2} - \frac{q\mu B}{V_b k_x} - \frac{\phi c_t \mu B}{5.615 k_x \Delta t} p_1^{n} \\[2ex] \frac{1}{144}\frac{g}{g_c}\rho\frac{G_1 - 2G_2 + G_3}{(\Delta x)^2} - \frac{q\mu B}{V_b k_x} - \frac{\phi c_t \mu B}{5.615 k_x \Delta t} p_2^{n} \\[2ex] \frac{1}{144}\frac{g}{g_c}\rho\frac{G_2 - G_3}{(\Delta x)^2} - \frac{q\mu B}{V_b k_x} - \frac{\phi c_t \mu B}{5.615 k_x \Delta t} p_3^{n} \end{bmatrix}.$$

$$\dots (7.319)$$

Problem 7.3.44

Consider the 2D, incompressible homogeneous, isotropic reservoir shown in **Fig. 7.40.** An infinitely large aquifer is connected to the reservoir through Block 4. The pressure at Block 4 is maintained at 3,000 psia by the aquifer. All other boundaries are completely sealed. Two identical wells are completed in Blocks 1 and 3. Both wells are producing at a sandface pressure of 100 psia. Assume that depth gradients can be ignored.

1. Write the PSOR iterative equations for this system using the reservoir properties found in **Table 7.25.**
2. Solve for the pressure distribution using the PSOR equations written in Part 1. Use a convergence criterion of 1 psia and a ω_{opt} of 1.1.
3. Calculate the production rates at wells in Blocks 1 and 3.
4. Perform an MBC.

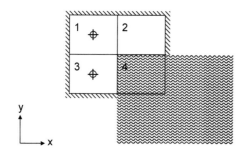

Fig. 7.40—Reservoir system for Problem 7.3.44.

Table 7.25—Reservoir properties for Problem 7.3.44.
h = 50 ft
Δx = 100 ft
Δy = 50 ft
k = 35 md
μ = 1.1 cp
B = 1.0 res bbl/STB
r_w = 0.2 ft
s = 0

Solution to Problem 7.3.44

The governing PDE for a 2D incompressible, homogeneous and isotropic reservoir is

$$\frac{\partial^2 p}{\partial x^2} + \frac{\partial^2 p}{\partial y^2} + \frac{q\mu B}{V_b k} = 0. \quad\dotfill (7.320)$$

The finite-difference approximation of the fluid flow equation is $(k_x = k_y = k)$

$$\frac{p_{i-1,j} - 2p_{i,j} + p_{i+1,j}}{(\Delta x)^2} + \frac{p_{i,j-1} - 2p_{i,j} + p_{i,j+1}}{(\Delta y)^2} + \frac{q\mu B}{V_b k}\bigg|_{i,j} = 0. \quad\dotfill (7.321)$$

Substituting the flow-rate term with the Peaceman wellbore model,

$$q_{i,j} = -\Omega_{i,j}(p_{i,j} - p_{sfi,j}), \quad\dotfill (7.322)$$

yields,

$$\frac{p_{i-1,j} - 2p_{i,j} + p_{i+1,j}}{(\Delta x)^2} + \frac{p_{i,j-1} - 2p_{i,j} + p_{i,j+1}}{(\Delta y)^2} - \frac{\Omega p \mu B}{V_b k}\bigg|_{i,j} + \frac{\Omega p_{sf} \mu B}{V_b k}\bigg|_{i,j} = 0. \quad\dotfill (7.323)$$

1. The PSOR iterative equations are as follows.

   - Block 1: Knowing that

$$r_e = 0.14\left[(\Delta x)^2 + (\Delta y)^2\right]^{\frac{1}{2}} = 0.14\left[(100)^2 + (50)^2\right]^{\frac{1}{2}} = 15.652 \text{ ft}, \quad\dotfill (7.324)$$

and

$$\Omega_1 = \frac{2\pi \bar{k} h}{\mu B\left[\ln\left(\dfrac{r_e}{r_w}\right) + s\right]} = \frac{2(3.14)(35)(50)}{(1.1)(1)\left[\ln\left(\dfrac{15.652}{0.2}\right) + 0\right]} = 2.5838 \text{ STB}/(\text{D}-\text{psia}), \quad\dotfill (7.325)$$

Eq. 7.321 becomes

$$\frac{p_1 - 2p_1 + p_2}{(100)^2} + \frac{p_3 - 2p_1 + p_1}{(50)^2} - \frac{2.5838(1.1)(1)p_1}{(100)(50)(0.001127)(35)} + \frac{2.5838(1.1)(1)(100)}{(100)(50)(0.001127)(35)} = 0, \quad\dots (7.326)$$

which simplifies to

$$-19{,}705.377p_1 + 2{,}500p_2 + 10{,}000p_3 = -720{,}537.7. \quad\dotfill (7.327)$$

   - Block 2:

$$\frac{p_1 - 2p_2 + p_2}{(100)^2} + \frac{3{,}000 - 2p_2 + p_2}{(50)^2} = 0, \quad\dotfill (7.328)$$

$$2{,}500p_1 - 12{,}500p_2 = -30{,}000{,}000. \quad\dotfill (7.329)$$

   - Block 3:

$$\Omega_3 = \Omega_1,$$

$$\frac{p_3 - 2p_3 + 3000}{(100)^2} + \frac{p_3 - 2p_3 + p_1}{(50)^2} - \frac{2.5838(1.1)(1)p_3}{(100)(50)(0.001127)(35)} + \frac{2.5838(1.1)(1)(100)}{(100)(50)(0.001127)(35)} = 0, \quad\dots (7.330)$$

$$-19{,}705.377p_3 + 10{,}000p_1 = -8{,}220{,}538. \quad\dotfill (7.331)$$

Rearranging the system of equations in a matrix form yields

$$\begin{bmatrix} -19{,}705.377 & 2{,}500 & 10{,}000 \\ 2{,}500 & -12{,}500 & 0 \\ 10{,}000 & 0 & -19{,}705.377 \end{bmatrix} \begin{bmatrix} p_1 \\ p_2 \\ p_3 \end{bmatrix} = \begin{bmatrix} -720{,}537.7 \\ -30{,}000{,}000 \\ -8{,}220{,}538 \end{bmatrix}. \quad\dotfill (7.332)$$

2. **Table 7.26** shows the results obtained at each PSOR iteration level (note that $\omega_{opt} = 1.1$).

Iteration level	p_1	p_2	p_3	Δp_{max}
0	3,000	3,000	3,000	0
1	1,833.559	2,743.383	1,182.425	1,817.575
2	899.7784	2,563.613	842.9243	933.7809
3	778.5513	2,554.92	809.2026	121.2271
4	770.6366	2,554.048	808.1566	7.914685
5	770.7225	2,554.154	808.3091	0.152544
6	770.8139	2,554.164	808.3449	0.09137
7	770.826	2,554.165	808.3481	0.012144
8	770.8268	2,554.165	808.3482	0.000815

Table 7.26—Results obtained at each PSOR iteration level in Part 2 of the Solution to Problem 7.3.44.

3. Production rates at Wells 1 and 3 are as follows:

  • Flow rate from Well 1:

$$q_1 = -\Omega_1\left(p_1 - p_{sf1}\right) = 2.5838\left(770.8 - 100\right) = -1,733.21 \text{ STB/D}. \dots\dots\dots\dots\dots (7.333)$$

  • Flow rate from Well 3:

$$q_3 = -\Omega_3\left(p_3 - p_{sf3}\right) = 2.5838\left(808.3 - 100\right) = -1,830.11 \text{ STB/D}. \dots\dots\dots\dots\dots (7.334)$$

4. The MBC is as follows:

$$q_{out} = q_1 + q_3 = -1,733.21 - 1,830.11 = -3,563.32 \text{ STB/D}, \dots\dots\dots\dots\dots\dots\dots (7.335)$$

$$q_{in} = q_2' + q_3' = \frac{kA_y}{\mu B}\frac{p_4 - p_2}{\Delta y} + \frac{kA_x}{\mu B}\frac{p_4 - p_3}{\Delta x}. \dots\dots\dots\dots\dots\dots\dots\dots (7.336)$$

Thus,

$$q_{in} = \frac{0.001127(35)(50)(100)}{(1.1)(1)}\frac{(3,000 - 2,554.2)}{50} + \frac{0.001127(35)(50)(50)}{(1.1)(1)}\frac{(3,000 - 808.3)}{100} \dots\dots (7.337)$$
$$= 3523.41 \text{ STB/D}.$$

$$\text{MBC} = \left|\frac{\text{Total } q_{out}}{\text{Total } q_{in}}\right| = \left|\frac{-3,563.32}{3,563.41}\right| = 0.99997\cdot \dots\dots\dots\dots\dots\dots\dots\dots\dots\dots\dots (7.338)$$

Problem 7.3.45

Can you solve the system in **Fig. 7.41** using the LSOR technique and calling for the Thomas algorithm only twice during one complete LSOR iteration? Provide the structure of the coefficients matrix for each line. If you cannot solve it by making only two calls for the Thomas algorithm, explain why not.

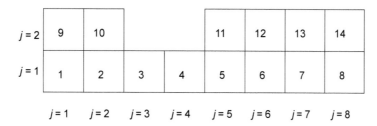

Fig. 7.41—Reservoir system for Problem 7.3.45.

Solution to Problem 7.3.45

You can apply Thomas' algorithm for the first row, because the structure of the resulting coefficient matrix is tridiagonal (the pressure at Blocks 9, 10, 13, 14, 15, and 16 are assumed to be one iteration level behind k).

For the second call for Thomas' algorithm to solve the unknowns of the second row, the solution will be determined by how one builds the coefficient matrix. One possibility is using the equations for the nodes that have active blocks. Another choice uses artificial equations for the inactive (non-existing) blocks. However, these superficial equations should not be coupled to the other equations.

The following matrix shows what the structure looks like when it is going to be solved using Thomas' algorithm:

- First call for Thomas' algorithm (for line $j = 1$); unknowns are p_1, p_2, p_3, p_4, p_5, p_6, p_7, and p_8: The structure of the coefficient matrix is

$$
\begin{bmatrix}
x & x & 0 & 0 & 0 & 0 & 0 & 0 \\
x & x & x & 0 & 0 & 0 & 0 & 0 \\
0 & x & x & x & 0 & 0 & 0 & 0 \\
0 & 0 & x & x & x & 0 & 0 & 0 \\
0 & 0 & 0 & x & x & x & 0 & 0 \\
0 & 0 & 0 & 0 & x & x & x & 0 \\
0 & 0 & 0 & 0 & 0 & x & x & x \\
0 & 0 & 0 & 0 & 0 & 0 & x & x
\end{bmatrix}_{8*8}
\begin{bmatrix}
p_1 \\ p_2 \\ p_3 \\ p_4 \\ p_5 \\ p_6 \\ p_7 \\ p_8
\end{bmatrix}
=
\begin{bmatrix}
c \\ c \\ c \\ c \\ c \\ c \\ c \\ c
\end{bmatrix}
. \qquad (7.339)
$$

The previous matrix has a tridiagonal structure.

- Second call for the Thomas' algorithm; unknowns are p_9, p_{10}, p_{13}, p_{14}, p_{15}, and p_{16}: The matrix is

$$
\begin{bmatrix}
x & x & 0 & 0 & 0 & 0 \\
x & x & 0 & 0 & 0 & 0 \\
0 & 0 & x & x & 0 & 0 \\
0 & 0 & x & x & x & 0 \\
0 & 0 & 0 & x & x & x \\
0 & 0 & 0 & 0 & x & x
\end{bmatrix}
\begin{bmatrix}
p_9 \\ p_{10} \\ p_{13} \\ p_{14} \\ p_{15} \\ p_{16}
\end{bmatrix}
=
\begin{bmatrix}
c \\ c \\ c \\ c \\ c \\ c
\end{bmatrix}
. \qquad (7.340)
$$

Again, the previous matrix has a tridiagonal structure.

The idea of introducing artificial equations is shown in the following matrix construction:

$$
\begin{bmatrix}
x & x & 0 & 0 & 0 & 0 & 0 & 0 \\
x & x & 0 & 0 & 0 & 0 & 0 & 0 \\
0 & 0 & 1 & 0 & 0 & 0 & 0 & 0 \\
0 & 0 & 0 & 1 & 0 & 0 & 0 & 0 \\
0 & 0 & 0 & 0 & x & x & 0 & 0 \\
0 & 0 & 0 & 0 & x & x & x & 0 \\
0 & 0 & 0 & 0 & 0 & x & x & x \\
0 & 0 & 0 & 0 & 0 & 0 & x & x
\end{bmatrix}
\begin{bmatrix}
p_9 \\ p_{10} \\ p_{art11} \\ p_{art12} \\ p_{13} \\ p_{14} \\ p_{15} \\ p_{16}
\end{bmatrix}
=
\begin{bmatrix}
c \\ c \\ c \\ c \\ c \\ c \\ c \\ c
\end{bmatrix}
. \qquad (7.341)
$$

Note that the artificial equation written for Blocks 11 and 12 have only zeros on the off-diagonal entries. This guarantees that they are not coupled to the other equations.

Another possible option is to solve the row $j = 2$ in two parts, but this will require a total of three calls for Thomas' algorithm:

- Solve for p_9 and p_{10}.
- Solve for p_{13}, p_{14}, p_{15}, and p_{16}.

Nomenclature

a_i = i^{th} element of vector a

$a_{i,j}$ = element (i,j) of matrix $[A]$

A = cross-sectional area perpendicular to the flow direction, L^2, ft^3

$[A]$ = coefficient matrix of a matrix equation

$[A']$ = upper triangular coefficient matrix

$A_{i,j,k}$ = transmissibility coefficient using SIP notation between block (i, j, k) and $(i, j, k + 1)$, L^4t/m, STB/(D-psia) for liquid, scf/(D-psia) for gas

b = constant

$[b]$ = right-hand-side vector of a matrix equation

b_i = i^{th} element of vector b

$b_{i,j}$ = element (i, j) of vector b in SIP factorization algorithm

B = formation volume factor, L^3/L^3, res bbl/STB for liquid and res bbl/scf for gas

$B_{i,j,k}$ = transmissibility coefficient using SIP notation between block (i, j, k) and $(i, j, k - 1)$, L^4t/m, STB/(D-psia) for liquid, scf/(D-psia) for gas

c = compressibility, Lt^2/m, psi^{-1}

c_i = i^{th} element of vector c

$c_{i,j}$ = element (i, j) of vector c in SIP factorization algorithm

c_t = total compressibility, Lt^2/m, psi^{-1}

$C_{i,j,k}$ = transmissibility coefficient using SIP notation of central block (i, j, k), L^4t/m, STB/(D-psia) for liquid, scf/(D-psia) for gas

$[d]$ = right-hand-side vector of a matrix equation

d_{mn} = intermediate variable defined in the analytical solution of temperature distribution in Problem 7.3.18

$d_{i,}$ = i^{th} element of vector d

$d_{i,j}$ = element (i, j) of vector d in SIP factorization algorithm

D = a constant defined in Problem 7.3.22

$e_{i,j}$ = element (i, j) of vector e in SIP factorization algorithm

$E_{i,j,k}$ = transmissibility coefficient using SIP notation between block (i, j, k) and $(i + 1, j, k)$, L^4t/m, STB/(D-psia) for liquid, scf/(D-psia) for gas

E_n = intermediate variable defined in the analytical solution of temperature distribution in Problem 7.3.17

$f_{i,j}$ = element (i, j) of vector f in SIP factorization algorithm

g = local gravitational acceleration, L/t^2, ft/s^2

g_c = gravitational acceleration on the Equator, L/t^2, ft/s^2

h = thickness, L, ft

I = identity matrix

k = permeability, L^2, darcies or perms (1 darcy = 1.127 perms)

k_x = x-direction permeability, L^2, darcies or perms

k_y = y-direction permeability, L^2, darcies or perms

k_z = z-direction permeability, L^2, darcies or perms

$l_{i,j}$ = element (i,j) of matrix $[L]$

L = the length of the wellbore in the wellblock, L, ft

$[L]$ = lower triangular matrix

$[L']$ = lower triangular matrix approximating to $[A']$

m = mass per unit volume of porous media, m/L^3, lbm/ft^3

M = preconditioner matrix

$N_{i,j,k}$ = transmissibility coefficient using SIP notation between block (i, j, k) and $(i, j + 1, k)$, L^4t/m, STB/(D-psia) for liquid, scf/(D-psia) for gas

N_x = number of gridblocks along x-direction

N_y = number of gridblocks along y-direction

N_z = number of gridblocks along z-direction

p = pressure, m/Lt^2, psia

p_{art} = image (artificial node) pressure, m/Lt^2, psia

p_B = specified pressure on the boundary, m/Lt^2, psia

p_{c1} = specified constant pressure boundary (Fig. 7.22), m/Lt^2, psia

p_{c2} = specified constant pressure boundary (Fig. 7.22), m/Lt^2, psia

p_i = initial pressure, m/Lt^2, psia

$p_{i,j,k}$ = pressure of block i,j,k, m/Lt^2, psia

p_k = intermediate variable of conjugate gradient algorithm at iteration level k

p_{sc} = standard condition pressure, m/Lt^2, psia

p_{sf} = sandface pressure, m/Lt^2, psia

p_{sp} = specified block pressure, m/Lt^2, psia

p_w = specified wellblock pressure, m/Lt^2, psia

$[P]$ = unknown vector

q = flow rate, L^3/t, STB/D for liquid and scf/D for gas

q_g = gas flow rate, L^3/t, scf/D

q_{in} = flow rate entering the system, L^3/t, STB/D for liquid and scf/D for gas

q_{out}	=	flow rate exiting the system, L^3/t, STB/D for liquid and scf/D for gas
$q_{production}$	=	production flow rate, L^3/t, STB/D for liquid and scf/D for gas
$q_{injection}$	=	injection flow rate, L^3/t, STB/D for liquid and scf/D for gas
q_{sc}	=	well flow rate at standard conditions, L^3/t, STB/D for liquid and scf/D for gas
q_{spec}	=	specified flow rate, L^3/t, STB/D for liquid and scf/D for gas
q_{well}	=	production/injection rate of the well, L^3/t, STB/D for liquid and scf/D for gas
$Q_{i,j,k}$	=	right-hand-side coefficient of governing flow equation using SIP notation for block (i, j, k) L^3/t, STB/D for liquid, scf/D for gas
$[Q]$	=	right-hand-side vector
$[Q']$	=	transformed right-hand-side vector in Eq. 7.2
r	=	radial direction in both cylindrical and spherical coordinate systems, L, ft
$[r]$	=	intermediate vector used to decompose $[A]$
r_e	=	Peaceman equivalent wellblock radius, L, ft
r_k	=	residual vector of iteration level k
r_w	=	wellblock radius, L, ft
$R_{i,j}$	=	residual, L^3/t, STB/D for liquid, scf/D for gas
s	=	skin factor, dimensionless, also indicates the maximum iteration number with in a SIP cycle (Eq. 7.56)
s_k	=	intermediate variables in preconditioned biconjugate gradient-stabilized algorithms at iteration level k
$S_{i,j,k}$	=	transmissibility coefficient using SIP notation between block (i, j, k) and $(i, j - 1, k)$, L^4t/m, STB/(D-psia) for liquid, scf/(D-psia) for gas
t	=	time, t, days
T	=	absolute temperature, T, °R
$T_{,}$	=	transmissibility factor between adjacent blocks L^4t/m, STB/(D-psia) for liquid, scf/(D-psia) for gas, also the temperature at node (i, j)
T_n	=	intermediate variable defined in the analytical solution of temperature distribution in Problem 7.3.17
T_{sc}	=	standard condition temperature, T, °R
T_x	=	transmissibility factor along x-direction, L^4t/m, STB/(D-psia) for liquid, scf/(D-psia) for gas
T_y	=	transmissibility factor along y-direction, L^4t/m, STB/(D-psia) for liquid, scf/(D-psia) for gas
T'	=	modified transmissibility term

t_k	=	intermediate variables in preconditioned biconjugate gradient-stabilized algorithms at iteration level k
u	=	an arbitrary function
$u_{i,j}$	=	element (i,j) of matrix $[U]$
$u_{i,j,k}$	=	dependent variable used to describe the problem in Eq. 7.183
$[U]$	=	upper triangular matrix
$[U']$	=	upper triangular matrix approximating to $[A']$
V_b	=	bulk volume, L^3, ft^3
v_k	=	intermediate variables in preconditioned biconjugate gradient-stabilized algorithms at iteration level k
$v_{i,j}$	=	element (i,j) of vector v in SIP iteration
$W_{i,j,k}$	=	transmissibility coefficient using SIP notation between block (i, j, k) and $(i - 1, j, k)$, L^4t/m, STB/(D-psia) for liquid, scf/(D-psia) for gas
x	=	distance in x-direction in the Cartesian coordinate system, L, ft
x_i	=	i^{th} vector of intermediate vector in Crout reduction
$[x]$	=	unknown vector of a matrix function
y	=	distance in y-direction in the Cartesian coordinate system
y_i	=	i^{th} vector of intermediate vector in Crout reduction
z	=	distance in z-direction in the Cartesian coordinate system
z_g	=	gas compressibility factor, dimensionless
z_k	=	intermediate variables in conjugate gradient algorithms at iteration level k
α	=	iteration parameter
α_k	=	cyclic acceleration parameter of SIP iteration, also intermediate variables in conjugate gradient algorithms at iteration level k
α_m	=	acceleration parameter for ADIP iterations
α_{max}	=	upper limitation of the cyclic acceleration parameter of SIP iteration
α_{min}	=	lower limitation of the cyclic acceleration parameter of SIP iteration
α_{mn}	=	intermediate variable defined in the analytical solution of temperature distribution in Problem 7.3.18
β	=	ratio of the acceleration parameter in SIP iteration
β_k	=	intermediate variable of conjugate gradient algorithm at iteration level k
Γ	=	accumulation coefficient, L^3/t, ft^3/D
Δ	=	difference of a variable

Δx	=	control volume dimension along the x-direction, L, ft
Δy	=	control volume dimension along the y-direction, L, ft
Δz	=	control volume dimension along the z-direction, L, ft
δ	=	magnitude of the improvement on an unknown between two successive SIP iterations.
ε_p	=	convergence tolerance
μ	=	viscosity, m/Lt, cp
μ_g	=	gas viscosity, m/Lt, cp
ρ	=	density, m/L^3, lbm/ft^3, also the right-hand-side term of Poisson's equation
ρ_{GS}	=	spectral radius
ρ_k	=	intermediate variable of conjugate gradient algorithm at iteration level k
ϕ	=	porosity, fraction
Ω	=	well productivity/injectivity index, L^{4}t/m, STB/(D-psia) for liquid, scf/(D-psia) for gas
ω_k	=	intermediate variables in preconditioned biconjugate gradient-stabilized algorithms at iteration level k.
ω	=	over-relaxation (acceleration) parameter
ω_{opt}	=	optimum over-relaxation (acceleration) parameter

Superscripts

(k)	=	old iteration level
$(k+1)$	=	current iteration level
(m)	=	old iteration number
$(m+1)$	=	new iteration number
n	=	old timestep level
$n+1$	=	current timestep level
T	=	transpose of a matrix or a vector
0	=	reference condition
$*$	=	intermediate value in ADIP iteration

Subscripts

i	=	pertaining to central block i in a 1D representation (x-direction only)
j	=	pertaining to central block j in a 1D representation (y-direction only)
k	=	pertaining to central block k in a 1D representation (z-direction only)
i,j	=	pertaining to central block i,j in a 2D representation (x- and y-directions only)
i,j,k	=	pertaining to central block i,j,k in a 3D representation
x	=	indicator for the x-direction component
y	=	indicator for the y-direction component
z	=	indicator for the z-direction component

References

Barrett, R., Berry, M., Chan, T. F. et al. 1994. *Templates for the Solution of Linear Systems: Building Blocks for Iterative Methods*, 1st ed., Philadelphia, Pennsylvania: Society for Industrial and Applied Mathematics. 1994. http://dx.doi.org/10.2118/9781611971538.

Ertekin, T., Abou-Kassem, J. H., and King, G. R. 2001. *Basic Applied Reservoir Simulation*. Richardson, Texas, US: Society of Petroleum Engineers.

Hestenes, M.R. and Stiefel, E. 1952. Method of Conjugate Gradients for Solving Linear Systems. *J. of Research of the Natl. Bureau of Standards* **49**(06): 409. Paper No. 2379.

Knyazev, A.V. 1991. A Preconditioned Conjugate Gradient Method for Eigenvalue Problems and its Implementation in a Subspace. In *Numerical Treatment of Eigenvalue Problems*, J. Albrecht, L. Collatz, P. Hagedorn, and W. Velte, eds., Vol. 5, Basel: Birkhäuser. https://dx.doi.org/10.1007/978-3-0348-6332-2_11.

Price, H. S. and Coats, K. H. 1974. Direct Methods in Reservoir Simulation. *SPE J.* **June:** 295–306. SPE-4278-PA. http://dx.doi.org/10.2118/4278-PA.

Stone, H. L. 1968. Iterative Solution of Implicit Approximations of Multidimensional Partial Differential Equations. *SIAM J. of Numerical Analysis* **5**(03): 530–558. http://dx.doi.org/10.1137/0705044.

van der Vorst, H. A. 1992. Bi-CGSTAB: A Fast and Smoothly Converging Variant of Bi-CG for the Solution of Nonsymmetric Linear Systems. *SIAM J. Sci. Stat. Comput.* **13**(2): 631–644. http://dx.doi.org/10.1137/0913035.

Chapter 8

Numerical Solution of Single-Phase Flow Equations

In previous chapters, we developed the partial-differential equation (PDE) of flow in porous media for single-phase flow conditions. We learned how to represent the same equations in finite-difference form. We also learned how to generate a system of linear algebraic equations, as well as some direct- and iterative-solution algorithms to solve the systems of linear algebraic equations. In this chapter, we will learn how to calculate transmissibility terms, which in one form or another, constitute the coefficients of the systems of equations we solve to obtain the solution we are after.

8.1 Single-Phase Incompressible Flow

The general form of the PDE describing the single-phase incompressible flow, which includes gravitational forces, can be written as follows:

$$\frac{\partial}{\partial x}\left(\frac{A_x k_x}{\mu B}\frac{\partial p}{\partial x}\right)\Delta x + \frac{\partial}{\partial y}\left(\frac{A_y k_y}{\mu B}\frac{\partial p}{\partial y}\right)\Delta y + \frac{\partial}{\partial z}\left(\frac{A_z k_z}{\mu B}\frac{\partial p}{\partial z}\right)\Delta z$$

$$-\left[\frac{\partial}{\partial x}\left(\frac{1}{144}\frac{g}{g_c}\rho\frac{A_x k_x}{\mu B}\frac{\partial G}{\partial x}\right)\Delta x + \frac{\partial}{\partial y}\left(\frac{1}{144}\frac{g}{g_c}\rho\frac{A_y k_y}{\mu B}\frac{\partial G}{\partial y}\right)\Delta y + \frac{\partial}{\partial z}\left(\frac{1}{144}\frac{g}{g_c}\rho\frac{A_z k_z}{\mu B}\frac{\partial G}{\partial y}\right)\Delta z\right] + q_{\text{STB/D}} = 0 \quad \dots\dots\dots (8.1)$$

The string of terms that appears in the rectangular brackets represents the gravitational forces acting along the x-, y-, and z-directions; they are presented to the model as part of the data. In other words, they are known entries in Eq. 8.1, and they need to be carried to the right-hand side of the equation. The finite-difference representation of Eq. 8.1 then becomes

$$\left.\frac{A_x k_x}{\mu B\Delta x}\right|_{i+\frac{1}{2},j,k}\left(p_{i+1,j,k}-p_{i,j,k}\right)-\left.\frac{A_x k_x}{\mu B\Delta x}\right|_{i-\frac{1}{2},j,k}\left(p_{i,j,k}-p_{i-1,j,k}\right)$$

$$+\left.\frac{A_y k_y}{\mu B\Delta y}\right|_{i,j+\frac{1}{2},k}\left(p_{i,j+1,k}-p_{i,j,k}\right)-\left.\frac{A_y k_y}{\mu B\Delta y}\right|_{i,j-\frac{1}{2},k}\left(p_{i,j,k}-p_{i,j-1,k}\right)$$

$$+\left.\frac{A_z k_z}{\mu B\Delta z}\right|_{i,j,k+\frac{1}{2}}\left(p_{i,j,k+1}-p_{i,j,k}\right)-\left.\frac{A_z k_z}{\mu B\Delta z}\right|_{i,j,k-\frac{1}{2}}\left(p_{i,j,k}-p_{i,j,k-1}\right)+q_{i,j,k}$$

$$=\begin{bmatrix}\left.\dfrac{1}{144}\dfrac{g}{g_c}\rho\dfrac{A_x k_x}{\mu B\Delta x}\right|_{i+\frac{1}{2},j,k}\left(G_{i+1,j,k}-G_{i,j,k}\right)-\left.\dfrac{1}{144}\dfrac{g}{g_c}\rho\dfrac{A_x k_x}{\mu B\Delta x}\right|_{i-\frac{1}{2},j,k}\left(G_{i,j,k}-G_{i-1,j,k}\right) \\[2mm] +\left.\dfrac{1}{144}\dfrac{g}{g_c}\rho\dfrac{A_y k_y}{\mu B\Delta y}\right|_{i,j+\frac{1}{2},k}\left(G_{i,j+1,k}-G_{i,j,k}\right)-\left.\dfrac{1}{144}\dfrac{g}{g_c}\rho\dfrac{A_y k_y}{\mu B\Delta y}\right|_{i,j-\frac{1}{2},k}\left(G_{i,j,k}-G_{i,j-1,k}\right) \\[2mm] +\left.\dfrac{1}{144}\dfrac{g}{g_c}\rho\dfrac{A_z k_z}{\mu B\Delta z}\right|_{i,j,k+\frac{1}{2}}\left(G_{i,j,k+1}-G_{i,j,k}\right)-\left.\dfrac{1}{144}\dfrac{g}{g_c}\rho\dfrac{A_z k_z}{\mu B\Delta z}\right|_{i,j,k-\frac{1}{2}}\left(G_{i,j,k}-G_{i,j,k-1}\right)\end{bmatrix} \quad \dots\dots\dots\dots\dots (8.2)$$

The transmissibility term, $\left.\dfrac{A_x k_x}{\mu B \Delta x}\right|_{i+\frac{1}{2},j,k}$, can then be described as

$$T_{ix_{i+\frac{1}{2},j,k}} = \left.\frac{A_x k_x}{\mu B \Delta x}\right|_{i+\frac{1}{2},j,k} \quad \dots\dots\dots\dots\dots\dots\dots\dots\dots\dots\dots\dots\dots\dots\dots\dots\dots\dots\dots\dots\dots\dots\dots\dots\dots\dots\dots\dots\dots \text{(8.3)}$$

All similar terms in Eq. 8.2 are known as transmissibility terms, which scale the rate of flow between two neighboring blocks (in this case blocks i, j, k, and $i+1, j, k$) and they need to be calculated at the interface of these two blocks.

For an incompressible fluid, terms μ and B are not functions of pressure; therefore, even if the block pressure of blocks i, j, k, and $i+1, j, k$ are different, they will display the same viscosity and formation volume factors (FVFs). Then the question that remains to be answered is: What kind of averaging technique needs to be used at the interface of these two neighboring blocks of i, j, k, and $i+1, j, k$ for the other remaining entries of the transmissibility term? There are three different averaging techniques that we can consider:

- Arithmetic averaging:

$$A_{\text{avg}} = \frac{a_1 + a_2}{2}. \quad \dots\dots\dots\dots\dots\dots\dots\dots\dots\dots\dots\dots\dots\dots\dots\dots\dots\dots\dots\dots\dots\dots\dots \text{(8.4a)}$$

- Geometric Averaging:

$$G_{\text{avg}} = (a_1 a_2)^{\frac{1}{2}}. \quad \dots\dots\dots\dots\dots\dots\dots\dots\dots\dots\dots\dots\dots\dots\dots\dots\dots\dots\dots\dots\dots\dots\dots \text{(8.4b)}$$

- Harmonic Averaging:

$$\frac{1}{H_{\text{avg}}} = \frac{1}{2}\left(\frac{1}{a_1} + \frac{1}{a_2}\right). \quad \dots\dots\dots\dots\dots\dots\dots\dots\dots\dots\dots\dots\dots\dots\dots\dots\dots\dots\dots\dots \text{(8.4c)}$$

Recall that $A_{\text{avg}} \geq G_{\text{avg}} \geq H_{\text{avg}}$ (the equality signs for the relationships between A_{avg}, G_{avg}, and H_{avg} hold true only when $a_1 = a_2$).

In calculating the term, $\dfrac{A_x k_x}{\Delta x}$ (or any other similar terms along the x-, y-, and z-directions), harmonic averaging is used. This is not an arbitrary choice, but is the result of studying the serial flow that occurs between two adjacent blocks. If we let

$$H_{\text{avg}} = \left(\frac{A_x k_x}{\Delta x}\right)_{i+\frac{1}{2},j,k}, \quad \dots\dots\dots\dots\dots\dots\dots\dots\dots\dots\dots\dots\dots\dots\dots\dots\dots\dots\dots\dots\dots\dots\dots \text{(8.5a)}$$

then we can write

$$\frac{1}{\left(\dfrac{A_x k_x}{\Delta x}\right)_{i+\frac{1}{2},j,k}} = \frac{1}{2}\left[\frac{1}{\left(\dfrac{A_x k_x}{\Delta x}\right)_{i,j,k}} + \frac{1}{\left(\dfrac{A_x k_x}{\Delta x}\right)_{i+1,j,k}}\right]. \quad \dots\dots\dots\dots\dots\dots\dots\dots\dots \text{(8.5b)}$$

Solving Eq. 8.5b for $\left(\dfrac{A_x k_x}{\Delta x}\right)_{i+\frac{1}{2},j,k}$ yields,

$$\left(\frac{A_x k_x}{\Delta x}\right)_{i+\frac{1}{2},j,k} = \frac{2 A_{x_{i,j,k}} A_{x_{i+1,j,k}} k_{x_{i,j,k}} k_{x_{i+1,j,k}}}{A_{x_{i,j,k}} k_{x_{i,j,k}} \Delta x_{i+1,j,k} + A_{x_{i+1,j,k}} k_{x_{i+1,j,k}} \Delta x_{i,j,k}}. \quad \dots\dots\dots\dots\dots\dots\dots \text{(8.5c)}$$

Similar to Eq. 8.5c, other similar terms can be calculated; for example,

$$\left(\frac{A_z k_z}{\Delta z}\right)_{i,j,k-\frac{1}{2}} = \frac{2 A_{z_{i,j,k}} A_{z_{i,j,k-1}} k_{z_{i,j,k}} k_{z_{i,j,k-1}}}{A_{z_{i,j,k}} k_{z_{i,j,k}} \Delta z_{i,j,k-1} + A_{z_{i,j,k-1}} k_{z_{i,j,k-1}} \Delta z_{i,j,k}}. \quad \dots\dots\dots\dots\dots\dots\dots \text{(8.5d)}$$

The final form of the transmissibility term between blocks i, j, k, and $i+1, j, k$ (including the constant μ and B terms) can be written as

$$T_{ix_{i+\frac{1}{2},j,k}} = \frac{1}{\mu B} \frac{2 A_{x_{i,j,k}} A_{x_{i+1,j,k}} k_{x_{i,j,k}} k_{x_{i+1,j,k}}}{A_{x_{i,j,k}} k_{x_{i,j,k}} \Delta x_{i+1,j,k} + A_{x_{i+1,j,k}} k_{x_{i+1,j,k}} \Delta x_{i,j,k}}. \quad \dots\dots\dots\dots\dots\dots\dots \text{(8.6)}$$

Before we further proceed to write the finite-difference approximation using the strongly implicit procedure (SIP) notation, let us focus on the modified transmissibility terms that appear on the right-hand side of Eq. 8.2; for example,

$$T'_{ix_{i+\frac{1}{2},j,k}} = \frac{1}{144}\frac{g}{g_c}\rho\left.\frac{A_x k_x}{\mu B \Delta x}\right|_{i+\frac{1}{2},j,k}. \dotfill (8.7a)$$

In Eq. 8.7a, for most practical purposes $\frac{g}{g_c} \approx 1.0$ (it is reasonable to assume that local gravitational acceleration is not that much different from the gravitational acceleration measured on the equator). The term ρ is constant, similar to μ and B terms for an incompressible fluid. Eq. 8.7a then becomes

$$T'_{ix_{i+\frac{1}{2},j,k}} = \left(\frac{1}{144}\rho\right)T_{ix_{i+\frac{1}{2},j,k}}. \dotfill (8.7b)$$

Eq. 8.2 can be written using SIP notation as follows:

$$B_{i,j,k}p_{i,j,k} + S_{i,j,k}p_{i,j,k} + W_{i,j,k}p_{i,j,k} + C_{i,j,k}p_{i,j,k} + E_{i,j,k}p_{i,j,k} + N_{i,j,k}p_{i,j,k} + A_{i,j,k}p_{i,j,k} = Q_{i,j,k}. \dotfill (8.8)$$

Note that Eq. 8.8 is the same as Eq. 8.2. In view of the equivalency of these two equations, one can easily obtain the definitions of the coefficients of Eq. 8.8, as follows:

- For a block that does not have a well ($q_{i,j,k} = 0$):

$$B_{i,j,k} = T_{i,j,k-\frac{1}{2}}, \dotfill (8.9a)$$

$$S_{i,j,k} = T_{i,j-\frac{1}{2},k}, \dotfill (8.9b)$$

$$W_{i,j,k} = T_{i-\frac{1}{2},j,k}, \dotfill (8.9c)$$

$$E_{i,j,k} = T_{i+\frac{1}{2},j,k}, \dotfill (8.9d)$$

$$N_{i,j,k} = T_{i,j+\frac{1}{2},k}, \dotfill (8.9e)$$

$$A_{i,j,k} = T_{i,j,k+\frac{1}{2}}, \dotfill (8.9f)$$

$$C_{i,j,k} = -\left(B_{i,j,k} + S_{i,j,k} + W_{i,j,k} + E_{i,j,k} + N_{i,j,k} + A_{i,j,k}\right), \dotfill (8.9g)$$

$$Q_{i,j,k} = \begin{bmatrix} T'_{ix_{i,j,k-\frac{1}{2}}}G_{i,j,k-1} + T'_{ix_{i,j-\frac{1}{2},k}}G_{i,j-1,k} + T'_{ix_{i-\frac{1}{2},j,k}}G_{i-1,j,k} + T'_{ix_{i+\frac{1}{2},j,k}}G_{i+1,j,k} + T'_{ix_{i,j+\frac{1}{2},k}}G_{i,j+1,k} \\ + T'_{ix_{i,j,k+\frac{1}{2}}}G_{i,j,k+1} - \left(T'_{ix_{i,j,k-\frac{1}{2}}} + T'_{ix_{i,j-\frac{1}{2},k}} + T'_{ix_{i-\frac{1}{2},j,k}} + T'_{ix_{i+\frac{1}{2},j,k}} + T'_{ix_{i,j+\frac{1}{2},k}} + T'_{ix_{i,j,k+\frac{1}{2}}}\right)G_{i,j,k} \end{bmatrix}. \dotfill (8.9h)$$

- For a block that hosts a well with a flow rate specification ($q_{i,j,k} = q_{\text{spec}}$): All of the $B_{i,j,k}$, $S_{i,j,k}$, $W_{i,j,k}$, $E_{i,j,k}$, $N_{i,j,k}$, $A_{i,j,k}$, and $C_{i,j,k}$ will have the same definitions as those given in Eqs. 8.9a through 8.9g. The definition of $Q_{i,j,k}$ will be slightly modified from how it appears in Eq. 8.9h, such that

$$Q_{i,j,k} = -q_{\text{spec}} + \begin{bmatrix} T'_{ix_{i,j,k-\frac{1}{2}}}G_{i,j,k-1} + T'_{ix_{i,j-\frac{1}{2},k}}G_{i,j-1,k} + T'_{ix_{i-\frac{1}{2},j,k}}G_{i-1,j,k} + T'_{ix_{i+\frac{1}{2},j,k}}G_{i+1,j,k} + T'_{ix_{i,j+\frac{1}{2},k}}G_{i,j+1,k} \\ + T'_{ix_{i,j,k+\frac{1}{2}}}G_{i,j,k+1} - \left(T'_{ix_{i,j,k-\frac{1}{2}}} + T'_{ix_{i,j-\frac{1}{2},k}} + T'_{ix_{i-\frac{1}{2},j,k}} + T'_{ix_{i+\frac{1}{2},j,k}} + T'_{ix_{i,j+\frac{1}{2},k}} + T'_{ix_{i,j,k+\frac{1}{2}}}\right)G_{i,j,k} \end{bmatrix}. \dotfill (8.9i)$$

Remember that if the well is a producer, the specified rate is a negative number ($q_{\text{spec}} < 0$), and if it is an injector, the specified rate is treated as a positive number ($q_{\text{spec}} > 0$), according to the sign convention we have adopted in this book.

- For a block that hosts a well with a specification of flowing sandface pressure $\left(p_{sf_{i,j,k}} = p_{sf_{spec}}\right)$: Again, the transmissibility coefficients $B_{i,j,k}$, $S_{i,j,k}$, $W_{i,j,k}$, $E_{i,j,k}$, $N_{i,j,k}$, and $A_{i,j,k}$ will remain the same as described in Eqs. 8.9a through 8.9f. However, $C_{i,j,k}$ and $Q_{i,j,k}$ equations need to be revised by incorporating the sandface-pressure specification. Recall that the wellbore equation we are using is

$$q_{i,j,k_{STB/D}} = -\Omega_{i,j,k}\left(p_{i,j,k} - p_{sf_{i,j,k}}\right). \quad\dotfill (8.10)$$

Eq. 8.10 must be substituted into Eq. 8.2 to replace the $q_{i,j,k_{STB/D}}$ term. Then the definition of $C_{i,j,k}$ will take the following form:

$$C_{i,j,k} = -\left(B_{i,j,k} + S_{i,j,k} + W_{i,j,k} + E_{i,j,k} + N_{i,j,k} + A_{i,j,k} + \Omega_{i,j,k}\right). \quad\dotfill (8.11)$$

In addition, the $Q_{i,j,k}$ term will have a new description:

$$Q_{i,j,k} = -\Omega_{i,j,k} \, p_{sf_{spec_{i,j,k}}} + \begin{bmatrix} T'_{ix_{i,j,k-\frac{1}{2}}}G_{i,j,k-1} + T'_{ix_{i,j-\frac{1}{2},k}}G_{i,j-1,k} + T'_{ix_{i-\frac{1}{2},j,k}}G_{i-1,j,k} + T'_{ix_{i+\frac{1}{2},j,k}}G_{i+1,j,k} + T'_{ix_{i,j+\frac{1}{2},k}}G_{i,j+1,k} \\ + T'_{ix_{i,j,k+\frac{1}{2}}}G_{i,j,k+1} - \left(T'_{ix_{i,j,k-\frac{1}{2}}} + T'_{ix_{i,j-\frac{1}{2},k}} + T'_{ix_{i-\frac{1}{2},j,k}} + T'_{ix_{i+\frac{1}{2},j,k}} + T'_{ix_{i,j+\frac{1}{2},k}} + T'_{ix_{i,j,k+\frac{1}{2}}}\right)G_{i,j,k} \end{bmatrix}. \quad\dots (8.12)$$

Once all these coefficients are calculated, then a system of linear algebraic equations can be developed. This system of equations can be solved using one of the solution techniques discussed in Chapter 7 of this book. For flow-rate-specified wells, recall that $p_{sf_{i,j,k}}$ is the remaining unknown (after all block pressures are calculated), and it can be determined from

$$p_{sf_{i,j,k}} = \frac{q_{spec} + \Omega_{i,j,k}\,p_{i,j,k}}{\Omega_{i,j,k}}. \quad\dotfill (8.13)$$

On the other hand, if sandface pressure is specified at the wellblock, this time the remaining unknown in the block will be the flow rate $q_{i,j,k}$. Again, once all block pressures are obtained, then the flow rate $q_{i,j,k}$ at the sandface-pressure-specified well can be directly obtained using the wellbore equation (Eq. 8.10):

$$q_{i,j,k_{STB/D}} = -\Omega_{i,j,k}\left(p_{i,j,k} - p_{sf_{i,j,k}}\right). \quad\dotfill (8.14)$$

8.1.1 Material-Balance Check. In an incompressible fluid flow problem, mass content of the reservoir does not change. In other words,

$$\Sigma\,\text{flow in} = \Sigma\,\text{flow out.} \quad\dotfill (8.15)$$

In Eq. 8.15, under the summation terms, all the flow in and out of the system across all the external and internal boundaries (physical boundaries and wells) of the reservoir must be accounted for. This process is known as a material-balance check (MBC). The MBC is conducted to find the accuracy level of the solution over the entire reservoir. Another check conducted on a block-by-block basis is known as the residual check and is expressed as follows:

$$R_{i,j,k} = Q_{i,j,k} - \left(\begin{array}{l} B_{i,j,k}p_{i,j,k-1} + S_{i,j,k}p_{i,j-1,k} + W_{i,j,k}p_{i-1,j,k} + C_{i,j,k}p_{i,j,k} \\ + E_{i,j,k}p_{i+1,j,k} + N_{i,j,k}p_{i,j+1,k} + A_{i,j,k}p_{i,j,k+1} \end{array}\right). \quad\dotfill (8.16)$$

The residual check provides a good indication of the precision of the block pressures calculated using an iterative-solution procedure. In other words, an MBC certifies whether the pressure-convergence criterion used in the solver was fine enough; a coarse criterion would yield a coarse pressure distribution, and hence, coarse flow rates and subsequently an unsatisfactory MBC.

8.2 Single-Phase Slightly Compressible Flow
The general form of the governing PDE that describes the slightly compressible flow is as follows:

$$\frac{\partial}{\partial x}\left(\frac{A_x k_x}{\mu B}\frac{\partial p}{\partial x}\right)\Delta x + \frac{\partial}{\partial y}\left(\frac{A_y k_y}{\mu B}\frac{\partial p}{\partial y}\right)\Delta y + \frac{\partial}{\partial z}\left(\frac{A_z k_z}{\mu B}\frac{\partial p}{\partial z}\right)\Delta z$$

$$-\left[\frac{\partial}{\partial x}\left(\frac{1}{144}\frac{g}{g_c}\rho\frac{A_x k_x}{\mu B}\frac{\partial G}{\partial x}\right)\Delta x + \frac{\partial}{\partial y}\left(\frac{1}{144}\frac{g}{g_c}\rho\frac{A_y k_y}{\mu B}\frac{\partial G}{\partial y}\right)\Delta y + \frac{\partial}{\partial z}\left(\frac{1}{144}\frac{g}{g_c}\rho\frac{A_z k_z}{\mu B}\frac{\partial G}{\partial z}\right)\Delta z\right]. \quad\dotfill (8.17)$$

$$+q_{STB/D} = \frac{V_b \phi c}{5.615}\frac{\partial p}{\partial t}$$

The slightly compressible flow equation inherently exhibits rather weak nonlinearities. These weak nonlinearities are caused by pressure dependencies of properties, such as μ, B, ρ, and c. Because these properties are weak functions of pressure, we typically can let them present one timestep behind. Accordingly, if the superscript n represents the old timestep, then the superscript $n+1$ will represent the current timestep. And then, the finite-difference representation of Eq. 8.17 takes the following form:

$$
\left.\frac{A_x k_x}{\mu B \Delta x}\right|^n_{i+\frac{1}{2},j,k}\left(p^{n+1}_{i+1,j,k}-p^{n+1}_{i,j,k}\right)-\left.\frac{A_x k_x}{\mu B \Delta x}\right|^n_{i-\frac{1}{2},j,k}\left(p^{n+1}_{i,j,k}-p^{n+1}_{i-1,j,k}\right)
$$

$$
+\left.\frac{A_y k_y}{\mu B \Delta y}\right|^n_{i,j+\frac{1}{2},k}\left(p^{n+1}_{i,j+1,k}-p^{n+1}_{i,j,k}\right)-\left.\frac{A_y k_y}{\mu B \Delta y}\right|^n_{i,j-\frac{1}{2},k}\left(p^{n+1}_{i,j,k}-p^{n+1}_{i,j-1,k}\right)
$$

$$
+\left.\frac{A_z k_z}{\mu B \Delta z}\right|^n_{i,j,k+\frac{1}{2}}\left(p^{n+1}_{i,j,k+1}-p^{n+1}_{i,j,k}\right)-\left.\frac{A_z k_z}{\mu B \Delta z}\right|^n_{i,j,k-\frac{1}{2}}\left(p^{n+1}_{i,j,k}-p^{n+1}_{i,j,k-1}\right)+q^{n+1}_{i,j,k} \qquad\qquad\dots\dots\dots\dots\dots\dots\dots\dots (8.18)
$$

$$
=\left(\frac{V_b \phi c}{5.615}\right)^n_{i,j,k}\left(\frac{p^{n+1}_{i,j,k}-p^n_{i,j,k}}{\Delta t}\right)
$$

$$
+\begin{bmatrix}
\left.\dfrac{1}{144}\dfrac{g}{g_c}\rho\dfrac{A_x k_x}{\mu B \Delta x}\right|^n_{i+\frac{1}{2},j,k}\left(G_{i+1,j,k}-G_{i,j,k}\right)+\left.\dfrac{1}{144}\dfrac{g}{g_c}\rho\dfrac{A_x k_x}{\mu B \Delta x}\right|^n_{i-\frac{1}{2},j,k}\left(G_{i-1,j,k}-G_{i,j,k}\right) \\[2ex]
+\left.\dfrac{1}{144}\dfrac{g}{g_c}\rho\dfrac{A_y k_y}{\mu B \Delta y}\right|^n_{i,j+\frac{1}{2},k}\left(G_{i,j+1,k}-G_{i,j,k}\right)+\left.\dfrac{1}{144}\dfrac{g}{g_c}\rho\dfrac{A_y k_y}{\mu B \Delta y}\right|^n_{i,j-\frac{1}{2},k}\left(G_{i,j-1,k}-G_{i,j,k}\right) \\[2ex]
+\left.\dfrac{1}{144}\dfrac{g}{g_c}\rho\dfrac{A_z k_z}{\mu B \Delta z}\right|^n_{i,j,k+\frac{1}{2}}\left(G_{i,j,k+1}-G_{i,j,k}\right)+\left.\dfrac{1}{144}\dfrac{g}{g_c}\rho\dfrac{A_z k_z}{\mu B \Delta z}\right|^n_{i,j,k-\frac{1}{2}}\left(G_{i,j,k-1}-G_{i,j,k}\right)
\end{bmatrix}
$$

The transmissibility coefficients that appear in Eq. 8.18 exhibit a different behavior compared with the transmissibility coefficients of the incompressible flow problem. Now the values of μ, B, and ρ are different across the interface between the blocks, because pressures in the neighboring blocks are different. Therefore, in calculating the μ, B, and ρ terms at the interface, we use the arithmetic average of the pressures and use that average pressure to calculate values of the weakly nonlinear terms of μ, B, and ρ. This is reasonable because the pressure surface is rather smooth and does not move in the form of a shock front (like a saturation front). Accordingly, if we define

$$
p^n_{i+\frac{1}{2},j,k}=\frac{1}{2}\left(p^n_{i,j,k}+p^n_{i+1,j,k}\right), \quad\dots\dots\dots\dots\dots\dots\dots\dots\dots\dots\dots\dots\dots\dots\dots\dots\dots\dots\dots\dots\dots (8.19)
$$

then we can calculate the pressure-dependent terms of the transmissibility coefficients as follows:

$$
\mu^n_{i+\frac{1}{2},j,k}=\mu^n\left(p^n_{i+\frac{1}{2},j,k}\right), \quad\dots\dots\dots\dots\dots\dots\dots\dots\dots\dots\dots\dots\dots\dots\dots\dots\dots\dots\dots (8.20a)
$$

$$
B^n_{i+\frac{1}{2},j,k}=B^n\left(p^n_{i+\frac{1}{2},j,k}\right), \quad\dots\dots\dots\dots\dots\dots\dots\dots\dots\dots\dots\dots\dots\dots\dots\dots\dots\dots\dots (8.20b)
$$

$$
\rho^n_{i+\frac{1}{2},j,k}=\rho^n\left(p^n_{i+\frac{1}{2},j,k}\right). \quad\dots\dots\dots\dots\dots\dots\dots\dots\dots\dots\dots\dots\dots\dots\dots\dots\dots\dots\dots (8.20c)
$$

Also, in Eq. 8.21 we recognize a new group of terms:

$$
\Gamma^n_{i,j,k}=\left(\frac{V_b \phi c}{5.615 \Delta t}\right)^n_{i,j,k}. \quad\dots\dots\dots\dots\dots\dots\dots\dots\dots\dots\dots\dots\dots\dots\dots\dots\dots\dots\dots (8.21)
$$

Now, one can write Eq. 8.18 using the SIP notation:

$$
B^n_{i,j,k}p^{n+1}_{i,j,k-1}+S^n_{i,j,k}p^{n+1}_{i,j-1,k}+W^n_{i,j,k}p^{n+1}_{i-1,j,k}+C^n_{i,j,k}p^{n+1}_{i,j,k}+E^n_{i,j,k}p^{n+1}_{i+1,j,k}+N^n_{i,j,k}p^{n+1}_{i,j+1,k}+A^n_{i,j,k}p^{n+1}_{i,j,k+1}=Q^{n+1}_{i,j,k}. \quad\dots\dots\dots\dots (8.22)
$$

A typical SIP coefficient, for example $B^n_{i,j,k}$, is defined as

$$
B^n_{i,j,k}=T^n_{i,j,k-\frac{1}{2}}=\left.\frac{1}{\mu B}\right|^n_{i,j,k-\frac{1}{2}}\left.\frac{A_z k_z}{\Delta x}\right|_{i,j,k-\frac{1}{2}}. \quad\dots\dots\dots\dots\dots\dots\dots\dots\dots\dots\dots\dots\dots\dots\dots\dots\dots\dots (8.23)
$$

The constant part of the transmissibility coefficient, $\dfrac{A_z k_z}{\Delta x}$, at the interface $i,j,k-\dfrac{1}{2}$ is calculated as usual using harmonic averaging. As we did with the incompressible flow equation, it is necessary to consider three possible cases for the definitions of the $C_{i,j,k}$ and $Q_{i,j,k}$ coefficients.

- For a block that does not have a well, $\left(q_{i,j,k}^{n+1}=0\right)$:

$$C_{i,j,k}^{n} = -\left(B_{i,j,k}^{n}+S_{i,j,k}^{n}+W_{i,j,k}^{n}+E_{i,j,k}^{n}+N_{i,j,k}^{n}+A_{i,j,k}^{n}+\Gamma_{i,j,k}^{n}\right), \quad \dots \dots \dots \dots \dots \dots \dots \dots \text{(8.24a)}$$

and

$$Q_{i,j,k}^{n+1} = -\Gamma_{i,j,k}^{n}p_{i,j,k}^{n}+\left[\begin{array}{l}T'^{n}_{ix_{i,j,k-\frac{1}{2}}}G_{i,j,k-1}+T'^{n}_{ix_{i,j-\frac{1}{2},k}}G_{i,j-1,k}+T'^{n}_{ix_{i-\frac{1}{2},j,k}}G_{i-1,j,k}+T'^{n}_{ix_{i+\frac{1}{2},j,k}}G_{i+1,j,k}+T'^{n}_{ix_{i,j+\frac{1}{2},k}}G_{i,j+1,k}\\[2mm]+T'^{n}_{ix_{i,j,k+\frac{1}{2}}}G_{i,j,k+1}-\left(T'^{n}_{ix_{i,j,k-\frac{1}{2}}}+T'^{n}_{ix_{i,j-\frac{1}{2},k}}+T'^{n}_{ix_{i-\frac{1}{2},j,k}}+T'^{n}_{ix_{i+\frac{1}{2},j,k}}+T'^{n}_{ix_{i,j+\frac{1}{2},k}}+T'^{n}_{ix_{i,j,k+\frac{1}{2}}}\right)G_{i,j,k}\end{array}\right]. \quad \dots \dots \text{(8.24b)}$$

- For a block that hosts a well with its flow rate specified $\left(q_{i,j,k}^{n+1}=q_{\text{spec}}\right)$:

$$C_{i,j,k}^{n} = -\left(B_{i,j,k}^{n}+S_{i,j,k}^{n}+W_{i,j,k}^{n}+E_{i,j,k}^{n}+N_{i,j,k}^{n}+A_{i,j,k}^{n}+\Gamma_{i,j,k}^{n}\right), \quad \dots \dots \dots \dots \dots \dots \dots \dots \text{(8.25a)}$$

and

$$Q_{i,j,k}^{n+1} = -q_{\text{spec}}^{n+1}-\Gamma_{i,j,k}^{n}p_{i,j,k}^{n}+\left[\begin{array}{l}T'^{n}_{ix_{i,j,k-\frac{1}{2}}}G_{i,j,k-1}+T'^{n}_{ix_{i,j-\frac{1}{2},k}}G_{i,j-1,k}+T'^{n}_{ix_{i-\frac{1}{2},j,k}}G_{i-1,j,k}+T'^{n}_{ix_{i+\frac{1}{2},j,k}}G_{i+1,j,k}+T'^{n}_{ix_{i,j+\frac{1}{2},k}}G_{i,j+1,k}\\[2mm]+T'^{n}_{ix_{i,j,k+\frac{1}{2}}}G_{i,j,k+1}-\left(T'^{n}_{ix_{i,j,k-\frac{1}{2}}}+T'^{n}_{ix_{i,j-\frac{1}{2},k}}+T'^{n}_{ix_{i-\frac{1}{2},j,k}}+T'^{n}_{ix_{i+\frac{1}{2},j,k}}+T'^{n}_{ix_{i,j+\frac{1}{2},k}}+T'^{n}_{ix_{i,j,k+\frac{1}{2}}}\right)G_{i,j,k}\end{array}\right]. \quad \dots \text{(8.25b)}$$

- For a block that hosts a well with flowing-sandface-pressure specification $\left(p_{sf_{i,j,k}}^{n+1}=p_{sf_{\text{spec}}}\right)$: Again, the wellbore equation will be brought into action,

$$q_{i,j,k}^{n+1} = -\Omega_{i,j,k}^{n}\left(p_{i,j,k}^{n+1}-p_{sf_{i,j,k}}^{n+1}\right). \quad \dots \dots \dots \dots \dots \dots \dots \dots \dots \dots \dots \dots \dots \text{(8.26)}$$

Then, when Eq. 8.26 is placed into Eq. 8.18, the following definitions of $C_{i,j,k}$ and $Q_{i,j,k}$ will be obtained:

$$C_{i,j,k}^{n} = -\left(B_{i,j,k}^{n}+S_{i,j,k}^{n}+W_{i,j,k}^{n}+E_{i,j,k}^{n}+N_{i,j,k}^{n}+A_{i,j,k}^{n}+\Gamma_{i,j,k}^{n}+\Omega_{i,j,k}^{n}\right), \quad \dots \dots \dots \dots \dots \dots \text{(8.27a)}$$

$$Q_{i,j,k}^{n+1} = -\Omega_{i,j,k}^{n+1}p_{sf_{\text{spec}}}-\Gamma_{i,j,k}^{n}p_{i,j,k}^{n}+\left[\begin{array}{l}T'^{n}_{ix_{i,j,k-\frac{1}{2}}}G_{i,j,k-1}+T'^{n}_{ix_{i,j-\frac{1}{2},k}}G_{i,j-1,k}+T'^{n}_{ix_{i-\frac{1}{2},j,k}}G_{i-1,j,k}+T'^{n}_{ix_{i+\frac{1}{2},j,k}}G_{i+1,j,k}+T'^{n}_{ix_{i,j+\frac{1}{2},k}}G_{i,j+1,k}\\[2mm]+T'^{n}_{ix_{i,j,k+\frac{1}{2}}}G_{i,j,k+1}-\left(T'^{n}_{ix_{i,j,k-\frac{1}{2}}}+T'^{n}_{ix_{i,j-\frac{1}{2},k}}+T'^{n}_{ix_{i-\frac{1}{2},j,k}}+T'^{n}_{ix_{i+\frac{1}{2},j,k}}+T'^{n}_{ix_{i,j+\frac{1}{2},k}}+T'^{n}_{ix_{i,j,k+\frac{1}{2}}}\right)G_{i,j,k}\end{array}\right]. \quad \dots \text{(8.27b)}$$

Again, after solving for all block pressures, $p_{i,j,k}^{n+1}$, Eq. 8.26 will be used to calculate $p_{sf_{i,j,k}}^{n+1}$, when $q_{i,j,k}^{n+1}$ is specified, or to calculate $q_{i,j,k}^{n+1}$, if the sandface pressure $p_{sf_{i,j,k}}^{n+1}$ is specified.

8.2.1 MBCs. At the end of a timestep calculation, three different checks typically are implemented.

Incremental MBC (IMBC). This check provides a measure on the conservation of mass over the entire reservoir during a timestep:

$$\text{IMBC} = \left| \frac{\sum_{k=1}^{N_z}\sum_{j=1}^{N_y}\sum_{i=1}^{N_x}\left(\dfrac{V_b\phi}{5.615B}\right)_{i,j,k}^{n+1}-\sum_{k=1}^{N_z}\sum_{j=1}^{N_y}\sum_{i=1}^{N_x}\left(\dfrac{V_b\phi}{5.615B}\right)_{i,j,k}^{n}}{\sum_{m=1}^{N_w}q_m^{n+1}\Delta t} \right|. \quad \dots \dots \dots \dots \dots \dots \dots \dots \text{(8.28)}$$

In Eq. 8.28, the index N_w represents the number of existing wells in the reservoir. Eq. 8.28 is written for a reservoir with closed boundaries. A good IMBC is expected to generate values between 0.999999 and 1.00000.

Cumulative MBC (CMBC). This check gives a measure on the conservation of mass from the beginning of the simulation ($t=0$) until the end of the current timestep over the entire reservoir:

$$\text{CMBC} = \left| \frac{\sum_{k=1}^{N_z} \sum_{j=1}^{N_y} \sum_{i=1}^{N_x} \left(\frac{V_b \phi}{5.615B} \right)_{i,j,k}^{n+1} - \sum_{k=1}^{N_z} \sum_{j=1}^{N_y} \sum_{i=1}^{N_x} \left(\frac{V_b \phi}{5.615B} \right)_{i,j,k}^{t=0}}{\sum_{l=1}^{n+1} \sum_{m=1}^{N_w} q_m^l \Delta t_l} \right| . \quad \dots\dots\dots\dots\dots\dots\dots\dots\dots \text{(8.29)}$$

Again, in Eq. 8.29, the goal is to obtain a value close to 1 (within the same margin previously mentioned).

Residual Check. The third check provides precision control (in the range of 10^{-7} to 10^{-8}) at each individual block level; $R_{i,j,k}^{n+1}$ values for each block are expected to be very close to zero because they are calculated from the following equation:

$$R_{i,j,k}^{n+1} = Q_{i,j,k}^{n+1} - \left(B_{i,j,k}^n p_{i,j,k-1}^{n+1} + S_{i,j,k}^n p_{i,j-1,k}^{n+1} + W_{i,j,k}^n p_{i-1,j,k}^{n+1} + C_{i,j,k}^n p_{i,j,k}^{n+1} + E_{i,j,k}^n p_{i+1,j,k}^{n+1} + N_{i,j,k}^n p_{i,j+1,k}^{n+1} + A_{i,j,k}^n p_{i,j,k+1}^{n+1} \right). \quad \dots\dots \text{(8.30)}$$

8.3 Single-Phase Compressible Flow Equation

A single-phase compressible flow equation describes the flow of a real gas in porous media. During the solution process, the compressible flow equation poses more challenges than incompressible and slightly compressible flow equations. The principal reason, which makes the numerical solution more demanding, is the more-pronounced dependence of transmissibility coefficients on changes in pressure. Therefore, it is not feasible to allow these coefficients to follow one timestep behind (increased level of nonlinearity). To accordingly overcome this difficulty, an iterative scheme needs to be embedded into the solution protocol such that coefficients of (nonlinear) terms can follow the solution protocol one iteration level behind. This procedure is called the simple iteration method, and it represents a typical linearization process. Another important item to keep in mind is the simplification of the problem formulation and subsequent computations by ignoring depth gradients. This assumption is quite reasonable, simply because the hydrostatic head exerted by the gas column is not significant because of the low density of gas (in this way, gravitational forces acting on the system are ignored).

Again, we start with the 3D representation of the single-phase compressible flow problem:

$$\frac{\partial}{\partial x}\left(\frac{A_x k_x}{\mu_g B_g} \frac{\partial p}{\partial x} \right)\Delta x + \frac{\partial}{\partial y}\left(\frac{A_y k_y}{\mu_g B_g} \frac{\partial p}{\partial y} \right)\Delta y + \frac{\partial}{\partial z}\left(\frac{A_z k_z}{\mu_g B_g} \frac{\partial p}{\partial z} \right)\Delta z + q_{\text{scf/D}} = \frac{V_b \phi}{5.615} \frac{\partial}{\partial t}\left(\frac{1}{B_g} \right). \quad \dots\dots\dots\dots\dots\dots \text{(8.31a)}$$

Recalling the definition of B_g,

$$B_g = \frac{p_{sc} T z_g}{5.615 T_{sc} p} \text{ res bbl/scf,} \quad \dots\dots\dots\dots\dots\dots\dots\dots\dots\dots\dots\dots\dots\dots\dots\dots\dots\dots\dots\dots\dots\dots\dots \text{(8.31b)}$$

and substituting on the right-hand side of Eq. 8.31a, one obtains

$$\frac{\partial}{\partial x}\left(\frac{A_x k_x}{\mu_g B_g} \frac{\partial p}{\partial x} \right)\Delta x + \frac{\partial}{\partial y}\left(\frac{A_y k_y}{\mu_g B_g} \frac{\partial p}{\partial y} \right)\Delta y + \frac{\partial}{\partial z}\left(\frac{A_z k_z}{\mu_g B_g} \frac{\partial p}{\partial z} \right)\Delta z + q_{\text{scf/D}} = \frac{V_b \phi T_{sc}}{p_{sc} T} \frac{\partial}{\partial t}\left(\frac{p}{z_g} \right). \quad \dots\dots\dots\dots\dots\dots \text{(8.31c)}$$

Note that we substituted B_g only on the right-hand side of Eq. 8.31a, for two reasons:

- To ensure that the dependent-variable pressure, p, is common on both sides of the equation.
- To ensure the generic definition of transmissibility terms on the left-hand side of the equation remains intact.

The finite-difference representation of Eq. 8.31c then becomes

$$\frac{A_x k_x}{\mu_g B_g \Delta x}\bigg|_{i+\frac{1}{2},j,k}^{(k)\,n+1}\left(p_{i+1,j,k}^{(k+1)\,n+1} - p_{i,j,k}^{(k+1)\,n+1} \right) - \frac{A_x k_x}{\mu_g B_g \Delta x}\bigg|_{i-\frac{1}{2},j,k}^{(k)\,n+1}\left(p_{i,j,k}^{(k+1)\,n+1} - p_{i-1,j,k}^{(k+1)\,n+1} \right) + \frac{A_y k_y}{\mu_g B_g \Delta y}\bigg|_{i,j+\frac{1}{2},k}^{(k)\,n+1}\left(p_{i,j+1,k}^{(k+1)\,n+1} - p_{i,j,k}^{(k+1)\,n+1} \right)$$

$$- \frac{A_y k_y}{\mu_g B_g \Delta y}\bigg|_{i,j-\frac{1}{2},k}^{(k)\,n+1}\left(p_{i,j,k}^{(k+1)\,n+1} - p_{i,j-1,k}^{(k+1)\,n+1} \right) + \frac{A_z k_z}{\mu_g B_g \Delta z}\bigg|_{i,j,k+\frac{1}{2}}^{(k)\,n+1}\left(p_{i,j,k+1}^{(k+1)\,n+1} - p_{i,j,k}^{(k+1)\,n+1} \right) - \frac{A_z k_z}{\mu_g B_g \Delta z}\bigg|_{i,j,k-\frac{1}{2}}^{(k)\,n+1}\left(p_{i,j,k}^{(k+1)\,n+1} - p_{i,j,k-1}^{(k+1)\,n+1} \right)$$

$$+ q_{i,j,k}^{(k+1)\,n+1} = \left(\frac{V_b \phi T_{sc}}{p_{sc} T} \right)_{i,j,k}^{n+1} \frac{1}{\Delta t}\left(\frac{p_{i,j,k}^{n+1}}{z_{g\,i,j,k}^{(k+1)\,n+1}} - \frac{p_{i,j,k}^n}{z_{g\,i,j,k}^n} \right). \quad \dots\dots\dots\dots\dots \text{(8.32)}$$

Note that in Eq. 8.32 the iteration levels for coefficients are one iteration level behind those of the unknowns (pressures).

Again, in writing Eq. 8.32 in its corresponding SIP notation, we observe the following two groups of terms:

$$\Gamma_{i,j,k}^{(k)\,n+1} = \left(\frac{V_b \phi T_{sc}}{p_{sc} T z_g \Delta t} \right)_{i,j,k}^{(k)\,n+1}, \quad \dots\dots\dots\dots\dots\dots\dots\dots\dots\dots\dots\dots\dots\dots\dots\dots\dots\dots\dots\dots\dots \text{(8.33a)}$$

and

$$\Gamma_{i,j,k}^n = \left(\frac{V_b \phi T_{sc}}{p_{sc} T z_g \Delta t} \right)_{i,j,k}^n . \quad \dots\dots\dots\dots\dots\dots\dots\dots\dots\dots\dots\dots \quad (8.33b)$$

We again use Peaceman's wellbore equation (Peaceman 1983), this time for a gas well:

$$q_{i,j,k}^{n+1\,(k+1)} = -\Omega_{i,j,k}^{n+1\,(k)} \left(p_{i,j,k}^{n+1\,(k+1)} - p_{sf_{i,j,k}}^{n+1\,(k+1)} \right). \quad \dots\dots\dots\dots\dots\dots\dots\dots \quad (8.34)$$

Now, the finite-difference representation of the compressible flow equation (Eq. 8.32) can be rewritten using the SIP notation:

$$B_{i,j,k}^{n+1\,(k)} p_{i,j,k-1}^{n+1\,(k+1)} + S_{i,j,k}^{n+1\,(k)} p_{i,j-1,k}^{n+1\,(k+1)} + W_{i,j,k}^{n+1\,(k)} p_{i-1,j,k}^{n+1\,(k+1)} + C_{i,j,k}^{n+1\,(k)} p_{i,j,k}^{n+1\,(k+1)} + E_{i,j,k}^{n+1\,(k)} p_{i+1,j,k}^{n+1\,(k+1)} + N_{i,j,k}^{n+1\,(k)} p_{i,j+1,k}^{n+1\,(k+1)} + A_{i,j,k}^{n+1\,(k)} p_{i,j,k+1}^{n+1\,(k+1)} = Q_{i,j,k}^{n+1\,(k+1)}. \quad \dots\dots\dots\dots \quad (8.35)$$

Once again, a typical SIP coefficient (e.g., $B_{i,j,k}^{n+1\,(k)}$) is defined as follows (other transmissibility coefficients follow similar definitions):

$$B_{i,j,k}^{n+1\,(k)} = \frac{A_z k_z}{\mu_g B_g \Delta z} \Bigg|_{i,j,k-\frac{1}{2}}^{n+1\,(k)} . \quad \dots\dots\dots\dots\dots\dots\dots\dots\dots\dots\dots\dots\dots\dots \quad (8.36)$$

In Eq. 8.36, we again see two groups of terms. One group is $\left(\dfrac{1}{\mu_g B_g} \right)_{i,j,k-\frac{1}{2}}^{n+1\,(k)}$, which is calculated at the interface of blocks i, j,

k, and $i, j, k-1$ using the arithmetic average of pressures $p_{i,j,k}^{n+1\,(k)}$ and $p_{i,j,k-1}^{n+1\,(k)}$. The second group is $\dfrac{A_x k_x}{\Delta x} \Big|_{i,j,k-\frac{1}{2}}$; again, this constant

part of the transmissibility term is calculated at the interface of blocks i, j, k, and $i, j, k-1$ using the harmonic average of $\dfrac{A_x k_x}{\Delta x} \Big|_{i,j,k}$ and $\dfrac{A_x k_x}{\Delta x} \Big|_{i,j,k-1}$.

For the remaining two coefficients, $C_{i,j,k}^{n+1\,(k)}$ and $Q_{i,j,k}^{n+1\,(k)}$ of Eq. 8.35 can be defined in consideration of whether or not block i, j, k, has a well; if there is a well, what is specified at the well as a boundary condition is further explained:

- For a block that does not have a well $\left(q_{i,j,k}^{n+1\,(k+1)} = 0 \right)$:

$$C_{i,j,k}^{n+1\,(k)} = -\left(B_{i,j,k}^{n+1\,(k)} + S_{i,j,k}^{n+1\,(k)} + W_{i,j,k}^{n+1\,(k)} + E_{i,j,k}^{n+1\,(k)} + N_{i,j,k}^{n+1\,(k)} + A_{i,j,k}^{n+1\,(k)} + \Gamma_{i,j,k}^{n+1\,(k)} \right), \quad \dots\dots\dots\dots\dots\dots \quad (8.37)$$

$$Q_{i,j,k}^{n+1\,(k)} = -\Gamma_{i,j,k}^n p_{i,j,k}^n. \quad \dots\dots\dots\dots\dots\dots\dots\dots\dots\dots\dots\dots\dots\dots \quad (8.38)$$

- For a block that hosts a well with a flow-rate specification $\left(q_{i,j,k}^{n+1\,(k+1)} = q_{spec} \right)$:

$$C_{i,j,k}^{n+1\,(k)} = -\left(B_{i,j,k}^{n+1\,(k)} + S_{i,j,k}^{n+1\,(k)} + W_{i,j,k}^{n+1\,(k)} + E_{i,j,k}^{n+1\,(k)} + N_{i,j,k}^{n+1\,(k)} + A_{i,j,k}^{n+1\,(k)} + \Gamma_{i,j,k}^{n+1\,(k)} \right), \quad \dots\dots\dots\dots\dots\dots \quad (8.39)$$

$$Q_{i,j,k}^{n+1\,(k)} = -q_{spec} - \Gamma_{i,j,k}^n p_{i,j,k}^n. \quad \dots\dots\dots\dots\dots\dots\dots\dots\dots\dots\dots \quad (8.40)$$

- For a block that hosts a well specification of flowing-sandface pressure $\left(p_{sf_{i,j,k}}^{n+1\,(k+1)} = p_{sf_{spec}} \right)$:

$$C_{i,j,k}^{n+1\,(k)} = -\left(B_{i,j,k}^{n+1\,(k)} + S_{i,j,k}^{n+1\,(k)} + W_{i,j,k}^{n+1\,(k)} + E_{i,j,k}^{n+1\,(k)} + N_{i,j,k}^{n+1\,(k)} + A_{i,j,k}^{n+1\,(k)} + \Gamma_{i,j,k}^{n+1\,(k)} + \Omega_{i,j,k}^{n+1\,(k)} \right), \quad \dots\dots\dots \quad (8.41)$$

$$Q_{i,j,k}^{n+1\,(k)} = -\Omega_{i,j,k}^{n+1\,(k)} p_{sf_{spec}} - \Gamma_{i,j,k}^n p_{i,j,k}^n. \quad \dots\dots\dots\dots\dots\dots\dots\dots \quad (8.42)$$

In arriving at Eqs. 8.41 and 8.42, note that the $q_{i,j,k}^{n+1\,(k+1)}$ entry in Eq. 8.32 was replaced by the right-hand side of Eq. 8.34.

8.3.1 MBCs.
Similar to a slightly compressible flow equation at the end of each timestep, three different checks need to be conducted to examine the quality of the generated solution.

IMBC. For a compressible flow equation, an IMBC is formulated as follows:

$$\text{IMBC} = \frac{\left| \sum_{k=1}^{N_z} \sum_{j=1}^{N_y} \sum_{i=1}^{N_x} \left(\dfrac{V_b \phi}{5.615 B_g} \right)_{i,j,k}^{n+1} - \sum_{k=1}^{N_z} \sum_{j=1}^{N_y} \sum_{i=1}^{N_x} \left(\dfrac{V_b \phi}{5.615 B_g} \right)_{i,j,k}^{n} \right|}{\sum_{l=1}^{N_w} q_{g_l}^{n+1} \Delta t} . \quad \dots\dots\dots\dots\dots\dots\dots \quad (8.43)$$

CMBC. The CMBC will be expressed as

$$\text{CMBC} = \left| \frac{\sum_{k=1}^{N_z} \sum_{j=1}^{N_y} \sum_{i=1}^{N_x} \left(\frac{V_b \phi}{5.615 B_g} \right)_{i,j,k}^{n+1} - \sum_{k=1}^{N_z} \sum_{j=1}^{N_y} \sum_{i=1}^{N_x} \left(\frac{V_b \phi}{5.615 B_g} \right)_{i,j,k}^{t=0}}{\sum_{l=1}^{n+1} \sum_{l=1}^{N_w} q_{g_l}^l \Delta t_l} \right| \dots \dots \dots \dots (8.44)$$

Both IMBC and CMBC checks are expected to yield results around 1.000000. If they do not, then it is good practice to go to a finer convergence criterion (simply by decreasing the current criterion by 50%) and continue with the iterations until the solution converges within the bounds of the new criterion. After two such passes, solution precision typically will be improved and satisfactory IMBC and CMBC checks will be achieved.

Residual Check. Similar to previous residual checks we studied, the residual check for compressible flow requires the calculation of residual gas volumes (in scf/D) at each gridblock using the equation given:

$$R_{i,j,k}^{n+1} = Q_{i,j,k}^{n+1} - \left(\begin{array}{l} B_{i,j,k}^n p_{i,j,k-1}^{n+1} + S_{i,j,k}^n p_{i,j-1,k}^{n+1} + W_{i,j,k}^n p_{i-1,j,k}^{n+1} + C_{i,j,k}^n p_{i,j,k}^{n+1} \\ + E_{i,j,k}^n p_{i+1,j,k}^{n+1} + N_{i,j,k}^n p_{i,j+1,k}^{n+1} + A_{i,j,k}^n p_{i,j,k+1}^{n+1} \end{array} \right). \dots \dots \dots (8.45)$$

The compressible flow equation is computationally more demanding, because the solution protocol requires an iterative scheme. Recall that in solving the incompressible flow problem, a system of equations is constructed only once and is subsequently solved by completing the simulation study. In slightly compressible flow, at each timestep only one system of equations is structured and solved only once, and similar workflow is repeated at the subsequent timestep calculations. In the case of the compressible flow equation, while implementing the iterative-solution protocol described in this subsection, it is necessary to construct a system of equations at each iteration level of every timestep. This is why the computational overhead is, by far, the largest among the three classes of flow problems discussed in this chapter.

8.3.2 Generalized Newton-Raphson Protocol. Generalized Newton-Raphson (GNR) protocol is broadly used to solve a nonlinear system of equations like the single-phase flow gas equations in reservoir simulation problems. We introduce this solution protocol by considering two equations in two unknowns, x and y: $f(x, y) = 0$ and $g(x, y) = 0$.

Let us assume that (x_0, y_0) is an approximate solution of the equations with a root of $[\xi, \eta]$. Suppose that the approximate solution is approaching the real root by way of $(x_0 + \Delta x, y_0 + \Delta y)$, and then in the neighborhood of (x_0, y_0), a Taylor's expansion can be written:

$$f(x_0 + \Delta x, y_0 + \Delta y) = f(x_0, y_0) + \Delta x \frac{\partial f}{\partial x} \bigg|_{x_0, y_0} + \Delta y \frac{\partial f}{\partial y} \bigg|_{x_0, y_0}, \dots \dots \dots \dots (8.46)$$

$$g(x_0 + \Delta x, y_0 + \Delta y) = g(x_0, y_0) + \Delta x \frac{\partial g}{\partial x} \bigg|_{x_0, y_0} + \Delta y \frac{\partial g}{\partial y} \bigg|_{x_0, y_0}. \dots \dots \dots \dots (8.47)$$

Eqs. 8.46 and 8.47 can be rewritten as

$$\Delta x \frac{\partial f}{\partial x} \bigg|_{x_0, y_0} + \Delta y \frac{\partial f}{\partial y} \bigg|_{x_0, y_0} \approx -f(x_0, y_0), \dots \dots \dots \dots (8.48)$$

$$\Delta x \frac{\partial g}{\partial x} \bigg|_{x_0, y_0} + \Delta y \frac{\partial g}{\partial y} \bigg|_{x_0, y_0} \approx -g(x_0, y_0). \dots \dots \dots \dots (8.49)$$

After defining $f_x = \frac{\partial f}{\partial x}\bigg|_{x_0, y_0}$, $g_x = \frac{\partial g}{\partial x}\bigg|_{x_0, y_0}$, $f_y = \frac{\partial f}{\partial y}\bigg|_{x_0, y_0}$, $g_y = \frac{\partial g}{\partial y}\bigg|_{x_0, y_0}$, one can solve for Δx and Δy with the following results:

$$\Delta x = \left(\frac{fg_y - gf_y}{f_y g_x - g_y f_x} \right)_{x_0, y_0}, \dots \dots \dots \dots (8.50)$$

$$\Delta y = \left(\frac{gf_x - fg_x}{f_y g_x - g_y f_x} \right)_{x_0, y_0}. \dots \dots \dots \dots (8.51)$$

At this stage, to complete the first iteration, the solution of the equations can be updated:

$$x_1 = x_0 + \left(\frac{fg_y - gf_y}{f_y g_x - g_y f_x} \right)_{x_0, y_0}, \dots \dots \dots \dots (8.52)$$

$$y_1 = y_0 + \left(\frac{gf_x - fg_x}{f_y g_x - g_y f_x} \right)_{x_0, y_0} . \quad\text{..}\quad (8.53)$$

Defining the Jacobian matrix of the problem as

$$J(f,g) = f_x g_y - f_y g_x = \begin{bmatrix} f_x & f_y \\ g_x & g_y \end{bmatrix} = \begin{bmatrix} \dfrac{\partial f}{\partial x} & \dfrac{\partial f}{\partial y} \\ \dfrac{\partial g}{\partial x} & \dfrac{\partial g}{\partial y} \end{bmatrix}. \quad\text{......................................}\quad (8.54)$$

Eqs. 8.52 and 8.53 can be expressed in a general form:

$$x^{(k+1)} = x^{(k)} - \left\{ \frac{fg_y - gf_y}{\det[J(f,g)]} \right\}_{x_0, y_0} , \quad\text{..}\quad (8.55)$$

$$y^{(k+1)} = y^{(k)} - \left\{ \frac{gf_y - fg_x}{\det[J(f,g)]} \right\}_{x_0, y_0} . \quad\text{..}\quad (8.56)$$

When this iteration converges, it converges quadratically. A set of conditions sufficient to ensure convergence can be summarized as follows:

- Function f, g, and all derivatives through their second order are continuous and bound in a region containing $[\xi, \eta]$.
- The Jacobian matrix does not vanish in R.
- The initial approximation (x_0, y_0) is chosen sufficiently close to the root $[\xi, \eta]$.

The GNR protocol, obviously, can be applied to a system of n equations in n unknowns. At each iteration level, we will then have to evaluate n^2 partial derivatives and n functions, which requires considerable computational effort.

8.4 Problems and Solutions

Problem 8.4.1 (Ertekin et al. 2001)

Consider the 1D, single-phase, incompressible flow of oil taking place in the z-direction of the system in **Fig. 8.1.1,** with gridblock properties given in **Table 8.1.** The boundary conditions are as follows:

- Pressure of the production block is maintained at 1,000 psia (Block 4).
- No-flow boundary at the top of the system.
- Pressure of the block located at the constant-pressure boundary is maintained at 5,000 psia (Block 1).
- The fluid properties are $\mu_o = 2$ cp, $\rho_o = 50$ lbm/ft^3, and $B_o = 1$ res bbl/STB.

Using the appropriate PDE and its finite-difference approximation, calculate the pressure distribution in the system.

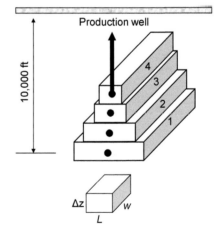

Fig. 8.1.1—Reservoir system for Problem 8.4.1.

Grid	Δz, ft	W, ft	L, ft	A_z, ft^2	k_z, md
1	400	1,200	200	240,000	100
2	600	1,200	180	216,000	100
3	600	1,200	160	192,000	100
4	200	1,200	140	168,000	100

Table 8.1—Gridblock properties for Problem 8.4.1.

Solution to Problem 8.4.1

The schematic representation of the system is displayed in **Fig. 8.1.2.** It should be noted that depths to the center of each gridblock are marked.

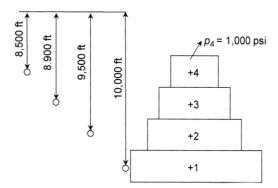

Fig. 8.1.2—Schematic representation of the reservoir system of Problem 8.4.1.

For a 1D reservoir with heterogeneous-thickness distribution and homogeneous permeability,

$$\frac{\partial}{\partial z}\left(\frac{A_z k_z}{\mu B}\frac{\partial p}{\partial y}\right)\Delta z + \frac{1}{144}\frac{g}{g_c}\rho\frac{\partial}{\partial z}\left(\frac{A_z k_z}{\mu B}\frac{\partial G}{\partial z}\right)\Delta z + q = 0. \quad\dots\dots\dots\dots\dots (8.57)$$

The transmissibility factors along the z-direction must be calculated:

$$T_{12} = \frac{2A_1 A_2 k_z}{\mu B\left(A_1 \Delta z_2 + A_2 \Delta z_1\right)} = \frac{2(240,000)(216,000)(0.1127)}{(2)(1)\left[(240,000)(600)+(216,000)(400)\right]} = 25.36 \text{ STB/D-psi}, \quad\dots\dots\dots\dots (8.58)$$

$$T_{23} = \frac{2A_3 A_2 k_z}{\mu B\left(A_3 \Delta z_2 + A_2 \Delta z_3\right)} = \frac{2(192,000)(216,000)(0.1127)}{(2)(1)\left[(192,000)(600)+(216,000)(600)\right]} = 19.09 \text{ STB/D-psi}, \quad\dots\dots\dots\dots (8.59)$$

$$T_{34} = \frac{2A_3 A_4 k_z}{\mu B\left(A_3 \Delta z_4 + A_4 \Delta z_3\right)} = \frac{2(192,000)(168,000)(0.1127)}{(2)(1)\left[(192,000)(200)+(168000)(600)\right]} = 26.12 \text{ STB/D-psi}. \quad\dots\dots\dots\dots (8.60)$$

The elevation of each block in Fig. 8.1.2 must be identified from the data shown in **Table 8.2** on the basis of thickness and reference depth.

The gravitational terms for Blocks 2 and 3 can be expressed in discretized form as

$$Q_2 = \frac{1}{144}\rho_f \frac{g}{g_c}[T_{12}G_1 - \left(T_{12}+T_{23}\right)G_2 + T_{23}G_3] = 424.7 \text{ STB/D}, \quad\dots\dots\dots (8.61)$$

$$Q_3 = \frac{1}{144}\rho_f \frac{g}{g_c}\left[T_{23}G_2 - \left(T_{34}+T_{23}\right)G_3 + T_{34}G_4\right] = 350.5 \text{ STB/D}. \quad\dots\dots\dots (8.62)$$

Grid	G, ft
1	10,000
2	9,500
3	8,900
4	8,500

Table 8.2—Elevation of each block in Problem 8.4.1.

Hence, the system of equations can be written as

- Block 2: $T_{12}p_1 - \left(T_{12}+T_{23}\right)p_2 + T_{23}p_3 = Q_2.$ $\dots\dots\dots\dots\dots\dots\dots\dots\dots\dots\dots(8.63)$

- Block 3: $T_{23}p_2 - \left(T_{34}+T_{23}\right)p_3 + T_{34}p_4 = Q_3.$ $\dots\dots\dots\dots\dots\dots\dots\dots\dots\dots\dots(8.64)$

Rewriting the system of equations using the calculated coefficients,

$$-44.45 p_2 + 19.09 p_3 = -126,363, \quad\dots\dots\dots\dots\dots\dots\dots\dots\dots\dots\dots\dots (8.65)$$

and

$$19.09 p_2 - 45.21 p_3 = -25,764.8, \quad\dots\dots\dots\dots\dots\dots\dots\dots\dots\dots\dots\dots (8.66)$$

and then solving for unknowns, $p_2 = 3,771.8$ psia and $p_3 = 2,162.87$ psia.

Problem 8.4.2 (Ertekin et al. 2001)

Consider the 1D, single-phase, steady-state flow of water taking place in the horizontal, homogeneous reservoir shown in **Fig. 8.2.** The block and fluid properties are $\Delta x = 400$ ft, $w = 200$ ft, $h = 80$ ft, $k_x = 60$ md, $\mu_w = 0.5$ cp, and $B_w = 1$ res bbl/STB. The boundary conditions are as follows:

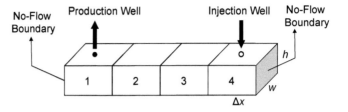

Fig. 8.2—Reservoir system for Problem 8.4.2.

- No-flow boundaries at the extreme ends of the 1D reservoir.
- Well in Block 1 produces at a rate of 400 B/D.
- Approximate pressure of Block 4 is 3,600 psia.

Calculate the pressure of all gridblocks using the pointwise-successive over-relaxation (PSOR) solution method, with $\omega = 1.5$ and a pressure convergence criterion of 3 psi. Use $p_1 = 3,000$ psia, $p_2 = 3,200$ psia, $p_3 = 3,400$ psia for initial guesses.

Solution to Problem 8.4.2

The governing flow equation for a 1D incompressible-reservoir flow system can be written as

$$\frac{\partial}{\partial x}\left(\frac{A_x k_x}{\mu B}\frac{\partial p}{\partial x}\right)\Delta x + q = 0, \quad\quad\quad\quad (8.67)$$

which can be simplified for a homogeneous system to

$$\frac{\partial^2 p}{\partial x^2} + \frac{q\mu B}{V_b k} = 0. \quad\quad\quad\quad (8.68)$$

The finite-difference representation of the previous equation is

$$\frac{p_{i-1} - 2p_i + p_{i+1}}{(\Delta x)^2} + \left(\frac{q\mu B}{\Delta x whk}\right)_i = 0 \quad\quad\quad\quad (8.69)$$

or

$$p_{i-1} - 2p_i + p_{i+1} + \frac{\Delta x}{w}\left(\frac{q\mu B}{hk}\right)_i = 0. \quad\quad\quad\quad (8.70)$$

- Block 1: $p_2 - p_1 + \frac{\Delta x}{w}\left(\frac{q\mu B}{kh}\right) = 0.$ $\quad\quad\quad\quad (8.71)$
- Block 2: $p_1 - 2p_2 + p_3 = 0.$ $\quad\quad\quad\quad (8.72)$
- Block 3: $p_2 - 2p_3 + p_4 = 0.$ $\quad\quad\quad\quad (8.73)$

Knowing that $p_4 = 3,600$ psia and rearranging the equations, we obtain

$$-p_1 + p_2 = 147.89, \quad\quad\quad\quad (8.74)$$

$$p_1 - 2p_2 + p_3 = 0, \qu\quad\quad\quad\quad (8.75)$$

$$p_2 - 2p_3 = -3,600. \quad\quad\quad\quad (8.76)$$

The iteration results using the PSOR method are listed in **Table 8.3.** With $\omega = 1.5$, it takes seven PSOR iterations to converge within a pressure tolerance of 3 psia. As seen in the table, the solution vector is $p_1 = 3,156.05$ psi, $p_2 = 3,303.67$ psi, and $p_3 = 3451.87$ psi.

PSOR Iteration Level	p_1	p_2	p_3
0	3,000	3,200	3,400
1	3,078.17	3,258.63	3,443.97
2	3,127.03	3,298.94	3,452.22
3	3,163.06	3,311.99	3,457.88
4	3,164.63	3,310.89	3,454.22
5	3,162.19	3,306.87	3,453.04
6	3,157.38	3,304.38	3,451.76
7	3,156.05	3,303.67	3,451.87

Table 8.3—Iteration results for Solution to Problem 8.4.2.

Problem 8.4.3 (Ertekin et al. 2001)

Consider the single-phase, incompressible flow of oil taking place in the 2D homogeneous, isotropic system of **Fig. 8.3.1.** Assume $\Delta x = \Delta y = 1{,}000$ ft, $h = 20$ ft, and $k_x = k_y = 100$ md. A production well in the center gridblock is produced at a rate of 12,000 STB/D ($\mu_o = 2$ cp). If pressure in the boundary gridblocks is kept at 2,000 psia, calculate the pressure distribution in the system and the flowing sandface pressure of the wellbore.

Solution to Problem 8.4.3

The governing flow equation for a 2D incompressible reservoir can be written as

$$\frac{\partial}{\partial x}\left(\frac{A_x k_x}{\mu B}\frac{\partial p}{\partial x}\right)\Delta x + \frac{\partial}{\partial y}\left(\frac{A_y k_y}{\mu B}\frac{\partial p}{\partial y}\right)\Delta y + q = 0, \dots\dots\dots\dots\dots\dots\dots\dots\dots\dots\dots\dots (8.77)$$

which can be simplified for a homogeneous system to

$$\frac{\partial^2 p}{\partial x^2} + \frac{\partial^2 p}{\partial y^2} + \frac{q\mu B}{V_b k} = 0. \dots\dots\dots\dots\dots\dots\dots\dots\dots\dots\dots\dots\dots\dots\dots\dots\dots\dots\dots (8.78)$$

The finite-difference representation of the previous equation is

$$\frac{p_{i-1,j} - 2p_{i,j} + p_{i+1,j}}{(\Delta x)^2} + \frac{p_{i,j-1} - 2p_{i,j} + p_{i,j+1}}{(\Delta y)^2} + \left(\frac{q\mu B}{\Delta x \Delta y h k}\right)_{i,j} = 0. \dots\dots\dots\dots\dots\dots (8.79)$$

Because $\Delta x = \Delta y$, the previous equation can be further simplified to

$$p_{i,j-1} + p_{i-1,j} - 4p_{i,j} + p_{i+1,j} + p_{i,j+1} = -\left(\frac{q\mu B}{hk}\right)_{i,j}. \dots\dots\dots\dots\dots\dots\dots\dots\dots\dots (8.80)$$

As displayed in **Fig. 8.3.2,** there are three unknown block pressures that need to be solved in the problem (in view of the existing symmetry). The blocks numbered 0 have a fixed pressure of 2,000 psia; hence, they are not part of the unknowns for this problem. The reservoir has homogeneous and isotropic permeability distribution and other rock properties are also homogeneous.

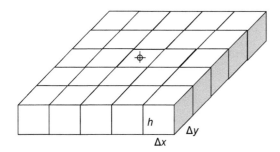

Fig. 8.3.1—Reservoir system for Problem 8.4.3.

0	0	0	0	0
0	1	2	1	0
0	2	3	2	0
0	1	2	1	0
0	0	0	0	0

Fig. 8.3.2—Numbering scheme of reservoir system for Problem 8.4.3.

Thus, the linear equations for each block can be written as follows:

- Block 1: $p_0 - 2p_1 + p_2 = 0.$ ....(8.81)
- Block 2: $p_0 - 2p_2 + p_3 + p_1 - 2p_2 + p_1 = 0.$ ....(8.82)
- Block 3: $(p_2 - 2p_3 + p_2) + (p_2 - 2p_3 + p_2) + \dfrac{q_3 \mu B}{kh} = 0$ ....(8.83)

or

$$2(p_2 - 2p_3 + p_2) + \frac{q_3 \mu B}{kh} = 0. \quad\quad\quad\quad\quad\quad\quad\quad\quad\quad\quad\quad (8.84)$$

Substituting $p_0 = 2{,}000$ psi and $q_3 = -12{,}000$ STB/D, and solving for p_1, p_2, and p_3 gives $p_1 = 1{,}733.8$ psia; $p_2 = 1{,}467.6$ psia; and $p_3 = 402.84$ psia.

An MBC can be carried out to validate the solution:

- Fluid flow from the boundary block into Block 1 is calculated to be

$$q_1 = 2 \times 1.127 \times 10^{-3} \times \frac{kh}{\mu B}(p_0 - p_1) = 1{,}500 \text{ STB/D}. \quad\quad\quad\quad\quad (8.85)$$

  The previous equation has a factor of 2, because the corner blocks receive flow across the two edges.
- Fluid flow from the boundary block to Block 2 is calculated to be

$$q_2 = 1.127 \times 10^{-3} \times \frac{kh}{\mu B}(p_0 - p_2) = 1{,}500 \text{ STB/D}. \quad\quad\quad\quad\quad\quad (8.86)$$

- Total fluid entering the reservoir is

$$q_{\text{total}} = 4q_1 + 4q_2 = 12{,}000 \text{ STB} / \text{D,} \quad\quad\quad\quad\quad\quad\quad\quad\quad\quad\quad (8.87)$$

which equals the production rate of the well, $q_3 = -12{,}000$ STB/D (a perfect MBC).

Problem 8.4.4 (Ertekin et al. 2001)

Consider the 1D, single-phase steady-state flow of oil in the horizontal, heterogeneous system shown in **Fig. 8.4. Table 8.4** gives the gridblock properties. The boundary conditions are prescribed as follows:

- Pressure in Block 1 is maintained at 3,000 psia.
- Production rate from Block 2 is specified as 1,000 STB/D.
- Pressure gradient at the extreme right of the system is given as 0.2 psi/ft.

The fluid properties are $\mu_o = 2$ cp and $B_o = 1$ res bbl/STB. With the appropriate PDE and its finite-difference approximation, calculate the pressure distribution in the system.

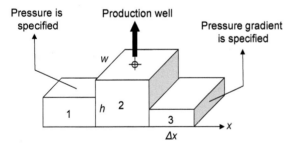

Fig. 8.4—Reservoir system for Problem 8.4.4.

Grid	Δx, ft	h, ft	w, ft	A_x, ft²	k_x, md
1	200	40	100	4,000	200
2	400	60	100	6,000	160
3	300	20	100	2,000	180

Table 8.4—Gridblock properties for Problem 8.4.4.

Solution to Problem 8.4.4

For a 1D reservoir with heterogeneous-permeability and thickness distributions, the flow equation is written as follows:

$$\frac{\partial}{\partial x}\left(\frac{A_x k_x}{\mu B} \frac{\partial p}{\partial x}\right)\Delta x + q = 0. \quad\quad\quad\quad\quad\quad\quad\quad\quad\quad\quad\quad (8.88)$$

The finite-difference approximation of the governing flow equation is

$$\left.\frac{A_x k_x}{\mu B \Delta x}\right|_{i+\frac{1}{2}} \left(p_{i+1} - p_i\right) + \left.\frac{A_x k_x}{\mu B \Delta x}\right|_{i-\frac{1}{2}} \left(p_i - p_{i-1}\right) + q_i = 0. \quad \dots\dots\dots\dots\dots\dots\dots\dots\dots\dots\dots\dots\dots \text{(8.89)}$$

The pressure of Block 1 is fixed at 3,000 psia. The transmissibility factors between blocks need to be calculated using the harmonic averaging technique:

$$\left.\frac{A_x k_x}{\mu B \Delta x}\right|_{1\frac{1}{2}} = T_{12} = \frac{2 A_1 A_2 k_1 k_2}{\mu B \left(A_1 k_1 \Delta x_2 + A_2 k_2 \Delta x_1\right)} = 1.690 \text{ STB/D-psi}, \quad \dots\dots\dots\dots\dots\dots\dots\dots\dots\dots \text{(8.90)}$$

$$\left.\frac{A_x k_x}{\mu B \Delta x}\right|_{2\frac{1}{2}} = T_{23} = \frac{2 A_3 A_2 k_3 k_2}{\mu B \left(A_3 k_3 \Delta x_2 + A_2 k_2 \Delta x_3\right)} = 0.902 \text{ STB/D-psi}, \quad \dots\dots\dots\dots\dots\dots\dots\dots\dots\dots \text{(8.91)}$$

$$T_{33'} = \frac{A_3 k_3}{\mu B \Delta x_3} = 0.68 \text{ STB/D-psi.} \quad \dots\dots\dots\dots\dots\dots\dots\dots\dots\dots\dots\dots\dots\dots\dots\dots\dots\dots\dots \text{(8.92)}$$

- For Block 2: $T_{12} p_1 - \left(T_{12} + T_{23}\right) p_2 + T_{23} p_3 = 1,000.$ $\quad \dots\dots\dots\dots\dots\dots\dots\dots\dots\dots\dots\dots\dots\dots \text{(8.93)}$
- For Block 3: $T_{23} p_2 - \left(T_{33'} + T_{23}\right) p_3 + T_{33'} p_3' = 0,$ $\quad \dots\dots\dots\dots\dots\dots\dots\dots\dots\dots\dots\dots\dots \text{(8.94)}$

and

$$\frac{p_3' - p_3}{\Delta x_3} = 0.2 \text{ psia/ft} \quad \dots\dots\dots\dots\dots\dots\dots\dots\dots\dots\dots\dots\dots\dots\dots\dots\dots\dots\dots\dots \text{(8.95)}$$

or

$$p_3' = 0.2 \Delta x_3 + p_3. \quad \dots\dots\dots\dots\dots\dots\dots\dots\dots\dots\dots\dots\dots\dots\dots\dots\dots\dots\dots\dots\dots\dots \text{(8.96)}$$

Rearranging,

$$-2.59 p_2 + 0.9 p_3 = -4,070, \quad \dots\dots\dots\dots\dots\dots\dots\dots\dots\dots\dots\dots\dots\dots\dots\dots\dots\dots\dots\dots \text{(8.97)}$$

$$0.9 p_2 - 0.9 p_3 = -40.6. \quad \dots\dots\dots\dots\dots\dots\dots\dots\dots\dots\dots\dots\dots\dots\dots\dots\dots\dots\dots\dots\dots \text{(8.98)}$$

Solving for unknowns, $p_2 = 2{,}432.46$ psia and $p_3 = 2{,}477.46$ psia.

Note: To validate the solution, an alternative method without solving the PDE can be used. Knowing the pressure gradient applied on the boundary of Block 3, the fluid flow from the external boundary to the reservoir can be calculated:

$$q_3 = 1.127 \times 10^{-3} \times \frac{Ak}{\mu B} 0.2 = 40.572 \text{ STB/D.} \quad \dots\dots\dots\dots\dots\dots\dots\dots\dots\dots\dots\dots\dots\dots \text{(8.99)}$$

With the 1,000 STB/D of production from the well at Block 2, the fluid flow from Block 1 to Block 2 is:

$$1{,}000 \text{ STB/D} - 40.572 \text{ STB/D} = 959.43 \text{ STB/D.} \quad \dots\dots\dots\dots\dots\dots\dots\dots\dots\dots\dots\dots \text{(8.100)}$$

Therefore,

$$T_{12}\left(p_1 - p_2\right) = 959.43 \text{ STB/D}, \quad \dots\dots\dots\dots\dots\dots\dots\dots\dots\dots\dots\dots\dots\dots\dots\dots\dots \text{(8.101)}$$

and $p_1 = 3{,}000$ psia; thus,

$$p_2 = p_1 - \frac{959.43}{T_{12}} = 2{,}432.46 \text{ psia.} \quad \dots\dots\dots\dots\dots\dots\dots\dots\dots\dots\dots\dots\dots\dots\dots\dots \text{(8.102)}$$

Then,

$$T_{23}\left(p_3 - p_2\right) = 40.572 \text{ STB/D}, \quad \dots\dots\dots\dots\dots\dots\dots\dots\dots\dots\dots\dots\dots\dots\dots\dots\dots \text{(8.103)}$$

$$p_3 = p_2 + \frac{40.572}{T_{23}} = 2{,}477.46 \text{ psia.} \quad \dots\dots\dots\dots\dots\dots\dots\dots\dots\dots\dots\dots\dots\dots\dots\dots \text{(8.104)}$$

Problem 8.4.5 (Ertekin et al. 2001)

Consider the incompressible fluid flow problem in the heterogeneous, anisotropic porous medium shown in **Fig. 8.5**. The fluid properties for this problem are $B = 1$ res bbl/STB, $\mu = 1$ cp, and $\rho = 62.4$ lbm/ft^3. **Table 8.5** shows the gridblock properties. Ignoring the potential gradient, calculate pressure distribution in the system and determine the nature of the boundary condition on the lowest z boundary of the system. Quantify the boundary conditions and provide the dimensionality of the problem; boundary conditions are as follows:

- All boundary planes perpendicular to the z-direction are no-flow boundaries, except the bottom plane of Block 1, which is the unknown of the problem.
- All boundary planes perpendicular to the y-direction, except the extreme right face of Block 4, are no-flow boundaries.
- Block 4 represents a constant-pressure block of 2,800 psia, and the well in Block 3 produces at a rate of 300 STB/D. Assume that $r_w = 0.25$ ft, $s = 0$, and the measured sandface pressure of the well is 144.13 psi.

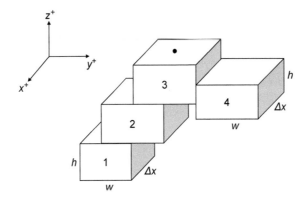

Fig. 8.5—Reservoir system for Problem 8.4.5.

Grid	Δx, ft	Δy, ft	Δz, ft	$k_x = k_y$, md	k_z, md	A_y, ft^2	A_z, ft^2
1	400	800	40	32	46	16,000	320,000
2	600	600	24	26	37	14,400	360,000
3	700	600	31	24	18	21,700	420,000
4	500	700	18	28	13	9,000	350,000

Table 8.5—Gridblock properties for Problem 8.4.5.

Solution to Problem 8.4.5

The governing flow equation for a 2D incompressible reservoir (ignoring the gravitational forces) can be written as

$$\frac{\partial}{\partial y}\left(\frac{A_y k_y}{\mu B}\frac{\partial p}{\partial y}\right)\Delta y + \frac{\partial}{\partial z}\left(\frac{A_z k_z}{\mu B}\frac{\partial p}{\partial z}\right)\Delta z + q = 0. \quad \dots \dots \dots \dots \dots \dots \dots \dots \dots \dots \dots \dots \dots \dots \dots \text{(8.105)}$$

The finite-difference approximation of the governing flow equation is

$$\frac{A_y k_y}{\mu B \Delta y}\bigg|_{j+\frac{1}{2},k}\left(p_{j+1,k}-p_{j,k}\right)+\frac{A_y k_y}{\mu B \Delta y}\bigg|_{j-\frac{1}{2},k}\left(p_{j,k}-p_{j-1,k}\right)+\frac{A_z k_z}{\mu B \Delta z}\bigg|_{j,k+\frac{1}{2}}\left(p_{j,k+1}-p_{j,k}\right)+$$

$$\frac{A_z k_z}{\mu B \Delta z}\bigg|_{j,k-\frac{1}{2}}\left(p_{j,k}-p_{j,k-1}\right)+q_{j,k}=0. \quad \dots \dots \dots \dots \dots \dots \dots \dots \dots \dots \dots \text{(8.106)}$$

The transmissibility coefficients are calculated as follows:

$$\frac{A_z k_z}{\mu B \Delta z}\bigg|_{1,1\frac{1}{2}}=T_{12}=\frac{2A_{z1}A_{z2}k_{z1}k_{z2}}{\mu B\left(A_{z1}k_{z1}\Delta z_2 + A_{z2}k_{z2}\Delta z_1\right)}=\frac{2(320,000)(360,000)(46)(37)(1.127\times10^{-3})}{(1)(1)\left[(320,000)(46)(24)+(360,000)(37)(40)\right]} \quad \dots \dots \text{(8.107)}$$

$$= 498.76 \text{ STB/D-psi,}$$

$$\left.\frac{A_z k_z}{\mu B \Delta z}\right|_{1,2\frac{1}{2}} = T_{23} = \frac{2A_{z3}A_{z2}k_{z3}k_{z2}}{\mu B\left(A_{z3}k_{z3}\Delta z_2 + A_{z2}k_{z2}\Delta z_3\right)} = \frac{2(420,000)(360,000)(18)(37)(1.127\times10^{-3})}{(1)(1)\left[(420,000)(18)(24)+(360,000)(37)(31)\right]} \dots \dots (8.108)$$

$$= 381.883 \text{ STB/D-psi,}$$

$$\left.\frac{A_y k_y}{\mu B \Delta y}\right|_{2,3\frac{1}{2}} = T_{34} = \frac{2A_{y3}A_{y4}k_{y3}k_{y4}}{\mu B\left(A_{y3}k_{y3}\Delta y_4 + A_{y4}k_{y4}\Delta y_3\right)} = \frac{2(21,700)(9,000)(28)(24)(1.127\times10^{-3})}{(1)(1)\left[(21,700)(24)(700)+(9,000)(28)(600)\right]} \dots \dots (8.109)$$

$$= 0.5736 \text{ STB/D-psi.}$$

Now we can calculate the productivity index of the well located in Block 3. Because the well is completed perpendicular to the x-y plane, the equivalent wellbore radius is

$$r_e = 0.14\sqrt{(\Delta x)^2 + (\Delta y)^2} = 0.14\sqrt{(700)^2 + (600)^2} = 129.07 \text{ ft.} \dots \dots \dots \dots \dots \dots (8.110)$$

The productivity index for the well is calculated using $\bar{k} = 24$ md:

$$\Omega = \frac{2\pi \bar{k} h}{\mu B\left[\ln\left(\dfrac{r_e}{r_w}\right)+s\right]} = \frac{2\pi(24)(1.127\times10^{-3})(31)}{(1)(1)\left[\ln\left(\dfrac{129.07}{0.25}\right)\right]} = 0.8430 \text{ STB/D-psi.} \dots \dots \dots \dots \dots (8.111)$$

Knowing that $q_3 = -300$ STB/D and $p_{sf3} = 144.13$ psi, the pressure of Block 3 can be calculated as

$$p_3 = p_{sf} - \frac{q_3}{\Omega_3} = 144.13 - \frac{-300}{0.8430} = 500 \text{ psi.} \dots \dots \dots \dots \dots \dots \dots \dots \dots (8.112)$$

The mass flow between Blocks 3 and 4,

$$q_{3-4} = T_{34}(p_4 - p_3) = 0.5736 \times (2,800 - 500) = 1,319.28 \text{ STB/D} > 300 \text{ STB/D.} \dots \dots \dots \dots (8.113)$$

For an incompressible system, the flow in must be equal to the flow out. Therefore, $q_{\text{boundary}} = 1,319.28 - 300 = 1,019.28$ STB/D, which indicates that $q_{\text{boundary}} = 1,019.28$ STB/D of fluid should leak across the boundary of Block 1. The boundary condition can be determined using Darcy's law (Darcy 1856), because the flow out is considered negative:

$$q_{\text{boundary}} = \frac{k_{z1}A_{z1}}{2}c = -1,019.28 \text{ STB/D.} \dots \dots \dots \dots \dots \dots \dots \dots \dots (8.114)$$

One can solve for the pressure gradient, c, across the bottom boundary of Block 1:

$$c = \frac{\mu B q_{\text{boundary}}}{k_{z1}A_{z1}} = \frac{(1)(1)(-1,019.28)}{(46)(1.127\times10^{-3})(320,000)} = -0.0614 \text{ psi/ft.} \dots \dots \dots \dots \dots (8.115)$$

The pressures of Blocks 1 and 2 can be calculated in a successive manner:

$$q_{2-3} = T_{2-3}(p_3 - p_2), \dots \dots \dots \dots \dots \dots \dots \dots \dots \dots \dots \dots \dots \dots \dots \dots \dots (8.116)$$

$$p_2 = p_3 - \frac{q_{2-3}}{T_{2-3}} = 497.33 \text{ psi,} \dots \dots \dots \dots \dots \dots \dots \dots \dots \dots \dots \dots \dots \dots \dots (8.117)$$

$$q_{1-2} = T_{1-2}(p_2 - p_1), \dots \dots \dots \dots \dots \dots \dots \dots \dots \dots \dots \dots \dots \dots \dots \dots \dots (8.118)$$

$$p_1 = p_2 - \frac{q_{1-2}}{T_{1-2}} = 495.28 \text{ psi.} \dots \dots \dots \dots \dots \dots \dots \dots \dots \dots \dots \dots \dots \dots \dots (8.119)$$

Problem 8.4.6 (Ertekin et al. 2001)

Consider the 2D single-phase flow of incompressible oil taking place in the horizontal, homogeneous reservoir of **Fig. 8.6.1.** The gridblock and fluid properties for this problem are $\Delta x = \Delta y = 400$ ft, $k_x = k_y = 200$ md, $h = 40$ ft, $\mu_o = 2$ cp, and $B_o = 1$ res bbl/STB. The boundary conditions are as follows:

- No-flow boundaries across all of the external flow planes.
- Pressure of each injection block is 5,000 psia.
- Pressure of production block is 2,000 psia.

Calculate the following:

- The pressure over the entire grid using a direct-solution method.
- The injection rate of each injection well.
- The production rate for the production well. *Hint:* Observe the symmetry.

● Production well

⌀ Injection well

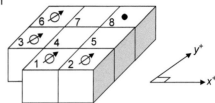

Fig. 8.6.1—Reservoir system for Problem 8.4.6.

Solution to Problem 8.4.6

According the symmetry of the system, by knowing the pressure of the injection blocks to be 5,000 psia and the production block to be 2,000 psia, there are two blocks with unknown pressures in the system, which are p_4 and p_5 (see **Fig. 8.6.2;** note that because of existing symmetry $p_7 = p_5$). The governing flow equation for a 2D incompressible reservoir can be written as

$$\frac{\partial}{\partial x}\left(\frac{A_x k_x}{\mu B}\frac{\partial p}{\partial x}\right)\Delta x + \frac{\partial}{\partial y}\left(\frac{A_y k_y}{\mu B}\frac{\partial p}{\partial y}\right)\Delta y + q = 0, \dots\dots\dots\dots\dots\dots (8.120)$$

which can be simplified for a homogeneous and isotropic system to

$$\frac{\partial^2 p}{\partial x^2} + \frac{\partial^2 p}{\partial y^2} + \frac{q\mu B}{V_b k} = 0, \dots\dots\dots\dots\dots\dots\dots\dots\dots\dots (8.121)$$

and then its finite-difference representation will be

$$\frac{p_{i-1,j} - 2p_{i,j} + p_{i+1,j}}{(\Delta x)^2} + \frac{p_{i,j-1} - 2p_{i,j} + p_{i,j+1}}{(\Delta y)^2} + \left(\frac{q\mu B}{\Delta x \Delta y h k}\right)_{i,j} = 0. \dots\dots\dots\dots\dots (8.122)$$

Because $\Delta x = \Delta y$, the previous equation can further be simplified to

$$p_{i,j-1} + p_{i-1,j} - 4p_{i,j} + p_{i+1,j} + p_{i,j+1} = -\left(\frac{q\mu B}{hk}\right)_{i,j}. \dots\dots\dots\dots\dots\dots (8.123)$$

One can generate the linear algebraic equations at the blocks with unknown pressures:

- Block 4: $(p_3 - 2p_4 + p_5) + (p_1 - 2p_4 + p_5) = 0.$ ....(8.124)
- Block 5: $(p_2 - 2p_5 + p_8) + (p_4 - p_5) = 0.$ ....(8.125)

Substituting for $p_1 = p_2 = p_3 = 5{,}000$ psi and $p_8 = 2{,}000$ psi, one obtains

$$-4p_4 + 2p_5 = -10{,}000 \dots\dots\dots\dots\dots\dots (8.126)$$

or

$$-2p_4 + p_5 = -5{,}000. \dots\dots\dots\dots\dots\dots (8.127)$$

Similarly,

$$-3p_5 + p_4 = -5{,}000 - 2{,}000 \dots\dots\dots\dots\dots (8.128)$$

or

$$p_4 - 3p_5 = -7{,}000. \dots\dots\dots\dots\dots\dots (8.129)$$

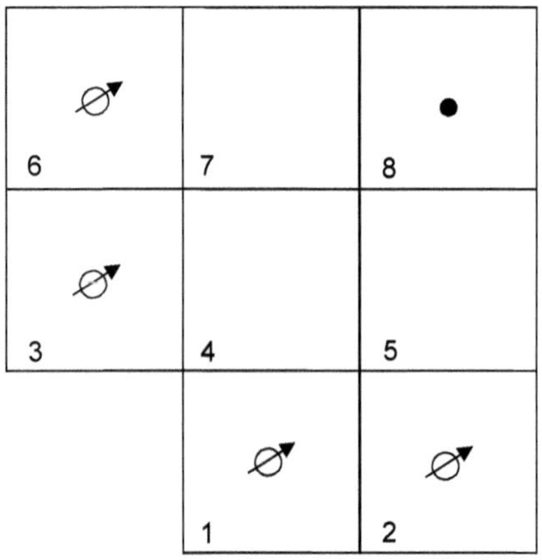

Fig. 8.6.2—Block numbering scheme of reservoir system for Problem 8.4.6.

The system of equations can be directly solved using the Gaussian elimination:

$$-p_4 + \frac{1}{2} p_5 = -2{,}500, \dots\dots\dots\dots\dots\dots\dots\dots\dots\dots\dots\dots\dots\dots\dots\dots\dots\dots \text{(8.130)}$$

$$-\frac{5}{2} p_5 = -9{,}500. \dots\dots\dots\dots\dots\dots\dots\dots\dots\dots\dots\dots\dots\dots\dots\dots\dots\dots\dots\dots \text{(8.131)}$$

Therefore, $p_5 = 3{,}800$ psia and $p_4 = 2{,}500 + \frac{1}{2} p_5 = 4{,}400$ psia.

To calculate the production rate of the producer, one needs to calculate the mass flow from Blocks 1 to 4 and from Blocks 2 to 5. Because $p_1 = p_2$, there is no flow across Blocks 1 and 2 or Blocks 3 and 6. Note that the flow rate between Blocks 1 and 4 and Blocks 2 and 5 needs to be multiplied by 2, because they are duplicated between Blocks 3 and 4 and Blocks 6 and 7, respectively:

$$q_{1,4} = 1.127 \times 10^{-3} \times \frac{hk}{\mu B}(5{,}000 - 4{,}400) = 5{,}409.6 \text{ STB/D}, \dots\dots\dots\dots\dots\dots\dots\dots\dots \text{(8.132)}$$

$$q_{2,5} = 1.127 \times 10^{-3} \times \frac{hk}{\mu B}(5{,}000 - 3{,}800) = 10{,}819.2 \text{ STB/D}. \dots\dots\dots\dots\dots\dots\dots\dots\dots \text{(8.133)}$$

Therefore, the flow rate of the producer is calculated as

$$q_8 = 2\left(q_{1,4} + q_{2,5}\right) = 2\left(5{,}409.6 + 10{,}819.2\right) = 32{,}457.6 \text{ STB/D}. \dots\dots\dots\dots\dots\dots\dots\dots \text{(8.134)}$$

Problem 8.4.7 (Ertekin et al. 2001)

Simulate the flow of incompressible oil in the horizontal reservoir represented by the grid shown in **Fig. 8.7.1. Table 8.6** displays gridblock permeability and thickness values. Fluid viscosity is 0.2 mPa·s. There are wells at Node (1,1), (2,3), and (3,2). The well at Node (2,3) produces at a rate of 103.68 std m³/D. The pressure at Node (1,1) and (3,2) are maintained at $p(1,1) = 3{,}500$ kPa and $p(3,2) = 100$ kPa.

Use the 1) direct-solution method, 2) Jacobi method, 3) Gauss-Seidel method, and 4) PSOR method to solve pressure distribution in the reservoir. Calculate well flow rate at Nodes (1,1) and (3,2), and check the material balance for each method. Use a tolerance of 1 kPa with iterative methods.

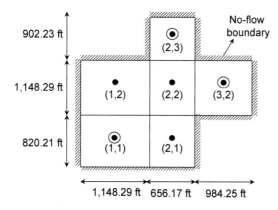

Fig. 8.7.1—Reservoir system for Problem 8.4.7.

Block	h, m	k_x, µm²	k_y, µm²
(1,1)	12	0.02	0.02
(1,2)	15	0.15	0.15
(2,1)	10	0.3	0.3
(2,2)	18	0.03	0.03
(2,3)	5	0.06	0.06
(3,2)	6	0.04	0.04

Table 8.6—Gridblock permeability and thickness values for Problem 8.4.7.

Solution to Problem 8.4.7

The problem is considered incompressible heterogeneous with isotropic permeability distribution. The governing equation can be written as

$$\frac{\partial}{\partial x}\left(\frac{A_x k_x}{\mu B} \frac{\partial p}{\partial x}\right)\Delta x + \frac{\partial}{\partial y}\left(\frac{A_y k_y}{\mu B} \frac{\partial p}{\partial y}\right)\Delta y + q = 0. \dots\dots\dots\dots\dots\dots\dots\dots\dots\dots\dots \text{(8.135)}$$

The PDE can be written in a discretized form:

$$T_{i,j-\frac{1}{2}} P_{i,j-1} + T_{i,j+\frac{1}{2}} P_{i,j+1} + T_{i-\frac{1}{2},j} P_{i-1,j} + T_{i+\frac{1}{2},j} P_{i+1,j} - \left(T_{i,j-\frac{1}{2}} + T_{i,j+\frac{1}{2}} + T_{i-\frac{1}{2},j} + T_{i+\frac{1}{2},j} \right) p_{i,j} + q = 0. \quad \dots\dots\dots\dots (8.136)$$

The transmissibility factor is calculated using the harmonic averaging technique

$$T_{i,j\pm\frac{1}{2}} = \frac{1}{\mu B} \frac{2 A_{i,j} A_{i,j\pm1} k_{y_{i,j}} k_{y_{i,j\pm1}}}{\left(A_{i,j} k_{y_{i,j}} \Delta y_{i,j\pm1} + A_{i,j\pm1} k_{y_{i,j\pm1}} \Delta y_{i,j} \right)}, \quad \dots\dots\dots\dots\dots\dots\dots\dots (8.137)$$

$$T_{i\pm\frac{1}{2},j} = \frac{1}{\mu B} \frac{2 A_{i,j} A_{i\pm1,j} k_{x_{i,j}} k_{x_{i\pm1,j}}}{\left(A_{i,j} k_{x_{i,j}} \Delta x_{i\pm1,j} + A_{i\pm1,j} k_{x_{i\pm1,j}} \Delta x_{i,j} \right)}. \quad \dots\dots\dots\dots\dots\dots\dots\dots (8.138)$$

Use the block numbering scheme found in **Fig. 8.7.2** and **Table 8.7.** The calculated transmissibility factors are listed in **Table 8.8.**

Because $p_1 = 3{,}500$ kPa and $p_6 = 100$ kPa, there are four unknowns in this problem, which are p_2, p_3, p_4, and p_5. The system of equations are written in matrix form:

$$\begin{bmatrix} -0.828 & 0 & 0.575 & 0 \\ 0 & -0.378 & 0.236 & 0 \\ 0.575 & 0.236 & -1.108 & 0.110 \\ 0 & 0 & 0.1104 & -0.110 \end{bmatrix} \begin{bmatrix} p_2 \\ p_3 \\ p_4 \\ p_5 \end{bmatrix} = \begin{bmatrix} -884.046 \\ -495.738 \\ -18.662 \\ 103.680 \end{bmatrix}. \quad \dots\dots\dots\dots\dots\dots (8.139)$$

1. Use Gaussian elimination, which is a direct solver:

$$\begin{bmatrix} 1 & 0 & -0.695 & 0 \\ 0 & -0.378 & 0.236 & 0 \\ 0 & 0.236 & -0.709 & 0.110 \\ 0 & 0 & 0.110 & -0.110 \end{bmatrix} \begin{bmatrix} p_2 \\ p_3 \\ p_4 \\ p_5 \end{bmatrix} = \begin{bmatrix} 1{,}068.243 \\ -495.738 \\ -632.887 \\ 103.680 \end{bmatrix}, \quad \dots\dots\dots\dots\dots\dots (8.140)$$

$$\begin{bmatrix} 1 & 0 & -0.695 & 0 \\ 0 & 1.000 & -0.625 & 0 \\ 0 & 0 & -0.561 & 0.110 \\ 0 & 0 & 0.110 & -0.110 \end{bmatrix} \begin{bmatrix} p_2 \\ p_3 \\ p_4 \\ p_5 \end{bmatrix} = \begin{bmatrix} 1{,}068.243 \\ 1{,}311.919 \\ -942.805 \\ 103.680 \end{bmatrix}, \quad \dots\dots\dots\dots\dots\dots (8.141)$$

$$\begin{bmatrix} 1 & 0 & -0.695 & 0 \\ 0 & 1.000 & -0.625 & 0 \\ 0 & 0 & 1 & -0.197 \\ 0 & 0 & 0 & -0.089 \end{bmatrix} \begin{bmatrix} p_2 \\ p_3 \\ p_4 \\ p_5 \end{bmatrix} = \begin{bmatrix} 1{,}068.243 \\ 1{,}311.919 \\ 1{,}680.301 \\ -81.873 \end{bmatrix}. \quad \dots\dots\dots\dots\dots\dots (8.142)$$

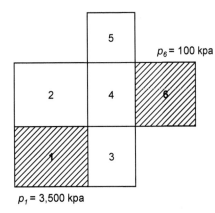

$p_6 = 100$ kpa

$p_1 = 3{,}500$ kpa

Fig. 8.7.2—Numbering of the blocks of the reservoir system as described in Problem 8.4.7.

Block	Δx, m	Δy, m	h, m	A_x, m²	A_y, m²	k_x, µm²	k_y, µm²
1	350	250	12	3,000	4,200	0.02	0.02
2	350	350	15	5,250	5,250	0.15	0.15
3	200	250	10	2,500	2,000	0.3	0.3
4	200	350	18	6,300	3,600	0.03	0.03
5	200	275	5	1,375	1,000	0.06	0.06
6	300	350	6	2,100	1,800	0.04	0.04

Table 8.7—Relevant reservoir and gridblock properties used in the solution to Problem 8.4.7.

Therefore,

- $p_5 = -81.873/-0.089 = 923.10$ kPa, ...(8.143)
- $p_4 = 1,680.301 + 0.197p_4 = 1,862.0$ kPa, ...(8.144)
- $p_3 = 1,311.92 + 0.625p_3 = 2,476.0$ kPa, ...(8.145)
- $p_2 = 1,068.24 + 0.695p_3 = 2,361.92$ kPa. ....(8.146)

The flow rate in Block (3,2) can be calculated the same as the flow from Blocks 4 to 6:

$$q_{well(3,2)} = q_{4-6} = T_{46}\left(p_4 - p_6\right) = 0.1866 \times \left(1,862.0 - 100\right) = 328.82 \, \text{std m}^3/\text{d}\left(\text{production}\right), \ldots \ldots (8.147)$$

$$q_{well(1,1)} = q_{1-2} + q_{1-3} = T_{12}\left(p_1 - p_2\right) + \left(p_1 - p_3\right)$$
$$= 0.2526 \times \left(3,500 - 2,361.92\right) + 0.1416 \times \left(3,500 - 2,475.96\right). \ldots \ldots (8.148)$$
$$= 432.51 \, \text{std m}^3/\text{d}\left(\text{injection}\right)$$

MBC—in this case, the production rate and the injection rate should be identical for the incompressible fluid system:

$$\text{MBC} = \left|\frac{q_{\text{injection}}}{q_{\text{production}}}\right| = \left|\frac{432.51}{432.51}\right| = 1.000, \ldots \ldots (8.149)$$

which results in a good MBC.

2. Jacobi method—It takes 28 iterations (see **Table 8.9**) to achieve the prescribed convergence tolerance of 1 kPa:

$$q_{well(3,2)} = q_{4-6} = T_{46}\left(p_4 - p_6\right) = 0.1866 \times \left(1,860.723 - 100\right) = 328.59 \, \text{std m}^3/\text{d}\left(\text{production}\right), \ldots \ldots (8.150)$$

$$q_{well(1,1)} = q_{1-2} + q_{1-3} = T_{12}\left(p_1 - p_2\right) + T_{13}\left(p_1 - p_3\right)$$
$$= 0.2526 \times \left(3,500 - 2,360.469\right) + 0.1416 \times \left(3,500 - 2,474.657\right) \ldots \ldots (8.151)$$
$$= 433.06 \, \text{std m}^3/\text{d}\left(\text{injection}\right),$$

$$\text{MBC} = \left|\frac{q_{\text{injection}}}{q_{\text{production}}}\right| = \left|\frac{432.27}{433.06}\right| = 0.9982, \ldots \ldots (8.152)$$

which is not as good as the result obtained from the direct solver, because of the relatively coarse nature of the convergence tolerance used.

3. Gauss-Seidel method—It takes 14 iterations (see **Table 8.10**) to achieve a prescribed convergence tolerance of 1 kPa:

$$q_{well(3,2)} = q_{4-6} = T_{46}\left(p_4 - p_6\right) = 0.1866 \times \left(1,860.829 - 100\right) = 328.61 \, \text{std m}^3/\text{d}\left(\text{production}\right), \ldots \ldots (8.153)$$

Transmissibility	std m³/D-kPa
T_{45}	0.110
T_{24}	0.575
T_{46}	0.187
T_{34}	0.236
T_{12}	0.253
T_{13}	0.142

Table 8.8—Calculated transmissibility coefficients for Solution to Problem 8.4.7.

Iteration Level	p_2	p_3	p_4	p_5
0	0	0	0	0
1	1,068.243	1,311.919	16.8392	−938.889
2	1,079.943	1,322.446	757.148	−922.05
3	1,594.3	1,785.262	767.1398	−181.741
4	1,601.242	1,791.509	1,206.411	−171.749
5	1,906.443	2,066.126	1,212.34	267.5223
...				
25	2,359.454	2,473.743	1,858.458	919.5366
26	2,359.476	2,473.763	1,859.868	919.5687
27	2,360.456	2,474.645	1,859.887	920.9788
28	2,360.469	2,474.657	1,860.723	920.9978

Table 8.9—Iterations using the Jacobi method in Part 2 of Solution to Problem 8.4.7.

Iteration level	p_2	p_3	p_4	p_5
0	0	0	0	0
1	1,068.243	1,311.919	850.6992	−88.1897
2	1,659.298	1,843.747	1,261.921	323.032
3	1,945.01	2,100.829	1,505.924	567.0354
4	2,114.541	2,253.372	1,650.707	711.8179
...				
11	2,355.512	2,470.196	1,856.501	917.6122
12	2,358.117	2,472.54	1,858.726	919.8369
13	2,359.663	2,473.931	1,860.046	921.157
14	2,360.58	2,474.756	1,860.829	921.9403

Table 8.10—Iterations using the Gauss-Seidel method in Part 3 of Solution to Problem 8.4.7.

$$q_{\text{well}(1,1)} = q_{1-2} + q_{1-3} = T_{12}\left(p_1 - p_2\right) + T_{13}\left(p_1 - p_3\right)$$
$$= 0.2526 \times \left(3,500 - 2,360.58\right) + 0.1416 \times \left(3,500 - 2,474.756\right), \quad \dotsb \quad (8.154)$$
$$= 433.01 \, \text{std m}^3/\text{d} \left(\text{injection}\right)$$

$$\text{MBC} = \left| \frac{q_{\text{injection}}}{q_{\text{production}}} \right| = \left| \frac{433.01}{433.29} \right| = 0.9983, \quad \dotsb \quad (8.155)$$

which is not as accurate as the result obtained from the direct solver, again because of the use of the coarse convergence tolerance.

4. PSOR method—Using $\omega_{\text{opt}} = 1.2212$, it takes eight iterations (see **Table 8.11**) to achieve a prescribed convergence tolerance of 1 kPa.

$$q_{\text{well}(3,2)} = q_{4-6} = T_{46}\left(p_4 - p_6\right) = 0.1866 \times \left(1,861.92 - 100\right) = 328.82 \, \text{std m}^3/\text{d} \left(\text{production}\right), \quad \dotsb \quad (8.156)$$

$$q_{\text{well}(1,1)} = q_{1-2} + q_{1-3} = T_{12}\left(p_1 - p_2\right) + T_{13}\left(p_1 - p_3\right)$$
$$= 0.2526 \times \left(3,500 - 2,361.83\right) + 0.1416 \times \left(3,500 - 2,475.879\right), \quad \dotsb \quad (8.157)$$
$$= 432.5 \, \text{std m}^3/\text{d} \left(\text{injection}\right)$$

$$\text{MBC} = \left| \frac{q_{\text{injection}}}{q_{\text{production}}} \right| = \left| \frac{433.5}{433.54} \right| = 0.9999. \quad \dotsb \quad (8.158)$$

At the convergence stage, PSOR with ω_{opt} yields the best MBC result compared with the other implemented iterative solvers. This is because of the reduced number of computations in which round-off error is decreased.

Iteration level	p_2	p_3	p_4	p_5
0	0	0	0	0
1	1,304.538	1,602.115	1,264.124	397.177
2	2,088.552	2,212.826	1,688.53	827.6066
3	2,275.226	2,401.751	1,814.477	886.2013
4	2,340.796	2,456.115	1,849.442	915.9396
5	2,355.959	2,470.784	1,858.751	920.7303
6	2,360.504	2,474.647	1,861.16	922.612
7	2,361.542	2,475.631	1,861.77	922.9411
8	2,361.83	2,475.879	1,861.923	923.0542

Table 8.11—Iterations using the PSOR method in Part 4 of Solution to Problem 8.4.7.

Problem 8.4.8 (Ertekin et al. 2001)

Consider the single-phase, steady-state flow in the horizontal porous medium with heterogeneous-permeability distribution. The boundary conditions are shown in **Fig. 8.8**.

- On the extreme right-hand side of the system, the pressure gradient is specified as −0.2 psi/ft.
- On the extreme-left-hand side of the system, the pressure gradient is specified as −0.8 psi/ft.
- The well in Block 4 produces at a constant block pressure of 100 psia.

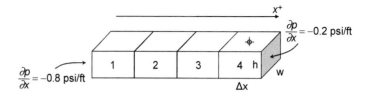

Fig. 8.8—Reservoir system for Problem 8.4.8.

The gridblock and fluid properties for this problem are $\Delta x = 600$ ft, $w = 100$ ft, $h = 80$ ft, $k_{x1} = 60$ md, $k_{x2} = 60$ md, $k_{x3} = 68$ md, $k_{x4} = 46$ md, $\mu = 1$ cp, and $B = 1$ res bbl/STB.

Solve for the steady-state pressure distribution in the system. In the solution of the system of equations generated by the characteristic finite-difference equations, use the Thomas' algorithm (Barrett et al. 1994). Calculate the production rate of the well.

Solution to Problem 8.4.8

For a 1D system with heterogeneous-permeability distribution, calculations of transmissibility factors are required. The SIP notation is employed for convenience:

$$W_i = \frac{2A_i A_{i-1} k_i k_{i-1}}{\mu B(A_i k_i \Delta x_{i-1} + A_{i-1} k_{i-1} \Delta x_i)}, \quad \ldots \ldots \ldots \ldots \ldots \ldots \ldots \ldots \ldots \ldots \ldots (8.159)$$

$$E_i = \frac{2A_i A_{i+1} k_i k_{i+1}}{\mu B(A_i k_i \Delta x_{i+1} + A_{i+1} k_{i+1} \Delta x_i)}, \quad \ldots \ldots \ldots \ldots \ldots \ldots \ldots \ldots \ldots \ldots \ldots (8.160)$$

$$C_i = -(W_i + E_i). \quad \ldots \ldots \ldots \ldots \ldots \ldots \ldots \ldots \ldots \ldots \ldots \ldots \ldots \ldots \ldots \ldots (8.161)$$

The calculated transmissibility factors are listed in **Table 8.12**.

One can then generate the system of equations:

- Block 1: $W_1 p_1' + C_1 p_1 + E_1 p_2 = 0.$ $\ldots \ldots \ldots \ldots \ldots \ldots \ldots \ldots \ldots \ldots \ldots \ldots \ldots (8.162)$

Applying the boundary condition between Block 1′ and Block 1, where Block 1′ is the reflection block for

Block 1: $\dfrac{p_1 - p_1'}{\Delta x} = -0.8$ psia / ft $\Rightarrow p_1' = p_1 + 480.$ $\ldots \ldots \ldots \ldots \ldots \ldots \ldots \ldots \ldots \ldots (8.163)$

Therefore, $-0.72 p_1 + 0.72 p_2 = -288.$ $\ldots \ldots \ldots \ldots \ldots \ldots \ldots \ldots \ldots \ldots \ldots \ldots (8.164)$

- Block 2: $W_2 p_1 + C_2 p_2 + E_2 p_3 = 0$ and $0.72 p_1 - 1.68 p_2 + 0.96 p_3 = 0.$ $\ldots \ldots \ldots \ldots \ldots \ldots \ldots (8.165)$

- Block 3: $W_3 p_2 + C_3 p_3 + E_3 p_4 = 0$, where $p_4 = 100$ psia and $0.96 p_2 - 1.78 p_3 = -82.5.$ $\ldots \ldots \ldots \ldots \ldots (8.166)$

Now the linear system of equations can be expressed in matrix form:

$$\begin{bmatrix} -0.72 & 0.72 & 0 \\ 0.72 & -1.68 & 0.96 \\ 0 & 0.96 & -1.78 \end{bmatrix} \begin{bmatrix} p_1 \\ p_2 \\ p_3 \end{bmatrix} = \begin{bmatrix} -288 \\ 0 \\ -82.5 \end{bmatrix}. \quad \ldots \ldots (8.167)$$

Block	W	C	E
1	0.60	−1.32	0.72
2	0.72	−1.68	0.96
3	0.96	−1.785	0.825
4	0.825	−1.515	0.69

Table 8.12—Calculated transmissibility factors from Solution to Problem 8.4.8.

Implementing the Thomas' algorithm:

- Forward solution:

$$w_1 = \frac{c_1}{b_1} = \frac{0.72}{-0.72} = -1, \quad \dots\dots\dots\dots\dots\dots\dots\dots\dots\dots \quad (8.168)$$

$$g_1 = \frac{d_1}{b_1} = -\frac{288}{-0.72} = 400, \quad \dots\dots\dots\dots\dots\dots\dots\dots\dots \quad (8.169)$$

$$w_2 = \frac{0.96}{-1.68 - 0.72 \times (-1)} = -1, \quad \dots\dots\dots\dots\dots\dots\dots\dots \quad (8.170)$$

$$g_2 = \frac{0 - 0.72 \times 400}{-1.68 - 0.72 \times (-1)} = 300, \quad \dots\dots\dots\dots\dots\dots\dots\dots \quad (8.171)$$

$$g_3 = \frac{-82.5 - 0.96 \times 300}{-1.78 - 0.96 \times (-1)} = 451.83 = p_3. \quad \dots\dots\dots\dots\dots\dots \quad (8.172)$$

- Backward solution:

$$p_2 = 300 - (-1) \times 451.83 = 751.83 \text{ psia}, \quad \dots\dots\dots\dots\dots\dots\dots \quad (8.173)$$

$$p_1 = 400 - (-1) \times 751.83 = 1151.83 \text{ psia}. \quad \dots\dots\dots\dots\dots\dots\dots \quad (8.174)$$

To calculate the production rate of the producer, the total mass transferred across the external boundaries needs to be calculated.

- Mass flux flow in from the left-extreme boundary:

$$q_{\text{left}} = -1.127 \times 10^{-3} \frac{Ak_1}{\mu B}(-0.8 \text{ psia} / \text{ft}) = 288.51 \text{ STB/D}. \quad \dots\dots\dots\dots \quad (8.175)$$

- Mass flux flow out through the right-extreme boundary:

$$q_{\text{right}} = 1.127 \times 10^{-3} \frac{Ak_4}{\mu B}(-0.2 \text{ psia} / \text{ft}) = 82.94 \text{ STB/D}. \quad \dots\dots\dots\dots \quad (8.176)$$

Hence, the production rate of the well is $q_{\text{well}} = q_{\text{left}} - q_{\text{right}} = 205.56$ STB/D. $\quad \dots\dots\dots\dots\dots \quad (8.177)$

Problem 8.4.9 (Ertekin et al. 2001)

Consider the 2D slightly compressible flow of oil in the porous medium shown in **Fig. 8.9.1**. Assume the system is horizontal. The block and fluid properties are $\phi = 20\%$, $k_x = k_y = 100$ md, $\Delta x = \Delta y = 600$ ft, $h = 40$ ft, $\mu = 4$ cp, $c = 5 \times 10^{-5}$ psi^{-1}, and $B = 1.2$ res bbl/STB. Assume that oil FVF, viscosity, and compressibility are not changing with pressure in the range in which you are working. The boundary conditions are as follows:

- The eastern, northern, and southern boundaries are no-flow boundaries.
- Across the western boundary, the pressure gradient is specified as $\partial p / \partial x = 0.1$ psi/ft.
- The production block is kept at 200 psia.

The initial pressure for all gridblocks is 3,600 psia at $t = 0$.

With the appropriate PDE and its backward finite-difference approximation, calculate the pressure distribution and production rate at the end of 15 and 30 days, respectively. Use a timestep of 15 days.

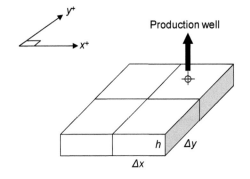

Fig. 8.9.1—Reservoir system for Problem 8.4.9.

Solution to Problem 8.4.9

For the 2D reservoir with homogeneous and isotropic properties shown in Fig. 8.9.1, the PDE describing the flow problem can be written as follows (note that μ, B, and c_t are treated as constants in the stated problem):

$$\frac{\partial^2 p}{\partial x^2} + \frac{\partial^2 p}{\partial y^2} + \frac{q\mu B}{k V_b} = \frac{\mu B c_t \phi}{5.615 k_x} \frac{\partial p}{\partial t}. \quad\dots\dots\dots\dots\dots\dots\dots\dots\dots\dots (8.178)$$

Using the numbering scheme shown in **Fig. 8.9.2,** the PDE can be written in a discretized form using the backward-difference (implicit) scheme on the temporal derivative for Block 1:

$$p_1^{'n+1} - 2p_1^{n+1} + p_4^{n+1} + p_1^{n+1} - 2p_1^{n+1} + p_2^{n+1} = \frac{\mu B c_t \phi \Delta x^2}{5.615 \Delta t k_x}(p_1^{n+1} - p_1^n). \quad\dots\dots\dots\dots\dots\dots\dots (8.179)$$

Applying the boundary condition,

$$\frac{p_1 - p_1'}{\Delta x} = 0.1 \text{ psi/ft} \quad\dots\dots\dots\dots\dots\dots\dots\dots\dots (8.180)$$

or

$$p_1' = p_1 - 0.1\Delta x. \quad\dots\dots\dots\dots\dots\dots\dots\dots\dots\dots (8.181)$$

Note that similar equations can be written between Blocks 2 and 2'.
Because the used timestep is 15 days,

$$\frac{\mu B c_t \phi \Delta x^2}{5.615 \Delta t k_x} = 1.82. \quad\dots\dots\dots\dots\dots\dots\dots\dots\dots (8.182)$$

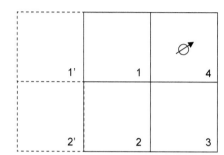

Fig. 8.9.2—Reservoir system for Problem 8.4.8.

Knowing that the pressure of the wellblock is kept at 200 psia, the characteristic linear equation for the blocks with unknown block pressures then can be rewritten:

- Block 1: $-3.82 p_1^{n+1} + p_2^{n+1} = -1.82 p_1^n - 140.$ $\dots\dots\dots\dots\dots\dots\dots\dots\dots\dots\dots (8.183)$

- Block 2: $p_1^{n+1} - 3.82 p_2^{n+1} + p_3^{n+1} = -1.82 p_2^n + 60.$ $\dots\dots\dots\dots\dots\dots\dots\dots\dots\dots (8.184)$

- Block 3: $p_2^{n+1} - 3.82 p_3^{n+1} = -1.82 p_3^n - 200.$ $\dots\dots\dots\dots\dots\dots\dots\dots\dots\dots\dots\dots (8.185)$

The previous three equations can be written in matrix form:

$$\begin{bmatrix} -3.82 & 1 & 0 \\ 1 & -3.82 & 1 \\ 0 & 1 & -3.82 \end{bmatrix} \begin{bmatrix} p_1^{n+1} \\ p_2^{n+1} \\ p_3^{n+1} \end{bmatrix} = \begin{bmatrix} -140 - 1.82 \times p_1^n \\ 60 - 1.82 \times p_2^n \\ -200 - 1.82 \times p_3^n \end{bmatrix}. \quad\dots\dots\dots\dots\dots\dots\dots (8.186)$$

Solving the linear system of equations expressed in matrix form (at the end of 15 days),

$$\begin{bmatrix} -3.82 & 1 & 0 \\ 1 & -3.82 & 1 \\ 0 & 1 & -3.82 \end{bmatrix} \begin{bmatrix} p_1^{n+1} \\ p_2^{n+1} \\ p_3^{n+1} \end{bmatrix} = \begin{bmatrix} -6,693.62 \\ -6,493.62 \\ -6,753.62 \end{bmatrix}, \quad\dots\dots\dots\dots\dots\dots\dots (8.187)$$

one obtains the pressure distribution: $p_1 = 2,547.0$ psia, $p_2 = 3,037.2$ psia, and $p_3 = 2,562.7$ psia.

- Fluid expansion of Block 1:

$$= (3,600 \text{ psia} - 2,547.0 \text{ psia}) \times 5 \times 10^{-5} \text{ psi}^{-1} \times \frac{600 \text{ ft} \times 600 \text{ ft} \times 40 \text{ ft} \times 0.2}{5.165} = 40,506.09 \text{ STB}. \dots (8.188)$$

- Fluid expansion of Block 2

$$= (3,600 \text{ psia} - 3,037.2 \text{ psia}) \times 5 \times 10^{-5} \text{ psi}^{-1} \times \frac{600 \text{ ft} \times 600 \text{ ft} \times 40 \text{ ft} \times 0.2}{5.165} = 21,650.89 \text{ STB}. \dots (8.189)$$

- Fluid expansion of Block 3

$$= \left(3,600 \text{ psia} - 2,562.7 \text{ psia}\right) \times 5 \times 10^{-5} \text{psi}^{-1} \times \frac{600 \text{ ft} \times 600 \text{ ft} \times 40 \text{ ft} \times 0.2}{5.165} = 39,901.95 \text{ STB}. \quad \dots \dots (8.190)$$

- The total fluid leakage from Blocks 1 and 2:

$$\text{Fluid leakage} = 2 \times 1.127 \times 10^{-3} \frac{kA}{\mu B} \frac{\partial p}{\partial x} \Delta t = 112.7 \text{ STB/D} \times 15 \text{ days} = 1,690.5 \text{ STB}. \quad \dots \dots \dots \dots (8.191)$$

- The total amount of fluid produced from the well during the first 15-day period:

$$Q_{t1} = \sum \text{fluid expansion} - \text{fluid leakage} = 102,058.9 \text{ STB} - 1,690.5 \text{ STB} = 100,368.4 \text{ STB}. \quad \dots \dots (8.192)$$

Then, the average production rate during the first 15 days is

$$q_1 = \frac{Q_{t1}}{15 \text{ days}} = 6,691.2 \text{ STB/D} \quad \dots \dots \dots \dots \dots \dots \dots \dots \dots \dots \dots \dots \dots (8.193)$$

Similarly, at the end of the second 15-day period (at the end of 30 days), the system of equations must be solved:

$$\begin{bmatrix} -3.82 & 1 & 0 \\ 1 & -3.82 & 1 \\ 0 & 1 & -3.82 \end{bmatrix} \begin{bmatrix} p_1^{n+1} \\ p_2^{n+1} \\ p_3^{n+1} \end{bmatrix} = \begin{bmatrix} -4,776.74 \\ -5,469.03 \\ -4,865.33 \end{bmatrix}, \quad \dots \dots \dots \dots \dots \dots \dots \dots \dots \dots (8.194)$$

which generates the following pressure distribution: $p_1 = 1,884.9$ psia, $p_2 = 2,424.3$ psia, and $p_3 = 1,908.1$ psia:

- Fluid expansion of Block 1:

$$= \left(2,547.0 - 1884.9 \text{ psia}\right) \times 5 \times 10^{-5} \text{psi}^{-1} \times \frac{600 \text{ ft} \times 600 \text{ ft} \times 40 \text{ ft} \times 0.2}{5.165} = 25,472.2 \text{ STB}. \quad \dots \dots (8.195)$$

- Fluid expansion of Block 2

$$= \left(3,037.2 \text{ psia} - 2,424.3 \text{ psia}\right) \times 5 \times 10^{-5} \text{psi}^{-1} \times \frac{600 \text{ ft} \times 600 \text{ ft} \times 40 \text{ ft} \times 0.2}{5.165} = 23,576.0 \text{ STB}. \quad \dots (8.196)$$

- Fluid expansion of Block 3

$$= \left(2,562.7 - 1,908.1 \text{ psia}\right) \times 5 \times 10^{-5} \text{psi}^{-1} \times \frac{600 \text{ ft} \times 600 \text{ ft} \times 40 \text{ ft} \times 0.2}{5.165} = 25,184.3 \text{ STB}. \quad \dots \dots (8.197)$$

- The total amount of fluid produced from the well:

$$Q_{t2} = \sum \text{fluid expansion} - \text{fluid leakage} = 74,232.6 \text{ STB} - 1,690.5 \text{ STB} = 72,542.1 \text{ STB}. \quad \dots \dots \dots (8.198)$$

The average production rate during the second 15-day production period is

$$q_2 = \frac{Q_{t2}}{15 \text{ days}} = 4,836.1 \text{ STB/D} \quad \dots \dots \dots \dots \dots \dots \dots \dots \dots \dots \dots \dots \dots \dots \dots (8.199)$$

Problem 8.4.10 (Ertekin et al. 2001)

Consider the 1D, single-phase, unsteady-state flow of oil in the system shown in **Fig. 8.10.** Uniform porosity, ϕ, and permeability, k, distributions exist in the system. Fluid properties that need to be considered in the formulation and the solution are compressibility, c_o, viscosity, μ_o, and FVF, B_o. The initial pressure distribution at $t = 0$ is uniform for all gridblocks and equals p_i.

1. State the PDE that describes the flow of oil in the described porous medium.
2. Using the backward finite-difference approximation, obtain the necessary system of equations to solve the pressure distribution in the system after a period of Δt.
3. Express the system of equations obtained in Part 2 in matrix form so that Thomas' algorithm is readily applicable.

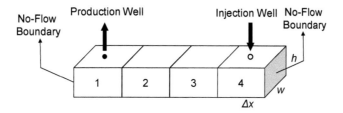

Fig. 8.10—Reservoir system for Problem 8.4.10.

Solution to Problem 8.4.10

1. For a 1D reservoir with homogeneous properties, the PDE describing the reservoir system can be written as

$$\frac{\partial}{\partial x}\left(\frac{A_x k_x}{\mu B}\frac{\partial p}{\partial x}\right)\Delta x + q = \frac{V_b \phi c_t}{5.615}\frac{\partial p}{\partial t}, \quad\dots\dots\dots\dots\dots\dots\dots\dots\dots\dots\dots\dots\dots\dots\dots\dots\dots\dots\dots (8.200)$$

which can be simplified for a homogeneous system to

$$\frac{\partial^2 p}{\partial x^2} + \frac{q\mu B}{kV_b} = \frac{\mu B c_t \phi}{5.615k}\frac{\partial p}{\partial t}. \quad\dots\dots\dots\dots\dots\dots\dots\dots\dots\dots\dots\dots\dots\dots\dots\dots\dots (8.201)$$

2. The PDE can be written in a discretized form using a backward finite-difference scheme. For an arbitrary Block i,

$$\frac{p_{i-1}^{n+1} - 2p_i^{n+1} + p_{i+1}^{n+1}}{(\Delta x)^2} + \left(\frac{q\mu B}{\Delta x whk}\right)_i = \frac{\mu B c_t \phi}{5.615k\Delta t}\left(p_i^{n+1} - p_i^n\right) \quad\dots\dots\dots\dots\dots\dots\dots\dots\dots\dots (8.202)$$

or

$$p_{i-1}^{n+1} - 2p_i^{n+1} + p_{i+1}^{n+1} + \left(\frac{\Delta x}{w}\right)\left(\frac{q\mu B}{hk}\right)_i = \frac{\mu B c_t \phi (\Delta x)^2}{5.615k\Delta t}\left(p_i^{n+1} - p_i^n\right). \quad\dots\dots\dots\dots\dots\dots\dots\dots (8.203)$$

Rearranging,

$$p_{i-1}^{n+1} - \left[2 + \frac{\mu B c_t \phi (\Delta x)^2}{5.615k\,\Delta t}\right]p_i^{n+1} + p_{i+1}^{n+1} = -\frac{\mu B c_t \phi (\Delta x)^2}{5.615k\,\Delta t}p_i^n - \left(\frac{\Delta x}{w}\right)\left(\frac{q\mu B}{hk}\right)_i. \quad\dots\dots\dots\dots\dots\dots\dots (8.204)$$

Note: q is negative for production, positive for injection, and zero if no well is prescribed in Block i.

3. The linear system of equations can be written in matrix form, which can be solved by Thomas' algorithm:

$$\begin{bmatrix} -1 - \dfrac{\mu B c_t \phi (\Delta x)^2}{5.165k\Delta t} & 1 & 0 \\[2em] 1 & -2 - \dfrac{\mu B c_t \phi (\Delta x)^2}{5.165k\Delta t} & 1 \\[2em] 0 & 1 & -1 - \dfrac{\mu B c_t \phi (\Delta x)^2}{5.165k\Delta t} \end{bmatrix}\begin{bmatrix} p_1^{n+1} \\[1em] p_2^{n+1} \\[1em] p_3^{n+1} \end{bmatrix} = \begin{bmatrix} -\left(\dfrac{\Delta x}{w}\right)\left(\dfrac{q\mu B}{kh}\right) - \dfrac{\mu B c_t \phi (\Delta x)^2}{5.165k\Delta t}\times p_1^n \\[1em] -\dfrac{\mu B c_t \phi (\Delta x)^2}{5.165k\Delta t}\times p_2^n \\[1em] -\dfrac{\mu B c_t \phi (\Delta x)^2}{5.165k\Delta t}\times p_3^n \end{bmatrix}.$$

$$\dots\dots\dots\dots\dots\dots\dots\dots\dots\dots\dots\dots\dots\dots\dots\dots\dots\dots\dots\dots\dots\dots\dots\dots\dots\dots\dots\dots\dots\dots\dots\dots\dots\dots\dots\dots\dots (8.205)$$

Problem 8.4.11 (Ertekin et al. 2001)

Consider the unsteady-state, single-phase flow of oil taking place in the 2D homogeneous, isotropic-horizontal reservoir system shown in **Fig. 8.11.** All boundaries are no-flow boundaries and production from the center gridblock is 400 STB/D. The initial condition for all gridblock pressures states that they are at 4,000 psia. The fluid properties are $\mu = 10$ cp, $c = 1 \times 10^{-5}$ psia^{-1}, and $B = 1$ res bbl/STB. The gridblock properties are $\Delta x = 400$ ft, $\Delta y = 400$ ft, $k_x = 88.7$ md, $k_y = 88.7$ md, $\phi = 20\%$, and $h = 100$ ft. Assume that FVF and viscosity are unchanged within the pressure range.

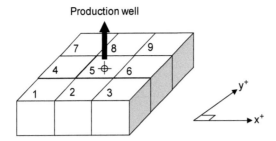

Production well

Fig. 8.11—Reservoir system for Problem 8.4.11.

Calculate the pressure distribution in the reservoir after 10 and 20 days, respectively. Use the appropriate PDE and its backward finite-difference approximation in the solution. Use a timestep of 10 days.

Solution to Problem 8.4.11

The governing flow equation for this homogeneous, 2D system can be written as

$$\frac{\partial^2 p}{\partial x^2} + \frac{\partial^2 p}{\partial y^2} + \frac{q\mu B}{kV_b} = \frac{\mu Bc_t\phi}{5.615k}\frac{\partial p}{\partial t}. \dots\dots\dots\dots\dots\dots\dots\dots\dots\dots\dots\dots \text{(8.206)}$$

The characteristic finite-difference equation can be written as

$$\frac{p_{i-1,j}^{n+1} - 2p_{i,j}^{n+1} + p_{i+1,j}^{n+1}}{(\Delta x)^2} + \frac{p_{i,j-1}^{n+1} - 2p_{i,j}^{n+1} + p_{i,j+1}^{n+1}}{(\Delta y)^2} + \left(\frac{q\mu B}{\Delta x \Delta y h k}\right)_{i,j} = \frac{\mu Bc_t\phi}{5.615k\Delta t}\left(p_{i,j}^{n+1} - p_{i,j}^n\right). \dots\dots\dots\dots \text{(8.207)}$$

Because $\Delta x = \Delta y$, the previous equation can be further simplified to

$$p_{i,j-1} + p_{i-1,j} - \left[4 - \frac{\mu Bc_t\phi(\Delta x)^2}{5.615k\Delta t}\right]p_{i,j} + p_{i+1,j} + p_{i,j+1} = -\left(\frac{q\mu B}{hk}\right)_{i,j} - \left[\frac{\mu Bc_t\phi(\Delta x)^2}{5.615k\Delta t}\right]p_{i,j}^n. \dots\dots\dots\dots\dots \text{(8.208)}$$

Considering the existing symmetry in the system, there are three unknowns that need to be solved in this problem—the block pressures of Blocks 2, 5, and 3. The system of equations can be written in matrix form:

$$\begin{bmatrix} -3.57 & 2 & 1 \\ 2 & -2.57 & 0 \\ 4 & 0 & -4.57 \end{bmatrix}\begin{bmatrix} p_2^{n+1} \\ p_3^{n+1} \\ p_5^{n+1} \end{bmatrix}\begin{bmatrix} -0.57 \times p_2^n \\ -0.57 \times p_3^n \\ 400.14 - 0.57 \times p_5^n \end{bmatrix}. \dots\dots\dots\dots\dots\dots\dots\dots\dots\dots\dots \text{(8.209)}$$

At the end of 10 days, the pressure distribution is solved as $p_2 = 3{,}923.11$ psia, $p_3 = 3{,}940.16$ psia and $p_5 = 3{,}845.17$ psia; and $p_2 = p_4 = p_6 = p_8$ and $p_1 = p_3 = p_7 = p_9$.

Similarly, at the end of 20 days, the pressure distribution is solved as $p_2 = 3{,}844.32$ psia, $p_3 = 3{,}865.58$ psia, and $p_5 = 3{,}756.90$ psia.

An MBC is conducted to check the accuracy of the solution. If the solved pressure distribution is correct, the material balance should satisfy (both at the end of 10 and 20 days):

$$\frac{[4 \times \text{fluid expansion at Block 2}] + [4 \times \text{fluid expansion at Block 3}] + [\text{fluid expansion at Block 5}]}{400 \text{ STB/D} \times 10 \text{ days}} \approx 1. \dots\dots \text{(8.210)}$$

Table 8.13 summarizes the MBC results.

Time/MBC	Block 2	Block 3	Block 5	MBC
10 days	438.21	341.02	882.38	0.999828
20 days	449.01	425.06	503.06	0.999828

Table 8.13—Results of the MBC in Solution to Problem 8.4.11.

Problem 8.4.12 (Ertekin et al. 2001)

Consider the 2D single-phase flow of slightly compressible oil taking place in the horizontal, homogeneous and isotropic reservoir shown in Fig. 8.11 from Problem 8.4.11. A production well located in the center (Block 5) is produced at a constant rate of 400 STB/D. No-flow boundaries are prescribed along all boundaries. **Table 8.14** gives the pressure distribution in the reservoir at 10 days. The gridblock properties are $\Delta x = 400$ ft, $\Delta y = 400$ ft, $k_x = 88.7$ md, $k_y = 88.7$ md, $h = 100$ ft, $\phi = 20\%$. The fluid properties are $\mu_o = 10$ cp, $B_o = 1$ res bbl/STB, and $c_o = 1 \times 10^{-5}$ psi^{-1}. Calculate the pressure distribution at $t = 20$ days by using the alternative direction implicit procedure (ADIP) solution method.

Gridblock	Pressure, psia
1	3,867
2	3,830
3	3,867
4	3,830
5	3,763
6	3,830
7	3,867
8	3,830
9	3,867

Table 8.14—Pressure distribution in the reservoir at 10 days.

Solution to Problem 8.4.12

The ADIP method can be implemented to solve the problem. For a 2D reservoir with homogeneous- and isotropic-property distribution, the transmissibility factor can be calculated as

$$T = \frac{kh}{\mu B} = 1.0 \text{ STB/D-psi.} \quad\quad\quad (8.211)$$

To solve the problem using the ADIP method, the iteration parameter needs to be calculated. For a 2D problem, $\omega_{max} = 2$, and

$$\omega_{min} = \min_{i,j}\left[\frac{\pi^2}{2N_x(1+\lambda_x)}, \frac{\pi^2}{2N_y(1+\lambda_y)}\right] = 0.0986. \quad\quad (8.212)$$

Also, n_k can be calculated as the minimum integer that satisfies

$$n_k \leq \frac{\log\left(\frac{\omega_{min}}{\omega_{max}}\right)}{\log(0.172)} + 1 = 2. \quad\quad\quad (8.213)$$

In this case, the number of iteration parameters included in one iteration cycle is 2.

Using a pressure tolerance of 1 psia, the problem can be solved using the ADIP iteration scheme. The intermediate solutions are displayed in **Table 8.15.**

Iteration 1					
p^*			$p^{(1)}$		
3,800.088	3,785.747	3,800.088	3,794.699	3,761.505	3,794.699
3,798.458	3,697.145	3,798.458	3,766.758	3,686.603	3,766.758
3,800.088	3,785.747	3,800.088	3,794.699	3,761.505	3,794.699
Iteration 2					
p^*			$p^{(2)}$		
3,791.635	3,764.353	3,791.635	3,790.053	3,765.461	3,790.053
3,767.759	3,682.396	3,767.759	3,767.141	3,680.756	3,767.141
3,791.635	3,764.353	3,791.635	3,790.053	3,765.461	3,790.053
Iteration 3					
p^*			$p^{(3)}$		
3,787.068	3,763.822	3,787.068	3,784.669	3,760.582	3,784.669
3,760.039	3,672.230	3,760.039	3,760.505	3,673.556	3,760.505
3,787.068	3,763.822	3,787.068	3,784.669	3,760.582	3,784.669
Iteration 4					
p^*			$p^{(4)}$		
3,784.439	3,760.612	3,784.439	3,784.332	3,760.615	3,784.332
3,760.625	3,673.465	3,760.625	3,760.602	3,673.420	3,760.602
3,784.439	3,760.612	3,784.439	3,784.332	3,760.615	3,784.332

Table 8.15—Intermediate solutions found in Solution to Problem 8.4.12.

Problem 8.4.13 (Ertekin et al. 2001)

Referring to the reservoir described in Problem 8.4.7, consider the flow of a slightly compressible oil with a compressibility of $7.3 \times 10^{-5}\,kPa^{-1}$, the porosity of Gridblocks (1,1), (2,1), (1,2), (2,2), and (3,2) are 0.2, 0.25, 0.18, 0.27, and 0.3, respectively. The porosity of Gridblock (2,3) is 0.4, because of the fracture. Initially, the pressure at all gridblocks is 3,500 kPa.

Calculate the pressure distribution of the reservoir at the end of 20 and 50 days with the following schemes:

1. Explicit formulation.
2. Implicit formulation, using a direct-solution scheme.
3. Implicit formulation, using the PSOR-solution method with a convergence criterion of 2 kPa.
4. Implicit formulation, using ADIP.

Obtain the material-balance error in each case.

Solution to Problem 8.4.13

The same numbering scheme as that in Problem 8.4.7 is used in the solution to this problem; see **Fig. 8.12.** For a heterogeneous and anisotropic system, the governing PDE can be written as

$$\frac{\partial}{\partial x}\left(\frac{A_x k_x}{\mu B}\frac{\partial p}{\partial x}\right)\Delta x + \frac{\partial}{\partial y}\left(\frac{A_y k_y}{\mu B}\frac{\partial p}{\partial y}\right)\Delta y + q = V_b \phi c_t \frac{\partial p}{\partial t}. \quad \dots (8.214)$$

First, transmissibility coefficients are calculated as shown in **Table 8.16.** For a slightly compressible system,

$$\Gamma_{i,j} = \frac{V_{bi,j}\phi_{i,j}c_t}{\Delta t}. \quad \dots (8.215)$$

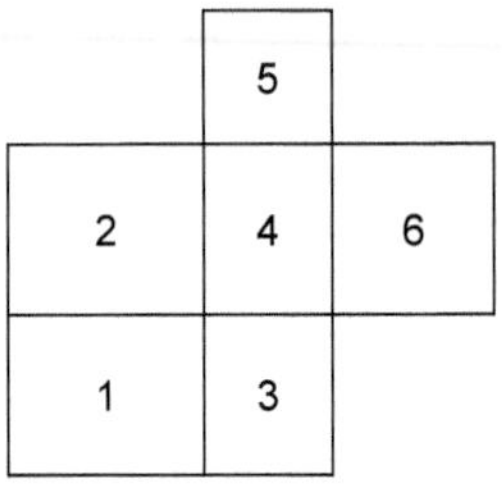

From days 0 through 20, $\Delta t = 20$ days, and from days 20 through 50, $\Delta t = 30$ days; see **Table 8.17.**
The PDE can be approximated using the finite-difference scheme:

- Implicit formulation:

$$T_{i,j-\frac{1}{2}}p_{i,j-1}^{n+1} + T_{i,j+\frac{1}{2}}p_{i,j+1}^{n+1} + T_{i-\frac{1}{2},j}p_{i-1,j}^{n+1} + T_{i+\frac{1}{2},j}p_{i+1,j}^{n+1} - \left(T_{i,j-\frac{1}{2}} + T_{i,j+\frac{1}{2}} + T_{i-\frac{1}{2},j} + T_{i+\frac{1}{2},j} + \Gamma_{i,j}\right)p_{i,j}^{n+1} = -\Gamma_{i,i}p_{i,i}^n - q. \quad \dots (8.216)$$

- Explicit formulation:

$$p_{i,j}^{n+1} = \frac{T_{i,j-\frac{1}{2}}p_{i,j-1}^n + T_{i,j+\frac{1}{2}}p_{i,j+1}^n + T_{i-\frac{1}{2},j}p_{i-1,j}^n + T_{i+\frac{1}{2},j}p_{i+1,j}^n - \left(T_{i,j-\frac{1}{2}} + T_{i,j+\frac{1}{2}} + T_{i-\frac{1}{2},j} + T_{i+\frac{1}{2},j} - \Gamma_{i,j}\right)p_{i,j}^n + q}{\Gamma_{i,j}}. \quad \dots (8.217)$$

The system of equations of the two time periods can be expressed in matrix forms:

- From $t = 0$ to 20 days:

$$\begin{bmatrix} -2.0348 & 0 & 0.575 & 0 \\ 0 & -0.8341 & 0.236 & 0 \\ 0.575 & 0.236 & -2.3500 & 0.110 \\ 0 & 0 & 0.1104 & -0.5119 \end{bmatrix}\begin{bmatrix} p_2^{n+1} \\ p_3^{n+1} \\ p_4^{n+1} \\ p_5^{n+1} \end{bmatrix} = \begin{bmatrix} -5,109.378 \\ -2,092.613 \\ -4,364.717 \\ -1,301.570 \end{bmatrix}. \quad \dots (8.218)$$

	Transmissibility
T_{45}	0.110
T_{24}	0.575
T_{46}	0.187
T_{34}	0.236
T_{12}	0.253
T_{13}	0.142

Table 8.16—Transmissibility coefficients for Solution to Problem 8.4.13.

Fig. 8.12—Numbering of the gridblocks of the reservoir system for Problem 8.4.13.

Block	Γ_1	Γ_2
1	0.7665	0.5110
2	1.2072	0.8048
3	0.4563	0.3042
4	1.2417	0.8278
5	0.4015	0.2677
6	0.6899	0.4599

Table 8.17—Solution of Problem 8.4.13.

- From $t = 20$ to 50 days:

$$\begin{bmatrix} -1.6324 & 0 & 0.575 & 0 \\ 0 & -0.6820 & 0.236 & 0 \\ 0.575 & 0.236 & -1.9361 & 0.110 \\ 0 & 0 & 0.1104 & -0.3781 \end{bmatrix} \begin{bmatrix} p_2^{n+1} \\ p_3^{n+1} \\ p_4^{n+1} \\ p_5^{n+1} \end{bmatrix} = \begin{bmatrix} -3,629.687 \\ -1,533.334 \\ -2,656.693 \\ -760.855 \end{bmatrix}. \quad \dots\dots\dots\dots\dots \quad (8.219)$$

1. Explicit formulation scheme. The problem can be solved using explicit solution:

- Between 0 and 20 days:

$$p_2^{n+1} = \frac{T_{12}p_1^n + T_{24}p_4^n - (T_{24} + T_{12} - \Gamma_2)p_2^n}{\Gamma_2} = 3,500 \text{ kPa}, \quad \dots\dots\dots\dots\dots\dots\dots \quad (8.220)$$

$$p_3^{n+1} = \frac{T_{13}p_1^n + T_{34}p_4^n - (T_{34} + T_{13} - \Gamma_3)p_3^n}{\Gamma_3} = 3,500 \text{ kPa}, \quad \dots\dots\dots\dots\dots\dots\dots \quad (8.221)$$

$$p_4^{n+1} = \frac{T_{24}p_2^n + T_{34}p_3^n + T_{45}p_5^n + T_{46}p_6^n - (T_{24} + T_{34} + T_{45} + T_{46} - \Gamma_4)p_4^n}{\Gamma_4} = 2,989 \text{ kPa}, \quad \dots\dots\dots \quad (8.222)$$

$$p_5^{n+1} = \frac{T_{45}p_4^n - (T_{45} - \Gamma_5)p_5^n + q}{\Gamma_5} = 3,241.77 \text{ kPa}. \quad \dots\dots\dots\dots\dots\dots\dots\dots\dots\dots \quad (8.223)$$

- Between 20 and 50 days:

$$p_2^{n+1} = \frac{T_{12}p_1^n + T_{24}p_4^n - (T_{24} + T_{12} - \Gamma_2)p_2^n}{\Gamma_2} = 3,134.93 \text{ kPa}, \quad \dots\dots\dots\dots\dots\dots\dots \quad (8.224)$$

$$p_3^{n+1} = \frac{T_{13}p_1^n + T_{34}p_4^n - (T_{34} + T_{13} - \Gamma_3)p_3^n}{\Gamma_3} = 3,103.13 \text{ kPa}, \quad \dots\dots\dots\dots\dots\dots\dots \quad (8.225)$$

$$p_4^{n+1} = \frac{T_{24}p_2^n + T_{34}p_3^n + T_{45}p_5^n + T_{46}p_6^n - (T_{24} + T_{34} + T_{45} + T_{46} - \Gamma_4)p_4^n}{\Gamma_4} = 2,872.17 \text{ kPa}, \quad \dots\dots\dots \quad (8.226)$$

$$p_5^{n+1} = \frac{T_{45}p_4^n - (T_{45} - \Gamma_5)p_5^n + q}{\Gamma_5} = 2,750.14 \text{ kPa}. \quad \dots\dots\dots\dots\dots\dots\dots\dots\dots\dots \quad (8.227)$$

Because the pressure of Blocks 1 and 6 are kept at 3,500 and 100 kPa, respectively, we recognize that there is no accumulation in Blocks 1 or 6. Therefore, the flow rate of the wells completed in Blocks 1 and 6 can be calculated using the mass flow to the neighboring blocks:

- Between 0 to 20 days:

$$q_{\text{well}(3,2)} = q_{4-6} = T_{46}(p_4 - p_6) = 0.1866 \times (2,989.00 - 100) = -539.157 \text{ std m}^3/\text{D (production)}, \quad \dots \quad (8.228)$$

$$q_{\text{well}(1,1)} = q_{1-2} + q_{1-3} = T_{12}(p_1 - p_2) + T_{13}(p_1 - p_3) = 0.2526 \times (3,500 - 3,500) + 0.1416$$
$$\times (3,500 - 3,500) = 0 \text{ std m}^3/\text{D (injection)}, \quad \dots\dots\dots \quad (8.229)$$

$$q_{\text{production}} = q_{\text{well}(3,2)} + q_{\text{well}(2,3)} = -642.837 \text{ std m}^3/\text{D}, \quad \dots\dots\dots\dots\dots\dots\dots\dots \quad (8.230)$$

$$q_{\text{injection}} = q_{\text{well}(1,1)} = 0 \text{ std m}^3/\text{D}. \quad \dots\dots\dots\dots\dots\dots\dots\dots\dots\dots\dots\dots\dots \quad (8.231)$$

Total fluid expansion can therefore be expressed as

$$\Delta V_{\text{total}} = c_t \sum_{i=1}^{5} (V_b \phi)_i (p_i^{n+1} - p_i^n) = 14,764.03 \text{ std m}^3, \quad \dots\dots\dots\dots\dots\dots\dots\dots \quad (8.232)$$

$$\text{MBC} = \left| \frac{(q_{\text{injection}} + q_{\text{production}})\Delta t}{\Delta V_{\text{total}}} \right| = \left| \frac{14,764.032}{12,856.74} \right| = 0.87. \quad \dots\dots\dots\dots\dots\dots\dots\dots \quad (8.233)$$

- Between 20 to 50 days:

$$q_{well(3,2)} = q_{4-6} = T_{46}(p_4 - p_6) = 0.1866 \times (2{,}872.17 - 100) = -517.354 \text{ std m}^3/\text{D (production)}, \ldots \text{ (8.234)}$$

$$q_{well(1,1)} = q_{1-2} + q_{1-3} = T_{12}(p_1 - p_2) + T_{13}(p_1 - p_3) = 0.2526 \times (3{,}500 - 3{,}134.93)$$
$$+ 0.1416 \times (3{,}500 - 3{,}103.13) = 148.42 \text{ std m}^3/\text{D (injection)}, \quad \ldots\ldots\ldots\ldots\ldots \text{ (8.235)}$$

$$q_{production} = q_{well(3,2)} + q_{well(2,3)} = -621.03 \text{ std m}^3/\text{D}, \ldots\ldots\ldots\ldots\ldots\ldots\ldots\ldots\ldots\ldots \text{ (8.236)}$$

$$q_{injection} = q_{well(1,1)} = 148.42 \text{ std m}^3/\text{D}. \ldots\ldots\ldots\ldots\ldots\ldots\ldots\ldots\ldots\ldots\ldots\ldots \text{ (8.237)}$$

Fluid expansion can therefore be expressed as

$$\Delta V_{total} = c_t \sum_{i=1}^{5}(V_b\phi)_i(p_i^{n+1} - p_i^n) = 19{,}285.11 \text{ std m}^3, \ldots\ldots\ldots\ldots\ldots\ldots\ldots\ldots \text{ (8.238)}$$

$$MBC = \left|\frac{(q_{injection} + q_{production})\Delta t}{\Delta V_{total}}\right| = \left|\frac{-14{,}178.32}{19{,}285.11}\right| = 0.7352. \ldots\ldots\ldots\ldots\ldots\ldots\ldots\ldots\ldots \text{ (8.239)}$$

2. Implicit formulation scheme using a direct solver. Employing Gaussian elimination and solving the system of equations, between 0 to 20 days:

$$\begin{bmatrix} 1 & 0 & -0.283 & 0 \\ 0 & -0.834 & 0.236 & 0 \\ 0 & 0.236 & -2.188 & 0.110 \\ 0 & 0 & 0.110 & -0.512 \end{bmatrix}\begin{bmatrix} p_2 \\ p_3 \\ p_4 \\ p_5 \end{bmatrix} = \begin{bmatrix} 2{,}510.998 \\ -2{,}092.61 \\ -5{,}808.51 \\ -1{,}301.57 \end{bmatrix}, \ldots\ldots\ldots\ldots\ldots\ldots\ldots\ldots \text{ (8.240)}$$

$$\begin{bmatrix} 1 & 0 & -0.283 & 0 \\ 0 & 1 & -0.283 & 0 \\ 0 & 0 & -2.121 & 0.110 \\ 0 & 0 & 0.110 & -0.512 \end{bmatrix}\begin{bmatrix} p_2 \\ p_3 \\ p_4 \\ p_5 \end{bmatrix} = \begin{bmatrix} 2{,}510.998 \\ 2{,}508.827 \\ -6{,}401.17 \\ -1{,}301.57 \end{bmatrix}, \ldots\ldots\ldots\ldots\ldots\ldots\ldots\ldots \text{ (8.241)}$$

$$\begin{bmatrix} 1 & 0 & -0.283 & 0 \\ 0 & 1 & -0.283 & 0 \\ 0 & 0 & 1 & -0.052 \\ 0 & 0 & 0 & -0.506 \end{bmatrix}\begin{bmatrix} p_2 \\ p_3 \\ p_4 \\ p_5 \end{bmatrix} = \begin{bmatrix} 2{,}510.998 \\ 2{,}508.827 \\ 3{,}018.543 \\ -1{,}634.9 \end{bmatrix}. \ldots\ldots\ldots\ldots\ldots\ldots\ldots\ldots \text{ (8.242)}$$

Therefore,

- $p_5 = -1{,}634.9/-0.506 = 3{,}229.89 \text{ kPa}, \ldots\ldots\ldots\ldots\ldots\ldots\ldots\ldots\ldots\ldots\ldots\ldots \text{ (8.243)}$
- $p_4 = 3{,}018.543 + 0.052 \times p_4 = 3{,}186.72 \text{ kPa}, \ldots\ldots\ldots\ldots\ldots\ldots\ldots\ldots\ldots\ldots \text{ (8.244)}$
- $p_3 = 2{,}508.83 + 0.283p_3 = 3{,}411.28 \text{ kPa}, \ldots\ldots\ldots\ldots\ldots\ldots\ldots\ldots\ldots\ldots\ldots \text{ (8.245)}$
- $p_2 = 2{,}510.998 + 0.283p_3 = 3{,}411.474 \text{ kPa}. \ldots\ldots\ldots\ldots\ldots\ldots\ldots\ldots\ldots\ldots \text{ (8.246)}$

Similarly, from $t = 20$ to $t = 50$ days: $p_2 = 3{,}243.271$ kPa; $p_3 = 3{,}250.901$ kPa; $p_4 = 2{,}895.054$ kPa; and $p_5 = 2{,}857.882$ kPa.

- Between 0 to 20 days:

$$q_{well(3,2)} = q_{4-6} = T_{46}(p_4 - p_6) = 0.1866 \times (3{,}186.72 - 100) = -576.056 \text{ std m}^3/\text{D (production)}, \ldots \text{ (8.247)}$$

$$q_{well(1,1)} = q_{1-2} + q_{1-3} = T_{12}(p_1 - p_2) + T_{13}(p_1 - p_3) = 0.2526 \times (3{,}500 - 3{,}411.47)$$
$$+ 0.1416 \times (3{,}500 - 3{,}411.28) = 34.93 \text{ std m}^3/\text{D (injection)}, \quad \ldots\ldots\ldots\ldots\ldots \text{ (8.248)}$$

$$q_{production} = q_{well(3,2)} + q_{well(2,3)} = -679.74 \text{ std m}^3/\text{D}, \cdots\cdots\cdots\cdots\cdots\cdots\cdots\cdots\cdots \text{ (8.249)}$$

$$q_{injection} = q_{well(1,1)} = 34.93 \text{ std m}^3/\text{D}. \ldots\ldots\ldots\ldots\ldots\ldots\ldots\ldots\ldots\ldots\ldots\ldots \text{ (8.250)}$$

Fluid expansion can therefore be expressed as

$$\Delta V_{\text{total}} = c_t \sum_{i=1}^{5} (V_b \phi)_i (p_i^{n+1} - p_i^n) = 12{,}896.2 \text{ std m}^3, \dots\dots\dots\dots\dots\dots\dots\dots\dots\dots\dots (8.251)$$

$$\text{MBC} = \left| \frac{(q_{\text{injection}} + q_{\text{production}})\Delta t}{\Delta V_{\text{total}}} \right| = \left| \frac{12{,}896.2}{12{,}896.2} \right| = 1.0000. \dots\dots\dots\dots\dots\dots\dots\dots\dots (8.252)$$

- Between 20 to 50 days:

$$q_{\text{well}(3,2)} = q_{4-6} = T_{46}(p_4 - p_6) = 0.1866 \times (2{,}895.05 - 100) = -521.624 \text{ std m}^3/\text{D (production)}, \dots (8.253)$$

$$q_{\text{well}(1,1)} = q_{1-2} + q_{1-3} = T_{12}(p_1 - p_2) + T_{13}(p_1 - p_3) = 0.2526 \times (3{,}500 - 3{,}243.27)$$
$$+ 0.1416 \times (3{,}500 - 3{,}250.90) = 100.128 \text{ std m}^3/\text{D (injection)}, \quad \dots\dots\dots\dots\dots (8.254)$$

$$q_{\text{production}} = q_{\text{well}(3,2)} + q_{\text{well}(2,3)} = -625.30 \text{ std m}^3/\text{D}, \dots\dots\dots\dots\dots\dots\dots\dots\dots (8.255)$$

$$q_{\text{injection}} = q_{\text{well}(1,1)} = 100.128 \text{ std m}^3/\text{D}. \dots\dots\dots\dots\dots\dots\dots\dots\dots\dots\dots\dots (8.256)$$

Fluid expansion can therefore be expressed as

$$\Delta V_{\text{total}} = c_t \sum_{i=1}^{5} (V_b \phi)_i (p_i^{n+1} - p_i^n) = 15{,}755.3 \text{ std m}^3, \dots\dots\dots\dots\dots\dots\dots\dots\dots (8.257)$$

$$\text{MBC} = \left| \frac{(q_{\text{injection}} + q_{\text{production}})\Delta t}{\Delta V_{\text{total}}} \right| = \left| \frac{15755.3}{15755.3} \right| = 1.0000. \dots\dots\dots\dots\dots\dots\dots\dots\dots (8.258)$$

3. Implicit formulation scheme solved by PSOR with $\omega_{\text{opt}} = 1.00$ and a convergence criterion of 2 kPa. The pressure distribution between 0 to 20 days, with iteration levels shown in **Table 8.18** is

$$q_{\text{well}(3,2)} = q_{4-6} = T_{46}(p_4 - p_6) = 0.1866 \times (3{,}186.799 - 100)$$
$$= -576.071 \text{ std m}^3/\text{D (production)}, \quad \dots\dots\dots\dots\dots\dots\dots\dots\dots\dots (8.259)$$

$$q_{\text{well}(1,1)} = q_{1-2} + q_{1-3} = T_{12}(p_1 - p_2) + T_{13}(p_1 - p_3) = 0.2526 \times (3{,}500 - 3{,}411.638)$$
$$+ 0.1416 \times (3{,}500 - 3{,}411.516) = 34.8517 \text{ std m}^3/\text{D (injection)}, \quad \dots\dots\dots\dots (8.260)$$

$$q_{\text{production}} = q_{\text{well}(3,2)} + q_{\text{well}(2,3)} = -679.751 \text{ std m}^3/\text{D}, \dots\dots\dots\dots\dots\dots\dots\dots\dots (8.261)$$

$$q_{\text{injection}} = q_{\text{well}(1,1)} = 34.8517 \text{ std m}^3/\text{D}. \dots\dots\dots\dots\dots\dots\dots\dots\dots\dots\dots\dots (8.262)$$

Fluid expansion can therefore be expressed as

$$\Delta V_{\text{total}} = c_t \sum_{i=1}^{5} (V_b \phi)_i (p_i^{n+1} - p_i^n) = 12886.493 \text{ std m}^3, \dots\dots\dots\dots\dots\dots\dots\dots (8.263)$$

$$\text{MBC} = \left| \frac{(q_{\text{injection}} + q_{\text{production}})\Delta t}{\Delta V_{\text{total}}} \right| = \left| \frac{12{,}886.493}{12{,}897.982} \right| = 1.00089. \dots\dots\dots\dots\dots\dots\dots (8.264)$$

Iteration Level	p_2	p_3	p_4	p_5
0	3,500	3,500	3,500	3,500
1	3,500.014	3,500.094	3,230.005	3,239.411
2	3,423.72	3,423.626	3,191.406	3,231.084
3	3,412.813	3,412.694	3,187.247	3,230.187
4	3,411.638	3,411.516	3,186.799	3,230.09

Table 8.18—Iteration levels for Part 3 of Solution to Problem 8.4.13.

The pressure distribution between 20 to 50 days, with iteration levels displayed in **Table 8.19:**

$$q_{well(3,2)} = q_{4-6} = T_{46}(p_4 - p_6) = 0.1866 \times (2,895.106 - 100) = -521.63 \text{ std m}^3/\text{D (production)}, \quad \ldots\ldots \quad (8.265)$$

$$q_{well(1,1)} = q_{1-2} + q_{1-3} = T_{12}(p_1 - p_2) + T_{13}(p_1 - p_3) = 0.2526 \times (3,500 - 3,243.364)$$
$$+ 0.1416 \times (3,500 - 3,251.186) = 100.0642 \text{ std m}^3/\text{D (injection)}, \quad \ldots\ldots\ldots\ldots \quad (8.266)$$

$$q_{production} = q_{well(3,2)} + q_{well(2,3)} = -625.314 \text{ std m}^3/\text{D}, \quad \ldots\ldots\ldots\ldots\ldots\ldots\ldots\ldots \quad (8.267)$$

$$q_{injection} = q_{well(1,1)} = 100.074 \text{ std m}^3/\text{D}. \quad \ldots\ldots\ldots\ldots\ldots\ldots\ldots\ldots\ldots\ldots\ldots \quad (8.268)$$

Fluid expansion can therefore be expressed as

$$\Delta V_{total} = c_t \sum_{i=1}^{5} (V_b \phi)_i (p_i^{n+1} - p_i^n) = 15,759.031 \text{ std m}^3, \quad \ldots\ldots\ldots\ldots\ldots\ldots\ldots \quad (8.269)$$

$$MBC = \left| \frac{(q_{injection} + q_{production})\Delta t}{\Delta V_{total}} \right| = \left| \frac{-15,757.4906}{15,759.03079} \right| = 0.9999. \quad \ldots\ldots\ldots\ldots\ldots\ldots \quad (8.270)$$

Iteration Level	p_2	p_3	p_4	p_5
0	3,411.638	3,411.516	3,186.799	3,230.09
1	3,346.025	3,352.142	2,959.14	2,876.562
2	3,265.836	3,273.285	2,905.539	2,860.907
3	3,246.956	3,254.718	2,896.774	2,858.347
4	3,243.869	3,251.682	2,895.341	2,857.929
5	3,243.364	3,251.186	2,895.106	2,857.86

Table 8.19—Iteration levels for Part 3 of Solution to Problem 8.4.13.

4. Implicit formulation scheme solved by ADIP. In this case, using the iteration parameters $\omega_{min} = 0.0602$ and $\omega_{max} = 2$:

$$n_k \leq \frac{\log\left(\dfrac{\omega_{min}}{\omega_{max}}\right)}{\log(0.172)} + 1 = 2. \quad \ldots\ldots\ldots\ldots\ldots\ldots\ldots\ldots\ldots\ldots\ldots\ldots\ldots \quad (8.271)$$

As shown in **Table 8.20,** with the pressure distribution between 0 and 20 days, it takes nine iterations to converge using a tolerance of 1×10^{-6} kPa:

$$q_{well(3,2)} = q_{4-6} = T_{46}(p_4 - p_6) = 0.1866 \times (3,186.72 - 100) = -576.056 \text{ std m}^3/\text{D (production)}, \quad \ldots\ldots \quad (8.272)$$

$$q_{well(1,1)} = q_{1-2} + q_{1-3} = T_{12}(p_1 - p_2) + T_{13}(p_1 - p_3) = 0.2526 \times (3,500 - 3,411.47)$$
$$+ 0.1416 \times (3,500 - 3,411.28) = 34.93 \text{ std m}^3/\text{D (injection)}, \quad \ldots\ldots\ldots\ldots \quad (8.273)$$

$$q_{production} = q_{well(3,2)} + q_{well(2,3)} = -679.74 \text{ std m}^3/\text{D}, \quad \ldots\ldots\ldots\ldots\ldots\ldots\ldots\ldots \quad (8.274)$$

$$q_{injection} = q_{well(1,1)} = 34.93 \text{ std m}^3/\text{D}. \quad \ldots\ldots\ldots\ldots\ldots\ldots\ldots\ldots\ldots\ldots\ldots \quad (8.275)$$

Fluid expansion can therefore be expressed as

$$\Delta V_{total} = c_t \sum_{i=1}^{5} (V_b \phi)_i (p_i^{n+1} - p_i^n) = 12,896.2 \text{ std m}^3, \quad \ldots\ldots\ldots\ldots\ldots\ldots\ldots \quad (8.276)$$

$$MBC = \left| \frac{(q_{injection} + q_{production})\Delta t}{\Delta V_{total}} \right| = \left| \frac{12,896.2}{12,896.2} \right| = 1.0000. \quad \ldots\ldots\ldots\ldots\ldots\ldots \quad (8.277)$$

Iteration 1

p^*			$p^{(1)}$		
0.000	3,245.973	0.000	0.000	3,258.118	0.000
3,923.243	3,311.036	100.000	3,280.788	3,318.295	100.000
3,500.000	3,500.000	0.000	0.000	3,439.984	0.000

Iteration 2

p^*			$p^{(2)}$		
0.000	3,258.248	0.000	0.000	3,240.787	0.000
3,367.288	3,241.347	100.000	3,406.236	3,202.299	100.000
3,500.000	3,445.256	0.000	3,500.000	3,428.226	0.000

....

Iteration 9

p^*			$p^{(9)}$		
0.000	3,229.894	0.000	0.000	3,229.894	0.000
3,411.475	3,186.720	100.000	3,411.475	3,186.720	100.000
3,500.000	3,411.275	0.000	3,500.000	3,411.275	0.000

Table 8.20—Iterations to convergence using a tolerance of 1×10^{-6} kPa, with pressure distribution between 0 and 20 days; from Solution to Problem 8.4.13, Part 4.

As shown in **Table 8.21,** with the pressure distribution between 20 to 50 days, it takes 19 convergence iterations using a tolerance of 1×10^{-6} kPa:

$$q_{\text{well}(3,2)} = q_{4-6} = T_{46}(p_4 - p_6) = 0.1866 \times (2{,}895.05 - 100) = -521.624 \text{ std m}^3/\text{D (production)}, \quad \dots \dots \quad (8.278)$$

$$q_{\text{well}(1,1)} = q_{1-2} + q_{1-3} = T_{12}(p_1 - p_2) + T_{13}(p_1 - p_3) = 0.2526 \times (3{,}500 - 3{,}243.27)$$
$$+ 0.1416 \times (3{,}500 - 3{,}250.90) = 100.128 \text{ std m}^3/\text{D (injection)}, \quad \dots \dots \dots \dots \quad (8.279)$$

$$q_{\text{production}} = q_{\text{well}(3,2)} + q_{\text{well}(2,3)} = -625.30 \text{ std m}^3/\text{D}, \dots \dots \dots \dots \dots \dots \dots \dots \dots \dots \dots \dots \dots \quad (8.280)$$

$$q_{\text{injection}} = q_{\text{well}(1,1)} = 100.128 \text{ std m}^3/\text{D}. \dots \dots \dots \dots \dots \dots \dots \dots \dots \dots \dots \dots \dots \dots \dots \quad (8.281)$$

Iteration 1

p^*			$p^{(1)}$		
0.000	2,834.551	0.000	0.000	2,907.491	0.000
3,830.847	3,097.429	100.000	3,069.666	3,069.301	100.000
3,500.000	3,324.880	0.000	0.000	3,308.377	0.000

Iteration 2

p^*			$p^{(2)}$		
0.000	2,908.484	0.000	0.000	2,881.467	0.000
3,178.351	2,975.861	100.000	3,234.549	2,921.775	100.000
3,500.000	3,310.010	0.000	3,500.000	3,283.776	0.000

....

Iteration 19

p^*			$p^{(19)}$		
0.000	2,857.882	0.000	0.000	2,857.882	0.000
3,243.271	2,895.054	100.000	3,243.271	2,895.054	100.000
3,500.000	3,250.901	0.000	3,500.000	3,250.901	0.000

Table 8.21—Iterations to convergence using a tolerance of 1×10^{-6} kPa, with the pressure distribution between 20 and 50 days; from Solution to Problem 8.4.13, Part 4.

Fluid expansion can therefore be expressed as

$$\Delta V_{\text{total}} = c_t \sum_{i=1}^{5} (V_b \phi)_i (p_i^{n+1} - p_i^n) = 15,755.3 \text{ std m}^3, \dots\dots\dots\dots\dots\dots\dots\dots\dots\dots \quad (8.282)$$

$$\text{MBC} = \left| \frac{(q_{\text{injection}} + q_{\text{production}})\Delta t}{\Delta V_{\text{total}}} \right| = \left| \frac{15,755.3}{15,755.3} \right| = 1.000. \dots\dots\dots\dots\dots\dots\dots\dots \quad (8.283)$$

Problem 8.4.14 (Ertekin et al. 2001)

Consider a 2D single-phase, slightly compressible flow problem simulated for the reservoir shown in **Fig. 8.13.** The two wellblocks are kept at 1,400 psia. The grid-block properties are $\Delta x = \Delta y = 600$ ft, $k_x = 100$ md, $k_y = 80$ md, $h = 70$ ft, and $\phi = 18\%$. The fluid properties are $c = 4.0 \times 10^{-7}$ psia^{-1}, $\mu = 1$ cp, and $B = 1$ res bbl/STB. The initial pressure of the system is 3,000 psia. Implement ADIP for at least one iteration to solve for the pressure distribution at $t = 30$ days. Start with an initial guess of 3,000 psia, and do not use iteration parameters.

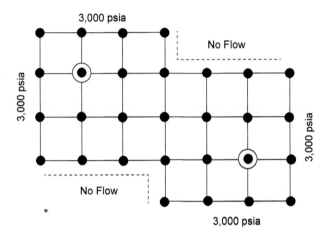

Fig. 8.13—Use of point-distributed grid for the reservoir system of Problem 8.4.14.

Solution to Problem 8.4.14

ADIP can be implemented to solve the 2D reservoir with homogeneous and anisotropic property distribution. For a homogeneous but anisotropic system:

$$W_{i,j} = E_{i,j} = k_x h = 7.8890 \text{ STB/D-psi}, \dots\dots\dots\dots\dots\dots\dots\dots\dots\dots \quad (8.284)$$

$$N_{i,j} = S_{i,j} = k_y h = 6.3112 \text{ STB/D-psi}, \dots\dots\dots\dots\dots\dots\dots\dots\dots\dots \quad (8.285)$$

$$\Gamma_{i,j} = \frac{V_{bi,j}\phi_{i,j}c_t}{5.615\Delta t} = 0.0108. \dots\dots\dots\dots\dots\dots\dots\dots\dots\dots\dots \quad (8.286)$$

Using a pressure tolerance of 1 psia, the problem can be solved using the ADIP iteration scheme with 18 iterations. The intermediate solutions are recorded **Table 8.22.**

Problem 8.4.15 (Ertekin et al. 2001)

Consider the flow of a compressible fluid in the 1D heterogeneous porous medium shown in **Fig. 8.14.** The initial pressure at every point in the system is 3,600 psia before the well is put on production. The boundary conditions are as follows:

- No-flow boundaries exist at the ends of the system.
- Production block (Block 2) is kept at 1,200 psia.

Table 8.23 shows the gridblock properties. The fluid properties at $T = 120°F$ are shown as Eqs. 8.287 and 8.288:

$$\mu_g = 5.306531 \times 10^{-6}p + 0.00829716 \dots\dots\dots\dots\dots\dots\dots\dots\dots\dots \quad (8.287)$$

and

$$z_g = 4.949892 \times 10^{-8}p^2 - 2.30415 \times 10^{-4}p + 1.00563, \dots\dots\dots\dots\dots\dots \quad (8.288)$$

where p is in psia and μ_g is in cp. The standard condition is $p_{sc} = 14.7$ psia and $T_{sc} = 520°R$.

Iteration 1

p*

3,000.00	3,000.00	3,000.00	3,000.00	0.00
1,400.00	3,230.04	3,488.52	3,106.36	3,022.15
3,207.00	3,039.47	3,014.09	3,039.47	3,207.00
3,022.15	3,106.36	3,488.52	3,230.04	1,400.00
0.00	3,000.00	3,000.00	3,000.00	3,000.00

$p^{(1)}$

3,000.00	3,000.00	3,000.00	3,000.00	0.00
1,400.00	3,228.78	3,089.89	3,179.13	3,061.67
2,866.56	2,995.46	2,824.48	2,995.46	2,866.56
3,061.67	3,179.13	3,089.89	3,228.78	1,400.00
0.00	3,000.00	3,000.00	3,000.00	3,000.00

Iteration 2

p*

3,000.00	3,000.00	3,000.00	3,000.00	0.00
1,400.00	2,685.36	3,055.13	3,093.96	2,993.19
3,095.01	2,706.95	2,988.44	2,706.95	3,095.01
2,993.19	3,093.96	3,055.13	2,685.36	1,400.00
0.00	3,000.00	3,000.00	3,000.00	3,000.00

$p^{(2)}$

3,000.00	3,000.00	3,000.00	3,000.00	0.00
1,400.00	2,590.58	3,010.41	2,977.56	3,019.71
2,898.46	2,946.72	2,654.26	2,946.72	2,898.46
3,019.71	2,977.56	3,010.41	2,590.58	1,400.00
0.00	3,000.00	3,000.00	3,000.00	3,000.00

......

Iteration 17

p*

3,000.00	3,000.00	3,000.00	3,000.00	0.00
1,400.00	2,326.61	2,596.42	2,596.14	2,527.94
2,471.82	2,430.04	2,693.38	2,430.04	2,471.82
2,527.94	2,596.14	2,596.42	2,326.61	1,400.00
0.00	3,000.00	3,000.00	3,000.00	3,000.00

$p^{(17)}$

3,000.00	3,000.00	3,000.00	3,000.00	0.00
1,400.00	2,326.35	2,595.66	2,596.06	2,527.44
2,471.39	2,429.83	2,693.09	2,429.83	2,471.39
2,527.44	2,596.06	2,595.66	2,326.35	1,400.00
0.00	3,000.00	3,000.00	3,000.00	3,000.00

Iteration 18

p*

3,000.00	3,000.00	3,000.00	3,000.00	0.00
1,400.00	2,326.15	2,595.73	2,595.33	2,527.01
2,471.03	2,429.63	2,692.87	2,429.63	2,471.03
2,527.01	2,595.33	2,595.73	2,326.15	1,400.00
0.00	3,000.00	3,000.00	3,000.00	3,000.00

$p^{(18)}$

3,000.00	3,000.00	3,000.00	3,000.00	0.00
1,400.00	2,325.98	2,595.00	2,595.49	2,526.67
2,470.73	2,429.49	2,692.68	2,429.49	2,470.73
2,526.67	2,595.49	2,595.00	2,325.98	1,400.00
0.00	3,000.00	3,000.00	3,000.00	3,000.00

Table 8.22—ADIP iteration scheme using a pressure tolerance of 1psia; from Solution to Problem 8.4.14.

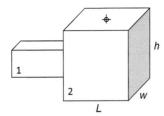

Fig. 8.14—Blocks 1 and 2 of the reservoir system described in Problem 8.4.15.

	Gridblock 1	Gridblock 2
w, ft	200	220
L, ft	400	400
h, ft	60	80
k, md	1	1.2
ϕ, %	7	8

Table 8.23—Gridblock properties for Problem 8.4.15.

Calculate the pressure distribution in the system at the end of 20 days of production. Also calculate the production rate of the well. Use the backward-difference scheme with a pressure-convergence tolerance of 10 psia. Use a timestep of 20 days.

Solution to Problem 8.4.15

There are two blocks in the system and the pressure of the production block is fixed at 1,200 psi. There is only one unknown to be solved in this problem, which is the pressure of Block 1. For Block 1, the discretized PDE can be written as

$$T_{12}(p_2 - p_1) = \left(\frac{\phi V_b T_{sc}}{p_{sc} T}\right)_1 \frac{\left[\left(\frac{p}{z_g}\right)_1^{n+1} - \left(\frac{p}{z_g}\right)_1^n\right]}{\Delta t}, \quad \dots \dots \dots \dots \dots \dots \dots \text{(8.289)}$$

while the transmissibility coefficient, T_{12}, can be calculated as

$$T_{12} = \frac{2A_1 A_2 k_1 k_2}{\mu_g B_g (k_1 A_1 \Delta x_2 + k_2 A_2 \Delta x_1)}. \quad \dots \dots \dots \dots \dots \dots \dots \text{(8.290)}$$

Note that the gas FVF needs to be calculated implementing the z_g factor:

$$B_g \left(\frac{\text{res bbl}}{\text{scf}}\right) = \frac{p_{sc} z_g T}{5.615 p T_{sc}}. \quad \dots \dots \dots \dots \dots \dots \dots \text{(8.291)}$$

The gas viscosity and the compressibility factor are both strong functions of pressure, as expressed in Eqs. 8.287 and 8.288, respectively. To solve the nonlinear system of equations, the GNR protocol scheme is implemented. The residual of Block 1 is

$$R(p) = T_{12}(p_2 - p_1) - \left(\frac{\phi V_b T_{sc}}{p_{sc} T}\right)_1 \frac{\left[\left(\frac{p}{z_g}\right)_1^{n+1} - \left(\frac{p}{z_g}\right)_1^n\right]}{\Delta t}. \quad \dots \dots \dots \dots \text{(8.292)}$$

Using $\varepsilon_p = 0.00001$ psia, the first-order derivative of the residual can be numerically calculated:

$$\frac{\partial R}{\partial p} = \frac{R(p + \varepsilon_p) - R(p)}{\varepsilon_p}. \quad \dots \dots \dots \dots \dots \dots \dots \text{(8.293)}$$

Then, the change of pressure is calculated as

$$\Delta p = -\frac{R}{\frac{\partial R}{\partial p}}. \quad \dots \dots \dots \dots \dots \dots \dots \text{(8.294)}$$

Table 8.24 summarizes the details of the GNR protocol. After three iterations, the prescribed pressure tolerance is achieved; the solution for Block 1 is 2,097.11 psia.

To calculate the production rate of the well over the 20 days of production, an MBC calculation can be implemented:

$$\frac{p}{z_g} = \frac{p_i}{z_{gi}}\left(1 - \frac{G_p}{G_i}\right). \quad \dots \dots \dots \dots \dots \dots \dots \text{(8.295)}$$

Iteration Level	Residual, scf/D	$\frac{\partial R}{\partial p}$, scf/D-psi	Δp, psi	p, psi
0	NA	NA	NA	3,600
1	–5,970,105	–4,072.34	–1,466.01	2,133.986
2	–135,951	–3,698.13	–36.7622	2,097.224
3	–434.463	–3,674.38	–0.11824	2,097.105

Table 8.24—Details of the GNR iterations from Solution to Problem 8.4.15.

Because the production block is kept at 1,200 psia, the fluid expansion occurs only in Block 1. The original gas in place is $G_i = \dfrac{\phi_1 V_{b1}}{5.615 B_{gi}} = 90$ MMscf, $z_{gi} = 0.8176$, and $z_g = 0.7401$:

$$G_p = G_i \left(1 - \frac{\dfrac{p}{z_g}}{\dfrac{p_i}{z_{gi}}} \right) = 32 \text{ MMscf.} \quad \dots\dots\dots\dots\dots\dots\dots\dots\dots\dots\dots\dots\dots\dots\dots\dots\dots\dots\dots\dots\dots\dots\dots(8.296)$$

The average production rate = G_p/20 days = 1.6 MMscf/D.

Problem 8.4.16 (Ertekin et al. 2001)

Consider the 2D horizontal, isotropic homogeneous porous medium shown in **Fig. 8.15.1**, in which incompressible single-phase fluid flow is taking place. Assume $\Delta x = \Delta y = 400$ ft, $\mu = 1$ cp, $\phi = 20\%$, $k_x = k_y = 200$ md, $h = 10$ ft, $B = 1$ res bbl/STB.

1. State the PDE that describes the fluid flow dynamics in the system.
2. Write the finite-difference approximation of the equation from Part 1.
3. If the pressure of the gridblocks located along the boundaries is specified as 1,000 psia and the production well is produced at a rate of 1,000 STB/D, generate the system of linear equations necessary to find the pressure distribution in the system. Write the linear system of equations in matrix form and solve the system of equations using the conjugate-gradient method.

Solution to Problem 8.4.16

1. The PDE that describes the system can be written as

$$\frac{\partial^2 p}{\partial x^2} + \frac{\partial^2 p}{\partial y^2} + \frac{q\mu B}{V_b k} = 0. \quad \dots\dots\dots\dots\dots\dots\dots\dots\dots\dots\dots\dots\dots\dots\dots\dots\dots\dots\dots\dots\dots\dots\dots\dots\dots\dots\dots\dots(8.297)$$

Considering the symmetry of the problem, there are six unknowns that need to be solved (see the numbering of the blocks with unknown pressures, as indicated in **Fig. 8.15.2**). The blocks numbered 0 are constant-pressure blocks.

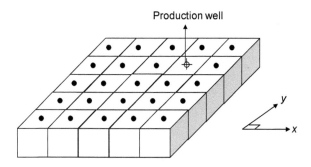

Production well

Fig. 8.15.1—Reservoir system for Problem 8.4.16.

0	0	0	0	0
0	5	4	1	0
0	6	2	4	0
0	3	6	5	0
0	0	0	0	0

Fig. 8.15.2—Numbering of the gridblocks of the reservoir system described in Problem 8.4.16.

2. For Block 1: $2(p_0 - 2p_1 + p_4) + \dfrac{q\mu B}{kh} = 0.$ . . . . (8.298)

For Block 2: $p_4 - 2p_2 + p_6 = 0.$ . . . . (8.299)

For Block 3: $p_0 - 2p_3 + p_6 = 0.$ . . . . (8.300)

For Block 4: $p_5 - 2p_4 + p_1 + p_0 - 2p_4 + p_2 = 0.$ . . . . (8.301)

For Block 5: $p_0 - 2p_5 + p_4 + p_0 - 2p_5 + p_6 = 0.$ . . . . (8.302)

For Block 6: $p_2 - 2p_6 + p_0 + p_3 - 2p_6 + p_5 = 0.$ . . . . (8.303)

3. Using $q = -1,000$ STB/D and $p_0 = 1,000$ psia, the linear system of equations can be written in matrix form:

$$\begin{bmatrix} -2 & 0 & 0 & 1 & 0 & 0 \\ 0 & -2 & 0 & 1 & 0 & 1 \\ 0 & 0 & -2 & 0 & 0 & 1 \\ 1 & 1 & 0 & -4 & 1 & 0 \\ 0 & 0 & 0 & 1 & -4 & 1 \\ 0 & 1 & 1 & 0 & 1 & -4 \end{bmatrix} \begin{bmatrix} p_1 \\ p_2 \\ p_3 \\ p_4 \\ p_5 \\ p_6 \end{bmatrix} = \begin{bmatrix} -778.17 \\ 0 \\ -1,000 \\ -1,000 \\ -2,000 \\ -1,000 \end{bmatrix}.$$ . . . . (8.304)

The linear equations can be solved using the conjugate-gradient method for a symmetrical matrix. It takes six iterations to obtain a convergence of 1 psia:

$$\vec{p} - \begin{bmatrix} 0 \\ 0 \\ 0 \\ 0 \\ 0 \\ 0 \end{bmatrix}^{(0)} = \begin{bmatrix} 378.06 \\ 0 \\ 485.83 \\ 485.83 \\ 971.66 \\ 485.83 \end{bmatrix}^{(1)} = \begin{bmatrix} 814.47 \\ 538.45 \\ 969.90 \\ 910.18 \\ 862.91 \\ 969.90 \end{bmatrix}^{(2)} = \begin{bmatrix} 920.69 \\ 851.53 \\ 1,078.80 \\ 893.10 \\ 946.98 \\ 944.51 \end{bmatrix}^{(3)} = \begin{bmatrix} 861.26 \\ 972.06 \\ 999.88 \\ 958.20 \\ 985.65 \\ 985.96 \end{bmatrix}^{(4)} = \begin{bmatrix} 864.54 \\ 972.47 \\ 996.16 \\ 954.94 \\ 986.17 \\ 989.55 \end{bmatrix}^{(5)} = \begin{bmatrix} 867.30 \\ 972.27 \\ 994.06 \\ 956.42 \\ 986.14 \\ 988.11 \end{bmatrix}^{(6)}.$$

. . . . (8.305)

To check the accuracy of the solution, an MBC is carried out:

- Mass flow entering Block 1:

$$q_1 = 2 \times -1.127 \times 10^{-3} \times \frac{kh}{\mu B}(p_1 - p_0) = 598.22 \text{ STB/D.}$$ . . . . (8.306)

- Mass flow entering Block 2: $q_2 = 0$ STB/D.

- Mass flow entering Block 3:

$$q_3 = 2 \times -1.127 \times 10^{-3} \times \frac{kh}{\mu B}(p_3 - p_0) = 26.79 \text{ STB/D.}$$ . . . . (8.307)

- Mass flow entering Block 4:

$$q_4 = 2 \times -1.127 \times 10^{-3} \times \frac{kh}{\mu B}(p_4 - p_0) = 196.43 \text{ STB/D.}$$ . . . . (8.308)

- Mass flow entering Block 5:

$$q_5 = 4 \times -1.127 \times 10^{-3} \times \frac{kh}{\mu B}(p_5 - p_0) = 125.0 \text{ STB/D.}$$ . . . . (8.309)

- Mass flow entering Block 6:

$$q_6 = 2 \times -1.127 \times 10^{-3} \times \frac{kh}{\mu B}(p_6 - p_0) = 53.57 \text{ STB/D.}$$ . . . . (8.310)

The total mass entering from the external boundaries is

$$q_{\text{total}} = q_1 + q_2 + q_3 + q_4 + q_5 = 1,000.00 \text{ STB/D} \approx q_{\text{production}}.$$ . . . . (8.311)

Such a result indicates a good material balance.

Problem 8.4.17 (Ertekin et al. 2001)

Consider the 3D system shown in **Fig. 8.16.1.** The steady-state flow of water is taking place in this completely closed system, and the system is known to be anisotropic and homogeneous. Note the existence of two wells in Gridblocks (1,3,2) and (4,1,2). These wells penetrate only through the top layer. Assume that potential gradients for this problem can be approximated by pressure gradients. Other relevant data are $\Delta x = 100$ ft, $\Delta y = 200$ ft, $\Delta z = 40$ ft, $\mu_w = 1$ cp, $B_w = 1$ res bbl/STB, $k_x = 400$ md, $k_y = 200$ md, and $k_z = 100$ md.

1. Write the PDE that describes the flow problem in the given system.
2. Give the finite-difference approximation for the PDE.
3. The production well in Gridblock (4,1,2) is maintained at atmospheric pressure, while 300 STB/D of water is injected through the injection well in Gridblock (1,3,2). Write the finite-difference equation for each gridblock, and generate the system of linear equations from which the pressure distribution in the system and the rate of production from the production well can be solved. Do not attempt to solve the equations.
4. If the pressure in the injection gridblock is specified as 1,500 psia and the production gridblock is kept at atmospheric pressure, how would you modify the equation in Part 3?

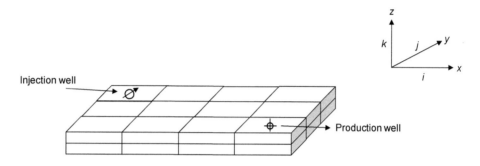

Fig. 8.16.1—Reservoir system for Problem 8.4.17.

Solution to Problem 8.4.17

1. For a single-phase, 3D incompressible flow in porous media, ignoring the depth gradients, the governing PDE can be written as

$$\frac{\partial}{\partial x}\left(\frac{A_x k_x}{\mu B}\frac{\partial p}{\partial x}\right)\Delta x + \frac{\partial}{\partial y}\left(\frac{A_y k_y}{\mu B}\frac{\partial p}{\partial y}\right)\Delta y + \frac{\partial}{\partial z}\left(\frac{A_z k_z}{\mu B}\frac{\partial p}{\partial z}\right)\Delta z + q = 0. \qquad (8.312)$$

The permeability distributions are homogeneous but anisotropic, so then, the governing equation can be simplified to

$$k_x \frac{\partial^2 p}{\partial x^2} + k_y \frac{\partial^2 p}{\partial y^2} + k_z \frac{\partial^2 p}{\partial z^2} + \frac{q\mu B}{V_b} = 0. \qquad (8.313)$$

2. The finite-difference approximation for the governing flow equation is

$$k_x\left(\frac{p_{i+1,j,k} + p_{i-1,j,k} - 2p_{i,j,k}}{\Delta x^2}\right) + k_y\left(\frac{p_{i,j+1,k} + p_{i,j-1,k} - 2p_{i,j,k}}{\Delta y^2}\right) +$$
$$k_z\left(\frac{p_{i,j,k+1} + p_{i,j,k-1} - 2p_{i,j,k}}{\Delta z^2}\right) + \frac{q\mu B}{V_b} = 0. \qquad (8.314)$$

Because $\Delta x = 100$ ft, $\Delta y = 200$ ft, $\Delta z = 40$ ft, $k_x = 400$ md, $k_y = 200$ md, and $k_z = 100$ md, the characteristic finite-difference equation will take the following form (note that the permeabilities are in perms):

$$8\left(p_{i+1,j,k} + p_{i-1,j,k} - 2p_{i,j,k}\right) + \left(p_{i,j+1,k} + p_{i,j-1,k} - 2p_{i,j,k}\right) + 12.5\left(p_{i,j,k+1} + p_{i,j,k-1} - 2p_{i,j,k}\right) +$$
$$\frac{200q\mu B}{1.127\times10^{-3}V_b} = 0. \qquad (8.315)$$

3. To list the equations to solve the problem, the grid system for both layers is numbered, as shown in **Fig. 8.16.2.**
   Case 1: The system of equations (24 equations in 24 unknowns) in matrix form for Case 1 is shown in **Fig. 8.16.3,** and the corresponding pressure surfaces calculated for the top layer and the bottom layer are shown in **Tables 8.25a and 8.25b,** respectively. If one calculates the flow rates entering Block 12 in Case 1, it is found that $q_x = 142.98$ STB/D, $q_y = 41.23$ STB/D, and $q_z = 115.79$ STB/D (totaling 300 STB/D).

4. Case 2: For the pure Dirichlet-type boundary conditions specified at the wellblocks, the system of equations and pressure surfaces are obtained. The system of equations (24 equations in 24 unknowns) in matrix form for Case 2 is shown in **Fig. 8.16.4**, and the pressure surfaces calculated for the top layer and the bottom layer are shown in **Tables 8.26a and 8.26b**, respectively, which follow the solutions presented for Case 1.

   The calculated flow rates in Case 2 from the injection block to the neighboring blocks are q_x = 9,413.33 STB/D, q_y= 2,714.13 STB/D, and q_z = 7,622.54 STB/D and, similarly, calculated flow rates from the neighboring blocks into the production block are q_x = −9,413.33 STB/D, q_y= −2,714.13 STB/D, and q_z = −7,622.54 STB/D (note that the injection-block rates and production-block rates have a combined total of 19,750 STB/D).

1 (Injection)	2	3	4
5	6	7	8
9	10	11	12 (Production)

Top Layer

13	14	15	16
17	18	19	20
21	22	23	24

Bottom Layer

Fig. 8.16.2—Numbering of gridblocks of the reservoir system of Problem 8.4.17.

−21.5	8	0	0	1	0	0	0	0	0	0	0	12.5	0	0	0	0	0	0	0	0	0	0	0		p_1		−66.55
8	−29.5	8	0	0	1	0	0	0	0	0	0	0	12.5	0	0	0	0	0	0	0	0	0	0		p_2		0
0	8	−29.5	8	0	0	1	0	0	0	0	0	0	0	12.5	0	0	0	0	0	0	0	0	0		p_3		0
0	0	8	−29.5	8	0	0	1	0	0	0	0	0	0	0	12.5	0	0	0	0	0	0	0	0		p_4		0
1	0	0	8	−30.5	8	0	0	1	0	0	0	0	0	0	0	12.5	0	0	0	0	0	0	0		p_5		0
0	1	0	0	8	−30.5	8	0	0	1	0	0	0	0	0	0	0	12.5	0	0	0	0	0	0		p_6		0
0	0	1	0	0	8	−30.5	8	0	0	1	0	0	0	0	0	0	0	12.5	0	0	0	0	0		p_7		0
0	0	0	1	0	0	8	−30.5	8	0	0	1	0	0	0	0	0	0	0	12.5	0	0	0	0		p_8		0
0	0	0	0	1	0	0	8	−29.5	8	0	0	0	0	0	0	0	0	0	0	12.5	0	0	0		p_9		0
0	0	0	0	0	1	0	0	8	−29.5	8	0	0	0	0	0	0	0	0	0	0	12.5	0	0		p_{10}		0
0	0	0	0	0	0	1	0	0	8	−29.5	8	0	0	0	0	0	0	0	0	0	0	12.5	0		p_{11}		0
0	0	0	0	0	0	0	0	0	0	0	1	0	0	0	0	0	0	0	0	0	0	0	0	×	p_{12}	=	14.7
12.5	0	0	0	0	0	0	0	0	0	0	0	−21.5	8	0	0	1	0	0	0	0	0	0	0		p_{13}		0
0	12.5	0	0	0	0	0	0	0	0	0	0	8	−29.5	8	0	0	1	0	0	0	0	0	0		p_{14}		0
0	0	12.5	0	0	0	0	0	0	0	0	0	0	8	−29.5	8	0	0	1	0	0	0	0	0		p_{15}		0
0	0	0	12.5	0	0	0	0	0	0	0	0	0	0	8	−29.5	8	0	0	1	0	0	0	0		p_{16}		0
0	0	0	0	12.5	0	0	0	0	0	0	0	1	0	0	8	−30.5	8	0	0	1	0	0	0		p_{17}		0
0	0	0	0	0	12.5	0	0	0	0	0	0	0	1	0	0	8	−30.5	8	0	0	1	0	0		p_{18}		0
0	0	0	0	0	0	12.5	0	0	0	0	0	0	0	1	0	0	8	−30.5	8	0	0	1	0		p_{19}		0
0	0	0	0	0	0	0	12.5	0	0	0	0	0	0	0	1	0	0	8	−30.5	8	0	0	1		p_{20}		0
0	0	0	0	0	0	0	0	12.5	0	0	0	0	0	0	0	1	0	0	8	−29.5	8	0	0		p_{21}		0
0	0	0	0	0	0	0	0	0	12.5	0	0	0	0	0	0	0	1	0	0	8	−29.5	8	0		p_{22}		0
0	0	0	0	0	0	0	0	0	0	12.5	0	0	0	0	0	0	0	1	0	0	8	−29.5	8		p_{23}		0
0	0	0	0	0	0	0	0	0	0	0	12.5	0	0	0	0	0	0	0	1	0	0	8	−21.5		p_{24}		0

Case 1

Fig. 8.16.3—Case 1 coefficient matrix, unknown vector and the right-hand-side vector.

37.26152	33.29671	30.78957	29.10485
28.11617	26.72706	25.23447	23.84535
22.85667	21.17196	18.66482	14.7

Table 8.25a—Case 1: Top-layer pressure distribution from Part 3 of Solution to Problem 8.4.17.

35.20677	32.88936	30.70763	29.08015
28.06159	26.71211	25.24941	23.89994
22.88138	21.2539	19.07216	16.75476

Table 8.25b—Case 1: Bottom-layer pressure distribution from Part 3 of Solution to Problem 8.4.17.

1	0	0	0	0	0	0	0	0	0	0	0	0	0	0	0	0	0	0	0	0	0	0	0		p_1		1,500
8	−29.5	8	0	0	1	0	0	0	0	0	0	0	12.5	0	0	0	0	0	0	0	0	0	0		p_2		0
0	8	−29.5	8	0	0	1	0	0	0	0	0	0	0	12.5	0	0	0	0	0	0	0	0	0		p_3		0
0	0	8	−29.5	8	0	0	1	0	0	0	0	0	0	0	12.5	0	0	0	0	0	0	0	0		p_4		0
1	0	0	8	−30.5	8	0	0	1	0	0	0	0	0	0	0	12.5	0	0	0	0	0	0	0		p_5		0
0	1	0	0	8	−30.5	8	0	0	1	0	0	0	0	0	0	0	12.5	0	0	0	0	0	0		p_6		0
0	0	1	0	0	8	−30.5	8	0	0	1	0	0	0	0	0	0	0	12.5	0	0	0	0	0		p_7		0
0	0	0	1	0	0	8	−30.5	8	0	0	1	0	0	0	0	0	0	0	12.5	0	0	0	0		p_8		0
0	0	0	0	1	0	0	8	−29.5	8	0	0	0	0	0	0	0	0	0	0	12.5	0	0	0		p_9		0
0	0	0	0	0	1	0	0	8	−29.5	8	0	0	0	0	0	0	0	0	0	0	12.5	0	0		p_{10}		0
0	0	0	0	0	0	1	0	0	8	−29.5	8	0	0	0	0	0	0	0	0	0	0	12.5	0		p_{11}		0
0	0	0	0	0	0	0	0	0	0	0	1	0	0	0	0	0	0	0	0	0	0	0	0	×	p_{12}	=	14.7
12.5	0	0	0	0	0	0	0	0	0	0	0	−21.5	8	0	0	1	0	0	0	0	0	0	0		p_{13}		0
0	12.5	0	0	0	0	0	0	0	0	0	0	8	−29.5	8	0	0	1	0	0	0	0	0	0		p_{14}		0
0	0	12.5	0	0	0	0	0	0	0	0	0	0	8	−29.5	8	0	0	1	0	0	0	0	0		p_{15}		0
0	0	0	12.5	0	0	0	0	0	0	0	0	0	0	8	−29.5	8	0	0	1	0	0	0	0		p_{16}		0
0	0	0	0	12.5	0	0	0	0	0	0	0	1	0	0	8	−30.5	8	0	0	1	0	0	0		p_{17}		0
0	0	0	0	0	12.5	0	0	0	0	0	0	0	1	0	0	8	−30.5	8	0	0	1	0	0		p_{18}		0
0	0	0	0	0	0	12.5	0	0	0	0	0	0	0	1	0	0	8	−30.5	8	0	0	1	0		p_{19}		0
0	0	0	0	0	0	0	12.5	0	0	0	0	0	0	0	1	0	0	8	−30.5	8	0	0	1		p_{20}		0
0	0	0	0	0	0	0	0	12.5	0	0	0	0	0	0	0	1	0	0	8	−29.5	8	0	0		p_{21}		0
0	0	0	0	0	0	0	0	0	12.5	0	0	0	0	0	0	0	1	0	0	8	−29.5	8	0		p_{22}		0
0	0	0	0	0	0	0	0	0	0	12.5	0	0	0	0	0	0	0	1	0	0	8	−29.5	8		p_{23}		0
0	0	0	0	0	0	0	0	0	0	0	12.5	0	0	0	0	0	0	0	1	0	0	8	−21.5		p_{24}		0

Case 2

Fig. 8.16.4—Case 2 coefficient matrix, unknown vector and the right-hand-side vector.

1,500	1,238.983	1,073.93	963.0192
897.9312	806.481	708.219	616.7688
551.6808	440.7703	275.7172	14.7

Table 8.26a—Case 2: Top layer pressure distribution from Part 3 of Solution to Problem 8.4.17.

1,364.729	1,212.166	1,068.535	961.3827
894.3376	805.4972	709.2028	620.3624
553.3073	446.1646	302.5341	149.9714

Table 8.26b—Case 2: Bottom layer pressure distribution from Part 3 of Solution to Problem 8.4.17.

Problem 8.4.18 (Ertekin et al. 2001)

Consider the incompressible flow of water in the 3D structure (Japanese crystal puzzle) shown in **Fig. 8.17.1.** Four injection wells are located in the outermost gridblocks of the largest *x*-*y* plane, and one production well is located in the central gridblock of the same *x*-*y* plane. The boundary conditions are as follows:

- The system is surrounded by no-flow boundaries.
- The injection rate at the injection well is 3,000 STB/D.
- The production gridblock is kept at a pressure of 100 psia.

The uniform gridblock properties are $\Delta x = \Delta y = \Delta z = 100$ ft and $k_x = k_y = k_z = 100$ md. The fluid properties are $\mu_w = 1$ cp and $B_w = 1$ res bbl/STB. Assume that the potential gradient for this problem can be approximated by pressure gradients.

Applying the finite-difference approximation to the governing flow equation, determine the steady-state pressure distribution and flow rate into the production well. Solve the linear equations of the finite-difference approximation by 1) the Jacobi iterative method, 2) the Gauss-Seidel method, and 3) the PSOR method with ω_{opt}. Use a pressure tolerance of 0.1 psia for each algorithm. Write the number of iterations with the solution obtained. Check for material balances.

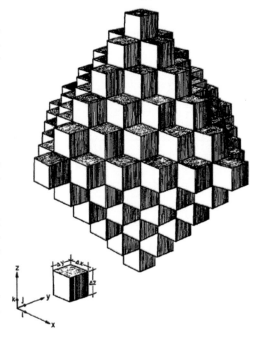

Fig. 8.17.1—Reservoir system for Problem 8.4.18.

Solution to Problem 8.4.18

For a 3D system with homogeneous and isotropic grid-property distributions, and in which the depth gradient is ignored, the PDE that characterizes the problem can be written as

$$\frac{\partial^2 p}{\partial x^2} + \frac{\partial^2 p}{\partial y^2} + \frac{\partial^2 p}{\partial z^2} + \frac{q\mu B}{kV_b} = 0. \quad\dots\dots\dots\dots\dots\dots\dots\dots\dots\dots\dots\dots\dots\dots\dots (8.316)$$

For the convenience of numbering the grids, the system is decomposed into six layers, shown in **Figs. 8.17.2 through 8.17.7.**

- For Block 1: $p_2 - p_1 = -\dfrac{q\mu B}{kh}$. $\dots\dots\dots\dots\dots\dots\dots\dots\dots\dots\dots\dots\dots\dots\dots\dots\dots (8.317)$
- For Block 2: $p_1 - 6p_2 + p_3 + 4p_7 = 0$. $\dots\dots\dots\dots\dots\dots\dots\dots\dots\dots\dots\dots\dots (8.318)$
- For Block 3: $p_2 + p_4 + 4p_8 - 6p_3 = 0$. $\dots\dots\dots\dots\dots\dots\dots\dots\dots\dots\dots\dots\dots (8.319)$
- For Block 4: $p_3 + p_5 + 4p_9 - 6p_4 = 0$. $\dots\dots\dots\dots\dots\dots\dots\dots\dots\dots\dots\dots\dots (8.320)$
- For Block 5: $p_4 + p_6 + 4p_{10} - 6p_5 = 0$. $\dots\dots\dots\dots\dots\dots\dots\dots\dots\dots\dots\dots\dots (8.321)$
- For Block 6: $p_6 = 100$ psia.
- For Block 7: $p_2 + p_8 - 2p_7 = 0$. $\dots\dots\dots\dots\dots\dots\dots\dots\dots\dots\dots\dots\dots\dots\dots (8.322)$
- For Block 8: $p_7 + p_9 + p_3 + p_{11} + 2p_{13} - 6p_8 = 0$. $\dots\dots\dots\dots\dots\dots\dots\dots\dots\dots (8.323)$
- For Block 9: $p_8 + p_4 + p_{10} + p_{12} + 2p_{15} - 6p_9 = 0$. $\dots\dots\dots\dots\dots\dots\dots\dots\dots (8.324)$
- For Block 10: $2p_5 + 2p_9 + 2p_{16} - 6p_{10} = 0$. $\dots\dots\dots\dots\dots\dots\dots\dots\dots\dots\dots\dots (8.325)$

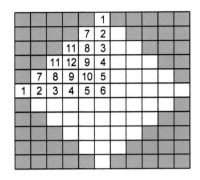

Fig. 8.17.2—Layer 1 (largest layer in the middle).

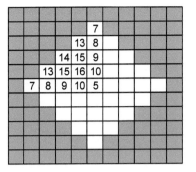

Fig. 8.17.3—Layer 2.

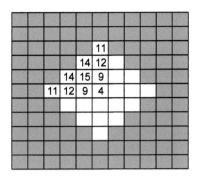

Fig. 8.17.4—Layer 3.

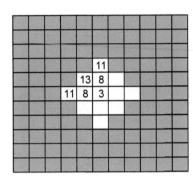

Fig. 8.17.5—Layer 4.

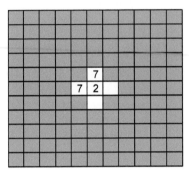

Fig. 8.17.6—Layer 5.

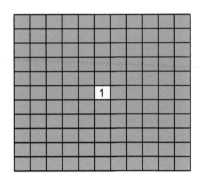

Fig. 8.17.7—Layer 6.

- For Block 11: $p_8 + p_{12} - 2p_{11} = 0.$ .... (8.326)
- For Block 12: $2p_{11} + 2p_9 + 2p_{14} - 6p_{12} = 0.$ .... (8.327)
- For Block 13: $2p_8 + p_{15} - 3p_{13} = 0.$ .... (8.328)
- For Block 14: $2p_{15} + p_{12} - 3p_{14} = 0.$ .... (8.329)
- For Block 15: $2p_9 + 2p_{14} + p_{13} + p_{16} - 6p_{15} = 0.$ .... (8.330)
- For Block 16: $p_{15} + p_{10} - 2p_{16} = 0.$ .... (8.331)

Eqs. 8.317 through 8.331 are written in matrix form:

$$
\begin{bmatrix}
-1 & 1 & 0 & 0 & 0 & 0 & 0 & 0 & 0 & 0 & 0 & 0 & 0 & 0 & 0 & 0 \\
1 & -6 & 1 & 0 & 0 & 0 & 4 & 0 & 0 & 0 & 0 & 0 & 0 & 0 & 0 & 0 \\
0 & 1 & -6 & 1 & 0 & 0 & 0 & 4 & 0 & 0 & 0 & 0 & 0 & 0 & 0 & 0 \\
0 & 0 & 1 & -6 & 1 & 0 & 0 & 0 & 4 & 0 & 0 & 0 & 0 & 0 & 0 & 0 \\
0 & 0 & 0 & 1 & -6 & 1 & 0 & 0 & 0 & 4 & 0 & 0 & 0 & 0 & 0 & 0 \\
0 & 0 & 0 & 0 & 0 & 1 & 0 & 0 & 0 & 0 & 0 & 0 & 0 & 0 & 0 & 0 \\
0 & 1 & 0 & 0 & 0 & 0 & -2 & 1 & 0 & 0 & 0 & 0 & 0 & 0 & 0 & 0 \\
0 & 0 & 0 & 0 & 0 & 0 & 1 & -6 & 1 & 0 & 1 & 0 & 2 & 0 & 0 & 0 \\
0 & 0 & 0 & 1 & 0 & 0 & 0 & 1 & -6 & 1 & 0 & 1 & 0 & 0 & 2 & 0 \\
0 & 0 & 0 & 0 & 2 & 0 & 0 & 2 & -6 & 0 & 0 & 0 & 0 & 0 & 2 \\
0 & 0 & 0 & 0 & 0 & 0 & 0 & 1 & 0 & 0 & -2 & 1 & 0 & 0 & 0 & 0 \\
0 & 0 & 0 & 0 & 0 & 0 & 0 & 0 & 2 & 0 & 2 & -6 & 0 & 2 & 0 & 0 \\
0 & 0 & 0 & 0 & 0 & 0 & 0 & 2 & 0 & 0 & 0 & 0 & -3 & 0 & 1 & 0 \\
0 & 0 & 0 & 0 & 0 & 0 & 0 & 0 & 0 & 0 & 0 & 1 & 0 & -3 & 2 & 0 \\
0 & 0 & 0 & 0 & 0 & 0 & 0 & 0 & 2 & 0 & 0 & 0 & 1 & 2 & -6 & 1 \\
0 & 0 & 0 & 0 & 0 & 0 & 0 & 0 & 0 & 1 & 0 & 0 & 0 & 0 & 1 & -2
\end{bmatrix}
\begin{bmatrix}
p_1 \\ p_2 \\ p_3 \\ p_4 \\ p_5 \\ p_6 \\ p_7 \\ p_8 \\ p_9 \\ p_{10} \\ p_{11} \\ p_{12} \\ p_{13} \\ p_{14} \\ p_{15} \\ p_{16}
\end{bmatrix}
=
\begin{bmatrix}
-266.2 \\ 0 \\ 0 \\ 0 \\ 0 \\ 100 \\ 0 \\ 0 \\ 0 \\ 0 \\ 0 \\ 0 \\ 0 \\ 0 \\ 0 \\ 0
\end{bmatrix}
\quad \cdots (8.332)
$$

Block/solved pressures	PSOR	Jacobian	Gauss-Seidel
1	831.37	821.17	821.26
2	565.24	555.07	555.16
3	482.89	472.82	473.01
4	438.55	429.05	429.23
5	365.85	358.17	358.39
6	100.00	100.00	100.00
7	519.33	509.11	509.30
8	473.43	463.34	463.53
9	445.68	435.83	436.11
10	414.16	405.10	405.36
11	463.06	452.91	453.18
12	452.70	442.66	442.92
13	464.85	454.70	454.99
14	449.32	439.26	439.57
15	447.65	437.70	437.99
16	430.93	421.31	421.68

Table 8.27—Solution vectors generated with different iterative
solutions from Solution to Problem 8.4.18.

Table 8.27 displays the solution vectors generated using different iterative procedures.

To achieve the pressure tolerance of 0.1 psia, the number of iterations carried out by using various iterative solvers is shown in **Fig. 8.17.8.**

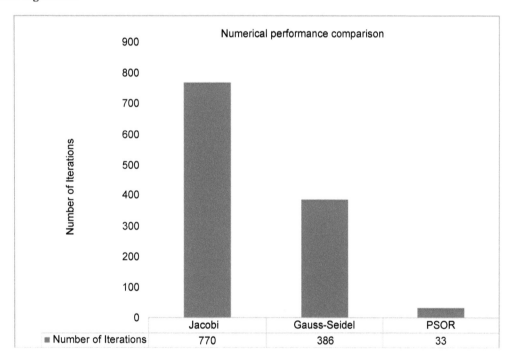

Fig. 8.17.8—Comparison of the numerical performance of different solvers.

By knowing the injection rate and the block pressure of the production well, an MBC can be carried out to validate the solutions, as shown in **Table 8.28.**

	PSOR	Jacobian	Gauss-Seidel
Calculated production rate, STB/D	11,984.52	11,638.40	11,648.40
$\dfrac{q_{producer}}{q_{injector}}$	0.998710	0.969867	0.970700

Table 8.28—MBC for the Solution to Problem 8.4.18.

Problem 8.4.19 (Ertekin et al. 2001)

Explain how to approximately solve Laplace's equation,

$$\frac{\partial^2 f}{\partial x^2} + \frac{\partial^2 f}{\partial y^2} + \frac{\partial^2 f}{\partial z^2} = 0,$$ in 3D inside a cube with an edge of unit length,

if $f = f(x,y,z)$ is equal to 1 on one face and 0 on the remaining faces. Illustrate your procedure by finding the approximate value of f at $x = 1/3$, $y = 1/3$, and $z = 1/3$. In the discretization process, use a 4×4×4 point-distributed grid system and base your finite-difference equations on the computational molecule shown in **Fig. 8.18.1.**

Hint: Consider the symmetry of the system—there are two unknowns to be solved: the node (p_1) close to the surface with f value of 1, and the node (p_2) away from the surface with f value of 0.

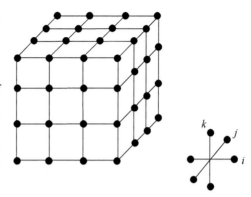

Fig. 8.18.1—Reservoir system for Problem 8.4.19.

Solution to Problem 8.4.19

Because the system is discretized by cubes with uniform edge lengths (see **Fig. 8.18.2**),

- For Node 1:

$$1 - 2f_1 + f_2 + 0 - 2f_1 + f_1 + 0 - 2f_1 + f_1 = 0. \quad (8.333)$$

- For Node 2:

$$0 - 2f_2 + f_1 + 0 - 2f_2 + f_2 + 0 - 2f_2 + f_2 = 0. \quad (8.334)$$

Rearranging,

$$1 - 4f_1 + f_2 = 0, \quad \dots \dots \dots \dots (8.335)$$

$$-4f_2 + f_1 = 0, \quad \dots \dots \dots \dots (8.336)$$

and solving for f_1 and f_2:

$$f_1 = \frac{4}{15}, \; f_2 = \frac{1}{15}. \quad \dots \dots \dots \dots \dots (8.337)$$

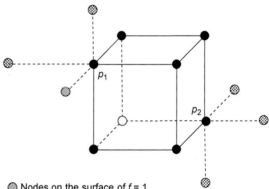

Nodes on the surface of $f = 1$
Nodes on the surface of $f = 0$

Fig. 8.18.2—Node representations for Problem 8.4.19.

Problem 8.4.20 (Ertekin et al. 2001)

Calculate the steady-state pressure distribution in the 3D porous medium shown in **Fig. 8.19.1,** in which a production well at Node (3,3,3) is completed. Calculate the flow rate of the well. Ignore gravitational effects. The boundary conditions are as follows:

- Nodes on the outer surface of the body are kept at 2,000 psia.
- Pressure of Gridblock (3,3,3) is kept at 400 psia.

The porous medium and fluid properties are $k_x = k_y = k_z = 8.88$ md, $\Delta x = \Delta y = \Delta z = 100$ ft (uniform), $\phi = 11\%$, $\mu = 10$ cp, and $B = 1$ res bbl/STB.

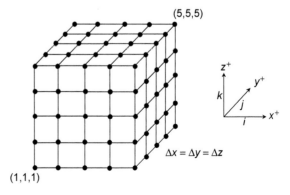

Fig. 8.19.1—Reservoir system for Problem 8.4.20.

Solution to Problem 8.4.20

The PDE that describes the flow system can be written as

$$\frac{\partial^2 p}{\partial x^2} + \frac{\partial^2 p}{\partial y^2} + \frac{\partial^2 p}{\partial z^2} + \frac{q\mu B}{V_b k} = 0. \quad \ldots\ldots\ldots\ldots\ldots\ldots\ldots\ldots\ldots \quad (8.338)$$

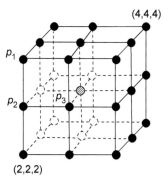

The unknown nodes of the system are illustrated in **Fig. 8.19.2.** The pressure of the well node is kept at 400 psia. Considering the existing symmetry in the system, there are only three unknowns to solve, which are the corner node (p_1), edge node (p_2), and surface node (p_3). Three linear equations must be solved to establish the pressure distribution throughout the reservoir:

- For the corner node (for example, Node 1):

$$-2p_1 + p_2 = -2{,}000. \quad \ldots\ldots\ldots\ldots\ldots\ldots\ldots\ldots\ldots \quad (8.339)$$

Fig. 8.19.2—Node representations for Problem 8.4.20.

- For the edge node (for example, Node 2):

$$p_1 - 3p_2 + p_3 = -2{,}000. \quad \ldots\ldots\ldots\ldots\ldots\ldots\ldots\ldots\ldots\ldots\ldots\ldots \quad (8.340)$$

- For surface node (for example, Node 3):

$$2p_2 - 3p_3 = -1{,}200. \quad \ldots\ldots\ldots\ldots\ldots\ldots\ldots\ldots\ldots\ldots\ldots\ldots\ldots\ldots\ldots \quad (8.341)$$

Solving for pressure values at these three nodes, one obtains $p_1 = 1{,}927.3$ psia, $p_2 = 1{,}854.6$ psia, and $p_3 = 1{,}636.4$ psia.

The conservation-of-mass principle requires that the mass flux that enters from the external boundaries should equal the production from the well; therefore,

- For the corner nodes:

$$q_c = 3 \times 1.127 \times 10^{-3} \times \frac{k\Delta x}{\mu B}(2{,}000 - 1{,}927.3) = 21.8 \ \text{STB/D}. \quad \ldots\ldots\ldots\ldots\ldots\ldots\ldots\ldots \quad (8.342)$$

- For the edge nodes:

$$q_s = 2 \times 1.127 \times 10^{-3} \times \frac{k\Delta x}{\mu B}(2{,}000 - 1{,}854.6) = 29.11 \ \text{STB/D}. \quad \ldots\ldots\ldots\ldots\ldots\ldots\ldots\ldots \quad (8.343)$$

- For the surface nodes:

$$q_f = 1.127 \times 10^{-3} \times \frac{k\Delta x}{\mu B}(2{,}000 - 1{,}636.4) = 36.4 \ \text{STB/D}. \quad \ldots\ldots\ldots\ldots\ldots\ldots\ldots\ldots \quad (8.344)$$

Therefore, the total mass flux that enters the cube can be calculated as

$$q_{\text{total}} = 8q_c + 12q_s + 6q_f = 742.4 \ \text{STB/D}. \quad \ldots\ldots\ldots\ldots\ldots\ldots\ldots\ldots\ldots\ldots\ldots\ldots \quad (8.345)$$

Hence, the production rate of the well must be equal to 742.4 STB/D.

Problem 8.4.21 (Ertekin et al. 2001)

What are the definition of $\Gamma_{g\ i,j,k}$, in terms of p/z_g of a compressible flow system when the definition of $B_{g\ i,j,k}$, as given by $B_g = \frac{\rho_{gsc}}{\alpha_c \rho_g} = \frac{p_{sc}}{\alpha_c \rho_{sc}} T \frac{z_g}{p}$, is used in $[(V_{b\ i,j,k}/\alpha_c \Delta t)(\phi^{n+1}/B_g^{\ n+1}) - (\phi^n/B_g^{\ n})]_{i,j,k}$?

Hint: Consider all the permutations of the time levels and the pressure-dependent terms ϕ, p, z_g.

Solution to Problem 8.4.21

Applying the definition of gas-FVF,

$$B_g\left(\frac{\text{res bbl}}{\text{scf}}\right)=\frac{p_{sc}z_gT}{\alpha_c pT_{sc}}. \quad\dots\dots\dots\dots\dots\dots\dots\dots\dots\dots\dots\dots\dots\dots\dots(8.346)$$

The right-hand side of the gas equation can be written as

$$\text{right}-\text{hand side}=\left(\frac{\phi V_b T_{sc}}{p_{sc}T}\right)_{i,j,k}\frac{\left[\left(\frac{p}{z_g}\right)^{n+1}_{i,j,k}-\left(\frac{p}{z_g}\right)^{n}_{i,j,k}\right]}{\Delta t}.\quad\dots\dots\dots\dots\dots(8.347)$$

Rearranging Eq. 8.347,

$$\Gamma^n_{i,j,k}=\left(\frac{\phi V_b T_{sc}}{p_{sc}T\Delta tz_g}\right)^n_{i,j,k}.\quad\dots\dots\dots\dots\dots\dots\dots\dots\dots\dots\dots\dots\dots(8.348)$$

Problem 8.4.22 (Ertekin et al. 2001)

Consider the steady-state flow of a single-phase incompressible fluid shown in the reservoir of **Fig. 8.20.** Permeability and thickness maps are provided. Well 1 produces at a flow rate of 600 STB/D and Well 2 produces at a rate of 400 STB/D. The known boundary conditions are shown along the three external boundaries. Calculate the value of the pressure gradient along the south boundary of the reservoir (note that the gradient along the south boundary is known to be uniform). Other relevant data are $\Delta x=\Delta y=1,000$ ft, $\mu=1$ cp, $B=1$ res bbl/STB, and $\phi=16\%$.

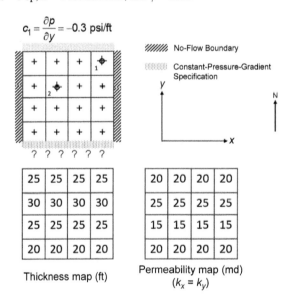

Fig. 8.20—Reservoir system for Problem 8.4.22.

Solution to Problem 8.4.22

The fluid leakage from the top boundary is

$$q_{\text{leakage}}=4\times1.127\times10^{-3}\times\frac{Ak}{\mu B}c_1=4\times1.127\times10^{-3}\times\frac{1,000\text{ ft}\times25\text{ ft}\times20\text{ md}}{1\text{ cp}\times1\text{ res bbl}/\text{STB}}\left(-\frac{0.3\text{ psi}}{\text{ft}}\right)\dots\dots(8.349)$$

$$=676.2\text{ STB}/\text{D}.$$

Total fluid loss from the system is

$$q_{\text{total}}=q_{\text{leakage}}+q_{\text{well 1}}+q_{\text{well 2}}=1,676.2\text{ STB}/\text{D}.\quad\dots\dots\dots\dots\dots\dots\dots\dots\dots\dots(8.350)$$

According to mass conservation, the fluid that enters the system by way of the bottom should be equal to the total fluid loss:

$$q_{total} = -4 \times 1.127 \times 10^{-3} \times \frac{Ak}{\mu B} c_x = -4 \times 1.127 \times 10^{-3} \times \frac{1{,}000 \text{ ft} \times 20 \text{ ft} \times 20 \text{ md}}{1 \text{ cp} \times 1 \text{ res bbl/STB}}(c_x) \quad \dots \dots \dots \dots \quad (8.351)$$

$$= 1{,}676.2 \text{ STB/D}.$$

Solving for c_x gives $c_x = -0.93$ psi/ft.

Problem 8.4.23 (Ertekin et al. 2001)

Consider the 2D heterogeneous anisotropic, volumetric single-phase gas reservoir shown in **Fig. 8.21.1.** Assume you are constructing a simulator using the GNR protocol. Use the notation introduced in this chapter to complete the following tasks:

1. Write a characteristic finite-difference equation in an explicit form that generates the system of nonlinear algebraic equations for this reservoir.
2. Write the characteristic linear equation that generates the coefficient matrix, unknown vector, and the right-hand-side vector.
3. Construct the coefficient matrix, unknown vector, and right-hand-side vector using a generic notation. Do not write the explicit description of the derivatives and the other entries.

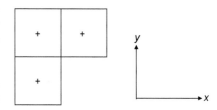

Fig. 8.21.1—Reservoir system for Problem 8.4.23.

Solution to Problem 8.4.23

1. The system can be numbered as shown in **Fig. 8.21.2.** For a 2D gas reservoir with heterogeneous- and anisotropic-property distribution,

Fig. 8.21.2—Numbering scheme of the reservoir system for Problem 8.4.23.

$$\frac{\partial}{\partial x}\left(\frac{A_x k_x}{\mu_g B_g}\frac{\partial p}{\partial x}\right)\Delta x + \frac{\partial}{\partial y}\left(\frac{A_y k_y}{\mu_g B_g}\frac{\partial p}{\partial y}\right)\Delta y = \left(\frac{\phi V_b T_{sc}}{p_{sc}T}\right)\frac{\partial\left(\frac{p}{z_g}\right)}{\partial t} \quad \dots \dots \quad (8.352)$$

- For Block 1:

$$T_{12}(p_2 - p_1) = \left(\frac{\phi V_b T_{sc}}{p_{sc}T}\right)_1 \frac{\left[\left(\frac{p}{z_g}\right)_1^{n+1} - \left(\frac{p}{z_g}\right)_1^{n}\right]}{\Delta t}. \quad \dots \dots \dots \dots \dots \dots \dots \dots \dots \dots \dots \quad (8.353)$$

- For Block 2:

$$T_{12}(p_1 - p_2) + T_{23}(p_3 - p_2) = \left(\frac{\phi V_b T_{sc}}{p_{sc}T}\right)_2 \frac{\left[\left(\frac{p}{z_g}\right)_2^{n+1} - \left(\frac{p}{z_g}\right)_2^{n}\right]}{\Delta t}. \quad \dots \dots \dots \dots \dots \dots \quad (8.354)$$

- For Block 3:

$$T_{23}(p_2 - p_3) = \left(\frac{\phi V_b T_{sc}}{p_{sc}T}\right)_3 \frac{\left[\left(\frac{p}{z_g}\right)_3^{n+1} - \left(\frac{p}{z_g}\right)_3^{n}\right]}{\Delta t}. \quad \dots \dots \dots \dots \dots \dots \dots \dots \dots \dots \quad (8.355)$$

2. The characteristic GNR equation is

$$\left(\frac{\partial R_i}{\partial p_{i-1}}\right)^{(k)} \Delta p_{i-1}^{(k+1)} + \left(\frac{\partial R_i}{\partial p_i}\right)^{(k)} \Delta p_i^{(k+1)} + \left(\frac{\partial R_i}{\partial p_{i+1}}\right)^{(k)} \Delta p_{i+1}^{(k+1)} = - R_i^{(k)}. \quad\dots\dots\dots\dots\dots\dots (8.356)$$

3. The system of equations can be expressed in the following matrix form:

$$\begin{bmatrix} \dfrac{\partial R_1}{\partial p_1} & \dfrac{\partial R_1}{\partial p_2} & 0 \\[2mm] \dfrac{\partial R_2}{\partial p_1} & \dfrac{\partial R_2}{\partial p_2} & \dfrac{\partial R_2}{\partial p_3} \\[2mm] 0 & \dfrac{\partial R_3}{\partial p_2} & \dfrac{\partial R_3}{\partial p_3} \end{bmatrix}^{(k)} \begin{bmatrix} \Delta p_1 \\ \Delta p_2 \\ \Delta p_3 \end{bmatrix}^{(k+1)} = \begin{bmatrix} -R_1 \\ -R_2 \\ -R_3 \end{bmatrix}^{(k)}. \quad\dots\dots\dots\dots\dots\dots (8.357)$$

Problem 8.4.24 (Ertekin et al. 2001)

Consider the flow of a slightly compressible fluid in a 2D homogeneous, isotropic porous medium shown in **Fig. 8.22**. The entire reservoir is surrounded by no-flow boundaries. Initial pressure of the reservoir is determined to be 2,800 psia. The well in Gridblock 2 produces at a constant rate of 400 STB/D.

Find the flowing sandface pressure of the well after 20 days of production. Use the ADIP solution technique. The uniform gridblock properties are $\Delta x = \Delta y = 600$ ft, $\phi = 24\%$, $k_x = k_y = 200$ md, $h = 60$ ft, and $r_w = 0.24$ ft. The fluid properties are $c = 8 \times 10^{-5}$ psi^{-1}, $B = 0.98$ res bbl/STB, and $\mu = 1.2$ cp ($s = 0$).

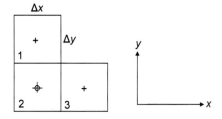

Fig. 8.22—Reservoir system for Problem 8.4.24.

Solution to Problem 8.4.24

The problem is a slightly compressible fluid flow in a reservoir with homogeneous- and isotropic-permeability distribution. The governing equation can be written as

$$\frac{\partial}{\partial x}\left(\frac{A_x k}{\mu B}\frac{\partial p}{\partial x}\right)\Delta x + \frac{\partial}{\partial y}\left(\frac{A_y k}{\mu B}\frac{\partial p}{\partial y}\right)\Delta y + q = \frac{V_b \phi c_t}{5.615}\frac{\partial p}{\partial t}. \quad\dots\dots\dots\dots\dots\dots (8.358)$$

The PDE can be written in a discretized form:

$$T_{i,j-\frac{1}{2}}p_{i,j-1}^{n+1} + T_{i,j+\frac{1}{2}}p_{i,j+1}^{n+1} + T_{i-\frac{1}{2},j}p_{i-1,j}^{n+1} + T_{i+\frac{1}{2},j}p_{i+1,j}^{n+1} - \left(T_{i,j-\frac{1}{2}} + T_{i,j+\frac{1}{2}} + T_{i-\frac{1}{2},j} + T_{i+\frac{1}{2},j} + \Gamma_{i,j}\right)p_{i,j}^{n+1} + q = -\Gamma_{i,j}p_{i,j}^{n}. \quad\dots\dots (8.359)$$

The transmissibility and storage coefficients are

$$T_{i,j\pm\frac{1}{2}} = T_{i\pm\frac{1}{2},j} = 11.5 \text{ STB/D-psi}, \quad\dots\dots\dots\dots\dots\dots (8.360)$$

$$\Gamma_{i,j} = \frac{V_b \phi c_t}{5.615\Delta t} = 3.6930 \text{ STB/D-psi}. \quad\dots\dots\dots\dots\dots\dots (8.361)$$

Using the calculated coefficients, the ADIP method can be implemented to solve the problem. In this case, we use $\omega = 1$ as the iteration parameter. As shown in **Table 8.29,** it takes 25 iterations to achieve a convergence tolerance of 1×10^{-3} psi. Performing an MBC, the fluid expansion is

$$\Delta V_{\text{total}} = c_t \sum_{i=1}^{3}\left(\frac{V_b \phi}{5.615}\right)_i \left(p_i^{n+1} - p_i^{n}\right) = 23{,}999.1 \text{ STB}. \quad\dots\dots\dots\dots\dots\dots (8.362)$$

Iteration 1			
p^*		$p^{(1)}$	
2,800.000	0.000	2,792.249	0.000
2,787.966	2,794.815	2,782.008	2,790.891
Iteration 2			
p^*		$p^{(2)}$	
2,786.382	0.000	2,782.733	0.000
2,777.701	2,786.469	2,773.539	2,783.121
...			
Iteration 25			
p^*		$p^{(25)}$	
2767.388	0.000	2767.388	0.000
2756.915	2767.388	2756.915	2767.388

Table 8.29—Iterations to achieve convergence tolerance of 1×10⁻³ psi using the ADIP method from the Solution to Problem 8.4.24.

$$\text{MBC}=\left|\frac{\left(q_{\text{production}}\right)\Delta t}{\Delta V_{\text{total}}}\right|=\left|\frac{24,000}{23,999.1}\right|=1.00004. \dots\dots\dots\dots\dots\dots\dots\dots\dots\dots\dots\dots\dots (8.363)$$

After calculating the productivity index, one can solve for the sandface pressure:

$$\Omega=\frac{2\pi \bar{k}h}{\mu B\left[\ln\left(\dfrac{r_e}{r_w}\right)+s\right]}=11.64 \text{ STB/D-psi}, \dots\dots\dots\dots\dots\dots\dots\dots\dots\dots\dots\dots(8.364)$$

$$p_{sf}=p+\frac{q}{\Omega}=2,756.915+\frac{-400}{11.64}=2,722.55\text{ psi}. \dots\dots\dots\dots\dots\dots\dots\dots\dots\dots\dots(8.365)$$

Problem 8.4.25 (Ertekin et al. 2001)

The reservoir in **Fig. 8.23** has homogeneous property distribution, but exhibits a pronounced permeability anisotropy. Initial reservoir pressure is 3,000 psia and depth gradients are ignored. The gridblock properties are $\Delta x = 1,000$ ft, $\Delta y = 800$ ft, $\phi = 18\%$, $k_x = 100$ md, $k_y = 50$ md, $h = 40$ ft, and $r_w = 0.2$ ft. The fluid properties are $c = 1 \times 10^{-6}$ psi⁻¹, $B = 1$ res bbl/STB, and $\mu = 0.87$ cp.

1. Write the finite-difference equations for Gridblocks 1, 18, and 25 at $t = 10$ days, if the injection rate of the well in Gridblock 18 is 1,000 STB/D.
2. Write the finite-difference equations for Gridblocks 8 and 18 at $t = 10$ days, if the sandface pressure of the well in Gridblock 8 is 4,100 psia.
3. What is the minimum number of unknowns that needs to be solved for Parts 1 and 2 of this problem, and what are they?

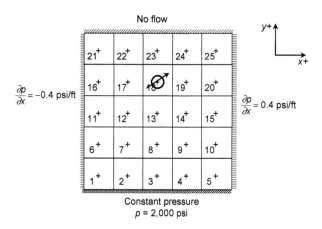

Fig. 8.23—Reservoir system for Problem 8.4.25.

Solution to Problem 8.4.25

The PDE that characterizes the slightly compressible flow in the reservoir described for this problem can be written as

$$k_x \frac{\partial^2 p}{\partial x^2} + k_y \frac{\partial^2 p}{\partial y^2} + \frac{q\mu B}{V_b} = \frac{\mu B c_t \phi}{5.615} \frac{\partial p}{\partial t}. \qquad (8.366)$$

Knowing that $k_x = 2k_y$,

$$2\frac{\partial^2 p}{\partial x^2} + \frac{\partial^2 p}{\partial y^2} + \frac{q\mu B}{k_y V_b} = \frac{\mu B c_t \phi}{k_y 5.615} \frac{\partial p}{\partial t}. \qquad (8.367)$$

Using finite-difference approximation,

$$2\left[\frac{p_{i-1,j}^{n+1} - 2p_{i,j}^{n+1} + p_{i+1,j}^{n+1}}{(\Delta x)^2}\right] + \frac{p_{i,j-1}^{n+1} - 2p_{i,j}^{n+1} + p_{i,j+1}^{n+1}}{(\Delta y)^2} + \left(\frac{q\mu B}{\Delta x \Delta y h k_y}\right)_{i,j} = \left(\frac{\mu B c_t \phi}{k_y 5.615 \Delta t}\right)\left(p_{i,j}^{n+1} - p_{i,j}^n\right). \qquad (8.368)$$

Because $\dfrac{\Delta y}{\Delta x} = \dfrac{4}{5}$, the previous equation can be further simplified to

$$2\left(p_{i-1,j}^{n+1} - 2p_{i,j}^{n+1} + p_{i+1,j}^{n+1}\right) + \left(\frac{\Delta x}{\Delta y}\right)^2 \left(p_{i,j-1}^{n+1} - 2p_{i,j}^{n+1} + p_{i,j+1}^{n+1}\right) + \left(\frac{\Delta x}{\Delta y}\right)\left(\frac{q\mu B}{h k_y}\right)_{i,j} = \left[\frac{\mu B c_t \phi (\Delta x)^2}{k_y 5.615 \Delta t}\right]\left(p_{i,j}^{n+1} - p_{i,j}^n\right), \qquad (8.369)$$

$$\frac{\mu B c_t \phi (\Delta x)^2}{k_y 5.615 \Delta t} = \frac{(0.87)(1)(1 \times 10^{-6})(0.18)(1,000)^2}{(50)(1.127 \times 10^{-3})(5.615)(10)} = 0.0495, \qquad (8.370)$$

and

$$\left(\frac{\Delta x}{\Delta y}\right)\left(\frac{\mu B}{h k_y}\right) = \frac{(1,000)}{(800)} \frac{(0.87)(1)}{(40)(50)(1.127 \times 10^{-3})} = 0.4825. \qquad (8.371)$$

Plugging in the preceding coefficients, the finite-difference approximation can be rewritten as

$$2\left(p_{i-1,j}^{n+1} - 2p_{i,j}^{n+1} + p_{i+1,j}^{n+1}\right) + 1.5625\left(p_{i,j-1}^{n+1} - 2p_{i,j}^{n+1} + p_{i,j+1}^{n+1}\right) + 0.4825 q_{i,j} = 0.0495\left(p_{i,j}^{n+1} - p_{i,j}^n\right). \qquad (8.372)$$

1.
- Block 1: The discretized finite-difference equation can be written as (p_1' represents the reflection block)

$$2\left(p_1' - 2p_1 + p_2\right) + 1.5625\left(2,000 - 2p_1 + p_6\right) = 0.0495\left(p_1 - 3,000\right). \qquad (8.373)$$

Applying the boundary condition,

$$\frac{p_1 - p_1'}{\Delta x} = -0.4 \text{ psi/ft}, \qquad (8.374)$$

gives

$$p_1' = p_1 + 0.4\Delta x, \qquad (8.375)$$

which yields the equation

$$2p_2 - 5.1745 p_1 + 1.563 p_6 = -4,073.48. \qquad (8.376)$$

- Block 18: Block 18 hosts an injection well with a specified rate of 1,000 STB/D:

$$2\left(p_{17} - 2p_{18} + p_{19}\right) + 1.5625\left(p_{13} - 2p_{18} + p_{23}\right) + 0.4825(1,000) = 0.0495\left(p_{18} - 3,000\right). \qquad (8.377)$$

Simplifying,

$$1.563 p_{13} + 2p_{17} - 7.1745 p_{18} + 2p_{19} + 1.563 p_{23} = -631.0. \qquad (8.378)$$

- Block 25:

$$2\left(p_{24} - 2p_{25} + p_{25}'\right) + 1.5625\left(p_{20} - p_{25}\right) = 0.0495\left(p_{25} - 3{,}000\right). \quad \dots\dots\dots\dots\dots\dots\dots \text{(8.379)}$$

Applying the boundary condition,

$$\frac{p_{25}' - p_{25}}{\Delta x} = 0.4 \text{ psi/ft}, \quad \dots\dots\dots\dots\dots\dots\dots\dots\dots\dots\dots\dots\dots\dots\dots\dots\dots\dots \text{(8.380)}$$

gives

$$p_{25}' = p_{25} + 0.4\Delta x, \quad \dots\dots\dots\dots\dots\dots\dots\dots\dots\dots\dots\dots\dots\dots\dots\dots\dots\dots\dots\dots \text{(8.381)}$$

$$2\left(p_{24} - p_{25} + 400\right) + 1.5625\left(p_{20} - p_{25}\right) = 0.0495\, p_{25} - 148.48. \quad \dots\dots\dots\dots\dots\dots\dots \text{(8.382)}$$

Simplifying,

$$1.5625\, p_{20} - 3.6125\, p_{25} + 2p_{24} = -948.52. \quad \dots\dots\dots\dots\dots\dots\dots\dots\dots\dots\dots\dots\dots \text{(8.383)}$$

2.

- Block 8:

$$2\left(p_{i-1,j}^{n+1} - 2p_{i,j}^{n+1} + p_{i+1,j}^{n+1}\right) + 1.5625\left(p_{i,j-1}^{n+1} - 2p_{i,j}^{n+1} + p_{i,j+1}^{n+1}\right) + 0.4825\, q_{i,j} = 0.0495\left(p_{i,j}^{n+1} - p_{i,j}^{n}\right), \quad \dots\dots \text{(8.384)}$$

$$2\left(p_{7} - 2p_{8} + p_{9}\right) + 1.5625\left(p_{3} - 2p_{8} + p_{13}\right) = 0.0495\left(p_{8} - 3{,}000\right). \quad \dots\dots\dots\dots\dots\dots\dots \text{(8.385)}$$

Simplifying,

$$1.563\, p_{3} + 2p_{7} - 7.1745\, p_{8} + 2p_{9} + 1.563\, p_{13} = -148.5. \quad \dots\dots\dots\dots\dots\dots\dots\dots\dots\dots\dots\dots \text{(8.386)}$$

- Block 18: Block 18 hosts an injection well with a sandface-pressure specification of 4,100 psi. The flow rate must be substituted by the wellbore model:

$$q_{18} = -\Omega\left(p_{18} - p_{sf_{spec}}\right). \quad \dots\dots\dots\dots\dots\dots\dots\dots\dots\dots\dots\dots\dots\dots\dots\dots\dots\dots \text{(8.387)}$$

The finite-difference equation can be rewritten as

$$2\left(p_{17} - 2p_{18} + p_{19}\right) + 1.5625\left(p_{13} - 2p_{18} + p_{23}\right) - 0.4825\,\Omega\left(p_{18} - p_{sf_{spec}}\right) = 0.0495\left(p_{18} - 3{,}000\right). \quad \dots\dots \text{(8.388)}$$

The well productivity index is calculated as

$$\Omega = \frac{2\pi \overline{k} h}{\mu B \ln\left(\dfrac{r_e}{r_w}\right)} = \frac{2\pi(70.71)(1.127 \times 10^{-3})(40)}{(1)(0.87)\ln\left(\dfrac{175.12}{0.2}\right)} = 3.4 \text{ STB/D-psi}, \quad \dots\dots\dots\dots\dots\dots \text{(8.389)}$$

where

$$r_e = \frac{0.28\sqrt{\left(\dfrac{k_x}{k_y}\right)^{\frac{1}{2}}(\Delta y)^2 + \left(\dfrac{k_y}{k_x}\right)^{\frac{1}{2}}(\Delta x)^2}}{\left(\dfrac{k_x}{k_y}\right)^{\frac{1}{4}} + \left(\dfrac{k_y}{k_x}\right)^{\frac{1}{4}}} = 175.12 \text{ ft} \quad \dots\dots\dots\dots\dots\dots\dots\dots\dots\dots \text{(8.390)}$$

and

$$\overline{k} = \sqrt{k_x k_y} = 70.71 \text{ md}. \quad \dots\dots\dots\dots\dots\dots\dots\dots\dots\dots\dots\dots\dots\dots\dots\dots\dots\dots\dots \text{(8.391)}$$

Simplifying,

$$1.5625\, p_{13} + 2p_{17} - 8.815\, p_{18} + 2p_{19} + 1.5625\, p_{23} = -6{,}874.55. \quad \dots\dots\dots\dots\dots\dots\dots\dots\dots\dots \text{(8.392)}$$

3. For Part 1,

- Equation of Block 1:

$$2p_2 - 5.1745p_1 + 1.563p_6 = -4,073.48. \quad\dots\dots\dots\dots\dots\dots\dots\dots\dots\dots\dots\text{(8.393)}$$

There are three unknowns, which are p_1, p_2, and p_6.

- Equation of Block 18:

$$1.563p_{13} + 2p_{17} - 7.1745p_{18} + 2p_{19} + 1.563p_{23} = -631.0. \quad\dots\dots\dots\dots\dots\dots\dots\text{(8.394)}$$

$p_{17} = p_{19}$ resulting from the symmetry; therefore, there are five unknowns: p_{13}, p_{17}, p_{18}, p_{23}, and p_{sf} of Block 18.

- Equation of Block 25:

$$1.5625p_{20} - 3.6125p_{25} + 2p_{24} = -948.52. \quad\dots\dots\dots\dots\dots\dots\dots\dots\dots\text{(8.395)}$$

Therefore, there are three unknowns (p_{20}, p_{25}, and p_{24}).

For Part 2:

- Block 8:

$$1.563p_3 + 2p_7 - 7.1745p_8 + 2p_9 + 1.563p_{13} = -148.5. \quad\dots\dots\dots\dots\dots\dots\dots\dots\text{(8.396)}$$

$p_7 = p_9$ resulting from the symmetry; therefore, there are four unknowns (p_7, p_3, p_8 and p_{13}).

- Block 18:

$$1.5625p_{13} + 2p_{17} - 8.815p_{18} + 2p_{19} + 1.5625p_{23} = -6,874.55. \quad\dots\dots\dots\dots\dots\text{(8.397)}$$

$p_{17} = p_{19}$ resulting from the symmetry; therefore, there are five unknowns p_{13}, p_{17}, p_{18}, p_{23}, and the flow rate of the well.

Problem 8.4.26 (Ertekin et al. 2001)

Use the GNR protocol to solve the following two equations in two unknowns; start with an initial guess of $x^0_1 = 1$ and $x_2{}^0 = 1$. In the solution, use a convergence tolerance of 10^{-4}.

- $x_1^2 + x_2^2 = 1.$ $\quad\dots\dots\dots\dots\dots\dots\dots\dots\dots\dots\dots\dots\dots\dots\dots\dots\dots\dots\text{(8.398)}$

- $(x_1 - 1)^2 + x_2^2 = 1.$ $\quad\dots\dots\dots\dots\dots\dots\dots\dots\dots\dots\dots\dots\dots\dots\dots\dots\text{(8.399)}$

Solution to Problem 8.4.26

Define two functions as f_1 and f_2 expressed as Eqs. 8.400 and 8.401, respectively:

$$f_1(x_1, x_2) = x_1^2 + x_2^2 - 1, \quad\dots\dots\dots\dots\dots\dots\dots\dots\dots\dots\dots\dots\dots\dots\text{(8.400)}$$

$$f_2(x_1, x_2) = (x_1 - 1)^2 + x_2^2 - 1. \quad\dots\dots\dots\dots\dots\dots\dots\dots\dots\dots\dots\dots\text{(8.401)}$$

The first-order derivatives of the functions are

$$\frac{\partial f_1}{\partial x_1} = 2x_1, \quad\dots\dots\dots\dots\dots\dots\dots\dots\dots\dots\dots\dots\dots\dots\dots\dots\dots\text{(8.402)}$$

$$\frac{\partial f_1}{\partial x_2} = 2x_2, \quad\dots\dots\dots\dots\dots\dots\dots\dots\dots\dots\dots\dots\dots\dots\dots\dots\dots\text{(8.403)}$$

$$\frac{\partial f_2}{\partial x_1} = 2(x_1 - 1), \quad\dots\dots\dots\dots\dots\dots\dots\dots\dots\dots\dots\dots\dots\dots\dots\text{(8.404)}$$

Iteration Level	Δx_1	Δx_2	x_1	x_2
0 (initial guess)	NA	NA	1.00	1.00
1	−0.50	0.00	0.50	1.00
2	0.000	−0.125	0.500	0.875
3	0.000	−0.009	0.500	0.866
4	0.000	0.000	0.500	0.866

Table 8.30—Iteration results from the Solution to Problem 8.4.26.

$$\frac{\partial f_2}{\partial x_2} = 2x_2. \quad\dots\dots\dots\dots\dots\dots\dots\dots\dots\dots\dots\dots\dots\dots\dots\dots\dots\dots\dots\dots\dots\dots\dots \text{(8.405)}$$

Constructing the Jacobian matrix,

$$J = \begin{bmatrix} 2x_1 & 2x_2 \\ 2(x_1-1) & 2x_2 \end{bmatrix}, \quad\dots\dots\dots\dots\dots\dots\dots\dots\dots\dots\dots\dots\dots\dots\dots\dots\dots\dots\dots\dots \text{(8.406)}$$

and implementing the GNR protocol,

$$\begin{bmatrix} 2x_1 & 2x_2 \\ 2(x_1-1) & 2x_2 \end{bmatrix}\begin{bmatrix} dx_1 \\ dx_2 \end{bmatrix} = \begin{bmatrix} x_1^2 + x_2^2 - 1 \\ (x_1-1)^2 + x_2^2 - 1 \end{bmatrix}, \quad\dots\dots\dots\dots\dots\dots\dots\dots\dots\dots\dots \text{(8.407)}$$

we observe the results in **Table 8.30.** After four GNR iterations, the solution converges to $x_1 = 0.5$ and $x_2 = 0.866$ within a convergence tolerance of 10^{-4}.

Problem 8.4.27 (Ertekin et al. 2001)

Consider the 2D incompressible fluid flow problem to be studied for the pressure distribution in the reservoir shown in **Fig. 8.24.1. Fig. 8.24.2** provides additional information. The gridblock properties are $\Delta x = 800$ ft, $\Delta y = 1{,}000$ ft, and $r_w = 0.2$ ft. The fluid properties are $\rho = 61$ lbm/ft³ and $\mu = 1.2$ cp.

1. Write the governing PDE that describes the flow problem in the reservoir. Also write the characteristic finite-difference approximation to the PDE and express it using SIP notation.
2. Write the finite-difference equation for Gridblock 7, and calculate the numerical values of all coefficients.
3. Set up the finite-difference equations in matrix form. Fill in the entries of the coefficient matrix, unknown vector, and the right-side vector using the SIP notation.

Solution to Problem 8.4.27

1. For a 2D incompressible fluid system with heterogeneous permeability, thickness, and grid dimensions, using existing wells and accounting for the depth gradients, the PDE can be written as

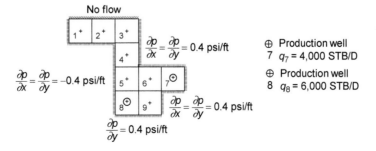

Fig. 8.24.1—Reservoir system for Problem 8.4.27.

Elevation of the Block Top, ft subsea

2,038	2,057	2,060		
		2,072		
		2,081	2,085	2,084
		2,090	2,098	

Block Thickness, ft

30	40	40	
		40	
		50	55	50
		60	60	

Gridblock Porosity, %

16	17	18	
		18	
		20	21	19
		20	20	

Gridblock permeability in x-direction, md

117	140	140	
		140	
		157	194	180
		176	178	

Gridblock permeability in y-direction, md

168	180	212	
		190	
		196	201	173
		182	197	

Fig. 8.24.2—Gridblock properties of the Reservoir system for Problem 8.4.24.

$$\frac{\partial}{\partial x}\left(\frac{A_x k_x}{\mu B}\frac{\partial p}{\partial x}\right)\Delta x + \frac{\partial}{\partial y}\left(\frac{A_y k_y}{\mu B}\frac{\partial p}{\partial y}\right)\Delta y - \left[\begin{array}{c} \dfrac{1}{144}\dfrac{g}{g_c}\rho\dfrac{\partial}{\partial x}\left(\dfrac{A_x k_x}{\mu B}\dfrac{\partial G}{\partial x}\right) \\[4pt] \Delta x + \dfrac{1}{144}\dfrac{g}{g_c}\rho\dfrac{\partial}{\partial y}\left(\dfrac{A_y k_y}{\mu B}\dfrac{\partial G}{\partial y}\right)\Delta y \end{array}\right] + q = 0. \quad \ldots\ldots\ldots\ldots(8.408)$$

The discretized PDE in a characteristic form using the SIP notation can be written as

$$S_{i,j}p_{i,j-1} + W_{i,j}p_{i-1,j} + C_{i,j}p_{i,j} + E_{i,j}p_{i+1,j} + N_{i,j}p_{i,j+1} = Q_{i,j}. \quad \ldots\ldots\ldots\ldots\ldots\ldots\ldots\ldots\ldots\ldots\ldots (8.409)$$

2. For Block 7, consider **Fig. 8.24.3.** Blocks $7'_{y+}$, $7'_{y-}$, and $7'_{x+}$ are employed to implement the boundary conditions by duplicating physical properties of Block 7:

$$S_7 = N_7 = \frac{k_{y7}h_7}{\mu B} = 8.12 \text{ STB/D-psi}, \quad \ldots\ldots\ldots\ldots\ldots\ldots (8.410)$$

$$W_7 = \frac{2A_{7x}A_{6x}k_{x7}k_{x6}}{\mu B(A_{7x}k_{7x}\Delta y_6 + A_{6x}k_{6x}\Delta y_7)} = 11.46 \text{ STB/D-psi}, \quad \ldots\ldots(8.411)$$

$$E_7 = \frac{k_{x7}h_7}{\mu B} = 8.45 \text{ STB/D-psi}. \quad \ldots\ldots\ldots\ldots\ldots\ldots\ldots\ldots(8.412)$$

The well is producing with a flow-rate specification; therefore,

$$C_7 = -(S_7 + W_7 + E_7 + N_7) = 36.15 \text{ STB/D-psi}. \quad \ldots\ldots\ldots\ldots(8.413)$$

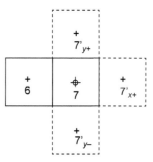

Fig. 8.24.3—Treatment of the boundary conditions of Block 7.

The gradient term is calculated as follows:

$$\frac{1}{144}\frac{g}{g_c}\rho W_7\left(G_6-G_7\right)=4.85 \text{ STB/D.} \quad\text{...(8.414)}$$

Then, $Q_7 = 4.85 + 4,000 = 4,004.85$ STB/D. $\quad$...(8.415)

The finite-difference equation for Block 7 can be expressed as

$$S_7 p_{7y-} + W_7 p_6 + C_7 p_7 + E_7 p_{7x+} + N_7 p_{7y+} = Q_7. \quad\text{...............................(8.416)}$$

Applying the boundary condition,

$$\frac{p_7-p_{7y-}}{\Delta y}=0.4 \text{ psi/ft} \Rightarrow p_{7y-}=p_7-400, \quad\text{.............................(8.417)}$$

$$\frac{p_{7y+}-p_7}{\Delta y}=0.4 \text{ psi/ft} \Rightarrow p_{7y+}=p_7+400, \quad\text{.............................(8.418)}$$

$$\frac{p_{7x+}-p_7}{\Delta x}=0.4 \text{ psi/ft} \Rightarrow p_{7x+}=p_7+320. \quad\text{.............................(8.419)}$$

Then, the finite-difference equation can be written as

$$W_7\left(p_6-p_7\right)=Q_7-320E_7-400N_7+400S_7, \quad\text{.............................(8.420)}$$

which reduces to

$$p_6-p_7=113.51. \quad\text{...............................(8.421)}$$

3. The system of linear equations in matrix form is

$$\begin{bmatrix} C_1 & E_1 & 0 & 0 & 0 & 0 & 0 & 0 & 0 \\ W_2 & C_2 & E_2 & 0 & 0 & 0 & 0 & 0 & 0 \\ 0 & W_3 & C_3 & S_3 & 0 & 0 & 0 & 0 & 0 \\ 0 & 0 & N_4 & C_4 & S_4 & 0 & 0 & 0 & 0 \\ 0 & 0 & 0 & N_5 & C_5 & E_5 & 0 & S_5 & 0 \\ 0 & 0 & 0 & 0 & W_6 & C_6 & E_6 & 0 & S_6 \\ 0 & 0 & 0 & 0 & 0 & W_7 & C_7 & 0 & 0 \\ 0 & 0 & 0 & 0 & N_8 & 0 & 0 & C_8 & E_8 \\ 0 & 0 & 0 & 0 & 0 & N_9 & 0 & 0 & C_9 \end{bmatrix} \begin{bmatrix} p_1 \\ p_2 \\ p_3 \\ p_4 \\ p_5 \\ p_6 \\ p_7 \\ p_8 \\ p_9 \end{bmatrix} = \begin{bmatrix} Q_1+320W_1+400S_1 \\ Q_1+400S_2 \\ Q_1-320E_3 \\ Q_4 \\ Q_5+320W_5 \\ Q_6-400N_6 \\ Q_7-320E_7 \\ Q_8+400S_8+320W_8 \\ Q_8+400S_9-320E_9 \end{bmatrix} \quad\text{..........(8.422)}$$

Problem 8.4.28 (Ertekin et al. 2001)

Find the pressure distribution in the reservoir presented in the 2D mesh-centered grid system shown in **Fig. 8.25.1.** Assume that the reservoir fluid is an incompressible liquid with a viscosity of 1 cp. Gridblocks at the outer boundaries of the reservoir represent a constant-pressure aquifer. The aquifer pressure is 2,000 psia. Both of the wells drilled in this reservoir are producing at a constant-sandface pressure of 1,650 psia.

1. If the coefficient matrix is structured to have the pentadiagonal structure shown in **Fig. 8.25.2,** how would you number the gridblocks?
2. Calculate the entries for the coefficient matrix and the right-hand-side vector.

Solution to Problem 8.4.28

1. To generate the matrix representation shown in Fig. 8.25.2, the gridblocks with unknown pressures should be numbered using natural ordering by row, as shown in **Fig. 8.25.3.**

p = 2,000 psia

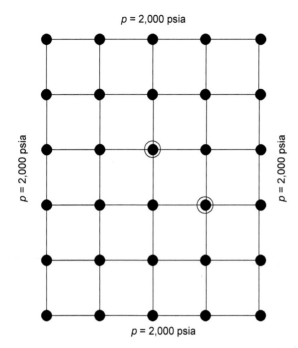

p = 2,000 psia

p = 2,000 psia

p = 2,000 psia

Fig. 8.25.1—Reservoir system of Problem 8.4.28.

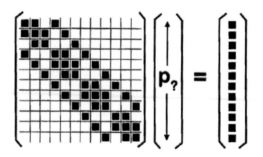

Fig. 8.25.2—Matrix structure of the reservoir system in Problem 8.4.28.

p = 2,000 psia

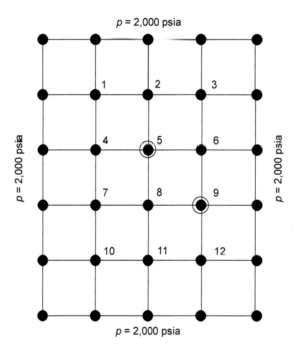

p = 2,000 psia

p = 2,000 psia

p = 2,000 psia

Fig. 8.25.3—Block numbering scheme of reservoir system in Problem 8.4.28.

2. Assuming that the system exhibits isotropic and homogeneous property distributions, the governing equation can be simplified to

$$\frac{\partial^2 p}{\partial x^2} + \frac{\partial^2 p}{\partial y^2} + \frac{\partial^2 p}{\partial z^2} + \frac{q\mu B}{V_b k} = 0 \ \dots\dots\dots\dots\dots\dots\dots\dots\dots\dots\dots\dots\dots\dots\dots\dots (8.423)$$

and

$$q_{i,j} = -\Omega_{i,j}\left(p_{i,j} - p_{sf}\right). \ \dots\dots\dots\dots\dots\dots\dots\dots\dots\dots\dots\dots\dots\dots\dots\dots (8.424)$$

In a finite-difference form (assuming $\Delta x = \Delta y$),

$$p_{i+1,j} + p_{i-1,j} + p_{i,j+1} + p_{i,j-1} - 4p_{i,j} + \frac{q\mu B}{kh} = 0. \quad\dots\dots\dots\dots\dots\dots\dots\dots\dots\dots\dots\dots\dots\dots (8.425)$$

Rewriting Eq. 8.425 using the wellbore model,

$$p_{i+1,j} + p_{i-1,j} + p_{i,j+1} + p_{i,j-1} - \left(4 + \frac{\Omega\mu B}{kh}\right) p_{i,j} + \frac{\Omega\mu B}{kh} p_{sf} = 0, \quad\dots\dots\dots\dots\dots\dots\dots\dots\dots\dots (8.426)$$

$$
\begin{bmatrix}
-4 & 1 & 0 & 1 & 0 & 0 & 0 & 0 & 0 & 0 & 0 & 0 \\
1 & -4 & 1 & 0 & 1 & 0 & 0 & 0 & 0 & 0 & 0 & 0 \\
0 & 1 & -4 & 0 & 0 & 1 & 0 & 0 & 0 & 0 & 0 & 0 \\
1 & 0 & 0 & -4 & 1 & 0 & 1 & 0 & 0 & 0 & 0 & 0 \\
0 & 1 & 0 & 1 & -4-\dfrac{\Omega\mu B}{kh} & 1 & 0 & 1 & 0 & 0 & 0 & 0 \\
0 & 0 & 1 & 0 & 1 & -4 & 0 & 0 & 1 & 0 & 0 & 0 \\
0 & 0 & 0 & 1 & 0 & 0 & -4 & 1 & 0 & 1 & 0 & 0 \\
0 & 0 & 0 & 0 & 1 & 0 & 1 & -4 & 1 & 0 & 1 & 0 \\
0 & 0 & 0 & 0 & 0 & 1 & 0 & 1 & -4-\dfrac{\Omega\mu B}{kh} & 0 & 0 & 1 \\
0 & 0 & 0 & 0 & 0 & 0 & 1 & 0 & 0 & -4 & 1 & 0 \\
0 & 0 & 0 & 0 & 0 & 0 & 0 & 1 & 0 & 1 & -4 & 1 \\
0 & 0 & 0 & 0 & 0 & 0 & 0 & 0 & 1 & 0 & 1 & -4
\end{bmatrix}
\begin{bmatrix}
p_1 \\ p_2 \\ p_3 \\ p_4 \\ p_5 \\ p_6 \\ p_7 \\ p_8 \\ p_9 \\ p_{10} \\ p_{11} \\ p_{12}
\end{bmatrix}
=
\begin{bmatrix}
-4,000 \\
-2,000 \\
-4,000 \\
-2,000 \\
-\dfrac{\Omega_5\mu B}{kh}1,650 \\
-2,000 \\
-2,000 \\
0 \\
-2,000 - \dfrac{\Omega_5\mu B}{kh}1,650 \\
-4,000 \\
-2,000 \\
-4,000
\end{bmatrix}
\quad..(8.427)
$$

Problem 8.4.29 (Ertekin et al. 2001)

Consider the 2D incompressible fluid problem in the homogeneous, anisotropic porous medium shown in **Fig. 8.26.** The gridblock properties are $\Delta x = \Delta y = 175$ ft, $k_x = 100$ md, $k_y = 150$ md, $h = 40$ ft, $r_w = 0.3$ ft, and $s = 0$. The fluid properties are $\mu = 1.2$ cp and $B = 1.05$ res bbl/STB. The northern and eastern boundaries of Gridblock 6 have pressure supports. The non-producing observation well in Gridblock 5 gives a pressure reading of 2,054 psia. Moreover, the well in Gridblock 2 is put on production with a sandface-pressure specification of 1,000 psia, leading to a production rate of 2,066 STB/D.

Determine the pressure distribution in the system and whether or not the south boundary of Gridblock 1 is sealed.

Solution to Problem 8.4.29

The flow rate entering across the northern and eastern boundaries of Gridblock 6 can be calculated as

$$q_x = 1.127 \times 10^{-3} \times \frac{A_x k_x}{\mu B} 0.6 \text{ psia/ft} = 375.67 \text{ STB/D}, \quad\dots\dots\dots\dots\dots\dots\dots\dots\dots\dots\dots (8.428)$$

$$q_y = 1.127 \times 10^{-3} \times \frac{A_y k_y}{\mu B} 0.6 \text{ psia/ft} = 751.33 \text{ STB/D}, \quad\dots\dots\dots\dots\dots\dots\dots\dots\dots\dots\dots (8.429)$$

$$q_{total} = q_x + q_y = 1,127 \text{ STB/D}. \quad\dots\dots\dots\dots\dots\dots\dots\dots\dots\dots\dots\dots\dots\dots\dots\dots\dots\dots (8.430)$$

In Block 5, $q_5 = 0$ (well is shut in); this condition implies that $p_5 = p_{sf_5} = 2,054$ psi. Knowing that $p_5 = 2,054$ psia, and fluid flowing out from Gridblock 6 can enter only Gridblock 5, one can calculate the flow between Gridblocks 5 and 6:

$$q_{56} = 1.127 \times 10^{-3} \times \frac{A_x k_x}{\mu B} \left(\frac{p_6 - p_5}{\Delta x}\right) = 1,127 \text{ STB/D}, \quad\dots\dots\dots\dots\dots\dots\dots\dots\dots\dots (8.431)$$

$$p_6 - p_5 = 315 \text{ psia}. \quad\dots\dots\dots\dots\dots\dots\dots\dots\dots\dots\dots\dots\dots\dots\dots\dots\dots\dots\dots\dots\dots (8.432)$$

Therefore, $p_6 = p_5 + 315 = 2,369$ psia. $\quad\dots\dots\dots\dots\dots\dots\dots\dots\dots\dots\dots\dots\dots\dots\dots\dots\dots (8.433)$

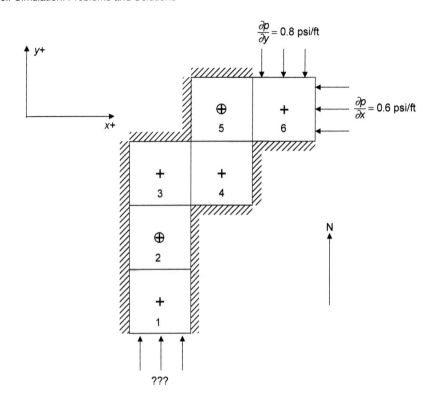

Fig. 8.26—Reservoir system of Problem 8.4.28.

Similarly, flow from Gridblock 5 to Gridblock 4 is

$$q_{45} = 1.127 \times 10^{-3} \times \frac{A_y k_y}{\mu B} (\frac{p_5 - p_4}{\Delta y}) = 1,127 \text{ STB/D}, \dots\dots\dots\dots\dots\dots\dots (8.434)$$

$$p_5 - p_4 - 210 \text{ psia.} \dots\dots\dots\dots\dots\dots\dots\dots\dots\dots\dots\dots\dots\dots\dots\dots (8.435)$$

Therefore, $p_4 = p_5 - 210 = 1,844$ psia. $\dots\dots\dots\dots\dots\dots\dots\dots\dots\dots\dots\dots\dots\dots (8.436)$

In a similar progressive manner, we can conclude that $p_4 - p_3 = 315$ and $p_3 = p_4 - 315 = 1,529$ psia, and finally, $p_3 - p_2 = 210$ psia and $p_2 = p_3 - 210 = 1,319$ psia.

Because total flow into the reservoir has to be equal to the total flow out of the reservoir, 2,066 STB/D − 1,127 STB/D = 939 STB/D must enter across the southern boundary of Gridblock 1. In other words,

$$q_y' = -1.127 \times 10^{-3} \times \frac{A_y k_y}{\mu B} c_1 \text{ psia/ft} = 939 \text{ STB/D}, \dots\dots\dots\dots\dots\dots\dots\dots (8.437)$$

which yields, $c_1 = -1$ psi/ft.

Problem 8.4.30 (Ertekin et al. 2001)

In constructing the Jacobian matrix for a 1D, single-phase compressible flow problem, an analytical expression representing the general form of the main diagonal entries for a wellblock with sandface-pressure specification is needed. Derive a generic expression that can be used for such entries.

Solution to Problem 8.4.30

The residual of a 1D single-phase gas flow problem can be expressed using SIP notation as follows:

$$R_i^{(k)\,n+1} = Q_i^{(k)\,n+1} - W_i^{(k)\,n+1} p_{i-1}^{(k)\,n+1} - E_i^{(k)\,n+1} p_{i+1}^{(k)\,n+1} - C_i^{(k)\,n+1} p_i^{(k)\,n+1}. \dots\dots\dots\dots\dots (8.438)$$

The diagonal entry of the Jacobian matrix is expressed as

$$\frac{\partial \overset{(k)}{R_i^{n+1}}}{\partial \overset{(k)}{p_i^{n+1}}} = \frac{\partial \overset{(k)}{Q_i^{n+1}}}{\partial \overset{(k)}{p_i^{n+1}}} - \frac{\partial \overset{(k)}{W_i^{n+1}}}{\partial \overset{(k)}{p_i^{n+1}}} \overset{(k)}{p_{i-1}^{n+1}} - \frac{\partial \overset{(k)}{E_i^{n+1}}}{\partial \overset{(k)}{p_i^{n+1}}} \overset{(k)}{p_{i+1}^{n+1}} - \frac{\partial \overset{(k)}{C_i^{n+1}}}{\partial \overset{(k)}{p_i^{n+1}}} \overset{(k)}{p_i^{n+1}} - \overset{(k)}{C_i^{n+1}}. \quad \ldots\ldots\ldots\ldots\ldots\ldots\ldots\ldots\ldots\ldots\ldots\ldots \quad (8.439)$$

Knowing that $\Gamma_i^n = \dfrac{V_b \phi T_{sc}}{p_{sc} T z_{gi}^n \Delta t}$ and $\overset{(k)}{\Gamma_i^{n+1}} = \dfrac{V_b \phi T_{sc}}{p_{sc} T \overset{(k)}{z_{gi}^{n+1}} \Delta t}$ $\ldots\ldots\ldots\ldots\ldots\ldots\ldots\ldots\ldots\ldots\ldots\ldots\ldots\ldots$ (8.440)

and

$$\overset{(k)}{Q_i^{n+1}} = -\Gamma_i^n p_i^n - \overset{(k)}{\Omega_i^{n+1}} p_{sfi} \quad \ldots\ldots\ldots\ldots\ldots\ldots\ldots\ldots\ldots\ldots\ldots\ldots\ldots\ldots\ldots \quad (8.441)$$

and

$$\overset{(k)}{C_i^{n+1}} = -\overset{(k)}{\Gamma_i^{n+1}} - \overset{(k)}{\Omega_i^{n+1}} - \overset{(k)}{W_i^{n+1}} - \overset{(k)}{E_i^{n+1}}, \quad \ldots\ldots\ldots\ldots\ldots\ldots\ldots\ldots\ldots\ldots\ldots\ldots \quad (8.442)$$

the original equation can be rewritten as

$$\frac{\partial \overset{(k)}{R_i^{n+1}}}{\partial \overset{(k)}{p_i^{n+1}}} = \frac{\partial \overset{(k)}{\Omega_i^{n+1}}}{\partial \overset{(k)}{p_i^{n+1}}}\left[\overset{(k)}{p_i^{n+1}} - p_{sf}\right] + \frac{\partial \overset{(k)}{W_i^{n+1}}}{\partial \overset{(k)}{p_i^{n+1}}}\left[\overset{(k)}{p_i^{n+1}} - \overset{(k)}{p_{i-1}^{n+1}}\right] + \frac{\partial \overset{(k)}{E_i^{n+1}}}{\partial \overset{(k)}{p_i^{n+1}}}\left[\overset{(k)}{p_i^{n+1}} - \overset{(k)}{p_{i+1}^{n+1}}\right] + \frac{\partial \overset{(k)}{\Gamma_i^{n+1}}}{\partial \overset{(k)}{p_i^{n+1}}}\overset{(k)}{p_i^{n+1}} - \overset{(k)}{C_i^{n+1}}. \quad \ldots\ldots\ldots\ldots \quad (8.443)$$

Problem 8.4.31 (Ertekin et al. 2001)

A 1D flow equation in the matrix notation for a compressible flow problem is given as

$$\overset{(v)}{W_i^{n+1}} \overset{(v+1)}{p_{i-1}^{n+1}} + \overset{(v)}{C_i^{n+1}} \overset{(v+1)}{p_i^{n+1}} + \overset{(v)}{E_i^{n+1}} \overset{(v+1)}{p_{i+1}^{n+1}} = Q_i^{(v)}. \quad \ldots\ldots\ldots\ldots\ldots\ldots\ldots\ldots\ldots\ldots\ldots\ldots\ldots \quad (8.444)$$

Write an algorithmic expression for implementation of the PSOR iterative-solution procedure to solve for pressures at time level $n + 1$ and iteration level $(v + 1)$.

Solution to Problem 8.4.31

1. Calculate the coefficients, $\overset{(v)}{W_i^{n+1}}$, $\overset{(v)}{C_i^{n+1}}$, $\overset{(v)}{E_i^{n+1}}$, and $Q_i^{(v)}$ using the pressure from the previous iteration level, v (at the first iteration level, pressure solved from previous timestep n can be used).

2. Update the pressure using the PSOR algorithm:

$$\overset{(v+1)}{p_i^{n+1}} = \overset{(v)}{p_i^{n+1}} + \frac{\omega_{opt}}{\overset{(v)}{C_i^{n+1}}}\left[Q_i^{(v)} - \overset{(v)}{W_i^{n+1}} \overset{(v+1)}{p_{i-1}^{n+1}} - \overset{(v)}{C_i^{n+1}} \overset{(v)}{p_i^{n+1}} - \overset{(v)}{E_i^{n+1}} \overset{(v)}{p_{i+1}^{n+1}}\right]. \quad \ldots\ldots\ldots\ldots\ldots\ldots\ldots \quad (8.445)$$

3. Check the difference between the solved pressures and the pressures of the previous iteration level.
4. If convergence is not achieved, update the coefficients using the most-recently obtained pressure values.
5. Repeat steps 2 and 3 until the maximum pressure difference between the last two iterations becomes smaller than the prescribed tolerance.

Problem 8.4.32 (Ertekin et al. 2001)

Consider the steady-state flow of a single-phase incompressible liquid in the reservoir shown in **Fig. 8.27.1.** The gridblock properties are $\Delta x = \Delta y = 400$ ft, $k_x = 60$ md, $k_y = 30$ md, and $h = 17$ ft. The fluid properties are $\mu = 1.2$ cp and $B = 1$ res bbl/STB.

The pressure along the external boundaries of the reservoir is 4,000 psia. Flowing sandface pressure at the well is controlled at 1,300 psia. What is the flow rate into the well? Perform an MBC to verify the accuracy of the solution. In the solution of the finite-difference equation, use the Gauss-Seidel method with a pressure tolerance of 1 psia.

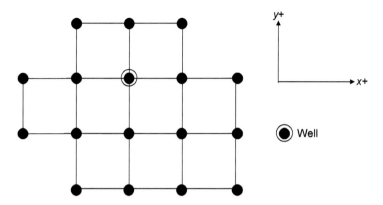

Fig. 8.27.1—Reservoir system of Problem 8.4.32.

Solution to Problem 8.4.32

For a single-phase 2D incompressible flow system with anisotropic but homogeneous permeability distribution, the PDE that describes the problem can be written as

$$k_x \frac{\partial^2 p}{\partial x^2} + k_y \frac{\partial^2 p}{\partial y^2} + \frac{q\mu B}{V_b} = 0. \quad\dotfill (8.446)$$

Because $k_x = 2k_y$,

$$2\frac{\partial^2 p}{\partial x^2} + \frac{\partial^2 p}{\partial y^2} + \frac{q\mu B}{k_y V_b} = 0. \quad\dotfill (8.447)$$

As displayed in **Fig. 8.27.2,** the external boundaries of the reservoir are kept at 4,000 psia. Hence, there are three pressure values that need to be solved in this problem. These blocks with unknown pressures are numbered as Nodes 1, 2, and 3, as shown in the figure.

- For Node 1,

$$2(4,000 - 2p_1 + 4,000) + (4,000 - 2p_1 + p_2) + \frac{\mu B}{hk_y}\Omega(p_{sf} - p_1) = 0, \quad\dotfill (8.448)$$

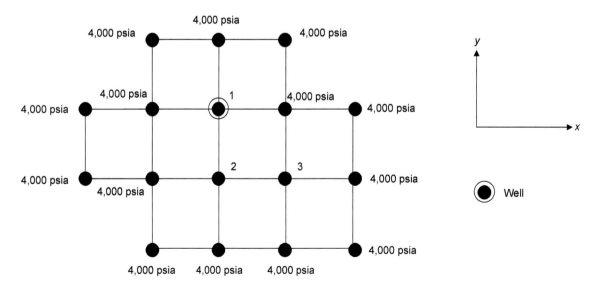

Fig. 8.27.2—Numbering scheme of reservoir system in Problem 8.4.32.

$$r_e = \frac{0.28\sqrt{\left(\frac{k_x}{k_y}\right)^{\frac{1}{2}}(\Delta y)^2 + \left(\frac{k_y}{k_x}\right)^{\frac{1}{2}}(\Delta x)^2}}{\left(\frac{k_x}{k_y}\right)^{\frac{1}{4}} + \left(\frac{k_y}{k_x}\right)^{\frac{1}{4}}} = 80.35 \text{ ft,} \dots\dots\dots\dots\dots\dots\dots(8.449)$$

and

$$\bar{k} = \sqrt{k_x k_y} = 42.43 \text{ md.} \dots\dots\dots\dots\dots\dots\dots (8.450)$$

Using Peaceman's wellbore equation,

$$\Omega = \frac{2\pi\bar{k}h}{\mu B \ln\left(\frac{r_e}{r_w}\right)} = 0.737 \text{ STB/D-psi,} \dots\dots\dots\dots\dots\dots (8.451)$$

with $p_{sf} = 1,300$ psia, the finite-difference equation at Node 1 can be rewritten as

$$-7.54 p_1 + p_2 = -22,000. \dots\dots\dots\dots\dots\dots(8.452)$$

- For Node 2, similarly,

$$2(4,000 - 2p_2 + p_3) + (4,000 - 2p_2 + p_1) = 0, \dots\dots\dots\dots\dots(8.453)$$

$$p_1 - 6p_2 + 2p_3 = -12,000. \dots\dots\dots\dots\dots\dots (8.454)$$

- For Node 3,

$$2(4,000 - 2p_3 + p_2) + (4,000 - 2p_3 + 4,000) = -16,000. \dots\dots\dots\dots\dots (8.455)$$

Collecting all the equations in a matrix form,

$$\begin{bmatrix} -7.54 & 1 & 0 \\ 1 & -6 & 2 \\ 0 & 2 & -6 \end{bmatrix} \begin{bmatrix} p_1 \\ p_2 \\ p_3 \end{bmatrix} = \begin{bmatrix} -22,000 \\ -12,000 \\ -16,000 \end{bmatrix}, \dots\dots\dots\dots\dots\dots(8.456)$$

and solving for pressures, one obtains $p_1 = 3,434.21$ psia, $p_2 = 3,893.9$ psia , and $p_3 = 3,964.6$ psia. Now, an MBC can be conducted.

- At Node 1, fluid flow into the node,

$$q_1 = 1.127 \times 10^{-3} \times \frac{hk_y}{\mu B}(4,000 - p_1) + 2 \times 1.127 \times 10^{-3} \times \frac{hk_x}{\mu B}(4,000 - p_1) = 1,355.1 \text{ STB/D.} \dots\dots\dots (8.457)$$

- At Node 2, fluid flow into the node,

$$q_2 = 1.127 \times 10^{-3} \times \frac{hk_y}{\mu B}(4,000 - p_2) + 1.127 \times 10^{-3} \times \frac{hk_x}{\mu B}(4,000 - p_2) = 152.45 \text{ STB/D.} \dots\dots\dots (8.458)$$

- At Node 3, fluid flow into the node,

$$q_3 = 2 \times 1.127 \times 10^{-3} \times \frac{hk_y}{\mu B}(4,000 - p_2) + 1.127 \times 10^{-3} \times \frac{hk_x}{\mu B}(4,000 - p_2) = 67.75 \text{ STB/D.} \dots\dots\dots (8.459)$$

Now the production from Node 1 is calculated using Peaceman's wellbore model:

$$q_w = \Omega(p_1 - p_{sf}) = 1,572.9 \text{ STB/D.} \dots\dots\dots\dots\dots\dots (8.460)$$

The MBC shows that the solution conserves the mass within an acceptable error margin:

$$\text{MBC} = \frac{1,355.1 \text{ STB/D} + 152.45 \text{ STB/D} + 67.75 \text{ STB/D}}{1,572.9 \text{ STB/D}} \approx 1.002. \dots\dots\dots\dots (8.461)$$

Problem 8.4.33 (Ertekin et al. 2001)

Consider the 2D gas reservoir shown in **Fig. 8.28.** The reservoir is surrounded by no-flow boundaries. Use the GNR protocol to formulate the solution for the pressure distribution in the reservoir. In the solution of the system of equations generated by the GNR protocol, use the line-successive over-relaxation (LSOR) technique with $\omega_{opt} = 1.00$.

1. Write a characteristic equation that will generate the system of nonlinear equations.
2. Write the system of nonlinear equations generated by the equation in Part 1 in a closed form.
3. Write a characteristic LSOR equation that will solve the system of equations generated by the GNR protocol, sweeping the system parallel to y-direction.

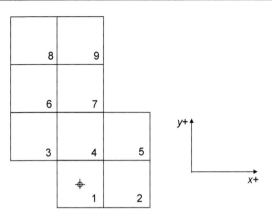

Fig. 8.28—Reservoir system of Problem 8.4.33.

Solution to Problem 8.4.33

1.

$$\frac{\partial \overset{(k)}{R_{i,j}^{n+1}}}{\partial \overset{(k)}{p_{i+1,j}^{n+1}}} \overset{(k+1)}{\Delta p_{i+1,j}^{n+1}} + \frac{\partial \overset{(k)}{R_{i,j}^{n+1}}}{\partial \overset{(k)}{p_{i-1,j}^{n+1}}} \overset{(k+1)}{\Delta p_{i-1,j}^{n+1}} + \frac{\partial \overset{(k)}{R_{i,j}^{n+1}}}{\partial \overset{(k)}{p_{i,j}^{n+1}}} \overset{(k+1)}{\Delta p_{i,j}^{n+1}} + \frac{\partial \overset{(k)}{R_{i,j}^{n+1}}}{\partial \overset{(k)}{p_{i,j+1}^{n+1}}} \overset{(k+1)}{\Delta p_{i,j+1}^{n+1}} + \frac{\partial \overset{(k)}{R_{i,j}^{n+1}}}{\partial \overset{(k)}{p_{i,j-1}^{n+1}}} \overset{(k+1)}{\Delta p_{i,j-1}^{n+1}} = -\overset{(k)}{R_{i,j}^{n+1}}. \quad \ldots \ldots \ldots \ldots (8.462)$$

Eq. 8.462 will generate a linear system of equations, as shown in the following matrix form:

$$\begin{bmatrix}
\frac{\partial R_1^{n+1}}{\partial p_1^{n+1}} & \frac{\partial R_1^{n+1}}{\partial p_2^{n+1}} & 0 & \frac{\partial R_1^{n+1}}{\partial p_4^{n+1}} & 0 & 0 & 0 & 0 & 0 \\
\frac{\partial R_2^{n+1}}{\partial p_1^{n+1}} & \frac{\partial R_2^{n+1}}{\partial p_2^{n+1}} & 0 & 0 & \frac{\partial R_2^{n+1}}{\partial p_5^{n+1}} & 0 & 0 & 0 & 0 \\
0 & 0 & \frac{\partial R_3^{n+1}}{\partial p_3^{n+1}} & \frac{\partial R_3^{n+1}}{\partial p_4^{n+1}} & 0 & \frac{\partial R_3^{n+1}}{\partial p_6^{n+1}} & 0 & 0 & 0 \\
\frac{\partial R_4^{n+1}}{\partial p_1^{n+1}} & 0 & \frac{\partial R_4^{n+1}}{\partial p_3^{n+1}} & \frac{\partial R_4^{n+1}}{\partial p_4^{n+1}} & \frac{\partial R_4^{n+1}}{\partial p_5^{n+1}} & 0 & \frac{\partial R_4^{n+1}}{\partial p_6^{n+1}} & 0 & 0 \\
0 & \frac{\partial R_5^{n+1}}{\partial p_2^{n+1}} & 0 & \frac{\partial R_5^{n+1}}{\partial p_4^{n+1}} & \frac{\partial R_5^{n+1}}{\partial p_5^{n+1}} & 0 & 0 & 0 & 0 \\
0 & 0 & \frac{\partial R_6^{n+1}}{\partial p_3^{n+1}} & 0 & 0 & \frac{\partial R_6^{n+1}}{\partial p_6^{n+1}} & \frac{\partial R_6^{n+1}}{\partial p_7^{n+1}} & \frac{\partial R_6^{n+1}}{\partial p_8^{n+1}} & 0 \\
0 & 0 & 0 & \frac{\partial R_7^{n+1}}{\partial p_4^{n+1}} & 0 & \frac{\partial R_7^{n+1}}{\partial p_6^{n+1}} & \frac{\partial R_7^{n+1}}{\partial p_7^{n+1}} & 0 & \frac{\partial R_7^{n+1}}{\partial p_9^{n+1}} \\
0 & 0 & 0 & 0 & 0 & \frac{\partial R_8^{n+1}}{\partial p_6^{n+1}} & 0 & \frac{\partial R_8^{n+1}}{\partial p_8^{n+1}} & \frac{\partial R_8^{n+1}}{\partial p_9^{n+1}} \\
0 & 0 & 0 & 0 & 0 & 0 & \frac{\partial R_9^{n+1}}{\partial p_7^{n+1}} & \frac{\partial R_9^{n+1}}{\partial p_8^{n+1}} & \frac{\partial R_9^{n+1}}{\partial p_9^{n+1}}
\end{bmatrix}^{(k)}
\begin{bmatrix}
\Delta p_1^{n+1} \\
\Delta p_2^{n+1} \\
\Delta p_3^{n+1} \\
\Delta p_4^{n+1} \\
\Delta p_5^{n+1} \\
\Delta p_6^{n+1} \\
\Delta p_7^{n+1} \\
\Delta p_8^{n+1} \\
\Delta p_9^{n+1}
\end{bmatrix}^{(k+1)}
= -
\begin{bmatrix}
R_1^{n+1} \\
R_2^{n+1} \\
R_3^{n+1} \\
R_4^{n+1} \\
R_5^{n+1} \\
R_6^{n+1} \\
R_7^{n+1} \\
R_8^{n+1} \\
R_9^{n+1}
\end{bmatrix}^{(k)} \quad \ldots (8.463)$$

2. The general form of the residual equation in a closed form can be written as

$$R_{i,j}^{n+1}\left(p_{i,j-1}^{n+1}, p_{i-1,j}^{n+1}, p_{i,j}^{n+1}, p_{i+1,j}^{n+1}, p_{i,j+1}^{n+1}\right) = 0, \quad \ldots \ldots \ldots \ldots \ldots \ldots \ldots \ldots \ldots (8.464)$$

for $i = 1$ to N_x, any $j = 1$ to N_y.

Eq. 8.464, when written at each block shown in Fig. 8.28, will generate the following system of equations (again written in a closed form):

- Block 1: $R_1^{n+1}\left(p_1^{n+1}, p_2^{n+1}, p_4^{n+1}\right) = 0.$ $\ldots \ldots \ldots \ldots \ldots \ldots \ldots \ldots \ldots \ldots \ldots \ldots (8.465)$

- Block 2: $R_2^{n+1}\left(p_1^{n+1}, p_2^{n+1}, p_5^{n+1}\right) = 0.$ .... (8.466)

- Block 3: $R_3^{n+1}\left(p_3^{n+1}, p_4^{n+1}, p_6^{n+1}\right) = 0.$ .... (8.467)

- Block 4: $R_4^{n+1}\left(p_1^{n+1}, p_3^{n+1}, p_4^{n+1}, p_5^{n+1}, p_7^{n+1}\right) = 0.$ .... (8.468)

- Block 5: $R_5^{n+1}\left(p_2^{n+1}, p_4^{n+1}, p_5^{n+1}\right) = 0.$ .... (8.469)

- Block 6: $R_6^{n+1}\left(p_3^{n+1}, p_6^{n+1}, p_7^{n+1}, p_8^{n+1}\right) = 0.$ .... (8.470)

- Block 7: $R_7^{n+1}\left(p_4^{n+1}, p_6^{n+1}, p_7^{n+1}, p_9^{n+1}\right) = 0.$ .... (8.471)

- Block 8: $R_8^{n+1}\left(p_6^{n+1}, p_8^{n+1}, p_9^{n+1}\right) = 0.$ .... (8.472)

- Block 9: $R_9^{n+1}\left(p_7^{n+1}, p_8^{n+1}, p_9^{n+1}\right) = 0.$ .... (8.473)

3. The algorithmic expression to implement the LSOR procedure sweeping along the y-direction can be written as

$$\frac{\partial R_{i,j}^{n+1}{}^{(k)}}{\partial p_{i+1,j}^{n+1}{}^{(k)}}\Delta p_{i+1,j}^{n+1}{}^{(m+1)}_{(k+1)} + \frac{\partial R_{i,j}^{n+1}{}^{(k)}}{\partial p_{i-1,j}^{n+1}{}^{(k)}}\Delta p_{i-1,j}^{n+1}{}^{(m+1)}_{(k+1)} + \frac{\partial R_{i,j}^{n+1}{}^{(k)}}{\partial p_{i,j}^{n+1}{}^{(k)}}\Delta p_{i,j}^{n+1}{}^{(m+1)}_{(k+1)} + R_{i,j}^{n+1}{}^{(k)} = -\frac{\partial R_{i,j}^{n+1}{}^{(k)}}{\partial p_{i,j+1}^{n+1}{}^{(k)}}\Delta p_{i,j+1}^{n+1}{}^{(m)}_{(k+1)} - \frac{\partial R_{i,j}^{n+1}{}^{(k)}}{\partial p_{i,j-1}^{n+1}{}^{(k)}}\Delta p_{i,j-1}^{n+1}{}^{(m+1)}_{(k+1)},$$ ....(8.474)

where superscript m is the iteration level of the LSOR iteration. In this implementation, the preceding system of equations will be generated for line $j = 1$ (Blocks 1 and 2), $j = 2$ (Blocks 3, 4, and 5), $j = 3$ (Blocks 6 and 7), and $j = 4$ (Blocks 8 and 9). Therefore, it will be necessary to call on the Thomas' algorithm four times during each LSOR iteration.

Problem 8.4.34 (Ertekin et al. 2001)

Consider the slightly compressible fluid flow through the 2D porous medium shown in **Fig. 8.29.** The porous medium has homogeneous but anisotropic property distributions and all external boundaries are designated as no-flow boundaries. The pressure at $t = 0$ for all grids is 4,000 psia.

Can the well sustain a flow rate of 500 STB/D for 5 days? Justify your answer. Use $h = 5$ ft, $k_x = 100$ md, $k_y = 150$ md, $\Delta x = \Delta y = 1,320$ ft, $\phi = 20\%$, $\mu = 1.2$ cp, $c = 1 \times 10^{-6}$ psi^{-1}, $B = 1$ res bbl/STB, $r_e = 263$ ft, $p_i = 4,000$ psi, $r_w = 0.2$ ft, and $s = 2$. In the solution of the finite-difference equation written for a timestep of five days, use the PSOR procedure with a pressure tolerance of 1 psi.

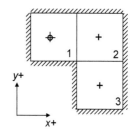

Fig. 8.29—Reservoir system of Problem 8.4.34.

Solution to Problem 8.4.34

The governing equation describing the flow problem in a homogeneous but anisotropic porous medium can be written as

$$k_x \frac{\partial^2 p}{\partial x^2} + k_y \frac{\partial^2 p}{\partial y^2} + \frac{q\mu B}{V_b} = \frac{\mu B c_t \phi}{5.615} \frac{\partial p}{\partial t}.$$ .... (8.475)

Because $k_y = 1.5 k_x$, the equation can be simplified to

$$\frac{\partial^2 p}{\partial x^2} + 1.5 \frac{\partial^2 p}{\partial y^2} + \frac{q\mu B}{k_x V_b} = \frac{\mu B c_t \phi}{5.615 k_x} \frac{\partial p}{\partial t}.$$ .... (8.476)

Furthermore, because $\Delta x = \Delta y$, the characteristic finite-difference equation can be recast as

$$(p_{i-1,j}^{n+1} - 2p_{i,j}^{n+1} + p_{i+1,j}^{n+1}) + 1.5(p_{i,j-1}^{n+1} - 2p_{i,j}^{n+1} + p_{i,j+1}^{n+1}) + \left(\frac{q\mu B}{hk_x}\right)_{i,j} = \left[\frac{(\Delta x)^2 \mu B c_t \phi}{k_x 5.615 \Delta t}\right]\left(p_{i,j}^{n+1} - p_{i,j}^{n}\right).$$ .... (8.477)

$$\text{Block 1: } p_2^{n+1} - p_1^{n+1} + \frac{q\mu B}{hk_x} = \frac{\mu B c_t \phi (\Delta x)^2}{5.615 \Delta t k_x}\left(p_1^{n+1} - p_1^{n}\right).$$ .... (8.478)

Knowing that $\Delta t = 5$ days and $q = -500$ STB/D, one can calculate the following coefficients:

$$\frac{\mu B c_t \phi (\Delta x)^2}{5.615 \Delta t k_x} = 0.1321 \quad \dots\dots\dots\dots\dots\dots\dots\dots\dots\dots\dots\dots\dots\dots\dots\dots\dots\dots\dots\dots \text{(8.479)}$$

and

$$\frac{q \mu B}{h k_x} = 1,064.77\,\text{psi.} \quad \dots\dots\dots\dots\dots\dots\dots\dots\dots\dots\dots\dots\dots\dots\dots\dots\dots\dots\dots\dots \text{(8.480)}$$

Therefore, the linear equation can be written as:

- Block 1: $-1.132 p_1 + p_2 = 536.115.$ $\dots\dots\dots\dots\dots\dots\dots\dots\dots\dots\dots\dots\dots\dots\dots\dots\dots$ (8.481)
- Block 2: $p_1 - 2.63 p_2 + 1.5 p_3 = 528.66.$ $\dots\dots\dots\dots\dots\dots\dots\dots\dots\dots\dots\dots\dots\dots\dots$ (8.482)
- Block 3: $1.5 p_2 - 1.63 p_3 = 528.66.$ $\dots\dots\dots\dots\dots\dots\dots\dots\dots\dots\dots\dots\dots\dots\dots\dots$ (8.483)

The PSOR procedure is employed to solve the system of equations with $\omega_{opt} = 1.23$. Iterations levels are shown in **Table 8.31.** An MBC is carried out to validate the solution using the results at the end of the 26th iteration:

- Fluid expansion of Block 1:

$$\Delta V_1 = (4,000\,\text{psia} - 827.389\,\text{psia}) \times 10^{-6}\,\text{psi}^{-1} \times \frac{1,320\,\text{ft} \times 1,320\,\text{ft} \times 5\,\text{ft} \times 0.2}{5.165} = 984.5\,\text{STB.} \quad \dots\dots\dots (8.484)$$

- Fluid expansion of Block 2:

$$\Delta V_2 = (4,000\,\text{psia} - 1,472.186\,\text{psia}) \times 10^{-6}\,\text{psi}^{-1} \times \frac{1,320\,\text{ft} \times 1,320\,\text{ft} \times 5\,\text{ft} \times 0.2}{5.165} = 788.0\,\text{STB.} \quad \dots\dots (8.485)$$

- Fluid expansion of Block 3:

$$\Delta V_3 = (4,000\,\text{psia} - 1,676.732\,\text{psia}) \times 10^{-6}\,\text{psi}^{-1} \times \frac{1,320\,\text{ft} \times 1,320\,\text{ft} \times 5\,\text{ft} \times 0.2}{5.165} = 724.2\,\text{STB.} \quad \dots\dots (8.486)$$

Total fluid expansion $= 984.5$ STB $+ 788.0$ STB $+ 724.42$ STB $= 2,496.7$ STB. $\dots\dots\dots\dots\dots\dots$ (8.487)

$$\frac{\text{Total fluid expansion}}{\text{Total fluid produced}} = \frac{2496.7\,\text{STB}}{500\,\text{STB/D} \times 5\,\text{days}} = 0.999 \approx 1. \quad \dots\dots\dots\dots\dots\dots\dots\dots\dots\dots\dots\dots \text{(8.488)}$$

Finally, at this stage one needs to calculate the productivity index to determine the sandface pressure:

$$\Omega = \frac{2\pi \bar{k} h_2}{\mu_o B_o \left[\ln\left(\dfrac{r_e}{r_w}\right) + s \right]} = 0.3934\,\text{STB/(D-psi),} \quad \dots\dots\dots\dots\dots\dots\dots\dots\dots\dots\dots\dots\dots \text{(8.489)}$$

$$p_{sf} = p_1 - \frac{500}{\Omega} = -454.43\,\text{psi.} \quad \dots\dots\dots\dots\dots\dots\dots\dots\dots\dots\dots\dots\dots\dots\dots\dots\dots\dots\dots\dots \text{(8.490)}$$

Iteration Level	p_1	p_2	p_3
0	4,000.000	4,000.000	4,000.000
1	2,839.170	3,459.268	3,386.620
2	2,521.640	3,005.136	3,015.189
3	2,100.938	2,652.620	2,702.337
4	1,815.190	2,380.870	2,467.378
5	1,585.875	2,171.514	2,284.946
...	...	...	...
22	833.239	1,477.576	1,681.420
23	831.165	1,475.665	1,679.758
24	829.568	1,474.193	1,678.478
25	828.337	1,473.059	1,677.491
26	827.389	1,472.186	1,676.732

Table 8.31—Iteration levels using the PSOR procedure to solve the system of equations with $\omega_{opt} = 1.23$.

Because the sandface pressure cannot fall below the atmospheric pressure, the well cannot sustain the specified production rate of 500 STB/D for five days. Because pressure at the sandface drops to atmospheric pressure, it implies that a downhole pump is placed across the sandface so that it does not allow fluid accumulation and a resulting back pressure at the wellbore across the sandface. In other words, when sandface pressure drops down to atmospheric pressure in practice, the choke is completely opened and the well switches to constant-sandface-pressure specification, during which time flow rate continually decreases from 500 STB/D.

Problem 8.4.35 (Ertekin et al. 2001)

Regardless of the outcome of the Solution to Problem 8.4.34, calculate the exact number of days Well 1 can sustain a specified flow rate of 500 STB/D.

Solution to Problem 8.4.35

To determine the exact days that the well can sustain the production rate of 500 STB/D, one needs to reduce the timestep from five days. In this case, a timestep of 1 day is used to solve the problem. The system of equations must be modified as well, in view of the reduced timestep:

$$\begin{bmatrix} -1.661 & 1 & 0 \\ 1 & -3.161 & 1.5 \\ 0 & 1.5 & -2.161 \end{bmatrix} \begin{bmatrix} p_1^{n+1} \\ p_2^{n+1} \\ p_3^{n+1} \end{bmatrix} = \begin{bmatrix} 1,064.77 - 0.661 \times p_1^n \\ -0.661 \times p_2^n \\ -0.661 \times p_3^n \end{bmatrix} \quad\ldots\ldots\ldots\ldots\ldots\ldots\ldots\ldots\ldots\ldots\ldots(8.491)$$

The PSOR with $\omega_{opt} = 1.234$ is employed to solve the linear system of equations of each timestep. The sandface pressure is also calculated at the end of each timestep. The total number of days that the well can sustain the production rate of 500 STB/D can be determined when the calculated sandface pressure becomes less than 14.7 psia.

In the aforementioned solution strategy, the calculated block pressure and sandface pressure of the wellblock is listed in **Table 8.32.** From the calculation result, one can tell that the well can sustain the production rate of 500 STB/D for four days, plus a few additional hours.

Time, days	Block Pressures			MBC	p_{sf}
	Block 1	Block 2	Block 3		
1	3,104.46	3,577.47	3,706.72	0.999	1,833.39
2	2,442.55	3,070.03	3,264.75	0.999	1,171.48
3	1,861.12	2,541.87	2,762.95	0.999	590.04
4	1,308.07	2,007.69	2,238.69	0.999	37.00
5	765.15	1,471.56	1,706.19	0.999	-505.92

Table 8.32—The calculated block pressure and sandface pressure of the wellblock from the Solution to Problem 8.4.35.

Problem 8.4.36 (Ertekin et al. 2001)

Consider the flow of a single-phase incompressible fluid in a reservoir represented by the body-centered grid in **Fig. 8.30.** The reservoir has homogeneous but anisotropic permeability distribution with $k_x = 100$ md and $k_y = 200$ md. The thickness of the formation is uniform and is equal to 60 ft. The grid dimension is $\Delta x = \Delta y = 800$ ft. The reservoir is surrounded by no-flow boundaries, with the exception of the two block boundaries indicated in Fig. 8.30. Flowing sandface pressure of the well in Gridblock 3 is kept at 600 psia. Assume that the entire top surface of the reservoir is 6,000 ft below the datum level. The wellbore radius is 0.24 ft and fluid viscosity is 2 cp. Use the LSOR procedure with $\omega_{opt} = 1$ to solve for the pressure distribution in the reservoir.

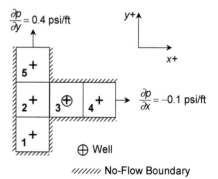

Fig. 8.30—Reservoir system of Problem 8.4.36.

Solution to Problem 8.4.36

For a single-phase 2D incompressible flow system with anisotropic and homogeneous permeability distributions, the PDE that describes the problem can be written as

$$k_x \frac{\partial^2 p}{\partial x^2} + k_y \frac{\partial^2 p}{\partial y^2} + \frac{q\mu B}{V_b} = 0. \dotfill (8.492)$$

Because $k_y = 2k_x$, the PDE can be further simplified to

$$\frac{\partial^2 p}{\partial x^2} + 2\frac{\partial^2 p}{\partial y^2} + \frac{q\mu B}{k_x V_b} = 0. \dotfill (8.493)$$

The characteristic finite-difference equation will be ($\Delta x = \Delta y$):

$$(p_{i-1,j} - 2p_{i,j} + p_{i+1,j}) + 2(p_{i,j-1} - 2p_{i,j} + p_{i,j+1}) + \left(\frac{q\mu B}{hk_x}\right)_{i,j} = 0. \dotfill (8.494)$$

At Block 5, the discretized equation in which p'_5 is a reflection block, can be written as

$$p'_5 = 2p_5 + p_2 = 0. \dotfill (8.495)$$

Applying the boundary condition at Block 5,

$$\frac{p'_5 - p_5}{\Delta y} = 0.4 \text{ psia/ft} \dotfill (8.496)$$

and

$$p'_5 = p_5 + 0.4\Delta y. \dotfill (8.497)$$

Thus, for Block 5,

$$p_2 + p_5 = -0.4\Delta y \text{ or } p_2 - p_5 = -320. \dotfill (8.498)$$

For Block 2,

$$(p_3 - p_2) + 2(p_5 - p_2) = 0 \dotfill (8.499)$$

(note that Block 1 is an inactive block with $p_1 = p_2$), or

$$p_3 - 3p_2 + 2p_5 = 0. \dotfill (8.500)$$

For Block 3,

$$p_2 - 2p_3 + p_4 + \frac{\mu B}{hk_x}\Omega\left(p_{sf} - p_3\right) = 0. \dotfill (8.501)$$

Calculating the equivalent wellbore radius,

$$r_e = \frac{0.28\sqrt{\left(\frac{k_x}{k_y}\right)^{\frac{1}{2}}(\Delta y)^2 + \left(\frac{k_y}{k_x}\right)^{\frac{1}{2}}(\Delta x)^2}}{\left(\frac{k_x}{k_y}\right)^{\frac{1}{4}} + \left(\frac{k_y}{k_x}\right)^{\frac{1}{4}}} = 160.7 \text{ ft}, \dotfill (8.502)$$

$$\bar{k} = \sqrt{k_x k_y} = 141.42 \text{ md.} \dotfill (8.503)$$

Using the Peaceman's wellbore equation,

$$\Omega = \frac{2\pi \bar{k}h}{\mu B \ln\left(\dfrac{r_e}{r_w}\right)} = 4.615 \text{ STB/D-psi,} \quad\quad\quad (8.504)$$

with $p_{sf} = 600$ psia.

Then,

$$p_2 - 2p_3 + p_4 + 1.365(600 - p_3) = 0 \quad\quad\quad (8.505)$$

or

$$p_2 - 3.365p_3 + p_4 = -819. \quad\quad\quad (8.506)$$

For Block 4,

$$p_3 - 2p_4 + p_4' = 0, \quad\quad\quad (8.507)$$

with consideration of the boundary condition (using the reflection Block 4'),

$$\frac{p_4' - p_4}{\Delta x} = -0.1 \text{ psia/ft} \quad\quad\quad (8.508)$$

or

$$p_4' = p_4 - 0.1\Delta x, \quad\quad\quad (8.509)$$

which gives

$$p_3 - p_4 = 0.1\Delta x = 80. \quad\quad\quad (8.510)$$

To solve the problem using LSOR, solve the linear equations row by row. To be more specific, pressure values of Blocks 2, 3, and 4 are to be solved using the pressure of Block 5 of the previous iteration. The solved pressures of Blocks 2, 3, and 4 can then be used to solve for the pressure of Block 5. This process can be done iteratively to solve for the pressure distribution.

For Blocks 2, 3, and 5, the line equations can be expressed in matrix form:

$$\begin{bmatrix} -3 & 1 & 0 \\ 1 & -3.365 & 1 \\ 0 & 1 & -1 \end{bmatrix} \begin{bmatrix} p_2^{(k+1)} \\ p_3^{(k+1)} \\ p_4^{(k+1)} \end{bmatrix} = \begin{bmatrix} 2p_5^{(k)} \\ -819 \\ 80 \end{bmatrix}. \quad\quad\quad (8.511)$$

Once $p_2^{(k+1)}$, $p_3^{(k+1)}$, and $p_4^{(k+1)}$ are solved, $p_5^{(k+1)}$ can be solved as

$$p_5^{(k+1)} = 320 + p_2^{(k+1)}. \quad\quad\quad (8.512)$$

Using an initial guess of 1,000 psia for all unknown block pressures, it takes 23 LSOR iterations to converge within a prescribed pressure tolerance of 1 psia, shown in **Table 8.33**.

An MBC can be implemented to validate the solution:

• Fluid entering into the reservoir across the northern boundary of Block 5:

$$q_5 = 1.127 \times 10^{-3} \times \frac{A_y k_y}{\mu B} 0.4 \frac{\text{psia}}{\text{ft}} = 2,163.84 \text{ STB/D.} \quad\quad\quad (8.513)$$

• Fluid leaving the reservoir across the eastern boundary of Block 4:

$$q_4 = 1.127 \times 10^{-3} \frac{A_x k_x}{\mu B} 0.1 \frac{\text{psia}}{\text{ft}} = 270.48 \text{ STB/D.} \quad\quad\quad (8.514)$$

• Fluid produced from the well:

$$q_3 = \Omega(p_3 - p_{sf}) = 4.61 \times (1,009.053 - 600) = 1,887.8 \text{ STB/D.} \quad\quad\quad (8.515)$$

Iteration Level/ Pressure, psia	$p_2^{(k+1)}$	$p_3^{(k+1)}$	$p_4^{(k+1)}$	$p_5^{(k+1)}$
0	1,000.000	1,000.000	1,000.000	1,000.000
1	897.293	691.879	611.879	1,217.293
2	1,065.922	763.181	683.181	1,385.922
3	1,196.786	818.514	738.514	1,516.786
4	1,298.343	861.456	781.456	1,618.343
5	1,377.155	894.780	814.780	1,697.155
...	...	...	...	...
19	1,642.409	1,006.938	926.938	1,962.409
20	1,644.166	1,007.681	927.681	1,964.166
21	1,645.530	1,008.258	928.258	1,965.530
22	1,646.589	1,008.706	928.706	1,966.589
23	1,647.410	1,009.053	929.053	1,967.410

Table 8.33—LSOR iterations to convergence within a prescribed pressure tolerance of 1 psia from the Solution to Problem 8.4.36.

Thus,

$$\mathrm{MBC} = \frac{270.48 \ \mathrm{STB/D} + 1{,}887.8 \ \mathrm{STB}}{2{,}163.84 \ \mathrm{STB/D}} \approx 1.003. \dots\dots\dots\dots\dots\dots\dots\dots\dots\dots\dots (8.516)$$

Problem 8.4.37 (Ertekin et al. 2001)

Use the following four equations in four unknowns to obtain an approximate coefficient matrix that is easy to factor:

- $x_1 + 2x_3 + 4x_4 = -8. \dots\dots\dots\dots\dots\dots\dots\dots\dots\dots\dots\dots\dots (8.517)$

- $3x_1 + 6x_2 + x_3 = 3. \dots\dots\dots\dots\dots\dots\dots\dots\dots\dots\dots\dots\dots (8.518)$

- $x_2 + 3x_3 + 2x_4 = 4. \dots\dots\dots\dots\dots\dots\dots\dots\dots\dots\dots\dots (8.519)$

- $4x_1 + 2x_2 + x_4 = -9. \dots\dots\dots\dots\dots\dots\dots\dots\dots\dots\dots\dots (8.520)$

Find the entries of the upper- and lower-tridiagonal matrices formed as a consequence of this factorization. Do not use an iteration parameter.

Solution to Problem 8.4.37

Using Crout reduction, one can obtain the following lower- and upper-triangular matrices, which satisfy Eq. 8.521:

$$\begin{bmatrix} l_{11} & 0 & 0 & 0 \\ l_{21} & l_{22} & 0 & 0 \\ l_{31} & l_{32} & l_{33} & 0 \\ l_{41} & l_{42} & l_{43} & l_{44} \end{bmatrix} \begin{bmatrix} u_{11} & u_{12} & u_{13} & u_{14} \\ 0 & u_{22} & u_{23} & u_{24} \\ 0 & 0 & u_{33} & u_{34} \\ 0 & 0 & 0 & u_{44} \end{bmatrix} = \begin{bmatrix} 1 & 0 & 2 & 4 \\ 3 & 6 & 1 & 0 \\ 0 & 1 & 3 & 2 \\ 4 & 2 & 0 & 1 \end{bmatrix}. \dots\dots\dots\dots\dots\dots (8.521)$$

For $j = 1, 2, \dots n$, perform the following two computational steps:

1. For $i \le j$, $u_{i,j} = a_{i,j}$ for $i = 1$;

$$u_{i,j} = a_{i,j} - \sum_{k=1}^{i-1} l_{i,k} u_{k,j}, \dots\dots\dots\dots\dots\dots\dots\dots\dots\dots\dots\dots (8.522)$$

for $i = 2, 3, \dots, j$.

2. For $i > j$, $l_{i,1} = \dfrac{a_{i,1}}{u_{11}}$ for $j = 1$;

$$l_{i,j} = \frac{1}{u_{i,j}}\left(a_{i,j} - \sum_{k=1}^{j-1} l_{i,k} u_{k,j} \right), \quad\ldots\ldots\ldots\ldots\ldots\ldots\ldots\ldots\ldots\ldots\ldots\ldots\ldots\ldots\ldots (8.523)$$

for $i = j + 1, j + 2, \ldots, n$, and for $j \geq 2$.

Finally, one obtains

$$[L] = \begin{bmatrix} 1 & 0 & 0 & 0 \\ 3 & 1 & 0 & 0 \\ 1 & 0 & 1 & 0 \\ 4 & 0.333 & -0.333 & 1 \end{bmatrix} \text{ and } [U] = \begin{bmatrix} 1 & 0 & 2 & 4 \\ 0 & 6 & -5 & -12 \\ 0 & 0 & 1 & -2 \\ 0 & 0 & 0 & -23.667 \end{bmatrix}. \quad\ldots\ldots\ldots\ldots\ldots\ldots (8.524)$$

Problem 8.4.38 (Ertekin et al. 2001)

Consider the 1D, single-phase, incompressible fluid flow in a reservoir, dipping at a constant angle of 10° from left to right (**Fig. 8.31**). Gridblocks 1 and 5 are located at the boundaries of the reservoir and maintained at 3,000 psia. A production well is located in the center of Gridblock 3 and produced at a rate of 2,000 STB/D. As Fig. 8.31 shows, all gridblocks have uniform dimensions: $\Delta x = 200$ ft, $\Delta y = 100$ ft, and $h = 50$ ft. The permeability distribution in the reservoir has a symmetry variation

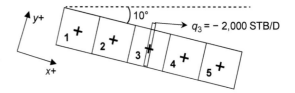

Fig. 8.31—Reservoir system of Problem 8.4.38.

of $k_{x1} = 88$ md, $k_{x2} = 178$ md, $k_{x3} = 355$ md, $k_{x4} = 178$ md, and $k_{x5} = 88$ md. With the appropriate PDE and the corresponding finite-difference approximation, calculate the pressure distribution in the system. Devise a procedure to check your solution. The fluid viscosity is 1 cp and FVF is 1 res bbl/STB. Use a fluid density of $\rho = 62.4$ lbm/ft^3 and assume that $g/g_c = 1$.

Solution to Problem 8.4.38

When a constant 10° dip exists in the system, the PDE should be modified in consideration of the depth gradient:

$$\frac{\partial}{\partial x}\left(\frac{A_x k_x}{\mu B} \frac{\partial p}{\partial x} \right)\Delta x - \frac{1}{144}\frac{g}{g_c}\rho\frac{\partial}{\partial x}\left(\frac{A_x k_x}{\mu B} \frac{\partial G}{\partial x} \right)\Delta x + q = 0. \quad\ldots\ldots\ldots\ldots\ldots\ldots\ldots\ldots\ldots (8.525)$$

In this system,

$$\frac{\partial G}{\partial x} = \sin(10°) = 0.174. \quad\ldots\ldots\ldots\ldots\ldots\ldots\ldots\ldots\ldots\ldots\ldots\ldots\ldots\ldots\ldots\ldots\ldots\ldots\ldots (8.526)$$

Therefore, the elevation variation of two adjacent blocks along the x-direction is constant:

$$\Delta G = 0.174\Delta x = 34.8 \text{ ft} \cdot \quad\ldots\ldots\ldots\ldots\ldots\ldots\ldots\ldots\ldots\ldots\ldots\ldots\ldots\ldots\ldots\ldots\ldots\ldots (8.527)$$

For a reservoir with heterogeneous-permeability distribution, the transmissibility factor must be calculated:

$$\left.\frac{A_x k_x}{\mu B \Delta x}\right|_{1\frac{1}{2}} = T_{12} = \frac{2A_1 A_2 k_1 k_2}{\mu B\left(A_1 k_1 \Delta x_2 + A_2 k_2 \Delta x_1\right)}. \quad\ldots\ldots\ldots\ldots\ldots\ldots\ldots\ldots\ldots\ldots\ldots (8.528)$$

Because $A_1 = A_2$ and the grid dimensions are uniform ($\Delta x_1 = \Delta x_2 = \Delta x_3 = \Delta x_4 = \Delta x_5$ and $\Delta y_1 = \Delta y_2 = \Delta y_3 = \Delta y_4 = \Delta y_5$), Eq. 8.528 can be simplified to

$$\left.\frac{A_x k_x}{\mu B \Delta x}\right|_{1\frac{1}{2}} = T_{12} = \frac{2\Delta y h k_1 k_2}{\Delta x \mu B\left(k_1 + k_2\right)} = \left(1.127\times10^{-3}\right)\frac{2(100)(50)(88)(178)}{(200)(1)(1)(88+178)} = 3.32 \text{ STB/D-psi}. \quad\ldots\ldots\ldots (8.529)$$

Because of the existing symmetry, $T_{12} = T_{45} = 3.32$ STB/D-psi,

$$\left.\frac{A_x k_x}{\mu B \Delta x}\right|_{2\frac{1}{2}} = T_{23} = \frac{2\Delta y h k_2 k_3}{\Delta x \mu B\left(k_2 + k_3\right)} = \left(1.127\times10^{-3}\right)\frac{2(100)(50)(355)(178)}{(200)(1)(1)(355+178)} = 6.68 \text{ STB / (D-psi)}. \quad\ldots\ldots\ldots (8.530)$$

Because of the existing symmetry, $T_{23} = T_{34} = 6.68$ STB/D-psi, Eq. 8.525 can be written using a finite-difference approximation:

$$T_{i+\frac{1}{2}}\left(p_{i+1} - p_i\right) - T_{i-\frac{1}{2}}\left(p_i - p_{i-1}\right) + \frac{1}{144}\frac{g}{g_c}\rho T_{i+\frac{1}{2}}\left(G_{i+1} - G_i\right) - \frac{1}{144}\frac{g}{g_c}\rho T_{i-\frac{1}{2}}\left(G_i - G_{i-1}\right) + q = 0. \quad\ldots\ldots\ldots\ldots(8.531)$$

- Block 2:

$$T_{23}\left(p_3 - p_2\right) - T_{12}\left(p_2 - p_1\right) + T_{23}'\left(G_3 - G_2\right) - T_{12}'\left(G_2 - G_1\right) = 0. \quad\ldots\ldots\ldots\ldots\ldots\ldots\ldots\ldots\ldots\ldots(8.532)$$

Note that

$$T_{23}' = T_{34}' = \frac{1}{144}\frac{g}{g_c}\rho T_{23} = 0.433 \times 6.68 = 2.89 \text{ STB}/(\text{D-psi}), \quad\ldots\ldots\ldots\ldots\ldots\ldots(8.533)$$

$$T_{12}' = T_{45}' = \frac{1}{144}\frac{g}{g_c}\rho T_{12} = 0.433 \times 3.32 = 1.44 \text{ STB}/(\text{D-psi}), \quad\ldots\ldots\ldots\ldots\ldots\ldots(8.534)$$

and $G_3 - G_2 = G_2 - G_1 = 34.8$ ft, $p_1 = 3{,}000$ psi.

Plugging in the coefficients and known block pressure, the finite-difference approximation of Block 2 can be written as

$$6.68\left(p_3 - p_2\right) - 3.32\left(p_2 - 3{,}000\right) + 1.45 \times 34.8 = 0, \quad\ldots\ldots\ldots\ldots\ldots\ldots\ldots\ldots(8.535)$$

which can be expressed as

$$-10 p_2 + 6.68 p_3 = -10{,}010.46. \quad\ldots\ldots\ldots\ldots\ldots\ldots\ldots\ldots\ldots\ldots\ldots(8.536)$$

- Block 3:

$$T_{34}\left(p_4 - p_3\right) - T_{23}\left(p_3 - p_2\right) + T_{34}'\left(G_4 - G_3\right) - T_{23}'\left(G_3 - G_2\right) + q_3 = 0. \quad\ldots\ldots\ldots\ldots\ldots(8.537)$$

Recall that $T_{23} = T_{34}$ and $T_{23}' = T_{34}'$, Eq. 8.537 can be simplified to

$$6.68 p_2 - 13.36 p_3 + 6.68 p_4 = 2{,}000. \quad\ldots\ldots\ldots\ldots\ldots\ldots\ldots\ldots\ldots\ldots(8.538)$$

- Block 4:

$$T_{45}\left(p_5 - p_4\right) - T_{34}\left(p_4 - p_3\right) + T_{45}'\left(G_5 - G_4\right) - T_{34}'\left(G_4 - G_3\right) = 0. \quad\ldots\ldots\ldots\ldots\ldots(8.539)$$

Plugging in the coefficients and known block-pressure value, the finite-difference approximation of Block 4 can be written as

$$3.32\left(3{,}000 - p_4\right) - 6.68\left(p_4 - p_3\right) - 1.45 \times 34.8 = 0, \quad\ldots\ldots\ldots\ldots\ldots\ldots\ldots(8.540)$$

$$6.68 p_3 - 10 p_4 = -9{,}909.54. \quad\ldots\ldots\ldots\ldots\ldots\ldots\ldots\ldots\ldots\ldots\ldots\ldots(8.541)$$

Rearranging Eqs. 8.536, 8.538, and 8.541, one will have

$$-10 p_2 + 6.68 p_3 = -10{,}010.46, \quad\ldots\ldots\ldots\ldots\ldots\ldots\ldots\ldots\ldots\ldots\ldots(8.542)$$

$$6.68 p_2 - 13.36 p_3 + 6.68 p_4 = 2{,}000, \quad\ldots\ldots\ldots\ldots\ldots\ldots\ldots\ldots\ldots\ldots(8.543)$$

$$6.68 p_3 - 10 p_4 = -9{,}909.54. \quad\ldots\ldots\ldots\ldots\ldots\ldots\ldots\ldots\ldots\ldots\ldots(8.544)$$

Solving for three unknowns in three equations, one obtains: $p_2 = 2{,}703.84$ psi, $p_3 = 2{,}549.09$ psi, and $p_4 = 2{,}693.75$ psi. To validate the solution, the mass flow from Block 1 to Block 2 is

$$q_{12} = T_{12}\left(p_1 - p_2\right) - T_{12}'\left(G_1 - G_2\right) = 3.32\left(3{,}000 - 2{,}703.84\right) + 1.44 \times 34.8 = 1{,}033.36 \text{ STB/D}. \quad\ldots\ldots\ldots(8.545)$$

Note that the gravitational force is stimulating the flow from Block 1 to Block 2.

The mass flow from Block 5 to Block 4 is

$$q_{54} = T_{45}(p_5 - p_4) - T'_{45}(G_5 - G_4) = 3.32(3,000 - 2,693.75) - 1.44 \times 34.8 = 966.64 \text{ STB/D}. \dots\dots (8.546)$$

Note that the gravitational force is reducing the flow from Block 5 to Block 4.

$$\text{MBC} = \frac{q_{54} + q_{12}}{q_{\text{well}}} = \frac{2,000 \text{ STB/D}}{2,000 \text{ STB/D}} = 1.0, \dots\dots\dots\dots\dots\dots\dots\dots\dots\dots\dots\dots\dots\dots\dots\dots (8.547)$$

which indicates a perfect material-balance check.

Problem 8.4.39

Consider the 1D compressible, homogeneous reservoir shown in **Fig. 8.32.** All structural boundaries of the reservoir are completely sealed. Gas density can be ignored. A well is drilled and completed through two layers. The flow rates from both layers are measured and known to be q_1 and q_2. The sandface pressures at both layers are also measured as $p_{sf,1}$ and $p_{sf,2}$ at the end of t days of production. In addition, it is found that the well is damaged in both layers and the skin values for both layers are unknown. Construct the Jacobian (coefficient) matrix by writing the generic forms of the entries for the matrix that can be used to simultaneously solve for all unknowns of the problem. In the solution, no numerical entries are needed.

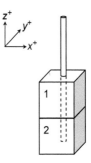

Fig. 8.32—Reservoir system of Problem 8.4.39.

Solution to Problem 8.4.39

There are four unknowns in this problem: p_1, p_2, S_1, and S_2; two residual equations are from flow equations (PDEs), and the remaining two residual equations are from Peaceman's wellbore equations with skin factor. The system of equations that must be solved is as follows:

$$\begin{bmatrix} \dfrac{\partial R_1}{\partial p_1} & \dfrac{\partial R_1}{\partial p_2} & \dfrac{\partial R_1}{\partial S_1} & 0 \\[2ex] \dfrac{\partial R_2}{\partial p_1} & \dfrac{\partial R_2}{\partial p_2} & 0 & \dfrac{\partial R_2}{\partial S_2} \\[2ex] \dfrac{\partial R_3}{\partial p_1} & 0 & \dfrac{\partial R_3}{\partial S_1} & 0 \\[2ex] 0 & \dfrac{\partial R_4}{\partial p_2} & 0 & \dfrac{\partial R_4}{\partial S_2} \end{bmatrix}_{n+1}^{(k)} \begin{bmatrix} \Delta p_1 \\[1ex] \Delta p_2 \\[1ex] \Delta S_1 \\[1ex] \Delta S_2 \end{bmatrix}_{n+1}^{(k+1)} = - \begin{bmatrix} R_1 \\[1ex] R_2 \\[1ex] R_3 \\[1ex] R_4 \end{bmatrix}_{n+1}^{(k)} \quad \dots\dots\dots\dots\dots\dots\dots\dots (8.548)$$

$$[\text{Jacobian}] \qquad [\text{Unknown vector}] [\text{Right-hand-side vector}]$$

Problem 8.4.40

Consider a 2D, single-phase compressible flow problem in the reservoir shown in **Fig. 8.33.** There is a well located in Block 3, and it is producing at a constant flow rate. Construct a Jacobian matrix system of equations to solve for pressure distribution (p_1, p_2, p_3, and p_4) and sandface pressure at Block 3 ($p_{sf,3}$), simultaneously using the GNR protocol. Write the system of equations that must be generated and express them in matrix form if you are asked to simultaneously solve all unknowns of the problem.

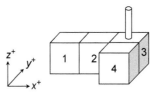

Fig. 8.33—Reservoir system of Problem 8.4.40.

Solution to Problem 8.4.40

$$
\begin{bmatrix}
\dfrac{\partial R_1}{\partial p_1} & \dfrac{\partial R_1}{\partial p_2} & 0 & 0 & 0 \\[2ex]
\dfrac{\partial R_2}{\partial p_1} & \dfrac{\partial R_2}{\partial p_2} & \dfrac{\partial R_2}{\partial p_3} & 0 & 0 \\[2ex]
0 & \dfrac{\partial R_3}{\partial p_2} & \dfrac{\partial R_3}{\partial p_3} & \dfrac{\partial R_3}{\partial p_4} & \dfrac{\partial R_3}{\partial p_{sf}} \\[2ex]
0 & 0 & \dfrac{\partial R_4}{\partial p_3} & \dfrac{\partial R_4}{\partial p_4} & 0 \\[2ex]
0 & 0 & \dfrac{\partial R_q}{\partial p_3} & 0 & \dfrac{\partial R_q}{\partial p_{sf}}
\end{bmatrix}^{(k)}_{n+1}
\begin{bmatrix}
\Delta p_1 \\[2ex] \Delta p_2 \\[2ex] \Delta p_3 \\[2ex] \Delta p_4 \\[2ex] \Delta p_{sf\,3}
\end{bmatrix}^{(k+1)}_{n+1}
= -
\begin{bmatrix}
R_1 \\[2ex] R_2 \\[2ex] R_3 \\[2ex] R_4 \\[2ex] R_5
\end{bmatrix}^{(k)}_{n+1}
\qquad \dots \dots \dots \dots \dots \dots \text{(8.549)}
$$

$[\text{Jacobian}]$ $[\text{Unknown vector}]\,[\text{Right-hand-side vector}]$

Problem 8.4.41

Consider the single-phase compressible fluid flow in the 2D reservoir shown in **Fig. 8.34.** The reservoir boundaries are completely sealed. There is a production well located at the center of the reservoir. Using the GNR protocol, construct the Jacobian matrix of the reservoir system.

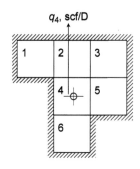

Fig. 8.34—Reservoir system of Problem 8.4.41.

Solution to Problem 8.4.41

Because there are six blocks in the reservoir for a single-phase flow problem, the Jacobian matrix will be 6×6. The first derivatives of the residuals at each block, with respect to block pressure and to the neighboring block pressures, need to be calculated. The final structure of the Jacobian matrix for the reservoir is shown:

$$
J =
\begin{bmatrix}
\dfrac{\partial R_1}{\partial p_1} & \dfrac{\partial R_1}{\partial p_2} & 0 & 0 & 0 & 0 \\[2ex]
\dfrac{\partial R_2}{\partial p_1} & \dfrac{\partial R_2}{\partial p_2} & \dfrac{\partial R_2}{\partial p_3} & \dfrac{\partial R_2}{\partial p_4} & 0 & 0 \\[2ex]
0 & \dfrac{\partial R_3}{\partial p_2} & \dfrac{\partial R_3}{\partial p_3} & \dfrac{\partial R_3}{\partial p_5} & 0 & 0 \\[2ex]
0 & \dfrac{\partial R_4}{\partial p_2} & 0 & \dfrac{\partial R_4}{\partial p_4} & \dfrac{\partial R_4}{\partial p_5} & \dfrac{\partial R_4}{\partial p_6} \\[2ex]
0 & 0 & \dfrac{\partial R_5}{\partial p_3} & \dfrac{\partial R_5}{\partial p_4} & \dfrac{\partial R_5}{\partial p_5} & 0 \\[2ex]
0 & 0 & 0 & \dfrac{\partial R_6}{\partial p_4} & 0 & \dfrac{\partial R_6}{\partial p_6}
\end{bmatrix}
\qquad \dots \dots \dots \dots \dots \dots \dots \dots \dots \text{(8.550)}
$$

Problem 8.4.42

Consider the incompressible flow of oil in the 2D homogeneous and anisotropic reservoir with no depth gradients shown in **Fig. 8.35.** The well is producing with a specified rate of 1,000 STB/D, and the external boundaries are completely sealed.

Calculate the sandface pressure of the well using the reservoir properties found in **Table 8.34.** Solve the system of equations using the Gauss-Seidel method. Use an initial guess of 3,000 psi for block pressures and 5 psi as the convergence criterion. Verify your results with an MBC.

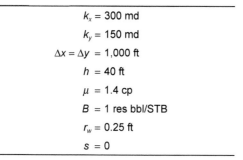

$$k_x = 300 \text{ md}$$
$$k_y = 150 \text{ md}$$
$$\Delta x = \Delta y = 1,000 \text{ ft}$$
$$h = 40 \text{ ft}$$
$$\mu = 1.4 \text{ cp}$$
$$B = 1 \text{ res bbl/STB}$$
$$r_w = 0.25 \text{ ft}$$
$$s = 0$$

Table 8.34—Reservoir properties for Problem 8.4.42.

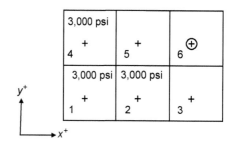

Fig. 8.35—Reservoir system of Problem 8.4.42.

Solution to Problem 8.4.42

For a 2D incompressible fluid reservoir, the governing PDE (ignoring the gravitational forces) is

$$\frac{\partial}{\partial x}\left(\frac{A_x k_x}{\mu B}\frac{\partial p}{\partial x}\right)\Delta x + \frac{\partial}{\partial y}\left(\frac{A_y k_y}{\mu B}\frac{\partial p}{\partial y}\right)\Delta y + q = 0. \qquad (8.551)$$

For a homogeneous and anisotropic system ($k_x = 2k_y$), Eq. 8.551 can be rewritten as

$$2\frac{\partial^2 p}{\partial x^2} + \frac{\partial^2 p}{\partial y^2} + \frac{q\mu B}{k_y V_b} = 0. \qquad (8.552)$$

The finite-difference approximation of the PDE can be expressed as (knowing $\Delta x = \Delta y$):

$$2\left(p_{i-1,j} + p_{i+1,j} - 2p_{i,j}\right) + \left(p_{i,j-1} + p_{i,j+1} - 2p_{i,j}\right) + \frac{q\mu B}{hk_y} = 0. \qquad (8.553)$$

- Block 3: $3p_3 - p_6 = 6,000.$ $\qquad (8.554)$
- Block 5: $5p_5 - 2p_6 = 9,000.$ $\qquad (8.555)$
- Block 6: $3p_6 - 2p_5 - p_3 = -207.04.$ $\qquad (8.556)$

Using the Gauss-Seidel method results at each iteration level are presented in **Table 8.35.**

Iteration Level	p_3	p_5	p_6
0	3,000	3,000	3,000
1	3,000	3,000	2,930.987
2	2,976.996	2,972.395	2,904.915
3	2,968.305	2,961.966	2,895.066
4	2,965.022	2,958.026	2,891.345

Table 8.35—Results encountered at each iteration level of the Gauss-Seidel method in the solution of Problem 8.4.42.

The productivity index of the well in Block 6 is

$$\Omega_6 = \frac{2\pi \bar{k} h}{\mu B \left[\ln\left(\frac{r_e}{r_w}\right) + s \right]} = \frac{2\pi(212.13)(1.127 \times 10^{-3})(40)}{(1)(1)\left(\ln\frac{200}{0.5} \right)} = 6.93 \text{ STB/D-psi}, \quad \dots\dots\dots\dots\dots\dots (8.557)$$

where $r_e = 0.2\Delta x = 200$ ft,

$$\bar{k} = \sqrt{k_x k_y} = \sqrt{(300)(150)} = 212.132 \text{ md.} \quad \dots\dots\dots\dots\dots\dots\dots\dots\dots\dots\dots\dots (8.558)$$

Then, the sandface pressure of the well can be calculated as

$$p_{sf} = p_6 + \frac{q_6}{\Omega_6} = 2747.1 \text{ psi.} \quad \dots\dots\dots\dots\dots\dots\dots\dots\dots\dots\dots\dots\dots\dots (8.559)$$

- Mass flow from Block 4 to Block 5:

$$q_{4 \text{ to } 5} = \frac{kh}{\mu B}(p_4 - p_5) = \frac{(300)(1.127 \times 10^{-3})(40)(3,000 - 2,958.026)}{(1)(1.4)} = 405.4688 \text{ STB/D.} \quad \dots\dots\dots (8.560)$$

- Mass flow from Block 2 to Block 5:

$$q_{2 \text{ to } 5} = \frac{kh}{\mu B}(p_2 - p_5) = \frac{(150)(1.127 \times 10^{-3})(40)(3,000 - 2,958.026)}{(1)(1.4)} = 202.7344 \text{ STB/D.} \quad \dots\dots\dots (8.561)$$

- Mass flow from Block 2 to Block 3:

$$q_{2 \text{ to } 3} = \frac{kh}{\mu B}(p_2 - p_3) = \frac{(300)(1.127 \times 10^{-3})(40)(3,000 - 2,965.022)}{(1)(1.4)} = 337.8875 \text{ STB/D}, \quad \dots\dots\dots (8.562)$$

$$\text{MBC} = \left| \frac{q_{\text{total}}}{q_{\text{well}}} \right| = \left| \frac{405.4688 + 202.7344 + 337.8875}{1,000} \right| = \left| \frac{946.1}{1,000} \right| = 0.9461. \quad \dots\dots\dots\dots\dots (8.563)$$

The relatively high error observed in the MBC is caused by the large tolerance used as the convergence criterion.

Problem 8.4.43

Consider the 2D single-phase incompressible fluid flow, homogeneous reservoir (with no depth gradient), as displayed in **Fig. 8.36.** The reservoir and fluid properties are shown in **Table 8.36.** The northern boundary of Block 5 and the eastern boundary of Block 4 are of the Dirichlet-type boundary, as displayed in Fig. 8.36. Block 9 is part of another large reservoir, and its pressure is kept at 3,000 psia. There is a no-flow boundary between Blocks 4 and 9. All other boundaries are closed. The injection well is operating at a fixed flow rate of 2,500 STB/D, and the production well is operating at a fixed sandface pressure of 250 psia.

Calculate the following SIP coefficients: C_2, Q_2, C_3, W_3, S_3, C_4, Q_4, C_5, Q_5, C_7, and Q_7 by using the rock and fluid properties in Table 8.36.

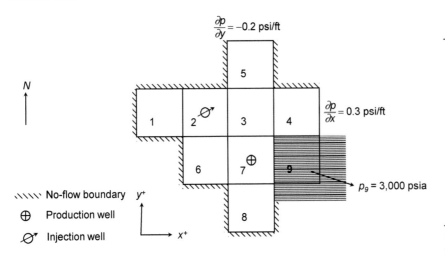

Table 8.36—Reservoir and fluid properties for Problem 8.4.43.	
Δx	= 200 ft
Δy	= 150 ft
k_y	= 10 md
k_x	= 15 md
μ	= 1.05 cp
B	= 1.0 res bbl/STB
h	= 50 ft
r_w	= 0.25 ft
r_e	= 34.17 ft
s	= 0.8

Fig. 8.36—Reservoir system of Problem 8.4.43.

Solution to Problem 8.4.43

All transmissibilities along the y-direction are the same:

$$N = S = \frac{\Delta x h k_y}{\Delta y \mu B} = \frac{(200)(50)(10)(0.001127)}{(150)(1.05)(1)} = 0.7156 \text{ STB/D-psi.} \quad \text{...........................} (8.564)$$

All transmissibilities along the x-direction are the same:

$$W = E = \frac{\Delta y h k_y}{\Delta x \mu B} = \frac{(150)(50)(15)(0.001127)}{(200)(1.05)(1)} = 0.6037 \text{ STB/D-psi,} \quad \text{...........................} (8.565)$$

$$C_2 = -(E_2 + W_2 + S_2) = -1.9231 \text{ STB/D-psi,} \quad \text{...............................} (8.566)$$

$$Q_2 = -2500 \text{ STB/D,} \quad \text{...} (8.567)$$

$$C_3 = -(E_3 + W_3 + S_3 + N_3) = -2.6386 \text{ STB/D-psi,} \quad \text{......................} (8.568)$$

$$W_3 = 0.6037 \text{ STB/D-psi,} \quad \text{...} (8.569)$$

$$S_3 = 0.7156 \text{ STB/D-psi,} \quad \text{...} (8.570)$$

$$C_4 = -W_4 = -0.6037 \text{ STB/D-psi,} \quad \text{....................................} (8.571)$$

$$Q_4 = -E_4 \Delta x \frac{\partial p}{\partial x} = -36.225 \text{ STB/D,} \quad \text{..............................} (8.572)$$

$$C_5 = -S_5 = -0.7156 \text{ STB/D-psi,} \quad \text{....................................} (8.573)$$

$$Q_5 = -N_5 \Delta y \frac{\partial p}{\partial y} = 21.467 \text{ STB/D-psi,} \quad \text{.............................} (8.574)$$

$$\Omega = \frac{2\pi \sqrt{k_x k_y} h}{\mu B \left(\ln \frac{r_e}{r_w} + S \right)} = \frac{2\pi \sqrt{(10)(15)}(1.127 \times 10^{-3})(50)}{(1.05)(1)\left(\ln \frac{34.17}{0.25} + 0.8 \right)} = 0.7223 \text{ STB/D-psi,} \quad \text{.....................} (8.575)$$

$$C_7 = -(N_7 + S_7 + W_7 + \Omega_7) = -(0.7156 + 0.7156 + 0.6037 + 0.722) = -3.3609 \text{ STB/D-psi,} \quad \text{.............} (8.576)$$

$$Q_7 = -\Omega p_{sf} = -3180.6 \text{ STB/D.} \quad \text{..........................} (8.577)$$

Problem 8.4.44

Consider the steady-state flow of a single-phase incompressible liquid in the reservoir shown in **Fig. 8.37.** Along the external boundaries of the system, pressure is specified as 3,500 psia, and the sandface pressure at the well is measured as 1,500 psia. Other reservoir properties are as listed in **Table 8.37.**

1. Calculate the well flow rate. Implement the Gauss-Seidel method to solve the system of equations using a pressure tolerance of 1 psi. Use 3,000 psia as your initial guess.
2. If the well were located at Node 11, would its production rate have been higher if the sandface pressure was specified as 1,500 psia?
3. If the southern boundary were a no-flow boundary, would the calculated pressure gradient between Nodes 5 and 6 have remained the same, as in the case of a constant pressure boundary of 3,500 psi?

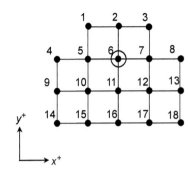

Fig. 8.37—Reservoir system of Problem 8.4.44.

$k_x = 20$ md	
$k_y = 10$ md	
$\Delta x = \Delta y = 500$ ft	
$h = 20$ ft	
$\mu = 1.3$ cp	
$B = 1$ res bbl/STB	
$r_w = 0.25$ ft	
$s = -0.5$	

Table 8.37—Reservoir properties for Problem 8.4.44.

Solution to Problem 8.4.44

1. The governing equation for a 2D incompressible system is

$$\frac{\partial}{\partial x}\left(\frac{A_x k_x}{\mu B}\frac{\partial p}{\partial x}\right)\Delta x + \frac{\partial}{\partial y}\left(\frac{A_y k_y}{\mu B}\frac{\partial p}{\partial y}\right)\Delta y + q = 0. \dots\dots\dots (8.578)$$

Because the reservoir is homogeneous and anisotropic in permeability distribution, the reduced form will be

$$k_x\frac{\partial^2 p}{\partial x^2} + k_y\frac{\partial^2 p}{\partial y^2} + \frac{q\mu B}{V_b} = 0. \dots\dots\dots (8.579)$$

Observing that $k_x = 2k_y$,

$$2\frac{\partial^2 p}{\partial x^2} + \frac{\partial^2 p}{\partial y^2} + \frac{q\mu B}{k_y V_b} = 0. \dots\dots\dots (8.580)$$

The finite-difference approximation of Eq. 8.580 can be written as

$$2\left[\frac{p_{i-1,j}+p_{i+1,j}-2p_{i,j}}{(\Delta x)^2}\right] + \left[\frac{p_{i,j-1}+p_{i,j+1}-2p_{i,j}}{(\Delta y)^2}\right] + \frac{q\mu B}{V_b k} = 0. \dots\dots\dots (8.581)$$

Because $\Delta x = \Delta y$, Eq. 8.581 can be simplified to

$$2\left(p_{i-1,j}+p_{i+1,j}-2p_{i,j}\right)+\left(p_{i,j-1}+p_{i,j+1}-2p_{i,j}\right)+\frac{q\mu B}{hk_y} = 0, \dots\dots\dots (8.582)$$

$$2p_{i-1,j}+2p_{i+1,j}+p_{i,j-1}+p_{i,j+1}-6p_{i,j}+\frac{q\mu B}{hk_y} = 0. \dots\dots\dots (8.583)$$

Employing the Peaceman's wellbore model to calculate the flow rate of the well located at Node 6,

$$q_6 = -\Omega_6\left(p_6 - p_{sf}\right), \dots\dots\dots (8.584)$$

and

$$\Omega_6 = \frac{2\pi \bar{k}h}{\mu B\left[\ln\left(\dfrac{r_e}{r_w}\right)+s\right]} = \frac{2\pi(14.14)(1.127\times10^{-3})(20)}{(1)(1.3)\left[\left(\ln\dfrac{100.45}{0.25}\right)-0.5\right]} = 0.28 \text{ STB/D-psi}, \dots\dots\dots (8.585)$$

where

$$r_e = \frac{0.28\left[\left(\dfrac{k_y}{k_x}\right)^{\frac{1}{2}}(\Delta x)^2 + \left(\dfrac{k_x}{k_y}\right)^{\frac{1}{2}}(\Delta y)^2\right]^{\frac{1}{2}}}{\left(\dfrac{k_y}{k_x}\right)^{\frac{1}{4}}+\left(\dfrac{k_x}{k_y}\right)^{\frac{1}{4}}} = 100.45 \text{ ft}, \dots\dots\dots (8.586)$$

$$\overline{k} = \sqrt{k_x k_y} = 14.14 \text{ md.} \quad \text{...} \quad (8.587)$$

- Block 6:

$$2p_5 + 2p_7 + p_2 + p_{11} - 6p_6 - \frac{\Omega \mu B}{k_y h}\left(p_6 - p_{sf}\right) = 0, \quad \text{.....................} \quad (8.588)$$

which reduces to

$$-7.6161 p_6 + p_{11} = -1,9924.15. \quad \text{........................} \quad (8.589)$$

- Block 10:

$$2p_9 + 2p_{11} + p_{15} + p_5 - 6p_{10} = 0, \quad \text{........................} \quad (8.590)$$

which reduces to

$$-6p_{10} + 2p_{11} = -14,000. \quad \text{...........................} \quad (8.591)$$

- Block 11:

$$2p_{10} + 2p_{12} + p_6 + p_{16} - 6p_{11} = 0 \text{ and } p_{10} = p_{12}, \quad \text{...................} \quad (8.592)$$

which again reduces to

$$p_6 + 4p_{10} - 6p_{11} = -3,500. \quad \text{.........................} \quad (8.593)$$

Solving for p_6, p_{10}, and p_{11} using the Gauss-Seidel method results at each iteration level are displayed in **Table 8.38.**

Iteration Level	p_6	p_{10}	p_{11}
0	3,000.000	3,000.000	3,000.000
1	3,009.959	3,333.333	3,307.215
2	3,050.297	3,435.738	3,382.208
3	3,060.143	3,460.736	3,400.515
4	3,062.547	3,466.838	3,404.983
5	3,063.134	3,468.328	3,406.074
6	3,063.277	3,468.691	3,406.34

Table 8.38—Calculated gridblock pressure values at the end of each iteration level of the Gauss-Seidel method used in the Solution to Problem 8.4.44.

The total flow from Nodes 2, 5, 7, and 11 to Node 6 should be equal to the production rate from the well:

$$T_x = \frac{k_x h}{\mu B} = 0.347 \text{ STB/D-psi,} \quad \text{.......................} \quad (8.594)$$

$$T_y = \frac{k_y h}{\mu B} = 0.173 \text{ STB/D-psi,} \quad \text{.......................} \quad (8.595)$$

$$\sum_{i=1}^{node} q_i = 2 \times 0.347(p_5 - p_6) + 0.173(p_2 + p_{11} - 2p_6) = 302.88 + 135.21 = 438.1 \text{ STB/D,} \quad \text{......} \quad (8.596)$$

$$q_6 = -\Omega(p_6 - p_{sf}) = -0.28(3,063.277 - 1,500) = 437.72 \text{ STB/D,} \quad \text{................} \quad (8.597)$$

$$\text{MBC} = \frac{\sum_{i=1}^{node} q_i}{q_6} = 1.0008. \quad \text{.......................} \quad (8.598)$$

2. If the well was located in Block 11,

- Block 6:

$$2p_5 + 2p_7 + p_2 + p_{11} - 6p_6 = 0, \quad \text{.......................} \quad (8.599)$$

which reduces to

$$-6p_6 + p_{11} = -17,500. \quad \dots\dots\dots\dots\dots\dots\dots\dots\dots\dots\dots\dots\dots\dots\dots\dots\dots\dots\dots(8.600)$$

- Block 10:

$$2p_9 + 2p_{11} + p_{15} + p_5 - 6p_{10} = 0, \quad \dots\dots\dots\dots\dots\dots\dots\dots\dots\dots\dots\dots\dots\dots\dots(8.601)$$

which reduces to

$$-6p_{10} + 2p_{11} = -14,000. \quad \dots\dots\dots\dots\dots\dots\dots\dots\dots\dots\dots\dots\dots\dots\dots\dots\dots(8.602)$$

- Block 11:

$$2p_{10} + 2p_{12} + p_6 + p_{16} - 6p_{11} - \frac{\Omega\mu B}{k_y h}\left(p_{11} - p_{sf}\right) = 0, \quad \dots\dots\dots\dots\dots\dots\dots\dots\dots(8.603)$$

and $p_{10} = p_{12}$, which again reduces to

$$p_6 + 4p_{10} - 7.6161p_{11} = -5,924.15. \quad \dots\dots\dots\dots\dots\dots\dots\dots\dots\dots\dots\dots\dots(8.604)$$

Solving simultaneously, $p_6 = 3,411.9$ psi, $p_{10} = 3,323.8$ psi, and $p_{11} = 2,971.5$ psi,

$$q_{11} = -\Omega(p_{11} - p_{sf}) = -0.28(2,971.5 - 1,500) = 412.2 \text{ STB/D}. \quad \dots\dots\dots\dots\dots\dots(8.605)$$

3. If the southern boundary is sealed,
   - Block 6:

$$2p_5 + 2p_7 + p_2 + p_{11} - 6p_6 - \frac{\Omega\mu B}{k_y h}(p_6 - p_{sf}) = 0, \quad \dots\dots\dots\dots\dots\dots\dots\dots\dots(8.606)$$

which reduces to

$$-7.6161p_6 + p_{11} = -19,924.15. \quad \dots\dots\dots\dots\dots\dots\dots\dots\dots\dots\dots\dots\dots\dots\dots(8.607)$$

- Block 10:

$$2p_9 + 2p_{11} + p_{10} + p_5 - 6p_{10} = 0, \quad \dots\dots\dots\dots\dots\dots\dots\dots\dots\dots\dots\dots\dots\dots(8.608)$$

which reduces to

$$-5p_{10} + 2p_{11} = -14,000. \quad \dots\dots\dots\dots\dots\dots\dots\dots\dots\dots\dots\dots\dots\dots\dots\dots(8.609)$$

- Block 11:

$$2p_{10} + 2p_{12} + p_6 + p_{11} - 6p_{11} = 0 \text{ and } p_{10} = p_{12}, \quad \dots\dots\dots\dots\dots\dots\dots\dots\dots\dots(8.610)$$

which again reduces to

$$p_6 + 4p_{10} - 5p_{11} = 0, \quad \dots\dots\dots\dots\dots\dots\dots\dots\dots\dots\dots\dots\dots\dots\dots\dots\dots(8.611)$$

$p_6 = 3,058.6$ psi, $p_{10} = 3,448.1$ psi, and $p_{11} = 3,370.2$ psi.

The pressure gradient between Nodes 5 and 6 is

$$\left.\frac{\partial p}{\partial x}\right|_{5-6} = \frac{3,500 - 3,058.6}{500} = 0.88 \text{ psi/ft.} \quad \dots\dots\dots\dots\dots\dots\dots\dots\dots\dots\dots(8.612)$$

For the case with a constant-pressure southern boundary,

$$\left.\frac{\partial p}{\partial x}\right|_{5-6} = \frac{3,500 - 3,063.3}{500} = 0.873 \text{ psi/ft.} \quad \dots\dots\dots\dots\dots\dots\dots\dots\dots\dots(8.613)$$

Problem 8.4.45

Consider the flow of a single-phase, incompressible fluid in a reservoir represented by the body-centered grid system shown in **Fig. 8.38.1,** with the boundary conditions of the reservoir as indicated. The well in Block 2 produces at a constant rate of 400 STB/D, and the sandface pressure of the well is measured to be 2,200 psi. A second well located in Block 4 is produced at a constant rate of 500 STB/D. Use the reservoir properties listed in **Table 8.39.**

1. Determine (qualitatively and quantitatively) the boundary condition across the eastern boundary of Block 4.
2. Calculate the sandface pressure of the well located in Block 4.
3. If the system is described as shown in **Fig. 8.38.2,** would the solution be the same?

k_x	= 30 md
k_y	= 20 md
$\Delta x = \Delta y$	= 600 ft
h	= 50 ft
μ	= 1 cp
B	= 1 res bbl/STB
r_w	= 0.25 ft
s	= +2.33

Table 8.39—Reservoir properties for Problem 8.4.45.

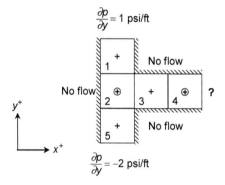

Fig. 8.38.1—Reservoir system of Problem 8.4.45.

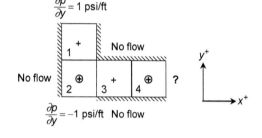

Fig. 8.38.2—Reservoir system of Problem 8.4.45, Part 3.

Solution to Problem 8.4.45

1. In this case, the boundary condition on the eastern boundary of Block 4 can be determined by mass conservation. Mass flux from the specified external boundaries:

- Across Block 1 into the reservoir:

$$q_1 = \frac{k_y A_y}{\mu B}\frac{\partial p}{\partial y} = \frac{(1.127 \times 10^{-3})(50)(600)(30)}{(1)(1.5)}[1] = 450.84 \text{ STB/D.} \dots\dots\dots\dots\dots\dots\dots (8.614)$$

- Across Block 5 into the reservoir:

$$q_5 = \frac{k_y A_y}{\mu B}\frac{\partial p}{\partial y} = \frac{(1.127 \times 10^{-3})(50)(600)(30)}{(1)(1.5)}[2] = 901.68 \text{ STB/D.} \dots\dots\dots\dots\dots\dots\dots (8.615)$$

The wells in Blocks 2 and 4 produce at rates of $q_{\text{well}_2} = -400$ STB/D and $q_{\text{well}_4} = -500$ STB/D, respectively. For an incompressible fluid system, the flow into the reservoir has to be equal to what exists in the reservoir; therefore,

$$q_1 + q_5 + q_{w2} + q_{w4} + q_{\text{unknown}} = 901.64 + 450.84 - 400 - 500 + q_{\text{unknown}} = 0, \dots\dots\dots\dots\dots\dots (8.616)$$

where q_{unknown} is the fluid flow across the unknown boundary, which can be calculated as

$$q_{\text{unknown}} = -(901.64 + 450.84 - 400 - 500) = -452.52 \text{ STB/D.} \dots\dots\dots\dots\dots\dots\dots\dots (8.617)$$

The negative sign indicates that 452.52 STB/D are exiting the reservoir across the eastern boundary of Block 4. Finally, the pressure gradient across the boundary is calculated:

$$q_{\text{unknown}} = \frac{k_x A_x}{\mu B}\frac{\partial p}{\partial x}, \dots\dots\dots\dots\dots\dots\dots\dots\dots\dots\dots\dots\dots\dots\dots\dots\dots\dots\dots\dots\dots\dots (8.618)$$

$$\frac{\partial p}{\partial x} = -0.67 \text{ psi/ft.} \dots\dots\dots\dots\dots\dots\dots\dots\dots\dots\dots\dots\dots\dots\dots\dots\dots\dots\dots\dots\dots (8.619)$$

The calculated pressure gradient of –0.67 psi/ft indicates that 452.52 STB/D will flow out of the reservoir across the eastern boundary of Block 4.

2. Knowing the sandface pressure of Block 2, $p_{sf2} = 2,200$ psi, the block pressure can be calculated using the wellbore equation:

$$q_2 = -\Omega_2(p_2 - p_{sf}) \dotfill (8.620)$$

and

$$\Omega_2 = \Omega_4 = \frac{2\pi \bar{k}h}{\mu B\left[\ln\left(\dfrac{r_e}{r_w}\right)+s\right]} = \frac{2\pi(24.5)(1.127\times10^{-3})(50)}{(1)(1)\left[\left(\ln\dfrac{100.45}{0.25}\right)+2.33\right]} = 1.0205 \text{ STB/D-psi}, \dotfill (8.621)$$

where

$$r_e = \frac{0.28\left[\left(\dfrac{k_y}{k_x}\right)^{\frac{1}{2}}(\Delta x)^2 + \left(\dfrac{k_x}{k_y}\right)^{\frac{1}{2}}(\Delta y)^2\right]^{\frac{1}{2}}}{\left(\dfrac{k_y}{k_x}\right)^{\frac{1}{4}} + \left(\dfrac{k_x}{k_y}\right)^{\frac{1}{4}}} = 119.40 \text{ ft}, \dotfill (8.622)$$

$$\bar{k} = \sqrt{k_x k_y} = 24.5 \text{ md}.$$

Therefore, $p_2 = p_{sf} - \dfrac{q_2}{\Omega_2} = 2,592.16$ psi. $\dotfill (8.623)$

- Block 1: $p_1 = \left[p_2 + (p_1 + c_1\Delta x)\right]/2 \rightarrow p_1 = 3,12.16$ psi. $\dotfill (8.624)$
- Block 5: $p_5 = \left[p_2 + (p_5 - c_3\Delta y)\right]/2 \rightarrow p_5 = 3,792.16$ psi. $\dotfill (8.625)$
- Block 2: $0.7514(p_1 - p_2) - 0.7514(p_2 - p_5) + 1.1271(p_3 - p_2) - 400 = 0 \rightarrow p_3 = 1,747.05$ psi. $\dots (8.626)$
- Block 3: $(p_4 - p_3) - (p_3 - p_2) = 0 \rightarrow p_4 = 901.95$ psi. $\dotfill (8.627)$

Thus, the sandface pressure of the well in Block 4 can be calculated employing the wellbore model:

$$q_4 = -\Omega_4(p_4 - p_{sf}), \dotfill (8.628)$$

which yields

$$p_{sf_4} = p_4 + \frac{q_4}{\Omega_4} = 901.95 + \frac{(-500)}{1.0205} = 412.0 \text{ psi}. \dotfill (8.629)$$

3. It would not be the same because the reflection node in Fig. 8.38.2 for Well 2 would have a different pressure from that solved in Fig. 8.38.1 for Block 5.

Problem 8.4.46

Consider the flow of a single-phase, 2D compressible fluid in a homogeneous and isotropic system with a sealed external boundary shown in **Fig. 8.39.1.**

1. Can you solve the problem as a 1D system? Explain.
2. Assume that you are using the GNR protocol to solve this problem. Give a generic expression that can be used in defining $\partial R_2 / \partial p_2$. Would the generic expression you suggested have been different if there were no wells in Block 2? Explain.

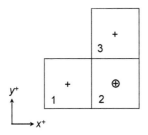

Fig. 8.39.1—Reservoir system of Problem 8.4.46.

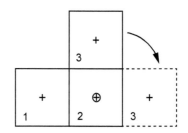

Fig. 8.39.2—1D representation of the reservoir system in Problem 8.4.46.

Solution to Problem 8.4.46

1. The reservoir system can be represented using the 1D grid (because it is homogeneous and isotropic) displayed in **Fig. 8.39.2.**

2.

$$R = A_x \left. \frac{k_x}{\Delta x} \right|_{i+\frac{1}{2}} \frac{\frac{1}{2}\left(p_i^{n+1} + p_{i+1}^{n+1}\right)}{\frac{1}{2}\left[\mu_g\left(p_i^{n+1}\right) + \mu_g\left(p_{i+1}^{n+1}\right)\right] \times \frac{1}{2}\left[\mu_g\left(p_i^{n+1}\right) + \mu_g\left(p_{i+1}^{n+1}\right)\right]} \left(p_{i+1}^{n+1} - p_i^{n+1}\right)$$

$$- A_x \left. \frac{k_x}{\Delta x} \right|_{i-\frac{1}{2}} \frac{\frac{1}{2}\left(p_i^{n+1} + p_{i-1}^{n+1}\right)}{\frac{1}{2}\left[\mu_g\left(p_i^{n+1}\right) + \mu_g\left(p_{i-1}^{n+1}\right)\right] \times \frac{1}{2}\left[\mu_g\left(p_i^{n+1}\right) + \mu_g\left(p_{i-1}^{n+1}\right)\right]} \left(p_i^{n+1} - p_{i-1}^{n+1}\right) + q_g \quad \dots\dots\dots\dots (8.630)$$

$$- \left(\frac{\phi V_b T_{sc}}{p_{sc} T}\right) \frac{1}{\Delta t}\left[\frac{p_i^{n+1}}{z_g\left(p_i^{n+1}\right)} - \frac{p_i^n}{z_g\left(p_i^n\right)}\right].$$

Using forward-difference approximation for the derivative,

$$\frac{\partial R_2}{\partial p_2} = \frac{R_2\left(p_1, p_2 + \varepsilon_p, p_3\right) - R_2\left(p_1, p_2, p_3\right)}{\varepsilon_p}, \quad \dots\dots\dots\dots\dots\dots\dots\dots\dots\dots\dots\dots\dots\dots\dots\dots\dots (8.631)$$

where ε_p is a small pressure used to perturb the residual equation to numerically calculate the partial derivatives (entries of the Jacobian matrix).

Problem 8.4.47

Consider the steady-state flow of a single-phase incompressible liquid in a horizontal reservoir with heterogeneous and aniso-tropic permeability distribution shown in **Fig. 8.40.** The reservoir is represented by a body-centered grid system. Boundary conditions of the reservoir are no flow, except with an aquifer charging from the west side of Block 1.

The strong aquifer charge maintains the pressure in Block 1 at 3,500 psia at all times. The well in Block 3 has been drilled, but has never been produced. On the other hand, the well at Node 4 was put on production upon its completion. A pressure survey conducted in the well located in Block 3 gives a pressure reading of 2,258 psia, while the well in Block 4 is producing.

Using the data provided and the reservoir properties in **Table 8.40,** determine the flow rate of the well and the pressure distribution in the system.

Fig. 8.40—Reservoir system of Problem 8.4.47.

	Table 8.40—Reservoir properties for Problem 8.4.47.
$\Delta x = \Delta y$	= 1,000 ft
h	= 100 ft
μ	= 2.0 cp
B	= 1 res bbl/STB
r_w	= 0.25 ft
s	= 0

Solution to Problem 8.4.47

For a 2D incompressible flow system,

$$\frac{\partial}{\partial x}\left(\frac{A_x k_x}{\mu B}\frac{\partial p}{\partial x}\right)\Delta x + \frac{\partial}{\partial y}\left(\frac{A_y k_y}{\mu B}\frac{\partial p}{\partial y}\right)\Delta y + q = 0, \quad\dotfill\text{(8.632)}$$

$$k_x \frac{\partial^2 p}{\partial x^2} + k_y \frac{\partial^2 p}{\partial y^2} + \frac{q\mu B}{V_b} = 0. \quad\dotfill\text{(8.633)}$$

Because the block pressures in Blocks 1 and 3 are known to be 3,500 and 2,250 psi, respectively, we can determine the pressure of Block 2:

$$k_{x1\text{to}2}\left(p_1 - p_2\right) + k_{y2\text{to}3}\left(p_3 - p_2\right) = 0, \quad\dotfill\text{(8.634)}$$

$$k_{x1\text{to}2} = \frac{2(100)(200)}{100+200} = 133.33 \text{ md}, \quad\dotfill\text{(8.635)}$$

$$k_{y2\text{to}3} = \frac{2(40)(10)}{40+10} = 16 \text{ md}, \quad\dotfill\text{(8.636)}$$

$$p_2 = \frac{133.33\ p_1 + 16 p_3}{133.33+16} = 3,366.93 \text{ psi}. \quad\dotfill\text{(8.637)}$$

For an incompressible system, the fluid flows from Blocks 1 to 2 and from Blocks 2 to 3 to be produced from Block 4. Fluid from 1 to 2,

$$q_{1\text{to}2} = \frac{k_x A_{x1\text{to}2}}{\mu B}\frac{\partial p}{\partial x} = \frac{(1.127\times10^{-3})(133.33)(1,000)(100)}{(1)(2)}\left[\frac{3,500 - 3,366.93}{\Delta x}\right] = 1,000 \text{ STB/D}. \quad\dotfill\text{(8.638)}$$

Therefore, the production rate from the well has to be 1,000 STB/D.

Calculating fluid flow from Block 3 to 4 as an additional check, we see that

$$k_{x3\text{to}4} = \frac{2(100)(150)}{100+150} = 120 \text{ md}, \quad\dotfill\text{(8.639)}$$

$$q_{3\text{to}4} = \frac{k_x A_{x3\text{to}4}}{\mu B}\frac{\partial p}{\partial x} = \frac{(1.127\times10^{-3})(120)(1,000)(100)}{(1)(2)}\left[\frac{2,258 - p_4}{\Delta x}\right] = 1,000 \text{ STB/D}. \quad\dotfill\text{(8.640)}$$

Finally, solving for p_4 yields $p_4 = 2,110.15$ psi.

Problem 8.4.48

Consider the single-phase incompressible flow in the reservoir shown in **Fig. 8.41**. There is a well completed only in Block 2 that produced at a constant-sandface pressure of 1,500 psia.

1. Calculate the SIP coefficients $B_{i,j,k}$, $S_{i,j,k}$, $W_{i,j,k}$, $C_{i,j,k}$, $E_{i,j,k}$, $N_{i,j,k}$, $A_{i,j,k}$, and $Q_{i,j,k}$ for Blocks 1 and 2.
2. Generate the coefficient matrix for this reservoir using the SIP notation (no numerical entries are requested) and the reservoir rock and fluid properties found in **Table 8.41**.

$\mu = 1.1$ cp

$B = 1.0$ res bbl/STB

$\rho = 62.4$ lbm/ft³

$\Delta x = \Delta y = 100$ ft (uniform)

$\Delta z = 80$ ft (uniform)

$r_w = 0.25$ ft

$s = 0$

Table 8.41—Reservoir rock and fluid properties for Problem 8.4.48.

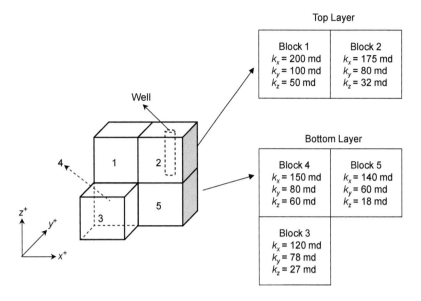

Fig. 8.41—Reservoir system of Problem 8.4.48.

Solution to Problem 8.4.48

1. For Block 1:

$$B_1 = \frac{\Delta x \Delta y}{\mu B \Delta z} \frac{2 k_{z1} k_{z4}}{k_{z1} + k_{z4}} = \frac{(100)(100)}{(1.1)(1)(80)} \frac{2(50)(60)}{(50+60)}(1.127 \times 10^{-3}) = 6.99 \text{ STB/D-psia},\ \ \ \dots\dots\dots\dots\dots (8.641)$$

$$E_1 = \frac{\Delta z}{\mu B} \frac{2 k_{x1} k_{x2}}{k_{x1} + k_{x2}} = \frac{80}{(1.1)(1)} \frac{2(200)(175)}{200+175}(1.127 \times 10^{-3}) = 15.30 \text{ STB/D-psia},\ \ \ \dots\dots\dots\dots (8.642)$$

$$C_1 = -(B_1 + E_1) = -(6.9855 + 15.30) = -22.29 \text{ STB/D-psia},\ \ \ \dots\dots\dots\dots\dots\dots\dots (8.643)$$

$$Q_1 = -\frac{\rho}{144} B_1 (G_1 - G_4) = -\frac{62.4}{144}(6.99)(-80) = 242.16 \text{ STB/D},\ \ \ \dots\dots\dots\dots\dots\dots\dots (8.644)$$

$$W_1 = N_1 = S_1 = A_1 = 0 \text{ STB/D-psia}.\ \ \ \dots\dots\dots\dots\dots\dots\dots\dots\dots\dots\dots\dots\dots\dots (8.645)$$

For Block 2:

$$B_2 = \frac{\Delta x \Delta y}{\mu B \Delta z} \frac{2 k_{z2} k_{z5}}{k_{z2} + k_{z5}} = \frac{(100)(100)}{(1.1)(1)(80)} \frac{2(32)(18)}{(32+18)}(1.127 \times 10^{-3}) = 2.95 \text{ STB/D-psia},\ \ \ \dots\dots\dots\dots (8.646)$$

$$W_2 = E_1 = 15.30 \text{ STB/D-psia},\ \ \ \dots\dots\dots\dots\dots\dots\dots\dots\dots\dots\dots\dots\dots\dots\dots\dots (8.647)$$

$$r_e = 0.28 \Delta x \frac{\left(\dfrac{k_x}{k_y}+1\right)^{\frac{1}{2}}}{\left(\dfrac{k_x}{k_y}\right)^{\frac{1}{2}}+1} = 0.28(100)\frac{\left(\dfrac{175}{80}+1\right)^{\frac{1}{2}}}{\left(\dfrac{175}{80}\right)^{\frac{1}{2}}+1} = 20.17 \text{ ft},\ \ \ \dots\dots\dots\dots\dots\dots\dots\dots\dots (8.648)$$

$$\Omega_2 = \frac{2\pi \sqrt{k_x k_y} \Delta z}{\mu B \left[\ln\left(\dfrac{r_e}{r_w}\right) + s\right]} = \frac{2(3.14)\sqrt{(175)(80)}(0.001127)(80)}{(1.1)(1)\left[\ln\left(\dfrac{20.1652}{0.25}\right) + 0\right]} = 13.87 \text{ STB/D-psia},\ \ \ \dots\dots\dots\dots (8.649)$$

$$C_2 = -(B_2 + W_2 + \Omega_2) = -(2.9507 + 15.30 + 13.8725) = -32.12 \text{ STB/D-psia},\ \ \ \dots\dots\dots\dots (8.650)$$

$$Q_2 = -\Omega_2 p_{sf} - \frac{\rho}{144} B_1 (G_2 - G_s) = -(13.87)(1,500) - \frac{62.4}{144}(6.99)(-80) \quad \dots\dots\dots\dots\dots\dots (8.651)$$

$$= -20,605.736 \text{ STB/D},$$

$$E_2 = N_2 = S_2 = A_2 = 0 \text{ STB/D-psia.} \quad \dots\dots\dots\dots\dots\dots\dots\dots\dots\dots (8.652)$$

2. The resulting coefficient matrix for this reservoir with five blocks will be

$$[A] = \begin{bmatrix} C_1 & E_1 & 0 & B_1 & 0 \\ W_2 & C_2 & 0 & 0 & B_2 \\ 0 & 0 & C_3 & N_3 & 0 \\ A_4 & 0 & S_4 & C_4 & E_4 \\ 0 & A_5 & 0 & W_5 & C_5 \end{bmatrix}. \quad \dots\dots\dots\dots\dots\dots\dots\dots (8.653)$$

Problem 8.4.49

Consider the single-phase incompressible fluid flow that is taking place in the 2D homogeneous and isotropic porous media shown in **Fig. 8.42**. The reservoir is surrounded by no-flow boundaries, with the exception of one side of Block 1 and the right-hand side of Block 2, which are connected to constant-pressure gradient boundaries. There is one well located in Block 2 with a constant sandface-pressure specification of 3,000 psia. Note that there is no depth gradient in the reservoir. Other reservoir properties are listed in **Table 8.42.**

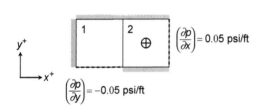

| $k_x = k_y =$ 100 md |
| $\Delta x = \Delta y$ = 1,000 ft |
| h = 50 ft |
| μ = 1 cp |
| B = 1 res bbl/STB |
| r_w = 0.3 ft |
| s = 0 |

Fig. 8.42—Reservoir system of Problem 8.4.49.   Table 8.42—Reservoir properties for Problem 8.4.49.

1. Write the differential equation governing the fluid flow dynamics in this reservoir.
2. Obtain the finite-difference approximation of the governing flow equation.
3. Calculate the pressure of each block using the PSOR iterative procedure with ω_{opt}. Begin the PSOR iterations with a uniform initial guess of 3,000 psia. Use a pressure tolerance of 1 psi for the convergence check. For spectral-radius calculations, use a tolerance of 0.01.
4. Calculate the flow rate from the well in Block 2 and describe the type of well (injector or producer).
5. Perform an MBC to verify your results.

Solution to Problem 8.4.49

1. The differential equation representing the fluid flow dynamics in this reservoir for a homogeneous and isotropic medium is

$$\frac{\partial^2 p}{\partial x^2} + \frac{q\mu B}{kV_b} = 0. \quad \dots\dots\dots\dots\dots\dots\dots\dots\dots\dots\dots (8.654)$$

2. The finite-difference approximation of the governing differential equation is

$$p_{i-1} - 2p_i + p_{i+1} + \left(\frac{q\mu B}{hk}\right)_i = 0. \quad \dots\dots\dots\dots\dots\dots\dots\dots (8.655)$$

3.
- Block 1:

$$p_1' - 2p_1 + p_2 = 0. \quad\dots\dots\dots\dots\dots\dots\dots\dots\dots\dots\dots\dots\dots\dots\dots\dots (8.656)$$

Applying the boundary condition,

$$\frac{p_1 - p_1'}{\Delta y} = -0.05 \quad\dots\dots\dots\dots\dots\dots\dots\dots\dots\dots\dots\dots\dots\dots\dots\dots (8.657)$$

and

$$p_1' = p_1 + 0.05\Delta y. \quad\dots\dots\dots\dots\dots\dots\dots\dots\dots\dots\dots\dots\dots\dots\dots\dots (8.658)$$

Therefore,

$$p_1 + 0.05\Delta y - 2p_1 + p_2 = 0, \quad\dots\dots\dots\dots\dots\dots\dots\dots\dots\dots\dots\dots\dots (8.659)$$

$$-p_1 + p_2 = -50. \quad\dots\dots\dots\dots\dots\dots\dots\dots\dots\dots\dots\dots\dots\dots\dots\dots (8.660)$$

- Block 2:

$$p_1 - 2p_2 + p_2' + q_2 = 0. \quad\dots\dots\dots\dots\dots\dots\dots\dots\dots\dots\dots\dots\dots\dots\dots (8.661)$$

Substituting the flow-rate term, employing the wellbore model,

$$q_2 = -\Omega_2(p_2 - p_{sf2}) \quad\dots\dots\dots\dots\dots\dots\dots\dots\dots\dots\dots\dots\dots\dots\dots (8.662)$$

and

$$\Omega_2 = \frac{2\pi \bar{k} h}{\mu B \left[\ln\left(\frac{r_e}{r_w}\right) + s \right]} = \frac{2\pi(100)(1.127 \times 10^{-3})(50)}{(1)(1)\left[\left(\ln\frac{200}{0.3}\right)\right]} = 5.45 \;\; \text{STB/D-psi}, \quad\dots\dots\dots\dots (8.663)$$

where $r_e = 0.2\Delta x = 200$ ft.
   Also, applying the boundary condition,

$$\frac{p_2' - p_2}{\Delta x} = 0.05 \quad\dots\dots\dots\dots\dots\dots\dots\dots\dots\dots\dots\dots\dots\dots\dots\dots (8.664)$$

and

$$p_2' = p_2 + 0.05\Delta x. \quad\dots\dots\dots\dots\dots\dots\dots\dots\dots\dots\dots\dots\dots\dots\dots\dots (8.665)$$

Substituting into the original finite-difference equation,

$$p_1 + 0.05\Delta x - p_2 - \frac{\mu B}{hk}\Omega_2(p_2 - p_{sf2}) = 0, \quad\dots\dots\dots\dots\dots\dots\dots\dots\dots\dots (8.666)$$

which reduces to

$$p_1 - 1.9678 p_2 = -2,953.4. \quad\dots\dots\dots\dots\dots\dots\dots\dots\dots\dots\dots\dots\dots\dots (8.667)$$

Implementing the PSOR iterative scheme,

$$p_1^{(k+1)} = p_1^{(k)} + \frac{\omega_{opt}}{-1}(-50 + p_1^{(k)} - p_2^{(k)}), \quad\dots\dots\dots\dots\dots\dots\dots\dots\dots\dots (8.668)$$

$$p_2^{(k+1)} = p_2^{(k)} + \frac{\omega_{opt}}{-1.9678}(-2,953.4 - p_1^{(k+1)} + 1.9678 p_2^{(k)}), \quad\dots\dots\dots\dots\dots\dots (8.669)$$

$$\omega_{opt} = \frac{2}{1 + \sqrt{1 - \rho_g}}. \quad\dots\dots\dots\dots\dots\dots\dots\dots\dots\dots\dots\dots\dots\dots\dots (8.670)$$

The iteration results are shown in **Table 8.43**. The solution is $p_1 = 3{,}153.24$ psi and $p_2 = 3{,}103.3$ psi.

k	p_1	p_2	p_S	Δp_{max}	ω_{opt}
0	3,000	3,000	–	–	–
1	3,050.00	3,050.82	0.00	50.82	1.00
2	3,100.82	3,076.64	0.00	50.82	1.00
3	3,126.64	3,089.77	0.508	25.82	1.00
4	3,139.77	3,096.44	0.508	13.12	1.1756
5	3,147.61	3,101.12	–	7.84	1.1756
6	3,151.74	3,102.76	–	4.13	1.1756
7	3,152.94	3,103.20	–	1.21	1.1756
8	3,153.24	3,103.30	–	0.30	1.1756

Table 8.43—Iteration results from the Solution to Problem 8.4.49.

4. Because the calculated p_2 is larger than the specified p_{sf}, which is 300 psia, the well is a producer; the production rate can be determined using Peaceman's wellbore model:

$$q_2 = -\frac{2\pi kh\left(p_2 - p_{sf}\right)}{\mu B\left[\ln\left(\dfrac{r_e}{r_w}\right) + s\right]} = -565 \text{ STB/D.} \quad\dotfill\quad (8.671)$$

5. MBC: Volumetric flow rate across the open boundaries,

$$q = \left(\frac{\partial p}{\partial x} + \frac{\partial p}{\partial y}\right)\left(\frac{kA}{\mu B}\right) = (0.05 + 0.05)\frac{(1,000)(50)(100)(1.127 \times 10^{-3})}{(1)(1)} = 563.5 \text{ STB/D.} \quad\dotfill\quad (8.672)$$

$$\text{MBC check: } \left|\frac{q_{in}}{q_{well}}\right| = \left|\frac{565}{563.5}\right| = 1.003. \quad\dotfill\quad (8.673)$$

Problem 8.4.50

Consider the 2D body-centered grid system without depth gradients shown in **Fig. 8.43.** The aquifer is always at constant pressure. All other boundaries are sealed.

1. Calculate the SIP coefficients: W_2, N_8, E_3, S_1, C_6, Q_5.
2. Show the complete SIP representation of the finite-difference equation when it is written at the production wellblock.

Use the rock and fluid properties in **Table 8.44** for the calculation.

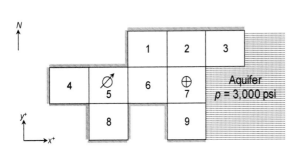

Fig. 8.43—Reservoir system of Problem 8.4.50.

$\Delta x = \Delta y$	= 200 ft
μ	= 1.01 cp
B	= 1 res bbl/STB
h	= 40 ft
r_w	= 0.25 ft
s	= –1.3
k_x	= 30 md
k_y	= 50 md
$p_{sf, injector}$	= 4,000 psia
$p_{sf, producer}$	= 1,500 psia
$p_{aquifer}$	= 3,000 psia
$\Omega_{injector} = \Omega_{producer}$	= 2.6262 STB/D-psi

Table 8.44—Rock and fluid properties for Problem 8.4.50.

Solution to Problem 8.4.50

1.

$$W_2 = \frac{A_x k_x}{\mu B \Delta x} = \frac{h k_x}{\mu B} = 40 \times 30 \times \frac{0.001127}{1.01} = 1.339 \text{ STB/D-psi,} \quad\quad\quad\quad\quad (8.674)$$

$$N_8 = \frac{A_y k_y}{\mu B \Delta x} = \frac{h k_y}{\mu B} = 40 \times 50 \times \frac{0.001127}{1.01} = 2.2317 \text{ STB/D-psi.} \quad\quad\quad\quad (8.675)$$

For a homogeneous system,

$$E_3 = W_2 = 1.339 \text{ STB/D-psi.} \quad\quad\quad\quad\quad\quad\quad\quad\quad\quad\quad\quad (8.676)$$

and

$$S_1 = N_6 = 2.2317 \text{ STB/D-psi,} \quad\quad\quad\quad\quad\quad\quad\quad\quad\quad\quad\quad (8.677)$$

$$C_6 = -(E_6 + W_6 + N_6) = -4.9097 \text{ STB/D,} \quad\quad\quad\quad\quad\quad\quad\quad (8.678)$$

$$Q_5 = -\Omega_5 p_{sf,5} = -10,505 \text{ STB/D.} \quad\quad\quad\quad\quad\quad\quad\quad\quad\quad (8.679)$$

2. At Block 7, which hosts the production well, the finite-difference analog of the governing equation can be written using SIP notation:

$$S_7 p_9 + W_7 p_6 + C_7 p_7 + E_7 p_a + N_7 p_2 = Q_7, \quad\quad\quad\quad\quad\quad\quad (8.680)$$

where $p_a = 3{,}000$ psi;

$$S_7 = N_7 = 2.2317 \text{ STB/D-psi,} \quad\quad\quad\quad\quad\quad\quad\quad\quad\quad\quad (8.681)$$

$$E_7 = W_7 = 1.339 \text{ STB/D-psi,} \quad\quad\quad\quad\quad\quad\quad\quad\quad\quad\quad (8.682)$$

$$C_7 = -(S_7 + W_7 + E_7 + N_7 + \Omega_7) = (2 \times 2.2317 + 2 \times 1.339 + 2.6262) = -9.7676 \text{ STB/D-psi,} \quad\quad (8.683)$$

$$Q_7 = -\Omega_7 p_{sf,7} - E_7 P_a = -2.6262 \times 1{,}500 - 1.339 \times 3{,}000 = -7{,}956.3 \text{ STB/D,} \quad\quad (8.684)$$

and

$$2.2317 p_9 + 1.339 p_6 - 9.7676 p_7 + 2.2317 p_2 = -7{,}956.3. \quad\quad\quad\quad (8.685)$$

Problem 8.4.51

Consider the heterogeneous reservoir shown in **Fig. 8.44.** Block 2 is connected to a large aquifer, which enables the pressure in it to be maintained at 2,500 psi. All other boundaries are known to be sealed. A single-phase, incompressible fluid is produced through a production well located in Block 1 at a rate of 1,000 STB/D. Using the reservoir properties in **Table 8.45,** determine the sandface pressure at the production well. Then, perform an MBC to verify the accuracy of the solution.

k_1 = 50 md	
k_2 = 100 md	
Δx = 300 ft	
Δy = 200 ft	
h_1 = 100 ft	
h_2 = 80 ft	
μ = 1.0 cp	
B = 1.0 res bbl/STB	
r_w = 0.25 ft	
ρ = 62.4 lbm/ft³	
s = 0	

Table 8.45—Reservoir properties for Problem 8.4.51.

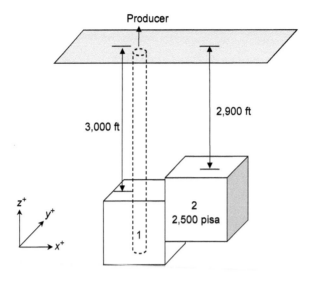

Fig. 8.44—Reservoir system of Problem 8.4.51.

Solution to Problem 8.4.51

For a 1D incompressible system, the governing flow equation can be written as

$$\frac{\partial}{\partial x}\left(\frac{A_x k_x}{\mu B}\frac{\partial p}{\partial x}\right)\Delta x - \frac{1}{144}\frac{g}{g_c}\rho\frac{\partial}{\partial x}\left(\frac{A_x k_x}{\mu B}\frac{\partial G}{\partial x}\right)\Delta x + q = 0. \dotfill (8.686)$$

The finite-difference approximation can be written as

$$\left.\frac{A_x k_x}{\mu B\Delta x}\right|_{i+\frac{1}{2}}\left(p_{i+1}-p_i\right)-\left.\frac{A_x k_x}{\mu B\Delta x}\right|_{i-\frac{1}{2}}\left(p_i-p_{i-1}\right)+\frac{1}{144}\frac{g}{g_c}\rho\left.\frac{A_x k_x}{\mu B\Delta x}\right|_{i+\frac{1}{2}}\left(G_{i+1}-G_i\right)-\frac{1}{144}\frac{g}{g_c}\rho\left.\frac{A_x k_x}{\mu B\Delta x}\right|_{i-\frac{1}{2}} \dotfill (8.687)$$

$$\left(G_i - G_{i-1}\right)+q = 0$$

or

$$T_{i+\frac{1}{2}}\left(p_{i+1}-p_i\right)-T_{i-\frac{1}{2}}\left(p_i-p_{i-1}\right)+\frac{1}{144}\frac{g}{g_c}\rho T_{i+\frac{1}{2}}\left(G_{i+1}-G_i\right)-\frac{1}{144}\frac{g}{g_c}\rho T_{i-\frac{1}{2}}\left(G_i-G_{i-1}\right)+q = 0. \dotfill (8.688)$$

In this case, the pressure in Block 2 is kept at 2,500 psia; therefore, the finite-difference equation can be expressed as

$$T_x\left(p_2 - p_1\right)=T_x'\left(G_2 - G_1\right)-q, \dotfill (8.689)$$

where

$$T_x = \frac{1}{\mu B}\left[\frac{2A_1 A_2 k_1 k_2}{\left(k_2 A_1\Delta x_1 + k_1 A_2\Delta x_2\right)}\right]=\frac{1}{(1)(1)}\left[\frac{2(16,000)(20,000)(100)(50)(1.127\times10^{-3})}{(20,000)(300)(100)+(16,000)(300)(50)}\right] \dotfill (8.690)$$
$$= 4.624 \text{ STB/D-psi},$$

and

$$T_x' = \frac{1}{144}\frac{g}{g_c}\rho T_x = \frac{1}{144}(62.4)(4.624)=2.004 \text{ STB/D-psi}. \dotfill (8.691)$$

Therefore,

$$-4.624\,p_1 + 4.624(2,500) = -2.004(3,050 - 2,940)+1,000. \dotfill (8.692)$$

Solving for p_1 yields $p_1 = 2,331.4$ psi.

The productivity index of the well hosted in Block 1 is

$$\Omega_1 = \frac{2\pi \bar{k} h}{\mu B \left[\ln\left(\frac{r_e}{r_w}\right) + s \right]} = \frac{2\pi(50)(1.127 \times 10^{-3})(100)}{(1)(1)\left(\ln\frac{50.48}{0.25} \right)} = 6.67 \text{ STB/D-psi}, \quad \dots \dots \dots \dots \dots \dots \text{(8.693)}$$

where

$$r_e = 0.14 \sqrt{(\Delta x)^2 + (\Delta y)^2} = 50.48 \text{ ft}. \quad \dots \dots \dots \dots \dots \dots \dots \dots \dots \dots \dots \dots \text{(8.694)}$$

Solving for p_{sf},

$$p_{sf} = p_1 + \frac{q}{\Omega_1} = 2,181.5 \text{ psi}. \quad \dots \dots \dots \dots \dots \dots \dots \dots \dots \dots \dots \dots \dots \text{(8.695)}$$

The fluid flow that comes from Block 2 to Block 1 is calculated as

$$q_{in} = T(\Delta\Phi) = 4.624 \left[\left(p_2 - \frac{1}{144}\frac{g}{g_c}\rho G_2 \right) - \left(p_1 - \frac{1}{144}\frac{g}{g_c}\rho G_1 \right) \right] = 1,000.02 \text{ STB/D}, \quad \dots \dots \dots \text{(8.696)}$$

which can be used to carry out an MBC of the solution:

$$\text{MBC} = \left| \frac{q_{in}}{q_{well}} \right| = \left| \frac{1,000.02}{1,000} \right| = 1.0002. \quad \dots \dots \dots \dots \dots \dots \dots \dots \dots \dots \dots \dots \dots \text{(8.697)}$$

The MBC indicates a good level of accuracy.

Problem 8.4.52

The grid system, depth, and pressure-distribution maps of a reservoir are given in **Fig. 8.45.** The reservoir is saturated by a slightly compressible fluid and is completely sealed. Using the reservoir properties found in **Table 8.46,**

1. Calculate the SIP coefficients C and Q for Block 6 during the time period $0 \le t \le 30$.
2. Recalculate the SIP coefficients C and Q for Block 6 during the time period $30 \le t \le 50$.

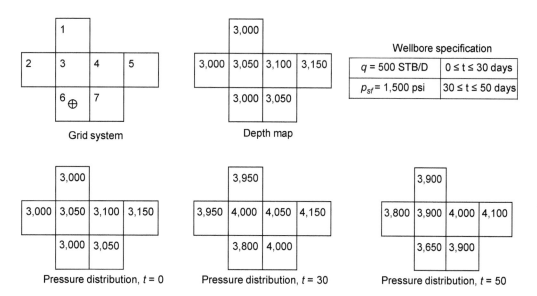

Fig. 8.45—Reservoir system of Problem 8.4.52.

$$k_x = k_y = 50 \text{ md}$$
$$\Delta x = \Delta y = 400 \text{ ft}$$
$$c = 1 \times 10^{-5} \text{ psi}^{-1}$$
$$\phi = 20\%$$
$$h = 100 \text{ ft}$$
$$\mu_{sc} = 0.9 \text{ cp}$$
$$p_{sc} = 14.7 \text{ psi}$$
$$\rho_{sc} = 62.4 \text{ lbm/ft}^3$$
$$\mu_o = \mu_{sc} + 0.00134 \ln(p)$$
$$B_o = [1 + c(p - p_{sc})]^{-1}$$
$$\rho_o = \rho_{sc}[1 + c(p - p_{sc})]$$
$$r_w = 0.25 \text{ ft}$$
$$s = 0$$

Table 8.46—Reservoir properties for Problem 8.4.52.

Solution to Problem 8.4.52

1. For $0 \le t \le 30$:

$$\mu_6 = \mu_{SC} + 0.00134 \ln(p) = 0.9 + 0.00134 \ln(4{,}000) = 0.911 \text{ cp}, \quad \dots\dots\dots\dots \quad (8.698)$$

$$\mu_7 = \mu_{SC} + 0.00134 \ln(p) = 0.9 + 0.00134 \ln(4{,}100) = 0.911 \text{ cp}, \quad \dots\dots\dots\dots \quad (8.699)$$

$$\bar{\mu}_{6-7} = \frac{\mu_6 + \mu_7}{2} = 0.911 \text{ cp}, \quad \dots\dots\dots\dots\dots\dots \quad (8.700)$$

$$B_6 = \left[1 + 1 \times 10^{-5} \times (4{,}000 - 14.7)\right]^{-1} = 0.9617 \text{ res bbl / STB}, \quad \dots\dots\dots \quad (8.701)$$

$$B_7 = \left[1 + 1 \times 10^{-5} \times (4{,}100 - 14.7)\right]^{1} = 0.9608 \text{ res bbl / STB}, \quad \dots\dots\dots \quad (8.702)$$

$$\bar{B}_{6-7} = \frac{B_6 + B_7}{2} = 0.96125 \text{ res bbl / STB}, \quad \dots\dots\dots\dots \quad (8.703)$$

$$\rho_6 = \rho_{sc}\left[1 + 1 \times 10^{-5} \times (4{,}000 - 14.7)\right] = 64.89 \text{ lbm/ft}^3, \quad \dots\dots\dots \quad (8.704)$$

$$\rho_7 = \rho_{sc}\left[1 + 1 \times 10^{-5} \times (4{,}100 - 14.7)\right] = 64.95 \text{ lbm/ft}^3, \quad \dots\dots\dots \quad (8.705)$$

$$\bar{\rho}_{6-7} = \frac{\rho_6 + \rho_7}{2} = 64.92 \text{ lbm/ft}^3, \quad \dots\dots\dots\dots \quad (8.706)$$

$$\Gamma_6 = \frac{V_b \phi c}{5.615 \Delta t} = \frac{(400)(400)(100)(0.2)(1 \times 10^{-5})}{5.615(30)} = 0.190 \text{ STB/D-psi}, \quad \dots\dots\dots \quad (8.707)$$

$$E_6 = N_6 = \frac{A_x k_x}{\bar{\mu}_{6-7} \bar{B}_{6-7} \Delta x} = \frac{(40{,}000)\left(50 \times 1.127 \times 10^{-3}\right)}{(0.96125 \times 0.911 \times 400)} = 6.435 \text{ STB/D-psi}, \quad \dots\dots \quad (8.708)$$

$$C_6 = -\left(E_6 + N_6 + \Gamma_6\right) = -13.06 \text{ STB/D-psi}, \quad \dots\dots\dots\dots \quad (8.709)$$

$$Q_6 = -\frac{1}{144}\frac{g}{g_c}\bar{\rho}_{6-7}\left(E_6 + N_6\right)G_6 + \frac{1}{144}\frac{g}{g_c}\bar{\rho}_{6-7}E_6 G_7 + \frac{1}{144}\frac{g}{g_c}\bar{\rho}_{6-7}N_6 G_3 - \Gamma_6 p_6 - q_6 \quad \dots\dots\dots \quad (8.710)$$
$$= 30 \text{ STB/D}.$$

2. For $30 \leq t \leq 50$:

$$\mu_6 = \mu_{sc} + 0.00134 \ln(p) = 0.9 + 0.00134 \ln(3,800) = 0.911 \text{ cp}, \dots\dots\dots\dots\dots\dots\dots (8.711)$$

$$\mu_7 = \mu_{sc} + 0.00134 \ln(p) = 0.9 + 0.00134 \ln(4,000) = 0.911 \text{ cp}, \dots\dots\dots\dots\dots\dots (8.712)$$

$$\bar{\mu}_{6-7} = \frac{\mu_6 + \mu_7}{2} = 0.911 \text{ cp}, \dots\dots\dots\dots\dots\dots\dots\dots\dots\dots\dots\dots\dots\dots (8.713)$$

$$B_6 = \left[1 + 1\times10^{-5} \times (3,800 - 14.7)\right]^{-1} = 0.9635 \text{ res bbl/STB}, \dots\dots\dots\dots\dots\dots\dots (8.714)$$

$$B_7 = \left[1 + 1\times10^{-5} \times (4,000 - 14.7)\right]^{-1} = 0.9617 \text{ res bbl/STB}, \dots\dots\dots\dots\dots\dots\dots (8.715)$$

$$\bar{B}_{6-7} = \frac{B_6 + B_7}{2} = 0.9626 \text{ res bbl/STB}, \dots\dots\dots\dots\dots\dots\dots\dots\dots\dots\dots\dots (8.716)$$

$$\rho_6 = \rho_{sc}\left[1 + 1\times10^{-5} \times (4,000 - 14.7)\right] = 64.76 \text{ lbm/ft}^3, \dots\dots\dots\dots\dots\dots\dots (8.717)$$

$$\rho_7 = \rho_{sc}\left[1 + 1\times10^{-5} \times (4,100 - 14.7)\right] = 64.89 \text{ lbm/ft}^3, \dots\dots\dots\dots\dots\dots\dots (8.718)$$

$$\bar{\rho}_{6-7} = \frac{\rho_6 + \rho_7}{2} = 64.825 \text{ lbm/ft}^3, \dots\dots\dots\dots\dots\dots\dots\dots\dots\dots\dots\dots\dots (8.719)$$

$$\Gamma_6 = \frac{V_b \phi c}{5.615 \Delta t} = \frac{(400)(400)(100)(0.2)(1\times10^{-5})}{5.615(20)} = 0.285 \text{ STB/D-psi}, \dots\dots\dots\dots\dots (8.720)$$

$$E_6 = N_6 = \frac{A_x k_x}{\bar{\mu}_{6-7}\bar{B}_{6-7}\Delta x} = \frac{(40,000)\left(50\times1.127\times10^{-3}\right)}{(0.9626\times0.911\times400)} = 6.426 \text{ STB/D-psi}, \dots\dots\dots (8.721)$$

$$\Omega_6 = \frac{2\pi\bar{k}h}{\mu B\left[\ln\left(\dfrac{r_e}{r_w}\right) + s\right]} = \frac{2\pi(50)(1.127\times10^{-3})(100)}{(0.911)(9,635)\left(\ln\dfrac{80}{0.25}\right)} = 6.99 \text{ STB/D-psi}, \dots\dots\dots (8.722)$$

$$C_6 = -\left(E_6 + N_6 + \Gamma_6 + \Omega_6\right) = -20.127 \text{ STB/D-psi}, \dots\dots\dots\dots\dots\dots\dots\dots (8.723)$$

$$Q_6 = -\frac{1}{144}\frac{g}{g_c}\bar{\rho}_{6-7}\left(E_6 + N_6\right)G_6 + \frac{1}{144}\frac{g}{g_c}\bar{\rho}_{6-7}E_6 G_7 + \frac{1}{144}\frac{g}{g_c}\bar{\rho}_{6-7}N_6 G_3 - \Gamma_6 p_6 - \Omega_6 p_{sf}$$
$$= -11,278.7 \text{ STB/D}. \dots\dots\dots (8.724)$$

Problem 8.4.53

Consider the 2D homogeneous isotropic reservoir saturated with incompressible fluid and with no depth gradients shown in **Fig. 8.46.** The reservoir boundaries are known to be sealed, except along the western boundary of Block 1 and the southern boundary of Block 3. A well is completed in Block 2, which is operated at a sandface pressure of 932 psi.

1. Determine the pressure distribution of the system using the reservoir properties found in **Table 8.47.**
2. Calculate the flow rate of the well in Block 2.
3. Perform an MBC to verify the accuracy of the results.

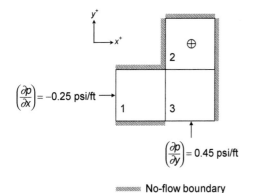

Fig. 8.46—Reservoir system of Problem 8.4.53.

$k_x = k_y$	= 80 md
$\Delta x = \Delta y$	= 200 ft
h	= 60 ft
μ	= 1 cp
B	= 1 res bbl/STB
r_w	= 0.25 ft
s	= 0

Table 8.47—Reservoir properties for Problem 8.4.53.

Solution to Problem 8.4.53

1. The PDE describing the flow is

$$\frac{\partial}{\partial x}\left(\frac{A_x k_x}{\mu B}\frac{\partial p}{\partial x}\right)\Delta x + \frac{\partial}{\partial y}\left(\frac{A_y k_y}{\mu B}\frac{\partial p}{\partial y}\right)\Delta y + q = 0, \quad\dots\dots\dots\dots\dots\dots\dots\dots\dots\dots\dots\dots \text{(8.725)}$$

which can be reduced to the following form for a homogeneous and isotropic system:

$$\frac{\partial^2 p}{\partial x^2} + \frac{\partial^2 p}{\partial y^2} + \frac{q\mu B}{kV_b} = 0. \quad\dots\dots\dots\dots\dots\dots\dots\dots\dots\dots\dots\dots\dots\dots\dots\dots \text{(8.726)}$$

The finite-difference approximation of the equation is

$$p_{i-1,j} + p_{i+1,j} - 4p_{i,j} + p_{i,j-1} + p_{i,j-1} + \left(\frac{q\mu B}{hk}\right)_{i,j} = 0. \quad\dots\dots\dots\dots\dots\dots\dots\dots\dots \text{(8.727)}$$

Across the western boundary,

$$\frac{p_1 - p_1'}{\Delta x} = -0.25 \quad\dots\dots\dots\dots\dots\dots\dots\dots\dots\dots\dots\dots\dots\dots\dots\dots\dots\dots\dots\dots \text{(8.728)}$$

or

$$p_1' = p_1 + 50. \quad\dots\dots\dots\dots\dots\dots\dots\dots\dots\dots\dots\dots\dots\dots\dots\dots\dots\dots\dots\dots\dots \text{(8.729)}$$

Across the southern boundary,

$$\frac{p_3 - p_3'}{\Delta y} = 0.45 \quad\dots\dots\dots\dots\dots\dots\dots\dots\dots\dots\dots\dots\dots\dots\dots\dots\dots\dots\dots\dots \text{(8.730)}$$

or

$$p_3' = p_3 - 90. \quad\dots\dots\dots\dots\dots\dots\dots\dots\dots\dots\dots\dots\dots\dots\dots\dots\dots\dots\dots\dots\dots \text{(8.731)}$$

- Block 1:

$$p_1' - 2p_1 + p_3 = 0 \quad\dots\dots\dots\dots\dots\dots\dots\dots\dots\dots\dots\dots\dots\dots\dots\dots\dots\dots\dots \text{(8.732)}$$

and

$$-p_1 + p_3 = -50. \quad\dots\dots\dots\dots\dots\dots\dots\dots\dots\dots\dots\dots\dots\dots\dots\dots\dots\dots\dots \text{(8.733)}$$

- Block 2:

$$p_3 - p_2 + \frac{\mu B}{kh}q_2 = 0. \quad\dots\dots\dots\dots\dots\dots\dots\dots\dots\dots\dots\dots\dots\dots\dots\dots\dots \text{(8.734)}$$

Substituting the flow-rate term using the wellbore model,

$$q_2 = -\Omega_2(p_2 - p_{sf2}), \quad\dots\dots\dots\dots\dots\dots\dots\dots\dots\dots\dots\dots\dots\dots\dots\dots\dots\dots \text{(8.735)}$$

with

$$\Omega_2 = \frac{2\pi kh}{\mu B\left[\ln\left(\dfrac{r_e}{r_w}\right)+s\right]} = \frac{2\pi(80)(1.127\times10^{-3})(60)}{(1)(1)\left(\ln\dfrac{40}{0.25}\right)} = 6.70 \text{ STB/D-psi}, \quad\ldots\ldots\ldots\ldots\ldots\ldots \text{(8.736)}$$

where $r_e = 0.2\Delta x = 40$ ft,

$$\frac{\mu B}{kh} = \frac{(1)(1)}{(80)(1.127\times10^{-3})(60)} = 0.185. \quad\ldots\ldots\ldots\ldots\ldots\ldots\ldots\ldots\ldots\ldots\ldots\ldots \text{(8.737)}$$

Now, substituting the calculated values into the finite-difference approximation,

$$p_3 - p_2 - \frac{\mu B}{kh}\Omega_2(p_2 - p_{sf2}) = 0, \quad\ldots\ldots\ldots\ldots\ldots\ldots\ldots\ldots\ldots\ldots\ldots\ldots\ldots \text{(8.738)}$$

one obtains

$$p_3 - 2.2380 p_2 = -1{,}153.8370. \quad\ldots\ldots\ldots\ldots\ldots\ldots\ldots\ldots\ldots\ldots\ldots\ldots\ldots \text{(8.739)}$$

- Block 3:

$$p_3' + p_1 - 3p_3 + p_2 = 0 \quad\ldots\ldots\ldots\ldots\ldots\ldots\ldots\ldots\ldots\ldots\ldots\ldots\ldots\ldots \text{(8.740)}$$

and

$$p_1 - 2p_3 + p_2 = 90. \quad\ldots\ldots\ldots\ldots\ldots\ldots\ldots\ldots\ldots\ldots\ldots\ldots\ldots\ldots \text{(8.741)}$$

Solving for pressures, $p_1 = 909.70$ psi, $p_2 = 899.70$ psi, and $p_3 = 859.70$ psi.

2. Flow-rate calculations:

$$\Omega = \frac{2\pi kh}{\mu B\left[\ln\left(\dfrac{r_e}{r_w}\right)+s\right]} = 6.71 \text{ STB/D-psi}, \quad\ldots\ldots\ldots\ldots\ldots\ldots\ldots\ldots\ldots\ldots \text{(8.742)}$$

and

$$q_2 = -\Omega\left(p_2 - p_{sf}\right) = -6.71(899.70 - 932) = 216.733 \text{ STB/D}. \quad\ldots\ldots\ldots\ldots\ldots\ldots\ldots \text{(8.743)}$$

The positive sign indicates that the well is an injector.

3. MBC:

$$q_{\text{south}} = \left(\frac{kA}{\mu B}\frac{\partial p}{\partial y}\right) = (-0.45)\frac{(200)(60)(80)(1.127\times10^{-3})}{(1)(1)} = -486.864 \text{ STB/D}, \quad\ldots\ldots\ldots\ldots \text{(8.744)}$$

$$q_{\text{west}} = \left(\frac{kA}{\mu B}\frac{\partial p}{\partial x}\right) = (0.25)\frac{(200)(60)(80)(1.127\times10^{-3})}{(1)(1)} = 270.48 \text{ STB/D}, \quad\ldots\ldots\ldots\ldots \text{(8.745)}$$

$$\left|\frac{q_{\text{in}}}{q_{\text{well}}}\right| = \left|\frac{270.48 + 216.733}{486.864}\right| = 1.0007. \quad\ldots\ldots\ldots\ldots\ldots\ldots\ldots\ldots\ldots\ldots\ldots \text{(8.746)}$$

Problem 8.4.54

A 2D homogeneous, isotropic horizontal reservoir is shown in **Fig. 8.47.** The reservoir fluid is assumed to be incompressible. The reservoir is known to be supported by a strong water drive, as shown in the figure. The aquifer is infinitely large so that it does not experience a measurable pressure drop. A production well is located in Block 2, which is producing as an artesian well from a depth of 7,788 ft (midpoint of the formation). Please note that the well is flowing into storage tanks that are kept at atmospheric pressure.

1. Solve for pressure distribution in the reservoir using the Gauss-Seidel method with a pressure convergence criterion of 2 psia. Use the average of the aquifer pressure and well sandface pressure as initial guess.
2. Calculate the flow rate at the production well using the reservoir properties in **Table 8.48.**
3. Perform an MBC and comment on the accuracy of the results.

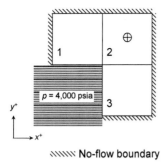

Fig. 8.47—Reservoir system of Problem 8.4.54.

$k_x = k_y$	= 50 md
$\Delta x = \Delta y$	= 400 ft
h	= 100 ft
μ	= 1 cp
B	= 1 res bbl/STB
r_w	= 0.25 ft
s	= 0

Table 8.48—Reservoir properties for Problem 8.4.54.

Solution to Problem 8.4.54

The PDE describing the flow is

$$\frac{\partial}{\partial x}\left(\frac{A_x k_x}{\mu B}\frac{\partial p}{\partial x}\right)\Delta x + \frac{\partial}{\partial y}\left(\frac{A_y k_y}{\mu B}\frac{\partial p}{\partial y}\right)\Delta y + q = 0, \dots\dots\dots\dots\dots\dots\dots\dots\dots\dots\dots\dots (8.747)$$

which can be reduced to the following form for a homogeneous and isotropic system:

$$\frac{\partial^2 p}{\partial x^2} + \frac{\partial^2 p}{\partial y^2} + \frac{q\mu B}{kV_b} = 0. \dots\dots\dots\dots\dots\dots\dots\dots\dots\dots\dots\dots\dots\dots\dots\dots\dots\dots (8.748)$$

The finite-difference approximation of the governing flow equation is

$$p_{i-1,j} + p_{i+1,j} - 4p_{i,j} + p_{i,j+1} + p_{i,j-1} + \left(\frac{q\mu B}{hk}\right)_{i,j} = 0. \dots\dots\dots\dots\dots\dots\dots (8.749)$$

Considering the symmetry of the system, $p_1 = p_3$; therefore,

- Block 1: $-2p_1 + p_2 = -4,000. \dots\dots\dots\dots\dots\dots\dots\dots\dots\dots\dots\dots\dots\dots\dots\dots\dots (8.750)$

- Block 2: $p_1 - 2p_2 + p_3 + \left(\frac{q\mu B}{kh}\right)_2 = 0. \dots\dots\dots\dots\dots\dots\dots\dots\dots\dots\dots\dots\dots (8.751)$

The sandface pressure of the production well can be calculated using the hydrostatic head exerted by the fluid column in the wellbore:

$$p_{sf} = 14.7 + 0.433G = 3,386.9 \text{ psi.} \dots\dots\dots\dots\dots\dots\dots\dots\dots\dots\dots\dots\dots\dots (8.752)$$

The productivity index also is calculated:

$$\Omega_2 = \frac{2\pi kh}{\mu B\left[\ln\left(\frac{r_e}{r_w}\right) + s\right]} = 6.135 \text{ STB/D-psi,} \dots\dots\dots\dots\dots\dots\dots\dots\dots\dots\dots (8.753)$$

where $r_e = 0.2\Delta x = 80$ ft.

Because the sandface pressure is specified for the well in Block 2, the wellbore model is employed:

$$q_2 = -\Omega_2\left(p_2 - p_{sf}\right). \dots\dots\dots\dots\dots\dots\dots\dots\dots\dots\dots\dots\dots\dots\dots\dots\dots\dots (8.754)$$

Employing the wellbore model to substitute q_2,

$$-2p_2 + 2p_1 - \frac{\mu B}{kh}\Omega_2(p_2 - p_{sf2}) = 0 \text{ (because } p_1 = p_3), \dots\dots\dots\dots\dots\dots\dots\dots (8.755)$$

which reduces to

$$-3.089p_2 + 2p_1 = -3,687.41. \dots\dots\dots\dots\dots\dots\dots\dots\dots\dots\dots\dots\dots\dots\dots\dots\dots (8.756)$$

Using the Gauss-Seidel iterations in **Table 8.49,**

$$p_1^{(k+1)} = -\frac{1}{2}\left[-4,000 - p_2^{(k)}\right] \quad \dots\dots\dots\dots\dots (8.757)$$

and

$$p_2^{(k+1)} = -\frac{1}{3.089}\left[-3,687.41 - 2p_1^{(k+1)}\right]. \quad \dots\dots\dots (8.758)$$

Iteration Level	p_1	p_2
0	3,693.45	3,693.45
1	3,846.725	3,684.322
2	3,842.161	3,681.367
3	3,840.683	3,680.41

Table 8.49—Gauss-Seidel iterations from Solution to Problem 8.4.54.

The initial guess = $(4,000 + 3,386.9)/2 = 3,693.45$ psi, $p_1 = p_3 = 3,840.68$ psi, and $p_2 = 3,680.41$ psi. Calculating the flow rate from the well in Block 2:

$$q_2 = -\Omega\left(p_2 - p_{sf}\right) = -6.135\left(3,680.41 - 3,386.9\right) = -1,800.68 \ \text{STB/D}. \quad \dots\dots\dots\dots\dots\dots\dots\dots\dots\dots\dots(8.759)$$

An MBC is carried out to validate the solution:

$$q_{in} = 2\left(\frac{kA}{\mu B}\right)\left(\frac{p_a - p_1}{\Delta x}\right) = 2\frac{(400)(50)(100)(1.127\times10^{-3})}{(1)(1)}\frac{(4,000 - 3,840.683)}{400} = 1,795.5 \ \text{STB/D}, \quad \dots\dots\dots(8.760)$$

$$\text{MBC} = \left|\frac{q_{in}}{q_{well}}\right| = \left|\frac{1,795.5}{1,800.68}\right| = 0.997. \quad \dots\dots\dots\dots\dots\dots\dots\dots\dots\dots\dots\dots\dots\dots\dots\dots\dots\dots\dots\dots(8.761)$$

Problem 8.4.55

A 2D homogeneous, isotropic, horizontal reservoir is shown in **Fig. 8.48.** The reservoir is known to be sealed, except on the east side of Block 3 and on the south side of Block 1, as shown in the figure. A single-phase, incompressible fluid is produced through a production well in Block 2. A downhole pump is installed at the well that maintains the sandface pressure at 14.7 psia.

1. Write the general flow equation for this reservoir using the reservoir properties shown in **Table 8.50.**
2. Generate the system of equations to solve for the pressure distribution in the reservoir.
3. Using the Gauss-Seidel method, determine the pressure distribution in the system. Use a convergence criterion of 1 psia and initial guesses for unknown pressures: $p_1 = 200$ psia, $p_2 = 275$ psi, and $p_3 = 195$ psi.
4. Calculate the flow rate from the well in Block 2.
5. Perform an MBC to verify the accuracy of your results.

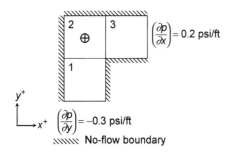

Fig. 8.48—Reservoir system of Problem 8.4.55.

$k_x = k_y$ =	50 md (uniform)
$\Delta x = \Delta y$ =	400 ft (uniform)
h =	100 ft (uniform)
μ =	1 cp
B =	1 res bbl/STB
r_w =	0.25 ft
ρ =	62.4 lbm/ft³
s =	0

Table 8.50—Reservoir properties for Problem 8.4.55.

Solution to Problem 8.4.55

1. The governing equation for a 2D incompressible system is

$$\frac{\partial}{\partial x}\left(\frac{A_x k_x}{\mu B}\frac{\partial p}{\partial x}\right)\Delta x + \frac{\partial}{\partial y}\left(\frac{A_y k_y}{\mu B}\frac{\partial p}{\partial y}\right)\Delta y + q = 0. \quad \dots\dots\dots\dots\dots\dots\dots\dots\dots\dots\dots\dots(8.762)$$

Simplifying for a homogeneous and isotropic reservoir,

$$\frac{\partial^2 p}{\partial x^2} + \frac{\partial^2 p}{\partial y^2} + \frac{q\mu B}{kV_b} = 0. \quad \dots\dots\dots\dots\dots\dots\dots\dots\dots\dots\dots\dots\dots\dots\dots\dots\dots\dots (8.763)$$

2. In the finite-difference form,

$$p_{i,j-1} + p_{i-1,j} - 4p_{i,j} + p_{i+1,j} + p_{i,j+1} + \frac{q\mu B}{hk} = 0. \quad \text{...} (8.764)$$

- Block 1:

$$\frac{p_1 - p_1'}{\Delta y} = -0.3 \quad \text{...} (8.765)$$

and

$$p_1' = p_1 + 0.3\Delta y; \quad \text{...} (8.766)$$

therefore, $-p_1 + p_2 = -120.$ ....(8.767)

- Block 2 (using two reflection blocks across the no-flow boundaries):

$$p_1 - 2p_2 + p_3 + \left(\frac{q\mu B}{hk}\right)_2 = 0. \quad \text{...}(8.768)$$

The wellbore model is employed to substitute the flow-rate term:

$$q_2 = -\Omega_2 \left(p_2 - p_{sf}\right), \quad \text{...}(8.769)$$

$$\Omega_2 = \frac{2\pi kh}{\mu B \left[\ln\left(\dfrac{r_e}{r_w}\right) + s\right]} = 6.135 \text{ STB/D-psi}, \quad \text{...........................}(8.770)$$

where $r_e = 0.2\Delta x = 80$ ft.

The original finite-difference equation can be rearranged as

$$p_1 - 2p_2 + p_3 - \left(\frac{\Omega_2 \mu B}{hk}\right)\left(p_2 - p_{sf}\right) = 0 \quad \text{...........................}(8.771)$$

and

$$\frac{\Omega_2 \mu B}{hk} = \frac{(6.135)(1)(1)}{(100)(50)(1.127\times 10^{-3})} = 1.09, \quad \text{..........................}(8.772)$$

because $p_{sf} = 14.7$ psi.

Thus,

$$p_1 - 2p_2 + p_3 - 1.09\left(p_2 - 14.7\right) = 0, \quad \text{............................}(8.773)$$

$$p_1 - 3.09p_2 + p_3 = -16. \quad \text{..}(8.774)$$

- Block 3:

$$\frac{p_3' - p_3}{\Delta x} = 0.2 \quad \text{...}(8.775)$$

and

$$p_3' = p_3 + 80, \quad \text{...}(8.776)$$

$$p_3' - 4p_3 + p_2 + 2p_3 = 0, \quad \text{..}(8.777)$$

$$p_2 - p_3 = -80. \quad \text{...}(8.778)$$

3. Implementation of the Gauss-Seidel method using the iteration levels in **Table 8.51:**

$$p_1 = p_2 + 120, \qquad (8.779)$$

$$p_2 = \frac{p_1 + p_3 + 16}{3.09}, \qquad (8.780)$$

$$p_3 = p_2 + 80. \qquad (8.781)$$

Iteration Level	p_1	p_2	p_3	Δp_{max}
0	200	275	195	–
1	395	196.12	276.11	195
2	316.12	196.84	276.84	78.88
3	316.84	197.31	277.31	0.72

Table 8.51—Gauss-Seidel iterations from Solution to Problem 8.4.55.

4. The flow rate can be calculated as

$$q_2 = -\Omega\left(p_2 - p_{sf}\right) = 1{,}120.527 \text{ STB/D}. \qquad (8.782)$$

5. Flow in across the boundaries,

$$q_{in} = \frac{kA}{\mu B}\left(\frac{\partial p}{\partial x} + \frac{\partial p}{\partial y}\right) = 1{,}127 \text{ STB/D}, \qquad (8.783)$$

$$\text{MBC} = \frac{q_{in}}{q_{out}} = \frac{1127}{1{,}120.527} = 1.0057. \qquad (8.784)$$

Problem 8.4.56

Consider the heterogeneous reservoir shown in **Fig. 8.49,** with the reservoir properties specified in **Table 8.52.** Block 2 is connected to an infinitely large aquifer (through Block 3), enabling the pressure at that boundary to be maintained at 3,000 psi when the system is in hydrodynamic equilibrium. All other boundaries are known to be sealed. Assume that reservoir fluid is incompressible.

1. If we have a pump with a maximum capacity of 2,000 psia, determine the depth at which the pump must be placed so that fluid can be transported to the surface.
2. What is the maximum-capacity flow rate the pump can deliver?

Note: ignore pressure losses in the production string.

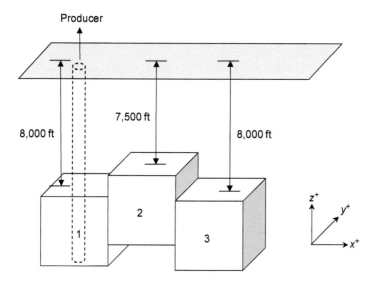

Fig. 8.49—Reservoir system of Problem 8.4.56.

h_1	= 150 ft
$h_2 = h_3$	= 100 ft
k_1	= 75 md
$k_2 = k_3$	= 100 md
μ	= 1.0 cp
B	= 1.0 res bbl/STB
r_w	= 0.25 ft
ρ	= 62.4 lbm/ft³
s	= 0
$\Delta x_1 = \Delta x_2 = \Delta x_3$	= 500 ft
$\Delta y_1 = \Delta y_2 = \Delta y_3$	= 400 ft

Table 8.52—Reservoir properties for Problem 8.4.56.

Solution to Problem 8.4.56

1. The fluid column that the pump can work against is calculated using hydrostatic pressure in the wellbore. Pressure gradient $\gamma = 62.4/144 = 0.433$ psi/ft, with a 2,000-psi limitation, the fluid column height above the intake valve of the pump is

$$h = \frac{2{,}000}{0.433} = 4{,}615.4 \text{ ft}. \qquad (8.785)$$

The pump needs to be installed at $8,000 - 4,615 = 3,385$ ft from the bottom of the well. The sandface pressure of the well can be calculated using

$$(8,000 - 4,615.4) \times 0.433 \text{ psi/ft} = 1,466.4 \text{ psia}. \dotfill (8.786)$$

2. For a 1D incompressible reservoir, the finite-difference approximation can be written using the SIP notation:

$$W_i p_{i-1} + C_i p_i + E_i p_{i+1} - \left(W_i' G_{i-1} + C_i' G_i + E_i' G_{i+1} \right) + q = 0. \dotfill (8.787)$$

*   At Block 1: $W_1 = W_1' = 0$,

$$E_1 = \frac{2A_1 k_1 A_2 k_2}{\mu B \left(\Delta x_1 A_2 k_2 + \Delta x_2 A_1 k_1 \right)} = \frac{2(60,000)\left(75 \times 1.127 \times 10^{-3}\right)(40,000)\left(100 \times 1.127 \times 10^{-3}\right)}{\left(500 \times 1.127 \times 10^{-3}\right)\left[(40,000)(100) + (60,000)(75)\right]} \dots (8.788)$$
$$= 9.55 \text{ STB/D-psi},$$

$$C_1 = -E_1 = -9.55 \text{STB/D-psi}, \dotfill (8.789)$$

$$E_1' = \frac{1}{144} \rho \frac{g}{g_c} E_1 = \frac{62.4}{144} \times 9.55 = 4.138 \text{ STB/D-ft}, \dotfill (8.790)$$

and

$$C_1' = -E_1' = -4.138 \text{ STB/D-ft}, \dotfill (8.791)$$

$$Q_1 = \left(W_i' G_{i-1} + C_i' G_i + E_i' G_{i+1} \right) - q = (-4.138)(-4.138)(8,000) \atop + (4.138)(7,500) - q = -2,069 - q \quad, \dotfill (8.792)$$

$$q = -\Omega \left(p_2 - p_{sf} \right), \dotfill (8.793)$$

$$\Omega = \frac{2\pi kh}{\mu B \left[\ln \left(\dfrac{r_e}{r_w} \right) + s \right]} = 13.55 \text{ STB/D-psi}, \dotfill (8.794)$$

where

$$r_e = 0.14 \sqrt{\left(\Delta x^2 + \Delta y^2 \right)} = 0.14 \sqrt{\left(500^2 + 400^2 \right)} = 89.64 \text{ STB/D-psi}. \dotfill (8.795)$$

Thus,

$$-23.1 p_1 + 9.55 p_2 = -21,938.7. \dotfill (8.796)$$

*   At Block 2:

$$E_1 = \frac{A_2 k_2}{\mu B \Delta x_2} = \frac{(40,000)\left(100 \times 1.127 \times 10^{-3}\right)}{500} = 9.016 \text{ STB/D-psi}, \dotfill (8.797)$$

$$W_2 = E_1 = 9.55 \text{ STB/D-psi}, \dotfill (8.798)$$

$$C_2 = -[E_2 + W_2] = -18.57 \text{ STB/D-psi}, \dotfill (8.799)$$

$$E_2' = \frac{1}{144} \rho \frac{g}{g_c} E_1 = \frac{62.4}{144} \times 9.016 = 3.907 \text{ STB/D-ft}, \dotfill (8.800)$$

$$W_2' = \frac{1}{144} \rho \frac{g}{g_c} E_1 = \frac{62.4}{144} \times 9.55 = 4.138 \text{ STB/D-ft}, \dotfill (8.801)$$

$$C_2' = -\left(E_2' + W_2' \right) = -8.045 \text{ STB/D-ft}. \dotfill (8.802)$$

Hence,

$$9.55 p_1 - 18.57 p_2 = -23,025.5. \dotfill (8.803)$$

Solving for $p_1 = 1,858.4$ psi and $p_2 = 2195.7$ psi,

$$q = -\Omega \left(p_1 - p_{sf} \right) = -5,311 \text{ STB/D}. \dotfill (8.804)$$

Problem 8.4.57

The system shown in **Fig. 8.50** is a 2D, single-phase flow with homogeneous and isotropic permeability distribution. There is a producer located in Block 5 of the reservoir. This production well is operating at a specified sandface pressure of 40 psia. There is no depth gradient in the system. The system boundaries are indicated in the figure. Blocks 12 and 15 are connected to an infinitely large aquifer and the pressure of the aquifer is 3,000 psi. Using SIP notation, write the simplest form of the finite-difference approximation of the governing flow equation for the following blocks (in your equations, numerical values for each coefficient should be included): Blocks 4, 5, 8, and 15. Use the rock and fluid properties in **Table 8.53** in your calculations.

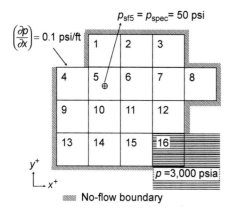

Fig. 8.50—Reservoir system of Problem 8.4.57.

$k_x = k_y = k =$	100 md
$\Delta x = \Delta y = \Delta L =$	1,000 ft
$h =$	50 ft
$\mu =$	1 cp
$B =$	1 res bbl/STB
$r_w =$	0.25 ft
$s =$	0
$r_e =$	198 ft

Table 8.53—Rock and fluid properties for Problem 8.4.57.

Solution to Problem 8.4.57

For a homogeneous and isotropic system, the transmissibility factor for each block and in each direction can be calculated from

$$T = \frac{Ak}{\mu B \Delta L}, \quad \dotfill \quad (8.805)$$

which reduces to

$$T = \frac{kh}{\mu B}, \quad \dotfill \quad (8.806)$$

because $A = \Delta L h$; therefore,

$$T = \frac{kh}{\mu B} = \frac{(1.127 \times 10^{-3})(100)(50)}{(1)(1)} = 5.635 \text{ STB/D-psi.} \quad \dotfill \quad (8.807)$$

The productivity index of the well will be

$$\Omega = \frac{2\pi kh}{\mu B \left[\ln\left(\frac{r_e}{r_w}\right) + s \right]} = \frac{2\pi (100)(1.127 \times 10^{-3})(50)}{(1)(1)\ln\left(\frac{198}{0.25}\right)} = 5.297 \text{ STB/D-psi.} \quad \dotfill \quad (8.808)$$

- Block 4:

$$W_4 = 0, E_4 = S_4 = N_4 = 5.635, \quad \dotfill \quad (8.809)$$

$$C_4 = -(E_4 + N_4 + S_4) = -16.905, \quad \dotfill \quad (8.810)$$

$$Q_4 = 0. \quad \dotfill \quad (8.811)$$

Boundary condition: $\dfrac{\partial p}{\partial y} = \dfrac{p_4' - p_4}{1,000} = 0.1 \rightarrow p_4' = p_4 + 100. \quad \dotfill \quad (8.812)$

Finite-difference equation in SIP notation:

$$5.635\,p_9 - 16.905\,p_4 + 5.635\,p_5 + 5.635(p_4 + 100) = 0, \quad\text{..................................}\quad (8.813)$$

or

$$-11.27\,p_4 + 5.635\,p_5 + 5.635\,p_9 = -563.5. \quad\text{..}\quad (8.814)$$

- Block 5:

$$W_5 = E_5 = S_5 = N_5 = 5.635, \quad\text{..}\quad (8.815)$$

$$C_5 = -\left(E_5 + N_5 + S_5 + W_5 + \Omega_5\right) = -\left(4 \times 5.635 + 5.635\right) = -27.837, \quad\text{.........................}\quad (8.816)$$

$$Q_5 = \Omega_{P_{sf}} = -5.297 \times 50 = -264.85, \quad\text{.......................................}\quad (8.817)$$

$$5.635\,p_1 + 5.635\,p_4 - 27.837\,p_5 + 5.635\,p_6 + 5.635\,p_{10} = -264.85. \quad\text{...........................}\quad (8.818)$$

- Block 8:

$$E_8 = S_8 = N_8 = 0, \quad\text{...}\quad (8.819)$$

$$W_8 = 5.635, \quad\text{...}\quad (8.820)$$

$$C_8 = -\left(E_8 + N_8 + S_8 + W_8\right) = -5.635, \quad\text{...................................}\quad (8.821)$$

$$Q_8 = 0, \quad\text{..}\quad (8.822)$$

$$5.635\,p_7 - 5.635\,p_8 = 0. \quad\text{..}\quad (8.823)$$

- Block 15:

$$S_{15} = 0,\; E_{15} = W_{15} = N_{15} = 5.635, \quad\text{..}\quad (8.824)$$

$$C_{15} = -\left(E_{15} + N_{15} + S_{15}\right) = -16.905, \quad\text{......................................}\quad (8.825)$$

$$Q_{15} = 0, \quad\text{...}\quad (8.826)$$

$$5.635\,p_{14} - 16.905\,p_{15} + 5.635\,p_{11} + 5.635 \times 3,000 = 0, \quad\text{................................}\quad (8.827)$$

$$-16.905\,p_{15} + 5.635\,p_{14} + 5.635\,p_{11} = -16,905. \quad\text{.....................................}\quad (8.828)$$

Problem 8.4.58

A 1D homogeneous, isotropic, horizontal reservoir is shown in **Fig. 8.51.** The reservoir boundaries are known to be sealed. A single-phase, slightly compressible fluid is produced through a production well in Block 3, which begins production at $t = 0$ at a rate of $-1,000$ STB/D.

Using the data provided in **Table 8.54,** determine the pressure distribution in the reservoir at $t = 30$ days in calculations using a timestep of 30 days.

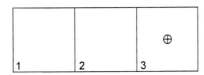

Fig. 8.51—Reservoir system of Problem 8.4.58.

Table 8.54—Reservoir properties for Problem 8.4.58.
$k_x = k_y = 10$ md
$\Delta x = \Delta y = 1,000$ ft
$h = 100$ ft
$\mu = 2$ cp
$B = 1$ res bbl/STB
$r_w = 0.25$ ft
$\rho = 62.4$ lbm/ft³
$\phi = 15\%$
$p_i = 3,000$ psi
$c = 5 \times 10^{-6}$ psi⁻¹
$s = 0$

Solution to Problem 8.4.58

The governing PDE of the system can be written as

$$\frac{\partial}{\partial x}\left(\frac{A_x k_x}{\mu B}\frac{\partial p}{\partial x}\right)\Delta x + q = \frac{V_b \phi c_t}{5.615}\left(\frac{\partial p}{\partial t}\right). \qquad (8.829)$$

For a homogeneous 1D reservoir,

$$\frac{\partial^2 p}{\partial x^2} + \frac{q\mu B}{V_b k} = \frac{\phi c_t \mu B}{5.615 k}\left(\frac{\partial p}{\partial t}\right), \qquad (8.830)$$

the characteristic finite-difference representation is

$$\left(p_{i-1}^{n+1} - 2p_i^{n+1} + p_{i+1}^{n+1}\right) + \left(\frac{q\mu B}{hk_x}\right)_i = \left[\frac{\mu B c_t \phi (\Delta x)^2}{k_x 5.615 \Delta t}\right]\left(p_i^{n+1} - p_i^n\right). \qquad (8.831)$$

The coefficient,

$$\frac{\mu B c_t \phi (\Delta x)^2}{k_x 5.615 \Delta t} = \frac{(2)(1)(0.15)(5\times10^{-6})(1,000)^2}{(5.615)(1.127\times10^{-3})(10)(30)} = 0.79, \qquad (8.832)$$

$$\frac{q\mu B}{hk_x} = \frac{(-1,000)(2)(1)}{(1.127\times10^{-3})(10)(100)} = -1,774.62. \qquad (8.833)$$

- Block 1: $-p_1^{n+1} + p_2^{n+1} = \dfrac{\mu B \phi c_t \Delta x^2}{5.615 k \Delta t}(p_1^{n+1} - p_1^n).$ \qquad (8.834)

- Block 2: $p_1^{n+1} - 2p_2^{n+1} + p_3^{n+1} = \dfrac{\mu B \phi c_t \Delta x^2}{5.615 k \Delta t}(p_2^{n+1} - p_2^n).$ \qquad (8.835)

- Block 3: $p_2^{n+1} - p_3^{n+1} + \dfrac{q\mu B}{hk} = \dfrac{\mu B \phi c_t \Delta x^2}{5.615 k \Delta t}(p_3^{n+1} - p_3^n).$ \qquad (8.836)

For the timestep between 0 and 30 days, rearranging the system of equations in a matrix form,

$$\begin{bmatrix} -1.790 & 1 & 0 \\ 1 & -2.790 & 1 \\ 0 & 1 & -1.790 \end{bmatrix} \begin{bmatrix} p_1^{n+1} \\ p_2^{n+1} \\ p_3^{n+1} \end{bmatrix} = \begin{bmatrix} -2370.378 \\ -2370.378 \\ -595.755 \end{bmatrix}. \qquad (8.837)$$

Solving the preceding system of equations yields $p_1^{n+1} = 2,668.97\,\text{psi}$, $p_2^{n+1} = 2,407.408\,\text{psi}$, $p_3^{n+1} = 1,677.626\,\text{psi}$. An MBC is carried out to validate the solution:

$$\text{Total fluid expansion} = \left[(3,000\,\text{psia} - 2,668.97\,\text{psia}) + (3,000\,\text{psia} - 2,407.77\,\text{psia}) + (3,000\,\text{psia} - 1,677.95\,\text{psia})\right]$$
$$\times 5\times10^{-6}\,\text{psi}^{-1} \times \frac{1,000\,\text{ft} \times 1,000\,\text{ft} \times 100\,\text{ft} \times 0.15}{5.165} = 29,999\ \text{STB}.$$
$$\qquad (8.838)$$

$$\text{Total production} = 30\,\text{days} \times 1,000\,\text{STB/D} = 30,000\,\text{STB}, \qquad (8.839)$$

$$\text{MBC} = \frac{\text{Total fluid expansion}}{\text{Total prodution}} = \frac{29,999}{30,000} = 0.999967. \qquad (8.840)$$

Problem 8.4.59

For the following polynomials,

$$f(x, y) = 4.5x^3 + 2y^2 - 54, \quad \dots\dots\dots\dots\dots\dots\dots\dots\dots\dots\dots\dots\dots \text{(8.841)}$$

$$g(x, y) = -1.5y^3 + 0.75x^2 + 37.5, \quad \dots\dots\dots\dots\dots\dots\dots\dots\dots\dots \text{(8.842)}$$

Implement the GNR protocol to solve for x and y expressed in the system of equations as

$$f(x, y) = 0, \quad \dots\dots\dots\dots\dots\dots\dots\dots\dots\dots\dots\dots\dots\dots\dots\dots \text{(8.843)}$$

$$g(x, y) = 0. \quad \dots\dots\dots\dots\dots\dots\dots\dots\dots\dots\dots\dots\dots\dots\dots\dots \text{(8.844)}$$

Use $x = 1.8$ and $y = 3.2$ as initial guesses.

Solution to Problem 8.4.59

The expressions of the entries of the Jacobian matrix are calculated as

$$\frac{\partial f}{\partial x} = 13.5x^2, \quad \dots\dots\dots\dots\dots\dots\dots\dots\dots\dots\dots\dots\dots\dots \text{(8.845)}$$

$$\frac{\partial f}{\partial g} = 4y, \quad \dots\dots\dots\dots\dots\dots\dots\dots\dots\dots\dots\dots\dots\dots\dots\dots \text{(8.846)}$$

$$\frac{\partial g}{\partial x} = 1.5x, \quad \dots\dots\dots\dots\dots\dots\dots\dots\dots\dots\dots\dots\dots\dots\dots \text{(8.847)}$$

$$\frac{\partial g}{\partial y} = -4.5y^2. \quad \dots\dots\dots\dots\dots\dots\dots\dots\dots\dots\dots\dots\dots\dots \text{(8.848)}$$

The Jacobian matrix is constructed as

$$J = \begin{bmatrix} \dfrac{\partial f}{\partial x} & \dfrac{\partial f}{\partial y} \\ \dfrac{\partial g}{\partial x} & \dfrac{\partial g}{\partial y} \end{bmatrix}. \quad \dots\dots\dots\dots\dots\dots\dots\dots\dots\dots \text{(8.849)}$$

The residual terms are defined as

$$R_1 = 4.5x^3 + 2y^2 - 54, \quad \dots\dots\dots\dots\dots\dots\dots\dots\dots\dots\dots\dots \text{(8.850)}$$

$$R_2 = -1.5y^3 + 0.75x^2 + 37.5. \quad \dots\dots\dots\dots\dots\dots\dots\dots\dots\dots \text{(8.851)}$$

The system of equations that needs to be solved is written as

$$\begin{bmatrix} 13.5x^2 & 4y \\ 1.5x & -4.5y^2 \end{bmatrix}^{(k)} \begin{bmatrix} \Delta x \\ \Delta y \end{bmatrix}^{(k+1)} = -\begin{bmatrix} 4.5x^3 + 2y^2 - 54 \\ -1.5y^3 + 0.75x^2 + 37.5 \end{bmatrix}^{(k)}. \quad \dots\dots \text{(8.852)}$$

At $x = 1.8$ and $y = 3.2$ and solving,

$$\begin{bmatrix} 43.74 & 12.8 \\ 2.7 & -46.08 \end{bmatrix}^{(0)} \begin{bmatrix} \Delta x \\ \Delta y \end{bmatrix}^{(1)} = \begin{bmatrix} 7.276 \\ 9.222 \end{bmatrix}^{(0)}. \quad \dots\dots\dots\dots\dots\dots \text{(8.853)}$$

Then, $\Delta x = 0.2211$ and $\Delta y = -0.1872$.
Updating,

$$x^{(1)} = x^{(0)} + \Delta x^{(1)}, \quad \dots\dots\dots\dots\dots\dots\dots\dots\dots\dots\dots\dots \text{(8.854)}$$

$$y^{(1)} = y^{(0)} + \Delta y^{(1)}, \quad \dots\dots\dots\dots\dots\dots\dots\dots\dots\dots\dots\dots \text{(8.855)}$$

yields $x = 2.021121$ and $y = 3.012826$.
  Solving,

$$\begin{bmatrix} 55.14655 & 12.0513 \\ 3.031681 & -40.847 \end{bmatrix}^{(1)} \begin{bmatrix} \Delta x \\ \Delta y \end{bmatrix}^{(2)} = \begin{bmatrix} -1.30685 \\ 0.457984 \end{bmatrix}^{(1)}. \quad \dots \dots \dots \dots \dots \dots \dots \dots \text{(8.856)}$$

Then, $\Delta x = -0.02091$ and $\Delta y = -0.01276$.
  Updating,

$$x^{(2)} = x^{(1)} + \Delta x^{(2)}, \quad \dots \dots \dots \dots \dots \dots \dots \dots \dots \dots \dots \dots \dots \dots \dots \text{(8.857)}$$

$$y^{(2)} = y^{(1)} + \Delta y^{(2)}, \quad \dots \dots \dots \dots \dots \dots \dots \dots \dots \dots \dots \dots \dots \dots \dots \text{(8.858)}$$

yields $x = 2.000212$, $y = 3.000062$, $R_1 = 0.012213$, and $R_2 = -0.00188$. Because the residuals are small, it can be stated that $x = 2.000212$ and $y = 3.000062$.

Problem 8.4.60

Consider the 2D homogeneous, isotropic, horizontal incompressible reservoir shown in **Fig. 8.52.** The reservoir boundaries are known to be sealed, except at the interface between Blocks 3 and 4. Block 4 is supported by a strong water drive so that its pressure is kept constant at 3,000 psi. A single-phase incompressible fluid is produced through a production well in Block 1, which is operated at a sandface pressure of 1,000 psia. Rock and fluid properties are given in **Table 8.55.**

1. Determine the pressure distribution in the system using the Gauss-Seidel method, with a pressure-convergence criterion of 1 psi. (*Note:* use an initial guess of 2,300 psi).
2. Calculate the flow rate of the well located in Block 1.
3. Perform an MBC to verify the accuracy of the results.
4. Consider the circumstance in which that the production well in Block 1 is suspected to be overloaded and thus, an additional well is planned to be drilled in Block 2. Note that the production rate of the well in Block 2 will not be allowed to exceed 50% of the original production rate of the well in Block 1. If the pressure gradient between the aquifer and Block 3 remains constant after drilling the second well, what is the sandface pressure of the new well in Block 2?

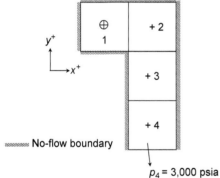

Fig. 8.52—Reservoir system of Problem 8.4.60.

$k_x = k_y =$ 80 md	
$\Delta x = \Delta y =$ 200 ft	
$h =$ 50 ft	
$\mu =$ 1 cp	
$B =$ 1 res bbl/STB	
$r_w =$ 0.3 ft	
$s =$ 0	

Table 8.55—Rock and fluid properties for Problem 8.4.60.

Solution to Problem 8.4.60

The governing flow equation for a 2D incompressible system is

$$\frac{\partial}{\partial x}\left(\frac{A_x k_x}{\mu B}\frac{\partial p}{\partial x}\right)\Delta x + \frac{\partial}{\partial y}\left(\frac{A_y k_y}{\mu B}\frac{\partial p}{\partial y}\right)\Delta y + q = 0. \quad \dots \dots \dots \dots \dots \dots \dots \dots \dots \text{(8.859)}$$

Simplifying for a homogeneous and isotropic reservoir,

$$\frac{\partial^2 p}{\partial x^2} + \frac{\partial^2 p}{\partial y^2} + \frac{q\mu B}{kV_b} = 0. \quad \dots \dots \dots \dots \dots \dots \dots \dots \dots \dots \dots \dots \dots \text{(8.860)}$$

In finite-difference form,

$$p_{i,j-1} + p_{i-1,j} - 4p_{i,j} + p_{i+1,j} + p_{i,j+1} + \frac{q\mu B}{hk} = 0. \qquad (8.861)$$

1.
- Block 1:

$$-p_1 + p_2 + \frac{q\mu B}{hk} = 0, \qquad (8.862)$$

and

$$q_1 = -\Omega_1 \left(p_1 - p_{sf} \right), \qquad (8.863)$$

$$\Omega = \frac{2\pi kh}{\mu B \left[\ln\left(\dfrac{r_e}{r_w}\right) + s \right]} = 5.79 \text{ STB/D-psi}, \qquad (8.864)$$

where $r_e = 0.2\Delta x = 40$ ft. Therefore,

$$-2.284 p_1 + p_2 = -1,284. \qquad (8.865)$$

- Block 2: $p_1 - 2p_2 + p_3 = 0.$ \qquad (8.866)

- Block 3: $p_1 - 2p_3 = -3,000$ \qquad (8.867)

Implementing the Gauss-Seidel method,

$$p_1^{(k+1)} = \frac{1,284 + p_2^{(k)}}{2.284}, \qquad (8.868)$$

$$p_2^{(k+1)} = \frac{p_3^{(k)} + p_1^{(k+1)}}{2}, \qquad (8.869)$$

$$p_3^{(k+1)} = \frac{3,000 + p_2^{(k+1)}}{2}. \qquad (8.870)$$

The calculation results are summarized in **Table 8.56.**
Therefore, $p_1 = 1,412$ psi, $p_2 = 1,941$ psi, $p_3 = 2,470$ psi.

Iteration Level	p_1	p_2	p_3	Δp_{max}
0	2,300	2,300	2,300	–
1	1,569.2	1,934.6	2,467.3	730.8
2	1,409.2	1,938.3	2,469.1	160
3	1410.8	1,940	2,470	1.7
4	1,411.6	1,940.8	2,470.4	0.8

Table 8.56—Calculation results from Block 3 of Part 1 from Solution to Problem 8.4.60.

2. Calculating q_1:

$$q_1 = -\Omega\left(p_1 - p_{sf}\right) = -5.79\left(1,412 - 1,000\right) = -2,385.5 \text{ STB/D}. \qquad (8.871)$$

MBC:

$$q_{in} = \frac{hk}{\mu B}\Delta p = 1.127 \times 10^{-3} \times \frac{50 \times 80 \text{ md}}{1 \times 1}\left(3,000 - 2,470\right) \text{ psi} = 2,389 \text{ STB/D}, \qquad (8.872)$$

$$\text{MBC} = \frac{q_{in}}{q_{out}} = \frac{2,389}{2,385.5} = 1.001. \qquad (8.873)$$

3.

$$\left.\frac{\partial p}{\partial y}\right|_{4\to3} = \frac{2,470-3,000}{200} = -2.65 \text{ psi/ft,} \dots\dots\dots\dots\dots\dots\dots\dots\dots\dots\dots\dots\dots\dots(8.874)$$

$$q_2 = q_1 = -2,385.5 \times 50\% = -1,192.75 \text{ STB/D.} \dots\dots\dots\dots\dots\dots\dots\dots\dots\dots\dots(8.875)$$

Thus,

$$\left.\frac{\partial p}{\partial y}\right|_{4\to3} = \left.\frac{\partial p}{\partial y}\right|_{3\to2} = -2.65 \text{ psi/ft} \dots\dots\dots\dots\dots\dots\dots\dots\dots\dots\dots\dots\dots\dots(8.876)$$

and

$$\left.\frac{\partial p}{\partial x}\right|_{2\to1} = \left.\frac{1}{2}\frac{\partial p}{\partial y}\right|_{3\to2} = 1.325\frac{\text{psi}}{\text{ft}}. \dots\dots\dots\dots\dots\dots\dots\dots\dots\dots\dots\dots\dots\dots(8.877)$$

- Find p_1:

$$p_1 = p_{sf} - \frac{q_1}{\Omega} = 1,000 - \frac{(-1,192.75)}{5.79} = 1,206 \text{ psi.} \dots\dots\dots\dots\dots\dots\dots\dots\dots\dots(8.878)$$

- Find p_2:

$$p_2 = \Delta x 1.325 + p_1 = 1,471 \text{ psi.} \dots\dots\dots\dots\dots\dots\dots\dots\dots\dots\dots\dots\dots\dots\dots\dots(8.879)$$

- Find p_{sf}:

$$p_{sf} = \frac{q_2}{\Omega} + p_2 = -\frac{1,192.75}{5.79} + 1,471 = 1,265 \text{ psi.} \dots\dots\dots\dots\dots\dots\dots\dots\dots\dots\dots(8.880)$$

Problem 8.4.61

The 1D homogeneous, horizontal reservoir shown in **Fig. 8.53** is 100% saturated with a slightly compressible fluid. The reservoir is surrounded by no-flow boundaries. Initial reservoir pressure is 3,000 psia. A production well in Block 2 is produced at a constant rate of 500 STB/D for 30 days. At the end of 30 days, the well is shut in for a period of 30 days. See the reservoir properties in **Table 8.57.** What is the average pressure in the reservoir at the end of the shut-in period? Using an MBC, check the accuracy of the pressure distribution obtained at the end of the shut-in period.

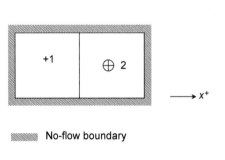

No-flow boundary

Fig. 8.53—Reservoir system of Problem 8.4.61.

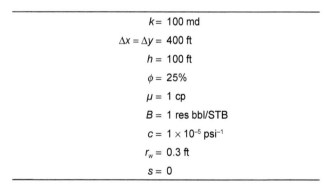

$k =$	100 md
$\Delta x = \Delta y =$	400 ft
$h =$	100 ft
$\phi =$	25%
$\mu =$	1 cp
$B =$	1 res bbl/STB
$c =$	1×10^{-5} psi^{-1}
$r_w =$	0.3 ft
$s =$	0

Table 8.57—Reservoir properties for Probem 8.4.61.

Solution to Problem 8.4.61

The governing PDE of the system can be written as

$$\frac{\partial}{\partial x}\left(\frac{A_x k_x}{\mu B}\frac{\partial p}{\partial x}\right)\Delta x + q = \frac{V_b \phi c_t}{5.615}\left(\frac{\partial p}{\partial t}\right). \dots\dots\dots\dots\dots\dots\dots\dots\dots\dots\dots(8.881)$$

For a homogeneous reservoir system,

$$\frac{\partial^2 p}{\partial x^2} + \frac{q\mu B}{V_b k} = \frac{\phi c_t \mu B}{5.615 k}\left(\frac{\partial p}{\partial t}\right). \dots\dots\dots\dots\dots\dots\dots\dots\dots\dots\dots\dots\dots\dots(8.882)$$

The characteristic finite-difference representation is

$$\left(p_{i-1,j}^{n+1} - 2p_{i,j}^{n+1} + p_{i+1,j}^{n+1}\right) + \left(\frac{q\mu B}{hk_x}\right)_{i,j} = \left[\frac{\mu Bc_t\phi(\Delta x)^2}{k_x 5.615\Delta t}\right]\left(p_{i,j}^{n+1} - p_{i,j}^n\right). \quad \dots \quad (8.883)$$

The coefficient is

$$\frac{\mu Bc_t\phi(\Delta x)^2}{k_x 5.615\Delta t} = \frac{(1)(1)(0.25)\left(1\times10^{-5}\right)(400)^2}{(5.165)\left(1.127\times10^{-3}\right)(100)(30)} = 0.021, \quad \dots \quad (8.884)$$

$$\frac{q\mu B}{hk_x} = \frac{(-500)(1)(1)}{\left(1.127\times10^{-3}\right)(100)(100)} = -44.37. \quad \dots \quad (8.885)$$

For Block 1,

$$p_2^{n+1} - 1.021p_1^{n+1} = 0.021(p_1^{n+1} - p_1^n). \quad \dots \quad (8.886)$$

For Block 2,

$$p_1^{n+1} - 1.021p_2^{n+1} = 0.021(p_2^{n+1} - p_2^n) + 44.37. \quad \dots \quad (8.887)$$

- Between 0 and 30 days: Rearranging the system of equations in a matrix form,

$$\begin{bmatrix} -1.021 & 1 \\ 1 & -1.021 \end{bmatrix}\begin{bmatrix} p_1^{n+1} \\ p_1^{n+1} \end{bmatrix} = \begin{bmatrix} -63 \\ -18.63 \end{bmatrix}. \quad \dots \quad (8.888)$$

Solving, $p_1^{n+1} = 1{,}954.55$ psi and $p_2^{n+1} = 1{,}932.59$ psi.

- Between 30 and 60 days: Rearranging the system of equations in a matrix form,

$$\begin{bmatrix} -1.021 & 1 \\ 1 & -1.021 \end{bmatrix}\begin{bmatrix} p_1^{n+1} \\ p_1^{n+1} \end{bmatrix} = \begin{bmatrix} -41.05 \\ -40.58 \end{bmatrix}.$$

Solving, $p_1^{n+1} = 1{,}943.7$ psi and $p_2^{n+1} = 1{,}943.5$ psi.

Use the cumulative MBC to explore the accuracy of these results.

- Total fluid expansion:

$$\Delta V = \Delta V_1 + \Delta V_2 = -cV_p\left(\Delta p_1 + \Delta p_2\right) = 1\times10^{-5}\frac{(400)(400)(100)(0.25)}{5.615}$$
$$\left[(1{,}943.7 - 3{,}000) + (1{,}943.5 - 3{,}000)\right] = 15{,}051.11 \text{ STB}. \quad \dots \quad (8.889)$$

- Total fluid produced:

$$Q_t = 30 \text{ days} \times 500 \text{ STB/D} = 15{,}000 \text{ STB}, \quad \dots \quad (8.890)$$

$$\text{MBC} = \frac{\Delta V}{Q_t} = 1.003. \quad \dots \quad (8.891)$$

Problem 8.4.62

Consider the incompressible flow system in **Fig. 8.54,** which is surrounded by no-flow boundaries except at Blocks 1 and 11. A pressure gradient at the eastern boundary of Block 1 is b psi/ft. Block 11 is charged by an aquifer so that the pressure of Block 11 is constant at p_{11} psia. Note that the production well in Block 5 is produced at a sandface pressure of p_{sf5} psia. The production well in Block 9 is produced at the rate of q_9 STB/D.

Give the definitions of the SIP coefficients $C_{i,j}$ and $Q_{i,j}$ for Blocks 1, 5, 6, 8, and 9.

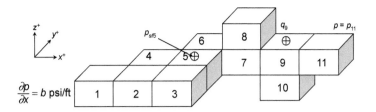

Fig. 8.54—Reservoir system of Problem 8.4.62.

Solution to Problem 8.4.62

Definitions for the incompressible flow system are as follows:

- C coefficient for a block with no well:

$$C_{i,j,k} = (B+S+W+E+N+A)_{i,j,k}. \dots\dots\dots\dots\dots\dots\dots\dots\dots\dots\dots\dots(8.892)$$

- C coefficient for a block with a flow-rate-specified well:

$$C_{i,j,k} = (B+S+W+E+N+A)_{i,j,k}. \dots\dots\dots\dots\dots\dots\dots\dots\dots\dots\dots(8.893)$$

- C coefficient for a block with a sandface-pressure-specified well:

$$C_{i,j,k} = (B+S+W+E+N+A+\Omega)_{i,j,k}. \dots\dots\dots\dots\dots\dots\dots\dots\dots\dots(8.894)$$

- Q coefficient for a block with no well:

$$Q_{i,j,k} = \text{gravitational terms.} \dots\dots\dots\dots\dots\dots\dots\dots\dots\dots\dots\dots(8.895)$$

- Q coefficient for a block with a flow-rate-specified well:

$$Q_{i,j,k} = \text{gravitational terms} - q_{i,j,k}. \dots\dots\dots\dots\dots\dots\dots\dots\dots\dots(8.896)$$

- Q coefficient for a block with a sandface-pressure-specified well:

$$C_{i,j,k} = \text{gravitational terms} - p_{sfi,j,k}\,\Omega_{i,j,k}. \dots\dots\dots\dots\dots\dots\dots\dots\dots(8.897)$$

Block 1: $C_1 = E_1$; $Q_1 = W_1 b \Delta x. \dots\dots\dots\dots\dots\dots\dots\dots\dots\dots\dots\dots\dots(8.898)$

Block 5: $C_5 = -\left(S_5 + W_5 + N_5 + \Omega_5\right)$; $Q_5 = -\Omega_5 p_{sf5}. \dots\dots\dots\dots\dots\dots\dots\dots(8.899)$

Block 6: $C_6 = -\left(S_6 + E_6\right)$; $Q_6 = 0. \dots\dots\dots\dots\dots\dots\dots\dots\dots\dots\dots\dots\dots(8.900)$

Block 8: $C_8 = -B_8$; $Q_8 = \dfrac{1}{144}\dfrac{g}{g_c}\rho B_g\left(G_8 - G_7\right). \dots\dots\dots\dots\dots\dots\dots\dots(8.901)$

Block 9: $C_9 = -\left(B_9 + W_9 + E_9\right)$; $Q_9 = \dfrac{1}{144}\dfrac{g}{g_c}\rho B_9\left(G_9 - G_{10}\right) - q_6 - E_9 p_{11}. \dots\dots\dots\dots(8.902)$

Problem 8.4.63

Fig. 8.55 represents the depth and thickness arrays of the 2D finite-difference grid of a homogeneous and isotropic reservoir in which a single-phase flow of incompressible fluid is taking place. There is one production well located in Block (5,3). Use $k_x = k_y = 100$ md, $\mu = 1$ cp, $B = 1$ res bbl/STB, $r_w = 0.3$ ft, and $s = 0$.

1. Calculate all SIP coefficients for Block (5,3), if a production rate of 600 STB/D is specified at the well.
2. Calculate all SIP coefficients for Block (5,3), if a sandface pressure of 1,100 psia is specified at the wellbore.
3. Calculate all SIP coefficients for Block (5,3), if the well is shut in.

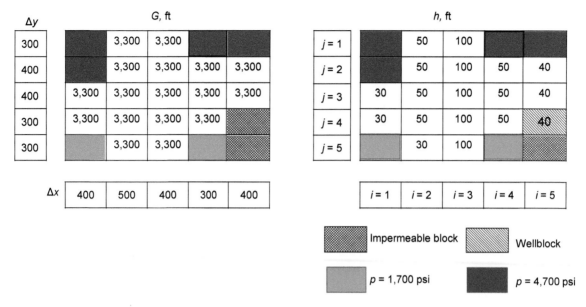

Fig. 8.55—Reservoir system of Problem 8.4.63.

Solution to Problem 8.4.63

1. For the well Block (5,3):

$$S = 0 \text{ STB/D-psi,} \qquad \qquad \qquad \qquad \qquad \qquad \qquad \qquad \qquad \text{(8.903)}$$

$$E = 0 \text{ STB/D-psi,} \qquad \qquad \qquad \qquad \qquad \qquad \qquad \qquad \qquad \text{(8.904)}$$

$$W = \frac{1}{\mu B}\left[\frac{2A_{x5,3}A_{x4,3}k}{\left(A_{x5,3}\Delta x_{4,3} + A_{x4,3}\Delta x_{5,3}\right)}\right] = \frac{1}{(1)(1)}\left[\frac{2(16,000)(20,000)(100)}{(20,000)(400) + (16,000)(300)}\right] \qquad \text{(8.905)}$$
$$= 5.635 \text{ STB/D-psi,}$$

$$N = \frac{1}{\mu B}\frac{A_{y\,5,3}k}{\Delta y_{5,3}} = 1.127 \times 10^{-3} \times \frac{1}{(1)(1)}\frac{(16,000)(100)}{400} = 4.508 \text{ STB/D-psi,} \qquad \text{(8.906)}$$

$$C = -\left(W + N + S + E\right) = -10.143 \text{ STB/D-psi,} \qquad \qquad \qquad \text{(8.907)}$$

$$Q = -q = 600 \text{ STB/D.} \qquad \qquad \qquad \qquad \qquad \qquad \qquad \text{(8.908)}$$

2. If sandface pressure is specified,

$$\Omega = \frac{2\pi kh}{\mu B\left[\ln\left(\dfrac{r_e}{r_w}\right) + s\right]} = 5.08 \text{ STB/D-psi,} \qquad \qquad \qquad \qquad \text{(8.909)}$$

$$C = -\left(W + N + S + E + \Omega\right) = -15.223 \text{ STB/D-psi,} \qquad \qquad \text{(8.910)}$$

$$Q = -\Omega p_{sf} = -5,587.8 \text{ STB/D.} \qquad \qquad \qquad \qquad \qquad \qquad \text{(8.911)}$$

3. If the well is shut in,

$$C = -\left(W + N + S + E\right) = -10.143 \text{ STB/D-psi,} \qquad \qquad \qquad \text{(8.912)}$$

$$Q = 0 \text{ STB/D-psi.} \qquad \qquad \qquad \qquad \qquad \qquad \qquad \qquad \text{(8.913)}$$

Problem 8.4.64

Consider the single-phase oil reservoir shown in **Fig. 8.56.1.** The permeability distribution is homogeneous and isotropic: $k_x = k_y = k_z = 30$ md. The permeability at Block 3 is found to be a strong function of pressure during the massive hydraulic-fracture treatment the well is exposed to: $k_3 = 50 + \sqrt{p}$ (unit: md).

Ignore all gravitational forces. Use the GNR protocol to solve the pressure distribution after 30 days, and calculate the flow rate from the well using the reservoir properties found in **Table 8.58** (in calculating the flow rate, use the first iteration result of the GNR protocol).

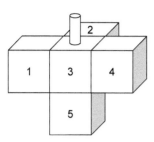

Fig. 8.56.1—Reservoir system of Problem 8.4.64.

$\Delta x = \Delta y = \Delta z = 600$ ft
Porosity = 18%
$r_w = 0.25$ ft
$c_o = 4 \times 10^{-5}$ psi^{-1}
$c_\phi = 0$ psi^{-1}
$B_o \approx$ const = 0.93 bbl/STB
$\mu_o \approx$ const = 0.8 cp
Initial pressure = 2,000 psi
Sandface pressure = 1,900 psi
$s = 0$

Table 8.58—Reservoir properties for Problem 8.4.64.

Solution to Problem 8.4.64

The entire system can be solved using only Blocks 1 and 3 (Block 1, 2, 4, and 5 are symmetrical with respect to each other), as shown in **Fig. 8.56.2.** The continuity equation could be written as

Fig. 8.56.2—Simplified reservoir system using asymmetry.

$$\frac{\partial}{\partial x}\left(\frac{A_x k_x}{\mu B}\frac{\partial p}{\partial x}\right)\Delta x + q = \frac{V_b \phi c_t}{5.615}\left(\frac{\partial p}{\partial t}\right). \quad (8.914)$$

Because sandface pressure is specified, and in consideration of the symmetry of the system shown in Fig. 8.56.2 ($p_1 = p_4 = p_5 = p_2$),

$$q_3' = -\frac{\Omega_3}{4}\left(p_3 - p_{sf}\right), \quad (8.915)$$

where

$$\frac{\Omega_3}{4} = \frac{2\pi k_3 h}{\mu B \ln\left(\frac{r_e}{r_w}\right)} \times \frac{1}{4} = \frac{2 \times 3.14159 \times 600 \text{ ft} \times 0.001127}{0.8 \text{ cp} \times 0.93 \times \ln(0.2 \times 600 \text{ ft}/0.25 \text{ ft})} \times \left(50 + \sqrt{p_3}\right) = 0.231\left(50 + \sqrt{p_3}\right), \quad (8.916)$$

and

$$\Gamma = \frac{V_b \phi c_t}{5.615 dt} = \frac{600 \text{ ft} \times 600 \text{ ft} \times 600 \text{ ft} \times 0.18 \times 4 \times 10^{-5}}{5.615 \times 30} = 9.2324 \text{ STB/D-psi.} \quad (8.917)$$

The transmissibility factor between Blocks 1 and 3 can be calculated as

$$T = \frac{2A_1 A_3 k_1 k_3}{(\Delta x_1 A_3 k_3 + \Delta x_3 A_1 k_1)\mu B}. \quad (8.918)$$

Because $\Delta x = \Delta y$ and the thickness of each block is uniform,

$$T = \frac{2hk_1 k_3}{\mu B(k_3 + k_1)} = \frac{2 \times 600 \text{ ft} \times 30 \text{ md} \times 0.001127 \times \left(50 + \sqrt{p_3}\right) \times 0.001127}{0.8 \text{ cp} \times 0.93 \times \left(50 + \sqrt{p_3} + 30\right) \times 0.001127}$$

$$= \frac{54.48\left(50 + \sqrt{p_3}\right)}{\left(80 + \sqrt{p_3}\right)} \text{ STB/D-psi.} \quad (8.919)$$

Writing the PDE in the finite-difference form using the SIP notation results in the following:

- At Block 1:

$$Tp_3 + C_1 p_1 = Q_1, \dots\dots\dots\dots\dots\dots\dots\dots\dots\dots\dots\dots\dots\dots\dots\dots\dots\dots (8.920)$$

$$C_1 = -(T + \Gamma), \dots\dots\dots\dots\dots\dots\dots\dots\dots\dots\dots\dots\dots\dots\dots\dots\dots\dots (8.921)$$

$$Q_1 = -\Gamma p_i. \dots\dots\dots\dots\dots\dots\dots\dots\dots\dots\dots\dots\dots\dots\dots\dots\dots\dots\dots (8.922)$$

The residual can be calculated as $R_1(p_1, p_3) = Q_1 - (Tp_3 + C_1 p_1)$. $\dots\dots\dots\dots\dots\dots\dots\dots (8.923)$

- At Block 3:

$$Tp_1 + C_3 p_3 = Q_3, \dots\dots\dots\dots\dots\dots\dots\dots\dots\dots\dots\dots\dots\dots\dots\dots\dots\dots (8.924)$$

$$C_3 = -(T + \Gamma + \Omega), \dots\dots\dots\dots\dots\dots\dots\dots\dots\dots\dots\dots\dots\dots\dots\dots\dots (8.925)$$

$$Q_3 = -\Gamma p_i - \Omega p_{sf}. \dots\dots\dots\dots\dots\dots\dots\dots\dots\dots\dots\dots\dots\dots\dots\dots\dots (8.926)$$

The residual can be calculated as $R_3(p_1, p_3) = Q_3 - (Tp_1 + C_3 p_3)$. $\dots\dots\dots\dots\dots\dots\dots\dots (8.927)$

Constructing the Jacobian matrix:

$$\frac{\partial R_1}{\partial p_1} = \frac{R_1(p_1 + \varepsilon, p_3) - R_1(p_1, p_3)}{\varepsilon} = 50.65, \dots\dots\dots\dots\dots\dots\dots\dots\dots\dots\dots (8.928)$$

$$\frac{\partial R_1}{\partial p_3} = \frac{R_1(p_1, p_3 + \varepsilon) - R_1(p_1, p_3)}{\varepsilon} = -41.42, \dots\dots\dots\dots\dots\dots\dots\dots\dots\dots (8.929)$$

$$\frac{\partial R_3}{\partial p_1} = \frac{R_3(p_1 + \varepsilon, p_3) - R_3(p_1, p_3)}{\varepsilon} = -41.42, \dots\dots\dots\dots\dots\dots\dots\dots\dots\dots (8.930)$$

$$\frac{\partial R_3}{\partial p_3} = \frac{R_3(p_1, p_3 + \varepsilon) - R_3(p_1, p_3)}{\varepsilon} = 72.81, \dots\dots\dots\dots\dots\dots\dots\dots\dots\dots (8.931)$$

$$R_1 = 0, \dots\dots\dots\dots\dots\dots\dots\dots\dots\dots\dots\dots\dots\dots\dots\dots\dots\dots\dots\dots\dots\dots\dots (8.932)$$

$$R_3 = -2,190.375. \dots\dots\dots\dots\dots\dots\dots\dots\dots\dots\dots\dots\dots\dots\dots\dots\dots\dots\dots\dots (8.933)$$

In matrix form,

$$\begin{bmatrix} \dfrac{\partial R_1}{\partial p_1} & \dfrac{\partial R_1}{\partial p_3} \\ \dfrac{\partial R_3}{\partial p_1} & \dfrac{\partial R_3}{\partial p_3} \end{bmatrix} \begin{bmatrix} \Delta p_1 \\ \Delta p_3 \end{bmatrix} = - \begin{bmatrix} R_1 \\ R_3 \end{bmatrix}. \dots\dots\dots\dots\dots\dots\dots\dots (8.934)$$

If one analytically calculates the Jacobian matrix, the results will be as follows:

$$\frac{\partial R_1}{\partial p_1} = \Gamma + T, \dots\dots\dots\dots\dots\dots\dots\dots\dots\dots\dots\dots\dots\dots\dots\dots\dots\dots\dots (8.935)$$

$$\frac{\partial R_1}{\partial p_3} = -T, \dots\dots\dots\dots\dots\dots\dots\dots\dots\dots\dots\dots\dots\dots\dots\dots\dots\dots\dots\dots (8.936)$$

$$\frac{\partial R_3}{\partial p_1} = -T, \dots\dots\dots\dots\dots\dots\dots\dots\dots\dots\dots\dots\dots\dots\dots\dots\dots\dots\dots\dots (8.937)$$

$$\frac{\partial R_3}{\partial p_3} = \Omega + \Gamma + T + \frac{\partial \Omega}{\partial p_3}(p_3 - p_{sf}). \dots\dots\dots\dots\dots\dots\dots\dots\dots\dots\dots\dots (8.938)$$

Solving the system of equations for Δp_1, Δp_3 results in $\Delta p_1 = -56.24$ psi and $\Delta p_3 = -45.99$ psi.

Stop the GNR protocol and update the pressure to result in $p_1 = 1,943.76$ psi and $p_3 = 1,954.01$ psi. Calculate the flow rate:

$$\Omega = 4 \times 0.231 \times \left(50 + \sqrt{p_3}\right) = 0.924 \times \left(50 + \sqrt{1,954.01}\right) = 86.94 \text{ STB/D-psi}, \dots\dots\dots\dots\dots\dots\dots (8.939)$$

$$q = -\Omega\left(p_3 - p_{sf}\right) = -86.94 \times \left(1,943.76 - 1,900\right) = -3,804.38 \text{ STB/D}. \dots\dots\dots\dots\dots (8.940)$$

Problem 8.4.65

Given a linear system of equations:

$$\begin{cases} a_{11}x_1 + a_{12}x_2 + a_{13}x_3 + \cdots + a_{1n}x_n = b_1 \\ a_{21}x_1 + a_{22}x_2 + a_{23}x_3 + \cdots + a_{2n}x_n = b_2 \\ a_{31}x_1 + a_{32}x_2 + a_{33}x_3 + \cdots + a_{3n}x_n = b_3 \\ \quad\vdots \\ a_{n1}x_1 + a_{n2}x_2 + a_{n3}x_3 + \cdots + a_{nn}x_n = b_n \end{cases}$$

If one were to solve the system of the equation using the GNR protocol and use $\left[x_1, x_2 \dots, x_n\right] = 0$ as an initial guess, is it correct to state that only one GNR protocol is required to obtain the solution?

Solution to Problem 8.4.65

The GNR protocol aims to update the dependent variable by solving a system of equations, which are arranged in the following form:

$$\begin{bmatrix} \dfrac{\partial R_1}{\partial x_1} & \dfrac{\partial R_1}{\partial x_2} & \cdots & \dfrac{\partial R_1}{\partial x_n} \\ \dfrac{\partial R_2}{\partial x_1} & \dfrac{\partial R_2}{\partial x_2} & \cdots & \dfrac{\partial R_2}{\partial x_n} \\ \vdots & \vdots & \ddots & \vdots \\ \dfrac{\partial R_n}{\partial x_1} & \dfrac{\partial R_n}{\partial x_2} & \cdots & \dfrac{\partial R_n}{\partial x_n} \end{bmatrix}^{(k)} \begin{bmatrix} \Delta x_1 \\ \Delta x_2 \\ \vdots \\ \Delta x_n \end{bmatrix}^{(k+1)} = - \begin{bmatrix} R_1 \\ R_2 \\ \vdots \\ R_n \end{bmatrix}^{(k)} \dots\dots\dots\dots\dots\dots\dots\dots\dots\dots (8.941)$$

Rewriting the original system of equations to calculate the residuals results in

$$\begin{cases} R_1 = -b_1 + a_{11}x_1 + a_{12}x_2 + a_{13}x_3 + \cdots + a_{1n}x_n \\ R_2 = -b_2 + a_{21}x_1 + a_{22}x_2 + a_{23}x_3 + \cdots + a_{2n}x_n \\ R_3 = -b_3 + a_{31}x_1 + a_{32}x_2 + a_{33}x_3 + \cdots + a_{3n}x_n \\ \quad\vdots \\ R_n = -b_n + a_{n1}x_1 + a_{n2}x_2 + a_{n3}x_3 + \cdots + a_{nn}x_n \end{cases} \dots\dots\dots\dots\dots\dots\dots (8.942)$$

The Jacobian matrix will turn out to be the coefficient matrix of the original system of equations, which is

$$J = \begin{bmatrix} a_{11} & a_{12} & \cdots & a_{1n} \\ a_{21} & a_{22} & \cdots & a_{2n} \\ \vdots & \vdots & \ddots & \vdots \\ a_{n1} & a_{n2} & \cdots & a_{nn} \end{bmatrix} . \dots\dots\dots\dots\dots\dots\dots\dots\dots\dots\dots\dots\dots\dots (8.943)$$

By taking the initial guess of zeros for $x_1, x_2 \dots x_n$, the residual will be calculated as

$$R = \begin{bmatrix} -b_1 \\ -b_2 \\ \vdots \\ -b_n \end{bmatrix} . \dots\dots\dots\dots\dots\dots\dots\dots\dots\dots\dots\dots\dots\dots\dots\dots\dots\dots (8.944)$$

Thus, the first GNR iteration solves the original system of equations, and the correct solution of the first GNR iteration yields the solution vector for the system of equations:

$$
\begin{bmatrix}
a_{11} & a_{12} & \cdots & a_{1n} \\
a_{21} & a_{22} & \cdots & a_{2n} \\
\vdots & \vdots & \ddots & \vdots \\
a_{n1} & a_{n2} & \cdots & a_{nn}
\end{bmatrix}
\begin{bmatrix}
\Delta x_1 \\
\Delta x_2 \\
\vdots \\
\Delta x_n
\end{bmatrix}
= -
\begin{bmatrix}
-b_1 \\
-b_2 \\
\vdots \\
-b_n
\end{bmatrix},
\quad \dots \dots \dots \dots \dots \dots \dots \dots \dots \dots \dots \dots \dots \dots \dots \dots (8.945)
$$

$$
\begin{cases}
x_1 = 0 + \Delta x_1 \\
x_2 = 0 + \Delta x_2 \\
\vdots \\
x_n = 0 + \Delta x_n
\end{cases}
\quad \dots \dots \dots \dots \dots \dots \dots \dots \dots \dots \dots \dots \dots \dots \dots \dots \dots \dots (8.946)
$$

Problem 8.4.66

Consider single-phase slightly compressible fluid flow of the reservoir shown in **Fig. 8.57.1.** All external boundaries are sealed. Initially the reservoir is at a pressure of 4,000 psia. The well in Block 2 is put on production at a constant-sandface pressure of 3,600 psia for a period of 10 days. The effect of the depth gradients can be ignored for this problem. The skin at the wellbore is dependent on the flow rate and is given by the following relationship:

$$ s = -\frac{q\,(\text{STB/D})}{100}. $$

Generate the necessary system of equations (Jacobian) to solve for the pressure distribution at the end of 10 days using the GNR protocol. (Use a timestep of $\Delta t = 10$ days and additional reservoir properties in **Table 8.59.**) *Note:* Do not attempt to solve the system of equations.

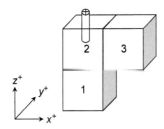

Fig. 8.57.1—Reservoir system
of Problem 8.4.66.

$k_x = k_y = 80$ md
$\Delta x = \Delta y = \Delta z = 200$ ft
$\phi = 15\%$
$c = 1 \times 10^{-6}$ psi^{-1}
$\mu = 1$ cp
$B = 1$ res bbl/STB
$r_w = 0.27$ ft

Table 8.59—Reservoir properties for Problem 8.4.66.

Solution to Problem 8.4.66

Because the given 2D reservoir is isotropic and homogeneous in terms of permeability, and the dimensions of the grid-blocks are uniform, the system could be simplified to a 1D reservoir system, shown in **Fig. 8.57.2** (recall that depth gradients are ignored).

The continuity equation for slightly compressible fluid flow in the given reservoir system shown in Fig. 8.57.2 could be written as

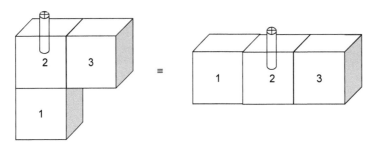

Fig. 8.57.2—Equivalent reservoir system.

$$\frac{\partial}{\partial x}\left(\frac{A_x k_x}{\mu B}\frac{\partial p}{\partial x}\right)\Delta x + q = \frac{V_b \phi c_t}{5.615}\left(\frac{\partial p}{\partial t}\right). \dots\dots\dots\dots\dots\dots\dots\dots\dots\dots (8.947)$$

Because the reservoir is homogeneous in permeability and area, Eq. 8.947 could be simplified to the following form:

$$\frac{\partial^2 p}{\partial x^2} + \frac{q\mu B}{V_b k} = \frac{\phi c_t}{5.615 k}\left(\frac{\partial p}{\partial t}\right). \dots\dots\dots\dots\dots\dots\dots\dots\dots\dots\dots\dots (8.948)$$

The finite-difference form of Eq. 8.948 could be written as

$$p_{i-1}^{n+1} - 2p_i^{n+1} + p_{i+1}^{n+1} + \frac{q\mu B}{hk} = \frac{\mu B \phi c_t \Delta x^2}{5.615 k \Delta t}(p_i^{n+1} - p_i^n). \dots\dots\dots\dots\dots\dots\dots (8.949)$$

Upon writing finite-difference equations at each block,

- Block 1 (see **Fig 8.57.3**):

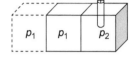

$$p_1^{n+1} - 2p_1^{n+1} + p_2^{n+1} = \frac{\mu B \phi c_t \Delta x^2}{5.615 k \Delta t}(p_1^{n+1} - p_1^n), \dots\dots\dots\dots (8.950)$$

$$-p_1^{n+1} + p_2^{n+1} = \frac{\mu B \phi c_t \Delta x^2}{5.615 k \Delta t}(p_1^{n+1} - p_1^n), \dots\dots\dots\dots (8.951)$$

Fig. 8.57.3—Treatment of boundary conditions of Block 1.

$$-p_1^{n+1} + p_2^{n+1} = \frac{(1\ \text{cp})\left(\dfrac{1\ \text{res bbl}}{\text{STB}}\right)(1 \times 10^{-5}\ \text{psi}^{-1})(200\ \text{ft})(200\ \text{ft})}{5.615(80\ \text{md})\left(0.001127\dfrac{\text{perms}}{\text{md}}\right)(10\ \text{days})}(p_1^{n+1} - 4,000), \dots\dots\dots\dots (8.952)$$

$$-p_1^{n+1} + p_2^{n+1} = 0.0118(p_1^{n+1} - 4,000), \dots\dots\dots\dots\dots\dots\dots\dots\dots\dots\dots\dots\dots (8.953)$$

$$R_1 = -1.0118 p_1^{n+1} + p_2^{n+1} + 47.2 = 0. \dots\dots\dots\dots\dots\dots\dots\dots\dots\dots\dots\dots\dots\dots (8.954)$$

- Block 2 (see **Fig. 8.57.4**): Because Blocks 3 and 1 are symmetrical with respect to Block 2 ($p_1 = p_3$),

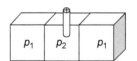

$$p_1^{n+1} - 2p_2^{n+1} + p_1^{n+1} + \frac{q_2^{n+1}\mu B}{hk} = \frac{\mu B \phi c_t \Delta x^2}{5.615 k \Delta t}(p_2^{n+1} - p_2^n), \dots\dots (8.955)$$

Fig. 8.57.4—Treatment of boundary conditions of Block 2.

$$2p_1^{n+1} - 2p_2^{n+1} + \frac{q_2^{n+1}(1)(1)}{(200)(80)(0.001127)} = 0.0118(p_2^{n+1} - 4,000),$$
$$\dots\dots\dots\dots\dots\dots\dots\dots\dots\dots\dots\dots\dots\dots\dots\dots\dots\dots (8.956)$$

$$R_2 = 2p_1^{n+1} - 2.0118 p_2^{n+1} + 0.055 q_2^{n+1} + 47.2 = 0. \dots\dots\dots\dots\dots\dots\dots\dots\dots (8.957)$$

Also at Block 2, the following equations can be written:

$$q_2^{n+1} = -\frac{2\pi kh}{\mu B\left[\ln\left(\dfrac{r_e}{r_w}\right) - \dfrac{q_2^{n+1}}{100}\right]}\left(p_2^{n+1} - p_{sf2}\right), \dots\dots\dots\dots\dots\dots\dots\dots\dots\dots (8.958)$$

$$q_2^{n+1} = -\frac{2(3.14)(80)(0.001127)(200)}{(1)(1)\left[\ln\left(\dfrac{0.197 \times 200}{0.27}\right) - \dfrac{q_2^{n+1}}{100}\right]}\left(p_2^{n+1} - 3,600\right), \dots\dots\dots\dots\dots\dots\dots\dots (8.959)$$

$$q_2^{n+1} = -\frac{2(3.14)(80)(0.001127)(200)}{(1)(1)\left[\ln\left(\dfrac{0.197 \times 200}{0.27}\right) - \dfrac{q_2^{n+1}}{100}\right]}\left(p_2^{n+1} - 3,600\right), \dots\dots\dots\dots\dots\dots\dots\dots (8.960)$$

$$q_2^{n+1}\left(5 - \frac{q_2^{n+1}}{100}\right) = -113.24\left(p_2^{n+1} - 3,600\right), \dots\dots\dots\dots\dots\dots\dots\dots\dots\dots\dots\dots (8.961)$$

$$R_3 = 113.24 p_2^{n+1} - \frac{(q_2^{n+1})^2}{100} + 5q_2^{n+1} - 407,667.5 = 0. \dots\dots\dots\dots\dots\dots\dots\dots\dots\dots (8.962)$$

Now, we have three equations and three unknowns to solve for: p_1^{n+1}, p_2^{n+1}, and q_2^{n+1}. The system of equations using the GNR protocol can be written as

$$
\begin{bmatrix}
\dfrac{\partial R_1}{\partial p_1^{n+1}} & \dfrac{\partial R_1}{\partial p_2^{n+1}} & 0 \\[2mm]
\dfrac{\partial R_2}{\partial p_1^{n+1}} & \dfrac{\partial R_2}{\partial p_2^{n+1}} & \dfrac{\partial R_2}{\partial q_2^{n+1}} \\[2mm]
0 & \dfrac{\partial R_3}{\partial p_2^{n+1}} & \dfrac{\partial R_3}{\partial q_2^{n+1}}
\end{bmatrix}^{(k)}
\begin{bmatrix}
\Delta p_1^{n+1} \\[2mm]
\Delta p_2^{n+1} \\[2mm]
\Delta q_2^{n+1}
\end{bmatrix}^{(k+1)}
= -
\begin{bmatrix}
R_1\left(p_1^{n+1},p_2^{n+1}\right) \\[2mm]
R_2\left(p_1^{n+1},p_2^{n+1},q_2^{n+1}\right) \\[2mm]
R_3\left(p_2^{n+1},q_2^{n+1}\right)
\end{bmatrix}^{(k)} \quad \dots\dots\dots\dots\dots\dots\dots\dots\dots (8.963)
$$

Substituting the values, the final form of the system of equations is obtained:

$$
\begin{bmatrix}
-1.0118 & 1 & 0 \\[2mm]
2 & -2.0118 & 0.055 \\[2mm]
0 & 113.24 & -\dfrac{q_2^{n+1}}{50}+5
\end{bmatrix}^{(k)}
\begin{bmatrix}
\Delta p_1^{n+1} \\[2mm]
\Delta p_2^{n+1} \\[2mm]
\Delta q_2^{n+1}
\end{bmatrix}^{(k+1)}
= -
\begin{bmatrix}
R_1\left(p_1^{n+1},p_2^{n+1}\right) \\[2mm]
R_2\left(p_1^{n+1},p_2^{n+1},q_2^{n+1}\right) \\[2mm]
R_3\left(p_2^{n+1},q_2^{n+1}\right)
\end{bmatrix}^{(k)} \quad \dots\dots\dots\dots\dots\dots\dots (8.964)
$$

Problem 8.4.67

Consider the homogeneous, isotropic reservoir shown in **Fig. 8.58.1,** which is saturated with a single-phase, slightly compressible fluid. All outer boundaries are sealed. Reservoir and fluid properties are given in **Table 8.60.** Initially, the reservoir has a uniform pressure distribution of 4,000 psia. The producer in Block 2 is producing at a constant flow rate of 1,200 STB/D. After 30 days, the downhole gauge shows that the sandface pressure of the producer is 1,500 psia. The well in Block 1 is an observation well, and its sandface pressure at the same time is recorded as 2,100 psia. Calculate the pressure of Block 2 at the end of 30 days and the permeability of the reservoir using the GNR protocol.

Note: Use $\Delta t = 30$ days, initial guesses of $p_2 = 2,000$ psia and $k = 15$ md, and convergence criteria of $\left|\Delta p_2\right| \leq 10$ psia and $\left|\Delta k\right| \leq 1$ md.

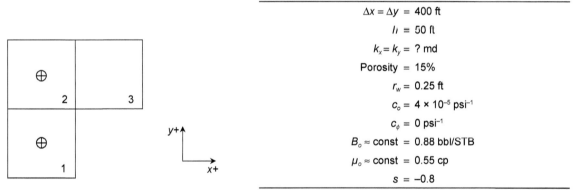

Fig. 8.58.1—Reservoir system of Problem 8.4.67.

Table 8.60—Reservoir and fluid properties for Problem 8.4.67.
$\Delta x = \Delta y = 400$ ft
$h = 50$ ft
$k_x = k_y = \;?$ md
Porosity $= 15\%$
$r_w = 0.25$ ft
$c_o = 4 \times 10^{-5}$ psi^{-1}
$c_\phi = 0$ psi^{-1}
$B_o \approx$ const $= 0.88$ bbl/STB
$\mu_o \approx$ const $= 0.55$ cp
$s = -0.8$

Solution to Problem 8.4.67

The 2D system can be reduced to a 1D system, as shown in **Fig. 8.58.2.** The governing flow equation for such a reservoir then can be written as

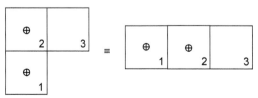

Fig. 8.58.2—Equivalent reservoir system.

$$\frac{\partial}{\partial x}\left(\frac{A_x k_x}{\mu B}\frac{\partial p}{\partial x}\right)\Delta x + q = \frac{V_b \phi c_t}{5.615}\left(\frac{\partial p}{\partial t}\right). \quad \dots\dots\dots\dots\dots\dots\dots\dots\dots\dots\dots\dots\text{(8.965)}$$

Because the reservoir is homogeneous in permeability and area, Eq. 8.965 could be simplified to the following form:

$$\frac{\partial^2 p}{\partial x^2} + \frac{q_{sc}\mu B}{V_b k} = \frac{\phi c_t}{5.615 k}\left(\frac{\partial p}{\partial t}\right). \quad \dots\dots\dots\dots\dots\dots\dots\dots\dots\dots\dots\dots\text{(8.966)}$$

The finite-difference form of Eq. 8.966 can be written as

$$p_{i-1}^{n+1} - 2p_i^{n+1} + p_{i+1}^{n+1} + \frac{q_{sc}^{n+1}\mu B}{hk} = \frac{\mu B \phi c_t \Delta x^2}{5.615 k \Delta t}(p_i^{n+1} - p_i^n). \quad \dots\dots\dots\dots\dots\dots\text{(8.967)}$$

- Block 1: $p_1 = p_{sf_1} = 2{,}100$ psia.

- Block 2:

$$\frac{hk}{\mu B}p_1 - \left(\frac{hk}{\mu B} + \frac{V_b \phi c_t}{5.615 \Delta t}\right)p_2 = -\frac{V_b \phi c_t}{5.615 \Delta t}p_i - q, \quad \dots\dots\dots\dots\dots\dots\dots\text{(8.968)}$$

$$q = -\frac{2\pi kh}{\mu B\left[\ln\left(\dfrac{r_e}{r_w}\right) + s\right]}(p_2 - p_{sf}), \quad \dots\dots\dots\dots\dots\dots\dots\dots\dots\text{(8.969)}$$

where $r_e = 0.2\,\Delta x = 90$ ft.

Now, the residual functions can be written as

$$R_1 = kp_1 - \left(k + \frac{\mu B \Delta x \Delta y \phi c_t}{5.615 \Delta t}\right)p_2 + \left[\frac{\mu B \Delta x \Delta y \phi c_t}{5.615 \Delta t}p_i + \frac{\mu B q}{h}\right], \quad \dots\dots\dots\text{(8.970)}$$

$$R_2 = \frac{q\mu B\left[\ln\left(\dfrac{r_e}{r_w}\right) + s\right]}{2\pi h} + k(p_2 - p_{sf}). \quad \dots\dots\dots\dots\dots\dots\dots\dots\text{(8.971)}$$

Terms A, B, and C are defined as constants to simplify the representation of the residual equations (Eqs. 8.970 and 8.971) as indicated below:

$$A = \frac{\mu B \Delta x \Delta y \phi c_t}{5.615 \Delta t}, \quad \dots\dots\dots\dots\dots\dots\dots\dots\dots\dots\dots\dots\dots\dots\dots\text{(8.972)}$$

$$B = \left(\frac{\mu B \Delta x \Delta y \phi c_t}{5.615 \Delta t}p_i + \frac{\mu B q}{h}\right), \quad \dots\dots\dots\dots\dots\dots\dots\dots\dots\dots\dots\text{(8.973)}$$

$$C = \frac{q\mu B\left[\ln\left(\dfrac{r_e}{r_w}\right) + s\right]}{2\pi h}. \quad \dots\dots\dots\dots\dots\dots\dots\dots\dots\dots\dots\dots\text{(8.974)}$$

Then,

$$R_1 = kp_1 - (k + A)p_2 + B, \quad \dots\dots\dots\dots\dots\dots\dots\dots\dots\dots\dots\dots\dots\text{(8.975)}$$

$$R_2 = C + k(p_2 - p_{sf}). \quad \dots\dots\dots\dots\dots\dots\dots\dots\dots\dots\dots\dots\dots\dots\dots\text{(8.976)}$$

The Jacobian matrix is

$$J = \begin{bmatrix} \dfrac{\partial R_1}{\partial k} & \dfrac{\partial R_1}{\partial p_2} \\[2mm] \dfrac{\partial R_2}{\partial k} & \dfrac{\partial R_2}{\partial p_2} \end{bmatrix} = \begin{bmatrix} p_1 - p_2 & k + A \\ p_2 - p_{sf} & k \end{bmatrix}, \quad \dots\dots\dots\dots\dots\dots\text{(8.977)}$$

$$\det(J) = \frac{\partial R_1}{\partial k}\frac{\partial R_2}{\partial p_2} - \frac{\partial R_1}{\partial p_2}\frac{\partial R_2}{\partial k}, \quad \dots\dots\dots\dots\dots\dots\dots\dots\dots\text{(8.978)}$$

$$\Delta k = \frac{\frac{\partial R_2}{\partial p_2} R_1 - \frac{\partial R_1}{\partial p_2} R_2}{\det(J)}, \quad \ldots \ldots \ldots \ldots \ldots \ldots \ldots \ldots \ldots \ldots \ldots \ldots (8.979)$$

$$\Delta p_2 = \frac{-\frac{\partial R_2}{\partial k} R_1 + \frac{\partial R_1}{\partial k} R_2}{\det(J)}. \quad \ldots \ldots \ldots \ldots \ldots \ldots \ldots \ldots \ldots \ldots \ldots \ldots (8.980)$$

- Iteration 1:

$$J = \begin{bmatrix} 100 & -0.0194 \\ 500 & 0.0169 \end{bmatrix}, \quad \ldots \ldots \ldots \ldots \ldots \ldots \ldots \ldots \ldots \ldots \ldots (8.981)$$

$\det(J) = 11.3968; \Delta k = 4.9247$ md; $\Delta p_2 = -169.9597$ psia; $k = 19.9247$ md; $p_2 = 1,830$ psia.

- Iteration 2:

$$J = \begin{bmatrix} 369.9597 & -0.025 \\ 330.0403 & 0.0225 \end{bmatrix}, \quad \ldots \ldots \ldots \ldots \ldots \ldots \ldots \ldots \ldots \ldots (8.982)$$

$\det(J) = 14.3006; \Delta k = 0.1468$ md; $\Delta p_2 = 39.5768$ psia; $k = 20.0714$ md; $p_2 = 1869.6$ psia.

- Iteration 3:

$$J = \begin{bmatrix} 230.3828 & 0.0251 \\ 369.6172 & 0.0226 \end{bmatrix}, \quad \ldots \ldots \ldots \ldots \ldots \ldots \ldots \ldots \ldots \ldots (8.983)$$

$\det(J) = 14.4991; \Delta k = -0.001$ md; $\Delta p_2 = -0.2709$ psia; $k = 20.0704$ md; $p_2 = 1869.3$ psia.

Problem 8.4.68

Consider the single-phase, slightly compressible reservoir shown in **Fig. 8.59.** A production well is completed at Block 1 and produces at a constant rate of –500 STB/D. The depth gradient is negligible and the external boundary is completely sealed. The permeability of the reservoir is known to be a function of pressure: $k \text{ (md)} = 2 \times 10^{-2} p$. Reservoir properties are given in **Table 8.61.**

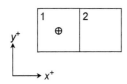

**Fig. 8.59—Reservoir
system of Problem 8.4.68.**

ϕ	= 0.25 fraction
h	= 100 ft
Δx	= 800 ft
Δy	= 800 ft
c_t	= 1×10⁻⁵ psi⁻¹
$\mu \approx$ const	= 0.95 cp
$B \approx$ const	= 0.98 res bbl/STB
p_i	= 1,000 psi
ρ	= 45 lbm/ft³
Δt	= 30 days
q	= –500 STB/D

Table 8.61—Reservoir properties for Problem 8.4.68.

1. List the system of equations in the matrix notation $[J] \times \vec{\Delta p} = -\vec{R}$, where J is the Jacobian matrix and $\vec{R}$ is the residual vector.

2. Solve for the pressure and permeability distribution in the reservoir system at the end of 30 days.

Solution to Problem 8.4.68

For a 1D slightly compressible flow system with no depth gradients,

$$\frac{\partial}{\partial x}\left(\frac{A_x k_x}{\mu B}\frac{\partial p}{\partial x}\right)\Delta x + q = \frac{V_b \phi c_t}{5.615}\left(\frac{\partial p}{\partial t}\right), \dots\dots\dots\dots\dots\dots\dots\dots\dots\dots\dots \text{(8.984)}$$

because

$$k = 2\times 10^{-2}\, p. \dots\dots\dots\dots\dots\dots\dots\dots\dots\dots\dots\dots\dots\dots\dots\dots\dots\dots \text{(8.985)}$$

The PDE can be simplified to

$$\frac{\partial}{\partial x}\left(p\frac{\partial p}{\partial x}\right)\Delta x + \frac{q\mu B}{V_b\left(1.127\times 10^{-3}\right)\left(2\times 10^{-2}\right)} = \frac{\mu B\phi c_t}{(2.254\times 10^{-5})\times 5.615}\left(\frac{\partial p}{\partial t}\right). \dots\dots\dots\dots \text{(8.986)}$$

In a discretized form,

$$\left[p_{i+\frac{1}{2}}\left(p_{i+1}-p_i\right)-p_{i-\frac{1}{2}}\left(p_i - p_{i-1}\right)\right]^{n+1} + \frac{q^{n+1}\mu B}{2.254\times 10^{-5} h} = \frac{\mu B\phi c_i \Delta x^2}{1.266\times 10^{-4}\Delta t}(p_i^{n+1}-p_i^n), \dots\dots\dots\dots \text{(8.987)}$$

$$\left[\frac{1}{2}(p_{i+1}+p_i)(p_{i+1}-p_i)-\frac{1}{2}(p_i+p_{i-1})(p_i-p_{i-1})\right]^{n+1} + \frac{q^{n+1}\mu B}{2.254\times 10^{-5} h} = \frac{\mu B\phi c_i \Delta x^2}{1.266\times 10^{-4}\Delta t}(p_i^{n+1}-p_i^n). \dots\dots \text{(8.988)}$$

Hence,

$$\left[\left(p_{i+1}^2 - p_i^2\right)-\left(p_i^2 - p_{i-1}^2\right)\right]^{n+1} + \frac{q^{n+1}\mu B}{1.127\times 10^{-5} h} = \frac{\mu B\phi c_i \Delta x^2}{6.328\times 10^{-5}\Delta t}(p_i^{n+1}-p_i^n). \dots\dots\dots\dots \text{(8.989)}$$

The residual equation is

$$R_i = \left[p_{i+1}^2 - 2p_i^2 - p_{i-1}^2\right]^{n+1} + \frac{q^{n+1}\mu B}{1.127\times 10^{-5} h} - \frac{\mu B\phi c_i \Delta x^2}{6.328\times 10^{-5}\Delta t}(p_i^{n+1}-p_i^n). \dots\dots\dots\dots \text{(8.990)}$$

At $t = 30$ days (using a timestep of 30 days),

$$\frac{\mu B\phi c_i \Delta x^2}{6.328\times 10^{-5}\Delta t} = 784.66. \dots\dots\dots\dots\dots\dots\dots\dots\dots\dots\dots\dots\dots\dots \text{(8.991)}$$

- At Block 1:

$$R_1 = \left[p_2^2 - p_1^2\right]^{n+1} - 784.66\, p_1^{n+1} + 2,725,596.52, \dots\dots\dots\dots\dots\dots\dots\dots\dots \text{(8.992)}$$

$$\frac{\partial R_1}{\partial p_1^{n+1}} = -2p_1^{n+1} - 784.66, \dots\dots\dots\dots\dots\dots\dots\dots\dots\dots\dots\dots\dots\dots \text{(8.993)}$$

$$\frac{\partial R_1}{\partial p_2^{n+1}} = 2p_2^{n+1}. \dots\dots\dots\dots\dots\dots\dots\dots\dots\dots\dots\dots\dots\dots\dots\dots\dots \text{(8.994)}$$

- At Block 2:

$$R_2 = \left[p_1^2 - p_2^2\right]^{n+1} - 784.66\, p_2^{n+1} + 3,738,640, \dots\dots\dots\dots\dots\dots\dots\dots\dots \text{(8.995)}$$

$$\frac{\partial R_2}{\partial p_1^{n+1}} = 2p_1^{n+1}, \dots\dots\dots\dots\dots\dots\dots\dots\dots\dots\dots\dots\dots\dots\dots\dots\dots \text{(8.996)}$$

$$\frac{\partial R_2}{\partial p_2^{n+1}} = -2p_2^{n+1} - 784.66. \dots\dots\dots\dots\dots\dots\dots\dots\dots\dots\dots\dots\dots\dots \text{(8.997)}$$

By implementing the GNR protocol,

$$J = \begin{bmatrix} \dfrac{\partial R_1}{\partial p_1} & \dfrac{\partial R_1}{\partial p_2} \\[2mm] \dfrac{\partial R_2}{\partial p_1} & \dfrac{\partial R_2}{\partial p_2} \end{bmatrix} = \begin{bmatrix} -2p_1^{n+1}-784.66 & 2p_2^{n+1} \\[2mm] 2p_1^{n+1} & -2p_2^{n+1}-784.66 \end{bmatrix} \dots\dots\dots\dots\dots \text{(8.998)}$$

and

$$
\begin{bmatrix} -2p_1^{n+1} - 784.66 & 2p_2^{n+1} \\ 2p_1^{n+1} & -2p_2^{n+1} - 784.66 \end{bmatrix}^{(k)} \begin{bmatrix} \Delta p_1 \\ \Delta p_2 \end{bmatrix}^{(k+1)} = \begin{bmatrix} -R_1 \\ -R_2 \end{bmatrix}^{(k+1)} \dots\dots\dots\dots\dots\dots\dots (8.999)
$$

Table 8.62 shows the iteration levels.

Iteration	p_1	p_2	Δp_1	Δp_2
0	4000	4000	−275.5	−250.89
1	3724.5	3749.11	0.826	0.819

Table 8.62—Results encountered at the completion of the first GNR iteration of Problem 8.4.68.

Therefore,

$$p_1^{(1)} = 3,724.5 \text{ psia and } k_1^{(1)} = 2\times10^{-2}\,p_1 = 74.5 \text{ md}, \dots\dots\dots\dots\dots\dots\dots\dots\dots\dots\dots (8.1000)$$

$$p_2^{(1)} = 3,749.11 \text{ psia and } k_2^{(1)} = 2\times10^{-2}\,p_2 = 75.0 \text{ md}. \dots\dots\dots\dots\dots\dots\dots\dots\dots\dots (8.1001)$$

Problem 8.4.69

Justify whether the following statements are TRUE or FALSE:

1. A reservoir engineer is simulating a single-phase, slightly compressible fluid reservoir using SIP; however, he updates the SIP coefficients within the SIP iterations similar to a gas reservoir. Therefore, the calculated pressure distribution would be totally wrong.
2. A single-phase, slightly compressible fluid reservoir is displayed in **Fig. 8.60.1**. An injection well is completed in Block 1 with a sandface-pressure specification to inject fluid for a certain period of time. Assume that the reservoir is characterized with homogeneous- and isotropic-property distributions. The pressure distribution would be the same as that solved for in a horizontal system (**Fig. 8.60.2**), if $G_1 - G_2 = G_3 - G_2$.

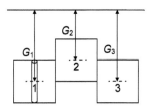

Fig. 8.60.1—Reservoir system of Problem 8.4.69, Part 2.

3. Consider two homogeneous reservoirs with slightly compressible fluid flow, as those shown in **Fig. 8.60.3**. All outer boundaries are sealed. All properties are the same for both reservoirs, with the exception of the permeability. The permeability of Reservoir 1 is less than that of Reservoir 2. The initial pressure for all blocks is 3,000 psia. The injectors and producers have the same constant flow rates of 100 STB/D. After 30 days, one can state the following:

   • $p_1 < p_4$.

   • $p_2 = p_5$.

   • $p_3 > p_6$.

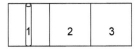

Fig. 8.60.2—Horizontal representation of the reservoir system in Problem 8.4.69, Part 2.

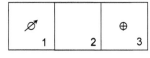

Reservoir 1 ($k_1 < k_2$)

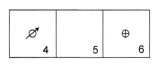

Reservoir 2 ($k_2 > k_1$)

Fig. 8.60.3—Reservoir system of Problem 8.4.69, Part 3.

4. To find either intersection points of the circle and line in **Fig. 8.60.4**, one can solve the following equation set using the GNR protocol:

$$
\begin{cases} x^2 + y^2 = 4 \\ \dfrac{1}{2}y - x = 0 \end{cases}.
$$

Note: Any point on the line or circle can be used as an initial guess.

5. For the single-phase, slightly compressible fluid flow in the homogeneous, isotropic reservoir with closed boundaries depicted in **Fig. 8.60.5**, it is impossible to obtain a solution in which $p_1 > p_3 > p_2 > p_4$. (The simulation starts with a homogenous pressure distribution; the well in Block 4 is a producer).

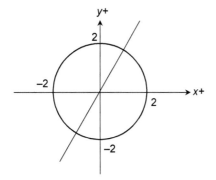

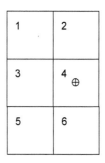

Fig. 8.60.5—Reservoir system
of Problem 8.4.69, Part 5.

Fig. 8.60.4—Geometric representation of
Problem 8.4.69, Part 4.

6. In a compressible fluid flow simulation, the iteration parameter α in SIP should be updated every timestep.
7. The GNR protocol can only be used to solve a system of nonlinear equations.
8. The p^2 approach is applicable if the gas reservoir is at initial pressure of 800 psi and only injectors are present.

Solution to Problem 8.4.69

1. FALSE. If a slightly compressible fluid reservoir is treated as a gas reservoir in the SIP solution process, the solved pressure distribution would still be correct, because the coefficients of the system of equations are slightly dependent on the pressure, which indicates that updating the coefficients brings only minor changes to the results. However, solving the problem in this way makes the solution scheme computationally more demanding.
2. FALSE. The finite-difference equations, when written for Block 2 of both systems, will be the same. However, when they are written for Blocks 1 and 3 of both systems, they will be different (on the right-hand side because of the depth gradients). This will yield two different systems of equations.
3. **Fig. 8.60.6** gives a qualitative description of the expected pressure profiles in both reservoirs at $t = 30$ days. In reference to this figure, the following explanations are provided:

   • FALSE. Injecting into a medium with smaller permeability requires higher injection pressure. Higher injection pressures will be encountered in Block 1 of Reservoir 1.
   • TRUE. Because the injection rate is equal to the production rate, the center block will stay at the initial pressure (the inflection point of the pressure profile at a constant pressure).
   • FALSE. A low-permeability system will experience a more-pronounced pressure drop compared with a high-permeability system.

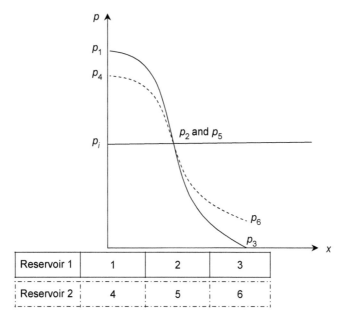

Fig. 8.60.6—Solution of Problem 8.4.69, Part 3.

4. FALSE. The Jacobian matrix can be written as

$$J = \begin{bmatrix} 2x & 2y \\ -1 & \dfrac{1}{2} \end{bmatrix}. \quad\dotfill\quad (8.1002)$$

For point $(0,0)$, the Jacobian matrix will have zero on the main diagonal, implying that the inverse of the Jacobian matrix does not exist.

5. TRUE. Blocks 2 and 3 are symmetrical with respect to Blocks 1 and 4. We would always obtain a solution with $p_1 > p_3 = p_2 > p_4$.

6. FALSE. In a compressible fluid flow problem, transmissibility terms are updated at every iteration, yielding a new system of equations. Accordingly, a new set of iteration parameters will be necessary at every iteration level when a new system of equations is generated.

7. FALSE. GNR protocol also can be used to solve a system of linear equations. In such an implementation, the solution will be obtained in only one iteration; however, the procedure collapses to a direct-solver application with excessive computational overhead.

8. FALSE. The p^2 approach is applied at low-pressure range ($< 1,000$ psi) where μz is constant; however, if the reservoir's initial pressure is at a low pressure (800 psi) and only when injectors are present (e.g., underground gas-storage injection cycle), the pressure could reach higher values where the p^2 approach is no longer applicable. Accordingly, in such an application, the ψ-approach (Ertekin et al. 2001) should be used.

Problem 8.4.70

Consider the 2D single-phase, slightly compressible fluid flow reservoir shown in **Fig. 8.61.** The reservoir boundaries are known to be sealed, with the exception of the southern boundary. A well is located in Block 2 and is produced at a rate of 1,500 STB/D. The permeability (md) in the x- and y-directions are shown in the figure. Use the reservoir properties in **Table 8.63** to do the following:

1. Calculate the SIP coefficients for Block 2 with a timestep of $\Delta t = 10$ days.
2. Write the finite-difference equation for Block 2.

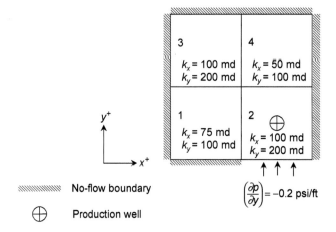

Fig. 8.61—Reservoir system of Problem 8.4.70.

h	= 50 ft
Δx	= 600 ft
Δy	= 500 ft
r_w	= 0.25 ft
ϕ	= 0.2
μ	= 2 cp
B	= 1.2 res bbl/STB
c	= 5×10^{-6} psi^{-1}
p_i	= 5,000 psi
s	= 0

Table 8.63—Reservoir properties for Problem 8.4.70.

Solution to Problem 8.4.70

1.

$$W_2 = \frac{h\Delta y}{\Delta x \mu B} \frac{2k_{x1}k_{x2}}{k_{x1}+k_{x2}} = \frac{2(50)(500)(1.127\times10^{-3})}{(600)(2)(1.2)} \frac{(100)(75)}{(100+75)} = 1.68 \text{ STB/D-psia}, \quad\ldots\ldots\ldots\ldots (8.1003)$$

$$N_2 = \frac{h\Delta x}{\Delta y \mu B} \frac{2k_{y1}k_{y2}}{k_{y1}+k_{y2}} = \frac{2(50)(600)(1.127\times10^{-3})}{(500)(2)(1.2)} \frac{(200)(100)}{(200+100)} = 3.76 \text{ STB/D-psia}, \quad\ldots\ldots\ldots\ldots (8.1004)$$

$$E_2 = 0 \text{ STB/D-psia}, \quad\ldots\ldots\ldots\ldots\ldots\ldots\ldots\ldots\ldots\ldots\ldots\ldots\ldots\ldots\ldots\ldots\ldots\ldots\ldots\ldots\ldots\ldots\ldots (8.1005)$$

$$S_2 = \frac{k_{y2}h\Delta x}{\Delta y \mu B} = \frac{(200)(1.127\times10^{-3})(50)(600)}{(500)(2)(1.2)} = 5.64 \text{ STB/D-psia}, \quad\ldots\ldots\ldots\ldots\ldots\ldots\ldots\ldots (8.1006)$$

$$\Gamma_2 = \frac{h\Delta x\Delta y\phi c}{5.615\Delta t} = \frac{(50)(600)(500)(0.2)(5\times10^{-6})}{(5.615)(10)} = 0.267 \text{ STB/D-psia}. \quad\ldots\ldots\ldots\ldots\ldots (8.1007)$$

Considering the condition at the southern boundary of Block 2,

$$\frac{p_2 - p_2'}{\Delta y} = -0.2\frac{\text{psia}}{\text{ft}} \Rightarrow p_2' = p_2 + 100. \quad\ldots\ldots\ldots\ldots\ldots\ldots\ldots\ldots\ldots\ldots\ldots\ldots\ldots\ldots\ldots (8.1008)$$

The characteristic equation for Block 2 can be written using the SIP notation:

$$S_2 p_2' + N_2 p_4 + W_2 p_1 + C_2 p_2 + q_2 = Q_2. \quad\ldots\ldots\ldots\ldots\ldots\ldots\ldots\ldots\ldots\ldots\ldots\ldots\ldots\ldots (8.1009)$$

Substituting the boundary condition, Eq. 8.1009 can be simplified to

$$W_2 p_1 + C_2 p_2 + N_2 p_4 + q_2 = Q_2, \quad\ldots\ldots\ldots\ldots\ldots\ldots\ldots\ldots\ldots\ldots\ldots\ldots\ldots\ldots\ldots\ldots (8.1010)$$

where,

$$C_2 = -(W_2 + N_2 + \Gamma_2) = -(1.68 + 3.76 + 0.267) = -5.707 \text{ STB/D-psia}, \quad\ldots\ldots\ldots\ldots\ldots\ldots (8.1011)$$

$$Q_2 = q_2 - 100S_2 - \Gamma_2 p_i = +2,500 - (3.76)(100) - (0.267)(5,000) = 789. \quad\ldots\ldots\ldots\ldots\ldots\ldots (8.1012)$$

2. The finite difference equation for Block 2 is

$$W_2 p_1 + C_2 p_2 + N_2 p_4 = Q_2. \quad\ldots\ldots\ldots\ldots\ldots\ldots\ldots\ldots\ldots\ldots\ldots\ldots\ldots\ldots\ldots\ldots\ldots\ldots (8.1013)$$

Substituting the numbers yields

$$1.68 p_1 - 5.707 p_2 + 3.76 p_4 = 789. \quad\ldots\ldots\ldots\ldots\ldots\ldots\ldots\ldots\ldots\ldots\ldots\ldots\ldots\ldots\ldots (8.1014)$$

Problem 8.4.71

A 1D homogeneous, horizontal reservoir is shown in **Fig. 8.62.** The reservoir boundaries are known to be sealed. A single-phase, slightly compressible fluid is produced through a production well in Block 3, which begins producing at $t = 0$, at a rate of 1,000 STB/D. Using the reservoir data provided in **Table 8.64,** determine the pressure distribution in the reservoir at $t = 30$ days.

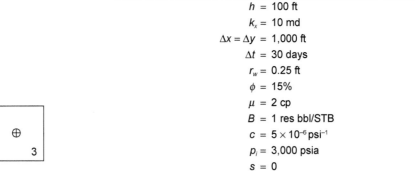

$$h = 100 \text{ ft}$$
$$k_x = 10 \text{ md}$$
$$\Delta x = \Delta y = 1{,}000 \text{ ft}$$
$$\Delta t = 30 \text{ days}$$
$$r_w = 0.25 \text{ ft}$$
$$\phi = 15\%$$
$$\mu = 2 \text{ cp}$$
$$B = 1 \text{ res bbl/STB}$$
$$c = 5 \times 10^{-6} \text{ psi}^{-1}$$
$$p_i = 3{,}000 \text{ psia}$$
$$s = 0$$

Fig. 8.62—Reservoir system of Problem 8.4.71.   **Table 8.64—Reservoir properties for Problem 8.4.71.**

Solution to Problem 8.4.71

This reservoir has uniform transmissibility coefficients; therefore,

$$W_i = E_i = T = \frac{h\Delta y k_x}{\Delta x \mu B} = \frac{(100)(1{,}000)(10)(1.127 \times 10^{-3})}{(1{,}000)(2)(1)} = 0.5635 \text{ STB/D-psia}, \quad \dots \dots \dots \dots \dots \dots \quad (8.1015)$$

$$\Gamma = \frac{h\Delta x \Delta y \phi c}{5.615\Delta t} = \frac{(100)(1{,}000)(1{,}000)(0.15)(5 \times 10^{-6})}{(5.615)(30)} = 0.445 \text{ STB/D-psia}. \quad \dots \dots \dots \dots \dots \quad (8.1016)$$

- Block 1: $T p_2 - (T + \Gamma) p_1 = -\Gamma p_i.$ $\dots \dots \dots \dots \dots \dots \dots \dots \dots \dots \dots \dots \dots \dots \dots$ (8.1017)
- Block 2: $T p_1 + T p_3 - (2T + \Gamma) p_2 = -\Gamma p_i.$ $\dots \dots \dots \dots \dots \dots \dots \dots \dots \dots \dots \dots \dots$ (8.1018)
- Block 3: $T p_2 - (T + \Gamma) p_3 = -\Gamma p_i + q_3.$ $\dots \dots \dots \dots \dots \dots \dots \dots \dots \dots \dots \dots \dots$ (8.1019)

Solving the matrix,

$$\begin{bmatrix} -(T+\Gamma) & T & 0 \\ T & -(2T+\Gamma) & T \\ 0 & T & -(T+\Gamma) \end{bmatrix} \begin{bmatrix} p_1 \\ p_2 \\ p_3 \end{bmatrix} = \begin{bmatrix} -\Gamma p_i \\ -\Gamma p_i \\ -\Gamma p_i + q_3 \end{bmatrix}, \quad \dots \dots \dots \dots \dots \dots \dots \dots \dots \quad (8.1020)$$

$$\begin{bmatrix} -1.0085 & 0.5635 & 0 \\ 0.5635 & -1.572 & 0.5635 \\ 0 & 0.5635 & -1.0085 \end{bmatrix} \begin{bmatrix} p_1 \\ p_2 \\ p_3 \end{bmatrix} = \begin{bmatrix} -1{,}335 \\ -1{,}335 \\ -335 \end{bmatrix}. \quad \dots \dots \dots \dots \dots \dots \dots \dots \dots \quad (8.1021)$$

Thus, $p_1 = 2{,}669.93 \text{ psi}$, $p_2 = 2{,}408.016 \text{ psia}$, and $p_3 = 1{,}678.36 \text{ psia}$.

An MBC is carried out to validate the result. The fluid expansion shows

$$\Delta V_{\text{total}} = c_t \sum_{i=1}^{3} \left(\frac{V_b \phi}{5.615} \right)_i \left(p_i^{n+1} \right) = 29{,}969.22 \text{ STB}, \quad \dots \dots \dots \dots \dots \dots \dots \dots \dots \quad (8.1022)$$

$$\text{MBC} = \left| \frac{(q_{\text{production}})\Delta t}{\Delta V_{\text{total}}} \right| = \left| \frac{30{,}000}{29{,}969.22} \right| = 1.00, \quad \dots \dots \dots \dots \dots \dots \dots \dots \dots \dots \dots \quad (8.1023)$$

which is a good MBC.

Problem 8.4.72

A 2D homogeneous, isotropic horizontal reservoir is shown in **Fig. 8.63.** The reservoir is known to be sealed, except across the eastern boundary of Block 3 and the southern boundary of Block 1, as shown in the figure. A single-phase incompressible fluid is produced through a production well in Block 2. A downhole pump is installed at the well, which maintains the sandface pressure at 14.7 psia. The reservoir has no depth gradients. Additional reservoir properties can be found in **Table 8.65.**

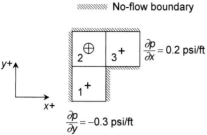

No-flow boundary

$\dfrac{\partial p}{\partial x} = 0.2$ psi/ft

$\dfrac{\partial p}{\partial y} = -0.3$ psi/ft

Fig. 8.63—Reservoir system of Problem 8.4.72.

h	$= 100$ ft
$k_x = k_y$	$= 50$ md
$\Delta x = \Delta y$	$= 400$ ft
r_w	$= 0.25$ ft
ρ	$= 62.4$ lbm/ft^3
μ	$= 1$ cp
B	$= 1$ res bbl/STB
s	$= 0$

Table 8.65—Reservoir properties for Problem 8.4.72.

1. Write the general flow equation for this reservoir.
2. Generate the system of equations to solve for the pressure distribution in the reservoir.
3. Using the Gauss-Seidel method, determine the pressure distribution in the reservoir. Use a convergence criterion of 1 psia and initial guesses for pressures as follows: $p_1 = 200$ psia, $p_2 = 275$ psia, and $p_3 = 195$ psia.
4. Calculate the flow rate at the well in Block 2.
5. Perform an MBC to verify the accuracy of your results.

Solution to Problem 8.4.72

1. For a 2D incompressible fluid reservoir, the governing PDE (ignoring the gravitational forces) can be written as

$$\frac{\partial}{\partial x}\left(\frac{A_x k_x}{\mu B}\frac{\partial p}{\partial x}\right)\Delta x + \frac{\partial}{\partial y}\left(\frac{A_y k_y}{\mu B}\frac{\partial p}{\partial y}\right)\Delta y + q = 0. \dotfill (8.1024)$$

For a homogeneous and isotropic system, Eq. 8.1024 can be reduced to

$$\frac{\partial^2 p}{\partial x^2} + \frac{\partial^2 p}{\partial y^2} + \frac{q\mu B}{k_y V_b} = 0. \dotfill (8.1025)$$

The finite-difference approximation of the PDE previously shown in Eq. 8.1025 (knowing that $\Delta x = \Delta y$) can be expressed as

$$p_{i,j-1} + p_{i-1,j} - 4p_{i,j} + p_{i+1,j} + p_{i,j+1} + \frac{q\mu B}{kh} = 0. \dotfill (8.1026)$$

2.

- Block 1: Applying the boundary condition across the southern boundary of Block 1,

$$\frac{p_1 - p_1'}{\Delta y} = -0.3\frac{\text{psia}}{\text{ft}} \Rightarrow p_1' = p_1 + 120. \dotfill (8.1027)$$

Writing the characteristic finite-difference equation for Block 1,

$$p_1' - 2p_1 + p_2 = 0, \dotfill (8.1028)$$

which (after applying the boundary condition) reduces to

$$p_2 - p_1 = -120. \dotfill (8.1029)$$

- Block 2:

$$p_1 + p_3 - \left(2 + \frac{2\pi}{\ln\frac{r_e}{r_w} + s}\right)p_2 = -\frac{2\pi}{\ln\frac{r_e}{r_w} + s}p_{sf}, \dotfill (8.1030)$$

where $r_e = 0.2\Delta x = 80$ ft.

So that,

$$\frac{2\pi}{\ln\left(\frac{r_e}{r_w}\right)+s} = \frac{(2)(3.14)}{\ln\left(\frac{80}{0.25}\right)+0} = 1.09 \quad \text{...(8.1031)}$$

and

$$p_1 + p_3 - 3.09 p_2 = -16. \quad \text{...(8.1032)}$$

- Block 3:

$$p_2 - p_3 = \Delta x \frac{\partial p}{\partial x} = -80. \quad \text{...(8.1033)}$$

3. Implementing the Gauss-Seidel method yields the iteration results as shown in **Table 8.66.**

Iteration level	p_1	p_2	p_3
0	200	275	195
1	395.00	196.17	276.17
2	316.17	196.92	276.92
3	316.92	197.41	277.41

Table 8.66—Iteration results found by implementing the
Gauss-Seidel iterative method in Solution to Problem 8.4.72.

4.

$$q = \frac{2\pi}{\left[\ln\left(\frac{r_e}{r_w}\right)+s\right]} \frac{kh}{\mu B}(p_2 - p_{sf}) = (1.09)\left(\frac{1.127\times10^{-3}\times50\times100}{1\times1}\right)(197.4128 - 14.7)$$
$$= 1,121.5 \text{ STB/D.} \quad \text{...........(8.1034)}$$

For Block 3, the fluid flows in across the east boundary:

$$q_x = \frac{k_x h \Delta x}{\mu B}\frac{\partial p}{\partial x} = \frac{(50)(1.127\times10^{-3})(100)(400)}{(1)(1)}(0.2) = 450 \text{ STB/D.} \quad \text{.....................(8.1035)}$$

For Block 1, the fluid flows in across the south boundary:

$$q_y = \frac{k_y h \Delta y}{\mu B}\frac{\partial p}{\partial y} = \frac{(50)(1.127\times10^{-3})(100)(400)}{(1)(1)}(0.3) = 676.2 \text{ STB/D.} \quad \text{.....................(8.1036)}$$

5. MBC shows a good degree of accuracy:

$$MBC = \frac{|\text{flow in}|}{|\text{flow out}|} = \frac{|q_x + q_y|}{|q_2|} = 1.0049. \quad \text{...(8.1037)}$$

Problem 8.4.73

Consider the incompressible fluid flow in the 1D reservoir of **Fig. 8.64.** It is known that the reservoir exhibits homogeneous permeability distribution. The thickness of the reservoir is uniform and the depth gradient of the system is negligible. The boundaries of the reservoir are shown in the figure and the reservoir properties are found in **Table 8.67.** It is known that Block 1 is connected to a large aquifer and pressure at this block is maintained at 2,500 psia.

1. Determine the production rate if the sandface pressure of the production well is specified at 2,300 psia.
2. Perform an MBC to verify your results.
3. What would you suggest doing to maximize the production rate of this system, and what is the maximum flow rate achievable from this system by way of a well in Block 2?

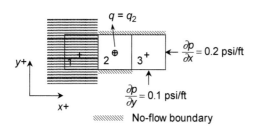

Fig. 8.64—Reservoir system of Problem 8.4.73.

$\Delta x = \Delta y$ = 1,000 ft	
h = 50 ft	
k_x = 100 md	
B = 1 res bbl/STB	
μ = 1 cp	
ϕ = 0.2 cp	
No depth gradient	
s = 0	
r_w = 0.25 ft	

Table 8.67—Reservoir properties for Problem 8.4.73.

Solution to Problem 8.4.73

1. For a 1D incompressible fluid flow system, the governing PDE (ignoring the gravitational forces) is

$$\frac{\partial}{\partial x}\left(\frac{A_x k_x}{\mu B}\frac{\partial p}{\partial x}\right)\Delta x + q = 0. \dotfill (8.1038)$$

This reservoir has uniform transmissibility coefficients along the x-direction:

$$T = \frac{A_x k_x}{\mu B \Delta x} = \frac{\Delta y h k_x}{\mu B \Delta x}. \dotfill (8.1039)$$

Because $\Delta x = \Delta y$,

$$T = \frac{h k_x}{\mu B} = \frac{(50)(100)(1.127 \times 10^{-3})}{(1)(1)} = 5.635 \text{ STB/D} - \text{psia.} \dotfill (8.1040)$$

The characteristic finite-difference equation can be written as $W_i p_{i-1} + C_i p_i + E_i p_{i+1} = Q_i$, where $W_i = E_i = T$.

- Block 2:

$$T p_1 + T p_3 - (2T + \Omega) p_2 = -\Omega p_{sf}, \dotfill (8.1041)$$

where $r_e = 0.2\Delta x = 200$ ft,

$$\Omega = \frac{2\pi kh}{\mu B\left[\ln\left(\dfrac{r_e}{r_w}\right) + s\right]} = \frac{(2)(3.14)(100)(1.127 \times 10^{-3})(50)}{(1)(1)\left[\ln\left(\dfrac{200}{0.25}\right) + 0\right]} = 5.294 \text{ STB/D} - \text{psia.} \dotfill (8.1042)$$

So that,

$$5.635 \times 2,500 + 5.635 p_3 - (2 \times 5.635 + 5.294) p_2 = -5.294 \times 2,300. \dotfill (8.1043)$$

- Block 3:

$$T p_2 - T p_3 - T \Delta y \frac{\partial p}{\partial y} + T \Delta x \frac{\partial p}{\partial x} = 0. \dotfill (8.1044)$$

Thus, $p_2 - p_3 + 100 = 0.$ $\dotfill (8.1045)$

Solving for p_2 and p_3, using the block equations derived above, $p_2 = 2,454.7$ psia, $p_3 = 2,554.7$ psia. So,

$$q_2 = -\Omega\left(p_2 - p_{sf}\right) = -5.294(2,454.7 - 2,300) = -818.98 \text{ STB/D.} \dotfill (8.1046)$$

2. MBC:

- Flow into the system:

$$q_{12} = T\left(p_1 - p_2\right) = 5.635\left(2,500 - 2,454.7\right) = 255.2655 \text{ STB/D}, \quad\dotfill (8.1047)$$

$$q_{3\text{in}} = T\Delta x \frac{\partial p}{\partial x} = 5.635\left(1,000\right)\left(0.2\right) = 1,127 \text{ STB/D}, \quad\dotfill (8.1048)$$

$$q_{\text{in}} = q_{12} + q_{3\text{in}} = 255.2655 + 1,127 = 1,382.2655 \text{ STB/D}. \quad\dotfill (8.1049)$$

- Flow out of the system:

$$q_2 = -818.98 \text{ STB/D}, \quad\dotfill (8.1050)$$

$$q_{3\text{out}} = -T\Delta y \frac{\partial p}{\partial y} = 5.635\left(1,000\right)\left(0.1\right) = -563.5 \text{ STB/D}, \quad\dotfill (8.1051)$$

$$q_{\text{out}} = q_2 + q_{3\text{out}} = -818.98 - 563.5 = -1,382.48 \text{ STB/D}. \quad\dotfill (8.1052)$$

Thus,

$$\text{MBC} = \frac{\left|q_{\text{in}}\right|}{\left|q_{\text{out}}\right|} = \frac{\left|1,382.2655\right|}{\left|-1,382.48\right|} = 0.9998. \quad\dotfill (8.1053)$$

3. To maximize the production rate, the sandface pressure of the well in Block 2 can be specified at 14.7 psia.

- At Block 2:

$$5.635 \times 2,500 + 5.635\, p_3 - \left(2 \times 5.635 + 5.294\right) p_2 = -5.294 \times 14.7. \quad\dotfill (8.1054)$$

- At Block 3:

$$p_2 - p_3 + 100 = 0. \quad\dotfill (8.1055)$$

Solving the system of equations, one obtains $p_2 = 1,347.68$ psia and $p_3 = 1,447.68$ psia. Then, the flow rate of the well in Block 2 is calculated as

$$q_2 = -\Omega\left(p_2 - p_{sf}\right) = -5.294\left(1,347.68 - 14.7\right) = -7,056.80 \text{ STB/D}. \quad\dotfill (8.1056)$$

Problem 8.4.74

State whether each of the following statements is TRUE or FALSE. Justify your answer.

1. Consider two different reservoirs with different property distributions. The pressure distributions of both reservoirs are obtained using the same iterative-solution technique. The pressure-convergence criteria used in obtaining both solutions are 0.01 psia. The levels of accuracy for solutions of both reservoirs are expected to be the same.

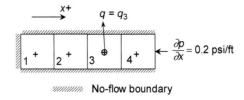

Fig. 8.65—Reservoir system of Problem 8.4.74, Part 2.

2. For the incompressible fluid system shown in **Fig. 8.65**, the production well located in Block 3 is producing with a flow-rate specification. It is possible to find the value of the sandface pressure of the production well, even though no block pressure in the system is given.

3. For a given reservoir, PSOR and Gauss-Seidel methods will generate solutions of the same level of accuracy, if the same number of iterations is performed with each protocol.

Solution to Problem 8.4.74

1. FALSE. The entries of the coefficient matrix would be different, thus the convergence paths also would be different, depending on the level of diagonal dominance (positive-definite characteristic) exhibited by the coefficient matrix.
2. FALSE. With the given production rate and the Neumann-type boundary condition, there would be an infinite number of pressure surfaces along the reservoir that are at different pressure levels but are parallel to each other (same pressure gradients); thus, it would be impossible to find a unique value of the sandface pressure for Well 3.
3. FALSE. Normally, the PSOR procedure would converge faster than the Gauss-Seidel method. Thus, at the same number of iterations, PSOR will bring the solution vector closer to the answer.

Problem 8.4.75

The system shown in **Fig. 8.66.1** is a 2D single-phase flow with homogeneous and isotropic permeability and area distributions. There is a horizontal well located in Block 2. This production well is operating at a specified sandface pressure of 20 psia. The system boundaries and imposed boundary conditions are highlighted in the figure. Reservoir properties are included in **Table 8.68.** Note that Block 4 is connected to a large aquifer that maintains the pressure of Block 4 at 700 psia. Depth gradients in the system may be ignored.

Generate a system of equations and solve for pressure distribution in the reservoir and flow rate into the horizontal well in Block 2.

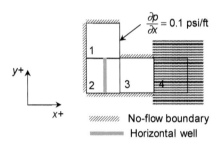

$\dfrac{\partial p}{\partial x} = 0.1$ psi/ft

/////// No-flow boundary
▬▬▬ Horizontal well

Fig. 8.66.1—Reservoir system of Problem 8.4.75.

$k_x = k_y = k_z$	= 500 md
$\Delta x = \Delta y$	= 1,000 ft
ϕ	= 20%
h	= 100 ft
μ	= 1 cp
B	= 1 res bbl/STB
r_w	= 0.25 ft
s	= 0

Table 8.68—Reservoir properties for Problem 8.4.75.

Solution to Problem 8.4.75

The horizontal well is completed parallel to the x-y plane, shown in **Fig. 8.66.2.** The productivity index of the horizontal well is calculated as

$$r_e = 0.14\sqrt{(\Delta x)^2 + (h)^2} = 0.14\sqrt{(1,000)^2 + (100)^2} = 140.7 \text{ ft}, \quad \ldots(8.1057)$$

$$\Omega = \frac{2\pi \overline{k}_{xz} \Delta y}{\mu B \left(\ln \dfrac{r_e}{r_w} + s \right)} = \frac{(2)(3.14)(500)(1.127\times10^{-3})(1,000)}{(1)(1)\left[\ln\left(\dfrac{140.7}{0.25}\right) + 0 \right]} \quad \ldots(8.1058)$$

$$= 558.79 \text{ STB/D} - \text{psia}.$$

Fig. 8.66.2—Schematic representation of the block hosting the horizontal well.

Block 2

For a homogeneous and isotropic system, the transmissibility coefficient is

$$T = \frac{kh}{\mu B} = \frac{(500)(1.127\times10^{-3})(100)}{(1)(1)} = 56.35 \text{ STB/D-psia.} \quad \ldots\ldots\ldots\ldots\ldots\ldots\ldots\ldots\ldots\ldots (8.1059)$$

- Block 1: Applying the boundary condition across the eastern edge of Block 1,

$$\frac{p_1' - p_1}{\Delta x} = 0.1\frac{\text{psia}}{\text{ft}} \implies p_1' = p_1 + 100, \quad \ldots\ldots\ldots\ldots\ldots\ldots\ldots\ldots\ldots\ldots\ldots\ldots (8.1060)$$

$$p_2 - p_1 + \Delta x \frac{\partial p}{\partial x} = 0, \quad \ldots\ldots\ldots\ldots\ldots\ldots\ldots\ldots\ldots\ldots\ldots\ldots\ldots\ldots\ldots\ldots\ldots (8.1061)$$

$$p_2 - p_1 = -100. \quad \ldots\ldots\ldots\ldots\ldots\ldots\ldots\ldots\ldots\ldots\ldots\ldots\ldots\ldots\ldots\ldots\ldots\ldots\ldots (8.1062)$$

- Block 2:

$$p_1 + p_3 - \left(2 + \frac{\Omega}{T}\right)p_2 = -\frac{\Omega p_{sf}}{T}, \qquad \text{(8.1063)}$$

$$p_1 + p_3 - 11.92 p_2 = -198.33. \qquad \text{(8.1064)}$$

- Block 3:

$$p_2 + p_4 - 2p_3 = 0, \qquad \text{(8.1065)}$$

$$p_2 - 2p_3 = -700. \qquad \text{(8.1066)}$$

Solving the system of equations, one obtains $p_1 = 162.22$ psia, $p_2 = 62.22$ psia, and $p_3 = 381.11$ psia.

The flow rate of the well is

$$q_2 = \Omega\left(p_2 - p_{sf2}\right) = -56.35\left(62.22 - 20\right) = -2,3592.02 \text{ STB/D}. \qquad \text{(8.1067)}$$

Fluid flow across the eastern boundary of Block 1 is

$$q_1 = \frac{k_x A_x}{\mu B}\frac{\partial p}{\partial x} = \frac{(500)\left(1.127 \times 10^{-3}\right)(1,000)(100)}{(1)(1)}(0.1) = 5,635 \text{ STB/D}. \qquad \text{(8.1068)}$$

Fluid flow from the aquifer (Block 4) to Block 3 is

$$q_{4\,3} = \frac{k_x h}{\mu B}\left(p_4 - p_3\right) = \frac{(500)\left(1.127 \times 10^{-3}\right)(100)}{(1)(1)}\left(700 - 381.11\right) = 17,969.46 \text{ STB/D}, \qquad \text{(8.1069)}$$

$$\text{MBC} = \frac{|\text{flow in}|}{|\text{flow out}|} = \frac{|q_1 + q_{4-3}|}{|q_2|} = \frac{|5,635 + 17,969.46|}{|23,592.02|} = 1.0005. \qquad \text{(8.1070)}$$

Problem 8.4.76

Consider the fluid flow in a 1D reservoir shown in **Fig. 8.67.** It is known that the reservoir has heterogeneous permeability distribution. The thickness of the reservoir is uniform and the depth gradient of the system is negligible. The boundaries of the reservoir are shown in the figure and the reservoir properties are outlined in **Table 8.69.** The production well in Block 2 is operating at a specified sandface pressure of 300 psia. Meanwhile, the production well in Block 1 is operating at a specified sandface pressure of 14.7 psia.

Use the Gauss-Seidel method, with $p_1 = 180$ psia and $p_2 = 380$ psia as initial guesses in your calculation of block pressures. Use a convergence criterion of 0.1 psia.

1. Determine the production rates from both production wells.
2. Perform an MBC to verify your results.

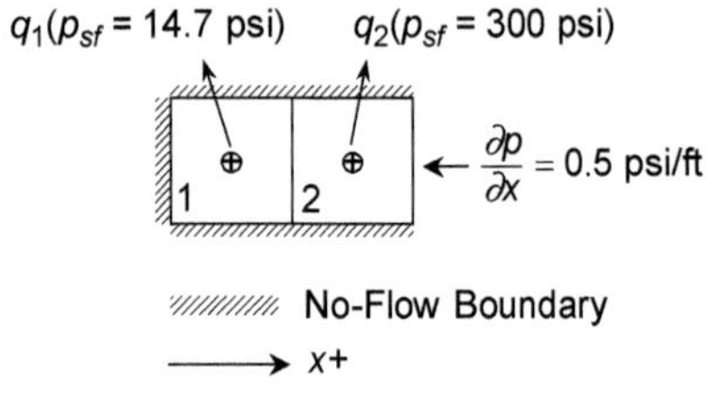

Fig. 8.67—Reservoir system of Problem 8.4.76.

Table 8.69—Reservoir properties for Problem 8.4.76.
$\Delta x = \Delta y = 500$ ft
$h = 50$ ft
$k_1 = 100$ md
$k_2 = 80$ md
$B = 1$ res bbl/STB
$\mu = 1$ cp
$\phi = 0.2$
$c_f = 0$ psi^{-1}
No depth gradient
$s = 0$
$r_w = 0.25$ ft

Solution to Problem 8.4.76

$$T_{12} = 2\frac{h}{\mu B}\left(\frac{k_1 k_2}{k_1 + k_2}\right) = 2\frac{50}{(1)(1)}\left[\frac{(100)(80)}{100 + 80}\right](1.127 \times 10^{-3}) = 5.01 \text{ STB/D-psia,} \quad \dots\dots\dots\dots\dots \quad (8.1071)$$

$$T_{22}' = \frac{hk_2}{\mu B} = \frac{(50)(80)(1.127 \times 10^{-3})}{(1)(1)} \text{ 4.508 STB/D-psia,} \quad \dots\dots\dots\dots\dots\dots\dots\dots \quad (8.1072)$$

where $r_e = 0.2\Delta x = 100$ ft,

$$\Omega_1 = \frac{2\pi k_1 h}{\mu B\left[\ln\left(\dfrac{r_e}{r_w}\right) + s\right]} = \frac{(2)(3.14)(100)(1.127 \times 10^{-3})(50)}{(1)(1)\left[\ln\left(\dfrac{100}{0.25}\right) + 0\right]} = 5.906 \text{ STB/D-psia,} \quad \dots\dots\dots\dots \quad (8.1073)$$

$$\Omega_2 = \frac{2\pi k_2 h}{\mu B\left[\ln\left(\dfrac{r_e}{r_w}\right) + s\right]} = \frac{(2)(3.14)(80)(1.127 \times 10^{-3}3)(50)}{(1)(1)\left[\ln\left(\dfrac{100}{0.25}\right) + 0\right]} = 4.725 \text{ STB/D-psia.} \quad \dots\dots\dots\dots \quad (8.1074)$$

Thus,

- Block 1:

$$T_{12}p_2 - (T_{12} + \Omega_1)p_1 = -\Omega_1 p_{sf1}, \quad \dots\dots\dots\dots\dots\dots\dots\dots\dots\dots\dots\dots\dots \quad (8.1075)$$

$$5.01p_2 - 10.916p_1 = -86.8182. \quad \dots\dots\dots\dots\dots\dots\dots\dots\dots\dots\dots\dots\dots \quad (8.1076)$$

- Block 2:

$$\frac{p_2' - p_2}{\Delta x} = 0.5\frac{\text{psia}}{\text{ft}} \Rightarrow p_2' = p_2 + 0.5\Delta x, \quad \dots\dots\dots\dots\dots\dots\dots\dots\dots\dots\dots \quad (8.1077)$$

$$T_{12}p_1 - (T_{12} + \Omega_2)p_2 + T_{22}'\Delta x\frac{\partial p}{\partial x} = -\Omega_2 p_{sf1}, \quad \dots\dots\dots\dots\dots\dots\dots\dots\dots \quad (8.1078)$$

$$5.01p_1 - 9.735p_2 = -2,544.5. \quad \dots\dots\dots\dots\dots\dots\dots\dots\dots\dots\dots\dots\dots \quad (8.1079)$$

Solving p_1 and p_2 using the Gauss-Seidel method results in iteration levels shown in **Table 8.70.** Accordingly, $p_1 = 167.4816$ psia and $p_2 = 347.5689$ psia.

Iteration level	p_1	p_2
0	180	380
1	182.3578	355.2247
2	170.987	349.3729
3	168.3012	347.9907
4	167.6669	347.6642
5	167.517	347.5871
6	167.4816	347.5689

Table 8.70—Results of the Gauss-Seidel iterative process from Solution to Problem 8.4.76.

MBC:

$$\text{MBC} = \frac{|\text{flow in}|}{|\text{flow out}|} = \frac{\left|T_2\Delta x\dfrac{\partial p}{\partial x}\right|}{\left|-\Omega_1\left(p_1 - p_{sf1}\right) - \Omega_2\left(p_2 - p_{sf2}\right)\right|} \quad \dots\dots\dots\dots\dots\dots \quad (8.1080)$$

$$= \frac{|(4.508)(500)(0.5)|}{|-(5.906)(167.4816 - 14.7) - 4.725(347.5689 - 300)|} = 0.999919.$$

Problem 8.4.77

Consider the incompressible fluid flow in the homogeneous reservoir shown in **Fig. 8.68,** with the reservoir properties displayed in **Table 8.71.** The reservoir has an aquifer support through the bottom of Block A; however, the other boundaries of the reservoir are completely sealed. The observation well located in Block E shows a pressure reading of 2,250 psia. The well in Block C is producing at a sandface pressure of 500 psia.

1. Provide the governing flow equation.
2. Determine the flow rate into the well located in Block C.
3. Determine the pressure of each block of the reservoir.

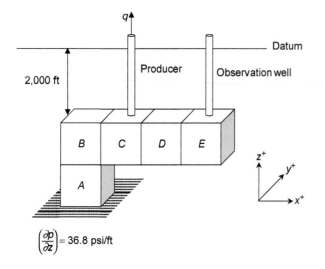

$$\left(\frac{\partial p}{\partial z}\right) = 36.8 \text{ psi/ft}$$

Fig. 8.68—Reservoir system of Problem 8.4.77.

$\mu =$	1.1 cp
$B =$	1.0 res bbl/STB
$\rho_w =$	62.43 lbm/ft³
$k_y =$	500 md (uniform)
$k_z =$	250 md (uniform)
$\Delta x =$	50 ft (uniform)
$\Delta y =$	50 ft (uniform)
$\Delta z =$	50 ft (uniform)
$\phi =$	0.20 (uniform)
$r_w =$	0.25 ft
$c =$	0 psi⁻¹

Table 8.71—Reservoir properties for Problem 8.4.77.

Solution to Problem 8.4.77

1. The governing PDE of a 2D incompressible reservoir flow system can be written as

$$\frac{\partial}{\partial y}\left(\frac{A_y k_y}{\mu B}\frac{\partial \Phi}{\partial y}\right)\Delta y + \frac{\partial}{\partial z}\left(\frac{A_z k_z}{\mu B}\frac{\partial \Phi}{\partial z}\right)\Delta z + q = 0. \quad\ldots\ldots\ldots\ldots\ldots\ldots\ldots\ldots\ldots\ldots\ldots(8.1081)$$

Knowing that $k_y = 2k_z$, Eq. 8.1081 can be reduced to (note that there is no depth gradient along the y-direction)

$$2\frac{\partial^2 p}{\partial y^2} + \frac{\partial^2 \Phi}{\partial z^2} + \frac{q\mu B}{k_z V_b} = 0. \quad\ldots\ldots\ldots\ldots\ldots\ldots\ldots\ldots\ldots\ldots\ldots\ldots\ldots\ldots\ldots\ldots(8.1082)$$

2. Blocks D and E are inactive blocks, so that $p_C = p_D = p_E = p_{sfE} = 2,250$ psia,

   where $r_e = 0.2\Delta x = 10$ ft,

$$\Omega = \frac{2\pi\sqrt{k_y k_x}\Delta z}{\mu B\left[\ln\left(\dfrac{r_e}{r_w}\right)+s\right]} = \frac{2(3.14)(500)(1.127\times10^{-3})(50)}{(1.1)(1)\left[\ln\left(\dfrac{10}{0.25}\right)+0\right]} = 43.6 \text{ STB/D-psia}, \quad\ldots\ldots\ldots\ldots(8.1083)$$

$$q_C = -\Omega\left(p_C - p_{sfC}\right) = -43.6(2,250-500) = -76,300 \text{ STB/D}, \quad\ldots\ldots\ldots\ldots\ldots\ldots\ldots(8.1084)$$

$$-\frac{A_y k_y}{\mu B \Delta y}\left(p_B - p_C\right) = q_C. \quad\ldots\ldots\ldots\ldots\ldots\ldots\ldots\ldots\ldots\ldots\ldots\ldots\ldots\ldots(8.1085)$$

3. So that,

$$p_B = -\frac{q\mu B}{\Delta z k_y} + p_C = -\frac{(-76,300)(1.1)(1)}{(50)(500)(1.127\times10^{-3})} + 2,250 = 5,228.88 \text{ psia}. \quad\ldots\ldots\ldots\ldots\ldots(8.1086)$$

Numerical Solution of Single-Phase Flow Equations   413

For Block A, applying the boundary condition supported by the aquifer,

$$\frac{p_A - p_A'}{\Delta z} = -36.8\,\frac{\text{psi}}{\text{ft}} \rightarrow p_A' = p_A + (36.8)(50) = p_A + 1,840. \quad\dots\dots\dots\dots\dots\dots\dots(8.1087)$$

Note that the pressure gradient between Blocks B and A must be equal to 36.8 psi/ft. Then,

$$\frac{p_A - p_B}{\Delta z} = -36.8\,\frac{\text{psi}}{\text{ft}} \rightarrow p_A = p_B + (36.8)(50), \quad\dots\dots\dots\dots\dots\dots\dots\dots(8.1088)$$

$$p_A = 5,228.88 + 1,840 = 7,068.88 \text{ psi.} \quad\dots\dots\dots\dots\dots\dots\dots\dots\dots\dots(8.1089)$$

Finally, the aquifer pressure is calculated from

$$p_A' = 7,068.88 + 1,840 = 8,908.88 \text{ psi.} \quad\dots\dots\dots\dots\dots\dots\dots\dots\dots(8.1090)$$

The calculated pressure distribution in the reservoir is summarized in **Table 8.72.**

Block	Pressure, psia
Aquifer	8,908.88
A	7,068.88
B	5,288.88
C	2,250
D	2,250
E	2,250

Table 8.72—Calculated-pressure distribution in the reservoir in Solution to Problem 8.4.77.

Problem 8.4.78

Consider a 2D single-phase incompressible flow in the reservoir shown in **Fig. 8.69.** The permeability distribution is heterogeneous and isotropic. All boundaries are completely sealed. The thickness of the reservoir is uniform and depth gradients are negligible. There is an injection well in Block 1 and a production well in Block 7. The injection well is operating at a rate of 650 STB/D, while the sandface pressure of the production well is maintained at 150 psia. Use the following properties in your calculations: $h = 40$ ft, $\Delta x = \Delta y = 50$ ft, $\mu = 1.2$ cp, $B = 1$ res bbl/STB, $r_w = 0.32$ ft, and $s = 0$.

1. Calculate the SIP coefficients S_{ij}, W_{ij}, C_{ij}, E_{ij}, N_{ij}, and Q_{ij} for Blocks 1, 2, 4, and 7.
2. Generate the coefficient matrix and the right-hand-side vector for this system using the SIP notation (provide the numerical value).

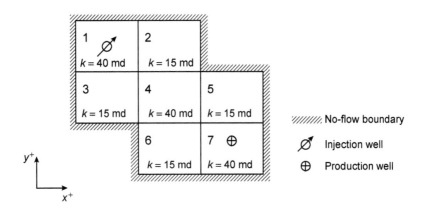

Fig. 8.69—Reservoir system of Problem 8.4.78.

1. Using the definitions of SIP coefficients, one can calculate the following:

   - Block 1:

$$S_1 = \frac{2A}{\mu B \Delta y}\left(\frac{k_1 k_3}{k_1 + k_3}\right)\frac{2(50)(40)(0.001127)}{(1.2)(1)(50)}\left[\frac{(40)(15)}{40+15}\right] = 0.8196 \text{ STB}/\text{D}-\text{psia}, \quad \ldots\ldots\ldots \quad (8.1091)$$

$$E_1 = \frac{2A}{\mu B \Delta x}\left(\frac{k_1 k_2}{k_1 + k_2}\right) = \frac{2(50)(40)(0.001127)}{(1.2)(1)(50)}\left[\frac{(40)(15)}{40+15}\right] = 0.820 \text{ STB}/\text{D}-\text{psia}, \quad \ldots\ldots\ldots \quad (8.1092)$$

$$W_1 = N_1 = 0, \quad \ldots\ldots\ldots\ldots\ldots\ldots\ldots\ldots\ldots\ldots\ldots\ldots\ldots\ldots\ldots\ldots\ldots\ldots \quad (8.1093)$$

$$C_1 = -\left(S_1 + E_1 + W_1 + N_1\right) = -\left(0.81963 + 0.81963\right) = -1.640 \text{ STB}/\text{D}-\text{psia}, \quad \ldots\ldots\ldots\ldots \quad (8.1094)$$

$$Q_1 = -650 \text{ STB}/\text{D}. \quad \ldots\ldots\ldots\ldots\ldots\ldots\ldots\ldots\ldots\ldots\ldots\ldots\ldots\ldots\ldots\ldots\ldots \quad (8.1095)$$

   - Block 2:

$$S_2 = \frac{2A}{\mu B \Delta y}\left(\frac{k_2 k_4}{k_2 + k_4}\right) = \frac{2(50)(40)(0.001127)}{(1.2)(1)(50)}\left[\frac{(40)(15)}{40+15}\right] = 0.820 \text{ STB}/\text{D}-\text{psia}, \quad \ldots\ldots\ldots \quad (8.1096)$$

$$W_2 = E_1 = 0.820 \text{ STB}/\text{D}-\text{psia}, \quad \ldots\ldots\ldots\ldots\ldots\ldots\ldots\ldots\ldots\ldots\ldots\ldots\ldots \quad (8.1097)$$

$$E_2 = N_2 = 0, \quad \ldots\ldots\ldots\ldots\ldots\ldots\ldots\ldots\ldots\ldots\ldots\ldots\ldots\ldots\ldots\ldots\ldots\ldots \quad (8.1098)$$

$$C_2 = -\left(0.81963 + 0.81963\right) = -1.64 \text{ STB}/\text{D}-\text{psia}, \quad \ldots\ldots\ldots\ldots\ldots\ldots\ldots\ldots \quad (8.1099)$$

$$Q_2 = 0 \text{ STB}/\text{D}. \quad \ldots\ldots\ldots\ldots\ldots\ldots\ldots\ldots\ldots\ldots\ldots\ldots\ldots\ldots\ldots\ldots\ldots \quad (8.1100)$$

   - Block 4:

$$E_4 = N_4 = S_4 = W_4 = S_2 = 0.82 \text{ STB}/\text{D}-\text{psia}, \quad \ldots\ldots\ldots\ldots\ldots\ldots\ldots\ldots\ldots \quad (8.1101)$$

$$C_4 = -\left(4 \times 0.81963\right) = -3.28 \text{ STB}/\text{D}-\text{psia}, \quad \ldots\ldots\ldots\ldots\ldots\ldots\ldots\ldots\ldots \quad (8.1102)$$

$$Q_4 = 0 \text{ STB}/\text{D}. \quad \ldots\ldots\ldots\ldots\ldots\ldots\ldots\ldots\ldots\ldots\ldots\ldots\ldots\ldots\ldots\ldots\ldots \quad (8.1103)$$

   - Block 7:

$$N_7 = S_5 = S_1 = 0.820 \text{ STB}/\text{D}-\text{psia}, \quad \ldots\ldots\ldots\ldots\ldots\ldots\ldots\ldots\ldots\ldots\ldots \quad (8.1104)$$

$$W_7 = E_1 = 0.820 \text{ STB}/\text{D}-\text{psia}, \quad \ldots\ldots\ldots\ldots\ldots\ldots\ldots\ldots\ldots\ldots\ldots\ldots\ldots \quad (8.1105)$$

$$S_7 = E_7 = 0 \text{ STB}/\text{D}-\text{psia}, \quad \ldots\ldots\ldots\ldots\ldots\ldots\ldots\ldots\ldots\ldots\ldots\ldots\ldots\ldots \quad (8.1106)$$

$$C_7 = -\left(0.81963 + 0.81963\right) = -1.6392 \text{ STB}/\text{D}-\text{psia}, \quad \ldots\ldots\ldots\ldots\ldots\ldots\ldots \quad (8.1107)$$

$$\Omega_7 = \frac{2\pi k h}{\mu B\left[\ln\left(\dfrac{r_e}{r_w}\right)+s\right]} = 2.74 \text{ STB}/\text{D}-\text{psi, where } r_e = 0.2\Delta x = 10 \text{ ft}, \quad \ldots\ldots\ldots\ldots \quad (8.1108)$$

$$Q_7 = -\Omega p_{sf7} = -411.45 \text{ STB}/\text{D}. \quad \ldots\ldots\ldots\ldots\ldots\ldots\ldots\ldots\ldots\ldots\ldots\ldots\ldots \quad (8.1109)$$

2.

$$\begin{bmatrix} -1.6392 & 0.8196 & 0.8196 & 0 & 0 & 0 & 0 \\ 0.8196 & -1.6392 & 0 & 0.8196 & 0 & 0 & 0 \\ 0.8196 & 0 & -1.6392 & 0.8196 & 0 & 0 & 0 \\ 0 & 0.8196 & 0.8196 & -3.2784 & 0.8196 & 0.8196 & 0 \\ 0 & 0 & 0 & 0.8196 & -1.6392 & 0 & 0.8196 \\ 0 & 0 & 0 & 0.8196 & 0 & -1.6392 & 0.8196 \\ 0 & 0 & 0 & 0 & 0.8196 & 0.8196 & -1.6392 \end{bmatrix} \begin{bmatrix} p_1 \\ p_2 \\ p_3 \\ p_4 \\ p_5 \\ p_6 \\ p_7 \end{bmatrix} = \begin{bmatrix} -650 \\ 0 \\ 0 \\ 0 \\ 0 \\ 0 \\ -411.45 \end{bmatrix}.$$

.... (8.1110)

The solution vector will be $p_1 = 1,973$ psia, $p_2 = p_3 = 1,576$ psia, $p_4 = 1,180$ psia, $p_5 = p_6 = 783$ psia, and $p_7 = 387$ psia.

Problem 8.4.79

As shown in **Fig. 8.70.1**, a 2×2 reservoir with an incompressible fluid is exposed to an infinitely large aquifer along the northern and eastern boundaries. The other boundaries are completely sealed. A production well is placed in Block 1, and wells in Blocks 2 and 3 are monitoring wells. Assume that water in the aquifer has the same properties as the reservoir fluid with $c = 0$. The aquifer is supporting a constant pressure of 1,000 psia. To solve the system of equations, use the Gauss-Seidel method, with 850 psia as the initial guess for each of the block pressures in your calculations. Use 2 psia as the convergence criterion and the reservoir properties found in **Table 8.73.** Check material balances to verify your results.

1. Determine the production rate if the well is producing at a constant sandface pressure of 100 psia.
2. A stimulation process is designed to achieve a 20% higher production rate with the same sandface pressure given in Part 1. If pressure values post-stimulation at the monitoring well of Blocks 2 and 3 are observed to be 749.3 and 803.1 psia, respectively, determine the skin factor.
3. Without stimulation, is it possible to achieve the same production rate from the pressure distribution observed in Part 2 by decreasing the sandface-pressure specification? Why or why not?

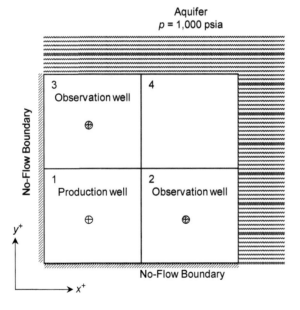

Fig. 8.70.1—Reservoir system of Problem 8.4.79.

Table 8.73—Reservoir properties for Problem 8.4.79.

$k_x =$	10 md
$k_y =$	5 md
$\Delta x = \Delta y =$	100 ft
$h =$	50 ft
$\mu =$	1.0 cp
$B =$	1.0 res bbl/STB
$r_w =$	0.25 ft
$s =$	0

Solution to Problem 8.4.79

1. There are four unknown pressures—p_1, p_2, p_3, p_4—that need to be solved. Therefore, four equations can be written at each of the four blocks. The governing flow equation for this problem states that

$$\frac{\partial}{\partial x}\left(\frac{A_x k_x}{\mu B}\frac{\partial p}{\partial x}\right)\Delta x + \frac{\partial}{\partial y}\left(\frac{A_y k_y}{\mu B}\frac{\partial p}{\partial y}\right)\Delta y + q = 0. \dots\dots\dots\dots\dots\dots\dots\dots\dots\dots (8.1111)$$

In this problem, μ is constant and $B = 1$, so that one can simplify Eq. 8.1111 to

$$k_x \frac{\partial^2 p}{\partial x^2} + k_y \frac{\partial^2 p}{\partial y^2} + \frac{q\mu B}{V_b} = 0. \quad \dots\dots\dots\dots\dots\dots\dots\dots\dots\dots\dots\dots\dots \quad (8.1112)$$

The finite-difference analog of the flow equation (because $k_y = 2k_x$ and $\Delta x = \Delta y$) is

$$2\left(p_{i+1,j} - 2p_{i,j} + p_{i-1,j}\right) + \left(p_{i,j+1} - 2p_{i,j} + p_{i,j-1}\right) + \frac{q\mu B (\Delta x)^2}{V_b k_y} = 0. \quad \dots\dots\dots\dots\dots\dots \quad (8.1113)$$

Rearranging,

$$2p_{i-1,j} + p_{i,j-1} - 6p_{i,j} + p_{i,j+1} + 2p_{i+1,j} + \frac{q\mu B (\Delta x)^2}{V_b k_y} = 0. \quad \dots\dots\dots\dots\dots\dots\dots\dots\dots \quad (8.1114)$$

At Block 1, Peaceman's wellbore model will be implemented:

$$q = \frac{2\pi \bar{k} h}{\mu B\left[\ln\left(\dfrac{r_e}{r_w}\right) + s\right]}\left(p_1 - p_{sf}\right), \quad \dots\dots\dots\dots\dots\dots\dots\dots\dots\dots\dots\dots\dots\dots\dots \quad (8.1115)$$

where $\bar{k} = \sqrt{k_x k_y} = \sqrt{10 \times 5} = 7.07$ md, and

$$r_e = 0.28 \frac{\left\{\left[\left(\dfrac{k_y}{k_x}\right)^{1/2}(\Delta x)^2\right] + \left[\left(\dfrac{k_x}{k_y}\right)^{1/2}(\Delta y)^2\right]\right\}^{1/2}}{\left(\dfrac{k_y}{k_x}\right)^{1/4} + \left(\dfrac{k_x}{k_y}\right)^{1/4}}, \quad \dots\dots\dots\dots\dots\dots\dots\dots \quad (8.1116)$$

$$r_e = 0.28 \frac{\left\{\left[\left(\dfrac{1}{2}\right)^{1/2}(100)^2\right] + \left[(2)^{1/2}(100)^2\right]\right\}^{1/2}}{\left(\dfrac{1}{2}\right)^{1/4} + (2)^{1/4}} = 20.09 \, \text{ft.} \quad \dots\dots\dots\dots\dots\dots\dots \quad (8.1117)$$

Substituting the Peaceman model, as well as reservoir/fluid property values into the flow equation, we obtain

$$-3p_1 + p_3 + 2p_2 - \frac{(1.0)(100)^2(1{,}000)}{(100)(100)(50)(5)(1.127)} \frac{2(3.14)(7.07)(1.127)(50)}{(1.0)\left(\ln\dfrac{20.09}{0.25} + 0\right)(1{,}000)}(p_1 - 100) = 0, \quad \dots\dots \quad (8.1118)$$

which reduces to

$$-5.024 p_1 + p_3 + 2p_2 + 202.4 = 0. \quad \dots\dots\dots\dots\dots\dots\dots\dots\dots\dots\dots\dots\dots\dots \quad (8.1119)$$

At Block 2, create the reflection Block 2′ along the $x+$-direction, shown in **Fig. 8.70.2,** which is at the same pressure as the aquifer:

$$2p_2 - 5p_2 + p_4 + 2p_1 = 0, \quad \dots\dots\dots\dots\dots\dots\dots\dots\dots\dots\dots\dots\dots\dots\dots\dots\dots \quad (8.1120)$$

$$-5p_2 + p_4 + 2p_1 = -2{,}000. \quad \dots\dots\dots\dots\dots\dots\dots\dots\dots\dots\dots\dots\dots\dots\dots\dots \quad (8.1121)$$

At Block 3, create the reflection Block 3′ along the $y+$-direction, shown in **Fig. 8.70.3:**

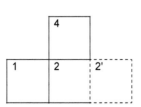

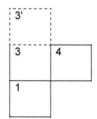

Fig. 8.70.2—Treatment of the boundary condition of Block 2.

Fig. 8.70.3—Treatment of the boundary condition of Block 3.

$$p_{3'} - 4p_3 + p_1 + 2p_4 = 0, \dots \dots \dots \dots \dots \dots \dots \dots \dots \dots (8.1122)$$

$$-4p_3 + p_1 + 2p_4 = -1,000. \dots \dots \dots \dots \dots \dots \dots \dots (8.1123)$$

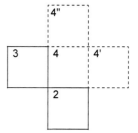

At Block 4, create the reflection Blocks 4′ and 4″ along the $x+$- and $y+$-directions, shown in **Fig. 8.70.4:**

$$2p_4 + p_4 - 6p_4 + p_2 + 2p_3 = 0, \dots \dots \dots \dots \dots \dots \dots (8.1124)$$

$$-6p_4 + p_2 + 2p3 = -3,000. \dots \dots \dots \dots \dots \dots \dots \dots (8.1125)$$

Fig. 8.70.4—Treatment of the boundary condition of Block 4.

Solve the four equations in four unknowns using the Gauss-Seidel method:

$$p_1^{(k+1)} = \frac{1}{5.204}\left[202.4 + 2p_2^{(k)} + p_2^{(k)}\right], \dots \dots \dots \dots \dots \dots \dots (8.1126)$$

$$p_2^{(k+1)} = \frac{1}{5}\left[2,000 + p_1^{(k+1)} + p_4^{(k)}\right], \dots \dots \dots \dots \dots \dots \dots (8.1127)$$

$$p_3^{(k+1)} = \frac{1}{4}\left[1,000 + p_1^{(k+1)} + 2p_4^{(k)}\right], \dots \dots \dots \dots \dots \dots \dots (8.1128)$$

$$p_4^{(k+1)} = \frac{1}{6}\left[3,000 + p_2^{(k+1)} + 2p_3^{(k+1)}\right]. \dots \dots \dots \dots \dots \dots \dots (8.1129)$$

Iteration results are displayed in **Table 8.74.**

Iteration level	p_1	p_2	p_3	p_4	|Max(Δp)|
0	850	850	850	850	–
1	547.85	789.14	811.93	902.18	302.15
2	516.05	786.86	830.1	907.84	31.8
3	518.75	789.07	833.61	909.38	3.51
4	520.33	790.01	834.77	909.93	1.58

Table 8.74—Iteration results from Part 1 of the Solution to Problem 8.4.79.

2. The production rate of the well in Block 1 is calculated using Peaceman's wellbore model:

$$q = \frac{2\pi \bar{k}h}{\mu B\left[\ln\left(\dfrac{r_e}{r_w}\right) + s\right]}(p_1 - p_{sf}) = \frac{2(3.14)(7.07)(1.127)(50)}{(1.0)\left(\ln\dfrac{20.09}{0.25} + 0\right)(1,000)}(p_1 - 100) = 239.74 \text{ STB/D.} \dots \dots (8.1130)$$

MBC:

- Mass flow across $y+$ boundary of Block 3:

$$q_{in} = \frac{A_y k_y}{\mu B}\frac{\partial p}{\partial y} = \frac{(100)(50)(5)(1.127)}{(1.0)(1,000)}\frac{1,000 - 834.77}{100} = 46.55 \text{ STB/D.} \dots \dots \dots \dots (8.1131)$$

- Mass flow across $y+$ boundary of Block 4:

$$q_{in} = \frac{A_y k_y}{\mu B}\frac{\partial p}{\partial y} = \frac{(100)(50)(5)(1.127)}{(1.0)(1,000)}\frac{(1,000 - 909.93)}{100} = 25.38 \text{ STB/D.} \dots \dots \dots \dots (8.1132)$$

- Mass flow across $x+$ boundary of Block 4:

$$q_{in} = \frac{A_x k_x}{\mu B}\frac{\partial p}{\partial y} = \frac{(100)(50)(10)(1.127)}{(1.0)(1,000)}\frac{(1,000 - 909.93)}{100} = 50.75 \text{ STB/D.} \dots \dots \dots \dots (8.1133)$$

- Mass flow across $x+$ boundary of Block 2:

$$q_{in} = \frac{A_x k_x}{\mu B}\frac{\partial p}{\partial x} = \frac{(100)(50)(10)(1.127)}{(1.0)(1,000)}\frac{(1,000 - 790.01)}{100} = 118.33 \text{ STB/D.} \dots \dots \dots \dots (8.1134)$$

$$\sum q_{in} = 46.55 + 25.38 + 50.75 + 118.33 = 241.01 \approx \sum q_{out} = 239.74, \quad \dots\dots\dots\dots\dots\dots\dots(8.1135)$$

which highlights a good MBC.

3. Production rate is to be increased to $239.74 \times 1.2 = 287.69$ STB/D.

At Block 1,

$$-3p_1 + p_3 + 2p_2 = \frac{(1.0)(100)^2(1,000)}{(100)(100)(50)(5)(1.127)}(287.69) = 1,021.1. \quad \dots\dots\dots\dots\dots\dots\dots(8.1136)$$

Because $p_2 = 749.3$ psia and $p_3 = 803.1$ psia, one can solve for $p_1 = 427.0$ psia.

Using the Peaceman model,

$$q = 287.69 = \frac{2(3.14)(7.07)(1.127)(50)}{(1.0)\left[\ln\dfrac{20.09}{0.25} + s\right](1,000)}(427 - 100), \text{ (skin factor } s = -1.54). \quad \dots\dots\dots\dots(8.1137)$$

4. With the same flow rate, the pressure distribution is the same as in Part 2.

Applying Peaceman's wellbore model,

$$q = \frac{2(3.14)(7.07)(1.127)(50)}{(1.0)\left[\left(\ln\dfrac{20.09}{0.25} + 0\right)\right](1,000)}(427 - p_{sf}) = 287.69 \text{ STB/D}. \quad \dots\dots\dots\dots\dots\dots(8.1138)$$

Solving for the sandface pressure yields $p_{sf} = -77.4$ psia.

Because a negative sandface pressure is encountered, it is not possible to achieve the flow rate by decreasing the sandface pressure (sandface pressure cannot be brought to less than 14.7 psia).

Problem 8.4.80

Consider the 2D body-centered grid with the boundary conditions indicated in **Fig. 8.71.** The well in Block 6 is producing at a constant sandface-pressure specification.

1. Show all reflection nodes and their respective equations needed to solve this problem.
2. Calculate these SIP coefficients: W_5, E_5, C_6, W_6, and N_7.

Use the following reservoir and fluid properties in your calculations: $\mu = 1.1$ cp, $B = 1$ res bbl/STB, $\Delta x = \Delta y = 100$ ft, $h = 50$ ft, $r_w = 0.25$ ft, and $s = 0$.

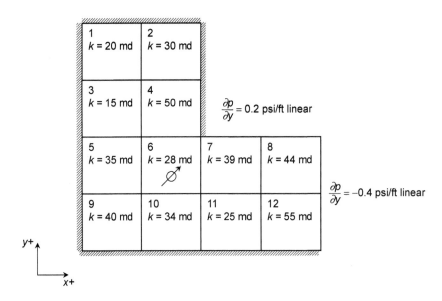

Fig. 8.71—Reservoir system of Problem 8.4.80.

Solution to Problem 8.4.80

1. Reflect Block 7 along the y+-direction. With specified-pressure gradient, we have

$$\frac{\partial p}{\partial y} = \frac{p_{7'} - p_7}{100} = 0.2, \dots \dots \dots \dots \dots \dots \dots \dots \dots \dots \dots \dots \dots \dots \dots \dots (8.1139)$$

$$p_{7'} = 20 + p_7. \dots \dots \dots \dots \dots \dots \dots \dots \dots \dots \dots \dots \dots \dots \dots \dots \dots \dots (8.1140)$$

Reflect Block 8 along the y+-direction. With specified-pressure gradient, we have

$$\frac{\partial p}{\partial y} = \frac{p_{8'} - p_8}{100} = 0.2, \dots \dots \dots \dots \dots \dots \dots \dots \dots \dots \dots \dots \dots \dots \dots \dots (8.1141)$$

$$p_{8'} = 20 + p_8. \dots \dots \dots \dots \dots \dots \dots \dots \dots \dots \dots \dots \dots \dots \dots \dots \dots \dots (8.1142)$$

Reflect Block 8 along the x+-direction. With specified-pressure gradient, we have

$$\frac{\partial p}{\partial x} = \frac{p_{8'} - p_8}{100} = -0.4, \dots \dots \dots \dots \dots \dots \dots \dots \dots \dots \dots \dots \dots \dots \dots (8.1143)$$

$$p_{8''} = p_8 - 40. \dots \dots \dots \dots \dots \dots \dots \dots \dots \dots \dots \dots \dots \dots \dots \dots \dots \dots (8.1144)$$

Reflect Block 12 along the x+-direction. With specified-pressure gradient, we have

$$\frac{\partial p}{\partial x} = \frac{p_{12'} - p_{12}}{100} = -0.4, \dots \dots \dots \dots \dots \dots \dots \dots \dots \dots \dots \dots \dots \dots (8.1145)$$

$$p_{12'} = p_{12} - 40. \dots \dots \dots \dots \dots \dots \dots \dots \dots \dots \dots \dots \dots \dots \dots \dots \dots (8.1146)$$

2. Block 5:

$$W_5 = 0, \dots \dots \dots \dots \dots \dots \dots \dots \dots \dots \dots \dots \dots \dots \dots \dots \dots \dots \dots \dots (8.1147)$$

$$E_5 = \frac{1}{\mu B} \frac{2 A_{5_x} A_{6_x} k_{5_x} k_{6_x}}{A_{5_x} k_{5_x} \Delta x_6 + A_{6_x} k_{6_x} \Delta x_5} = \frac{2(100)(50)(100)(50)(35)(0.001127)(28)(0.001127)}{(1.1)(100)(50)(0.001127)(100)(28 + 35)} = 1.59 \text{ STB/D-psi.}$$

$$\dots \dots \dots \dots \dots \dots \dots \dots \dots \dots \dots \dots \dots \dots \dots \dots \dots \dots \dots \dots (8.1148)$$

Block 6:

$$S_6 = \frac{1}{\mu B} \frac{2 A_{6_y} A_{6_y} k_{10_y} k_{10_y}}{A_{6_y} k_{6_y} \Delta y_{10} + A_{10_y} k_{10_y} \Delta y_6} = \frac{2(100)(50)(100)(50)(28)(0.001127)(34)(0.001127)}{(1.1)(100)(50)(0.001127)(100)(28 + 34)} = 1.57 \text{ STB/D-psi,}$$

$$\dots \dots \dots \dots \dots \dots \dots \dots \dots \dots \dots \dots \dots \dots \dots \dots \dots \dots \dots \dots (8.1149)$$

$$N_6 = \frac{1}{\mu B} \frac{2 A_{6_y} A_{6_y} k_{4_y} k_{4_y}}{A_{6_y} k_{6_y} \Delta y_4 + A_{4_y} k_{4_y} \Delta y_6} = \frac{2(100)(50)(100)(50)(28)(0.001127)(50)(0.001127)}{(1.1)(100)(50)(0.001127)(100)(28 + 50)} = 1.84 \text{ STB/D-psi,}$$

$$\dots \dots \dots \dots \dots \dots \dots \dots \dots \dots \dots \dots \dots \dots \dots \dots \dots \dots \dots \dots (8.1150)$$

$$E_6 = \frac{1}{\mu B} \frac{2 A_{7_x} A_{6_x} k_{7_x} k_{6_x}}{A_{7_x} k_{7_x} \Delta x_6 + A_{6_x} k_{6_x} \Delta x_7} = \frac{2(100)(50)(100)(50)(39)(0.001127)(28)(0.001127)}{(1.1)(100)(50)(0.001127)(100)(28 + 39)} = 1.67 \text{ STB/D-psi,}$$

$$\dots \dots \dots \dots \dots \dots \dots \dots \dots \dots \dots \dots \dots \dots \dots \dots \dots \dots \dots \dots (8.1151)$$

$$W_6 = E_5 = 1.59 \text{ STB/D-psi.} \dots \dots \dots \dots \dots \dots \dots \dots \dots \dots \dots \dots \dots \dots \dots (8.1152)$$

Using the Peaceman model:

$$r_e \approx 0.2\Delta x = 20 \text{ft and } \bar{k} = \sqrt{k_x k_y} = 28 \text{ md.} \dots \dots \dots \dots \dots \dots \dots \dots \dots (8.1153)$$

Productivity index:

$$\Omega = \frac{2\pi \bar{k} h}{\mu B \left[\ln\left(\frac{r_e}{r_w}\right) + s \right]} = \frac{2(3.14)(28)(1.127)(50)}{(1.1)\left(\ln\frac{20}{0.25} + 0 \right)(1,000)} = 2.06 \text{ STB/D-psi,} \dots \dots \dots \dots (8.1154)$$

$$C_6 = -(E_6 + W_6 + N_6 + S_6 + \Omega) = -(1.67 + 1.59 + 1.84 + 1.57 + 2.06) = -8.73 \text{ STB/D-psi.} \dots \dots (8.1155)$$

Block 7 has a constant-pressure-gradient boundary along its northern edge so that the northern transmissibility is

$$N_7 = \frac{1}{\mu B} \frac{A_{7y} k_{7y}}{\Delta y_7} = \frac{(100)(50)(39)(0.001127)}{(1.1)(100)} = 2.0 \text{ STB/D-psi.} \dots \dots \dots (8.1156)$$

Problem 8.4.81

Consider the following Dirichlet-type problem given in **Fig. 8.72,** with the reservoir properties given in **Table 8.75,** which describes the 2D flow of a single-phase incompressible liquid in an isotropic and homogeneous porous medium with no depth gradients. The well is producing at 300 STB/D.

1. Write the PDE that describes the problem and simplify it.
2. Write the characteristic finite-difference approximation.
3. Identify the unknowns of the problems, generate the system of equations, and solve for the pressure distribution.

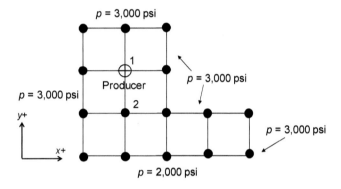

Fig. 8.72—Reservoir system of Problem 8.4.81.

Table 8.75—Reservoir properties for Problem 8.4.81.
$k_x = k_y$ = 100 md
$\Delta x = \Delta y$ = 400 ft
h = 40 ft
μ = 1.0 cp
B = 1.0 res bbl/STB

Solution to Problem 8.4.81

1. The governing flow equation of the problem states that

$$\frac{\partial}{\partial x}\left(\frac{A_x k_x}{\mu B} \frac{\partial p}{\partial x} \right) \Delta x + \frac{\partial}{\partial y}\left(\frac{A_y k_y}{\mu B} \frac{\partial p}{\partial y} \right) \Delta y + q = 0. \dots \dots \dots (8.1157)$$

Upon simplification for a homogeneous and isotropic reservoir,

$$\frac{\partial^2 p}{\partial x^2} + \frac{\partial^2 p}{\partial y^2} + \frac{q \mu B}{V_b k} = 0. \dots \dots \dots (8.1158)$$

By substituting values for the wellblock,

$$\frac{\partial^2 p}{\partial x^2} + \frac{\partial^2 p}{\partial y^2} + \frac{(-300)(1)(1)}{(400)(400)(40)(0.001127)(100)} = 0, \dots \dots \dots (8.1159)$$

one obtains

$$\frac{\partial^2 p}{\partial x^2} + \frac{\partial^2 p}{\partial y^2} - 0.000416 = 0. \dots \dots \dots (8.1160)$$

For a block without a well, the PDE is

$$\frac{\partial^2 p}{\partial x^2} + \frac{\partial^2 p}{\partial y^2} = 0. \dots \dots \dots (8.1161)$$

2. The finite-difference approximation of the governing flow equation is

$$(p_{i+1,j} - 2p_{i,j} + p_{i-1,j}) + (p_{i,j+1} - 2p_{i,j} + p_{i,j-1}) + \frac{q\mu(\Delta x)^2}{V_b k_x} = 0. \quad\dots\dots\dots\dots\dots\dots\dots\text{(8.1162)}$$

One may number the block hosting the well as Block 1 and the other block with unknown pressure as Block 2. At Block 1, the finite-difference approximation can be written as

$$p_{i-1,j} + p_{i,j-1} - 4p_{i,j} + p_{i,j+1} + p_{i+1,j} + \frac{(-300)(1)(1)}{(40)(0.001127)(100)} = 0, \quad\dots\dots\dots\dots\dots\text{(8.1163)}$$

$$p_{i-1,j} + p_{i,j-1} - 4p_{i,j} + p_{i,j+1} + p_{i+1,j} - 66.5 = 0. \quad\dots\dots\dots\dots\dots\dots\dots\dots\dots\text{(8.1164)}$$

Applying the boundary conditions,

$$p_{i-1,j} = p_{i,j+1} = p_{i+1,j} = 3,000, \quad\dots\dots\dots\dots\dots\dots\dots\dots\dots\dots\dots\dots\text{(8.1165)}$$

$$p_2 - 4p_1 + 8,933.5 = 0. \quad\dots\dots\dots\dots\dots\dots\dots\dots\dots\dots\dots\dots\dots\dots\text{(8.1166)}$$

At Block 2,

$$p_{i-1,j} + p_{i,j-1} + 4p_{i,j} + p_{i,j+1} + p_{i+1,j} = 0. \quad\dots\dots\dots\dots\dots\dots\dots\dots\dots\dots\text{(8.1167)}$$

Similarly, applying the boundary condition,

$$p_{i-1,j} = p_{i+1,j} = 3,000, \; p_{i,j-1} = 2,000, \quad\dots\dots\dots\dots\dots\dots\dots\dots\dots\dots\text{(8.1168)}$$

$$p_1 - 4p_2 + 8,000 = 0. \quad\dots\dots\dots\dots\dots\dots\dots\dots\dots\dots\dots\dots\dots\dots\text{(8.1169)}$$

3. The system of equations generated for this problem will be

$$p_2 - 4p_1 + 8,933.5 = 0, \quad\dots\dots\dots\dots\dots\dots\dots\dots\dots\dots\dots\dots\dots\text{(8.1170)}$$

$$p_1 - 4p_2 + 8,000 = 0. \quad\dots\dots\dots\dots\dots\dots\dots\dots\dots\dots\dots\dots\dots\dots\text{(8.1171)}$$

Solving the system of equations, one obtains $p_1 = 2,915.6$ psi and $p_2 = 2,782.9$ psi.

Problem 8.4.82

Generate the simplest form of the system of equations to find the pressure distribution in a reservoir for the 2D incompressible fluid flow problem in **Fig. 8.73,** with boundary conditions described in **Table 8.76.**

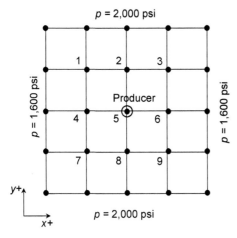

Fig. 8.73—Reservoir system of Problem 8.4.82.

$k_x = k_y =$	10 md
$\Delta x = \Delta y =$	400 ft
$h =$	50 ft
$\mu =$	1.0 cp
$B =$	1.0 res bbl/STB
$r_w =$	0.25 ft
$s =$	0
Production rate $=$	−500 STB/D

Table 8.76—Reservoir properties for Problem 8.4.82.

Solution to Problem 8.4.82

The blocks with unknown pressures can be numbered as shown in Fig. 8.73. The existing symmetry across the x and y axes calls $p_1 = p_3 = p_7 = p_9$; $p_2 = p_8$; and $p_4 = p_6$. Note that because of different boundary conditions along the boundaries, $p_2 \neq p_4$; $p_8 \neq p_4$; $p_6 \neq p_2$; and $p_6 \neq p_8$. Thus, unknowns of the problem are p_1, p_2, p_4, and p_5.

The characteristic finite-difference equation is (note that $\Delta x = \Delta y$ and $k_x = k_y$)

$$(p_{i+1,j} - 2p_{i,j} + p_{i-1,j}) + (p_{i,j+1} - 2p_{i,j} + p_{i,j-1}) + \frac{q\mu B(\Delta x)^2}{V_b k_x} = 0. \quad \text{(8.1172)}$$

- Node 1: $p_2 + p_4 + 1{,}600 + 2{,}000 - 4p_1 = 0.$ .... (8.1173)
- Node 2: $p_1 + p_1 + p_5 + 2{,}000 - 4p_2 = 0.$ .... (8.1174)
- Node 4: $p_1 + p_1 + p_5 + 1{,}600 - 4p_4 = 0.$ .... (8.1175)
- Node 5: $p_2 + p_4 + p_2 + p_4 - 4p_5 - 887.3 = 0.$ .... (8.1176)

The solutions to the system of equations are $p_1 = 1{,}752.2$ psi; $p_2 = 1{,}746.7$ psi; $p_3 = 1{,}646.7$ psi; and $p_4 = 1{,}474.9$ psi.

Problem 8.4.83

Consider the incompressible fluid flow through the 2D porous media shown in **Fig. 8.74,** with properties listed in **Table 8.77.** The reservoir has homogeneous property distribution with no-flow boundaries. Both wells are stimulated and hold a skin factor of –2. The shaded blocks are maintained at 3,000 psia by means of a strong edgewater encroachment. Injection rate is 600 STB/D. Sandface pressure at the production well is 1,000 psia.

1. Write the PDE describing the problem and identify the unknown blocks and other unknowns of the problem.
2. Write the finite-difference equations for Blocks 4, 6, 7, and 12 using the numerical values of the coefficients. Simplify them, if possible.

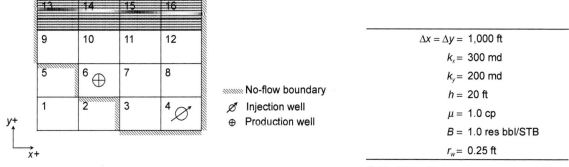

$\Delta x = \Delta y =$	1,000 ft
$k_x =$	300 md
$k_y =$	200 md
$h =$	20 ft
$\mu =$	1.0 cp
$B =$	1.0 res bbl/STB
$r_w =$	0.25 ft

Fig. 8.74—Reservoir system of Problem 8.4.83.

Table 8.77—Reservoir properties for Problem 8.4.83.

Solution to Problem 8.4.83

1. The generalized form of PDE for a 2D incompressible problem in a reservoir with no depth gradients is

$$\frac{\partial}{\partial x}\left(\frac{A_x k_x}{\mu B}\frac{\partial p}{\partial x}\right)\Delta x + \frac{\partial}{\partial y}\left(\frac{A_y k_y}{\mu B}\frac{\partial p}{\partial y}\right)\Delta y + q = 0. \quad \text{(8.1177)}$$

Simplifying Eq. 8.1177 for a homogenous reservoir with anisotropic permeability distribution,

$$k_x \frac{\partial^2 p}{\partial x^2} + k_y \frac{\partial^2 p}{\partial y^2} + \frac{q\mu B}{V_b} = 0. \quad \text{(8.1178)}$$

Unknowns of the problem are pressure values of Blocks 3, 4, 6, 7, 8, 9, 10, 11, and 12. The finite-difference analog of the flow equation for this reservoir is

$$(300)(0.001127)(p_{i+1,j} - 2p_{i,j} + p_{i-1,j}) + (200)(0.001127)(p_{i,j+1} - 2p_{i,j} + p_{i,j-1}) + \frac{q\mu B}{h} = 0. \quad \ldots\ldots (8.1179)$$

At Block 4, $q_4 = 600$ STB/D:

$$300(-p_4 + p_3) + 200(p_8 - p_4) + \frac{600}{20(0.001127)} = 0, \quad \ldots\ldots\ldots\ldots\ldots\ldots\ldots\ldots\ldots\ldots\ldots\ldots\ldots\ldots (8.1180)$$

$$3p_3 + 2p_8 - 5p_4 + 266.2 = 0. \quad \ldots\ldots\ldots\ldots\ldots\ldots\ldots\ldots\ldots\ldots\ldots\ldots\ldots\ldots\ldots\ldots\ldots\ldots (8.1181)$$

In Block 6, applying the Peaceman model:

$$q = -\frac{2\pi \bar{k} h}{\mu B\left[\ln\left(\dfrac{r_e}{r_w}\right) + s\right]}(p_1 - p_{sf}), \quad \ldots\ldots\ldots\ldots\ldots\ldots\ldots\ldots\ldots\ldots\ldots\ldots\ldots\ldots\ldots (8.1182)$$

$$\bar{k} = \sqrt{k_x k_y} = \sqrt{300 \times 200} = 245 \text{md},$$

$$r_e = 0.28 \frac{\left\{\left[\left(\dfrac{k_y}{k_x}\right)^{1/2}(\Delta x)^2\right] + \left[\left(\dfrac{k_x}{k_y}\right)^{1/2}(\Delta y)^2\right]\right\}^{1/2}}{\left(\dfrac{k_y}{k_x}\right)^{1/4} + \left(\dfrac{k_x}{k_y}\right)^{1/4}}$$

$$= 0.28 \frac{\left\{\left[\left(\dfrac{2}{3}\right)^{1/2}(1,000)^2\right] + \left[\left(\dfrac{3}{2}\right)^{1/2}(1,000)^2\right]\right\}^{1/2}}{\left(\dfrac{2}{3}\right)^{1/4} + \left(\dfrac{3}{2}\right)^{1/4}} = 199\text{ft}, \quad \ldots\ldots\ldots\ldots\ldots\ldots\ldots\ldots\ldots\ldots (8.1183)$$

$$\Omega = \frac{2\pi \bar{k} h}{\mu B\left[\ln\dfrac{r_e}{r_w} + s\right]} = \frac{2(3.14)(245)(1.127)(20)}{(1.0)(\ln\dfrac{199}{0.25} - 2)(1,000)} = 7.41 \text{ STB/D-psi}. \quad \ldots\ldots\ldots\ldots\ldots (8.1184)$$

The finite-difference equation is

$$300(-p_6 + p_7) + 200(p_{10} - p_6) + \frac{-\Omega(p_6 - p_{sf})}{20(0.001127)} = 0, \text{ (where } p_{sf} = 1,000 \text{ psi)}, \quad \ldots\ldots\ldots\ldots (8.1185)$$

$$3p_7 + 2p_{10} - 8.3p_6 + 3,288 = 0. \quad \ldots\ldots\ldots\ldots\ldots\ldots\ldots\ldots\ldots\ldots\ldots\ldots\ldots\ldots\ldots (8.1186)$$

At Block 7:

$$300(p_8 - 2p_7 + p_6) + 200(p_{11} - 2p_7 + p_3) = 0, \quad \ldots\ldots\ldots\ldots\ldots\ldots\ldots\ldots\ldots\ldots\ldots\ldots\ldots (8.1187)$$

$$3p_8 + 3p_6 - 10p_7 + 2p_{11} + 2p_3 = 0. \quad \ldots\ldots\ldots\ldots\ldots\ldots\ldots\ldots\ldots\ldots\ldots\ldots\ldots\ldots\ldots (8.1188)$$

At Block 12:

$$300(p_{11} - p_{12}) + 200(p_8 - 2p_{12} + p_{16}) = 0. \quad \ldots\ldots\ldots\ldots\ldots\ldots\ldots\ldots\ldots\ldots\ldots\ldots\ldots\ldots (8.1189)$$

Because $p_{16} = 3,000$ psia,

$$3p_{11} - 7p_{12} + 2p_8 + 6,000 = 0. \quad \ldots\ldots\ldots\ldots\ldots\ldots\ldots\ldots\ldots\ldots\ldots\ldots\ldots\ldots\ldots (8.1190)$$

Problem 8.4.84

Consider a single-phase reservoir with slightly compressible fluid flow. There is only one production well located in Block 2, which has been put on production for 10 days at a flow rate of –3,000 STB/D. Earlier investigation indicates that all boundaries, except the southwest boundary, are completely sealed (shown as a dashed line in **Fig. 8.75**). At day 10, a measurement in the observation well in Block 3 indicates that pressure is 2,750 psia. Using the data in **Table 8.78,** analyze and determine the nature of the boundary condition at the southwest boundary.

Note: Use $\Delta t = 10$ days in your calculations.

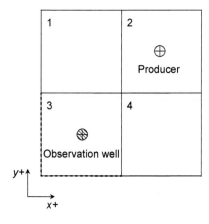

Fig. 8.75—Reservoir system of Problem 8.4.84.

$$k_x = k_y = 100 \text{ md}$$
$$\phi = 0.1, \text{ constant}$$
$$p_{initial} = 3{,}000 \text{ psia}$$
$$r_w = 0.25 \text{ ft}$$
$$\Delta x = \Delta y = 500 \text{ ft}$$
$$s = 0$$
$$\rho = 62 \text{ lbm/ft}^3$$
$$c = 5 \times 10^{-5} \text{ psia}^{-1}, \text{ constant}$$
$$h = 100 \text{ ft}$$
$$\mu \approx 1.1 \text{ cp}, \text{ constant}$$
$$B \approx 0.9 \text{ res bbl/STB}, \text{ constant}$$

Depth gradients can be ignored

Table 8.78—Reservoir properties for Problem 8.4.84.

Solution to Problem 8.4.84

The reservoir-diffusivity equation could be written as

$$\frac{\partial}{\partial x}\left(\frac{A_x k_x}{\mu B}\frac{\partial p}{\partial x}\right)\Delta x + \frac{\partial}{\partial y}\left(\frac{A_y k_y}{\mu B}\frac{\partial p}{\partial y}\right)\Delta y + q = \frac{V_b \phi c_t}{5.615}\frac{\partial p}{\partial t}. \quad \dots \dots \dots \dots \dots \dots \dots \text{(8.1191)}$$

Because the reservoir is homogeneous and isotropic in terms of permeability and other reservoir properties,

$$\frac{A_x k_x}{\mu B}\frac{\partial^2 p}{\partial x^2}\Delta x + \frac{A_y k_y}{\mu B}\frac{\partial^2 p}{\partial y^2}\Delta y + q = \frac{V_b \phi c_t}{5.615}\frac{\partial p}{\partial t}. \quad \dots \dots \dots \dots \dots \dots \dots \dots \dots \text{(8.1192)}$$

Multiplying on both sides with $\frac{\mu B}{V_b k}$, where $k_x = k_y = k$,

$$\frac{\partial^2 p}{\partial x^2} + \frac{\partial^2 p}{\partial y^2} + \frac{q\mu B}{V_b k} = \frac{\mu B \phi c_t}{5.615 k}\frac{\partial p}{\partial t}. \quad \dots \dots \dots \dots \dots \dots \dots \dots \dots \dots \dots \text{(8.1193)}$$

The finite-difference approximation can be written as

$$\frac{(p_{i+1,j}^{n+1} - 2p_{i,j}^{n+1} + p_{i-1,j}^{n+1})}{(\Delta x)^2} + \frac{(p_{i,j+1}^{n+1} - 2p_{i,j}^{n+1} + p_{i,j-1}^{n+1})}{(\Delta y)^2} + \frac{q\mu B}{V_b k}\bigg|_{i,j} = \frac{\mu B \phi c_t}{5.615 k}\frac{\left(p_{i,j}^{n+1} - p_{i,j}^{n}\right)}{\Delta t}. \quad \dots \dots \dots \text{(8.1194)}$$

One can then multiply the entire equation with Δx^2:

$$(p_{i+1,j}^{n+1} - 2p_{i,j}^{n+1} + p_{i-1,j}^{n+1}) + (p_{i,j+1}^{n+1} - 2p_{i,j}^{n+1} + p_{i,j-1}^{n+1}) + \left.\frac{q\mu B}{hk}\right|_{i,j} = \frac{\Delta x^2 \mu B\phi c_t}{5.615k} \frac{\left(p_{i,j}^{n+1} - p_{i,j}^n\right)}{\Delta t}. \quad \ldots\ldots\ldots\ldots\ldots\ldots (8.1195)$$

Assuming that the southwest boundary is completely sealed, the problem can be solved in the following way: Because Blocks 1 and 4 are symmetrical with respect to each other, we can write $p_1 = p_4$. On writing the finite-difference approximations at each of the reservoir blocks, we find

- Block 1:

$$p_3^{n+1} + p_2^{n+1} - 2p_1^{n+1} = \frac{(500)^2 \left(5 \times 10^{-5}\right)(0.1)(1.1)(0.9)}{(5.615)(100 \times 0.001127)} \frac{\left(p_1^{n+1} - 3,000\right)}{10}, \quad \ldots\ldots\ldots\ldots\ldots\ldots\ldots (8.1196)$$

$$p_3^{n+1} + p_2^{n+1} - 2p_1^{n+1} = 0.196\left(p_1^{n+1} - 3,000\right), \quad \ldots\ldots\ldots\ldots\ldots\ldots\ldots\ldots\ldots\ldots\ldots (8.1197)$$

$$-2.196 p_1^{n+1} + p_2^{n+1} + p_3^{n+1} = -586.7. \quad \ldots\ldots\ldots\ldots\ldots\ldots\ldots\ldots\ldots\ldots\ldots\ldots (8.1198)$$

- Block 2:

$$2p_1^{n+1} - 2p_2^{n+1} + \frac{(1.1)(0.9)(-3,000)}{(100)(100 \times 0.001127)} = 0.196\left(p_2^{n+1} - 3,000\right), \quad \ldots\ldots\ldots\ldots\ldots\ldots\ldots\ldots\ldots (8.1199)$$

$$2p_1^{n+1} - 2.196 p_2^{n+1} = -323.17. \quad \ldots\ldots\ldots\ldots\ldots\ldots\ldots\ldots\ldots\ldots\ldots\ldots (8.1200)$$

- Block 3:

$$2p_1^{n+1} + 2p_3^{n+1} = 0.196\left(p_3^{n+1} - 3,000\right), \quad \ldots\ldots\ldots\ldots\ldots\ldots\ldots\ldots\ldots\ldots\ldots\ldots (8.1201)$$

$$2p_1^{n+1} - 2.196 p_3^{n+1} = -586.7, \quad \ldots\ldots\ldots\ldots\ldots\ldots\ldots\ldots\ldots\ldots\ldots\ldots\ldots (8.1202)$$

$$\begin{bmatrix} -2.196 & 1 & 1 \\ 2 & -2.196 & 0 \\ 2 & 0 & -2.196 \end{bmatrix} \begin{bmatrix} p_1^{n+1} \\ p_2^{n+1} \\ p_3^{n+1} \end{bmatrix} = - \begin{bmatrix} 586.7 \\ 323.2 \\ 586.7 \end{bmatrix}. \quad \ldots\ldots\ldots\ldots\ldots\ldots\ldots\ldots\ldots (8.1203)$$

On solving the preceding matrix, the solution for pressure gives pressure values at $t = 10$ days: $p_1 = p_4 = 2,672.9$ psia; $p_2 = 2,581.5$ psia; and $p_3 = 2,701.5$ psia. Because p_3 calculated is less than p_3 observed, it can be concluded that fluid is entering through the external boundaries of Block 3, and hence, our assumption of a sealed boundary condition is invalid.

Problem 8.4.85

A slightly compressible fluid reservoir is shown in **Fig. 8.76.** A production well has been placed in Block 2. Assume that the well produces at a constant sandface pressure. Reservoir properties can be found in **Table 8.79.**

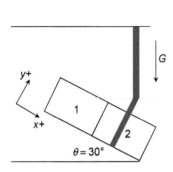

Fig. 8.76—Reservoir system and well configuration described in Problem 8.4.85.

$\Delta x = \Delta z = 600$ ft
(z is perpendicular to x-y plane)
$\Delta y = 40$ ft
$\phi = 11\%$
$k_x = k_y = k_z = 42$ md
$r_w = 0.24$ ft
$\mu = 3.4$ cp
$c = 4\times10^{-6}$ psia^{-1}
$B = 0.941$ res bbl/STB
$p_{sf} = 1,400$ psia
$s = 0$
$p_i = 3,000$ psia
$\rho = 52$ lbm/ft^3

Table 8.79—Reservoir properties for Problem 8.4.85.

1. Write the PDE representing the problem and simplify it.
2. Construct the Jacobian matrix.
3. Obtain pressures of Blocks 1 and 2 and the flow rate of the well at the end of 10 days.

Solution to Problem 8.4.85

1. For a 1D slightly compressible flow, the PDE can be written as

$$\frac{\partial}{\partial x}\left(\frac{A_x k_x}{\mu B}\frac{\partial \Phi}{\partial x}\right)\Delta x + q = \frac{V_b c_t \phi}{5.615}\frac{\partial p}{\partial t}. \quad\dots\dots\dots\dots\dots\dots\dots\dots\dots\dots\dots\dots\dots(8.1204)$$

Because μ, B, and ϕ are treated as constants, Eq. 8.1204 collapses to

$$\frac{\partial^2 p}{\partial x^2} - \frac{1}{144}\frac{g}{g_c}\rho\frac{\partial^2 G}{\partial x^2} + \frac{\mu B q}{V_b k} = \frac{\mu B}{k}\frac{c\phi}{5.615}\frac{\partial p}{\partial t}. \quad\dots\dots\dots\dots\dots\dots\dots\dots\dots\dots\dots(8.1205)$$

The finite-difference equivalent of Eq. 8.1205 is

$$p_{i-1}^{n+1} - 2p_i^{n+1} + 2p_{i+1}^{n+1} + \frac{\mu B q}{hk} = \frac{\mu B}{k}\frac{(\Delta x)^2 c\phi}{5.615\Delta t}\left(p_i^{n+1} - p_i^n\right) + \frac{1}{144}\frac{g}{g_c}\rho\left(G_{i+1} - 2G_i + G_{i-1}\right). \quad\dots\dots\dots(8.1206)$$

In consideration of the dipping nature of the reservoir,

$$\frac{G_2 - G_1}{\Delta x} = \sin\theta = \frac{1}{2}, \quad\dots\dots\dots\dots\dots\dots\dots\dots\dots\dots\dots\dots\dots\dots\dots\dots\dots\dots\dots\dots\dots\dots(8.1207)$$

$$G_2 - G_1 = \frac{1}{2}(600) = 300. \quad\dots\dots\dots\dots\dots\dots\dots\dots\dots\dots\dots\dots\dots\dots\dots\dots\dots\dots\dots(8.1208)$$

- At Block 1:

$$-p_1^{n+1} + p_2^{n+1} = \frac{\mu B}{k}\frac{(\Delta x)^2 c\phi}{5.615\Delta t}\left(p_1^{n+1} - p_i^n\right) + \frac{1}{144}\frac{g}{g_c}\rho\left(G_2 - G_1\right), \quad\dots\dots\dots\dots\dots\dots\dots\dots(8.1209)$$

$$-p_1^{n+1} + p_2^{n+1} = \frac{(3.4)(0.941)}{(42)(0.001127)} \frac{(600)^2 (4 \times 10^{-6})(0.11)}{5.615(10)} \left(p_1^{n+1} - p_1^n \right) + \frac{1}{144}(52)(300), \quad \ldots\ldots(8.1210)$$

$$-p_1^{n+1} + p_2^{n+1} = 0.19\left(p_1^{n+1} - p_1^n \right) + 108.3, \quad \ldots\ldots\ldots\ldots\ldots\ldots\ldots\ldots\ldots\ldots(8.1211)$$

$$R_1 = -1.19 p_1^{n+1} + p_2^{n+1} = 463.7 = 0. \quad \ldots\ldots\ldots\ldots\ldots\ldots\ldots\ldots\ldots\ldots\ldots(8.1212)$$

- At Block 2:

$$p_1^{n+1} - p_2^{n+1} + \frac{\mu B q}{kh} = \frac{\mu B}{k} \frac{(\Delta x)^2 c\phi}{5.615\Delta t}\left(p_2^{n+1} - p_2^n \right) + \frac{1}{144}\frac{g}{g_c}\rho\left(G_1 + G_2 \right), \quad \ldots\ldots\ldots\ldots(8.1213)$$

$$p_1^{n+1} - p_2^{n+1} + \frac{(3.4)(0.941)q}{(42)(0.001127)(40)} = \frac{(3.4)(0.941)}{(42)(0.001127)} \frac{(600)^2 (4 \times 10^{-6})(0.11)}{5.615(10)}\left(p_2^{n+1} - p_2^n \right) + \frac{1}{144}(52)(-300).$$
$$\ldots\ldots\ldots\ldots\ldots\ldots\ldots\ldots\ldots\ldots\ldots\ldots\ldots\ldots\ldots\ldots\ldots\ldots\ldots\ldots\ldots\ldots\ldots\ldots\ldots\ldots\ldots(8.1214)$$

Applying Peaceman's wellbore model:

$$q = -\frac{2\pi \bar{k}\Delta y}{\mu B\left(\ln \frac{r_{eq}}{r_w} + s \right)}\left(p_2 - p_{sf} \right) = -\frac{2(3.14)(42)(0.001127)(600)}{(3.4)(0.941)\left[\ln \frac{(0.2)(600)}{(0.25)} \right]}\left(p_2 - 1,400 \right) = -9.03\left(p_2 - 1,400 \right).$$
$$\ldots\ldots\ldots\ldots\ldots\ldots\ldots\ldots\ldots\ldots\ldots\ldots\ldots\ldots\ldots\ldots\ldots\ldots\ldots\ldots\ldots\ldots\ldots\ldots\ldots\ldots(8.1215)$$

where $r_e = 0.2\Delta x = 120$ ft.

Substituting q into the equation at Block 2,

$$p_1^{n+1} - p_2^{n+1} - 15.28 p_2^{n+1} = 0.19\left(p_2^{n+1} - p_2^n \right) - 108.3 - 21,386.1, \quad \ldots\ldots\ldots\ldots\ldots\ldots(8.1216)$$

$$R_2 = p_1^{n+1} - 16.45 p_2^{n+1} + 22,041.7 = 0. \quad \ldots\ldots\ldots\ldots\ldots\ldots\ldots\ldots\ldots\ldots\ldots(8.1217)$$

2. Entries for Jacobian:

$$\left(\frac{\partial R_1}{\partial p_1} \right) = -1.19, \left(\frac{\partial R_1}{\partial p_2} \right) = 1, \left(\frac{\partial R_2}{\partial p_1} \right) = 1, \left(\frac{\partial R_2}{\partial p_2} \right) = -16.45. \quad \ldots\ldots\ldots\ldots\ldots\ldots\ldots\ldots(8.1218)$$

Using the initial condition as an initial guess for unknown pressures, $p_1 = p_2 = 3,000$,

$$R_1^{(0)} = -108.3, \quad R_2^{(0)} = -24,308.3. \quad \ldots\ldots\ldots\ldots\ldots\ldots\ldots\ldots\ldots\ldots\ldots\ldots(8.1219)$$

3. Applying the GNR protocol:

$$\left(\frac{\partial R_1}{\partial p_1} \right)^{(0)} \Delta p_1^{(1)} + \left(\frac{\partial R_1}{\partial p_2} \right)^{(0)} \Delta p_2^{(1)} = -R_1^{(0)}, \quad \ldots\ldots\ldots\ldots\ldots\ldots\ldots\ldots\ldots\ldots(8.1220)$$

$$\left(\frac{\partial R_2}{\partial p_1} \right)^{(0)} \Delta p_1^{(1)} + \left(\frac{\partial R_2}{\partial p_2} \right)^{(0)} \Delta p_2^{(1)} = -R_2^{(0)}, \quad \ldots\ldots\ldots\ldots\ldots\ldots\ldots\ldots\ldots\ldots(8.1221)$$

$$\begin{bmatrix} \left(\frac{\partial R_1}{\partial p_1} \right) \left(\frac{\partial R_1}{\partial p_2} \right) \\ \left(\frac{\partial R_2}{\partial p_1} \right) \left(\frac{\partial R_2}{\partial p_2} \right) \end{bmatrix}^{(0)} \begin{bmatrix} \Delta p_1 \\ \Delta p_2 \end{bmatrix}^{(1)} = -\begin{bmatrix} R_1 \\ R_2 \end{bmatrix}^{(0)}, \quad \ldots\ldots\ldots\ldots\ldots\ldots\ldots\ldots(8.1222)$$

$$\begin{bmatrix} -1.19 & 1 \\ 1 & -16.45 \end{bmatrix} \begin{bmatrix} \Delta p_1 \\ \Delta p_2 \end{bmatrix}^{(1)} = -\begin{bmatrix} -108.3 \\ -24,308.3 \end{bmatrix}. \quad \ldots\ldots\ldots\ldots\ldots\ldots\ldots\ldots\ldots(8.1223)$$

Solving the previous system, one obtains the following:

$$\Delta p_1^{(1)} = -1,404.5 \rightarrow p_1^{(1)} = p_1^{(0)} - 1,404.5, \quad \ldots\ldots\ldots\ldots\ldots\ldots\ldots\ldots\ldots\ldots(8.1224)$$

$$\Delta p_2^{(1)} = -1,563.1 \rightarrow p_2^{(1)} = p_2^{(0)} - 1,563.1, \quad \dots\dots\dots\dots\dots\dots\dots\dots\dots\dots\dots\dots\dots (8.1225)$$

with the resulting pressure values for Blocks 1 and 2, $p_1^{(1)} = 1,595.5$ psia and $p_2^{(1)} = 1,436.9$ psia.

New residuals, $R_1^{(1)} = 0.00$ and $R_2^{(1)} = 0.00$, indicates that this is the final solution for this timestep, because the assumptions that constant μ, B, and c generate a linear system of equations, and hence, GNR protocol converges in one iteration.

Problem 8.4.86

As a reservoir engineer, you are asked to study the slightly compressible oil reservoir shown in **Fig. 8.77,** with reservoir properties shown in **Table 8.80.** There is one well completed in Block 2, which has been producing for 30 days at a constant flow rate of $q = 200$ STB/D. After 30 days of production, the well was shut in for another 30 days. One of the reservoir engineers claims that the average pressure of the reservoir at the end of the 60 days is within 1% of the sandface pressure recorded at the well. Is this view correct? Justify your analysis with relevant calculations.

Note: Use $\Delta t = 30$ days in your calculations.

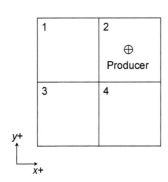

Fig. 8.77—Reservoir system of Problem 8.4.86.

$k_x = k_y = 1$ md	
$\phi = 0.1$, constant	
$p_{initial} = 3,000$ psia	
$r_w = 0.25$ ft	
$\Delta x = \Delta y = 500$ ft	
$s = 0$	
$\rho = 62$ lbm/ft³	
$c = 5\times10^{-5}$ psia⁻¹, constant	
$h = 100$ ft	
$\mu \approx 1.1$ cp, constant	
$B \approx 0.9$ res bbl/STB, constant	
All external boundaries are sealed	

Table 8.80   Reservoir properties for Problem 8.4.86.

Solution to Problem 8.4.86

The governing equation for slightly compressible flow is

$$\frac{\partial}{\partial x}\left(\frac{A_x k_x}{\mu B}\frac{\partial \Phi}{\partial x}\right)\Delta x + \frac{\partial}{\partial y}\left(\frac{A_y k_y}{\mu B}\frac{\partial \Phi}{\partial y}\right)\Delta y + q = \frac{V_b c_t \phi}{5.615}\frac{\partial p}{\partial t}. \quad \dots\dots\dots\dots\dots\dots\dots\dots\dots\dots (8.1226)$$

Because μ, B, and ϕ are treated as constants and there are no depth gradients, the previous equation collapses to

$$\frac{\partial^2 p}{\partial x^2} + \frac{\partial^2 p}{\partial y^2} + \frac{\mu B q}{V_b k} = \frac{\mu B}{k}\frac{c\phi}{5.615}\frac{\partial p}{\partial t}. \quad \dots\dots\dots\dots\dots\dots\dots\dots\dots\dots\dots\dots (8.1227)$$

The finite-difference form of the preceding PDE is

$$p_{i,j-1}^{n+1} + p_{i-1,j}^{n+1} - 4p_{i,j}^{n+1} + p_{i+1,j}^{n+1} + p_{i,j+1}^{n+1} + \frac{\mu B q}{hk} = \frac{\mu B}{k}\frac{(\Delta x)^2}{5.615\Delta t}c\phi\left(p_i^{n+1} - p_i^n\right). \quad \dots\dots\dots\dots\dots (8.1228)$$

- Block 1:

$$p_3^{n+1} - 2p_1^{n+1} + p_2^{n+1} = \frac{\mu B}{k}\frac{\Delta x \Delta y c\phi}{5.615\Delta t}\left(p_1^{n+1} - p_1^n\right). \quad \dots\dots\dots\dots\dots\dots\dots\dots\dots (8.1229)$$

- Block 2:

$$p_1^{n+1} - 2p_2^{n+1} + p_1^{n+1} + \frac{\mu B q}{hk} = \frac{\mu B}{k} \frac{\Delta x \Delta y c \phi}{5.615 \Delta t} \left(p_2^{n+1} - p_2^n \right). \quad \dots\dots\dots\dots\dots\dots\dots\dots\dots\dots\dots(8.1230)$$

- Block 3:

$$p_1^{n+1} - 2p_3^{n+1} + p_1^{n+1} = \frac{\mu B}{k} \frac{\Delta x \Delta y c \phi}{5.615 \Delta t} \left(p_3^{n+1} - p_3^n \right), \quad \dots\dots\dots\dots\dots\dots\dots\dots\dots\dots(8.1231)$$

where

$$\frac{\mu B}{k} \frac{\Delta x \Delta y c \phi}{5.615 \Delta t} = \frac{(1.1)(0.9)(500)(500)(5 \times 10^{-5})(0.1)}{(0.001127 \times 1)(5.615)(30)} = 6.52. \quad \dots\dots\dots\dots\dots\dots\dots\dots(8.1232)$$

For the first timestep, $\Delta t = 30$ days, $q = -200$ STB/D:

$$\begin{cases} p_2 - 8.52 p_1 + p_3 = -19,555.6 \\ p_1 - 4.26 p_2 = -8,902 \\ p_1 - 4.26 p_3 = -9,780 \end{cases}. \quad \dots\dots\dots\dots\dots\dots\dots\dots\dots\dots\dots\dots\dots\dots\dots\dots(8.1233)$$

Solving the preceding system of equations, one obtains the following:

$$\begin{cases} p_1 = p_4 = 2,973.86 \\ p_2 = 2,787.76 \\ p_3 = 2,993.86 \end{cases}. \quad \dots\dots\dots\dots\dots\dots\dots\dots\dots\dots\dots\dots\dots\dots\dots\dots\dots(8.1234)$$

For the second timestep, $\Delta t = 30$ days and $q = 0$,

$$\begin{cases} p_2 - 8.52 p_1 + p_3 = -19,389.6 \\ p_1 - 4.26 p_2 = -9,088 \\ p_1 - 4.26 p_3 = -9,760 \end{cases}. \quad \dots\dots\dots\dots\dots\dots\dots\dots\dots\dots\dots\dots\dots\dots\dots\dots(8.1235)$$

Solving the preceding system of equations yields,

$$\begin{cases} p_1 = p_4 = 2,958.08 \\ p_2 = 2,827.72 \\ p_3 = 2,985.46 \end{cases}, \quad \dots\dots\dots\dots\dots\dots\dots\dots\dots\dots\dots\dots\dots\dots\dots\dots\dots(8.1236)$$

because $q = 0$, $p_{sf2} = p_2 = 2,827.72$ psia.

Considering the compressibility as a constant, we have

$$C = -\frac{1}{V_p} \left(\frac{\partial V}{\partial p} \right) = -\frac{1}{V_{pinitial}} \left(\frac{\Delta V}{\Delta p} \right), \quad \dots\dots\dots\dots\dots\dots\dots\dots\dots\dots\dots\dots\dots\dots(8.1237)$$

where

$$\Delta V_p = q \times \Delta t. \quad \dots\dots\dots\dots\dots\dots\dots\dots\dots\dots\dots\dots\dots\dots\dots\dots\dots\dots\dots\dots(8.1238)$$

Then, the average pressure drop is

$$\overline{\Delta p} = -\frac{\Delta V}{C V_{initial}} = \frac{(200)(30)(5.615)}{(5 \times 10^{-5})(500)^2 (100)(4)} = 67.38 \text{ psia.} \quad \dots\dots\dots\dots\dots\dots\dots(8.1239)$$

Yielding an average reservoir pressure of

$$\overline{p} = p_{initial} - \overline{p} = 3,000 - 67.38 = 2,932.62 \text{ psia.} \quad \dots\dots\dots\dots\dots\dots\dots\dots\dots\dots\dots(8.1240)$$

Comparing the two results, we find

$$\frac{\overline{p} - p_{sf}}{p_{sf}} = \frac{2,932.62 - 2,827.72}{2,827.72} = 3.7\% > 1\%. \quad \dots\dots\dots\dots\dots\dots\dots\dots\dots\dots\dots\dots(8.1241)$$

It is concluded that the statement is not correct.

Problem 8.4.87

Consider the 3D, single-phase compressible gas flow system shown in **Fig. 8.78**. A vertical well is completed only in Block 3, with a specified flow rate of q. All external boundaries are no-flow boundaries. Provide answers to the following:

1. A PDE describing the flow.
2. A residual equation in finite-difference form only for Block 3, as applicable to the GNR protocol.
3. Using SIP notation, write the GNR equation only for Block 4. Give the definition of SIP coefficients.
4. Construct the Jacobian matrix.

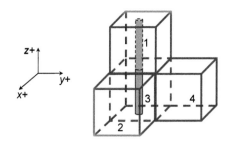

Fig. 8.78—Reservoir system of Problem 8.4.87.

Note: Block size Δx, Δy, and Δz are constants. Permeability is homogeneous and anisotropic. All external boundaries are sealed. You may take x^+-direction as East and y^+-direction as North.

Solution to Problem 8.4.87

1. For compressible fluid flow, gravitational forces are negligible. Therefore, the governing flow equation is as follows:

$$\frac{\partial}{\partial x}\left(\frac{A_x p}{\mu_g z_g}\frac{\partial p}{\partial x}\right)k_x \Delta x + \frac{\partial}{\partial y}\left(\frac{A_y p}{\mu_g z_g}\frac{\partial p}{\partial y}\right)k_y \Delta y + \frac{\partial}{\partial z}\left(\frac{A_z p}{\mu_g z_g}\frac{\partial p}{\partial z}\right)k_z \Delta z + q_g \frac{p_{sc}T}{5.615 T_{sc}} = \frac{V_b \phi}{5.615 \Delta t}\frac{\partial}{\partial t}\left(\frac{p}{z_g}\right). \quad \ldots (8.1242)$$

2. The finite-difference approximation form of the equation in Block 3 is

$$k_x \frac{2A_x}{\Delta x}\Big|_3 \frac{\left(p_2^{n+1}\right)^2 - \left(p_3^{n+1}\right)^2}{\left(\mu_{g2}^{n+1}+\mu_{g3}^{n+1}\right)\left(z_{g2}^{n+1}+z_{g3}^{n+1}\right)} + k_y \frac{2A_y}{\Delta y}\Big|_3 \frac{\left(p_4^{n+1}\right)^2 - \left(p_3^{n+1}\right)^2}{\left(\mu_{g4}^{n+1}+\mu_{g3}^{n+1}\right)\left(z_{g4}^{n+1}+z_{g3}^{n+1}\right)}$$

$$+ k_z \frac{2A_z}{\Delta z}\Big|_3 \frac{\left(p_1^{n+1}\right)^2 - \left(p_3^{n+1}\right)^2}{\left(\mu_{g1}^{n+1}+\mu_{g3}^{n+1}\right)\left(z_{g1}^{n+1}+z_{g3}^{n+1}\right)} + \frac{p_{sc}T}{5.615 T_{sc}} q_{scf/D}^{n+1} = \frac{V_b \phi}{5.615 \Delta t}\Big|_3 \frac{p_3^{n+1}}{z_{g3}^{n+1}} - \frac{V_b \phi}{5.615 \Delta t}\Big|_3 \frac{p_3^{n}}{z_{g3}^{n}}. \quad \ldots (8.1243)$$

Thus, the preceding finite-difference equation can be expressed in a residual form for Block 3:

$$R_3 = k_x \frac{2A_x}{\Delta x}\Big|_3 \frac{\left(p_2^{n+1}\right)^2 - \left(p_3^{n+1}\right)^2}{\left(\mu_{g2}^{n+1}+\mu_{g3}^{n+1}\right)\left(z_{g2}^{n+1}+z_{g3}^{n+1}\right)} + k_y \frac{2A_y}{\Delta y}\Big|_3 \frac{\left(p_4^{n+1}\right)^2 - \left(p_3^{n+1}\right)^2}{\left(\mu_{g4}^{n+1}+\mu_{g3}^{n+1}\right)\left(z_{g4}^{n+1}+z_{g3}^{n+1}\right)}$$

$$+ k_z \frac{2A_z}{\Delta z}\Big|_3 \frac{\left(p_1^{n+1}\right)^2 - \left(p_3^{n+1}\right)^2}{\left(\mu_{g1}^{n+1}+\mu_{g3}^{n+1}\right)\left(z_{g1}^{n+1}+z_{g3}^{n+1}\right)} + \frac{p_{sc}T}{5.615 T_{sc}} q_{scf/D}^{n+1} - \frac{V_b \phi}{5.615 \Delta t}\Big|_3 \frac{p_3^{n+1}}{z_{g3}^{n+1}} + \frac{V_b \phi}{5.615 \Delta t}\Big|_3 \frac{p_3^{n}}{z_{g3}^{n}}. \quad \ldots (8.1244)$$

3. In a generic form, the GNR equation for Block 4 can be written as

$$S_4^{n+1}{}^{(k)} \Delta p_3^{n+1}{}^{(k+1)} + C_4^{n+1}{}^{(k)} \Delta p_4^{n+1}{}^{(k+1)} = -R_4^{n+1}{}^{(k)}, \quad \ldots \ldots \ldots \ldots \ldots (8.1245)$$

where $S_4 = \dfrac{\partial R_4}{\partial p_3}$ and $C_4 = \dfrac{\partial R_4}{\partial p_4}$.

4. Jacobian matrix for the problem is

$$\begin{bmatrix} \dfrac{\partial R_1}{\partial p_1} & 0 & \dfrac{\partial R_1}{\partial p_3} & 0 \\[2mm] 0 & \dfrac{\partial R_2}{\partial p_2} & \dfrac{\partial R_2}{\partial p_3} & 0 \\[2mm] \dfrac{\partial R_3}{\partial p_1} & \dfrac{\partial R_3}{\partial p_2} & \dfrac{\partial R_3}{\partial p_3} & \dfrac{\partial R_3}{\partial p_4} \\[2mm] 0 & 0 & \dfrac{\partial R_4}{\partial p_3} & \dfrac{\partial R_4}{\partial p_4} \end{bmatrix}^{(k)}_{n+1} \begin{bmatrix} \Delta p_1 \\ \Delta p_2 \\ \Delta p_3 \\ \Delta p_4 \end{bmatrix}^{(k+1)}_{n+1} = - \begin{bmatrix} R_1 \\ R_2 \\ R_3 \\ R_4 \end{bmatrix}^{(k)}_{n+1}. \quad \ldots \ldots \ldots \ldots (8.1246)$$

Problem 8.4.88

Do the following for the two linear equations given:

$$\begin{cases} 2x + y = 3 \\ 4x - 2y = 5. \end{cases}$$

1. Rewrite the equations in the form of a matrix and solve them.
2. Rewrite the equations with the GNR protocol and solve them.
3. Comment on the two solution techniques.

Solution to Problem 8.4.88

1. The matrix form is

$$\begin{bmatrix} 2 & 1 \\ 4 & -2 \end{bmatrix} \begin{bmatrix} x \\ y \end{bmatrix} = \begin{bmatrix} 3 \\ 5 \end{bmatrix}. \quad \dots \dots \dots \dots \dots \dots \dots \dots \dots (8.1247)$$

Solve the equations by Gaussian elimination (or any other solver you choose): $x = 1.4$ and $y = 0.2$.

2. Residual equations are

$$R_1 = 2x + y - 3, \quad \dots \dots \dots \dots \dots \dots \dots \dots (8.1248)$$

$$R_2 = 4x - 2y - 5. \quad \dots \dots \dots \dots \dots \dots \dots \dots (8.1249)$$

For the GNR protocol, we have

$$\left(\frac{\partial R_1}{\partial x}\right)\Delta x + \left(\frac{\partial R_1}{\partial y}\right)\Delta y = -R_1, \quad \dots \dots \dots \dots \dots (8.1250)$$

$$\left(\frac{\partial R_2}{\partial x}\right)\Delta x + \left(\frac{\partial R_2}{\partial y}\right)\Delta y = -R_2. \quad \dots \dots \dots \dots (8.1251)$$

Take $x_{\text{initial}} = 0$, $y_{\text{initial}} = 0$ as an initial guess:

$$2\Delta x + \Delta y = -3, \quad \dots \dots \dots \dots \dots \dots \dots (8.1252)$$

$$4\Delta x - 2\Delta y = -5. \quad \dots \dots \dots \dots \dots \dots \dots (8.1253)$$

The matrix equation for the GNR protocol is

$$\begin{bmatrix} 2 & 1 \\ 4 & -2 \end{bmatrix} \begin{bmatrix} \Delta x \\ \Delta y \end{bmatrix} = -\begin{bmatrix} -3 \\ -5 \end{bmatrix}, \quad \dots \dots \dots \dots \dots (8.1254)$$

which collapses to the original linear system of equations.

The GNR takes one iteration to converge, with the result $\Delta x = 1.4$ and $\Delta y = 0.2$. This yields the solution of $x = x_{\text{initial}} + \Delta x = 1.4$ and $y = y_{\text{initial}} + \Delta y = 0.2$.

3. GNR protocol is specifically designed for systems of nonlinear equations, which cannot be solved by linear solvers. The overall idea of the GNR protocol is to generate a system of linear equations based on residual equations. If we apply the GNR protocol to solve linear equations, it will generate the same linear equations as the original, if the initial guess is ($x_{\text{initial}} = 0$ and $y_{\text{initial}} = 0$). Different initial guesses will still generate the same results and will take only one iteration to converge.

Problem 8.4.89

Consider the incompressible fluid flow in a 2D reservoir with no depth gradients shown in **Fig. 8.79.1,** with reservoir properties displayed in **Table 8.81.** The reservoir is known to be homogeneous in thickness, depth, and permeability: A uniform body-centered grid with nine blocks placed over the reservoir. Fluid is produced through wells located in Blocks 1, 3, 7, and 9 at a sandface pressure of 500 psia at each of the wells. Water is injected through the injector located in Block 5 at a sandface pressure of 3,500 psia.

1. Determine the pressure distribution in the reservoir, starting with the PDE describing the flow dynamics.
2. Determine the flow rates of the wells.
3. Perform an MBC to verify the accuracy of your results.

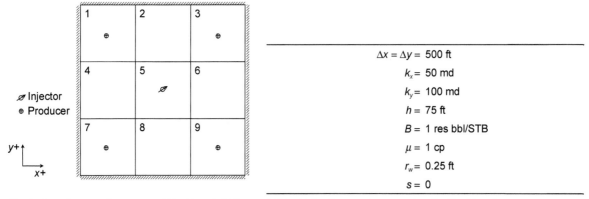

Fig. 8.79.1—Reservoir system of Problem 8.4.89.

Table 8.81—Reservoir properties for Problem 8.4.89.

$\Delta x = \Delta y = 500$ ft

$k_x = 50$ md

$k_y = 100$ md

$h = 75$ ft

$B = 1$ res bbl/STB

$\mu = 1$ cp

$r_w = 0.25$ ft

$s = 0$

In the solution of the system of equations, use the Gauss-Seidel method, with 1,400 psia as the initial guesses for all unknowns in your calculations. Use 1 psia as the convergence criterion.

Solution to Problem 8.4.89

We can simplify the problem by use of the existing symmetry in the reservoir. The given reservoir is homogeneous and anisotropic. In addition, all producers are producing at a sandface pressure of 500 psia, and the injector located in the center of the reservoir is injecting at 3,500 psia. By incorporating the symmetry, we are stating that the pressure of Block 1 will be the same as the pressure in Blocks 3, 7, and 9. Furthermore, pressure in Block 2 is the same as in Block 8, and pressure in Block 4 is the same as in Block 6. Therefore, only the pressures of Blocks 1, 2, 4, and 5 need to be estimated as shown in **Fig. 8.79.2.**

Writing a general PDE for the 2D homogeneous anisotropic ($k_y = 2k_x$) reservoir with wells in the system, for an incompressible flow problem with no depth gradients, we obtain

$$\frac{A_x k_x \Delta x}{\mu B}\frac{\partial^2 p}{\partial x^2}+\frac{A_y 2k_x \Delta y}{\mu B}\frac{\partial^2 p}{\partial y^2}+q=0. \quad\quad (8.1255)$$

Simplifying,

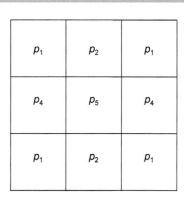

Fig. 8.79.2—Numbering scheme of the reservoir system in Problem 8.4.89.

$$\frac{\partial^2 p}{\partial x^2}+2\frac{\partial^2 p}{\partial y^2}+\frac{\mu B q}{A_x k_x \Delta x}=0. \quad\quad (8.1256)$$

Writing a finite-difference analog for Eq. 8.1256,

$$\frac{p_{i-1,j}-2p_{i,j}+p_{i+1,j}}{\Delta x^2}+2\frac{p_{i,j-1}-2p_{i,j}+p_{i,j+1}}{\Delta y^2}+\frac{\mu B q}{\Delta x \Delta y h k_x}=0. \quad\quad (8.1257)$$

Because $\Delta x = \Delta y$, one can multiply the entire equation with Δx^2 to obtain

$$p_{i-1,j}+2p_{i,j-1}-6p_{i,j}+p_{i+1,j}+2p_{i,j+1}+\frac{\mu B q}{h k_x}=0. \quad\quad (8.1258)$$

Now, calculating the productivity index for all the wells of the system (all of the productivity indices will be the same as all the reservoir properties, which are the same at the wellblocks). The productivity index is defined as

$$\Omega=\frac{2\pi \bar{k} h}{\mu B\left(\ln\frac{r_e}{r_w}+s\right)}, \quad\quad (8.1259)$$

where

$$\bar{k}=\sqrt{k_x k_y}=\sqrt{50\times 100}=70.71\ \text{md}, \quad\quad (8.1260)$$

$$\bar{k}=70.71\times 0.001127\,\text{perms}=0.0797\ \text{perms}, \quad\quad (8.1261)$$

and r_e is

$$r_e = 0.28 \frac{\left[\left(\frac{k_y}{k_z}\right)^{0.5}(\Delta z)^2 + \left(\frac{k_z}{k_y}\right)^{0.5}(\Delta y)^2\right]^{0.5}}{\left(\frac{k_y}{k_z}\right)^{0.25} + \left(\frac{k_z}{k_y}\right)^{0.25}} = 0.28 \frac{\left[\left(\frac{100}{50}\right)^{0.5}(500)^2 + \left(\frac{50}{100}\right)^{0.5}(500)^2\right]^{0.5}}{\left(\frac{100}{50}\right)^{0.25} + \left(\frac{50}{100}\right)^{0.25}} = 100.42 \text{ ft}, \ \ldots (8.1262)$$

$$\Omega = \frac{2\pi \times 0.07997 \times 75}{1 \times 1 \times \left(\ln\frac{100.42}{0.25} + 0\right)} = 6.264 \text{ STB/D-psi.} \ \ldots\ldots\ldots\ldots\ldots\ldots\ldots\ldots\ldots\ldots\ldots\ldots\ldots\ldots\ldots (8.1263)$$

Now, writing the finite-difference equation for Block 1 (**Fig. 8.79.3**),

$$p_1 + 2p_1 - 6p_1 + p_2 + 2p_4 - \frac{\mu B}{hk_x}\Omega(p_1 - 500) = 0, \ \ldots\ldots\ldots\ldots\ldots\ldots\ldots\ldots\ldots\ldots\ldots\ldots\ldots (8.1264)$$

or

$$-3p_1 + p_2 + 2p_4 - \frac{6.264}{4.22625}(p_1 - 500) = 0, \ \ldots\ldots\ldots\ldots\ldots\ldots\ldots\ldots\ldots\ldots\ldots\ldots (8.1265)$$

$$-4.4822 p_1 + p_2 + 2p_4 = -741.08. \ \ldots\ldots\ldots\ldots\ldots\ldots\ldots\ldots\ldots\ldots\ldots\ldots\ldots\ldots (8.1266)$$

Applying the Gauss-Seidel method, the Block 1 equation will take the following form:

$$p_1^{(k+1)} = \frac{p_2^{(k)} + 2p_4^{(k)} + 741.08}{4.4822}. \ \ldots\ldots\ldots\ldots\ldots\ldots\ldots\ldots\ldots\ldots\ldots\ldots\ldots\ldots (8.1267)$$

At Block 2 (**Fig. 8.79.4**),

$$p_1 + 2p_2 - 6p_2 + p_1 + 2p_5 = 0 \ \ldots\ldots\ldots\ldots\ldots\ldots\ldots\ldots\ldots\ldots\ldots\ldots\ldots\ldots\ldots (8.1268)$$

or

$$-4p_2 + 2p_1 + 2p_5 = 0. \ \ldots\ldots\ldots\ldots\ldots\ldots\ldots\ldots\ldots\ldots\ldots\ldots\ldots\ldots\ldots\ldots\ldots (8.1269)$$

Rearranging for the Gauss-Seidel equation, the Block 2 equation becomes

$$p_2^{(k+1)} = \frac{p_1^{(k+1)} + p_5^{(k)}}{2}. \ \ldots\ldots\ldots\ldots\ldots\ldots\ldots\ldots\ldots\ldots\ldots\ldots\ldots\ldots\ldots\ldots\ldots (8.1270)$$

Now, writing the finite-difference equation for Block 4 (**Fig. 8.79.5**),

$$p_4 + 2p_1 - 6p_4 + 2p_1 + p_5 = 0. \ \ldots\ldots\ldots\ldots\ldots\ldots\ldots\ldots\ldots\ldots\ldots\ldots\ldots\ldots (8.1271)$$

Rearranging for the Gauss-Seidel equation, we have the Block 4 equation:

$$p_4^{(k)} = \frac{4p_1^{(k+1)} + p_5^{(k)}}{5}. \ \ldots\ldots\ldots\ldots\ldots\ldots\ldots\ldots\ldots\ldots\ldots\ldots\ldots\ldots\ldots\ldots (8.1272)$$

Now, writing the finite-difference equation for Block 5 (**Fig. 8.79.6**),

$$p_4 + 2p_2 - 6p_5 + 2p_2 + p_4 - \frac{6.264}{4.22625}(p_5 - 3,500) = 0. \ \ldots\ldots\ldots\ldots\ldots\ldots\ldots\ldots (8.1273)$$

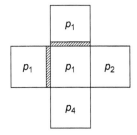

Fig. 8.79.3—Treatment of the boundary condition of Block 1.

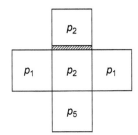

Fig. 8.79.4—Treatment of the boundary condition of Block 2.

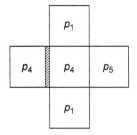

Fig. 8.79.5—Treatment of the boundary condition of Block 4.

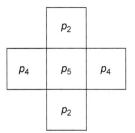

Fig. 8.79.6—Treatment of the boundary condition of Block 5.

Finally, rearranging for the Gauss-Seidel equation, Block 5 yields

$$p_5^{(k+1)} = \frac{4p_2^{(k+1)} + 2p_4^{(k+1)} + 5,187.6}{7.4822}. \quad\dots\dots\dots\dots\dots\dots\dots\dots\dots\dots\dots\dots\dots\dots(8.1274)$$

1. At this stage, we have four equations and four unknowns. Solving the linear system of equations using the Gauss-Seidel method with an initial guess of 1,400 psia for all the blocks appearing in these equations, and using a convergence criterion of 1 psia, **Table 8.82** is generated.

2. $q_1 = q_3 = q_7 = q_9 = -6.264 \times (951.53 - 500) = -2828.4$ STB/D, $\dots\dots\dots\dots\dots\dots\dots\dots\dots\dots\dots(8.1275)$

   $q_5 = -6.264 \times (1,694.73 - 3,500) = 11,308.2$ STB/D. $\dots\dots\dots\dots\dots\dots\dots\dots\dots\dots\dots(8.1276)$

3. Total production $= Q_t = 4 \times 2,828.4 = 11,313.6$ STB/D, $\dots\dots\dots\dots\dots\dots\dots\dots\dots\dots\dots(8.1277)$

   Total injection $= Q_{inj} = 11,308.2$ STB/D, $\dots\dots\dots\dots\dots\dots\dots\dots\dots\dots\dots\dots\dots(8.1278)$

   $$MBC = \frac{Q_t}{Q_{inj}} = \frac{11,313.6}{11,308.2} = 1.00048. \quad\dots\dots\dots\dots\dots\dots\dots\dots\dots\dots\dots\dots\dots(8.1279)$$

k	p_1	p_2	p_4	p_5	Max Improvement
0	1,400	1,400	1,400	1,400	–
1	1,102.378	1,251.189	1,161.903	1,672.792	297.6217
2	962.9366	1,317.864	1,104.908	1,693.201	139.4417
3	952.3804	1,322.791	1,100.545	1,694.669	10.55621
4	951.5328	1,323.101	1,100.16	1,694.732	0.847634

Table 8.82—Results of using the Gauss-Seidel method to solve the linear equation in Part 1 of the Solution to Problem 8.4.89.

Problem 8.4.90

An oil company has acquired property-contour maps of a reservoir. The system is a 2D single-phase system saturated with an incompressible fluid. The maps have been discretized, and the corresponding properties are assigned to each block, shown in **Table 8.83.** The reservoir fluid viscosity is 1.1 cp, density is 65 lbm/ft³, and FVF is 1 res bbl/STB. A horizontal well is drilled along the x-direction across the full length of Blocks 7 and 8. The well has a radius of 0.25 ft and is producing at a sandface pressure of 200 psia. The well is stimulated and has a skin factor of –1. The permeability values in the y- and z-directions can be taken as 60% and 30%, respectively, of those of x-direction.

$k_x = 100$ md	$k_x = 90$ md	$k_x = 85$ md	$k_x = 80$ md
$\phi = 10\%$	$\phi = 12\%$	$\phi = 9\%$	$\phi = 15\%$
$h = 70$ ft	$h = 70$ ft	$h = 70$ ft	$h = 70$ ft
$G = 3,100$ ft	$G = 3,100$ ft	$G = 3,100$ ft	$G = 3,100$ ft
$k_x = 120$ md	$k_x = 100$ md	$k_x = 90$ md	$k_x = 95$ md
$\phi = 20\%$	$\phi = 22\%$	$\phi = 18\%$	$\phi = 12\%$
$h = 80$ ft	$h = 85$ ft	$h = 80$ ft	$h = 80$ ft
$G = 3,000$ ft	$G = 3,000$ ft	$G = 3,000$ ft	$G = 3,000$ ft
$k_x = 120$ md	$k_x = 120$ md	$k_x = 100$ md	$k_x = 90$ md
$\phi = 25\%$	$\phi = 18\%$	$\phi = 10\%$	$\phi = 15\%$
$h = 90$ ft	$h = 90$ ft	$h = 90$ ft	$h = 90$ ft
$G = 2,900$ ft	$G = 2,900$ ft	$G = 2,900$ ft	$G = 2,900$ ft
$k_x = 110$ md	$k_x = 110$ md	$k_x = 90$ md	$k_x = 90$ md
$\phi = 18\%$	$\phi = 20\%$	$\phi = 16\%$	$\phi = 28\%$
$h = 100$ ft	$h = 100$ ft	$h = 100$ ft	$h = 100$ ft
$G = 2,800$ ft	$G = 2,900$ ft	$G = 2,800$ ft	$G = 2,800$ ft

Table 8.83—Corresponding properties assigned to each block in discretized maps.

1. Write the governing flow equation for the system shown in **Fig. 8.80.1.**
2. Write the finite-difference representation of the equation given in Part 1, using the SIP notation for Blocks 7 and 10 (no numerical values are requested).
3. Calculate the following SIP coefficients by stating their units:
   - S_3.
   - E_6.
   - N_7.
   - S_{15}.
   - Q_3.
   - W_9.
   - C_7.
4. Which portion of the horizontal well will produce more (base your answer on a qualitative analysis)?

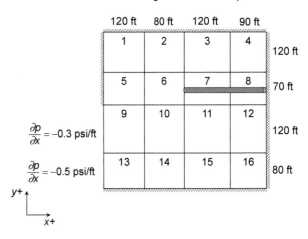

Fig. 8.80.1—Reservoir system of Problem 8.4.90.

Solution to Problem 8.4.90

1. Starting with the general form of single-phase incompressible flow,

$$\frac{\partial}{\partial x}\left(\frac{A_x k_x}{\mu B}\frac{\partial p}{\partial x}\right)\Delta x + \frac{\partial}{\partial y}\left(\frac{A_y k_y}{\mu B}\frac{\partial p}{\partial y}\right)\Delta y + \frac{\partial}{\partial z}\left(\frac{A_z k_z}{\mu B}\frac{\partial p}{\partial x}\right)\Delta z$$
$$-\left[\frac{\partial}{\partial x}\left(\frac{1}{144}\frac{\rho g}{g_c}\frac{A_x k_x}{\mu B}\frac{\partial G}{\partial x}\right)\Delta x + \frac{\partial}{\partial y}\left(\frac{1}{144}\frac{\rho g}{g_c}\frac{A_y k_y}{\mu B}\frac{\partial G}{\partial y}\right)\Delta y + \frac{\partial}{\partial z}\left(\frac{1}{144}\frac{\rho g}{g_c}\frac{A_z k_z}{\mu B}\frac{\partial G}{\partial x}\right)\Delta z\right] + q = 0. \quad \ldots (8.1280)$$

Relevant to this problem, the following observations are made:

- It is a 2D incompressible flow system.
- Depth gradients cannot be ignored.
- It has anisotropic permeability distribution.
- It has heterogeneous properties.
- It is a nonuniform grid.
- There is a production well.

The general governing flow equation for this reservoir system can be written as

$$\frac{\partial}{\partial x}\left(\frac{A_x k_x}{\mu B}\frac{\partial p}{\partial x}\right)\Delta x + \frac{\partial}{\partial y}\left(\frac{A_y k_y}{\mu B}\frac{\partial p}{\partial y}\right)\Delta y$$
$$-\left[\frac{\partial}{\partial x}\left(\frac{1}{144}\frac{\rho g}{g_c}\frac{A_x k_x}{\mu B}\frac{\partial G}{\partial x}\right)\Delta x + \frac{\partial}{\partial y}\left(\frac{1}{144}\frac{\rho g}{g_c}\frac{A_y k_y}{\mu B}\frac{\partial G}{\partial y}\right)\Delta y\right] + q = 0.$$
$$\ldots\ldots\ldots\ldots\ldots\ldots\ldots\ldots\ldots\ldots\ldots\ldots\ldots\ldots\ldots\ldots\ldots\ldots(8.1281)$$

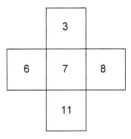

Fig. 8.80.2—Treatment of the boundary condition in Block 7.

2. The general form of the finite-difference equation using SIP notation is

$$S_{i,j}p_{i-1,j} + W_{i,j}p_{i,j-1} + C_{i,j}p_{i,j} + E_{i,j}p_{i+1,j} + N_{i,j}p_{i,j+1} = Q_{i,j}. \quad \ldots(8.1282)$$

- Block 7 is depicted in **Fig. 8.80.2.** The equation, using the SIP notation, can be written as

$$S_7 p_{11} + W_7 p_6 + C_7 p_7 + E_7 p_8 + N_7 p_3 = Q_7. \quad \ldots\ldots\ldots\ldots (8.1283)$$

- Block 10 is depicted in **Fig. 8.80.3.** The equation, using the SIP notation, can be written as

$$S_{10} p_{14} + W_{10} p_9 + C_{10} p_{10} + E_{10} p_{11} + N_{10} p_6 = Q_{10}. \quad \ldots\ldots\ldots(8.1284)$$

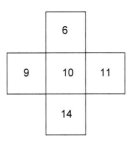

Fig. 8.80.3—Treatment of the boundary condition in Block 10.

3.

- Determine S_3:

$$S_{i,j} = \frac{A_y k_y}{\mu B \Delta y}\bigg|_{i,j-\frac{1}{2}}, \quad \dots \dots \dots \dots \dots \dots \dots \dots \dots \dots \dots \dots \dots (8.1285)$$

$$S_3 = \frac{1}{\mu B}\left(\frac{2A_{y,3}A_{y,7}k_{y,3}k_{y,7}}{A_{y,3}k_{y,3}\Delta_{y7} + A_{y,7}k_{y,7}\Delta y_3}\right), \quad \dots \dots \dots \dots \dots \dots \dots \dots \dots (8.1286)$$

$$S_3 = \frac{1}{1.1 \times 1}\left[\frac{2(120 \times 70)(120 \times 80)(51 \times 0.001127)(54 \times 0.001127)}{(120 \times 70)(51 \times 0.001127)(70) + (120 \times 80)(54 \times 0.001127)(120)}\right] \dots \dots \dots (8.1287)$$
$$= 4.936 \text{ STB/D-psi.}$$

- Determine E_6:

$$E_{i,j} = \frac{A_x k_x}{\mu B \Delta x}\bigg|_{i+\frac{1}{2},j}, \quad \dots \dots \dots \dots \dots \dots \dots \dots \dots \dots \dots \dots \dots (8.1288)$$

$$E_6 = \frac{1}{\mu B}\left(\frac{2A_{y,7}A_{x,6}k_{x,7}k_{x,6}}{A_{x,7}k_{x,7}\Delta x_6 + A_{x,6}k_{x,6}\Delta x_7}\right), \quad \dots \dots \dots \dots \dots \dots \dots \dots (8.1289)$$

$$E_6 = \frac{1}{1.1 \times 1}\left[\frac{2(80 \times 70)(85 \times 70)(90 \times 0.001127)(100 \times 0.001127)}{(80 \times 70)(90 \times 0.001127)(80) + (70 \times 85)(100 \times 0.001127)(120)}\right] \dots \dots \dots (8.1290)$$
$$= 5.50 \text{ STB/D-psi.}$$

- Determine N_7:

$$N_7 = S_3 = 4.936 \text{ STB/D-psi.} \quad \dots \dots \dots \dots \dots \dots \dots \dots \dots \dots \dots (8.1291)$$

- Determine S_{15}: Because there is no block on the southern boundary of Block 15, $S_{15} = 0$.
- Determine Q_3: Q_3 contains only the existing depth gradients between Block 3 and 7.

$$Q_3 = \left[\frac{1}{144}\frac{g}{g_c}\rho\frac{A_y k_y}{\mu B \Delta y}\bigg|_{i,j-\frac{1}{2}}\left(G_{i,j} - G_{i,j-1}\right)\right], \quad \dots \dots \dots \dots \dots \dots \dots (8.1292)$$

$$Q_3 = \left[\frac{1}{144}\left(\frac{32.174}{32.174}\right)65 \times 4.9358 \times (3,100 - 3,000)\right] = 222.796 \text{ STB/D.} \quad \dots \dots \dots \dots 8.1293)$$

- Determine W_9:

$$W_9 = \frac{A_x k_x}{\mu B \Delta x}, \quad \dots \dots \dots \dots \dots \dots \dots \dots \dots \dots \dots \dots \dots (8.1294)$$

$$E_6 = \frac{(120 \times 90)(120 \times 0.001127)}{1.1 \times 1} = 11.065 \text{ STB/D.} \quad \dots \dots \dots \dots \dots \dots \dots (8.1295)$$

- Determine C_7:

$$C_7 = -\left(S_7 + W_7 + E_7 + N_7 + \Omega_7\right). \quad \dots \dots \dots \dots \dots \dots \dots \dots \dots (8.1296)$$

- Calculate E_7:

$$E_{i,j} = \frac{A_x k_x}{\mu B \Delta x}\bigg|_{i+-\frac{1}{2},j}, \quad \dots \dots \dots \dots \dots \dots \dots \dots \dots \dots \dots (8.1297)$$

$$E_7 = \frac{1}{\mu B}\left(\frac{2A_{x,7}A_{x,8}k_{x,7}k_{x,8}}{A_{x,7}k_{x,7}\Delta x_8 + A_{x,8}k_{x,8}\Delta x_7}\right), \quad \dots \dots \dots \dots \dots \dots \dots \dots (8.1298)$$

$$E_6 = \frac{1}{1.1 \times 1}\left[\frac{2(80 \times 70)(80 \times 70)(90 \times 0.001127)(95 \times 0.001127)}{(80 \times 70)(90 \times 0.001127)(90) + (70 \times 80)(95 \times 0.001127)(120)}\right] \dots \dots \dots (8.1299)$$
$$= 5.0313 \text{ STB/D-psi.}$$

- Calculate S_7:

$$S_{i,j} = \frac{A_y k_y}{\mu B \Delta y}\bigg|_{i,j-\frac{1}{2}}, \quad \dots \dots \dots \dots \dots \dots \dots \dots \dots \dots \dots \dots \dots (8.1300)$$

$$S_7 = \frac{1}{\mu B}\left(\frac{2A_{y,11}A_{y,7}k_{x,11}k_{x,7}}{A_{y,11}k_{y,11}\Delta y_7 + A_{y,7}k_{y,7}\Delta y_{11}}\right), \quad\dots\dots\dots\dots\dots\dots\dots\dots\dots\dots\dots(8.1301)$$

$$S_7 = \frac{1}{1.1\times 1}\left[\frac{2(120\times 90)(120\times 80)(60\times 0.001127)(54\times 0.001127)}{(120\times 90)(60\times 0.001127)(70)+(120\times 80)(54\times 0.001127)(120)}\right] \quad\dots\dots\dots\dots (8.1302)$$

$$= 6.40 \text{ STB/D-psi.}$$

- Calculate Ω_7:

$$r_{e,7} = 0.28\frac{\left\{\left[\left(\frac{k_y}{k_x}\right)^{0.5}(\Delta z)^2\right]+\left[\left(\frac{k_z}{k_y}\right)^{0.5}(\Delta y)^2\right]\right\}^{0.5}}{\left(\frac{k_y}{k_z}\right)^{0.25}+\left(\frac{k_z}{k_y}\right)^{0.25}} = 0.28\frac{\left\{\left[\left(\frac{54}{27}\right)^{0.5}(80)^2\right]+\left[\left(\frac{27}{54}\right)^{0.5}(70)^2\right]\right\}^{0.5}}{\left(\frac{54}{27}\right)^{0.25}+\left(\frac{27}{54}\right)^{0.25}} = 15.43 \text{ ft.}$$

$$\dots\dots\dots\dots\dots\dots\dots\dots\dots\dots\dots\dots\dots\dots\dots\dots\dots\dots\dots\dots\dots\dots\dots (8.1303)$$

$$\Omega_1 = \frac{2\pi \bar{k}h}{\mu B}\left[\frac{1}{\ln\frac{r_e}{r_w}+s}\right] = \frac{2\pi\sqrt{54\times 27}\times(0.001127)^2(120)}{(1.1)(1)}\left[\frac{1}{\ln\frac{15.4301}{0.25}-1}\right] \quad\dots\dots\dots\dots\dots(8.1304)$$

$$= 9.45 \text{ STB/D-psi.}$$

Therefore,

$$C_7 = -(S_7 + W_7 + E_7 + N_7 + \Omega_7), \quad\dots\dots\dots\dots\dots\dots\dots\dots\dots\dots\dots\dots\dots (8.1305)$$

$$C_7 = -(6.40 + 5.50 + 5.0313 + 4.9358 + 9.4461), \quad\dots\dots\dots\dots\dots\dots\dots\dots\dots (8.1306)$$

$$C_7 = -31.3125 \text{ STB/D-psi.} \quad\dots\dots\dots\dots\dots\dots\dots\dots\dots\dots\dots\dots\dots (8.1307)$$

4. Fluid encroachment into Block 9:

$$q_9 = -\frac{kA_x}{\mu B}\frac{\partial p_9}{\partial x}, \quad\dots\dots\dots\dots\dots\dots\dots\dots\dots\dots\dots\dots\dots\dots\dots\dots (8.1308)$$

$$q_9 = \frac{(120\times 0.001127)(120\times 90)}{(1.1)(1)}(-0.3) = 398.343 \text{ STB/D.} \quad\dots\dots\dots\dots\dots\dots (8.1309)$$

Fluid encroachment into Block 13:

$$q_{13} = -\frac{kA_x}{\mu B}\frac{\partial p_{13}}{\partial x}, \quad\dots\dots\dots\dots\dots\dots\dots\dots\dots\dots\dots\dots\dots\dots\dots (8.1310)$$

$$q_{13} = \frac{(110\times 0.001127)(80\times 100)}{(1.1)(1)}(-0.5) = 450.80 \text{ STB/D.} \quad\dots\dots\dots\dots\dots\dots\dots (8.1311)$$

Thus, the total encroachment is $q_9 + q_{13} = 849.143$ STB/D. Because the system is incompressible, the total production from the horizontal well has to be equal to 849.143 STB/D. For the horizontal well section in Block 7,

$$q_7 = -\Omega_7\left(p_7 - p_{sf}\right). \quad\dots\dots\dots\dots\dots\dots\dots\dots\dots\dots\dots\dots\dots\dots\dots (8.1312)$$

For the horizontal well section in Block 8,

$$q_8 = -\Omega_8\left(p_8 - p_{sf}\right), \quad\dots\dots\dots\dots\dots\dots\dots\dots\dots\dots\dots\dots\dots\dots\dots (8.1313)$$

$$q_{total} = q_7 + q_8 = -\Omega_7\left(p_7 - p_{sf}\right) - \Omega_8\left(p_8 - p_{sf}\right). \quad\dots\dots\dots\dots\dots\dots\dots\dots (8.1314)$$

Therefore, we cannot determine which portion of the horizontal well will produce more, because we have one equation but two unknowns, unless p_7 and p_8 values are determined first by solving for the pressure distribution in the reservoir.

Problem 8.4.91

Draw the approximate streamlines for the system described in **Fig. 8.81.1.**

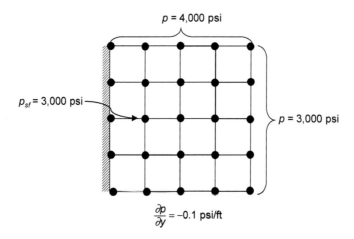

Fig. 8.81.1—Reservoir system of Problem 8.4.91.

Solution to Problem 8.4.91

With the given constraints in the system, the expected streamlines can be drawn as shown in **Fig. 8.81.2.**

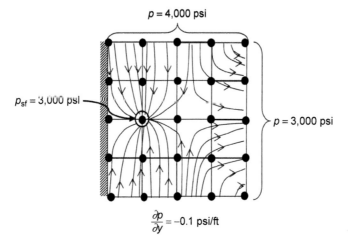

Fig. 8.81.2—Streamline of reservoir system of Problem 8.4.90

Problem 8.4.92

Consider the 1D homogeneous, horizontal reservoir with sealed boundaries shown in **Fig. 8.82.1,** with reservoir properties shown in **Table 8.84.** A single-phase, slightly compressible fluid is produced at a constant rate of 800 STB/D from a production well drilled in Block 1. Pressure is monitored at an observation well located in Block 3. At the end of 30 days, pressure-gauge readings show that $p_{sf1} = 2,855$ psia and $p_{sf3} = 3,560$ psi.

State the number of unknowns of the problem. If you are asked to solve this problem using the GNR protocol with a 3×3 Jacobian matrix, develop the necessary system of equations to solve the problem, and then do the following:

1. Give a map of the Jacobian, with only definitions of its entries.
2. What are the analytical expressions that define the main diagonal entries of the Jacobian (no numerical values are required)?
3. In this formulation, what is the last unknown that will need to be solved, and how will you solve it? (Please note that no numerical calculations are required.)

⊕ Observation well

⊕ Producer

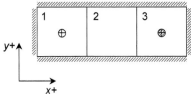

Fig. 8.82.1—Reservoir system of Problem 8.4.92.

$p_i =$	4,000 psia
$\Delta x = \Delta y =$	300 ft
$\phi =$	30%
$c =$	1×10^{-5} psia^{-1} (constant)
$s =$	?
$k_x =$	?
$h =$	80 ft
$r_w =$	0.25 ft
$B =$	1.1 res bbl/STB (constant)
$\mu =$	0.9 cp (constant)

Table 8.84—Reservoir properties for Problem 8.4.92.

Solution to Problem 8.4.92

For a 1D slightly compressible fluid in a homogeneous and horizontal reservoir, one can write

$$\frac{A_x k_x \Delta x}{\mu B} \frac{\partial^2 p}{\partial x^2} + q = \frac{V_b c_t \phi}{5.615} \frac{\partial p}{\partial t}. \quad \text{...} (8.1315)$$

Simplifying that to

$$\frac{\partial^2 p}{\partial x^2} + \frac{q \mu B}{A_x k_x \Delta x} = \frac{c_t \phi}{5.615} \frac{\mu B}{k_x} \frac{\partial p}{\partial t}. \quad \text{...} (8.1316)$$

Writing the finite-difference analog of the equation,

$$\frac{(p_{i-1}^{n+1} - 2p_i^{n+1} + p_{i+1}^{n+1})}{(\Delta x)^2} + \frac{\mu B q}{\Delta y h k_x \Delta x} = \frac{c_t \phi}{5.615} \frac{\mu B}{k_x} \frac{(p_i^{n+1} - p_i^n)}{\Delta t}, \quad \text{...................................} (8.1317)$$

and simplifying,

$$p_{i-1}^{n+1} - 2p_i^{n+1} + p_{i+1}^{n+1} + \frac{\mu B q}{h k_x} = \frac{\Delta x^2 c_t \phi}{5.615} \frac{\mu B}{k_x} \frac{(p_i^{n+1} - p_i^n)}{\Delta t}. \quad \text{...................................} (8.1318)$$

The unknowns of the problem are: p_1, p_2, k, and s.

Writing the wellbore equation at the producing well,

$$q = -\frac{2\pi \bar{k} h}{\mu B \left(\ln \dfrac{r_e}{r_w} + s \right)} \left(p - p_{sf} \right). \quad \text{...} (8.1319)$$

Substituting the data,

$$-800 = -\frac{2\pi \times 0.001127 \times k \times 80}{1.1 \times 0.9 \left(\ln \dfrac{0.2 \times 300}{0.25} + s \right)} \left(p_1 - 2855 \right), \quad \text{...................................} (8.1320)$$

one obtains

$$800 = \frac{0.5722k}{(5.48 + s)} (p_1 - 2,855). \quad \text{...} (8.1321)$$

Rearranging Eq. 8.1321,

$$5.48 + s = 7.15 \times 10^{-4} k(p_1 - 2,855), \quad \text{...} (8.1322)$$

one can obtain,

$$s = 7.15 \times 10^{-4} k p_1 - 7.522. \quad \text{...} (8.1323)$$

This equation states that skin is a dependent variable and can be solved once we have solved for k and p_i. Now, writing the equation at Block 1 using the finite-difference approximation shown in **Fig. 8.82.2,**

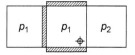

$$p_1^{n+1} - 2p_1^{n+1} + 2p_2^{n+1} + \frac{\mu B q}{h k_x} = \frac{\Delta x^2 c_t \phi}{5.615} \frac{\mu B}{k_x \Delta t}(p_1^{n+1} - 4{,}000). \qquad (8.1324)$$

Fig. 8.82.2—Treatment of the boundary condition in Block 1.

Simplifying to

$$-p_1^{n+1} + 2p_2^{n+1} - \frac{0.9 \times 1.1 \times 800}{80 \times 0.001127 k} = \frac{300^2 \times 10^{-5} \times 0.3}{5.615} \frac{1.1 \times 0.9}{0.001127 \times k \times 30}\left(p_1^n - 4{,}000\right) \qquad (8.1325)$$

or

$$-p_1^{n+1}k + p_2^{n+1}k - 8{,}784.4 = 1.408(p_1^n - 4{,}000). \qquad (8.1326)$$

Rearranging to,

$$R_1^{n+1} = -p_1^{n+1}k + p_2^{n+1}k - 1.408 p_1^{n+1} - 3{,}152.4. \qquad (8.1327)$$

- At Block 2 (see **Fig. 8.82.3),**

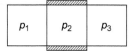

$$p_1^{n+1} - 2p_2^{n+1} + p_3^{n+1} = \frac{\Delta x^2 c_t \phi}{5.615} \frac{\mu B}{k_x \Delta t}(p_2^{n+1} - 4{,}000). \qquad (8.1328)$$

Fig. 8.82.3—Treatment of the boundary condition in Block 2.

Substituting the values,

$$-p_1^{n+1} + 2p_2^{n+1} + 3{,}560 = \frac{300^2 \times 10^{-5} \times 0.3}{5.615} \frac{1.1 \times 0.9}{0.001127 k \times 30}\left(p_2^{n+1} - 4{,}000\right). \qquad (8.1329)$$

Simplifying,

$$-p_1^{n+1}k + 2p_2^{n+1}k + 3{,}560k = 1.408(p_2^{n+1} - 4{,}000). \qquad (8.1330)$$

Rearranging,

$$R_2^{n+1} = -p_1^{n+1}k + 2p_2^{n+1}k + 3{,}560k - 1.408 p_2^{n+1} + 5{,}632. \qquad (8.1331)$$

- At Block 3 (see **Fig. 8.82.4),**

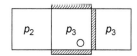

$$p_2^{n+1} - 2p_2^{n+1} + p_3^{n+1} = \frac{\Delta x^2 c_t \phi}{5.615} \frac{\mu B}{k_x \Delta t}(p_3^{n+1} - 4{,}000). \qquad (8.1332)$$

Fig. 8.82.4—Treatment of the boundary condition in Block 3.

Substituting the values,

$$p_2^{n+1} - 3{,}560 = \frac{300^2 \times 10^{-5} \times 0.3}{5.615} \frac{1.1 \times 0.9}{0.001127 k \times 30}\left(p_3^{n+1} - 4{,}000\right), \qquad (8.1333)$$

and simplifying further,

$$p_2^{n+1}k - 3{,}560k = 1.408(p_3^{n+1} - 4{,}000), \qquad (8.1334)$$

results in the following expression:

$$R_3^{n+1} = p_2^{n+1}k - 3{,}560k - 619.52. \qquad (8.1335)$$

From the preceding formulation, we can solve the system of nonlinear equations using the GNR protocol for a 3×3 system. Therefore, one needs to solve for p_1, p_2, and k using the following residual equations:

$$R_1^{n+1} = -p_1^{n+1}k + p_2^{n+1}k - 1.408 p_1^{n+1} - 3{,}152.4, \qquad (8.1336)$$

$$R_2^{n+1} = -p_1^{n+1}k + 2p_2^{n+1}k + 3{,}560k - 1.408 p_2^{n+1} + 5{,}632, \qquad (8.1337)$$

$$R_3^{n+1} = -p_2^{n+1}k - 3{,}560k - 619.52. \qquad (8.1338)$$

1. Jacobian matrix of the previously described system can be written as

$$
\begin{bmatrix}
\dfrac{\partial R_1}{\partial p_1} & \dfrac{\partial R_1}{\partial p_2} & \dfrac{\partial R_1}{\partial k} \\[2ex]
\dfrac{\partial R_2}{\partial p_1} & \dfrac{\partial R_2}{\partial p_2} & \dfrac{\partial R_2}{\partial k} \\[2ex]
0 & \dfrac{\partial R_3}{\partial p_2} & \dfrac{\partial R_3}{\partial k}
\end{bmatrix}^{(k)}_{n+1}
\begin{bmatrix}
\Delta p_1 \\[1ex]
\Delta p_2 \\[1ex]
\Delta p_3
\end{bmatrix}^{(k+1)}_{n+1}
= -
\begin{bmatrix}
R_1 \\[1ex]
R_2 \\[1ex]
R_3
\end{bmatrix}^{(k)}_{n+1}
. \quad\dots\dots\dots\dots\dots\dots\dots\dots\dots \quad (8.1339)
$$

2. The main diagonal entries are

$$
\left(\frac{\partial R_1}{\partial p_1}\right)^{n+1} = -k - 1.408, \quad \dots\dots\dots\dots\dots\dots\dots\dots\dots\dots\dots\dots\dots\dots \quad (8.1340)
$$

$$
\left(\frac{\partial R_2}{\partial p_2}\right)^{n+1} = 2k - 1.408, \quad \dots\dots\dots\dots\dots\dots\dots\dots\dots\dots\dots\dots\dots\dots\dots \quad (8.1341)
$$

and

$$
\left(\frac{\partial R_3}{\partial k}\right)^{n+1} = p_2 - 3,560. \quad \dots\dots\dots\dots\dots\dots\dots\dots\dots\dots\dots\dots\dots\dots \quad (8.1342)
$$

3. It will be necessary to first solve for p_1, p_2, and k by implementing the following calculation steps:

   1) Once a Jacobian matrix is created, calculate the entries of the matrix and the residual vector at the initial guess.
   2) Solve for $\Delta p_1^{(k+1)}$, $\Delta p_2^{(k+1)}$, and $\Delta k^{(k+1)}$.
   3) Update unknown variables as $p_1^{n+1\,(k+1)} = p_1^{n+1\,(k)} + \Delta p_1^{n+1\,(k+1)}$, $p_2^{n+1\,(k+1)} = p_2^{n+1\,(k)} + \Delta p_2^{n+1\,(k+1)}$, and $k^{n+1\,(k+1)} = k^{n+1\,(k)} + \Delta k^{n+1\,(k+1)}$.

   Repeat Steps 1, 2, and 3 until convergence is achieved (i.e., when all improvements of $\Delta p_1^{(k+1)}$, $\Delta p_2^{(k+1)}$, and $\Delta k^{(k+1)}$ become negligibly small). Once solutions for p_1, p_2, and k are generated, then the skin factor can be calculated:

$$
s = 7.15 \times 10^{-4}\, kp_1 - 7.522. \quad \dots\dots\dots\dots\dots\dots\dots\dots\dots\dots\dots\dots\dots \quad (8.1343)
$$

Problem 8.4.93

Consider the 2D reservoir system shown in **Fig. 8.83.1.** The reservoir is homogeneous and isotropic in terms of all the reservoir characteristics. A producing well is completed at the location shown in the figure, and the sandface pressure at the well is maintained at 1,000 psia. Note that all external boundaries of the system are represented by a constant pressure of 2,000 psia.

1. Using the data provided in **Table 8.85,** calculate the pressure at each meshpoint using the Gauss-Seidel method (suggested $|\varepsilon_p| = 0.1$ psia and initial guess 2,000 psia).
2. Calculate the flow rate for the well.
3. Does your solution conserve mass?

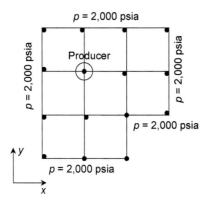

Fig. 8.83.1—Reservoir system of Problem 8.4.93.

Table 8.85—Reservoir properties for Problem 8.4.93.	
μ	= 1 cp
B	= 1 res bbl/STB
$k_x = k_y$	= 60 md
$\Delta x = \Delta y$	= 200 ft
h	= 200 ft
r_w	= 0.25 ft
ϕ	= 0.3, fraction
c	= 0 psi^{-1}
p_i	= 2,000 psia
G	= 2,000 ft (uniform)
ρ	= 48 lbm/ft
s	= 0

Solution to Problem 8.4.93

1. Using the existing symmetry, the system can be reduced to two unknown block pressures, $p_1 = p_2$, as explained in the schematics in **Fig. 8.83.2.**

    Writing the finite-difference approximation for an incompressible fluid flow in a homogenous and isotropic system, we have

   - Node 1:

   $$-4p_1 + p_2 + 6,000 = 0, \quad \dots\dots\dots\dots\dots (8.1344)$$

   $$-4p_1 + p_2 = -6,000. \quad \dots\dots\dots\dots\dots (8.1345)$$

   - Node 2:

   $$2p_1 - 4p_2 + \frac{q\mu B}{kh} + 4,000 = 0. \quad \dots\dots\dots\dots (8.1346)$$

Substituting q using the Peaceman's wellbore equation:

$$2p_1 - 4p_2 - \frac{2\pi kh\mu B}{\left[\ln\left(r_e/r_w\right) - s\right]\mu Bkh}\left(p_2 - p_{sf}\right) + 4,000 = 0, \quad \dots\dots\dots\dots\dots\dots\dots (8.1347)$$

$$2p_1 - 4p_2 - \frac{2\pi}{\left[\ln\left(r_e/r_w\right) - s\right]}\left(p_2 - p_{sf}\right) + 4,000, \quad \dots\dots\dots\dots\dots\dots\dots\dots (8.1348)$$

$$2p_1 - 4p_2 - \frac{2\pi}{\left[\ln\left(r_e/r_w\right) - s\right]}p_2 = -\frac{2\pi}{\left[\ln\left(r_e/r_w\right) - s\right]}p_{sf} - 4,000, \quad \dots\dots\dots\dots\dots (8.1349)$$

where $r_e = 0.2 \times \Delta x = 0.2 \times 800 \text{ ft} = 160 \text{ ft}$,

$$2p_1 - 4,9724p_2 = -4,972.4083. \quad \dots\dots\dots\dots\dots\dots\dots\dots\dots\dots\dots\dots\dots (8.1350)$$

The system of equations to be solved is given by

$$-4p_1 + p_2 = -6,000, \quad \dots\dots\dots\dots\dots\dots\dots\dots\dots\dots\dots\dots\dots\dots (8.1351)$$

$$2p_1 - 4,9724p_2 = -4,972.4083. \quad \dots\dots\dots\dots\dots\dots\dots\dots\dots\dots\dots\dots\dots (8.1352)$$

Eqs. 8.1351 and 8.1352 can be simplified in the following ways:

$$p_1 - 0.25p_2 = 1,500, \quad \dots\dots\dots\dots\dots\dots (8.1353)$$

$$-0.40222p_1 + p_2 = 1,000. \quad \dots\dots\dots\dots\dots (8.1354)$$

Using the Gauss-Seidel method,

$$p_1^{(k+1)} = 1,500 + 0.25 \times p_2^{(k)}, \quad \dots\dots\dots\dots\dots (8.1355)$$

$$p_2^{(k+1)} = 1,000 + 0.4022 \times p_1^{(k+1)}. \quad \dots\dots\dots\dots (8.1356)$$

Iterations can be found in **Table 8.86.** The final pressures are $p_1 = 1,945.64$ psia and $p_2 = 1,782.54$ psia.

2. The flow rate is given by

$$q - \frac{2\pi \bar{k}h}{\left[\ln\left(r_e/r_w\right) + s\right]\mu B}\left(p_2 - p_{sf}\right) = -\frac{2\pi 0.060 \times 1.127 \times 200}{\left[\ln\left(160/0.25\right) + 0\right] \times 1 \times 1}\left(1,782.54 - 1,000\right) \quad \dots\dots\dots\dots (8.1357)$$

$$= 1,0291.066 \text{ STB/D.}$$

3. To verify the material balance, we must calculate the flow entering the reservoir. There is a total of six nodes with 2,000 psia connecting to the reservoir with p_1, and two 2,000-psi nodes connecting to the reservoir with p_2. Therefore, the total mass flow into the reservoir can be calculated as

Fig. 8.83.2—Reduced system in Problem 8.4.93.

k	p_1 (psia)	p_2 (psia)	|Maximum Error|
0	2,000	2,000	
1	2,000	1,804.56	195.44
2	1,951.14	1,784.74	48.86
3	1,946.18	1,782.76	4.96
4	1,945.69	1,782.56	0.49
5	1,945.64	1,782.54	0.02

Table 8.86—Iterations using the Gauss-Seidel method in Solution to Problem 8.4.93.

$$q_{\text{in}} = \frac{kA}{\mu B \Delta x} \Big[6 \times \big(2,000 - p_1 \big) + 2 \big(2,000 - p_2 \big) \Big]. \quad \dots\dots\dots\dots\dots\dots\dots\dots\dots\dots\dots\dots\dots\dots\dots (8.1358)$$

Substituting the data,

$$q_{\text{in}} - \frac{0.06 \times 1.127 \times 200 \times 200}{1 \times 1 \times 200} \Big[6 \times (2,000 - 1,945.64) + 2(2,000 - 1,782.54) \Big] \quad \dots\dots\dots\dots\dots (8.1359)$$
$$= 1,0292.84 \text{ STB/D}.$$

We then have

$$\left| \frac{q_{\text{in}}}{q_{\text{well}}} \right| = \left| \frac{10,292.84}{10,291.066} \right| = 1.00017, \quad \dots\dots\dots\dots\dots\dots\dots\dots\dots\dots\dots\dots\dots\dots\dots\dots\dots (8.1360)$$

which indicates a good MBC.

Problem 8.4.94

Consider the homogenous and slightly compressible system (1D and single phase) shown in **Fig. 8.84.** The well located in Block 3 produces at a flow rate of 400 STB/D. At the end of 30 days, flowing-sandface pressure in the production well is measured at 800 psia. The entire reservoir has sealed boundaries. A pressure survey in Block 1 records 2,500 psia at 30 days (in the observation well). From seismic studies, it is suggested that the average formation thickness is 50 ft. However, the reservoir and production engineers dispute this value of thickness. From that pressure and the production data provided in **Table 8.87,** can you assign a representative thickness to this system? Write the equations and structure for the coefficient matrix to solve the system using the GNR protocol (use a convergence criterion of 1 for all unknowns).

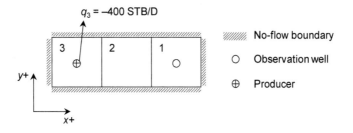

Fig. 8.84—Reservoir system of Problem 8.4.94.

B = 1 res bbl/STB
k_x = 50 md
$\Delta x = \Delta y$ = 200 ft
r_w = 0.25 ft
ϕ = 0.2
c = 5×10^{-5} psi^{-1}
p_i = 3,000 psia (at t = 0 days)
μ = 1.2 cp

Table 8.87—Reservoir properties for Problem 8.4.94.

Solution to Problem 8.4.94

The problem can be solved using 1D single-phase slightly compressible flow equation with no depth gradient:

$$\frac{\partial}{\partial x} \left(\frac{A_x k_x}{\mu B} \frac{\partial p}{\partial x} \right) \Delta x + q = \frac{V_b \phi c}{5.615} \frac{\partial p}{\partial t}. \quad \dots\dots\dots\dots\dots\dots\dots\dots\dots\dots\dots\dots\dots\dots\dots\dots (8.1361)$$

A homogenous system,

$$\frac{A_x k_x \Delta x}{\mu B} \frac{\partial^2 p}{\partial x^2} + q = \frac{V_b \phi c}{5.615} \frac{\partial p}{\partial t}. \quad \dots\dots\dots\dots\dots\dots\dots\dots\dots\dots\dots\dots\dots\dots\dots\dots\dots\dots (8.1362)$$

$$\frac{\partial^2 p}{\partial x^2} + \frac{q\mu B}{V_b k} = \frac{\mu B c \phi}{5.615k}\frac{\partial p}{\partial t}. \quad\dots\dots\dots\dots\dots\dots\dots\dots\dots\dots\dots\dots (8.1363)$$

Using the finite-difference approximation,

$$p_{i+1}^{n+1} - 2p_i^{n+1} + p_{i-1}^{n+1} + \frac{q\mu B}{hk} = \frac{\mu\phi c B \Delta x^2}{5.615k\Delta t}(p_i^{n+1} - p_i^n), \quad\dots\dots\dots\dots\dots (8.1364)$$

$$p_{i+1}^{n+1} - \left(\frac{\mu\phi c B \Delta x^2}{5.615k\Delta t} + 2\right)p_i^{n+1} + p_{i-1}^{n+1} + \frac{q\mu B}{hk} = -\left(\frac{\mu\phi c B \Delta x^2}{5.615k\Delta t}\right)p_i^n. \quad\dots\dots (8.1365)$$

Calculating the coefficients,

$$\left[\frac{\mu\phi c B \Delta x^2}{5.615k\Delta t}\right] = \frac{1.2\times0.2\times5\times10^{-5}\times1\times200^2}{5.615\times1.127\times10^{-3}\times50\times30} = 0.051, \quad\dots\dots\dots\dots\dots (8.1366)$$

$$\left[\frac{\mu\phi c B \Delta x^2}{5.615k\Delta t} + 2\right] = 0.051 + 2 = 2.051. \quad\dots\dots\dots\dots\dots\dots\dots\dots\dots (8.1367)$$

Writing the equations for each block and well:

- At Block 2:

$$2500 - 2.051p_2 + p_3 = -0.051\times(3,000), \quad\dots\dots\dots\dots\dots\dots\dots\dots\dots (8.1368)$$

$$-2.051p_2 + p_3 = -2,651.70. \quad\dots\dots\dots\dots\dots\dots\dots\dots\dots\dots\dots (8.1369)$$

- At Block 3:

$$-1.051p_3 + p_2 - \frac{400\times1.2\times1}{50\times1.127\times10^{-3}h} = -151.704, \quad\dots\dots\dots\dots\dots (8.1370)$$

$$-1.051p_3 + p_2 - \frac{8518.19}{h} = -151.704. \quad\dots\dots\dots\dots\dots\dots\dots\dots (8.1371)$$

For the well located in Block 3:

$$q = \frac{-2\pi\overline{k}h}{\mu B\left[\ln\left(\frac{r_e}{r_w}\right) + s\right]}\left(p_3 - p_{sf}\right), \quad\dots\dots\dots\dots\dots\dots\dots\dots (8.1372)$$

$$400 = -\frac{2\pi\times50\times1.127\times10^{-3}\times h}{1.2\times1\left(\ln\frac{0.2\times200}{0.25} + 0\right)}\left(p_3 - 800\right), \quad\dots\dots\dots\dots\dots (8.1373)$$

$$0.0581\times h\left(p_3 - 800\right) - 400 = 0. \quad\dots\dots\dots\dots\dots\dots\dots\dots\dots (8.1374)$$

To solve the system of equations using GNR protocol, the residual equations are calculated as

$$R_1 = 2,651.704 - 2.051p_2 + p_3, \quad\dots\dots\dots\dots\dots\dots\dots\dots\dots\dots (8.1375)$$

$$R_2 = 1.051hp_3 + hp_2 - 8,518.19 + 151.704h = 0, \quad\dots\dots\dots\dots\dots\dots (8.1376)$$

$$R_3 = 0.058hp_3 - 46.508h - 400 = 0. \quad\dots\dots\dots\dots\dots\dots\dots\dots\dots (8.1377)$$

The Jacobian entries are also calculated as

$$\frac{\partial R_1}{\partial p_2} = -2.051, \quad\dots\dots\dots\dots\dots\dots\dots\dots\dots\dots\dots\dots\dots\dots (8.1378)$$

$$\frac{\partial R_1}{\partial p_3} = 1, \quad\dots\dots\dots\dots\dots\dots\dots\dots\dots\dots\dots\dots\dots\dots\dots\dots (8.1379)$$

$$\frac{\partial R_1}{\partial h} = 0, \quad\dots\dots\dots\dots\dots\dots\dots\dots\dots\dots\dots\dots\dots\dots\dots\dots (8.1380)$$

$$\frac{\partial R_2}{\partial p_2} = h, \dotfill (8.1381)$$

$$\frac{\partial R_2}{\partial p_3} = -1.051h, \dotfill (8.1382)$$

$$\frac{\partial R_2}{\partial h} = p_2 - 1.051p_3 + 151.704, \dotfill (8.1383)$$

$$\frac{\partial R_3}{\partial p_2} = 0, \dotfill (8.1384)$$

$$\frac{\partial R_3}{\partial p_3} = 0.0581h, \dotfill (8.1385)$$

$$\frac{\partial R_3}{\partial h} = 0.0581p_3 - 46.5084. \dotfill (8.1386)$$

In the matrix form, the GNR equations can be written as

$$\begin{bmatrix} -2.051 & 1 & 0 \\ h & -1.021h & p_2 - 1.051p_3 + 151.704 \\ 0 & 0.0581h & 0.05815p_3 - 46.51 \end{bmatrix}^{(k)} \begin{bmatrix} \Delta p_2 \\ \Delta p_3 \\ \Delta h \end{bmatrix}^{(k+1)} = - \begin{bmatrix} -R_1 \\ -R_2 \\ -R_3 \end{bmatrix}^{(k)}. \dotfill (8.1387)$$

- First iteration ($k = 0$): Assuming 2,500 psi as initial guess for pressure and 50 ft for thickness, we will have

$$\begin{bmatrix} -2.050568 & 1 & 0 \\ 50 & -52.5284 & 25.284 \\ 0 & 2.90678 & 98.83035 \end{bmatrix} \begin{bmatrix} \Delta p_2 \\ \Delta p_3 \\ \Delta h \end{bmatrix} = \begin{bmatrix} -25.284 \\ 7,253.99 \\ -4,541.5175 \end{bmatrix}, \dotfill (8.1388)$$

$\Delta p_2 = -119.331$ psi, $p_2 = 2,380.669$ psi, $\Delta p_3 = -269.98$ psi, $p_3 = 2,230.02$ psi, and $\Delta h = -38.012$ ft, $h = 11.988$ ft.

- Second iteration ($k = 1$):

$$\begin{bmatrix} -2.050568 & 1 & 0 \\ 11.988 & -12.5942 & 189.5853 \\ 0 & 0.696928 & 83.13493 \end{bmatrix} \begin{bmatrix} \Delta p_2 \\ \Delta p_3 \\ \Delta h \end{bmatrix} = \begin{bmatrix} -0.00033001 \\ 6245.4408 \\ -596.621513 \end{bmatrix}, \dotfill (8.1389)$$

$\Delta p_2 = -444.89$ psi, $p_2 = 1,935.779$ psi, $\Delta p_3 = -912.28$ psi, $p_3 = 1,317.74$ psi, and $\Delta h = 0.47$ ft, $h = 12.458$ ft.

- Third iteration ($k = 2$):

$$\begin{bmatrix} -2.050568 & 1 & 0 \\ 12.458 & -13.088 & 703.1075 \\ 0 & 0.724252 & 30.09907 \end{bmatrix} \begin{bmatrix} \Delta p_2 \\ \Delta p_3 \\ \Delta h \end{bmatrix} = \begin{bmatrix} -0.0024725 \\ -241.12353 \\ 25.025739 \end{bmatrix}, \dotfill (8.1390)$$

$\Delta p_2 = 16.825$ psi, $p_2 = 1,952.604$ psi, $\Delta p_3 = 34.504$ psi, $p_3 = 1,352.244$ psi, and $\Delta h = 0.00121$ ft, $h = 12.459$ ft.

- Fourth iteration ($k = 3$):

$$\begin{bmatrix} -2.050568 & 1 & 0 \\ 12.459 & -13.088 & 683.6827 \\ 0 & 0.72431 & 32.10504 \end{bmatrix} \begin{bmatrix} \Delta p_2 \\ \Delta p_3 \\ \Delta h \end{bmatrix} = \begin{bmatrix} -0.00172093 \\ 0.1875542 \\ 0.0033166 \end{bmatrix}, \dotfill (8.1391)$$

$\Delta p_2 = -0.0014$ psi, $p_2 = 1,952.603$ psi, $\Delta p_3 = -0.0005$ psi, $p_3 = 1,352.244$ psi, and $\Delta h = 0.0002$ ft, $h = 12.459$ ft.

The final pressure is 1,952.6 psia for Block 2 and 1,352.2 psia for Block 3. The thickness for the reservoir is 12.5 ft, which is less than the data reported by the petrophysics department.

Here, we implement an MBC to validate the solution. Total fluid expansion from Blocks 2 and 3:

$$\text{fluid expansion} = \left(V_{p2} + V_{p3}\right) \times c_t \times \left(\Delta p_2 + \Delta p_3\right)$$

$$= 0.2 \times \left[\left(200 \times 200 \times 12.46\right) + \left(200 \times 200 \times 12.46\right)\right] \times \left(5 \times 10^{-6}\right) \quad\dots\dots\dots\dots\dots\dots\dots (8.1392)$$

$$\times \left[\left(3,000 - 1,952.6\right) + \left(3,000 - 1,352.2\right)\right] = 2,392.29 \text{ STB}.$$

Mass flow from the constant pressure block (Block 1) to Block 2 is

$$q_{1-2} = \frac{kh}{\mu B}\left(p_1 - p_2\right) = \frac{(50)(1.127 \times 10^{-3})(12.46)}{(1.2)(1)}(2,500 - 1,952.6) = 9,607.99 \text{ STB}. \quad\dots\dots\dots\dots\dots\dots (8.1393)$$

Thus,

$$\text{MBC} = \left|\frac{400 \times 30}{q_{1-2} + \text{fluid expansion}}\right| = \left|\frac{12,000}{9,607.99 + 2,392.29}\right| = 1,000, \quad\dots\dots\dots\dots\dots\dots\dots\dots (8.1394)$$

which gives a perfect MBC.

Nomenclature

a	=	a constant
$a_{i,j}$	=	element (i,j) of matrix $[A]$
A	=	cross-sectional area, L^2, ft^2
A_{avg}	=	arithmetic averaging
$A_{i,j,k}$	=	transmissibility coefficient using SIP notation between block (i, j, k) and $(i, j, k–1)$, L^{4}t/m, STB/D–psia for liquid, scf/D–psia for gas
A_x	=	cross-sectional area along x-direction, L^2, ft^2
A_y	=	cross-sectional area along y-direction, L^2, ft^2
A_z	=	cross-sectional area along z-direction, L^2, ft^2
$[A]$	=	coefficient matrix of a matrix equation
b	=	a constant
B	=	FVF, L^3/L^3, res bbl/STB for liquid and res bbl/scf for gas
B_g	=	gas FVF, L^3/L^3, res bbl/scf
$B_{i,j,k}$	=	transmissibility coefficient using SIP notation between block (i, j, k) and $(i, j, k+1)$, L^{4}t/m, STB/D–psia for liquid, scf/D–psia for gas
B_o	=	oil FVF, L^3/L^3, res bbl/STB
B_w	=	water FVF, L^3/L^3, res bbl/STB
c	=	a constant,
c_o	=	oil compressibility, Lt^2/m, psi^{-1}
c_t	=	total compressibility, Lt^2/m, psi^{-1}
c_ϕ	=	porosity compressibility, Lt^2/m, psi^{-1}
C	=	transmissibility coefficient using SIP notation of central block (i, j, k), L^{4}t/m, STB/D–psia for liquid, scf/D–psia for gas
d	=	an intermediate variable of Thomas' algorithm
E	=	transmissibility coefficient using SIP notation between block (i, j, k) and $(i+1, j, k)$, L^{4}t/m, STB/D–psia for liquid, scf/D–psia for gas
f	=	function
f_x	=	first-order derivative of function $f(x, y)$ in terms of x

f_y	=	first-order derivative of function $f(x, y)$ in terms of y
g	=	acceleration of gravity, L/t^2, ft/s^2, also an intermediate variable of Thomas' algorithm
g_c	=	unit conversion factor in Newton's Law
g_x	=	first-order derivative of function $g(x, y)$ in terms of x
g_y	=	first-order derivative of function $g(x, y)$ in terms of y
G	=	elevation with respect to absolute datum positive downward, L, ft
G_{avg}	=	geometric averaging
G_i	=	original gas in place, L^3, scf
G_p	=	cumulative gas production, L^3, scf
h	=	thickness, L, ft
H_{avg}	=	harmonic averaging
i	=	dummy index, row i, Gridblock i
j	=	dummy index, row j, Gridblock j
J	=	Jacobian matrix of GNR protocol
k	=	permeability, L^2, darcies
$\bar{k}$	=	the geometric mean of the directional permeabilities in the plane that the well is orthogonal, L^2, darcies
k_x	=	permeability along x-direction, L^2, darcies or perms (1 darcy = 1.127 perms)
k_y	=	permeability along y-direction, L^2, darcies or perms (1 darcy = 1.127 perms)
k_z	=	permeability along z-direction, L^2, darcies or perms (1 darcy = 1.127 perms)
$l_{i,j}$	=	lower triangular element (i,j) of matrix $[L]$
L	=	the length of the wellbore in the wellblock, L, ft
$[L]$	=	lower triangular matrix
m	=	mass per unit volume of porous media, m/L^3, lbm/ft^3

n	=	number of functions or equations	
n_k	=	number of integer that satisfies the iteration criteria of ADIP	
N	=	transmissibility coefficient using SIP notation between block (i, j, k) and $(i, j+1, k)$, $L^4 t/m$, STB/D–psia for liquid, scf/D–psia for gas	
N_w	=	number of existing wells in a reservoir system	
N_x	=	number of gridblocks along x-direction	
N_y	=	number of gridblocks along y-direction	
N_z	=	number of gridblocks along z-direction	
p	=	pressure, m/Lt^2, psia	
p'	=	pressure of an image block, m/Lt^2, psia	
p_a	=	aquifer pressure, m/Lt^2, psia	
$p_{initial}$	=	initial pressure, m/Lt^2, psia	
p_{sc}	=	standard condition pressure, m/Lt^2, psia	
p_{sf}	=	sandface pressure, m/Lt^2, psia	
$p_{sf_{spec}}$	=	specified sandface pressure, m/Lt^2, psia	
q	=	production or injection flow rate, L^3/t, STB/D	
$q_{boundary}$	=	flow rate across the boundary, L^3/t, STB/D	
q_c	=	flow rate from the corner node, L^3/t, STB/D	
q_f	=	flow rate from the surface node, L^3/t, STB/D	
q_g	=	gas production or flow rate, L^3/t, scf/D	
q_{gm}	=	gas production or flow rate of Well m, L^3/t, scf/D	
q_{in}	=	flow rate entering the system, L^3/t, STB/D for liquid and scf/D for gas	
$q_{injection}$	=	injection rate, L^3/t, STB/D for liquid and scf/D for gas	
$q_{leakage}$	=	flow leaking from the system, L^3/t, STB/D for liquid and scf/D for gas	
q_{left}	=	flow rate across the left boundary, L^3/t, STB/D for liquid and scf/D for gas	
q_m	=	flow rate of Well m in the MBC, scf/D for gas and STB/D for liquid	
$q_{production}$	=	production rate, STB/D for liquid and scf/D for gas	
q_{out}	=	flow rate existing the system, L^3/t, STB/D for liquid and scf/D for gas	
q_{right}	=	flow rate across the right boundary, L^3/t, STB/D for liquid and scf/D for gas	
q_s	=	flow rate from the side node, L^3/t, STB/D	
q_{south}	=	flow rate across the southern boundary, L^3/t, STB/D for liquid and scf/D for gas	
q_{spec}	=	specified production or flow rate, L^3/t, STB/D for liquid and scf/D for gas	
q_{total}	=	total flow rate, L^3/t, STB/D for liquid and scf/D for gas	
$q_{unknown}$	=	flow rate across the unknown boundary, L^3/t, STB/D	

q_{well}	=	well production rate, L^3/t, STB/D for liquid and scf/D for gas
q_{west}	=	flow rate across the western boundary, L^3/t, STB/D for liquid and scf/D for gas
q_x	=	flow rate across the x–direction, L^3/t, STB/D
q_y	=	flow rate across the y–direction, L^3/t, STB/D
q_z	=	flow rate across the z–direction, L^3/t, STB/D
Q	=	right-hand-side coefficient of governing flow equation using SIP notation for block (i, j, k) L^3/t, STB/D for liquid, scf/D for gas
Q_{inj}	=	total fluid injected, L^3, STB
Q_t	=	total fluid produced, L^3, STB
r_e	=	Peaceman's equivalent wellblock radius, L, ft
r_w	=	wellblock radius, L, ft
R	=	universal gas constant, °R
$R_{i,j,k}$	=	residual of block i,j,k, L^3/t , STB/D for liquid, scf/D for gas
s	=	skin factor, dimensionless
S	=	transmissibility coefficient using SIP notation between block (i, j, k) and $(i, j–1, k)$, $L^4 t/m$, STB/D–psia for liquid, scf/D–psia for gas
t	=	time, t, days
T_{sc}	=	temperature at standard conditions, T, °R
T	=	transmissibility coefficient between two adjacent blocks $L^4 t/m$, STB/D–psia for liquid, scf/D–psia for gas
T'	=	gravitational transmissibility factor between adjacent block i and j, $L^4 t/m$, STB/D–psia for liquid, scf/D–psia for gas
T_c	=	critical temperature, T, °R
T_{ix}	=	transmissibility coefficient of block i along x-direction, $L^4 t/m$, STB/D–psia for liquid, scf/D–psia for gas
T_{sc}	=	standard condition temperature, T, °R
$u_{i,j}$	=	upper triangular element (i,j) of matrix $[U]$
U	=	upper triangular matrix
V_b	=	bulk volume, L^3, ft^3
V_p	=	pore volume, L^3, ft^3
w	=	width, L, ft.
W	=	transmissibility coefficient using SIP notation between block (i, j, k) and $(i–1, j, k)$, $L^4 t/m$, STB/D–psia for liquid, scf/D–psia for gas
x	=	distance in x-direction in the Cartesian coordinate system, L, ft
y	=	distance in y-direction in the Cartesian coordinate system, also an unknown variable in system of equations
z	=	distance in z-direction in the Cartesian coordinate system
z_g	=	gas-compressibility factor, dimensionless

z_{gi}	=	initial gas-compressibility factor, dimensionless
α	=	iteration parameter of SIP method
α_c	=	unit conversion factor between bbl and ft^3 (5.615 ft^3/bbl)
Δ	=	change in a variable or property
Γ	=	accumulation coefficient, L^3/t, ft^3/D
γ	=	pressure gradient, m/L^2 t^2, psi/ft
ΔV	=	change of volume, L^3, ft^3
ΔV_{total}	=	total change of volume, L^3, ft^3
Δx	=	control volume dimension along the x-direction, L, ft
Δy	=	control volume dimension along the y-direction, L, ft
Δz	=	control volume dimension along the z-direction, L, ft
ε	=	tolerance
ε_p	=	pressure tolerance applied to calculate the numerical derivatives
η	=	root of an arbitrary equation
λ_x	=	iteration variable used in ADIP algorithm along x-direction
λ_y	=	iteration variable used in ADIP algorithm along y-direction
μ	=	viscosity, cp
μ_g	=	gas viscosity, cp
μ_o	=	oil viscosity, cp
μ_w	=	water viscosity, cp
ξ	=	root of an arbitrary equation
ρ	=	density, m/L^3, lbm/ft^3
ρ_o	=	oil density, m/L^3, lbm/ft^3
ρ_S	=	spectral radius
ϕ	=	porosity, fraction
ψ	=	defined function
Ω	=	well productivity/injectivity index, L^{4}t/m, STB/D–psia for liquid, scf/D–psia for gas
ω	=	iteration parameter of ADIP and over-relaxation algorithms
ω_{max}	=	maximum iteration parameter of ADIP iteration
ω_{min}	=	minimum iteration parameter of ADIP iteration
ω_{opt}	=	optimum over-relaxation parameter

Superscripts

m	=	previous iteration level of outer iteration loop
$m+1$	=	current iteration level of outer iteration loop
n	=	previous time level
$n+1$	=	current time level
k	=	previous iteration level
$k+1$	=	current iteration level

Subscripts

aquifer	=	property related to aquifer
c	=	component
f	=	property related to fluid
g	=	related to gas phase
i,j,k	=	related to block i,j,k in a 3D representation
i,j	=	related to block i,j in a 2D representation
i	=	related to block i in a 2D representation
initial	=	related to initial condition
injector	=	property or variable related to injection wells
j	=	block address along the y-direction
j,k	=	related to block j,k in a 2D representation
k	=	block address along the z-direction
l	=	counter for time levels (Eq. 8.29)
n	=	equation number of a gridblock
m	=	counter for wells (Eq. 8.29)
max	=	maximum value
o	=	related to oil phase
producer	=	property or variable related to production well
s	=	spatial directions of the Cartesian coordinate system
sc	=	standard conditions
sf	=	sandface
spec	=	specified
t	=	implies total of a quantity
w	=	related to water phase
well	=	related to well
x	=	related to x-direction
y	=	related to y-direction
z	=	related to z-direction

References

Barrett, R., Berry, M., Chan, T. F. et al. 1994. *Templates for the Solution of Linear Systems: Building Blocks for Iterative Methods*, 1st ed., Philadelphia, Pennsylvania: Society for Industrial and Applied Mathematics. 1994. http://dx.doi.org/10.2118/9781611971538.

Darcy, H. 1856. *Les Fontaines Publiques de la Ville de Dijon*, first edition. Paris: Victor Dalmont, ed; Libraire des Corps Impériaux des Ponts et Chaussées et des Mines.

Ertekin, T., Abou-Kassem, J. H., and King, G. R. 2001. *Basic Applied Reservoir Simulation*. Richardson, Texas, US: Society of Petroleum Engineers.

Peaceman, D.W. 1983. Interpretation of Well-Block Pressures in Numerical Reservoir Simulation with Nonsquare Grid Blocks and Anisotropic Permeability. *SPE J* **23**(3): 531–543. SPE-10528-PA. https://dx.doi.org/10.2118/10528-PA.

Chapter 9

Numerical Solution of Multiphase-Flow Equations in Porous Media

In this chapter, we focus on the formulation of the multiphase-flow equations and the solution techniques widely implemented to solve the systems of partial-differential equations (PDEs). So far in this book, we have worked on single-phase flow formalisms and their respective solutions. In this chapter, we start with the generalized form of the multiphase-flow equation and then proceed with a review of the solution protocols applicable to systems of the PDEs.

9.1 Multiphase-Flow Equations

In multiphase-flow formulations, Darcy's law for multiphase flow becomes the constituent expression substituted into the continuity equation through the velocity term. Darcy's law for multiphase flow is written in the following form:

$$u_{ls} = -\frac{A_s k_s k_{rl}}{\mu_l B_l}\frac{\partial \Phi}{\partial s}. \quad \dots\dots\dots\dots\dots\dots\dots\dots\dots\dots\dots\dots\dots\dots\dots\dots\dots\dots \quad (9.1)$$

In Eq. 9.1, $l = o,w,g$, and $s = x,y,z$. Black-oil formulation (Eqs. 9.2 through 9.4) is broadly employed to describe a three-phase-flow system in porous media, which allows transfer of gas from gaseous phase to oleic and aqueous phases and vice versa, assuming no compositional changes are taking place. In other words, during these interphase mass transfers, the composition of gas that goes into the solution and comes out of the solution remains the same.

The oil equation is as follows:

$$
\frac{\partial}{\partial x}\left(\frac{A_x k_x k_{ro}}{\mu_o B_o}\frac{\partial p_o}{\partial x}\right)\Delta x + \frac{\partial}{\partial y}\left(\frac{A_y k_y k_{ro}}{\mu_o B_o}\frac{\partial p_o}{\partial y}\right)\Delta y + \frac{\partial}{\partial z}\left(\frac{A_z k_z k_{ro}}{\mu_o B_o}\frac{\partial p_o}{\partial z}\right)\Delta z -
$$

$$
\begin{bmatrix}
\dfrac{\partial}{\partial x}\left(\dfrac{1}{144}\dfrac{g}{g_c}\rho_o\dfrac{A_x k_x k_{ro}}{\mu_o B_o}\dfrac{\partial G}{\partial x}\right)\Delta x \\[2ex]
+\dfrac{\partial}{\partial y}\left(\dfrac{1}{144}\dfrac{g}{g_c}\rho_o\dfrac{A_y k_y k_{ro}}{\mu_o B_o}\dfrac{\partial G}{\partial y}\right)\Delta y \\[2ex]
+\dfrac{\partial}{\partial z}\left(\dfrac{1}{144}\dfrac{g}{g_c}\rho_o\dfrac{A_z k_z k_{ro}}{\mu_o B_o}\dfrac{\partial G}{\partial z}\right)\Delta z
\end{bmatrix}
+ q_{o\,\text{STB/D}} = \frac{V_b}{5.615}\frac{\partial}{\partial t}\left(\frac{\phi S_o}{B_o}\right). \quad \dots\dots\dots\dots\dots\dots\dots\dots\dots \quad (9.2)
$$

The water equation is as follows:

$$
\frac{\partial}{\partial x}\left(\frac{A_x k_x k_{rw}}{\mu_w B_w}\frac{\partial p_w}{\partial x}\right)\Delta x + \frac{\partial}{\partial y}\left(\frac{A_y k_y k_{rw}}{\mu_w B_w}\frac{\partial p_w}{\partial y}\right)\Delta y + \frac{\partial}{\partial z}\left(\frac{A_z k_z k_{rw}}{\mu_w B_w}\frac{\partial p_w}{\partial z}\right)\Delta z -
$$

$$
\begin{bmatrix}
\dfrac{\partial}{\partial x}\left(\dfrac{1}{144}\dfrac{g}{g_c}\rho_w\dfrac{A_x k_x k_{rw}}{\mu_w B_w}\dfrac{\partial G}{\partial x}\right)\Delta x \\[2ex]
+\dfrac{\partial}{\partial y}\left(\dfrac{1}{144}\dfrac{g}{g_c}\rho_w\dfrac{A_y k_y k_{rw}}{\mu_w B_w}\dfrac{\partial G}{\partial y}\right)\Delta y \\[2ex]
+\dfrac{\partial}{\partial z}\left(\dfrac{1}{144}\dfrac{g}{g_c}\rho_w\dfrac{A_z k_z k_{rw}}{\mu_w B_w}\dfrac{\partial G}{\partial z}\right)\Delta z
\end{bmatrix}
+ q_{w\,\text{STB/D}} = \frac{V_b}{5.615}\frac{\partial}{\partial t}\left(\frac{\phi S_w}{B_w}\right). \quad \dots\dots\dots\dots\dots\dots\dots\dots \quad (9.3)
$$

The gas equation is as follows:

$$\frac{\partial}{\partial x}\left(\frac{A_x k_x k_{rg}}{\mu_g B_g}\frac{\partial p_g}{\partial x}+R_{so}\frac{A_x k_x k_{ro}}{\mu_o B_o}\frac{\partial p_o}{\partial x}+R_{sw}\frac{A_x k_x k_{rw}}{\mu_w B_w}\frac{\partial p_w}{\partial x}\right)\Delta x$$

$$+\frac{\partial}{\partial y}\left(\frac{A_y k_y k_{rg}}{\mu_g B_g}\frac{\partial p_g}{\partial y}+R_{so}\frac{A_y k_y k_{ro}}{\mu_o B_o}\frac{\partial p_o}{\partial y}+R_{sw}\frac{A_y k_y k_{rw}}{\mu_w B_w}\frac{\partial p_w}{\partial y}\right)\Delta y$$

$$+\frac{\partial}{\partial z}\left(\frac{A_z k_z k_{rg}}{\mu_g B_g}\frac{\partial p_g}{\partial z}+R_{so}\frac{A_z k_z k_{ro}}{\mu_o B_o}\frac{\partial p_o}{\partial z}+R_{sw}\frac{A_z k_z k_{rw}}{\mu_w B_w}\frac{\partial p_w}{\partial z}\right)\Delta z-\begin{bmatrix}\dfrac{\partial}{\partial x}\left(\dfrac{1}{144}\dfrac{g}{g_c}\rho_g\dfrac{A_x k_x k_{rg}}{\mu_g B_g}\dfrac{\partial G}{\partial x}\right)\Delta x\\[2mm]+\dfrac{\partial}{\partial y}\left(\dfrac{1}{144}\dfrac{g}{g_c}\rho_g\dfrac{A_y k_y k_{rg}}{\mu_g B_g}\dfrac{\partial G}{\partial y}\right)\Delta y\\[2mm]+\dfrac{\partial}{\partial z}\left(\dfrac{1}{144}\dfrac{g}{g_c}\rho_g\dfrac{A_z k_z k_{rg}}{\mu_g B_g}\dfrac{\partial G}{\partial z}\right)\Delta z\end{bmatrix}$$

$$-\begin{bmatrix}\dfrac{\partial}{\partial x}\left(\dfrac{1}{144}\dfrac{g}{g_c}\rho_o R_{so}\dfrac{A_x k_x k_{ro}}{\mu_o B_o}\dfrac{\partial G}{\partial x}\right)\Delta x\\[2mm]+\dfrac{\partial}{\partial y}\left(\dfrac{1}{144}\dfrac{g}{g_c}\rho_o R_{so}\dfrac{A_y k_y k_{ro}}{\mu_o B_o}\dfrac{\partial G}{\partial y}\right)\Delta y\\[2mm]+\dfrac{\partial}{\partial z}\left(\dfrac{1}{144}\dfrac{g}{g_c}\rho_o R_{so}\dfrac{A_z k_z k_{ro}}{\mu_o B_o}\dfrac{\partial G}{\partial z}\right)\Delta z\end{bmatrix}-\begin{bmatrix}\dfrac{\partial}{\partial x}\left(\dfrac{1}{144}\dfrac{g}{g_c}\rho_w R_{sw}\dfrac{A_x k_x k_{rw}}{\mu_w B_w}\dfrac{\partial G}{\partial x}\right)\Delta x\\[2mm]+\dfrac{\partial}{\partial y}\left(\dfrac{1}{144}\dfrac{g}{g_c}\rho_w R_{sw}\dfrac{A_y k_y k_{rw}}{\mu_w B_w}\dfrac{\partial G}{\partial y}\right)\Delta y\\[2mm]+\dfrac{\partial}{\partial z}\left(\dfrac{1}{144}\dfrac{g}{g_c}\rho_w R_{sw}\dfrac{A_z k_z k_{rw}}{\mu_w B_w}\dfrac{\partial G}{\partial z}\right)\Delta z\end{bmatrix}+q_{g(\text{free})\text{scf/D}}+R_{so}q_o\text{ STB/D}+R_{sw}q_w\text{ STB/D}$$

$$=\frac{V_b}{5.615}\frac{\partial}{\partial t}\left(\frac{\phi S_g}{B_g}+R_{sw}\frac{\phi S_w}{B_w}+R_{so}\frac{\phi S_o}{B_o}\right).$$

.... (9.4)

Eq. 9.4 is much more elaborate than Eqs. 9.2 or 9.3. This is expected, because the gas phase can be transported, stored, and produced in three different modes, including as a free phase in solution with oil and in solution with water. For most practical purposes in black-oil formulation, gas transport, gas production, and gas storage throughout the water phase can be ignored; this is easily achieved by setting $R_{sw}=0$ in Eq. 9.4.

9.1.1 Inventory of Equations and Principal Unknowns in Black-Oil Formulation. A careful review of Eqs. 9.2 through 9.4 shows that there are six principal unknowns in a three-phase black-oil formulation. Three of these unknowns are phase pressures, and the remaining three are the phase saturations. Accordingly, the six principal unknowns are p_o, p_w, and p_g, (phase pressures) and S_o, S_w, and S_g (phase saturations). Therefore, in any gridblock during the solution process, we will work with six unknowns. It is necessary to have six equations available in each block so that the problem description becomes well posed. **Table 9.1** provides an inventory check on the equations available to close the formulation and gives an effective summary of the three-phase black-oil formulation. It is easy to see in Table 9.1 how some subsets of black-oil formulation can be identified. For example, a two-phase oil/gas black-oil formulation will have four equations (two mass balances, one capillary pressure relationship, and one saturation relationship) with four unknowns (p_o, p_w, S_o, and S_w). Other expressions of two-phase formulations and three relevant unknowns also can easily be similarly identified.

9.2 Solution Techniques for Solving Multiphase-Flow Equations of Porous Media

The two most-popular methods for solving multiphase-flow equations are known as the simultaneous solution (SS) and the implicit pressure/explicit saturation (IMPES).

The SS method has two different types of implementations. Both implementations involve eliminating three of the principal unknowns using the auxiliary equations given in Table 9.1 and leaving three principal unknowns simultaneously solved. After three auxiliary equations are embedded into three PDEs, the next stage involves writing the finite-difference representations of each of the three PDEs, which now have three principal unknowns. In the SS method, the three principal unknowns that are simultaneously solved are often p_o, S_w, and S_g.

In the IMPES method, five out of six unknowns are eliminated and water-phase pressure usually remains as the only unknown solved implicitly. Then, using the water equation, the water saturation is solved explicitly. Once the p_w and S_w values are obtained, by use of the capillary pressure relationship, one can obtain the values of p_o, which then can be used in explicitly obtaining the S_o values from the oil equation. In the final stage, through use of the saturation relationship, one can solve for the gas-saturation values, and once the gas-saturation values are obtained using the capillary pressure relationship between the oil and gas phases, the last unknown p_g can be determined. In the next subsection, we will review the mathematical manipulations necessary for these two solution methods, both at the PDE level and, subsequently, at the finite-difference-equation level.

Name of Equation	Type of Equation	Equation	Principal Unknowns
Oil-phase mass balance	Partial-differential equation	Eq. 9.2	
Water-phase mass balance	Partial-differential equation	Eq. 9.3	
Gas-phase mass balance	Partial-differential equation	Eq. 9.4	
Capillary pressure relationship between oil and water phase	Algebraic equation	$P_{cow}(S_w) = p_o - p_w$	$p_o, p_w, p_g, S_o, S_w, S_g$
Capillary pressure relationship between oil and gas phase	Algebraic equation	$P_{cgo}(S_g) = p_g - p_o$	
Saturation relationship	Algebraic equation	$S_o + S_w + S_g = 1$	
Total number of equations = 6			Total number of unknowns = 6

Table 9.1—Inventory of three-phase black-oil formulation equations and unknowns.

9.2.1 The SS Method as Applied to Two-Phase, Oil/Water Black-Oil Formulation. *Fully Implicit Formulation.* In this case, we choose p_o and p_w as the principal unknowns, and convert $\frac{\partial S_o}{\partial t}$ and $\frac{\partial S_w}{\partial t}$ terms into pressure derivatives for both oil and water equations. This results in two equations in two unknowns per block (p_o and p_w).

Consider the flow of oil and water in two dimensions:

- Oil equation:

$$\frac{\partial}{\partial x}\left(\frac{A_x k_x k_{ro}}{\mu_o B_o}\frac{\partial \Phi_o}{\partial x}\right)\Delta x + \frac{\partial}{\partial y}\left(\frac{A_y k_y k_{ro}}{\mu_o B_o}\frac{\partial \Phi_o}{\partial y}\right)\Delta y + \frac{\partial}{\partial z}\left(\frac{A_z k_z k_{ro}}{\mu_o B_o}\frac{\partial \Phi_o}{\partial z}\right)\Delta z + q_{o\text{ STB/D}} = \frac{V_b}{5.615}\frac{\partial}{\partial t}\left(\frac{\phi S_o}{B_o}\right). \quad (9.5)$$

- Water equation:

$$\frac{\partial}{\partial x}\left(\frac{A_x k_x k_{rw}}{\mu_w B_w}\frac{\partial \Phi_w}{\partial x}\right)\Delta x + \frac{\partial}{\partial y}\left(\frac{A_y k_y k_{rw}}{\mu_w B_w}\frac{\partial \Phi_w}{\partial y}\right)\Delta y + \frac{\partial}{\partial z}\left(\frac{A_z k_z k_{rw}}{\mu_w B_w}\frac{\partial \Phi_w}{\partial z}\right)\Delta z + q_{w\text{ STB/D}} = \frac{V_b}{5.615}\frac{\partial}{\partial t}\left(\frac{\phi S_w}{B_w}\right). \quad (9.6)$$

As previously stated, we now focus on the right-hand side of Eqs. 9.5 and 9.6, where $\frac{\partial S_o}{\partial t}$ and $\frac{\partial S_w}{\partial t}$ terms appear. The right-hand side of Eq. 9.5 can be expanded:

$$\frac{\partial}{\partial t}\left(\frac{\phi S_o}{B_o}\right) = \left(\frac{S_o}{B_o}\frac{\partial \phi}{\partial p_o} - \frac{\phi S_o}{B_o^2}\frac{\partial B_o}{\partial p_o}\right)\frac{\partial p_o}{\partial t} + \frac{\phi}{B_o^2}\frac{\partial S_o}{\partial t}. \quad (9.7)$$

In reaching Eq. 9.7, we assume that $\phi = \phi(p_o)$. Because $S_o + S_w = 1$, then we can state that $-\frac{\partial S_w}{\partial t} = \frac{\partial S_o}{\partial t}$, and $P_{cow}(S_w) = p_o - p_w$ implies $S_w = S_w(P_{cow})$, which is known as the inverted capillary pressure relationship. Because

$$\frac{\partial P_{cow}}{\partial t} = \frac{\partial p_o}{\partial t} - \frac{\partial p_w}{\partial t}, \quad (9.8)$$

one can state that

$$\frac{\partial S_w}{\partial t} = \frac{\partial S_w}{\partial P_{cow}}\frac{\partial P_{cow}}{\partial t} \quad (9.9)$$

or

$$\frac{\partial S_o}{\partial t} = -\frac{\partial S_w}{\partial P_{cow}}\left(\frac{\partial p_o}{\partial t} - \frac{\partial p_w}{\partial t}\right). \quad (9.10)$$

Accordingly, the final form of the right-hand side of the oil equation (Eq. 9.5) is then

$$\frac{V_b}{5.615}\frac{\partial}{\partial t}\left(\frac{\phi S_o}{B_o}\right) = \frac{V_b}{5.615}\left[\left(\frac{S_o}{B_o}\frac{\partial \phi}{\partial p_o} - \frac{\phi S_o}{B_o^2}\frac{\partial B_o}{\partial p_o} - \frac{\phi}{B_o}\frac{\partial S_w}{\partial P_{cow}}\right)\frac{\partial p_o}{\partial t} + \frac{\phi}{B_o}\frac{\partial S_w}{\partial P_{cow}}\frac{\partial p_w}{\partial t}\right]. \quad (9.11)$$

In a similar way, the right-hand side of the water equation (Eq. 9.6) can be expressed as

$$\frac{V_b}{5.615}\frac{\partial}{\partial t}\left(\frac{\phi S_w}{B_w}\right) = \frac{V_b}{5.615}\left[\left(\frac{S_w}{B_w}\frac{\partial \phi}{\partial p_o} + \frac{\phi}{B_w}\frac{\partial S_w}{\partial P_{cow}}\right)\frac{\partial p_o}{\partial t} - \left(\frac{\phi}{B_w}\frac{\partial S_w}{\partial P_{cow}} + \frac{\phi S_w}{B_w^2}\frac{\partial B_w}{\partial P_{cow}}\right)\frac{\partial p_w}{\partial t}\right]. \quad (9.12)$$

In Eqs. 9.11 and 9.12, there are no saturation derivatives (yes, there are saturations and saturation- and pressure-dependent functions, but they are in the form of coefficients to $\dfrac{\partial p_o}{\partial t}$ and $\dfrac{\partial p_w}{\partial t}$ terms, and they can be calculated at one iteration level behind). Before we write the finite-difference approximation of the oil and water equation, we introduce the $\Gamma_{i,j}$ notation, so that

$$\frac{V_b}{5.615}\frac{\partial}{\partial t}\left(\frac{\phi S_o}{B_o}\right)=\frac{V_b}{5.615}\left(\Gamma_{1o}\frac{\partial p_o}{\partial t}+\Gamma_{1w}\frac{\partial p_w}{\partial t}\right), \dots\dots\dots\dots\dots\dots\dots (9.13)$$

and

$$\frac{V_b}{5.615}\frac{\partial}{\partial t}\left(\frac{\phi S_w}{B_w}\right)=\frac{V_b}{5.615}\left(\Gamma_{2o}\frac{\partial p_o}{\partial t}+\Gamma_{2w}\frac{\partial p_w}{\partial t}\right). \dots\dots\dots\dots\dots\dots\dots (9.14)$$

Note that in Eqs. 9.13 and 9.14, the gamma terms are defined as

$$\Gamma_{1o}=\frac{S_o}{B_o}\frac{\partial \phi}{\partial p_o}-\frac{\phi S_o}{B_o^2}\frac{\partial B_o}{\partial p_o}-\frac{\phi}{B_o}\frac{\partial S_w}{\partial P_{cow}}, \dots\dots\dots\dots\dots\dots\dots (9.15a)$$

$$\Gamma_{1w}=\frac{\phi}{B_o}\frac{\partial S_w}{\partial P_{cow}}, \dots\dots\dots\dots\dots\dots\dots\dots\dots\dots\dots (9.15b)$$

$$\Gamma_{2o}=\frac{S_w}{B_w}\frac{\partial \phi}{\partial p_o}+\frac{\phi}{B_w}\frac{\partial S_w}{\partial P_{cow}}, \dots\dots\dots\dots\dots\dots\dots\dots\dots (9.15c)$$

and

$$\Gamma_{2w}=-\left(\frac{\phi}{B_w}\frac{\partial S_w}{\partial P_{cow}}+\frac{\phi S_w}{B_w^2}\frac{\partial B_w}{\partial P_{cow}}\right). \dots\dots\dots\dots\dots\dots\dots (9.15d)$$

Also, the subscript 1 refers to the first equation (oil phase →1), and the subscript 2 refers to the second equation (water phase →2). At this stage of development, we also focus on typical transmissibility terms that originate from the left-hand side of the oil and water equations. The typical transmissibility terms for the oil and water equations will be in the following forms:

$$T_{o,i,j-\frac{1}{2}}=\left(\frac{A_y k_y}{\Delta y}\right)\Bigg|_{i,j-\frac{1}{2}}\left(\frac{1}{\mu_o B_o}\right)\Bigg|_{i,j-\frac{1}{2}}(k_{ro})\big|_{i,j-\frac{1}{2}}, \dots\dots\dots\dots\dots\dots\dots (9.16a)$$

and

$$T_{w,i,j-\frac{1}{2}}=\left(\frac{A_y k_y}{\Delta y}\right)\Bigg|_{i,j-\frac{1}{2}}\left(\frac{1}{\mu_w B_w}\right)\Bigg|_{i,j-\frac{1}{2}}(k_{rw})\big|_{i,j-\frac{1}{2}}. \dots\dots\dots\dots\dots\dots\dots (9.16b)$$

As seen in the previous transmissibility terms, a new term, k_{ro} (or k_{rw}), appears. The first group of terms is known as the constant part of the transmissibility and, as usual, it is calculated using harmonic averaging of the values between the two adjacent blocks. The second group of terms has two pressure-dependent terms (viscosity and formation volume factor); they represent the weak nonlinearities of the transmissibility group, and can be calculated using the arithmetic average of the pressures of two adjacent blocks. The last term of Eqs. 9.16a and 9.16b represent the relative permeabilities; they pose a significant challenge in terms of the strong nonlinearities they introduce into the finite-difference equations. As we know, relative-permeability terms are functions only of saturations. The saturation front moves in the form of a shock front; therefore, instead of using an averaging scheme, it is much more accurate to use an upstream-weighting method. There are two commonly used upstream-weighting techniques for relative permeability values:

- Single-point upstream weighting:

$$k_{rli+\frac{1}{2}}=\begin{cases}k_{rl}(S_{wi}), & \text{if } \Phi_{li}>\Phi_{li+1}\\ k_{rl}(S_{wi+1}), & \text{if } \Phi_{li+1}>\Phi_{li}\end{cases}. \dots\dots\dots\dots\dots\dots\dots (9.17)$$

- Two-point upstream weighting: This technique uses the concept of linear extrapolation of k_{rl} values using the two upstream points:

$$k_{rli+\frac{1}{2}}=\begin{cases}\frac{1}{2}\left[3k_{rl}(S_{wi})-k_{rl}(S_{wi-1})\right], & \text{if } \Phi_{li}>\Phi_{li+1}\\ \frac{1}{2}\left[3k_{rl}(S_{wi+1})-k_{rl}(S_{wi+2})\right], & \text{if } \Phi_{li+1}>\Phi_{li}\end{cases} \dots\dots\dots\dots\dots\dots\dots (9.18)$$

It should be noted that Eq. 9.18 is correct for uniform-block spacing. If the dimension of the blocks along the flow direction is changing, then block dimensions must be accordingly included in the extrapolation; a more-general equation for two-point upstream weighting for nonuniform block dimensions can be written as

$$k_{rli\mp\frac{1}{2}} = k_{rup1} + \left[\frac{\left(k_{rup1} - k_{rup2}\right)}{\sum\limits_{j=1}^{2} \Delta x_{upj}} \right] \left(\Delta x_{up1}\right). \quad\ldots\ldots\ldots\ldots\ldots\ldots\ldots\ldots\ldots\ldots\ldots\ldots\ldots\ldots\ldots (9.19)$$

Note that Eq. 9.19 collapses to Eq. 9.18 if Δx values are uniform. Now, one can write the finite-difference approximations to the oil and water equations (using Eqs. 9.13 and 9.14 as right-hand sides of the corresponding PDEs, respectively):

- Oil equation in finite-difference form:

$$S_{i,j}^{1o\ (k)}\ p_{oi,j-1}^{n+1\ (k+1)} + W_{i,j}^{1o\ (k)}\ p_{oi-1,j}^{n+1\ (k+1)} + C_{i,j}^{1o\ (k)}\ p_{oi,j}^{n+1\ (k+1)} + C_{i,j}^{1w\ (k)}\ p_{wi,j}^{n+1\ (k+1)} + E_{i,j}^{1o\ (k)}\ p_{oi+1,j}^{n+1\ (k+1)} + N_{i,j}^{1o\ (k)}\ p_{oi,j+1}^{n+1\ (k+1)} = Q_{i,j}^{1o\ (k)}. \quad\ldots\ldots\ldots\ldots\ldots (9.20)$$

- Water equation in finite-difference form:

$$S_{i,j}^{2w\ (k)}\ p_{wi,j-1}^{n+1\ (k+1)} + W_{i,j}^{2w\ (k)}\ p_{wi-1,j}^{n+1\ (k+1)} + C_{i,j}^{2o\ (k)}\ p_{oi,j}^{n+1\ (k+1)} + C_{i,j}^{2w\ (k)}\ p_{wi,j}^{n+1\ (k+1)} + E_{i,j}^{2w\ (k)}\ p_{wi+1,j}^{n+1\ (k+1)} + N_{i,j}^{2w\ (k)}\ p_{wi,j+1}^{n+1\ (k+1)} = Q_{i,j}^{2w\ (k)}. \quad\ldots\ldots\ldots\ldots\ldots (9.21)$$

In Eq. 20, we observe the following coefficients:

$$S_{i,j}^{1o} = \frac{A_y k_y k_{ro}}{\mu_o B_o \Delta y}\bigg|_{i,j-\frac{1}{2}}, \quad\ldots\ldots\ldots\ldots\ldots\ldots\ldots\ldots\ldots\ldots\ldots\ldots\ldots\ldots\ldots\ldots\ldots (9.22a)$$

$$W_{i,j}^{1o} = \frac{A_x k_x k_{ro}}{\mu_o B_o \Delta x}\bigg|_{i-\frac{1}{2},j}, \quad\ldots\ldots\ldots\ldots\ldots\ldots\ldots\ldots\ldots\ldots\ldots\ldots\ldots\ldots\ldots\ldots (9.22b)$$

$$E_{i,j}^{1o} = \frac{A_x k_x k_{ro}}{\mu_o B_o \Delta x}\bigg|_{i+\frac{1}{2},j}, \quad\ldots\ldots\ldots\ldots\ldots\ldots\ldots\ldots\ldots\ldots\ldots\ldots\ldots\ldots\ldots\ldots (9.22c)$$

$$N_{i,j}^{1o} = \frac{A_y k_y k_{ro}}{\mu_o B_o \Delta y}\bigg|_{i,j+\frac{1}{2}}, \quad\ldots\ldots\ldots\ldots\ldots\ldots\ldots\ldots\ldots\ldots\ldots\ldots\ldots\ldots\ldots\ldots (9.22d)$$

$$C_{i,j}^{1o} = -\left[S_{i,j}^{1o} + W_{i,j}^{1o} + E_{i,j}^{1o} + N_{i,j}^{1o} + \left(\frac{V_b}{5.615}\right)_{i,j} \frac{1}{\Delta t} \Gamma_{1oi,j} \right], \quad\ldots\ldots\ldots\ldots\ldots\ldots\ldots\ldots\ldots\ldots (9.22e)$$

$$C_{i,j}^{1w} = -\left(\frac{V_b}{5.615}\right)_{i,j} \frac{1}{\Delta t} \Gamma_{1wi,j}, \quad\ldots\ldots\ldots\ldots\ldots\ldots\ldots\ldots\ldots\ldots\ldots\ldots\ldots\ldots\ldots (9.22f)$$

and

$$Q_{i,j}^{1o} = -q_{oi,j} + E_{i,j}^{'1o}\left(G_{i+1,j} - G_{i,j}\right) - W_{i,j}^{'1o}\left(G_{i,j} - G_{i-1,j}\right) + N_{i,j}^{'1o}\left(G_{i,j+1} - G_{i,j}\right) - S_{i,j}^{'1o}\left(G_{i,j} - G_{i,j-1}\right)$$
$$\qquad\qquad - \left(\frac{V_b}{5.615}\right)_{i,j} \frac{1}{\Delta t} \Gamma_{1oi,j} p_{oi.j}^n - \left(\frac{V_b}{5.615}\right)_{i,j} \frac{1}{\Delta t} \Gamma_{1wi,j} p_{wi.j}^n. \qquad\qquad\ldots\ldots\ldots\ldots\ldots\ldots (9.22g)$$

Similarly, for the water equation (Eq. 9.21), the definitions of coefficients are as follows:

$$S_{i,j}^{2w} = \frac{A_y k_y k_{rw}}{\mu_w B_w \Delta y}\bigg|_{i,j-\frac{1}{2}}, \quad\ldots\ldots\ldots\ldots\ldots\ldots\ldots\ldots\ldots\ldots\ldots\ldots\ldots\ldots\ldots\ldots (9.23a)$$

$$W_{i,j}^{2w} = \frac{A_x k_x k_{rw}}{\mu_w B_w \Delta x}\bigg|_{i-\frac{1}{2},j}, \quad\ldots\ldots\ldots\ldots\ldots\ldots\ldots\ldots\ldots\ldots\ldots\ldots\ldots\ldots\ldots\ldots (9.23b)$$

$$E_{i,j}^{2w} = \frac{A_x k_x k_{rw}}{\mu_w B_w \Delta x}\bigg|_{i+\frac{1}{2},j}, \quad\ldots\ldots\ldots\ldots\ldots\ldots\ldots\ldots\ldots\ldots\ldots\ldots\ldots\ldots\ldots\ldots (9.23c)$$

$$N_{i,j}^{2w} = \frac{A_y k_y k_{rw}}{\mu_w B_w \Delta y}\bigg|_{i,j+\frac{1}{2}}, \quad\ldots\ldots\ldots\ldots\ldots\ldots\ldots\ldots\ldots\ldots\ldots\ldots\ldots\ldots\ldots\ldots (9.23d)$$

$$C_{i,j}^{2w}=-\left[S_{i,j}^{2w}+W_{i,j}^{2w}+E_{i,j}^{2w}+N_{i,j}^{2w}+\left(\frac{V_b}{5.615}\right)_{i,j}\frac{1}{\Delta t}\Gamma_{2wi,j}\right], \quad (9.23e)$$

$$C_{i,j}^{2o}=-\left(\frac{V_b}{5.615}\right)_{i,j}\frac{1}{\Delta t}\Gamma_{2oi,j}, \quad (9.23f)$$

and

$$Q_{i,j}^{2w}=-q_{wi,j}+\overset{2w}{E_{i,j}'}\left(G_{i+1,j}-G_{i,j}\right)-\overset{2w}{W_{i,j}'}\left(G_{i,j}-G_{i-1,j}\right)+\overset{2w}{N_{i,j}'}\left(G_{i,j+1}-G_{i,j}\right)-\overset{2w}{S_{i,j}'}\left(G_{i,j}-G_{i,j-1}\right). \quad (9.23g)$$

Note that coefficients such as $E_{i,j}'$, $W_{i,j}'$, $N_{i,j}'$, and $S_{i,j}'$ represent the modified transmissibility coefficients, and they arise from gravitational terms; for example,

$$\overset{2w}{E_{i,j}'}=\frac{1}{144}\frac{g}{g_c}\rho_w\left.\frac{A_xk_xk_{rw}}{\mu_wB_w\Delta x}\right|_{i+\frac{1}{2},j} \text{ and } \overset{1o}{W_{i,j}'}=\frac{1}{144}\frac{g}{g_c}\rho_o\left.\frac{A_xk_xk_{ro}}{\mu_oB_o\Delta x}\right|_{i,j+\frac{1}{2}}.$$

At this stage, in each block we have two simultaneous equations with two principal unknowns, $p_{oi,j}^{n+1}$ and $p_{wi,j}^{n+1}$ (Eqs. 9.20 and 9.21). When these two equations are implemented, for example, for a 3x3 reservoir system, the coefficient matrix will be 18x18. **Fig. 9.1** shows the numbering scheme for the 3x3 example reservoir, and **Fig. 9.2** shows the resulting system of equations. It is clear from Fig. 9.2 that the resulting coefficient matrix is a pentadiagonal (2x2) block matrix.

(1,3)	(2,3)	(3,3)
(1,2)	(2,2)	(3,2)
(1,1)	(2,1)	(3,1)

Fig. 9.1—The node numbering scheme for the 3×3 sample reservoir.

Coefficients **of Unknowns**

Equations	$p_{o1,1}$	$p_{w1,1}$	$p_{o2,1}$	$p_{w2,1}$	$p_{o3,1}$	$p_{w3,1}$	$p_{o1,2}$	$p_{w1,2}$	$p_{o2,2}$	$p_{w2,2}$	$p_{o3,2}$	$p_{w3,2}$	$p_{o3,1}$	$p_{w3,1}$	$p_{o3,2}$	$p_{w3,2}$	$p_{o3,3}$	$p_{w3,3}$			
Block (1,1) Oil	$C_{1,1}^{1o}$	$C_{1,1}^{1w}$	$E_{1,1}^{1o}$	0	0	0	$N_{1,1}^{1o}$	0	0	0	0	0	0	0	0	0	0	0	$p_{o1,1}$		$Q_{1,1}^1$
Block (1,1) water	$C_{1,1}^{2o}$	$C_{1,1}^{2w}$	0	$E_{1,1}^{2w}$	0	0	0	$N_{1,1}^{2w}$	0	0	0	0	0	0	0	0	0	0	$p_{w1,1}$		$Q_{1,1}^2$
Block (2,1) Oil	$W_{2,1}^{1o}$	0	$C_{2,1}^{1o}$	$C_{2,1}^{1w}$	$E_{2,1}^{1o}$	0	0	0	$N_{2,1}^{1o}$	0	0	0	0	0	0	0	0	0	$p_{o2,1}$		$Q_{2,1}^1$
Block (2,1) water	0	$W_{2,1}^{2w}$	$C_{2,1}^{2o}$	$C_{2,1}^{2w}$	0	$E_{2,1}^{2w}$	0	0	0	$N_{2,1}^{2w}$	0	0	0	0	0	0	0	0	$p_{w2,1}$		$Q_{2,1}^2$
Block (3,1) Oil	0	0	$W_{3,1}^{1o}$	0	$C_{3,1}^{1o}$	$C_{3,1}^{1w}$	0	0	0	0	$N_{3,1}^{1o}$	0	0	0	0	0	0	0	$p_{o3,1}$		$Q_{3,2}^1$
Block (3,1) water	0	0	0	$W_{3,1}^{2w}$	$C_{3,1}^{2o}$	$C_{3,1}^{2w}$	0	0	0	0	0	$N_{3,1}^{2w}$	0	0	0	0	0	0	$p_{w3,1}$		$Q_{3,2}^2$
Block (1,2) Oil	$S_{1,2}^{1o}$	0	0	0	0	0	$C_{1,2}^{1o}$	$C_{1,2}^{1w}$	$E_{2,1}^{1o}$	0	0	0	$N_{1,2}^{1o}$	0	0	0	0	0	$p_{o1,2}$		$Q_{1,2}^1$
Block (1,2) water	0	$S_{1,2}^{2w}$	0	0	0	0	$C_{1,2}^{2o}$	$C_{1,2}^{2w}$	0	$E_{1,2}^{2w}$	0	0	0	$N_{1,2}^{2w}$	0	0	0	0	$p_{w1,2}$		$Q_{1,2}^2$
Block (2,2) Oil	0	0	$S_{2,2}^{1o}$	0	0	0	$W_{2,2}^{1o}$	0	$C_{2,2}^{1o}$	$C_{2,2}^{1w}$	$E_{2,2}^{1o}$	0	0	0	$N_{2,2}^{1o}$	0	0	0	$p_{o2,2}$	=	$Q_{2,2}^1$
Block (2,2) water	0	0	0	$S_{2,2}^{2w}$	0	0	0	$W_{2,2}^{2w}$	$C_{2,2}^{2o}$	$C_{2,2}^{2w}$	0	$E_{2,w}^{2w}$	0	0	0	$N_{2,2}^{2w}$	0	0	$p_{w2,2}$		$Q_{2,2}^2$
Block (3,2) Oil	0	0	0	0	$S_{3,2}^{1o}$	0	0	0	$W_{3,2}^{1o}$	0	$C_{3,2}^{1o}$	$C_{3,2}^{1w}$	0	0	0	0	$N_{3,2}^{1o}$	0	$p_{o3,2}$		$Q_{3,3}^1$
Block (3,2) water	0	0	0	0	0	$S_{3,2}^{2w}$	0	0	0	$W_{3,2}^{2w}$	$C_{3,2}^{2o}$	$C_{3,2}^{2w}$	0	0	0	0	0	$N_{3,2}^{2w}$	$p_{w3,2}$		$Q_{3,3}^2$
Block (1,3) Oil	0	0	0	0	0	0	$S_{1,3}^{1o}$	0	0	0	0	0	$C_{1,3}^{1o}$	$C_{1,3}^{1w}$	$E_{1,3}^{1o}$	0	0	0	$p_{o1,3}$		$Q_{1,3}^1$
Block (1,3) water	0	0	0	0	0	0	0	$S_{1,3}^{2w}$	0	0	0	0	$C_{1,3}^{2o}$	$C_{1,3}^{2w}$	0	$E_{1,3}^{2w}$	0	0	$p_{w1,3}$		$Q_{1,3}^2$
Block (2,3) Oil	0	0	0	0	0	0	0	0	$S_{2,3}^{1o}$	0	0	0	$W_{2,3}^{1o}$	0	$C_{2,3}^{1o}$	$C_{2,3}^{1w}$	$E_{2,3}^{1o}$	0	$p_{o2,3}$		$Q_{2,3}^1$
Block (2,3) water	0	0	0	0	0	0	0	0	0	$S_{2,3}^{2w}$	0	0	0	$W_{2,3}^{2w}$	$C_{2,3}^{2o}$	$C_{2,3}^{2w}$	0	$E_{2,3}^{2w}$	$p_{w2,3}$		$Q_{2,3}^2$
Block (3,3) Oil	0	0	0	0	0	0	0	0	0	0	$S_{3,3}^{1o}$	0	0	0	$W_{3,3}^{1o}$	0	$C_{3,3}^{1o}$	$C_{3,3}^{1w}$	$p_{o3,3}$		$Q_{3,3}^1$
Block (3,3) water	0	0	0	0	0	0	0	0	0	0	0	$S_{3,3}^{2w}$	0	0	0	$W_{3,3}^{2w}$	$C_{3,3}^{2o}$	$C_{3,3}^{2w}$	$p_{w3,3}$		$Q_{3,3}^2$

Fig. 9. 2 —The coefficient matrix, the unknown vector, and the right-hand-side vector of the reservoir system of Fig. 9.1.

9.2.2 Generalized Newton-Raphson Formulation of the Fully Implicit Procedure. In Chapter 8, we discussed a summary of the generalized Newton-Raphson (GNR) formulation for single-phase flow problems. The same protocol can be extended to multiphase-flow problems. In this discussion, we show an implementation of GNR applicable to two-phase (oil/water) problems. In this application, there are two residual equations in each block representing each phase. In other words, for a 1D flow problem, these two equations are in the generic forms shown in Eqs. 9.24 and 9.25:

$$R_{oi}^{n+1} = 0, \dotfill (9.24)$$

and

$$R_{wi}^{n+1} = 0. \dotfill (9.25)$$

Now, we can implement the GNR in two-phase (oil/water) problems. We consider a 1D problem and develop a system of equations, this time using the principal unknowns p_o and S_w (in other words, p_w and S_o will be eliminated). Then, 1D simultaneous flow of oil and water can be expressed with the help of the following four equations:

- Oil equation:

$$\frac{\partial}{\partial x}\left(\frac{A_x k_x k_{ro}}{\mu_o B_o}\frac{\partial p_o}{\partial x}\right)\Delta x + q_{o\,\text{STB/D}} = \frac{V_b}{5.615}\frac{\partial}{\partial t}\left(\frac{\phi S_o}{B_o}\right). \dotfill (9.26)$$

- Water equation:

$$\frac{\partial}{\partial x}\left(\frac{A_x k_x k_{rw}}{\mu_w B_w}\frac{\partial p_w}{\partial x}\right)\Delta x + q_{w\,\text{STB/D}} = \frac{V_b}{5.615}\frac{\partial}{\partial t}\left(\frac{\phi S_w}{B_w}\right). \dotfill (9.27)$$

- Auxiliary equations:

$$S_w + S_o = 1, \dotfill (9.28)$$

$$P_{cow}\left(S_w\right) = p_o - p_w. \dotfill (9.29)$$

The finite-difference approximations to Eqs. 9.24 and 9.25 can be written in the forms that follow [note that S_o is eliminated by setting $S_o = 1 - S_w$ in the oil equation, and p_w is eliminated by setting $p_w = p_o - P_{cow}(S_w)$ in the water equation]. When these substitutions are made, the following residual form of the finite-difference approximations will be obtained:

- Residual equation for oil phase:

$$R_o\left(p_o^{n+1}, S_w^{n+1}\right) = \frac{A_x k_x}{\Delta x}\Bigg|_{i+\frac{1}{2}}\frac{1}{\mu_o B_o}\Bigg|_{i+\frac{1}{2}}^{n+1} k_{ro}\Big|_{i+\frac{1}{2}}^{n+1}\left(p_{oi+1}^{n+1} - p_{oi}^{n+1}\right) - \frac{A_x k_x}{\Delta x}\Bigg|_{i-\frac{1}{2}}\frac{1}{\mu_o B_o}\Bigg|_{i-\frac{1}{2}}^{n+1} k_{ro}\Big|_{i-\frac{1}{2}}^{n+1}\left(p_{oi}^{n+1} - p_{oi-1}^{n+1}\right)$$

$$+ q_{oi\,\text{STB/D}}^{n+1} - \left(\frac{V_b}{5.615}\right)_i\frac{1}{\Delta t}\left(\frac{\phi - \phi S_w}{B_o}\right)_i^{n+1} + \left(\frac{V_b}{5.615}\right)_i\frac{1}{\Delta t}\left(\frac{\phi - \phi S_w}{B_o}\right)_i^{n}. \dotfill (9.30)$$

- Residual equation for water phase:

$$R_w\left(p_o^{n+1}, S_w^{n+1}\right) = \frac{A_x k_x}{\Delta x}\Bigg|_{i+\frac{1}{2}}\frac{1}{\mu_w B_w}\Bigg|_{i+\frac{1}{2}}^{n+1} k_{rw}\Big|_{i+\frac{1}{2}}^{n+1}\left(p_{oi+1}^{n+1} - p_{oi}^{n+1}\right) - \frac{A_x k_x}{\Delta x}\Bigg|_{i-\frac{1}{2}}\frac{1}{\mu_w B_w}\Bigg|_{i-\frac{1}{2}}^{n+1} k_{rw}\Big|_{i-\frac{1}{2}}^{n+1}\left(p_{oi}^{n+1} - p_{oi-1}^{n+1}\right)$$

$$- \frac{A_x k_x}{\Delta x}\Bigg|_{i+\frac{1}{2}}\frac{1}{\mu_w B_w}\Bigg|_{i+\frac{1}{2}}^{n+1} k_{rw}\Big|_{i+\frac{1}{2}}^{n+1}\left(P_{cow+1}^{n+1} - P_{cowi}^{n+1}\right) + \frac{A_x k_x}{\Delta x}\Bigg|_{i-\frac{1}{2}}\frac{1}{\mu_w B_w}\Bigg|_{i-\frac{1}{2}}^{n+1} k_{rw}\Big|_{i-\frac{1}{2}}^{n+1}\left(P_{cowi}^{n+1} - P_{cowi-1}^{n+1}\right) \dotfill (9.31)$$

$$+ q_{wi\,\text{STB/D}}^{n+1} - \left(\frac{V_b}{5.615}\right)_i\frac{1}{\Delta t}\left(\frac{\phi S_w}{B_w}\right)_i^{n+1} + \left(\frac{V_b}{5.615}\right)_i\frac{1}{\Delta t}\left(\frac{\phi S_w}{B_w}\right)_i^{n}.$$

Eqs. 9.30 and 9.31 are the characteristic equations for the system of equations and $(i = 1, 2,..., \underline{NX})\times 2$ nonlinear equations need to be solved for $p_{o1}^{n+1}, p_{o2}^{n+1}, \dots p_{oNX}^{n+1}$, and $S_{w1}^{n+1}, S_{w2}^{n+1} \dots S_{wNX}^{n+1}$. In shorthand, Eqs. 9.30 and 9.31 can be written as

$$R_o\left(p_{oi-1}^{n+1}, S_{wi-1}^{n+1}, p_{oi}^{n+1}, S_{wi}^{n+1}, p_{oi+1}^{n+1}, S_{wi+1}^{n+1}\right) = 0, \dotfill (9.32)$$

and

$$R_w\left(p_{oi-1}^{n+1}, S_{wi-1}^{n+1}, p_{oi}^{n+1}, S_{wi}^{n+1}, p_{oi+1}^{n+1}, S_{wi+1}^{n+1}\right) = 0. \dotfill (9.33)$$

Then, the generalized linear equations will be

- Oil equation:

$$\frac{\partial R_{oi}^{n+1}}{\partial p_{oi-1}^{n+1}}\bigg|^{(k)} \left[p_{oi-1}^{n+1\,(k+1)} - p_{oi-1}^{n+1\,(k)} \right] + \frac{\partial R_{oi}^{n+1}}{\partial S_{wi-1}^{n+1}}\bigg|^{(k)} \left[S_{wi-1}^{n+1\,(k+1)} - S_{wi-1}^{n+1\,(k)} \right] + \frac{\partial R_{oi}^{n+1}}{\partial p_{oi}^{n+1}}\bigg|^{(k)} \left[p_{oi}^{n+1\,(k+1)} - p_{oi}^{n+1\,(k)} \right] + \frac{\partial R_{oi}^{n+1}}{\partial S_{wi}^{n+1}}\bigg|^{(k)} \left[S_{wi}^{n+1\,(k+1)} - S_{wi}^{n+1\,(k)} \right]$$
$$+ \frac{\partial R_{oi}^{n+1}}{\partial p_{oi+1}^{n+1}}\bigg|^{(k)} \left[p_{oi+1}^{n+1\,(k+1)} - p_{oi+1}^{n+1\,(k)} \right] + \frac{\partial R_{oi}^{n+1}}{\partial S_{wi+1}^{n+1}}\bigg|^{(k)} \left[S_{wi+1}^{n+1\,(k+1)} - S_{wi+1}^{n+1\,(k)} \right] = -R_{oi}^{n+1\,(k)}. \qquad \dots (9.34)$$

- Water equation:

$$\frac{\partial R_{wi}^{n+1}}{\partial p_{wi-1}^{n+1}}\bigg|^{(k)} \left(p_{oi-1}^{n+1\,(k+1)} - p_{oi-1}^{n+1\,(k)} \right) + \frac{\partial R_{wi}^{n+1}}{\partial S_{wi-1}^{n+1}}\bigg|^{(k)} \left(S_{wi-1}^{n+1\,(k+1)} - S_{wi-1}^{n+1\,(k)} \right) + \frac{\partial R_{wi}^{n+1}}{\partial p_{oi}^{n+1}}\bigg|^{(k)} \left(p_{oi}^{n+1\,(k+1)} - p_{oi}^{n+1\,(k)} \right) + \frac{\partial R_{wi}^{n+1}}{\partial S_{wi}^{n+1}}\bigg|^{(k)} \left(S_{wi}^{n+1\,(k+1)} - S_{wi}^{n+1\,(k)} \right)$$
$$+ \frac{\partial R_{wi}^{n+1}}{\partial p_{oi+1}^{n+1}}\bigg|^{(k)} \left(p_{oi+1}^{n+1\,(k+1)} - p_{oi+1}^{n+1\,(k)} \right) + \frac{\partial R_{wi}^{n+1}}{\partial S_{wi+1}^{n+1}}\bigg|^{(k)} \left(S_{wi+1}^{n+1\,(k+1)} - S_{wi+1}^{n+1\,(k)} \right) = -R_{wi}^{n+1\,(k)}. \qquad \dots (9.35)$$

A simple numerical-differentiation protocol is implemented to evaluate the partial derivatives that appear in the residual equations. For example, using a forward-difference scheme, one obtains

$$\frac{\partial R_{oi}^{n+1}}{\partial p_{oi}^{n+1}}\bigg|^{(k)} = \frac{R_o^{(k)}\left(p_{oi-1}^{n+1}, S_{wi-1}^{n+1}, p_{oi}^{n+1}+\varepsilon_p, S_{wi}^{n+1}, p_{oi+1}^{n+1}, S_{wi+1}^{n+1} \right) - R_o^{(k)}\left(p_{oi-1}^{n+1}, S_{wi-1}^{n+1}, p_{oi}^{n+1}, S_{wi}^{n+1}, p_{oi+1}^{n+1}, S_{wi+1}^{n+1} \right)}{\varepsilon_p}. \qquad \dots (9.36)$$

It is also possible to use a more-accurate central-difference approximation for the first-order derivative of the residual equations. However, the use of the central-difference approximation will require calculation of the residual two times (in both forward and backward directions). This will significantly increase the computational time. By keeping ε_p in Eq. 9.36 small (typically in the order of 10^{-7} to 10^{-8}), the accuracy of the forward-difference scheme is assured at a lower computational overhead. Other partial derivatives are also calculated using a similar procedure.

Consider the simple 1D reservoir with four blocks shown in **Fig. 9.3.** When Eqs. 9.34 and 9.35 are written for $i =1$, 2, 3, 4 for the reservoir system shown in Fig. 9.3, a system of equations, as shown in **Fig. 9.4,** will be generated. The coefficient matrix shown in Fig. 9.4 is called the Jacobian of the linearized system of equations.

	$i=1$	$i=2$	$i=3$	$i=4$
Unknowns	Δp_{o1}^{n+1} ΔS_{w1}^{n+1}	Δp_{o2}^{n+1} ΔS_{w2}^{n+1}	Δp_{o3}^{n+1} ΔS_{w3}^{n+1}	Δp_{o4}^{n+1} ΔS_{w4}^{n+1}

Fig. 9.3—A 1D, four-block reservoir system. The unknowns in each block are identified.

Coefficients of Unknowns

Fig. 9.4—System of equations generated from the implementation of the Newton-Raphson protocol to the reservoir system of Fig. 9.3.

9.2.3 IMPES Method. The basic idea of the IMPES method is to obtain an equation that contains the pressure of only one phase as the principal unknown. This equation is obtained by combining the flow equations demonstrated for the two-phase (oil/water) flow problem, as explained in this subsection. Later in this subsection, it is shown that once the pressure equation is implicitly solved, saturation of the same phase can be explicitly solved.

In applying the IMPES method to the two-phase (oil/water) problem in the first step, we need to expand the right-hand side of both equations:

- Oil equation:

$$\frac{\partial}{\partial x}\left(\frac{A_x k_x k_{ro}}{\mu_o B_o}\frac{\partial \Phi_o}{\partial x}\right)\Delta x + \frac{\partial}{\partial y}\left(\frac{A_y k_y k_{ro}}{\mu_o B_o}\frac{\partial \Phi_o}{\partial y}\right)\Delta y + q_{o\,\text{STB/D}} = \frac{V_b}{5.615}\frac{\partial}{\partial t}\left(\frac{\phi S_o}{B_o}\right). \quad \dots \dots \dots (9.37)$$

- Water equation:

$$\frac{\partial}{\partial x}\left(\frac{A_x k_x k_{rw}}{\mu_w B_w}\frac{\partial \Phi_w}{\partial x}\right)\Delta x + \frac{\partial}{\partial y}\left(\frac{A_y k_y k_{rw}}{\mu_w B_w}\frac{\partial \Phi_w}{\partial y}\right)\Delta y + q_{w\,\text{STB/D}} = \frac{V_b}{5.615}\frac{\partial}{\partial t}\left(\frac{\phi S_w}{B_w}\right). \quad \dots \dots \dots (9.38)$$

Consider the temporal derivative of Eq. 9.37, which can be expanded to

$$\frac{\partial}{\partial t}\left(\frac{\phi S_o}{B_o}\right) = \frac{\phi}{B_o}\frac{\partial S_o}{\partial t} + \left(\frac{S_o}{B_o}\frac{\partial \phi}{\partial p_o} - \frac{\phi S_o}{B_o^2}\frac{\partial B_o}{\partial p_o}\right)\frac{\partial p_o}{\partial t}. \quad \dots \dots \dots (9.39)$$

Similarly, the right-hand side of the water equation (Eq. 9.38) can be expanded as follows:

$$\frac{\partial}{\partial t}\left(\frac{\phi S_w}{B_w}\right) = -\frac{\phi S_w}{B_w^2}\frac{\partial B_w}{\partial p_w}\frac{\partial p_w}{\partial t} + \frac{\phi}{B_w}\frac{\partial S_w}{\partial t} + \frac{S_w}{B_w}\frac{\partial \phi}{\partial p_o}\frac{\partial p_o}{\partial t}. \quad \dots \dots \dots (9.40)$$

Then, the addition of the right-hand sides of Eqs. 9.37 and 9.38 gives

$$\frac{V_b}{5.615}\left[\frac{\partial}{\partial t}\left(\frac{\phi S_w}{B_w}\right) + \frac{\partial}{\partial t}\left(\frac{\phi S_o}{B_o}\right)\right]$$

$$= \frac{V_b}{5.615}\left[\left(\frac{S_o}{B_o}\frac{\partial \phi}{\partial p_o} - \frac{\phi S_o}{B_o^2}\frac{\partial B_o}{\partial p_o}\right)\frac{\partial p_o}{\partial t} + \left(\frac{S_w}{B_w}\frac{\partial \phi}{\partial p_o}\frac{\partial p_o}{\partial t} - \frac{\phi S_w}{B_w^2}\frac{\partial B_w}{\partial p_w}\frac{\partial p_w}{\partial t}\right) + \left(\frac{\phi}{B_w}\frac{\partial S_w}{\partial t} + \frac{\phi}{B_o}\frac{\partial S_o}{\partial t}\right)\right]. \quad \dots \dots (9.41)$$

Our goal is to eliminate the saturation derivatives $\dfrac{\partial S_o}{\partial t}$ and $\dfrac{\partial S_w}{\partial t}$ in Eq. 9.41. This can be achieved by introducing a constant, A, such that the entire oil equation is multiplied by A on both sides and added side by side with the water equation. Our aim is to select A, such that

$$A\left(\frac{\phi}{B_o}\frac{\partial S_o}{\partial t}\right) + \frac{\phi}{B_w}\frac{\partial S_w}{\partial t} = 0, \quad \dots \dots \dots (9.42)$$

because $\dfrac{\partial S_o}{\partial t} = -\dfrac{\partial S_w}{\partial t}$, then

$$\left(\frac{1}{B_w} - \frac{A}{B_o}\right)\phi\frac{\partial S_w}{\partial t} = 0. \quad \dots \dots \dots (9.43)$$

Again, because $\dfrac{\partial S_w}{\partial t} \neq 0$, then we should put the constraint that

$$\left(\frac{1}{B_w} - \frac{A}{B_o}\right) = 0, \quad \dots \dots \dots (9.44)$$

which results in the definition of A as

$$A = \frac{B_o}{B_w}. \quad \dots \dots \dots (9.45)$$

When the oil equation is multiplied by $\dfrac{B_o}{B_w}$, then the linear combination of oil and water equations will result in the last component

Eq. 9.41; $\left(\dfrac{\phi}{B_w}\dfrac{\partial S_w}{\partial t} + \dfrac{\phi}{B_o}\dfrac{\partial S_o}{\partial t}\right)$ will vanish. Hence, the resulting equation from the addition of oil and water equations will be

$$\frac{B_o}{B_w}\left[\frac{\partial}{\partial x}\left(\frac{A_x k_x k_{ro}}{\mu_o B_o}\frac{\partial \Phi_o}{\partial x}\right)\Delta x + \frac{\partial}{\partial y}\left(\frac{A_y k_y k_{ro}}{\mu_o B_o}\frac{\partial \Phi_o}{\partial y}\right)\Delta y\right] + \left[\frac{\partial}{\partial x}\left(\frac{A_x k_x k_{rw}}{\mu_w B_w}\frac{\partial \Phi_w}{\partial x}\right)\Delta x + \frac{\partial}{\partial y}\left(\frac{A_y k_y k_{rw}}{\mu_w B_w}\frac{\partial \Phi_w}{\partial y}\right)\Delta y\right]$$
.........(9.46)

$$+\frac{B_o}{B_w}q_{o\,\text{STB/D}} + q_{w\,\text{STB/D}} = \frac{V_b}{5.615}\left[\left(\frac{S_o}{B_w}\frac{\partial \phi}{\partial p_o} - \frac{\phi S_o}{B_w B_o}\frac{\partial B_o}{\partial p_o}\right)\frac{\partial p_o}{\partial t} + \left(\frac{S_w}{B_w}\frac{\partial \phi}{\partial p_o}\frac{\partial p_o}{\partial t} - \frac{\phi S_w}{B_w^2}\frac{\partial B_w}{\partial p_w}\frac{\partial p_w}{\partial t}\right)\right].$$

To eliminate p_o from Eq. 9.46, we use the capillary pressure relationship between the oil and water phases:

$$p_o = P_{cow}(S_w) + p_w,\; \dotfill \;(9.47\text{a})$$

$$\frac{p_o}{\partial t} = \frac{P_{cow}(S_w)}{\partial t} + \frac{p_w}{\partial t}.\; \dotfill \;(9.47\text{b})$$

Similarly,

$$\frac{p_o}{\partial x} = \frac{P_{cow}(S_w)}{\partial x} + \frac{p_w}{\partial x},\; \dotfill \;(9.47\text{c})$$

and

$$\frac{p_o}{\partial y} = \frac{P_{cow}(S_w)}{\partial y} + \frac{p_w}{\partial y}.\; \dotfill \;(9.47\text{d})$$

When Eqs. 9.47b, 9.47c, and 9.47d are substituted into Eq. 9.46, we obtain the following final form of the combined oil and water equation with p_w as the only principal unknown:

$$\frac{B_o}{B_w}\left\{\begin{array}{l}\frac{\partial}{\partial x}\left[\frac{A_x k_x k_{ro}}{\mu_o B_o}\left(\frac{\partial p_w}{\partial x} + \frac{\partial P_{cow}}{\partial x} - \frac{1}{144}\frac{g}{g_c}\rho_o\frac{\partial G}{\partial x}\right)\right]\Delta x\\ +\frac{\partial}{\partial y}\left[\frac{A_y k_y k_{ro}}{\mu_o B_o}\left(\frac{\partial p_w}{\partial y} + \frac{\partial P_{cow}}{\partial y} - \frac{1}{144}\frac{g}{g_c}\rho_o\frac{\partial G}{\partial y}\right)\right]\Delta y\end{array}\right\} + \left\{\begin{array}{l}\frac{\partial}{\partial x}\left[\frac{A_x k_x k_{rw}}{\mu_w B_w}\left(\frac{\partial p_w}{\partial x} - \frac{1}{144}\frac{g}{g_c}\rho_w\frac{\partial G}{\partial x}\right)\right]\Delta x\\ +\frac{\partial}{\partial y}\left[\frac{A_y k_y k_{rw}}{\mu_w B_w}\left(\frac{\partial p_w}{\partial y} - \frac{1}{144}\frac{g}{g_c}\rho_w\frac{\partial G}{\partial y}\right)\right]\Delta y\end{array}\right\}$$
......(9.48)

$$+\frac{B_o}{B_w}q_{o\,\text{STB/D}} + q_{w\,\text{STB/D}} = \frac{V_b}{5.615}\left[\left(\frac{1}{B_w}\frac{\partial \phi}{\partial p_o} - \frac{\phi S_o}{B_w B_o}\frac{\partial B_o}{\partial p_o} - \frac{\phi S_w}{B_w^2}\frac{\partial B_w}{\partial p_w}\right)\frac{\partial p_w}{\partial t} + \left(\frac{1}{B_w}\frac{\partial \phi}{\partial p_o} - \frac{\phi S_o}{B_w B_o}\frac{\partial B_o}{\partial p_o}\right)\frac{\partial P_{cow}}{\partial t}\right].$$

Eq. 9.48 can be written in finite-difference form using the strongly implicit procedure (SIP) notation:

$$S_{i,j}^{n+1}p_{i,j-1}^{(k+1)} + W_{i,j}^{n+1}p_{i-1,j}^{(k+1)} + C_{i,j}^{(k)}p_{i,j}^{(k+1)} + E_{i,j}^{n+1}p_{i+1,j}^{(k+1)} + N_{i,j}^{n+1}p_{i,j+1}^{(k+1)} - Q_{i,j}^{n+1}\; \dotfill \;(9.49)$$

The coefficients of Eq. 9.49 can be identified as follows:

$$S_{i,j} = \left(\frac{B_o}{B_w}\right)_{i,j}\left.\frac{A_y k_y k_{ro}}{\mu_o B_o \Delta y}\right|_{i,j-\frac{1}{2}} + \left.\frac{A_y k_y k_{rw}}{\mu_w B_w \Delta y}\right|_{i,j-\frac{1}{2}},\; \dotfill \;(9.50\text{a})$$

$$W_{i,j} = \left(\frac{B_o}{B_w}\right)_{i,j}\left.\frac{A_y k_y k_{ro}}{\mu_o B_o \Delta y}\right|_{i-\frac{1}{2},j} + \left.\frac{A_y k_y k_{rw}}{\mu_w B_w \Delta y}\right|_{i-\frac{1}{2},j},\; \dotfill \;(9.50\text{b})$$

$$E_{i,j} = \left(\frac{B_o}{B_w}\right)_{i,j}\left.\frac{A_y k_y k_{ro}}{\mu_o B_o \Delta y}\right|_{i+\frac{1}{2},j} + \left.\frac{A_y k_y k_{rw}}{\mu_w B_w \Delta y}\right|_{i+\frac{1}{2},j},\; \dotfill \;(9.50\text{c})$$

$$N_{i,j} = \left(\frac{B_o}{B_w}\right)_{i,j}\left.\frac{A_y k_y k_{ro}}{\mu_o B_o \Delta y}\right|_{i,j+\frac{1}{2}} + \left.\frac{A_y k_y k_{rw}}{\mu_w B_w \Delta y}\right|_{i,j+\frac{1}{2}}.\; \dotfill \;(9.50\text{d})$$

Also, if we define $\Gamma_{i,j}$ as

$$\Gamma_{i,j} = \left[\frac{V_b}{5.615\Delta t}\left(\frac{1}{B_w}\frac{\partial \phi}{\partial p_o} - \frac{\phi S_o}{B_w B_o}\frac{\partial B_o}{\partial p_o} - \frac{\phi S_w}{B_w^2}\frac{\partial B_w}{\partial p_w}\right)\right]_{i,j},\; \dotfill \;(9.50\text{e})$$

then we define $C_{i,j}$ and $Q_{i,j}$ coefficients as

$$C_{i,j} = -\left(S_{i,j} + W_{i,j} + E_{i,j} + N_{i,j} + \Gamma_{i,j}\right),\; \dotfill \;(9.50\text{f})$$

$$Q_{i,j} = -\left(\frac{B_o}{B_w}\right)_{i,j} q_{oi,j} - q_{wi,j}$$

$$-\left(\frac{B_o}{B_w}\right)_{i,j}\left[\begin{array}{l} T_{oxi+\frac{1}{2},j}\left(P_{cowi+1,j} - P_{cowi,j}\right) - T_{oxi-\frac{1}{2},j}\left(P_{cowi,j} - P_{cowi-1,j}\right) \\ +T_{oyi,j+\frac{1}{2}}\left(P_{cowi,j+1} - P_{cowi,j}\right) - T_{oyi,j-\frac{1}{2}}\left(P_{cowi,j} - P_{cowi,j-1}\right) \end{array}\right]$$

$$+\left(\frac{B_o}{B_w}\right)_{i,j}\left[\begin{array}{l} T'_{oxi+\frac{1}{2},j}\left(G_{i+1,j} - G_{i,j}\right) - T'_{oxi-\frac{1}{2},j}\left(G_{i,j} - G_{i-1,j}\right) \\ +T'_{oyi,j+\frac{1}{2}}\left(G_{i,j+1} - G_{i,j}\right) - T'_{oyi,j-\frac{1}{2}}\left(G_{i,j} - G_{i,j-1}\right) \end{array}\right] \quad \dots\dots\dots\dots\dots\dots\dots (9.51)$$

$$+\left[\begin{array}{l} T'_{wxi+\frac{1}{2},j}\left(G_{i+1,j} - G_{i,j}\right) - T'_{wxi-\frac{1}{2},j}\left(G_{i,j} - G_{i-1,j}\right) \\ +T'_{wyi,j+\frac{1}{2}}\left(G_{i,j+1} - G_{i,j}\right) - T'_{wyi,j-\frac{1}{2}}\left(G_{i,j} - G_{i,j-1}\right) \end{array}\right]$$

$$-\left[\frac{V_b}{5.615\Delta t}\left(\frac{1}{B_w}\frac{\partial\phi}{\partial p_o} - \frac{\phi S_o}{B_w B_o}\frac{\partial B_o}{\partial p_o} - \frac{\phi S_w}{B_w^2}\frac{\partial B_w}{\partial p_w}\right)\right]p_{wi,j}^n$$

$$+\left[\frac{V_b}{5.615\Delta t}\left(\frac{1}{B_w}\frac{\partial\phi}{\partial p_o} - \frac{\phi S_o}{B_w B_o}\frac{\partial B_o}{\partial p_o}\right)\right]\left(P_{cowi,j}^n - P_{cowi,j}^{n-1}\right).$$

In Eq. 9.51, it is noted that the $\dfrac{\partial P_{cow}}{\partial t}$ derivative is approximated using backward difference in time. We can do this on the basis of the observation that the middle portion of the capillary pressure curve exhibits a linear relationship with almost a constant slope, so a backwards-linear extrapolation will work effectively. Now, if Eq. 9.49 is applied to the reservoir of Fig. 9.1, this time for the implicit-pressure solution part of the IMPES method, the coefficient matrix shown in **Fig. 9.5** will be obtained.

The system of equation that generated from Eq. 9.51 can be solved for $p_{i,j}^{n+1^{(k+1)}}$ at every block. After that, we are ready to move forward to calculate water-saturation values, $S_{wi,j}^{n+1^{(k+1)}}$ at every block. This is the explicit-calculation part of the IMPES method. After writing the finite-difference approximation of the water equation and rearranging it to solve for $S_{wi,j}^{n+1^{(k+1)}}$, we end up with the following equation:

$$S_{wi,j}^{n+1^{(k+1)}} = S_{wi,j}^n + \left(\frac{5.615\Delta t}{V_b\phi}\right)B_{wi,j}^{n+1^{(k+1)}}\left\{\begin{array}{l} T_{wxi+\frac{1}{2},j}^{n+1^{(k)}}\left(p_{wi+1,j}^{n+1^{(k+1)}} - p_{wi,j}^{n+1^{(k+1)}}\right) - T_{wxi-\frac{1}{2},j}^{n+1^{(k)}}\left(p_{wi,j}^{n+1^{(k+1)}} - p_{wi-1,j}^{n+1^{(k+1)}}\right) \\ +T_{wyi,j+\frac{1}{2}}^{n+1^{(k)}}\left(p_{wi,j+1}^{n+1^{(k+1)}} - p_{wi,j}^{n+1^{(k+1)}}\right) - T_{wyi,j-\frac{1}{2}}^{n+1^{(k)}}\left(p_{wi,j}^{n+1^{(k+1)}} - p_{wi,j-1}^{n+1^{(k+1)}}\right) \\ -\left[\begin{array}{l} T'_{wxi+\frac{1}{2},j}^{(n+1)^{(k)}}\left(G_{i+1,j} - G_{i,j}\right) - T'_{wxi-\frac{1}{2},j}^{(n+1)^{(k)}}\left(G_{i,j} - G_{i-1,j}\right) \\ +T'_{wxi,j+\frac{1}{2}}^{(n+1)^{(k)}}\left(G_{i,j+1} - G_{i,j}\right) - T'_{wxi,j-\frac{1}{2}}^{(n+1)^{(k)}}\left(G_{i,j} - G_{i,j-1}\right) \end{array}\right] + q_{wi,j}^{n+1} \text{ STB/D} \end{array}\right\} \quad \dots\dots\dots\dots (9.52)$$

$$+\left(\frac{S_w}{B_w}\frac{\partial B_w}{\partial p_w}\right)_{i,i}^{n+1^{(k)}}\left(p_{wi,j}^{n+1^{(k+1)}} - p_{wi,j}^n\right) - \left(\frac{S_w}{\phi}\frac{\partial\phi}{\partial p_o}\right)_{i,i}^{n+1^{(k)}}\left(p_{wi,j}^{n+1^{(k+1)}} - p_{wi,j}^n\right) - \left(\frac{S_w}{\phi}\frac{\partial\phi}{\partial p_o}\right)_{i,i}^{n+1^{(k)}}\left(P_{cowi,j}^n - P_{cowi,j}^{n-1}\right).$$

After solving for $S_{wi,j}^{n+1^{(k+1)}}$ at every block, then the two remaining principal unknowns can be solved to complete the iteration:

$$S_{oi,j}^{n+1^{(k+1)}} = 1 - S_{wi,j}^{n+1^{(k+1)}}, \quad \dots\dots\dots\dots\dots\dots\dots\dots\dots\dots\dots\dots\dots\dots\dots\dots\dots\dots\dots\dots\dots (9.53)$$

and

$$p_{oi,j}^{n+1^{(k+1)}} = p_{wi,j}^{n+1^{(k+1)}} + P_{cowi,j}\left[S_{wi,j}^{n+1^{(k+1)}}\right]. \quad \dots\dots\dots\dots\dots\dots\dots\dots\dots\dots\dots\dots\dots\dots\dots\dots\dots (9.54)$$

Coefficients of Unknowns

$$
\begin{array}{c}
\begin{array}{ccccccccc}
p^{n+1}_{w1,1} & p^{n+1}_{w2,1} & p^{n+1}_{w3,1} & p^{n+1}_{w1,2} & p^{n+1}_{w2,2} & p^{n+1}_{w3,2} & p^{n+1}_{w1,3} & p^{n+1}_{w2,3} & p^{n+1}_{w3,3}
\end{array}
\end{array}
$$

	$p^{n+1}_{w1,1}$	$p^{n+1}_{w2,1}$	$p^{n+1}_{w3,1}$	$p^{n+1}_{w1,2}$	$p^{n+1}_{w2,2}$	$p^{n+1}_{w3,2}$	$p^{n+1}_{w1,3}$	$p^{n+1}_{w2,3}$	$p^{n+1}_{w3,3}$			
Block (1,1)	$C_{1,1}$	$E_{1,1}$	0	$N_{1,1}$	0	0	0	0	0	$p^{n+1}_{W1,1}$		$Q_{1,1}$
Block (2,1)	$W_{2,1}$	$C_{2,1}$	$E_{2,1}$	0	$N_{2,1}$	0	0	0	0	$p^{n+1}_{W2,1}$		$Q_{2,1}$
Block (3,1)	0	$W_{3,1}$	$C_{3,1}$	0	0	$N_{3,1}$	0	0	0	$p^{n+1}_{W3,1}$		$Q_{3,1}$
Block (1,2)	$S_{1,2}$	0	0	$C_{1,2}$	$E_{1,2}$	0	$N_{1,2}$	0	0	$p^{n+1}_{W1,2}$		$Q_{1,2}$
Block (2,2)	0	$S_{2,2}$	0	$W_{2,2}$	$C_{2,2}$	$E_{2,2}$	0	$N_{2,2}$	0	$p^{n+1}_{W2,2}$	=	$Q_{2,2}$
Block (3,2)	0	0	$S_{3,2}$	0	$W_{3,2}$	$C_{3,2}$	0	0	$N_{3,2}$	$p^{n+1}_{W3,2}$		$Q_{3,2}$
Block (1,3)	0	0	0	$S_{1,3}$	0	0	$C_{1,3}$	$E_{1,3}$	0	$p^{n+1}_{W1,3}$		$Q_{1,3}$
Block (2,3)	0	0	0	0	$S_{2,3}$	0	$W_{2,3}$	$C_{2,3}$	$E_{2,3}$	$p^{n+1}_{W2,3}$		$Q_{2,3}$
Block (3,3)	0	0	0	0	0	$S_{3,3}$	0	$W_{3,3}$	$C_{3,3}$	$p^{n+1}_{W3,3}$		$Q_{3,3}$

(left label: Equations)

Fig. 9.5—The coefficient matrix, unknown vector, and the right-hand-side vector of the implicit part of the IMPES procedure when implemented to the reservoir of Fig. 9.1.

Once the convergence criteria are satisfied for each principal unknown at every block, then timestep calculations are completed and material-balance checks and residual checks on each phase are conducted.

In **Fig. 9.6,** we summarize the basic steps of the SS method for the three-phase flow (oil/water/gas) problems.

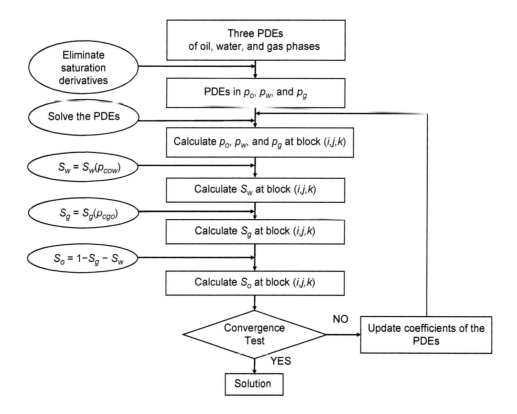

Fig. 9.6—Flow chart summarizing the basic steps of simultaneous solution process for a three-phase system.

9.3 Problems and Solutions

Problem 9.3.1 (Ertekin et al. 2001)

Apply the general conservative-expansion formula given by

$$\Delta_t (UVXY) = (VXY)^n \Delta_t U + U^{n+1}(XY)^n \Delta_t V + (UV)^{n+1} Y^n \Delta_t X + (UVX)^{n+1} \Delta_t Y \dots \dots (9.55)$$

to obtain

$$\Delta_t\left[\left(\frac{\phi R_s}{B_o}\right)S_o\right] = \left\{\left[\frac{\phi'}{B_o^n} + \phi^{n+1}\left(\frac{1}{B_o}\right)'\right]R_s^n + \left(\frac{\phi}{B_o}\right)^{n+1}R_s'\right\} \times S_o^n \Delta_t p_o + \left(\frac{\phi}{B_o}\right)^{n+1}R_s^{n+1}\Delta_t S_o, \dots \dots (9.56)$$

where superscript $n+1$ represents the current time level, and n represents the previous time level. Variables with a superscript prime (for example ϕ') are the derivatives with respect to p_o. *Hint:* let $U = \phi$, $V = \frac{1}{B_o}$, $X = R_s$, and $Y = S_o$.

Solution to Problem 9.3.1

Applying the differential operator to the variables, one obtains

$$\Delta_t U = \Delta_t \phi = \phi'\Delta_t p_o. \dots \dots (9.57)$$

Note that the chain rule is also employed in Eq. 9.57 by assuming $\phi = \phi(p_o)$. Similarly, $\Delta_t V = \Delta_t\left(\frac{1}{B_o}\right) = \left(\frac{1}{B_o}\right)'\Delta_t p_o$, $\Delta_t X = \Delta_t R_s = R_s'\Delta_t p_o$, and $\Delta_t Y = \Delta_t S_o$. Substituting into the original conservative expansion,

$$\Delta_t\left[\left(\frac{\phi R_s}{B_o}\right)S_o\right] = \left[\left(\frac{R_s}{B_o}\right)S_o\right]^n \phi'\Delta_t p_o + \phi^{n+1}\left(R_s S_o\right)^n\left(\frac{1}{B_o}\right)'\Delta_t p_o + \left(\frac{\phi}{B_o}\right)^{n+1}S_o^n R_s'\Delta_t p_o + \left(\frac{\phi}{B_o}\right)^{n+1}R_s^{n+1}\Delta_t S_o. \dots \dots (9.58)$$

Rearranging the previous equation, we have

$$\Delta_t\left[\left(\frac{\phi R_s}{B_o}\right)S_o\right] = \left\{\left[\frac{\phi'}{B_o^n} + \phi^{n+1}\left(\frac{1}{B_o}\right)'\right]R_s^n + \left(\frac{\phi}{B_o}\right)^{n+1}R_s'\right\} \times S_o^n \Delta_t p_o + \left(\frac{\phi}{B_o}\right)^{n+1}R_s^{n+1}\Delta_t S_o. \dots \dots (9.59)$$

Problem 9.3.2 (Ertekin et al. 2001)

Demonstrate that the 1D finite-difference equation for gas, shown next, can be derived by combining mass balance, mass flux (product of density and Darcy velocity), and the external-forcing function (sink/source):

$$\Delta_x\left[\beta_c \frac{k_x A_x}{\Delta x}\frac{k_{rg}}{\mu_g B_g}\left(\Delta_x p_g - \gamma_g \Delta_x G\right)\right]_i + \Delta_x\left[\beta_c \frac{k_x A_x}{\Delta x}\frac{k_{ro}R_s}{\mu_o B_o}\left(\Delta_x p_o - \gamma_o \Delta_x G\right)\right]_i$$

$$= \frac{V_{b_i}}{\alpha_c \Delta t}\left\{\left[\frac{\phi S_g}{B_g} + \frac{\phi R_s\left(1 - S_g\right)}{B_o}\right]_i^{n+1} - \left[\frac{\phi S_g}{B_g} + \frac{\phi R_s\left(1 - S_g\right)}{B_o}\right]_i^n\right\} - \left(q_{fgsc_i} + R_s q_{osc_i}\right) \dots \dots (9.60)$$

Solution to Problem 9.3.2

The mass-balance equation is as follows:

$$-\left[\left(\dot{m}_{cx}A_x\right)_{x-\frac{\Delta x}{2}} - \left(\dot{m}_{cx}A_x\right)_{x+\frac{\Delta x}{2}}\right]\Delta t + \left(q_{m_{t_c}} + q_{m_c}\right)\Delta t = V_b\left[\left(m_{vc}\right)_{t+\Delta t} - \left(m_{vc}\right)_t\right]. \dots \dots (9.61)$$

The mass-flux equations are as follows:

$$\dot{m}_{cx} = \begin{cases} \alpha_c \rho_c u_{cx}, \text{when } c = o, w, \text{ and free gas} \\ \left(\rho_{gsc}\dfrac{R_{so}}{B_o}\right)u_{ox}, \text{ when } c = \text{solution gas}, \end{cases} \dots \dots (9.62)$$

$$\dot{m}_{vc} = \begin{cases} \phi \rho_c S_c \text{, when } c = o, w, \text{ and } g \\ \dfrac{\phi}{\alpha_c}\left(\rho_{gsc}\dfrac{R_{so}}{B_o}\right)S_o \text{, when } c = \text{solution gas.} \end{cases} \quad\dotfill\quad (9.63)$$

The external sink/source equations are as follows:

$$q_{mc} = \begin{cases} \alpha_c \rho_c q_c \text{, when } c = o, w, \text{ and free gas} \\ \dot{m}_{cx} = \left(\rho_{gsc}\dfrac{R_{so}}{B_o}\right)q_o \text{, when } c = \text{solution gas,} \end{cases} \quad\dotfill\quad (9.64)$$

$$q_{mt_c} = \begin{cases} 0 \text{, when } c = o, w \\ q_{mt_{sg}} \text{, when } c = \text{solution gas} \\ -q_{mt_{sg}} \text{, when } c = \text{free gas.} \end{cases} \quad\dotfill\quad (9.65)$$

Substituting the mass flux and sink/source equations into the mass-balance equation, one obtains

$$\frac{\partial}{\partial x}\left[\beta_c k_x A_x \frac{k_{rg}}{\mu_g B_g}\left(\frac{\partial p_g}{\partial x} - \gamma_g \frac{\partial G}{\partial x}\right) + \beta_c k_x A_x \frac{k_{ro}R_{so}}{\mu_o B_o}\left(\frac{\partial p_o}{\partial x} - \gamma_o \frac{\partial G}{\partial x}\right)\right] = \frac{V_b}{\alpha_c \Delta t}\frac{\partial}{\partial t}\left[\frac{\phi S_g}{B_g} + \frac{\phi R_{so}\left(1 - S_g\right)}{B_o}\right]$$
$$- \left(q_{fgsc} + R_{so}q_{osc}\right). \quad\dotfill\quad (9.66)$$

Using a finite-difference analog form, one can write

$$\Delta_x\left[\beta_c \frac{k_x A_x}{\Delta x}\frac{k_{rg}}{\mu_g B_g}\left(\Delta_x p_g - \gamma_g \Delta_x G\right)\right]_i + \Delta_x\left[\beta_c \frac{k_x A_x}{\Delta x}\frac{k_{ro}R_{so}}{\mu_o B_o}\left(\Delta_x p_o - \gamma_o \Delta_x G\right)\right]_i$$
$$= \frac{V_{b_i}}{\alpha_c \Delta t}\left\{\left[\frac{\phi S_g}{B_g} + \frac{\phi R_{so}\left(1 - S_g\right)}{B_o}\right]_i^{n+1} - \left[\frac{\phi S_g}{B_g} + \frac{\phi R_{so}\left(1 - S_g\right)}{B_o}\right]_i^{n}\right\} - \left(q_{fgsc_i} + R_{so}q_{osc_i}\right). \quad\dotfill\quad (9.67)$$

Note that unit conversion constant $\beta_c = 1.127$, and k is expressed in darcies.

Problem 9.3.3 (Ertekin et al. 2001)

Consider the oil reservoir described by Odeh (1981) and shown in **Fig. 9.7.** It is one-quarter of a five-spot pattern (with no-flow boundaries). The reservoir consists of three layers showing vertical heterogeneity. It is discretized into 10×10×3 equal gridblocks with a block-centered grid. **Table 9.2** shows gridblock properties, dimensions, and initial conditions. Initially, the reservoir is undersaturated with $R_{si} = 1,270$ scf/STB. Initial fluid saturations are $S_{oi} = 0.88$ and $S_{wi} = S_{iw} = 0.12$. **Tables 9.3 and 9.4** give pressure-volume-temperature (PVT) data and relative permeability data, respectively.

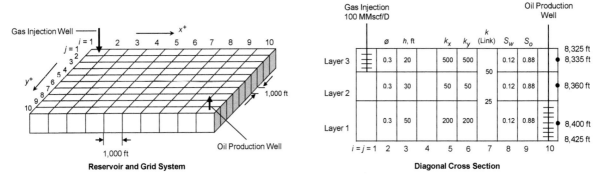

Fig. 9.7—Reservoir system of Problem 9.3.3 (redrawn after Odeh 1981).

Layer	Δx (ft)	Δy (ft)	Δz (ft)	k_x (md)	k_y (md)	k_z (md)	G (ft)	ϕ	p_i (psia)
3 (top)	1,000	1,000	20	500	500	50	8,335	0.30	4,783.5
2	1,000	1,000	30	50	50	50	8,360	0.30	4,789.8
1 (bottom)	1,000	1,000	50	200	200	19.23	8,400	0.30	4,800.0

Table 9.2—Gridblock properties, dimensions, and initial conditions (Problem 9.3.3).

p (psia)	R_s (scf/STB)	B_o (res bbl/STB)	μ_o (cp)	ρ_o (lbm/ft^3)	B_g (res bbl/scf)	μ_g (cp)	ρ_g (lbm/ft^3)
3,014.7	930	1.5650	0.5940	37.781	0.00108	0.0228	9.984
4,014.7	1,270	1.6950	0.5100	37.046	0.00081	0.0268	13.295
5,014.7	1,618	1.8270	0.4490	36.424	0.00065	0.0309	16.614

Table 9.3—Fluid PVT data in pressure range of interest (Problem 9.3.3).

S_g	k_{rg}	k_{ro}	P_{cgo}
0.000	0.0000	1.0000	0.0
0.001	0.0000	1.0000	0.0
0.020	0.0000	0.9970	0.0
0.050	0.0050	0.9800	0.0
0.120	0.0250	0.7000	0.0
0.200	0.0750	0.3500	0.0
0.250	0.1250	0.2000	0.0
0.300	0.1900	0.0900	0.0
0.400	0.4100	0.0210	0.0
0.450	0.6000	0.0100	0.0
0.500	0.7200	0.0010	0.0
0.600	0.8700	0.0001	0.0
0.700	0.9400	0.0000	0.0
0.850	1.0000	0.0000	0.0

Table 9.4—Two-phase oil/gas relative permeability and capillary pressure data (Problem 9.3.3).

As shown in Fig. 9.7, gas is injected into the upper-corner Gridblock (1,1,3) at a rate of 100 MMscf/D, and oil is produced from the lower-corner Gridblock (10,10,1) at a rate of 20 MSTB/D. As long as $p_{wf} \geq 1,000$ psia, the production well will be operated with constant-bottomhole pressure of $p_{wf} = 1,000$ psia until $t = 10$ years. Table 9.5 displays the oil pressure and fluid saturations for production wellblock (10,10,1) and its neighboring gridblocks, Gridblocks (9,10,1), (10,9,1), (10,10,2), (8,10,1), and (10,10,3) at $t = 10$ years.

Given the 3D oil reservoir described in this problem, calculate the constant part of the transmissibility (G), saturation-dependent (f_s), and pressure-dependent (f_g) components for the $T_{gx_{9\frac{1}{2},10,1}}$ and $T_{gz_{10,10,1\frac{1}{2}}}$ using appropriate weighting techniques.

Gridblock (i,j,k)	p_o (psia)	S_o	S_g	S_w
(10,10,1)	3,315.7	0.6406	0.2391	0.1203
(9,10,1)	3,533.9	0.8520	0.0267	0.1212
(10,9,1)	3,533.9	0.8520	0.0267	0.1212
(10,10,2)	3,325.4	0.5258	0.3541	0.1200
(8,10,1)	3,661.5	0.8528	0.0259	0.1213
(10,10,3)	3,325.3	0.3963	0.4837	0.1200

Table 9.5—Oil pressure and fluid saturations at the wellbock and its neighboring gridblocks at t =10 years (Problem 9.3.3).

Solution to Problem 9.3.3

The constant part of the transmissibility G term is the same for all phases:

$$G_{9\frac{1}{2},10,1} = \beta_c \left(\frac{k_x A_x}{\Delta x}\right)_{9\frac{1}{2},10,1} = \beta_c \frac{k_x A_x}{\Delta x} = \frac{(1.127)(0.2)(1000)(50)}{1000} = 11.27 \frac{\text{cp}-\text{res bbl}}{\text{D}-\text{psi}}, \quad \dots\dots\dots\dots \quad (9.68)$$

$$G_{10,10,1\frac{1}{2}} = \beta_c \left(\frac{k_z A_z}{\Delta z}\right)_{10,10,1\frac{1}{2}} = 2\beta_c A_z \left(\frac{k_{z_{10,10,1}} k_{z_{10,10,2}}}{k_{z_{10,10,1}} z_{10,10,1} + k_{z_{10,10,2}} \Delta z_{10,10,2}}\right) \quad \dots\dots\dots\dots\dots\dots\dots\dots \quad (9.69)$$

$$= 2\frac{(1.127)(10^6)(0.01923)(0.05)}{(0.01923)(30)+(0.05)(50)} = 704.35 \frac{\text{cp}-\text{res bbl}}{\text{D}-\text{psi}}.$$

At $t = 10$ years, using the data provided in Tables 9.3 and 9.4, one can calculate gas viscosity (μ_g), formation volume factor (B_g), density (ρ_g), solution gas/oil ratio (R_s), gas saturation (S_g), and relative permeability (k_{rg}) by way of linear interpolation, shown in **Table 9.6.**

Gridblock (i,j,k)	p_g (psia)	μ_g (cp)	B_g (res bbl/scf)	ρ_g (lbm/ft³)	R_s (scf/STB)	S_g	k_{rg}
10,10,1	3,315.7	0.024004	0.000999	10.980611	1032.340	0.2391	0.114100
9,10,1	3,533.9	0.024877	0.000940	11.703071	1106.528	0.0267	0.001117
8,10,1	3,661.5	0.025387	0.000905	12.125555	1149.912	0.0259	0.000983
10,10,2	3,325.4	0.024043	0.000996	11.012728	1035.638	0.3541	0.309020
10,10,3	3,325.3	0.024042	0.000996	11.012397	1035.604	0.4837	0.680880

Table 9.6—Calculation of gas viscosity (μ_g), formation volume factor (B_g), density (ρ_g), solution gas/oil ratio (R_s), gas saturation (S_g), and relative permeability (k_{rg}) by way of linear interpolation from the Solution to Problem 9.3.3.

For upstream weighting, determine the upstream gridblock for the gas phase, between Gridblocks (10,10,1) and (9,10,1) of the reservoir shown in Fig.9.7:

$$\Delta\Phi_{g_{9\frac{1}{2},10,1}} = \left(p_{g_{9,10,1}} - p_{g_{10,10,1}}\right) - \gamma_{g_{9\frac{1}{2},10,1}} \left(G_{9,10,1} - G_{10,10,1}\right) = (3,533.9 - 3,315.7)$$

$$-\frac{11.703071+10.980611}{2(144)}(8,400-8,400) = 218.2 \text{ psi.} \quad \dots\dots\dots\dots\dots\dots \quad (9.70)$$

For upstream weighting, determine the upstream gridblock for the gas phase, between Gridblocks (10,10,1) and (10,10,2):

$$\Delta\Phi_{g_{10,10,1\frac{1}{2}}} = \left(p_{g_{10,10,2}} - p_{g_{10,10,1}}\right) - \gamma_{g_{10,10,1\frac{1}{2}}} \left(G_{10,10,2} - G_{10,10,1}\right) = (3,325.4 - 3,315.7)$$

$$-\frac{11.012728+10.980611}{2(144)}(8,360-8,400) = 12.75 \text{ psi.} \quad \dots\dots\dots\dots\dots\dots \quad (9.71)$$

Hence, for upstream weighting,

$$f_{p_{9\frac{1}{2},10,1}} = \left(\frac{1}{\mu_g B_g}\right)_{9,10,1} = 42,772.3 \text{ STB}/\text{res bbl}-\text{cp}, \quad \dots\dots\dots\dots\dots\dots \quad (9.72)$$

$$f_{p_{10,10,1\frac{1}{2}}} = \left(\frac{1}{\mu_g B_g}\right)_{10,10,2} = 41,754.9 \text{ STB}/\text{res bbl}-\text{cp}. \quad \dots\dots\dots\dots\dots\dots \quad (9.73)$$

For midpoint weighting,

$$f_{p_{9\frac{1}{2},10,1}} = \frac{1}{2}\left[\left(\frac{1}{\mu_g B_g}\right)_{10,10,1} + \left(\frac{1}{\mu_g B_g}\right)_{9,10,1}\right] = 42,242.5 \text{ STB}/\text{res bbl}-\text{cp}, \quad \dots\dots\dots\dots\dots\dots \quad (9.74)$$

$$f_{p_{10,10,1\frac{1}{2}}} = \frac{1}{2}\left[\left(\frac{1}{\mu_g B_g}\right)_{10,10,1} + \left(\frac{1}{\mu_g B_g}\right)_{10,10,2}\right] = 41,733.8 \text{ STB/res bbl} - \text{cp.} \dots\dots\dots\dots\dots\dots\dots \tag{9.75}$$

Or, using average pressures,

$$f_{p_{9\frac{1}{2},10,1}} = \left[\frac{1}{\mu_g\left(\overline{p}_g\right)B_o\left(\overline{p}_g\right)}\right] = 42,212.9 \text{ STB/res bbl} - \text{cp,} \dots\dots\dots\dots\dots\dots\dots\dots\dots \tag{9.76}$$

$$f_{p_{10,10,1\frac{1}{2}}} = \left[\frac{1}{\mu_g\left(\overline{p}_g\right)B_o\left(\overline{p}_g\right)}\right] = 41,733.7 \text{ STB/res bbl} - \text{cp.} \dots\dots\dots\dots\dots\dots\dots\dots\dots \tag{9.77}$$

Or,

$$f_{p_{9\frac{1}{2},10,1}} = \left[\frac{1}{\overline{\mu}_g \overline{B}_g}\right] = 42,212.9 \text{ STB/res bbl} - \text{cp,} \dots\dots\dots\dots\dots\dots\dots\dots\dots\dots\dots \tag{9.78}$$

$$f_{p_{10,10,1\frac{1}{2}}} = \left[\frac{1}{\overline{\mu}_g \overline{B}_g}\right] = 41,733.7 \text{ STB/res bbl} - \text{cp.} \dots\dots\dots\dots\dots\dots\dots\dots\dots\dots\dots \tag{9.79}$$

To calculate strongly nonlinear relative permeability terms using single-point upstream weighting,

$$f_{s_{9\frac{1}{2},10,1}} = \left(k_{rg}\right)_{9,10,1} = 0.001117, \dots\dots\dots\dots\dots\dots\dots\dots\dots\dots\dots\dots\dots\dots \tag{9.80}$$

$$f_{s_{10,10,1\frac{1}{2}}} = \left(k_{rg}\right)_{10,10,2} = 0.30902. \dots\dots\dots\dots\dots\dots\dots\dots\dots\dots\dots\dots\dots\dots \tag{9.81}$$

Similarly, using midpoint weighting,

$$f_{s_{9\frac{1}{2},10,1}} = \frac{1}{2}\left[\left(k_{rg}\right)_{10,10,1} + \left(k_{rg}\right)_{9,10,1}\right] = 0.057608, \dots\dots\dots\dots\dots\dots\dots\dots\dots\dots \tag{9.82}$$

$$f_{s_{10,10,1\frac{1}{2}}} = \frac{1}{2}\left[\left(k_{rg}\right)_{10,10,1} + \left(k_{rg}\right)_{10,10,2}\right] = 0.21156. \dots\dots\dots\dots\dots\dots\dots\dots\dots\dots \tag{9.83}$$

And finally, for two-point weighting,

$$\begin{aligned} f_{s_{9\frac{1}{2},10,1}} &= \frac{\delta_{9,10,1}^- + \delta_{9,10,1}^+ + \delta_{8,10,1}^-}{\delta_{9,10,1}^+ + \delta_{8,10,1}^-}\left(k_{rg}\right)_{9,10,1} - \frac{\delta_{9,10,1}^-}{\delta_{9,10,1}^+ + \delta_{8,10,1}^-}\left(k_{rg}\right)_{8,10,1} \\ &= \frac{500+500+500}{500+500}0.001117 - \frac{500}{500+500}0.000983 = 0.001184, \end{aligned} \dots\dots\dots\dots\dots \tag{9.84}$$

$$\begin{aligned} f_{s_{10,10,1\frac{1}{2}}} &= \frac{d_{10,10,2}^- + d_{10,10,2}^+ + d_{10,10,3}^-}{d_{10,10,2}^+ + d_{10,10,3}^-}\left(k_{rg}\right)_{10,10,2} - \frac{d_{10,10,2}^-}{d_{10,10,2}^+ + d_{10,10,3}^-}\left(k_{rg}\right)_{10,10,3} \\ &= \frac{15+15+10}{15+10}0.309020 - \frac{15}{15+10}0.680880 = 0.085904. \end{aligned} \dots\dots\dots\dots\dots \tag{9.85}$$

Problem 9.3.4 (Ertekin et al. 2001)

Given the 3D oil reservoir described in Problem 9.3.3, calculate the interblock-gas flow rate components in the x-, y-, and z-directions between production Gridblock (10,10,1) in Fig. 9.7 and its immediate neighboring gridblocks at 10 years. Use one-point upstream weighting for the approximation of the interblock transmissibilities.

Solution to Problem 9.3.4

For production Gridblock (10,10,1), there are three no-flow boundaries in the negative direction of the z axis and the positive directions of the x and y axes. Applying single-point upstream weighting,

$$T_{gx_{9\frac{1}{2},10,1}}^{n+1} = G_{9\frac{1}{2},10,1}f_{p_{9\frac{1}{2},10,1}}f_{s_{9\frac{1}{2},10,1}} = 11.27 \times 42,772.3 \times 0.001117 = 538.443 \text{ scf/D-psi,} \dots\dots\dots \tag{9.86}$$

$$T^{n+1}_{gx_{10,10,1\frac{1}{2}}} = G_{10,10,1\frac{1}{2}} f_{p_{10,10,1\frac{1}{2}}} f_{s_{10,10,1\frac{1}{2}}} = 704.35 \times 41,754.9 \times 0.30902 = 9.088 \times 10^6 \text{scf/D-psi}, \quad\quad\quad\quad\quad\quad\quad\quad (9.87)$$

$$R^{n+1}_{sx_{9\frac{1}{2},10,1}} = R_{s_{9,10,1}} = 1,106.528 \text{ scf/STB}, \quad\quad\quad\quad\quad\quad\quad\quad\quad\quad\quad\quad\quad\quad\quad\quad\quad\quad\quad\quad (9.88)$$

$$R^{n+1}_{sx_{10,10,1\frac{1}{2}}} = R_{s_{10,10,2}} = 1,035.638 \text{ scf/STB}, \quad\quad\quad\quad\quad\quad\quad\quad\quad\quad\quad\quad\quad\quad\quad\quad\quad\quad (9.89)$$

$$q^{n+1}_{gx_{9\frac{1}{2},10,1}} = T^{n+1}_{gx_{9\frac{1}{2},10,1}} \left(p_{g_{9,10,1}} - p_{g_{10,10,1}} \right) + R^{n+1}_{sx_{9\frac{1}{2},10,1}} q^{n+1}_{ox_{9\frac{1}{2},10,1}} = 538.443(3,533.9 - 3,315.7) + 1,106.528 \times 2,718.3$$
$$= 3.125 \times 10^6 \text{scf/D}, \quad\quad\quad\quad\quad\quad\quad (9.90)$$

$$q^{n+1}_{gy_{9\frac{1}{2},10,1}} = q^{n+1}_{gx_{9\frac{1}{2},10,1}}, \quad\quad\quad\quad\quad\quad\quad\quad\quad\quad\quad\quad\quad\quad\quad\quad\quad\quad\quad\quad\quad\quad\quad\quad\quad\quad (9.91)$$

$$q^{n+1}_{gx_{10,10,1\frac{1}{2}}} = T^{n+1}_{gx_{10,10,1\frac{1}{2}}} \left(p_{g_{10,10,2}} - p_{g_{10,10,1}} \right) + R^{n+1}_{sx_{10,10,1\frac{1}{2}}} q^{n+1}_{ox_{10,10,1\frac{1}{2}}} = 9.088 \times 10^6 \left(3,325.4 - 3,315.7 \right)$$
$$+ 1,035.638 \times 820.12 = 8.9 \times 10^7 \text{ scf / D}. \quad\quad\quad\quad\quad\quad\quad\quad\quad\quad (9.92)$$

Problem 9.3.5 (Ertekin et al. 2001)

Write the flow equations for a solution-gas-drive reservoir with immobile water phase. The producing formation is relatively thin and has a large areal extension with several wells drilled and on production. Also, write the finite-difference equations for Gridblock n for (i,j) in the 2D x-y space.

Solution to Problem 9.3.5

Oil-flow equation for a solution-gas-drive reservoir with immobile water phase:

$$\frac{\partial}{\partial x}\left[k_x A_x \frac{k_{ro}}{\mu_o B_o} \left(\frac{\partial p_o}{\partial x} - \gamma_o \frac{\partial G}{\partial x} \right) \right] \Delta x + \frac{\partial}{\partial y}\left[k_y A_y \frac{k_{ro}}{\mu_o B_o} \left(\frac{\partial p_o}{\partial y} - \gamma_o \frac{\partial G}{\partial y} \right) \right] \Delta y = \frac{V_{b_{i,j}}}{\alpha_c \Delta t} \frac{\partial}{\partial t}\left[\frac{\phi(1-S_g)}{B_o} \right] - q_{osc}. \quad\quad (9.93)$$

Gas-flow equation for a solution-gas-drive reservoir with immobile water phase:

$$\frac{\partial}{\partial x}\left[\begin{matrix} k_x A_x \frac{k_{rg}}{\mu_g B_g} \left(\frac{\partial p_o}{\partial x} + \frac{\partial P_{cgo}}{\partial x} \right) \\ + k_x A_x \frac{k_{ro} R_s}{\mu_o B_o} \left(\frac{\partial p_o}{\partial x} - \gamma_g \frac{\partial G}{\partial x} \right) \end{matrix} \right] \Delta x + \frac{\partial}{\partial y}\left[\begin{matrix} k_y A_y \frac{k_{rg}}{\mu_g B_g} \left(\frac{\partial p_o}{\partial y} + \frac{\partial P_{cgo}}{\partial y} \right) \\ + k_y A_y \frac{k_{ro} R_s}{\mu_o B_o} \left(\frac{\partial p_o}{\partial y} - \gamma_o \frac{\partial G}{\partial y} \right) \end{matrix} \right] \quad\quad\quad\quad (9.94)$$
$$\Delta y = \frac{V_{b_{i,j}}}{\alpha_c \Delta t} \frac{\partial}{\partial t}\left[\frac{\phi S_g}{B_o} + \frac{\phi R_s (1-S_g)}{B_o} \right] - \left(q_{gsc} + R_s q_{osc} \right).$$

The finite-difference representation of the oil equation will be

$$\Delta_x\left[\frac{k_x A_x}{\Delta x} \frac{k_{ro}}{\mu_o B_o} \left(\Delta_x p_o - \gamma_o \Delta_x G \right) \right]_{i,j} + \Delta_y\left[\frac{k_y A_y}{\Delta y} \frac{k_{ro}}{\mu_o B_o} \left(\Delta_y p_o - \gamma_o \Delta_y G \right) \right]_{i,j}$$
$$= \frac{V_{b_{i,j}}}{\alpha_c \Delta t}\left\{ \left[\frac{\phi(1-S_g)}{B_o} \right]^{n+1}_{i,j} - \left[\frac{\phi(1-S_g)}{B_o} \right]^{n}_{i,j} \right\} - q_{osc_{i,j}}, \quad\quad\quad\quad\quad\quad\quad\quad\quad\quad (9.95)$$

resulting in an expanded form:

$$\frac{A_x k_x}{\Delta x}\left.\frac{k_{ro}}{\mu_o B_o}\right|_{i+\frac{1}{2},j}^{n+1}\left(p_{o_{i+1,j}}^{n+1}-p_{o_{i,j}}^{n+1}\right)-\frac{A_x k_x}{\Delta x}\left.\frac{k_{ro}}{\mu_o B_o}\right|_{i-\frac{1}{2},j}^{n+1}\left(p_{o_{i,j}}^{n+1}-p_{o_{i-1,j}}^{n+1}\right)$$

$$+\frac{A_y k_y}{\Delta y}\left.\frac{k_{ro}}{\mu_o B_o}\right|_{i,j+\frac{1}{2}}^{n+1}\left(p_{o_{i,j+1}}^{n+1}-p_{o_{i,j}}^{n+1}\right)-\frac{A_y k_y}{\Delta y}\left.\frac{k_{ro}}{\mu_o B_o}\right|_{i,j-\frac{1}{2}}^{n+1}\left(p_{o_{i,j}}^{n+1}-p_{o_{i,j-1}}^{n+1}\right)$$

$$-\left[\begin{array}{l}\gamma_o\dfrac{A_x k_x}{\Delta x}\left.\dfrac{k_{ro}}{\mu_o B_o}\right|_{i+\frac{1}{2},j}^{n+1}\left(G_{i+1,j}-G_{i,j}\right)-\gamma_o\dfrac{A_x k_x}{\Delta x}\left.\dfrac{k_{ro}}{\mu_o B_o}\right|_{i-\frac{1}{2},j}^{n+1}\left(G_{i,j}-G_{i-1,j}\right)\\[2ex]+\gamma_o\dfrac{A_y k_y}{\Delta y}\left.\dfrac{k_{ro}}{\mu_o B_o}\right|_{i,j+\frac{1}{2}}^{n+1}\left(G_{i,j+1}-G_{i,j}\right)-\gamma_o\dfrac{A_y k_y}{\Delta y}\left.\dfrac{k_{ro}}{\mu_o B_o}\right|_{i,j-\frac{1}{2}}^{n+1}\left(G_{i,j}-G_{i,j-1}\right)\end{array}\right]\ \dots\dots\dots\ (9.96)$$

$$=\frac{V_{b_{i,j}}}{\Delta t}\left\{\left[\frac{\phi\left(1-S_g\right)}{B_o}\right]_{i,j}^{n+1}-\left[\frac{\phi\left(1-S_g\right)}{B_o}\right]_{i,j}^{n}\right\}-q_{osc_{i,j}}.$$

Gas-flow equation in finite-difference form:

$$\Delta_x\left[\frac{k_x A_x}{\Delta x}\frac{k_{rg}}{\mu_g B_g}\left(\Delta_x p_o+\Delta_x P_{cgo}\right)\right]_{i,j}+\Delta_x\left[\frac{k_x A_x}{\Delta x}\frac{k_{ro}R_s}{\mu_o B_o}\left(\Delta_x p_o-\gamma_o\Delta_x G\right)\right]_{i,j}$$

$$+\Delta_y\left[\frac{k_y A_y}{\Delta y}\frac{k_{rg}}{\mu_g B_g}\left(\Delta_y p_o+\Delta_y P_{cgo}\right)\right]_{i,j}+\Delta_y\left[\frac{k_y A_y}{\Delta y}\frac{k_{ro}R_s}{\mu_o B_o}\left(\Delta_y p_o-\gamma_o\Delta_y G\right)\right]_{i,j}\ \dots\dots\dots\dots\ (9.97)$$

$$=\frac{V_{b_{i,j}}}{\alpha_c\Delta t}\left\{\left[\frac{\phi S_g}{B_g}+\frac{\phi R_s\left(1-S_g\right)}{B_o}\right]_{i,j}^{n+1}-\left[\frac{\phi S_g}{B_g}+\frac{\phi R_s\left(1-S_g\right)}{B_o}\right]_{i,j}^{n}\right\}-\left(q_{fgsc_{i,j}}+R_s q_{osc_{i,j}}\right).$$

Recall that the spatial derivative of capillary pressure needs be written in terms of gas saturation; for instance, along the *x*-direction using the chain rule,

$$\Delta_x\left(P_{cgo}\right)=\frac{\partial P_{cgo}}{\partial x}=\frac{\partial P_{cgo}}{\partial S_g}\frac{\partial S_g}{\partial x};\ \dots\dots\dots\dots\dots\dots\dots\dots\dots\dots\dots\dots\dots\dots\dots\dots\ (9.98)$$

therefore, the equation for the gas phase becomes (in finite-difference form)

$$\left(R_s\frac{A_x k_x}{\Delta x}\left.\frac{k_{ro}}{\mu_o B_o}\right|_{i+\frac{1}{2},j}^{n+1}+\frac{A_x k_x}{\Delta x}\left.\frac{k_{rg}}{\mu_g B_g}\right|_{i+\frac{1}{2},j}^{n+1}\right)\left(p_{o_{i+1,j}}^{n+1}-p_{o_{i,j}}^{n+1}\right)-\left(R_s\frac{A_x k_x}{\Delta x}\left.\frac{k_{ro}}{\mu_o B_o}\right|_{i-\frac{1}{2},j}^{n+1}+\frac{A_x k_x}{\Delta x}\left.\frac{k_{rg}}{\mu_g B_g}\right|_{i-\frac{1}{2},j}^{n+1}\right)\left(p_{o_{i,j}}^{n+1}-p_{o_{i-1,j}}^{n+1}\right)$$

$$+\left(R_s\frac{A_x k_x}{\Delta x}\left.\frac{k_{ro}}{\mu_o B_o}\right|_{i,j+\frac{1}{2}}^{n+1}+\frac{A_x k_x}{\Delta x}\left.\frac{k_{rg}}{\mu_g B_g}\right|_{i,j+\frac{1}{2}}^{n+1}\right)\left(p_{o_{i,j+1}}^{n+1}-p_{o_{i,j}}^{n+1}\right)-\left(R_s\frac{A_x k_x}{\Delta x}\left.\frac{k_{ro}}{\mu_o B_o}\right|_{i,j-\frac{1}{2}}^{n+1}+\frac{A_x k_x}{\Delta x}\left.\frac{k_{rg}}{\mu_g B_g}\right|_{i,j-\frac{1}{2}}^{n+1}\right)\left(p_{o_{i,j}}^{n+1}-p_{o_{i,j-1}}^{n+1}\right)$$

$$\left(\frac{A_x k_x}{\Delta x}\left.\frac{k_{rg}}{\mu_g B_g}\right|_{i+\frac{1}{2},j}^{n+1}\left.\frac{\partial P_{cgo}}{\partial S_g}\right|_{i+\frac{1}{2},j}^{n+1}\right)\left(S_{g_{i+1,j}}^{n+1}-S_{g_{i,j}}^{n+1}\right)-\left(\frac{A_x k_x}{\Delta x}\left.\frac{k_{rg}}{\mu_g B_g}\right|_{i-\frac{1}{2},j}^{n+1}\left.\frac{\partial P_{cgo}}{\partial S_g}\right|_{i-\frac{1}{2},j}^{n+1}\right)\left(S_{g_{i,j}}^{n+1}-S_{g_{i-1,j}}^{n+1}\right)$$

$$+\left(\frac{A_x k_x}{\Delta x}\left.\frac{k_{rg}}{\mu_g B_g}\right|_{i,j+\frac{1}{2}}^{n+1}\left.\frac{\partial P_{cgo}}{\partial S_g}\right|_{i,j+\frac{1}{2}}^{n+1}\right)\left(S_{g_{i,j+1}}^{n+1}-S_{g_{i,j}}^{n+1}\right)-\left(\frac{A_x k_x}{\Delta x}\left.\frac{k_{rg}}{\mu_g B_g}\right|_{i,j-\frac{1}{2}}^{n+1}\left.\frac{\partial P_{cgo}}{\partial S_g}\right|_{i,j-\frac{1}{2}}^{n+1}\right)\left(S_{g_{i,j}}^{n+1}-S_{g_{i,j-1}}^{n+1}\right)$$

$$-\left[\begin{array}{l}\gamma_o R_s\dfrac{A_x k_x}{\Delta x}\left.\dfrac{k_{ro}}{\mu_o B_o}\right|_{i+\frac{1}{2},j}^{n+1}\left(G_{i+1,j}-G_{i,j}\right)-\gamma_o R_s\dfrac{A_x k_x}{\Delta x}\left.\dfrac{k_{ro}}{\mu_o B_o}\right|_{i-\frac{1}{2},j}^{n+1}\left(G_{i,j}-G_{i-1,j}\right)\\[2ex]+\gamma_o R_s\dfrac{A_y k_y}{\Delta y}\left.\dfrac{k_{ro}}{\mu_o B_o}\right|_{i,j+\frac{1}{2}}^{n+1}\left(G_{i,j+1}-G_{i,j}\right)-\gamma_o R_s\dfrac{A_y k_y}{\Delta y}\left.\dfrac{k_{ro}}{\mu_o B_o}\right|_{i,j-\frac{1}{2}}^{n+1}\left(G_{i,j}-G_{i,j-1}\right)\end{array}\right]$$

$$=\frac{V_{b_{i,j}}}{\Delta t}\left\{\left[\frac{\phi\left(1-S_g\right)}{B_o}\right]_{i,j}^{n+1}-\left[\frac{\phi\left(1-S_g\right)}{B_o}\right]_{i,j}^{n}\right\}-q_{osc_{i,j}}.$$

$$\dots\dots\dots\dots\dots\dots\dots\dots\dots\dots\dots\dots\dots\dots\dots\dots\dots\dots\dots\dots\dots\dots\dots\dots\dots\dots\ (9.99)$$

Note that the derivatives of capillary pressures, with respect to gas saturation, need to be evaluated using single-point upstream weighting.

Problem 9.3.6 (Ertekin et al. 2001)

Write the implicit finite-difference equations for two-phase flow of water and gas in a 1D horizontal reservoir. Assume negligible gas solution in water and ignore the capillary pressure.

Solution to Problem 9.3.6

A 1D form of the water equation is

$$\frac{\partial}{\partial x}\left(\frac{A_x k_x k_{rw}}{\mu_w B_w}\frac{\partial p}{\partial x}\right)\Delta x + q_w = \frac{V_b}{5.615}\frac{\partial}{\partial t}\left(\frac{\phi S_w}{B_w}\right). \quad\dots\dots\dots\dots\dots\dots\dots\dots\dots\dots\dots\dots\dots\dots\dots\dots\dots\dots\dots(9.100)$$

A 1D form of the gas equation, assuming negligible gas solution in water, is

$$\frac{\partial}{\partial x}\left(\frac{A_x k_x k_{rg}}{\mu_g B_g}\frac{\partial p}{\partial x}\right)\Delta x + q_g = \frac{V_b}{5.615}\frac{\partial}{\partial t}\left(\frac{\phi S_g}{B_g}\right). \quad\dots\dots\dots\dots\dots\dots\dots\dots\dots\dots\dots\dots\dots\dots\dots\dots\dots(9.101)$$

In the previous equation, p_w is set to be equal to p_g (no capillary pressure between the water and gas phase). This is why the dependent variable (pressure) appears with no phase subscripts. The implicit form of the finite-difference equations for the water and gas phase will be (assuming porosity is not a function of pressure) as follows:

- Water equation:

$$\left.\frac{A_x k_x}{\Delta x}\frac{k_{rw}}{\mu_w B_w}\right|_{i+\frac{1}{2}}^{n+1}\left(p_{i+1}^{n+1}-p_i^{n+1}\right) - \left.\frac{A_x k_x}{\Delta x}\frac{k_{rw}}{\mu_w B_w}\right|_{i-\frac{1}{2}}^{n+1}\left(p_{i,j}^{n+1}-p_{i-1,j}^{n+1}\right) + q_{wsc_i}^{n+1}$$

$$= \left(\frac{V_b\phi}{5.615\Delta t}\right)_i\left[\left(\frac{S_w}{B_w}\right)_i^{n+1}-\left(\frac{S_w}{B_w}\right)_i^n\right]. \quad\dots\dots\dots\dots\dots\dots\dots\dots\dots(9.102)$$

- Gas equation:

$$\left.\frac{A_x k_x}{\Delta x}\frac{k_{rg}}{\mu_g B_g}\right|_{i+\frac{1}{2}}^{n+1}\left(p_{i+1}^{n+1}-p_i^{n+1}\right) - \left.\frac{A_x k_x}{\Delta x}\frac{k_{rg}}{\mu_g B_g}\right|_{i-\frac{1}{2}}^{n+1}\left(p_{i,j}^{n+1}-p_{i-1,j}^{n+1}\right) + q_{gsc_i}^{n+1} = \left(\frac{V_b\phi}{5.615\Delta t}\right)_i\left[\left(\frac{S_g}{B_g}\right)_i^{n+1}-\left(\frac{S_g}{B_g}\right)_i^n\right]. \quad..(9.103)$$

Problem 9.3.7 (Ertekin et al. 2001)

Derive the linearized-implicit approximations for the difference operator $\Delta_x\left(T_{wx}\Delta_x P_{cow}\right)$.

Solution to Problem 9.3.7

Option 1 is as follows:

$$\Delta_x\left(T_{wx}\Delta_x P_{cow}\right) \approx T_{wx_+}^n\left(P_{cow_+}'\right)^n\left(S_{w_{i+1}}^{n+1}-S_{w_i}^n\right) + T_{wx_-}^n\left(P_{cow_-}'\right)^n\left(S_{w_{i-1}}^{n+1}-S_{w_i}^n\right)$$

$$+ \left(\frac{\partial T_{wx}}{\partial S_w}\right)_+^n\left(P_{cow_{i+1}}^n-P_{cow_i}^n\right)\left(S_w^{n+1}-S_w^n\right)_+ + \left(\frac{\partial T_{wx}}{\partial S_w}\right)_-^n\left(P_{cow_{i-1}}^n-P_{cow_i}^n\right)\left(S_w^{n+1}-S_w^n\right)_-. \quad\dots\dots\dots\dots(9.104)$$

Option 2 is as follows:

$$\Delta_x\left(T_{wx}\Delta_x P_{cow}\right) \approx T_{wx_+}^n\left(P_{cow_{i+1}}^n-P_{cow_i}^n\right) + T_{wx_-}^n\left(P_{cow_{i-1}}^n-P_{cow_i}^n\right) + T_{wx_+}^n\left(P_{cow_{i+1}}'\right)^n\left(S_{w_{i+1}}^{n+1}-S_{w_i}^n\right)$$

$$+ \left(\frac{\partial T_{wx}}{\partial S_w}\right)_+^n\left(P_{cow_{i+1}}^n-P_{cow_i}^n\right)\left(S_w^{n+1}-S_w^n\right)_+ + \left(\frac{\partial T_{wx}}{\partial S_w}\right)_-^n\left(P_{cow_{i-1}}^n-P_{cow_i}^n\right)\left(S_w^{n+1}-S_w^n\right)_- \quad\dots\dots\dots\dots(9.105)$$

$$+ T_{wx_-}^n\left(P_{cow_{i-1}}'\right)^n\left(S_{w_{i-1}}^{n+1}-S_{w_i}^n\right) - \left(T_{wx_+}^n+T_{wx_-}^n\right)\left(P_{cow_i}'\right)^n\left(S_{w_i}^{n+1}-S_{w_i}^n\right),$$

where

$$\left(P_{cow_i}'\right)^n = \left(\frac{\partial P_{cow}}{\partial S_w}\right)^n. \quad\dots\dots\dots\dots\dots\dots\dots\dots\dots\dots\dots\dots\dots\dots\dots\dots\dots\dots\dots\dots\dots\dots(9.106)$$

Problem 9.3.8 (Ertekin et al. 2001)

Show that the fully implicit approximation for the difference operator $\Delta_x\left(T_{px}\Delta_x p_o\right)_i^{n+1}$,

$$\Delta_x\left(T_{px}\Delta_x p_o\right)_i^{n+1} \approx T_{px_+}^{(v)}\left[p_{o_{i+1}}^{(v)} - p_{o_i}^{(v)}\right] + T_{px_-}^{(v)}\left[p_{o_{i-1}}^{(v)} - p_{o_i}^{(v)}\right] + T_{px_+}^{(v)}\left[\delta p_{o_{i+1}}^{(v)} - \delta p_{o_i}^{(v)}\right] + T_{px_-}^{(v)}\left[\delta p_{o_{i-1}}^{(v)} - \delta p_{o_i}^{(v)}\right]$$

$$+ \sum_{l=w,g}\left[p_{o_{i+1}}^{(v)} - p_{o_i}^{(v)}\right]\left(\frac{\partial T_{px}}{\partial S_l}\right)_+^{(v)}(\delta S_l)_+ + \sum_{l=w,g}\left[p_{o_{i-1}}^{(v)} - p_{o_i}^{(v)}\right]\left(\frac{\partial T_{px}}{\partial S_l}\right)_-^{(v)}(\delta S_l)_- \qquad \dots\dots (9.107)$$

$$+ \left[p_{o_{i+1}}^{(v)} - p_{o_i}^{(v)}\right]\left(\frac{\partial T_{px}}{\partial S_l}\right)_+^{(v)}(\delta p_o)_+ + \left[p_{o_{i-1}}^{(v)} - p_{o_i}^{(v)}\right]\left(\frac{\partial T_{px}}{\partial S_l}\right)_-^{(v)}(\delta p_o)_-,$$

can be reduced to the semi-implicit approximation of Nolen and Berry (1972), as given by

$$\Delta_x\left(T_{px}\Delta_x p_o\right)_i^{n+1} \approx \left\{T_{px+}^{\ n} + \sum_{l=w,g}\left(\frac{\partial T_{px}}{\partial S_l}\right)_+^n\left[S_l^{(v)} - S_l^n\right]_+\right\}\left[p_{o_{i+1}}^{(v)} - p_{o_i}^{(v)}\right]$$

$$+ \left\{T_{px-}^{\ n} + \sum_{l=w,g}\left(\frac{\partial T_{px}}{\partial S_l}\right)_-^n\left[S_l^{(v)} - S_l^n\right]_-\right\}\left[p_{o_{i-1}}^{(v)} - p_{o_i}^{(v)}\right]$$

$$+ \left\{T_{px+}^{\ n} + \sum_{l=w,g}\left(\frac{\partial T_{px}}{\partial S_l}\right)_+^n\left[S_l^{(v)} - S_l^n\right]_+\right\}\left(\delta p_{o_{i+1}} - \delta p_{o_i}\right) \qquad \dots\dots\dots\dots (9.108)$$

$$+ \left\{T_{px-}^{\ n} + \sum_{l=w,g}\left(\frac{\partial T_{px}}{\partial S_l}\right)_-^n\left[S_l^{(v)} - S_l^n\right]_-\right\}\left(\delta p_{o_{i-1}} - \delta p_{o_i}\right)$$

$$+ \sum_{l=w,g}\left(\frac{\partial T_{px}}{\partial S_l}\right)_+^n\left[p_{o_{i+1}}^{(v)} - p_{o_i}^{(v)}\right](\delta S_l)_+ + \sum_{l=w,g}\left(\frac{\partial T_{px}}{\partial S_l}\right)_-^n\left[p_{o_{i-1}}^{(v)} - p_{o_i}^{(v)}\right](\delta S_l)_-.$$

State and implement all necessary assumptions and perform all manipulations. Note that the superscript v is the iteration level and n is the timestep level.

Solution to Problem 9.3.8

The difference operator, $\Delta_x\left(T_{px}\Delta_x p_o\right)_i^{n+1}$, will yield

$$\Delta_x\left(T_{px}\Delta_x p_o\right)_i^{n+1} \approx T_{px_+}^{(v)}\left[p_{o_{i+1}}^{(v)} - p_{o_i}^{(v)}\right] + T_{px_-}^{(v)}\left[p_{o_{i-1}}^{(v)} - p_{o_i}^{(v)}\right] + T_{px_+}^{(v)}\left[\delta p_{o_{i+1}}^{(v)} - \delta p_{o_i}^{(v)}\right] + T_{px_-}^{(v)}\left[\delta p_{o_{i-1}}^{(v)} - \delta p_{o_i}^{(v)}\right]$$

$$+ \sum_{l=w,g}\left[p_{o_{i+1}}^{(v)} - p_{o_i}^{(v)}\right]\left(\frac{\partial T_{px}}{\partial S_l}\right)_+^{(v)}(\delta S_l)_+ + \sum_{l=w,g}\left[p_{o_{i-1}}^{(v)} - p_{o_i}^{(v)}\right]\left(\frac{\partial T_{px}}{\partial S_l}\right)_-^{(v)}(\delta S_l)_- \qquad \dots\dots\dots (9.109)$$

$$+ \left[p_{o_{i+1}}^{(v)} - p_{o_i}^{(v)}\right]\left(\frac{\partial T_{px}}{\partial S_l}\right)_+^{(v)}(\delta p_o)_+ + \left[p_{o_{i-1}}^{(v)} - p_{o_i}^{(v)}\right]\left(\frac{\partial T_{px}}{\partial S_l}\right)_-^{(v)}(\delta p_o)_-.$$

The fully implicit equation, Eq. 9.109, can be reduced to the semi-implicit approximation of Nolen and Berry, if $f_p^{n+1} \approx f_p^n$ and f_s is a straight line. In such a case, we have

$$\delta S_l = S_l^{(v+1)} - S_l^{(v)} \approx S_l^{(v)} - S_l^n. \dots\dots\dots\dots\dots\dots\dots\dots\dots\dots\dots\dots\dots\dots\dots (9.110)$$

The final result will be

$$\Delta_x\left(T_{px}\Delta_x p_o\right)_i^{n+1} \approx T_{px_+}^{(v)}\left[p_{o_{i+1}}^{(v)} - p_{o_i}^{(v)}\right] + T_{px_-}^{(v)}\left[p_{o_{i-1}}^{(v)} - p_{o_i}^{(v)}\right] + T_{px_+}^{(v)}\left[\delta p_{o_{i+1}}^{(v)} - \delta p_{o_i}^{(v)}\right]$$

$$+ T_{px_-}^{(v)}\left[\delta p_{o_{i-1}}^{(v)} - \delta p_{o_i}^{(v)}\right] + \sum_{l=w,g}\left[p_{o_{i+1}}^{(v)} - p_{o_i}^{(v)}\right]\left(\frac{\partial T_{px}}{\partial S_l}\right)_+^{(v)}\left[S_l^{(v)} - S_l^n\right]_+ \qquad \dots\dots\dots\dots (9.111)$$

$$+ \sum_{l=w,g}\left[p_{o_{i-1}}^{(v)} - p_{o_i}^{(v)}\right]\left(\frac{\partial T_{px}}{\partial S_l}\right)_-^{(v)}\left[S_l^{(v)} - S_l^n\right]_- + \left[p_{o_{i+1}}^{(v)} - p_{o_i}^{(v)}\right]\left(\frac{\partial T_{px}}{\partial S_l}\right)_+^{(v)}(\delta p_o)_+,$$

which can be expressed as

$$
\Delta_x \left(T_{px} \Delta_x p_o \right)_i^{n+1} \approx T_{px_+}^{(v)} \left[p_{o_{i+1}}^{(v)} - p_{o_i}^{(v)} \right] + T_{px_-}^{(v)} \left[p_{o_{i-1}}^{(v)} - p_{o_i}^{(v)} \right] + T_{px_+}^{(v)} \left[\delta p_{o_{i+1}}^{(v)} - \delta p_{o_i}^{(v)} \right] + T_{px_-}^{(v)} \left[\delta p_{o_{i-1}}^{(v)} - \delta p_{o_i}^{(v)} \right]
$$

$$
+ \sum_{l=w,g} \left[p_{o_{i+1}}^{(v)} - p_{o_i}^{(v)} \right] \left(\frac{\partial T_{px}}{\partial S_l} \right)_+^{(v)} (\delta S_l)_+ + \sum_{l=w,g} \left[p_{o_{i-1}}^{(v)} - p_{o_i}^{(v)} \right] \left(\frac{\partial T_{px}}{\partial S_l} \right)_-^{(v)} (\delta S_l)_- \qquad \ldots\ldots\ldots (9.112)
$$

$$
+ \left[p_{o_{i+1}}^{(v)} - p_{o_i}^{(v)} \right] \left(\frac{\partial T_{px}}{\partial S_l} \right)_+^{(v)} (\delta p_o)_+ + \left[p_{o_{i-1}}^{(v)} - p_{o_i}^{(v)} \right] \left(\frac{\partial T_{px}}{\partial S_l} \right)_-^{(v)} (\delta p_o)_- \;.
$$

The fully implicit approximation can be reduced to Nolen and Berry's semi-explicit approximation by considering the assumptions and arguments that follow. The following terms that appear in Nolen and Berry's approximation are expected to be small compared with $T_{px_+}^{\,n}$ and $T_{px_-}^{\,n}$ terms that appear outside of the summation signs:

$$
\sum_{l=w,g} \left(\frac{\partial T_{px}}{\partial S_l} \right)_+^{(v)} \left[S_l^{(v)} - S_l^n \right]_+ . \ldots\ldots\ldots\ldots\ldots\ldots\ldots\ldots\ldots\ldots\ldots\ldots\ldots\ldots\ldots\ldots\ldots\ldots\ldots\ldots (9.113)
$$

and

$$
\sum_{l=w,g} \left(\frac{\partial T_{px}}{\partial S_l} \right)_-^{(v)} \left[S_l^{(v)} - S_l^n \right]_- . \ldots\ldots\ldots\ldots\ldots\ldots\ldots\ldots\ldots\ldots\ldots\ldots\ldots\ldots\ldots\ldots\ldots\ldots\ldots\ldots (9.114)
$$

This is expected because even at the later stages of the convergence, the magnitude of $\left[S_l^{(v)} - S_l^n \right]$ is very small because changes in saturations from one timestep to the other are expected to be small. Similarly, the partial derivatives, with respect to saturations, will be small as well. And finally, the product of two small numbers will result in a smaller value.

Problem 9.3.9 (Ertekin et al. 2001)

Using upstream weighting and knowing that flow is in the direction of increasing i in a 1D reservoir, write the finite-difference equations for Gridblock i in an oil/water flow system using the following methods:

1. Simple iteration method.
2. Linearized-implicit method.
3. Semi-implicit method of Nolen and Berry.

Solution to Problem 9.3.9

1. Simple iteration method.
   - Oil equation:

$$
\left(T_{ox_{i+\frac{1}{2},j,k}}^{n+1} \right)^{(v)} \left[\left(p_{oi+1,j,k}^{n+1} \right)^{(v+1)} - \left(p_{oi,j,k}^{n+1} \right)^{(v+1)} \right] + \left(T_{ox_{i-\frac{1}{2},j,k}}^{n+1} \right)^{(v)} \left[\left(p_{oi-1,j,k}^{n+1} \right)^{(v+1)} - \left(p_{oi,j,k}^{n+1} \right)^{(v+1)} \right]
$$

$$
= \frac{V_{b_i}}{\Delta t} \left\{ \left[\frac{\phi S_o}{B_o} \right]_i^{n+1} - \left[\frac{\phi S_o}{B_o} \right]_i^n \right\} - q_{osc_i} . \qquad \ldots\ldots\ldots (9.115)
$$

   - Water equation:

$$
\left(T_{wx_{i+\frac{1}{2},j,k}}^{n+1} \right)^{(v)} \left[\left(p_{wi+1,j,k}^{n+1} \right)^{(v+1)} - \left(p_{wi,j,k}^{n+1} \right)^{(v+1)} \right] + \left(T_{wx_{i-\frac{1}{2},j,k}}^{n+1} \right)^{(v)} \left[\left(p_{wi-1,j,k}^{n+1} \right)^{(v+1)} - \left(p_{wi,j,k}^{n+1} \right)^{(v+1)} \right]
$$

$$
= \frac{V_{b_i}}{\Delta t} \left\{ \left[\frac{\phi S_w}{B_w} \right]_i^{n+1} - \left[\frac{\phi S_w}{B_w} \right]_i^n \right\} - q_{wsc_i} . \qquad \ldots\ldots\ldots (9.116)
$$

2. Linearized-implicit method.
- Oil equation:

$$T_{px_+}^n \left(p_{o_{i+1}}^{n+1} - p_{o_i}^{n+1} \right) + T_{px_-}^n \left(p_{o_{i-1}}^{n+1} - p_{o_i}^{n+1} \right) +$$

$$\left(\frac{\partial T_{px}}{\partial S_w} \right)_+^n \left(p_{o_{i+1}}^{n+1} - p_{o_i}^{n+1} \right)\left(S_w^{n+1} - S_w^n \right)_+ + \left(\frac{\partial T_{px}}{\partial S_w} \right)_-^n \left(p_{o_{i-1}}^{n+1} - p_{o_i}^{n+1} \right)\left(S_w^{n+1} - S_w^n \right)_-$$

$$+ T_{wx_+}^n \left(P_{cow_+}' \right)^n \left(S_{w_{i+1}}^{n+1} - S_{w_i}^n \right) + T_{wx_-}^n \left(P_{cow_-}' \right)^n \left(S_{w_{i-1}}^{n+1} - S_{w_i}^n \right) + \left(\frac{\partial T_{wx}}{\partial S_w} \right)_+^n$$

$$\left(P_{cow_{i+1}}^n - P_{cow_i}^n \right)\left(S_w^{n+1} - S_w^n \right)_+ + \left(\frac{\partial T_{wx}}{\partial S_w} \right)_-^n \left(P_{cow_{i-1}}^n - P_{cow_i}^n \right)\left(S_w^{n+1} - S_w^n \right)_- \qquad \ldots\ldots\ldots\ldots\ldots (9.117)$$

$$= \frac{V_{b_i}}{\Delta t} \left\{ \left[\frac{\phi S_o}{B_o} \right]_i^{n+1} - \left[\frac{\phi S_o}{B_o} \right]_i^n \right\} - q_{osc_i}.$$

- Water equation:

$$T_{px_+}^n \left(p_{w_{i+1}}^{n+1} - p_{w_i}^{n+1} \right) + T_{px_-}^n \left(p_{w_{i-1}}^{n+1} - p_{w_i}^{n+1} \right) = \frac{V_{b_i}}{\Delta t} \left[\left(\frac{\phi S_w}{B_w} \right)_i^{n+1} - \left(\frac{\phi S_w}{B_w} \right)_i^n \right] - q_{wsc_i}. \qquad \ldots\ldots\ldots\ldots (9.118)$$

3. Semi-implicit method of Nolen and Berry.
- Oil equation:

$$\left\{ T_{px+}^n + \left(\frac{\partial T_{px}}{\partial S_w} \right)_+^n \left[S_w^{(v)} - S_w^n \right]_+ \right\} \left[p_{o_{i+1}}^{(v)} - p_{o_i}^{(v)} \right] + \left\{ T_{px-}^n + \left(\frac{\partial T_{px}}{\partial S_w} \right)_-^n \left[S_w^{(v)} - S_w^n \right]_- \right\}$$

$$\left[p_{o_{i-1}}^{(v)} - p_{o_i}^{(v)} \right] + \left\{ T_{px+}^n + \left(\frac{\partial T_{px}}{\partial S_w} \right)_+^n \left[S_w^{(v)} - S_w^n \right]_+ \right\} \left(\delta p_{o_{i+1}} - \delta p_{o_i} \right)$$

$$+ \left\{ T_{px-}^n + \left(\frac{\partial T_{px}}{\partial S_w} \right)_-^n \left[S_w^{(v)} - S_w^n \right]_- \right\} \left(\delta p_{o_{i-1}} - \delta p_{o_i} \right) \qquad \ldots\ldots\ldots\ldots\ldots (9.119)$$

$$+ \left(\frac{\partial T_{px}}{\partial S_w} \right)_+^n \left[p_{o_{i+1}}^{(v)} - p_{o_i}^{(v)} \right]\left(\delta S_w \right)_+ + \left(\frac{\partial T_{px}}{\partial S_w} \right)_-^n \left[p_{o_{i-1}}^{(v)} - p_{o_i}^{(v)} \right]\left(\delta S_w \right)_-$$

$$= \frac{V_{b_i}}{\Delta t} \left\{ \left[\frac{\phi S_o}{B_o} \right]_i^{n+1} - \left[\frac{\phi S_o}{B_o} \right]_i^n \right\} - q_{osc_i}.$$

- Water equation:

$$T_{px+}^n \left[p_{w_{i+1}}^{(v)} - p_{w_i}^{(v)} \right] + T_{px-}^n \left[p_{w_{i-1}}^{(v)} - p_{w_i}^{(v)} \right] + T_{px+}^n \left(\delta p_{w_{i+1}} - \delta p_{w_i} \right) + T_{px-}^n \left(\delta p_{w_{i-1}} - \delta p_{w_i} \right)$$

$$= \frac{V_{b_i}}{\Delta t} \left\{ \left[\frac{\phi S_w}{B_w} \right]_i^{n+1} - \left[\frac{\phi S_w}{B_w} \right]_i^n \right\} - q_{wsc_i}. \qquad \ldots\ldots\ldots\ldots (9.120)$$

Problem 9.3.10 (Ertekin et al. 2001)

Write the equations of the SS method for the explicitly linearized oil and gas flow model in a 1D reservoir.

Solution to Problem 9.3.10

Consider the unknowns in a vector form of representation for a 1D reservoir with N blocks:

$$\bar{X} = \left(\bar{X}_1, \bar{X}_2, \ldots, \bar{X}_N \right)^T, \qquad \ldots\ldots\ldots\ldots\ldots\ldots\ldots\ldots\ldots\ldots\ldots\ldots\ldots\ldots\ldots\ldots\ldots\ldots\ldots\ldots\ldots\ldots (9.121)$$

$$\bar{X}_n = \left(S_{g_n}, P_{o_n} \right)^T, \qquad \ldots\ldots\ldots\ldots\ldots\ldots\ldots\ldots\ldots\ldots\ldots\ldots\ldots\ldots\ldots\ldots\ldots\ldots\ldots\ldots\ldots\ldots (9.122)$$

where $n = 1, 2, 3, \ldots$, The finite-difference equations, written for all gridblocks, may be expressed in a matrix form as

$$\left([T]-[C]\right)\vec{X}^{n+1} = -[C]\vec{X}^{n} - \vec{Q} + \vec{G}, \dotfill (9.123)$$

where $[T]$ = transmissibility matrix, $[C]$ = accumulation matrix, $\vec{Q}$ = source vector, and $\vec{G}$ = vector of gravitational terms. The definition of these vectors and matrices follow:

$$\vec{Q} = \left(\vec{Q}_1, \vec{Q}_2, \vec{Q}_3, \ldots, \vec{Q}_N\right)^T, \dotfill (9.124)$$

$$\vec{Q}_n = \begin{bmatrix} q^n_{gsc_n} \\ q^n_{osc_n} \end{bmatrix}, \dotfill (9.125)$$

where $n = 1, 2, 3, \ldots, N$;

$$\vec{G} = \left(\vec{G}_1, \vec{G}_2, \vec{G}_3, \ldots, \vec{G}_N\right)^T, \dotfill (9.126)$$

$$\vec{G}_n = \begin{bmatrix} \sum_{m \in \Psi_n} \left(T_g \bar{\gamma}_g + T_o R_s \bar{\gamma}_o\right)^n_{n,m} \Delta_m \\ \sum_{m \in \Psi_n} \left(T_o \bar{\gamma}_o\right)^n_{n,m} \Delta_m G \end{bmatrix}, \dotfill (9.127)$$

where $n = 1, 2, 3, \ldots, N$;

$$[C] = \begin{bmatrix} [C]_1 & & & & \\ & [C]_2 & & & \\ & & [C]_3 & & \\ & & & \ddots & \\ & & & & [C]_N \end{bmatrix}, \dotfill (9.128)$$

$$[C]_n = \begin{bmatrix} C_{gg_n} & C_{gp_n} \\ C_{og_n} & C_{op_n} \end{bmatrix}, \dotfill (9.129)$$

where $n = 1, 2, 3, \ldots, N$.

In the SS method for multiphase flow, the transmissibility matrix $[T]$ has a block structure. That is, the elements of the matrix are submatrices. For the nth-block row of the matrix $[T]$, the definitions of the submatrices are

$$[T]_{n,m} = \begin{bmatrix} \left(T_g P'_{cgo}\right)^n_{n,m} & \left(T_g + T_o R_s\right)^n_{n,m} \\ 0 & T^n_{o_{n,m}} \end{bmatrix}, \dotfill (9.130)$$

where $m, \in, \Psi_n$, and

$$[T]_{n,n} = \begin{bmatrix} -\left(T_g P'_{cgo}\right)^n_n & -\left(T_g + T_o R_s\right)^n_{n,m} \\ 0 & -T^n_{o_{n,m}} \end{bmatrix}. \dotfill (9.131)$$

For the explicit method of linearization, the elements of the submatrices are evaluated at time level n.

Problem 9.3.11 (Ertekin et al. 2001)

Write the equations of the SS method for the explicitly linearized oil and water flow model in a 2D horizontal reservoir. Use the natural ordering of the gridblocks.

Solution to Problem 9.3.11

Consider the unknowns in a vector form of representation for a 2D reservoir with $N = N_x \times N_y$ blocks:

$$\bar{X} = \left(\bar{X}_1, \bar{X}_2, \ldots, \bar{X}_N \right)^T, \dotfill (9.132)$$

$$\bar{X}_n = \left(S_{w_n}, p_{o_n} \right)^T, \dotfill (9.133)$$

where $n = 1, 2, 3, \ldots, N$. The finite-difference equations written for all gridblocks may be expressed in matrix form:

$$\left([T] - [C] \right) \bar{X}^{n+1} = -[C] \bar{X}^n - \bar{Q} + \bar{G}, \dotfill (9.134)$$

where $[T] =$ transmissibility matrix, $[C] =$ accumulation matrix, $\bar{Q} =$ source vector, and $\overline{G} =$ vector of gravitational terms. The definition of these vectors and matrices follow:

$$\bar{Q} = \left(\bar{Q}_1, \bar{Q}_2, \bar{Q}_3, \ldots, \bar{Q}_N \right)^T, \dotfill (9.135)$$

$$\bar{Q}_n = \begin{bmatrix} q^n_{wsc_n} \\ q^n_{osc_n} \end{bmatrix}, \dotfill (9.136)$$

where $n = 1, 2, 3, \ldots, N$;

$$\bar{G} = \left(\bar{G}_1, \bar{G}_2, \bar{G}_3, \ldots, \bar{G}_N \right)^T, \dotfill (9.137)$$

$$\bar{G}_n = \begin{bmatrix} \sum_{m \in \Psi_n} \left(T_w \bar{\gamma}_w \right)^n_{n,m} \Delta_m G \\ \sum_{m \in \Psi_n} \left(T_o \bar{\gamma}_o \right)^n_{n,m} \Delta_m G \end{bmatrix}, \dotfill (9.138)$$

where $n = 1, 2, 3, \ldots, N$;

$$[C] = \begin{bmatrix} [C]_1 & & & & \\ & [C]_2 & & & \\ & & [C]_3 & & \\ & & & \ddots & \\ & & & & [C]_N \end{bmatrix}, \dotfill (9.139)$$

$$[C]_n = \begin{bmatrix} C_{ww_n} & C_{wp_n} \\ C_{ow_n} & C_{op_n} \end{bmatrix}, \dotfill (9.140)$$

where $n = 1, 2, 3, \ldots, N$.

In the SS method for multiphase flow, the transmissibility matrix $[T]$ has a block structure. That is, the elements of the matrix are submatrices. For the nth-block row of the matrix $[T]$, the definitions of the submatrices are

$$[T]_{n,m} = \begin{bmatrix} -\left(T_w P'_{cow} \right)^n_{n,m} & \left(T_w \right)^n_{n,m} \\ 0 & T^n_{o_{n,m}} \end{bmatrix}, \dotfill (9.141)$$

where $m = x$, and

$$[T]_{n,n} = -\sum_{m \in \Psi_n} [T]_{n,m}. \dotfill (9.142)$$

For the explicit method of linearization, the elements of the submatrices are evaluated at time level n.

Problem 9.3.12 (Ertekin et al. 2001)

A homogeneous, 1D horizontal oil reservoir is 1,000-ft long with a cross-sectional area of 10,000 ft² (**Fig. 9.8**). It is discretized into four equal gridblocks. Initially, $S_{wi} = S_{wirr} = 0.160$ and $p_i = 1,000$ psia everywhere. Water is injected at $x = 0$ at a rate of 75.96 res bbl/D at the standard condition, and oil is produced at $x = 1,000$ ft at the same rate. The gridblock dimensions and properties are $\Delta x = 250$ ft, $A_x = 10,000$ ft², $k_x = 300$ md, and $\phi = 0.20$. The reservoir fluids are incompressible with $B_w = B_o = 1$ res bbl/STB and $\mu_o = \mu_w = 1$ cp. The oil/water capillary pressure is zero. **Table 9.7** gives the relative permeability data for this 1D reservoir.

Using the IMPES method, find the pressure and saturation distributions at 100 and 300 days. Use single-point upstream weighting for relative permeabilities. The time-stability limit resulting from explicit transmissibilities for this problem is $\Delta t < \Delta x / 0.4453$ or $\Delta t < 561$ days.

Using the pressure and saturation distribution at $t = 300$ days from **Table 9.8**, find the pressure and saturation distributions at $t = 600$ and 800 days. Use single timesteps to progress simulation from 300 to 600 and from 600 to 800 days. Check incremental material balance for both oil and water at the end of each timestep.

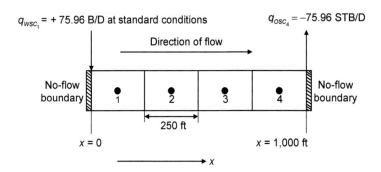

Fig. 9.8—Reservoir system of Problem 9.3.12.

S_w	k_{rw}	k_{ro}
0.16	0.000	1.000
0.20	0.010	0.700
0.30	0.035	0.325
0.40	0.060	0.140
0.50	0.110	0.045
0.60	0.160	31.000
0.70	0.240	0.015
0.80	0.420	0.000

Table 9.7—Relative permeability data for Fig. 9.8 (Problem 9.3.12).

Gridblock	p_o (psia)	p_w (psia)	S_w	S_o
1	1,000	1,000	0.409311	0.590689
2	989.8145	989.8146	0.166598	0.833402
3	984.1978	984.1978	0.16	0.84
4	978.5811	978.5811	0.16	0.84

Table 9.8—Pressure and saturation distribution at $t = 300$ days for Fig. 9.8 (Problem 9.3.12).

Solution to Problem 9.3.12

At $t = 600$ days, $\Delta t = 600 - 300 = 300$ days,

$$\frac{V_b \phi}{\alpha_c \Delta t} = \frac{\left(2.5 \times 10^6\right) \times 0.20}{(5.615)(200)} = 296.824 \text{ STB/D}. \dotfill (9.143)$$

Using the data provided at $t = 300$ days, we can get the relative permeability values shown in **Table 9.9**. For $t = 600$ days, the coefficients of the fluid flow equation are shown in **Table 9.10**. As $p_{o_1}^{n+1} = 1,000$ psia, the equations become those displayed in **Table 9.11**. Results of solving the system of equations are shown in **Table 9.12**. Then, the saturation distribution can be explicitly obtained; the results of this timestep at $t = 600$ days are shown in **Table 9.13**.

At $t = 800$ days, $\Delta t = 800 - 600 = 200$ days,

$$\frac{V_b \phi}{\alpha_c \Delta t} = \frac{\left(2.5 \times 10^6\right) \times 0.20}{(5.615)(200)} = 445.236 \text{ STB/D}. \dotfill (9.144)$$

Gridblock	S_w	k_{rw}	k_{ro}
1	0.409311	0.062328	0.131154
2	0.166598	0.001649	0.825258
3	0.16	0	1
4	0.16	0	1

Table 9.9—Relative permeability values at t = 300 days from Solution to Problem 9.3.12.

Gridblock	$p_{o_1}^{n+1}$	$p_{o_2}^{n+1}$	$p_{o_3}^{n+1}$	$p_{o_4}^{n+1}$	Right-Hand Side (RHS)
1	−2.61665	2.616651	0	0	−75.96
2	2.616651	−13.7997	11.1831	0	0
3	0	11.1831	−24.7071	13.524	0
4	0	0	13.524	−13.524	75.96

Table 9.10—Coefficients of the fluid-flow equations for t = 600 days from Solution to Problem 9.3.12.

Gridblock	$p_{o_2}^{n+1}$	$p_{o_3}^{n+1}$	$p_{o_4}^{n+1}$	RHS
2	−13.7997	11.1831	0	−2,616.65
3	11.1831	−24.7071	13.524	0
4	0	13.524	−13.524	75.96

Table 9.11—Equations as they become for $p_{o_1}^{n+1}$ = 1,000 psia from Solution to Problem 9.3.12.

Gridblock	p_o^{n+1} (psia)	p_w^{n+1} (psia)
1	1,000	1,000
2	970.9705	970.9705
3	964.1781	964.1781
4	958.5615	958.5615

Table 9.12—Results of Solving the systems of equations from Solution to Problem 9.3.12.

Gridblock	p_o (psia)	p_w (psia)	S_w	S_o	k_{rw}	k_{ro}
1	1,000	1,000	0.582783	0.417217	0.105696	−0.03364
2	970.9705	970.9705	0.248525	0.751475	0.022131	0.51803
3	964.1781	964.1781	0.16017	0.83983	4.25E-05	1
4	958.5615	958.5615	0.16	0.84	0	1

Table 9.13—Saturation distribution explicitly obtained in Solution to Problem 9.3.12.

The coefficients of the fluid flow equation are shown in **Table 9.14.** As $p_{o_1}^{n+1}$ = 1,000 psia, the equations become what is shown in **Table 9.15.** The results of solving the system of equations are shown in **Table 9.16.** Then, saturation values are again explicitly solved and the results of this timestep (t = 800 days) are shown in **Table 9.17.**

Gridblock	$p_{o_1}^{n+1}$	$p_{o_2}^{n+1}$	$p_{o_3}^{n+1}$	$p_{o_4}^{n+1}$	RHS
1	-0.97443	0.974433	0	0	-75.96
2	0.974433	-8.27957	7.30514	0	0
3	0	7.30514	-20.8297	13.52458	0
4	0	0	13.52458	-13.5246	75.96

Table 9.14—Coefficients of the fluid flow equation from Solution to Problem 9.3.12.

Gridblock	$p_{o_2}^{n+1}$	$p_{o_3}^{n+1}$	$p_{o_4}^{n+1}$	RHS
2	-8.27957	7.30514	0	-974.433
3	7.30514	-20.8297	13.52458	0
4	0	13.52458	-13.5246	75.96

Table 9.15—Equations as they become for $p_{o_1}^{n+1} = 1{,}000$ psia from Solution to Problem 9.3.12.

Gridblock	p_o^{n+1} (psia)	p_w^{n+1} (psia)
1	1,000	1,000
2	922.047	922.047
3	911.6488	911.6488
4	906.0324	906.0324

Table 9.16—Results of solving the systems of equations from Solution to Problem 9.3.12.

Gridblock	p_o (psia)	p_w (psia)	S_w	S_o	k_{rw}	k_{ro}
1	1,000	1,000	0.503121	0.496879	0.08578	0.042035
2	922.047	922.047	0.491803	0.508197	0.082951	-0.39426
3	011.6488	911.6488	0.163662	0.836338	0.000915	1
4	906.0324	906.0324	0.160007	0.839993	1.81E-06	1

Table 9.17—Saturation values as they are explicitly solved for $t = 800$ days in Solution to Problem 9.3.12.

Problem 9.3.13 (Ertekin et al. 2001)

Given the oil reservoir and data described in Problem 9.3.12, the capillary pressure data in **Table 9.18,** and fixing the oil pressure for Gridblock 1 at 1,000 psia, find the pressure and saturation distributions at 100 and 300 days using the IMPES method. Compare your simulation results with those obtained in the Solution to Problem 9.3.12.

S_w	P_{cow} (psi)
0.16	0.90
0.20	0.70
0.30	0.50
0.40	0.40
0.50	0.30
0.60	0.22
0.70	0.17
0.80	-0.20

Table 9.18—Capillary pressure data from Problem 9.3.13.

Solution to Problem 9.3.13

The solution to this question is similar to the solution we developed for Problem 9.3.12. In this problem, $P_{cow} \neq 0$:

- $i = 1$:

$$\left[\left(T_{w_{1,2}}^n\right)_1 + \left(T_{o_{1,2}}^n\right)_1\right]p_{o_2}^{n+1} - \left[\left(T_{w_{1,2}}^n\right)_1 + \left(T_{o_{1,2}}^n\right)_1\right]p_{o_1}^{n+1} - \left(T_{w_{1,2}}^n\right)_1\left(P_{cow_2}^{n+1} - P_{cow_1}^{n+1}\right) = -q_{wsc_1}^n. \quad \dots \dots \dots \dots \dots \quad (9.145)$$

- $i = 2$:

$$\left[\left(T_{w_{2,1}}^n\right)_1 + \left(T_{o_{2,1}}^n\right)_1\right]p_{o_1}^{n+1} + \left[\left(T_{w_{2,3}}^n\right)_2 + \left(T_{o_{2,3}}^n\right)_2\right]p_{o_3}^{n+1} - \left\{\left[\left(T_{w_{2,1}}^n\right)_1 + \left(T_{o_{2,1}}^n\right)_1\right] + \left[\left(T_{w_{2,3}}^n\right)_2 + \left(T_{o_{2,3}}^n\right)_2\right]\right\}$$
$$p_{o_2}^{n+1} - \left[\left(T_{w_{2,1}}^n\right)_1 p_{cow_1}^{n+1} + \left(T_{w_{2,3}}^n\right)_2 P_{cow_3}^{n+1} - \left[\left(T_{w_{2,1}}^n\right)_1 + \left(T_{w_{2,3}}^n\right)_2\right]P_{cow_2}^{n+1}\right] = 0. \quad \dots \dots \dots \quad (9.146)$$

- $i = 3$:

$$\left[\left(T_{w_{3,2}}^n\right)_2 + \left(T_{o_{3,2}}^n\right)_2\right]p_{o_2}^{n+1} + \left[\left(T_{w_{3,4}}^n\right)_3 + \left(T_{o_{3,4}}^n\right)_3\right]p_{o_4}^{n+1} - \left\{\left[\left(T_{w_{3,2}}^n\right)_2 + \left(T_{o_{3,2}}^n\right)_2\right] + \left[\left(T_{w_{3,4}}^n\right)_3 + \left(T_{o_{3,4}}^n\right)_3\right]\right\}$$
$$p_{o_3}^{n+1} - \left[\left(T_{w_{3,2}}^n\right)_2 P_{cow_2}^{n+1} + \left(T_{w_{3,4}}^n\right)_3 P_{cow_4}^{n+1} - \left[\left(T_{w_{3,2}}^n\right)_2 + \left(T_{w_{3,4}}^n\right)_3\right]P_{cow_3}^{n+1}\right] = 0. \quad \dots \dots \dots \quad (9.147)$$

- $i = 4$:

$$\left[\left(T_{w_{4,3}}^n\right)_3 + \left(+T_{o_{4,3}}^n\right)_3\right]p_{o_3}^{n+1} - \left[\left(T_{w_{4,3}}^n\right)_3 + \left(+T_{o_{4,3}}^n\right)_3\right]p_{o_4}^{n+1} - \left(T_{w_{4,3}}^n\right)_3\left(P_{cow_3}^{n+1} - P_{cow_4}^{n+1}\right) = -q_{osc_4}^n. \quad \dots \dots \dots \dots \quad (9.148)$$

As $p_{o_1}^{n+1} = 1{,}000$ psia, the coefficient matrix to be solved is

$$\begin{bmatrix} -\left\{\left[\left(T_{w_{2,1}}^n\right)_1 + \left(T_{o_{2,1}}^n\right)_1\right] + \left[\left(T_{w_{2,3}}^n\right)_2 + \left(T_{o_{2,3}}^n\right)_2\right]\right\} & \left[\left(T_{w_{2,3}}^n\right)_2 + \left(T_{o_{2,3}}^n\right)_2\right] & 0 \\ \left[\left(T_{w_{3,2}}^n\right)_2 + \left(T_{o_{3,2}}^n\right)_2\right] & -\left\{\left[\left(T_{w_{3,2}}^n\right)_2 + \left(T_{o_{3,2}}^n\right)_2\right] + \left[\left(T_{w_{3,4}}^n\right)_3 + \left(T_{o_{3,4}}^n\right)_3\right]\right\} & \left[\left(T_{w_{3,4}}^n\right)_3 + \left(T_{o_{3,4}}^n\right)_3\right] \\ 0 & \left[\left(T_{w_{4,3}}^n\right)_3 + \left(+T_{o_{4,3}}^n\right)_3\right] & -\left[\left(T_{w_{4,3}}^n\right)_3 + \left(+T_{o_{4,3}}^n\right)_3\right] \end{bmatrix} \begin{bmatrix} p_{o_2}^{n+1} \\ p_{o_3}^{n+1} \\ p_{o_4}^{n+1} \end{bmatrix}$$

$$= \begin{bmatrix} -\left[\left(T_{w_{2,1}}^n\right)_1 + \left(T_{o_{2,1}}^n\right)_1\right]p_{o_1}^{n+1} + \left[\left(T_{w_{2,1}}^n\right)_1 P_{cow_1}^{n+1} + \left(T_{w_{2,3}}^n\right)_2 P_{cow_3}^{n+1} - \left[\left(T_{w_{2,1}}^n\right)_1 + \left(T_{w_{2,3}}^n\right)_2\right]P_{cow_2}^{n+1}\right] - q_{wsc_1}^n \\ \left[\left(T_{w_{3,2}}^n\right)_2 P_{cow_2}^{n+1} + \left(T_{w_{3,4}}^n\right)_3 P_{cow_4}^{n+1} - \left[\left(T_{w_{3,2}}^n\right)_2 + \left(T_{w_{3,4}}^n\right)_3\right]P_{cow_3}^{n+1}\right] \\ \left(T_{w_{4,3}}^n\right)_3\left(P_{cow_3}^{n+1} - P_{cow_4}^{n+1}\right) - q_{osc_4}^n \end{bmatrix}.$$

$$\dots \dots \dots \dots \dots \dots \dots \dots \dots \dots \dots \dots \dots \dots \dots \dots \dots \dots \dots \dots \dots \dots \dots \dots \dots \dots \dots \dots \dots \dots \dots \dots \dots \dots \dots \dots \quad (9.149)$$

And the saturation equations to be solved explicitly are

- $i = 1$:

$$S_{w_1}^{n+1} = S_{w_1}^n + \frac{\alpha_c \Delta t}{V_b \phi}\left[\left(T_{w_{1,2}}^n\right)_1\left(p_{w_2}^{n+1} - p_{w_1}^{n+1}\right) + q_{wsc_1}^n\right]. \quad \dots \dots \dots \dots \dots \dots \dots \dots \dots \dots \dots \dots \dots \dots \quad (9.150)$$

- $i = 2$:

$$S_{w_2}^{n+1} = S_{w_2}^n + \frac{\alpha_c \Delta t}{V_b \phi}\left[\left(T_{w_{2,1}}^n\right)_1\left(p_{w_1}^{n+1} - p_{w_2}^{n+1}\right) + \left(T_{w_{2,3}}^n\right)_2\left(p_{w_3}^{n+1} - p_{w_2}^{n+1}\right)\right]. \quad \dots \dots \dots \dots \dots \dots \dots \dots \dots \quad (9.151)$$

- $i = 3$:

$$S_{w_3}^{n+1} = S_{w_3}^n + \frac{\alpha_c \Delta t}{V_b \phi}\left[\left(T_{w_{3,2}}^n\right)_2\left(p_{w_2}^{n+1} - p_{w_3}^{n+1}\right) + \left(T_{w_{3,4}}^n\right)_3\left(p_{w_4}^{n+1} - p_{w_3}^{n+1}\right)\right]. \quad \dots \dots \dots \dots \dots \dots \dots \dots \dots \quad (9.152)$$

- $i = 4$:

$$S_{w_4}^{n+1} = S_{w_4}^n + \frac{\alpha_c \Delta t}{V_b \phi}\left[\left(T_{w_{4,3}}^n\right)_3\left(p_{w_3}^{n+1} - p_{w_4}^{n+1}\right)\right]. \quad \dots \dots \dots \dots \dots \dots \dots \dots \dots \dots \dots \dots \dots \dots \dots \dots \quad (9.153)$$

At $t = 100$ days, $\Delta t = 100 - 0 = 100$ days,

$$\frac{V_b \phi}{\alpha_c \Delta t} = \frac{\left(2.5 \times 10^6\right) \times 0.20}{(5.615)(100)} = 296.824 \text{ STB/D}. \quad \dots \dots \dots \dots \dots \dots \dots \dots \dots \dots \dots \dots \dots \dots \dots \dots \dots \dots \quad (9.154)$$

From the data at $t = 100$ days, we can get the relative permeability values shown in **Table 9.19.** For $t = 100$ days, the coefficients of the fluid flow equations are shown in **Table 9.20.** As $p_{o_1}^{n+1} = 1,000$ psia, the equations become those displayed in **Table 9.21.** Solving the system of equations, (note that $p_w = p_o - P_{cow}$) one obtains the results shown in **Table 9.22.** Then S_w can be solved; the results of this timestep are shown in **Table 9.23.**

At $t = 300$ days, $\Delta t = 300 - 100 = 200$ days,

$$\frac{V_b\phi}{\alpha_c\Delta t} = \frac{(2.5\times10^6)\times0.20}{(5.615)(200)} = 445.236 \text{ STB/D.} \dots\dots\dots\dots\dots\dots\dots\dots\dots\dots\dots\dots\dots\dots\dots\dots\dots\dots (9.155)$$

Gridblock	p_o (psia)	p_w (psia)	S_w	S_o	k_{rw}	k_{ro}
1	1,000	999.1	0.16	0.84	0	1
2	1,000	999.1	0.16	0.84	0	1
3	1,000	999.1	0.16	0.84	0	1
4	1,000	999.1	0.16	0.84	0	1

Table 9.19—Relative permeability values from the data at $t = 0$ days.

Gridblock	$p_{o_1}^{n+1}$	$p_{o_2}^{n+1}$	$p_{o_3}^{n+1}$	$p_{o_4}^{n+1}$	RHS
1	−13.524	13.524	0	0	−75.96
2	13.524	−27.048	13.524	0	0
3	0	13.524	−27.048	13.524	0
4	0	0	13.524	−13.524	75.96

Table 9.20—Coefficients of the fluid flow equations for $t = 100$ days.

Gridblock	$p_{o_2}^{n+1}$	$p_{o_3}^{n+1}$	$p_{o_4}^{n+1}$	RHS
2	−27.048	13.524	0	−13,524
3	13,524	−27.048	13.524	0
4	0	13.524	−13.524	75.96

Table 9.21—System of equations that need to be solved for $p + = 1,000$ psia.

Gridblock	p_o^{n+1} (psia)	p_w^{n+1} (psia)
1	1,000	999.3906
2	994.3833	993.4833
3	988.7666	987.8666
4	983.15	982.25

Table 9.22—Solution of the systems of equations.

Gridblock	p_o (psia)	p_w (psia)	S_w	S_o	k_{rw}	k_{ro}
1	1,000	999.3906	0.245303	0.754697	0.021326	0.530113
2	994.3833	993.4833	0.16	0.84	0	1
3	988.7666	987.8666	0.16	0.84	0	1
4	983.15	982.25	0.16	0.84	0	1

Table 9.23—Results of solving for S_w from Solution to Problem 9.3.13.

The coefficients of the fluid flow equation are shown in **Table 9.24.** As $p_{o_1}^{n+1} = 1{,}000$ psia, the equations become those shown in **Table 9.25.** Solving the system of equations (p_w values are obtained from the capillary pressure relationship), one gets the results displayed in **Table 9.26.** Then S_w can be solved and the results of this timestep are shown in **Table 9.27.**

Gridblock	$p_{o_1}^{n+1}$	$p_{o_2}^{n+1}$	$p_{o_3}^{n+1}$	$p_{o_4}^{n+1}$	RHS
1	−7.45766	7.457664	0	0	−75.96
2	7.457664	−20.9817	13.524	0	0
3	0	13.524	−27.048	13.524	0
4	0	0	13.524	−13.524	75.96

Table 9.24—Coefficients of the fluid flow equation in Solution to Problem 9.3.13.

Gridblock	$p_{o_2}^{n+1}$	$p_{o_3}^{n+1}$	$p_{o_4}^{n+1}$	RHS
2	−20.9817	13.524	0	−7,457.75
3	13.524	−27.048	13.524	0
4	0	13.524	−13.524	75.96

Table 9.25—Equations as they become for $p_{o_1}^{n+1} = 1{,}000$ psia from Solution to Problem 9.3.13.

Gridblock	p_o^{n+1} (psia)	p_w^{n+1} (psia)
1	1,000	999.3906
2	989.8257	988.9257
3	984.2091	983.3091
4	978.5924	977.6924

Table 9.26—Results of solving the systems of equations from Solution to Problem 9.3.13.

Gridblock	p_o (psia)	p_w (psia)	S_w	S_o
1	1,000	999.3906	0.40913	0.59087
2	989.8257	988.9257	0.166779	0.833221
3	984.2091	983.3091	0.16	0.84
4	978.5924	977.6924	0.16	0.84

Table 9.27—Result of solving for S_w from Solution to Problem 9.3.13.

Problem 9.3.14 (Ertekin et al. 2001)

Write the pressure equation and implicit water-saturation equation of the sequential solution (SEQ) method for oil and water flow in a 1D reservoir by reducing the general equations for the black-oil model. Neglect gravity forces.

Note: The objective of the SEQ method is to improve the stability of the IMPES method. The SEQ method, therefore, consists of two consecutive steps. This is accomplished by implicitly solving for pressure, followed by explicitly solving for pressure in a subsequent step.

Solution to Problem 9.3.14

The implicit water-saturation equation is as follows:

$$\sum_{m \in \Psi_n} \left[\left(\Delta_m p_o^n - \Delta_m P_{cow}^n \right) \left(\frac{\partial T_{w_{n,m}}}{\partial S_{w_m}} \right)^n - T_{w_{n,m}}^n P_{cow}'^n \right] \Delta_t S_{w_m}$$

$$+ \left\{ \sum_{m \in \Psi_n} \left[\left(\Delta_m p_o^n - \Delta_m P_{cow}^n \right) \left(\frac{\partial T_{w_{n,m}}}{\partial S_{w_m}} \right)^n - T_{w_{n,m}}^n P_{cow}'^n \right] - \left(C_{ww_n} - q_{wwsc_n}'^n \right) \right\} \quad \dots \dots \dots \dots \dots \dots \dots \dots \dots (9.156)$$

$$\Delta_t S_{w_n} = -C_{ww_n} \left(S_{w_n}^* - S_{w_n}^n \right).$$

The implicit-pressure equation is as follows:

$$\sum_{m \in \Psi_n} \left[\left(B_o \right)_n^{n+1} T_{o,m}^n + B_{w_n}^{n+1} T_{w_{n,m}}^n \right] p_{o_m}^{n+1} - \left\{ \begin{array}{l} \left(B_{o_n}^{n+1} C_{op_n} + B_{w_n}^{n+1} C_{wp_n} \right) - \left(B_{o_n}^{n+1} q'^n_{opsc_n} + B_{w_n}^{n+1} q'^n_{wpsc_n} \right) \\ + \sum_{m \in \Psi_n} \left[\left(B_o \right)_n^{n+1} T_{o,m}^n + B_{w_n}^{n+1} T_{w_{n,m}}^n \right] p_{o_m}^{n+1} \end{array} \right\}$$

$$p_{o_m}^{n+1} = - \left\{ \begin{array}{l} \left(B_{o_n}^{n+1} C_{op_n} + B_{w_n}^{n+1} C_{wp_n} \right) - \left(B_{o_n}^{n+1} q'^n_{opsc_n} + B_{w_n}^{n+1} q'^n_{wpsc_n} \right) \\ + \sum_{m \in \Psi_n} \left[\left(B_o \right)_n^{n+1} T_{o,m}^n + B_{w_n}^{n+1} T_{w_{n,m}}^n \right] p_{o_m}^{n+1} \end{array} \right\} \quad\quad\quad \dots\dots\dots\dots(9.157)$$

$$p_{o_m}^n - \left(B_{o_n}^{n+1} q^n_{opsc_n} + B_{w_n}^{n+1} q^n_{wpsc_n} \right) + \sum_{m \in \Psi_n} B_{w_n}^{n+1} T_{w_{n,m}}^n \Delta_m P_{cow}^n.$$

Problem 9.3.15 (Ertekin et al. 2001)

Prove that material balance is satisfied for the IMPES method. Use a 1D reservoir with no-flow boundaries.

Solution to Problem 9.3.15

$$I_{MB_w} = \frac{\sum_n \dfrac{V_b \phi}{\alpha_c \Delta t} \left(S_{w_n}^{n+1} - S_{w_n}^n \right)}{\sum q_w}$$

$$= \frac{\sum_n \left(\dfrac{V_b \phi}{\alpha_c \Delta t} \dfrac{1}{C_{ww_n}} \left\{ \sum_{m \in \psi_n} \left[T_{w,m}^n \left(\Delta_m p_w^{n+1} - \overline{\gamma}_{w_{n,m}}^n \Delta_m G \right) \right] - C_{wp_n} \left(p_{o_n}^{n+1} - p_{o_n}^n \right) + q_{wsc_n}^n \right\} \right)}{\sum q_{wsc_n}^n} \quad \dots\dots\dots\dots (9.158)$$

$$= \frac{\sum_n \left\{ \sum_{m \in \psi_n} \left[T_{w,m}^n \left(\Delta_m p_w^{n+1} - \overline{\gamma}_{w_{n,m}}^n \Delta_m G \right) \right] - C_{wp_n} \left(p_{o_n}^{n+1} - p_{o_n}^n \right) + q_{wsc_n}^n \right\}}{\sum q_{wsc_n}^n}.$$

Because the boundary is closed,

$$\sum_n \sum_{m \in \psi_n} \left[T_{w,m}^n \left(\Delta_m p_w^{n+1} - \overline{\gamma}_{w_{n,m}}^n \Delta_m G \right) \right] = \sum_n C_{wp_n} \left(p_{o_n}^{n+1} - p_{o_n}^n \right), \quad \dots\dots\dots\dots\dots\dots\dots (9.159)$$

so that

$$I_{MB_w} = \frac{\sum q_{wsc_n}^n}{\sum q_{wsc_n}^n} = 1. \quad \dots\dots\dots\dots\dots\dots\dots\dots\dots\dots\dots\dots\dots\dots\dots\dots\dots\dots (9.160)$$

Similar derivations can be made for oil and gas phases.

Problem 9.3.16 (Ertekin et al. 2001)

For two-phase flow of oil and water, derive the equation for total velocity and fractional-flow equations in the x-direction. Check your equations with those obtained from the corresponding equation in three-phase flow.

Solution to Problem 9.3.16

The definition of superficial velocity, in terms of total flow and for the water and oil phases, are as follows:
- Total flow:

$$\vec{u}_t = -\lambda_t \left[\vec{\nabla} p_o - f_w \left(\vec{\nabla} p_w + \Delta \gamma_{ow} \vec{\nabla} G \right) - \gamma_o \vec{\nabla} G \right]. \quad \dots\dots\dots\dots\dots\dots\dots\dots\dots\dots\dots\dots (9.161)$$

- Water flow:

$$\vec{u}_w = f_w \vec{u}_t + \beta_c k \zeta_w \left(\vec{\nabla} P_{cow} + \Delta \gamma_{ow} \vec{\nabla} G \right). \quad \dots\dots\dots\dots\dots\dots\dots\dots\dots\dots\dots\dots (9.162)$$

- Oil flow:

$$\vec{u}_o = f_o \vec{u}_t - \beta_c k \zeta_w \left(\vec{\nabla} P_{cow} + \Delta \gamma_{ow} \vec{\nabla} G \right); \dots \dots \dots \dots \dots \dots \dots \dots \dots (9.163)$$

where $\lambda_t = \lambda_o + \lambda_w$,

$$\lambda_l = \frac{\beta_c k k_{rl}}{\mu_l}; \dots \dots \dots \dots \dots \dots \dots \dots \dots \dots \dots \dots \dots \dots \dots (9.164)$$

where $l = o, w$,

$$f_l = \frac{\lambda_l}{\lambda_t}; \dots \dots \dots \dots \dots \dots \dots \dots \dots \dots \dots \dots \dots \dots \dots \dots (9.165)$$

where $l = o, w$,

$$f_w + f_o = 1, \dots \dots \dots \dots \dots \dots \dots \dots \dots \dots \dots \dots \dots \dots \dots \dots (9.166)$$

$$\zeta_w = \frac{k_{ro}}{\mu_o} f_w = \frac{k_{rw}}{\mu_w} f_o, \dots \dots \dots \dots \dots \dots \dots \dots \dots \dots \dots \dots \dots (9.167)$$

$$\Delta \gamma_{ow} = \gamma_w - \gamma_o. \dots \dots \dots \dots \dots \dots \dots \dots \dots \dots \dots \dots \dots \dots (9.168)$$

Problem 9.3.17 (Ertekin et al. 2001)

The fractional flow coefficients $f_w, f_o,$ and f_g are defined by $f_l = \frac{\lambda_l}{\lambda_t}$, where $\lambda_l = \frac{k k_{rl}}{\mu_l}$ (for $l = o$, w, and g). Derive the expressions

for $\dfrac{\partial f_o}{\partial S_w}, \dfrac{\partial f_o}{\partial S_g}, \dfrac{\partial f_w}{\partial S_w}, \dfrac{\partial f_w}{\partial S_g}, \dfrac{\partial f_g}{\partial S_w}$, and $\dfrac{\partial f_g}{\partial S_g}$ in terms of $\dfrac{\partial k_{ro}}{\partial S_w}, \dfrac{\partial k_{ro}}{\partial S_g}, \dfrac{\partial k_{rw}}{\partial S_w}$, and $\dfrac{\partial k_{rw}}{\partial S_g}$, which should be used in

$$\sum_{m \in \psi_n} \left(\frac{q_t}{B_o^n} \right)_{n,m} \left(\frac{\partial f_{o_{n,m}}}{\partial S_{w_m}} \right)^n \Delta_t S_{w_m} + \sum_{m \in \psi_n} \left(\frac{q_t}{B_o^n} \right)_{n,m} \left(\frac{\partial f_{o_{n,m}}}{\partial S_{g_m}} \right)^n \Delta_t S_{g_m}$$

$$+ \left[-\left(C_{ow_n} - q_{owsc_n}^{'n} \right) + \sum_{m \in \psi_n} \left(\frac{q_t}{B_o^n} \right)_{n,m} \left(\frac{\partial f_{o_{n,m}}}{\partial S_{w_n}} \right)^n \right] \Delta_t S_{w_n} + \left[-\left(C_{og_n} - q_{owsc_n}^{'n} \right) + \sum_{m \in \psi_n} \left(\frac{q_t}{B_o^n} \right)_{n,m} \left(\frac{\partial f_{o_{n,m}}}{\partial S_{g_n}} \right)^n \right] \dots \dots (9.169)$$

$$\Delta_t S_{g_n} = -C_{ow_n} \left(S_{w_n}^* - S_{w_n}^n \right) - C_{og_n} \left(S_{g_n}^* - S_{g_n}^n \right),$$

and

$$\sum_{m \in \psi_n} \left(\frac{q_t}{B_w^n} \right)_{n,m} \left(\frac{\partial f_{w_{n,m}}}{\partial S_{w_m}} \right)^n \Delta_t S_{w_m} + \sum_{m \in \psi_n} \left(\frac{q_t}{B_w^n} \right)_{n,m} \left(\frac{\partial f_{w_{n,m}}}{\partial S_{g_m}} \right)^n \Delta_t S_{g_m}$$

$$+ \left[-\left(C_{ww_n} - q_{owsc_n}^{'n} \right) + \sum_{m \in \psi_n} \left(\frac{q_t}{B_w^n} \right)_{n,m} \left(\frac{\partial f_{w_{n,m}}}{\partial S_{w_n}} \right)^n \right] \Delta_t S_{w_n} + \left[-\left(C_{wg_n} - q_{owsc_n}^{'n} \right) + \sum_{m \in \psi_n} \left(\frac{q_t}{B_w^n} \right)_{n,m} \left(\frac{\partial f_{w_{n,m}}}{\partial S_{g_n}} \right)^n \right] \dots \dots (9.170)$$

$$\Delta_t S_{g_n} = -C_{ww_n} \left(S_{w_n}^* - S_{w_n}^n \right).$$

Solution to Problem 9.3.17

The general form of the derivative of the fractional flow term, with respect to the relative permeability, is

$$\frac{\partial f_l}{\partial k_{rl}} = \frac{\partial}{\partial k_{rl}} \left(\frac{\lambda_l}{\lambda_t} \right) = \frac{\partial}{\partial k_{rl}} \left(\frac{1}{1 + \dfrac{\mu_l}{\mu_a} \dfrac{k_{ra}}{k_{rl}} + \dfrac{\mu_l}{\mu_b} \dfrac{k_{rb}}{k_{rl}}} \right) = - \frac{\dfrac{\partial}{\partial k_{rl}} \left(\dfrac{\mu_l}{\mu_a} \dfrac{k_{ra}}{k_{rl}} \right) + \dfrac{\partial}{\partial k_{rl}} \left(\dfrac{\mu_l}{\mu_b} \dfrac{k_{rb}}{k_{rl}} \right)}{\left(1 + \dfrac{\mu_l}{\mu_a} \dfrac{k_{ra}}{k_{rl}} + \dfrac{\mu_l}{\mu_b} \dfrac{k_{rb}}{k_{rl}} \right)^2}, \dots \dots \dots (9.171)$$

where $l = o, w, g$.

For term $\dfrac{\partial f_o}{\partial S_w}$,

$$\frac{\partial f_o}{\partial S_w} = \frac{\partial f_o}{\partial k_{ro}} \frac{\partial k_{ro}}{\partial S_w} = -\frac{\dfrac{\partial}{\partial k_{ro}}\left(\dfrac{\mu_o}{\mu_w}\dfrac{k_{rw}}{k_{ro}}\right) + \dfrac{\partial}{\partial k_{ro}}\left(\dfrac{\mu_o}{\mu_g}\dfrac{k_{rg}}{k_{ro}}\right)}{\left(1 + \dfrac{\mu_o}{\mu_w}\dfrac{k_{rw}}{k_{ro}} + \dfrac{\mu_o}{\mu_g}\dfrac{k_{rg}}{k_{ro}}\right)^2}\frac{\partial k_{ro}}{\partial S_w}, \quad\text{.....................(9.172)}$$

$$\frac{\partial f_o}{\partial S_w} = \frac{\partial f_o}{\partial k_{rw}} \frac{\partial k_{rw}}{\partial S_w} = -\frac{\dfrac{\partial}{\partial k_{rw}}\left(\dfrac{\mu_o}{\mu_w}\dfrac{k_{rw}}{k_{ro}}\right) + \dfrac{\partial}{\partial k_{rw}}\left(\dfrac{\mu_o}{\mu_g}\dfrac{k_{rg}}{k_{ro}}\right)}{\left(1 + \dfrac{\mu_o}{\mu_w}\dfrac{k_{rw}}{k_{ro}} + \dfrac{\mu_o}{\mu_g}\dfrac{k_{rg}}{k_{ro}}\right)^2}\frac{\partial k_{rw}}{\partial S_w}. \quad\text{.....................(9.173)}$$

For term $\dfrac{\partial f_o}{\partial S_g}$,

$$\frac{\partial f_o}{\partial S_g} = \frac{\partial f_o}{\partial k_{ro}} \frac{\partial k_{ro}}{\partial S_g} = -\frac{\dfrac{\partial}{\partial k_{ro}}\left(\dfrac{\mu_o}{\mu_w}\dfrac{k_{rw}}{k_{ro}}\right) + \dfrac{\partial}{\partial k_{ro}}\left(\dfrac{\mu_o}{\mu_g}\dfrac{k_{rg}}{k_{ro}}\right)}{\left(1 + \dfrac{\mu_o}{\mu_w}\dfrac{k_{rw}}{k_{ro}} + \dfrac{\mu_o}{\mu_g}\dfrac{k_{rg}}{k_{ro}}\right)^2}\frac{\partial k_{ro}}{\partial S_g}, \quad\text{.....................(9.174)}$$

$$\frac{\partial f_o}{\partial S_g} - \frac{\partial f_o}{\partial k_{rw}} \frac{\partial k_{rw}}{\partial S_g} = -\frac{\dfrac{\partial}{\partial k_{rw}}\left(\dfrac{\mu_o}{\mu_w}\dfrac{k_{rw}}{k_{ro}}\right) + \dfrac{\partial}{\partial k_{rw}}\left(\dfrac{\mu_o}{\mu_g}\dfrac{k_{rg}}{k_{ro}}\right)}{\left(1 + \dfrac{\mu_o}{\mu_w}\dfrac{k_{rw}}{k_{ro}} + \dfrac{\mu_o}{\mu_g}\dfrac{k_{rg}}{k_{ro}}\right)^2}\frac{\partial k_{rw}}{\partial S_g}. \quad\text{.....................(9.175)}$$

For term $\dfrac{\partial f_w}{\partial S_w}$,

$$\frac{\partial f_w}{\partial S_w} = \frac{\partial f_w}{\partial k_{ro}} \frac{\partial k_{ro}}{\partial S_w} = -\frac{\dfrac{\partial}{\partial k_{ro}}\left(\dfrac{\mu_w}{\mu_o}\dfrac{k_{ro}}{k_{rw}}\right) + \dfrac{\partial}{\partial k_{ro}}\left(\dfrac{\mu_w}{\mu_g}\dfrac{k_{rg}}{k_{rw}}\right)}{\left(1 + \dfrac{\mu_w}{\mu_o}\dfrac{k_{ro}}{k_{rw}} + \dfrac{\mu_w}{\mu_g}\dfrac{k_{rg}}{k_{rw}}\right)^2}\frac{\partial k_{ro}}{\partial S_w}, \quad\text{.....................(9.176)}$$

$$\frac{\partial f_w}{\partial S_w} = \frac{\partial f_w}{\partial k_{rw}} \frac{\partial k_{rw}}{\partial S_w} = -\frac{\dfrac{\partial}{\partial k_{rw}}\left(\dfrac{\mu_w}{\mu_o}\dfrac{k_{ro}}{k_{rw}}\right) + \dfrac{\partial}{\partial k_{rw}}\left(\dfrac{\mu_w}{\mu_g}\dfrac{k_{rg}}{k_{rw}}\right)}{\left(1 + \dfrac{\mu_w}{\mu_o}\dfrac{k_{ro}}{k_{rw}} + \dfrac{\mu_w}{\mu_g}\dfrac{k_{rg}}{k_{rw}}\right)^2}\frac{\partial k_{rw}}{\partial S_w} \qquad \frac{\partial f_w}{\partial S_g} = \frac{\partial f_w}{\partial k_{ro}} \frac{\partial k_{ro}}{\partial S_g}$$

$$\text{.....................(9.177)}$$

$$= -\frac{\dfrac{\partial}{\partial k_{ro}}\left(\dfrac{\mu_w}{\mu_o}\dfrac{k_{ro}}{k_{rw}}\right) + \dfrac{\partial}{\partial k_{ro}}\left(\dfrac{\mu_w}{\mu_g}\dfrac{k_{rg}}{k_{rw}}\right)}{\left(1 + \dfrac{\mu_w}{\mu_o}\dfrac{k_{ro}}{k_{rw}} + \dfrac{\mu_w}{\mu_g}\dfrac{k_{rg}}{k_{rw}}\right)^2}\frac{\partial k_{ro}}{\partial S_g}.$$

For term $\dfrac{\partial f_w}{\partial S_g}$,

$$\frac{\partial f_w}{\partial S_g} = \frac{\partial f_w}{\partial k_{ro}} \frac{\partial k_{ro}}{\partial S_g} = -\frac{\dfrac{\partial}{\partial k_{ro}}\left(\dfrac{\mu_w}{\mu_o}\dfrac{k_{ro}}{k_{rw}}\right) + \dfrac{\partial}{\partial k_{ro}}\left(\dfrac{\mu_w}{\mu_g}\dfrac{k_{rg}}{k_{rw}}\right)}{\left(1 + \dfrac{\mu_w}{\mu_o}\dfrac{k_{ro}}{k_{rw}} + \dfrac{\mu_w}{\mu_g}\dfrac{k_{rg}}{k_{rw}}\right)^2}\frac{\partial k_{ro}}{\partial S_g}, \quad\text{.....................(9.178)}$$

$$\frac{\partial f_w}{\partial S_g} = \frac{\partial f_w}{\partial k_{rw}}\frac{\partial k_{rw}}{\partial S_g} = -\frac{\dfrac{\partial}{\partial k_{rw}}\left(\dfrac{\mu_w}{\mu_o}\dfrac{k_{ro}}{k_{rw}}\right) + \dfrac{\partial}{\partial k_{rw}}\left(\dfrac{\mu_w}{\mu_g}\dfrac{k_{rg}}{k_{rw}}\right)}{\left(1 + \dfrac{\mu_w}{\mu_o}\dfrac{k_{ro}}{k_{rw}} + \dfrac{\mu_w}{\mu_g}\dfrac{k_{rg}}{k_{rw}}\right)^2}\frac{\partial k_{rw}}{\partial S_g}. \quad \dots\dots (9.179)$$

For term $\dfrac{\partial f_g}{\partial S_w}$,

$$\frac{\partial f_g}{\partial S_w} = \frac{\partial f_g}{\partial k_{ro}}\frac{\partial k_{ro}}{\partial S_w} = -\frac{\dfrac{\partial}{\partial k_{ro}}\left(\dfrac{\mu_g}{\mu_o}\dfrac{k_{ro}}{k_{rg}}\right) + \dfrac{\partial}{\partial k_{ro}}\left(\dfrac{\mu_g}{\mu_w}\dfrac{k_{rw}}{k_{rg}}\right)}{\left(1 + \dfrac{\mu_g}{\mu_o}\dfrac{k_{ro}}{k_{rg}} + \dfrac{\mu_g}{\mu_w}\dfrac{k_{rw}}{k_{rg}}\right)^2}\frac{\partial k_{ro}}{\partial S_w}, \quad \dots\dots (9.180)$$

$$\frac{\partial f_g}{\partial S_w} = \frac{\partial f_g}{\partial k_{rw}}\frac{\partial k_{rw}}{\partial S_w} = -\frac{\dfrac{\partial}{\partial k_{rw}}\left(\dfrac{\mu_g}{\mu_o}\dfrac{k_{ro}}{k_{rg}}\right) + \dfrac{\partial}{\partial k_{rw}}\left(\dfrac{\mu_g}{\mu_w}\dfrac{k_{rw}}{k_{rg}}\right)}{\left(1 + \dfrac{\mu_g}{\mu_o}\dfrac{k_{ro}}{k_{rg}} + \dfrac{\mu_g}{\mu_w}\dfrac{k_{rw}}{k_{rg}}\right)^2}\frac{\partial k_{rw}}{\partial S_w}. \quad \dots\dots (9.181)$$

For term $\dfrac{\partial f_g}{\partial S_g}$,

$$\frac{\partial f_g}{\partial S_g} = \frac{\partial f_g}{\partial k_{ro}}\frac{\partial k_{ro}}{\partial S_g} = -\frac{\dfrac{\partial}{\partial k_{ro}}\left(\dfrac{\mu_g}{\mu_o}\dfrac{k_{ro}}{k_{rg}}\right) + \dfrac{\partial}{\partial k_{ro}}\left(\dfrac{\mu_g}{\mu_w}\dfrac{k_{rw}}{k_{rg}}\right)}{\left(1 + \dfrac{\mu_g}{\mu_o}\dfrac{k_{ro}}{k_{rg}} + \dfrac{\mu_g}{\mu_w}\dfrac{k_{rw}}{k_{rg}}\right)^2}\frac{\partial k_{ro}}{\partial S_g}, \quad \dots\dots (9.182)$$

$$\frac{\partial f_g}{\partial S_g} = \frac{\partial f_g}{\partial k_{rw}}\frac{\partial k_{rw}}{\partial S_g} = -\frac{\dfrac{\partial}{\partial k_{rw}}\left(\dfrac{\mu_g}{\mu_o}\dfrac{k_{ro}}{k_{rg}}\right) + \dfrac{\partial}{\partial k_{rw}}\left(\dfrac{\mu_g}{\mu_w}\dfrac{k_{rw}}{k_{rg}}\right)}{\left(1 + \dfrac{\mu_g}{\mu_o}\dfrac{k_{ro}}{k_{rg}} + \dfrac{\mu_g}{\mu_w}\dfrac{k_{rw}}{k_{rg}}\right)^2}\frac{\partial k_{rw}}{\partial S_g}. \quad \dots\dots (9.183)$$

Problem 9.3.18 (Ertekin et al. 2001)

Using the expressions obtained in the Solution to Problem 9.3.17, together with $q_l = f_l\, q_t$, prove that equations $q_o + q_w + q_g = \left(f_o + f_w + f_g\right)q_t = q_t$, $\dfrac{\partial q_o}{\partial S_w} + \dfrac{\partial q_w}{\partial S_w} + \dfrac{\partial q_g}{\partial S_w} = 0$, and $\dfrac{\partial q_o}{\partial S_g} + \dfrac{\partial q_w}{\partial S_g} + \dfrac{\partial q_g}{\partial S_g} = 0$ are satisfied.

Solution to Problem 9.3.18

$$f_o + f_w + f_g = \frac{\lambda_o + \lambda_w + \lambda_g}{\lambda_t} = 1. \quad \dots\dots (9.184)$$

Because $q_l = f_l\, q_t$ for $l = o$, w, and g yields

$$q_o = f_o q_t, \quad \dots\dots (9.185)$$

$$q_w = f_w q_t, \quad \dots\dots (9.186)$$

$$q_g = f_g q_t. \quad \dots\dots (9.187)$$

Thus, we have

$$q_o + q_w + q_g = \left(f_o + f_w + f_g\right)q_t = q_t. \quad \dots\dots (9.188)$$

As $\Delta_t q_t = 0$,

$$\frac{\partial q_o}{\partial S_w} + \frac{\partial q_w}{\partial S_w} + \frac{\partial q_g}{\partial S_w} = \frac{\partial}{\partial S_w}\left[\left(f_o + f_w + f_g\right)q_t\right] = \frac{\partial q_t}{\partial S_w} = 0, \dots\dots\dots\dots\dots\dots\dots\dots \text{(9.189)}$$

$$\frac{\partial q_o}{\partial S_g} + \frac{\partial q_w}{\partial S_g} + \frac{\partial q_g}{\partial S_g} = \frac{\partial}{\partial S_g}\left[\left(f_o + f_w + f_g\right)q_t\right] = \frac{\partial q_t}{\partial S_g} = 0. \dots\dots\dots\dots\dots\dots\dots \text{(9.190)}$$

Problem 9.3.19 (Ertekin et al. 2001)

Using the equations,

$$q_{osc_n} = \frac{f_{o_n}}{B_{o_n}^{n+1}}q_{t_n}, \dots\dots\dots\dots\dots\dots\dots\dots\dots\dots\dots\dots\dots\dots\dots\dots\dots\dots\dots\dots \text{(9.191)}$$

$$q_{wsc_n} = \frac{f_{w_n}}{B_{w_n}^{n+1}}q_{t_n}, \dots\dots\dots\dots\dots\dots\dots\dots\dots\dots\dots\dots\dots\dots\dots\dots\dots\dots\dots\dots \text{(9.192)}$$

$$q_{gsc_n} = \left(\frac{f_{g_n}}{B_{g_n}^{n+1}} + R_{s_n}^{n+1}\frac{f_{o_n}}{B_{o_n}^{n+1}}\right)q_{t_n}, \dots\dots\dots\dots\dots\dots\dots\dots\dots\dots\dots\dots\dots\dots\dots \text{(9.193)}$$

and the expressions obtained in Problem 9.3.17, derive equations for the partial derivatives of q_{oscn} and q_{wscn} with respect to S_{wn} and S_{gn}. Note that these partial derivatives can be used in the second step of the SEQ method.

Solution to Problem 9.3.19

Equations for the partial derivatives of q_{oscn} and q_{wscn} with respect to S_{wn} and S_{gn} are as follows:

$$\frac{\partial q_{osc_n}}{\partial S_{w_n}} = \frac{\partial}{\partial S_{w_n}}\left(\frac{f_{o_n}}{B_{o_n}^{n+1}}q_{t_n}\right) = \frac{q_{t_n}}{B_{o_n}^{n+1}}\frac{\partial f_{o_n}}{\partial S_{w_n}} = -\frac{q_{t_n}}{B_{o_n}^{n+1}}\frac{\dfrac{\partial}{\partial k_{ro}}\left(\dfrac{\mu_o}{\mu_w}\dfrac{k_{rw}}{k_{ro}}\right) + \dfrac{\partial}{\partial k_{ro}}\left(\dfrac{\mu_o}{\mu_g}\dfrac{k_{rg}}{k_{ro}}\right)}{\left(1 + \dfrac{\mu_o}{\mu_w}\dfrac{k_{rw}}{k_{ro}} + \dfrac{\mu_o}{\mu_g}\dfrac{k_{rg}}{k_{ro}}\right)^2}\frac{\partial k_{ro}}{\partial S_w}, \dots\dots \text{(9.194)}$$

$$\frac{\partial q_{osc_n}}{\partial S_{w_n}} = \frac{\partial}{\partial S_{w_n}}\left(\frac{f_{o_n}}{B_{o_n}^{n+1}}q_{t_n}\right) = \frac{q_{t_n}}{B_{o_n}^{n+1}}\frac{\partial f_{o_n}}{\partial S_{w_n}} = -\frac{q_{t_n}}{B_{o_n}^{n+1}}\frac{\dfrac{\partial}{\partial k_{rw}}\left(\dfrac{\mu_o}{\mu_w}\dfrac{k_{rw}}{k_{ro}}\right) + \dfrac{\partial}{\partial k_{rw}}\left(\dfrac{\mu_o}{\mu_g}\dfrac{k_{rg}}{k_{ro}}\right)}{\left(1 + \dfrac{\mu_o}{\mu_w}\dfrac{k_{rw}}{k_{ro}} + \dfrac{\mu_o}{\mu_g}\dfrac{k_{rg}}{k_{ro}}\right)^2}\frac{\partial k_{rw}}{\partial S_w}, \dots\dots \text{(9.195)}$$

$$\frac{\partial q_{osc_n}}{\partial S_{g_n}} = \frac{\partial}{\partial S_{g_n}}\left(\frac{f_{o_n}}{B_{o_n}^{n+1}}q_{t_n}\right) = \frac{q_{t_n}}{B_{o_n}^{n+1}}\frac{\partial f_{o_n}}{\partial S_{g_n}} = -\frac{q_{t_n}}{B_{o_n}^{n+1}}\frac{\dfrac{\partial}{\partial k_{ro}}\left(\dfrac{\mu_o}{\mu_w}\dfrac{k_{rw}}{k_{ro}}\right) + \dfrac{\partial}{\partial k_{ro}}\left(\dfrac{\mu_o}{\mu_g}\dfrac{k_{rg}}{k_{ro}}\right)}{\left(1 + \dfrac{\mu_o}{\mu_w}\dfrac{k_{rw}}{k_{ro}} + \dfrac{\mu_o}{\mu_g}\dfrac{k_{rg}}{k_{ro}}\right)^2}\frac{\partial k_{ro}}{\partial S_g}, \dots\dots \text{(9.196)}$$

$$\frac{\partial q_{osc_n}}{\partial S_{g_n}} = \frac{\partial}{\partial S_{g_n}}\left(\frac{f_{o_n}}{B_{o_n}^{n+1}}q_{t_n}\right) = \frac{q_{t_n}}{B_{o_n}^{n+1}}\frac{\partial f_{o_n}}{\partial S_{g_n}} = -\frac{q_{t_n}}{B_{o_n}^{n+1}}\frac{\dfrac{\partial}{\partial k_{rw}}\left(\dfrac{\mu_o}{\mu_w}\dfrac{k_{rw}}{k_{ro}}\right) + \dfrac{\partial}{\partial k_{rw}}\left(\dfrac{\mu_o}{\mu_g}\dfrac{k_{rg}}{k_{ro}}\right)}{\left(1 + \dfrac{\mu_o}{\mu_w}\dfrac{k_{rw}}{k_{ro}} + \dfrac{\mu_o}{\mu_g}\dfrac{k_{rg}}{k_{ro}}\right)^2}\frac{\partial k_{rw}}{\partial S_g}, \dots\dots \text{(9.197)}$$

$$\frac{\partial q_{wsc_n}}{\partial S_{w_n}} = \frac{\partial}{\partial S_{w_n}}\left(\frac{f_{w_n}}{B_{w_n}^{n+1}}q_{t_n}\right) = \frac{q_{t_n}}{B_{w_n}^{n+1}}\frac{\partial f_{w_n}}{\partial S_{w_n}} = -\frac{q_{t_n}}{B_{w_n}^{n+1}}\frac{\dfrac{\partial}{\partial k_{ro}}\left(\dfrac{\mu_w}{\mu_o}\dfrac{k_{ro}}{k_{rw}}\right) + \dfrac{\partial}{\partial k_{ro}}\left(\dfrac{\mu_w}{\mu_g}\dfrac{k_{rg}}{k_{rw}}\right)}{\left(1 + \dfrac{\mu_w}{\mu_o}\dfrac{k_{ro}}{k_{rw}} + \dfrac{\mu_w}{\mu_g}\dfrac{k_{rg}}{k_{rw}}\right)^2}\frac{\partial k_{ro}}{\partial S_w}, \dots\dots \text{(9.198)}$$

$$\frac{\partial q_{wsc_n}}{\partial S_{w_n}} = \frac{\partial}{\partial S_{w_n}}\left(\frac{f_{w_n}}{B_{w_n}^{n+1}}q_{t_n}\right) = \frac{q_{t_n}}{B_{w_n}^{n+1}}\frac{\partial f_{w_n}}{\partial S_{w_n}} = -\frac{q_{t_n}}{B_{w_n}^{n+1}}\frac{\dfrac{\partial}{\partial k_{rw}}\left(\dfrac{\mu_w}{\mu_o}\dfrac{k_{ro}}{k_{rw}}\right) + \dfrac{\partial}{\partial k_{rw}}\left(\dfrac{\mu_w}{\mu_g}\dfrac{k_{rg}}{k_{rw}}\right)}{\left(1 + \dfrac{\mu_w}{\mu_o}\dfrac{k_{ro}}{k_{rw}} + \dfrac{\mu_w}{\mu_g}\dfrac{k_{rg}}{k_{rw}}\right)^2}\frac{\partial k_{rw}}{\partial S_w}, \dots\dots \text{(9.199)}$$

$$\frac{\partial q_{wsc_n}}{\partial S_{g_n}} = \frac{\partial}{\partial S_{g_n}}\left(\frac{f_{w_n}}{B_{w_n}^{n+1}}q_{t_n}\right) = \frac{q_{t_n}}{B_{w_n}^{n+1}}\frac{\partial f_{w_n}}{\partial S_{g_n}} = -\frac{q_{t_n}}{B_{w_n}^{n+1}}\frac{\dfrac{\partial}{\partial k_{ro}}\left(\dfrac{\mu_w}{\mu_o}\dfrac{k_{ro}}{k_{rw}}\right)+\dfrac{\partial}{\partial k_{ro}}\left(\dfrac{\mu_w}{\mu_g}\dfrac{k_{rg}}{k_{rw}}\right)}{\left(1+\dfrac{\mu_w}{\mu_o}\dfrac{k_{ro}}{k_{rw}}+\dfrac{\mu_w}{\mu_g}\dfrac{k_{rg}}{k_{rw}}\right)^2}\frac{\partial k_{ro}}{\partial S_g},\quad\ldots\ldots\ldots\ldots\ldots\ldots(9.200)$$

$$\frac{\partial q_{wsc_n}}{\partial S_{g_n}} = \frac{\partial}{\partial S_{g_n}}\left(\frac{f_{w_n}}{B_{w_n}^{n+1}}q_{t_n}\right) = \frac{q_{t_n}}{B_{w_n}^{n+1}}\frac{\partial f_{w_n}}{\partial S_{g_n}} = -\frac{q_{t_n}}{B_{w_n}^{n+1}}\frac{\dfrac{\partial}{\partial k_{rw}}\left(\dfrac{\mu_w}{\mu_o}\dfrac{k_{ro}}{k_{rw}}\right)+\dfrac{\partial}{\partial k_{rw}}\left(\dfrac{\mu_w}{\mu_g}\dfrac{k_{rg}}{k_{rw}}\right)}{\left(1+\dfrac{\mu_w}{\mu_o}\dfrac{k_{ro}}{k_{rw}}+\dfrac{\mu_w}{\mu_g}\dfrac{k_{rg}}{k_{rw}}\right)^2}\frac{\partial k_{rw}}{\partial S_g},\quad\ldots\ldots\ldots\ldots\ldots\ldots(9.201)$$

$$\frac{\partial q_{gsc_n}}{\partial S_{w_n}} = \frac{\partial}{\partial S_{w_n}}\left[\left(\frac{f_{g_n}}{B_{g_n}^{n+1}}+R_{s_n}^{n+1}\frac{f_{o_n}}{B_{o_n}^{n+1}}\right)q_{t_n}\right] = \frac{q_{t_n}}{B_{g_n}^{n+1}}\frac{\partial f_{g_n}}{\partial S_{w_n}}$$

$$+R_{s_n}^{n+1}\frac{q_{t_n}}{B_{o_n}^{n+1}}\frac{\partial f_{o_n}}{\partial S_{w_n}} = -\frac{q_{t_n}}{B_{g_n}^{n+1}}\frac{\dfrac{\partial}{\partial k_{ro}}\left(\dfrac{\mu_g}{\mu_o}\dfrac{k_{ro}}{k_{rg}}\right)+\dfrac{\partial}{\partial k_{ro}}\left(\dfrac{\mu_g}{\mu_w}\dfrac{k_{rw}}{k_{rg}}\right)}{\left(1+\dfrac{\mu_g}{\mu_o}\dfrac{k_{ro}}{k_{rg}}+\dfrac{\mu_g}{\mu_w}\dfrac{k_{rw}}{k_{rg}}\right)^2}\frac{\partial k_{ro}}{\partial S_w}\quad\ldots\ldots\ldots\ldots\ldots\ldots\ldots(9.202)$$

$$-R_{s_n}^{n+1}\frac{q_{t_n}}{B_{o_n}^{n+1}}\frac{\dfrac{\partial}{\partial k_{ro}}\left(\dfrac{\mu_o}{\mu_w}\dfrac{k_{rw}}{k_{ro}}\right)+\dfrac{\partial}{\partial k_{ro}}\left(\dfrac{\mu_o}{\mu_g}\dfrac{k_{rg}}{k_{ro}}\right)}{\left(1+\dfrac{\mu_o}{\mu_w}\dfrac{k_{rw}}{k_{ro}}+\dfrac{\mu_o}{\mu_g}\dfrac{k_{rg}}{k_{ro}}\right)^2}\frac{\partial k_{ro}}{\partial S_w},$$

$$\frac{\partial q_{gsc_n}}{\partial S_{w_n}} = \frac{\partial}{\partial S_{w_n}}\left[\left(\frac{f_{g_n}}{B_{g_n}^{n+1}}+R_{s_n}^{n+1}\frac{f_{o_n}}{B_{o_n}^{n+1}}\right)q_{t_n}\right] = \frac{q_{t_n}}{B_{g_n}^{n+1}}\frac{\partial f_{g_n}}{\partial S_{w_n}}$$

$$+R_{s_n}^{n+1}\frac{q_{t_n}}{B_{o_n}^{n+1}}\frac{\partial f_{o_n}}{\partial S_{w_n}} = -\frac{q_{t_n}}{B_{g_n}^{n+1}}\frac{\dfrac{\partial}{\partial k_{rw}}\left(\dfrac{\mu_g}{\mu_o}\dfrac{k_{ro}}{k_{rg}}\right)+\dfrac{\partial}{\partial k_{rw}}\left(\dfrac{\mu_g}{\mu_w}\dfrac{k_{rw}}{k_{rg}}\right)}{\left(1+\dfrac{\mu_g}{\mu_o}\dfrac{k_{ro}}{k_{rg}}+\dfrac{\mu_g}{\mu_w}\dfrac{k_{rw}}{k_{rg}}\right)^2}\frac{\partial k_{rw}}{\partial S_w}\quad\ldots\ldots\ldots\ldots\ldots\ldots\ldots(9.203)$$

$$-R_{s_n}^{n+1}\frac{q_{t_n}}{B_{o_n}^{n+1}}\frac{\dfrac{\partial}{\partial k_{rw}}\left(\dfrac{\mu_o}{\mu_w}\dfrac{k_{rw}}{k_{ro}}\right)+\dfrac{\partial}{\partial k_{rw}}\left(\dfrac{\mu_o}{\mu_g}\dfrac{k_{rg}}{k_{ro}}\right)}{\left(1+\dfrac{\mu_o}{\mu_w}\dfrac{k_{rw}}{k_{ro}}+\dfrac{\mu_o}{\mu_g}\dfrac{k_{rg}}{k_{ro}}\right)^2}\frac{\partial k_{rw}}{\partial S_w},$$

$$\frac{\partial q_{gsc_n}}{\partial S_{g_n}} = \frac{\partial}{\partial S_{w_n}}\left[\left(\frac{f_{g_n}}{B_{g_n}^{n+1}}+R_{s_n}^{n+1}\frac{f_{o_n}}{B_{o_n}^{n+1}}\right)q_{t_n}\right] = \frac{q_{t_n}}{B_{g_n}^{n+1}}\frac{\partial f_{g_n}}{\partial S_{w_n}}+R_{s_n}^{n+1}\frac{q_{t_n}}{B_{o_n}^{n+1}}\frac{\partial f_{o_n}}{\partial S_{w_n}}$$

$$=-\frac{q_{t_n}}{B_{g_n}^{n+1}}\frac{\dfrac{\partial}{\partial k_{ro}}\left(\dfrac{\mu_g}{\mu_o}\dfrac{k_{ro}}{k_{rg}}\right)+\dfrac{\partial}{\partial k_{ro}}\left(\dfrac{\mu_g}{\mu_w}\dfrac{k_{rw}}{k_{rg}}\right)}{\left(1+\dfrac{\mu_g}{\mu_o}\dfrac{k_{ro}}{k_{rg}}+\dfrac{\mu_g}{\mu_w}\dfrac{k_{rw}}{k_{rg}}\right)^2}\frac{\partial k_{ro}}{\partial S_g}-R_{s_n}^{n+1}\frac{q_{t_n}}{B_{o_n}^{n+1}}\frac{\dfrac{\partial}{\partial k_{ro}}\left(\dfrac{\mu_o}{\mu_w}\dfrac{k_{rw}}{k_{ro}}\right)+\dfrac{\partial}{\partial k_{ro}}\left(\dfrac{\mu_o}{\mu_g}\dfrac{k_{rg}}{k_{ro}}\right)}{\left(1+\dfrac{\mu_o}{\mu_w}\dfrac{k_{rw}}{k_{ro}}+\dfrac{\mu_o}{\mu_g}\dfrac{k_{rg}}{k_{ro}}\right)^2}\frac{\partial k_{ro}}{\partial S_g},\quad\ldots\ldots\ldots(9.204)$$

$$\frac{\partial q_{gsc_n}}{\partial S_{g_n}} = \frac{\partial}{\partial S_{w_n}}\left[\left(\frac{f_{g_n}}{B_{g_n}^{n+1}}+R_{s_n}^{n+1}\frac{f_{o_n}}{B_{o_n}^{n+1}}\right)q_{t_n}\right] = \frac{q_{t_n}}{B_{g_n}^{n+1}}\frac{\partial f_{g_n}}{\partial S_{w_n}}+R_{s_n}^{n+1}\frac{q_{t_n}}{B_{o_n}^{n+1}}\frac{\partial f_{o_n}}{\partial S_{w_n}}$$

$$=-\frac{q_{t_n}}{B_{g_n}^{n+1}}\frac{\dfrac{\partial}{\partial k_{rw}}\left(\dfrac{\mu_g}{\mu_o}\dfrac{k_{ro}}{k_{rg}}\right)+\dfrac{\partial}{\partial k_{rw}}\left(\dfrac{\mu_g}{\mu_w}\dfrac{k_{rw}}{k_{rg}}\right)}{\left(1+\dfrac{\mu_g}{\mu_o}\dfrac{k_{ro}}{k_{rg}}+\dfrac{\mu_g}{\mu_w}\dfrac{k_{rw}}{k_{rg}}\right)^2}\frac{\partial k_{rw}}{\partial S_g}-R_{s_n}^{n+1}\frac{q_{t_n}}{B_{o_n}^{n+1}}\frac{\dfrac{\partial}{\partial k_{rw}}\left(\dfrac{\mu_o}{\mu_w}\dfrac{k_{rw}}{k_{ro}}\right)+\dfrac{\partial}{\partial k_{rw}}\left(\dfrac{\mu_o}{\mu_g}\dfrac{k_{rg}}{k_{ro}}\right)}{\left(1+\dfrac{\mu_o}{\mu_w}\dfrac{k_{rw}}{k_{ro}}+\dfrac{\mu_o}{\mu_g}\dfrac{k_{rg}}{k_{ro}}\right)^2}\frac{\partial k_{rw}}{\partial S_g}.\quad\ldots\ldots\ldots(9.205)$$

Problem 9.3.20 (Ertekin et al. 2001)

Write the implicit water-saturation equation for Gridblock n for incompressible two-phase flow of oil and water with negligible capillary pressure in a 2D horizontal reservoir. In your derivation, assume that each Gridblock $n = i, j, k$ may have a production well in it, and use the following equations as your starting points:

1. Implicit-saturation step of the SEQ method (the implicit-saturation equation):

$$\sum_{m \in \psi_n} \left[\left(\Delta_m p_o^n - \Delta_m P_{cow}^n - \overline{\gamma}_{w_{n,m}}^n \Delta_m G \right) \left(\frac{\partial T_{w_{n,m}}}{\partial S_{w_m}} \right)^n - T_{w_{n,m}}^n P_{cow_m}^{'\,n} \right]$$

$$\Delta_t S_{w_m} + \left\{ \sum_{m \in \psi_n} \left[\left(\Delta_m p_o^n - \Delta_m P_{cow}^n - \overline{\gamma}_{w_{n,m}}^n \Delta_m G \right) \left(\frac{\partial T_{w_{n,m}}}{\partial S_{w_m}} \right)^n - T_{w_{n,m}}^n P_{cow_m}^{'\,n} \right] - \left(C_{ww_n} - q_{wwsc_n}^{'\,n} \right) \right\} \quad \dots\dots\dots (9.206)$$

$$\Delta_t S_{w_m} + q_{wgsc_n}^{'\,n} \Delta_t S_{g_n} = -C_{ww_n} \left(S_{w_n}^* - S_{w_n}^n \right).$$

2. Implicit-saturation equation of the SEQ method for two-phase flow (oil and water):

$$\sum_{m \in \psi_n} \left[\left(\frac{f_w}{B_w} \right)_{n,m}^n \Delta_t q_{t_{n,m}} \right] + \sum_{m \in \psi_n} \left[\begin{array}{c} \left(\frac{q_t}{B_w^n} \right)_{n,m} \left(\frac{\partial f_{w_{n,m}}}{\partial S_{w_m}} \right)^n \\ + \frac{G_{n,m}}{B_{w_{n,m}}^n} \left\{ \begin{array}{c} -\left[\Delta_m P_{cow}^n + \overline{(\Delta\gamma)}_{ow_{n,m}}^n \Delta_m G \right] \times \left[\left(\frac{\partial Y_{g_{n,m}}}{\partial S_{w_m}} \right)^n + \left(\frac{\zeta_{w_{n,m}}}{\partial S_{w_m}} \right)^n \right] - \\ \left(Y_g^n + \zeta_w^n \right)_{n,m} P_{cow_m}^{'\,n} - \left[\Delta_m P_{cgo}^n + \overline{(\Delta\gamma)}_{og_{n,m}}^n \Delta_m G \right] \times \left(\frac{\partial Y_{g_{n,m}}}{\partial S_{w_m}} \right)^n \end{array} \right\} \end{array} \right] \Delta_t S_{w_m}$$

$$+ \sum_{m \in \psi_n} \left(\left(\frac{q_t}{D_w^n} \right)_{n,m} \left(\frac{\partial f_{w_{n,m}}}{\partial S_{g_m}} \right)^n + \frac{G_{n,m}}{B_{w_{n,m}}^n} \left\{ \begin{array}{c} -\left[\Delta_m P_{cow}^n + \overline{(\Delta\gamma)}_{ow_{n,m}}^n \Delta_m G \right] \times \left[\left(\frac{\partial Y_{g_{n,m}}}{\partial S_{g_m}} \right)^n + \left(\frac{\zeta_{w_{n,m}}}{\partial S_{g_m}} \right)^n \right] \\ -\left(Y_g^n \right)_{n,m} P_{cgo_m}^{'\,n} - \left[\Delta_m P_{cgo}^n + \overline{(\Delta\gamma)}_{og_{n,m}}^n \Delta_m G \right] \times \left(\frac{\partial Y_{g_{n,m}}}{\partial S_{g_m}} \right)^n \end{array} \right\} \right) \Delta_t S_{g_m}$$

$$+ \left[-\left(C_{ww_n} - q_{wwsc_n}^{'\,n} \right) + \sum_{m \in \psi_n} \left(\begin{array}{c} \left(\frac{q_t}{B_w^n} \right)_{n,m} \left(\frac{\partial f_{w_{n,m}}}{\partial S_{w_m}} \right)^n + \frac{G_{n,m}}{B_{w_{n,m}}^n} \\ \left\{ \begin{array}{c} -\left(\Delta_m P_{cow}^n + \overline{(\Delta\gamma)}_{ow_{n,m}}^n \Delta_m G \right) \times \left[\left(\frac{\partial Y_{g_{n,m}}}{\partial S_{w_m}} \right)^n + \left(\frac{\zeta_{w_{n,m}}}{\partial S_{w_m}} \right)^n \right] \\ +\left(Y_g^n + \zeta_w^n \right)_{n,m} P_{cow_m}^{'\,n} - \left(\Delta_m P_{cgo}^n + \overline{(\Delta\gamma)}_{og_{n,m}}^n \Delta_m G \right) \times \left(\frac{\partial Y_{g_{n,m}}}{\partial S_{w_m}} \right)^n \end{array} \right\} \end{array} \right) \right] \Delta_t S_{w_n}$$

$$+ \left[q_{wgsc_n}^{'\,n} + \sum_{m \in \psi_n} \left(\begin{array}{c} \left(\frac{q_t}{B_w^n} \right)_{n,m} \left(\frac{\partial f_{w_{n,m}}}{\partial S_{g_m}} \right)^n + \frac{G_{n,m}}{B_{w_{n,m}}^n} \\ \left\{ \begin{array}{c} -\left(\Delta_m P_{cow}^n + \overline{(\Delta\gamma)}_{ow_{n,m}}^n \Delta_m G \right) \times \left[\left(\frac{\partial Y_{g_{n,m}}}{\partial S_{g_m}} \right)^n + \left(\frac{\zeta_{w_{n,m}}}{\partial S_{g_m}} \right)^n \right] \\ +\left(Y_g^n \right)_{n,m} P_{cgo_m}^{'\,n} - \left(\Delta_m P_{cgo}^n + \overline{(\Delta\gamma)}_{og_{n,m}}^n \Delta_m G \right) \times \left(\frac{\partial Y_{g_{n,m}}}{\partial S_{g_m}} \right)^n \end{array} \right\} \end{array} \right) \right]$$

$$\Delta_t S_{g_n} = -C_{ww_n} \left(S_{w_n}^* - S_{w_n}^n \right).$$

$$\dots\dots\dots\dots\dots\dots\dots\dots\dots\dots\dots\dots\dots\dots\dots\dots\dots\dots\dots\dots\dots\dots\dots\dots\dots\dots\dots (9.207)$$

3. The implicit approximation of the saturation equation, if capillary and gravitational forces are neglected is

$$\sum_{m\in\psi_n}\left(\frac{q_t}{B_w^n}\right)_{n,m}\left(\frac{\partial f_{w_{n,m}}}{\partial S_{w_m}}\right)^n\Delta_t S_{w_m}+\sum_{m\in\psi_n}\left(\frac{q_t}{B_w^n}\right)_{n,m}\left(\frac{\partial f_{w_{n,m}}}{\partial S_{g_m}}\right)^n\Delta_t S_{g_m}$$

$$+\left[-\left(C_{ww_n}-q_{wwsc_n}^{'n}\right)+\sum_{m\in\psi_n}\left(\frac{q_t}{B_w^n}\right)_{n,m}\left(\frac{\partial f_{w_{n,n}}}{\partial S_{w_n}}\right)^n\right] \quad\ldots\ldots\ldots\ldots\ldots\ldots (9.208)$$

$$\Delta_t S_{w_n}+\left[-\left(C_{wg_n}-q_{wgsc_n}^{'n}\right)+\sum_{m\in\psi_n}\left(\frac{q_t}{B_w^n}\right)_{n,m}\left(\frac{\partial f_{w_{n,m}}}{\partial S_{g_n}}\right)^n\right]\Delta_t S_{g_n}=-C_{ww_n}\left(S_{w_n}^*-S_{w_n}^n\right).$$

Solution to Problem 9.3.20

1. $$\sum_{m\in\psi_n}\left[\left(\Delta_m p_o^n\right)\left(\frac{\partial T_{w_{n,m}}}{\partial S_{w_m}}\right)^n\right]\Delta_t S_{w_m}$$

$$+\left\{\sum_{m\in\psi_n}\left[\left(\Delta_m p_o^n\right)\left(\frac{\partial T_{w_{n,m}}}{\partial S_{w_m}}\right)^n\right]-\left(C_{ww_n}-q_{wwsc_n}^{'n}\right)\right\}\Delta_t S_{w_n}=-C_{ww_n}\left(S_{w_n}^*-S_{w_n}^n\right). \quad\ldots\ldots\ldots\ldots\ldots (9.209)$$

2. $$\sum_{m\in\psi_n}\left[\left(\frac{f_w}{B_w}\right)_{n,m}^n\Delta_t q_{t_{n,m}}\right]+\sum_{m\in\psi_n}\left[\left(\frac{q_t}{B_w^n}\right)_{n,m}\left(\frac{\partial f_{w_{n,n}}}{\partial S_{w_n}}\right)^n\right]\Delta_t S_{w_m}$$

$$+\left[-\left(C_{ww_n}-q_{wwsc_n}^{'n}\right)+\sum_{m\in\psi_n}\left(\frac{q_t}{B_w^n}\right)_{n,m}\left(\frac{\partial f_{w_{n,n}}}{\partial S_{w_n}}\right)^n\right]\Delta_t S_{w_m}=-C_{ww_n}\left(S_{w_n}^*-S_{w_n}^n\right). \quad\ldots\ldots\ldots\ldots\ldots (9.210)$$

3. $$\sum_{m\in\psi_n}\left[\left(\frac{q_t}{B_w^n}\right)_{n,m}\left(\frac{\partial f_{w_{n,n}}}{\partial S_{w_n}}\right)^n\right]\Delta_t S_{w_m}+\left[-\left(C_{ww_n}-q_{wwsc_n}^{'n}\right)+\sum_{m\in\psi_n}\left(\frac{q_t}{B_w^n}\right)_{n,m}\left(\frac{\partial f_{w_{n,n}}}{\partial S_{w_n}}\right)^n\right] \quad\ldots\ldots\ldots\ldots\ldots (9.211)$$

$$\Delta_t S_{w_m}=-C_{ww_n}\left(S_{w_n}^*-S_{w_n}^n\right).$$

Problem 9.3.21 (Ertekin et al. 2001)

Give a complete formulation of two-phase gas/water flow applicable to a 2D domain with no depth gradients. Assume the solubility of gas in water and the capillary pressure between the two phases to be negligible. Treat the porous medium as homogeneous and isotropic. Reduce the equations to their simplest forms and identify the principal unknowns and equations necessary to solve the problem.

Solution to Problem 9.3.21

Two-phase gas/water flow equations for the reservoir described are

$$\frac{\partial}{\partial x}\left(\frac{A_x k_x k_{rw}}{\mu_w B_w}\frac{\partial p_w}{\partial x}\right)\Delta x+\frac{\partial}{\partial y}\left(\frac{A_y k_y k_{rw}}{\mu_w B_w}\frac{\partial p_w}{\partial y}\right)\Delta y+q_w=\frac{V_b}{5.615}\frac{\partial}{\partial t}\left(\frac{\phi S_w}{B_w}\right), \quad\ldots\ldots\ldots\ldots\ldots\ldots\ldots\ldots\ldots\ldots (9.212)$$

$$\frac{\partial}{\partial x}\left(\frac{A_x k_x k_{rg}}{\mu_g B_g}\frac{\partial p_g}{\partial x}\right)\Delta x+\frac{\partial}{\partial y}\left(\frac{A_y k_y k_{rg}}{\mu_g B_g}\frac{\partial p_g}{\partial y}\right)\Delta y+q_g=\frac{V_b}{5.615}\frac{\partial}{\partial t}\left(\frac{\phi S_g}{B_g}\right). \quad\ldots\ldots\ldots\ldots\ldots\ldots\ldots\ldots\ldots\ldots (9.213)$$

For a homogeneous and isotropic ($k_x = k_y = k$) reservoir with no capillary pressure ($p_g = p_w = p$), water and gas equations can be reduced to the following forms:

$$\frac{\partial}{\partial x}\left(\frac{k_{rw}}{\mu_w B_w}\frac{\partial p_w}{\partial x}\right)+\frac{\partial}{\partial y}\left(\frac{k_{rw}}{\mu_w B_w}\frac{\partial p_w}{\partial y}\right)+\frac{q_w}{V_b k}=\frac{1}{k5.615}\frac{\partial}{\partial t}\left(\frac{\phi S_w}{B_w}\right),\dotfill (9.214)$$

$$\frac{\partial}{\partial x}\left(\frac{k_{rg}}{\mu_g B_g}\frac{\partial p_g}{\partial x}\right)+\frac{\partial}{\partial y}\left(\frac{k_{rg}}{\mu_g B_g}\frac{\partial p_g}{\partial y}\right)+\frac{q_g}{V_b k}=\frac{1}{k5.615}\frac{\partial}{\partial t}\left(\frac{\phi S_g}{B_g}\right).\dotfill (9.215)$$

Auxiliary equation is the saturation relationship: $S_w + S_g = 1$.

The principal unknowns of the problem are p and S_w and S_g; with the three equation as described, it is obvious that the problem is well posed.

Problem 9.3.22 (Ertekin et al. 2001)

1. After examining the formulations that follow, describe the fluid flow problem and the porous media to the fullest extent:

$$\frac{\partial}{\partial x}\left(\frac{k_x k_{ro}}{\mu_o B_o}\frac{\partial p}{\partial x}\right)+\frac{q_o}{V_b}=\frac{1}{5.615}\frac{\partial}{\partial t}\left(\frac{\phi S_o}{B_o}\right)+\frac{\partial}{\partial x}\left(\gamma_c \rho_o g\frac{k_x k_{ro}}{\mu_o B_o}\frac{\partial G}{\partial x}\right),\dotfill (9.216)$$

$$\frac{\partial}{\partial x}\left(\frac{k_x k_{rw}}{\mu_w B_w}\frac{\partial p}{\partial x}\right)+\frac{q_w}{V_b}=\frac{1}{5.615}\frac{\partial}{\partial t}\left(\frac{\phi S_w}{B_w}\right)+\frac{\partial}{\partial x}\left(\gamma_c \rho_w g\frac{k_x k_{rw}}{\mu_w B_w}\frac{\partial G}{\partial x}\right).\dotfill (9.217)$$

2. What are the principal unknowns of the problem? Provide the necessary auxiliary equations to solve the problem.

Solution to Problem 9.3.22

1. It is a 1D two-phase (oil/water) reservoir with wells; depth gradient exists and is not negligible. Rock is compressible. The reservoir is heterogeneous with respect to permeability distribution. There is no capillary pressure between phases.
2. Principle unknowns of the problem are p, S_w, and S_o. Auxiliary equation is $S_w + S_o = 1$.

Problem 9.3.23

Consider the two-phase flow problem in the reservoir shown in **Fig. 9.9**. Give a complete formulation of the flow equations to the fullest extent using the information provided in the figure.

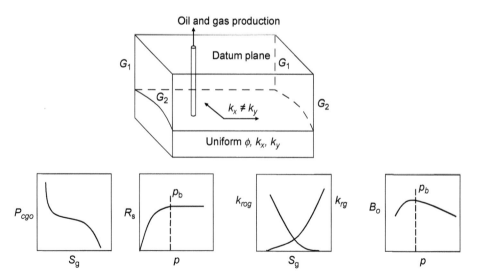

Fig. 9.9—Description of the reservoir and fluid properties of Problem 9.3.23.

Solution to Problem 9.3.23

Saturated conditions: $S_o + S_g = 1$.

- Oil equation:

$$k_x \frac{\partial}{\partial x}\left(\frac{k_{ro}}{\mu_o B_o}\frac{\partial \Phi_o}{\partial x}\right)\Delta x + k_y \frac{\partial}{\partial y}\left(\frac{k_{ro}}{\mu_o B_o}\frac{\partial \Phi_o}{\partial y}\right)\Delta y + k_z \frac{\partial}{\partial z}\left(\frac{k_{ro}}{\mu_o B_o}\frac{\partial \Phi_o}{\partial z}\right)\Delta z + q_o = \frac{V_b \phi}{5.615}\frac{\partial}{\partial t}\left(\frac{S_o}{B_o}\right). \quad\dots\dots\dots (9.218)$$

- Gas equation:

$$k_x \frac{\partial}{\partial x}\left(\frac{A_x k_{rg}}{\mu_g B_g}\frac{\partial p_g}{\partial x} + R_{so}\frac{A_x k_{ro}}{\mu_o B_o}\frac{\partial \Phi_o}{\partial x}\right)\Delta x + k_y \frac{\partial}{\partial y}\left(\frac{A_y k_{rg}}{\mu_g B_g}\frac{\partial p_g}{\partial y} + R_{so}\frac{A_y k_{ro}}{\mu_o B_o}\frac{\partial \Phi_o}{\partial y}\right)\Delta y$$

$$+ k_z \frac{\partial}{\partial z}\left(\frac{A_z k_{rg}}{\mu_g B_g}\frac{\partial p_g}{\partial z} + R_{so}\frac{A_z k_{ro}}{\mu_o B_o}\frac{\partial \Phi_o}{\partial z}\right)\Delta z + q_g + R_{so} q_o = \frac{V_b \phi}{5.615}\frac{\partial}{\partial t}\left(\frac{S_g}{B_g} + R_{so}\frac{S_o}{B_o}\right). \quad\dots\dots\dots\dots (9.219)$$

Undersaturated conditions: $S_o = 1$ and $S_g = 0$.

- Oil equation:

$$k_x \frac{\partial}{\partial x}\left(\frac{k_{ro}}{\mu_o B_o}\frac{\partial \Phi_o}{\partial x}\right)\Delta x + k_y \frac{\partial}{\partial y}\left(\frac{k_{ro}}{\mu_o B_o}\frac{\partial \Phi_o}{\partial y}\right)\Delta y + k_z \frac{\partial}{\partial z}\left(\frac{k_{ro}}{\mu_o B_o}\frac{\partial \Phi_o}{\partial z}\right)\Delta z + q_o = \frac{V_b \phi}{5.615}\frac{\partial}{\partial t}\left(\frac{S_o}{B_o}\right). \quad\dots\dots\dots (9.220)$$

- Gas equation:

$$k_x \frac{\partial}{\partial x}\left(R_{so}\frac{A_x k_{ro}}{\mu_o B_o}\frac{\partial \Phi_o}{\partial x}\right)\Delta x + k_y \frac{\partial}{\partial y}\left(R_{so}\frac{A_y k_{ro}}{\mu_o B_o}\frac{\partial \Phi_o}{\partial y}\right)$$

$$\Delta y + k_z \frac{\partial}{\partial z}\left(R_{so}\frac{A_z k_{ro}}{\mu_o B_o}\frac{\partial \Phi_o}{\partial z}\right)\Delta z + R_{so} q_o = \frac{V_b \phi}{5.615}\frac{\partial}{\partial t}\left(R_{so}\frac{S_o}{B_o}\right). \quad\dots\dots\dots (9.221)$$

Problem 9.3.24

Give a complete description of the multiphase-flow problem expressed by the given equations:

$$\frac{\partial}{\partial x}\left(\frac{k_{ro}}{\mu_o B_o}\frac{\partial p}{\partial x}\right)\Delta x + \frac{q_o}{V_b k_x} = \frac{\phi}{5.615 k_x}\frac{\partial}{\partial t}\left(\frac{S_g}{B_g}\right), \quad\dots\dots\dots\dots\dots\dots\dots\dots\dots\dots (9.222)$$

$$\frac{\partial}{\partial x}\left(k_{rw}\frac{\partial p}{\partial x}\right)\Delta x + \frac{q_w \mu_w}{V_b k_x} = \frac{\phi \mu_w}{5.615 k_x}\frac{\partial S_w}{\partial t}, \quad\dots\dots\dots\dots\dots\dots\dots\dots\dots\dots (9.223)$$

$$\frac{\partial}{\partial x}\left(\frac{R_{so} k_{ro}}{\mu_o B_o}\frac{\partial p}{\partial x}\right)\Delta x + R_{so}\frac{q_o}{V_b k_x} = \frac{\phi}{5.615 k_x}\frac{\partial}{\partial t}\left(R_{so}\frac{S_g}{B_g}\right), \quad\dots\dots\dots\dots\dots\dots (9.224)$$

$$S_o + S_w + S_g = 1. \quad\dots\dots\dots\dots\dots\dots\dots\dots\dots\dots\dots\dots\dots\dots\dots\dots\dots (9.225)$$

Are there any redundant equations? Why?

Solution to Problem 9.3.24

- 1D flow along the x-direction.
- Homogeneous permeability distribution.
- Incompressible pore space.
- Horizontal reservoir.
- Three-phase flow.
- Reservoir is undersaturated.
- Water phase is treated as incompressible.
- R_{sw} is assumed to be equal to zero.
- Capillary pressure is ignored.
- There is (are) well(s) in the flow domain.

The third equation is the same as the first equation; therefore, it is redundant. Undersaturated conditions indicate that R_{so} is constant. If the third equation is divided by R_{so}, the first equation will be obtained.

Problem 9.3.25

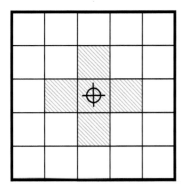

Fig. 9.10—Reservoir system of Problem 9.3.25.

Fig. 9.10 shows the plan view of a 2D geopressured aquifer. A geopressured aquifer is defined as an ultrahigh-pressure gas reservoir in which the gas phase, originally in its entirety, is dissolved in the water phase.

The crosshatched blocks in Fig. 9.10 represent the portion of the reservoir in which the pressure dropped below the bubblepoint as a result of the production well. Give two sets of equations that can be used in both regions of the reservoir. Assume that the reservoir has homogeneous, but anisotropic, property distribution.

Solution to Problem 9.3.25

PDEs:

- Water equation (inner and outer regions):

$$k_x \frac{\partial}{\partial x}\left(\frac{k_{rw}}{\mu_w B_w} \frac{\partial \Phi_w}{\partial x} \right) + k_y \frac{\partial}{\partial y}\left(\frac{k_{rw}}{\mu_w B_w} \frac{\partial \Phi_w}{\partial y} \right) + \frac{q_w}{V_b} = \frac{1}{V_b} \frac{\partial}{\partial t}\left(\frac{\phi S_w}{B_w} \right). \quad \dots \dots \dots (9.226)$$

- Gas equation (inner region):

$$k_x \frac{\partial}{\partial x}\left(\frac{k_{rg}}{\mu_g B_g} \frac{\partial p_g}{\partial x} + R_{sw} \frac{k_{rw}}{\mu_w B_w} \frac{\partial \Phi_w}{\partial x} \right) + k_y \frac{\partial}{\partial y}\left(\frac{k_{rg}}{\mu_g B_g} \frac{\partial p_g}{\partial y} + R_{sw} \frac{k_{rw}}{\mu_w B_w} \frac{\partial \Phi_w}{\partial y} \right)$$
$$+ \frac{q_g + R_{sw}q_w}{V_b} = \frac{1}{V_b} \frac{\partial}{\partial t}\left(\frac{\phi S_g}{B_g} + R_{sw} \frac{\phi S_w}{B_w} \right). \quad \dots \dots \dots (9.227)$$

- Gas equation (outer region):

$$k_x \frac{\partial}{\partial x}\left(R_{sw} \frac{k_{rw}}{\mu_w B_w} \frac{\partial \Phi_w}{\partial x} \right) + k_y \frac{\partial}{\partial y}\left(R_{sw} \frac{k_{rw}}{\mu_w B_w} \frac{\partial \Phi_w}{\partial y} \right) + \frac{R_{sw}q_w}{V_b} = \frac{1}{V_b} \frac{\partial}{\partial t}\left(R_{sw} \frac{\phi S_w}{B_w} \right). \quad \dots \dots \dots (9.228)$$

Capillary pressure relationships:

- $P_{cgw} = p_g - p_w$ (inner region).

- $P_{cgw} = 0$ (outer region).

Saturation relationships:

- $S_g + S_w = 1$ (inner region).

- $S_g = 0$ and $S_w = 1$ (outer region).

Problem 9.3.26

Consider a two-phase, saturated-oil/-gas reservoir with significant capillary pressure difference between the phases, give a complete mathematical formulation of this system for 2D flow (x-y directions only) and reduce the equations to their simplest forms. The reservoir has homogeneous and isotropic permeability distribution and is a water-wet reservoir. The top and cross-sectional view of the reservoir is shown in **Fig. 9.11** (parts a and b, respectively).

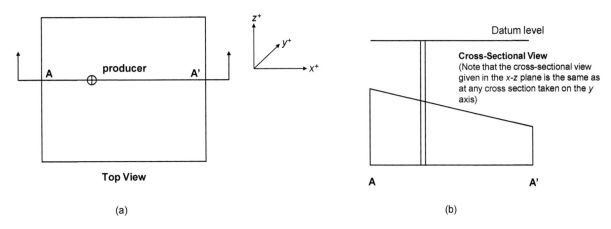

Fig. 9.11—Reservoir system of Problem 9.3.26.

Solution to Problem 9.3.26

Gas equation is as follows:

$$\frac{\partial}{\partial x}\left\{A_x\left[\frac{k_{rog}}{\mu_g B_g}\frac{\partial p_g}{\partial x}+R_{so}\left(\frac{k_{ro}}{\mu_o B_o}\frac{\partial p_o}{\partial x}-\frac{1}{144}\frac{g}{g_c}\rho_o\frac{\partial G}{\partial x}\right)\right]\right\}$$

$$\Delta x+V_b\frac{\partial}{\partial y}\left[\left(\frac{k_{rog}}{\mu_g B_g}\frac{\partial p_g}{\partial y}+R_{so}\frac{k_{ro}}{\mu_o B_o}\frac{\partial p_o}{\partial y}\right)\right]+\frac{q_g+R_{so}q_o}{k}=\frac{V_b}{5.615k}\frac{\partial}{\partial t}\left(\frac{\phi S_g}{B_g}+R_{so}\frac{\phi S_o}{B_o}\right). \quad \ldots\ldots\ldots\ldots\ldots (9.229)$$

Oil equation is as follows:

$$\frac{\partial}{\partial x}\left[A_x k_{ro}\left(R_v\frac{1}{\mu_g B_g}\frac{\partial p_g}{\partial x}+\frac{1}{\mu_o B_o}\frac{\partial p_o}{\partial x}-\frac{1}{144}\frac{g}{g_c}\rho_o\frac{\partial G}{\partial x}\right)\right]\Delta x$$

$$+V_b\frac{\partial}{\partial y}\left[\left(R_v\frac{k_{ro}}{\mu_g B_g}\frac{\partial p_g}{\partial y}+\frac{k_{ro}}{\mu_o B_o}\frac{\partial p_o}{\partial y}\right)\right]+\frac{R_v q_g+q_o}{k}=\frac{V_b}{5.615k}\frac{\partial}{\partial t}\left(R_v\frac{\phi S_g}{B_g}+\frac{\phi S_o}{B_o}\right). \quad \ldots\ldots\ldots\ldots\ldots (9.230)$$

Capillary pressure relationship is as follows:

$$P_{cgo}\left(S_g\right)=p_g-p_o. \quad \ldots\ldots\ldots\ldots\ldots\ldots\ldots\ldots\ldots\ldots\ldots\ldots\ldots\ldots\ldots\ldots\ldots\ldots\ldots\ldots (9.231)$$

Saturation relationship is as follows:

$$S_o+S_g=1. \quad \ldots\ldots\ldots\ldots\ldots\ldots\ldots\ldots\ldots\ldots\ldots\ldots\ldots\ldots\ldots\ldots\ldots\ldots\ldots\ldots\ldots\ldots\ldots (9.232)$$

Problem 9.3.27

Give a complete mathematical formulation of the following system:
- Thick and layered reservoir.
- $k_x = k_y \neq k_z$.
- Homogeneous rock properties.
- Two-phase flow (water and oil are slightly compressible).
- Negligible capillary pressure effects.
- Production wells only.
- Compressible pore volume.

Solution to Problem 9.3.27

$$\frac{\partial}{\partial x}\left(\frac{k_{ro}}{\mu_o B_o}\frac{\partial \Phi_o}{\partial x}\right)+\frac{\partial}{\partial y}\left(\frac{k_{ro}}{\mu_o B_o}\frac{\partial \Phi_o}{\partial y}\right)+\frac{k_z}{k_x}\frac{\partial}{\partial z}\left(\frac{k_{ro}}{\mu_o B_o}\frac{\partial \Phi_o}{\partial z}\right)+\frac{q_o}{k_x V_b}=\frac{1}{5.615 k_x}\frac{\partial}{\partial t}\left(\frac{\phi S_o}{B_o}\right), \quad \ldots\ldots\ldots\ldots (9.233)$$

$$\frac{\partial}{\partial x}\left(\frac{k_{rw}}{\mu_w B_w}\frac{\partial \Phi_w}{\partial x}\right) + \frac{\partial}{\partial y}\left(\frac{k_{rw}}{\mu_w B_w}\frac{\partial \Phi_w}{\partial y}\right) + \frac{k_z}{k_x}\frac{\partial}{\partial z}\left(\frac{k_{rw}}{\mu_w B_w}\frac{\partial \Phi_w}{\partial z}\right) + \frac{q_w}{k_x V_b} = \frac{1}{5.615 k_x}\frac{\partial}{\partial t}\left(\frac{\phi S_w}{B_w}\right), \quad \ldots\ldots\ldots\ldots\ldots(9.234)$$

where, $\Phi_o = p - \frac{1}{144}\frac{g}{g_c}\rho_o G$ and $\Phi_w = p - \frac{1}{144}\frac{g}{g_c}\rho_w G$.

Problem 9.3.28

Give the governing equations describing a 2D gas-condensate reservoir, in which the water-saturation level is at its irreducible saturation. The boundary of the reservoir is completely sealed. The reservoir is heterogeneous and anisotropic in permeability and thickness distributions. There are production wells in the reservoir.

Solution to Problem 9.3.28

The governing equation, when reservoir pressure is greater than dewpoint pressure, is

$$\frac{\partial}{\partial x}\left(\frac{A_x k_x k_{rg}}{\mu_g B_g}\frac{\partial \Phi_g}{\partial x}\right)\Delta x + \frac{\partial}{\partial y}\left(\frac{A_y k_y k_{rg}}{\mu_g B_g}\frac{\partial \Phi_g}{\partial y}\right)\Delta y + q_g = \frac{V_b \phi}{5.615}\frac{\partial}{\partial t}\left(\frac{S_g}{B_g}\right). \quad \ldots\ldots\ldots\ldots\ldots\ldots\ldots\ldots (9.235)$$

The governing equations, when reservoir pressure is less than the dewpoint pressure, are

- Gas equation:

$$\frac{\partial}{\partial x}\left(\frac{A_x k_x k_{rg}}{\mu_g B_g}\frac{\partial \Phi_g}{\partial x} + R_{so}\frac{A_x k_x k_{ro}}{\mu_o B_o}\frac{\partial \Phi_o}{\partial x}\right)\Delta x + \frac{\partial}{\partial y}\left(\frac{A_y k_y k_{rg}}{\mu_g B_g}\frac{\partial \Phi_g}{\partial y} + R_{so}\frac{A_y k_y k_{ro}}{\mu_o B_o}\frac{\partial \Phi_o}{\partial y}\right)\Delta y$$
$$+ q_g + R_{so}q_o = \frac{V_b \phi}{5.615}\frac{\partial}{\partial t}\left(\frac{S_g}{B_g} + R_{so}\frac{S_o}{B_o}\right). \quad \ldots\ldots\ldots\ldots (9.236)$$

- Oil equation:

$$\frac{\partial}{\partial x}\left(\frac{A_x k_x k_{ro}}{\mu_o B_o}\frac{\partial \Phi_o}{\partial x} + R_v\frac{A_x k_x k_{rg}}{\mu_g B_g}\frac{\partial \Phi_g}{\partial x}\right)\Delta x + \frac{\partial}{\partial y}\left(\frac{A_y k_y k_{ro}}{\mu_o B_o}\frac{\partial \Phi_o}{\partial y} + R_v\frac{A_y k_y k_{rg}}{\mu_g B_g}\frac{\partial \Phi_g}{\partial y}\right)\Delta y$$
$$+ q_o + R_v q_g = \frac{V_b \phi}{5.615}\frac{\partial}{\partial t}\left(\frac{S_o}{B_o} + R_v\frac{S_g}{B_g}\right). \quad \ldots\ldots\ldots\ldots\ldots(9.237)$$

Problem 9.3.29

For the following reservoir fluid system, provide the flow equations. Also, give an inventory of the unknowns and equations available.

Two-phase (oil/gas) flow in a 2D system. The gas is soluble in oil. The permeability distribution is homogeneous and isotropic. The reservoir is horizontal, has homogeneous thickness, and is made of incompressible rock; water saturation is at its irreducible value. A gas-recycling process is being held in the reservoir, which implies that there are gas injectors and producers. (Include sets of equations for both producer and injector blocks.)

Solution to Problem 9.3.29

Equations for the production block are as follows:
- Gas equation:

$$A_x k \frac{\partial}{\partial x}\left[\left(\frac{k_{rog}}{\mu_g B_g}\frac{\partial p_g}{\partial x} + R_{so}\frac{k_{ro}}{\mu_o B_o}\frac{\partial p_o}{\partial x}\right)\right]\Delta x + A_y k \frac{\partial}{\partial y}\left[\left(\frac{k_{rog}}{\mu_g B_g}\frac{\partial p_g}{\partial y} + R_{so}\frac{k_{ro}}{\mu_o B_o}\frac{\partial p_o}{\partial y}\right)\right]\Delta y + q_g + R_{so}q_o$$
$$= \frac{V_b}{5.615}\frac{\partial}{\partial t}\left(\frac{\phi S_g}{B_g} + R_{so}\frac{\phi S_o}{B_o}\right). \quad \ldots (9.238)$$

- Oil equation:

$$A_x k \frac{\partial}{\partial x}\left[\left(R_v \frac{k_{rog}}{\mu_g B_g}\frac{\partial p_g}{\partial x} + \frac{k_{ro}}{\mu_o B_o}\frac{\partial p_o}{\partial x}\right)\right]\Delta x + A_y k \frac{\partial}{\partial y}\left[\left(R_v \frac{k_{rog}}{\mu_g B_g}\frac{\partial p_g}{\partial y} + \frac{k_{ro}}{\mu_o B_o}\frac{\partial p_o}{\partial y}\right)\right]\Delta y + R_v q_g + q_o$$

$$= \frac{V_b}{5.615}\frac{\partial}{\partial t}\left(R_v \frac{\phi S_g}{B_g} + \frac{\phi S_o}{B_o}\right). \qquad \ldots\ldots(9.239)$$

Equations for the injection block are as follows:
- Gas equation:

$$A_x k \frac{\partial}{\partial x}\left[\left(\frac{k_{rog}}{\mu_g B_g}\frac{\partial p_g}{\partial x} + R_{so}\frac{k_{ro}}{\mu_o B_o}\frac{\partial p_o}{\partial x}\right)\right]\Delta x + A_y k \frac{\partial}{\partial y}\left[\left(\frac{k_{rog}}{\mu_g B_g}\frac{\partial p_g}{\partial y} + R_{so}\frac{k_{ro}}{\mu_o B_o}\frac{\partial p_o}{\partial y}\right)\right]\Delta y + q_g$$

$$= \frac{V_b}{5.615}\frac{\partial}{\partial t}\left(\frac{\phi S_g}{B_g} + R_{so}\frac{\phi S_o}{B_o}\right). \qquad \ldots\ldots\ldots(9.240)$$

- Oil equation:

$$A_x k \frac{\partial}{\partial x}\left[\left(R_v \frac{k_{rog}}{\mu_g B_g}\frac{\partial p_g}{\partial x} + \frac{k_{ro}}{\mu_o B_o}\frac{\partial p_o}{\partial x}\right)\right]\Delta x + A_y k \frac{\partial}{\partial y}\left[\left(R_v \frac{k_{rog}}{\mu_g B_g}\frac{\partial p_g}{\partial y} + \frac{k_{ro}}{\mu_o B_o}\frac{\partial p_o}{\partial y}\right)\right]\Delta y$$

$$= \frac{V_b}{5.615}\frac{\partial}{\partial t}\left(R_v \frac{\phi S_g}{B_g} + \frac{\phi S_o}{B_o}\right). \qquad \ldots\ldots\ldots(9.241)$$

There are four unknowns: p_o, p_g, S_o, S_g.

The equations are as follows: two PDEs, two auxiliary equations, which are $S_g + S_o = 1$, and $P_{cgo} = p_g - p_o$.

Problem 9.3.30

For the following reservoir fluid systems, write flow equations describing the flow dynamics in the reservoir. Also, provide an inventory of the unknowns and the equations available.

1. Two-phase (gas/water), 2D horizontal reservoir with isotropic but heterogeneous permeability characteristics, homogeneous thickness distribution, and with production wells (assume $\Delta x = \Delta y$).
2. Single-phase gas reservoir in 3D with an isotropic and homogeneous permeability distribution and with varying thickness. No well exists in the system.

Solution to Problem 9.3.30

1. Knowing that $k_x = k_y = k$ and permeability is characterized with heterogeneous distribution,
   - Water equation:

$$\frac{\partial}{\partial x}\left(\frac{k k_{rw}}{\mu_w B_w}\frac{\partial p_w}{\partial x}\right) + \frac{\partial}{\partial y}\left(\frac{k k_{rw}}{\mu_w B_w}\frac{\partial p_w}{\partial y}\right) + \frac{q_w}{V_b} = \frac{\phi}{5.615}\frac{\partial}{\partial t}\left(\frac{S_w}{B_w}\right). \qquad \ldots\ldots\ldots\ldots(9.242)$$

   - Gas equation:

$$\frac{\partial}{\partial x}\left(\frac{k k_{rg}}{\mu_g B_g}\frac{\partial p_g}{\partial x} + R_{sw}\frac{k k_{rw}}{\mu_w B_w}\frac{\partial \Phi_w}{\partial x}\right) + \frac{\partial}{\partial y}\left(\frac{k k_{rg}}{\mu_g B_g}\frac{\partial p_g}{\partial y} + R_{sw}\frac{k k_{rw}}{\mu_w B_w}\frac{\partial p_o}{\partial y}\right)$$

$$+ \frac{q_g + R_{sw}q_w}{V_b} = \frac{\phi}{5.615}\frac{\partial}{\partial t}\left(\frac{S_g}{B_g} + R_{sw}\frac{S_w}{B_w}\right). \qquad \ldots\ldots\ldots(9.243)$$

2. Because $k_x = k_y = k_z = k$ and permeability is homogeneously distributed,

$$\frac{\partial}{\partial x}\left(\frac{A_x}{\mu_g B_g}\frac{\partial p}{\partial x}\right)\Delta x + \frac{\partial}{\partial y}\left(\frac{A_y}{\mu_g B_g}\frac{\partial p}{\partial y}\right)\Delta y + \frac{\partial}{\partial z}\left(\frac{A_z}{\mu_g B_g}\frac{\partial p}{\partial z}\right)\Delta z = \frac{T_{sc}V_b}{kp_{sc}T}\frac{\partial}{\partial t}\left(\frac{\phi p}{z_g}\right). \qquad \ldots\ldots\ldots(9.244)$$

Problem 9.3.31

Give full descriptions of the two reservoir systems, which are represented by the following governing flow equations:

1.

$$\frac{\partial}{\partial x}\left(\frac{k_x}{\mu B}\frac{\partial \Phi}{\partial x}\right) + \frac{\partial}{\partial y}\left(\frac{k_y}{\mu B}\frac{\partial \Phi}{\partial y}\right) + \frac{q}{V_b} = \frac{\phi c}{5.615}\frac{\partial p}{\partial t}. \dots \dots (9.245)$$

2.

$$\frac{\partial}{\partial x}\left(\frac{A_x k_{rg}}{\mu_g B_g}\frac{\partial p_g}{\partial x} + R_{so}\frac{k_{ro}}{\mu_o B_o}\frac{\partial p_o}{\partial x}\right) + \frac{\partial}{\partial y}\left(\frac{A_y k_{rg}}{\mu_g B_g}\frac{\partial p_g}{\partial y} + R_{so}\frac{k_{ro}}{\mu_o B_o}\frac{\partial p_o}{\partial y}\right)$$

$$+\frac{q_g + R_{so}q_o}{k\Delta x} = \frac{A\phi}{5.615k}\frac{\partial}{\partial t}\left(\frac{S_g}{B_g} + R_{so}\frac{S_o}{B_o}\right), \dots \dots (9.246)$$

$$\frac{\partial}{\partial x}\left(\frac{A_x k_{ro}}{\mu_o B_o}\frac{\partial p_o}{\partial x}\right) + \frac{\partial}{\partial y}\left(\frac{A_y k_{ro}}{\mu_o B_o}\frac{\partial p_o}{\partial y}\right) + \frac{q_o}{k\Delta x} = \frac{A\phi}{5.615k}\frac{\partial}{\partial t}\left(\frac{S_o}{B_o}\right). \dots (9.247)$$

Solution to Problem 9.3.31

1.
- Single-phase, slightly compressible flow in a 2D system.
- Anisotropic and heterogeneous permeability distribution.
- Reservoir has depth gradients in the x- and y-directions.
- Homogeneous in thickness.
- Reservoir has wells.
- Incompressible pore space.

2.
- Two-phase (oil/gas) flow in 2D.
- Gas is soluble in oil.
- Uniform grid dimensions along the x- and y-directions and $\Delta x = \Delta y$.
- Homogeneous and isotropic permeability distribution.
- No depth gradients along the x- and y-directions.
- There are wells in the reservoir.
- Incompressible pore space.
- Pronounced capillary pressure between the oil and gas phases.

Problem 9.3.32

Give a full description of the system described by the following flow equations:

1.

$$k_x \frac{\partial}{\partial x}\left(\frac{A_x k_{rg}}{\mu_g B_g}\frac{\partial p_g}{\partial x} + \frac{R_{so}A_x k_{ro}}{\mu_o B_o}\frac{\partial p_o}{\partial x} + \frac{R_{sw}A_x k_{rw}}{\mu_w B_w}\frac{\partial p_w}{\partial x}\right)\Delta x$$

$$+k_y \frac{\partial}{\partial y}\left(\frac{A_y k_{rg}}{\mu_g B_g}\frac{\partial p_g}{\partial y} + \frac{R_{so}A_x k_{ro}}{\mu_o B_o}\frac{\partial p_o}{\partial y} + \frac{R_{sw}A_y k_{rw}}{\mu_w B_w}\frac{\partial p_w}{\partial y}\right)\Delta y \dots \dots (9.248)$$

$$+q_{g,\text{free}} + R_{so}q + R_{sw}q_w = \frac{V_b\phi}{5.615}\frac{\partial}{\partial t}\left(\frac{S_g}{B_g} + R_{so}\frac{S_o}{B_o} + R_{sw}\frac{S_w}{B_w}\right),$$

$$k_x \frac{\partial}{\partial x}\left(\frac{A_x k_{ro}}{\mu_o B_o}\frac{\partial p_o}{\partial x}\right)\Delta x + k_y \frac{\partial}{\partial y}\left(\frac{A_y k_{ro}}{\mu_o B_o}\frac{\partial p_o}{\partial y}\right)\Delta y + q_{o(\text{STB/D})} = \frac{V_b\phi}{5.615}\frac{\partial}{\partial t}\left(\frac{S_o}{B_o}\right), \dots \dots (9.249)$$

$$k_x \frac{\partial}{\partial x}\left(\frac{A_x k_{rw}}{\mu_w B_w}\frac{\partial p_w}{\partial x}\right)\Delta x + k_y \frac{\partial}{\partial y}\left(\frac{A_y k_{rw}}{\mu_w B_w}\frac{\partial p_w}{\partial y}\right)\Delta y + q_{w(\text{STB/D})} = \frac{V_b\phi}{5.615}\frac{\partial}{\partial t}\left(\frac{S_w}{B_w}\right). \dots \dots (9.250)$$

2.

$$\frac{\partial}{\partial x}\left(\frac{k_x}{\mu B}\frac{\partial p}{\partial x}\right)+\frac{\partial}{\partial y}\left(\frac{k_y}{\mu B}\frac{\partial p}{\partial y}\right)+\frac{\partial}{\partial z}\left(\frac{k_z}{\mu B}\frac{\partial p}{\partial z}\right)+\frac{q_o}{V_b}=\frac{\partial}{\partial x}\left(\frac{k_x}{\mu B}\frac{1}{144}\frac{g}{g_c}\rho\frac{\partial G}{\partial x}\right)$$

$$+\frac{\partial}{\partial y}\left(\frac{k_y}{\mu B}\frac{1}{144}\frac{g}{g_c}\rho\frac{\partial G}{\partial y}\right)+\frac{\partial}{\partial z}\left(\frac{k_z}{\mu B}\frac{1}{144}\frac{g}{g_c}\rho\frac{\partial G}{\partial z}\right). \quad\quad\quad\quad\quad\quad (9.251)$$

Solution to Problem 9.3.32

1.
- Three-phase, 2D reservoir flow system.
- Reservoir does not have depth gradients in either direction.
- Homogeneous but anisotropic permeability distribution ($k_x \neq k_y$).
- Solution gas in oil and water phases.
- Nonuniform reservoir thickness.
- There are wells in the reservoir.
- $\Delta x \neq \Delta y$.
- Incompressible pore space.

2.
- Single-phase incompressible fluid in a 3D reservoir system.
- Depth gradients exist in each direction.
- Heterogeneous, anisotropic permeability distribution.
- Uniform thickness distribution.
- Reservoir has wells.

Problem 9.3.33

Give a full description of the system with the following governing equations. In each case, give an inventory of the unknowns to be solved and complete the set of equations, if necessary.

1.

$$\frac{\partial}{\partial x}\left(\frac{A_x k_x k_{ro}}{\mu_o B_o}\frac{\partial p}{\partial x}\right)\Delta x+\frac{\partial}{\partial y}\left(\frac{A_y k_y k_{ro}}{\mu_o B_o}\frac{\partial p}{\partial y}\right)\Delta y+k_z\frac{\partial}{\partial z}\left(\frac{A_z k_{ro}}{\mu_o B_o}\frac{\partial p}{\partial z}\right)\Delta z+q_o=\frac{V_b}{5.615}\frac{\partial}{\partial t}\left(\frac{\phi S_o}{B_o}\right), \quad\quad (9.252)$$

$$\frac{\partial}{\partial x}\left(\frac{A_x k_x k_{rw}}{\mu_w B_w}\frac{\partial p}{\partial x}\right)\Delta x+\frac{\partial}{\partial y}\left(\frac{A_y k_y k_{rw}}{\mu_w B_w}\frac{\partial p}{\partial y}\right)\Delta y+k_z\frac{\partial}{\partial z}\left(\frac{A_z k_{rw}}{\mu_w B_w}\frac{\partial p}{\partial z}\right)\Delta z+q_w=\frac{V_b}{5.615}\frac{\partial}{\partial t}\left(\frac{\phi S_w}{B_w}\right). \quad\quad (9.253)$$

2.

$$\frac{\partial}{\partial x}\left(\frac{k}{\mu B}\frac{\partial \Phi}{\partial x}\right)+\frac{\partial}{\partial y}\left(\frac{k}{\mu B}\frac{\partial \Phi}{\partial y}\right)+\frac{q_o}{V_b}=\frac{c}{5.615}\frac{\partial(\phi p)}{\partial t}. \quad\quad\quad\quad\quad\quad (9.254)$$

Solution to Problem 9.3.33

1.
- Two-phase (oil/water) immiscible flow in a 3D reservoir system.
- Anisotropic and heterogeneous permeability distribution.
- Varying thickness in each direction.
- Heterogeneous in k_x and k_y, and homogeneous in k_z.
- Capillary pressure between oil and water phases is ignored.
- Depth gradients either do not exist or are ignored.
- Reservoir has wells.
- Pore space is compressible.

Three unknowns are present: p, S_o, and S_w. It comprises three equations: oil mass-balance equation, water mass-balance equation, and saturation relationship ($S_o + S_w = 1$).

2.
- Single-phase slightly compressible flow in a 2D reservoir system.
- Isotropic but heterogeneous permeability distribution.
- Homogeneous thickness distribution.
- Reservoir has depth gradients in both *x*- and *y*-directions.
- Reservoir has wells.
- Compressible pore space.

One unknown exists: *p*. It comprises one equation: oil mass-balance equation.

Problem 9.3.34

1. Write the flow equations for the following type of reservoir:
   - One dimensional.
   - Two-phase flow of oil and water.
   - Heterogeneous permeability distribution.
   - Homogeneous thickness and reservoir width.
   - Inclined reservoir.
   - Slightly compressible fluid.
   - Formation porosity is a function of pressure.
   - Capillary pressure between the two phases is considered negligible.
2. What are the principal unknowns in these equations?
3. What additional equations are required to solve for these unknowns?

Solution to Problem 9.3.34

1. Oil equation:

$$\frac{\partial}{\partial x}\left(\frac{k_{ro}k}{\mu_o B_o}\frac{\partial p}{\partial x}\right) - \frac{1}{144}\frac{\partial}{\partial x}\left(\frac{g}{g_C}\rho_o\frac{k_{ro}k}{\mu_o B_o}\frac{\partial G}{\partial x}\right) + \frac{q_o}{V_b} = \frac{1}{5.615}\frac{\partial}{\partial t}\left(\frac{\phi S_o}{B_o}\right). \quad\quad (9.255)$$

Water equation:

$$\frac{\partial}{\partial x}\left(\frac{k_{rw}k}{\mu_w B_w}\frac{\partial p}{\partial x}\right) - \frac{1}{144}\frac{\partial}{\partial x}\left(\frac{g}{g_C}\rho_w\frac{k_{rw}k}{\mu_w B_w}\frac{\partial G}{\partial x}\right) + \frac{q_w}{V_b} = \frac{1}{5.615}\frac{\partial}{\partial t}\left(\frac{\phi S_w}{B_w}\right). \quad\quad (9.256)$$

2. Three unknowns, which are p, S_o, and S_w ($p_o = p_w$), because capillary pressure is ignored.
3. Saturation relationship: $S_w + S_o = 1$.

Problem 9.3.35

Give the governing flow equations for the gas-condensate reservoir shown in **Fig. 9.12**. The boundaries of the reservoir are completely sealed. The reservoir is isotropic and heterogeneous in permeability and has varying thickness. There are producers and injectors in the system. This reservoir has no depth gradients.

1. Give the (2D) governing equation of the system at the initial reservoir condition marked at Point a in the given $\angle pT$ diagram. State any assumptions you are making.
2. After one year of operation, the reservoir condition changes to Point b in the diagram. Give the (2D) governing flow equation of the system. State any assumptions you are making.

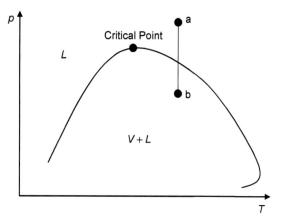

Fig. 9.12—Phase envelope of the reservoir system of Problem 9.3.35.

Solution to Problem 9.3.35

1. Assuming that no water is presented and porosity is not changing with pressure,

$$\frac{\partial}{\partial x}\left(\frac{A_x k}{\mu_o B_o}\frac{\partial p_o}{\partial x}\right)\Delta x + \frac{\partial}{\partial y}\left(\frac{A_y k}{\mu_o B_o}\frac{\partial p_o}{\partial y}\right)\Delta y + q_o = \frac{V_b \phi c_t}{5.615}\frac{\partial p}{\partial t}. \quad\dots\dots\dots\dots\dots\dots\dots\dots (9.257)$$

2. For the gas phase,

$$\frac{\partial}{\partial x}\left[A_x k\left(\frac{k_{rog}}{\mu_g B_g}\frac{\partial p_g}{\partial x} + R_{so}\frac{k_{ro}}{\mu_o B_o}\frac{\partial p_o}{\partial x}\right)\right]\Delta x + \frac{\partial}{\partial y}\left[A_y k\left(\frac{k_{rog}}{\mu_g B_g}\frac{\partial p_g}{\partial y} + R_{so}\frac{k_{ro}}{\mu_o B_o}\frac{\partial p_o}{\partial y}\right)\right]\Delta y + q_g + R_{so}q_o$$

$$= \frac{V_b}{5.615}\frac{\partial}{\partial t}\left(\frac{\phi S_g}{B_g} + R_{so}\frac{\phi S_o}{B_o}\right). \quad\dots (9.258)$$

3. For the oil phase,

$$\frac{\partial}{\partial x}\left[A_x k\left(R_v\frac{k_{rog}}{\mu_g B_g}\frac{\partial p_g}{\partial x} + \frac{k_{ro}}{\mu_o B_o}\frac{\partial p_o}{\partial x}\right)\right]\Delta x + \frac{\partial}{\partial y}\left[A_y k\left(R_v\frac{k_{rog}}{\mu_g B_g}\frac{\partial p_g}{\partial y} + \frac{k_{ro}}{\mu_o B_o}\frac{\partial p_o}{\partial y}\right)\right]\Delta y + R_v q_g +$$

$$= \frac{V_b}{5.615}\frac{\partial}{\partial t}\left(R_v\frac{\phi S_g}{B_g} + \frac{\phi S_o}{B_o}\right). \quad\dots (9.259)$$

Problem 9.3.36

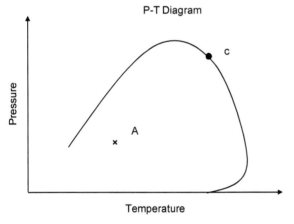

P-T Diagram

Consider the P-T diagram presented in **Fig. 9.13.** Provide the governing flow equation(s) describing the fluid flow at P-T condition of Point A in a 2D reservoir with homogeneous and isotropic permeability distribution. Porosity remains the same throughout the production period. All boundaries are sealed. Depth gradient along the x-direction can be ignored and $S_w = S_{wirr}$. State all principle unknowns and equations of this reservoir system.

Fig. 9.13—Phase envelope of the reservoir system of Problem 9.3.36.

Solution to Problem 9.3.36

At Point A, there are two phases and, hence, two-phase equations. The principal equations (oil and gas mass balances) are as follows:

$$\frac{\partial}{\partial x}\left(\frac{A_x k_{ro}}{\mu_o B_o}\frac{\partial p_o}{\partial x}\right)\Delta x + \frac{\partial}{\partial y}\left(\frac{A_y k_{ro}}{\mu_o B_o}\frac{\partial \Phi_o}{\partial y}\right)\Delta y + \frac{q_{osc}}{k} = \frac{V_b \phi}{k5.615}\frac{\partial}{\partial t}\left(\frac{S_o}{B_o}\right), \quad\dots\dots\dots\dots\dots\dots\dots (9.260)$$

$$\frac{\partial}{\partial x}\left(\frac{A_x}{\mu_g B_g}\frac{\partial p_g}{\partial x} + R_{so}\frac{A_x k_{ro}}{\mu_o B_o}\frac{\partial p_o}{\partial x}\right)\Delta x + \frac{\partial}{\partial y}\left(\frac{A_y}{\mu_g B_g}\frac{\partial p_g}{\partial y} + R_{so}\frac{A_y k_{ro}}{\mu_o B_o}\frac{\partial \Phi_o}{\partial y}\right)$$

$$\Delta y + \frac{q_{gsc} + R_{so}q_{osc}}{k} = \frac{V_b \phi}{k5.615}\frac{\partial}{\partial t}\left(\frac{S_g}{B_g} + R_{so}\frac{S_o}{B_o}\right). \quad\dots\dots\dots\dots\dots\dots\dots\dots\dots\dots\dots\dots (9.261)$$

Auxiliary equations are $P_{cgo}\left(S_g\right) = p_g - p_o$ and $S_o + S_g = 1 - S_{wirr}$. The principal unknowns are p_o, p_g, S_o, and S_g.

Problem 9.3.37

Consider the undersaturated oil reservoir with homogenous and isotropic permeability distribution shown in **Fig. 9.14.** There exists an infinitely large aquifer at the bottom of the reservoir. The aquifer provides pressure support to the reservoir in such a way that the reservoir pressure will not decrease below bubblepoint pressure.

1. Given that Δx, Δy, and Δz are nonuniform, provide a full description of the characteristics of the system.
2. Write the governing flow equations of the system in their simplest forms. State all unknowns and all equations needed to solve for the unknowns.

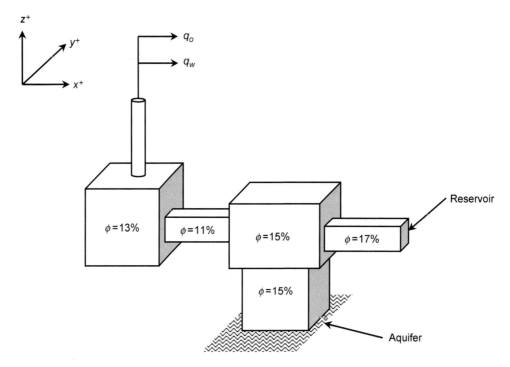

Fig. 9.14—Schematic representation of the reservoir system of Problem 9.3.37.

Solution to Problem 9.3.37

1. It has 2D rectangular coordinates (y-z), two-phase (water and oil) fluid flow, and a nonuniform grid system. There is a well in the system, and the reservoir has a heterogeneous porosity distribution. The problem states that the reservoir also has homogeneous and isotropic permeability distribution.
2. Water equation:

$$\frac{\partial}{\partial y}\left(\frac{A_y k_{rw}}{\mu_w B_w}\frac{\partial D_w}{\partial y}\right)\Delta y + \frac{\partial}{\partial z}\left(\frac{A_z k_{rw}}{\mu_w B_w}\frac{\partial \Phi_w}{\partial z}\right)\Delta z + \frac{q_w}{k} = \frac{V_b \phi}{5.615k}\frac{\partial}{\partial t}\left(\frac{S_w}{B_w}\right). \qquad (9.262)$$

Oil equation:

$$\frac{\partial}{\partial y}\left(\frac{A_y k_{ro}}{\mu_o B_o}\frac{\partial \Phi_o}{\partial y}\right)\Delta y + \frac{\partial}{\partial z}\left(\frac{A_z k_{ro}}{\mu_o B_o}\frac{\partial \Phi_o}{\partial z}\right)\Delta z + \frac{q_o}{k} = \frac{V_b \phi}{5.615k}\frac{\partial}{\partial t}\left(\frac{S_o}{B_o}\right). \qquad (9.263)$$

Auxiliary equations are $P_{cow}(S_w) = p_o - p_w$ and $S_o + S_w = 1$. The principal unknowns are p_o, p_w, S_o, and S_w.

Problem 9.3.38

Write the governing flow equation(s) for the 2D, two-phase (gas/water) flow reservoir with heterogeneous but isotropic permeability distribution. There is depth gradient in the system; however, gravitational forces for the gas phase can be ignored. Porosity is changing throughout the production period and the solution of gas in water exists. Also, there is one producing well in the system. Capillary pressure between gas and water can be expressed as $P_{cgw} = p_g - p_w$.

Express the flow equation(s) in terms of the principal unknowns p_g and S_w.

Solution to Problem 9.3.38

From the capillary pressure relationship, one can write

$$\frac{\partial p_w}{\partial x} = \frac{\partial p_g}{\partial x} - \frac{\partial P_{cgw}}{\partial x}, \quad\dotsfill (9.264)$$

and

$$\frac{\partial p_w}{\partial y} = \frac{\partial p_g}{\partial y} - \frac{\partial P_{cgw}}{\partial y}. \quad\dotsfill (9.265)$$

Using these two equations, the $\dfrac{\partial p_w}{\partial x}$ and $\dfrac{\partial p_w}{\partial x}$ terms can be eliminated, as shown next (note that $S_o = 1 - S_w$):

- Gas equation:

$$\frac{\partial}{\partial x}\left[R_{sw}\frac{A_x k k_{rw}}{\mu_w B_w}\left(\frac{\partial p_g}{\partial x} - \frac{\partial P_{cgw}}{\partial x}\right) + \frac{A_x k k_{rg}}{\mu_g B_g}\frac{\partial p_g}{\partial x}\right]\Delta x + \frac{\partial}{\partial y}\left[R_{sw}\frac{A_y k k_{rw}}{\mu_w B_w}\left(\frac{\partial p_g}{\partial y} - \frac{\partial P_{cgw}}{\partial y}\right) + \frac{A_x k k_{rg}}{\mu_g B_g}\frac{\partial p_g}{\partial y}\right]\Delta y$$

$$-\left[\frac{\partial}{\partial x}\left(R_{sw}\frac{A_x k k_{rw}}{\mu_w B_w}\frac{1}{144}\frac{g}{g_c}\rho_w\frac{\partial G}{\partial x}\right)\Delta x + \frac{\partial}{\partial y}\left(R_{sw}\frac{A_y k k_{rw}}{\mu_w B_w}\frac{1}{144}\frac{g}{g_c}\rho_w\frac{\partial G}{\partial y}\right)\Delta y\right] \quad\dots (9.266)$$

$$+ q_g + R_{sw}q_w = \frac{V_b}{5.615}\frac{\partial}{\partial t}\left(R_{sw}\frac{\phi S_w}{B_w} + \frac{\phi(1 - S_w)}{B_g}\right).$$

- Water equation:

$$\frac{\partial}{\partial x}\left[\frac{A_x k k_{rw}}{\mu_w B_w}\left(\frac{\partial p_g}{\partial x} - \frac{\partial P_{cgw}}{\partial x}\right)\right]\Delta x + \frac{\partial}{\partial y}\left[\frac{A_y k k_{rw}}{\mu_w B_w}\left(\frac{\partial p_g}{\partial y} - \frac{\partial P_{cgw}}{\partial y}\right)\right]\Delta y$$

$$-\left[\frac{\partial}{\partial x}\left(\frac{A_x k k_{rw}}{\mu_w B_w}\frac{1}{144}\frac{g}{g_c}\rho_w\frac{\partial G}{\partial x}\right)\Delta x + \frac{\partial}{\partial y}\left(\frac{A_y k k_{rw}}{\mu_w B_w}\frac{1}{144}\frac{g}{g_c}\rho_w\frac{\partial G}{\partial y}\right)\Delta y\right] + q_w = \frac{V_b}{5.615}\frac{\partial}{\partial t}\left(\frac{\phi S_w}{B_w}\right). \quad\dotsfill (9.267)$$

Eqs. 9.266 and 9.267 contain two principal unknowns: p_g and S_w.

Problem 9.3.39

You are performing a two-phase relative permeability experiment by injecting gas into a horizontal, homogeneous linear core that is 100% saturated with water.

Write the equations of the mathematical model describing this experiment. Assume no solubility of gas in water and treat the water phase as a slightly compressible fluid.

Solution to Problem 9.3.39

Water equation:

$$\frac{\partial}{\partial x}\left(\frac{A_x k_x k_{rw}}{\mu_w B_w}\frac{\partial p_w}{\partial x}\right)\Delta x + q_w = \frac{V_b}{5.615}\frac{\partial}{\partial t}(\frac{\phi S_w}{B_w}). \quad\dotsfill (9.268)$$

Water is treated as a slightly compressible fluid with $B_w \approx 1$ and $\mu_w \approx$ constant. Throughout the core, A_x and k_x are assumed to be constant. At the downstream end of the core, water is produced with gas (only gas is injected at the upstream end). Inserting these conditions into Eq. 9.268, the water equation becomes

$$\frac{\partial}{\partial x}\left(k_{rw}\frac{\partial p_w}{\partial x}\right) + \frac{q_w \mu_w B_w}{V_b k_x} = \frac{\mu_w B_w}{5.615 k_x}\frac{\partial}{\partial t}(\phi S_w). \quad\dotsfill (9.269)$$

The gas equation is as follows:

$$\frac{\partial}{\partial x}\left(\frac{A_x k_x k_{rg}}{\mu_g B_g}\frac{\partial p_g}{\partial x}\right)\Delta x + q_g = \frac{V_b}{5.615}\frac{\partial}{\partial t}\left(\frac{\phi S_g}{B_g}\right), \dots \dots \dots \dots \dots \dots (9.270)$$

which can be further simplified to the following form:

$$\frac{\partial}{\partial x}\left(\frac{k_{rg}}{\mu_g B_g}\frac{\partial p_g}{\partial x}\right)\Delta x + \frac{q_g}{V_b k_x} = \frac{1}{5.615 k_x}\frac{\partial}{\partial t}\left(\frac{\phi S_g}{B_g}\right). \dots \dots \dots \dots \dots (9.271)$$

Capillary pressure relation and the saturation relationship shown next provide the auxiliary equations:

$$P_{cgw}\left(S_g\right) = p_g - p_w \dots \dots \dots \dots \dots \dots \dots \dots \dots \dots \dots \dots \dots \dots \dots (9.272)$$

and

$$S_g + S_w = 1. \dots \dots \dots \dots \dots \dots \dots \dots \dots \dots \dots \dots (9.273)$$

Nomenclature

A = cross-sectional area, L², ft², and coefficient of Eq. 9.42

A_s = cross-sectional area along s-direction ($s = x, y, z$), L², ft²

A_x = cross-sectional area along x-direction, L², ft²

A_y = cross-sectional area along y-direction, L², ft²

A_z = cross-sectional area along z-direction, L², ft²

B = formation volume factor, L³/L³, res bbl/scf for liquid and res bbl/STB for gas, and coefficient in Eq. 9.43

B_g = gas formation volume factor, L³/L³, res bbl/scf

B_l = formation volume factor of phase l, L³/L³, res bbl/STB for liquid and res bbl/scf for gas

B_o = oil formation volume factor, L³/L³, res bbl/STB

B_w = water formation volume factor, L³/L³, res bbl/STB

c = constant

c_t = total compressibility, Lt²/m, psi⁻¹

$[C]$ = accumulation matrix

C_{gg} = coefficient of $\Delta_t S_g$ in the expansion of gas accumulation, L³/t, STB/D

C_{gp} = coefficient of $\Delta_t p_o$ in the expansion of gas accumulation, L⁴t/m, STB/(D–psi)

$C_{i,j,k}$ = transmissibility coefficient using SIP notation of central block (i, j, k), L⁴t/m, STB/(D–psia) for liquid, scf/(D–psia) for gas

C_{og} = coefficient of $\Delta_t S_g$ in the expansion of oil accumulation, L³/t, STB/D

C_{op} = coefficient of $\Delta_t p_o$ in the expansion of oil accumulation, L⁴t/m, STB/(D–psi)

C_{ow} = coefficient of $\Delta_t S_w$ in the expansion of oil accumulation, L³/t, STB/D

C_{wg} = coefficient of $\Delta_t S_g$ in the expansion of water accumulation, L³/t, STB/D

C_{wp} = coefficient of $\Delta_t p_o$ in the expansion of water accumulation, L⁴t/m, STB/(D–psi)

C_{ww} = coefficient of $\Delta_t S_w$ in the expansion of water accumulation, L³/t, STB/D

E = transmissibility coefficient using SIP notation between block (i, j, k) and ($i+1, j, k$), L⁴t/m, STB/(D–psia) for liquid, scf/(D–psia) for gas

f = function

f_g = fractional flow term for the gas phase, dimensionless

f_l = fractional flow term for the phase l ($l = o, w,$ and g), dimensionless

f_o = fractional flow term for the oil phase, dimensionless

f_p = weak nonlinear function and pressure-dependent term of transmissibility

f_s = strong nonlinear function and saturation-dependent term of transmissibility

f_w = fractional flow term for the water phase, dimensionless

g = acceleration of gravity, L/t², ft/s²

g_c = unit conversion factor in Newton's law

G = depth to the center of the block from datum, L, ft

$\overline{G}$ = vector of gravitational terms

H = thickness, L, ft

I_{MB} = incremental material balance, dimensionless

k = permeability, L², darcies or perms (1 darcies = 1.127 perms)

k_{rg} = relative permeability to gas, dimensionless

k_{ri} = relative permeability to the ith phase, dimensionless

k_{rl} = relative permeability to phase l, dimensionless (l = oil, water and gas)

k_{ro} = relative permeability to oil, dimensionless

k_{rog} = relative permeability to oil in oil/gas system, dimensionless

k_{rup} = relative permeability of a upstream node

k_{rw} = relative permeability to water, dimensionless

k_s = permeability along s-direction ($s = x, y, z$), L^2, darcies or perms

k_x = permeability along x-direction, L^2, darcies or perms

k_y = permeability along y-direction, L^2, darcies or perms

k_z = permeability along z-direction, L^2, darcies or perms

L = the length of the wellbore in the wellblock, L, ft

M = mass per unit volume of porous media, m/L^3, lbm/ft^3

$\dot{m}_{cx}$ = mass flux, m/L^2t, lbm/ft^2–D

N = transmissibility coefficient using SIP notation between block (i, j, k) and ($i, j+1, k$), L^4t/m, STB/(D–psia) for liquid, scf/(D–psia) for gas

p = pressure, m/Lt^2, psia

p_b = bubblepoint pressure, m/Lt^2, psia

P_{cgo} = gas/oil capillary pressure, m/Lt^2, psia

P_{cow} = oil/water capillary pressure, m/Lt^2, psia

P_{cgw} = gas/water capillary pressure, m/Lt^2, psia

p_g = gas pressure, m/Lt^2, psia

p_i = initial pressure, m/Lt^2, psia

p_o = oil pressure, m/Lt^2, psia

p_{sc} = standard condition pressure, m/Lt^2, psia

p_w = water pressure, m/Lt^2, psia

p_{wf} = well-bottomhole flowing pressure, m/Lt^2, psia

Q_t = production or flow rate, L^3/t, scf/D for gas, STB/D for liquid

q = fluid injection/production rate, L^3/t, scf/D for gas and STB/D for liquid

q_c = fluid injection/production rate for component c, L^3/t, scf/D for gas and STB/D for liquid

q_g = gas production or flow rate, L^3/t, scf/D

q_{mc} = rate of mass depletion of component c through well, m/t, lbm/D

q_{mt_c} = rate of mass transfer of component c between phases, m/t, lbm/D

$q_{mt_{sg}}$ = rate of mass transfer of solution gases between phases, m/t, lbm/D

q_o = oil production or flow rate, L^3/t, STB/D

q_t = total flow rate, L^3/t, STB/D for oil and scf/D for gas

q_w = water production or flow rate, L^3/t, STB/D

Q = right-hand-side coefficient of governing flow equation using SIP notation for block (i,j,k) L^3/t, STB/D for liquid, scf/D for gas

$\overline{Q}$ = source vector

r_w = well radius, L, ft

R_o = residual of the oil equation, L^3/t, STB/D

R_s = solution gas ratio, L^3/L^3, scf/STB

R_{so} = solution gas/oil ratio, L^3/L^3, scf/STB

R_{sw} = solution gas/water ratio, L^3/L^3, scf/STB

R_v = volatile oil/gas ratio, L^3/L^3, STB/scf

R_w = residual of the water equation, L^3/t, STB/D

s = skin factor, dimensionless; an arbitrary direction ($s = x, y, z$)

S = transmissibility coefficient using SIP notation between block (i,j,k) and ($i, j-1, k$), L^4t/m, STB/(D–psia) for liquid, scf/(D–psia) for gas

S_g = gas saturation, fraction

S_{gn} = normalized gas saturation, fraction

S_l = saturation of phase l

S_{wirr} = irreducible water saturation, fraction

S_o = oil saturation, fraction

S_{oi} = initial oil saturation, fraction

S_w = water saturation, fraction

S_{wi} = initial water saturation, fraction

t = time, t, days

T = generic representation of the transmissibility term, L^4t/m, STB/(D–psia) or absolute temperature, T, °R

$[\boldsymbol{T}]$ = transmissibility matrix

$T_{o,i,j}$ = oil transmissibility factor between adjacent block i and j, L^4t/m, STB/(D–psia)

T_{ox} = oil transmissibility factor along x-direction, L^4t/m, STB/(D–psia)

T_{oy} = oil transmissibility factor along y-direction, L^4t/m, STB/(D–psia)

T_{px} = pressure-dependent transmissibility factor along x-direction

T_{sc} = standard condition temperature, T, °R

$T_{w,i,j}$ = water-transmissibility factor between adjacent block i and j, L^4t/m, STB/(D–psia)

T_{wx} = water-transmissibility factor along x-direction, L^4t/m, STB/(D–psia)

T_{wy} = water-transmissibility factor along y-direction, L^4t/m, STB/(D–psia)

u_{ls} = superficial velocity of phase l (l = water, oil and gas) along direction s ($s = x, y, z$), L/t, res bbl/(D–ft^2)

u_{cs} = superficial velocity of component c along direction s ($s = x, y, z$), L/t, res bbl/(D–ft^2)

u_o = superficial velocity of the oil flow, L/t, res bbl/(D–ft^2)

u_t = superficial velocity of the total flow, L/t, res bbl/(D–ft^2)

u_w = superficial velocity of the water flow, L/t, res bbl/(D–ft^2)

U = arbitrary dependent variable

V = arbitrary dependent variable

V_b = bulk volume, L^3, ft^3

W = transmissibility coefficient using SIP notation between block (i, j, k) and (i-$1, j, k$), L^{4}t/m, STB/(D–psia) for liquid, scf/(D–psia) for gas

x = distance in x-direction in the Cartesian coordinate system, L, ft

X = arbitrary dependent variable

$\overline{X}$ = unknown vector

y = distance in y-direction in the Cartesian coordinate system

Y = arbitrary dependent variable

z = distance or depth in z-direction in the Cartesian coordinate system, L, ft

z_g = gas compressibility factor, dimensionless

α_c = volumetric conversion factor with a magnitude of 5.615 scf/STB

β_c = transmissibility conversion factor with a magnitude of 1.127

ε_p = small pressure differential, m/Lt^2, psi

Γ = accumulation coefficient, L^3/t, ft^3/D

Γ_{1o} = accumulation coefficient of oil component in oleic equation, L^3/t, ft^3/D

Γ_{1w} = accumulation coefficient of water component in oleic phase, L^3/t, ft^3/D

Γ_{2o} = accumulation coefficient of oil component in aqueous equation, L^3/t, ft^3/D

Γ_{2w} = accumulation coefficient of oil component in aqueous equation, L^3/t, ft^3/D

γ_g = gas gravity, m/L^{2}t^2, psi/ft

γ_o = oil gravity, m/L^{2}t^2, psi/ft

γ_w = water gravity, m/L^{2}t^2, psi/ft

$\bar{\nabla}$ = differential operator

Δ = difference operator and difference of a variable

Δ_x = difference operator along x-direction

Δx = control-volume dimension along the x-direction, L, ft

Δy = control-volume dimension along the y-direction, L, ft

δz = control-volume dimension along the z-direction, L, ft

Δ = central difference operator

ζ_w = function defined in the Solution to Problem 9.3.16

λ_g = gas-mobility term, L^{3}t/m, res bbl/(D–psi-ft)

λ_l = fluid-mobility term of phase l ($l = o$, w, and g), L^{3}t/m, res bbl/(D–psi-ft)

λ_o = oil-mobility term, L^{3}t/m, res bbl/(D–psi-ft)

λ_t = total-mobility term, L^{3}t/m, res bbl/(D–psi-ft)

λ_w = water-mobility term, L^{3}t/m, res bbl/(D–psi-ft)

μ = viscosity, m/Lt, cp

μ_g = gas viscosity, m/Lt, cp

μ_l = viscosity of phase l ($l =$ oil, water, and gas), m/Lt, cp

μ_o = oil viscosity, m/Lt, cp

μ_w = water viscosity, m/Lt, cp

ρ = density, m/L^3, lbm/ft^3

ρ_c = density of component c, m/L^3, lbm/ft^3

ρ_g = gas density, m/L^3, lbm/ft^3

ρ_l = fluid density of phase l, m/L^3, lbm/ft^3

ρ_o = oil density, m/L^3, lbm/ft^3

ρ_w = water density, m/L^3, lbm/ft^3

Φ = flow potential, m/Lt^2, psia

Φ_g = gas-flow potential, m/Lt^2, psia

Φ_o = oil-flow potential, m/Lt^2, psia

Φ_w = water-flow potential, m/Lt^2, psia

ϕ = porosity, fraction

Ψ_n = set of neighborhood blocks of Block n

Superscripts

$n+1$ = current time level

n = previous time level

$k+1$ = current iteration level

k = previous iteration level

T = transpose of a matrix or a vector

(v) = iteration level (old)

$(v+1)$ = iteration level (new)

$1o$ = oil-component term of the first equation

$1w$ = water-component term of the first equation

$2o$ = oil-component term of the second equation

$2w$ = water-component term of the second equation

$'$ = prime used to identify the modified transmissibility coefficients

$\rightarrow$ = vector notation

Subscripts

c = component

f or free = related to free gas

g = related to gas phase

i,j,k = related to block i,j,k in a 3D representation

i,j = related to block i,j in a 2D representation

i	=	block address along the x-direction
j	=	block address along the y-direction
k	=	block address along the z-direction
l	=	phases of the fluid (l = water, oil, and gas)
n	=	gridblock number
N	=	phase number
$\underline{NX}$	=	related to block NX^{th} (last block) along the x-direction
m	=	number of the neighborhood block to block n
o	=	related to oil phase

s	=	spatial directions of the Cartesian coordinate system
sc	=	standard conditions
t	=	related to time dimension or total
$up1$	=	first upstream block
$up2$	=	second upstream block
w	=	related to water phase
x	=	related to x-direction
y	=	related to y-direction
z	=	related to z-direction

References

Ertekin, T., Abou-Kassem, J. H., and King, G. R. 2001. *Basic Applied Reservoir Simulation*. Richardson, Texas, US: Society of Petroleum Engineers.

Nolen, J. S. and Berry, D. W. 1972. Tests of the Stability and Time-Step Sensitivity of Semi-Implicit Reservoir Simulation Techniques. *SPE J* **12**(03): 253–266. SPE-2981-PA. http://dx.doi.org/10.2118/2981-PA.

Odeh, A. S. 1981. Comparison of Solutions to a Three-Dimensional Black-Oil Reservoir Simulation Problem. *J Pet Technol* **33**(01): 13-25. SPE-9723-PA. http://dx.doi.org/10.2118/9723-PA.

Chapter 10

Reservoir Simulation in Practice

So far in this book, we have discussed the building blocks of reservoir simulation and focused on the solution techniques available to solve flow equations. In this chapter, we focus more on the use of reservoir simulators in practice. In general, we can state that a reservoir simulation study involves some important stages, all of which require significant planning. These stages include

- Defining the problem and setting the objectives.
- Selecting the formalism (model) to be used.
- Deciding the data needed and exploring their availability.
- Finding the availability of expertise and manpower.
- Finding the availability of computational platforms.
- Planning and designing computer runs.
- Analyzing the output files generated and checking their validity.
- Making necessary inferences and deductions and assembling the results into one coherent document.

The successful planning and implementation of every step outlined plays a critical role in the successful completion of a simulation study.

10.1 Mathematical Models of A Reservoir

As shown in **Fig. 10.1,** there are different mathematical models of hydrocarbon reservoirs used in petroleum engineering. A detailed description of the geological area results in a geologic model. A geologic model, when constructed, will not only describe the physical limits of the reservoir, but also characterize the reservoir in terms of spatial distribution of the reservoir parameters. A detailed description of the area of interest is required to reduce the model to a simplified reservoir flow simulation model.

The reservoir flow simulation models have become known as the most important tool in the hands of a reservoir engineer to conduct reservoir-management studies. Fig. 10.1 also shows that a material-balance model is a simplified version of a full-fledge reservoir simulation model.

Fig. 10.2 shows the dynamic nature of a simulation cycle. The first step of the cycle is the data-acquisition cycle. Then a history-matching study is conducted to tune the parameters so that the simulator can accurately mimic the history. In the next stage, certain performance studies, including production predictions, are conducted for reservoir-management purposes. Over the course of years, as new production data become available, further history-matching studies are carried out, which will result in updated reservoir models through this kind of fine tuning of the reservoir parameters.

For the flow model to be effectively accurate, the geologic model must capture features that define and control in-situ flow. So, the geologic model must do the following:

- Identify reservoir stratification and the degree of vertical communication within the various stratigraphic zones.
- Define what constitutes the net reservoir rock.
- Establish areal connectivity by identifying the sealing faults and natural fractures of the reservoir and the variation on reservoir quality.
- Identify constructing lithologic zones.
- Identify reservoir boundaries and existing boundary conditions.
- Distinguish between local and regional geologic features.

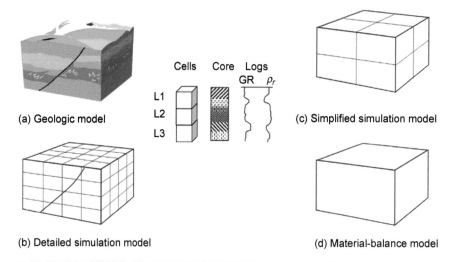

Fig. 10. 1—Various mathematical models of a reservoir.

10.2 Uncertainty in Reservoir Simulation

Throughout this book we have reiterated that reservoir simulation formalism is a statement of certain assumptions in mathematical terms. Typically, five fundamental principles, when assembled together, form a backbone of how a reservoir model will be structured. These five fundamental principles include

- Conservation of mass.
- Conservation of energy.
- Conservation of momentum (essentially this principle constitutes a force balance).
- Constitutive equation.
- Equation of state.

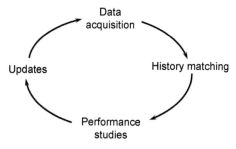

Fig. 10. 2—A typical reservoir simulation cycle.

Because there is a good degree of uncertainty in data that are assembled for a reservoir model, there will be a good degree of uncertainty encountered in the predictions generated by the model. It is normal to expect an increase in the band of uncertainty in predictive studies, as shown in **Fig. 10.3.**

As the band of uncertainty grows, it is necessary to have good checks and balances in place to reduce (or at least to keep under control) the size of the uncertainty band. In reservoir simulation, the uncertainty sources may include items such as

- Data quantity and quality.
- Geology.
- Scale up.
- Mathematical.

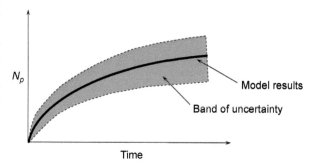

Fig. 10. 3—Uncertainty of reservoir simulation.

Among the four uncertainty sources listed, only mathematical uncertainty can be formally computed. This is possible because a reservoir engineer should understand the error introduced by the discretization process (truncation error) and linearization of the flow equations (conservation equations). Furthermore, it is necessary to have an idea about the magnitude of numerical dispersion, stability, and truncation error, because they are all caused by several mathematical assumptions and mathematical manipulations used in building a reservoir simulation code. Recall that several input options, such as convergence criteria and the selection of a linear-equation solver, all directly operate at the mathematical level.

Because of existing uncertainties, it is necessary to view the predictions as probabilistic, because of the lack of engineering data, sparse geologic data, constraints introduced by model selection, and inherent uncertainties in reservoir characterization.

10.3 Incentives for Carrying Out a Reservoir-Simulation Study

Reservoir simulation is a process that offers many incentives to the reservoir management engineer. These incentives are mainly based on the economics of the project being considered, but also go well beyond economics. Below is a list of the incentives that immediately surface:

- Scoping studies.
- Green field-development plans.
- Reservoir injectivity and deliverability studies.
- Alternative development plans on operating strategies and recovery processes with the goal of improving capital efficiency.
- Decision support and sensitivity analysis.
- Integration of data and physics.
- Credibility and reliability, because the study promises unbiased calculations.
- Performance monitoring through well-test analysis and decline-curve analysis.
- Educational (pedagogical) tools for parametric studies, research programs, and training programs.

10.4 Objectives of a Simulation Study and Prioritization of Objectives

It is of the foremost importance to set clear objectives and priorities. Objectives of a typical simulation study may include

- Pressure, injection, and production forecasts.
- Calibrating to historical reservoir performance.
- Assessment of critical gas- and water-coning rates.
- Assisting in timing and sizing of the surface facilities.
- Evaluating infill-drilling options.
- Evaluating workover potential.
- Mitigation of lease-line migration.

Once a good understanding of the entries of the preceding list is made, the reservoir engineer and geoscientist will be able to find answers to critical questions, such as

- How are the fluids moving in the reservoir?
- What are the optimum injection and production rates?
- What is the optimum well spacing?
- Can better well-completion methods be implemented?
- When will artificial lift be needed?
- What kind of artificial lift should be used?
- What is the dominant recovery mechanism?
- Are enhanced recovery methods needed?
- What type and size of surface facilities will be needed?

10.5 Problems and Solutions

Problem 10.5.1 (Ertekin et al. 2001)

Comment on the following statement regarding **Fig. 10.4** and the conditions listed in **Table 10.1.** Show the details of your analysis.

If Condition 1 is satisfied, it is possible to solve the flow problem as if it were a 1D problem. However, if Condition 2 applies, we cannot solve this system with a 1D fluid flow formulation.

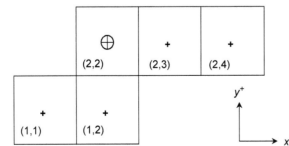

Condition 1	Condition 2
$\Delta x = \Delta y$	$\Delta x \neq \Delta y$
$k_x = k_y$	$k_x \neq k_y$

Table 10.1—Reservoir conditions for Fig. 10.4.

Fig. 10.4—Schematic representation of the reservoir system of Problem 10.5.1.

Solution to Problem 10.5.1

Condition 1 cannot be treated as a 1D problem, unless the thickness distribution is uniform between Blocks (1,2) and (2,2).

If the transmissibility factor between (1,1) and (1,2) and the transmissibility factor between (1,2) and (2,2) are the same, Condition 2 can be treated as a 1D system.

Problem 10.5.2 (Ertekin et al. 2001)

Fig. 10.5 represents a homogeneous, isotropic reservoir with uniform thickness. The depth at the top surface of the formation is 4,000 ft at every point, and the reservoir is 100% saturated with a slightly compressible liquid. The well in Gridblock 1 is put on production at a rate of 6,000 STB/D and remains at this constant rate for 20 days. A pressure survey in the wellbore at the end of 20 days reveals the sandface pressure to be 3,590 psi, which represents a pressure drop and brings into question the validity of the no-flow conditions imposed on two of the boundaries shown in Fig. 10.5. From seismic studies, it is certain that the other two boundaries are no-flow boundaries. Use the data provided in **Table 10.2** to resolve the question regarding the validity of the no-flow boundaries.

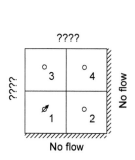

Fig. 10.5—Schematic of the reservoir system of Problem 10.5.2.

$k_x = k_y$ = 140 md
Δx = 800 ft
Δy = 800 ft
h = 80 ft
μ = 0.93 cp
B = 1 res bbl/STB
r_w = 0.24 ft
ϕ = 18%
s = 0
c = 6×10^{-5} psi^{-1}

Table 10.2—Reservoir data for Fig. 10.5.

Solution to Problem 10.5.2

To understand the nature of the boundary, we can solve the problem by assuming the boundaries are sealed. In view of the existing symmetry in the system, the pressure of Block 2 is expected to be equal to that of Block 3. Therefore, the primary unknowns of the problem are p_1, p_2, and p_4, with $p_3 = p_2$.

For an isotropic and homogeneous system, the governing partial-differential equation (PDE) for slightly compressible flow can be written as:

$$\frac{\partial^2 p}{\partial x^2} + \frac{\partial^2 p}{\partial y^2} + \frac{q\mu B}{kV_b} = \frac{\mu B \phi c_t}{5.615} \frac{\partial p}{\partial t}. \dots\dots\dots\dots\dots\dots\dots\dots\dots\dots\dots\dots (10.1)$$

In finite-difference form,

$$p_{i,j-1}^{n+1} + p_{i-1,j}^{n+1} - 4p_{i,j}^{n+1} + p_{i+1,j}^{n+1} + p_{i,j+1}^{n+1} + \frac{q\mu B}{kh} = \frac{(\Delta x)^2 \mu B \phi c_t}{5.615 \Delta t k} \left(p_{i,j}^{n+1} - p_{i,j}^n \right) \dots\dots\dots\dots (10.2)$$

and

$$\frac{(\Delta x)^2 \mu B \phi c_t}{5.615 k \Delta t} = \frac{(800)^2 (0.93)(1)(0.18)(6 \times 10^{-5})}{(1.127 \times 10^{-3})(140)(5.615)(20)} = 0.36279. \dots\dots\dots\dots\dots\dots\dots\dots (10.3)$$

Because $q = -6,000$ STB/D,

$$\frac{q\mu B}{kh} = \frac{(-6,000)(0.93)(1)}{(1.127 \times 10^{-3})(140)(80)} = -442.071. \dots\dots\dots\dots\dots\dots\dots\dots\dots\dots (10.4)$$

- Block 1:

$$2p_2^{n+1} - 2p_1^{n+1} + \frac{q\mu B}{kh} = \frac{(\Delta x)^2 \mu B \phi c_t}{k5.615\Delta t} \left(p_1^{n+1} - p_1^n \right). \dots\dots\dots\dots\dots\dots\dots\dots (10.5)$$

- Block 2:

$$p_1^{n+1} - 2p_1^{n+1} + p_4^{n+1} = \frac{(\Delta x)^2 \mu B \phi c_t}{k 5.615 \Delta t}\left(p_2^{n+1} - p_2^n\right). \quad \dots \dots \dots \dots \dots \dots \dots \dots \dots \dots \dots \dots \dots \dots \quad (10.6)$$

- Block 4:

$$2p_2^{n+1} - 2p_4^{n+1} = \frac{(\Delta x)^2 \mu B \phi c_t}{k 5.615 \Delta t}\left(p_4^{n+1} - p_4^n\right). \quad \dots \dots \dots \dots \dots \dots \dots \dots \dots \dots \dots \dots \dots \dots \quad (10.7)$$

From 0 to 20 days, rearranging the system of equations into a matrix form,

$$\begin{bmatrix} -2.363 & 2 & 0 \\ 1 & -2.363 & 1 \\ 0 & 2 & -2.363 \end{bmatrix}\begin{bmatrix} p_1^{n+1} \\ p_2^{n+1} \\ p_3^{n+1} \end{bmatrix} = \begin{bmatrix} -1,009.09 \\ -1,451.16 \\ -1,451.16 \end{bmatrix}. \quad \dots \dots \dots \dots \dots \dots \dots \dots \dots \dots \dots \quad (10.8)$$

Solving, $p_1^{n+1} = 3,576.486\,\text{psi}$, $p_2^{n+1} = p_3^{n+1} = 3,720.699\,\text{psi}$, $p_4^{n+1} = 3,763.583\,\text{psi}$,

$$\Omega = \frac{2\pi kh}{\mu B\left[\ln\left(\dfrac{r_e}{r_w}\right) + s\right]} = 13.14\,\text{STB/psi-D}, \quad \dots \dots \dots \dots \dots \dots \dots \dots \dots \dots \dots \dots \dots \quad (10.9)$$

$$p_{sf} = p + \frac{q}{\Omega} = 3,119.706\,\text{psi} < 3,590\,\text{psi}\;\; p_{sf}. \quad \dots \dots \dots \dots \dots \dots \dots \dots \dots \dots \dots \dots \dots \dots \dots \quad (10.10)$$

Implement a material-balance check (MBC) to validate the results.
Fluid expansion:

$$\Delta V_p = cV_p \sum_{i=1}^{4} \Delta p_i = 6\times10^{-5}\times\left(6,565,272\,\text{bbl}\right)\times\left(423.514 + 279.301 + 279.301 + 236.417\right) = 120,000\,\text{STB}, \quad (10.11)$$

$$\text{MBC} = \left|\frac{\Delta tq}{\Delta V_p}\right| = \left|\frac{120,000\,\text{STB}}{120,000\,\text{STB}}\right| = 1. \quad \dots \dots \dots \dots \dots \dots \dots \dots \dots \dots \dots \dots \dots \dots \dots \dots \quad (10.12)$$

Calculations show that if the boundaries were all sealed, the calculated sandface pressure would have been much lower than that of the reported value, which indicates that there is external pressure support coming in from the investigated boundaries.

Problem 10.5.3 (Ertekin et al. 2001)

Consider the flow of a slightly compressible liquid in a 2D reservoir with no depth gradients (**Fig. 10.6**). The reservoir has homogeneous and isotropic property distribution and is discovered at an initial pressure of 4,000 psi. **Table 10.3** provides the related reservoir rock and fluid properties. The well in Gridblock 2 is produced at a rate of 4,000 STB/D. You consider

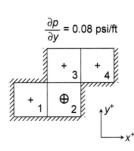

Fig. 10.6—Reservoir system of Problem 10.5.3.

Table 10.3—Reservoir rock and fluid properties for Fig. 10.6.	
$k_x = k_y =$	80 md
$\Delta x =$	1,000 ft
$\Delta y =$	1,000 ft
$h =$	100 ft
$\mu =$	1.3 cp
$B =$	1 res bbl/STB
$r_w =$	0.24 ft
$\phi =$	20%
$s =$	0
$c =$	1×10^{-5} psi^{-1}

drilling a second production well in the center of Gridblock 4, if the average pressure in Gridblock 4 has not dropped more than 10% of the initial pressure when the well in Gridblock 2 has produced for 20 days.

1. Will you be able to drill the new well in Gridblock 4 on the basis of the proposed criterion?
2. Perform an MBC on the solution to verify the validity of your computation. (*Hint:* Because $B = 1$ res bbl/STB is assumed to remain constant, the conventional material-balance equation will not work; however, you can devise another approach to check the material balance.)

Solution to Problem 10.5.3

1. For a 2D homogeneous and isotropic, slightly compressible fluid flow system, the governing flow equation will be

$$\frac{\partial^2 p}{\partial x^2} + \frac{\partial^2 p}{\partial y^2} + \frac{q\mu B}{kV_b} = \frac{\mu B\phi c_t}{k5.615}\frac{\partial p}{\partial t}. \quad\dotfill \text{(10.13)}$$

The PDE can be written in finite-difference form:

$$p_{i,j-1}^{n+1} + p_{i-1,j}^{n+1} - 4p_{i,j}^{n+1} + p_{i+1,j}^{n+1} + p_{i,j+1}^{n+1} + \left(\frac{q\mu B}{kh}\right)_{i,j} = \frac{(\Delta x)^2 \mu B\phi c_t}{5.615\Delta t}\left(p_{i,j}^{n+1} - p_{i,j}^n\right), \quad\dotfill \text{(10.14)}$$

where

$$\frac{(\Delta x)^2 \mu B\phi c_t}{5.615k\Delta t} = \frac{(1,000)^2 (1.3)(1)(0.2)(1\times10^{-5})}{(1.127\times10^{-3})(80)(5.615)(20)} = 0.257. \quad\dotfill\text{(10.15)}$$

Because $q = -4,000$ STB/D,

$$\frac{q\mu B}{kh} = \frac{(-4,000)(1.3)(1)}{(1.127\times10^{-3})(80)(100)} = -576.75. \quad\dotfill \text{(10.16)}$$

The linear system of equations generated from the characteristic finite-difference approximation can be assembled in a matrix form:

$$\begin{bmatrix} -1.257 & 1 & 0 & 0 \\ 1 & -2.257 & 1 & 0 \\ 0 & 1 & -2.257 & 1 \\ 0 & 0 & 1 & -1.257 \end{bmatrix} \begin{bmatrix} p_1^{n+1} \\ p_2^{n+1} \\ p_3^{n+1} \\ p_4^{n+1} \end{bmatrix} = \begin{bmatrix} -1,027.16 \\ -450.41 \\ -1,107.16 \\ -1,107.16 \end{bmatrix}. \quad\dotfill \text{(10.17)}$$

Solving the system of equations in Eq. 10.17 for pressures will yield pressure distribution:

- $p_1^{n+1} = 3,509.88\,\text{psi}$,
- $p_2^{n+1} = 3,384.02\,\text{psi}$,
- $p_3^{n+1} = 3,676.74\,\text{psi}$,
- $p_4^{n+1} = 3,806.44\,\text{psi}$,

because $p_4 = 3,806.44$ psi $> 90\%$ p_i (3,600 psi). The infill-drilling program should proceed as planned.

2. Implement an MBC to validate the results.

- Fluid expansion:

$$\Delta V = cV_p\sum_{i=1}^{4}\Delta p_i = 1\times10^{-5}\times(3,561,887.8\,\text{bbl})\times(490.12+615.98+323.26+193.56) \quad\dotfill\text{(10.18)}$$
$$= 57,807\,\text{STB}.$$

- Flow coming in across the boundary:

$$q_{in} = \Delta t \times \left(\frac{kA}{\mu B} c \right) = 20 \times 1,109.7 \text{ STB/D} = 22,193 \text{ STB}, \dots\dots\dots\dots\dots\dots\dots\dots(10.19)$$

$$\text{MBC} = \left| \frac{\Delta t q}{q_{in} + \Delta V} \right| = \left| \frac{80,000}{57,807 + 22,193} \right| = 1.00. \dots\dots\dots\dots\dots\dots\dots\dots\dots (10.20)$$

Problem 10.5.4 (Ertekin et al. 2001)

Consider the 2D homogeneous and isotropic reservoir shown in **Fig. 10.7.** The well in Gridblock 4 was put on production 30 days before the well in Gridblock 5 was brought on line. Both wells were produced at a constant rate of 300 STB/D. **Table 10.4** contains available well pressure-survey data. Other pertinent data are $k_x = k_y = 100$ md, $\Delta x = \Delta y = 1,320$ ft, $h = 40$ ft, $\phi = 20\%$, $\mu_w = 1.0$ cp, $B_w = 1.0$ res bbl/STB, $c_w = 8 \times 10^{-7}$ psi^{-1}, and $r_w = 0.25$ ft (for both wells).

1. Find the nature of the boundary along line ABC.
2. Find the pressure distribution in the reservoir at $t = 30$ and $t = 60$ days.

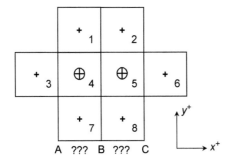

Fig. 10.7—Reservoir system of Problem 10.5.4.

t (days)	p_{wf}, psi	
	Well 4	Well 5
0	5,000	N/A
30 days	3,697	3,783.5

Table 10.4—Available well pressure survey data for the wells depicted in Fig. 10.7.

Solution to Problem 10.5.4

1. For a 2D homogeneous and isotropic, slightly compressible system,

$$\frac{\partial^2 p}{\partial x^2} + \frac{\partial^2 p}{\partial y^2} + \frac{q\mu B}{kV_b} = \frac{\mu B \phi c_t}{5.615} \frac{\partial p}{\partial t}. \dots\dots\dots\dots\dots\dots\dots\dots\dots (10.21)$$

In a finite-difference form,

$$p_{i,j-1}^{n+1} + p_{i-1,j}^{n+1} - 4 p_{i,j}^{n+1} + p_{i+1,j}^{n+1} + p_{i,j+1}^{n+1} + \left(\frac{q\mu B}{kh} \right)_{i,j} = \left[\frac{(\Delta x)^2 \mu B \phi c_t}{5.615 k \Delta t} \right]_{i,j} \left(p_{i,j}^{n+1} - p_{i,j}^n \right), \dots\dots\dots\dots(10.22)$$

$$\frac{(\Delta x)^2 \mu B \phi c_t}{5.615 k \Delta t} = \frac{(1,320)^2 (1)(1)(0.2)(8 \times 10^{-7})}{(1.127 \times 10^{-3})(100)(5.615)(40)} = 0.014685. \dots\dots\dots\dots\dots\dots (10.23)$$

In the solution, $p_{initial} = 5,000$ psi is uniform throughout the reservoir. With the flow rate and sandface pressure specified, the pressure of Block 2 can be calculated using the wellbore-flow equation:

$$p_4 = p_{sf_4} - \left(\frac{q}{\Omega} \right)_4 = 3,770.63 \text{ psi}, \dots\dots\dots\dots\dots\dots\dots\dots\dots (10.24)$$

where

$$\Omega = \frac{2\pi kh}{\mu B \left[\ln\left(\frac{r_e}{r_w} \right) + s \right]} = 4.074 \text{ STB/psi-D}. \dots\dots\dots\dots\dots\dots\dots\dots (10.25)$$

Note that in Eq. 10.25 r_e is calculated using the wellblock equivalent radius concept for a square gridblock and $s = 0$. With the finite-difference equation, the pressure of each block can be calculated in a propagating manner:

- Block 3:

$$p_4 - \left[\frac{(\Delta x)^2 \, \mu B \phi c_t}{5.615 \Delta t} + 1\right] p_4 = -\frac{(\Delta x)^2 \, \mu B \phi c_t}{5.615 \Delta t} p_i, \text{ solving for } p_3 = 3{,}788.426 \text{ psi.} \quad \dots\dots\dots\dots\dots \text{(10.26)}$$

- Block 6:

$$p_5 - \left[\frac{(\Delta x)^2 \, \mu B \phi c_t}{5.615 \Delta t} + 1\right] p_6 = -\frac{(\Delta x)^2 \, \mu B \phi c_t}{5.615 \Delta t} p_i, \text{ solving for } p_6 = 3{,}801.10572 \text{ psi.} \quad \dots\dots\dots\dots\dots \text{(10.27)}$$

- Block 1:

$$p_4 + p_2 - \left[\frac{(\Delta x)^2 \, \mu B \phi c_t}{5.615 \Delta t} + 2\right] p_1 = -\frac{(\Delta x)^2 \, \mu B \phi c_t}{5.615 \Delta t} p_i. \quad \dots\dots\dots\dots\dots\dots\dots\dots \text{(10.28)}$$

- Block 2:

$$p_1 + p_5 - \left[\frac{(\Delta x)^2 \, \mu B \phi c_t}{5.615 \Delta t} + 2\right] p_2 = -\frac{(\Delta x)^2 \, \mu B \phi c_t}{5.615 \Delta t} p_i. \quad \dots\dots\dots\dots\dots\dots\dots\dots \text{(10.29)}$$

Solving the equations of Block 1 and 2 simultaneously: $p_1 = 3{,}792.63$ psi and $p_2 = 3{,}796.90$ psi.

- Block 4:

$$p_1 + p_3 + p_7 + p_5 - \left[\frac{(\Delta x)^2 \, \mu B \phi c_t}{5.615 \Delta t} + 4\right] p_4 + \frac{q \mu B}{kh} = -\frac{(\Delta x)^2 \, \mu B \phi c_t}{5.615 \Delta t} p_i, \text{ solving for}$$

$$p_7 = 3{,}766.4743673 \text{ psi.} \quad \dots\dots\dots\dots\dots\dots\dots\dots\dots\dots\dots\dots\dots\dots\dots\dots\dots \text{(10.30)}$$

- Block 5:

$$p_2 + p_4 + p_6 + p_8 - \left[\frac{(\Delta x)^2 \, \mu B \phi c_t}{5.615 \Delta t} + 4\right] p_5 + \frac{q \mu B}{kh} = -\frac{(\Delta x)^2 \, \mu B \phi c_t}{5.615 \Delta t} p_i, \text{ solving for}$$

$$p_8 = 3{,}747.496104 \text{ psi.} \quad \dots\dots\dots\dots\dots\dots\dots\dots\dots\dots\dots\dots\dots\dots\dots\dots\dots\dots \text{(10.31)}$$

- At Block 7, if a Neumann-type boundary condition is applied to segment AB, with a magnitude of $+c_1$ psi/ft,

$$\frac{p_7 - p_7'}{\Delta y} = c_1, \ p_7' = p_7 - c_1 \Delta y. \quad \dots\dots\dots\dots\dots\dots\dots\dots\dots\dots\dots\dots\dots\dots \text{(10.32)}$$

For Block 7:

$$p_4 + p_8 - \left[\frac{(\Delta x)^2 \, \mu B \phi c_t}{5.615 \Delta t} + 2\right] p_7 - c_1 \Delta y = -\frac{(\Delta x)^2 \, \mu B \phi c_t}{5.615 \Delta t} p_i, \text{ solving for } c_1 = 0.00249 \text{ psi/ft.} \quad \dots\dots \text{(10.33)}$$

- For Block 8, if a Neumann-type boundary condition is applied to segment AB, with a magnitude of $+c_2$ psi/ft,

$$\frac{p_8 - p_8'}{\Delta y} = c_2, \ p_8' = p_8 - c_2 \Delta y. \quad \dots\dots\dots\dots\dots\dots\dots\dots\dots\dots\dots\dots\dots\dots \text{(10.34)}$$

For Block 8:

$$p_7 + p_5 - \left[\frac{(\Delta x)^2 \, \mu B \phi c_t}{5.615 \Delta t} + 2\right] p_8 - c_2 \Delta y = -\frac{(\Delta x)^2 \, \mu B \phi c_t}{5.615 \Delta t} p_i, \text{ solving for } c_2 = 0.0556 \text{ psi/ft.} \quad \dots\dots \text{(10.35)}$$

Implementing an MBC to validate the results:

- Fluid expansion:

$$\Delta V = c V_p \sum_{i=1}^{8} \Delta p_i = 8 \times 10^{-7} \times (2{,}482{,}493.32 \text{ bbl}) \times (9{,}752.8) = 19{,}369.07319 \text{ STB.} \quad \dots\dots\dots\dots \text{(10.36)}$$

- Flow out from the boundary:

$$Q_{out} = \Delta t \times \left[\frac{kh}{\mu B}(c_1 + c_2) \right] = 30 \times 345.64 \text{ STB/D} = 10{,}369.07319 \text{ STB}, \quad \text{.......................} \quad (10.37)$$

$$\text{MBC} = \left| \frac{\Delta t q + Q_{out}}{\Delta V} \right| = \left| \frac{19{,}369.07319}{19{,}369.07319} \right| = 1. \quad \text{...} \quad (10.38)$$

2. Using the solved pressure distribution of the previous step, the well in Block 5 starts to produce. A system of equations can be written and expressed in the following matrix form:

$$
\begin{bmatrix}
-2.015 & 1 & 0 & 1 & 0 & 0 & 0 & 0 \\
1 & -2.015 & 0 & 0 & 1 & 0 & 0 & 0 \\
0 & 0 & -1.015 & 1 & 0 & 0 & 0 & 0 \\
0 & 0 & 1 & -4.015 & 1 & 0 & 1 & 0 \\
0 & 0 & 0 & 1 & -4.015 & 1 & 0 & 1 \\
0 & 0 & 0 & 0 & 1 & -1.015 & 0 & 0 \\
0 & 0 & 0 & 1 & 0 & 0 & -2.015 & 1 \\
0 & 0 & 0 & 0 & 1 & 0 & 1 & -2.015
\end{bmatrix}
\begin{bmatrix}
p_1^{n+1} \\
p_2^{n+1} \\
p_3^{n+1} \\
p_4^{n+1} \\
p_5^{n+1} \\
p_6^{n+1} \\
p_7^{n+1} \\
p_8^{n+1}
\end{bmatrix}
=
\begin{bmatrix}
-55.70 \\
-55.76 \\
-55.63 \\
11.18 \\
10.99 \\
-55.82 \\
-52.01 \\
18.34
\end{bmatrix}.
$$

$$\text{...} \quad (10.39)$$

Solving for pressure at the end of 60 days, $p_1 = 2{,}012.57$ psi, $p_2 = 2{,}010.33$ psi, $p_3 = 2{,}014.72$ psi, $p_4 = 1{,}988.67$ psi, and $p_5 = 1{,}981.86$ psi.

Problem 10.5.5 (Ertekin et al. 2001)

Consider the 1D, slightly compressible fluid flow problem shown in **Fig. 10.8.** As shown in the figure, Block 2 hosts a production well at a constant rate of 1,400 STB/D. Blocks 1 and 3 have observation wells, which are not put on production. The sandface pressure logs after 30 days are shown in **Table 10.5.**

Time	Well 1	Well 2	Well 3
30 days	1,514.3 psi	1,000 psi	1,436.5 psi

Table 10.5—Sandface pressure logs for the wells shown in Fig. 10.8.

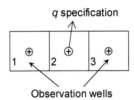

q specification

Observation wells

Fig. 10.8—Reservoir system of Problem 10.5.5.

$k_1 = 100$ md
$\Delta x = 1{,}000$ ft
$\Delta y = 1{,}000$ ft
$h = 100$ ft
$\mu = 2$ cp
$B = 1$ res bbl/STB
$r_w = 0.25$ ft
$s = 0$
$c_t = 1 \times 10^{-5}$ psi^{-1}
Ignore depth gradient

Table 10.6—Reservoir data for Fig. 10.8.

Using the information given in **Table 10.6,** estimate the permeability values of Blocks 1 and 3 (from a pressure-transient analysis, the permeability of Block 2 is found to be 100 md). The initial reservoir pressure is 2,000 psi.

Solution to Problem 10.5.5

The productivity index for the well in Block 2 is

$$\Omega_2 = \frac{2\pi kh}{\mu B \left[\ln\left(\dfrac{r_e}{r_w} \right) + s \right]} = 5.3019 \text{ STB/D-psi}, \quad \text{...} \quad (10.40)$$

where $r_e = 0.2\Delta x = 200$ ft;

$$q = -\Omega\left(p_2 - p_{sf} \right) = -1{,}400 \text{ STB/D}. \quad \text{...} \quad (10.41)$$

Solving for p_2,

$$p_2 = p_{sf_2} - \frac{q_2}{\Omega_2} = 1,000 - \left(\frac{-1,400}{5.3019}\right) = 1,264.1 \text{ psi.} \quad \text{..(10.42)}$$

The fluid expansion in Block 1 is

$$\Delta V_1 = c_t \frac{V_b \phi}{5.615}(p_i - p_1) = 20,760.1 \text{ STB.} \quad \text{..(10.43)}$$

The fluid expansion in Block 2 is

$$\Delta V_2 = c_t \frac{V_b \phi}{5.615}(p_i - p_1) = 31,454.3 \text{ STB.} \quad \text{..(10.44)}$$

The fluid expansion in Block 3 is

$$\Delta V_3 = c_t \frac{V_b \phi}{5.615}(p_i - p_1) = 24,085.5 \text{ STB.} \quad \text{..(10.45)}$$

Flow between Blocks 1 and 2 is

$$q_{1to2} = \frac{\bar{k}h}{\mu B}(p_1 - p_2) = \frac{\Delta V_1}{30} = 692 \text{ STB/D,} \quad \text{..(10.46)}$$

$$\bar{k} = \frac{2k_1 k_2}{k_1 + k_2} = 49.10 \text{ md;} \quad \text{..(10.47)}$$

solving for k_1, $k_1 = 32.523$ md.

Mass flux from Block 3 to Block 2 is

$$q_{1to2} = \frac{\bar{k}h}{\mu B}(p_3 - p_2) = \frac{\Delta V_2}{30} = 802 \text{ STB/D,} \quad \text{..(10.48)}$$

$$\bar{k} = \frac{2k_3 k_2}{k_3 + k_2} = 82.64 \text{ md;} \quad \text{..(10.49)}$$

solving for k_3: $k_3 = 70.92$ md.

Problem 10.5.6 (Ertekin et al. 2001)

Consider the reservoir system shown in **Fig. 10.9,** in which the permeability, thickness, and width of the gridblocks are uniform throughout. The sandface pressure in Gridblock 4 is 3,000 psi. A pressure gradient of –0.6 psia is specified at the bottom of the reservoir. All other boundaries are considered to be completely sealed. The observed flow rate from the well in Gridblock 4 is measured as 2,255 STB/D. After conducting a simulation study, you realize this rate does not match the simulation result. Going back to the geological records of the reservoir, you find there is a possibility that the boundary of Block 3 is active. Neglecting the depth gradients, assuming incompressible flow, and using the data provided in **Table 10.7,** answer the following questions:

1. What is the nature of the boundary condition at the right boundary of Block 3?
2. Assume that the well is completed in Blocks 2 and 4. What is the flow rate from the well if the same sandface pressure is specified? Find the pressure values in each gridblock, and comment on the results.

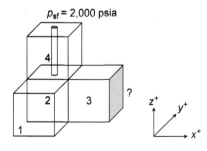

Fig. 10.9—Reservoir system of Problem 10.5.6.

$k_x = k_y = k_z$ = 100 md
Δx = 100 ft
Δy = 100 ft
Δz = 100 ft
μ = 1 cp
B = 1 res bbl/STB
r_w = 0.3 ft
s = 0
Ignore depth gradient

Table 10.7—Reservoir data for Fig. 10.9.

Solution to Problem 10.5.6

1. The flow in from the bottom of the formation can be calculated as

$$q_{in} = 3 \times \left(\frac{kA}{\mu B} c \right) = \left(\frac{1.127 \times 10^{-3} \times 100 \times 100}{1 \times 1} \right)(0.6 \text{ psi/ft}) = 2,028.6 \text{ STB/D}. \dots \dots \dots (10.50)$$

The measured flow rate is 2,255 STB/D. Therefore, the right boundary of Block 3 needs to compensate with a flow rate of 2,255 − 2,028.6 = 226.4 STB/D. Thus, the pressure gradient imposed on the right-most surface of Block 3 can be calculated:

$$\frac{\partial p}{\partial x} = q \left(\frac{\mu B}{kA} \right) = \left(\frac{226.4}{1.127 \times 10^{-3} \times 100 \times 100 \times 100} \right) = +0.2 \text{ psi/ft}. \dots \dots \dots (10.51)$$

The positive sign indicates that flow is entering the reservoir across the eastern face of Block 3.

2. For an incompressible flow system, the flow rate when the well is completed at both Blocks 2 and 4 should be 2,255 STB/D, because flow in must be equal to flow out. Recalling that there is flow in of 2,028.6 / 3 = 676.2 STB/D to each of Blocks 1, 2, and 3 from the bottom, we list the finite-difference equations:

- Block 1: The volume of 676.2 STB/D entering into Block 1 must be transmitted to Block 2:

$$p_2 - p_1 + \frac{672.2 \mu B}{kh} = 0. \dots \dots \dots (10.52)$$

- Block 2:

$$p_1 + p_3 + p_4 - 3p_2 + \left[-\frac{\Omega(p_2 - p_{sf})\mu B}{kh} \right] + \frac{672.2 \mu B}{kh} = 0. \dots \dots \dots (10.53)$$

- Block 3:

$$p_2 - p_3 + 0.2 \Delta x + \frac{676.2 \mu B}{kh} = 0. \dots \dots \dots (10.54)$$

- Block 4:

$$p_2 - p_4 + \left[-\frac{\Omega(p_4 - p_{sf})\mu B}{kh} \right] = 0. \dots \dots \dots (10.55)$$

In matrix form,

$$\begin{bmatrix} -1 & 1 & 0 & 0 \\ 1 & -4.5 & 1 & 1 \\ 0 & 1 & -1 & 1 \\ 0 & 1 & 0 & -2.5 \end{bmatrix} \begin{bmatrix} p_1 \\ p_2 \\ p_3 \\ p_4 \end{bmatrix} = \begin{bmatrix} -60 \\ -3,060 \\ -80 \\ -3,000 \end{bmatrix}. \dots \dots \dots (10.56)$$

Solving for pressures in psi, $p_1 = 2,155.3$ psi, $p_2 = 2,095.3$ psi, $p_3 = 2,175.3$ psi, and $p_4 = 2,038.0$ psi. Checking flow rates,

$$q_2 = -\Omega(p_2 - p_{sf}) = -1,611.13 \text{ STB/D}, \dots \dots \dots (10.57)$$

$$q_4 = -\Omega(p_4 - p_{sf}) = -643.8 \text{ STB/D}, \dots \dots \dots (10.58)$$

$$q_{total} = q_2 + q_4 = -2255.0 \text{ STB/D}. \dots \dots \dots (10.59)$$

Problem 10.5.7 (Ertekin et al. 2001)

Consider the slightly compressible fluid flow through the 2D homogeneous and isotropic porous medium shown in **Fig. 10.10.1.** All boundaries are known to be no-flow boundaries, except the right boundary of Gridblock 3. The well in Gridblock 2 has a flow-rate specification of 500 STB/D. The sandface pressure on the same gridblock is measured as 2,028 psia at the end of five days of production. Initial pressure of the reservoir is known to be 3,000 psia.

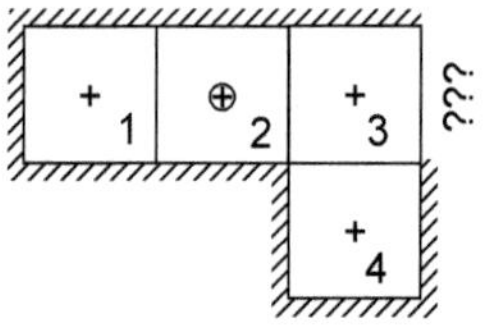

Fig. 10.10.1—Reservoir system of Problem 10.5.7.

1. What is the boundary condition specified at the right boundary of Gridblock 3?
2. Obtain the pressure distribution of the system at the end of a five-day production period. Perform a material balance on your solution to check the accuracy of the computations.
3. Assume that the boundary conditions specified are maintained, and implement the line-successive over-relaxation-solution (LSOR) technique to find the pressure distribution at the end of 10 days. Use an initial guess of 1,680 psia for the LSOR iteration. Use $\omega_{opt} = 1.1$ and a convergence criterion of 2 psi.

$k_x = k_y =$ 75 md
$\Delta x =$ 200 ft
$\Delta y =$ 200 ft
$h =$ 40 ft
$\mu =$ 1.3 cp
$B =$ 1 res bbl/STB
$r_w =$ 0.23 ft
$\phi =$ 25%
$s =$ +2
$c =$ 1×10⁻⁵ psi⁻¹

Table 10.8—Reservoir data for Fig. 10.10.1.

Use the data presented in **Table 10.8** for the analysis.

Solution to Problem 10.5.7

Answer to questions 1 and 2: For a 2D homogeneous and isotropic, slightly compressible system,

$$\frac{\partial^2 p}{\partial x^2} + \frac{\partial^2 p}{\partial y^2} + \frac{q\mu B}{kV_b} = \frac{\mu B \phi c_t}{5.615k}\frac{\partial p}{\partial t}. \quad \dots \dots \dots \dots \dots \dots \dots \dots \dots \dots \dots \dots \dots \dots \dots \dots \dots \dots \dots \dots \dots \dots (10.60)$$

In finite-difference form,

$$p_{i,j-1}^{n+1} + p_{i-1,j}^{n+1} - 4p_{i,j}^{n+1} + p_{i+1,j}^{n+1} + p_{i,j+1}^{n+1} + \left(\frac{q\mu B}{kh}\right)_{i,j} = \left[\frac{(\Delta x)^2 \mu B \phi c_t}{5.615k\Delta t}\right]_{i,j} \left(p_{i,j}^{n+1} - p_{i,j}^{n}\right), \quad \dots \dots \dots (10.61)$$

where

$$\frac{(\Delta x)^2 \mu B \phi c_t}{5.615k\Delta t} = \frac{(200)^2 (1.3)(1)(0.25)(1\times 10^{-5})}{(1.127\times 10^{-3})(75)(5.615)(5)} = 0.054782. \quad \dots \dots \dots \dots \dots \dots \dots \dots \dots (10.62)$$

With flow rate specified and sandface pressure measured at the bottom of the well in Gridblock 2, the pressure of Gridblock 2 can be calculated:

$$p_2 = p_{sf_2} - \frac{q_2}{\Omega_2} = 2,246.73\,\text{psi}, \quad \dots \dots \dots \dots \dots \dots \dots \dots \dots \dots \dots \dots \dots \dots \dots \dots \dots \dots \dots (10.63)$$

where

$$\Omega_2 = \frac{2\pi kh}{\mu B\left[\ln\left(\frac{r_e}{r_w}\right) + s\right]} = 2.286 \text{ STB/psi-D}. \quad \dots \dots \dots \dots \dots \dots \dots \dots \dots \dots \dots \dots \dots \dots (10.64)$$

With the finite-difference equation, the pressure of each block can be calculated in a propagating manner:

- At Block 1:

$$p_2 - \left[\frac{(\Delta x)^2 \mu B \phi c_t}{5.615\Delta t} + 1\right]p_1 = -\frac{(\Delta x)^2 \mu B \phi c_t}{5.615\Delta t}p_i, \text{ solving for } p_1 = 2,285.85 \text{ psi}. \quad \dots \dots \dots \dots \dots (10.65)$$

- At Block 2:

$$p_1 + p_3 - \left[\frac{(\Delta x)^2 \mu B \phi c_t}{5.615\Delta t} + 2\right]p_2 = \left(\frac{q\mu B}{kh}\right)_2 - \frac{(\Delta x)^2 \mu B \phi c_t}{5.615\Delta t}p_i, \text{ solving for } p_3 = 2,358.60 \text{ psi}. \quad \dots \dots (10.66)$$

- At Block 4:

$$p_3 - \left[\frac{(\Delta x)^2 \mu B \phi c_t}{5.615\Delta t} + 1\right]p_4 = -\frac{(\Delta x)^2 \mu B \phi c_t}{5.615\Delta t}p_i, \text{ solving for } p_4 = 2,391.91 \text{ psi}. \quad \dots \dots \dots \dots \dots \dots (10.67)$$

- At Block 3, if a Neumann-type boundary condition is applied with magnitude of $+c$ psi/ft, then

$$\frac{p_3' - p_3}{\Delta x} = c, \ p_3' = p_3 + c\Delta x. \dotfill (10.68)$$

For Block 3:

$$p_2 + p_4 - \left[\frac{(\Delta x)^2 \ \mu B\phi c_t}{5.615\Delta t} + 2\right]p_3 + c\Delta x = -\frac{(\Delta x)^2 \ \mu B\phi c_t}{5.615\Delta t}p_i, \ \text{solving for } c = 0.217 \text{ psi/ft.} \dotfill (10.69)$$

Implementing an MBC to validate the results,

- Fluid expansion:

$$\Delta V = cV_p \sum_{i=1}^{4} \Delta p_i = 1\times10^{-5} \times (71,237.76 \text{ bbl}) \times (714.1498 + 753.2724 + 641.41 + 608.0972) \dotfill (10.70)$$
$$= 1,935.48 \text{ STB}.$$

- Flow in across the boundary:

$$q_{\text{in}} = \Delta t \times \left(\frac{kh}{\mu B}c\right) = 5\times112.90 \text{ STB/D} = 564.52 \text{ STB}, \dotfill (10.71)$$

$$\text{MBC} = \left|\frac{\Delta t q}{q_{\text{in}} + \Delta V}\right| = \left|\frac{2,500 \text{ STB}}{564.52 \text{ STB} + 1,1935.45 \text{ STB}}\right| = 1.0000. \dotfill (10.72)$$

3. The solution process continues from $t = 5$ to $t = 10$ days (using a timestep size of 5 days). The original system can be simplified to a 1D system shown in **Fig. 10.10.2**. Therefore, there is only one line equation to be solved in the system.

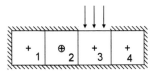

Fig. 10.10.2—Representation of the equivalent reservoir system.

The system of equations can be listed in matrix form. Because there is only one line, we are simultaneously solving all four unknowns. This is why there are no LSOR iterations.

$$\begin{bmatrix} -1.055 & 1 & 0 & 0 \\ 1 & -2.055 & 1 & 0 \\ 0 & 1 & -2.055 & 1 \\ 0 & 0 & 1 & -1.055 \end{bmatrix} \begin{bmatrix} p_1^{n+1} \\ p_2^{n+1} \\ p_3^{n+1} \\ p_4^{n+1} \end{bmatrix} = \begin{bmatrix} -125.22 \\ -69.17 \\ -172.62 \\ -131.03 \end{bmatrix}. \dotfill (10.73)$$

Solving for the pressures, $p_1^{n+1} = 1,601.73 \text{ psia}$, $p_2^{n+1} = 1,564.25 \text{ psia}$, $p_3^{n+1} = 1,681.64 \text{ psia}$, and $p_4^{n+1} = 1,718.53 \text{ psia}$. Implementing an MBC to validate the results:

- Fluid expansion:

$$\Delta V = cV_p \sum_{i=1}^{4} \Delta p_i = 1\times10^{-5} \times (71,237.76 \text{ bbl}) \times (684.12 + 682.48 + 679.96 + 673.88) \dotfill (10.74)$$
$$= 1,935.48 \text{ STB}.$$

- Flow in across the boundary:

$$q_{\text{in}} = \Delta t \times \left(\frac{kh}{\mu B}c\right) = 5\times112.90 \text{ STB/D} = 564.52 \text{ STB}, \dotfill (10.75)$$

$$\text{MBC} = \left|\frac{\Delta t q}{q_{\text{in}} + \Delta V}\right| = \left|\frac{2,500 \text{ STB}}{564.52 \text{ STB} + 1,935.45 \text{ STB}}\right| = 1.000. \dotfill (10.76)$$

The agreement between Parts 1 and 2 implies that the system has entered a pseudosteady-state condition and eventually will enter into steady-state condition.

Problem 10.5.8 (Ertekin et al. 2001)

Consider the 1D reservoir with no-depth gradients shown in **Fig. 10.11.** Assume single-phase, slightly compressible fluid flow dynamics. The known reservoir characteristics are listed in **Table 10.9. Table 10.10** displays the pressure measurements recorded at Wells 1 and 2 at $t = 10$ and $t = 30$ days, respectively. Formal seismic studies concluded that the western boundary of the reservoir is completely sealed. There is disagreement on the nature of the eastern boundary. Using the provided pressure and production data, respond to the following:

1. Determine the nature of the boundary condition along the eastern boundary.
2. Drilling a third well at $t = 30$ days is being considered. Your engineering manager indicates that the third well should be drilled only if the pressure at the location of the third well is not less than 50% of the initial formation pressure at $t = 60$ days. What would you recommend to your manager? Justify your recommendation. Use the implicit finite-difference approximation in generating the system of equations.

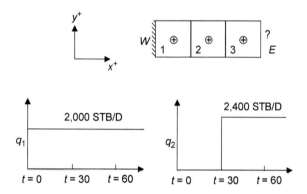

Fig. 10.11—Reservoir system of Problem 10.5.8.

$k_x = k_y =$ 180 md	
$\Delta x =$ 4,000 ft	
$\Delta y =$ 4,000 ft	
$h =$ 100 ft	
$\mu =$ 1 cp	
$B =$ 1 res bbl/STB	
$r_w =$ 0.2 ft	
$\phi =$ 17%	
$s = 0$	
$c = 6 \times 10^{-7}$ psi^{-1}	

Table 10.9—Known reservoir characteristics for Fig. 10.11.

	Well 1	Well 2
$t = 0$ days	$p_{sf} =$ 4,000 psi	—
$t = 30$ days	$p_{sf} =$ 3,129.40 psi	$p_{sf} =$ 3,322.67 psi

Table 10.10—Pressure measurements recorded at Wells 1 and 2.

Solution to Problem 10.5.8

1. For a 1D slightly compressible system, the governing flow equation is

$$\frac{\partial}{\partial x}\left(\frac{A_x k_x}{\mu B} \frac{\partial p}{\partial x} \right) \Delta x + q = \frac{V_b \phi c_t}{5.615}\left(\frac{\partial p}{\partial t} \right), \quad \dots \dots \dots \dots \dots \dots \dots \dots \dots \dots \dots \dots (10.77)$$

For the homogeneous reservoir considered in this problem,

$$\frac{\partial^2 p}{\partial x^2} + \frac{q \mu B}{V_b k} = \frac{\phi c_t \mu B}{5.615 k}\left(\frac{\partial p}{\partial t} \right), \quad \dots \dots \dots \dots \dots \dots \dots \dots \dots \dots \dots \dots (10.78)$$

the intermediate variables are $\dfrac{\mu B}{hk} = 0.049$ and $\dfrac{\mu B \phi c_t \Delta x^2}{5.615 k \Delta t} = 0.0478$.

Employing the Peaceman model (Peaceman 1983), the productivity index is calculated as

$$\Omega = \frac{2\pi k h}{\mu B \left[\ln\left(\dfrac{r_e}{r_w} \right) + s \right]} = 15.38 \text{ STB/D-psi.} \quad \dots \dots \dots \dots \dots \dots \dots \dots \dots \dots (10.79)$$

In Eq. 10.79, Peaceman's equivalent wellblock-radius equation is used: $r_e = 0.2 \Delta x = 800$ ft.
The governing flow equation in finite-difference form is written as

$$p_{i-1}^{n+1} - 2.0478 p_i^{n+1} + p_{i+1}^{n+1} = -0.049 q - 0.0478 p_i^n. \quad \dots \dots \dots \dots \dots \dots \dots \dots (10.80)$$

From 0 to 30 days, the pressure of Block 1 can be calculated:

$$p_1 = \frac{q}{-\Omega} + p_{sf} = 3,259.44 \text{ psi}; \dots\dots\dots\dots\dots\dots\dots\dots\dots\dots\dots\dots\dots(10.81)$$

also, $p_2 = 3,322.67$ psi.

We can assume that the eastern boundary of the system is sealed,

$$-1.0478 p_3^{n+1} + p_2^{n+1} = -0.0478 p_i^n, \dots\dots\dots\dots\dots\dots\dots\dots\dots\dots\dots\dots(10.82)$$

which gives $p_3 = 3,353.57$ psi.

An MBC can be performed:

- Total fluid volume expansion:

$$\Delta V = \Delta V_1 + \Delta V_2 + \Delta V_3 = -cV_b \left(\Delta p_1 + \Delta p_2 + \Delta p_3\right)$$

$$= 6 \times 10^{-7} \frac{(4,000)(4,000)(100)(0.17)}{5.615}\Big[(4,000 - 3,259.44) + (4,000 - 3,322.67) + (4,000 - 3,353.57)\Big]$$

$$= 60,000 \text{ STB}.$$

$$\dots\dots\dots\dots\dots\dots\dots\dots\dots\dots\dots\dots\dots\dots\dots\dots\dots\dots\dots(10.83)$$

- Total production:

$$Q_{\text{total}} = 2,000 \text{ STB/D} \times 30 \text{ days} = 60,000 \text{ STB}. \dots\dots\dots\dots\dots\dots\dots\dots\dots\dots(10.84)$$

Because the MBC yields excellent results, the assumption of a sealed eastern boundary is valid.

2. From 30 to 60 days, expressing the system of equations in matrix form:

$$\begin{bmatrix} -1.04776 & 1 & 0 \\ 1 & -2.04776 & 1 \\ 0 & 1 & -1.04776 \end{bmatrix} \begin{bmatrix} p_1^{n+1} \\ p_2^{n+1} \\ p_3^{n+1} \end{bmatrix} = \begin{bmatrix} -57.0769 \\ -40.3783 \\ -160.161 \end{bmatrix}, \dots\dots\dots\dots\dots\dots(10.85)$$

and solving for block pressure at $t = 60$ days, gives $p_1 = 1,756.31$ psia, $p_2 = 1,783.11$ psia, and $p_3 = 1,854.70$ psia. Because $p_3 < 50\% \, p_{\text{initial}}$, it is not recommended to drill a new well in Gridblock 3.

Problem 10.5.9

Consider the 2D reservoir shown in **Fig. 10.12.** The well in Block 2 is producing water at a specified sandface pressure of 450 psia. The well in Block 8 is shut in and its permanently installed downhole pressure gauge reads 2,974 psia. Block 1 is connected to constant-pressure Aquifer A, and Block 3 is connected with another constant-pressure Aquifer B, with $p_A = 4,000$ psia and $p_B = 3,000$ psia, respectively. All other boundaries are closed. The laboratory-analysis results show that Aquifer A has a certain uniform molar concentration of Na^+, but no K^+; while Aquifer B has the same uniform molar concentration of K^+, but no Na^+. Calculate the molar ratio of Na^+ over K^+ in the water produced from the well. In your calculation, assume that any kind of chemical activity is negligible.

Note: The fluid and reservoir properties are listed in **Table 10.11** (depth gradient is negligible). Use the Gauss-Seidel method (detailed more fully in Chapter 7 of this manuscript) to solve the system of equations. Use 1,900 psia as the initial guess for all unknowns. The convergence criterion to be used is $\varepsilon < 5$ psi. Check the validity of your calculations by performing an MBC.

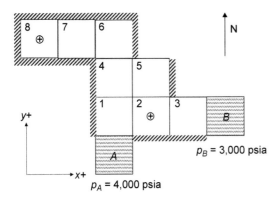

Fig. 10.12—Reservoir system of Problem 10.5.9.

$c = 0$
$\mu = 0.95$ cp
$B = 1.0$ res bbl/STB
$h = 40$ ft
$r_w = 0.25$ ft
$s = -1.5$
$\Delta x = \Delta y = 200$ ft
$k_x = 5$ md
$k_y = 20$ md

Table 10.11—Fluid and reservoir data for Fig. 10.12.

Solution to Problem 10.5.9

Blocks 6, 7, and 8 are all inactive (no flow from Block 4 into Block 6, because reservoir fluid is incompressible), so their pressures are the same as those of Block 4 (i.e., $p_4 = p_6 = p_7 = p_8 = 2{,}974$ psia). The system of equations will have p_1, p_2, p_3, and p_5 as unknowns:

- At Node 1: $k_y\left(p_A - p_1\right) + k_y\left(p_4 - p_1\right) + k_x\left(p_2 - p_1\right) = 0.$ .... (10.86)

- At Node 2: $k_x\left(p_1 - p_2\right) + k_y\left(p_5 - p_2\right) + k_x\left(p_3 - p_2\right) - \dfrac{2\pi\sqrt{k_x k_y}}{\ln\dfrac{r_e}{r_w} + s}\left(p_2 - p_{wf}\right) = 0.$ .... (10.87)

- At Node 3: $k_x\left(p_B - p_3\right) + k_x\left(p_2 - p_3\right) = 0.$ .... (10.88)

- At Node 5: $k_x\left(p_4 - P_5\right) + k_y\left(P_2 - P_5\right) = 0,$ .... (10.89)
  where

$$r_e = \frac{0.28\sqrt{\left(\dfrac{k_x}{k_y}\right)^{\frac{1}{2}}(\Delta y)^2 + \left(\dfrac{k_y}{k_x}\right)^{\frac{1}{2}}(\Delta x)^2}}{\left(\dfrac{k_x}{k_y}\right)^{\frac{1}{4}} + \left(\dfrac{k_y}{k_x}\right)^{\frac{1}{4}}} = 41.74 \text{ ft.} \quad\quad\quad (10.90)$$

Move all constant terms to the right-hand side to end up with four equations in four unknowns (the unknowns are p_1, p_2, p_3, and p_5):

$$-\left(2k_y + k_x\right)p_1 + k_x p_2 = -k_y\left(p_A + p_4\right), \quad\quad\quad\quad\quad\quad\quad (10.91)$$

$$k_x p_1 - \left(2k_x + k_y + \frac{2\pi\sqrt{k_x k_y}}{\ln\dfrac{r_e}{r_w} + s}\right)p_2 + k_x p_3 + k_y p_5 = -\frac{2\pi\sqrt{k_x k_y}}{\ln\dfrac{r_e}{r_w} + s}p_{wf}, \quad\quad (10.92)$$

$$k_x p_2 - 2k_x p_3 = -k_x p_B, \quad\quad\quad\quad\quad\quad\quad\quad\quad\quad (10.93)$$

$$k_y p_2 - \left(k_x + k_y\right)p_5 = -k_x p_4. \quad\quad\quad\quad\quad\quad\quad\quad\quad (10.94)$$

Divide both sides of the equations by k_x and rearrange them for the Gauss-Seidel-iteration implementation:

$$p_1 = \frac{1}{2r+1}\left[r\left(p_A + p_4\right) + p_2\right], \quad\quad\quad\quad\quad\quad\quad\quad (10.95)$$

$$p_2 = \frac{1}{2 + r + \dfrac{2\pi\sqrt{r}}{\ln\dfrac{r_e}{r_w} + s}}\left(p_1 + p_3 + rp_5 + \frac{2\pi\sqrt{r}}{\ln\dfrac{r_e}{r_w} + s}p_{wf}\right), \quad\quad (10.96)$$

$$p_3 = \frac{1}{2}\left(p_B + p_2\right), \quad\quad\quad\quad\quad\quad\quad\quad\quad\quad (10.97)$$

and

$$p_5 = \frac{1}{1+r}\left(rp_2 + p_4\right), \quad\quad\quad\quad\quad\quad\quad\quad\quad (10.98)$$

where

$$r = \frac{k_y}{k_x}. \quad\quad\quad\quad\quad\quad\quad\quad\quad\quad\quad\quad\quad (10.99)$$

After substituting the given data,

$$\frac{2\pi\sqrt{r}}{\ln\dfrac{r_e}{r_w} + s} = 3.4735, \quad\quad\quad\quad\quad\quad\quad\quad\quad (10.100)$$

$$p_1 = \frac{1}{9}\left(28{,}328 + p_2\right), \quad\quad\quad\quad\quad\quad\quad\quad\quad (10.101)$$

$$p_2 = \frac{1}{9.4735}\left(p_1 + p_3 + 4p_5 + 1{,}563.1\right), \quad\text{...} \quad (10.102)$$

$$p_3 = \frac{1}{2}\left(3{,}000 + p_2\right), \quad\text{...} \quad (10.103)$$

$$p_5 = \frac{1}{5}\left(4p_2 + 2{,}974\right). \quad\text{..} \quad (10.104)$$

Results of the first Gauss-Seidel iterations from Eqs. 10.101 through 10.104 are shown in **Table 10.12.** The max difference is 2.1014 < 5. The molar ratio of Na^+/K^+ is

$$q_A = 1.127\times10^{-3}\times\frac{A_y k_y}{\Delta y}\left(p_A - p_1\right) = 695.1251\ \text{STB/D}, \quad\text{.....................................} \quad (10.105)$$

$$q_B = 1.127\times10^{-3}\times\frac{A_x k_x}{\Delta x}\left(p_B - p_3\right) = 176.7694\ \text{STB/D}, \quad\text{.....................................} \quad (10.106)$$

$$\frac{Na^+}{K^+} = \frac{q_A}{q_B} = 3.9324. \quad\text{...} \quad (10.107)$$

	p_1	p_2	p_3	p_5
0	1,900	1,900	1,900	1,900
1	3,310.667	1,517.254	2,258.627	1,808.604
2	3,268.139	1,512.031	2,256.015	1,804.425
3	3,267.559	1,509.929	2,254.965	1,802.744

Table 10.12—Gauss-Seidel iteration for the Solution to Problem 10.5.9.

MBC:

$$q_{\text{in}} = q_A + q_B = 871.8945\ \text{STB/D}, \quad\text{...} \quad (10.108)$$

$$q_{\text{out}} = 873.5309\ \text{STB/D}, \quad\text{..} \quad (10.109)$$

$$\text{MBC} = \frac{q_{\text{in}}}{q_{\text{out}}} = 0.99813. \quad\text{..} \quad (10.110)$$

Problem 10.5.10

Fig. 10.13 shows a mesh-centered grid imposed on a 2D reservoir experiencing two-phase (oil/water) flow conditions. External boundaries of the reservoir are exposed to an aquifer able to sustain a constant pressure of p_e all around the system. Furthermore, you can assume that on the external boundaries of the reservoir $S_w = 100\%$. The well at Node 13 is produced at a constant sandface pressure.

Assuming both phases are incompressible, and ignoring the capillary pressure between the phases, respond to the following:

1. Give a complete formulation of the problem, including the description of the boundary conditions.
2. Identify the principal unknowns of the problem for a typical meshpoint. What are the equations available to solve for these unknowns?
3. Give a comprehensive list of all unknowns needed to be determined using the node numbers.
4. Is this a steady-state problem? Why or why not?

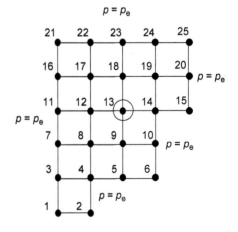

Fig. 10.13—Reservoir system of Problem 10.5.10.

Solution to Problem 10.5.10

1. The governing oil and water equations are given:

$$\frac{\partial}{\partial x}\left(\frac{A_x k_x k_{ro}}{\mu_o B_o}\frac{\partial \Phi_o}{\partial x}\right)\Delta x + \frac{\partial}{\partial y}\left(\frac{A_y k_y k_{ro}}{\mu_o B_o}\frac{\partial \Phi_o}{\partial_y}\right)\Delta y + q_o = \frac{\partial}{\partial t}\left(\frac{\phi S_o}{B_o}\right), \dots\dots\dots\dots\dots\dots\dots (10.111)$$

$$\frac{\partial}{\partial x}\left(\frac{A_x k_x k_{rw}}{\mu_o B_w}\frac{\partial \Phi_w}{\partial x}\right)\Delta x + \frac{\partial}{\partial y}\left(\frac{A_y k_y k_{rw}}{\mu_w B_w}\frac{\partial \Phi_w}{\partial_y}\right)\Delta y + q_w = \frac{\partial}{\partial t}\left(\frac{\phi S_w}{B_w}\right). \dots\dots\dots\dots\dots\dots\dots (10.112)$$

External boundary: constant-pressure specification.
Internal boundary: constant sandface-pressure specification.

2. Principal unknowns are p, S_o, and S_w.
Equations are as follows:

- Oil- and water-transport equations (PDEs).
- $S_o + S_w = 1$.
- Well equation for oil and water.

3. Comprehensive list of unknowns:

- p_8, p_9, p_{12}, p_{13}, p_{17}, p_{18}, and p_{19}.
- S_{o8}, S_{o9}, S_{o12}, S_{o13}, S_{o17}, S_{o18}, and S_{o19}.
- S_{w8}, S_{w9}, S_{w12}, S_{w13}, S_{w17}, S_{w18}, and S_{w19}.
- q_{o13} and q_{w13}.

4. This is not a steady-state problem; S_o and S_w are dependent variables, and they are changing with time (right-hand side of the PDEs are not equal to zero).

Problem 10.5.11

A plan view of a horizontal, homogeneous, anisotropic reservoir and its grid discretization are shown in **Fig. 10.14.** There is an aquifer charging from the southern end of the reservoir, and the geologist is uncertain whether the northern end of the reservoir is sealed. The steady-state flow of a single-phase incompressible liquid occurs in the reservoir. The strong aquifer charge retains the constant block pressure in Node 1 at 3,500 psia. The well located at Node 4 produces 130.5 STB/D of liquid, and the sandface pressure of the well is specified as 1,200 psia. The wellbore radius is 0.25 ft. Additional reservoir data are provided in **Table 10.13.**

1. Give the PDE that describes the flow problem.
2. Write the characteristic finite-difference approximation to the PDE of Part 1.
3. Determine the qualitative and quantitative nature of the boundary condition across the northern boundary of Block 5.

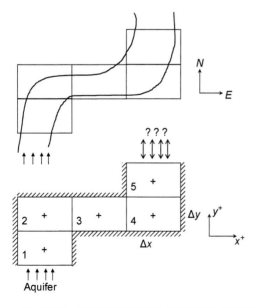

Fig. 10.14—Reservoir system of Problem 10.5.11.

Table 10.13—Reservoir data for Fig. 10.14.	
k_x =	20 md
k_y =	10 md
Δx =	500 ft
Δy =	1,000 ft
h =	20 ft
μ =	2.0 cp
B =	1 res bbl/STB
r_w =	0.25 ft
s =	0

Solution to Problem 10.5.11

1. The governing equation for a 2D incompressible fluid flow problem is

$$\frac{\partial}{\partial x}\left(\frac{A_x k_x}{\mu B}\frac{\partial p}{\partial x}\right)\Delta x + \frac{\partial}{\partial y}\left(\frac{A_y k_y}{\mu B}\frac{\partial p}{\partial y}\right)\Delta y + q = 0. \quad \dots\dots\dots\dots\dots\dots\dots\dots\dots \quad (10.113)$$

For a homogeneous but anisotropic system, and because $k_x = 2k_y$,

$$2\frac{\partial^2 p}{\partial x^2} + \frac{\partial^2 p}{\partial y^2} + \frac{q\mu B}{V_b k_y} = 0. \quad \dots\dots\dots\dots\dots\dots\dots\dots\dots\dots\dots\dots\dots\dots \quad (10.114)$$

2. The infinite-difference form is

$$2\left[\frac{p_{i-1,j} - 2p_{i,j} + p_{i+1,j}}{(\Delta x)^2}\right] + \left[\frac{p_{i,j-1} - 2p_{i,j} + p_{i,j-1}}{(\Delta y)^2}\right] + \frac{q\mu B}{V_b k_y} = 0. \quad \dots\dots\dots\dots\dots \quad (10.115)$$

Because $2\Delta x = \Delta y$,

$$2\left[\frac{p_{i-1,j} - 2p_{i,j} + p_{i+1,j}}{(500)^2}\right] + \left[\frac{p_{i,j-1} - 2p_{i,j} + p_{i,j-1}}{(1,000)^2}\right] + \frac{q\mu B}{V_b k_y} = 0, \quad \dots\dots\dots\dots \quad (10.116)$$

which results in

$$8\left(p_{i-1,j} - 2p_{i,j} + p_{i+1,j}\right) + \left(p_{i,j-1} - 2p_{i,j} + p_{i,j-1}\right) + \frac{2q\mu B}{hk} = 0. \quad \dots\dots\dots\dots \quad (10.117)$$

3. Applying Peaceman's well model,

$$r_e = \frac{0.28\left[\left(\frac{k_y}{k_x}\right)^{\frac{1}{2}}(\Delta x)^2 + \left(\frac{k_x}{k_y}\right)^{\frac{1}{2}}(\Delta y)^2\right]^{\frac{1}{2}}}{\left(\frac{k_y}{k_x}\right)^{\frac{1}{4}} + \left(\frac{k_x}{k_y}\right)^{\frac{1}{4}}} = 173.97\text{ft}, \quad \dots\dots\dots\dots\dots\dots\dots \quad (10.118)$$

where $\bar{k} = \sqrt{k_x k_y} = 14.14\text{md}$. $\dots\dots\dots\dots\dots\dots\dots\dots\dots\dots\dots\dots\dots\dots\dots$ (10.119)

The productivity index is

$$\Omega = \frac{2\pi \bar{k} h}{\mu B\left[\ln\left(\frac{r_e}{r_w}\right) + s\right]} = \frac{2\pi(14.14)(1.127\times 10^{-3})(20)}{(1)(2)\left(\ln\frac{173.97}{0.25}\right)} = 0.153\text{ STB}/\text{D-psi.} \quad \dots\dots\dots\dots \quad (10.120)$$

The flow rate of the well is

$$q_4 = -\Omega\left(p_4 - p_{sf}\right). \quad \dots\dots\dots\dots\dots\dots\dots\dots\dots\dots\dots\dots\dots\dots\dots\dots\dots \quad (10.121)$$

Thus,

$$p_4 = -\frac{q_4}{\Omega} + p_{sf} = -\frac{-130}{0.153} + 1,200\text{ psi} = 2,052.94\text{ psi.} \quad \dots\dots\dots\dots\dots\dots\dots\dots \quad (10\ 122)$$

In this case, the pressure in Blocks 2 and 3 can be solved:

- Block 2:

$$p_1 - p_2 + 8(p_3 - p_2) = 0, \quad \dots\dots\dots\dots\dots\dots\dots\dots\dots\dots\dots\dots\dots\dots\dots\dots \quad (10.123)$$

rearranged to get

$$-9p_2 + 8p_3 = -3,500. \quad \dots\dots\dots\dots\dots\dots\dots\dots\dots\dots\dots\dots\dots\dots\dots\dots\dots \quad (10.124)$$

- Block 3:

$$p_2 - 2p_3 + p_4 = 0, \quad \dots\dots\dots\dots\dots\dots\dots\dots\dots\dots\dots\dots\dots\dots\dots\dots\dots\dots \quad (10.125)$$

rearranged to get

$$p_2 - 2p_3 = -2{,}052.94. \dotfill (10.126)$$

Solving, $p_2 = 2{,}342.35$ psia and $p_3 = 2{,}197.65$ psi.

Mass transfer from Block 3 to Block 4:

$$q_{3to4} = \frac{k_x A_x}{\mu B}\frac{\partial p}{\partial x} = \frac{\left(1.127\times10^{-3}\right)(20)(1{,}000)(20)}{(1)(2)}\left[\frac{p_4 - p_3}{\Delta x}\right] = 225.4\left[\frac{2{,}197.65 - 2{,}052.94}{500}\right] \dotfill (10.127)$$
$$= 65.23 \text{ STB/D}.$$

Therefore, the external boundary needs to support $130.5 - 65.23 = 65.26$ STB/D:

$$q_{boundary} = \frac{k_y A_y}{\mu B}\frac{\partial p}{\partial y} = \frac{\left(1.127\times10^{-3}\right)(10)(500)(20)}{(1)(2)}\left(\frac{\partial p}{\partial y}\right) = 56.35\left(\frac{\partial p}{\partial y}\right) = 65.26 \text{ STB/D}, \dotfill (10.128)$$

$$\frac{\partial p}{\partial y} = +1.158 \text{ psi/ft}. \dotfill (10.129)$$

Problem 10.5.12

Consider the homogeneous reservoir shown in **Fig. 10.15.** Reservoir Block 1 is connected to a large aquifer so that pressure in Block 1 is maintained at 3,000 psia. The east boundary of Block 4 has a Neumann-type boundary condition that gives a pressure gradient of –0.4 psi/ft. All other boundaries are believed to be sealed; however, it is suspected that the bottom plane of Block 4 might not be completely sealed. A single-phase, incompressible fluid is produced through a production well located in Block 4 at a rate of 1,000 STB/D. The measured sandface pressure in the well is 1,567 psia. As a reservoir engineer, your task is to investigate whether there is leak across the bottom plane of Block 4. Use the properties listed in **Table 10.14** to determine your answer.

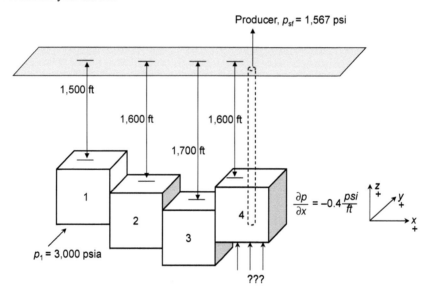

Fig. 10.15—Reservoir system of Problem 10.5.12.

$k =$	50 md
$\Delta x =$	500 ft
$\Delta y =$	500 ft
$\phi =$	25%
$h =$	80 ft
$\mu =$	1 cp
$B =$	1.0 res bbl/STB
$r_w =$	0.25 ft
$s =$	0

Table 10.14—Reservoir properties for the reservoir shown in Fig. 10.15.

Solution to Problem 10.5.12

The governing equation for 1D incompressible fluid flow in a reservoir with depth gradients is

$$\frac{\partial}{\partial x}\left(\frac{A_x k_x}{\mu B}\frac{\partial p}{\partial x}\right)\Delta x + q = \frac{\partial}{\partial x}\left(\frac{1}{144}\frac{g}{g_c}\rho\frac{A_x k_x}{\mu B}\frac{\partial G}{\partial x}\right)\Delta x. \dots\dots\dots\dots\dots\dots (10.130)$$

For a homogeneous 1D reservoir system,

$$\frac{\partial^2 p}{\partial x^2} + \frac{q\mu B}{V_b k} = \frac{1}{144}\frac{g}{g_c}\rho\frac{\partial^2 G}{\partial x^2}. \dots\dots\dots\dots\dots\dots\dots\dots (10.131)$$

In finite-difference form,

$$p_{i-1} - 2p_i + p_{i+1} + \left(\frac{q\mu B}{hk}\right)_i = \frac{1}{144}\frac{g}{g_c}\rho\left(G_{i-1} - 2G + G_{i+1i}\right). \dots\dots\dots\dots\dots (10.132)$$

Block pressure, p_4, can be directly calculated from Peaceman's equation (p_{sf4} and q_4 are known):

$$q_4 = -\Omega 4\left(p_4 - p_{sf4}\right), \dots\dots\dots\dots\dots\dots\dots\dots\dots\dots\dots\dots (10.133)$$

$$\Omega_4 = \frac{2\pi kh}{\mu B\left[\ln\left(\dfrac{r_e}{r_w}\right) + s\right]} = \frac{2\pi(80)(1.127\times10^{-3})(50)}{(1)(1)\left(\ln\dfrac{100}{0.25}\right)} = 4.73\ \text{STB/D-psi}. \dots\dots\dots (10.134)$$

Thus,

$$p_4 = -\frac{q_4}{\Omega_4} + p_{sf4} = -\frac{(-1,000)}{4.73} + 1,567 = 1,778\ \text{psia}. \dots\dots\dots\dots\dots\dots (10.135)$$

- Block 2:

$$3,000 - 2p_2 + p_3 = \frac{1}{144}(62.4)(1,540 - 2\times1,640 + 1,740), \dots\dots\dots\dots\dots\dots (10\ 136)$$

$$-2p_2 + p_3 = -3,000. \dots\dots\dots\dots\dots\dots\dots\dots\dots\dots\dots\dots\dots (10.137)$$

- Block 3:

$$p_2 - 2p_3 + 1,778 = \frac{1}{144}(62.4)(1,640 - 2\times1,740 + 1,640), \dots\dots\dots\dots\dots\dots (10.138)$$

$$-2p_2 + p_3 = -1,864.5. \dots\dots\dots\dots\dots\dots\dots\dots\dots\dots\dots\dots\dots (10.139)$$

Solving for unknown pressures: $p_2 = 2,621.5$ psia and $p_3 = 2,243.0$ psia.
MBC:
- Flow in from Block 1 to 2:

$$q_{in} = \frac{kA}{\mu B}\frac{\partial\Phi}{\partial x} = \frac{kA}{\mu B}\left(\frac{\partial p}{\partial x} - \frac{1}{144}\frac{g}{g_c}\rho\frac{\partial G}{\partial x}\right) = \frac{(0.05635)(500)(80)}{(1)(1)}\left[\frac{p_1 - p_2}{\Delta x} - \frac{1}{144}\frac{g}{g_c}\rho\frac{G_1 - G_2}{\Delta x}\right] \dots (10.140)$$

$$= 2,254\left[\frac{3,000 - 2,621.5}{500} - \frac{1}{144}(62.4)\frac{1,540 - 1,640}{500}\right] = 1,901.3\ \text{STB/D}.$$

- Flow out (including production from the well and leakage across the boundary of Block 4):

$$q_{boundary} = \frac{kA}{\mu B}\frac{\partial p}{\partial x} = \frac{(0.05635)(500)(80)}{(1)(1)}(0.4) = 901.6\ \text{STB/D}, \dots\dots\dots\dots\dots\dots (10.141)$$

$$q_{out} = q_{boundary} + q_{well} = 1,901.6\ \text{STB/D}, \dots\dots\dots\dots\dots\dots\dots\dots (10.142)$$

$$\text{MBC} = \left|\frac{q_{in}}{q_{out}}\right| = \left|\frac{1,901.5}{1,901.6}\right| = 0.9998. \dots\dots\dots\dots\dots\dots\dots\dots\dots (10.143)$$

Because the MBC is close to 1, the reservoir is sealed, and there is no leak across the bottom plane of Block 4.

Problem 10.5.13

Consider the 1D homogeneous and isotropic, horizontal reservoir shown in **Fig. 10.16.** The west and east boundaries are known to be open. A producer well is located in Block 5, which is producing 1,000 STB/D of water (incompressible liquid). What is the pressure gradient across the west boundary of Block 1? Use the properties listed in **Table 10.15** for the analysis.

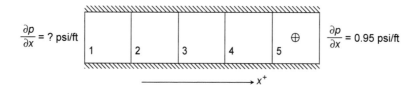

Fig. 10.16—Reservoir system of Problem 10.5.13.

$k_x = k_y =$	25 md
$\Delta x =$	100 ft
$\Delta y =$	200
$h =$	80 ft
$\mu =$	1 cp
$B =$	1 res bbl/STB
$r_w =$	0.35 ft
$s =$	0

Table 10.15—Reservoir properties for Fig. 10.16.

Solution to Problem 10.5.13

Flow across the eastern boundary (into the reservoir):

$$q_{east} = \left(\frac{kA}{\mu B} \frac{\partial p}{\partial x} \right) = (0.95) \frac{(200)(80)(25)(1.127 \times 10^{-3})}{(1)(1)} = 428.26 \text{ STB/D.} \quad \dots \dots \dots \dots \dots \dots \quad (10.144)$$

For an incompressible reservoir, production from the well must be compensated for by the fluid entering the reservoir across the boundaries; therefore,

$$q_{west} = 1,000 - 426.26 = 571.74 \text{ STB/D (into the reservoir).} \quad \dots \dots \dots \dots \dots \dots \dots \dots \dots \dots \dots \quad (10.145)$$

The pressure gradient is calculated as

$$\left. \frac{\partial p}{\partial x} \right|_{west} = -\frac{q_{west} \mu B}{kA} = -\frac{(1)(1)(571.74)}{(25)(1.127 \times 10^{-3})(200)(80)} = -1.27 \text{ psi/ft.} \quad \dots \dots \dots \dots \dots \dots \dots \dots \dots \quad (10.146)$$

The negative sign indicates the fluid is entering the reservoir across the western boundary.

Problem 10.5.14

Consider the slightly compressible fluid flow in a 1D reservoir shown in **Fig. 10.17.** This reservoir is homogeneous with uniform thickness. It is surrounded with no-flow boundaries. The depth gradient in the system can be ignored. The initial pressure of the system was 2,000 psi. A production well located in Block 2 was put on production at a constant rate of 100 STB/D. It is found that the production well is damaged. The sandface pressure of the well in Block 2 at the end of 30 days was measured as 1,741.51 psia. In the calculation, assume B and μ are constant and use the data provided in **Table 10.16** to respond to the following:

1. Determine the skin factor of the producer located in Block 2.
2. Perform the incremental MBC for the solution at the end of 30 days.
3. If we continue producing at this rate, what will be the value of the sandface pressure of the production well at the end of 60 days?

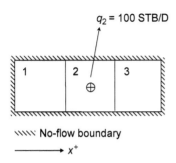

<table>
<tr><td>$k_x = k_y =$ 100 md</td></tr>
</table>

$k_x = k_y =$ 100 md

$\Delta x = \Delta y =$ 500 ft

$h =$ 50 ft

$\mu =$ 1 cp

$B =$ 1 res bbl/STB

$c_f =$ 1×10⁻⁵ psi⁻¹

$\phi =$ 0.2

$r_w =$ 0.25 ft

$s =$ 0

$\Delta t =$ 30 days

Fig. 10.17—Reservoir system of Problem 10.5.14.

Table 10.16—Reservoir properties for Fig. 10.17.

Solution to Problem 10.5.14

For a 1D slightly compressible reservoir with homogeneous property distributions, the governing equation can be written as:

$$\frac{\partial}{\partial x}\left(\frac{A_x k_x}{\mu B}\frac{\partial p}{\partial x}\right)\Delta x + q = \frac{V_b \phi c_t}{5.615}\frac{\partial p}{\partial t}. \dots\dots\dots\dots\dots\dots\dots\dots\dots\dots\dots\dots\dots\dots (10.147)$$

Simplifying for a homogeneous reservoir and treating μ and B as constants,

$$\frac{\partial^2 p}{\partial x^2} + \frac{q\mu B}{kV_b} = \frac{\mu B \phi c_t}{5.615k}\frac{\partial p}{\partial t}. \dots\dots\dots\dots\dots\dots\dots\dots\dots\dots\dots\dots\dots\dots\dots\dots (10.148)$$

Writing the finite-difference approximation,

$$p_{i-1}^{n+1} - 2p_i^{n+1} + p_{i+1}^{n+1} + \left(\frac{q\mu B}{hk}\right)_i = \frac{\mu B \phi c_t (\Delta x)^2}{5.615k\Delta t}\left(p_i^{n+1} - p_i^n\right). \dots\dots\dots\dots\dots\dots\dots\dots (10.149)$$

Observing the existing symmetry in the reservoir, $p_1 = p_3$, one can write the system of equations using the coefficient calculated next:

$$\frac{\mu B \phi c_t (\Delta x)^2}{5.615k\Delta t} = \frac{(1)(1)(0.2)(1\times 10^{-5})(500)^2}{5.615(100\times 1.127\times 10^{-3})30} = 0.0263. \dots\dots\dots\dots\dots\dots\dots\dots\dots (10.150)$$

At $t = 30$ days ($p_1^n = 2,000$ psia and $p_2^n = 2,000$ psia),

- Block 1:

$$p_2^{n+1} - p_1^{n+1} = \frac{(1)(1)(0.2)\left(1\times 10^{-5}\right)\left(500\right)^2}{5.615(100\times 1.127\times 10^{-3})30}\left(p_1^{n+1} - 2,000\right), \dots\dots\dots\dots\dots\dots\dots\dots (10.151)$$

$$-1.02634 p_1^{n+1} + p_2^{n+1} = -52.68. \dots\dots\dots\dots\dots\dots\dots\dots\dots\dots\dots\dots\dots\dots (10.152)$$

- Block 2:

$$p_1^{n+1} - 2p_2^{n+1} + p_1^{n+1} + = \frac{(-100)}{\left(100\times 1.127\times 10^{-3}\right)(50)} = 0.02634\left(p_2^{n+1} - 2,000\right), \dots\dots\dots\dots\dots (10.153)$$

$$2p_1^{n+1} - 2.02634 p_2^{n+1} = -34.93. \dots\dots\dots\dots\dots\dots\dots\dots\dots\dots\dots\dots\dots\dots\dots (10.154)$$

Solving for p_1 and p_2 and recalling $p_1 = p_3$ because of the existing symmetry, $p_1 = p_3 = 1,777.38$ psi and $p_2 = 1,771.51$ psi, because

$$q_2 = -\Omega_2\left(p_2 - p_{sf2}\right), \dots\dots\dots\dots\dots\dots\dots\dots\dots\dots\dots\dots\dots\dots\dots\dots\dots\dots (10.155)$$

$$\Omega_2 = -\frac{q_2}{\left(p_2 - p_{sf2}\right)} = 3.33 \text{ STB/D-psi}. \dots\dots\dots\dots\dots\dots\dots\dots\dots\dots\dots\dots\dots\dots (10.156)$$

Employing the definition of productivity index,

$$\Omega = \frac{2\pi kh}{\mu B\left[\ln\left(\dfrac{r_e}{r_w}\right)+s\right]} = 3.33 \text{ STB/D-psi}, \dots\dots\dots\dots\dots\dots\dots\dots\dots\dots \text{(10.157)}$$

one can solve for skin factor, which is found to be $s = 4.63$.

MBC:

$$\Delta V = \Delta V_1 + \Delta V_2 + \Delta V_3 = -cV_p(\Delta p_1 + \Delta p_2 + \Delta p_3)$$

$$= -1\times10^{-5}\frac{(50)(500)(500)(0.2)}{5.615}\left[3\times2,000-(1,777.38\times2+1,771.51)\right]=-2,999.69 \text{ STB}, \dots\dots\dots \text{(10.158)}$$

$$\text{MBC} = \frac{\Delta V}{Q_{\text{total}}} = 0.99999. \dots\dots\dots\dots\dots\dots\dots\dots\dots\dots\dots\dots\dots \text{(10.159)}$$

At $t = 60$ days ($p_1^n = 1,777.38$ psia and $p_2^n = 1,771.51$ psia),

- Block 1:

$$p_2^{n+1} - p_1^{n+1} = \frac{(1)(1)(0.2)(1\times10^{-5})(500)^2}{5.615(100\times1.127\times10^{-3})30}\left(p_1^{n+1}-p_1^n\right), \dots\dots\dots\dots\dots\dots \text{(10.160)}$$

$$-1.02634\,p_1^{n+1} + p_2^{n+1} = -46.8162. \dots\dots\dots\dots\dots\dots\dots\dots\dots \text{(10.161)}$$

- Block 2:

$$p_1^{n+1} - p_2^{n+1} + \frac{(-100)}{(100\times1.127\times10^{-3})(50)} = 0.2634\left(p_2^{n+1}-p_2^n\right), \dots\dots\dots\dots\dots \text{(10.162)}$$

$$p_1^{n+1} - 1.01317\,p_2^{n+1} = 14.45767, \dots\dots\dots\dots\dots\dots\dots\dots\dots\dots \text{(10.163)}$$

$p_1 = p_3 = 1,552.82$ psi and $p_2 = 1,546.90$ psi; $q = -\Omega(p_2 - p_{sf})$ and $p_{sf} = 1,516.90$ psi.

Problem 10.5.15

Fig. 10.18 shows a representation of a brine-disposal reservoir. The reservoir is homogeneous and isotropic and has no depth gradients. A brine-disposal well is located at Block 1 and is operated at a constant injection rate of 1,000 STB/D. Three observation wells are completed in Blocks 2, 3, and 4. The reservoir is known to be completely sealed. However, there is a freshwater well located 1 mile northwest of the reservoir. The well owner claims that the injection brine from the brine-disposal well contaminated the water in his well. To investigate this issue, the sandface pressure of the injection well and the observation wells are recorded. The initial reservoir pressure is 1,000 psia (uniform). As a reservoir engineer, you are asked to quantitatively analyze the possibility that the brine in the reservoir would contaminate the fresh water in that well using the pressure-log data **(Table 10.17)**. In your analysis, assume that $r_w = 0.25$ ft and skin factor = 0. Other related reservoir properties are listed in **Table 10.18**.

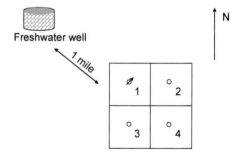

Fig. 10.18—Reservoir system of Problem 10.5.15.

$\phi =$	0.2 fraction
$k_x = k_y = k_z =$	150 md
$h =$	100 ft
$\Delta x =$	500 ft
$\Delta y =$	500 ft
$c_t =$	5×10^{-5} psi^{-1}
$\mu \approx$ const $=$	1.01 cp
$B \approx$ const $=$	0.8 res bbl/STB
$p_{\text{initial}} =$	800 psi
$\Delta t =$	30 days
$q =$	1,000 STB/D

Table 10.18—Reservoir data for Fig. 10.18.

Time, days	p_{sf1}, psia	p_{sf2}, psia	p_{sf3}, psia	p_{sf4}, psia
30	1,029	–	–	960
60	–	1,033	–	–

Table 10.17—Pressure-log data for the reservoir shown in Fig. 10.18.

Solution to Problem 10.5.15

The existing symmetry plane in the reservoir system implies that $p_2 = p_3$. Therefore, the primary unknowns of the problem are p_1, p_2, and p_4. The governing flow equation is

$$\frac{\partial}{\partial x}\left(\frac{A_x k_x}{\mu B}\frac{\partial P}{\partial x}\right)\Delta x + \frac{\partial}{\partial y}\left(\frac{A_y k_y}{\mu B}\frac{\partial p}{\partial y}\right)\Delta y + q = \frac{V_b \phi c_t}{5.615}\left(\frac{\partial p}{\partial t}\right). \quad\text{......................................} \quad (10.164)$$

For an isotropic and homogeneous 2D reservoir system ($k_x = k_y = k$), the preceding PDE can be simplified to

$$\frac{\partial^2 p}{\partial x^2} + \frac{\partial^2 p}{\partial y^2} = \frac{q\mu B}{V_b k} = \frac{\phi c_t \mu B}{5.615 k}\left(\frac{\partial p}{\partial t}\right). \quad\text{..} \quad (10.165)$$

The characteristic finite-difference equation is ($\Delta x = \Delta y$):

$$\frac{p_{i-1,j}^{n+1} - 2p_{i,j}^{n+1} + p_{i+1,j}^{n+1}}{(\Delta x)^2} + \frac{p_{i,j-1}^{n+1} - 2p_{i,j}^{n+1} + p_{i,j+1}^{n+1}}{(\Delta y)^2} + \left(\frac{q\mu B}{kV_b}\right)_{i,j} = \left[\frac{\mu B \phi c_t}{5.615 k \Delta t}\right]_{i,j}\left(p_{i,j}^{n+1} - p_{i,j}^{n}\right). \quad\text{................} \quad (10.166)$$

Because $\Delta x = \Delta y$, the previous equation can be simplified to

$$p_{i,j-1}^{n+1} + p_{i-1,j}^{n+1} - 4p_{i,j}^{n+1} + p_{i+1,j}^{n+1} + p_{i,j+1}^{n+1}\left(\frac{q\mu B}{kh}\right)_{i,j} = \left[\frac{(\Delta x)^2 \mu B \phi c_t}{5.615 k \Delta t}\right]_{i,j}\left(p_{i,j}^{n+1} - p_{i,j}^{n}\right), \quad\text{......................} \quad (10.167)$$

where

$$\frac{(\Delta x)^2 \mu B \phi c_t}{5.615 k \Delta t} = 0.07094. \quad\text{...} \quad (10.168)$$

- Block 1:

$$-2p_1^{n+1} + 2p_2^{n+1} + \left(\frac{q\mu B}{hk}\right)_1 = \frac{\mu B \phi c_t \Delta x^2}{5.615 k \Delta t}(p_1^{n+1} - p_1^{n}), \text{ with }\left(\frac{q\mu B}{hk}\right)_1 = 47.79. \quad\text{......................} \quad (10.169)$$

- Block 2:

$$p_1^{n+1} - 2p_2^{n+1} + p_4^{n+1} = \frac{\mu B \phi c_t \Delta x^2}{5.615 k \Delta t}\left(p_2^{n+1} - p_2^{n}\right). \quad\text{...} \quad (10.170)$$

- Block 4:

$$2p_2^{n+1} - 2p_4^{n+1} = \frac{\mu B \phi c_t \Delta x^2}{5.615 k \Delta t}\left(p_4^{n+1} - p_4^{n}\right). \quad\text{...} \quad (10.171)$$

From 0 to 30 days, with $p_1^n = p_2^n = p_3^n = 800$ psia, after rearranging the system of equations in a matrix form,

$$\begin{bmatrix} -2.071 & 2 & 0 \\ 1 & -2.071 & 1 \\ 0 & 2 & -2.071 \end{bmatrix}\begin{bmatrix} p_1^{n+1} \\ p_2^{n+1} \\ p_3^{n+1} \end{bmatrix} = \begin{bmatrix} -104.545 \\ -56.7486 \\ -56.7486 \end{bmatrix}. \quad\text{.....................................} \quad (10.172)$$

Solving the system of equations,

$$p_1^{n+1} = 982.93 \text{ psia}, p_2^{n+1} = p_3^{n+1} = 965.51 \text{ psia, and } p_4^{n+1} = 959.85 \text{ psia}, \quad\text{..............................} \quad (10.173)$$

$$r_e = 0.2\Delta x = 100 \text{ ft}, \quad\text{...} \quad (10.174)$$

$$\Omega = \frac{2\pi kh}{\mu B\left[\ln\left(\dfrac{r_e}{r_w}\right) + s\right]} = 21.94 \text{ STB/D-psi}, \quad\text{............................} \quad (10.175)$$

and $p_{sf} = 1,028.50$ psi. The calculated pressures are in agreement with the pressure log. Accordingly, there is no leakage observed during the first 30 days.

From 30 to 60 days, again rearranging the system of equations in a matrix form,

$$\begin{bmatrix} -2.071 & 2 & 0 \\ 1 & -2.071 & 1 \\ 0 & 2 & -2.071 \end{bmatrix} \begin{bmatrix} p_1^{n+1} \\ p_2^{n+1} \\ p_3^{n+1} \end{bmatrix} = \begin{bmatrix} -117.521 \\ -68.4895 \\ -68.0874 \end{bmatrix}, \quad \dots \dots \dots \dots \dots \dots \dots \dots \dots \dots \dots \dots \dots (10.176)$$

and solving for pressures, one obtains

$$p_1^{n+1} = 1{,}151.821 \text{ psi}, \; p_2^{n+1} = p_3^{n+1} = 1{,}133.914 \text{ psi}, \text{ and } p_4^{n+1} = 1{,}127.951 \text{ psi}. \quad \dots \dots \dots \dots \dots \dots (10.177)$$

The calculated pressure for Block 2 is 100 psi higher than the pressure measured. This observation indicates that the reservoir possibly is leaking. This may be indicative of potential contamination of the freshwater zone.

Problem 10.5.16

A single-phase incompressible fluid flow reservoir with anisotropic and homogeneous permeability distribution is displayed in **Fig. 10.19.** Critical reservoir rock and fluid properties are listed in **Table 10.19.** A well completed in Block 6 is producing at a constant rate of 1,000 STB/D. In the nonproducing observation well located in Block 2, the pressure of the sandface is measured to be 2,000 psia. Neumann-type boundary conditions are applied on the north and east surfaces of Block 1, with pressure-gradient values of 1.2 psi/ft and 0.75 psi/ft, respectively.

1. Calculate the pressure of Block 7.
2. Assuming Neumann-type boundary condition across the eastern boundary of Block 8, calculate the expected active pressure gradient.

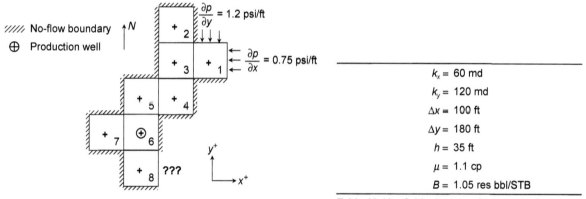

Fig. 10.19—Reservoir system of Problem 10.5.16.

Table 10.19—Critical reservoir rock and fluid properties for the reservoir displayed in Fig. 10.19.

k_x = 60 md
k_y = 120 md
Δx = 100 ft
Δy = 180 ft
h = 35 ft
μ = 1.1 cp
B = 1.05 res bbl/STB

Solution to Problem 10.5.16

The mass flux through Block 1 can be calculated as

$$q_x = 1.127 \times 10^{-3} \times \frac{A_x k_x}{\mu B} 0.75 \frac{\text{psia}}{\text{ft}} = 276.63 \text{ STB/D}, \quad \dots \dots \dots \dots \dots \dots \dots \dots \dots \dots \dots \dots \dots (10.178)$$

$$q_y = 1.127 \times 10^{-3} \times \frac{A_y k_y}{\mu B} 1.2 \frac{\text{psia}}{\text{ft}} = 491.78 \text{ STB/D}, \quad \dots \dots \dots \dots \dots \dots \dots \dots \dots \dots \dots \dots \dots (10.179)$$

$$q_{\text{total}} = q_x + q_y = 768.40 \text{ STB/D}. \quad \dots \dots \dots \dots \dots \dots \dots \dots \dots \dots \dots \dots \dots \dots \dots \dots (10.180)$$

Knowing that p_2 = 2,000 psia (shut-in well, $p_2 = p_{sf2}$) and $p_3 = p_2$; therefore, p_3 = 2,000 psia. One can calculate the transmissibilities along the x- and y-directions:

$$T_x = \frac{k_x h \Delta y}{\mu B \Delta x} = \frac{1.127 \times 10^{-3} \times 60 \text{ md} \times 35 \text{ ft} \times 180 \text{ ft}}{1.1 \text{ cp} \times 1.05 \text{ res bbl / STB} \times 100 \text{ ft}} = 3.69 \text{ STB/D-psi}, \quad \dots \dots \dots \dots \dots \dots (10.181)$$

$$T_y = \frac{k_y h \Delta x}{\mu B \Delta y} = \frac{1.127 \times 10^{-3} \times 120\,\text{md} \times 35\,\text{ft} \times 100\,\text{ft}}{1.1\,\text{cp} \times 1.05\,\dfrac{\text{res bbl}}{\text{STB}} \times 180\,\text{ft}} = 2.78\ \text{STB/D-psi.} \quad \dotfill \text{(10.182)}$$

Block pressures can be calculated on the basis of the material balance:

$$p_4 = p_3 - \frac{q_{\text{total}}}{T_y} = 2{,}000 - \frac{768.4}{2.78} = 1{,}662.5\,\text{psia,} \quad \dotfill \text{(10.183)}$$

$$p_5 = p_4 - \frac{q_{\text{total}}}{T_x} = 1{,}662.5 - \frac{768.4}{3.69} = 1{,}454.2\,\text{psia,} \quad \dotfill \text{(10.184)}$$

$$p_6 = p_5 - \frac{q_{\text{total}}}{T_y} = 1{,}454.2 - \frac{768.4}{2.78} = 1{,}116.7\,\text{psia,} \quad \dotfill \text{(10.185)}$$

and $p_7 = p_6 = 1{,}116.7$ psia.

The well in Block 6 is producing at 1,000 STB/D. Thus, the fluid flow from Block 8 to Block 6 is calculated as

$$q_{8-6} = 1{,}000 - 768.4 = 231.6\ \text{STB/D.} \quad \dotfill \text{(10.186)}$$

Now, one can calculate pressure of Block 8:

$$p_8 = p_6 + \frac{q_{\text{boundary}}}{T_y} = 1{,}454.2 + \frac{231.6}{2.78} = 1{,}218.4\,\text{psia.} \quad \dotfill \text{(10.187)}$$

The Neumann-type boundary condition is applied to Block 8, with $q_{\text{boundary}} = 231.6$ STB/D, $\Delta x = 100$ ft, and $T_x = 2.78$ STB/D–psi, which gives:

$$q_{\text{boundary}} = \frac{A_x k_x}{\mu B} \frac{\partial p}{\partial x}, \quad \dotfill \text{(10.188)}$$

or

$$\frac{\partial p}{\partial x} = \frac{q_{\text{boundary}} \mu B}{A_x k_x} = \frac{(231.6)(1.1)(1.05)}{(180)(35)(1.127 \times 10^{-3})(60)} = 0.63\,\text{psi/ft.} \quad \dotfill \text{(10.189)}$$

Problem 10.5.17

Consider a slightly compressible fluid reservoir (**Fig. 10.20**) with properties listed in **Table 10.20.** A well is completed in Block 2 and all boundaries are sealed. The well is put on production at a constant sandface pressure of 500 psia. A stimulation job is carried out around the well, which makes the productivity index of the well a strong function of pressure: $\Omega = 0.2 \ln(p)$. Ignore all gravitational forces. As a reservoir engineer, you are asked to do the following:

1. Calculate the pressure distribution of the system at 30 days of production ($\Delta t = 30$ days). Use the generalized Newton-Raphson (GNR) protocol to solve the nonlinear system of equations with a pressure tolerance of 5 pisa. Clearly list calculations of the entries of the Jacobian matrix, residuals, and pressure change of each GNR iteration. Use 1,000 psia as your initial guess.
2. How would you evaluate the stimulation job using the results of Part 1?

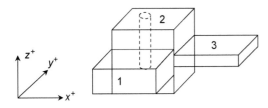

Fig. 10.20—Reservoir system of Problem 10.5.17.

$h = h_1 = 10,\ h_2 = 20,\ h_3 = 4$ ft	
$\phi = \phi_1 = 0.1,\ \phi_2 = 0.2,\ \phi_3 = 0.25$ fraction	
$\Delta x = 120$ ft	
$\Delta y = 240$ ft	
$k_x = 15$ md	
$k_y = 30$ md	
$B_o \approx \text{const} = 1$ res bbl/STB	
$\mu_o \approx \text{const} = 0.8$ cp	
$c_o = c_t = 1 \times 10^{-4}$ psia^{-1}	
$r_w = 0.25$ ft	
$p_{sf} = 500$ psia	
$p_i = 1{,}000$ psia	

Table 10.20—Reservoir properties for Fig. 10.20.

1. The transmissibility between Blocks 1 and 2 is calculated using harmonic averaging:

$$T_{1y} = \frac{2A_{1y}A_{2y}k_yk_y}{\Delta y\left(A_{1y}k_y + A_{2y}k_y\right)\mu_oB_o}, \quad \dots\dots\dots\dots\dots\dots\dots\dots\dots\dots\dots (10.190)$$

and

$$A_{21} = h_2\Delta x; A_{12} = h_1\Delta x. \quad \dots\dots\dots\dots\dots\dots\dots\dots\dots\dots\dots\dots\dots\dots (10.191)$$

Hence,

$$T_{1y} = \frac{2h_1h_2\Delta x k_y}{\Delta y\left(h_1 + h_2\right)\mu_oB_o} = \frac{\left(2\times1.127\times10^{-3}\dfrac{\text{perm}}{\text{md}}\times10\,\text{ft}\times20\,\text{ft}\times120\,\text{ft}\times30\,\text{md}\right)}{240\,\text{ft}\times\left(10\,\text{ft} + 20\,\text{ft}\right)\times0.8\,\text{cp}\times1\,\text{res bbl/STB}} \quad \dots\dots\dots\dots (10.192)$$
$$= 0.282\ \text{STB/D-psia.}$$

Similarly, the transmissibility between Blocks 2 and 3 is calculated as

$$T_{23} = \frac{2h_2h_3\Delta y k_x}{\Delta x\left(h_2 + h_3\right)\mu_oB_o} = \frac{\left(2\times1.127\times10^{-3}\dfrac{\text{perm}}{\text{md}}\times20\,\text{ft}\times4\,\text{ft}\times240\,\text{ft}\times15\,\text{md}\right)}{120\,\text{ft}\times\left(20\,\text{ft} + 4\,\text{ft}\right)\times0.8\,\text{cp}\times1\,\text{res bbl/STB}} \quad \dots\dots\dots\dots (10.193)$$
$$= 0.282\ \text{STB/D-psia.}$$

In implementing the strong implicit procedure (SIP) notation to generate the system of equations, Γ terms first must be calculated:

$$\Gamma_1 = \frac{V_{b1}\phi_1c_o}{5.615\Delta t} = \frac{120\,\text{ft}\times240\,\text{ft}\times10\,\text{ft}\times0.1\times10^{-4}\,\text{psia}^{-1}}{5.615\times30\,\text{days}} = 0.0171\ \text{STB/D-psia,} \quad \dots\dots\dots\dots\dots (10.194)$$

$$\Gamma_2 = \frac{V_{b2}\phi_2c_o}{5.615\Delta t} = \frac{120\,\text{ft}\times240\,\text{ft}\times20\,\text{ft}\times0.2\times10^{-4}\,\text{psia}^{-1}}{5.615\times30\,\text{days}} = 0.0684\ \text{STB/D-psia,} \quad \dots\dots\dots\dots\dots (10.195)$$

$$\Gamma_3 = \frac{V_{b3}\phi_3c_o}{5.615\Delta t} = \frac{120\,\text{ft}\times240\,\text{ft}\times4\,\text{ft}\times0.25\times10^{-4}\,\text{psi}^{-1}}{5.615\times30\,\text{days}} = 0.0171\ \text{STB/D-psia.} \quad \dots\dots\dots\dots\dots (10.196)$$

Note that h_1 is 2.5X larger than h_3, whereas ϕ_3 is 2.5X larger than ϕ_1; therefore, the storage (depletion) capacities of these blocks (Blocks 1 and 3) will be the same. This implies that $\Gamma_1 = \Gamma_3$. Furthermore, it is stated that $T_{12} = T_{23}$. These observations hint at an existing symmetry, so that $p_1 = p_3$. Thus, it is necessary to calculate only the pressures of Blocks 1 and 2:

- Block 1: $T_{12}p_{12} - \left(T_{12} + \Gamma_1\right)p_1 = -\Gamma_1p_i.$ $\dots\dots\dots\dots\dots\dots\dots\dots\dots\dots\dots\dots\dots\dots\dots (10.197)$

- Block 2: $2T_{12}p_1 - \left(2T_{12} + \Gamma_2 + \Omega_2\right)p_2 = -\Gamma_2p_i - \Omega_2p_{sf}.$ $\dots\dots\dots\dots\dots\dots\dots\dots\dots\dots (10.198)$

In the form of residual equations,

$$R_1 = T_{12}p_{12} - \left(T_{12} + \Gamma_1\right)p_1 + \Gamma_1p_i, \quad \dots\dots\dots\dots\dots\dots\dots\dots\dots\dots\dots\dots\dots\dots (10.199)$$

$$R_2 = 2T_{12}p_1 - \left(2T_{12} + \Gamma_2 + \Omega_2\right)p_2 + \Gamma_2p_i + \Omega_2p_{sf}. \quad \dots\dots\dots\dots\dots\dots\dots\dots\dots\dots (10.200)$$

Substitute the terms with the numbers,

$$R_1 = 0.282p_2 - 0.2991p_1 + 17.1, \quad \dots\dots\dots\dots\dots\dots\dots\dots\dots\dots\dots\dots\dots\dots\dots (10.201)$$

$$R_2 = 0.564p_1 - 0.6324p_2 - 0.2p_2\ln\left(p_2\right) + 68.4 + 100\ln\left(p_2\right). \quad \dots\dots\dots\dots\dots\dots\dots (10.202)$$

For the Jacobian matrix,

$$J = \begin{bmatrix} \dfrac{\partial R_1}{\partial p_1} = -0.2991 & \dfrac{\partial R_1}{\partial p_2} = 0.282 \\[3ex] \dfrac{\partial R_2}{\partial p_1} = 0.564 & \dfrac{\partial R_2}{\partial p_2} = -0.6324 - 0.2 - 0.2\ln\left(p_2\right) + \dfrac{100}{p_2} \end{bmatrix}. \quad \dots\dots\dots\dots\dots\dots (10.203)$$

Using 1,000 psia as an initial guess,

- Iteration 1:

$$\begin{bmatrix} -0.2991 & 0.282 \\ 0.564 & -2.114 \end{bmatrix} \begin{bmatrix} \Delta p_1 \\ \Delta p_2 \end{bmatrix} = -\begin{bmatrix} 0 \\ -690.78 \end{bmatrix}, \dots\dots\dots\dots\dots\dots\dots\dots\dots\dots\dots\dots (10.204)$$

$$\begin{cases} \Delta p_1 = -411.6 \, \text{psia} \\ \Delta p_2 = -436.6 \, \text{psia} \end{cases} , \rightarrow \begin{cases} p_1 = -588.4 \, \text{psia} \\ p_2 = -563.4 \, \text{psia} \end{cases}. \dots\dots\dots\dots\dots\dots\dots (10.205)$$

- Iteration 2:

$$\begin{bmatrix} -0.2991 & 0.282 \\ 0.564 & -1.922 \end{bmatrix} \begin{bmatrix} \Delta p_1 \\ \Delta p_2 \end{bmatrix} = -\begin{bmatrix} 0.01164 \\ -36.35 \end{bmatrix}, \dots\dots\dots\dots\dots\dots\dots\dots\dots\dots (10.206)$$

$$\begin{cases} \Delta p_1 = -24.607 \, \text{psia} \\ \Delta p_2 = -26.13 \, \text{psia} \end{cases} , \rightarrow \begin{cases} p_1 = -563.7 \, \text{psia} \\ p_2 = -537.2 \, \text{psia} \end{cases}. \dots\dots\dots\dots\dots\dots (10.207)$$

- Iteration 3:

$$\begin{bmatrix} -0.2991 & 0.282 \\ 0.564 & -1.90 \end{bmatrix} \begin{bmatrix} \Delta p_1 \\ \Delta p_2 \end{bmatrix} = -\begin{bmatrix} -9.9 \times 10^{-4} \\ -0.25 \end{bmatrix}, \dots\dots\dots\dots\dots\dots\dots\dots (10.208)$$

$$\begin{cases} \Delta p_1 = -0.18 \, \text{psia} \\ \Delta p_2 = -0.18 \, \text{psia} \end{cases}. \dots\dots\dots\dots\dots\dots\dots\dots\dots\dots\dots\dots\dots\dots (10.209)$$

So that, $p_1 = 563.5$ psia, $p_2 = 537.1$ psia, and $p_3 = 563.5$ psia.

Convergence is achieved after three iterations:

$$\Omega = 0.2 \ln(p_2) = 1.26 \, \text{STB/D-psia}. \dots\dots\dots\dots\dots\dots\dots\dots\dots\dots\dots\dots (10.210)$$

MBC:
- Fluid expansion of Block 1:

$$(1,000 \, \text{psia} - 563.5 \, \text{psia}) \times 10^{-4} \, \text{psi}^{-1} \times \frac{120 \, \text{ft} \times 240 \, \text{ft} \times 10 \times 0.1}{5.165} = 223.88 \, \text{STB}. \dots\dots\dots (10.211)$$

- Fluid expansion of Block 2:

$$(1,000 \, \text{psia} - 537.1 \, \text{psia}) \times 10^{-4} \, \text{psi}^{-1} \times \frac{120 \, \text{ft} \times 240 \, \text{ft} \times 10 \times 0.1}{5.165} = 949.71 \, \text{STB}. \dots\dots\dots (10.212)$$

- Total fluid expansion: $223.88 + 223.88 + 949.71 = 1397.5$ STB. $\dots\dots\dots\dots\dots\dots (10.213)$

- Total fluid produced: $\Omega(p_2 - p_{sf}) \times 30 \, \text{days} = 1,400.1$ STB. $\dots\dots\dots\dots\dots\dots\dots (10.214)$

- $\text{MBC} = \left| \dfrac{\text{Total fluid expansion}}{\text{Total fluid produced}} \right| = \left| \dfrac{1,397.5}{1,400.1} \right| \approx 0.998. \dots\dots\dots\dots\dots\dots\dots (10.215)$

2. Using the results of Part 1:

$$r_e = \frac{\sqrt{0.5^{\frac{1}{2}}(240)^2 + 2^{\frac{1}{2}}120^2}}{0.5^{\frac{1}{4}} + 2^{\frac{1}{4}}} = 34.09 \, \text{ft}, \dots\dots\dots\dots\dots\dots\dots\dots\dots\dots\dots (10.216)$$

$$\bar{k} = \sqrt{k_x k_y} = 21.21 \, \text{md}, \dots\dots\dots\dots\dots\dots\dots\dots\dots\dots\dots\dots\dots\dots (10.217)$$

$$\Omega = \frac{2\pi \bar{k} h_2}{\mu_o B_o \left[\ln\left(\frac{r_e}{r_w}\right) + s \right]}, \dots\dots\dots\dots\dots\dots\dots\dots\dots\dots\dots\dots\dots (10.218)$$

$$\left[\ln\left(\frac{r_e}{r_w}\right) + s \right] = 3.00. \dots\dots\dots\dots\dots\dots\dots\dots\dots\dots\dots\dots\dots (10.219)$$

$s = -1.9$, which indicates that the stimulation is still effective at a value of $s = -1.9$ at 30 days.

Problem 10.5.18

The reservoir shown in **Fig. 10.21** has been uniformly contaminated. A long-term water-injection project is being considered to clean it. The injection well is completed only in Block 4, and the two production wells are located in Blocks 2 and 7. All external boundaries are sealed. Consider incompressible flow conditions. Answer the following questions using the data provided in **Table 10.21**:

1. List the system of equations necessary for calculating all flow rates.
2. Rewrite the equations in Part 1 in matrix form.
3. Find the pressure distribution in the reservoir and the flow rates of the wells by solving the system of equations using the Gauss-Seidel procedure (initial guess: 2,500 psia; pressure tolerance of the convergence is 5 psia).
4. If time is not limited, will it be possible to flush out all of the contamination in this reservoir? Explain your reasoning.

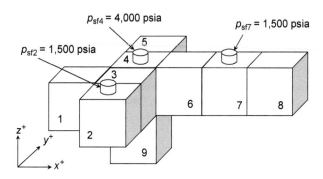

Fig. 10.21—Reservoir system of Problem 10.5.18.

Table 10.21—Reservoir data for Fig. 10.21.	
μ = 0.98 cp	
B = 1 res bbl/STB	
Hydraulic gradient = 0.433 psi/ft	
h = 50 ft	
r_w (injection) = 0.25 ft	
r_w (production) = 0.32 ft	
s (injection) = −1.5	
s (production) = 1.0	
$\Delta x = \Delta y$ = 500 ft	
$k_x = k_y = k_z$ = 40 md	

Solution to Problem 10.5.18

Only Blocks 2, 3, 4, 6, and 7 are active. Therefore, the problem can be reduced into a 2D flow domain (*x-y* plane). The flow paths 4-3-2 and 4-6-7 are symmetrical with respect to each other. Thus, the problem can be simplified by solving the pressures of Blocks 4, 6, and 7 or Blocks 4, 3, and 2 ($p_7 = p_2$, $p_3 = p_6$).

The governing flow equation for a 2D incompressible-reservoir-flow system can be written as

$$\frac{\partial}{\partial x}\left(\frac{A_x k_x}{\mu B}\frac{\partial p}{\partial x}\right)\Delta x + \frac{\partial}{\partial y}\left(\frac{A_y k_y}{\mu B}\frac{\partial p}{\partial x}\right)\Delta y + q = 0, \quad \dots\dots\dots\dots\dots\dots\dots\dots\dots\dots\dots \text{(10.220)}$$

which can be simplified for a homogeneous and isotropic reservoir to

$$\frac{\partial^2 p}{\partial x^2} + \frac{\partial^2 p}{\partial y^2} + \frac{q\mu B}{V_b k} = 0. \quad \dots\dots\dots\dots\dots\dots\dots\dots\dots\dots\dots\dots\dots\dots\dots\dots \text{(10.221)}$$

The finite-difference representation of Eq. 10.221 is

$$\frac{p_{i-1,j}-2p_{i,j}+p_{i+1,j}}{(\Delta x)^2} + \frac{p_{i,j-1}-2p_{i,j}+p_{i,j+1}}{(\Delta y)^2} + \left(\frac{q\mu B}{\Delta x \Delta y h k}\right)_{i,j} = 0. \quad \dots\dots\dots\dots\dots\dots\dots \text{(10.222)}$$

Eq. 10.222 can be rewritten by considering $\Delta x = \Delta y$:

$$p_{i,j-1} + p_{i-1,j} - 4p_{i,j} + p_{i+1,j} + p_{i,j+1} + \left(\frac{q\mu B}{hk}\right)_{i,j} = 0. \quad \dots\dots\dots\dots\dots\dots\dots\dots \text{(10.223)}$$

The Peaceman model is employed to substitute the flow rate (q) term in Eq. 10.223 for the sandface-pressure-specified wells:

$$q_{i,j} = -\Omega_{i,j}\left(p_{i,j} - p_{sf_{i,j}}\right), \quad \dots\dots\dots\dots\dots\dots\dots\dots\dots\dots\dots\dots\dots\dots\dots\dots \text{(10.224)}$$

which yields

$$p_{i,j-1} + p_{i-1,j} - 4p_{i,j} + p_{i+1,j} + p_{i,j+1} + \left(\frac{\Omega\mu B}{hk}\right)_{i,j}\left(p_{i,j} - p_{sf_{i,j}}\right) = 0. \quad \dots\dots\dots\dots\dots \text{(10.225)}$$

We can calculate the productivity indices of the wells in Blocks 2, 4, and 7:

$$\Omega_4 = \frac{2\pi kh}{\mu B \left[\ln\left(\frac{r_e}{r_{w_4}}\right) + s_4 \right]} = 3.4 \text{ STB/D-psi}, \quad\text{..} \quad (10.226)$$

$$\Omega_2 = \Omega_7 = \frac{2\pi kh}{\mu B \left[\ln\left(\frac{r_e}{r_{w_2}}\right) + s_4 \right]} = 2.041 \text{ STB/D-psi}, \quad\text{.......................................} \quad (10.227)$$

where $r_e = 0.2\Delta x = 100$ ft.

Working with Blocks 4, 6, and 7:

- Block 4:

$$p_6 + p_4 - 4p_4 + p_6 + p_4 + \frac{\mu B}{hk}\left[-\Omega_4 \left(p_4 - 4,000\right)\right] = 0. \quad\text{.....................................} \quad (10.228)$$

- Block 6:

$$p_4 - 2p_6 + p_7 = 0. \quad\text{...} \quad (10.229)$$

- Block 7:

$$p_6 + p_7 - 4p_7 + p_7 + p_7 + \frac{\mu B}{hk}\left[-\Omega_7 \left(p_7 - 1,500\right)\right] = 0. \quad\text{.............................} \quad (10.230)$$

The matrix form of the system equations is

$$\begin{bmatrix} -3.4966 & 2.3 & 0 \\ 1 & -2 & 1 \\ 0 & 2.3 & -4.3408 \end{bmatrix} \begin{bmatrix} p_4 \\ p_6 \\ p_7 \end{bmatrix} = \begin{bmatrix} -5,986.5 \\ 0 \\ -3,061.3 \end{bmatrix}. \quad\text{.................................} \quad (10.231)$$

The progression of Gauss-Seidel iterations is summarized in **Table 10.22**. Thus, $p_5 = p_4 = 3,181.77$ psia, $p_1 = p_3 = p_6 = 2,645.38$ psia, and $p_2 = p_7 = p_8 = 2,106.88$ psia.

The pressure of Block 9 can be solved using hydrostatic head:

$$p_9 = p_4 + 0.433h = 3,181.77 + 0.433 \times 50 = 3203.4 \text{ psia}. \quad\text{........................} \quad (10.232)$$

Iteration level	p_4	p_6	p_7
0	2,500	2,500	2,500
1	3,091.3	2,795.65	2,186.5
2	3,270.41	2,728.45	2,150.9
3	3,229.7	2,690.3	2,130.68
4	3,206.58	2,668.63	2,119.2
5	3,193.46	2,656.33	2,112.68
6	3,186.01	2,649.34	2,108.98
7	3,181.77	2,645.38	2,106.88

Table 10.22—Progression of Gauss-Seidel iterations found in Solution to Problem 10.5.18.

The flow rates of the wells are

$$q_4 = -\Omega_4 \left(p_4 - p_{sf4}\right) = 2,462.8 \text{ STB/D}, \quad\text{...} \quad (10.233)$$

$$q_2 = q_7 = -\Omega_7 \left(p_7 - p_{sf7}\right) = -1,233.9 \text{ STB/D}. \quad\text{.......................................} \quad (10.234)$$

An MBC is carried out to validate the result:

$$q_{in} = q_4 = 2,462.8 \text{ STB/D}, \quad\text{...} \quad (10.235)$$

$$q_{out} = q_2 + q_7 = 2,467.8 \text{ STB/D}, \dots\dots\dots\dots\dots\dots\dots\dots\dots\dots\dots\dots\dots\dots\dots\dots\dots (10.236)$$

$$\text{MBC} = \left|\frac{q_{in}}{q_{out}}\right| = 1.002, \dots\dots\dots\dots\dots\dots\dots\dots\dots\dots\dots\dots\dots\dots\dots\dots\dots\dots (10.237)$$

which is a good MBC.

Problem 10.5.19

The reservoir represented in **Fig. 10.22** is known to be isotropic, homogeneous, and horizontal, with reservoir properties listed in **Table 10.23.** It contains a single-phase, slightly compressible fluid. All boundaries are sealed. Two wells are drilled at the beginning, but during the first five days, only the well in Block 2 was produced at a constant rate of 100 STB/D. At day five, the well was shut in, and on the tenth day, the following pressures were recorded: $p_{sf_1} = 3{,}798.173$ psi and $p_{sf_2} = 3{,}798.08$ psi. Unfortunately, initial pressure of the reservoir was not recorded when the wells were completed.

1. Calculate the sandface pressure of Well 2 at five days.
2. Assuming a different scenario, if both wells were put on production at a constant rate of 100 STB/D from the beginning, what would the sandface pressure of Well 2 be after five days of production?

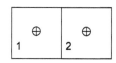

Fig. 10.22—Reservoir system of Problem 10.5.19.

Table 10.23—Reservoir data for Fig. 10.22.
$\Delta x = \Delta y = 200$ ft
$k = 100$ md
$h = 50$ ft
$\phi = 0.2$
$\mu = 1.1$ cp
$c = 5\times10^{-6}$ psi^{-1}
$B = 1.2$ res bbl/STB
$r_w = 0.25$ ft
$s = 0$

Solution to Problem 10.5.19

1. The governing PDE of the 1D, slightly compressible fluid problem is

$$\frac{\partial}{\partial x}\left(\frac{A_x k_x}{\mu B}\frac{\partial p}{\partial x}\right)\Delta x + q = \frac{V_b \phi c_t}{5.615}\left(\frac{\partial p}{\partial t}\right). \dots\dots\dots\dots\dots\dots\dots\dots\dots\dots\dots\dots\dots\dots (10.238)$$

The finite-difference analog of the flow equation is

$$p_{i-1}^{n+1} - 2p_i^{n+1} + p_{i+1}^{n+1} + \frac{q\mu B}{hk} = \frac{(\Delta x)^2 \phi c \mu B}{5.615 \Delta t k}\left(p_i^{n+1} - p_i^n\right). \dots\dots\dots\dots\dots\dots\dots\dots\dots\dots (10.239)$$

The coefficient on the right-hand side is calculated:

$$\frac{(\Delta x)^2 \phi c \mu B}{5.615 \Delta t k} = \frac{(200)^2 (0.2)(5\times10^{-6})(1.2)(1.1)}{(5.615)(100)(1.127\times10^{-3})(5)} = 0.01669. \dots\dots\dots\dots\dots\dots\dots\dots\dots (10.240)$$

The group of terms with the flow rate is

$$\frac{q\mu B}{hk} = \frac{(-100)(1.1)(1.2)}{(50)(1.127\times10^{-3})(100)} = -23.425. \dots\dots\dots\dots\dots\dots\dots\dots\dots\dots\dots\dots (10.241)$$

The well productivity index is

$$\Omega = \frac{2\pi kh}{\mu B\left[\ln\left(\dfrac{r_e}{r_w}\right) + s\right]} = 5.2824 \text{ STB/D-psi}, \dots\dots\dots\dots\dots\dots\dots\dots\dots\dots\dots\dots (10.242)$$

with $r_e = 0.2\Delta x = (0.2)(200) = 40$ ft.

Because the pressure at the end of the tenth day is recorded from $t = 5$ to 10 (with $n \to t = 5$ days and $n + 1 \to t = 10$ days):

- For Block 1:

$$p_2^{n+1} - p_1^{n+1} = (0.01669)\left(p_1^{n+1} - p_1^n\right), \dots\dots\dots\dots\dots\dots\dots\dots\dots\dots\dots\dots\dots\dots (10.243)$$

$$p_1^{t=5} = p_1^{t=10} - \frac{p_2^{t=10} - p_1^{t=10}}{0.01669} = 3,803.93 \text{ psi.} \quad\dots\dots\dots\dots\dots\dots\dots\dots\dots\dots\dots\dots \quad (10.244)$$

- For Block 2:

$$p_1^{n+1} - p_2^{n+1} = (0.01669)\left(p_2^{n+1} - p_2^{n}\right), \quad\dots\dots\dots\dots\dots\dots\dots\dots\dots\dots\dots\dots\dots \quad (10.245)$$

$$p_2^{t=5} = p_2^{t=10} - \frac{p_1^{t=10} - p_2^{t=10}}{0.01669} = 3,792.32 \text{ psi.} \quad\dots\dots\dots\dots\dots\dots\dots\dots\dots\dots\dots\dots \quad (10.246)$$

The value of p_{sf_2} at $t = 5$ days can be calculated using the wellbore model:

$$p_{sf_2}^{t=5} = p_2^{t=5} + \frac{q}{\Omega} = 3,792.32 + \frac{-100}{5.2824} = 3,773.39 \text{ psi.} \quad\dots\dots\dots\dots\dots\dots\dots\dots \quad (10.247)$$

2. From $t = 0$ to 5 (with $n \rightarrow t = 0$ days and $n + 1 \rightarrow t = 5$ days):

- For Block 1:

$$p_2^{n+1} - p_1^{n+1} = (0.01669)\left(p_1^{n+1} - p_1^{n}\right), \quad\dots\dots\dots\dots\dots\dots\dots\dots\dots\dots\dots\dots\dots \quad (10.248)$$

$$p_1^{t=0} = p_1^{t=5} - \frac{p_2^{t=5} - p_1^{t=5}}{0.01669} = 4,500.00 \text{ psi.} \quad\dots\dots\dots\dots\dots\dots\dots\dots\dots\dots\dots \quad (10.249)$$

- For Block 2:

$$p_1^{n+1} - p_2^{n+1} + (-23.425) = (0.01669)\left(p_2^{n+1} - p_2^{n}\right), \quad\dots\dots\dots\dots\dots\dots\dots\dots\dots\dots \quad (10.250)$$

$$p_2^{t=0} = p_2^{t=5} - \frac{p_1^{t=5} - p_2^{t=5} - 23.425}{0.01669} = 4,500.03 \text{ psi.} \quad\dots\dots\dots\dots\dots\dots\dots\dots \quad (10.251)$$

3. If both wells are producing,

- For Block 1:

$$-1.016686\,p_1^{n+1} + p_2^{n+1} = 51.669. \quad\dots\dots\dots\dots\dots\dots\dots\dots\dots\dots\dots\dots\dots\dots \quad (10.252)$$

- For Block 2:

$$p_1^{n+1} - 1.016686\,p_2^{n+1} = 51.669, \quad\dots\dots\dots\dots\dots\dots\dots\dots\dots\dots\dots\dots\dots\dots \quad (10.253)$$

$$p_1^{n+1} = p_2^{n+1} = 3,096.2 \text{ psi,} \quad\dots\dots\dots\dots\dots\dots\dots\dots\dots\dots\dots\dots\dots\dots\dots\dots \quad (10.254)$$

$$p_{sf_2} = p_{sf_1} = p_2 + \frac{q}{\Omega} = 3,096.2 + \frac{(-100)}{5.2824} = 3,077.3 \text{ psi.} \quad\dots\dots\dots\dots\dots\dots\dots \quad (10.255)$$

Problem 10.5.20

Consider the single-phase slightly compressible oil reservoir shown in **Fig. 10.23.1.** A permanent downhole pressure gauge was installed in the well located in Block 1. Initially the pressure of the reservoir is uniformly distributed. After 30 days of production, the flow rate and bottomhole pressure at the end of the 30-day period is recorded. Your advisor suspects that the eastern boundary of Block 3 may not be sealed (all other boundaries have proved to be sealed through previous tests). Your task as a reservoir engineer is to provide an analysis on the nature of the eastern boundary of Block 3. Justify your answer with detailed calculations and explanations using the data in **Table 10.24.**

Note: Depth gradients can be ignored. In solving this problem, it is suggested to use the SIP notation.

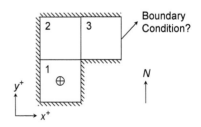

Fig. 10.23.1—Reservoir system of Problem 10.5.20.

$$\Delta x = \Delta y = 500 \text{ ft}$$
$$h = 100 \text{ ft}$$
$$k_x = k_y = 50 \text{ md}$$
$$\phi = 20\%$$
$$r_w = 0.25 \text{ ft}$$
$$c_o = 5 \times 10^{-5} \text{ psi}^{-1}$$
$$c_\phi = 0 \text{ psi}^{-1}$$
$$B_o = 0.85 \text{ res bbl/STB}$$
$$\mu_o = 0.75 \text{ cp}$$
$$\text{Initial pressure} = 2,000 \text{ psi}$$
$$\text{Sandface pressure} = 1,600 \text{ psi}$$
$$\text{Flow rate} = -2,000 \text{ STB/D}$$
$$\text{Skin} = +1.5$$

Table 10.24—Reservoir data for Fig. 10.23.1.

Solution to Problem 10.5.20

For a 2D slightly compressible system, the governing flow equation can be written as

$$\frac{\partial}{\partial x}\left(\frac{A_x k_x}{\mu B}\frac{\partial p}{\partial x}\right)\Delta x+\frac{\partial}{\partial y}\left(\frac{A_y k_y}{\mu B}\frac{\partial p}{\partial y}\right)\Delta y+q=\frac{V_b \phi c_t}{5.615}\frac{\partial p}{\partial t}. \quad\dots\dots\dots\dots\dots\dots\dots\dots (10.256)$$

The finite-difference approximation for the governing flow equation is

$$\left.\frac{A_x k_x}{\mu B \Delta x}\right|_{i+\frac{1}{2},j}^{n}\left(p_{i+1,j}^{n+1}-p_{i,j}^{n+1}\right)-\left.\frac{A_x k_x}{\mu B \Delta x}\right|_{i-\frac{1}{2},j}^{n}\left(p_{i,j}^{n+1}-p_{i-1,j}^{n+1}\right)+\left.\frac{A_x k_x}{\mu B \Delta x}\right|_{i,j+\frac{1}{2}}^{n}\left(p_{i,j+1}^{n+1}-p_{i,j}^{n+1}\right)-\left.\frac{A_x k_x}{\mu B \Delta x}\right|_{i,j-\frac{1}{2}}^{n}\left(p_{i,j}^{n+1}-p_{i,j-1}^{n+1}\right)+q$$

$$=\frac{V_b \phi c_t}{5.615 \Delta t}\left(p_{i,j}^{n+1}-p_{i,j}^{n}\right). \quad\dots\dots\dots\dots\dots\dots\dots\dots\dots\dots\dots\dots\dots\dots\dots\dots\dots\dots\dots\dots\dots\dots (10.257)$$

The finite-difference approximation, as shown in Eq. 10.257, can be written using SIP notation, as follows:

$$S_{i,j,k}^{n}p_{i,j-1}^{n+1}+W_{i,j,k}^{n}p_{i-1,j}^{n+1}+C_{i,j,k}^{n}p_{i,j}^{n+1}+E_{i,j,k}^{n}p_{i+1,j}^{n+1}+N_{i,j,k}^{n}p_{i,j+1}^{n+1}=Q_{i,j,k}^{n+1}. \quad\dots\dots\dots\dots\dots\dots (10.258)$$

Because the reservoir is homogeneous and isotropic ($k_x = k_y = k$, $A_x = A_y$), and also because $\Delta x = \Delta y$,

$$N_{i,j}^{n}=S_{i,j}^{n}=W_{i,j}^{n}=E_{i,j}^{n}=\left(\frac{hk_y}{\mu_o B_o}\right)_{i,j}, \quad \Gamma_{i,j}=\left(\frac{\Delta x \Delta y h \phi c_o}{5.615 \Delta t}\right)_{i,j}, \quad\dots\dots\dots\dots\dots\dots\dots (10.259)$$

$$\Omega_{i,j}=\frac{2\pi h\sqrt{k_x k_y}}{\mu_o B_o\left(\ln\frac{r_e}{r_w}+s\right)}. \quad\dots\dots\dots\dots\dots\dots\dots\dots\dots\dots\dots\dots\dots\dots\dots\dots (10.260)$$

For a sandface-pressure-specified wellblock, the $C_{i,j}$ and $Q_{i,j}$ coefficients are defined as follows:

$$C_{i,j}^{n}=-\left(S_{i,j}^{n}+W_{i,j}^{n}+E_{i,j}^{n}+N_{i,j}^{n}+\Gamma_{i,j}^{n}+\Omega_{i,j}^{n}\right), \quad\dots\dots\dots\dots\dots\dots\dots\dots\dots\dots\dots (10.261)$$

$$Q_{i,j}^{n+1}=-\Omega_{i,j}^{n+1}p_{sf_{spec}}-\Gamma_{i,j}^{n}p_{i,j}^{n}. \quad\dots\dots\dots\dots\dots\dots\dots\dots\dots\dots\dots\dots\dots\dots (10.262)$$

There are two possible types of boundary conditions that need to be considered on the eastern boundary of Block 3 (Dirichlet-type and Neumann-type). First, SIP coefficients are calculated:

$$\Omega=\frac{2\pi h\sqrt{k_x k_y}}{\mu_o B_o\left(\ln\frac{r_e}{r_w}+s\right)}=7.4136 \text{ STB/D-psi}, \quad\dots\dots\dots\dots\dots\dots\dots\dots\dots\dots\dots\dots (10.263)$$

$$N = S = W = E = \frac{hk_y}{\mu_o B_o} = 8.8392 \text{ STB/D-psi,} \quad\dotfill \quad (10.264)$$

$$\Gamma = \frac{\Delta x \Delta y h \phi c_o}{5.615 \, \Delta t} = 1.4841 \text{ STB/D-psi.} \quad\dotfill \quad (10.265)$$

- Block 1: Because both flow rate and bottomhole-pressure values are known, one can calculate p_1 using the wellbore equation:

$$p_1 = \frac{q_{osc}}{\Omega} + p_{sf} = 1,869.8 \text{ psia.} \quad\dotfill \quad (10.266)$$

The finite-difference equation written using SIP notation for Block 1 is

$$N_1 p_2 + C_1 p_1 = Q_1, \quad\dotfill \quad (10.267)$$

where

$$C_1 = -(N_1 + \Gamma_1 + \Omega) = -17.7369 \text{ STB/D-psia,} \quad\dotfill \quad (10.268)$$

$$Q_1 = -\Omega p_{sf} - \Gamma_1 p_i = -14,830 \text{ STB/D.} \quad\dotfill \quad (10.269)$$

Using the Block 1 equation, one can solve for p_2:

$$p_2 = \frac{Q_1 - C_1 p_1}{N_1} = 2,074.2 \text{ psia.} \quad\dotfill \quad (10.270)$$

- Block 2: The finite-difference equation written for Block 2 using SIP notation is

$$S_2 p_1 + C_2 p_2 + E_2 p_3 = Q_2, \quad\dotfill \quad (10.271)$$

where

$$C_2 = -(S_2 + E_2 + \Gamma_2) = -19.1626 \text{ STB/D-psia,} \quad\dotfill \quad (10.272)$$

$$Q_2 = -\Gamma_2 p_i = -2,968 \text{ STB/D.} \quad\dotfill \quad (10.273)$$

Again, this time we can solve for p_3:

$$p_3 = \frac{Q_2 - S_2 p_1 - C_2 p_2}{E_2} = 2,291 \text{ psia.} \quad\dotfill \quad (10.274)$$

- Block 3: The image block across the eastern boundary of Block 3 is depicted in **Fig. 10.23.2.**

Fig. 10.23.2—Treatment of the boundary condition of Block 3.

The finite-difference equation written for Block 3 is

$$W_3 p_2 + C_3 p_3 + E_3 p_3' = Q_3, \quad\dotfill \quad (10.275)$$

where

$$C_3 = -(W_3 + E_3 + \Gamma_3) = -19.1626 \text{ STB/D-psia,} \quad\dotfill \quad (10.276)$$

and

$$Q_3 = Q_2. \quad\dotfill \quad (10.277)$$

If Block 3 has Dirichlet-type boundary conditions, then one can solve for p_3':

$$p_3' = \frac{Q_3 - W_3 p_2 - C_3 p_3}{E_3} = 2,556.7 \text{ psia.} \quad\dotfill \quad (10.278)$$

If Block 3 has Neumann-type boundary conditions,

$$\frac{p_3' - p_3}{\Delta x} = \frac{\partial p}{\partial x} \rightarrow p_3' = \frac{\partial p}{\partial x} \Delta x + p_3. \quad \dots \dots \dots \dots \dots \dots \dots \dots \dots \dots \dots \dots (10.279)$$

The finite-difference equation for Block 3 can be modified as follows:

$$W_3 p_2 + C_3 p_3 + E_3 \left(\frac{\partial p}{\partial x} \Delta x + p_3 \right) = Q_3. \quad \dots \dots \dots \dots \dots \dots \dots \dots \dots \dots \dots (10.280)$$

Now, we can solve for $\dfrac{\partial p}{\partial x}$:

$$\frac{\partial p}{\partial x} = \frac{Q_3 - W_3 p_2 - (C_3 + E_3) p_3}{E_3 \Delta x} = 0.5314 \text{ psia/ft.} \quad \dots \dots \dots \dots \dots \dots \dots \dots \dots (10.281)$$

To summarize, if the eastern boundary of Block 3 is not sealed, it is not possible to conclude whether it had a Dirichlet- or Neumann-type boundary condition.

Problem 10.5.21

Two companies, A and B, have shares of the single-phase oil reservoir presented in **Fig. 10.24.** Company A is responsible for operating the well located in Block 2. An agreement was signed between the two parties to produce the well at a rate of 1,200 STB/D. After 30 days of production, however, Company B suspects that Company A has breached the agreement and produced more than 1,200 B/D. To verify the actual production rate, the well was shut in for 10 days after 30 days of production to allow the reservoir pressure to build up. The bottomhole pressure at the end of 10 days of shut-in time is measured as 2,322 psia. As a technical expert, analyze the data to decide whether Company A or B is right in its claim. Use the related petrophysical properties listed in **Table 10.25.**

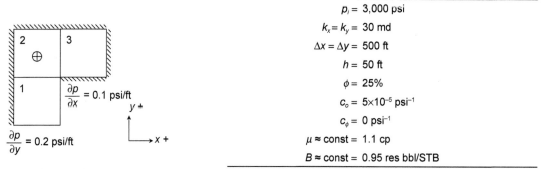

Fig. 10.24—Reservoir system of Problem 10.5.21.   Table 10.25—Petrophysical properties of the reservoir in Fig. 10.24.

Table 10.25 values:
$p_i = 3{,}000$ psi
$k_x = k_y = 30$ md
$\Delta x = \Delta y = 500$ ft
$h = 50$ ft
$\phi = 25\%$
$c_o = 5 \times 10^{-5}$ psi^{-1}
$c_\phi = 0$ psi^{-1}
$\mu \approx \text{const} = 1.1$ cp
$B \approx \text{const} = 0.95$ res bbl/STB

Solution to Problem 10.5.21

The governing flow equation for the reservoir system studied in this problem is

$$\frac{\partial}{\partial x} \left(\frac{A_x k_x}{\mu B} \frac{\partial p}{\partial x} \right) \Delta x + \frac{\partial}{\partial y} \left(\frac{A_y k_y}{\mu B} \frac{\partial p}{\partial y} \right) \Delta y + q = \frac{V_b \, \phi c_t}{5.615} \left(\frac{\partial p}{\partial t} \right). \quad \dots \dots \dots \dots \dots (10.282)$$

For an isotropic and homogeneous system, the PDE can be reduced to

$$\frac{\partial^2 p}{\partial x^2} + \frac{\partial^2 p}{\partial y^2} + \frac{q \mu B}{V_b k} = \frac{\phi c_t \mu B}{5.615 k} \left(\frac{\partial p}{\partial t} \right). \quad \dots \dots \dots \dots \dots \dots \dots \dots (10.283)$$

The finite-difference representation of the PDE is

$$p_{i,j-1}^{n+1} + p_{i-1,j}^{n+1} - 4 p_{i,j}^{n+1} + p_{i+1,j}^{n+1} + p_{i,j+1}^{n+1} + \frac{q^{n+1} \mu B}{hk} = \frac{\mu B \phi c_t \Delta x^2}{5.615 k \Delta t} (p_{i,j}^{n+1} - p_{i,j}^n). \quad \dots \dots \dots (10.284)$$

Now, one can calculate the pressure distribution at $t = 30$ days and at $t = 40$ days using 30-day and, subsequently, 10-day timesteps.

Timestep I ($t = 30$ days with $\Delta t = 30$ days):

- Block 1:

$$p_{1x}^{n+1} - 2p_1^{n+1} + p_1^{n+1} + p_2^{n+1} - 2p_1^{n+1} + p_{1y}^{n+1} = \left(\frac{\mu B \phi\, c_t \Delta x^2}{5.615 k \Delta t}\right)(p_1^{n+1} - p_1^n). \quad\quad\quad (10.285)$$

p_{1x} and p_{1y} are the reflection-block pressures across the eastern and southern boundaries, respectively. In other words,

$$p_{1x}^{n+1} = p_1^{n+1} + \frac{\partial p}{\partial x}\Delta x, \quad\quad\quad\quad\quad\quad\quad\quad\quad\quad\quad (10.286)$$

and

$$p_{1y}^{n+1} = p_1^{n+1} - \frac{\partial p}{\partial x}\Delta y. \quad\quad\quad\quad\quad\quad\quad\quad\quad\quad\quad (10.287)$$

Knowing that,

$$\frac{\mu B \phi\, c_t \Delta x^2}{5.615 k \Delta t} = \frac{1.1\,\text{cp} \times \dfrac{0.95\,\text{bbl}}{\text{STB}} \times 0.25 \times (5 \times 10^{-5}) \times 500\,\text{ft} \times 500\,\text{ft}}{5.615 \times 30\,\text{md} \times 0.001127\,\dfrac{\text{perm}}{\text{md}} \times 30\,\text{days}} = 0.573. \quad\quad (10.288)$$

The equation for Block 1 will take the following form (after implementing the boundary conditions):

$$-1.573 p_1 + p_2 = -1{,}870.17. \quad\quad\quad\quad\quad\quad\quad\quad\quad\quad (10.289)$$

- Block 2: The finite-difference equation can be written as

$$p_3^{n+1} - 2p_2^{n+1} + p_2^{n+1} + p_1^{n+1} - 2p_2^{n+1} + p_2^{n+1} + \frac{q^{n+1}\mu B}{hk} = \frac{\mu B \phi\, c_t \Delta x^2}{5.615 k \Delta t}(p_2^{n+1} - p_2^n). \quad\quad (10.290)$$

Now, the $\dfrac{q^{n+1}\mu B}{hk}$ group on the left side of the equal sign is calculated:

$$\frac{q^{n+1}\mu B}{hk} = \frac{-1{,}200\,\text{STB/D} \times 1.1\,\text{cp} \times 0.95\,\text{res bbl/STB}}{50\,\text{ft} \times 30\,\text{md} \times 0.001127\,\dfrac{\text{perm}}{\text{md}}} = -741.79. \quad\quad\quad\quad (10.291)$$

After making the substitutions for the two coefficients, the equation for Block 2 becomes

$$p_1 = -2.573 p_2 + p_3 = -978.38. \quad\quad\quad\quad\quad\quad\quad\quad\quad (10.292)$$

- Block 3: Finally, the finite-difference equation for Block 3 is written as

$$p_2 = -1.573 p_3 = -1{,}720.17. \quad\quad\quad\quad\quad\quad\quad\quad\quad\quad (10.293)$$

Now, we are set to solve three equations in three unknowns:

$$-1.573 p_1 + p_2 = -1{,}870.17, \quad\quad\quad\quad\quad\quad\quad\quad\quad\quad (10.294)$$

$$p_1 - 2.573 p_2 + p_3 = -978.38, \quad\quad\quad\quad\quad\quad\quad\quad\quad\quad (10.295)$$

$$p_2 - 1.573 p_3 = -1{,}720.17. \quad\quad\quad\quad\quad\quad\quad\quad\quad\quad (10.296)$$

The solution to the system of equations will give the pressure distribution at the end of the 30th day of production: $p_1 = 2{,}779.8$ psia, $p_2 = 2{,}503.6$ psia, and $p_3 = 2{,}684.5$ psia.

Timestep II ($t = 40$ days with $\Delta t = 10$ days):
Equations for Blocks 1, 2, and 3 will be as follows:

$$-2.72 p_1 + p_2 = -4{,}931.78, \quad\quad\quad\quad\quad\quad\quad\quad\quad\quad (10.297)$$

$$p_1 - 3.72 p_2 + p_3 = -4{,}306.59, \quad\quad\quad\quad\quad\quad\quad\quad\quad\quad (10.298)$$

$$p_2 - 2.72 p_3 = -4{,}617.78. \quad\quad\quad\quad\quad\quad\quad\quad\quad\quad (10.299)$$

Solving the system of equations, one will obtain the following pressure distribution: $p_1 = 2,775.81$ psia, $p_2 = 2,618.91$ psia, and $p_3 = 2,660.38$ psia

As a result of these calculations, one can reach the following conclusion: If the well, indeed, were produced at a constant rate of 1,200 STB/D, the sandface pressure of Well 2 would have been 2,618.91 psia. However, the sandface pressure of Well 2 was measured to be 2,322 psia. This additional pressure drop of almost 300 psia indicates that during the first 30 days, Well 2 was produced at a higher rate than 1,200 STB/D.

Problem 10.5.22

Consider a reservoir saturated with a single-phase slightly compressible fluid (**Fig. 10.25**). The reservoir has sealed northern, western, and southern boundaries. The eastern reservoir boundary is a constant-pressure gradient boundary. Well 1 has been producing under a constant flow rate of 4,500 STB/D for seven days. On the seventh day, pressure in the observation well (Well 2) located in Block 2 is observed to be 3,709.4 psia. Determine the value of the constant-pressure gradient across the eastern reservoir boundary using the data presented in **Table 10.26**.

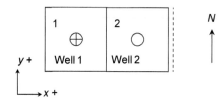

Fig. 10.25—Reservoir system of Problem 10.5.22.

$k_x = k_y = 80$ md
$\phi = 0.13$
$p_{initial} = 4,000$ psi
$r_w = 0.32$ ft
$\Delta x = \Delta y = 550$ ft
$s = 0$
$\rho = 62$ lbm/ft^3
$c = 5 \times 10^{-5}$ psi^{-1}
$h = 90$ ft
$\mu \approx 1.1$ cp
$B \approx 0.9$ res bbl/STB
$\Delta t = 7$ days

Table 10.26—Reservoir data for Fig. 10.25

Solution to Problem 10.5.22

The governing equation for a 1D reservoir containing slightly compressible fluid is

$$\frac{\partial}{\partial x}\left(\frac{A_x k_x}{\mu B}\frac{\partial p}{\partial x}\right)\Delta x + q = \frac{V_b \phi c_t}{5.615}\left(\frac{\partial p}{\partial t}\right). \dots\dots\dots\dots (10.300)$$

Eq. 10.300 can be simplified by treating μ and B as constants. Furthermore, because the reservoir has homogeneous property distributions, the reduced form of the equation will be

$$\frac{\partial^2 p}{\partial x^2} + \frac{q\mu B}{V_b k} = \frac{\phi c_t \mu B}{5.615 k}\left(\frac{\partial p}{\partial t}\right). \dots\dots\dots\dots (10.301)$$

The characteristic finite-difference equation is written as follows:

$$\frac{p_{i-1}^{n+1} - 2p_i^{n+1} + p_{i+1}^{n+1}}{(\Delta x)^2} + \left(\frac{q\mu B}{kV_b}\right)_i = \left[\frac{\mu B \phi c_t}{5.615 k \Delta t}\right](p_i^{n+1} - p_i^n), \dots\dots\dots\dots (10.302)$$

which reduces to the following form (because $\Delta x = \Delta y$):

$$p_{i-1}^{n+1} - 2p_i^{n+1} + p_{i+1}^{n+1} + \left(\frac{q\mu B}{kh}\right)_i = \left[\frac{(\Delta x)^2 \mu B \phi c_t}{5.615 k \Delta t}\right](p_i^{n+1} - p_i^n). \dots\dots\dots\dots (10.303)$$

Method 1:
- Block 1 ($p_1^n = 4,000$ psia):

$$-p_1^{n+1} + p_2^{n+1} + \left(\frac{q\mu B}{hk}\right)_1 = \frac{(\Delta x)^2 \mu B \phi c_t}{5.615 k \Delta t}(p_1^{n+1} - p_1^n), \dots\dots\dots\dots (10.304)$$

$$-p_1 + 3,709.4 - \frac{4,500 \times 1.1 \times 0.9}{90 \times 80 \times 0.001127} = \frac{550 \times 550 \times 1.1 \times 0.9 \times 0.13 \times 5 \times 10^{-5}}{5.615 \times 80 \times 0.001127 \times 7}(p_1 - 4,000). \dots\dots (10.305)$$

One can solve for p_1:

$$p_1 = 3,458.1 \text{ psi.} \dotfill (10.306)$$

- Block 2 ($p_2^n = 4,000$ psia):

$$p_1^{n+1} - 2p_2^{n+1} + p_2'^{n+1} = \frac{(\Delta x)^2 \mu B \phi c_t}{5.615 k \Delta t}(p_2^{n+1} - p_2^n). \dotfill (10.307)$$

Implementing the pressure gradient across the boundary, the following equations can be obtained:

$$\frac{p_2' - p_2}{\Delta x} = \alpha, \dotfill (10.308)$$

$$p_2' = \alpha \Delta x + p_2. \dotfill (10.309)$$

The preceding definition of p_2' is substituted into the Block 2 equation to solve for α:

$$p_1^{n+1} - p_2^{n+1} + \alpha \Delta x = \frac{(\Delta x)^2 \mu B \phi c_t}{5.615 k \Delta t}(p_2^{n+1} - p_2^n), \dotfill (10.310)$$

$$3,458.1 - 3,709.4 + 550\alpha = \frac{550 \times 550 \times 1.1 \times 0.9 \times 0.13 \times 5 \times 10^{-5}}{5.615 \times 80 \times 0.001127 \times 7}(3,709.4 - 4,000), \dotfill (10.311)$$

$$\alpha = 0.167 \text{ psi/ft.} \dotfill (10.312)$$

Method 2:
- Block 1:
  Pore volume of Block 1 is

$$V_1 = 550 \text{ ft} \times 550 \text{ ft} \times 90 \text{ ft} \times 0.13 = 3.53925 \times 10^6 \text{ ft}^3. \dotfill (10.313)$$

To calculate the fluid expansion, we employ the definition of compressibility:

$$c = -\frac{1}{V}\frac{\partial V}{\partial p}, \dotfill (10.314)$$

$$5 \times 10^{-5} = \frac{1}{3.53925 \times 10^6}\frac{\Delta V_1}{(4,000 - 3,458.1)}, \dotfill (10.315)$$

$$\Delta V_1 = -95,896 \text{ ft}^3 \times \frac{1}{5.615}\frac{\text{STB}}{\text{ft}^3} = -17,078.5 \text{ STB.} \dotfill (10.316)$$

- Block 2:
  Pore volume of Block 2 is

$$V_2 = 550 \text{ ft} \times 550 \text{ ft} \times 90 \text{ ft} \times 0.13 = 3.53925 \times 10^6 \text{ ft}^3. \dotfill (10.317)$$

Similarly, the fluid-volume expansion in Block 2 is calculated:

$$5 \times 10^{-5} = \frac{1}{3.53925 \times 10^6}\frac{\Delta V_2}{(4,000 - 3,709.4)}, \dotfill (10.318)$$

$$\Delta V_2 = -51,425 \text{ ft}^3 \times \frac{1}{5.615}\frac{\text{STB}}{\text{ft}^3} = -9,158.6 \text{ STB.} \dotfill (10.319)$$

The total fluid volume expansion is

$$\Delta V_1 + \Delta V_2 = -17,078.5 \text{ STB} - 9,158.6 \text{ STB} = -26,237.1 \text{ STB.} \dotfill (10.320)$$

The total volume produced from the well is

$$\Delta V_{\text{well}} = 7 \text{ days} \times (-4,500) \text{ STB/D} = -31,500 \text{ STB.} \dotfill (10.321)$$

Because the total fluid expansion is less than the total production $(|-26,237 \text{ STB}| < |-31,500 \text{ STB}|)$, it is clear that 31,500 STB – 26,237 STB = 5,263 STB of fluid has entered the reservoir across the eastern boundary. The pressure gradient that will give that volume (5,263 STB) can be calculated as shown:

$$q = -\frac{A \times k}{\mu \times B}\frac{\partial p}{\partial x} \rightarrow \frac{\partial p}{\partial x} = -\frac{q \times \mu \times B}{A \times k}, \dotfill (10.322)$$

$$q = -\frac{5,263 \text{ STB}}{7 \text{ days}} = -752.9 \text{ STB/D}, \dotfill (10.323)$$

$$\frac{\partial p}{\partial x} = \frac{753 \times 1.1 \times 0.9}{550 \times 90 \times 80 \times 0.001127} = 0.167 \text{ psi/ft.} \dotfill (10.324)$$

Problem 10.5.23

Are the following statements TRUE or FALSE? Please provide an explanation for your choice.

1. Two 3×3 slightly compressible reservoirs are shown in **Fig. 10.26.1,** with Well 1 as an injector, Well 3 as a producer, and Well 2 as an observation well. The initial pressure in both reservoirs is 2,000 psia. A reservoir engineer makes an operational plan and specifies the sandface pressure at the injector to be 2,500 psia and at the producer to be 14.7 psia. The engineer predicts that after operating for 30 days, the pressure of Block 2 in Reservoir 1 will be less than that of Reservoir 2. Using the reservoir data in **Table 10.27,** complete the analysis; what do you think?

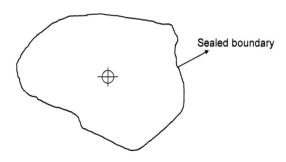

Fig. 10.26.1—Reservoir system of Part 1 of Problem 10.5.23.

Reservoir 1		Reservoir 2	
$\Delta x = \Delta y =$	500 ft	$\Delta x = \Delta y =$	500 ft
$h =$	40 ft	$h =$	20 ft
$k_x = k_y =$	30 md	$k_x = k_y =$	60 md
$\phi =$	0.2	$\phi =$	0.4

Table 10.27—Reservoir data for wells found in Fig. 10.26.1.

2. A reservoir engineer reports that a well is producing natural gas from a reservoir (**Fig. 10.26.2**) at an average gas-flow rate of 0.9 MMscf/D over the past 30 days. Given the flowing reservoir properties, quantitatively justify whether or not the engineer's report is correct (use the properties given in **Table 10.28**).

Sealed boundary

Fig. 10.26.2—Reservoir system of Part 2 Problem 10.5.23.

Pressure at $t = 40 =$	3,000 psi
Pressure at $t = 30 =$	2,800 psi
z–factor at $t = 0 =$	0.96
z–factor at $t = 30 =$	0.95
$\phi =$	20%
$T =$	150°F
$T_{sc} =$	60°F
$p_{sc} =$	14.7 psi
$h =$	30 ft
$A =$	10 acres

Table 10.28—Properties for the reservoir given in Fig. 10.26.2.

3. Consider the reservoir shown in **Fig. 10.26.3.** All boundaries are sealed. The northern part (the shaded region) of the reservoir has $k_x = k_y = 50$ md; the southern part of the reservoir has $k_x = k_y = 100$ md. All other reservoir and fluid parameters are homogeneous and isotropic. After a period of production through the well in the center of the reservoir (under constant flow-rate production),
    a) $p_A < p_B$.
    b) $p_C < p_D$.

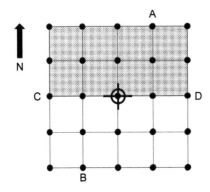

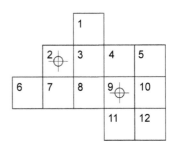

Fig. 10.26.3—Reservoir system in Part 3 of Problem 10.5.23.   **Fig. 10.26.4—Reservoir system in Part 4 of Problem 10.5.23.**

4. Consider the single-phase gas reservoir shown in **Fig. 10.26.4.** All outer boundaries are entirely sealed. Two production wells are located in the centers of Block 2 and 9. All reservoir parameters are homogeneous and isotropic. Depth gradient is ignored. Initially the reservoir pressure ($p_{initial}$) is uniformly distributed. After a period of time of production,
   a) $p_{11} < p_{12}$.
   b) $p_{10} < p_{12}$.

5. An injection well is injecting at a constant flow rate ($q_{inj} = n$ lbm/D) into a reservoir with sealed boundaries. Fluid in the reservoir is the same as the injected fluid. The p_{frac} is the constraint on the sandface pressure and once $p_{sf} = p_{frac}$, the injection stops. Qualitatively draw the sandface pressure as a function of time if the injected fluid is (i) incompressible fluid; (ii) slightly compressible fluid; and (iii) compressible fluid (please label your curves). Provide a brief explanation of your answer.

Solution to Problem 10.5.23

1. FALSE. The 2D slightly compressible flow equation is

$$\frac{\partial}{\partial x}\left(\frac{A_x k_x}{\mu B}\frac{\partial p}{\partial x}\right)\Delta x + \frac{\partial}{\partial y}\left(\frac{A_y k_y}{\mu B}\frac{\partial p}{\partial y}\right)\Delta y + q_{sc} = \frac{V_b \phi c_t}{5.615}\left(\frac{\partial p}{\partial t}\right). \quad \dots \dots \dots \dots \dots \dots \dots \dots (10.325)$$

For homogeneous and isotropic systems, Eq. 10.325 reduces to

$$\frac{\partial^2 p}{\partial x^2} + \frac{\partial^2 p}{\partial y^2} + \frac{q\mu B}{V_b k} = \frac{\phi c_t \mu B}{5.615 k}\left(\frac{\partial p}{\partial t}\right). \quad \dots \dots \dots \dots \dots \dots \dots \dots \dots \dots (10.326)$$

- For Reservoir 1:

$$V_b k = 500 \text{ ft} \times 500 \text{ ft} \times 40 \text{ ft} \times 30 \text{ md} \times 0.001127 = 338{,}100, \quad \dots \dots \dots \dots \dots \dots (10.327)$$

$$\frac{\phi}{k} = \frac{0.2}{30} = \frac{1}{150}. \quad \dots \dots \dots \dots \dots \dots \dots \dots \dots \dots \dots \dots \dots \dots \dots \dots (10.328)$$

- For Reservoir 2:

$$V_b k = 500 \text{ ft} \times 500 \text{ ft} \times 20 \text{ ft} \times 60 \text{ md} \times 0.001127 = 338{,}100, \quad \dots \dots \dots \dots \dots \dots (10.329)$$

$$\frac{\phi}{k} = \frac{0.4}{60} = \frac{1}{150}. \quad \dots \dots \dots \dots \dots \dots \dots \dots \dots \dots \dots \dots \dots \dots \dots \dots (10.330)$$

In this case, the continuity equations for Reservoir 1 and 2 are the same. Thus, pressure at Block 2 is the same for both reservoirs.

2. TRUE.

$$B_{g1} = \frac{p_{sc} z_{g1} T_1}{5.615 p_1 T_{sc}} = \frac{14.7 \text{ psi} \times 0.96 \times (150 + 460)}{5{,}615 \times 3{,}000 \times (60 + 460)} = 9.83 \times 10^{-4} \text{ res bbl/scf}, \quad \dots \dots \dots \dots (10.331)$$

$$B_{g2} = \frac{p_{sc} z_{g2} T_2}{5.615 p_2 T_{sc}} = \frac{14.7 \text{ psi} \times 0.95 \times (150 + 460)}{5{,}615 \times 2{,}800 \times (60 + 460)} = 1.04 \times 10^{-3} \text{ res bbl/scf}. \quad \dots \dots \dots \dots (10.332)$$

Total gas produced is

$$\frac{V_b\phi}{5.615}\times\left(\frac{1}{B_{g2}}-\frac{1}{B_{g1}}\right)=26{,}978{,}130.34 \text{ scf.} \quad\dots\dots\dots\dots\dots\dots\dots\dots\dots\dots\dots\dots\dots\dots\dots \text{(10.333)}$$

The average production rate over a period of 30 days is

$$q=\frac{26{,}978{,}130.34 \text{ scf}}{30 \text{ days}}=0.9 \text{ MMscf/D,} \quad\dots\dots\dots\dots\dots\dots\dots\dots\dots\dots\dots\dots\dots\dots \text{(10.334)}$$

which agrees with the value reported by the reservoir engineer.

3. (a) FALSE. Because the permeability of the southern part of the reservoir is larger, more fluid will be extracted from this part, and thus, although the location of Points A and B are symmetrical, the pressure of Point B will be lower.
   (b) FALSE. Because each parameter of Points C and D are the same and their locations are symmetrical, then we expect to have $p_C < p_D$.
4. (a) FALSE. The fluid should flow from Block 12 to Block 11 because $p_{11} < p_{12}$.
   (b) TRUE. The fluid should flow from Block 12 to Block 10.
5. As shown in **Fig. 10.26.5**, for a closed incompressible fluid reservoir, the injection cannot be carried out, thus there is no representation for the incompressible fluid. For a slightly compressible system, the initial sandface pressure is higher than that of a compressible fluid system to inject the same amount of fluid. The sandface pressure increases faster and reaches to p_{frac} earlier compared with the injection operation of a compressible fluid.

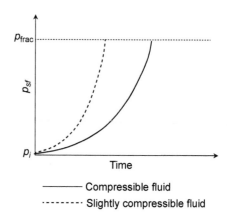

Fig. 10.26.5—Sandface pressure vs. time.

Problem 10.5.24

Consider the slightly compressible fluid flow in the 1D reservoir shown in **Fig. 10.27.** The reservoir exhibits homogeneous properties. Reservoir rock and fluid properties are listed in **Table 10.29.** The depth to the top surface of the formation is 3,000 ft at every point. The reservoir boundaries are completely sealed. The initial pressure of the reservoir is 4,000 psia (uniform). The well located in Block 2 is put on production at a constant flow rate of q STB/D. The well in Block 3 is not put on line. A pressure survey conducted the at the observation well located in Block 3 at the end of 30 days reveals a sandface pressure of 3,850 psia.

1. What is the magnitude of flow rate q?
2. What is the sandface pressure of the producer at the end of 30 days?
3. Perform an MBC to verify the accuracy of your results.

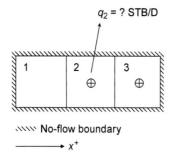

Fig. 10.27—Reservoir system of Problem 10.5.24.

Table 10.29—Reservoir rock and fluid properties for Fig. 10.27.
$k = 100$ md
$\Delta x = \Delta y = 500$ ft
$\phi = 30\%$
$c = 1\times10^{-5}$ psi^{-1}
$r_w = 0.25$ ft
$h = 100$ ft
$\rho = 62.4$ lbm/ft^3
$\mu \approx \text{const} = 1$ cp
$B \approx \text{const} = 1$ res bbl/STB
$s = 0$
$\Delta t = 30$ days

Solution to Problem 10.5.24

1. The governing flow equation for the 1D slightly compressible reservoir is

$$\frac{\partial}{\partial x}\left(\frac{A_x k_x}{\mu B}\frac{\partial p}{\partial x}\right)\Delta x + q = \frac{V_b \phi c_t}{5.615}\left(\frac{\partial p}{\partial t}\right). \quad \dotfill \text{(10.335)}$$

The finite-difference approximation can be written as

$$\left.\frac{A_x k_x}{\mu B \Delta x}\right|_{i+\frac{1}{2}}^{n}\left(p_{i+1}^{n+1} - p_i^{n+1}\right) - \left.\frac{A_x k_x}{\mu B \Delta x}\right|_{i-\frac{1}{2}}^{n}\left(p_i^{n+1} - p_{i-1}^{n+1}\right) + q = \frac{V_b \phi c_t}{5.615 \Delta t}\left(p_i^{n+1} - p_i^{n}\right). \quad \dotfill \text{(10.336)}$$

The transmissibility terms are calculated as

$$\left.\frac{A_x k_x}{\mu B \Delta x}\right|_{i+\frac{1}{2}}^{n} = \left.\frac{A_x k_x}{\mu B \Delta x}\right|_{i-\frac{1}{2}}^{n} = T = \frac{hk}{\mu B} = \frac{(100)(100)(1.127\times10^{-3})}{(1)(1)} = 11.27 \text{ STB/D-psia}, \quad \dotfill \text{(10.337)}$$

and

$$\Gamma = \frac{\Delta x \Delta y h \phi c}{5.615 \Delta t} = \frac{(500)(500)(100)(0.3)(1\times10^{-5})}{(5.615)(30)} = 0.4452 \text{ STB/D-psia}. \quad \dotfill \text{(10.338)}$$

• Block 3:

$$T p_2 - (T+\Gamma) p_3 = -\Gamma p_i, \quad \dotfill \text{(10.339)}$$

$$11.27 p_2 - 11.7152 p_3 = -1,780.8. \quad \dotfill \text{(10.340)}$$

As we know, $p_3 = 3,850$ psia,

$$p_2 = \frac{-1,780.8 + 11.7152 + 3,850}{11.27} = 3,844.1 \text{ psia}. \quad \dotfill \text{(10.341)}$$

• Block 1:

$$T p_2 - (T+\Gamma) p_1 = -\Gamma p_i. \quad \dotfill \text{(10.342)}$$

Thus,

$$11.27 p_2 - 11.715 p_1 = -1,780.8, \quad \dotfill \text{(10.343)}$$

$$p_1 = \frac{-1,780.8 - 11.27\times3,844.1}{-11.715} = 3,850.1 \text{ psia}. \quad \dotfill \text{(10.344)}$$

• Block 2:

$$T p_1 - (2T+\Gamma) p_2 + T p_3 + q = -\Gamma p_i, \quad \dotfill \text{(10.345)}$$

$$11.27 p_1 - 22.9852 p_2 + 11.27 p_3 + q = -1,780.8, \quad \dotfill \text{(10.346)}$$

$$q = -1,780.8 - (11.27\times3,850.1 - 22.9852\times3,844.1 + 11.27\times3,850) = -203.5 \text{ STB/D}. \quad \dotfill \text{(10.347)}$$

2. Using the wellbore model, we can calculate the sandface pressure of the well located in Block 2:

$$\Omega_4 = \frac{2\pi k h}{\mu B\left[\ln\left(\frac{r_e}{r_w}\right) + s\right]} = \frac{(2)(3.14)(100)(1.127\times10^{-3})(100)}{(1)(1)\left[\ln\left(\frac{0.2\times500}{0.25}\right) + 0\right]} = 11.8127 \text{ STB/D-psi}, \quad \dotfill \text{(10.348)}$$

$$p_{sf} = p_2 + \frac{q}{\Omega} = 3,844.1 + \frac{(-203.5)}{11.8127} = 3,826.87 \text{ psia}. \quad \dotfill \text{(10.349)}$$

3. An MBC shows that

$$MB = \frac{\left|\Gamma\left(p_1 + p_2 + p_3 - 3p_i\right)\right|}{|q|} = \frac{\left|0.4452\left(3,850.1 + 3,844.1 + 3,850 - 3\times4,000\right)\right|}{|-203.5|} = 0.9972. \quad \ldots\ldots\ldots (10.350)$$

Problem 10.5.25

Consider incompressible fluid flow in the 1D reservoir shown in **Fig. 10.28.1.** Both outer reservoir boundaries are maintained at a constant pressure of 2,500 psi. There is an existing vertical well in Block 2. The plan is to stimulate the well to increase its production. Using the data in **Table 10.30,** you can either a) Acidize the well to achieve a skin factor of –4 (the well will be produced under a constant p_{wf} = 200 psi), or b) Hydraulically fracture the well to create a stimulated reservoir volume (SRV) with properties specified in the table. Assume that all flow goes to the SRV, and the pressure in it (p_{frac}) is 200 psi. Which option are you going to choose and why? Please show all your work.

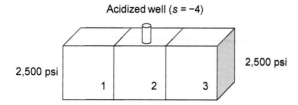

Acidized well (s = –4)

2,500 psi   2,500 psi

1   2   3

Hydraulically fractured well

SRV

2,500 psi   2,500 psi

1   2'   3

p_{frac} = 200 psi

Fig. 10.28.1—Reservoir system of Problem 10.5.25.

	Reservoir properties		SRV properties
k_x =	20 md	k_x =	400 md
$\Delta x = \Delta y$ =	150 ft	Δw_f =	0.3 ft
h =	80 ft	Δy =	150 ft
μ =	1 cp	h =	80 ft
B =	1 res bbl/STB		
r_w =	0.32 ft		
s =	–4		
ρ =	64 lbm/ft³		
p_{wf} =	200 psi		

Table 10.30—Reservoir and SRV properties for Fig. 10.28.1.

Solution to Problem 10.5.25

Because the given reservoir flow is 1D, the governing flow equation for the incompressible fluid flow can be written as

$$\frac{\partial}{\partial x}\left(\frac{A_x k_x}{\mu B}\frac{\partial p}{\partial x}\right)\Delta x + q_{sc} = 0. \quad \ldots\ldots\ldots\ldots\ldots\ldots\ldots\ldots\ldots\ldots\ldots\ldots\ldots\ldots\ldots\ldots\ldots (10.351)$$

a) Acidizing to stimulate the well: Homogeneous in area and permeability, μB = 1; therefore, the equation could be simplified to

$$\frac{\partial^2 p}{\partial x^2} + \frac{q_{sc}}{V_b k} = 0. \quad \ldots\ldots\ldots\ldots\ldots\ldots\ldots\ldots\ldots\ldots\ldots\ldots\ldots\ldots\ldots\ldots\ldots\ldots\ldots (10.352)$$

Expressing Eq. 10.352 in finite-difference form, one obtains

$$p_{i-1} - 2p_i + p_{i+1} + \frac{q_{sc}}{hk} = 0. \quad \ldots\ldots\ldots\ldots\ldots\ldots\ldots\ldots\ldots\ldots\ldots\ldots\ldots\ldots\ldots\ldots (10.353)$$

Because Blocks 1 and 3 are symmetrically located from Block 2 and external pressure conditions, $p_1 = p_3$, for Block 2,

$$\Omega = \frac{2\pi \bar{k} h}{\mu B\left[\ln\left(\frac{r_e}{r_w}\right) + s\right]} = \frac{2\times3.14\times20\ mD\times0.001127\left(\frac{perms}{mD}\right)\times80\ ft}{1\times1\times\left[\ln\left(\frac{0.2\times\Delta x}{0.32}\right) - 4\right]} = \frac{11.324}{0.541} = 20.94. \quad \ldots\ldots (10.354)$$

Writing finite-difference equations in each block,

- Block 1, as seen in **Fig. 10.28.2:**

$$2,500 - 2p_1 + p_2 = 0, \dots\dots\dots\dots\dots\dots\dots\dots\dots\dots\dots\dots\dots\dots (10.355)$$

$$-2p_1 + p_2 = -2,500. \dots\dots\dots\dots\dots\dots\dots\dots\dots\dots\dots\dots\dots\dots (10.356)$$

- Block 2, as seen in **Fig. 10.28.3:**

$$p_1 - 2p_2 + p_1 + \frac{-20.94(p_2 - p_{sf})}{80 \text{ ft} \times 20(\text{mD}) \times 0.001127} = 0, \dots\dots\dots\dots\dots\dots\dots\dots\dots (10.357)$$

$$2p_1 - 2p_2 - 11.61(p_2 - 200) = 0, \dots\dots\dots\dots\dots\dots\dots\dots\dots\dots\dots (10.358)$$

$$2p_1 - 13.61p_2 = -2,322. \dots\dots\dots\dots\dots\dots\dots\dots\dots\dots\dots\dots\dots (10.359)$$

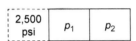

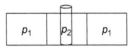

Fig. 10.28.2—Treatment of Block 1.   **Fig. 10.28.3—Treatment of Block 2.**

Solving for p_1 and p_2 simultaneously, $p_2 = 382.4$ psi and $p_1 = 1,441$ psi. Therefore, the flow rate into the wellbore in Block 2 is calculated as

$$q_2 = -\Omega(p_2 - p_{sf}) = -20.94(382.4 - 200) = -3,819.5 \text{ STB/D}. \dots\dots\dots\dots\dots\dots (10.360)$$

b) Hydraulic fracturing to stimulate the well,

$$\frac{\partial}{\partial x}\left(\frac{A_x k_x}{\mu B}\frac{\partial p}{\partial x}\right)\Delta x + q_{sc} = 0. \dots\dots\dots\dots\dots\dots\dots\dots\dots\dots\dots\dots (10.361)$$

Because k_x and Δx change in this problem (note that Δx is equal to Δw_f for the SRV), Eq. 10.361 needs to be written in the following form,

$$\left.\frac{A_x k_x}{\Delta x}\right|_{i+\frac{1}{2}}(p_{i+1} - p_i) - \left.\frac{A_x k_x}{\Delta x}\right|_{i-\frac{1}{2}}(p_i - p_{i-1}) = 0. \dots\dots\dots\dots\dots\dots\dots\dots\dots (10.362)$$

For this problem, because all the flow goes into the SRV, the SRV can be modeled as an additional block with $\Delta x = 0.3$ ft and $k_x = 400$ md. Hence, the well term could be ignored. One would not need to use Peaceman's wellbore model because the sandface and SRV pressures are identical. This simply implies that the volume of fluid entering into the SRV volume is immediately produced by the well (See **Fig. 10.28.4**).

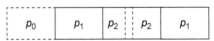

Fig. 10.28.4—Treatment of the SRV.

Block 1:

- Transmissibility between Blocks 0 and 1:

$$T_{0-1} = \left.\frac{A_x k_x}{\Delta x}\right|_{0+\frac{1}{2}} = \frac{A_{x1} k_{x1}}{\Delta x_1} = 1.8 \text{ STB/psia-D}. \dots\dots\dots\dots\dots\dots\dots\dots\dots (10.363)$$

- Transmissibility between Blocks 1 and 2:

$$T_{1-2} = \left.\frac{A_x k_x}{\Delta x}\right|_{1+\frac{1}{2}} = \frac{2A_{x1} k_{x1} A_{x2} k_{x2}}{A_{x2} k_{x2}\Delta x_1 + A_{x1} k_{x1}\Delta x_2}$$

$$= \frac{2(80)(150)(20\times 0.001127)(80)(150)(20\times 0.001127)}{(80)(150)(20\times 0.001127)\times 150 + (80)(150)(20\times 0.001127)\times 74.85} = 2.4 \text{ STB/D-psi.} \dots\dots (10.364)$$

- Transmissibility between Blocks 2 and 3:

$$T_{2-3}\left|\frac{A_x k_x}{\Delta x}\right|_{2+\frac{1}{2}} = \frac{2 A_{x2} k_{x2} A_{x3} k_{x3}}{A_{x3} k_{x3} \Delta x_2 + A_{x2} k_{x2} \Delta x_3}$$

$$= \frac{2(80)(150)(20 \times 0.001127)(80)(150)(400 \times 0.001127)}{(80)(150)(400 \times 0.001127) \times 74.85 + (80)(150)(20 \times 0.001127) \times 0.3} = 7.2 \text{ STB/D-psi.} \quad \ldots \ldots (10.365)$$

Now, one can write the finite-difference approximation at each block,

- At Block 1:

$$2.4(p_2 - p_1) - 1.8(p_1 - 2,500) = 0, \quad \ldots \ldots \ldots \ldots \ldots \ldots \ldots \ldots \ldots \ldots \ldots \ldots \ldots \ldots \ldots (10.366)$$

$$2.4 p_2 - 4.2 p_1 = -4,500. \quad \ldots \ldots \ldots \ldots \ldots \ldots \ldots \ldots \ldots \ldots \ldots \ldots \ldots \ldots \ldots \ldots (10.367)$$

- At Block 2:

$$7.2(p_3 - p_2) - 2.4(p_2 - p_1) = 0, \quad \ldots \ldots \ldots \ldots \ldots \ldots \ldots \ldots \ldots \ldots \ldots \ldots \ldots \ldots (10.368)$$

where $p_3 = 200$ psi $\left(\text{SRV } p_{\text{frac}}\right)$. Therefore, Eq. 10.368 could be written as

$$7.2(200 - p_2) - 2.4(p_2 - p_1) = 0, \quad \ldots \ldots \ldots \ldots \ldots \ldots \ldots \ldots \ldots \ldots \ldots \ldots \ldots (10.369)$$

$$1,440 - 9.6 p_2 + 2.4 p_1 = 0. \quad \ldots \ldots \ldots \ldots \ldots \ldots \ldots \ldots \ldots \ldots \ldots \ldots \ldots \ldots \ldots (10.370)$$

Solving for p_1 and p_2: $p_2 = 487.5$ psia and $p_1 = 1,350$ psia.
Therefore, the flow into Block 3 can be calculated as:

$$q_3 = 2 \times 7.2(p_2 - 200) = 2 \times 7.2 \times (487.5 - 200) = 4,140 \text{ STB/D.} \quad \ldots \ldots \ldots \ldots \ldots \ldots (10.371)$$

The production from the SRV (Block 3) is equal to –4,140 STB/D. The flow rate obtained in Case (b) = –4,140 STB/D, which is higher than Case (a) = –3,819 STB/D. Hence, it is better to hydraulically fracture the well than to acidize it.

Problem 10.5.26

Consider the 1D, single-phase flow problem shown in **Fig. 10.29.** All boundaries are sealed. The well is set to produce at 1,650 STB/D; however, it is reported that there was an error with the flowmeter at the wellhead. A pressure survey was conducted at the well at the end of 30 days of production, and sandface pressure was measured at 50 psia. Your task as a reservoir engineer is to verify whether the flowmeter needs to be recalibrated. The initial pressure of Block 1 is known to be 3,000 psia and other reservoir properties are listed in **Table 10.31.**

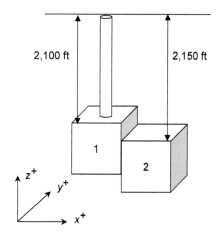

Fig. 10.29—Reservoir system of Problem 10.5.26.

$k =$ 50 md	
$\Delta x = \Delta y =$ 200 ft	
$\phi =$ 12%	
$c =$ 5×10⁻⁵ psi⁻¹	
$r_w =$ 0.25 ft	
$h =$ 200 ft	
$\rho =$ 45 lbm/ft³	
$\mu \approx$ const = 1.2 cp	
$B \approx$ const = 1 res bbl/STB	
$s =$ 0	
$\Delta t =$ 30 days	

Table 10.31—Reservoir properties for Fig. 10.29.

Solution to Problem 10.5.26

The governing equation for the 1D slightly compressible reservoir with depth gradients is

$$\frac{\partial}{\partial x}\left(\frac{A_x k_x}{\mu B}\frac{\partial p}{\partial x}\right)\Delta x - \frac{\partial}{\partial x}\left(\frac{1}{144}\frac{g}{g_c}\rho\frac{A_x k_x}{\mu B}\frac{\partial G}{\partial x}\right)\Delta x + q = \frac{V_b \phi c_t}{5.615}\left(\frac{\partial p}{\partial t}\right).\qquad(10.371)$$

The finite-difference approximation can be written as

$$\left.\frac{A_x k_x}{\mu B \Delta x}\right|_{i+\frac{1}{2}}^{n}\left(p_{i+1}^{n+1} - p_i^{n+1}\right) - \left.\frac{A_x k_x}{\mu B \Delta x}\right|_{i-\frac{1}{2}}^{n}\left(p_i^{n+1} - p_{i-1}^{n+1}\right) - \frac{1}{144}\frac{g}{g_c}\rho\left[\left.\frac{A_x k_x}{\mu B \Delta x}\right|_{i+\frac{1}{2}}^{n}\left(G_{i+1} - G_i\right) - \left.\frac{A_x k_x}{\mu B \Delta x}\right|_{i-\frac{1}{2}}^{n}\left(G_i - G_{i-1}\right)\right] \qquad(10.372)$$
$$+ q = \frac{V_b \phi c_t}{5.615 \Delta t}\left(p_i^{n+1} - p_i^{n}\right).$$

The transmissibility factors are calculated as

$$\left.\frac{A_x k_x}{\mu B \Delta x}\right|_{i+\frac{1}{2}}^{n} = \left.\frac{A_x k_x}{\mu B \Delta x}\right|_{i-\frac{1}{2}}^{n} = T = \frac{hk}{\mu B} = \frac{(200)(50)(1.127\times10^{-3})}{(1.2)(1)} = 9.39 \text{ STB/D-psi},\qquad(10.373)$$

and coefficient Γ is

$$\Gamma = \frac{V_b \phi c}{5.615 \Delta t} = \frac{(200)(200)(200)(0.12)(5\times10^{-5})}{(5.615)(30)} = 0.285 \text{ STB/D-psi},\qquad(10.374)$$

$$\gamma = \frac{\rho}{144} = \frac{45}{144} = 0.3125 \text{ psi/ft}.\qquad(10.375)$$

Using the Peaceman model, $r_e = 0.2\Delta x = 40$ ft, and the productivity index,

$$\Omega = \frac{2\pi k h}{\mu B\left[\ln\left(\frac{r_e}{r_w}\right)+s\right]} = \frac{2(3.14)(50)(1.127\times10^{-3})(200)}{(1.2)(1)\left[\ln\left(\frac{40}{0.25}\right)+0\right]} = 11.62 \text{ STB/D-psi}.\qquad(10.376)$$

- Block 1:

$$Tp_2 - (T + \Gamma + \Omega)p_1 = -\Gamma p_{i1} - \Omega p_{sf} - T\gamma(G_1 - G_2),\qquad(10.377)$$

$$9.39 p_2 - (9.39 + 0.285 + 11.62)p_1 = -0.285\times3,000 - 11.62\times50 - 9.39\times0.3215(2,100 - 2,150),\quad(10.378)$$

$$9.39 p_2 - 21.295 p_1 = -1,289.28.\qquad(10.379)$$

- Block 2:

$$Tp_1 - (T + \Gamma)p_2 = -\Gamma p_{i2} - T\gamma(G_2 - G_1),\qquad(10.380)$$

$$9.39 p_1 - 9.675 p_2 = -0.285(3,000 + 0.3215\times50) - 9.39\times0.3215(2,150 - 2,100),\qquad(10.381)$$

$$9.39 p_1 - 9.675 p_2 = -1,006.17.\qquad(10.382)$$

Solving for pressures, $p_1 = 186$ psia and $p_2 = 284.53$ psi, the calculated flow rate would be

$$q_1 = -\Omega\left(p_1 - p_{sf}\right) = -11.621(186 - 50) = -1,580.32 \text{ STB/D}.\qquad(10.383)$$

Thus, the flowmeter was wrong and needs to be recalibrated.

Problem 10.5.27

Consider the fluid flow in a 1D reservoir shown in **Fig. 10.30.** The boundaries of the system are completely sealed. The reservoir has homogeneous permeability distribution. Depth gradients in the system cannot be ignored. The injection well located in Block 2 is operating at a constant injection rate of 100 STB/D. Using the additional reservoir properties found in **Table 10.32,** determine the sandface pressure at the injection well so that it is possible to produce fluid from the production well in Block 1 without installing any downhole pumps.

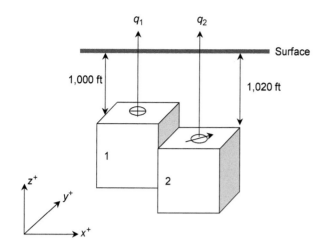

Table 10.32—Reservoir properties for Fig. 10.30.

$\Delta x = \Delta y = 500$ ft
$h = 50$ ft
$k_1 = k_2 = 100$ md
$B = 1$ res bbl/STB
$\mu = 1$ cp
$\phi = 0.2$
$c_f = 0$ psi^{-1}
$\rho = 62.4$ lbm/ft^3
Skin factor $= 0$
$r_w = 0.25$ ft

Fig. 10.30—Reservoir system of Problem 10.5.27.

Solution to Problem 10.5.27

Calculating the hydrostatic gradient in the wellbore,

$$\gamma = \frac{\rho}{144} = 0.433 \text{ psi / ft}, \quad\dots\dots\dots\dots\dots\dots\dots\dots\dots\dots\dots\dots\dots\dots\dots\dots\dots \text{(10.384)}$$

and calculating the sandface pressure of the well in Block 1 (when the wellbore is completely filled with the reservoir fluid),

$$p_{sf1} = p_{surface} + \gamma G_1 = 14.7 + (0.433)\left(1{,}000 + \frac{50}{2}\right) = 458.525 \text{ psi}. \quad\dots\dots\dots\dots\dots\dots\dots\dots\dots \text{(10.385)}$$

Now, one can calculate the block pressures:

$$r_e = 0.2\Delta x = 100 \text{ ft}, \quad\dots\dots\dots\dots\dots\dots\dots\dots\dots\dots\dots\dots\dots\dots\dots\dots\dots\dots\dots\dots\dots \text{(10.386)}$$

$$\Omega = \frac{2\pi kh}{\mu B\left[\ln\left(\dfrac{r_e}{r_w}\right) + s\right]} = \frac{(2)(3.14)(100)\left(1.127 \times 10^{-3}\right)(50)}{(1)(1)\left[\ln\left(\dfrac{100}{0.25}\right) + 0\right]} = 5.906 \text{ STB/D} - \text{psi}, \quad\dots\dots\dots\dots\dots \text{(10.387)}$$

$$T = \frac{hk}{\mu B} = \frac{(50)(100)\left(1.127 \times 10^{-3}\right)}{(1)(1)} = 5.635 \text{ STB/D} - \text{psi}. \quad\dots\dots\dots\dots\dots\dots\dots\dots\dots\dots \text{(10.388)}$$

- At Block 1:

$$p_2 - \gamma G_1 - \left(1 + \frac{\Omega}{T}\right)p_1 + \gamma G_2 = -\frac{\Omega p_{sf1}}{T}, \quad\dots\dots\dots\dots\dots\dots\dots\dots\dots\dots\dots\dots\dots \text{(10.389)}$$

$$p_2 - \left(1 + \frac{\Omega}{T}\right)p_1 = \gamma\left(G_1 - G_2\right) - \frac{\Omega p_{sf1}}{T}, \quad\dots\dots\dots\dots\dots\dots\dots\dots\dots\dots\dots\dots\dots \text{(10.390)}$$

$$p_2 - 2.05 p_1 = -471.92. \quad\dots\dots\dots\dots\dots\dots\dots\dots\dots\dots\dots\dots\dots\dots\dots\dots\dots\dots\dots \text{(10.391)}$$

- At Block 2:

$$p_1 - p_2 = \gamma\left(G_2 - G_1\right) - \frac{q_2}{T}, \quad\dots\dots\dots\dots\dots\dots\dots\dots\dots\dots\dots\dots\dots\dots\dots\dots\dots \text{(10.392)}$$

$$p_1 - p_2 = -26.406. \quad\dots\dots\dots\dots\dots\dots\dots\dots\dots\dots\dots\dots\dots\dots\dots\dots\dots\dots\dots\dots \text{(10.393)}$$

Solving the two equation for the two unknowns, $p_1 = 474.6$ psi and $p_2 = 501$ psi. Calculate the sandface pressure of the well in Block 2:

$$p_{sf2} = p_2 + \frac{q}{\Omega} = 501 + \frac{100}{5.906} = 517.9 \text{ psi}. \quad\dots\dots\dots\dots\dots\dots\dots\dots\dots\dots\dots\dots\dots \text{(10.394)}$$

Problem 10.5.28

Consider a single-phase slightly compressible fluid flow in the reservoir shown in **Fig. 10.31.** All boundaries are closed. The permeability distribution of the reservoir is homogeneous. The initial pressure is known to be 1,000 psia. There are plans to install a pump in the wellbore for testing purposes for a period of one week. The flow rate will be specified as 2,000 STB/D. A production engineer plans to install the pump in the wellbore at a depth of 2,000 ft from the surface; however, another production engineer disagrees and proposes to install the pump at a depth of 2,700 ft or deeper. As a reservoir engineer, what would be your suggestion on the installation depth of the pump? Use the reservoir properties found in **Table 10.33** to determine your answer.

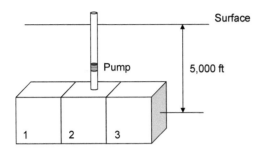

Fig. 10.31—Reservoir system of Problem 10.5.28.

$\mu = 1.2$ cp
$B = 1.0$ res bbl/STB
$k = 500$ md (uniform)
$\Delta x = \Delta y = \Delta z = 200$ ft (uniform)
$\phi = 0.20$ (uniform)
$r_w = 0.25$ ft
$c = 5 \times 10^{-5}$ psi^{-1}
$\rho = 40$ lbm/ft^3

Table 10.33—Reservoir properties for Fig. 10.31.

Solution to Problem 10.5.28

The governing equation describing the flow problem in a homogeneous porous medium can be written as

$$\frac{\partial^2 p}{\partial x^2} + \frac{q\mu B}{k_x V_b} = \frac{\mu B c_t \phi}{5.615 k_x} \frac{\partial p}{\partial t}. \qquad (10.395)$$

The characteristic finite-difference equation can be written as

$$p_{i-1}^{n+1} - 2p_i^{n+1} + p_{i+1}^{n+1} + \left(\frac{q\mu B}{hk_x}\right)_i = \left[\frac{\mu B c_t \phi (\Delta x)^2}{k_x 5.615 \Delta t}\right]\left(p_i^{n+1} - p_i^{n}\right). \qquad (10.396)$$

The coefficient on the right-hand side of the equation is calculated as

$$\Gamma = \frac{\mu B (\Delta x)^2 \phi c}{5.615 \Delta t k} = \frac{(1.2)(1)(200)(200)(0.2)\left(5 \times 10^{-5}\right)}{5.615(7)(500)\left(1.127 \times 10^{-3}\right)} = 0.0217. \qquad (10.397)$$

Because Blocks 1 and 3 are symmetrical, $p_1 = p_3$,

- At Block 1:

$$p_2 - (1+\Gamma)p_1 = -\Gamma p_i, \qquad (10.398)$$

$$p_2 - 1.0217 p_1 = -21.7. \qquad (10.399)$$

- At Block 2:

$$2p_1 - (2+\Gamma)p_2 = -\Gamma p_i - \frac{q\mu B}{hk}, \qquad (10.400)$$

$$2p_1 - 2.0217 p_2 = -0.4045. \qquad (10.401)$$

Solving for pressure distribution, $p_1 = p_3 = 675.23$ psi and $p_2 = 668.18$ psi,

$$r_e = 0.2\Delta x = 40 \text{ ft}, \qquad (10.402)$$

$$\Omega = \frac{2\pi kh}{\mu B\left[\ln\left(\dfrac{r_e}{r_w}\right) + s\right]} = \frac{2(3.14)(200)(500)\left(1.127 \times 10^{-3}\right)}{(1.2)(1)\left[\ln\left(\dfrac{40}{0.25}\right) + 0\right]} = 116.212 \text{ STB/D} - \text{psi}, \qquad (10.403)$$

$$p_{sf} = p_2 - \frac{q}{-\Omega} = 668.18 + \frac{-2,000}{116.212} = 650.97 \text{ psi.} \quad \dots\dots\dots\dots\dots\dots\dots\dots\dots\dots \quad (10.404)$$

The height of the fluid column is

$$h = \frac{p_{sf} - 14.7}{\dfrac{\rho}{144}} = \frac{650.97 - 14.7}{\dfrac{40}{144}} = 2,290.57 \text{ ft.} \quad \dots\dots\dots\dots\dots\dots\dots\dots\dots\dots \quad (10.405)$$

The depth from the surface to install the pump is

$$G = 5,000 - h = (5,000 - 2,290.572) = 2,709.43 \text{ ft.} \quad \dots\dots\dots\dots\dots\dots\dots\dots\dots \quad (10.406)$$

To produce the well at the desired rate of 2,000 STB/D or more, a pump should be installed below this depth. Thus, the second engineer was right in his recommendation.

Problem 10.5.29

Consider a 1D single-phase, slightly compressible flow in the reservoir shown in **Fig. 10.32,** with reservoir properties detailed in **Table 10.34.** The permeability distribution is homogeneous. All boundaries are completely sealed. The thickness of the reservoir is uniform, and there is no depth gradient. The initial pressure is known to be 3,500 psia. A well was drilled and completed in Block 1. The well was put on production at a constant sandface pressure of 14.7 psia for 30 days.

1. According to the development plan, drilling a second well in Block 2 is being considered, if the block pressure is larger than 90% of the initial pressure after 30 days of production from the well located in Block 1. Provide a quantitative analysis on whether or not the second well should be drilled in Block 2.
2. Consider the possibility that the second well is drilled in Block 2 and both wells are put on production for another 30 days. Conduct an economical analysis to determine whether or not it would be possible to recover the cost of the second well after 30 days of production. Would this option be more profitable than producing only from the well in Block 1 (without drilling the second well)? A sandface pressure of 14.7 psia is maintained in both wells.

Assume that a drilling and completion cost for the second well in Block 2 is $500,000 and the oil price is $80/bbl. Both wells are identical. Base your economical analysis on using the stabilized flow rate encountered at the end of 60 days.

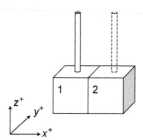

Fig. 10.32—Reservoir system of Problem 10.5.29.

$h = 200$ ft
$\Delta x = \Delta y = 200$ ft
$\mu = 1.2$ cp
$k = 0.55$ md
$\phi = 10\%$
$c_t = 5 \times 10^{-5}$ psi^{-1}
$B = 1.1$ res bbl/STB
$r_w = 0.32$ ft
$s = 0$

Table 10.34—Reservoir properties for Fig. 10.32.

Solution to Problem 10.5.29

1. The governing equation describing the flow problem in a homogeneous porous medium can be written as

$$\frac{\partial^2 p}{\partial x^2} + \frac{q\mu B}{k_x V_b} = \frac{\mu B c_t \phi}{5.615 k_x} \frac{\partial p}{\partial t} \quad \dots\dots\dots\dots\dots\dots\dots\dots\dots\dots\dots\dots\dots\dots\dots\dots \quad (10.407)$$

The characteristic finite-difference equation can be written as

$$p_{i-1}^{n+1} - 2p_i^{n+1} + p_{i+1}^{n+1} + \left(\frac{q\mu B}{hk_x}\right)_i = \left(\frac{\mu B c_t \phi (\Delta x)^2}{k_x 5.615 \Delta t}\right)(p_i^{n+1} - p_i^n). \quad \dots\dots\dots\dots\dots \quad (10.408)$$

The equivalent wellblock radius is calculated as

$$r_e = 0.2\Delta x = 40 \text{ ft,} \quad \dots\dots\dots\dots\dots\dots\dots\dots\dots\dots\dots\dots\dots\dots\dots\dots\dots\dots\dots\dots \quad (10.409)$$

$$\Omega' = \Omega\left(\frac{\mu B}{hk_x}\right) = \frac{2\pi}{\left[\ln\left(\dfrac{r_e}{r_w}\right) + s\right]} = \frac{2(3.14)}{\left[\ln\left(\dfrac{40}{0.32}\right) + 0\right]} = 1.3, \quad\dotfill \quad (10.410)$$

$$\Gamma = \frac{\mu B}{k}\frac{\Delta x \Delta y \phi c_t}{5.615\Delta t} = \frac{(1.2)(1.1)(200)(200)(0.1)\left(5\times10^{-5}\right)}{(0.55)(1.127\times10^{-3})(5.615)(30)} = 2.5284. \quad\dotfill \quad (10.411)$$

- At Block 1:

$$p_2 - \left(1 + \Gamma + \Omega'\right)p_1 = -\Gamma p_i - \Omega' p_{sf1}, \quad\dotfill \quad (10.412)$$

$$p_2 - 4.8284p_1 = -8,868.51. \quad\dotfill \quad (10.413)$$

- At Block 2:

$$p_1 - \left(1 + \Gamma\right)p_2 = -\Gamma p_i, \quad\dotfill \quad (10.414)$$

$$p_1 - 3.5284p_2 = -8,849.4. \quad\dotfill \quad (10.415)$$

Solving two equations in two unknowns, $p_1 = 2,503.1$ psi and $p_2 = 3,217.5$ psi.
Because,

$$\frac{p_2}{p_i} = \frac{3,217.5}{3,500} = 0.919 > 90\%; \quad\dotfill \quad (10.416)$$

therefore, the second well can be drilled in Block 2.
2. If the well in Block 2 is drilled and put on production at the end of 30 days, we can calculate the new pressure distribution at the end of 60 days.

- At Block 1:

$$p_2 - \left(1 + \Gamma + \Omega'\right)p_1 = -\Gamma p_1^n - \Omega' p_{sf1}, \quad\dotfill \quad (10.417)$$

$$p_2 - 4.8284p_1 = -6,347.95. \quad\dotfill \quad (10.418)$$

- At Block 2

$$p_1 - \left(1 + \Gamma + \Omega\right)p_2 = -\Gamma p_2^n - \Omega p_{sf2}, \quad\dotfill \quad (10.419)$$

$$p_1 - 4.8284p_2 = -8,154.237. \quad\dotfill \quad (10.420)$$

Solving the system of equations, one obtains $p_1 = 1,739.1$ psi and $p_2 = 2,049$ psi,

$$\Omega = \frac{\Omega'kh}{\mu B} = \frac{(1.3)(0.55)\left(1.127\times10^{-3}\right)(200)}{(1.2)(1.1)} = 0.122 \text{ STB/D} - \text{psi}, \quad\dotfill \quad (10.421)$$

$$\text{Well 2 Production} = \Omega\left(p_2^{n+1} - p_{sf}\right)\Delta t = (0.122)(2,049 - 14.7)(30) = 7,745.538 \text{ STB}, \quad\dotfill \quad (10.422)$$

$$\text{Revenue} = 80 \times \$7,745.538 = \$619,643.04 > \$500,000. \quad\dotfill \quad (10.423)$$

Accordingly, Well 2 can cover the cost of the drilling in 30 days. Then, one can calculate the total production rate during the period of the first 60 days of production from the reservoir:

- From 0 to 30 days (Well 2 is not producing):

Well 1 production,

$$q_{0-30} = q_1 = -\Omega_1\left(p_1^{30} - p_{sf_1}^{30}\right) = -0.122 \times (2,503.1 - 14.7) = -303.58 \text{ STB/D}, \quad\dotfill \quad (10.424)$$

$$Q_{\text{total}_1}^{30} = 303.58 \times 30 = 9,107.54 \text{ STB}. \quad\dotfill \quad (10.425)$$

- From 30 to 60 days (if Well 2 is put on production):

$$q_1^{60} = -\Omega_1\left(p_1^{60} - p_{sf_1}^{60}\right) = 210.38 \text{ STB/D}, \dots\dots\dots\dots\dots\dots\dots\dots\dots\dots (10.426)$$

$$q_2^{60} = -\Omega_2\left(p_2^{60} - p_{sf_2}^{60}\right) = 248.18 \text{ STB/D}, \dots\dots\dots\dots\dots\dots\dots\dots\dots\dots (10.427)$$

$$Q_{total_2}^{30-60} = (q_1^{60} + q_2^{60}) \times 30 = 13,756.84 \text{ STB}. \dots\dots\dots\dots\dots\dots\dots\dots\dots (10.428)$$

Therefore, the total production from the reservoir is

$$Q_{total} = Q_{total_1}^{0-30} + Q_{total_2}^{30-60} = 22,864.39 \text{ STB}. \dots\dots\dots\dots\dots\dots\dots\dots\dots (10.429)$$

To the calculate the total revenue at the end of 60 days, it is necessary to subtract the cost of drilling the second well:

$$\text{Total Revenue} = \$22,864.39 \times 80 - 500,000 = \$1,329,151. \dots\dots\dots\dots\dots\dots\dots\dots (10.430)$$

- From 30 to 60 days (if Well 2 is not drilled):
  For another 30 days, if the well in Block 2 is not drilled,

  ○ At Block 1:

$$p_2 - (1 + \Gamma + \Omega')p_1 = -\Gamma p_1^n - \Omega' p_{sf}, \dots\dots\dots\dots\dots\dots\dots\dots\dots\dots (10.431)$$

$$p_2 - 4.8284 p_1 = -6,347.95. \dots\dots\dots\dots\dots\dots\dots\dots\dots\dots\dots\dots (10.432)$$

  ○ At Block 2:

$$p_1 - (1 + \Gamma)p_2 = -\Gamma p_2^n, \dots\dots\dots\dots\dots\dots\dots\dots\dots\dots\dots\dots (10.433)$$

$$p_1 - 3.5284 p_2 = -8,135.127. \dots\dots\dots\dots\dots\dots\dots\dots\dots\dots\dots\dots (10.434)$$

Solving for the unknown block pressures, $p_1 = 1,904$ psi and $p_2 = 2,845$ psi:

$$q_1^{60} = -\Omega_1\left(p_1^{60} - p_{sf_1}^{60}\right) = 230.49 \text{ STB/D}, \dots\dots\dots\dots\dots\dots\dots\dots\dots\dots (10.435)$$

$$Q_{total_1}^{30-60} = q_1^{60} \times 30 = 6,914.84 \text{ STB}, \dots\dots\dots\dots\dots\dots\dots\dots\dots\dots\dots (10.436)$$

$$Q_{total} = Q_{total_1}^{0-30} + Q_{total_1}^{30-60} = 16,022.38 \text{ STB}, \dots\dots\dots\dots\dots\dots\dots\dots\dots (10.437)$$

$$\text{Total Revenue} = 16,022.382 \times 80 = \$1,281,790.56. \dots\dots\dots\dots\dots\dots\dots\dots\dots (10.438)$$

Therefore, drilling Well 2 is economically feasible.

Problem 10.5.30

You are working as a reservoir engineer at an exploration and production company. Your company is planning to acquire a new small field that has been on production for only 30 days. The current owner of the field claims that the field contains a crude with 38 °API and produced at 500 STB/D for the entire month from a well located in Block 3. During that period, the well was flowing as an artesian well. Storage-inventory stubs show that the field has indeed produced for 30 days at 500 STB/D. Your company surveyed the wellhead pressure at the producing well and found it to be 700 psia at the end of 30 days. In addition to the producing well, this field also has an observation well located in Block 2. Pressure in the well located in Block 2 after 30 days was measured as 2,800 psia, again by your company personnel.

Your company looks at a nearby crude-oil refinery as a potential buyer of the crude from this field to minimize the associated transportation costs. Upon investigation, your company found that the refinery can process the crude oil within a gravity range of 35 to 40 °API. Your company has determined that the field will be a good investment, provided all the data represented by the current owner is accurate. As a reservoir engineer, perform an analysis to ensure the feasibility of the project and make a final recommendation towards the purchase of the field. Use **Fig. 10.33.1** and **Table 10.35** to make an informed decision.

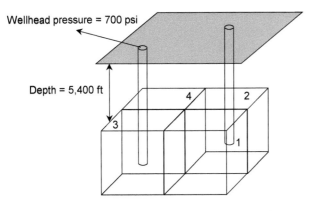

Wellhead pressure = 700 psi

Depth = 5,400 ft

Fig. 10.33.1—Reservoir system of Problem 10.5.30.

$p_i =$	2,900 psi
$k_x = k_y =$	100 md
$\Delta x = \Delta y =$	500 ft
$\phi =$	25%
$c =$	2×10^{-5} psi^{-1} (const)
$s =$	0
$h =$	100 ft
$r_w =$	0.25 ft
$B =$	1 res bbl/STB (const)
$\mu =$	1 cp (const)

Table 10.35—Reservoir properties for the small field-production depicted in Fig. 10.33.1.

Solution to Problem 10.5.30

Your company measures wellhead pressure (700 psia) for the producing well and the observation wellblock pressure in Block 2 is 2,800 psia at the end of 30 days. Flow-rate records indicate that the field has actually produced at a rate of 500 STB/D. Initial reservoir pressure is known to be 2,900 psia. Let us calculate the bottomhole pressure and °API gravity for the given reservoir system and make a decision on that basis. In this system, pressure in Block 1 will be the same as pressure in Block 4 because of existing symmetry.

First, let us represent our system with a PDE. It is a 2D reservoir with no-flow boundaries and no depth gradients, a homogeneous and isotropic system. Therefore, the flow equation can be written as

$$\frac{\partial^2 p}{\partial x^2} + \frac{\partial^2 p}{\partial y^2} + \frac{q\mu B}{A_x k_x \Delta x} = \frac{c_t \phi}{5.615} \frac{\mu B}{k_x} \frac{\partial p}{\partial t}. \quad \dots \quad (10.439)$$

Writing the finite-difference analog of the PDE we will have

$$\frac{p_{i-1,j}^{n+1} - 2p_{i,j}^{n+1} + p_{i+1,j}^{n+1}}{\Delta x^2} + \frac{p_{i,j-1}^{n+1} - 2p_{i,j}^{n+1} + p_{i,j+1}^{n+1}}{\Delta y^2} + \frac{\mu B}{\Delta y} \frac{q}{h\,k_x \Delta x} = \frac{c_t \phi}{5.615} \frac{\mu B}{k_x} \frac{\left(p_{i,j}^{n+1} - p_{i,j}^{n}\right)}{\Delta t}. \quad \dots \quad (10.440)$$

Simplifying the equation for a homogeneous and isotropic reservoir system represented with uniform blocks ($\Delta x = \Delta y$),

$$p_{i-1,j}^{n+1} + p_{i,j-1}^{n+1} - 4p_{i,j}^{n+1} + p_{i,j+1}^{n+1} + p_{i+1,j}^{n+1} + \frac{q\mu B}{hk_x} = \frac{\Delta x^2 c_t \phi}{5.615} \frac{\mu B}{k_x} \frac{\left(p_{i,j}^{n+1} - p_{i,j}^{n}\right)}{\Delta t}. \quad \dots \quad (10.441)$$

- At Block 1, shown in **Fig. 10.33.2:**

$$p_1 + p_1 - 4p_1 + p_2 + p_3 = \frac{500^2 \times 2 \times 10^{-5} \times 0.25 \times 1 \times 1}{5.615 \times 0.001127 \times 100 \times 30}\left(p_1 - 2{,}900\right). \quad \dots \quad (10.442)$$

Note that from the observation well measurement, $p_2 = p_{sf2}$:

$$-2.066\ p_1 + p_3 = -2{,}990.95. \quad \dots \quad (10.443)$$

- At Block 3 (we already know the pressure in Block 2), shown in **Fig. 10.33.3:**

$$p_3 + p_3 - 4p_3 + p_1 + p_1 - \frac{1 \times 1 \times 500}{0.001127 \times 100 \times 100} = 0.06584\left(p_3 - 2{,}900\right), \quad \dots \quad (10.444)$$

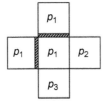

p_1

p_1 p_1 p_2

p_3

Fig. 10.33.2—Treatment of boundary conditions of Block 1.

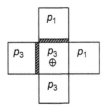

p_1

p_3 p_3 p_1

p_3

Fig. 10.33.3—Treatment of boundary conditions of Block 3.

which is

$$2p_1 - 2.066p_3 = -147.034. \quad \dots \dots \dots \dots \dots \dots \dots \dots \dots \dots \dots \dots \dots \dots (10.445)$$

Now, solve for p_1 and p_3 from the preceding two equations. These two equations in two unknowns, will yield $p_1 = 2,788.9$ psi and $p_3 = 2,771.0$ psi. Now, one can calculate the sandface pressure for the well located in Block 3 using the wellbore equation:

$$q_3 = -\frac{2\pi \bar{k} h}{\mu B \left(\ln \frac{r_e}{r_w} + s \right)} \left(p - p_{sf} \right). \quad \dots \dots \dots \dots \dots \dots \dots \dots \dots \dots \dots \dots (10.446)$$

Substituting the data for Block 3,

$$-500 = -\frac{2\pi \times 0.001127 \times 100 \times 100}{1 \times 1 \times \left(\ln \frac{0.2 \times 500}{0.25} + 0 \right)} \left(2,771 - p_{3,sf} \right). \quad \dots \dots \dots \dots \dots \dots \dots (10.447)$$

The sandface pressure of the well can be solved: $42.30 = 2,771 - p_{3,sf}$. Resulting in $p_{3,sf} = 2,728.7$ psi. Now, we know that

$$p_{\text{wellhead}} + \frac{\rho_o}{144} \frac{g}{g_c} h = p_{3,sf}, \quad \dots \dots \dots \dots \dots \dots \dots \dots \dots \dots \dots \dots (10.448)$$

or

$$700 + \frac{\rho_o}{144} 5,400 = 2,728.7, \quad \dots \dots \dots \dots \dots \dots \dots \dots \dots \dots \dots \dots (10.449)$$

or

$$\rho_o = \frac{2,728.7 - 700}{5,400} 144 = 54.1 \text{ lbm/ft}^3, \quad \dots \dots \dots \dots \dots \dots \dots \dots \dots (10.450)$$

$$\gamma_o = \frac{54.1}{62.4} = 0.827. \quad \dots \dots \dots \dots \dots \dots \dots \dots \dots \dots \dots \dots (10.451)$$

Calculating °API gravity:

$$°\text{API} = \frac{141.5}{\gamma_o} - 131.5 = \frac{141.5}{0.867} - 131.5 = 31.7. \quad \dots \dots \dots \dots \dots \dots \dots (10.452)$$

Accordingly, it would not be prudent to recommend your company invest in this project, because the °API gravity of the crude oil is lower than the suggested limit of 35 to 40 °API.

Problem 10.5.31

A hydraulic fracturing project is planned for the single-phase incompressible fluid flow homogeneous reservoir with wells shown in **Fig. 10.34.** The horizontal well is perforated in Block 3. The vertical well is perforated *only* in Block 4 to collect the injected fluid (assuming 100% flowback). The wellbore of the vertical well is open to atmosphere. All outer boundaries are closed. On the basis of the petrophysical report provided by the laboratory, the fracturing pressure of the reservoir rock is 4,000 psia. The current goal of the project is to fracture the rock *only* in Block 3. The fracturing fluid used in this project is $5/STB. As a reservoir engineer on the project, please report on the following items with your detailed analysis and accompanying calculations:

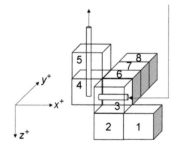

Fig. 10.34—Reservoir system for Problem 10.5.31.

1. The minimum fracturing fluid cost of the project.
2. Pressure distribution in all blocks (using the minimum amount of fracturing fluid).
3. The sandface pressure of the horizontal well (using the minimum amount of fracturing fluid).

The reservoir and fluid properties are shown in **Table 10.36** (assume the properties of the reservoir fluid are the same as the injected fracturing fluid). Assume North is the same as the $y+$-direction shown in Fig. 10.34.

Solution to Problem 10.5.31

All blocks other than 3, 6, and 4 are inactive blocks. Assuming the flow rate is q,

- For Block 3: $N_3 p_6 - N_3 p_3 + q = 0.$ ....(10.453)
- For Block 4: $E_4 p_6 - (E_4 + \Omega_4) p_4 = -\Omega_4 p_{sf4}.$ ....(10.454)
- For Block 6: $S_6 p_3 + W_6 p_4 - (S_6 + W_6) p_6 = 0,$ ....(10.455)

where

$$W = E = 0.7245 \text{ STB}/\text{D}-\text{psi}, \dots\dots\dots\dots\dots (10.456)$$

$$N = S = 3.22 \text{ STB}/\text{D}-\text{psi}, \dots\dots\dots\dots\dots (10.457)$$

For the vertical well,

$$r_{e4} = \frac{0.28\sqrt{\left(\frac{k_x}{k_y}\right)^{\frac{1}{2}}(\Delta y)^2 + \left(\frac{k_y}{k_x}\right)^{\frac{1}{2}}(\Delta x)^2}}{\left(\frac{k_x}{k_y}\right)^{\frac{1}{4}} + \left(\frac{k_y}{k_x}\right)^{\frac{1}{4}}} = 75.9355 \text{ ft}, \dots\dots (10.458)$$

$$\Omega_4 = \frac{2\pi\sqrt{k_x k_y}\,h}{\mu B\left(\ln\frac{r_{e4}}{r_w} + s\right)} = 1.3876 \text{ STB/D}-\text{psi}. \dots\dots\dots\dots\dots\dots\dots\dots\dots\dots\dots (10.459)$$

μ = 0.7 cp	
B = 1.0 res bbl/STB	
ρ = 65.88 lbm/ft³	
h = 60 ft	
r_w = 0.25 ft	
s = 1.2	
Δx = 400 ft	
Δy = 300 ft	
k_y = 25 md	
k_x = 10 md	
k_z = 15 md	

Table 10.36—Reservoir and fluid properties for Fig. 10.34.

Because the sandface pressure of the vertical well is $p_{sf} = 14.7$ psi, solving the system of equations one obtains $q = 1,652.6$ STB/D, $p_6 = 3,486.8$ psi, and $p_4 = 1,205.7$ psi. Thus, the cost is

$$\text{Cost} = q \times \$5 \text{ STB/D} = \$ 8,263.1. \dots\dots\dots\dots\dots\dots\dots\dots\dots\dots (10.460)$$

The static hydraulic pressure gradient is

$$\gamma = \frac{\rho}{144} = \frac{65.88}{144} = 0.4575 \text{ psi/ft}. \dots\dots\dots\dots\dots\dots\dots\dots\dots\dots\dots (10.461)$$

And the pressure for each block is

$$p_2 = p_3 + \gamma h = 4,027.5 \text{ psi}, \dots\dots\dots\dots\dots\dots\dots\dots\dots (10.462)$$

$$p_1 = p_2 = 4,027.5 \text{ psi}, \dots\dots\dots\dots\dots\dots\dots\dots\dots\dots (10.463)$$

$$p_5 = p_4 - \gamma h = 1,178.3 \text{ psi}, \dots\dots\dots\dots\dots\dots\dots\dots\dots (10.464)$$

$$p_7 = p_6 = 3,486.8 \text{ psi}, \dots\dots\dots\dots\dots\dots\dots\dots\dots\dots (10.465)$$

$$p_8 = p_7 = 3,486.8 \text{ psi}. \dots\dots\dots\dots\dots\dots\dots\dots\dots\dots (10.466)$$

For the horizontal well,

$$r_{e3} = \frac{0.28\sqrt{\left(\frac{k_x}{k_y}\right)^{\frac{1}{2}}(\Delta y)^2 + \left(\frac{k_y}{k_x}\right)^{\frac{1}{2}}(\Delta x)^2}}{\left(\frac{k_x}{k_y}\right)^{\frac{1}{4}} + \left(\frac{k_y}{k_x}\right)^{\frac{1}{4}}} = 80.5137 \text{ ft}, \dots\dots\dots\dots\dots (10.467)$$

$$\Omega_3 = \frac{2\pi\sqrt{k_x k_y}\,h}{\mu B \left(\ln \frac{r_{e3}}{r_w} + s \right)} = 50.1977 \text{ STB/D} - \text{psi}. \dots\dots\dots\dots\dots\dots\dots\dots\dots\dots\dots\dots\dots (10.468)$$

Thus, the sandface pressure of the horizontal injector is $p_{sf3} = 4{,}032.9$ psi.

Problem 10.5.32

Incompressible Flow Simulation Project. A wastewater-filtration project is planned to be implemented in a reservoir with ten existing wells. The spatial distributions of the reservoir properties are shown in **Figs. 10.35.1 through 10.35.3** Note that the permeability along the North/East direction is 2X larger than that of the permeability along the South/West direction.

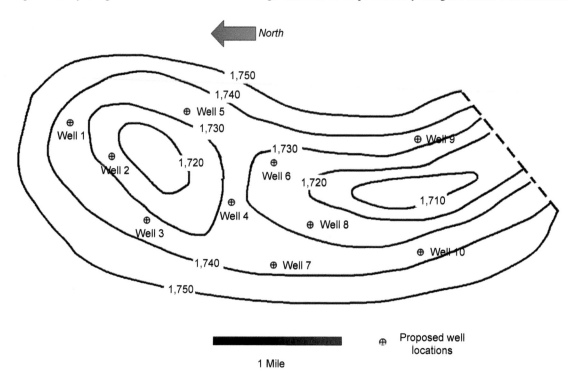

Fig. 10.35.1—Structure map (formation top), depth is positive from the datum level, ft.

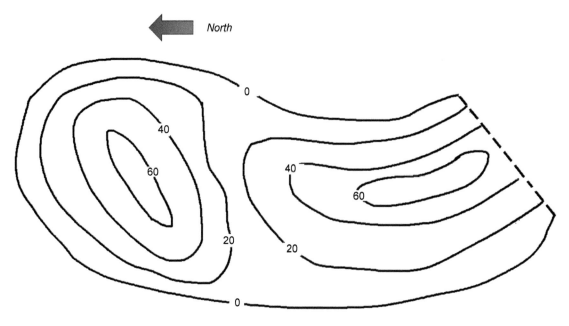

Fig. 10.35.2—Isopach map, ft.

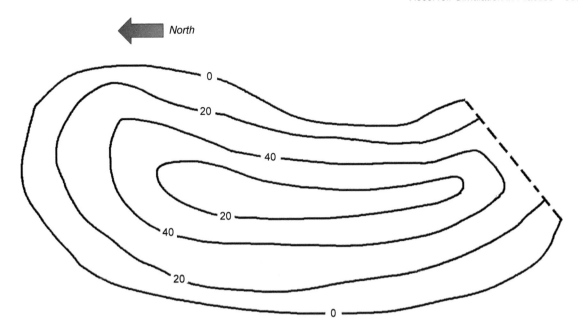

Fig. 10.35.3—Permeability contours (North/East), md.

The implementation plan for the project is

1. Four of the wells are selected to inject contaminated brine.
2. The dissolved toxic material will be adsorbed by the formation rock, and the other five wells are used to produce the water, which is cleaned from contamination (reservoir is expected to filter the contaminants).

In designing the project, the following operational constraints are in place:

- For injection operations: Formation-fracturing pressure is known to be 4,500 psi. To prevent the fracturing of the reservoir, do not exceed this pressure at any point in the reservoir.
- For production operations: Capacity of the surface facilities is 10,000 STB/D. Accordingly, the total field-production flow rate cannot exceed this limit.

Other fluid and wellbore properties are summarized in **Table 10.37.** Your assignment is to do the following:

1. Structure a computer code implementing the LSOR routine as a solver and simulating the brine-filtration problem. Use Thomas' algorithm (see Ertekin et al. 2001, for a more-detailed explanation) to solve the line equations of the LSOR protocol.
2. Use a convergence tolerance less than 1×10^{-4} psi as the convergence criterion.
3. Show the computed pressure distribution after the convergence is achieved.
4. Implement material-balance and residual checks to highlight the accuracy of the solution.
5. Complete the entries of the well and wellblock properties in **Table 10.38.1.**

Fluid compressibility	0 psi^{-1}
Viscosity (cp)	0.98
Density (lbm/ft^3)	62.4
Formation Volume Factor (res bbl/STB)	1.0
r_w (ft)	0.25
Skin of the wells	−1.5

Table 10.37—Fluid and wellbore properties for the Incompressible Flow Simulation Project.

Solution to Problem 10.5.32

Summary of the Overall Approach:

1. Structure a Cartesian grid system and assign readings from the contour maps (elevation, thickness, and permeability). Capture the reservoir boundaries (irregular boundaries). It is recommended to employ 2D arrays to store the input data. In the provided solution, a 29×19 grid system is employed (29 gridblocks along the x-direction and 19 gridblocks along the y-direction).
2. Design an injection/production scheme on the basis of the existing wells. Assign internal boundary conditions to each well. In the solution provided, Wells 1, 2, 3, 8, 9, and 10 are designated as injectors, with sandface-pressure specifications of 4,500 psia; and Wells 4, 5, 6, and 7 are assigned as producers (at each of these wells, sandface pressure is specified as 14.7 psia).

Well #	Well Type	Flow Rate (STB/D)	Productivity Index, Ω (STB/D–psi)	p_{sf} (psi)	Wellblock Pressure (psi)
1					
2					
3					
4					
5					
6					
7					
8					
9					
10					

Table 10.38.1—Wells and wellblock entries of the simulation study.

3. Calculate the SIP coefficients for each block. Note that in this incompressible fluid flow study, the SIP coefficients will not change during the computations.
4. Establish line equations along the *x*-direction using the SIP coefficients calculated in Step 3. Note that each line equation aims to solve unknown block pressures along a certain line, which can be achieved by calling on Thomas' algorithm.
5. Sweep the entire reservoir to solve the line equations along the *y*-direction. In the provided solution, Thomas' algorithm is called for 19 times during one completed LSOR iteration.
6. Compare the pressure-distribution results against that of the previous iteration level. Find the maximum pressure improvement and compare against the prescribed tolerance. More importantly, update the ω value from the spectral radius of the coefficient matrix. When the spectral radius is stabilized, calculate the ω_{opt}.
7. Repeat Step 4 to continue the LSOR iteration until the maximum pressure improvement at every gridblock in the reservoir is smaller than the prescribed tolerance (1×10^{-4} psi).

Once convergence is achieved, it will be necessary to output and save the pressure values of each block. More importantly, the injection rate and production rate of each well will need to be calculated. Last, but not the least, an MBC must be carried out to highlight the accuracy of the simulation results.

Results. **Table 10.38.2** shows results of the simulation study. In conducting the LSOR iterations, the value of the ω_{opt} is found to be 1.6735.

Well #	Well Type	Flow Rate (STB/D)	Productivity Index, Ω (STB/D–psi)	p_{sf} (psi)	Wellblock Pressure (psi)
1	Injector	2,105.48	1.67941	4,500	3,246.3
2	Injector	2,018.64	1.36558	4,500	3,021.77
3	Injector	946.62	1.44978	4,500	3,847.06
4	Producer	–3,098.83	4.00491	14.7	788.456
5	Producer	–2,003.13	1.84705	14.7	1,099.2
6	Producer	–710.829	0.684419	14.7	1,053.29
7	Producer	–2,045.8	4.51711	14.7	467.599
8	Injector	1,463.91	1.92781	4,500	3,740.63
9	Injector	649.105	1.21394	4,500	3,965.29
10	Injector	674.821	0.546347	4,500	3,264.85

Table 10.38.2—Results of the simulation study.

MBC results:

- Total injection rate = 7,858.58 STB/D.
- Total production rate = 7,858.58 STB/D.

Fig. 10.35.4 displays the residual-check results. One can compare the solved pressure distribution shown in **Fig. 10.35.5** to validate the result.

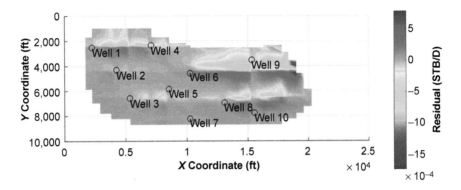

Fig. 10.35.4—Residual surface.

·	2146	2089	2030	2021	2025	2052	2093	2176	2215	·	·	·	·	·	·	·	·	·	·	·	·	·	·	·	·	·	·	·	·
2285	2212	2118	2022	2003	2008	2047	2106	2226	2305	2356	2348	·	·	·	·	·	·	·	·	·	·	·	·	·	·	·	·	·	·
2368	2380	2163	1981	1944	1953	2016	2099	2319	2572	2409	2314	2203	2139	·	·	·	·	·	·	·	·	·	3603	·	·	·	·	·	·
2390	2692	2172	1919	1886	1899	1970	2065	2348	3246	2393	2244	2138	2088	2088	·	·	·	·	·	·	·	·	3634	3603	3588	·	·	·	·
2370	3022	2150	1878	1846	1861	1935	2032	2280	2671	2325	2198	2109	2071	2088	2167	2369	2595	2826	3140	3505	3653	3656	3639	3603	3586	·	·	·	·
2242	2382	2017	1805	1774	1789	1863	1955	2151	2294	2212	2128	2067	2047	2079	2169	2373	2595	2826	3140	3508	3676	3877	3645	3603	3584	3677	·	·	·
1927	1860	1735	1615	1582	1605	1673	1755	1891	1963	2014	1994	1980	1990	2040	2144	2361	2591	2824	3135	3518	3847	3764	3650	3596	3578	3572	3568	·	·
1669	1580	1485	1365	1249	1342	1426	1519	1645	1709	1813	1845	1853	1866	1921	2059	2329	2583	2823	3113	3388	3505	3527	3550	3560	3563	3563	3581	·	·
1521	1425	1333	1188	788	1151	1265	1361	1472	1532	1630	1702	1732	1709	1699	1970	2314	2591	2836	3117	3362	3443	3471	3513	3540	3554	3558	3558	·	·
1434	1334	1257	1147	990	1114	1194	1271	1366	1420	1493	1584	1638	1557	1099	1893	2324	2616	2866	3135	3343	3422	3448	3494	3529	3548	3554	3556	3556	·
1322	1234	1183	1115	1067	1084	1122	1167	1244	1296	1363	1436	1567	1703	1812	2073	2416	2695	2941	3178	3350	3416	3440	3482	3519	3542	3551	3554	3555	·
1235	1149	1110	1062	1027	1014	1023	1051	1124	1174	1231	1053	1489	1862	2106	2330	2628	2877	3095	3276	3409	3457	3474	3505	3534	3549	3554	3557	3558	3559
·	1075	1048	1007	954	910	868	940	1069	1157	1290	1422	1680	2035	2293	2540	2845	3100	3313	3415	3495	3523	3532	3550	3561	3564	3564	3564	3564	·
·	1056	1025	985	923	844	468	872	1070	1178	1335	1525	1791	2137	2393	2648	2965	3246	3544	3518	3547	3569	3575	3586	3593	3576	3574	3571	3570	3568
·	1032	1003	967	911	859	801	918	1085	1197	1364	1579	1844	2193	2446	2709	3028	3322	3741	3562	3584	3608	3613	3617	3606	3592	3584	3679	3575	3572
·	·	987	959	922	899	919	978	1120	1231	1398	1632	1911	2270	2515	2767	3087	3326	3579	3571	3626	3677	3689	3671	3639	3615	3597	3589	3581	3575
·	·	971	959	955	966	1020	1068	1181	1283	1453	1716	1996	2371	2634	2860	3113	3315	3477	3571	3697	3863	3965	3794	3702	3650	3624	3607	3596	·
·	·	·	968	987	1015	1092	1143	1241	1337	1512	1796	2076	2582	3265	3077	3178	3320	3439	3559	3689	3765	3793	3770	3724	3677	3647	3625	·	·
·	·	·	·	·	1180	1203	1283	1378	1639	1797	2081	2621	2870	3042	3193	3322	3429	3547	3670	3724	3738	3750	3727	3693	·	·	·	·	·

Fig. 10.35.5—Simulated pressure distribution of the incompressible-fluid-flow reservoir.

Problem 10.5.33

Slightly Compressible Flow Simulation Project. A reservoir has been put on production for a period of 2 years, under a constant-sandface pressure of 1,000 psia (note that all of the wells were brought online at the same time). The spatial-property distributions can be found in **Figs. 10.36.1 through 10.36.3**. It was found that because of an error encountered in the original permeability values, the reservoir permeability distribution of the existing model needs to be recalibrated by way of multiplying the existing values by a certain factor. As shown in **Fig. 10.36.4**, there are four zones separated by dashed lines: Region I (northeastern region), Region II (southeastern region), Region III (southwestern region), and Region IV (northwestern region). Each of these four zones are expected to have their own permeability-calibration factor. As a new reservoir engineer in the field, you are asked to develop a single-phase slightly compressible fluid flow simulator. Conduct a history-matching study to find a corrected permeability distribution using the reservoir well and fluid properties presented in **Table 10.39** during your computations.

With the corrected permeability distribution, you should obtain a reasonable match of the field-production data displayed in **Table 10.40.** You can safely assume that all other properties of the field are the same as those provided in the maps (Figs. 10.36.1 through 10.36.3).

Implement the SIP to solve the system of equations generated during each timestep. A maximum pressure tolerance of 0.0001 psia and a maximum residual tolerance of 0.0001 STB/D can be used for convergence. *Note:* Use 30 days as one month. The maximum timestep size you can use is 30 days.

You are asked to do the following:

- Develop a numerical model to simulate the single-phase slightly compressible problem.
- Find out the permeability multipliers for each zone that will accurately history match the field-production data.
- Plot the residual surface at the end of each year.
- Plot the MBC results for validation purposes.
- Plot the pressure surfaces at the end of each year.
- Plot the wellblock pressure profiles for each well.

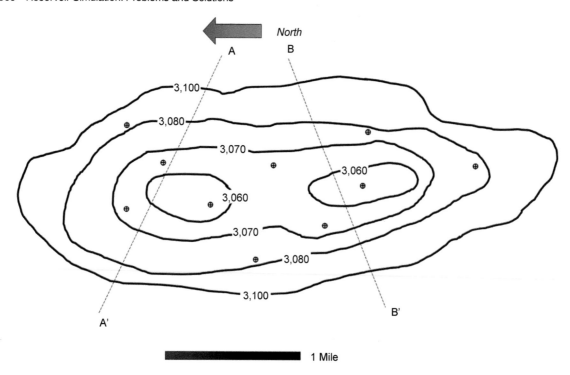

Fig. 10.36.1—Structure map (formation top); depth is positive from the datum level, ft.

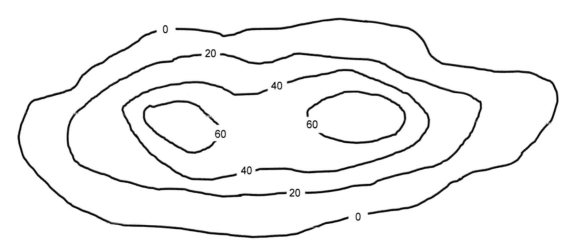

Fig. 10.36.2—Isopach map (isopach contours are in ft).

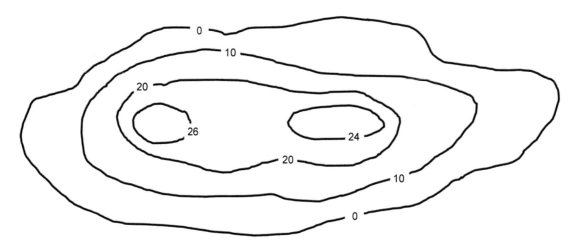

Fig. 10.36.3—Porosity contours, %.

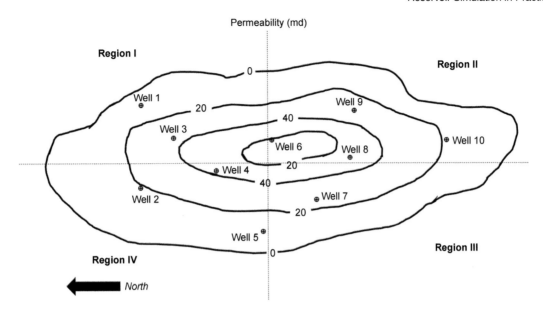

Fig. 10.36.4—Permeability distribution in North/East direction and well locations.

Parameter	Unit	Parameter	Unit
$\mu_o = \mu_{sc} + 0.00134\ln(p)$	cp	$p_{sc} = 14.7$	psi
$\mu_{sc} = 0.6$	cp	$r_w = 0.25$	ft
$B_o = \left[1 + c_o\left(p - p_{sc}\right)\right]^{-1}$	res bbl/STB	$p_{initial} = 5,000$	psi
$\rho_o = \rho_{sc}\left[1 + c_o\left(p - p_{sc}\right)\right]^{-1}$	lbm/ft³	$c_\phi = 0$	psi⁻¹
$\rho_{sc} = 42$	lbm/ft³	Skin = 0 for all wells	
$c_o = 3 \times 10^{-5}$	psi⁻¹	$p_{sf} = 1,000$	psi

Table 10.39—Critical input data of the problem.

Time, days	Well 1	Well 2	Well 3	Well 4	Well 5	Well 6	Well 7	Well 8	Well 9	Well 10
30	1,040	3,228	9,309	14,900	1,619	5,078	3,493	6,342	2,775	1,434
60	757	3,779	7,504	12,481	1,233	4,317	2,995	5,154	2,117	1,159
90	822	2,593	6,462	8,678	858	3,513	2,359	4,448	2,003	810
120	636	2,168	6,539	8,445	1,004	2,637	2,541	3,684	1,741	921
150	464	1,817	5,412	5,928	920	2,492	2,245	4,012	1,464	651
180	472	1,807	4,835	7,087	800	2,383	1,889	2,714	1,928	755
210	396	1,422	4,667	6,295	516	2,223	1,877	3,491	1,681	558
240	376	1,338	3,806	4,948	458	1,508	1,466	2,536	1,687	649
270	236	985	2,790	4,135	508	1,554	1,534	2,572	1,357	568
300	252	1,128	2,193	2,992	369	1,227	1,308	2,118	1,028	572
330	186	923	2,159	3,718	440	1,308	1,190	2,341	1,057	476
360	211	863	2,168	2,817	360	1,015	772	2,221	1,097	453
390	149	554	2,156	2,211	245	944	749	2,120	775	342
420	168	519	1,550	1,874	242	868	921	1,688	977	382
450	115	479	1,636	1,653	219	817	579	1,777	614	338
480	107	482	1,502	1,698	226	612	576	1,340	615	315
510	121	380	917	1,798	161	492	540	1,391	550	287
540	110	413	883	1,589	144	614	475	1,019	619	280
570	89	337	943	1,444	142	519	486	1,134	438	255
600	89	351	762	1,381	134	430	362	826	393	242
630	58	263	629	1,221	127	468	428	742	420	176
660	67	201	766	1,040	113	309	393	747	397	195
690	49	231	556	1,000	94	373	353	737	340	163
720	53	206	552	727	106	247	355	680	340	132

Table 10.40–Field-production data in STB/D (refer to well locations in Fig. 10.36.4).

Solution to Problem 10.5.33

Summary of the Overall Approach: To solve the history-match problem, it is necessary to develop a numerical simulator capable of mimicking a 2D single-phase slightly compressible fluid flow reservoir. Here is a summary of the recommended steps to be taken during such a development:

1. Structure a Cartesian grid system (body centered), and assign readings from the contour maps (Figs. 10.36.1 through Figs. 10.36.3) (elevation, thickness, permeability, and porosity). Try to capture irregularities along the reservoir boundaries using nonuniform grid dimensions. It is recommended that you employ 2D arrays to store the reservoir data. In the provided solution, a 33×16 grid system along the x- and y-directions was overlaid.
2. Assign uniform boundary conditions of $p_{sf} = 1,000$ psia at each well location.
3. At the beginning of each timestep, calculate (update) the SIP coefficients of each block. Because the nonlinearities are not pronounced, use the pressure values of the previous timestep to update the pressure-dependent fluid properties (μ_o and B_o).
4. Employ the SIP algorithm to solve the pressure distribution using the coefficients calculated in Step 3.
5. Once the pressure distribution is solved at the current timestep level, calculate the production rate from each well.
6. Conduct cumulative and incremental MBCs at the conclusion of each timestep computation.
7. Repeat Step 3 until the cumulative simulation time reaches 2 years (730 days) of production.

Using the single-phase slightly compressible simulator developed as part of this project, one can carry out an iterative history-matching process to fine-tune the permeability multipliers for each zone:

1. Use a permeability multiplier of 1 for all regions as a starting point. Run the simulator, and compare the simulated production profile of each well against the field data provided (Table 10.40).
2. In any of the four regions studied, if the simulated production data show higher rates during the early times of the simulation and deplete faster than the field history, it is understood that the permeability multiplier is larger than the correct value. Then it will be necessary to reduce the multiplier accordingly (this can be done using a linear ratio). In the event of the contrary observation, increase the multipliers. Note that changing the permeability of one of the regions potentially will influence the simulation results of the other zones as well.
3. Repeat the process described in Step 2 until satisfactory matches are obtained for all of the producers.

Results. **Table 10.41** displays the permeability multipliers found for each of the zones marked in Fig. 10.36.4. Using the multipliers shown in Table 10.41, history-matching results in terms of production from each well, as shown in **Fig. 10.36.5,** were obtained. In addition, the calculated production rates (**Table 10.42**), calculated well-block pressures for each well (**Table 10.43**), simulated pressure distribution over the entire reservoir (**Figs. 10.36.6 and 10.36.7),** and residual-check results (**Fig. 10.36.8**) are displayed.

Region I	2
Region II	0.3
Region III	1
Region IV	3.0

Table 10.41—Permeability multipliers found for each of the zones marked in Fig. 10.36.4.

Time	Well 1	Well 2	Well 3	Well 4	Well 5	Well 6	Well 7	Well 8	Well 9	Well 10
30	1,105	4,031	11,270	14,211	1,410	4,638	3,174	5,289	2,714	1,204
60	868	3,238	8,827	11,297	1,164	3,763	2,701	4,603	2,333	1,035
90	727	2,719	7,332	9,538	1,010	3,229	2,414	4,205	2,131	942
120	615	2,307	6,192	8,163	882	2,799	2,168	3,865	1,962	868
150	525	1,972	5,284	7,044	773	2,441	1,950	3,553	1,808	802
180	451	1,697	4,545	6,118	679	2,138	1,757	3,265	1,665	741
210	389	1,468	3,933	5,340	598	1,879	1,584	2,998	1,532	685
240	338	1,276	3,422	4,681	529	1,657	1,429	2,751	1,407	633
270	295	1,114	2,991	4,119	469	1,466	1,291	2,522	1,292	584
300	258	977	2,624	3,637	416	1,300	1,167	2,311	1,185	539
330	227	859	2,310	3,220	371	1,156	1,055	2,116	1,086	497
360	200	759	2,041	2,859	331	1,030	955	1,937	994	458
390	177	672	1,808	2,545	296	920	866	1,772	910	421
420	157	596	1,606	2,270	265	823	785	1,621	833	388
450	139	530	1,430	2,029	238	737	712	1,482	761	356
480	124	473	1,275	1,817	214	662	646	1,355	696	328
510	111	423	1,140	1,630	193	595	587	1,239	637	301
540	99	378	1,021	1,464	174	535	533	1,132	582	276
570	89	339	916	1,318	157	482	485	1,034	532	253
600	80	305	823	1,187	142	435	441	945	486	232
630	72	274	740	1,070	128	393	401	864	444	213
660	65	247	667	966	116	355	365	789	406	195
690	58	223	601	873	105	322	332	721	371	179
720	52	201	543	790	95	291	302	659	339	164

Table 10.42—Calculated production rates, STB/D, for the wells in Fig. 10.36.4.

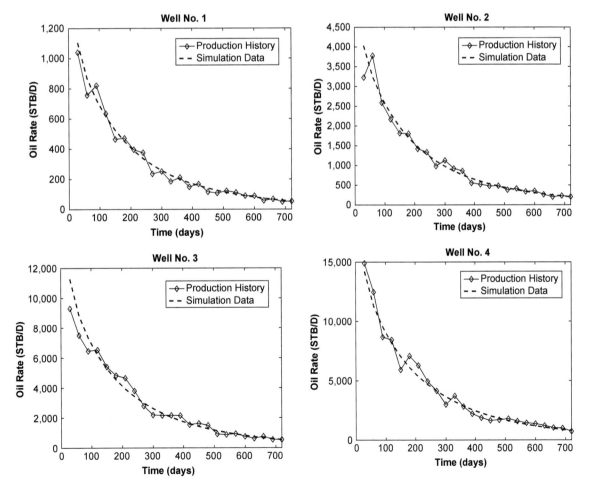

Fig. 10.36.5 (a) —History-match results using the established permeability multipliers.

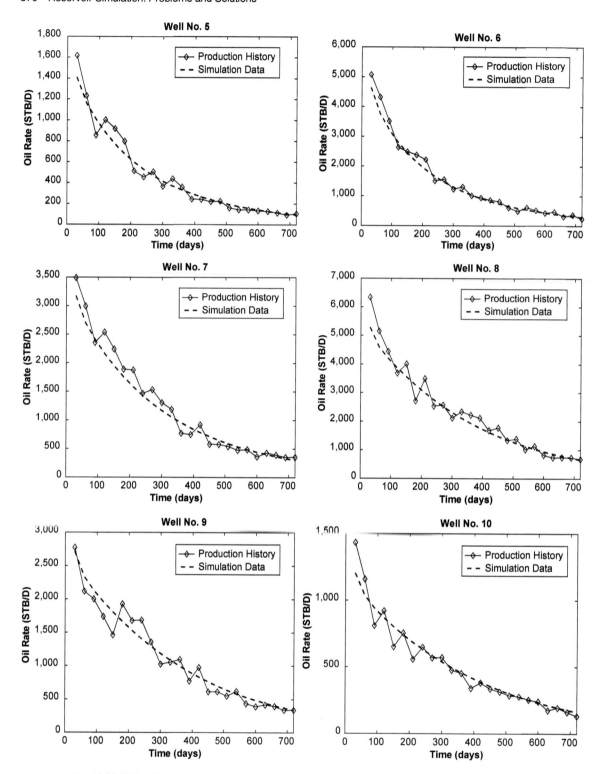

Fig. 10.36.5(b)—History-match results using the established permeability multipliers (continued).

Time	Well 1	Well 2	Well 3	Well 4	Well 5	Well 6	Well 7	Well 8	Well 9	Well 10
30	3,253	3,618	3,385	3,186	3,375	3,356	3,493	3,749	3,472	3,679
60	2,873	3,184	2,950	2,821	3,070	3,009	3,236	3,527	3,263	3,444
90	2,566	2,837	2,630	2,546	2,801	2,734	3,011	3,336	3,096	3,261
120	2,322	2,557	2,379	2,325	2,569	2,505	2,810	3,157	2,942	3,097
150	2,125	2,329	2,178	2,143	2,371	2,311	2,629	2,989	2,796	2,945
180	1,964	2,142	2,012	1,992	2,202	2,147	2,467	2,831	2,656	2,803
210	1,831	1,986	1,875	1,864	2,057	2,007	2,322	2,682	2,524	2,670
240	1,720	1,856	1,761	1,757	1,933	1,886	2,192	2,543	2,399	2,544
270	1,627	1,747	1,664	1,665	1,825	1,783	2,075	2,414	2,283	2,427
300	1,548	1,654	1,582	1,586	1,732	1,693	1,970	2,295	2,175	2,316
330	1,481	1,575	1,511	1,518	1,651	1,615	1,877	2,184	2,074	2,213
360	1,424	1,507	1,451	1,459	1,580	1,547	1,792	2,082	1,982	2,117
390	1,374	1,448	1,399	1,408	1,518	1,487	1,716	1,988	1,897	2,026
420	1,331	1,397	1,354	1,363	1,464	1,435	1,648	1,902	1,818	1,943
450	1,294	1,353	1,315	1,324	1,415	1,389	1,587	1,823	1,747	1,865
480	1,262	1,314	1,280	1,290	1,373	1,349	1,531	1,751	1,681	1,793
510	1,233	1,281	1,250	1,259	1,335	1,313	1,481	1,684	1,621	1,726
540	1,208	1,251	1,224	1,233	1,301	1,281	1,436	1,624	1,566	1,665
570	1,186	1,225	1,200	1,209	1,271	1,253	1,396	1,568	1,515	1,608
600	1,167	1,201	1,180	1,188	1,245	1,228	1,359	1,518	1,469	1,556
630	1,150	1,181	1,161	1,169	1,221	1,205	1,325	1,472	1,428	1,508
660	1,135	1,163	1,145	1,152	1,199	1,185	1,295	1,430	1,389	1,464
690	1,121	1,146	1,131	1,137	1,180	1,167	1,268	1,391	1,354	1,423
720	1,109	1,132	1,118	1,124	1,163	1,151	1,243	1,356	1,323	1,386

Table 10.43—Calculated wellblock pressures, psia, for the wells in Fig. 10.36.4.

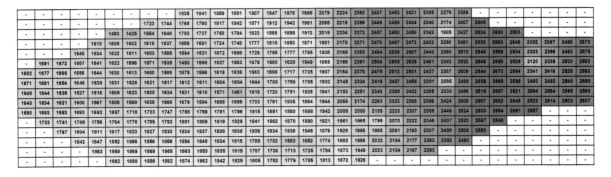

Fig. 10.36.6—Simulated pressure distribution at the end of 360 days.

1	2	3	4	5	6	7	8	9	10	11	12	13	14	15	16	17	18	19	20	21	22	23	24	25	26	27	28	29	30	31	32	33
·	·	·	·	·	·	·	·	1219	1221	1227	1235	1244	1258	1268	1275	1306	1360	1413	1446	1458	1454	1438	1418	1467	·	·	·	·	·	·	·	·
·	·	·	·	·	·	·	1187	1193	1200	1207	1216	1224	1234	1248	1259	1265	1299	1358	1414	1449	1462	1458	1434	1386	1470	1544	·	·	·	·	·	·
·	·	·	·	·	·	1126	1109	1143	1168	1183	1193	1203	1211	1221	1235	1246	1252	1288	1355	1417	1452	1467	1464	1435	1324	1477	1551	1569	1546	·	·	·
·	·	·	1158	1155	1155	1159	1165	1171	1181	1191	1199	1209	1223	1235	1241	1278	1347	1416	1457	1473	1472	1456	1425	1496	1554	1570	1534	1479	1487	1525	1563	·
·	·	1164	1162	1159	1157	1155	1154	1155	1165	1178	1186	1195	1205	1213	1219	1266	1343	1420	1465	1482	1483	1470	1459	1506	1558	1571	1529	1452	1476	1523	1554	·
·	1174	1170	1167	1163	1159	1153	1147	1139	1118	1149	1166	1174	1182	1184	1172	1151	1248	1344	1422	1468	1486	1489	1476	1460	1510	1559	1571	1525	1387	1465	1525	1556
1173	1171	1169	1167	1165	1162	1158	1156	1152	1150	1157	1164	1170	1178	1189	1197	1204	1259	1344	1416	1459	1477	1481	1464	1425	1505	1567	1571	1538	1463	1491	1633	1559
1170	1168	1167	1165	1164	1162	1162	1161	1161	1161	1161	1162	1174	1194	1209	1219	1267	1339	1402	1441	1460	1466	1449	1367	1498	1563	1569	1549	1510	1517	1540	1560	·
1166	1165	1163	1161	1159	1158	1162	1164	1166	1166	1164	1153	1124	1168	1199	1218	1233	1271	1327	1380	1417	1437	1448	1448	1436	1495	1545	1564	1555	1538	1536	1644	1567
1164	1163	1160	1155	1147	1132	1157	1168	1180	1183	1186	1190	1201	1220	1235	1245	1273	1314	1356	1391	1416	1435	1452	1465	1500	1540	1561	1561	1556	1555	1564	1569	·
1178	1179	1179	1180	1180	1182	1189	1194	1198	1202	1207	1212	1218	1225	1238	1249	1255	1271	1297	1326	1350	1383	1415	1447	1472	1503	1536	1559	1564	1565	·	·	·
·	1190	1192	1195	1197	1200	1204	1207	1210	1214	1218	1222	1227	1232	1240	1247	1252	1265	1282	1290	1244	1334	1387	1432	1467	1502	1532	1556	·	·	·	·	·
·	·	1208	1210	1212	1214	1217	1219	1220	1223	1225	1227	1229	1230	1232	1235	1239	1252	1272	1289	1301	1340	1376	1418	1458	1496	1526	·	·	·	·	·	·
·	·	·	1220	1222	1224	1225	1226	1227	1228	1228	1226	1222	1216	1195	1163	1184	1222	1258	1286	1312	1342	1369	1411	1449	1488	·	·	·	·	·	·	·
·	·	·	1226	1228	1229	1230	1230	1230	1229	1226	1222	1216	1204	1200	1204	1224	1255	1282	1311	1341	1364	1409	·	·	·	·	·	·	·	·	·	·
·	·	·	·	1235	1234	1234	1234	1233	1231	1228	1224	1221	1217	1218	1220	1231	1253	1272	·	·	·	·	·	·	·	·	·	·	·	·	·	·

Fig. 10.36.7—Simulated pressure distribution at the end of 720 days.

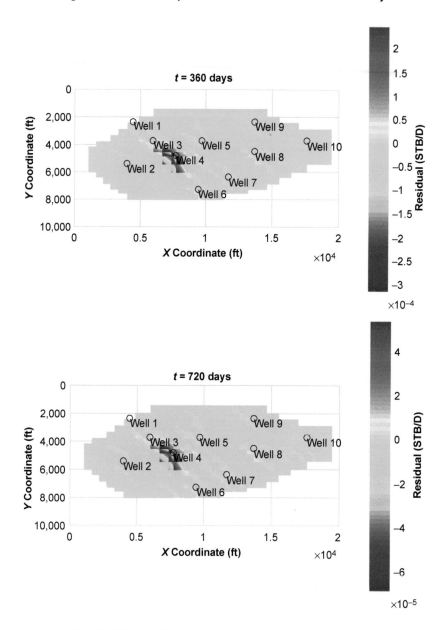

Fig. 10.36.8—Residual checks at the end of 360 and 720 days.

Problem 10.5.34

Compressible Flow Simulation Project: The depleted gas reservoir shown in **Fig. 10.37.1** is used as a gas-storage facility for City A. The structural, isopach, porosity, and permeability maps are given in **Figs. 10.37.2 through 10.37.5** Gas purchased from a nearby larger gas field will be injected into this storage reservoir. Gas from the storage field will be used by the communities in City A. The natural gas demand per month over the course of a typical year for each household is shown in **Table 10.44.** Wells can be used as either injectors or producers, depending on the demand. Note that the bursting pressure of the reservoir is 5,000 psia. The pressure should be maintained above 1,000 psia at all times to avoid the closure of pores/fractures in the reservoir. The transmission line from the gas field to the storage reservoir is limited to a working capacity of 100 MMscf/D (maximum total injection rate into the storage reservoir). Reservoir boundaries are completely sealed, and the initial pressure and temperature of the reservoir are recorded as 4,000 psia and 150°F, respectively. Within the reservoir, isothermal flow conditions exist. In your calculations, use 30 days for one month.

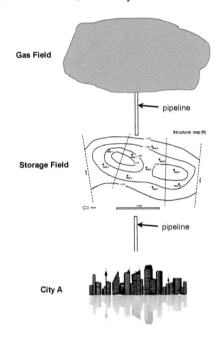

Fig. 10.37.1—Illustration of the gas storage project.

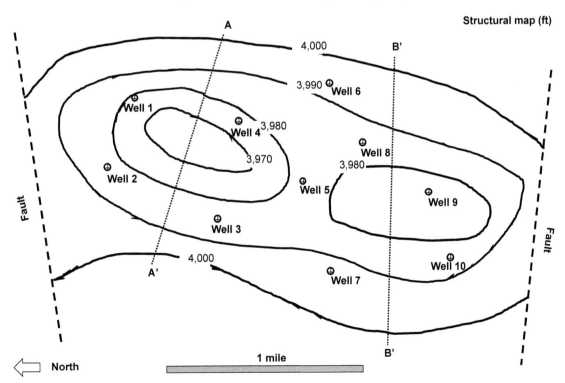

Fig. 10.37.2—Structure map (formation top), depth is positive from the datum level, ft.

Thickness (ft)

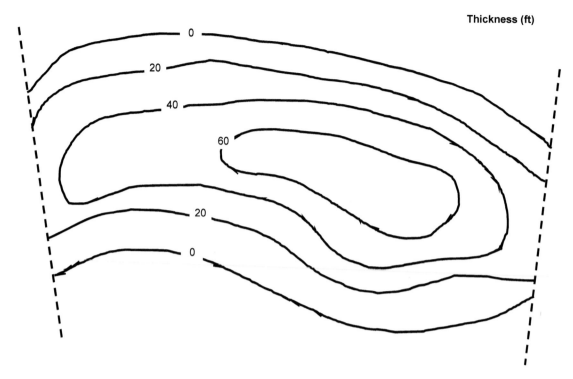

Fig. 10.37.3—Isopach map (contours are in ft).

Porosity, fraction

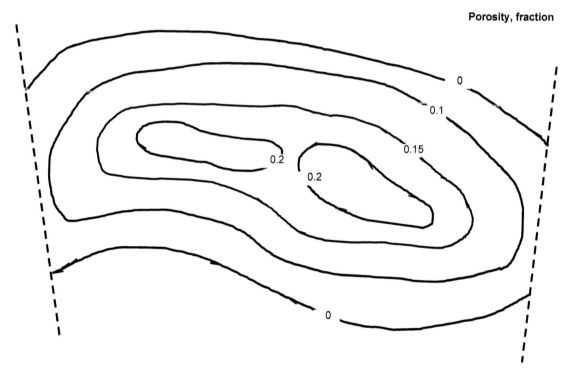

Fig. 10.37.4—Porosity contours, fraction.

Permeability (NE), md

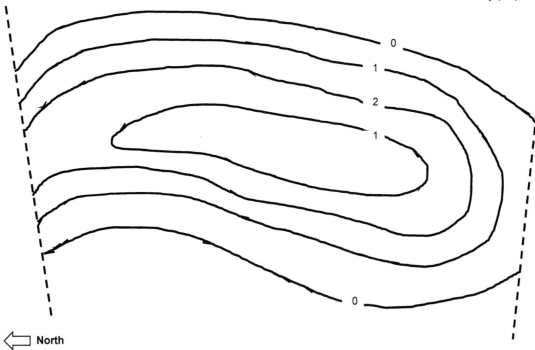

⟵ **North**

Fig. 10.37.5—Permeability distribution in North/East direction.

Gas Demand (Mscf/month/household)		Gas Demand (Mscf/month/household)	
Jan	140	Jul	28
Feb	130	Aug	25
Mar	115	Sep	45
Apr	90	Oct	65
May	60	Nov	85
Jun	40	Dec	120

Table 10.44—Natural gas demand per month over the course of a typical year for each household in City A.

The specific gravity of the natural gas in the reservoir is 0.6. The critical pressure and critical temperature for the gas are 680 psia and 385 °R, respectively. The gas viscosity can be estimated using the correlation of Lee, Gonzalez, and Eakin (Lee et al. 1966) as presented in **Table 10.45.** The gas-compressibility factor can be estimated using Dranchuk and Abou-Kassem's method (Dranchuk and Abou-Kassem 1975).

$$\mu_o = 10^{-4} K \exp[X(\frac{\rho}{62.4})^Y]$$

$$X = 3.448 + \frac{986.4}{T} + 0.01009MW$$

$$Y = 2.447 - 0.2224X$$

$$K = \frac{(9.379 + 0.01607MW)T^{1.5}}{(209.2 + 19.26MW + T)}$$

MW = molecular weight (lbm/lbmole)

T = reservoir temperature (°R)

μ = gas viscosity (cp)

ρ = gas density (lbm/ft³)

R = 10.731 psia–ft³/(lbmole–°R)

Table 10.45—Lee, Gonzalez, and Eakin correlation (using data from Lee et al. 1966).

Conduct a simulation study for one year (from January 1 to December 30). Determine an injection/production schedule for the storage field that will maximize the number of houses the storage reservoir would be able to meet to fulfill the city's natural gas need over a period of one year. Use the SIP to solve the generated system of equations. A maximum pressure tolerance of 0.01 psia can be used for convergence.

Solution to Problem 10.5.34

Summary of the Overall Approach: To maximize the number of households to be supported by the gas-storage field, it is necessary to first program a numerical model applicable to 2D single-phase gas reservoirs. The recommended steps for the development of the model are as follows:

1. Structure a Cartesian grid system and assign data to each block using the contour maps (elevation, thickness, permeability, and porosity) depicted in Figs. 10.37.2 through 10.37.5. Accurately capture the reservoir boundaries (it is recommended to use irregular grid spaces). Employ 2D arrays to store reservoir data. In this case, a 28×20 grid system was deployed.
2. Prepare subroutines that calculate the z-factor and gas viscosity employing the Dranchuk and Abou-Kassem (1975) protocol and the Lee, Gonzalez, and Eakin (Lee et al. 1966) correlations, respectively.
3. Assign the internal boundary conditions at the wellpoints ($p_{sf} = 1,000$ psia for producers and $p_{sf} = 5,000$ psia for injectors).
4. During the calculations, at a given timestep, use the SIP solver to solve for the pressure distribution. The SIP coefficients that are strong functions of pressure should be updated after each SIP iteration.
5. With the pressure distribution solved at a current timestep level, calculate the production rate of each well.
6. Conduct cumulative and incremental MBCs at the conclusion of each timestep calculation.
7. Repeat Step 3 until the cumulative simulation time reaches one year (360 days) of production.

It is worth emphasizing that the gas-reservoir simulator under development must be competent enough to change the specifications of the wells at the end of the storage/supply cycles. To maximize the number of households supplied by the gas-storage project, it is necessary to run the simulator multiple times to test various operational strategies. It is recommended to assign more wells as producer during the winter months and assign more wells as injectors during the summer months, when the demand is less.

Results. The operational schedule to meet the maximum number of household needs shown in **Table 10.46** was established as a result of the study. During the simulation studies, constant sandface pressure specifications of 1,000 and 5,000 psi are implemented at the producers and injectors, respectively. The number of each producer and injector is shown in **Table 10.47.**

	Well 1	Well 2	Well 3	Well 4	Well 5
Jan	Producer	Producer	Producer	Producer	Producer
Feb	Producer	Producer	Producer	Producer	Producer
Mar	Injector	Producer	Producer	Producer	Producer
Apr	Injector	Injector	Producer	Producer	Producer
May	Injector	Injector	Injector	Producer	Injector
Jun	Injector	Injector	Injector	Producer	Injector
Jul	Producer	Injector	Injector	Producer	Injector
Aug	Producer	Injector	Injector	Producer	Injector
Sep	Injector	Injector	Injector	Producer	Injector
Oct	Injector	Injector	Injector	Producer	Injector
Nov	Injector	Injector	Producer	Producer	Producer
Dec	Producer	Producer	Producer	Producer	Producer

	Well 6	Well 7	Well 8	Well 9	Well 10
Jan	Producer	Producer	Producer	Producer	Producer
Feb	Producer	Producer	Producer	Producer	Producer
Mar	Producer	Producer	Producer	Producer	Injector
Apr	Producer	Producer	Injector	Producer	Injector
May	Producer	Producer	Injector	Producer	Injector
Jun	Injector	Injector	Injector	Producer	Producer
Jul	Injector	Injector	Injector	Injector	Producer
Aug	Injector	Injector	Injector	Injector	Producer
Sep	Producer	Injector	Producer	Injector	Producer
Oct	Producer	Producer	Injector	Producer	Injector
Nov	Producer	Producer	Injector	Producer	Injector
Dec	Producer	Producer	Producer	Producer	Producer

Table 10.46—Operational schedule of the wells in the Compressible Flow Simulation Project.

	Number of Producers	Number of Injectors
Jan	10	0
Feb	10	0
Mar	8	2
Apr	6	4
May	4	6
Jun	3	7
Jul	3	7
Aug	3	7
Sep	4	6
Oct	4	6
Nov	6	4
Dec	10	0

Table 10.47— Number of injectors and producers shown in the Compressible Flow Simulation Project.

The simulated gas-production and -injection rates (MMscf/D; negative numbers are production rates) and the wellblock pressures are provided in **Tables 10.48 and 10.49.** As an example, we see that the maximum number of households supplied by the field during the month of October is 18,464, based on the seasonal household-supply data displayed in **Table 10.50.** The accuracy levels of the simulation results are captured by the residual surface maps (**Fig. 10.37.6**) provided at the end of January, April, August, and December. Figures showing the pressure distributions at the end of January (**Fig. 10.37.7**), June (**Fig. 10.37.8**), and December (**Fig. 10.37.9**) are also provided.

	Well 1	Well 2	Well 3	Well 4	Well 5	Well 6	Well 7	Well 8	Well 9	Well 10
Jan	−3.12	−8.52	−9.40	−14.34	−8.83	−9.57	−9.88	−6.15	−12.39	−4.62
Feb	−2.88	−7.83	−8.78	−13.20	−8.26	−8.93	−9.26	−5.72	−11.35	−4.31
Mar	1.17	−7.35	−8.36	−12.45	−7.90	−8.53	−8.83	−5.45	−10.65	1.66
Apr	1.11	3.64	−8.02	−11.87	−7.60	−8.18	−8.45	2.41	−10.10	1.50
May	1.14	3.30	3.86	−11.38	3.57	−7.88	−8.11	2.14	−9.64	1.48
Jun	1.16	3.11	3.48	−10.95	3.18	4.14	4.35	2.06	−9.26	−4.32
Jul	−2.72	2.95	3.34	−10.60	3.02	3.71	4.04	2.03	6.17	−3.92
Aug	−2.46	2.81	3.27	−10.31	2.90	3.56	3.93	2.01	5.43	−3.71
Sep	1.47	2.70	3.23	−10.08	2.80	−8.48	3.85	−5.76	5.04	−3.59
Oct	1.33	2.60	3.18	−9.90	2.72	−7.76	−8.49	2.39	−10.69	2.15
Nov	1.30	2.52	−8.67	−9.73	−8.48	−7.38	−7.79	2.19	−9.65	1.91
Dec	−2.57	−8.29	−8.02	−9.55	−7.89	−7.13	−7.37	−5.59	−9.05	−3.93

Table 10.48—Simulated gas-production (MMscf/D) rate of the wells.

	Well 1	Well 2	Well 3	Well 4	Well 5	Well 6	Well 7	Well 8	Well 9	Well 10
Jan	3,059	2,749	2,950	3,095	3,033	2,952	3,089	2,902	2,920	2,979
Feb	2,884	2,594	2,809	2,910	2,889	2,808	2,943	2,755	2,743	2,829
Mar	3,919	2,486	2,714	2,791	2,798	2,717	2,843	2,665	2,626	4,009
Apr	3,979	3,940	2,637	2,698	2,722	2,640	2,756	3,953	2,533	4,104
May	3,955	4,044	3,875	2,621	3,848	2,575	2,677	4,073	2,457	4,116
Jun	3,928	4,099	3,988	2,554	3,977	3,811	3,708	4,107	2,392	2,835
Jul	2,766	4,146	4,031	2,499	4,032	3,938	3,803	4,121	3,638	2,648
Aug	2,586	4,186	4,052	2,455	4,070	3,982	3,838	4,131	3,813	2,552
Sep	3,631	4,221	4,064	2,418	4,102	2,707	3,862	2,770	3,900	2,496
Oct	3,763	4,250	4,076	2,389	4,130	2,547	2,766	3,960	2,631	3,699
Nov	3,800	4,272	2,784	2,363	2,944	2,466	2,606	4,050	2,458	3,853
Dec	2,665	2,698	2,636	2,335	2,795	2,410	2,510	2,712	2,357	2,654

Table 10.49—Simulated wellblock pressures, psia.

	Households Supplied		Households Supplied
Jan	20,624	Jul	19,269
Feb	19,206	Aug	20,174
Mar	18,596	Sep	20,210
Apr	18,464	Oct	18,464
May	18,879	Nov	19,424
Jun	19,172	Dec	18,424

Table 10.50—Number of households supplied.

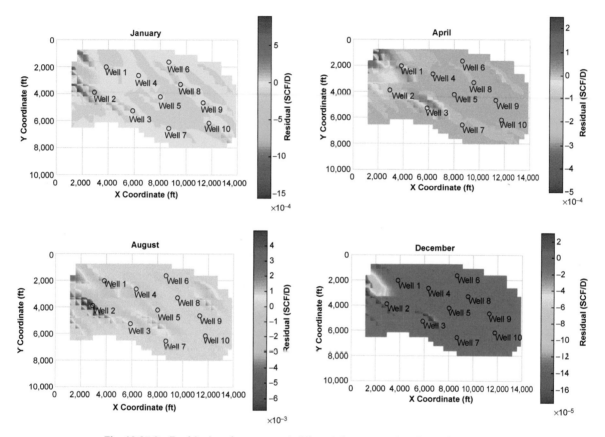

Fig. 10.37.6—Residual surface maps at different times over the simulation period.

-	3988	3965	3916	3857	3815	3815	3864	3906	3944	3974	3980	3996	3987	3976	3947	3867	3937	-	-	-	-	-	-	-	-	-
-	3961	3923	3895	3819	3755	3716	3788	3872	3928	3958	3961	3959	3942	3895	3789	3444	3699	3877	3950	3981	3994	-	-	-	-	-
3970	3937	3910	3858	3766	3636	3522	3670	3815	3894	3930	3926	3927	3912	3847	3682	3059	3602	3823	3914	3959	3991	3996	3999	-	-	-
3968	3947	3916	3848	3704	3407	2749	3443	3740	3847	3860	3809	3851	3895	3904	3882	3813	3861	3894	3924	3950	3969	3981	3989	3993	3996	-
3974	3957	3934	3888	3799	3646	3455	3658	3811	3859	3793	3571	3750	3870	3913	3919	3889	3862	3839	3873	3918	3952	3974	3985	3991	3995	-
3976	3963	3947	3919	3877	3826	3790	3844	3898	3884	3712	2950	3612	3850	3922	3939	3908	3819	3732	3802	3882	3935	3965	3961	3989	3994	3997	3999
3972	3961	3943	3914	3888	3921	3953	3974	3983	3977	3941	3863	3920	3961	3969	3956	3890	3609	3095	3572	3805	3906	3951	3971	3982	3991	3995	3998	3998
3962	3937	3885	3761	3556	3789	3925	3975	3990	3989	3976	3953	3965	3964	3936	3899	3913	3795	3607	3770	3883	3929	3943	3956	3965	3980	3990	3995	3996
3958	3924	3845	3627	3033	3659	3990	3970	3990	3991	3987	3983	3978	3954	3878	3783	3908	3922	3897	3924	3946	3939	3923	3936	3954	3974	3987	3994	3995
3967	3946	3905	3821	3717	3839	3928	3966	3971	3960	3946	3962	3955	3884	3633	2952	3784	3940	3966	3966	3940	3856	3741	3840	3909	3953	3979	3990	3994
3978	3968	3951	3924	3902	3925	3950	3955	3933	3882	3817	3895	3929	3925	3855	3764	3918	3975	3986	3972	3906	3684	3089	3679	3855	3934	3972	3989	3994
0006	3676	3969	3954	3940	3948	3956	3942	3881	3740	3527	3865	3924	3924	3909	3963	3987	3992	3982	3943	3832	3653	3788	3877	3935	3970	3987	3993
3991	3990	3987	3982	3975	3973	3961	3915	3787	3499	2902	3592	3812	3917	3950	3962	3977	3989	3993	3989	3971	3926	3858	3847	3884	3928	3965	3984	3992
3998	3996	3995	3994	3991	3985	3968	3916	3797	3606	3409	3714	3860	3929	3950	3950	3942	3964	3977	3977	3957	3902	3800	3687	3771	3871	3940	3974	3987
3999	3999	3999	3999	3999	3998	3995	3984	3942	3814	3773	3895	3935	3941	3914	3866	3792	3892	3938	3944	3906	3787	3521	2920	3469	3758	3898	3957	3975
3999	4000	4000	4000	4000				3975	3938	3915	3937	3938	3906	3798	3580	2979	3699	3887	3931	3915	3843	3714	3561	3683	3838	3930	3966	3980
-	-	-	-	-	-	-	-	-	3963	3928	3885	3808	3690	3491	3792	3906	3944	3942	3904	3842	3787	3826	3900	3950	3973	3982
-	-	-	-	-	-	-	-	-	3965	3929	3848	3787	3733	3882	3941	3963	3963	3946	3916	3893	3905	3939	3964	3983	3996
-	-	-	-	-	-	-	-	-	-	3968	3938	3874	3943	3967	3978	3977	3968	3952	3943	3943	3962	3991	3998	-
-	-	-	-	-	-	-	-	-	-	3978	3962	3985	3991	3991	3987	3992	3991	3991	3995	3999	-	-

Fig. 10.37.7—Pressure distribution at the end of January.

.	3580	3615	3662	3699	3718	3711	3679	3636	3598	3564	3559	3570	3609	3660	3714	3790	3759	.	.	.	.	.	.	.	.	.	.	.
.	3623	3658	3684	3720	3747	3754	3710	3653	3607	3573	3569	3576	3597	3633	3692	3842	3755	3657	3603	3584	3600	.	.	.	.	.	.	.
3618	3644	3662	3693	3738	3792	3830	3760	3683	3628	3592	3588	3590	3603	3635	3702	3928	3707	3601	3543	3518	3532	3522	3545	3565	3576	.	.	.
3613	3629	3649	3686	3753	3870	4099	3848	3721	3659	3634	3646	3623	3598	3582	3571	3555	3471	3412	3393	3402	3424	3455	3485	3510	3540	.	.	.
3607	3620	3635	3663	3709	3777	3852	3772	3700	3664	3672	3746	3667	3604	3563	3526	3463	3358	3290	3298	3340	3390	3437	3475	3501	3534	.	.	.
3603	3612	3624	3643	3671	3703	3725	3705	3675	3667	3719	3988	3728	3611	3548	3491	3398	3256	3161	3204	3287	3362	3420	3465	3498	3531	3576	3617	.
3603	3609	3617	3633	3651	3649	3652	3660	3665	3664	3664	3674	3624	3569	3513	3442	3299	2991	2654	2969	3201	3329	3404	3451	3488	3534	3568	3594	3605
3616	3625	3647	3699	3783	3704	3666	3668	3681	3686	3681	3657	3608	3545	3487	3437	3349	3192	3027	3178	3302	3378	3424	3453	3475	3507	3543	3570	3579
3624	3637	3670	3757	3977	3759	3682	3676	3692	3703	3697	3657	3601	3530	3478	3462	3397	3361	3335	3366	3400	3420	3440	3453	3468	3496	3533	3560	3570
3633	3642	3659	3694	3738	3699	3678	3685	3706	3723	3726	3672	3605	3537	3556	3811	3491	3482	3487	3478	3456	3445	3477	3445	3444	3465	3505	3540	3556
3644	3647	3653	3664	3676	3674	3678	3696	3725	3768	3781	3721	3647	3571	3532	3545	3532	3570	3576	3544	3490	3491	3708	3456	3415	3430	3474	3518	3541
3656	3654	3657	3661	3667	3671	3682	3708	3752	3817	3890	3792	3709	3641	3602	3592	3605	3629	3622	3576	3502	3437	3438	3362	3354	3384	3436	3487	3513
3667	3667	3666	3666	3667	3673	3692	3731	3803	3916	4107	3857	3756	3693	3664	3653	3662	3672	3652	3598	3511	3406	3321	3269	3282	3331	3396	3452	3488
3686	3688	3688	3688	3685	3687	3705	3748	3817	3898	3962	3833	3761	3718	3696	3681	3669	3668	3640	3576	3478	3339	3185	3056	3121	3225	3325	3400	3439
3721	3725	3737	3761	3763	3761	3744	3730	3746	3822	3843	3785	3755	3730	3695	3647	3565	3612	3596	3532	3416	3226	2930	2392	2825	3078	3239	3334	3373
3737	3753	3788	3818	3835	.	.	.	3700	3741	3774	3766	3746	3717	3623	3422	2835	3447	3554	3520	3421	3272	3088	2910	2988	3133	3255	3324	3358
.	.	.	.	.	.	.	.	.	.	.	3765	3746	3718	3661	3554	3341	3543	3577	3535	3453	3337	3209	3108	3118	3192	3271	3326	3353
.	.	.	.	.	.	.	.	.	.	.	.	3782	3767	3706	3653	3570	3624	3807	3555	3481	3392	3298	3224	3206	3238	3283	3363	3485
.	.	.	.	.	.	.	.	.	.	.	.	.	3789	3770	3679	3664	3624	3566	3504	3433	3358	3302	3275	3299	3467	3564	.	.
.	.	.	.	.	.	.	.	.	.	.	.	.	3754	3692	3643	3598	3547	3501	3505	3496	3502	3552	3667	.	.	.	.	.

Fig. 10.37.8—Pressure distribution at the end of June.

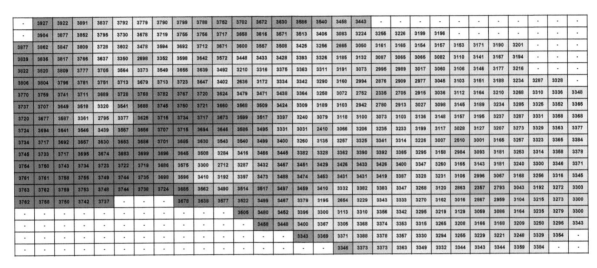

Fig. 10.37.9—Pressure distribution at the end of December.

Problem 10.5.35

Three-Phase Flow Simulation Project: Consider the three-phase reservoir shown in **Fig. 10.38.1.** Complete the following:

1. Code a computer program that constructs the Jacobian matrix and the residual vector that can be employed by the GNR method. Use the fully implicit protocol to formulate the reservoir fluid flow dynamics.
2. Solve for the pressure and saturation distribution for one timestep ($\Delta t = 5$ days).
3. Carry out MBCs on each phase to check the accuracy of the results.

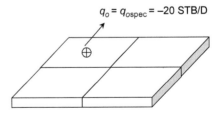

$q_o = q_{ospec} = -20$ STB/D

Fig. 10.38.1—Reservoir system for Problem 10.5.35.

Use the reservoir rock and fluid properties shown in **Tables 10.51 through 10.53** in your computations. Note that the well in the reservoir is produced at a rate of 20 STB/D.

$$k_x = k_y = 100 \text{ md}$$
$$\Delta x = \Delta y = 500 \text{ ft}$$
$$h = 50 \text{ ft}$$
$$\phi = 0.2$$
$$c_r = 3.00 \times 10^{-6} \text{ psi}^{-1}$$
$$p_i = 4,800 \text{ psi}$$
$$S_{wi} = 0.3$$
$$S_{gi} = 0.1$$
$$\text{skin} = 0$$
$$\Delta t = 5 \text{ days}$$

Table 10.51–Reservoir data used for Fig. 10.38.1.

Oil Properties

p	ρ_o	B_o	μ_o	R_{so}
1,500	49.0113	1.20413	1.7356	292.75
2,000	48.5879	1.2321	1.5562	368
2,500	48.1774	1.26054	1.4015	443.75
3,000	47.6939	1.29208	1.2516	522.71
3,500	47.1788	1.32933	1.1024	619
4,000	46.5899	1.37193	0.9647	724.92
4,500	45.5756	1.42596	0.918	818.6
5,000	45.1925	1.46387	0.92	923.12
5,500	45.4413	1.44983	0.9243	965.28
6,000	45.7426	1.43831	0.9372	966.32

Table 10.52.1–PVT oil-properties data used for Fig. 10.38.1.

Water Properties

p	ρ_w	B_w	μ_w	R_{sw}
1,500	62.228	1.0253	0.52	0
2,000	62.413	1.0222	0.52	0
2,500	62.597	1.0192	0.52	0
3,000	62.782	1.0162	0.52	0
3,500	62.968	1.0132	0.52	0
4,000	63.153	1.0102	0.52	0
4,500	63.337	1.0073	0.52	0
5,000	63.523	1.0051	0.52	0
5,500	63.708	1.0017	0.52	0
6,000	63.893	0.9986	0.52	0

Gas Properties

p	ρ_g	B_g	μ_g
1,500	5.8267	0.0018	0.015
2,000	8.0573	0.00133	0.0167
2,500	10.228	0.00105	0.0185
3,000	12.208	0.00088	0.0204
3,500	13.942	0.00077	0.0222
4,000	15.431	0.00069	0.0241
4,500	16.705	0.00064	0.026
5,000	17.799	0.0006	0.0278
5,500	18.748	0.00057	0.0296
6,000	19.577	0.00055	0.0313

Table 10.52.2–PVT water- and gas-properties data used for Fig. 10.38.1.

S_w	k_{rw}	k_{row}	P_{cow}	S_g	k_{rg}	k_{rog}	P_{cgo}
0.18	0	1	9	0	0	1	0
0.21	0	0.92692	7.26	0.04	0.01103	0.70778	0.01
0.24	2.00E–05	0.85441	5.04	0.08	0.02912	0.55844	0.062
0.27	0.00014	0.79288	3.78	0.12	0.05138	0.4454	0.114
0.3	0.00045	0.71312	3	0.16	0.07687	0.35562	0.166
0.33	0.00111	0.64526	2.634	0.2	0.10506	0.28302	0.218
0.36	0.00232	0.5798	2.268	0.24	0.13561	0.22392	0.27
0.39	0.0043	0.51709	1.902	0.28	0.16827	0.17574	0.366
0.42	0.00733	0.45744	1.666	0.32	0.20286	0.13656	0.462
0.45	0.01175	0.4011	1.495	0.36	0.23923	0.10485	0.580667
0.48	0.01791	0.34831	1.324	0.4	0.27725	0.07938	0.722
0.51	0.02623	0.29924	1.168	0.44	0.31683	0.05912	0.863333
0.54	0.03714	0.25403	1.042	0.48	0.35788	0.04319	1.004667
0.57	0.05116	0.21278	0.916	0.52	0.40031	0.03084	1.175
0.6	0.06882	0.17552	0.79	0.56	0.44408	0.02143	1.355
0.63	0.09069	0.14228	0.682	0.6	0.48911	0.01442	1.563
0.66	0.11741	0.11301	0.574	0.64	0.53536	0.00933	1.855
0.69	0.14963	0.08763	0.466	0.68	0.58279	0.00574	2.147
0.72	0.18807	0.06603	0.364	0.72	0.63134	0.00332	2.652
0.75	0.23347	0.04803	0.265				
0.78	0.28664	0.03344	0.166				
0.81	0.34842	0.02199	0.09				
0.84	0.41968	0.0134	0.06				
0.87	0.50135	0.00733	0.03				
0.9	0.59439	0.0034	0				

Table 10.53–Relative permeability data and capillary pressure data used for the reservoir shown in Fig. 10.38.1.

Solution to Problem 10.5.35

Summary of the Overall Approach: In this exercise, a three-phase black simulator will be developed. It will then be used to solve for pressure and saturation distributions by employing the GNR protocol. In the development of the three-phase simulator, the following steps must be completed:

1. Prepare subroutines capable of reading pressure- and saturation-dependent fluid properties from the given PVT and relative permeability tables using linear interpolation.
2. Formulate the residual of the oil/gas/water equation of each block at the initial condition.
3. Start the GNR iterations by constructing the Jacobian matrix that serves the GNR protocol in solving for pressure and saturation changes in a given timestep.
4. Use the changes calculated in Step 3 to update pressure and saturation values for each block.
5. Update the residuals of the oil, gas, and water equations.
6. Update the Jacobian matrix.
7. Repeat Step 3 until the maximum magnitude of the calculated residuals is smaller than a prescribed tolerance.
8. Conduct MBCs for oil, gas, and water phases.

Results. The calculated pressure, water-saturation, and gas-saturation distributions for each of the four blocks in Fig. 10.38.1 are shown **Table 10.54.** These results were obtained in three iterations. Furthermore, the computed Jacobian and residual entries after conducting three GNR iterations are displayed in **Fig. 10.38.2.** The mobility-ratio relationships were used to calculate accompanying water- and gas-production rates for the oil-flow rate specified ($q_{ospec} = -20$ STB/D):

$$q_w = q_{ospec}\left[\frac{k_{rw}}{\mu_w B_w}\frac{\mu_o B_o}{k_{ro}}\right] = 0.0664 \text{ STB/D}, \dots\dots\dots\dots\dots\dots\dots\dots\dots\dots\dots\dots\dots\dots\dots(10.469)$$

$$q_g = q_{ospec}\left[\frac{k_{rg}}{\mu_g B_g}\frac{\mu_o B_o}{k_{ro}} + R_{so}\right] + R_{sw}q_w = 204,937.88 \text{ Scf/D}. \dots\dots\dots\dots\dots\dots\dots\dots\dots\dots\dots (10.470)$$

Pressure Distribution		Water Saturation		Gas Saturation	
4,785.941	4,790.462	0.300031	0.300021	0.100495	0.100337
4,790.462	4,791.819	0.300021	0.300018	0.100337	0.100289

Table 10.54—Calculated pressure, water-saturation, and gas-saturation distribution for each of the four blocks in Fig. 10.38.1.

Jacobian Matrix (Iteration level 1) — Residual

												Residual
-1.110230745	61466.71237	61466.71237	1.464876654	0	0	1.464876654	0	0	0	0	0	-20
-0.205514042	-88521.70396	-0.415997701	0.004847393	0.059141061	0	0.004847393	0.059141061	0	0	0	0	-0.06618
-37815.98201	53508304.63	-94158691.34	14887.61781	0	17675.82272	14887.61781	0	17675.82272	0	0	0	-203261
1.464876656	0	0	-1.110230743	61466.71237	61466.71237	0	0	0	1.464877	0	0	0
0.004847393	0.059141061	0	-0.20551	-88518.23201	0	0	0	0	0.004847	0.059141	0	0
14887.61781	0	17675.82272	-37801.84714	54171351.21	-90425227.95	0	0	0	14887.62	0	17675.82272	0
1.464876656	0	0	0	0	0	-1.110230743	61466.71237	61466.71237	1.464877	0	0	0
0.004847393	0.059141061	0	0	0	0	-0.20551	-88518.23201	0	0.004847	0.059141	0	0
14887.61781	0	17675.82272	0	0	0	-37801.84714	54171351.21	-90425227.95	14887.62	0	17675.82272	0
0	0	0	1.464876656	0	0	1.464876656	0	0	-1.11023	61466.71	61466.71237	0
0	0	0	0.004847393	0.059141061	0	0.004847393	0.059141061	0	-0.20551	-88518.2	0	0
0	0	0	14887.61781	0	17675.82272	14887.61781	0	17675.82272	-37801.8	54171351	-90425227.95	0

Jacobian Matrix (Iteration level 2) — Residual

												Residual
-1.105049649	61509.43352	61509.43352	1.462453646	-23.70907746	-41.77956208	1.462453646	-23.70907746	0	0	0	0	-0.01674
-0.205520296	-88512.52534	-0.418932923	0.004852151	1.133748713	0	0.004852151	1.133748713	0	0	0	0	9.64E05
-37937.28657	53357962.26	-94053308.85	14946.37132	-20836.56308	833491.2998	14946.37132	-20836.56305	833491.2998	0	0	0	591.3413
1.462829851	0	0	-1.105114165	61519.3741	61537.44458	0	0	0	1.462799	-7.11449	0	0.009043
0.004852055	0.059195055	0	-0.205515721	-88513.07851	0	0	0	0	0.004851	0.38163	0	-3.82E06
14945.53061	0	17758.80595	-37914.44013	54095034.14	-91155214.25	0	0	0	14937.99	-6256.9	262511.2015	-164.921
1.462829851	0	0	0	0	0	-1.105114165	61519.3741	61537.44458	1.462799	-7.11449	-12.53458582	0.009043
0.004852055	0.059195055	0	0	0	0	-0.205515721	-88513.07851	0	0.004851	0.38163	0	-3.82E06
14945.53061	0	17758.80595	0	0	0	-37914.44012	54095034.14	-91155214.25	14937.99	-6256.9	262511.2015	-164.921
0	0	0	1.462912196	0	0	1.462912196	0	0	-1.10544	61505.76	61516.60402	0.006714
0	0	0	0.004851444	0.059187598	0	0.004851444	0.059187598	0	-0.2055	-88513.5	0	-7.18E06
0	0	0	14937.73692	0	17746.77747	14937.73692	0	17746.77747	-37905.8	54100570	-90841213.99	-113.779

Jacobian Matrix (Iteration level 3) — Residual

												Residual
-1.105062388	61509.37915	61509.37915	1.46245805	-23.63839174	-41.65491449	1.462458046	-23.63839247	0	0	0	0	2.13E-05
-0.205511024	-88512.53703	-0.41893473	0.004852149	1.130543523	0	0.004852149	1.130543555	0	0	0	0	9.15E-08
-37937.10616	53358137.91	-94053491.93	14946.27999	-20774.49646	831057.5273	14946.28002	-20774.49707	831057.5395	0	0	0	0.002673
1.462833129	0	0	-1.105114751	61519.29013	61537.30666	0	0	0	1.462802	-7.09762	0	3.76E-06
0.004852053	0.059195033	0	-0.205512629	-88513.07816	0	0	0	0	0.004851	0.380865	0	1.67E-08
14945.44186	0	17758.67969	-37914.28722	54095016.65	-91152819.67	0	0	0	14937.94	-6242.08	261930.677	-0.00107
1.462833133	0	0	0	0	0	-1.10511475	61519.29013	61537.30666	1.462802	-7.09762	-12.50485631	7.17E-08
0.004852053	0.059195033	0	0	0	0	-0.20550954	-88513.07816	0	0.004851	0.380865	0	1.67E-08
14945.44184	0	17758.67966	0	0	0	-37914.28405	54095016.65	-91152819.67	14937.94	-6242.08	261930.6524	-0.00107

Fig. 10.38.2—Jacobian and residual entries encountered in the calculations.

An incremental MBC is implemented to verify the accuracy of the results at the end of five days of simulation. Because the solution for only one timestep is computed, the cumulative material balance and the incremental MBC will be the same.

$$\text{Oil phase MBC} = V_b \left[\frac{\left(\frac{\phi S_o}{B_o}\right)^{t=5} - \left(\frac{\phi S_o}{B_o}\right)^{t=0}}{\Delta t \times q_o} \right] = \frac{100.00 \text{ STB}}{100 \text{ STB}} = 1.000, \quad \dots \dots \dots \dots \dots \dots \dots (10.471)$$

$$\text{Water phase MBC} = V_b \left[\frac{\left(\frac{\phi S_w}{B_w}\right)^{t=5} - \left(\frac{\phi S_w}{B_w}\right)^{t=0}}{\Delta t \times q_w} \right] = \frac{0.3322 \text{ STB}}{0.3322 \text{ STB}} = 1.000, \quad \dots \dots \dots \dots \dots \dots (10.472)$$

$$\text{Gas phase MBC} = V_b \left\{ \frac{\left[\left(\frac{\phi S_g}{B_g}\right)^{t=5} - \left(\frac{\phi S_g}{B_g}\right)^{t=0}\right] + R_{so}\left[\left(\frac{\phi S_o}{B_o}\right)^{t=5} - \left(\frac{\phi S_o}{B_o}\right)^{t=0}\right] + R_{sw}\left[\left(\frac{\phi S_w}{B_w}\right)^{t=5} - \left(\frac{\phi S_w}{B_w}\right)^{t=0}\right]}{\Delta t \times \left(q_g + R_{so}q_o + R_{sw}q_w\right)} \right\}. \quad \dots (10.473)$$

$$= \frac{1,024,689 \text{ scf}}{1,024,685 \text{ scf}} = 0.999996$$

Problem 10.5.36

Waterflooding Simulation Project. A waterflooding project is carried out in the reservoir shown in Figs. 10.36.1 through 10.36.4 in Problem 10.5.33. Wells 3 and 5 are deployed as water-injection wells with sandface-pressure specifications of 5,000 psia. The other wells in the reservoir are production wells producing at a sandface-pressure specification of 1,000 psia. Implement a black-oil reservoir simulator to predict the project response over a period of 360 days. Use $p_i = 2,600$ and $S_{wi} = 0.3$ to initialize the simulation model. The PVT properties for oil, water, and gas phases are given in **Tables 10.55 through 10.57.** Relative permeability data is given in **Table 10.58.**

	p (psia)	ρ_o (lbm/ft^3)	B_o (res bbl/STB)	μ_o (cp)	R_{so} (scf/STB)
Saturated	14.7	45.36	1.062	1.04	1
	270	44.08	1.15	0.975	90.5
	520	42.93	1.207	0.91	180
	1,015	41	1.295	0.83	371
	2,015	39.04	1.435	0.695	636
	2,515 (p_b)	38.52	1.5	0.641	775
	3,015	37.55	1.565	0.594	930
	4,015	36.81	1.695	0.51	1270
	5,015	36.05	1.827	0.449	1618
	9,015	34.4	2.357	0.203	2984
Undersaturated	2,515	38.52	1.5	0.641	775
	9,015	41.35	1.397	0.93	775

Table 10.55–Oil PVT properties for the Waterflooding Simulation Project.

	p (psia)	ρ_w (lbm/ft^3)	B_w (res bbl/STB)	μ_w (cp)	R_{sw} (scf/STB)
Saturated	14.7	62.24	1.041	0.31	0
	270	62.28	1.0403	0.31	0
	520	62.33	1.0395	0.31	0
	1,015	62.42	1.038	0.31	0
	2,015	62.6	1.035	0.31	0
	2,515	62.69	1.0335	0.31	0
	3,015	62.78	1.032	0.31	0
	4,015	62.96	1.029	0.31	0
	5,015	63.16	1.0258	0.31	0
	9,015	63.96	1.013	0.31	0
Undersaturated	2,515	62.69	1.0335	0.31	0
	9,015	63.96	1.013	0.31	0

Table 10.56–Water PVT properties for the Waterflooding Simulation Project.

p (psia)	ρ_g (lbm/ft^3)	B_g (res bbl/scf)	μ_g (cp)
14.7	0.0647	0.166666	0.008
270	0.8916	0.012093	0.0096
520	1.7185	0.006274	0.0112
1,015	3.3727	0.003197	0.014
2,015	6.6806	0.001614	0.0189
2,515	8.3326	0.001294	0.0208
3,015	9.9837	0.00108	0.0228
4,015	13.2952	0.000811	0.0268
5,015	16.6139	0.000649	0.0309
9,015	27.9483	0.000386	0.047

Table 10.57–Gas PVT properties for the Waterflooding Simulation Project.

Water-Relative Permeability and Capillary Pressure Data				Gas-Relative Permeability and Capillary Pressure Data			
S_w	k_{rw}	k_{row}	P_{cow}	S_g	k_{rg}	k_{rog}	P_{cgo}
0.22	0	1	7	0	0	1	0
0.3	0.08	0.41	4	0.04	0.005	0.602	0.21
0.4	0.17	0.128	3	0.1	0.022	0.333	0.55
0.5	0.26	0.0645	2.5	0.2	0.1	0.104	1.03
0.6	0.35	0.0045	2	0.3	0.24	0.021	1.54
0.8	0.68	0	1	0.4	0.34	0	2.09
0.9	0.85	0	0.5	0.5	0.42	0	2.51
1	1	0	0	0.6	0.5	0	3.05
				0.7	0.8125	0	3.5
				0.78	1	0	3.93

Table 10.58–Relative permeability data for the Waterflooding Simulation Project.

Solution to Problem 10.5.36

Summary of the Overall Approach. In this exercise, a black-oil simulator capable of implementing variable bubblepoint formulations needs to be developed. The same property-distribution maps given in Problem 10.5.34 are to be used. In developing the simulator, the following steps will be taken:

1. Prepare subroutines capable of reading pressure- and saturation-dependent fluid properties from the given PVT and relative permeability tables using linear interpolation.
2. Given that the initial pressure of the system, $p_{initial} = 2,600$ psia $> p_{bubble} = 2,516$ psia, indicates that the reservoir is initially undersaturated, calculate the residuals of the oil, gas, and water equations of each block at the initial condition with respect to the primary unknowns, p_o, S_w, and R_{so}.
3. Begin the GNR iteration during any timestep calculation by constructing the Jacobian matrix that serves the GNR protocol, and solve for changes in the primary unknowns.
4. Using the changes calculated in Step 3, update the primary unknowns for each block.
5. Implement the treatment of a variable bubblepoint, if necessary. When the calculated R_{so} is larger than that of the R_{so} value on a saturated curve, it is understood that the gas is coming out of solution and forming a free phase. The primary unknown of the gas equation shifts from R_{so} to S_g starting with the next iteration within the same timestep. On the other hand, when the calculated S_g value is less than zero, the primary unknown of the gas equation shifts from S_g to R_{so} for the next iteration. Recall that these shifts in unknowns, and hence in Jacobian entries, need to be implemented only in the blocks in which such observations are made.
6. At the completion of an iteration, update the residuals of the oil, gas, and water equations.
7. Update the Jacobian matrix, and solve for the new system of equations.
8. Repeat Step 3 until the maximum magnitude of the residual becomes smaller than a prescribed tolerance.
9. Carry out incremental and cumulative MBCs for oil, gas, and water phases.
10. Repeat the same simulation process for the next timestep until the total simulation time reaches 360 days.

Results. See the entries in **Tables 10.59 through 10.61** for the calculated oil-, gas-, and water-production rates and **Table 10.62** for the wellblock pressure results. The pressure- and saturation-distribution maps are also provided in **Figs. 10.39.1 through 10.39.6.**

	Well 1	Well 2	Well 3	Well 4	Well 5	Well 6	Well 7	Well 8	Well 9	Well 10
30	−128.629	−169.099	0	−104.489	0	−67.1528	−248.632	−96.6174	−202.7	−195.718
60	−119.896	−146.059	0	−93.1304	0	−59.9063	−211.12	−84.8003	−177.321	−169.236
90	−118.922	−140.281	0	−92.3135	0	−59.6826	−198.873	−82.958	−171.338	−162.807
120	−117.657	−139.246	0	−93.2881	0	−60.7749	−195.316	−83.2842	−169.906	−160.16
150	−115.912	−138.914	0	−94.0615	0	−61.6689	−194.68	−83.5439	−168.686	−157.279
180	−113.973	−138.221	0	−94.3887	0	−62.0283	−194.422	−83.3945	−166.719	−153.596
210	−112.125	−137.059	0	−94.3633	0	−61.9639	−193.711	−82.9531	−164.082	−149.379
240	−110.439	−135.535	0	−94.1074	0	−61.6572	−192.488	−82.293	−161.047	−144.945
270	−108.877	−133.781	0	−93.7285	0	−61.2207	−190.992	−81.4766	−157.801	−140.539
300	−107.398	−131.895	0	−93.3047	0	−60.7344	−189.301	−80.584	−154.488	−137.457
330	−106.059	−129.941	0	−92.9141	0	−60.2734	−187.5	−79.6699	−151.176	−134.16
360	−104.82	−127.969	0	−92.5703	0	−59.8203	−185.633	−78.7539	−147.926	−131.012

Table 10.59—Oil-production rates (STB/D) results for the Waterflooding Simulation Project.

	Well 1	Well 2	Well 3	Well 4	Well 5	Well 6	Well 7	Well 8	Well 9	Well 10
30	−138,279	−175,670	0	−109,552	0	−67,805.5	−248,030	−102,537	−209,478	−210,895
60	−129,894	−174,946	0	−107,122	0	−67,453.5	−251,755	−101,208	−207,603	−207,978
90	−122,826	−167,294	0	−101,558	0	−64,640.5	−245,700	−95,930	−198,302	−197,142
120	−119,035	−159,690	0	−97,242	0	−62,007.5	−236,470	−91,250	−189,688	−188,246
150	−116,934	−154,078	0	−94,368	0	−60,080	−227,880	−87,982	−183,624	−182,666
180	−115,624	−150,248	0	−92,428	0	−58,721	−221,120	−85,778	−179,724	−179,568
210	−114,654	−147,664	0	−91,028	0	−57,720	−216,096	−84,250	−177,300	−178,020
240	−113,848	−145,892	0	−89,954	0	−56,953	−212,404	−83,166	−175,820	−177,396
270	−113,156	−144,640	0	−89,076	0	−56,348	−209,608	−82,370	−174,952	−177,276
300	−112,512	−143,740	0	−88,324	0	−55,856	−207,456	−81,766	−174,468	−179,024
330	−111,888	−143,080	0	−87,650	0	−55,448	−205,760	−81,280	−174,236	−181,092
360	−111,348	−142,576	0	−87,044	0	−55,108	−204,376	−80,884	−174,164	−182,680

Table 10.60—Gas-production rates (scf/D) results for the Waterflooding Simulation Project.

	Well 1	Well 2	Well 3	Well 4	Well 5	Well 6	Well 7	Well 8	Well 9	Well 10
30	−89.1929	−116.817	362.1895	−72.3008	404.834	−45.9536	−169.854	−67.2087	−139.637	−136.873
60	−82.792	−105.067	339.5605	−65.9282	382.8594	−42.0007	−152.357	−60.7002	−126.296	−122.845
90	−80.8242	−100.49	327.5488	−64.0532	372.0625	−40.8679	−144.635	−58.4336	−121.209	−117.492
120	−79.4756	−98.4268	320.4727	−63.3359	365.5781	−40.5244	−140.785	−57.498	−118.796	−114.655
150	−78.2344	−97.1709	316.0938	−62.8633	361.2188	−40.3506	−138.625	−56.9082	−117.143	−112.432
180	−77.0479	−96.123	318.5195	−62.4219	360.7813	−40.1489	−137.076	−56.3896	−115.617	−110.305
210	−75.959	−95.1152	327.1016	−61.9795	370.7656	−39.8975	−135.703	−55.8955	−114.08	−108.184
240	−74.9746	−94.0996	334.9219	−61.542	379.7891	−39.6191	−134.375	−55.3955	−112.523	−106.102
270	−74.0762	−93.084	342.1172	−61.1211	388.1016	−39.3301	−133.09	−54.8887	−110.965	−104.086
300	−73.2578	−92.0781	348.7969	−60.7344	395.8594	−39.043	−131.844	−54.3867	−109.426	−102.133
330	−72.5273	−91.0898	355.0469	−60.3848	403.1641	−38.7734	−130.633	−53.9004	−107.918	−100.297
360	−71.8418	−90.125	360.9297	−60.0801	410.0156	−38.5127	−129.449	−53.4238	−106.461	−98.7266

Table 10.61—Water-production rates (STB/D) results for the Waterflooding Simulation Project.

	Well 1	Well 2	Well 3	Well 4	Well 5	Well 6	Well 7	Well 8	Well 9	Well 10
30	2,191.459	2,291.555	3,053.691	2,266.865	3,053.263	2,314.952	2,336.065	2,275.822	2,291.192	2,265.42
60	2,126.034	2,217.085	3,080.273	2,198.032	3,073.144	2,253.685	2,264.375	2,203.181	2,219.961	2,186.707
90	2,095.284	2,174.046	3,086.64	2,164.86	3,073.362	2,225.142	2,220.84	2,163.69	2,179.057	2,141.565
120	2,076.02	2,147.199	3,084.59	2,146.33	3,067.493	2,208.774	2,191.598	2,139.761	2,152.847	2,112.266
150	2,061.457	2,128.667	3,078.785	2,133.828	3,059.836	2,196.964	2,170.885	2,123.009	2,133.965	2,090.678
180	2,049.169	2,114.455	3,072.199	2,124.054	3,051.918	2,186.622	2,155.003	2,109.938	2,118.839	2,073.026
210	2,038.433	2,102.637	3,068.998	2,115.707	3,047.682	2,176.966	2,141.907	2,099.155	2,105.837	2,057.584
240	2,028.857	2,092.248	3,067.134	2,108.24	3,045.191	2,168.006	2,130.601	2,089.729	2,094.164	2,043.546
270	2,020.143	2,082.777	3,066.222	2,101.425	3,043.697	2,159.733	2,120.661	2,081.16	2,083.38	2,030.521
300	2,012.026	2,073.951	3,065.911	2,095.231	3,042.809	2,152.177	2,111.725	2,073.241	2,073.231	2,017.697
330	2,004.545	2,065.615	3,065.944	2,089.641	3,042.256	2,145.44	2,103.521	2,065.854	2,063.573	2,004.046
360	1,997.677	2,057.677	3,066.141	2,084.633	3,042.113	2,139.307	2,095.885	2,058.902	2,054.321	1,991.236

Table 10.62—Wellblock pressure (psia) results for the Waterflooding Simulation Project.

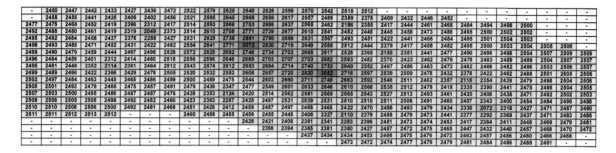

Fig. 10.39.1—Simulated pressure distribution at *t* = 180 days.

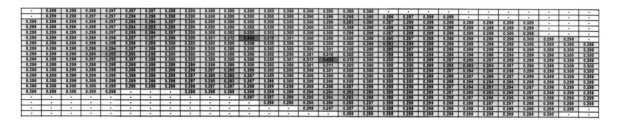

Fig. 10.39.2—Simulated water saturation at *t* = 180 days.

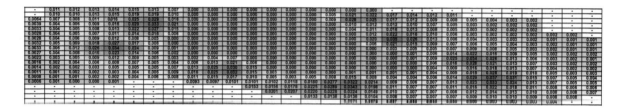

Fig. 10.39.3—Simulated gas saturation at *t* = 180 days.

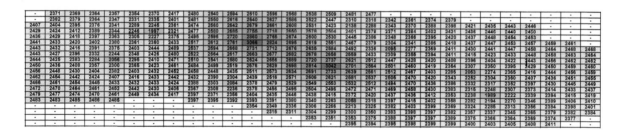

Fig. 10.39.4—Simulated pressure distribution at *t* = 365 days.

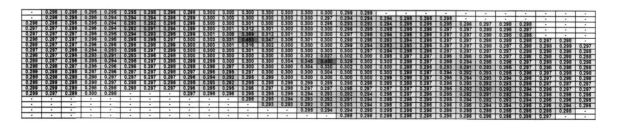

Fig. 10.39.5—Simulated water saturation at *t* = 365 days.

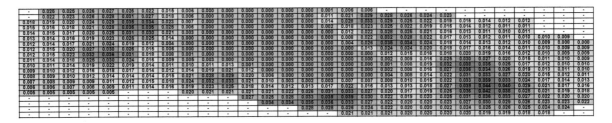

Fig. 10.39.6—Simulated gas saturation at t = 365 days.

Nomenclature

A	=	cross-sectional area, L^2, ft³
A_x	=	cross-sectional area along x-direction, L^2, ft²
A_y	=	cross-sectional area along y-direction, L^2, ft²
B	=	formation volume factor, L^3/L^3, res bbl/scf for liquid and res bbl/STB for gas
B_g	=	gas formation volume factor, L^3/L^3, RB/scf
B_o	=	oil formation volume factor, L^3/L^3, res bbl/STB
B_w	=	water formation volume factor, L^3/L^3, res bbl/STB
c_f	=	fluid compressibility, Lt^2/m, psi⁻¹
c_o	=	oil compressibility, Lt^2/m, psi⁻¹
c_t	=	total compressibility, Lt^2/m, psi⁻¹
c_w	=	water compressibility, Lt^2/m, psi⁻¹
c_ϕ	=	porosity compressibility, Lt^2/m, psi⁻¹
C	=	transmissibility coefficient using SIP notation of central block (i,j,k), L^4t/m, STB/(D–psia) for liquid, scf/(D–psia) for gas
E	=	transmissibility coefficient using SIP notation between block (i,j,k) and $(i+1, j, k)$, L^4t/m, STB/(D–psia) for liquid, scf/(D–psia) for gas
g	=	acceleration of gravity, L/t^2, ft/s²
g_c	=	unit conversion factor in Newton's law
G	=	elevation with respect to absolute datum being positive upward, L, ft
h	=	thickness, L, ft
J	=	Jacobian matrix
k	=	permeability, L^2, darcies
$\bar{k}$	=	average permeability, L^2, darcies or perms
k_{rg}	=	relative permeability to gaseous phase, fraction
k_{ro}	=	relative permeability to oil phase, fraction
k_{rog}	=	relative permeability to oil phase in the presence of gas phase, fraction
k_{row}	=	relative permeability to oil in the presence of water phase, fraction
k_{rw}	=	relative permeability to water phase, fraction

k_x	=	permeability along x-direction, L^2, darcies or perms
k_y	=	permeability along y-direction, L^2, darcies or perms
k_z	=	permeability along z-direction, L^2, darcies or perms
N	=	transmissibility coefficient using SIP notation between block (i,j,k) and $(i, j+1, k)$, L^4t/m, STB/(D–psia) for liquid, scf/(D–psia) for gas
N_p	=	cumulative production, L^3, STB for liquid and scf for gas
p	=	pressure, m/Lt^2, psia
p_{bubble}	=	bubblepoint pressure, m/Lt^2, psia
p_e	=	aquifer pressure, m/Lt^2, psia
p_{frac}	=	formation fracking pressure, m/Lt^2, psia
p_{intial}	=	initial pressure, m/Lt^2, psia
p_{sc}	=	standard condition pressure, m/Lt^2, psia
p_{sf}	=	sandface pressure, m/Lt^2, psia
$p_{surface}$	=	surface pressure
p_{wf}	=	well-bottomhole flowing pressure, m/Lt^2, psia
P_{cgo}	=	gas liquid capillary pressure, m/Lt^2, psia
P_{cow}	=	oil water capillary pressure, m/Lt^2, psia
q	=	production or flow rate, L^3/t, scf/D for gas, STB/D for liquid
$q_{boundary}$	=	flow rate supported by external boundary, L^3/t, STB/D for liquid and scf/D for gas
q_{east}	=	flow rate supported by eastern boundary, L^3/t, STB/D for liquid and scf/D for gas
q_{in}	=	flow rate entering the system, L^3/t, STB/D for liquid and scf/D for gas
q_{osc}	=	oil production or flow rate, L^3/t, STB/D
q_{out}	=	flow rate existing the system, L^3/t, STB/D for liquid and scf/D for gas
q_{total}	=	total production or flow rate, L^3/t, scf/D for gas, STB/D for liquid

q_{well} = production/injection rate of the well, L^3/t, STB/D for liquid and scf/D for gas

q_{west} = flow rate supported by western boundary, L^3/t, STB/D for liquid and scf/D for gas

q_x = x-direction component of the flow rate, L^3/t, scf/D for gas, STB/D for liquid

q_y = y-direction component of the flow rate, L^3/t, scf/D for gas, STB/D for liquid

Q = right-hand-side coefficient of governing flow equation using SIP notation for block (i,j,k) L^3/t, STB/D for liquid, scf/D for gas

Q_{out} = volumetric flow rate leaving the system, L^3/t, STB/D for liquid, scf/D for gas

Q_{total} = cumulative production, scf for gas and STB for liquid

r = radial distance in both cylindrical and spherical coordinator system, L, ft

r_e = Peaceman's equivalent wellblock radius, L, ft

r_w = wellblock radius, L, ft

R = universal gas constant

R_{so} = solution gas/oil ratio, L^3/L^3, scf/STB

R_{sw} = solution gas/water, ratio, L^3/L^3, scf/STB

s = skin factor, dimensionless

S = transmissibility coefficient using SIP notation between block (i,j,k) and $(i, j\text{-}1, k)$, L^4t/m, STB/(D–psia) for liquid, scf/(D–psia) for gas

S_g = gas saturation, fraction

S_o = oil saturation, fraction

S_w = water saturation, fraction

S_{wi} = initial water saturation, fraction

S_{gi} = initial gas saturation, fraction

t = time, t, days

T = absolute temperature, T, °R

T_{sc} = standard condition temperature, T, °R

T_x = transmissibility along x-direction, L^4t/m, scf/(D–psia) for gas and STB/(D–psia) for liquid

T_y = transmissibility along y-direction, L^4t/m, scf/(D–psia) for gas and STB/(D–psia) for liquid

V_b = bulk volume, L^3, ft^3

V_p = pore volume, L^3, ft^3

V_{well} = total volume produced through the well, L^3, scf for gas and STB for liquid

$W_{i,j,k}$ = transmissibility coefficient using SIP notation between block (i,j,k) and $(i\text{-}1, j, k)$, L^4t/m, STB/(D–psia) for liquid, scf/(D–psia) for gas

w_f = width of the fracture, L, ft

x = distance in x-direction in the Cartesian coordinate system, L, ft

y = distance in y-direction in the Cartesian coordinate system

z = gas compressibility factor, dimensionless

Γ = accumulation coefficient, L^3/t, ft^3/D

γ = hydrostatic gradient , m/L^2t^2, psi/ft

Δ = small change of a parameter

Δx = control volume dimension along the x-direction, L, ft

Δy = control volume dimension along the y-direction, L, ft

Δz = control volume dimension along the z-direction, L, ft

μ = viscosity, m/Lt, cp

μ_g = gas viscosity, m/Lt, cp

μ_o = oil viscosity, m/Lt, cp

μ_w = water viscosity, m/Lt, cp

ρ = density, m/L^3, lbm/ft^3

Φ = flow potential, m/Lt^2, psia

ϕ = porosity, fraction

ω_{opt} = acceleration parameter

Ω = well productivity/injectivity index, L^4t/m, STB/(D–psia) for liquid, scf/(D–psia) for gas

Superscripts

n = previous time level

$n+1$ = current time level

Subscripts

g = relevant to gas phase

i = address of a block located in a 1D system at position i

i,j = address of a block located in a 2D system at position i,j

i,j,k = address of a block located in a 3D system at position i,j,k

initial = refers to initial state

o = relevant to oil phase

p = pore-related property

sc = standard conditions

spec = specified

t = total

w = relevant to water phase

wellhead = any parameter expressed at the wellhead conditions

References

Dranchuk, P. M. and Abou-Kassem, H. 1975. Calculation of Z Factor for Natural Gases Using Equation of State. *J Can Pet Technol*, **14**(3): 34. PETSOC-75-03-03. http://dx.doi.org/10.2118/75-03-03.

Ertekin, T., Abou-Kassem, J. H., and King, G. R. 2001. *Basic Applied Reservoir Simulation*. Richardson, Texas, US: Society of Petroleum Engineers.

Lee, A. B., Gonzalez, M. H., Eakin, B. E. 1966. The Viscosity of Natural Gases. *J of Pet Tech* **18**(08): 997–1,000. SPE-1340-PA. http://dx.doi.org/10.2118/1340-PA.

Peaceman, D.W. 1983. Interpretation of Well-Block Pressures in Numerical Reservoir Simulation with Nonsquare Grid Blocks and Anisotropic Permeability. *SPE J* **23**(3): 531–543. SPE-10528-PA. https://dx.doi.org/10.2118/10528-PA

Chapter 11

Classical Reservoir Engineering Protocols as Subsets of Reservoir Simulation

There is ongoing debate among reservoir simulation engineers and petroleum engineers who are more attuned to the use of classical reservoir engineering analysis tools. The goal of this chapter is to settle this debate by showing that most classical reservoir engineering tools are simplified forms of reservoir simulation formulations.

11.1 Reservoir Engineering Analysis Tools: Numerical vs. Analytical

A numerical reservoir model provides opportunities to incorporate effects of heterogeneities, anisotropy, and existing non-linearities more effectively so that it is possible to stay closer to the exact representation of the problem. In solving reservoir simulation equations, one realizes that conducting a reservoir simulation is equivalent to carrying out a material-balance analysis, analyzing a water-encroachment study, finding the size of the gas cap, performing a pressure-transient analysis, and making a Buckley and Leverett (1942) analysis and decline-curve analysis in a concurrent manner. Accordingly, one can comfortably state that classical reservoir engineering analysis tools simply are subsets of reservoir simulation equations studied in this book.

Most of the analytical and empirical models previously mentioned are built into reservoir simulation equations. A broad classification of the functionalities of the analytical and numerical tools can be described as follows:

- Resources—estimation of original oil in place (OOIP) and original gas in place (OGIP). This can be achieved using a simplistic approach, such as a material-balance equation, but lacks accurate calculation of saturations and is unsuitable for time scheduling.
- Front tracking—most of the time this is tied to well-coning and pattern-flood studies. Water-encroachment models, while providing a simplistic approach, require fine grid and are unsuitable for larger-scale field studies.
- Resources and front tracking—this study results in pressure and production forecasting, which is used at the end for field development. While it is a complex approach and results in accurate pressure and saturation distributions, it also allows for time scheduling.
- Fully integrated reservoir modeling—in fully integrated reservoir modeling, we are referring to not only the integration of the geological and geomechanical features of the reservoir to the reservoir flow equations, but also the integration of the wellbore hydraulics and operational conditions of the surface facilities. In such an integration, while one boundary condition is at the physical limits of the reservoir, the other boundary condition can be specified at the separator outlet, as shown in **Fig. 11.1.** What makes a fully integrated reservoir-model development more challenging is the definite need to ensure the synchronous flow of information in this class of models. **Fig. 11.2** shows the synchronous flow of information in an integrated reservoir model and schematically depicts the types of concurrent calculations needed to ensure isochronicity.

11.2 Problems and Solutions

Problem 11.2.1

Buckley-Leverett Model and Numerical Simulation. Consider the system shown in **Fig. 11.3.** Calculate the water-saturation distribution along the injection path using the Buckley-Leverett model (Buckley and Leverett 1942) and numerical-simulation protocol, and then compare the results. Use the input data listed in **Tables 11.1 and 11.2.**

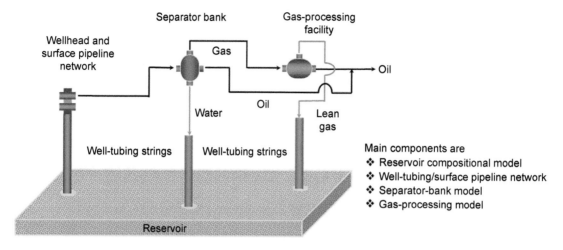

Fig. 11.1—Schematic representation of an integrated reservoir-surface network simulation.

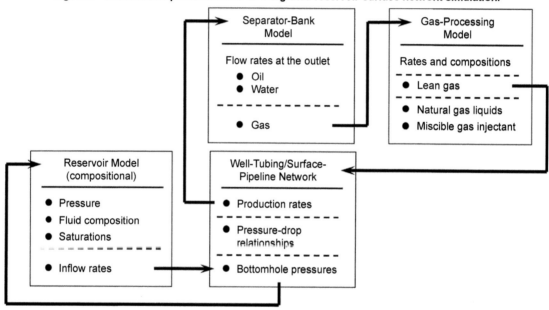

Fig. 11.2—Concurrent calculations taking place in an integrated reservoir model.

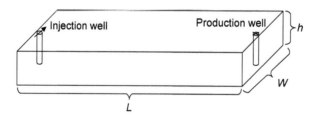

Fig. 11.3—Description of the 1D waterflooding project described in Problem 11.2.1.

$W = 500$ ft	$k_{rwro} = 0.4$
$L = 500$ ft	$k_{roq} = 0.8$
$h = 50$ ft	$S_{iw} = 0.15$
$\phi = 0.2$ fraction	$S_{orw} = 0.15$
$\mu_o = 2$ cp	$n_w = 2$
$\mu_w = 2$ cp	$n_{ow} = 2$
$q_{injection} = 1,000$ STB/D	

Table 11.1—Reservoir properties from Problem 11.2.1.

Table 11.2—Coefficients to be used in Part 1 of Problem 11.2.1.

1. Generate a relative permeability curve using the Corey model (Corey 1954):

$$k_{rw} = k_{rwro}\left(\frac{S_w - S_{iw}}{1 - S_{iw} - S_{orw}}\right)^{n_w}, \dots\dots\dots\dots\dots\dots\dots\dots\dots\dots\dots\dots\dots\dots\dots\dots\dots (11.1)$$

$$k_{row} = k_{roiw}\left(\frac{1 - S_w - S_{orw}}{1 - S_{iw} - S_{orw}}\right)^{n_{ow}}. \dots\dots\dots\dots\dots\dots\dots\dots\dots\dots\dots\dots\dots\dots\dots (11.2)$$

2. Calculate the water-saturation front at $t = 100$ days and $t = 275$ days using the Buckley-Leverett model.
3. Repeat Part 2 using the numerical-simulation protocol and compare the results.

Solution to Problem 11.2.1

1. **Fig. 11.4** shows the relative permeability curves generated using the Corey model. **Table 11.3** shows the results for reference.
2. The Buckley-Leverett model expresses the saturation front as

$$x_f = \frac{5.615qt}{A\phi}\left(\frac{df_w}{dS_w}\right). \dots\dots\dots\dots (11.3)$$

Fig. 11.5 shows the calculation results of the fractional flow curve and the derivative of the fractional flow with respect to water saturation. **Fig. 11.6** displays the saturation front calculated using the solution of the Buckley-Leverett equation. **Table 11.4** shows the results for reference.

3. The following partial-differential equations (PDEs) are to be solved employing the numerical-simulation model:

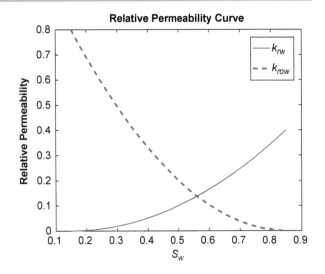

Fig. 11.4—Relative permeability curves for Problem 11.2.1.

- Water Equation: $\dfrac{\partial}{\partial x}\left(\dfrac{A_x k_x k_{rw}}{\mu_w B_w}\dfrac{\partial p}{\partial x}\right)\Delta x + q_w = \dfrac{V_b\phi}{5.615}\dfrac{\partial S_w}{\partial t}. \dots\dots\dots\dots\dots\dots\dots\dots\dots\dots\dots\dots\dots (11.4)$

- Oil Equation: $\dfrac{\partial}{\partial x}\left(\dfrac{A_x k_x k_{ro}}{\mu_o B_o}\dfrac{\partial p}{\partial x}\right)\Delta x + q_o = \dfrac{V_b\phi}{5.615}\dfrac{\partial S_o}{\partial t}. \dots\dots\dots\dots\dots\dots\dots\dots\dots\dots\dots\dots\dots (11.5)$

For a homogeneous system, Eqs. 11.4 and 11.5 can be simplified to

- Water Equation: $\dfrac{\partial}{\partial x}\left(k_{rw}\dfrac{\partial p}{\partial x}\right) + \dfrac{\mu_w B_w q_w}{k V_b} = \dfrac{\mu_w B_w \phi}{5.615k}\dfrac{\partial S_w}{\partial t}. \dots\dots\dots\dots\dots\dots\dots\dots\dots\dots\dots\dots\dots (11.6)$

- Oil Equation: $\dfrac{\partial}{\partial x}\left(k_{ro}\dfrac{\partial p}{\partial x}\right) + \dfrac{\mu_o B_o q_o}{k V_b} = \dfrac{\mu_o B_o \phi}{5.615k}\dfrac{\partial(1 - S_w)}{\partial t}. \dots\dots\dots\dots\dots\dots\dots\dots\dots\dots\dots\dots\dots (11.7)$

Suppose that the system is discretized using uniform Δx. The PDE can be written in a finite-difference form (note that $V_b = \Delta x \Delta y h$):

- Water Equation: $k_{rw}\big|_{i+\frac{1}{2}}\left(p_{i+1}^{n+1} - p_i^{n+1}\right) - k_{rw}\big|_{i-\frac{1}{2}}\left(p_i^{n+1} - p_{i-1}^{n+1}\right) + \dfrac{\Delta x \mu_w B_w q_w}{k \Delta y h} = \dfrac{(\Delta x)^2 \mu_w B_w \phi}{5.615k\Delta t}\left(S_w^{n+1} - S_w^n\right). \dots (11.8)$

- Oil Equation: $k_{ro}\big|_{i+\frac{1}{2}}\left(p_{i+1}^{n+1} - p_i^{n+1}\right) - k_{ro}\big|_{i-\frac{1}{2}}\left(p_i^{n+1} - p_{i-1}^{n+1}\right) + \dfrac{\Delta x \mu_o B_o q_o}{k \Delta y h} = \dfrac{(\Delta x)^2 \mu_o B_o \phi}{5.615k\Delta t}\left(S_w^n - S_w^{n+1}\right). \dots\dots (11.9)$

S_w	k_{rw}	k_{row}
0.1500	0.0000	0.8000
0.2901	0.0160	0.5119
0.4301	0.0641	0.2878
0.5702	0.1441	0.1278
0.7103	0.2563	0.0319
0.8500	0.4000	0.0000

S_w	f_w	df_w/dS_w	x_f ($t = 100$ days)	x_f ($t = 275$ days)
0.1500	0.0000	0.0007	0.0803	0.220921
0.2901	0.0589	0.9912	111.3064	306.0926
0.4301	0.3080	2.5386	285.0880	783.992
0.5702	0.6928	2.5326	284.4120	782.1329
0.7103	0.9415	0.9840	110.4982	303.87
0.8500	1.0000	0.0000	0.0000	0

Table 11.3—Results for Part 1 of the Solution to Problem 11.2.1.

Table 11.4—Results for Part 2 of the Solution to Problem 11.2.1.

Two unknowns, S_w and p, are to be solved from the PDEs. For the purposes of validation, the reservoir is discretized using a 5×1 grid system and a timestep of $\Delta t = 10$ days. Each block has dimensions of 100 ft × 500 ft × 50 ft. **Fig. 11.7** provides sample results.

Please note that to implement the numerical-solution protocol, one needs to assume the initial pressure, permeability, and production-well internal-boundary condition of the system. However, those values have no impact on the results of the S_w front calculation.

Fig. 11.8 shows the comparison of the saturation front using the Buckley-Leverett model and the numerical-simulation method. To guarantee the matching quality, the system needs to be discretized using a finer-grid dimension. We recommend the Δx value to be less than 2 ft.

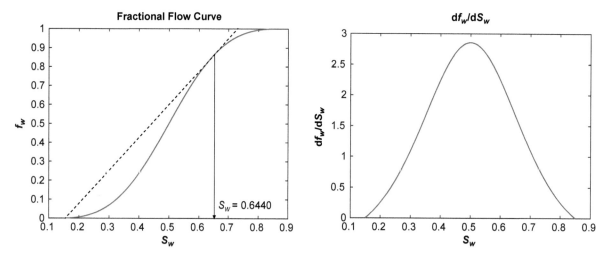

Fig. 11.5–Fractional flow curves of Problem 11.2.1.

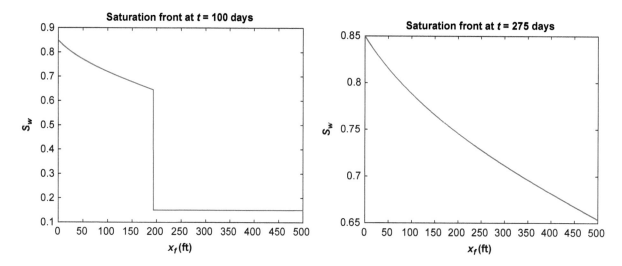

Fig. 11.6—Saturation profile before water breakthrough (at $t = 100$ days) and after water breakthrough ($t = 275$ days).

Sample results:

	Block 2	Block 3	Block 4	
S_w	0.4042	0.2275	0.1551	0.1500	0.1500

Block 1
(Injection well)

Block 5
(Production well)

Fig. 11.7—Sample numerical simulation results (Problem 11.2.1, Part 3).

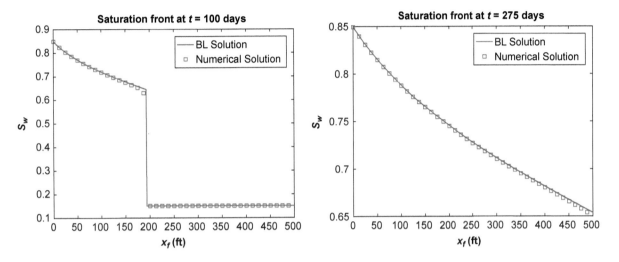

Fig. 11.8—Validation with Buckley-Leverett model (Problem 11.2.1).

Problem 11.2.2

Material Balance Equation and Numerical Simulation. The production data in **Table 11.5** is generated using a numerical reservoir simulation model with the given pressure-volume-temperature (PVT) data (**Table 11.6**). Knowing that the bubblepoint pressure of the reservoir is 2,800 psi, and the initial reservoir pressure is 2,850 psi, the water saturation of the system is lower than the irreducible saturation. Use the material-balance-equation approach to calculate the OOIP and compare the reported value against the numerical model, which is 3.06 MMSTB.

$\bar{p}$, psi	N_p, STB	G_p, MMscf
2,500.43	93,015.93	52.51
2,148.70	222,070.52	150.53
1,890.71	308,312.31	266.51
1,655.49	366,858.63	407.70
1,443.15	406,699.78	556.44
1,255.20	433,946.41	700.23
764.32	486,207.13	1,103.76
665.71	494,549.03	1,185.86
558.78	503,384.41	1,275.92

Table 11.5—Average reservoir pressure vs. cumulative production data from Problem 11.2.2.

p, psi	R_s, scf/STB	B_o, res bbl/STB	B_g, res bbl/scf
14.696	3.2563	1.02599	0.197669
386.07	53.7206	1.04285	0.007205
571.757	84.1373	1.05338	0.004763
757.444	116.585	1.06488	0.003522
943.131	150.662	1.07724	0.002773
1,128.82	186.112	1.09036	0.002275
1,314.5	222.753	1.10419	0.001922
1,500.19	260.453	1.11869	0.001659
1,685.88	299.106	1.13381	0.001459
1,871.57	338.63	1.14953	0.001302
2,057.25	378.955	1.16581	0.001178
2,242.94	420.024	1.18264	0.001077
2,428.63	461.788	1.19999	0.000994
2,614.31	504.205	1.21784	0.000925
2,800	547.238	1.23618	0.000868
2,840	556.586	1.24019	0.000857
2,880	565.96	1.24423	0.000847
3,000	594.24	1.25646	0.000817

Table 11.6—PVT data for Problem 11.2.2.

Solution to Problem 11.2.2

For an undersaturated reservoir with no mobile water ($S_w <$ critical water saturation), the material-balance equation can be written as Eq. 11.10:

$$N\left(B_o - B_{oi}\right) + N\left(R_{si} - R_s\right)B_g = N_p B_o + N_p\left(R_p - R_r\right)B_g. \quad\quad\quad (11.10)$$

Defining

$$F = N_p B_o + N_p\left(R_p - R_s\right)B_g \quad\quad\quad\quad\quad (11.11)$$

and

$$E = \left(B_o - B_{oi}\right) + \left(R_{si} - R_s\right)B_g, \quad\quad\quad\quad\quad (11.12)$$

the material-balance equation can be rewritten as $F = NE$.

If one plots E vs. F, as shown in **Fig. 11.9,** the slope indicates the OOIP. The slope is calculated as 3.02 MMSTB, which is 1.3% off compared with the reported value from the numerical model (3.06 MMSTB). **Table 11.7** shows the calculation results for reference.

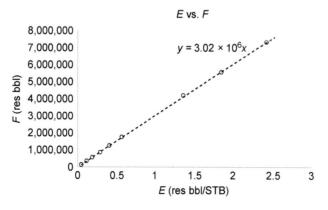

Fig. 11.9—Result of material balance analysis (Problem 11.2.2).

$\bar{p}$, psi	R_p, scf/STB	B_o, RB/STB	B_g, RB/scf	R_s, scf/STB	F, RB	E, RB/STB
2,500.429	564.5274	1.207046	0.000972	478.3909	120,060.7	0.044116
2,148.699	677.8364	1.174238	0.001134	399.369	330,880.5	0.113959
1,890.713	864.4098	1.151237	0.001291	342.8256	562,508.2	0.188977
1,655.489	1,111.333	1.131586	0.00151	293.1441	868,423.3	0.29177
1,443.146	1,368.194	1.114338	0.001754	249.034	1,251,576	0.416714
1,255.203	1,613.627	1.100001	0.002066	211.4114	1,734,218	0.576633
764.3215	2,270.142	1.065352	0.003499	117.8726	4,179,390	1.367371
665.7068	2,397.861	1.059438	0.004432	101.0562	5,558,031	1.847487
558.7815	2,534.68	1.053095	0.005492	82.95637	7,308,618	2.42615

Table 11.7—Calculation results for reference for Problem 11.2.2

Problem 11.2.3

Decline Curve and Numerical Simulation. Consider the production data in **Table 11.8,** which is generated using a numerical-simulation model for a single-phase gas reservoir. Estimate the OGIP using Arps' decline curve (Arps 1945), and compare the result with the reported value of 4.22 Bcf from the numerical model.

Solution to Problem 11.2.3

The Arps decline curve can be written as

$$q_g = \frac{q_i}{\left(1+bD_i t\right)^{1/b}}. \qquad (11.13)$$

The production data must be fitted to the decline curve (**Fig. 11.10**) for decline parameters of q_i, b, and D_i.

$$OGIP = \frac{q_i}{(1-b)D_i} = 4.021 \text{ Bcf}, \qquad (11.14)$$

which is 4.74% off from the value reported by the numerical model.

t, days	q_g, MMscf/D
100	3.801311
200	3.224789
400	2.386424
600	1.821354
800	1.425438
1,000	1.139078
1,200	0.926382
1,400	0.764802
1,600	0.639656
1,800	0.541086
2,000	0.462297

Table 11.8—Production rate data for Problem 11.2.3.

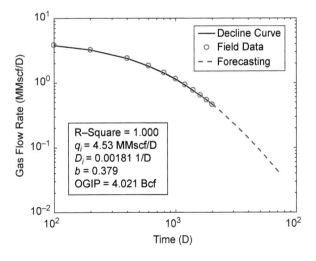

Fig. 11.10—Production analysis using Arp's Model for Problem 11.2.3.

Nomenclature

A_x	=	cross-sectional area perpendicular to flow along the x-direction, L^2, ft^2
b	=	decline exponent of the Arp model
B_g	=	gas formation volume factor, L^3/L^3, res bbl/scf
B_o	=	oil formation volume factor, L^3/L^3, res bbl/STB
B_{oi}	=	initial oil formation volume factor, L^3/L^3, res bbl/STB
B_w	=	water formation volume factor, L^3/L^3, res bbl/STB
D_i	=	initial decline rate, 1/t, D^{-1}
E	=	parameter defined in material balance equation, L^3/L^3, res bbl/STB
f_w	=	fractional flow ratio, dimensionless
F	=	parameter defined in material balance equation, L^3, RB
G_p	=	total gas produced, L^3, scf
h	=	thickness, L, ft
k	=	permeability, L^2, darcies or perms
k_{ro}	=	relative permeability to oil, dimensionless
k_{rog}	=	relative permeability to oil in oil/gas system, dimensionless
k_{roiw}	=	relative permeability to gas at irreducible water saturation, dimensionless
k_{row}	=	relative permeability to oil in oil/water system, dimensionless
k_{rw}	=	relative permeability to water, dimensionless
k_{rwro}	=	relative permeability to water at residual oil saturation, dimensionless
k_x	=	permeability along x-direction, L^2, darcies
L	=	the length of the wellbore in the wellblock, L, ft
n_w	=	exponent of water relative permeability in oil/water systems
n_{ow}	=	exponent of oil relative permeability in oil/water systems
N	=	original oil in place, L^3, STB
N_p	=	cumulative oil production, L^3, STB

p	=	pressure, m/Lt^2, psia
p_i	=	initial pressure, m/Lt^2, psia
q	=	production or flow rate, L^3/t, STB/D
q_g	=	gas flow rate, L^3/t, scf/D
q_i	=	initial flow rate, L^3/t, STB/D for liquid and scf/D for gas
$q_{injection}$	=	injection rate, L^3/t, STB/D for liquid and scf/D for gas
R_p	=	producing gas/oil ratio, L^3/L^3, scf/STB
R_s	=	solution gas/oil ratio, L^3/L^3, scf/STB
R_{si}	=	solution gas/oil ratio at initial condition, L^3/L^3, scf/STB
S_{iw}	=	irreducible water saturation, fraction
S_o	=	oil saturation, fraction
S_{orw}	=	residual oil saturation in oil/water system, fraction
S_w	=	water saturation, fraction
t	=	time, t, days
V_b	=	bulk volume, L^3, ft^3
W	=	width, L, ft.
x	=	distance in x-direction in the Cartesian coordinate system, L, ft
x_f	=	the saturation front defined by the Buckley-Leverett model
y	=	distance in y-direction in the Cartesian coordinate system
Δx	=	control volume dimension along the x-direction, L, ft
Δy	=	control volume dimension along the y-direction, L, ft
μ_o	=	oil viscosity, m/Lt, cp
μ_w	=	water viscosity, m/Lt, cp
ϕ	=	porosity, fraction

Superscripts

n	=	old timestep level
$n+1$	=	current time level

Subscripts

i	=	index of block/node

References

Arps, J.J. 1945. Analysis of Decline Curves. *Trans. of the AIME* **160**(01): 228–247. SPE-945228-G. http://dx.doi.org/10.2118/945228-G.

Buckley, S.E. and Leverett, M.C. 1942. Mechanism of Fluid Displacements in Sands. *Trans. of the AIME* **146**(01): 107–116. SPE-942107-G. http://dx.doi.org/10.2118/942107-G.

Corey, A. T. 1954. The Interrelation Between Gas and Oil Relative Permeabilities. *Producer Monthly*, **19**(November): 38–41. https://www.discovery-group.com/pdfs/Corey_1954.pdf.

INDEX